Algorithms and The
Computation Handbook

Second Edition

Special Topics
and Techniques

Edited by

Mikhail J. Atallah
Marina Blanton

CRC Press
Taylor & Francis Group
Boca Raton London New York

CRC Press is an imprint of the
Taylor & Francis Group, an **informa** business

A CHAPMAN & HALL BOOK

Chapman & Hall/CRC
Applied Algorithms and Data Structures Series

Series Editor

Samir Khuller
University of Maryland

Aims and Scopes

The design and analysis of algorithms and data structures form the foundation of computer science. As current algorithms and data structures are improved and new methods are introduced, it becomes increasingly important to present the latest research and applications to professionals in the field.

This series aims to capture new developments and applications in the design and analysis of algorithms and data structures through the publication of a broad range of textbooks, reference works, and handbooks. We are looking for single authored works and edited compilations that will:

- Appeal to students and professionals by providing introductory as well as advanced material on mathematical, statistical, and computational methods and techniques

- Present researchers with the latest theories and experimentation

- Supply information to interdisciplinary researchers and practitioners who use algorithms and data structures but may not have advanced computer science backgrounds

The inclusion of concrete examples and applications is highly encouraged. The scope of the series includes, but is not limited to, titles in the areas of parallel algorithms, approximation algorithms, randomized algorithms, graph algorithms, search algorithms, machine learning algorithms, medical algorithms, data structures, graph structures, tree data structures, and more. We are willing to consider other relevant topics that might be proposed by potential contributors.

Proposals for the series may be submitted to the series editor or directly to:

Randi Cohen

Acquisitions Editor

Chapman & Hall/CRC Press

6000 Broken Sound Parkway NW, Suite 300

Boca Raton, FL 33487

Algorithms and Theory of Computation Handbook, Second Edition

Algorithms and Theory of Computation Handbook, Second Edition: General Concepts and Techniques
Algorithms and Theory of Computation Handbook, Second Edition: Special Topics and Techniques

CRC Press
Taylor & Francis Group
6000 Broken Sound Parkway NW, Suite 300
Boca Raton, FL 33487-2742

First issued in paperback 2019

ISBN-13: 978-1-58488-820-8 (hbk)
ISBN-13: 978-0-367-38484-5 (pbk)

Library of Congress Cataloging-in-Publication Data

Algorithms and theory of computation handbook. Special topics and techniques / editors, Mikhail J. Atallah and Marina Blanton. -- 2nd ed.
    p. cm. -- (Chapman & Hall/CRC applied algorithms and data structures series)
    Includes bibliographical references and index.
    ISBN 978-1-58488-820-8 (alk. paper)
    1. Computer algorithms. 2. Computer science. 3. Computational complexity. I. Atallah, Mikhail J. II. Blanton, Marina. III. Title. IV. Series.

QA76.9.A43A433 2009
005.1--dc22
                                              2009017978

Visit the Taylor & Francis Web site at
http://www.taylorandfrancis.com

and the CRC Press Web site at
http://www.crcpress.com

Contents

Preface

This handbook aims to provide a comprehensive coverage of algorithms and theoretical computer science for computer scientists, engineers, and other professionals in related scientific and engineering disciplines. Its focus is to provide a compendium of fundamental topics and techniques for professionals, including practicing engineers, students, and researchers. The handbook is organized along the main subject areas of the discipline, and also contains chapters from application areas that illustrate how the fundamental concepts and techniques come together to provide efficient solutions to important practical problems.

Thecontents of each chapter were chosen in such a manner as to help the computer professional and the engineer in finding significant information on a topic of his or her interest. While the reader may not find all the specialized topics in a given chapter, nor will the coverage of each topic be exhaustive, the reader should be able to find sufficient information for initial inquiries and a number of references to the current in-depth literature. In addition to defining terminology and presenting the basic results and techniques for their respective topics, the chapters also provide a glimpse of the major research issues concerning the relevant topics.

Compared to the first edition, this edition contains 21 new chapters, and therefore provides a significantly broader coverage of the field and its application areas. This, together with the updating and revision of many of the chapters from the first edition, has made it necessary to move into a two-volume format.

It is a pleasure to extend our thanks to the people and organizations who made this handbook possible. First and foremost the chapter authors, whose dedication and expertise are at the core of this handbook; the universities and research laboratories with which the authors are affiliated for providing the computing and communication facilities and the intellectual environment for this project; Randi Cohen and her colleagues at Taylor & Francis for perfect organization and logistics that spared us the tedious aspects of such a project and enabled us to focus on its scholarly side; and, last but not least, our spouses and families, who provided moral support and encouragement.

Mikhail Atallah
Marina Blanton

Editors

Mikhail Atallah obtained his PhD from The Johns Hopkins University in 1982 and immediately thereafter joined the computer science department at Purdue University, Klest Lafayette Indiana, where he currently holds the rank of distinguished professor of computer science. His research interests include information security, distributed computing, algorithms, and computational geometry. A fellow of both the ACM and the IEEE, Dr. Atallah has served on the editorial boards of top journals and on the program committees of top conferences and workshops. He was a keynote and invited speaker at many national and international meetings, and a speaker in the Distinguished Colloquium Series of top computer science departments on nine occasions. In 1999, he was selected as one of the best teachers in the history of Purdue and was included in a permanent wall display of Purdue's best teachers, past and present.

Marina Blanton is an assistant professor in the computer science and engineering department at the University of Notre Dame, Notre Dame, Indiana. She holds a PhD from Purdue University. Her research interests focus on information security, privacy, and applied cryptography, and, in particular, span across areas such as privacy-preserving computation, authentication, anonymity, and key management. Dr. Blanton has numerous publications at top venues and is actively involved in program committee work.

Contributors

Srinivas Aluru
Department of Electrical and
  Computer Engineering
Iowa State University
Ames, Iowa

and

Department of Computer Science
  and Engineering
Indian Institute of Technology
Bombay, India

Guy E. Blelloch
School of Computer Science
Carnegie Mellon University
Pittsburgh, Pennsylvania

Chris Charnes
Institute of Applied Physics
  and CASED Technical
  University Darmstadt
Darmstadt, Germany

Danny Z. Chen
Department of Computer Science
  and Engineering
University of Notre Dame
Notre Dame, Indiana

Valentina Ciriani
Dipartimento di Technologie
  dell'Informazione
Università degli Studi di Milano
Crema, Italy

Vincent Conitzer
Department of Computer Science
  and Department of Economics
Duke University
Durham, North Carolina

Sabrina De Capitani di Vimercati
Dipartimento di Technologie
  dell'Informazione
Università degli Studi di Milano
Crema, Italy

Yvo Desmedt
Department of Computer Science
University College London
London, United Kingdom

Wenliang Du
Department of Electrical Engineering
  and Computer Science
Syracuse University
Syracuse, New York

Peter Eades
School of Information Technologies
University of Sydney
Sydney, New South Wales, Australia

Sara Foresti
Dipartimento di Technologie
  dell'Informazione
Università degli Studi di Milano
Crema, Italy

Keith B. Frikken
Department of Computer Science
  and Software Engineering
Miami University
Oxford, Ohio

Concettina Guerra
College of Computing
Georgia Institute of Technology
Atlanta, Georgia

Carsten Gutwenger
Department of Computer Science
Dortmund University of Technology
Dortmund, Germany

Dan Halperin
School of Computer Science
Tel-Aviv University
Tel-Aviv, Israel

Seok-Hee Hong
School of Information Technologies
University of Sydney
Sydney, New South Wales, Australia

Jing Jia
Department of Electrical Engineering
  and Computer Science
Syracuse University
Syracuse, New York

David Karger
Computer Science and Artificial
  Intelligence Laboratory
Massachusetts Institute of Technology
Cambridge, Massachusetts

Lila Kari
Department of Computer Science
University of Western Ontario
London, Ontario, Canada

Lydia Kavraki
Department of Computer Science
Rice University
Houston, Texas

Maleq Khan
Department of Computer Science
Purdue University
West Lafayette, Indiana

Andrew Klapper
Department of Computer Science
University of Kentucky
Lexington, Kentucky

Richard E. Korf
Computer Science Department
University of California
Los Angeles, California

Andrea S. LaPaugh
Department of Computer Science
Princeton University
Princeton, New Jersey

Jean-Claude Latombe
Computer Science Department
Stanford University
Stanford, California

D.T. Lee
Institute of Information Science and Research
  Center for Information Technology
  Innovation
Academia Sinica
Taipei, Taiwan

Bruce M. Maggs
Department of Computer Science
Duke University
Durham, North Carolina

and

Akamai Technologies
Cambridge, Massachusetts

Kalpana Mahalingam
Department of Mathematics
Indian Institute of Technology
Guindy, Chennai, India

Russ Miller
Department of Computer Science
  and Engineering
State University of New York
Buffalo, New York

T.M. Murali
Department of Computer Science
Virginia Polytechnic Institute and
  State University
Blacksburg, Virginia

Mummoorthy Murugesan
Department of Electrical Engineering
  and Computer Science
Syracuse University
Syracuse, New York

Petra Mutzel
Department of Computer Science
Dortmund University of Technology
Dortmund, Germany

Mark-Jan Nederhof
School of Computer Science
University of St Andrews
St. Andrews, Scotland, United Kingdom

Panagiota N. Panagopoulou
Research Academic Computer Technology
  Institute
University Campus, Rion
Patras, Greece

and

Department of Computer Engineering
  and Informatics
Patras University
Patras, Greece

Gopal Pandurangan
Department of Computer Science
Purdue University
West Lafayette, Indiana

Josef Pieprzyk
Department of Computer Science
Macquarie University
Sydney, New South Wales, Australia

Yves Robert
Laboratoire del'Informatique
  du Parallélisme
Ecole Normale Supérieure de Lyon
Lyon, France

Rei Safavi-Naini
Department of Computer Science
University of Calgary
Calgary, Alberta, Canada

Pierangela Samarati
Dipartimento di Technologie
  dell'Informazione
Università degli Studi di Milano
Crema, Italy

Giorgio Satta
Department of Information Engineering
University of Padua
Padua, Italy

Berry Schoenmakers
Department of Mathematics and
  Computer Science
Technical University of Eindhoven
Eindhoven, the Netherlands

Nicole Schweikardt
Institut für Informatik
Goethe-Universität Frankfurt am Main
Frankfurt, Germany

Thomas Schwentick
Fakultät für Informatik
Technische Universität Dortmund
Dortmund, Germany

Jennifer Seberry
Centre for Computer Security Research
University of Wollongong
Wollongong, New South Wales, Australia

Luc Segoufin
LSV
INRIA and ENS Cachan
Cachan, France

Paul G. Spirakis
Research Academic Computer Technology
  Institute
University Campus, Rion
Patras, Greece

and

Department of Computer Engineering
  and Informatics
Patras University
Patras, Greece

Cliff Stein
Department of Computer Science
Dartmouth College
Hanover, New Hampshire

Quentin F. Stout
Department of Electrical Engineering
  and Computer Science
University of Michigan
Ann Arbor, Michigan

Ruppa K. Thulasiram
Department of Computer Science
University of Manitoba
Winnipeg, Manitoba, Canada

Parimala Thulasiraman
Department of Computer Science
University of Manitoba
Winnipeg, Manitoba, Canada

Sébastien Tixeuil
Laboratory LIP6–CNRS UMR 7606
Université Pierre et Marie Curie
Paris, France

Konstantinos Tsianos
Department of Computer Science
Rice University
Houston, Texas

George Varghese
Department of Computer Science
University of California
San Diego, California

Frédéric Vivien
Laboratoire del'Informatique du
  Parallélisme (INRIA)
Ecole Normale Supérieure de Lyon
Lyon, France

Samuel S. Wagstaff, Jr.
Department of Computer Science
Purdue University
West Lafayette, Indiana

Chao Wang
Department of Computer Science
  and Engineering
University of Notre Dame
Notre Dame, Indiana

Joel Wein
Department of Computer and
  Information Science
Polytechnic Institute of New York University
Brooklyn, New York

Afra Zomorodian
Department of Computer Science
Dartmouth College
Hanover, New Hampshire

1

Computational Geometry I

D.T. Lee
Academia Sinica

1.1 Introduction

Computational geometry, since its inception [66] in 1975,has received a great deal of attention from researchers in the area of design and analysis of algorithms. It has evolved into a discipline of its own. It is concerned with the computational complexity of geometric problems that arise in various disciplines such as pattern recognition, computer graphics, geographical information system, computer vision, **CAD/CAM**, robotics, VLSI layout, operations research, and statistics. In contrast with the classical approach to proving mathematical theorems about geometry-related problems, this discipline emphasizes the computational aspect of these problems and attempts to exploit the underlying geometric properties possible, e.g., the metric space, to derive efficient algorithmic solutions.

An objective of this discipline in the theoretical context is to study the computational complexity (giving lower bounds) of geometric problems, and to devise efficient algorithms (giving upper bounds) whose complexity preferably matches the lower bounds. That is, not only are we interested in the intrinsic difficulty of geometric computational problems under a certain computation model, but we are also concerned with the algorithmic solutions that are efficient or provably optimal in the worst or average case. In this regard, the **asymptotic time** (or **space**) **complexity** of an algorithm, i.e., the behavior of an algorithm, as the input size approaches infinity, is of interest. Due to its applications to various science and engineering related disciplines, researchers in this field have begun to address the *efficacy* of the algorithms, the issues concerning *robustness* and *numerical stability* [33,82], and the actual running times of their implementations. In order to value and get better understanding of the geometric algorithms in action, a computational problem solving environment has been developed at the Institute of Information Science and the Research Center for Information Technology Innovation, Academia Sinica, Taiwan. Actual implementations of several geometric algorithms have been incorporated into a Java-based algorithm visualization and debugging software system, dubbed *GeoBuilder* (http://webcollab.iis.sinica.edu.tw/Components/GeoBuilder/), which supports remote compilation, visualization of intermediate execution results, and other run-time features, e.g., visual debugging, etc. This system facilitates geometric algorithmic researchers in testing their ideas and demonstrating their findings in computational geometry. GeoBuilder system is embedded into a knowledge portal [51], called OpenCPS (Open Computational Problem Solving), (http://www.opencps.org/) and possesses three important features. First, it is a platform-independent software system based on Java's promise of portability, and can be invoked by Sun's Java Web Start technology in any browser-enabled environment. Second, it has the collaboration capability for multiple users to concurrently develop programs, manipulate geometric objects, and control the camera. Finally, its three-dimensional (3D) geometric drawing bean provides an optional function that can automatically position the camera to track 3D objects during algorithm visualization [79]. GeoBuilder develops its rich client platform based on Eclipse RCP and has already built in certain functionalities such as remote addition, deletion, and saving of files as well as remote compiling, and execution of LEDA C/C++ programs, etc., based on a multipage editor. Other notable geometric software projects include, among others, CGAL (http://www.cgal.org/) [32] and LEDA (http://www.algorithmic-solutions.com/leda/about/index.htm.) [60].

In this and the following chapter (Chapter 2) we concentrate mostly on the theoretical development of this field in the context of sequential computation, and discuss a number of typical topics and the algorithmic approaches. We will adopt the *real* RAM (random access machine) model of computation in which all arithmetic operations, comparisons, kth root, exponential, or logarithmic functions take unit time.

1.2   Convex Hull

The convex hull of a set of points in $\Re^k$ is the most fundamental problem in computational geometry. Given is a set of points in $\Re^k$, and we are interested in computing its convex hull, which is defined to be the smallest convex set containing these points. There are two ways to represent a convex hull. An implicit representation is to list all the **extreme points**, whereas an explicit representation is to list all the extreme d-faces of dimensions $d = 0, 1, \ldots, k - 1$. Thus, the complexity of any convex hull algorithm would have two parts, computation part and the output part. An algorithm is said to be **output-sensitive** if its complexity depends on the size of the output.

1.2.1   Convex Hulls in Two and Three Dimensions

For an arbitrary set of n points in two and three dimensions, we can compute its convex hull using the *Graham scan*, *gift-wrapping* method, or *divide-and-conquer* paradigm, which are briefly described below.

Note that the convex hull of an arbitrary set S of points in two dimensions is a convex polygon. We'll describe algorithms that compute the *upper hull* of S, since the convex hull is just the union of the upper and lower hulls. Let v_0 denote the point with minimum x-coordinate; if there are more than one, pick the one with the maximum y-coordinate. Let v_{n-1} be similarly defined except that it denotes the point with the maximum x-coordinate. In two dimensions, the upper hull consists of two vertical lines passing through v_0 and v_{n-1}, respectively and a sequence of edges, known as a *polygonal chain*, $\mathcal{C} = \{\overline{v_{j_{i-1}}, v_{j_i}} \mid i = 1, 2, \ldots, k\}$, where $v_{j_0} = v_0$ and $v_{j_k} = v_{n-1}$, such that the entire set S of points lies on one side of or *below* the lines $\mathcal{L}_i$ containing each edge $\overline{v_{j_{i-1}}, v_{j_i}}$. See Figure 1.1a for an illustration of the upper hull. The lower hull is similarly defined.

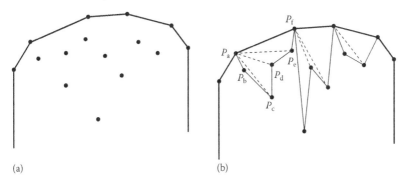

FIGURE 1.1 The upper hull of a set of points (a) and illustration of the Graham scan (b).

The *Graham scan* computes the convex hull by (1) sorting the input set of points in ascending order of their x-coordinates (in case of ties, in ascending order of their y-coordinates), (2) connecting these points into a polygonal chain P stored as a doubly linked list L, and (3) performing a linear scan to compute the upper hull of the polygon [66].

The triple (v_i, v_j, v_k) of points is said to form a *right* turn if and only if the determinant

$$\begin{vmatrix} x_i & y_i & 1 \\ x_j & y_j & 1 \\ x_k & y_k & 1 \end{vmatrix} < 0,$$

where (x_i, y_i) are the x- and y-coordinates of v_i. If the determinant is positive, then the triple (v_i, v_j, v_k) of points is said to form a *left* turn. The points v_i, v_j, and v_k are collinear if the determinant is zero. This is also known as the **side test**, determining on which side of the line defined by points v_i and v_j the point v_k lies.

It is obvious that when we scan points in L in ascending order of x-coordinate, the middle point of a triple (v_i, v_j, v_k) that does not form a right turn is not on the upper hull and can be deleted. The following is the algorithm.

ALGORITHM GRAHAM_SCAN

Input: A set S of points sorted in lexicographically ascending order of their (x, y)-coordinate values.

Output: A sorted list L of points in ascending x-coordinates.

begin
    if $(|S| == 2)$ **return** $(\overline{v_0, v_{n-1}})$;
    $i = 0$; $v_{n-1} = next(v_{n-1})$; /* set sentinel */
    $p_a = v_0$; $p_b = next(p_a)$, $p_c = next(p_b)$;
    while $(p_b \neq v_{n-1})$ **do**
        if (p_a, p_b, p_c) forms a right turn
        then begin /* *advance* */
                $p_a = p_b$; $p_b = p_c$;
                $p_c = next(p_b)$;
        end
        else begin /* *backtrack* */
                delete p_b;

$$\textbf{if } (p_a \neq v_0)$$
$$\textbf{then } p_a = prev(p_a);$$
$$p_b = next(p_a); p_c = next(p_b);$$
$$\textbf{end}$$
$$p_t = next(v_0);$$
$$L = \{\overline{v_0, p_t}\};$$
$$\textbf{while } (p_t \neq v_{n-1}) \textbf{ do}$$
$$\textbf{begin}$$
$$p_u = next(p_t);$$
$$L = L \cup \{\overline{p_t, p_u}\};$$
$$p_t = p_u;$$
$$\textbf{end;}$$
$$\textbf{return } (L);$$
$$\textbf{end.}$$

Step (i) being the dominating step, ALGORITHM GRAHAM_SCAN, takes $O(n \log n)$ time. Figure 1.1b shows the initial list L and vertices not on the upper hull are removed from L. For example, p_b is removed since (p_a, p_b, p_c) forms a left turn; p_c is removed since (p_a, p_c, p_d) forms a left turn; p_d, and p_e are removed for the same reason.

One can also use the *gift-wrapping* technique to compute the upper hull. Starting with a vertex that is known to be on the upper hull, say the point $v_0 = v_{i_0}$. We sweep clockwise the half-line emanating from v_0 in the direction of the positive y-axis. The first point v_{i_1} this half-line hits will be the next point on the upper hull. We then march to v_{i_1}, repeat the same process by sweeping clockwise the half-line emanating from v_{i_1} in the direction from v_{i_0} to v_{i_1}, and find the next vertex v_{i_2}. This process terminates when we reach v_{n-1}. This is similar to wrapping an object with a *rope*. Finding the next vertex takes time proportional to the number of points not yet known to be on the upper hull. Thus, the total time spent is $O(n\mathcal{H})$, where $\mathcal{H}$ denotes the number of points on the upper hull. The gift-wrapping algorithm is output-sensitive, and is more efficient than the ALGORITHM GRAHAM_SCAN if the number of points on the upper hull is small, i.e., $O(\log n)$.

One can also compute the upper hull recursively by divide-and-conquer. This method is more amenable to parallelization. The divide-and-conquer paradigm consists of the following steps.

ALGORITHM UPPER_HULL_D&C (*2d-Point S*)

Input: A set S of points.

Output: A sorted list L of points in ascending x-coordinates.

1. If $|S| \leq 3$, compute the upper hull UH(S) explicitly and **return** (UH(S)).
2. Divide S by a vertical line $\mathcal{L}$ into two approximately equal subsets S_l and S_r such that S_l and S_r lie, respectively, to the left and to the right of $\mathcal{L}$.
3. UH(S_l) = Upper_Hull_D&C(S_l).
4. UH(S_r) = Upper_Hull_D&C(S_r).
5. UH(S) = **Merge**(UH(S_l), UH(S_r)).
6. **return** (UH(S)).

The key step is the **Merge** of two upper hulls, each of which is the solution to a subproblem derived from the recursive step. These two upper hulls are separated by a vertical line $\mathcal{L}$. The **Merge** step basically calls for computation of a common tangent, called *bridge* over line $\mathcal{L}$, of these two upper hulls (Figure 1.2).

The computation of the bridge begins with a segment connecting the rightmost point l of the left upper hull to the leftmost point r of the right upper hull, resulting in a sorted list L. Using the Graham

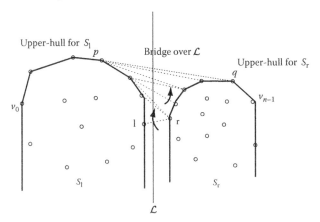

FIGURE 1.2 The bridge $\overline{p,q}$ over the vertical line $\mathcal{L}$.

scan one can obtain in linear time the two endpoints of the bridge, ($\overline{p,q}$ shown in Figure 1.2), such that the entire set of points lies on one side of the line, called *supporting line*, containing the bridge. The running time of the divide-and-conquer algorithm is easily shown to be $O(n \log n)$ since the **Merge** step can be done in $O(n)$ time.

A more sophisticated output-sensitive and *optimal* algorithm which runs in $O(n \log \mathcal{H})$ time has been developed by Kirkpatrick and Seidel [48]. It is based on a variation of the divide-and-conquer paradigm, called **divide-and-marriage-before-conquest** method. It has been shown to be asymptotically optimal; a lower bound proof of $\Omega(n \log \mathcal{H})$ can be found in [48]. The main idea in achieving the optimal result is that of eliminating redundant computations. Observe that in the divide-and-conquer approach after the bridge is obtained, some vertices belonging to the left and right upper hulls that are below the bridge are deleted. Had we known that these vertices are not on the final hull, we could have saved time without computing them. Kirkpatrick and Seidel capitalized on this concept and introduced the marriage-before-conquest principle putting **Merge** step before the two recursive calls.

The divide-and-conquer scheme can be easily generalized to three dimensions. The Merge step in this case calls for computing common supporting faces that wrap two recursively computed convex polyhedra. It is observed by Preparata and Shamos [66] that the common supporting faces are computed from connecting two cyclic sequences of edges, one on each polyhedron (Figure 1.3). See [3] for a characterization of the two cycles of seam edges. The computation of these supporting

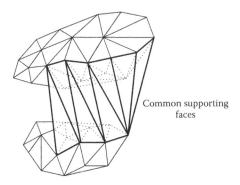

FIGURE 1.3 Common supporting faces of two disjoint convex polyhedra.

faces can be accomplished in linear time, giving rise to an $O(n \log n)$ time algorithm. By applying the marriage-before-conquest principle Edelsbrunner and Shi [28] obtained an $O(n \log^2 \mathcal{H})$ algorithm.

The gift-wrapping approach for computing the convex hull in three dimensions would mimic the process of wrapping a gift with a piece of paper. One starts with a plane supporting S, i.e., a plane determined by three points of S such that the entire set of points lie on one side. In general, the supporting face is a triangle $\Delta(a, b, c)$. Pivoting at an edge, say (a, b) of this triangle, one rotates the plane in space until it hits a third point v, thereby determining another supporting face $\Delta(a, b, v)$. This process repeats until the entire set of points are wrapped by a collection of supporting faces. These supporting faces are called 2-faces, the edges common to two supporting faces, 1-faces, and the vertices (or extreme points) common to 2-faces and 1-faces are called 0-faces. The gift-wrapping method has a running time of $O(n\mathcal{H})$, where $\mathcal{H}$ is the total number of i-faces, $i = 0, 1, 2$.

The following optimal output-sensitive algorithm that runs in $O(n \log \mathcal{H})$ time in two and three dimensions is due to Chan [13]. It is a modification of the *gift-wrapping* method (also known as the Jarvis' March method) and uses a *grouping* technique.

ALGORITHM 2DHULL (S)

1. For $i = 1, 2, \ldots$ do
2. $P \leftarrow$ HULL2D ($S, \mathcal{H}_0, \mathcal{H}_0$), where $\mathcal{H}_0 = \min\{2^{2^i}, n\}$
3. If $P \neq nil$ then return P.

FUNCTION HULL2D ($S, m, \mathcal{H}_0$)

1. Partition S into subsets $S_1, S_2, \ldots, S_{\lceil \frac{n}{m} \rceil}$, each of size at most m
2. For $i = 1, 2, \ldots, \lceil \frac{n}{m} \rceil$ do
3. Compute $CH(S_i)$ and preprocess it in a suitable data structure
4. $p_0 \leftarrow (0, -\infty), p_1 \leftarrow$ the rightmost point of S
5. For $j = 1, 2, \ldots, \mathcal{H}_0$ do
6. For $i = 1, 2, \ldots, \lceil \frac{n}{m} \rceil$ do
7. Compute a point $q_i \in S_i$ that maximizes $\cap p_{j-1}p_jq_i$
8. $p_{j+1} \leftarrow$ a point q from $\{q_1, \ldots, q_{\lceil \frac{n}{m} \rceil}\}$ maximizing $\cap p_{j-1}p_jq$
9. If $p_{j+1} = p_1$ then return list $(p_1, \ldots, p_j)$
10. return nil

Let us analyze the complexity of the algorithm. In Step 2, we use an $O(m \log m)$ time algorithm for computing the convex hull for each subset of m points, e.g., Graham's scan for S in two dimensions, and Preparata–Hong algorithm for S in three dimensions. Thus, it takes $O((\frac{n}{m})m \log m) = O(n \log m)$ time. In Step 5 we build a suitable data structure that supports the computation of the supporting vertex or supporting face in logarithmic time. In two dimensions we can use an array that stores the vertices on the convex hull in say, clockwise order. In three dimensions we use Dobkin–Kirkpatrick hierarchical representation of the faces of the convex hull [24]. Thus, Step 5 takes $\mathcal{H}_0(\frac{n}{m})O(\log m)$ time. Setting $m = \mathcal{H}_0$ gives an $O(n \log \mathcal{H}_0)$ time. Note that setting $m = 1$ we have the Jarvis' March, and setting $m = n$ the two-dimensional (2D) convex hull algorithm degenerates to the Graham's scan. Since we do not know $\mathcal{H}$ in advance, we use in Step 2 of ALGORITHM 2DHULL(S) a sequence $\mathcal{H}_i = 2^{2^i}$ such that $\mathcal{H}_1 + \cdots + \mathcal{H}_{k-1} < \mathcal{H} \leq \mathcal{H}_1 + \cdots + \mathcal{H}_k$ to guess it. The total running time is

$$O\left(\sum_{i=1}^{k} n \log \mathcal{H}_i\right) = O\left(\sum_{i=1}^{\lceil \log \log \mathcal{H} \rceil} n 2^i\right) = O\left(n \log \mathcal{H}\right)$$

1.2.2 Convex Hulls in k Dimensions, $k > 3$

For convex hulls of higher dimensions, Chazelle [16] showed that the convex hull can be computed in time $O(n \log n + n^{\lfloor k/2 \rfloor})$, which is optimal in all dimensions $k \geq 2$ in the worst case. But this result is insensitive to the output size. The gift-wrapping approach generalizes to higher dimensions and yields an output-sensitive solution with running time $O(n\mathcal{H})$, where $\mathcal{H}$ is the total number of i-faces, $i = 0, 1, \ldots, k - 1$ and $\mathcal{H} = O(n^{\lfloor k/2 \rfloor})$ [27]. One can also use *beneath–beyond* method [66] of adding points one at a time in ascending order along one of the coordinate axis.* We compute the convex hull $CH(S_{i-1})$ for points $S_{i-1} = \{p_1, p_2, \ldots, p_{i-1}\}$. For each added point p_i we update $CH(S_{i-1})$ to get $CH(S_i)$ for $i = 2, 3, \ldots, n$ by deleting those t-faces, $t = 0, 1, \ldots, k - 1$, that are internal to $CH(S_{i-1} \cup \{p_i\})$. It has been shown by Seidel [27] that $O(n^2 + \mathcal{H} \log h)$ time is sufficient, where h is the number of extreme points. Later Chan [13] obtained an algorithm based on gift-wrapping method using the data structures for ray-shooting queries in polytopes developed by Agarwal and Matoušek [1] and refined by Matoušek and Schwarzkopf [58], that runs in $O(n \log \mathcal{H} + (n\mathcal{H})^{1-1/(\lfloor k/2 \rfloor + 1)} \log^{O(1)} n)$ time. Note that the algorithm is optimal when $k = 2, 3$. In particular, it is optimal when $\mathcal{H} = O(n^{1/(\lfloor k/2 \rfloor)} / \log^\delta n)$ for a sufficiently large δ.

We conclude this section with the following theorem [13].

THEOREM 1.1 *The convex hull of a set S of n points in $\mathfrak{R}^k$ can be computed in $O(n \log \mathcal{H})$ time for $k = 2$ or $k = 3$, and in $O(n \log \mathcal{H} + (n\mathcal{H})^{1-1/(\lfloor k/2 \rfloor + 1)} \log^{O(1)} n)$ time for $k > 3$, where $\mathcal{H}$ is the number of i-faces, $i = 0, 1, \ldots, k - 1$.*

1.2.3 Convex Layers of a Planar Set

The convex layers $\mathcal{C}(S)$ of a set S of n points in the Euclidean plane is obtained by a process, known as *onion peeling*, i.e., compute the convex hull of S and remove its vertices from S, until S becomes empty. Figure 1.4 shows the convex layer of a point set. This onion peeling process of a point set is central in the study of robust estimators in statistics, in which the outliers, points lying on the outermost convex layers, should be removed. In this section we describe an efficient algorithm due to Chazelle [14] that runs in optimal $O(n \log n)$ time.

As described in Section 1.2.1, each convex layer of $\mathcal{C}(S)$ can be decomposed into two convex polygonal chains, called upper and lower hulls (Figure 1.5).

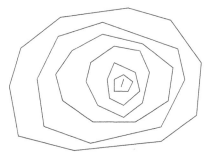

FIGURE 1.4 Convex layers of a point set.

---

* If the points of S are not given *a priori*, the algorithm can be made **on-line** by adding an extra step of checking if the newly added point is internal or external to the current convex hull. If it is internal, just discard it.

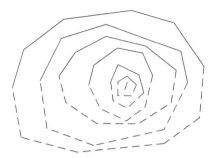

FIGURE 1.5 Decomposition of each convex layer into upper and lower hulls.

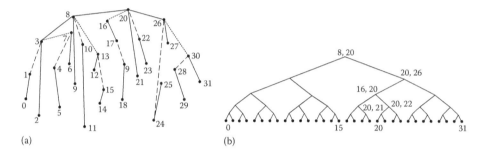

FIGURE 1.6 The hull graph of upper hull (a) and a complete binary tree representation (b).

Let l and r denote the points with the minimum and maximum x-coordinate, respectively, in a convex layer. The upper (respectively, lower) hull of this layer runs clockwise (respectively, counterclockwise) from l to r. The upper and lower hulls are the same if the convex layer has one or two points. Assume that the set S of points $p_0, p_1, \ldots, p_{n-1}$ are ordered in nondescending order of their x-coordinates. We shall concentrate on the computation of upper hulls of $\mathcal{C}(S)$; the other case is symmetric. Consider the complete binary tree $T(S)$ with leaves $p_0, p_1, \ldots, p_{n-1}$ from left to right. Let $S(v)$ denote the set of points stored at the leaves of the subtree rooted at node v of T and let $U(v)$ denote its upper hull of the convex hull of $S(v)$. Thus, $U(\rho)$, where ρ denotes the root of T, is the upper hull of the convex hull of S in the outermost layer. The union of all the upper hulls $U(v)$ for all nodes v is a tree, called *hull graph* [14]. (A similar graph is also computed for the lower hull of the convex hull.) To minimize the amount of space, at each internal node v we store the bridge (common tangent) connecting a point in $U(v_l)$ and a point in $U(v_r)$, where v_l and v_r are the left and right children of node v, respectively. Figure 1.6a and b illustrates the binary tree T and the corresponding hull graph, respectively.

Computation of the hull graph proceeds from bottom up. Computing the bridge at each node takes time linear in the number of vertices on the respective upper hulls in the left and right subtrees. Thus, the total time needed to compute the hull graph is $O(n \log n)$. The bridges computed at each node v which are incident upon a vertex p_k are naturally separated into two subsets divided by the vertical line $\mathcal{L}(p_k)$ passing through p_k. Those on the left are arranged in a list $L(p_k)$ in counterclockwise order from the positive y direction of $\mathcal{L}(p_k)$, and those on the right are arranged in a list $R(p_k)$ in clockwise order. This adjacency list at each vertex in the hull graph can be maintained fairly easily. Suppose the bridge at node v connects vertex p_j in the left subtree and vertex p_k in the right subtree. The edge $\overline{p_j, p_k}$ will be inserted at the *first* position in the current lists $R(p_j)$ and $L(p_k)$. That is, edge

$\overline{p_j, p_k}$ is the *top* edge in both lists $R(p_j)$ and $L(p_k)$. It is easy to retrieve the vertices on the upper hull of the outermost layer from the hull graph beginning at the root node of $\mathcal{T}$.

To compute the upper hull of the next convex layer, one needs to remove those vertices on the first layer (including those vertices in the lower hull). Thus, update of the hull graph includes deletion of vertices on both upper hull and lower hull. Deletions of vertices on the upper hull can be performed in an arbitrary order. But if deletions of vertices on the lower hull from the hull graph are done in say clockwise order, then the update of the adjacency list of each vertex p_k can be made easy, e.g., $R(p_k) = \emptyset$. The deletion of a vertex p_k on the upper hull entails removal of edges incident on p_k in the hull graph. Let $v_1, v_2, \ldots, v_l$ be the list of internal nodes on the leaf-to-root path from p_k. The edges in $L(p_k)$ and $R(p_k)$ are deleted from bottom up in $O(1)$ time each, i.e., the top edge in each list gets deleted last. Figure 1.6b shows the leaf-to-root path when vertex p_{20} is deleted. Figure 1.7a–f shows the updates of bridges when p_{20} is deleted and Figure 1.7g is the final upper hull after the update is finished. It can be shown that the overall time for deletions can be done in $O(n \log n)$ time [14].

THEOREM 1.2 *The convex layers of a set of n points in the plane can be computed in $O(n \log n)$ time.*

Nielsen [64] considered the problem of computing the first k layers of convex hull of a planar point set S that arises in statistics and pattern recognition, and gave an $O(n \log \mathcal{H}_k)$ time algorithm, where $\mathcal{H}_k$ denotes the number of points on the first k layers of convex hull of S, using the grouping scheme of Chan [13].

1.2.4 Applications of Convex Hulls

Convex hulls have applications in clustering, linear regression, and Voronoi diagrams (see Chapter 2). The following problems have solutions derived from the convex hull.

Problem C1 (Set Diameter) Given a set S of n points, find the two points that are the farthest apart, i.e., find $p_i, p_j \in S$ such that $d(p_i, p_j) = \max\{d(p_k, p_l)\} \ \forall p_k, p_l \in S$, where $d(p, q)$ denotes the Euclidean distance between p and q.

In two dimensions $O(n \log n)$ time is both sufficient and necessary in the worst case [66]. It is easy to see that the farthest pair must be extreme points of the convex hull of S. Once the convex hull is computed, the farthest pair in two dimensions can be found in linear time by observing that it admits a pair of parallel supporting lines. Various attempts, including geometric sampling and parametric search method, have been made to solve this problem in three dimensions. See e.g., [59].

Clarkson and Shor [20] gave a randomized algorithm with an optimal expected $O(n \log n)$ time. Later Ramos [67] gave an optimal deterministic algorithm, based on a simplification of the randomization scheme of Clarkson and Shor [20], and derandomization making use of the efficient construction of ϵ-nets by Matoušek [57].

Problem C2 (Smallest enclosing rectangle) Given a set S of n points, find the smallest rectangle that encloses the set.

Problem C3 (Regression line) Given a set S of n points, find a line such that the maximum distance from S to the line is minimized.

These two problems can be solved in optimal time $O(n \log n)$ using the convex hull of S [54] in two dimensions. In k dimensions Houle et al. [43] gave an $O(n^{\lfloor k/2+1 \rfloor})$ time and $O(n^{\lfloor (k+1)/2 \rfloor})$ space algorithm. The time complexity is essentially that of computing the convex hull of the point set.

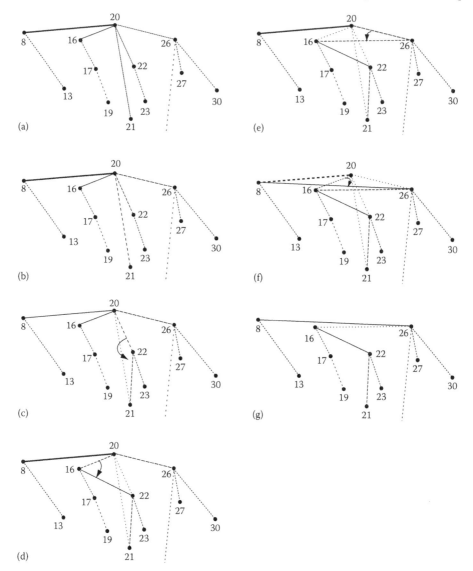

FIGURE 1.7 Update of hull graph.

1.3 Maxima Finding

In this section we discuss a problem concerned with the *extremes* of a point set which is somewhat related to that of convex hull problems. Consider a set S of n points in $\mathfrak{R}^k$ in the Cartesian coordinate system. Let $(x_1(p), x_2(p), \ldots, x_k(p))$ denote the coordinates of point $p \in \mathfrak{R}^k$. Point p is said to *dominate* point q, denoted $p \succeq q$, (or q is dominated by p, denoted $q \preceq p$) if $x_i(p) \geq x_i(q)$ for all $1 \leq i \leq k$. A point p is said to be *maximal* (or a *maximum*) in S if no point in S dominates p. The maxima-finding problem is that of finding the set $\mathcal{M}(S)$ of maximal elements for a set S of points in $\mathfrak{R}^k$.

1.3.1 Maxima in Two and Three Dimensions

In two dimensions the problem can be done fairly easily by a plane-sweep technique. (For a more detailed description of plane-sweep technique, see, e.g., [50] or Section 1.5.1) Assume that the set S of points $p_1, p_2, \ldots, p_n$ are ordered in nondescending order of their x-coordinates, i.e., $x(p_1) \leq x(p_2) \leq \cdots \leq x(p_n)$.

We shall scan the points from right to left. The point p_n is necessarily a maximal element. As we scan the points, we maintain the maximum y-coordinate among those that have been scanned so far. Initially, $\max_y = y(p_n)$. The next point p_i is a maximal element if and only if $y(p_i) > \max_y$. If $y(p_i) > \max_y$, then $p_i \in \mathcal{M}(S)$, and $\max_y$ is set to $y(p_i)$, and we continue. Otherwise $p_i \preceq p_j$ for some $j > i$. Thus, after the initial sorting, the set of maxima can be computed in linear time. Note that the set of maximal elements satisfies the property that their x- and y-coordinates are totally ordered: If they are ordered in strictly ascending x-coordinate, their y-coordinates are ordered in strictly descending order.

In three dimensions we can use the same strategy. We will scan the set in descending order of the x-coordinate by a plane $\mathcal{P}$ orthogonal to the x-axis. Point p_n as before is a maximal element. Suppose we have computed $\mathcal{M}(S_{i+1})$, where $S_{i+1} = \{p_{i+1}, \ldots, p_n\}$, and we are scanning point p_i. Consider the orthogonal projection S_{i+1}^x of the points in S_{i+1} to $\mathcal{P}$ with $x = x(p_i)$. We now have an instance of an *on-line* 2D maximal problem, i.e., for point p_i, if $p_i^x \preceq p_j^x$ for some $p_j^x \in S_{i+1}^x$, then it is not a maximal element, otherwise it is (p_i^x denotes the projection of p_i onto $\mathcal{P}$). If we maintain the points in $\mathcal{M}(S_{i+1}^x)$ as a **height-balanced binary search tree** in either y- or z-coordinate, then testing whether p_i is maximal or not can be done in logarithmic time. If it is dominated by some point in $\mathcal{M}(S_{i+1}^x)$, then it is ignored. Otherwise, it is in $\mathcal{M}(S_{i+1}^x)$ (and also in $\mathcal{M}(S_{i+1})$); $\mathcal{M}(S_{i+1}^x)$ will then be updated to be $\mathcal{M}(S_i^x)$ accordingly. The update may involve deleting points in $\mathcal{M}(S_{i+1}^x)$ that are no longer maximal because they are dominated by p_i^x. Figure 1.8 shows the effect of adding a maximal element p_i^x to the set $\mathcal{M}(S_{i+1}^x)$ of maximal elements. Points in the shaded area will be deleted. Thus, after the initial sorting, the set of maxima in three dimensions can be computed in $O(n \log \mathcal{H})$ time, as the *on-line* 2D maximal problem takes $O(\log \mathcal{H})$ time to maintain $\mathcal{M}(S_i^x)$ for each point p_i, where $\mathcal{H}$ denotes the size of $\mathcal{M}(S)$.

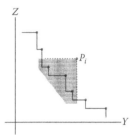

FIGURE 1.8   Update of maximal elements.

Since the total number of points deleted is at most n, we conclude the following.

LEMMA 1.1   Given a set of n points in two and three dimensions, the set of maxima can be computed in $O(n \log n)$ time.

For two and three dimensions one can solve the problem in optimal time $O(n \log \mathcal{H})$, where $\mathcal{H}$ denotes the size of $\mathcal{M}(S)$. The key observation is that we need not *sort* S in its entirety. For instance, in two dimensions one can solve the problem by divide-and-marriage-before-conquest paradigm. We first use a linear time median finding algorithm to divide the set into two halves L and R with points in R having larger x-coordinate values than those of points in L. We then recursively compute $\mathcal{M}(R)$. Before we recursively compute $\mathcal{M}(L)$ we note that points in L that are dominated by points in $\mathcal{M}(R)$ can be eliminated from consideration. We trim L before we invoke the algorithm recursively. That is, we compute $\mathcal{M}(L')$ recursively, where $L' \subseteq L$ consists of points $q \npreceq p$ for all $p \in \mathcal{M}(R)$. A careful analysis of the running time shows that the complexity of this algorithm is $O(n \log \mathcal{H})$. For three dimensions we note that other than the initial sorting step, the subsequent plane-sweep step takes $O(n \log \mathcal{H})$ time. It turns out that one can replace the full-fledged $O(n \log n)$ sorting step with a so-called *lazy sorting* of S using a technique similar to those described in Section 1.2.1 to derive an output-sensitive algorithm.

THEOREM 1.3   *Given a set S of n points in two and three dimensions, the set $\mathcal{M}(S)$ of maxima can be computed in $O(n \log \mathcal{H})$ time, where $\mathcal{H}$ is the size of $\mathcal{M}(S)$.*

1.3.2 Maxima in Higher Dimensions

The set of maximal elements in $\mathfrak{R}^k, k \geq 4$, can be solved by a generalization of plane-sweep method to higher dimensions. We just need to maintain a data structure for $\mathcal{M}(S_{i+1})$, where $S_{i+1} = \{p_{i+1}, \ldots, p_n\}$, and test for each point p_i if it is a maximal element in S_{i+1}, reducing the problem to one dimension lower, assuming that the points in S are sorted and scanned in descending lexicographical order. Thus, in a straightforward manner we can compute $\mathcal{M}(S)$ in $O(n^{k-2} \log n)$ time. However, we shall show below that one can compute the set of maxima in $O(n \log^{k-2} n)$ time, for $k > 3$ by, divide-and-conquer. Gabow et al. [36] gave an algorithm which improved the time by a $O\left(\frac{\log n}{\log \log n}\right)$ factor to $O(n \log \log n \log^{k-3} n)$.

Let us first consider a *bichromatic maxima-finding* problem. Consider a set of n red and a set of m blue points, denoted R and B, respectively. The bichromatic maxima-finding problem is to find a subset of points in R that are not dominated by any points in B and vice versa. That is, find $\mathcal{M}(R, B) = \{r | r \not\preceq b, b \in B\}$ and $\mathcal{M}(B, R) = \{b | b \not\preceq r, r \in R\}$.

In three dimensions, this problem can be solved by plane-sweep method in a manner similar to the maxima-finding problem as follows. As before, the sets R and B are sorted in nondescending order of x-coordinates and we maintain two subsets of points $\mathcal{M}\left(R_{i+1}^x\right)$ and $\mathcal{M}\left(B_{j+1}^x\right)$, which are the maxima of the projections of R_{i+1} and B_{j+1} onto the yz-plane for $R_{i+1} = \{r_{i+1}, \ldots, r_n\} \subseteq R$ and $B_{j+1} = \{b_{j+1}, \ldots, b_m\} \subseteq B$, respectively. When the next point $r_i \in R$ is scanned, we test if r_i^x is dominated by any points in $\mathcal{M}\left(B_{j+1}^x\right)$. The point $r_i \in \mathcal{M}(R, B)$, if r_i^x is not dominated by any points in $\mathcal{M}\left(B_{j+1}^x\right)$. We then update the set of maxima for $R_i^x = R_{i+1}^x \cup \{r_i^x\}$. That is, if $r_i^x \preceq q$ for $q \in \mathcal{M}\left(R_{i+1}^x\right)$, then $\mathcal{M}\left(R_i^x\right) = \mathcal{M}\left(R_{i+1}^x\right)$. Otherwise, the subset of $\mathcal{M}\left(R_{i+1}^x\right)$ dominated by r_i^x is removed, and r_i^x is included in $\mathcal{M}\left(R_i^x\right)$. If the next point scanned is $b_j \in B$, we perform similar operations. Thus, for each point scanned we spend $O(\log n + \log m)$ time.

LEMMA 1.2   The bichromatic maxima-finding problem for a set of n red and m blue points in three dimensions can be solved in $O(N \log N)$ time, where $N = m + n$.

Using Lemma 1.2 as basis, one can solve the bichromatic maxima-finding problem in $\mathfrak{R}^k$ in $O(N \log^{k-2} N)$ time for $k \geq 3$ using multidimensional divide-and-conquer.

LEMMA 1.3   The bichromatic maxima-finding problem for a set of n red and m blue points in $\mathfrak{R}^k$ can be solved in $O(N \log^{k-2} N)$ time, where $N = m + n$, and $k \geq 3$.

Let us now turn to the maxima-finding problem in $\mathfrak{R}^k$. We shall use an ordinary divide-and-conquer method to solve the maxima-finding problem. Assume that the points in $S \subseteq \mathfrak{R}^k$ have been sorted in all dimensions. Let L_x denote the median of all the x-coordinate values. We first divide S into two subsets S_1 and S_2, each of size approximately $|S|/2$ such that the points in S_1 have x-coordinates larger than L_x and those of points in S_2 are less than L_x. We then recursively compute $\mathcal{M}(S_1)$ and $\mathcal{M}(S_2)$. It is clear that $\mathcal{M}(S_1) \subseteq \mathcal{M}(S)$. However, some points in $\mathcal{M}(S_2)$ may be dominated by points in $\mathcal{M}(S_1)$, and hence, are not in $\mathcal{M}(S)$. We then project points in S onto the hyperplane $\mathcal{P}: x = L_x$. The problem now reduces to the bichromatic maxima-finding problem in $\mathfrak{R}^{k-1}$, i.e., finding among $\mathcal{M}(S_2)$ those that are maxima with respect to $\mathcal{M}(S_1)$. By Lemma 1.3 we know that this bichromatic maxima-finding problem can be solved in $O(n \log^{k-3} n)$ time. Since the merge step takes $O(n \log^{k-3} n)$ time, we conclude the following.

THEOREM 1.4 *The maxima-finding problem for a set of n points in $\mathfrak{R}^k$ can be solved in $O(n \log^{k-2} n)$ time, for $k \geq 3$.*

We note here also that if we apply the *trimming* operation of S_2 with $\mathcal{M}(S_1)$, i.e., removing points in S_2 that are dominated by points in $\mathcal{M}(S_1)$, before recursion, one can compute $\mathcal{M}(S)$ more efficiently as stated in the following theorem.

THEOREM 1.5 *The maxima-finding problem for a set S of n points in $\mathfrak{R}^k$, $k \geq 4$, can be solved in $O(n \log^{k-2} \mathcal{H})$ time, where $\mathcal{H}$ is the number of maxima in S.*

1.3.3 Maximal Layers of a Planar Set

The maximal layers of a set of points in the plane can be obtained by a process similar to that of convex layers discussed in Section 1.2.3. A brute-force method would yield an $O(\delta \cdot n \log \mathcal{H})$ time, where δ is the number of layers and $\mathcal{H}$ is the maximum number of maximal elements in any layer. In this section we shall present an algorithm due to Atallah and Kosaraju [7] for computing not only the maximal layers, but also some other functions associated with dominance relation.

Consider a set S of n points. As in the previous section, let $\mathcal{D}_S(p)$ denote the set of points in S dominated by p, i.e., $\mathcal{D}_S(p) = \{q \in S | q \preceq p\}$. Since p is always dominated by itself, we shall assume $\mathcal{D}_S(p)$ does not include p, when $p \in S$. The first subproblem we consider is the *maxdominance problem*, which is defined as follows: for each $p \in S$, find $\mathcal{M}(\mathcal{D}_S(p))$. That is, for each $p \in S$ we are interested in computing the set of maximal elements among those points that are dominated by p. Another related problem is to compute the labels of each point p from the labels of those points in $\mathcal{M}(\mathcal{D}_S(p))$. More specifically, suppose each point is associated with a weight $w(p)$. The label $l_S(p)$ is defined to be $w(p)$ if $\mathcal{D}_S(p) = \emptyset$ and is $w(p) + \max\{l_S(q), q \in \mathcal{M}(\mathcal{D}_S(p))\}$. The max function can be replaced with min or any other associative functional operation. In other words, $l_S(p)$ is equal to the maximum among the labels of all the points dominated by p. Suppose we let $w(p) = 1$ for all $p \in S$. Then those points with labels equal to 1 are points that do not dominate any points. These points can be thought of as *minimal* points in S. That a point p_i has label λ implies there exists a sequence of λ points $p_{j_1}, p_{j_2}, \ldots, p_{j_\lambda} = p_i$, such that $p_{j_1} \preceq p_{j_2} \preceq \cdots \preceq p_{j_\lambda} = p_i$. In general, points with label λ are on the λth minimal layer and the maximum label gives the number of minimal layers. If we modify the definition of domination to be p dominates q if and only if $x(p) \leq x(q)$ and $y(p) \leq y(q)$, then the minimal layers obtained using the method to be described below correspond to the maximal layers.

Let us now discuss the labeling problem defined earlier. We recall a few terms as used in [7].*

Let L and R denote two subsets of points of S separated by a vertical line, such that $x(l) \leq x(r)$ for all $l \in L$ and $r \in R$. leader$_R(p), p \in R$ is the point $\mathcal{H}_p$ in $\mathcal{D}_R(p)$ with the largest y-coordinate. Strip$_L(p, R), p \in R$ is the subset of points of $\mathcal{D}_L(p)$ dominated by p but with y-coordinates greater than leader$_R(p)$, i.e., Strip$_L(p, R) = \{q \in \mathcal{D}_L(p) | y(q) > y(\mathcal{H}_p)\}$ for $p \in R$. Left$_L(p, R), p \in R$, is defined to be the largest $l_S(q)$ over all $q \in$ Strip$_L(p, R)$ if Strip$_L(p, R)$ is nonempty, and $-\infty$ otherwise.

Observe that for each $p \in R \, \mathcal{M}(\mathcal{D}_S(p))$ is the concatenation of $\mathcal{M}(\mathcal{D}_R(p))$ and Strip$_L(p, R)$. Assume that the points in $S = \{p_1, p_2, \ldots, p_n\}$ have been sorted as $x(p_1) < x(p_2) < \cdots < x(p_n)$. We shall present a divide-and-conquer algorithm that can be called with $R = S$ and Left$_\emptyset(p, S) = -\infty$ for all $p \in S$ to compute $l_S(p)$ for all $p \in S$. The correctness of the algorithm hinges on the following lemma.

---

* Some of the notations are slightly modified. In [7] min is used in the *label* function, instead of max. See [7] for details.

LEMMA 1.4   For any point $p \in R$, if $\mathcal{D}_S(p) \neq \emptyset$, then $l_S(p) = w(p) + \max\{Left_L(p, R), \max\{l_S(q), q \in \mathcal{M}(\mathcal{D}_R(p))\}\}$.

ALGORITHM MAXDOM_LABEL(R)

Input: A consecutive sequence of m points of S, i.e., $R = \{p_r, p_{r+1}, \ldots, p_{r+m-1}\}$ and for each $p \in R$, $Left_L(p, R)$, where $L = \{p_1, p_2, \ldots, p_{r-1}\}$. Assume a list Q_R of points of R sorted by increasing y-coordinate.

Output: The labels $l_S(q), q \in R$.

1. If $m = 1$ then we set $l_S(p_r)$ to $w(p_r) + Left_L(p_r, R)$, if $Left_L(p_r, R) \neq -\infty$ and to $w(p_r)$ if $Left_L(p_r, R) = -\infty$, and **return**.

2. Partition R by a vertical line $\mathcal{V}$ into subsets R_1 and R_2 such that $|R_1| = |R_2| = m/2$ and R_1 is to the left of R_2. Extract from Q_R the lists Q_{R_1} and Q_{R_2}.

3. Call MAXDOM_LABEL(R_1). Since $Left_L(p, R_1)$ equals $Left_L(p, R)$, this call will return the labels for all $q \in R_1$ which are the final labels for $q \in R$.

4. Compute $Left_{R_1}(p, R_2)$.

5. Compute $Left_{L \cup R_1}(p, R_2)$, given $Left_{R_1}(p, R_2)$ and $Left_L(p, R)$. That is, for each $p \in R_2$, set $Left_{L \cup R_1}(p, R_2)$ to be $\max\{Left_{R_1}(p, R_2), Left_L(p, R)\}$.

6. Call MAXDOM_LABEL(R_2). This will return the labels for all $q \in R_2$ which are the final labels for $q \in R$.

All steps other than Step 4 are self-explanatory. Steps 4 and 5 are needed in order to set up the correct invariant condition for the second recursive call. The computation of $Left_{R_1}(p, R_2)$ and its complexity is the key to the correctness and time complexity of the algorithm MAXDOM_LABEL(R). We briefly discuss this problem and show that this step can be done in $O(m)$ time. Since all other steps take linear time, the overall time complexity is $O(m \log m)$.

Consider in general two subsets L and R of points separated by a vertical line $\mathcal{V}$, with L lying to the left of R and points in $L \cup R$ are sorted in ascending y-coordinate (Figure 1.9). Suppose we have computed the labels $l_L(p), p \in L$. We compute $Left_L(p, R)$ by using a plane-sweep technique scanning

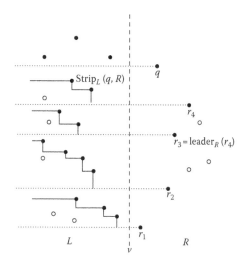

FIGURE 1.9 Computation of $Left_L(p, R)$.

points in $L \cup R$ in ascending y-coordinate. We will maintain for each point $r \in R$ $\text{Strip}_L(r, R)$ along with the highest and rightmost points in the subset, denoted $1\text{st}_L(r, R)$ and $\text{last}_L(r, R)$, respectively, and $\text{leader}_R(r)$. For each point $p \in L \cap \text{Strip}_L(r, R)$ for some $r \in R$ we maintain a label $\text{max\_}l(p)$, which is equal to $\max\{l_L(q)|q \in \text{Strip}_L(r, R) \text{ and } y(q) < y(p)\}$.

A stack ST_R will be used to store $\text{leader}_R(r_i)$ of $r_i \in R$ such that any element r_t in ST_R is $\text{leader}_R(r_{t+1})$ for point r_{t+1} above r_t, and the top element r_i of ST_R is the last scanned point in R. For instance in Figure 1.9 ST_R contains r_4, r_3, r_2, and r_1 when r_4 is scanned. Another stack ST_L is used to store $\text{Strip}_L(r, R)$ for a yet-to-be-scanned point $r \in R$. (The staircase above r_4 in Figure 1.9 is stored in ST_L. The solid staircases indicate $\text{Strip}_L(r_i, R)$ for r_i, $i = 2, 3, 4$.)

Let the next point scanned be denoted q. If $q \in L$, we pop off the stack ST_L all points that are dominated by q until q'. And we compute $\text{max\_}l(q)$ to be the larger of $l_L(q)$ and $\text{max\_}l(q')$. We then push q onto ST_L. That is, we update ST_L to make sure that all the points in ST_L are maximal.

Suppose $q \in R$. Then $\text{Strip}_L(q, R)$ is initialized to be the entire contents of ST_L and let $1\text{st}_L(q, R)$ be the top element of ST_L and $\text{last}_L(q, R)$ be the bottom element of ST_L.

If the top element of ST_R is equal to $\text{leader}_R(q)$, we set $\text{Left}_L(q, R)$ to $\text{max\_}l(q')$, where q' is $1\text{st}_L(q, R)$, initialize ST_L to be empty, and continue to scan the next point. Otherwise we need to pop off the stack ST_R all points that are *not* dominated by q, until q', which is $\text{leader}_R(q)$. As shown in Figure 1.9, r_i, $i = 4, 3, 2$ will be popped off ST_R when q is scanned. As point r_i is popped off ST_R, $\text{Strip}_L(r_i, R)$ is concatenated with $\text{Strip}_L(q, R)$ to maintain its maximality. That is, the points in $\text{Strip}_L(r_i, R)$ are scanned from $1\text{st}_L(r_i, R)$ to $\text{last}_L(r_i, R)$ until a point, if any, α_i is encountered such that $x(\alpha_i) > x(\text{last}_L(q, R))$. $\text{max\_}l(q')$, $q' = 1\text{st}_L(q, R)$, is set to be the larger of $\text{max\_}l(q')$ and $\text{max\_}l(\alpha)$, and $\text{last}_L(q, R)$ is temporarily set to be $\text{last}_L(r_i, R)$. If no such α_i exists, then the entire $\text{Strip}_L(r_i, R)$ is ignored. This process repeats until $\text{leader}_R(q)$ of q is on top of ST_R. At that point, we would have computed $\text{Strip}_L(q, R)$ and $\text{Left}_L(q, R)$ is $\text{max\_}l(q')$, where $q' = 1\text{st}_L(q, R)$. We initialize ST_L to be empty and continue to scan the next point.

It has been shown in [7] that this scanning operation takes linear time (with path compression), so the overall algorithm takes $O(m \log m)$ time.

THEOREM 1.6 *Given a set S of n points with weights $w(p_i), p_i \in S, i = 1, 2, \ldots, n$,* ALGORITHM MAXDOM_LABEL(S) *returns $l_S(p)$ for each point $p \in S$ in $O(n \log n)$ time.*

Now let us briefly describe the algorithm for the *maxdominance problem*. That is to find for each $p \in S$, $\mathcal{M}(\mathcal{D}_S(p))$.

ALGORITHM MAXDOM_LIST(S)

Input: A sorted sequence of n points of S, i.e., $S = \{p_1, p_2, \ldots, p_n\}$, where $x(p_1) < x(p_2) < \cdots < x(p_n)$.

Output: $\mathcal{M}(\mathcal{D}_S(p))$ for each $p \in S$ and the list Q_S containing the points of S in ascending y-coordinates.

1. If $n = 1$ then we set $\mathcal{M}(\mathcal{D}_S(p_1)) = \emptyset$ and **return**.

2. Call ALGORITHM MAXDOM_LIST(L), where $L = \{p_1, p_2, \ldots, p_{n/2}\}$. This call returns $\mathcal{M}(\mathcal{D}_S(p))$ for each $p \in L$ and the list Q_L.

3. Call ALGORITHM MAXDOM_LIST(R), where $R = \{p_{n/2+1}, \ldots, p_n\}$. This call returns $\mathcal{M}(\mathcal{D}_R(p))$ for each $p \in R$ and the list Q_R.

4. Compute for each $r \in R$ $\text{Strip}_L(r, R)$ using the algorithm described in Step 4 of ALGORITHM MAXDOM_LABEL(R).

5. For every $r \in R$ compute $\mathcal{M}(\mathcal{D}_S(p))$ by concatenating $\text{Strip}_L(r, R)$ and $\mathcal{M}(\mathcal{D}_R(p))$.

6. Merge Q_L and Q_R into Q_S and return.

Since Steps 4, 5, and 6, excluding the output time, can be done in linear time, we have the following.

THEOREM 1.7   *The maxdominance problem of a set S of n points in the plane can be solved in $O(n \log n + \mathcal{F})$ time, where $\mathcal{F} = \sum_{p \in S} |\mathcal{M}(\mathcal{D}_S(p))|$.*

Note that the problem of computing the layers of maxima in three dimensions can be solved in $O(n \log n)$ time [12].

1.4 Row Maxima Searching in Monotone Matrices

The *row maxima-searching problem* in a matrix is that given an $n \times m$ matrix M of real entries, find the leftmost maximum entry in each row.

A matrix is said to be *monotone*, if $i_1 > i_2$ implies that $j(i_1) \geq j(i_2)$, where $j(i)$ is the index of the leftmost column containing the maximum in row i. It is totally monotone if all of its submatrices are monotone.

In fact if every 2×2 submatrix $M[i, j; k, l]$ with $i < j$ and $k < l$ is monotone, then the matrix is totally monotone. Or equivalently if $M(i, k) < M(i, l)$ implies $M(j, k) < M(j, l)$ for any $i < j$ and $k < l$, then M is totally monotone.

The algorithm for solving the row maxima-searching problem is due to Aggarwal et al. [2], and is commonly referred to as the SMAWK algorithm. Specifically the following results were obtained: $O(m \log n)$ time for an $n \times m$ monotone matrix, and $\theta(m)$ time, $m \geq n$, and $\theta(m(1 + \log(n/m)))$ time, $m < n$, if the matrix is totally monotone.

We use as an example the distance matrix between pairs of vertices of a convex n-gon P, represented as a sequence of vertices $p_1, p_2, \ldots, p_n$ in counterclockwise order. For an integer j, let $*j$ denote $((j - 1) \bmod n) + 1$. Let M be an $n \times (2n - 1)$ matrix defined as follows. If $i < j \leq i + n - 1$ then $M[i, j] = d(p_i, p_{*j})$, where $d(p_i, p_j)$ denotes the Euclidean distance between two vertices p_i and p_j. If $j \leq i$ then $M[i, j] = j - i$, and if $j \geq i + n$ then $M[i, j] = -1$. The problem of computing for each vertex its farthest neighbor is now the same as the row maxima-searching problem.

Consider submatrix $M[i, j; k, l]$, with $i < j$ and $k < l$, that has only positive entries, i.e., $i < j < k < l < i + n$. In this case vertices p_i, p_j, p_{*k}, and p_{*l} are in counterclockwise order around the polygon. From the triangle inequality we have $d(p_i, p_{*k}) + d(p_j, p_{*l}) \geq d(p_i, p_{*l}) + d(p_j, p_{*k})$. Thus, $M[i, j; k, l]$ is monotone. The nonpositive entries ensure that all other 2×2 submatrices are monotone. We will show below that the all farthest neighbor problem for each vertex of a convex n-gon can be solved in $O(n)$ time.

A straightforward divide-and-conquer algorithm for the row maxima-searching problem in monotone matrices is as follows.

ALGORITHM MAXIMUM_D&C

1. Find the maximum entry $j = j(i)$, in the ith row, where $i = \lceil \frac{n}{2} \rceil$.
2. Recursively solve the row maxima-searching problem for the submatrices $M[1, \ldots, i - 1; 1, \ldots, j]$ when $i, j > 1$ and $M[i + 1, \ldots, n; j, \ldots, m]$ when $i < n$ and $j < m$.

The time complexity required by the algorithm is given by the recurrence

$$f(n, m) \leq m + \max_{1 \leq j \leq m} (f(\lceil n/2 \rceil - 1, j) + f(\lfloor n/2 \rfloor, m - j + 1))$$

with $f(0, m) = f(n, 1) = $ constant. We have $f(n, m) = O(m \log n)$.

Now let us consider the case when the matrix is totally monotone. We distinguish two cases: (a) $m \geq n$ and (b) $m < n$.

Case (a): Wide matrix $m \geq n$.

An entry $M[i, j]$ is bad if $j \neq j(i)$, i.e., column j is not a solution to row i. Column j, $M[*, j]$ is bad if all $M[i, j]$, $1 \leq i \leq n$ are bad.

LEMMA 1.5 For $j_1 < j_2$ if $M[r, j_1] \geq M[r, j_2]$, then $M[i, j_2]$, $1 \leq i \leq r$, are bad; otherwise $M[i, j_1]$, $r \leq i \leq n$, are bad.

Consider an $n \times n$ matrix C, the *index* of C is defined to be the largest k such that $C[i, j]$, $1 \leq i < j$, $1 \leq j \leq k$ are bad.

The following algorithm REDUCE reduces in $O(m)$ time a totally monotone $m \times n$ matrix M to an $n \times n$ matrix C, a submatrix of M, such that for $1 \leq i \leq n$ it contains column $M^{j(i)}$. That is, bad columns of M (which are known not to contain solutions) are eliminated.

ALGORITHM REDUCE(M)

    1. $C \leftarrow M$; $k \leftarrow 1$;

    2. **while** C has more than n columns **do**

            case $C(k, k) \geq C(k, k + 1)$ and $k < n$: $k \leftarrow k + 1$;
                $C(k, k) \geq C(k, k + 1)$ and $k = n$: Delete column C^{k+1};
                $C(k, k) < C(k, k + 1)$:   Delete column C^k; **if** $k > 1$ **then** $k \leftarrow k - 1$
            end case

    3. **return**(C)

The following algorithm solves the maxima-searching problem in an $n \times m$ totally monotone matrix, where $m \geq n$.

ALGORITHM MAX_COMPUTE(M)

    1. $B \leftarrow$ REDUCE(M);

    2. **if** $n = 1$ **then** output the maximum and **return**;

    3. $C \leftarrow B[2, 4, \ldots, 2\lfloor n/2 \rfloor; 1, 2, \ldots, n]$;

    4. Call MAX_COMPUTE(C);

    5. From the known positions of the maxima in the even rows of B, find the maxima in the odd rows.

The time complexity of this algorithm is determined by the following recurrence:

$$f(n, m) \leq c_1 n + c_2 m + f(n/2, n)$$

with $f(0, m) = f(n, 1) = $ constant. We therefore have $f(n, m) = O(m)$.

Case (b): Narrow matrix $m < n$.

In this case we decompose the problem into m subproblems each of size $\lfloor n/m \rfloor \times m$ as follows. Let $r_i = \lfloor in/m \rfloor$, for $0 \leq i \leq m$. Apply MAX_COMPUTE to the $m \times m$ submatrix $M[r_1, r_2, \ldots, r_m; 1, 2, \ldots, m]$ to get $c_1, c_2, \ldots, c_m$, where $c_i = j(r_i)$. This takes $O(m)$ time. Let $c_0 = 1$. Consider submatrices $B_i = M[r_{i-1}+1, r_{i-1}+2, \ldots, r_i-1; c_{i-1}, c_{i-1}+1, \ldots, c_i]$ for $1 \leq i \leq m$ and $r_{i-1} \leq r_i - 2$. Applying the straightforward divide-and-conquer algorithm to the submatrices, B_i, we obtain the column positions of the maxima for all remaining rows. Since each submatrix has at most $\lfloor n/m \rfloor$ rows, the time for finding the maxima is at most $c(p_i - p_{i-1} + 1)\log(n/m)$ for some constant c. Summing over all $1 \leq i \leq m$ we get the total time, which is $O(m(1 + \log(n/m)))$. The bound can be shown to be tight [2].

The applications of the matrix-searching algorithm include the problems of finding all farthest neighbors for all vertices of a convex n-gon ($O(n)$ time), and finding the extremal (maximum perimeter or area) polygons (inscribed k-gons) of a convex n-gon ($O(kn + n\log n)$). If one adopts the algorithm by Hershberger and Suri [42] the above problems can be solved in $O(n)$ time. It is also used in solving the Largest Empty Rectangle Problem discussed in Section 2.3.6 of this book.

1.5 Decomposition

Polygon decomposition arises in pattern recognition [77] in which recognition of a shape is facilitated by first decomposing it into simpler components, called primitives, and comparing them to templates previously stored in a library via some similarity measure. This class of decomposition is called *component-directed decomposition*. The primitives are often convex.

1.5.1 Trapezoidalization

We will consider first *trapezoidalization* of a polygon P with n vertices, i.e., decomposition of the interior of a polygon into a collection of *trapezoids* with two horizontal sides, one of which may degenerate into a point, reducing a trapezoid to a triangle. Without loss of generality let us assume that no edge of P is horizontal. For each vertex v let us consider the horizontal line passing through v, denoted $\mathcal{H}_v$. The vertices of P are classified into three types. A vertex v is *regular* if the other two vertices adjacent to v lie on different sides of $\mathcal{H}_v$. A vertex v is a V-*cusp* if the two vertices adjacent to v are above $\mathcal{H}_v$, and is a $\wedge$-*cusp* if the two vertices adjacent to v are below $\mathcal{H}_v$. In general the intersection of $\mathcal{H}_v$ and the interior of P consists of a number of horizontal segments, one of which contains v. Let this segment be denoted $\overline{v_\ell, v_r}$, where v_ℓ and v_r are called the left and right projections of v on the boundary of P, denoted ∂P, respectively. If v is regular, either $\overline{v, v_\ell}$ or $\overline{v, v_r}$ lies totally in the interior of P. If v is a V-cusp or $\wedge$-cusp, then $\overline{v_\ell, v_r}$ either lies totally in the interior of P or degenerates to v itself.

Consider only the segments $\overline{v_\ell, v_r}$ that are *nondegenerate*. These segments collectively partition the interior of P into a collection of trapezoids, each of which contains no vertex of P in its interior (Figure 1.10a).

The trapezoidalization can be generalized to a **planar straight-line graph** $G(V, E)$, where the entire plane is decomposed into trapezoids, some of which are unbounded. This trapezoidalization is sometimes referred to as horizontal **visibility map** of the edges, as the horizontal segments connect two edges of G that are *visible* (horizontally) (Figure 1.10b). The trapezoidalization of a planar straight-line graph $G(V, E)$ can be computed by plane-sweep technique in $O(n\log n)$ time, where $n = |V|$ [66], while the trapezoidalization of a simple polygon can be found in linear time [15].

The plane-sweep algorithm works as follows. The vertices of the graph $G(V, E)$ are sorted in descending y-coordinates. We will sweep the plane by a horizontal sweep-line from top down. Associated with this approach there are two basic data structures containing all relevant information that should be maintained.

1. Sweep-line status, which records the information of the geometric structure that is being swept. In this example the *sweep-line status* keeps track of the set of edges intersecting the current sweep-line.

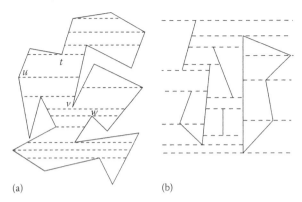

(a)   (b)

FIGURE 1.10 Trapezoidalization of a polygon (a) and horizontal visibility map of a planar straight line graph (b).

2. *Event schedule,* which defines a sequence of *event points* that the sweep-line status will change. In this example, the sweep-line status will change only at the vertices.

The event schedule is normally represented by a data structure, called *priority queue.* The content of the queue may not be available entirely at the start of the plane-sweep process. Instead, the list of events may change dynamically. In this case, the events are static; they are the y-coordinates of the vertices. The sweep-line status is represented by a suitable data structure that supports insertions, deletions, and computation of the left and right projections, v_l and v_r, of each vertex v. In this example a *red-black tree* or any *height-balanced binary search tree* is sufficient for storing the edges that intersect the sweep-line according to the x-coordinates of the intersections. Suppose at event point v_{i-1} we maintain a list of edges intersecting the sweep-line from left to right. Analogous to the trapezoidalization of a polygon, we say that a vertex v is regular if there are edges incident on v that lie on different sides of $\mathcal{H}_v$; a vertex v is a V-cusp if all the vertices adjacent to v are above $\mathcal{H}_v$; v is a $\wedge$-cusp if all the vertices adjacent to v are below $\mathcal{H}_v$. For each event point v_i we do the following.

1. v_i is regular. Let the leftmost and rightmost edges that are incident on v_i and above $\mathcal{H}_{v_i}$ are $E_\ell(v_i)$ and $E_r(v_i)$, respectively. The left projection $v_{i\ell}$ of v_i is the intersection of $\mathcal{H}_{v_i}$ and the edge to the left of $E_\ell(v_i)$ in the sweep-line status. Similarly the right projection v_{ir} of v_i is the intersection of $\mathcal{H}_{v_i}$ and the edge to the right of $E_r(v_i)$ in the sweep-line status. All the edges between $E_\ell(v_i)$ and $E_r(v_i)$ in the sweep-line status are replaced in an order-preserving manner by the edges incident on v_i that are below $\mathcal{H}_{v_i}$.
2. v_i is a V-cusp. The left and right projections of v_i are computed in the same manner as in Step 1 above. All the edges incident on v_i are then deleted from the sweep-line status.
3. v_i is a $\wedge$-cusp. We use binary search in the sweep-line status to look for the two adjacent edges $E_\ell(v_i)$ and $E_r(v_i)$ such that v_i lies in between. The left projection $v_{i\ell}$ of v_i is the intersection of $\mathcal{H}_{v_i}$ and $E_\ell(v_i)$ and the right projection v_{ir} of v_i is the intersection of $\mathcal{H}_{v_i}$ and $E_r(v_i)$. All the edges incident on v_i are then inserted in an order-preserving manner between $E_\ell(v_i)$ and $E_r(v_i)$ in the sweep-line status.

Figure 1.11 illustrates these three cases. Since the update of the sweep-line status for each event point takes $O(\log n)$ time, the total amount of time needed is $O(n \log n)$.

THEOREM 1.8   *Given a planar straight-line graph $G(V, E)$, the horizontal visibility map of G can be computed in $O(n \log n)$ time, where $n = |V|$. However, if G is a simple polygon then the horizontal visibility map can be computed in linear time.*

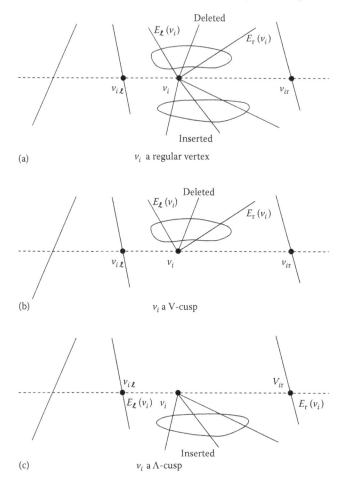

(a) v_i a regular vertex

(b) v_i a V-cusp

(c) v_i a Λ-cusp

FIGURE 1.11 Updates of sweep-line status. (a) v_i is regular (b) v_i is a V-cusp and (c) v_i is a Λ-cusp.

1.5.2 Triangulation

In this section we consider triangulating a planar straight-line graph by introducing noncrossing edges so that each face in the final graph is a triangle and the outermost boundary of the graph forms a convex polygon. Triangulation of a set of (discrete) points in the plane is a special case. This is a fundamental problem that arises in computer graphics, geographical information systems, and finite element methods. Let us start with the simplest case.

1.5.2.1 Polygon Triangulation

Consider a simple polygon P with n vertices. It is obvious that to triangulate the interior of P (into $n - 2$ triangles) one needs to introduce at most $n - 3$ diagonals. A pioneering work is due to Garey et al. [37] who gave an $O(n \log n)$ algorithm and a linear algorithm if the polygon is *monotone*. A polygon is monotone if there exists a straight line $\mathcal{L}$ such that the intersection of ∂P and any line orthogonal to $\mathcal{L}$ consists of no more than two points. The shaded area in Figure 1.12 is a monotone polygon.

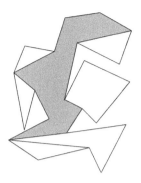

FIGURE 1.12 Decomposition of a simple polygon into monotone subpolygons.

The $O(n \log n)$ time algorithm can be illustrated by the following two-step procedure.

1. Decompose P into a collection of monotone subpolygons with respect to the y-axis in time $O(n \log n)$.
2. Triangulate each monotone subpolygons in linear time.

To find a decomposition of P into a collection of monotone polygons we first obtain the horizontal visibility map described in Section 1.5.1. In particular we obtain for each cusp v the left and right projections and the associated trapezoid below $\mathcal{H}_v$ if v is a V-cusp, and above $\mathcal{H}_v$ if v is a Λ-cusp. (Recall that $\mathcal{H}_v$ is the horizontal line passing through v.) For each V-cusp v we introduce an edge $\overline{v, w}$ where w is the vertex through which the other base of the trapezoid below passes. $\overline{t, u}$ and $\overline{v, w}$ in Figure 1.10a illustrate these two possibilities, respectively. For each Λ-cusp we do the same thing. In this manner we convert each vertex into a regular vertex, except the cusps v for which $\overline{v_l, v_r}$ lies totally outside of P, where v_l and v_r are the left and right projections of v in the horizontal visibility map. This process is called *regularization* [66]. Figure 1.12 shows a decomposition of the simple polygon in Figure 1.10a into a collection of monotone polygons.

We now describe an algorithm that triangulates a monotone polygon P in linear time. Assume that the monotone polygon has v_0 as the topmost vertex and v_{n-1} as the lowest vertex. We have two polygonal chains from v_0 to v_{n-1}, denoted $\mathcal{L}$ and $\mathcal{R}$, that define the left and right boundary of P, respectively. Note that vertices on these two polygonal chains are already sorted in descending order of their y-coordinates. The algorithm is based on a *greedy* method, i.e., whenever a triangle can be formed by connecting vertices either on the same chain or on opposite chains, we do so immediately. We shall examine the vertices in order, and maintain a polygonal chain $\mathcal{C}$ consisting of vertices whose internal angles are greater than π. Initially $\mathcal{C}$ consists of two vertices v_0 and v_1 that define an edge $\overline{v_0, v_1}$. Suppose $\mathcal{C}$ consists of vertices $v_{i_0}, v_{i_1}, \ldots, v_{i_k}, k \geq 1$. We distinguish two cases for each vertex v_ℓ examined, $l < n - 1$. Without loss of generality we assume $\mathcal{C}$ is a left chain, i.e., $v_{i_k} \in \mathcal{L}$. The other case is treated symmetrically.

1. $v_\ell \in \mathcal{L}$. Let v_{i_j} be the last vertex on $\mathcal{C}$ that is visible from v_ℓ. That is, the internal angle $\cap(v_{i_j}, v_{i_{j'}}, v_\ell)$, where $j < j' \leq k$, is less than π, and either $v_{i_j} = v_{i_0}$ or the internal angle $\cap(v_{i_{j-1}}, v_{i_j}, v_\ell)$ is greater than π. Add diagonals $\overline{v_\ell, v_{i_{j'}}}$, for $j \leq j' < k$. Update $\mathcal{C}$ to be composed of vertices $v_{i_0}, v_{i_1}, \ldots, v_{i_j}, v_\ell$.
2. $v_\ell \in \mathcal{R}$. In this case we add diagonals $\overline{v_\ell, v_{i_{j'}}}$, for $0 \leq j' \leq k$. $\mathcal{C}$ is updated to be composed of v_{i_k} and v_ℓ and it becomes a right chain.

Figure 1.13a and b illustrates these two cases, respectively, in which the shaded portion has been triangulated.

Fournier and Montuno [34] and independently Chazelle and Incerpi [17] showed that triangulation of a polygon is linear-time equivalent to computing the horizontal visibility map. Based on this result Tarjan and Van Wyk [76] first devised an $O(n \log \log n)$ time algorithm that computes the horizontal visibility map and hence, an $O(n \log \log n)$ time algorithm for triangulating a simple polygon. A breakthrough result of Chazelle [15] finally settled the longstanding open problem, i.e., triangulating a simple polygon in linear time. But the method is quite involved. As a result of this linear triangulation algorithm, a number of problems can be solved asymptotically in linear time. Note that if the polygons have holes, the problem of triangulating the interior is shown to require $\Omega(n \log n)$ time [5].

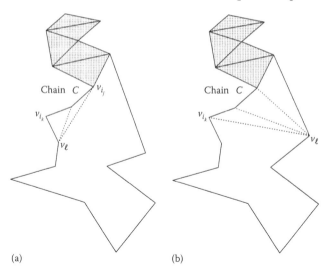

FIGURE 1.13 Triangulation of a monotone polygon. (a) v_ℓ on the left chain and (b) v_ℓ on the right chain.

1.5.2.2 Planar Straight-Line Graph Triangulation

The triangulation is also known as the *constrained triangulation*. This problem includes the triangulation of a set of points. A triangulation of a given planar straight-line graph $G(V, E)$ with $n = |V|$ vertices is a planar graph $\mathcal{G}(V, \mathcal{E})$ such that $E \subseteq \mathcal{E}$ and each face is a triangle, except the exterior one, which is unbounded. A constrained triangulation $\mathcal{G}(V, \mathcal{E})$ of a $G(V, E)$ can be obtained as follows.

1. Compute the convex hull of the set of vertices, ignoring all the edges. Those edges that belong to the convex hull are necessarily in the constrained triangulation. They define the boundary of the exterior face.
2. Compute the horizontalvisibility map for the graph, $G' = G \cup CH(V)$, where $CH(V)$ denotes the convex hull of V, i.e., for each vertex, its left and right projections are calculated, and a collection of trapezoids are obtained.
3. Apply the *regularization* process by introducing edges to vertices in the graph G' that are not regular. An isolated vertex requires two edges, one from above and one from below. Regularization will yield a collection of *monotone* polygons that comprise collectively the interior of $CH(V)$.
4. Triangulate each monotone subpolygon.

It is easily seen that the algorithm runs in time $O(n \log n)$, which is asymptotically optimal. (This is because the problem of sorting is linearly reducible to the problem of constrained triangulation.)

1.5.2.3 Delaunay and Other Special Triangulations

Sometimes we want to look for quality triangulation, instead of just an arbitrary one. For instance, triangles with large or small angles is not desirable. The Delaunay triangulation of a set of points in the plane is a triangulation that satisfies the *empty circumcircle property*, i.e., the circumcircle of each triangle does not contain any other points in its interior. It is well-known that the Delaunay triangulation of points in general position is unique and it will maximize the minimum angle. In fact, the **characteristic angle vector** of the Delaunay triangulation of a set of points is *lexicographically maximum* [49]. The notion of Delaunay triangulation of a set of points can be generalized to a planar straight-line graph $G(V, E)$. That is, we would like to have G as a subgraph of a triangulation $\mathcal{G}(V, \mathcal{E}'), E \subseteq \mathcal{E}'$, such that each triangle satisfies the empty circumcircle property: no vertex *visible* from the vertices of triangle is contained in the interior of the circle. This *generalized* Delaunay triangulation is thus a constrained triangulation that maximizes the minimum angle. The generalized Delaunay triangulation was first introduced by the author and an $O(n^2)$ (respectively, $O(n \log n)$) algorithm for constructing the generalized triangulation of a planar graph (respectively a simple polygon) with n vertices was given in [52]. As the generalized Delaunay triangulation (also known as *constrained* Delaunay triangulation) is of fundamental importance, we describe in Section 1.5.2.4 an optimal algorithm due to Chew [19] that computes the constrained Delaunay triangulation for a planar straight-line graph $G(V, E)$ with n vertices in $O(n \log n)$ time. Triangulations that minimize the maximum angle or maximum edge length [29] were also studied. But if the constraints are on the measure of the triangles, for instance, each triangle in the triangulation must be nonobtuse, then **Steiner points** must be introduced. See Bern and Eppstein (in [26, pp. 23–90]) for a survey of triangulations satisfying different criteria and discussions of triangulations in two and three dimensions. Bern and Eppstein gave an $O(n \log n + \mathcal{F})$ algorithm for constructing a nonobtuse triangulation of polygons using $\mathcal{F}$ triangles. Bern et al. [11] showed that $\mathcal{F}$ is $O(n)$ and gave an $O(n \log n)$ time algorithm for simple polygons without holes, and an $O(n \log^2 n)$ time algorithm for polygons with holes. For more results about acute triangulations of polygons and special surfaces see [56,83,84] and the references therein. The problem of finding a triangulation of a set of points in the plane whose total edge length is minimized, known as the *minimum weight triangulation*, is listed as an *open* problem (called *minimum length triangulation*) in Johnson's NP-complete column [45]. On the assumption that this problem is NP-hard, many researchers have obtained polynomial-time approximation algorithms for it. See Bern and Eppstein [10] for a survey of approximation algorithms. Only recently this problem was settled in the affirmative by Mulzer and Rote [63].

The problem of triangulating a set P of points in $\Re^k$, $k \geq 3$, is less studied. In this case the convex hull of P is to be partitioned into $\mathcal{F}$ nonoverlapping simplices, the vertices of which are points in P. A simplex in k-dimensions consists of exactly $k + 1$ points, all of which are extreme points. In $\Re^3$ $O(n \log n + \mathcal{F})$ time suffices, where $\mathcal{F}$ is linear if no three points are collinear, and $O(n^2)$ otherwise. Recently Saraf [71] gave a simpler proof that acute triangulations for general polyhedral surfaces exist and also showed that it is possible to obtain a nonobtuse triangulation of a general polyhedral surface. See [26] for more references on 3D triangulations and Delaunay triangulations in higher dimensions.

1.5.2.4 Constrained Delaunay Triangulation

Consider a **planar straight-line graph** $G(V, E)$, where V is a set of points in the plane, and edges in E are nonintersecting except possibly at the endpoints. Let $n = |V|$. Without loss of generality we assume that the edges on the convex hull CH(V) are all in E. These edges, if not present, can be computed in $O(n \log n)$ time (cf. Section 1.2.1).

In the constrained Delaunay triangulation $G_{DT}(V, \mathcal{E})$ the edges in $\mathcal{E} \setminus E$ are called *Delaunay* edges. It can be shown that two points $p, q \in V$ define a Delaunay edge if there exists a circle $\mathcal{K}$ passing through p and q which does not contain in its interior any other point visible from p and from q.

Let us assume without loss of generality that the points are in general position that no two have the same x-coordinate, and no four are cocircular. Let the points in V be sorted by ascending order of x-coordinate so that $x(p_i) < x(p_j)$ for $i < j$. Let us associate this set V (and graph $G(V, E)$) with it a bounding rectangle R_V with diagonal points $U(u.x, u.y), L(l.x, l.y)$, where $u.x = x(p_n), u.y = \max y(p_i), l.x = x(p_1), l.y = \min y(p_i)$. That is, L is at the lower left corner with x- and y-coordinates equal to the minimum of the x- and y-coordinates of all the points in V, and U is at the upper right corner. Given an edge $\overline{p_i, p_j}, i < j$, its x-interval is the interval $(x(p_i), x(p_j))$. The x-interval of the

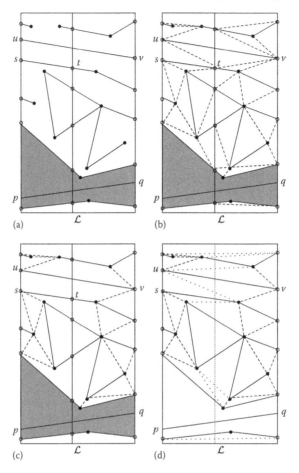

FIGURE 1.14 Computation of constrained Delaunay triangulation for subgraphs in adjacent bounding rectangles.

bounding rectangle R_V, denoted by $\mathcal{X}_V$, is the interval $(x(p_1), x(p_n))$. The set V will be recursively divided by vertical lines $\mathcal{L}$'s and so will be the bounding rectangles. We first divide V into two halves V_1 and V_r by a line $\mathcal{L}(X = m)$, where $m = \frac{1}{2}(x(p_{\lfloor n/2 \rfloor}) + x(p_{\lfloor n/2 \rfloor+1}))$. The edges in E that lie totally to the left and to the right of $\mathcal{L}$ are assigned respectively to the left graph $G_\ell(V_\ell, E_\ell)$ and to the right graph $G_r(V_r, E_r)$, which are associated respectively with bounding rectangle R_{V_ℓ} and R_{V_r} whose x-intervals are $\mathcal{X}_{V_\ell} = (x(p_1), m)$ and $\mathcal{X}_{V_r} = (m, x(p_n))$, respectively. The edges $\overline{p, q} \in E$ that are intersected by the dividing line and that do not *span** the associated x-interval $\mathcal{X}_V$ will each get cut into *two edges* and a *pseudo* point $\epsilon(p, q)$ on the edge is introduced. Two edges, called *half-edges*, $\overline{p, \epsilon(p, q)} \in E_\ell$ and $\overline{\epsilon(p, q), q} \in E_r$ are created. Figure 1.14a shows the creation of half-edges with pseudo points shown in hollow circles. Note that the edge $\overline{p, q}$ in the shaded area is not considered "present" in the associated bounding rectangle, $\overline{u, v}$ spans the x-interval for which no pseudo point is created, and $\overline{s, t}$ is a half-edge that spans the x-interval of the bounding rectangle to the left of the dividing line ($\mathcal{L}$).

---

* An edge $\overline{p, q}, x(p) < x(q)$ is said to span an x-interval (a, b), if $x(p) < a$ and $x(q) > b$.

It can be shown that for each edge $\overline{p,q} \in E$, the number of half-edges so created is at most $O(\log n)$. Within each bounding rectangle the edges that span its x-interval will divide the bounding rectangle into various parts. The constrained Delaunay triangulation gets computed for each part recursively. At the bottom of recursion each bounding rectangle contains at most three vertices of V, the edges incident on them, plus a number of half-edges spanning the x-interval including the pseudo endpoints of half-edges. Figure 1.14b illustrates an example of the constrained Delaunay triangulation at some intermediate step. No edges shall intersect one another, except at the endpoints.

As is usually the case for divide-and-conquer paradigm, the Merge step is the key to the method. We describe below the Merge step that combines constrained Delaunay triangulations in adjacent bounding rectangles that share a common dividing vertical edge $\mathcal{L}$.

1. Eliminate the pseudo points along the boundary edge $\mathcal{L}$, including the Delaunay edges incident on those pseudo points. This results in a partial constrained Delaunay triangulation within the union of these two bounding rectangles. Figure 1.14c illustrates the partial constrained Delaunay triangulation as a result of the removal of the Delaunay edges incident on the pseudo points on the border of the constrained Delaunay triangulation shown in Figure 1.14b.

2. Compute new Delaunay edges that cross $\mathcal{L}$ as follows. Let A and B be the two endpoints of an edge that crosses $\mathcal{L}$ with A on the left and B on the right of $\mathcal{L}$, and $\overline{A,B}$ is known to be part of the desired constrained Delaunay triangulation. That is, either $\overline{A,B}$ is an edge of E or a Delaunay edge just created. Either A or B may be a pseudo point. Let $\overline{A,C}$ be the first edge counterclockwise from edge $\overline{A,B}$. To decide if $\overline{A,C}$ remains to be a Delaunay edge in the desired constrained Delaunay triangulation, we consider the next edge $\overline{A,C_1}$, if it exists counterclockwise from $\overline{A,C}$. If $\overline{A,C_1}$ does not exist, or $\overline{A,C}$ is in E, $\overline{A,C}$ will remain. Otherwise we test if the circumcircle $\mathcal{K}(A,C,C_1)$ contains B in its interior. If so, $\overline{A,C}$ is eliminated, and the test continues on $\overline{A,C_1}$. Otherwise, $\overline{A,C}$ stays. We do the same thing to test edges incident on B, except that we consider edges incident on B in clockwise direction from $\overline{B,A}$. Assume now we have determined that both $\overline{A,C}$ and $\overline{B,D}$ remain. The next thing to do is to decide which of edge $\overline{B,C}$ and $\overline{A,D}$ should belong to the desired constrained Delaunay triangulation. We apply the circle test: test if circle $\mathcal{K}(A,B,C)$ contains D in the interior. If not, $\overline{B,C}$ is the desired Delaunay edge. Otherwise $\overline{A,D}$ is. We then repeat this step.

Step 2 of the merge step is similar to the method of constructing unconstrained Delaunay triangulation given in [52] and can be accomplished in linear time in the number of edges in the combined bounding rectangle. The dotted lines in Figure 1.14d are the Delaunay edges introduced in Step 2. We therefore conclude with the following theorem.

THEOREM 1.9 *Given a planar straight-line graph $G(V, E)$ with n vertices, the constrained Delaunay triangulation of G can be computed in $O(n \log n)$ time, which is asymptotically optimal.*

An implementation of this algorithm can be found in http://www.opencps.org/Members/ta/ Constrained Delaunay Triangulation/cpstype_view

A plane-sweep method, for constructing the constrained Delaunay triangulation was also presented by De Floriani and Puppo [23] and by Domiter and Žalik [25]. The constrained Delaunay triangulation has been used as a basis for the so-called *Delaunay refinement* [31,40,61,68] for generating triangular meshes satisfying various *angle conditions* suitable for use in interpolation and the finite element method. Shewchuk [75] gave a good account of this subject and has also developed a tool, called Triangle: A Two-Dimensional Quality Mesh Generator and Delaunay Triangulator [74]. See http://www.cs.cmu.edu/~quake/triangle.html for more details.

1.5.3 Other Decompositions

Partitioning a simple polygon into shapes such as convex polygons, **star-shaped polygons**, spiral polygons, etc., has also been investigated. After a polygon has been triangulated one can partition the polygon into star-shaped polygons in linear time. This algorithm provided a very simple proof of the traditional art gallery problem originally posed by Klee, i.e., $\lfloor n/3 \rfloor$ vertex guards are always sufficient to see the entire region of a simple polygon with n vertices. But if a partition of a simple polygon into a minimum number of star-shaped polygons is desired, Keil [47] gives an $O(n^5 N^2 \log n)$ time, where N denotes the number of reflex vertices. However, the problem of decomposing a simple polygon into a minimum number of star-shaped parts that may overlap is shown to be **NP-hard** [53,65]. This problem sometimes is referred to as the *covering* problem, in contrast to the *partitioning* problem, in which the components are not allowed to overlap. The problem of partitioning a polygon into a minimum number of convex parts can be solved in $O(N^2 n \log n)$ time [47]. It is interesting to note that it may not be possible to partition a simple polygon into convex quadrilaterals, but it is always possible for rectilinear polygons. The problem of determining if a convex quadrilateralization of a polygonal region (with holes) exists is NP-complete. It is interesting to note that $\lfloor n/4 \rfloor$ vertex guards are always sufficient for the art gallery problem in a rectilinear polygon. An $O(n \log n)$ algorithm for computing a convex quadrilateralization or positioning at most $\lfloor n/4 \rfloor$ guards is known (see [65] for more information). Modifying the proof of Lee and Lin [53], Schuchardt and Hecker [72] showed that the minimum covering problem by star-shaped polygons for rectilinear polygons is also NP-hard. Most recently Katz and Roisman [46] proved that even guarding the vertices of a rectilinear polygon with minimum number of guards is NP-hard. However, the minimum covering problem by **r-star-shaped polygons** or by **s-star-shaped polygons** for rectilinear polygons can be solved in polynomial time [62,80]. When the rectilinear polygon to be covered is monotone or *uni-s-star-shaped*,* then the minimum covering by r-star-shaped polygons can be found in linear time [38,55].

For variations and results of art gallery problems the reader is referred to [65,73,78]. Polynomial time algorithms for computing the minimum partition of a simple polygon into simpler parts while allowing Steiner points can be found in [5].

The minimum partition problem for simple polygons becomes NP-hard when the polygons are allowed to have *holes* [47]. Asano et al. [4] showed that the problem of partitioning a simple polygon with h holes into a minimum number of trapezoids with two horizontal sides can be solved in $O(n^{h+2})$ time, and that the problem is NP-complete if h is part of the input. An $O(n \log n)$ time 3-approximation algorithm was presented.

The problem of partitioning a rectilinear polygon with holes into a minimum number of rectangles (allowing Steiner points) arises in VLSI artwork data. Imai and Asano [44] gave an $O(n^{3/2} \log n)$ time and $O(n \log n)$ space algorithm for partitioning a rectilinear polygon with holes into a minimum number of rectangles (allowing Steiner points). The problem of covering a rectilinear polygon (without holes) with a minimum number of rectangles, however, is NP-hard [21,41].

Given a polyhedron with n vertices and r notches (features causing nonconvexity), $\Omega(r^2)$ convex components are required for a complete convex decomposition in the worst case. Chazelle and Palios [18] gave an $O((n + r^2) \log r)$ time $O(n + r^2)$ space algorithm for this problem. Bajaj and Dey addressed a more general problem where the polyhedron may have holes and internal voids [9]. The problem of minimum partition into convex parts and the problem of determining if a nonconvex polyhedron can be partitioned into tetrahedra without introducing Steiner points are NP-hard [69].

---

* Recall that a rectilinear polygon P is s-star-shaped, if there exists a point p in P such that all the points of P are *s-visible* from p. When the staircase paths connecting p and all other *s-visible* points q in P are of the same (uniform) orientation, then the rectilinear polygon is *uni-s-star-shaped*.

1.6   Research Issues and Summary

We have covered in this chapter a number of topics in computational geometry, including convex hulls, maximal-finding problems, decomposition, and maxima searching in monotone matrices. Results, some of which are classical, and some which represent the state of the art of this field, were presented. More topics will be covered in Chapter 2.

In Section 1.2.3 an optimal algorithm for computing the layers of planar convex hulls is presented. It shows an interesting fact: within the same time as computing the convex hull (the outermost layer) of a point set in two dimensions, one can compute *all* layers of convex hull. Whether or not one can do the same for higher dimensions $\Re^k, k > 2$, remains to be seen. Although the triangulation problem of a simple polygon has been solved by Chazelle [15], the algorithm is far from being practical. As this problem is at the heart of this field, a simpler and more practical algorithm is of great interest. Recall that the following two minimum covering problems have been shown to be NP-hard: the minimum covering of rectilinear polygons by *rectangles* [21,41], and the minimum covering of rectilinear polygons by star-shaped polygons [46,72]. However, the minimum covering of rectilinear polygons by r-star-shaped polygons or by s-star-shaped polygons, as described above, can be found in polynomial time. There seems to be quite a complexity gap. Note that the components in the decomposition in these two cases, rectangles versus r-star-shaped polygons and star-shape polygons versus s-star-shaped polygons, are related in the sense that the former one is a special case of the latter. That is, a rectangle is an r-star-shaped polygon and a star-shaped primitive of a rectilinear polygon is s-star-shaped. But the converse is not true. When the primitives used in the covering problem are more restricted, they seem to make the minimum covering problem harder computationally. But when and if the fact that the minimum covering of a rectilinear polygon by s-star-shaped polygons solvable in polynomial time can be used to approximate the minimum covering by primitives of more restricted types is not clear. The following results, just to give some examples, were known. Eidenbenz et al. [30] presented inapproximability results of some art gallery problems and Fragoudakis et al. [35] showed certain maximization problems of guarded boundary of an art gallery to be APX-complete. There is a good wealth of problems related to art gallery or polygon decomposition that are worth further investigation.

1.7   Further Information

For some problems we present efficient algorithms in pseudocode and for others that are of more theoretical interest we only give a sketch of the algorithms and refer the reader to the original articles. The text book by de Berg et al. [22] contains a very nice treatment of this topic. The reader who is interested in *parallel* computational geometry is referred to [6]. For current research results, the reader may consult the *Proceedings of the Annual ACM Symposium on Computational Geometry*, *Proceedings of the Annual Canadian Conference on Computational Geometry*, and the following three journals, *Discrete & Computational Geometry*, *International Journal of Computational Geometry & Applications*, and *Computational Geometry: Theory and Applications*. More references can be found in [39,50,70,81]. The ftp site <ftp://ftp.cs.usask.ca/pub/geometry/geombib.tar.gz> contains close to 14,000 entries of bibliography in this field.

David Avis announced a convex hull/vertex enumeration code, *lrs,* based on reverse search and made it available. It finds all vertices and rays of a polyhedron in $\Re^k$ for any k, defined by a system of inequalities, and finds a system of inequalities describing the convex hull of a set of vertices and rays. More details can be found at this site http://cgm.cs.mcgill.ca/~avis/C/lrs.html. See Avis et al. [8] for more information about other convex hull codes. Those who are interested in the implementations or would like to have more information about other software available can consult http://www.geom.umn.edu/software/cglist/.

The following WWW page on *Geometry in Action* maintained by David Eppstein at http://www.ics.uci.edu/~eppstein/geom.html and the Computational Geometry Pages by J. Erickson at http://compgeom.cs. uiuc.edu/~jeffe/compgeom/ give a comprehensive description of research activities of computational geometry.

Defining Terms

Asymptotic time or space complexity: Asymptotic behavior of the time (or space) complexity of an algorithm when the size of the problem approaches infinity. This is usually denoted in big-Oh notation of a function of input size. A time or space complexity $T(n)$ is $O(f(n))$ means that there exists a constant $c > 0$ such that $T(n) \leq c \cdot f(n)$ for sufficiently large n, i.e., $n > n_0$, for some n_0.

CAD/CAM: Computer-aided design and computer-aided manufacturing, a discipline that concerns itself with the design and manufacturing of products aided by a computer.

Characteristic angle vector: A vector of minimum angles of each triangle in a triangulation arranged in nondescending order. For a given point set the number of triangles is the same for all triangulations, and therefore each of these triangulation has a characteristic angle vector.

Divide-and-marriage-before-conquest: A problem-solving paradigm derived from divide-and-conquer. A term coined by the Kirkpatrick and Seidel [48], authors of this method. After the divide step in a divide-and-conquer paradigm, instead of conquering the subproblems by recursively solving them, a merge operation is performed first on the subproblems. This method is proven more effective than conventional divide-and-conquer for some applications.

Extreme point: A point in S is an extreme point if it cannot be expressed as a convex combination of other points in S. A convex combination of points $p_1, p_2, \ldots, p_n$ is $\Sigma_{i=1}^{n} \alpha_i p_i$, where $\alpha_i, \forall i$ is nonnegative, and $\Sigma_{i=1}^{n} \alpha_i = 1$.

Geometric duality: A transform between a point and a hyperplane in $\Re^k$ that preserves incidence and order relation. For a point $p = (\mu_1, \mu_2, \ldots, \mu_k)$, its dual $\mathcal{D}(p)$ is a hyperplane denoted by $x_k = \Sigma_{j=1}^{k-1} \mu_j x_j - \mu_k$; for a hyperplane $\mathcal{H} : x_k = \Sigma_{j=1}^{k-1} \mu_j x_j + \mu_k$, its dual $\mathcal{D}(\mathcal{H})$ is a point denoted by $(\mu_1, \mu_2, \ldots, -\mu_k)$. See [22,27] for more information.

Height-balanced binary search tree: A data structure used to support membership, insert/delete operations each in time logarithmic in the size of the tree. A typical example is the *AVL* tree or *red-black* tree.

NP-hard problem: A complexity class of problems that are intrinsically *harder* than those that can be solved by a Turing machine in nondeterministic polynomial time. When a decision version of a combinatorial optimization problem is proven to belong to the class of NP-complete problems, which includes well-known problems such as satisfiability, traveling salesman problem, etc., an optimization version is NP-hard. For example, to decide if there exist k star-shaped polygons whose union is equal to a given simple polygon, for some parameter k, is NP-complete. The optimization version, i.e., finding a minimum number of star-shaped polygons whose union is equal to a given simple polygon, is NP-hard.

On-line algorithm: An algorithm is said to be online if the input to the algorithm is given one at a time. This is in contrast to the off-line case where the input is known in advance. The algorithm that works online is similar to the off-line algorithms that work incrementally, i.e., it computes a partial solution by considering input data one at a time.

Planar straight-line graph: A graph that can be embedded in the plane without crossings in which every edge in the graph is a straight line segment. It is sometimes referred to as planar subdivision or map.

Star-shaped polygon: A polygon P in which there exists an interior point p such that all the points of P are visible from p. That is, for any point q on the boundary of P, the intersection of the line segment $\overline{p,q}$ with the boundary of P is the point q itself.

r-Star-shaped polygon: A (rectilinear) polygon P in which there exists an interior point p such that all the points of P are r-visible from p. Two points in P are said to be r-visible if there exists a rectangle in P that totally contains these two points.

s-Star-shaped polygon: A (rectilinear) polygon P in which there exists an interior point p such that all the points of P are s-visible from p. Two points in P are said to be s-visible if there exists a staircase path in P that connects these two points. Recall that a staircase path is a rectilinear path that is monotone in both x- and y-directions, i.e., its intersection with every horizontal or vertical line is either empty, a point, or a line segment.

Steiner point: A point that is not part of the input set. It is derived from the notion of Steiner tree. Consider a set of three points determining a triangle $\Delta(a, b, c)$ all of whose angles are smaller than $120°$, in the Euclidean plane. Finding a shortest tree interconnecting these three points is known to require a fourth point s in the interior such that each side of $\Delta(a, b, c)$ subtends the angle at s equal to $120°$. The optimal tree is called the Steiner tree of the three points, and the fourth point is called the Steiner point.

Visibility map: A planar subdivision that encodes the visibility information. Two points p and q are visible if the straight line segment $\overline{p,q}$ does not intersect any other object. A horizontal (or vertical) visibility map of a planar straight-line graph is a partition of the plane into regions by drawing a horizontal (or vertical) straight line through each vertex p until it intersects an edge e of the graph or extends to infinity. The edge e is said to be horizontally (or vertically) visible from p.

References

1. Agarwal, P.K. and Matoušek, J., Ray shooting and parametric search, *SIAM J. Comput.*, 22, 794–806, 1993.
2. Aggarwal, A., Klawe, M.M., Moran, S., Shor, P., and Wilber, R., Geometric applications of a matrix-searching algorithm, *Algorithmica*, 2(2), 195–208, 1987.
3. Amato, N. and Preparata, F.P., The parallel 3D convex hull problem revisited, *Int. J. Comput. Geom. Appl.*, 2(2), 163–173, June 1992.
4. Asano, Ta., Asano, Te., and Imai, H., Partitioning a polygonal region into trapezoids, *J. Assoc. Comput. Mach.*, 33(2), 290–312, April 1986.
5. Asano, Ta., Asano, Te., and Pinter, R.Y., Polygon triangulation: Efficiency and minimality, *J. Algorithms*, 7, 221–231, 1986.
6. Atallah, M.J., Parallel techniques for computational geometry, *Proc. IEEE*, 80(9), 1435–1448, September 1992.
7. Atallah, M.J. and Kosaraju, S.R., An efficient algorithm for maxdominance with applications, *Algorithmica*, 4, 221–236, 1989.
8. Avis, D., Bremner, D., and Seidel, R., How good are convex hull algorithms, *Comput. Geom. Theory Appl.*, 7(5/6), 265–301, April 1997.
9. Bajaj, C. and Dey, T.K., Convex decomposition of polyhedra and robustness, *SIAM J. Comput.*, 21, 339–364, 1992.
10. Bern, M. and Eppstein, D., Approximation algorithms for geometric problems, in *Approximation Algorithms for NP-Hard Problems*, Hochbaum, D. (Ed.), PWS Publishing, Boston, MA, pp. 296–345, 1996.
11. Bern, M., Mitchell, S., and Ruppert, J., Linear-size nonobtuse triangulation of polygons, *Discrete Comput. Geom.*, 14, 411–428, 1995.

12. Buchsbaum, A.L. and Goodrich, M.T., Three dimensional layers of maxima, *Algorithmica*, 39(4), 275–286, May 2004.

13. Chan, T.M., Output-sensitive results on convex hulls, extreme points, and related problems, *Discrete Comput. Geom.*, 16(4), 369–387, April 1996.

14. Chazelle, B., On the convex layers of a planar set, *IEEE Trans. Inf. Theory*, IT-31, 509–517, 1985.

15. Chazelle, B., Triangulating a simple polygon in linear time, *Discrete Comput. Geom.*, 6, 485–524, 1991.

16. Chazelle, B., An optimal convex hull algorithm for point sets in any fixed dimension, *Discrete Comput. Geom.*, 8, 145–158, 1993.

17. Chazelle, B. and Incerpi, J., Triangulation and shape-complexity, *ACM Trans. Graph.*, 3(2), 135–152, 1984.

18. Chazelle, B. and Palios, L., Triangulating a non-convex polytope, *Discrete Comput. Geom.*, 5, 505–526, 1990.

19. Chew, L.P., Constrained Delaunay triangulations, *Algorithmica*, 4(1), 97–108, 1989.

20. Clarkson, K.L. and Shor, P.W., Applications of random sampling in computational geometry, II, *Discrete Comput. Geom.*, 4, 387–421, 1989.

21. Culberson, J.C. and Reckhow, R.A., Covering polygons is hard, *J. Algorithms*, 17(1), 2–44, July 1994.

22. de Berg, M., Cheong, O., van Kreveld, M., and Overmars, M., *Computational Geometry: Algorithms and Applications*, Springer-Verlag, Berlin, Germany, 386 pp., 2008.

23. De Floriani, L. and Puppo, E., An on-line algorithm for constrained Delaunay triangulation, *CVGIP Graph. Models Image Process.*, 54(4), 290–300, July 1992.

24. Dobkin, D.P. and Kirkpatrick, D.G., Fast detection of polyhedral intersection, *Theore. Comput. Sci.*, 27, 241–253, 1983.

25. Domiter, V. and Žalik, B., Sweep-line algorithm for constrained Delaunay triangulation, *Int. J. Geogr. Inf. Sci.*, 22(4), 449–462, 2008.

26. Du, D.Z. and Hwang, F.K. (Eds.), *Computing in Euclidean Geometry*, World Scientific, Singapore, 1992.

27. Edelsbrunner, H., *Algorithms in Combinatorial Geometry*, Springer-Verlag, Berlin, Germany, 1987.

28. Edelsbrunner, H. and Shi, W., An $O(n \log^2 h)$ time algorithm for the three-dimensional convex hull problem, *SIAM J. Comput.*, 20(2), 259–269, April 1991.

29. Edelsbrunner, H. and Tan, T.S., A quadratic time algorithm for the minmax length triangulation, *SIAM J. Comput.*, 22, 527–551, 1993.

30. Eidenbenz, S., Stamm, C., and Widmayer, P., Inapproximability results for guarding polygons and terrains, *Algorithmica*, 31(1), 79–113, 2001.

31. Erten, H. and Ügör, A., Triangulations with locally optimal Steiner points, *Proceedings of the 5th Eurographics Symposium on Geometry Processing*, Barcelona, Spain, pp. 143–152, 2007.

32. Fabri, A., Giezeman, G., Kettner, L., Schirra, S., and Schönherr, S., On the design of CGAL a computational geometry algorithms library, *Softw. Prac. Exp.*, 30(11), 1167–1202, Aug. 2000. http://www.cgal.org/

33. Fortune, S., Progress in computational geometry, in *Directions in Computational Geometry*, Martin, R. (Ed.), Information Geometers, Winchester, UK, pp. 81–128, 1993.

34. Fournier, A. and Montuno, D.Y., Triangulating simple polygons and equivalent problems, *ACM Trans. Graph.*, 3(2), 153–174, 1984.

35. Fragoudakis, C., Markou, E., and Zachos, S., Maximizing the guarded boundary of an art gallery is APX-complete, *Comput. Geom. Theory Appl.*, 38(3), 170–180, October 2007.

36. Gabow, H.N., Bentley, J.L., and Tarjan, R.E., Scaling and related techniques for geometry problems, *Proceedings of the 16th ACM Symposium on Theory of Computing*, New York, pp. 135–143, 1984.

37. Garey, M.R., Johnson, D.S., Preparata, F.P., and Tarjan, R.E., Triangulating a simple polygon, *Inf. Proc. Lett.*, 7, 175–179, 1978.

38. Gewali, L., Keil, M., and Ntafos, S., On covering orthogonal polygons with star-shaped polygons, *Inf. Sci.,* 65(1–2), 45–63, November 1992.

39. Goodman, J.E. and O'Rourke, J. (Eds)., *The Handbook of Discrete and Computational Geometry,* CRC Press LLC, Boca Raton, FL, 1997.

40. Har-Peled, S. and Üngör, A., A time-optimal Delaunay refinement algorithm in two dimensions, *Proceedings of the 21st ACM Symposium on Computational Geometry,* ACM, New York, 228–236, June 2005.

41. Heinrich-Litan, L. and Lübbecke, M.E., Rectangle covers revisited computationally, *J. Exp. Algorithmics,* 11, 1–21, 2006.

42. Hershberger, J. and Suri, S., Matrix searching with the shortest path metric, *SIAM J. Comput.,* 26(6), 1612–1634, December 1997.

43. Houle, M.E., Imai, H., Imai, K., Robert, J.-M., and Yamamoto, P., Orthogonal weighted linear L_1 and L_∞ approximation and applications, *Discrete Appl. Math.,* 43, 217–232, 1993.

44. Imai, H. and Asano, Ta., Efficient algorithms for geometric graph search problems, *SIAM J. Comput.,* 15(2), 478–494, May 1986.

45. Johnson, D.S., The NP-completeness column, *ACM Trans. Algorithms,* 1(1), 160–176, July 2005.

46. Katz, M. and Roisman, G.S., On guarding the vertices of rectilinear domains, *Comput. Geom. Theory Appl.,* 39(3), 219–228, April 2008.

47. Keil, J.M., Decomposing a polygon into simpler components, *SIAM J. Comput.,* 14(4), 799–817, 1985.

48. Kirkpatrick, D.G. and Seidel, R., The ultimate planar convex Hull algorithm? *SIAM J. Comput.,* 15(1), 287–299, February 1986.

49. Lee, D.T., Proximity and reachability in the plane, PhD thesis, Tech. Rep. R-831, Coordinated Science Laboratory, University of Illinois at Urbana-Champaign, Urbana-Champaign, IL, 1978.

50. Lee, D.T., Computational geometry, in *Computer Science and Engineering Handbook,* Tucker, A. (Ed)., CRC Press, Boca Raton, FL, pp. 111–140, 1996.

51. Lee, D.T., Lee, G.C., and Huang, Y.W., Knowledge management for computational problem solving, *J. Universal Comput. Sci.,* 9(6), 563–570, 2003.

52. Lee, D.T. and Lin, A.K., Generalized Delaunay triangulation for planar graphs, *Discrete Comput. Geom.,* 1, 201–217, 1986.

53. Lee, D.T. and Lin, A.K., Computational complexity of art gallery problems, *IEEE Trans. Inf. Theory,* IT-32, 276–282, 1986.

54. Lee, D.T. and Wu, Y.F., Geometric complexity of some location problems, *Algorithmica,* 1(1), 193–211, 1986.

55. Lingas, A., Wasylewicz, A., and Zylinski, P., Note on covering monotone orthogonal polygons with star-shaped polygons, *Inf. Proc. Lett.,* 104(6), 220–227, December 2007.

56. Maehara, H., Acute triangulations of polygons, *Eur. J. Combin.,* 23, 45–55, 2002.

57. Matoušek, J., Efficient partition trees, *Discrete Comput. Geom.,* 8(1), 315–334, 1992.

58. Matoušek, J. and Schwarzkopf, O., On ray shooting on convex polytopes, *Discrete Comput. Geom.,* 10, 215–232, 1993.

59. Matoušek, J. and Schwarzkopf, O., A deterministic algorithm for the three-dimensional diameter problem, *Comput. Geom. Theory Appl.,* 6(4), 253–262, July 1996.

60. Mehlhorn, K. and Näher, S., *LEDA, A Platform for Combinatorial and Geometric Computing,* Cambridge University Press, Cambridge, U.K., 1999.

61. Miller, G.L., A time efficient Delaunay refinement algorithm, *Proceedings of the 15th ACM-SIAM Symposium on Discrete Algorithms,* SIAM, Philadelphia, PA, 400–409, 2004.

62. Motwani, R., Raghunathan, A., and Saran, H., Covering orthogonal polygons with star polygons: The perfect graph approach, *J. Comput. Syst. Sci.,* 40(1), 19–48, 1990.

63. Mulzer, W. and Rote, G., Minimum weight triangulation is NP-Hard, *Proceedings of the 22nd ACM Symposium on Computational Geometry,* Sedona, AZ, 1–10, June 2006. Revised version to appear in *J. Assoc. Comput. Mach.,* 55(2), May 2008.

64. Nielsen, F., Output-sensitive peeling of convex and maximal layers, *Inf. Proc. Lett.,* 59(5), 255–259, September 1996.

65. O'Rourke, J., *Art Gallery Theorems and Algorithms,* Oxford University Press, New York, 1987.

66. Preparata, F.P. and Shamos, M.I., *Computational Geometry: An Introduction,* Springer-Verlag, New York, 1988.

67. Ramos, E.A., An optimal deterministic algorithm for computing the diameter of a three-dimensional point set, *Discrete Comput. Geom.,* 26(2), 233–244, 2001.

68. Ruppert, J., A Delaunay refinement algorithm for quality 2-dimensional mesh generation, *J. Algorithms,* 18(3), 548–585, May 1995.

69. Ruppert, J. and Seidel, R., On the difficulty of triangulating three-dimensional non-convex polyhedra, *Discrete Comput. Geom.,* 7, 227–253, 1992.

70. Sack, J. and Urrutia, J. (Eds.), *Handbook of Computational Geometry,* Elsevier Science Publishers, B.V. North-Holland, Amsterdam, the Netherlands, 2000.

71. Saraf, S., Acute and nonobtuse triangulations of polyhedral surfaces, *Eur. J. Combin.,* on line September 2008, doi:10.1016/j.ejc.2008.08.004.

72. Schuchardt, D. and Hecker, H.D., Two NP-hard art-gallery problems for ortho-polygons, *Math. Log. Q.,* 41, 261–267, 1995.

73. Shermer, T.C., Recent results in art galleries, *Proc. IEEE,* 80(9), 1384–1399, September 1992.

74. Shewchuk, J.R., Triangle: Engineering a 2D quality mesh generator and Delaunay triangulator, in *Applied Computational Geometry: Towards Geometric Engineering,* Lin, M.C. and Manocha, D. (Eds.), Springer-Verlag, Berlin, Germany, pp. 203–222, May 1996.

75. Shewchuk, J.R., Delaunay refinement algorithms for triangular mesh generation, *Comput. Geom. Theory Appl.,* 22(1–3), 21–74, May 2002.

76. Tarjan, R.E. and Van Wyk, C.J., An $O(n \log \log n)$-time algorithm for triangulating a simple polygon, *SIAM J. Comput.,* 17(1), 143–178, February 1988. Erratum: 17(5), 1061, 1988.

77. Toussaint, G.T., New results in computational geometry relevant to pattern recognition in practice, in *Pattern Recognition in Practice II,* Gelsema, E.S. and Kanal, L.N. (Eds.), North-Holland, Amsterdam, the Netherlands, pp. 135–146, 1986.

78. Urrutia, J., Art gallery and illumination problems, in *Handbook on Computational Geometry,* Sack, J.R. and Urrutia, J. (Eds.), Elsevier Science Publishers, Amsterdam, the Netherlands, pp. 973–1026, 2000.

79. Wei, J.D., Tsai, M.H., Lee, G.C., Huang, J.H., and Lee, D.T., GeoBuilder: A geometric algorithm visualization and debugging system for 2D and 3D geometric computing, *IEEE Trans. Vis. Comput. Graph.,* 15(2), 234–248, March 2009.

80. Worman, C. and Keil, J.M., Polygon decomposition and the orthogonal art gallery problem, *Int. J. Comput. Geom. Appl.,* 17(2), 105–138, April 2007.

81. Yao, F.F., Computational geometry, in *Handbook of Theoretical Computer Science, Vol. A: Algorithms and Complexity,* van Leeuwen, J. (Ed.), MIT Press, Cambridge, MA, pp. 343–389, 1994.

82. Yap, C., Towards exact geometric computation, *Comput. Geom. Theory Appl.,* 7(3), 3–23, February 1997.

83. Yuan, L., Acute triangulations of polygons, *Discrete Comput. Geom.,* 34, 697–706, 2005.

84. Yuan, L. and Zamfirescu, T., Acute triangulations of flat möbius strips, *Discrete Comput. Geom.,* 37, 671–676, 2007.

2

Computational Geometry II

D.T. Lee
Academia Sinica

2.1 Introduction

This chapter is a follow-up of Chapter 1, which deals with geometric problems and their efficient solutions. The classes of problems that we address in this chapter include proximity, optimization, intersection, searching, point location, and some discussions of geometric software that has been developed.

2.2 Proximity

Geometric problems abound pertaining to the questions of how close two geometric entities are among a collection of objects or how similar two geometric patterns match each other. For example, in pattern classification and clustering, features that are similar according to some metric are to be clustered in a group. The two aircrafts that are the closest at any time instant in the air space will have the largest likelihood of collision with each other. In some cases one may be interested in how far apart or how dissimilar the objects are. Some of these proximity-related problems will be addressed in this section.

2.2.1 Closest Pair

Consider a set S of n points in $\mathfrak{R}^k$. The closest pair problem is to find in S a pair of points whose distance is the minimum, i.e., find p_i and p_j, such that $d(p_i, p_j) = \min_{k \neq l}\{d(p_k, p_l)$, for all points $p_k, p_l \in S\}$, where $d(a, b)$ denotes the Euclidean distance between a and b. (The result below holds for any distance metric in Minkowski's norm.) Enumerating all pairs of distances to find the pair with the minimum distance would take $O(k \cdot n^2)$ time. As is well-known, in one dimension one can solve the problem much more efficiently: Since the closest pair of points must occur consecutively on the real line, one can sort these points and then scan them in order to solve the closest pair problem in $O(n \log n)$ time. The time complexity turns out to be the best possible, since the problem has a lower bound of $\Omega(n \log n)$, following from a linear time transformation from the *element uniqueness problem* [90].

But unfortunately there is no total ordering for points in $\mathfrak{R}^k$ for $k \geq 2$, and thus, sorting is not applicable. We will show that by using divide-and-conquer approach, one can solve this problem in $O(n \log n)$ optimal time. Let us consider the case when $k = 2$. In the following, we only compute the minimum distance between the closest pair; the actual identity of the closest pair that realizes the minimum distance can be found easily by some straightforward bookkeeping operations. Consider a vertical separating line $\mathcal{V}$ that divides S into S_1 and S_2 such that $|S_1| = |S_2| = n/2$. Let δ_i denote the minimum distance defined by the closest pair of points in S_i, $i = 1, 2$. Observe that the minimum distance defined by the closest pair of points in S is either δ_1, δ_2, or $d(p, q)$ for some $p \in S_1$ and $q \in S_2$. In the former case, we are done. In the latter, points p and q must lie in the vertical strip of width $\delta = \min\{\delta_1, \delta_2\}$ on each side of the separating line $\mathcal{V}$ (Figure 2.1). The problem now reduces to that of finding the closest pair between points in S_1 and S_2 that lie inside the strip $\mathcal{L}$ of width 2δ. This subset of points $\mathcal{L}$ possesses a special property, known as *sparsity*, i.e., for each square box* of length 2δ the number of points in $\mathcal{L}$ is bounded by a constant $c = 4 \cdot 3^{k-1}$, since in each set S_i, there exists no point that lies in the interior of the δ-ball centered at each point in S_i, $i = 1, 2$ [90] (Figure 2.2). It is this sparsity property that enables us to solve the *bichromatic closest pair problem* in $O(n)$ time.

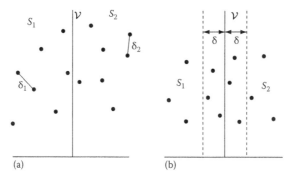

FIGURE 2.1 Divide-and-conquer scheme for closest pair problem. Solutions to subproblems S_1 and S_2 (a) and candidates must lie in the vertical strip of width δ on each side of $\mathcal{V}$ (b).

---

* A box is a hypercube in higher dimensions.

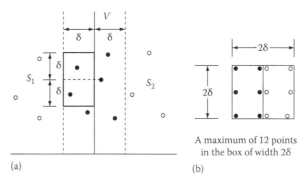

A maximum of 12 points
in the box of width 2δ

(a) (b)

FIGURE 2.2 The box of width 2δ dissected by the separating line has at most 12 points; each point in S_2 needs to examine at most 6 points in S_1 to find its closest neighbor (a) and the box of width 2δ dissected by the separating line has at most 12 points (b).

The *bichromatic closest pair problem* is defined as follows. Given two sets of red and blue points, denoted R and B, find the closest pair $r \in R$ and $b \in B$, such that $d(r,b)$ is minimum among all possible distances $d(u,v)$, $u \in R$, $v \in B$. Let $\overline{S}_i \subseteq S_i$ denote the set of points that lie in the vertical strip. In two dimensions, the sparsity property ensures that for each point $p \in \overline{S}_1$ the number of candidate points $q \in \overline{S}_2$ for the closest pair is at most six (Figure 2.2). We, therefore, can scan these points $\overline{S}_1 \cup \overline{S}_2$ in order along the separating line V and compute the distance between each point in $\overline{S}_1$ (respectively, $\overline{S}_2$) scanned and its six candidate points in $\overline{S}_2$ (respectively, $\overline{S}_1$). The pair that gives the minimum distance δ_3 is the bichromatic closest pair. The minimum distance of all pairs of points in S is then equal to $\delta_S = \min\{\delta_1, \delta_2, \delta_3\}$.

Since the merge step takes linear time, the entire algorithm takes $O(n \log n)$ time. This idea generalizes to higher dimensions, except that to ensure the sparsity property of the set $\mathcal{L}$, the separating hyperplane should be appropriately chosen so as to obtain an $O(n \log n)$-time algorithm [90], which is asymptotically optimal.

We note that the bichromatic closest pair problem is in general more difficult than the closest pair problem. Edelsbrunner and Sharir [46] showed that in three dimensions the number of possible closest pairs is $O((|R| \cdot |B|)^{2/3} + |R| + |B|)$. Agarwal et al. [3] gave a randomized algorithm with an expected running time of $O((mn \log m \log n)^{2/3} + m \log^2 n + n \log^2 m)$ in $\mathfrak{R}^3$ and $O((mn)^{1-1/(\lceil k/2 \rceil + 1) + \epsilon} + m \log n + n \log m)$ in $\mathfrak{R}^k$, $k \geq 4$, where $m = |R|$ and $n = |B|$. Only when the two sets possess the sparsity property defined above can the problem be solved in $O(n \log n)$ time, where $n = |R| + |B|$. A more general problem, known as the fixed radius all nearest neighbor problem in a sparse set [90], i.e., given a set M of points in $\mathfrak{R}^k$ that satisfies the sparsity condition, finds that all pairs of points whose distance is less than a given parameter δ can be solved in $O(|M| \log |M|)$ time [90].

The closest pair of vertices u and v of a simple polygon P such that $\overline{u,v}$ lies totally within P can be found in linear time [55]; $\overline{u,v}$ is also known as a diagonal of P.

2.2.2 Voronoi Diagrams

The Voronoi diagram $\mathcal{V}(S)$ of a set S of points, called *sites*, $S = \{p_1, p_2, \ldots, p_n\}$ in $\mathfrak{R}^k$ is a partition of $\mathfrak{R}^k$ into Voronoi cells $V(p_i)$, $i = 1, 2, \ldots, n$, such that each cell contains points that are closer to site p_i than to any other site $p_j, j \neq i$, i.e.,

$$V(p_i) = \left\{ x \in R^k \,\middle|\, d(x, p_i) \leq d(x, p_j) \,\forall p_j \in R^k, j \neq i \right\}.$$

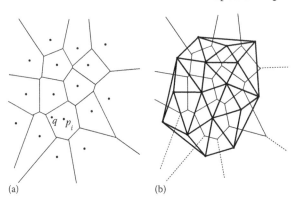

FIGURE 2.3 The Voronoi diagram of a set of 16 points in the plane (a) and its dual graph a Delaunay triangulation (b).

In two dimensions, $\mathcal{V}(S)$ is a planar graph and is of size linear in $|S|$. In dimensions $k \geq 2$, the total number of d-faces of dimensions $d = 0, 1, \ldots, k - 1$, in $\mathcal{V}(S)$ is $O(n^{\lceil k/2 \rceil})$.

Figure 2.3a shows the Voronoi diagram of 16 point sites in two dimensions. Figure 2.3b shows the straight-line dual graph of the Voronoi diagram, which is called the Delaunay triangulation (cf. Section 1.5.2). In this triangulation, the vertices are the sites, and the two vertices are connected by an edge, if their Voronoi cells are adjacent.

2.2.2.1 Construction of Voronoi Diagrams in Two Dimensions

The Voronoi diagram possesses many proximity properties. For instance, for each site p_i, the closest site must be among those whose Voronoi cells are adjacent to $V(p_i)$. Thus, the closest pair problem for S in $\mathfrak{R}^2$ can be solved in linear time after the Voronoi diagram has been computed. Since this pair of points must be adjacent in the Delaunay triangulation, all one has to do is to examine all the adjacent pairs of points and report the pair with the smallest distance. A divide-and-conquer algorithm to compute the Voronoi diagram of a set of points in the L_p-metric for all $1 \leq p \leq \infty$ is known [62]. There is a rich body of literature concerning the Voronoi diagram. The interested reader is referred to the surveys [41,93].

We give below a brief description of a plane-sweep algorithm, known as the *wavefront approach*, due to Dehne and Klein [38]. Let $S = \{p_1, p_2, \ldots, p_n\}$ be a set of point sites in $\mathfrak{R}^2$ sorted in ascending x-coordinate value, i.e., $x(p_1) < x(p_2) < \cdots < x(p_n)$. Consider that we sweep a vertical line $\mathcal{L}$ from left to right and as we sweep $\mathcal{L}$, we compute the Voronoi diagram $\mathcal{V}(S_t)$, where

$$S_t = \{p_i \in S \,|\, x\,(p_i) \,<\, t\} \cup \{\mathcal{L}_t\}.$$

Here $\mathcal{L}_t$ denotes the vertical line whose x-coordinate equals t. As is well-known, $\mathcal{V}(S_t)$ will contain not only straight-line segments, which are portions of perpendicular bisectors of two-point sites, but also parabolic curve segments, which are portions of *bisectors* of one-point site and $\mathcal{L}_t$. The wavefront W_t, consisting of a sequence of parabolae, called *waves*, is the boundary of the Voronoi cell $V(\mathcal{L}_t)$ with respect to S_t. Figure 2.4a and b illustrate two instances, $\mathcal{V}(S_t)$ and $\mathcal{V}(S_{t'})$. Those Voronoi cells that do not contribute to the wavefront are final, whereas those that do will change as $\mathcal{L}$ moves to the right. There are two possible events at which the wavefront needs an update. One, called site event, is when a site is hit by $\mathcal{L}$ and a new wave appears. The other, called spike event, is when an old wave disappears. Let p_i and p_j be two sites such that the associated waves are adjacent in W_t. The bisector of p_i and p_j defines an edge of $\mathcal{V}(S_t)$ to the left of W_t. Its extension into the cell $V(\mathcal{L}_t)$ is called a spike. The spikes can be viewed as tracks along which two neighboring waves travel.

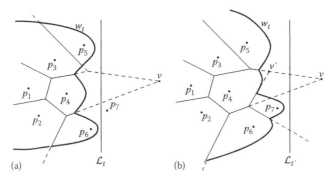

FIGURE 2.4 The Voronoi diagrams of (a) $\mathcal{V}(S_t)$ and (b) $\mathcal{V}(S_{t'})$.

A wave disappears from W_t, once it has reached the point where its two neighboring spikes intersect. In Figure 2.4a dashed lines are spikes and v is a potential spike event point. Without p_7 the wave of p_3 would disappear first and then the wave of p_4. After p_7, a site event, has occurred, a new point v' will be created and it defines an earlier spike event than v. v' will be a spike event point at which the wave of p_4 disappears and waves of p_5 and p_7 become adjacent. Note that the spike event corresponding to v' does not occur at $\mathcal{L}_t$, when $t = x(v')$. Instead, it occurs at $\mathcal{L}_t$, when $t = x(v') + d(v', p_4)$. If there is no site event between $\mathcal{L}_{x(p_7)}$ and $\mathcal{L}_t$, then the wave of p_4 will disappear. It is not difficult to see that after all site events and spike events have been processed at time τ, $\mathcal{V}(S)$ is identical to $\mathcal{V}(S_\tau)$ with the wavefront removed.

Since the waves in W_t can be stored in a *height-balanced binary search tree* and the site events and spike events can be maintained as a *priority queue*, the overall time and space needed are $O(n \log n)$ and $O(n)$, respectively.

Although $\Omega(n \log n)$ is the lower bound for computing the Voronoi diagram for an arbitrary set of n sites, this lower bound does not apply to special cases, e.g., when the sites are on the vertices of a convex polygon. In fact, the Voronoi diagram of a convex polygon can be computed in linear time [5]. This demonstrates further that additional properties of the input can sometimes help reduce the complexity of the problem.

2.2.2.2 Construction of Voronoi Diagrams in Higher Dimensions

The Voronoi diagrams in $\mathfrak{R}^k$ are related to the convex hulls $\mathfrak{R}^{k+1}$ via a **geometric duality** transformation. Consider a set S of n sites in $\mathfrak{R}^k$, which is the hyperplane $\mathcal{H}^0$ in $\mathfrak{R}^{k+1}$ such that $x_{k+1} = 0$, and a paraboloid $\mathcal{P}$ in $\mathfrak{R}^{k+1}$ represented as $x_{k+1} = x_1^2 + x_2^2 + \cdots + x_k^2$. Each site $p_i = (\mu_1, \mu_2, \dots, \mu_k)$

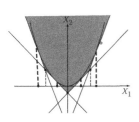

is transformed into a hyperplane $\mathcal{H}(p_i)$ in $\mathfrak{R}^{k+1}$ denoted as $x_{k+1} = 2\sum_{j=1}^{k} \mu_j x_j - (\sum_{j=1}^{k} \mu_j^2)$. That is, $\mathcal{H}(p_i)$ is tangent to the paraboloid $\mathcal{P}$ at point $\mathcal{P}(p_i) = (\mu_1, \mu_2, \dots, \mu_k, \mu_1^2 + \mu_2^2 + \cdots + \mu_k^2)$, which is just the vertical projection of site p_i onto the paraboloid $\mathcal{P}$. See Figure 2.5 for an illustration of the transformation in one dimension. The half-space defined by $\mathcal{H}(p_i)$ and containing the paraboloid $\mathcal{P}$ is denoted as $\mathcal{H}^+(p_i)$. The intersection of all half-spaces $\bigcap_{i=1}^{n} \mathcal{H}^+(p_i)$ is a convex body and the boundary of the convex body is denoted by $CH(\mathcal{H}(S))$. Any point $q \in \mathfrak{R}^k$ lies in the Voronoi cell $V(p_i)$, if the vertical projection of q onto $CH(\mathcal{H}(S))$ is contained in $\mathcal{H}(p_i)$. The distance between point q and its closest site p_i can be shown to be equal to the square root of the vertical distance between its vertical projection $\mathcal{P}(q)$ on the paraboloid $\mathcal{P}$ and

FIGURE 2.5 The paraboloid transformation of a site in one dimension to a line tangent to a parabola.

on $CH(\mathcal{H}(S))$. Moreover every κ-face of $CH(\mathcal{H}(S))$ has a vertical projection on the hyperplane $\mathcal{H}^0$ equal to the κ-face of the Voronoi diagram of S in $\mathcal{H}^0$.

We thus obtain the result which follows from the theorem for the convex hull in Chapter 1.

THEOREM 2.1 *The Voronoi diagram of a set S of n points in $\Re^k$, $k \geq 2$ can be computed in $O(n \log \mathcal{H})$ time for $k = 2$, and in $O(n \log \mathcal{H} + (n\mathcal{H})^{1-1/(\lfloor (k+1)/2 \rfloor + 1)} \log^{O(1)} n)$ time for $k > 2$, where $\mathcal{H}$ is the number of i-faces, $i = 0, 1, \ldots, k$.*

It has been shown that the Voronoi diagram in $\Re^k$, for $k = 3, 4$, can be computed in $O((n + \mathcal{H}) \log^{k-1} \mathcal{H})$ time [12].

2.2.2.3 Farthest Neighbor Voronoi Diagram

The Voronoi diagram defined in Section 2.2.2 is also known as the *nearest neighbor Voronoi diagram*. The nearest neighbor Voronoi diagram partitions the space into cells such that each site has its own cell, which contains all the points that are closer to this site than to any other site. A variation of this partitioning concept is a partition of the space into cells, each of which is associated with a site, and contains all the points that are farther from the site than from any other site. This diagram is called the *farthest neighbor Voronoi diagram*. Unlike the nearest neighbor Voronoi diagram, the farthest neighbor Voronoi diagram only has a subset of sites which have a Voronoi cell associated with them. Those sites that have a nonempty Voronoi cell are those that lie on the convex hull of S. A similar partitioning of the space is known as the *order κ-nearest neighbor Voronoi diagram*, in which each Voronoi cell is associated with a subset of κ sites in S for some fixed integer κ such that these κ sites are the closest among all other sites. For $\kappa = 1$, we have the nearest neighbor Voronoi diagram, and for $\kappa = n - 1$, we have the farthest neighbor Voronoi diagram. The construction of the order κ-nearest neighbor Voronoi diagram in the plane can be found in, e.g., [90]. *The order κ Voronoi diagrams in $\Re^k$ are related to the levels of hyperplane arrangements in $\Re^{k+1}$ using the paraboloid transformation discussed in Section 2.2.2.2. See, e.g., [2] for details. Below is a discussion of the farthest neighbor Voronoi diagram in two dimensions.

Given a set S of sites $s_1, s_2, \ldots, s_n$, the f-neighbor Voronoi cell of site s_i is the locus of points that are farther from s_i than from any other site s_j, $i \neq j$, i.e.,

$$f\_V(s_i) = \left\{ p \in \Re^2 \mid d(p, s_i) \geq d(p, s_j), s_i \neq s_j \right\}.$$

The union of these f-neighbor Voronoi cells is called the farthest neighbor Voronoi diagram of S. Figure 2.6 shows the farthest neighbor Voronoi diagram for a set of 16 sites. Note that only sites that are on the convex hull $CH(S)$ will have a nonempty f-neighbor Voronoi cell [90] and that all the f-neighbor Voronoi cells are unbounded.

Since the farthest neighbor Voronoi diagram in the plane is related to the convex hull of the set of sites, one can use the divide-and-marriage-before-conquest paradigm to compute the farthest neighbor Voronoi diagram of S in two dimensions in time $O(n \log \mathcal{H})$, where $\mathcal{H}$ is the number of sites on the convex hull. Once the convex hull is available, the linear time algorithm [5] for computing the Voronoi diagram for a convex polygon can be applied.

2.2.2.4 Weighted Voronoi Diagrams

When the sites have weights such that the distance from a point to the sites is weighted, the structure of the Voronoi diagram can be drastically different than the unweighted case. We consider a few examples.

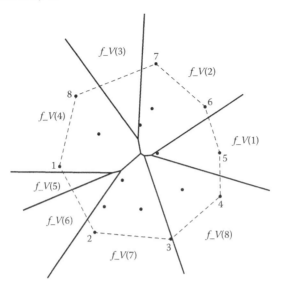

FIGURE 2.6 The farthest neighbor Voronoi diagram of a set of 16 sites in the plane.

Example 2.1: Power Diagrams

Suppose each site s in $\Re^k$ is associated with a nonnegative weight, w_s. For an arbitrary point p in $\Re^k$ the weighted distance from p to s is defined as

$$\delta(s,p) = d(s,p)^2 - w_s^2.$$

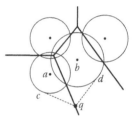

If w_s is positive, and if $d(s,p) \geq w_s$, then $\sqrt{\delta(s,p)}$ is the length of the tangent of p to the ball, $b(s)$, of radius w_s and centered at s. $\delta(s,p)$ is also called the *power of p* with respect to the ball $b(s)$. The locus of points p equidistant from two sites $s \neq t$ of equal weight will be a hyperplane called the *chordale* of s and t (see Figure 2.7). Point q is equidistant to sites a and b, and the distance is the length of the tangent line $\overline{q,c} = \overline{q,d}$.

FIGURE 2.7 The power diagram in two dimensions. $\delta(q,a) = \delta(q,b) = $ length of $\overline{q,c}$.

The power diagram in two dimensions can be used to compute the contour of the union of n disks, and the connected components of n disks in $O(n \log n)$ time, and in higher dimensions, it can be used to compute the union or intersection of n axis-parallel cones in $\Re^k$ with apexes in a common hyperplane in time $O(CH_{k+1}(n))$, the multiplicative-weighted nearest neighbor Voronoi diagram (defined below) for n points in $\Re^k$ in time $O(CH_{k+2}(n))$, and the Voronoi diagrams for n spheres in $\Re^k$ in time $O(CH_{k+2}(n))$, where $CH_\ell(n)$ denotes the time for constructing the convex hull of n points in $\Re^\ell$ [93]. For the best time bound for $CH_\ell(n)$, See Section 1.1. For more results on the union of spheres and the volumes see [45].

Example 2.2: Multiplicative-Weighted Voronoi Diagrams

Suppose each site $s \in \Re^k$ is associated with a positive weight w_s. The distance from a point $p \in \Re^k$ to s is defined as

$$\delta_{multi-w}(s,p) = d(p,s)/w_s.$$

In two dimensions, the locus of points equidistant to two sites $s \neq t$ is a disk, if $w_s \neq w_t$, and a perpendicular bisector of line segment $\overline{s,t}$, if $w_s = w_t$. Each cell associated with a site s consists of all points closer to s than to any other site and may be disconnected. In the worst case, the multiplicative-weighted nearest neighbor Voronoi diagram of a set S of n points in two dimensions can have $O(n^2)$ regions and can be computed in $O(n^2)$ time. But in one dimension, the diagram can be computed optimally in $O(n \log n)$ time. On the other hand, the multiplicative-weighted farthest neighbor Voronoi diagram has a very different characteristic. Each Voronoi cell associated with a site remains connected, and the size of the diagram is still linear in the number of sites. An $O(n \log^2 n)$-time algorithm for constructing such a diagram is given in [66]. See [80] for more applications of the diagram.

Example 2.3: Additive-Weighted Voronoi Diagrams

Suppose each site $s \in \mathfrak{R}^k$ is associated with a positive weight w_s. The distance of a point $p \in \mathfrak{R}^k$ to a site s is defined as

$$\delta_{add-w}(s,p) = d(p,s) - w_s.$$

In two dimensions, the locus of points equidistant to two sites $s \neq t$ is a branch of a hyperbola, if $w_s \neq w_t$, and a perpendicular bisector of line segment $\overline{s,t}$, if $w_s = w_t$. The Voronoi diagram has properties similar to the ordinary unweighted diagram. For example, each cell is still connected and the size of the diagram is linear. If the weights are positive, the diagram is the same as the Voronoi diagram of a set of spheres centered at site s and of radius w_s, and in two dimensions, this diagram for n disks can be computed in $O(n \log n)$ time [15,81], and in $k \geq 3$, one can use the notion of power diagram (cf. Example 2.1) to compute the diagram [93].

2.2.2.5   Generalizations of Voronoi Diagrams

We consider two variations of Voronoi diagrams that are of interest and have applications.

Example 2.4: Geodesic Voronoi Diagrams

The nearest neighbor geodesic Voronoi diagram is a Voronoi diagram of sites in the presence of obstacles. The distance from point p to a site s, called the *geodesic distance* between p and s, is the length of the shortest path from p to s avoiding all the obstacles (cf. Section 2.6.1). The locus of points equidistant to two sites s and t is in general a collection of hyperbolic segments. The cell associated with a site is the locus of points whose geodesic distance to the site is shorter than to any other site [86]. The farthest neighbor geodesic Voronoi diagram can be similarly defined. Efficient algorithms for computing either kind of geodesic Voronoi diagram for k point sites in an n-sided simple polygon

FIGURE 2.8   The geodesic Voronoi diagram within a simple polygon.

in $O((n+k) \log(n+k))$ time can be found in [86]. Figure 2.8 illustrates the geodesic Voronoi diagram of a set of point sites within a simple polygon; the whole shaded region is $V(s_i)$.

Example 2.5: Skew Voronoi Diagrams

A *directional distance* function between two points in the plane is introduced in Aichholzer et al. [8] that models a more realistic distance measure. The distance, called *skew distance*, from point p to point q is defined as

$$\tilde{d}(p,q) = d(p,q) + k \cdot d_y(p,q),$$

where $d_y(p,q) = y(q) - y(p)$, and $k \geq 0$ is a parameter. This distance function is *asymmetric* and satisfies $\tilde{d}(p,q) + \tilde{d}(q,p) = 2d(p,q)$, and the triangle inequality. Imagine we have a tilted plane $\mathcal{T}$

obtained by rotating the *xy*-plane by an angle α about the *x*-axis. The height (*z*-coordinate) $h(p)$ of a point *p* on $\mathcal{T}$ is related to its *y*-coordinate by $h(p) = y(p) \cdot \sin \alpha$.

The distance function defined above reflects the cost that is proportional to the difference of their heights; the distance is *smaller* going *downhill* than going *uphill*. That is, the distance from *p* to *q* defined as $\tilde{d}(p,q) = d(p,q) + \kappa \cdot (h(q) - h(p))$ for $\kappa > 0$ serves this purpose; $\tilde{d}(p,q)$ is less than $\tilde{d}(q,p)$, if $h(q)$ is smaller than $h(p)$.

Because the distance is directional, one can define two kinds of Voronoi diagrams defined by the set of sites. A skew Voronoi cell from a site *p*, $\mathcal{V}_{\text{from}}(p)$, is defined as the set of points that are closest to *p* than to any other site. That is,

$$\mathcal{V}_{\text{from}}(p) = \left\{ x \,\middle|\, \tilde{d}(p,x) \leq \tilde{d}(q,x) \right\}$$

for all $q \neq p$. Similarly one can define a skew Voronoi cell to a site *p* as follows:

$$\mathcal{V}_{\text{to}}(p) = \left\{ x \,\middle|\, \tilde{d}(x,p) \leq \tilde{d}(x,q) \right\}$$

for all $q \neq p$.

The collection of these Voronoi cells for all sites is called the skew (or directional) Voronoi diagram.

For each site *p*, we define an *r*-disk centered at *p*, denoted from$_r(p)$ to be the set of points to which the skew distance from *p* is *r*. That is, from$_r(p) = \{x | \tilde{d}(p,x) = r\}$. Symmetrically, we can also define an *r*-disk centered at *p*, denoted to$_r(p)$ to be the set of points from which the skew distance to *p* is *r*. That is, to$_r(p) = \{x | \tilde{d}(x,p) = r\}$. The subscript *r* is omitted, when $r = 1$. It can be shown that to$_r(p)$ is just a mirror reflection of from$_r(p)$ about the horizontal line passing through *p*. We shall consider only the skew Voronoi diagram which is the collection of the cells $\mathcal{V}_{\text{from}}(p)$ for all $p \in S$.

LEMMA 2.1   For $k > 0$, the unit disk *from(p)* is a conic with focus *p*, directrix the horizontal line at *y*-distance $1/k$ above *p*, and eccentricity *k*. Thus, *from(p)* is an ellipse for $k < 1$, a parabola for $k = 1$, and a hyperbola for $k > 1$. For $k = 0$, *from(p)* is a disk with center *p* (which can be regarded as an ellipse of eccentricity zero).

Note that when *k* equals 0, the skew Voronoi diagram reduces to the ordinary nearest neighbor Voronoi diagram. When $k < 1$, it leads to known structures: By Lemma 2.1, the skew distance $\tilde{d}$ is a *convex distance function* and the Voronoi diagrams for convex distance functions are well studied (see, e.g., [93]). They consist of $O(n)$ edges and vertices, and can be constructed in time $O(n \log n)$ by divide-and-conquer.

When $k \geq 1$, since the unit disks are no longer bounded, the skew Voronoi diagrams have different behavior from the ordinary ones. As it turns out, some of the sites do not have nonempty skew Voronoi cells in this case. In this regard, it looks like ordinary farthest neighbor Voronoi diagram discussed earlier.

Let $L_0(p,k)$ denote the locus of points *x* such that $\tilde{d}(p,x) = 0$. It can be shown that for $k = 1$, $L_0(p,k)$ is a vertical line emanating downwards from *p*; and for $k > 1$, it consists of two rays, emanating from, and extending below, *p*, with slopes $1/(\sqrt{k^2 - 1})$ and $-1/(\sqrt{k^2 - 1})$, respectively. Let $N(p,k)$ denote the area below $L_0(p,k)$ (for $k > 1$). Let the *0-envelope*, $E_0(S)$, be the upper boundary of the union of all $N(p,k)$ for $p \in S$. $E_0(S)$ is the upper envelope of the graphs of all $L_0(p,k)$, when being seen as functions of the *x*-coordinate. For each point *u* lying above $E_0(S)$, we have $\tilde{d}(p,u) > 0$ for all $p \in S$, and for each point *v* lying below $E_0(S)$, there is at least one $p \in S$ with $\tilde{d}(p,v) < 0$. See Figure 2.9 for an example of a 0-envelope (shown as the dashed polygonal line) and the corresponding skew Voronoi diagram. Note that the skew Voronoi cells associated with sites *q* and *t* are empty. The following results are obtained [8].

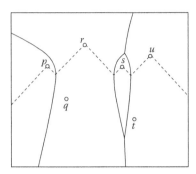

FIGURE 2.9 The 0-envelope and the skew Voronoi diagram when $k = 1.5$.

LEMMA 2.2   For $k > 1$, the 0-envelope $E_0(S)$ of a set S of n sites can be computed in $O(n \log \mathcal{H})$ time and $O(n)$ space, where $\mathcal{H}$ is the number of edges of $E_0(S)$.

LEMMA 2.3   Let $p \in S$ and $k > 1$, then $\mathcal{V}_{\text{from}}(p) \neq \emptyset$, if and only if $p \in E_0(S)$. $\mathcal{V}_{\text{from}}(p)$ is unbounded if and only if p lies on the upper hull of $E_0(S)$. For $k = 1$, $\mathcal{V}_{\text{from}}(p)$ is unbounded for all p.

THEOREM 2.2   *For any $k \geq 0$, the skew Voronoi diagram for n sites can be computed in $O(n \log \mathcal{H})$ time and $O(n)$ space, where $\mathcal{H}$ is the number of nonempty skew Voronoi cells in the resulting Voronoi diagram.*

The sites mentioned so far are point sites. They can be of different shapes. For instance, they can be line segments or polygonal objects. The Voronoi diagram for the edges of a simple polygon P that divides the interior of P into Voronoi cells is also known as the *medial axis* or *skeleton* of P [90]. The distance function used can also be the convex distance function or other norms.

2.3 Optimization

The geometric optimization problems arise in operations research, Very Large-Scale Integrated Circuit (VLSI) layout, and other engineering disciplines. We give a brief description of a few problems in this category that have been studied in the past.

2.3.1 Minimum Cost Spanning Tree

The minimum (cost) spanning tree (MST) of an undirected, weighted graph $G(V, E)$, in which each edge has a nonnegative weight, is a well-studied problem in graph theory and can be solved in $O(|E| \log |V|)$ time [90]. When cast in the Euclidean or other L_p-metric plane in which the input consists of a set S of n points, the complexity of this problem becomes different. Instead of constructing a complete graph with edge weight being the distance between its two endpoints, from which to extract an MST, a sparse graph, known as the *Delaunay triangulation* of the point set, is computed. The Delaunay triangulation of S, which is a planar graph, is the straight-line dual of the Voronoi diagram of S. That is, two points are connected by an edge, if and only if the Voronoi cells of these two sites share an edge. (cf. Section 1.5.2.3). It can be shown that the MST of S is a subgraph of the Delaunay triangulation. Since the MST of a planar graph can be found in linear time [90], the problem can be solved in $O(n \log n)$ time. In fact, this is asymptotically optimal, as the closest pair of the set of points must define an edge in the MST, and the closest pair problem is known to have an $\Omega(n \log n)$ lower bound, as mentioned in Section 2.2.1.

This problem in dimensions three or higher can be solved in subquadratic time. Agarwal et al. [3] showed that the Euclidean MST problem for a set of N points in $\Re^k$ can be solved in time $O(T_k(N,N) \log^k N)$, where $T_k(m,n)$ denotes the time required to compute a bichromatic closest pair among m red and n blue points in $\Re^k$. If $T_k(N,N) = \Omega(N^{1+\epsilon})$, for some fixed $\epsilon > 0$, then the running time improves to be $O(T_k(N,N))$. They also gave a randomized algorithm with an expected time of $O((N \log N)^{4/3})$ in three dimensions and of $O(N^{2(1-1/(\lceil k/2 \rceil+1))+\epsilon})$ for any positive ϵ in $k \geq 4$ dimensions [3]. Interestingly enough, if we want to find an MST that spans at least k nodes in a planar graph (or in the Euclidean plane), for some parameter $k \leq n$, then the problem, called k-MST problem, is NP-hard [92]. Approximation algorithms for the k-MST problem can be found in [18,75].

2.3.2 Steiner Minimum Tree

The *Steiner minimum tree* (SMT) of a set of vertices $S \subseteq V$ in an undirected weighted graph $G(V, E)$ is a spanning tree of $S \cup Q$ for some $Q \subseteq V$ such that the total weight of the spanning tree is minimum. This problem differs from MST in that we need to identify a set $Q \subseteq V$ of *Steiner vertices* so that the total cost of the spanning tree is minimized. Of course, if $S = V$, SMT is the same as MST. It is the identification of the Steiner vertices that makes this problem intractable. In the plane, we are given a set S of points and are to find the shortest tree interconnecting points in S, while additional Steiner points are allowed. Both Euclidean and rectilinear (L_1-metric) SMT problems are known to be NP-hard. In the geometric setting, the rectilinear SMT problem arises mostly in VLSI net routing, in which a number of terminals need to be interconnected using horizontal and vertical wire segments using the shortest wire length. As this problem is intractable, heuristics are proposed. For more information, the reader is referred to a special issue of *Algorithmica* on Steiner trees, edited by Hwang [57]. Most heuristics for the L_1 SMT problem are based on a classical theorem, known as the *Hanan grid theorem*, which states that the Steiner points of an SMT must be at the grid defined by drawing horizontal and vertical lines through each of the given points. However, when the number of orientations permitted for routing is greater than 2, the Hanan grid theorem no longer holds true. Lee and Shen [65] established a *multi-level grid* theorem, which states that the Steiner points of an SMT for n points must be at the grid defined by drawing λ lines in the feasible orientation recursively for up to $n - 2$ levels, where λ denotes the number of orientations of the wires allowed in routing. That is, the given points are assumed to be at the *0th level*. At each level, λ lines in the feasible orientations are drawn through each new grid point created at the previous level. In this λ-geometry plane, feasible orientations are assumed to make an angle $i\pi/\lambda$ with the positive x-axis. For the rectilinear case, $\lambda = 2$, Figure 2.10 shows that Hanan grid is insufficient for determining a Steiner SMT for $\lambda = 3$. Steiner point s_3 does not lie on the Hanan grid.

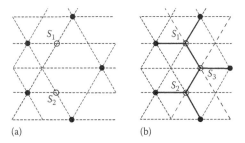

(a) (b)

FIGURE 2.10 Hanan grid theorem fails for $\lambda = 3$. Steiner point s_3 does not lie on the Hanan grid (a), but they line on a second-level grid (b).

DEFINITION 2.1   The performance ratio of any approximation $\mathcal{A}$ in metric space $\mathcal{M}$ is defined as

$$\rho_{\mathcal{M}}(\mathcal{A}) = \inf_{P \in \mathcal{M}} \frac{L_s(P)}{L_{\mathcal{A}}(P)}$$

where $L_s(P)$ and $L_{\mathcal{A}}(P)$ denote, respectively, the lengths of a Steiner minimum tree and of the approximation $\mathcal{A}$ on P in space $\mathcal{M}$. When the MST is the approximation, the performance ratio is known as the Steiner ratio, denoted simply as ρ.

It is well-known that the Steiner ratios for the Euclidean and rectilinear SMTs are $\frac{\sqrt{3}}{2}$ and $\frac{2}{3}$, respectively [57]. The λ-*Steiner ratio** for the λ-geometry SMTs is no less than $\frac{\sqrt{3}\cos(\pi/2\lambda)}{2}$. The following interesting result regarding Steiner ratio is reported in [65], which shows that the Steiner ratio is not an increasing function from $\frac{2}{3}$ to $\frac{\sqrt{3}}{2}$, as λ varies from 2 to ∞.

THEOREM 2.3   *The λ-Steiner ratio is $\frac{\sqrt{3}}{2}$, when λ is a multiple of 6, and $\frac{\sqrt{3}\cos(\pi/2\lambda)}{2}$ when λ is a multiple of 3 but not a multiple of 6.*

2.3.3 Minimum Diameter Spanning Tree

The *minimum diameter spanning tree* (MDST) of an undirected weighted graph $G(V, E)$ is a spanning tree such that its **diameter**, i.e., total weight of the longest path in the tree, is minimum. This arises in applications to communication network where a tree is sought such that the maximum delay, instead of the total cost, is to be minimized. Using a graph-theoretic approach one can solve this problem in $O(|E||V|\log|V|)$ time. However, by the triangle inequality one can show that there exists an MDST such that the longest path in the tree consists of no more than *three segments* [56]. Based on this an $O(n^3)$-time algorithm was obtained.

THEOREM 2.4   *Given a set S of n points, the minimum diameter spanning tree for S can be found in $\Theta(n^3)$ time and $O(n)$ space.*

We remark that the problem of finding a spanning tree whose total cost and the diameter are both bounded is NP-complete [56]. In [92], the problem of finding a *minimum diameter cost spanning tree* is studied. In this problem for each pair of vertices v_i and v_j there is a weighting function $w_{i,j}$ and the diameter *cost* of a spanning tree is defined to be the maximum over $w_{i,j} * d_{i,j}$, where $d_{i,j}$ denotes the distance between vertices v_i and v_j. To find a spanning tree with minimum diameter cost as defined above is shown to be NP-hard [92].

Another similar problem that arises in VLSI clock tree routing is to find a tree from a source to multiple sinks such that every source-to-sink path is the shortest rectilinear path and the total wire length is to be minimized. This problem, also known as *rectilinear Steiner arborescence problem* (see [57]), has been shown to be NP-complete [98]. A polynomial time approximation scheme of approximation ratio $(1 + 1/\epsilon)$ in time $O(n^{O(\epsilon)}\log n)$ was given by Lu and Ruan [70]. Later a simple 2-approximation algorithm in time $O(n\log n)$ was provided by Ranmath [91]. The problem of finding a minimum spanning tree such that the longest source-to-sink path is bounded by a given parameter is shown also to be NP-complete [96].

---

* The λ-Steiner ratio is defined as the greatest lower bound of the length of SMT over the length of MST in the λ-geometry plane.

2.3.4 Minimum Enclosing Circle

Given a set S of points the problem is to find the smallest disk enclosing the set. This problem is also known as the (unweighted) one-center problem. That is, find a center such that the maximum distance from the center to the points in S is minimized. More formally, we need to find the center $c \in \Re^2$ such that $\max_{p_j \in S} d(c, p_j)$ is minimized. The weighted one-center problem, in which the distance function $d(c, p_j)$ is multiplied by the weight w_j, is a well-known *min–max problem*, also referred to as the *emergency center problem* in operations research. In two dimensions, the one-center problem can be solved in $O(n)$ time. The minimum enclosing ball problem in higher dimensions is also solved by using linear programming technique [103]. The general p-center problem, i.e., finding p circles whose union contains S such that the maximum radius is minimized, is known to be NP-hard. For a special case when $p = 2$, Eppstein [48] gave an $O(n \log^2 n)$ randomized algorithm based on parametric search technique, and Chan [22] gave a deterministic algorithm with a slightly worse running time. For the problem of finding a minimum enclosing ellipsoid for a point set in $\Re^k$ and other types of geometric location problem see, e.g., [47,103].

2.3.5 Largest Empty Circle

This problem, in contrast to the minimum enclosing circle problem, is to find a circle centered in the interior of the convex hull of the set S of points that does not contain any given point and the radius of the circle is to be maximized. This is mathematically formalized as a *max–min problem*, the minimum distance from the center to the set is maximized. The weighted version is also known as the *obnoxious center problem* in facility location. For the unweighted version, the center must be either at a vertex of the Voronoi diagram for S in the convex hull or at the intersection of a Voronoi edge and the boundary of the convex hull. $O(n \log n)$ time is sufficient for this problem. Following the same strategy one can solve the largest empty square problem for S in $O(n \log n)$ time as well, using the Voronoi diagram in the L_∞-metric [62]. The time complexity of the algorithm is asymptotically optimal, as the maximum gap problem, i.e., finding the maximum gap between two consecutive numbers on the real line, which requires $\Omega(n \log n)$ time, is reducible to this problem [90]. In contrast to the minimum enclosing ellipsoid problem is the largest empty ellipsoid problem, which has also been studied [43].

2.3.6 Largest Empty Rectangle

In Section 1.2.4, we mentioned the smallest enclosing rectangle problem. Here, we look at the problem of finding the largest rectangle that is empty. Mukhopadhyay and Rao [78] gave an $O(n^3)$-time and $O(n^2)$-space algorithm for finding the largest empty arbitrarily oriented rectangle of a set of n points. A special case of this problem is to find the largest area *restricted* rectangle with sides parallel to those of the original rectangle containing a given set S of n points, whose interior contains no points from S. The problem arises in document analysis of printed-page layout in which white space in the black-and-white image of the form of a maximal empty rectangle is to be recognized. A related problem, called the *largest empty corner rectangle problem*, is that given two subsets S_l and S_r of S separated by a vertical line, find the largest rectangle containing no other points in S such that the lower left corner and the upper right corner of the rectangle are in S_l and S_r, respectively. This problem can be solved in $O(n \log n)$ time, where $n = |S|$, using fast matrix searching technique (cf. Section 1.4). With this as a subroutine, one can solve the largest empty restricted rectangle problem in $O(n \log^2 n)$ time. When the points define a rectilinear polygon that is orthogonally convex, the largest empty restricted rectangle that can fit inside the polygon can be found in $O(n\alpha(n))$ time, where $\alpha(n)$ is the slowly growing inverse of Ackermann's function using a result of Klawe and Kleitman [61]. When the polygon P is arbitrary and may contain holes, Daniels et al. [36] gave an

$O(n \log^2 n)$ algorithm, for finding the largest empty restricted rectangle in P. Orlowski [82] gave an $O(n \log n + s)$ algorithm, where s is the number of restricted rectangles that are possible candidates, and showed that s is $O(n^2)$ in the worst case and the expected value is $O(n \log n)$.

2.3.7   Minimum-Width Annulus

Given a set S of n points find an annulus (defined by two concentric circles) whose center lies internal to the convex hull of S such that the width of the annulus is minimized. This problem arises in dimensional tolerancing and metrology which deals with the specification and measurement of error tolerances in geometric shapes. To measure if a manufactured circular part is round, an American National Standards Institute (**ANSI**) standard is to use the width of an annulus covering the set of points obtained from a number of measurements. This is known as the roundness problem [51,99,100]. It can be shown that the center of the annulus can be located at the intersection of the nearest neighbor and the farthest neighbor Voronoi diagrams, as discussed in Section 2.2.2. The center can be computed in $O(n \log n)$ time [51]. If the input is defined by a simple polygon P with n vertices, then the problem is to find a minimum-width annulus that contains the boundary of P. The center of the smallest annulus can be located at the medial axis of P [100]. In particular, the problem can be solved in $O(n \log n + k)$, where k denotes the number of intersection points of the medial axis of the simple polygon and the farthest neighbor Voronoi diagram of the vertices of P. In [100], k is shown to be $\theta(n^2)$. However, if the polygon is convex, one can solve this problem in linear time [100]. Note that the minimum-width annulus problem is equivalent to the *best circle approximation problem*, in which a circle approximating a given shape (or a set of points) is sought such that the error is minimized. The error of the approximating circle is defined to be the maximum over all distances between points in the set and the approximating circle. To be more precise, the error is equal to one half of the width of the smallest annulus. See Figure 2.11.

If the center of the smallest annulus of a point set can be arbitrarily placed, the center may lie at infinity and the annulus degenerates to a pair of parallel lines enclosing the set of points. When the center is to be located at infinity, the problem becomes the well-known *minimum-width problem*, which is to find a pair of parallel lines enclosing the set such that the distance between them is minimized. The width of a set of n points can be computed in $O(n \log n)$ time, which is optimal [67]. In three dimensions the width of a set is also used as a measure for flatness of a plate, a flatness problem in computational metrology. Chazelle et al. [28] gave an $O(n^{8/5+\epsilon})$-time algorithm for this problem, improving over a previously known algorithm that runs in $O(n^2)$ time.

Shermer and Yap [97] introduced the notion of *relative roundness*, where one wants to minimize the ratio of the annulus width and the radius of the inner circle. An $O(n^2)$ algorithm was presented. Duncan et al. [42] define another notion of roundness, called *referenced roundness*, which becomes equivalent to the flatness problem when the radius of the reference circle is set to infinity. Specifically given a reference radius ρ of an annulus A that contains S, i.e., ρ is the mean of the two concentric

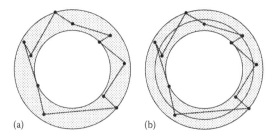

(a)   (b)

FIGURE 2.11   Minimum-width annulus (a) and the best circle approximation (b).

circles defining the annulus, find an annulus of a minimum width among all annuli with radius ρ containing S, or for a given $\epsilon > 0$, find an annulus containing S whose width is upper bounded by ϵ. They presented an $O(n \log n)$ algorithm for two dimensions and a near quadratic-time algorithm for three dimensions. In contrast to the minimum-width annulus problem is the *largest empty annulus problem*, in which we want to find the largest-width annulus that contains *no* points. The problem is much harder and can be solved in $O(n^3 \log n)$ time [39].

2.4   Geometric Matching

Matching in general graphs is one of the classical subjects in combinatorial optimization and has applications in operations research, pattern recognition, and VLSI design. Only geometric versions of the matching problem are discussed here. For graph-theoretic matching problems, see [83].

Given a weighted undirected complete graph on a set of $2n$ vertices, a *complete matching* is a set of n edges such that each vertex has exactly one edge incident on it. The weight of a matching is the sum of the weights of the edges in the matching. In a metric space, the vertices are points in the plane and the weight of an edge between two points is the distance between them. The *Euclidean minimum weight matching problem* is that given $2n$ points, find n matching pairs of points (p_i, q_i) such that $\Sigma d(p_i, q_i)$ is minimized.

It was not known if geometric properties can be exploited to obtain an algorithm that is faster than the $\theta(n^3)$ algorithm for general graphs (see [83]). Vaidya [101] settled this question in the affirmative. His algorithm is based on a well-studied primal-dual algorithm for weighted matching. Making use of additive-weighted Voronoi diagram discussed in Section 2.2.2.4 and the range search tree structure (see Section 2.7.1), Vaidya solved the problem in $O(n^{2.5} \log^4 n)$ time. This algorithm also generalizes to $\Re^k$ but the complexity is increased by a $\log^k n$ factor.

The *bipartite minimum weight matching* problem is defined similarly, except that we are given a set of red points $R = \{r_1, r_2, \ldots, r_n\}$ and a set of blue points $B = \{b_1, b_2, \ldots, b_n\}$ in the plane, and look for n matching pairs of points $(r, b) \in R \times B$ with minimum cost. In [101] Vaidya gave an $O(n^{2.5} \log n)$-time algorithm for Euclidean metric and an $O(n^2 \log^3 n)$ algorithm for L_1-metric. Approximation algorithms for this problem can be found in [7] and in [58].

If these $2n$ points are given as vertices of a polygon, the problems of minimum weight matching and bipartite matching can be solved in $O(n \log n)$ time if the polygon is convex and in $O(n \log^2 n)$ time if the polygon is simple. In this case, the weight of each matching pair of vertices is defined to be the geodesic distance between them [71]. However, if a *maximum weight matching* is sought, a $\log n$ factor can be shaved off [71].

Because of the triangle inequality, one can easily show that in a minimum weight matching, the line segments defined by the matched pairs of points cannot intersect one another. Generalizing this *nonintersecting property* the following *geodesic minimum matching problem* in the presence of obstacles can be formulated. Given $2m$ points and polygonal obstacles in the plane, find a matching of these $2m$ points such that the sum of the geodesic distances between the matched pairs is minimized. These m paths must not cross each other (they may have portions of the paths overlapping each other). There is no efficient algorithm known to date, except for the obvious method of reducing it to a minimum matching of a complete graph, in which the weight of an edge connecting any two points is the geodesic distance between them. Note that finding a geodesic matching without optimization is trivial, since these m noncrossing paths can always be found. This geodesic minimum matching problem in the general polygonal domain seems nontrivial. The *noncrossing* constraint and the optimization objective function (minimizing total weight) makes the problem hard.

When the matching of these $2m$ points is given *a priori*, finding m noncrossing paths minimizing the total weight seems very difficult. This resembles global routing problem in VLSI for which m two-terminal nets are given, and a routing is sought that optimizes a certain objective function,

including total wire length, subject to some capacity constraints. The
noncrossing requirement is needed when single-layer routing or planar
routing model is used. Global routing problems in general are NP-
hard. Since the paths defined by matching pairs in an optimal routing
cannot cross each other, paths obtained by earlier matched pairs become
obstacles for subsequently matched pairs. Thus, the *sequence* in which
the pairs of points are matched is very crucial. In fact, the path defined
by a matched pair of points need not be the shortest. Thus, to route
the matched pairs in a greedy manner sequentially does not give an
optimal routing. Consider the configuration shown in Figure 2.12 in
which $R = X, Y, Z$, $B = x, y, z$, and points (X, x), (Y, y), and (Z, z) are
to be matched. Note that in this optimal routing, none of the matched
pairs is realized by the shortest path, i.e., a straight line. This problem

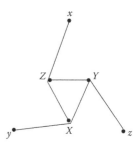

FIGURE 2.12   An instance
of three noncrossing pair
matching problem.

is referred to as the *shortest k-pair noncrossing path problem*. However, if the m matching pairs of
points are on the boundary of a simple polygon, and the path must be confined to the interior of the
polygon, Papadopoulou [84] gave an $O(n+m)$ algorithm for finding an optimal set of m noncrossing
paths, if a solution exists, where n is the number of vertices of the polygon.

Atallah and Chen [14] considered the following bipartite matching problem: Given n red and n blue
disjoint isothetic rectangles in the plane, find a matching of these n red–blue pairs of rectangles such
that the rectilinear paths connecting the matched pairs are noncrossing and monotone. Surprisingly
enough, they showed that such a matching satisfying the constraints always exists and gave an
asymptotically optimal $O(n \log n)$ algorithm for finding such a matching.

To conclude this section we remark that the min–max versions of the general matching or bipartite
matching problems are open. In the red-blue matching, if one of the sets is allowed to translate,
rotate, or scale, we have a different matching problem. In this setting, we often look for the *best
match* according to min–max criterion, i.e., the maximum error in the matching is to be minimized.
A dual problem can also be defined, i.e., given a maximum error bound, determine if a matching
exists, and if so, what kind of *motions* are needed. For more information about various types of
matching, see a survey by Alt and Guibas [9].

2.5   Planar Point Location

Planar point location is a fundamental problem in computational geometry. Given a planar subdivision, and a query point, we want to find the region that contains the query point. Figure 2.13
shows an example of a planar subdivision. This problem arises in geographic information systems,
in which one often is interested in locating, for example, a certain facility in a map. Consider the
skew Voronoi diagram, discussed earlier in Section 2.2.2.5, for a set S of emergency dispatchers.
Suppose an emergency situation arises at a location q and that the nearest dispatcher p is to be called
so that the distance $\bar{d}(p, q)$ is the smallest among all distances $\bar{d}(r, q)$, for $r \in S$. This is equivalent to
locating q in the Voronoi cell $V_{\text{from}}(p)$ of the skew Voronoi diagram that contains q. In situations like
this, it is vital that the nearest dispatcher be located quickly. We therefore address the point location
problem under the assumption that the underlying planar map is fixed and the main objective is
to have a fast response time to each query. Toward this end we preprocess the planar map into a
suitable structure so that it would facilitate the point location task.

An earlier preprocessing scheme is based on the *slab method* [90], in which parallel lines are drawn
through each vertex, thus, partitioning the plane into parallel slabs. Each parallel slab is further
divided into subregions by the edges of the subdivision that can be linearly ordered. Any given query
point q can thus be located by two binary searches; one to locate among the $n + 1$ horizontal slabs the

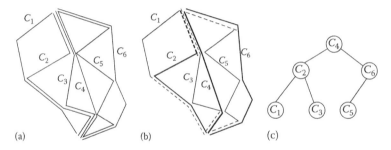

FIGURE 2.13 Chain decomposition method for a planar subdivision.

slab containing q, and followed by another to locate the region defined by a pair of consecutive edges which are ordered from left to right. We use a three tuple, $(P(n), S(n), Q(n)) =$ (preprocessing time, space requirement, query time) to denote the performance of the search strategy. The slab method gives an $(O(n^2), O(n^2), O(\log n))$ algorithm. Since preprocessing time is only performed once, the time requirement is not as critical as the space requirement, which is permanently engaged. The primary goal of any search strategy is to minimize the query time and the space required. Lee and Preparata [90] first proposed a *chain decomposition* method to decompose a monotone planar subdivision with n points into a collection of $m \leq n$ monotone chains organized in a complete binary tree. Each node in the binary tree is associated with a monotone chain of at most n edges, ordered in y-coordinate. This set of monotone chains forms a totally ordered set partitioning the plane into collections of regions. In particular, between two adjacent chains, there are a number of disjoint regions. The point location process begins with the root node of the complete binary tree. When visiting a node, the query point is compared with the node, hence, the associated chain, to decide on which side of the chain the query point lies. Each chain comparison takes $O(\log n)$ time, and the total number of nodes visited is $O(\log m)$. The search on the binary tree will lead to two adjacent chains, and, hence, identify a region that contains the point. Thus, the query time is $O(\log m \log n) = O(\log^2 n)$. Unlike the slab method in which each edge may be stored as many as $O(n)$ times, resulting in $O(n^2)$ space, it can be shown that with an appropriate chain assignment scheme, each edge in the planar subdivision is stored only once. Thus, the space requirement is $O(n)$. For example, in Figure 2.13, the edges shared by the root chain C_4 and its descendant chains are assigned to the root chain; in general, any edge shared by two nodes on the same root-to-leaf path will be assigned to the node that is an ancestor of the other node. The chain decomposition scheme gives rise to an $(O(n \log n), O(n), O(\log^2 n))$ algorithm. The binary search on the chains is not efficient enough. Recall that after each chain comparison, we will move down the binary search tree to perform the next chain comparison and start over another binary search on the same y-coordinate of the query point to find an edge of the chain, against which a comparison is made to decide if the point lies to the left or right of the chain. A more efficient scheme is to be able to perform a binary search of the y-coordinate at the root node and to spend only $O(1)$ time per node as we go down the chain tree, shaving off an $O(\log n)$ factor from the query time. This scheme is similar to the ones adopted by Chazelle and Guibas [37,90] in fractional cascading search paradigm and by Willard [37] in his range tree-search method. With the linear time algorithm for triangulating a simple polygon (cf. Section 1.5), we conclude with the following optimal search structure for planar point location.

THEOREM 2.5   *Given a planar subdivision of n vertices, one can preprocess the subdivision in linear time and space such that each point location query can be answered in $O(\log n)$ time.*

The point location problem in arrangements of hyperplanes is also of significant interest. See, e.g., [30]. **Dynamic** versions of the point location problem, where the underlying planar subdivision is subject to changes (insertions and deletions of vertices or edges), have also been investigated. See [34] for a survey of dynamic computational geometry.

2.6   Path Planning

This class of problems is mostly cast in the following setting. Given is a set of obstacles O, an object, called *robot*, and an initial and final position, called source and destination, respectively. We wish to find a path for the robot to move from the source to the destination avoiding all the obstacles. This problem arises in several contexts. For instance, in robotics this is referred to as the *piano movers' problem* or *collision avoidance problem*, and in VLSI design this is the routing problem for two-terminal nets. In most applications, we are searching for a collision avoidance path that has the shortest length, where the distance measure is based on the Euclidean or L_1-metric. For more information regarding motion planning see, e.g., [11,94].

2.6.1   Shortest Paths in Two Dimensions

In two dimensions, the Euclidean shortest path problem in which the robot is a point, and the obstacles are simple polygons, is well studied. A most fundamental approach is by using the notion of the *visibility graph*. Since the shortest path must make turns at polygonal vertices, it is sufficient to construct a graph whose vertices include the vertices of the polygonal obstacles, the source and the destination, and whose edges are determined by vertices that are mutually visible, i.e., the segment connecting the two vertices does not intersect the interior of any obstacle. Once the visibility graph is constructed with edge weight equal to the Euclidean distance between the two vertices, one can then apply the Dijkstra's shortest path algorithms [90] to find the shortest path between the source and destination. The Euclidean shortest path between two points is referred to as the *geodesic path* and the distance as the *geodesic distance*. The visibility graph for a set of polygonal obstacles with a total of n vertices can be computed trivially in $O(n^3)$ time. The computation of the visibility graph is the dominating factor for the complexity of any visibility graph-based shortest path algorithm. Research results aiming at more efficient algorithms for computing the visibility graph and the geodesic path in time proportional to the size of the graph have been obtained. For example, in [89] Pocchiola and Vetger gave an optimal output-sensitive algorithm that runs in $O(\mathcal{F} + n \log n)$ time and $O(n)$ space for computing the visibility graph, where $\mathcal{F}$ denotes the number of edges in the graph.

Mitchell [74] used the *continuous Dijkstra* wavefront approach to the problem for general polygonal domain of n obstacle vertices and obtained an $O(n^{3/2+\epsilon})$-time algorithm. He constructed the *shortest path map* that partitions the plane into regions such that all points q that lie in the same region have the same vertex sequence in the shortest path from the given source to q. The shortest path map takes $O(n)$ space and enables us to perform the shortest path queries, i.e., find the shortest path from the given source to any query points, in $O(\log n)$ time. Hershberger and Suri [54] on the other hand used plane subdivision approach and presented an $O(n \log^2 n)$-time and $O(n \log n)$-space algorithm to compute the shortest path map of a given source point. They later improved the time bound to $O(n \log n)$. It remains an open problem if there exists an algorithm for finding the Eucliedan shortest path in a general polygonal domain of h obstacles and n vertices in $O(n + h \log h)$ time and $O(n)$ space. If the source-destination path is confined in a simple polygon with n vertices, the shortest path can be found in $O(n)$ time [37].

In the context of VLSI routing, one is mostly interested in rectilinear paths (L_1-metric) whose edges are either horizontal or vertical. As the paths are restricted to be rectilinear, the shortest path problem can be solved more easily. Lee et al. [68] gave a survey on this topic.

In a two-layer routing model, the number of segments in a rectilinear path reflects the number of *vias*, where the wire segments change layers, which is a factor that governs the fabrication cost. In robotics, a straight-line motion is not as costly as *making turns*. Thus, the number of segments (or turns) has also become an objective function. This motivates the study of the problem of finding a path with the least number of segments, called the *minimum link path problem* [76].

These two cost measures, length and number of links, are in conflict with each other. That is, the shortest path may have far too many links, whereas a minimum link path may be arbitrarily long compared with the shortest path. A path that is optimal in both criteria is called the *smallest path*. In fact it can be easily shown that in a general polygonal domain, the smallest path does not exist. However, the smallest rectilinear path in a simple rectilinear polygon exists, and can be found in linear time. Instead of optimizing both measures *simultaneously* one can either seek a path that optimizes a linear function of both length and the number of links, known as the *combined* metric [104], or optimizes them in a lexicographical order. For example, we optimize the length first, and then the number of links, i.e., among those paths that have the same shortest length, find one whose number of links is the smallest and vice versa. In the rectilinear case see, e.g., [104]. Mitchell [76] gave a comprehensive treatment of the geometric shortest path and optimization problems.

A generalization of the collision avoidance problem is to allow collision with a cost. Suppose each obstacle has a weight which represents the cost if the obstacle is *penetrated*. Lee et al. [68] studied this problem in the rectilinear case. They showed that the shortest rectilinear path between two given points in the presence of weighted rectilinear polygons can be found in $O(n \log^{3/2} n)$ time and space. Chen et al. [31] showed that a data structure can be constructed in $O(n \log^{3/2} n)$ time and $O(n \log n)$ space that enables one to find the shortest path from a given source to any query point in $O(\log n + \mathcal{H})$ time, where $\mathcal{H}$ is the number of links in the path. Another generalization is to include in the set of obstacles some subset $F \subset O$ of obstacles, whose vertices are *forbidden* for the solution path to make turns. Of course, when the weight of obstacles is set to be ∞, or the forbidden set $F = \emptyset$, these generalizations reduce to the ordinary collision avoidance problem.

2.6.2   Shortest Paths in Three Dimensions

The Euclidean shortest path problem between two points in a three-dimensional polyhedral environment turns out to be much harder than its two-dimensional counterpart. Consider a convex polyhedron P with n vertices in three dimensions and two points s and d on the surface of P. The shortest path from s to d on the surface will cross a sequence of edges, denoted by $\xi(s, d)$. $\xi(s, d)$ is called the *shortest path edge sequence* induced by s and d and consists of distinct edges. For given s and d, the shortest path from s to d is not unique. However, $\xi(s, d)$ is unique. If $\xi(s, d)$ is known, the shortest path between s and d can be computed by a planar unfolding procedure so that these faces crossed by the path lie in a common plane and the path becomes a straight-line segment.

The shortest paths on the surface of a convex polyhedron P possess the following topological properties: (1) they do not pass through the vertices of P and do not cross an edge of P more than once, (2) they do not intersect themselves, i.e., they must be simple, and (3) except for the case of the two shortest paths sharing a common subpath, they intersect transversely in at most one point, i.e., they cross each other. If the shortest paths are grouped into equivalent classes according to the sequences of edges that they cross, then the number of such equivalent classes denoted by $|\xi(P)|$ is $\theta(n^4)$, where n is the number of vertices of P. These equivalent classes can be computed in $O(|\xi(P)|n^3 \log n)$ time. Chen and Han [32] gave an $O(n^2)$ algorithm for finding the shortest path between a fixed source s and any destination d, where n is the number of vertices and edges of the polyhedron, which may or may not be convex. If s and d lie on the surface of two different polyhedra, $O(N^{O(k)})$ time suffices for computing the shortest path between them amidst a set of k polyhedra, where N denotes the total number of vertices of these obstacles.

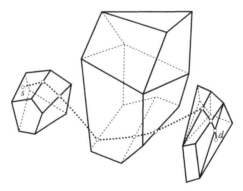

FIGURE 2.14 The possible shortest path edge sequence between points s and d.

The crux of the problem lies in the fact that the number of possible edge sequences may be exponential in the number of obstacles, if s and d lie on the surface of different polyhedra. It was established that the problem of determining the shortest path edge sequence is indeed NP-hard [11]. Figure 2.14 shows an example of the possible shortest path edge sequence induced by s and d in the presence of three convex polyhedra. Approximation algorithms for this problem can be found in, e.g., Choi et al. [35] and a recent article [76] for a survey.

2.7 Searching

This class of problems is cast in the form of query-answering. Given a collection of objects, with preprocessing allowed, one is to find the objects that satisfy the queries. The problem can be **static** or dynamic, depending on the fact that if the database to be searched is allowed to change over the course of query-answering sessions, and it is studied mostly in two modes, *count mode* and *report mode*. In the former case, only the *number* of objects satisfying the query is to be answered, whereas in the latter, the actual identity of the objects is to be reported. In the report mode, the query time of the algorithm consists of two components, the *search time* and the *retrieval time*, and expressed as $Q_A(n) = O(f(n) + \mathcal{F})$, where n denotes the size of the database, $f(n)$ a function of n, and $\mathcal{F}$ the size of output. Sometimes we may need to perform some semigroup operations to those objects that satisfy the query. For instance, we may have weights $w(v)$ assigned to each object v, and we want to compute $\Sigma w(v)$ for all $v \cap q \neq \emptyset$. This is referred to as *semigroup range searching*. The semigroup range searching problem is the most general form: if the semigroup operation is set union, we get report-mode range searching problem and if the semigroup operation is just addition (of uniform weight), we have the count-mode range searching problem. We will not discuss the semigroup range searching here. It is obvious that algorithms that handle the report-mode queries can also handle the count-mode queries ($\mathcal{F}$ is the answer). It seems natural to expect that the algorithms for count-mode queries would be more efficient (in terms of the order of magnitude of the space required and query time), as they need not search for the objects. However, it was argued that in the report-mode range searching, one could take advantage of the fact that since reporting takes time, the more to report, the sloppier the search can be. For example, if we were to know that the ratio $n/\mathcal{F}$ is $O(1)$, we could use a sequential search on a linear list. This notion is known as the *filtering search* [90]. In essence, more objects than necessary are identified by the search mechanism, followed by a filtering process leaving out unwanted objects. As indicated below, the count-mode range searching problem is harder than the report-mode counterpart.

2.7.1 Range Searching

This is a fundamental problem in database applications. We will discuss this problem and the algorithm in the two-dimensional space. The generalization to higher dimensions is straightforward using a known technique, called multidimensional divide-and-conquer [90]. Given is a set of n points in the plane, and the ranges are specified by a product $(l_1, u_1) \times (l_2, u_2)$. We would like to find points $p = (x, y)$ such that $l_1 \leq x \leq u_1$ and $l_2 \leq y \leq u_2$. Intuitively we want to find those points that lie inside a query rectangle specified in the range. This is called *orthogonal range searching*, as opposed to other kinds of range searching problems discussed below, e.g., half-space range searching, and simplex range searching, etc. Unless otherwise specified, a range refers to an orthogonal range. We discuss the static case, as this belongs to the class of **decomposable searching problems**, the **dynamization transformation** techniques can be applied. We note that the range tree structure mentioned below can be made dynamic by using a weight-balanced tree, called $BB(\alpha)$ tree.

For count-mode queries, it can be solved by using the locus method as follows. In two dimensions, we can divide the plane into $O(n^2)$ cells by drawing a horizontal and a vertical line through each point. The answer to the query q, i.e., find the number of points dominated by q (those points whose x- and y-coordinates are both no greater than those of q), can be found by locating the cell containing q. Let it be denoted by $\text{Dom}(q)$. Thus, the answer to the count-mode range queries can be obtained by some simple arithmetic operations of $\text{Dom}(q_i)$ for the four corners, q_1, q_2, q_3, q_4, of the query rectangle. Let q_4 be the northeast corner and q_2 be the southwest corner. The answer will be $\text{Dom}(q_4) - \text{Dom}(q_1) - \text{Dom}(q_3) + \text{Dom}(q_2)$. Thus, in $\Re^k$, we have $Q(k, n) = O(k \log n)$, $S(k, n) = P(k, n) = O(n^k)$. To reduce space requirement at the expense of query time has been a goal of further research on this topic. Bentley introduced a data structure called the *range trees* [90]. Using this structure the following results were obtained: for $k \geq 2$, $Q(k, n) = O(\log^{k-1} n)$, $S(k, n) = P(k, n) = O(n \log^{k-1} n)$. See [1,4] for more recent results.

For report-mode queries, by using filtering search technique the space requirement can be further reduced by a $\log \log n$ factor. If the range satisfies additional conditions, e.g., grounded in one of the coordinates, say $l_1 = 0$, or the aspect ratio of the intervals specifying the range is fixed, less space is needed. For instance, in two dimensions, the space required is linear (a saving of $\log n / \log \log n$ factor) for these two cases. By using the so-called *functional approach* to data structures, Chazelle developed a compression scheme to reduce further the space requirement. Thus, in k-dimensions, $k \geq 2$, for the count-mode range queries, we have $Q(k, n) = O(\log^{k-1} n)$ and $S(k, n) = O(n \log^{k-2} n)$ and for report-mode range queries $Q(k, n) = O(\log^{k-1} n + \mathcal{F})$, and $S(k, n) = O(n \log^{k-2+\epsilon} n)$ for some $0 < \epsilon < 1$ [1,4].

As regards the lower bound of range searching in terms of space–time trade-offs, Chazelle [23] showed that in k-dimensions, if the query time is $O(\log^c n + \mathcal{F})$ for any constant c, the space required is $\Omega(n(\log n / \log \log n)^{k-1})$ for the pointer machine models and the bound is tight for any $c \geq 1$, if $k = 2$, and any $c \geq k - 1 + \epsilon$ (for any $\epsilon > 0$), if $k > 2$. See also [1,4] for more lower-bound results related to orthogonal range searching problems.

2.7.2 Other Range Searching Problems

There are other range searching problems called *simplex range searching problem* and the *half-space range searching problem* that have been well studied. A simplex range in $\Re^k$ is a range whose boundary is specified by $k + 1$ hyperplanes. In two dimensions it is a triangle. For this problem, there is a lower bound on the query time for simplex range queries: let m denote the space required, $Q(k, n) = \Omega((n / \log n) m^{1/k}), k > 2$, and $Q(2, n) = \Omega(n / \sqrt{m})$ [37].

The report-mode half-space range searching problem in the plane can be solved optimally in $Q(n) = O(\log n + \mathcal{F})$ time and $S(n) = O(n)$ space, using geometric duality transform [37]. But this method does not generalize to higher dimensions. In [6], Agarwal and Matoušek obtained a general result for this problem: for $n \leq m \leq n^{\lfloor k/2 \rfloor}$, with $O(m^{1+\epsilon})$ space and preprocessing, $Q(k, n) = O((n/m^{1/\lfloor k/2 \rfloor}) \log n + \mathcal{F})$. As half-space range searching problem is also decomposable, standard dynamization techniques can be applied.

A general method for simplex range searching is to use the notion of *partition tree*. The search space is partitioned in a hierarchical manner using cutting hyperplanes [25] and a search structure is built in a tree structure. Using a **cutting theorem** of hyperplanes, Matoušek [72] showed that for k-dimensions, there is a linear space search structure for the simplex range searching problem with query time $O(n^{1-1/k})$, which is optimal in two dimensions and within $O(\log n)$ factor of being optimal for $k > 2$. For more detailed information regarding geometric range searching, see [72].

The above discussion is restricted to the case in which the database is a collection of points. One may consider also other kinds of objects, such as line segments, rectangles, triangles, etc., whatever applications may take. The inverse of the orthogonal range searching problem is that of *point enclosure searching problem*. Consider a collection of isothetic rectangles. The point enclosure searching is to find all rectangles that contain the given query point q. We can cast these problems as *intersection searching problem*, i.e., given a set S of objects, and a query object q, find a subset $\mathcal{F}$ of S such that for any $f \in \mathcal{F}, f \cap q \neq \emptyset$. We have then the rectangle enclosure searching problem, rectangle containment problem, segment intersection searching problem, etc. Janardan and Lopez [59] generalized the intersection searching in the following manner. The database is a collection of groups of objects, and the problem is to find all the groups of objects intersecting a query object. A group is considered to be intersecting the query object if any object in the group intersects the query object. When each group has only one object, this reduces to the ordinary searching problems.

2.8 Intersection

This class of problems arises in, for example, architectural design, computer graphics, etc. In an architectural design, no two objects can share a common region. In computer graphics, the well-known hidden-line or hidden-surface elimination problems [40] are examples of intersection problems. This class encompasses two types of problems, *intersection detection* and *intersection computation*.

2.8.1 Intersection Detection

The intersection detection problem is of the form: Given a set of objects, do any two intersect? For instance, given n line segments in the plane, are there two that intersect? The intersection detection problem has a lower bound of $\Omega(n \log n)$ [90].

In two dimensions, the problem of detecting if two polygons of r and b vertices intersect was easily solved in $O(n \log n)$ time, where $n = r + b$ using the red–blue segment intersection algorithm [29]. However, this problem can be reduced in linear time to the problem of detecting the self-intersection of a polygonal curve. The latter problem is known as the *simplicity test* and can be solved optimally in linear time by Chazelle's linear time triangulation algorithm (cf. Section 1.5). If the two polygons are convex, then $O(\log n)$ suffices in detecting if they intersect [26]. Note that although detecting if two convex polygons intersect can be done in logarithmic time, detecting if the boundary of the two convex polygons intersect requires $\Omega(n)$ time [26]. Mount [77] investigated the intersection detection of two simple polygons and computed a separator of m links in $O(m \log^2 n)$ time if they do not intersect.

In three dimensions, detecting if two convex polyhedra intersect can be solved in linear time [26] by using a hierarchical representation of the convex polyhedron or by formulating it as a linear programming problem in three variables.

2.8.2 Intersection Reporting/Counting

One of the simplest of such intersecting reporting problems is that of reporting pairwise intersection, e.g., intersecting pairs of line segments in the plane. An earlier result due to Bentley and Ottmann [90] used the plane-sweep technique that takes $O((n + \mathcal{F}) \log n)$ time, where $\mathcal{F}$ is the output size. This is based on the observation that the line segments intersected by a vertical sweep-line can be ordered according to the y-coordinates of their intersection with the sweep-line, and the sweep-line status can be maintained in logarithmic time per event point, which is either an endpoint of a line segment or the intersection of two line segments. It is not difficult to see that the lower bound for this problem is $\Omega(n \log n + \mathcal{F})$; thus, the above algorithm is $O(\log n)$ factor from the optimal. This segment intersection reporting problem has been solved optimally by Chazelle and Edelsbrunner [27], who used several important algorithm design and data structuring techniques, as well as some crucial combinatorial analysis. In contrast to this asymptotically time-optimal deterministic algorithm, a simpler randomized algorithm was obtained [37] for this problem which is both time- and space-optimal. That is, it requires only $O(n)$ space (instead of $O(n + \mathcal{F})$ as reported in [27]). Balaban [16] gave a deterministic algorithm that solves this problem optimally both in time and space.

On a separate front, the problem of finding intersecting pairs of segments from two different sets was considered. This is called *bichromatic line segment intersection problem.*

Chazelle et al. [29] used *hereditary segment trees* structure and fractional cascading and solved both segment intersection reporting and counting problems optimally in $O(n \log n)$ time and $O(n)$ space. (The term $\mathcal{F}$ should be included in case of reporting.) If the two sets of line segments form connected subdivisions, then merging or overlaying these two subdivisions can be computed in $O(n + \mathcal{F})$ [50]. Boissonnat and Snoeyink [20] gave yet another optimal algorithm for the bichromatic line segment intersection problem, taking into account the notion of algebraic degree proposed by Liotta et al. [69].

The *rectangle intersection* reporting *problem* arises in the design of VLSI, in which each rectangle is used to model a certain circuitry component. These rectangles are isothetic, i.e., their sides are all parallel to the coordinate axes. This is a well-studied classical problem, and optimal algorithms ($O(n \log n + \mathcal{F})$ time) have been reported. See [4] for more information. The k-dimensional hyperrectangle intersection reporting (respectively, counting) problem can be solved in $O(n^{k-2} \log n + \mathcal{F})$ time and $O(n)$ space (respectively, in time $O(n^{k-1} \log n)$ and space $O(n^{k-2} \log n)$). Gupta et al. [53] gave an $O(n \log n \log \log n + \mathcal{F} \log \log n)$-time and linear-space algorithm for the *rectangle enclosure reporting problem* that calls for finding all the enclosing pairs of rectangles.

2.8.3 Intersection Computation

Computing the actual intersection is a basic problem, whose efficient solutions often lead to better algorithms for many other problems.

Consider the problem of computing the common intersection of half-planes by divide-and-conquer. Efficient computation of the intersection of the two convex polygons is required during the merge step. The intersection of the two convex polygons can be solved very efficiently by plane-sweep in linear time, taking advantage of the fact that the edges of the input polygons are ordered. Observe that in each vertical strip defined by the two consecutive sweep-lines, we only need to compute the intersection of two trapezoids, one derived from each polygon [90].

The problem of the intersecting two convex polyhedra was first studied by Muller and Preparata [90], who gave an $O(n \log n)$ algorithm by reducing the problem to the problems of intersection detection and convex hull computation. Following this result one can easily derive an $O(n \log^2 n)$ algorithm for computing the common intersection of n half-spaces in three dimensions by divide-and-conquer. However, using geometric duality and the concept of separating plane, Preparata and Muller [90] obtained an $O(n \log n)$ algorithm for computing the common intersection of n half-spaces, which is asymptotically optimal. There appears to be a difference in the approach to solving the common intersection problem of half-spaces in two and three dimensions. In the latter, we resorted to geometric duality instead of divide-and-conquer. This inconsistency was later resolved. Chazelle [24] combined the hierarchical representation of convex polyhedra, geometric duality, and other ingenious techniques to obtain a linear time algorithm for computing the intersection of two convex polyhedra. From this result, several problems can be solved optimally: (1) the common intersection of half-spaces in three dimensions can now be solved by divide-and-conquer optimally, (2) the merging of two Voronoi diagrams in the plane can be done in linear time by observing the relationship between the Voronoi diagram in two dimensions and the convex hull in three dimensions (cf. Section 1.2), and (3) the medial axis of a simple polygon or the Voronoi diagram of vertices of a convex polygon [5] can be solved in linear time.

2.9   Research Issues and Summary

We have covered in this chapter a number of topics in computational geometry, including proximity, optimization, planar point location, geometric matching, path planning, searching, and intersection. These topics discussed here and in the previous chapter are not meant to be exhaustive. New topics arise as the field continues to flourish.

In Section 2.3, we discussed the problems of the smallest enclosing circle and the largest empty circle. These are the two extremes: either the circle is empty or it contains all the points. The problem of finding the smallest (respectively, largest) circle containing at least (respectively, at most) k points for some integer $0 \le k \le n$ is also a problem of interest. Moreover, the shape of the object is not limited to circles. A number of open problems remain. Given two points s and t in a simple polygon, is it NP-complete to decide whether there exists a path with at most k links and of length at most L? What is the complexity of the shortest k-pair noncrossing path problem discussed in Section 2.4? Does there exist an algorithm for finding the Euclidean shortest path in a general polygonal demain of h obstacles and n vertices in $O(n + h \log h)$ time and $O(n)$ space? How fast can one solve the geodesic minimum matching problem for $2m$ points in the presence of polygonal obstacles? Can one solve the largest empty restricted rectangle problem for a rectilinear polygon in $O(n \log n)$ time? The best known algorithm to date runs in $O(n \log^2 n)$ time [36]. Is $\Omega(n \log n)$ a lower bound of the minimum-width annulus problem? Can the technique used in [51] be applied to the polygonal case to yield an $O(n \log n)$-time algorithm for the minimum-width annulus problem? In [76] Mitchell listed a few open problems worth studying. Although the minimum spanning tree problem for general graph with m edges and n vertices can be solved optimally in $O(T^*(m, n))$ time, where T^* is the minimum number of edge-weight comparisons and $T^*(m, n) = \Omega(m)$ and $T^*(m, n) = O(m \cdot \alpha(m, n))$, where $\alpha(m, n)$ is the inverse of Ackermann's function [88], an optimal algorithm for the Euclidean minimum spanning tree is still open. The problem of finding a spanning tree of bounded cost and bounded diameter is known to be NP-complete [96], but there is no known approximation algorithm to date.

Researchers in computational geometry have begun to address the issues concerning the actual running times of the algorithms and their robustness when the computations in their implementations are not exact [106]. It is understood that the real-RAM computation model with an implicit infinite-precision arithmetic is unrealistic in practice. In addition to the robustness issue concerning the accuracy of the output of an algorithm, one needs to find a new cost measure to evaluate the efficiency of an algorithm. In the infinite-precision model, the asymptotic time complexity was accepted as an adequate cost measure. However, when the input data have a finite-precision representation and computation time varies with the precision required, an alternative cost measure is warranted. The notion of the **degree** of a geometric algorithm could be an important cost measure for comparing the efficiency of the algorithms when they are actually implemented [69]. For example, Chen et al. [33] showed that the Voronoi diagram of a set of arbitrarily oriented segments can be constructed by a plane-sweep algorithm with degree 14 for certain regular k-gon metrics. This notion of the degree of robustness could play a similar role as the asymptotic time complexity has in the past for the real-RAM computation model.

On the applied side, there are efforts put into development of geometric software. A project known as the Computational Geometry Algorithms Library (CGAL) project (http://www.cgal.org/) [49] is an ongoing collaborative effort of researchers in Europe to organize a system library containing primitive geometric abstract data types useful for geometric algorithm developers. This is concurrent with the Library of Efficient Data Types and Algorithms (LEDA) project [73] which was initiated at the Max-Planck-Institut für Informatik, Saarbrücken, Germany, and now maintained by Algorithmic Solutions Software GmbH (http://www.algorithmic-solutions.com/leda/about/index.htm.) LEDA is a C++ class library for efficient data types and algorithms, and provides algorithmic in-depth knowledge in geometric computing, combinatorial optimization, etc. In Asia, a web-collaboratory project was initiated at the Institute of Information Science and the Research Center for Information Technology Innovation, Academia Sinica, from which a geometric algorithm visualization and debugging system, GeoBuilder, for 2D and 3D geometric computing has been developed (http://webcollab.iis.sinica.edu.tw/Components/GeoBuilder/). This system facilitates geometric algorithmic researchers in not only testing their ideas and demonstrating their findings, but also teaching algorithm design in the classroom. GeoBuilder is embedded into a knowledge portal [64], called OpenCPS (Open Computational Problem Solving) (http://www.opencps.org/), as a practice platform for a course on geometric computing and algorithm visualization. The GeoBuilder system possesses three important features: First, it is a platform-independent software system based on Java's promise of portability, and can be invoked by Sun's Java Web Start technology in any browser-enabled environment. Second, it has the collaboration capability for multiple users to concurrently develop programs, manipulate geometric objects and control the camera. Finally, its 3D geometric drawing bean provides an optional function that can automatically position the camera to track 3D objects during algorithm visualization [102]. GeoBuilder develops its rich client platform based on Eclipse RCP and has already built in certain functionalities such as remote addition, deletion, and saving of files as well as remote compiling, and execution of LEDA C/C++ programs, etc. based on a multipage editor. Other projects related to the efforts of building geometric software or problem-solving environment, include GeoLab, developed at the Institute of Computing at UNICAMP (State University of Campinas) as a programming environment for implementation, testing, and animation of geometric algorithms (http://www.dcc.unicamp.br/~rezende/GeoLab.htm), and *XYZ GeoBench*, developed at Zürich, as a programming environment on Macintosh computers for geometric algorithms, providing an interactive user interface similar to a drawing program which can be used to create and manipulate geometric objects such as points, line segments, polygons, etc. (http://www.schorn.ch/geobench/XYZGeoBench.html).

2.10 Further Information

Additional references about various variations of closest pair problems can be found in [17,21,60,95]. For additional results concerning the Voronoi diagrams in higher dimensions and the duality transformation see [15]. For more information about Voronoi diagrams for sites other than points, in various distance functions or norms, see [10,19,81,85,87,93]. A recent textbook by de Berg et al. [37] contains a very nice treatment of computational geometry in general. The book by Narasimhan and Smid [79] covers topics pertaining to geometric spanner networks. More information can be found in [52,63,93,105]. The reader who is interested in parallel computational geometry is referred to [13]. For current research activities and results, the reader may consult the *Proceedings of the Annual ACM Symposium on Computational Geometry, Proceedings of the Annual Canadian Conference* on *Computational Geometry* and the following three journals: *Discrete & Computational Geometry, International Journal of Computational Geometry & Applications,* and *Computational Geometry: Theory and Applications.* The following site http://compgeom.cs.uiuc.edu/~jeffe/compgeom/biblios.html contains close to 14,000 entries of bibliography in this field.

Those who are interested in the implementations or would like to have more information about available software can consult http://www.cgal.org/.

The following WWW page on *Geometry in Action* maintained by David Eppstein at http://www.ics.uci.edu/~eppstein/geom.html and the Computational Geometry Page by J. Erickson at http://compgeom.cs.uiuc.edu/~jeffe/compgeom give a comprehensive description of research activities of computational geometry.

Defining Terms

ANSI: American National Standards Institute.

Bisector: A bisector of two elements e_i and e_j is defined to be the locus of points equidistant from both e_i and e_j. That is, $\{p|d(p, e_i) = d(p, e_j)\}$. For instance, if e_i and e_j are two points in the Euclidean plane, the bisector of e_i and e_j is the perpendicular bisector to the line segment $\overline{e_i, e_j}$.

Cutting theorem: This theorem [25] states that for any set $\mathcal{H}$ of n hyperplanes in $\mathfrak{R}^k$, and any parameter r, $1 \leq r \leq n$, there always exists a $(1/r)$-cutting of size $O(r^k)$. In two dimensions, a $(1/r)$-cutting of size s is a partition of the plane into s disjoint triangles, some of which are unbounded, such that no triangle in the partition intersects more than n/r lines in $\mathcal{H}$. In $\mathfrak{R}^k$, triangles are replaced by simplices. Such a cutting can be computed in $O(nr^{k-1})$ time.

Decomposable searching problems: A searching problem with query Q is decomposable if there exists an efficiently computable associative, and commutative binary operator @ satisfying the condition: $Q(x, A \cup B) = @(Q(x, A), Q(x, B))$. In other words, one can decompose the searched domain into subsets, find answers to the query from these subsets, and combine these answers to form the solution to the original problem.

Degree of an algorithm or problem: Assume that each input variable is of arithmetic degree 1 and that the arithmetic degree of a polynomial is the common arithmetic degree of its monomials, whose degree is defined to be the sum of the arithmetic degrees of its variables. An algorithm has degree d, if its test computation involves evaluation of multivariate polynomials of arithmetic degree d. A problem has degree d, if any algorithm that solves it has degree at least d [69].

Diameter of a graph: The distance between two vertices u and v in a graph is the sum of weights of the edges of the shortest path between them. (For an unweighted graph, it is the number of edges of the shortest path.) The diameter of a graph is the maximum among all the distances between all possible pairs of vertices.

Dynamic versus static: This refers to cases when the underlying problem domain can be subject to updates, i.e., insertions and deletions. If no updates are permitted, the problem or data structure is said to be static; otherwise, it is said to be dynamic.

Dynamization transformation: A data structuring technique can transform a static data structure into a dynamic one. In so doing, the performance of the dynamic structure will exhibit certain space-time trade-offs. See, e.g., [63,90] for more references.

Geometric duality: A transform between a point and a hyperplane in $\mathfrak{R}^k$, that preserves incidence and order relation. For a point $p = (\mu_1, \mu_2, \ldots, \mu_k)$, its dual $\mathcal{D}(p)$ is a hyperplane denoted by $x_k = \sum_{j=1}^{k-1} \mu_j x_j - \mu_k$; for a hyperplane $\mathcal{H} : x_k = \sum_{j=1}^{k-1} \mu_j x_j + \mu_k$, its dual $\mathcal{D}(\mathcal{H})$ is a point denoted by $(\mu_1, \mu_2, \ldots, -\mu_k)$. There are other duality transformations. What is described in the text is called the paraboloid transform. See [37,44,90] for more information.

Height-balanced binary search tree: A data structure used to support membership, insert/delete operations each in time logarithmic in the size of the tree. A typical example is the *AVL* tree or red-black tree.

Orthogonally convex rectilinear polygon: A rectilinear polygon *P* is orthogonally convex if every horizontal or vertical segment connecting two points in *P* lies totally within *P*.

Priority queue: A data structure used to support insert and delete operations in time logarithmic in the size of the queue. The elements in the queue are arranged so that the element of the minimum priority is always at one end of the queue, readily available for delete operation. Deletions only take place at that end of the queue. Each delete operation can be done in constant time. However, since restoring the above property after each deletion takes logarithmic time, we often say that each delete operation takes logarithmic time. A heap is a well-known priority queue.

References

1. Agarwal, P.K., Range Searching, in *Handbook of Discrete and Computational Geometry*, Goodman, J.E. and O'Rourke, J. (Eds.), CRC Press LLC, Boca Raton, FL, 2004.
2. Agarwal, P.K., de Berg, M., Matoušek, J., and Schwarzkopf, O., Constructing levels in arrangements and higher order Voronoi diagrams, *SIAM J. Comput.*, 27, 654–667, 1998.
3. Agarwal, P.K., Edelsbrunner, H., Schwarzkopf, O., and Welzl, E., Euclidean minimum spanning trees and bichromatic closest pairs, *Discrete Comput. Geom.*, 6(5), 407–422, 1991.
4. Agarwal, P.K. and Erickson, J., Geometric range searching and its relatives, in *Advances in Discrete and Computational Geometry*, Chazelle, B., Goodman, J.E., and Pollack, R. (Eds.), AMS Press, Providence, RI, pp. 1–56, 1999.
5. Aggarwal, A., Guibas, L.J., Saxe, J., and Shor, P.W., A linear-time algorithm for computing the Voronoi diagram of a convex polygon, *Discrete Comput. Geom.*, 4(6), 591–604, 1989.
6. Agarwal, P.K. and Matoušek, J., Dynamic half-space range reporting and its applications, *Algorithmica*, 13(4) 325–345, Apr. 1995.
7. Agarwal, P. and Varadarajan, K., A near-linear constant-factor approximation for Euclidean bipartite matching? *Proceedings of the 20th Symposium on Computational Geometry*, ACM, New York, pp. 247–252, 2004.
8. Aichholzer, O., Aurenhammer, F., Chen, D.Z., Lee, D.T., Mukhopadhyay, A., and Papadopoulou, E., Skew Voronoi diagrams, *Int. J. Comput. Geom. Appl.*, 9(3), 235–247, June 1999.
9. Alt, H. and Guibas, L.J., Discrete geometric shapes: Matching, interpolation, and approximation, in *Handbook of Computational Geometry*, Sack, J.R. and Urrutia, J. (Eds.), Elsevier Science Publishers, B.V. North-Holland, Amsterdam, the Netherlands, pp. 121–153, 2000.
10. Alt, H. and Schwarzkopf, O., The Voronoi diagram of curved objects, *Discrete Comput. Geom.*, 34(3), 439–453, Sept. 2005.
11. Alt, H. and Yap, C.K., Algorithmic aspect of motion planning: A tutorial, Part 1 & 2, *Algorithms Rev.*, 1, 43–77, 1990.
12. Amato, N.M. and Ramos, E.A., On computing Voronoi diagrams by divide-prune-and-conquer, *Proceedings of the 12th ACM Symposium on Computational Geometry*, ACM, New York, pp. 166–175, 1996.
13. Atallah, M.J., Parallel techniques for computational geometry, *Proc. IEEE*, 80(9), 1435–1448, Sept. 1992.
14. Atallah, M.J. and Chen, D.Z., On connecting red and blue rectilinear polygonal obstacles with nonintersecting monotone rectilinear paths, *Int. J. Comput. Geom. Appl.*, 11(4), 373–400, Aug. 2001.

15. Aurenhammer, F. and Klein, R., Voronoi diagrams, in *Handbook of Computational Geometry*, Sack, J.R. and Urrutia, J. (Eds.), Elsevier Science Publishers, B.V. North-Holland, Amsterdam, the Netherlands, pp. 201–290, 2000.

16. Balaban, I.J., An optimal algorithm for finding segments intersections, *Proceedings of the 11th Symposium on Computational Geometry*, Vancouver, BC, Canada, pp. 211–219, June 1995.

17. Bespamyatnikh, S.N., An optimal algorithm for closest pair maintenance, *Discrete Comput. Geom.*, 19(2), 175–195, 1998.

18. Blum, A., Ravi, R., and Vempala, S., A Constant-factor approximation algorithm for the k-MST problem, *J. Comput. Syst. Sci.*, 58(1), 101–108, Feb. 1999.

19. Boissonnat, J.-D., Sharir, M., Tagansky, B., and Yvinec, M., Voronoi diagrams in higher dimensions under certain polyhedra distance functions, *Discrete Comput. Geom.*, 19(4), 473–484, 1998.

20. Boissonnat, J.-D. and Snoeyink, J., Efficient algorithms for line and curve segment intersection using restricted predicates, *Comp. Geom. Theor. Appl.*, 19(1), 35–52, May 2000.

21. Callahan, P. and Kosaraju, S.R., A decomposition of multidimensional point sets with applications to k-nearest-neighbors and n-body potential fields, *J. Assoc. Comput. Mach.*, 42(1), 67–90, Jan. 1995.

22. Chan, T.M., More planar 2-center algorithms, *Comp. Geom. Theor. Appl.*, 13(3), 189–198, Sept. 1999.

23. Chazelle, B., Lower bounds for orthogonal range searching. I. The reporting case, *J. Assoc. Comput. Mach.*, 37(2), 200–212, Apr. 1990.

24. Chazelle, B., An optimal algorithm for intersecting three-dimensional convex polyhedra, *SIAM J. Comput.*, 21(4), 671–696, 1992.

25. Chazelle, B., Cutting hyperplanes for divide-and-conquer, *Discrete Comput. Geom.*, 9(2), 145–158, 1993.

26. Chazelle, B. and Dobkin, D.P., Intersection of convex objects in two and three dimensions, *J. Assoc. Comput. Mach.*, 34(1), 1–27, 1987.

27. Chazelle, B. and Edelsbrunner, H., An optimal algorithm for intersecting line segments in the plane, *J. Assoc. Comput. Mach.*, 39(1), 1–54, 1992.

28. Chazelle, B., Edelsbrunner, H., Guibas, L.J., and Sharir, M., Diameter, width, closest line pair, and parametric searching, *Discrete Comput. Geom.*, 8, 183–196, 1993.

29. Chazelle, B., Edelsbrunner, H., Guibas, L.J., and Sharir, M., Algorithms for bichromatic line-segment problems and polyhedral terrains, *Algorithmica*, 11(2), 116–132, Feb. 1994.

30. Chazelle, B. and Friedman, J., Point location among hyperplanes and unidirectional ray-shooting, *Comp. Geom. Theor. Appl.*, 4, 53–62, 1994.

31. Chen, D., Klenk, K.S., and Tu, H.-Y.T., Shortest path queries among weighted obstacles in the rectilinear plane, *SIAM J. Comput.*, 29(4), 1223–1246, Feb. 2000.

32. Chen, J. and Han, Y., Shortest paths on a polyhedron, Part I: Computing shortest paths, *Int. J. Comput. Geom. Appl.*, 6(2), 127–144, 1996.

33. Chen, Z. Papadopoulou, E., and Xu J., Robustness of k-gon Voronoi diagram construction, *Inf. Process. Lett.*, 97(4), 138–145, Feb. 2006.

34. Chiang, Y.-J. and Tamassia, R., Dynamic algorithms in computational geometry, *Proc. IEEE*, 80(9), 1412–1434, Sept. 1992.

35. Choi, J., Sellen, J., and Yap, C.K., Approximate Euclidean shortest path in 3-space, *Int. J. Comput. Geom. Appl.*, 7(4), 271–295, 1997.

36. Daniels, K., Milenkovic, V., and Roth, D., Finding the largest area axis-parallel rectangle in a polygon, *Comp. Geom. Theor. Appl.*, 7, 125–148, Jan. 1997.

37. de Berg, M., Cheong, O., van Kreveld, M. and Overmars, M., *Computational Geometry: Algorithms and Applications*, Springer-Verlag, Berlin, Germany, p. 386, 2008.

38. Dehne, F. and Klein, R., The Big Sweep: On the power of the wavefront approach to Voronoi diagrams, *Algorithmica*, 17(1), 19–32, Jan. 1997.

39. Diaz-Banez, J.M., Hurtado, F., Meijer, H., Rappaport, D., and Sellares, J.A., The largest empty annulus problem, *Int. J. Comput. Geom. Appl.*, 13(4), 317–325, Aug. 2003.

40. Dorward, S.E., A survey of object-space hidden surface removal, *Int. J. Comput. Geom. Appl.*, 4(3), 325–362, Sept. 1994.

41. Du, D.Z. and Hwang, F.K., Eds., *Computing in Euclidean Geometry*, World Scientific, Singapore, 1992.

42. Duncan, C.A., Goodrich, M.T., and Ramos, E.A., Efficient approximation and optimization algorithms for computational metrology, *Proceedings of the 8th annual ACM-SIAM Symposium on Discrete Algorithms*, SIAM, Philadelphia, PA, pp. 121–130, 1997.

43. Dwyer, R.A. and Eddy, W.F., Maximal empty ellipsoids, *Int. J. Comput. Geom. Appl.*, 6(2), 169–186, 1996.

44. Edelsbrunner, H., *Algorithms in Combinatorial Geometry*, Springer-Verlag, Berlin, Germany, 1987.

45. Edelsbrunner, H., The union of balls and its dual shape, *Discrete Comput. Geom.*, 13(1), 415–440, 1995.

46. Edelsbrunner, H. and Sharir, M., A hyperplane incidence problem with applications to counting distances, in *Applied Geometry and Discrete Mathematics. The Victor Klee Festschrift*, Gritzmann, P. and Sturmfels, B. (Eds.), *DIMACS Series in Discrete Mathematics and Theoretical Computer Science*, AMS Press, Providence, RI, pp. 253–263, 1991.

47. Efrat, A. and Sharir, M., A near-linear algorithm for the planar segment center problem, *Discrete Comput. Geom.*, 16(3), 239–257, 1996.

48. Eppstein, D., Fast construction of planar two-centers, *Proceedings of the 8th annual ACM-SIAM Symposium on Discrete Algorithms*, SIAM, Philadelphia, PA, pp. 131–138, 1997.

49. Fabri A., Giezeman G., Kettner L., Schirra, S., and Schonherr S., On the design of CGAL a computational geometry algorithms library, *Software Pract. Ex.*, 30(11), 1167–1202, Aug. 2000. http://www.cgal.org/

50. Finkle, U. and Hinrichs, K., Overlaying simply connected planar subdivision in linear time, *Proceedings of the 11th annual ACM Symposium on Computational Geometry*, ACM, New York, pp. 119–126, 1995.

51. Garcia-Lopez, J., Ramos, P.A., and Snoeyink, J., Fitting a set of points by a circle, *Discrete Comput. Geom.*, 20, 389–402, 1998.

52. Goodman, J.E. and O'Rourke, J., Eds., *The Handbook of Discrete and Computational Geometry*, CRC Press LLC, Boca Raton, FL, 2004.

53. Gupta, P., Janardan, R., Smid, M., and Dasgupta, B., The rectangle enclosure and point-dominance problems revisited, *Int. J. Comput. Geom. Appl.*, 7(5), 437–455, May 1997.

54. Hershberger, J. and Suri, S., Efficient computation of Euclidean shortest paths in the plane, *Proceedings of the 34th Symposium on Foundations of Computer Science*, Palo Alto, CA, 508–517, Nov. 1993.

55. Hershberger, J. and Suri, S., Finding a shortest diagonal of a simple polygon in linear time, *Comput. Geom. Theor. Appl.*, 7(3), 149–160, Feb. 1997.

56. Ho, J.M., Chang, C.H., Lee, D.T., and Wong, C.K., Minimum diameter spanning tree and related problems, *SIAM J. Comput.*, 20(5), 987–997, Oct. 1991.

57. Hwang, F.K., Foreword, *Algorithmica*, 7(2/3), 119–120, 1992.

58. Indyk, P. A near linear time constant factor approximation for Euclidean bichromatic matching (Cost), *Proceedings of the 18th Annual ACM-SIAM Symposium on Discrete Algorithms*, SIAM, Philadelphia, PA, pp. 39–42, 2007.

59. Janardan, R. and Lopez, M., Generalized intersection searching problems, *Int. J. Comput. Geom. Appl.*, 3(1), 39–69, Mar. 1993.

60. Kapoor, S. and Smid, M., New techniques for exact and approximate dynamic closest-point problems, *SIAM J. Comput.*, 25(4), 775–796, Aug. 1996.

61. Klawe, M.M. and Kleitman, D.J., An almost linear time algorithm for generalized matrix searching, *SIAM J. Discrete Math.,* 3(1), 81–97, Feb. 1990.

62. Lee, D.T., Two dimensional voronoi diagrams in the L_p-metric, *J. Assoc. Comput. Mach.,* 27, 604–618, 1980.

63. Lee, D.T., Computational geometry, in *Computer Science and Engineering Handbook,* Tucker, A. (Ed.), CRC Press, Boca Raton, FL, pp. 111–140, 1997.

64. Lee, D.T., Lee, G.C., and Huang, Y.W., Knowledge management for computational problem solving, *J. Universal Comput. Sci.,* 9(6), 563–570, 2003.

65. Lee, D.T. and Shen, C.F., The Steiner minimal tree problem in the λ-geometry plane, *Proceedings of the 7th International Symposium on Algorithms and Computation,* Asano, T., Igarashi, Y., Nag-amochi, H., Miyano, S., and Suri, S. (Eds.), LNCS, Vol. 1173, Springer-Verlag, Berlin, Germany, pp. 247–255, Dec. 1996.

66. Lee, D.T. and Wu, V.B., Multiplicative weighted farthest neighbor Voronoi diagrams in the plane, *Proceedings of International Workshop on Discrete Mathematics and Algorithms,* Hong Kong, pp. 154–168, Dec. 1993.

67. Lee, D.T. and Wu, Y.F., Geometric complexity of some location problems, *Algorithmica,* 1(1), 193–211, 1986.

68. Lee, D.T., Yang, C.D., and Wong, C.K., Rectilinear paths among rectilinear obstacles, *Perspect. Discrete Appl. Math.,* Bogart, K. (Ed.), 70, 185–215, 1996.

69. Liotta, G., Preparata, F.P., and Tamassia, R., Robust proximity queries: An illustration of degree-driven algorithm design, *SIAM J. Comput.,* 28(3), 864–889, Feb. 1998.

70. Lu, B. and Ruan, L., Polynomial time approximation scheme for the rectilinear Steiner arborescence problem, *J. Comb. Optim.,* 4(3), 357–363, 2000.

71. Marcotte, O. and Suri, S., Fast matching algorithms for points on a polygon, *SIAM J. Comput.,* 20(3), 405–422, June 1991.

72. Matoušek, J., Geometric range searching, *ACM Comput. Surv.,* 26(4), 421–461, 1994.

73. Mehlhorn, K. and Näher, S., *LEDA, A Platform for Combinatorial and Geometric Computing,* Cambridge University Press, Cambridge, U.K., 1999.

74. Mitchell, J.S.B., Shortest paths among obstacles in the plane, *Int. J. Comput. Geom. Appl.,* 6(3), 309–332, Sept. 1996.

75. Mitchell, J.S.B., Guillotine subdivisions approximate polygonal subdivisions: A simple polynomial-time approximation scheme for geometric TSP, k-MST, and Related Problems, *SIAM J. Comput.,* 28(4), 1298–1309, Aug. 1999.

76. Mitchell, J.S.B., Geometric shortest paths and network optimization, in *Handbook of Computational Geometry,* Sack, J.R. and Urrutia, J. (Eds.), Elsevier Science Publishers, B.V. North-Holland, Amsterdam, the Netherlands, 633–701, 2000.

77. Mount, D.M., Intersection detection and separators for simple polygons, *Proceedings of the 8th Annual ACM Symposium on Computational Geometry,* ACM, New York, pp. 303–311, 1992.

78. Mukhopadhyay, A. and Rao, S.V., Efficient algorithm for computing a largest empty arbitrarily oriented rectangle, *Int. J. Comput. Geom. Appl.,* 13(3), 257–272, June 2003.

79. Narasimhan, G. and Smid, M. *Geometric Spanner Networks*, Cambridge University Press, Cambridge, U.K., 2007.

80. Nielsen, F., Boissonnat, J.-D., and Nock, R., On bregman Voronoi diagrams, *Proceedings of the 18th Annual ACM-SIAM Symposium on Discrete Algorithms,* SIAM, Philadelphia, PA, pp. 746–755, 2007.

81. Okabe, A., Boots, B., Sugihara, K., and Chiu, S.N., *Spatial Tessellations: Concepts and Applications of Voronoi Diagrams,* (2nd ed.), John Wiley & Sons, Chichester, U.K., 2000.

82. Orlowski, M., A new algorithm for the largest empty rectangle, *Algorithmica,* 5(1), 65–73, 1990.

83. Papadimitriou, C.H. and Steiglitz, K., *Combinatorial Optimization: Algorithms and Complexity,* Dover Publications Inc., Mineola, NY, 1998.

84. Papadopoulou, E., *k*-Pairs non-crossing shortest paths in a simple polygon, *Int. J. Comput. Geom. Appl.*, 9(6), 533–552, Dec. 1999.
85. Papadopoulou, E., The hausdorff Voronoi diagram of point clusters in the plane, *Algorithmica*, 40(2), 63–82, July 2004.
86. Papadopoulou, E. and Lee, D.T., A new approach for the geodesic voronoi diagram of points in a simple polygon and other restricted polygonal domains, *Algorithmica*, 20(4), 319–352, Apr. 1998.
87. Papadopoulou, E. and Lee, D.T., The Hausdorff Voronoi diagram of polygonal objects: A divide and conquer approach, *Int. J. Comput. Geom. Appl.*, 14(6), 421–452, Dec. 2004.
88. Pettie, S. and Ramachandran, V., An optimal minimum spanning tree algorithm, *J. Assoc. Comput. Mach.*, 49(1), 16–34, Jan. 2002.
89. Pocchiola, M. and Vegter, G., Topologically sweeping visibility complexes via pseudotriangulations, *Discrete Comput. Geom.*, 16, 419–453, Dec. 1996.
90. Preparata, F.P. and Shamos, M.I., *Computational Geometry: An Introduction,* Springer-Verlag, Berlin, Germany, 1988.
91. Ranmath, S., New Approximations for the rectilinear Steiner arborescence problem, *IEEE Trans. Comput. Aided Design.*, 22(7), 859–869, July 2003.
92. Ravi, R., Sundaram, R., Marathe, M.V., Rosenkrantz, D.J., and Ravi, S.S., Spanning trees short or small, *SIAM J. Discrete Math.*, 9(2), 178–200, May 1996.
93. Sack, J. and Urrutia, J., Eds., *Handbook of Computational Geometry,* Elsevier Science Publishers, B.V. North-Holland, Amsterdam, the Netherlands, 2000.
94. Sharir, M., Algorithmic motion planning, in *Handbook of Discrete and Computational Geometry,* Goodman, J.E. and O'Rourke, J. (Eds.) CRC Press LLC, Boca Raton, FL, 2004.
95. Schwartz, C., Smid, M., and Snoeyink, J., An optimal algorithm for the on-line closest-pair problem, *Algorithmica,* 12(1), 18–29, July 1994.
96. Seo, D.Y., Lee, D.T., and Lin, T.C., Geometric minimum diameter minimum cost spanning tree problem, *Algorithmica,* submitted for publication, 2009.
97. Shermer, T. and Yap, C., Probing near centers and estimating relative roundness, *Proceedings of the ASME Workshop on Tolerancing and Metrology,* University of North Carolina, Charlotte, NC, 1995.
98. Shi, W. and Chen S., The rectilinear Steiner arborescence problem is NP-complete, *SIAM J. Comput.*, 35(3), 729–740, 2005.
99. Smid, M. and Janardan, R., On the width and roundness of a set of points in the plane, *Comput. Geom. Theor. Appl.*, 9(1), 97–108, Feb. 1999.
100. Swanson, K., Lee, D.T., and Wu, V.L., An Optimal algorithm for roundness determination on convex polygons, *Comput. Geom. Theor. Appl.*, 5(4), 225–235, Nov. 1995.
101. Vaidya, P.M., Geometry helps in matching, *SIAM J. Comput.*, 18(6), 1201–1225, Dec. 1989.
102. Wei, J.D., Tsai, M.H., Lee, G.C., Huang, J.H., and Lee, D.T., GeoBuilder: A geometric algorithm visualization and debugging system for 2D and 3D geometric computing, *IEEE Trans. Vis. Comput. Graph*, 15(2), 234–248, Mar. 2009.
103. Welzl, E., Smallest enclosing disks, balls and ellipsoids, in *New Results and New Trends in Computer Science,* Maurer, H.A. (Ed.), LNCS, Vol. 555, Springer-Verlag, Berlin, Germany, pp. 359–370, 1991.
104. Yang, C.D., Lee, D.T., and Wong, C.K., Rectilinear path problems among rectilinear obstacles revisited, *SIAM J. Comput.*, 24(3), 457–472, June 1995.
105. Yao, F.F., Computational geometry, in *Handbook of Theoretical Computer Science, Vol. A, Algorithms and Complexity,* van Leeuwen, J. (Ed.), MIT Press, Cambridge, MA, pp. 343–389, 1994.
106. Yap, C., Exact computational geometry and tolerancing, in *Snapshots of Computational and Discrete Geometry,* Avis, D. and Bose, J. (Eds.), School of Computer Science, McGill University, Montreal, QC, Canada, 1995.

3

Computational Topology

Afra Zomorodian
Dartmouth College

3.1 Introduction

According to the *Oxford English Dictionary*, the word *topology* is derived from *topos* (τόπος) meaning *place*, and *-logy* (λογια), a variant of the verb λέγειν, meaning *to speak*. As such, topology speaks about places: how local neighborhoods connect to each other to form a space. Computational topology, in turn, undertakes the challenge of studying topology using a computer.

The field of geometry studies intrinsic properties that are invariant under rigid motion, such as the curvature of a surface. In contrast, topology studies invariants under continuous deformations. The larger set of transformations enables topology to extract more qualitative information about a space, such as the number of connected components or tunnels. Computational topology has theoretical and practical goals. Theoretically, we look at the tractability and complexity of each problem, as well as the design of efficient data structures and algorithms. Practically, we are interested in heuristics and fast software for solving problems that arise in diverse disciplines. Our input is often a finite discrete set of noisy samples from some underlying space. This type of input has renewed interest in combinatorial and algebraic topology, areas that had been overshadowed by point set topology in the last one hundred years.

Computational topology developed in response to topological impediments emerging from within geometric problems. In computer graphics, researchers encountered the problem of connectivity in reconstructing *watertight* surfaces from point sets. Often, their heuristics resulted in surfaces with extraneous holes or tunnels that had to be detected and removed before proper geometric processing was feasible. Researchers in computational geometry provided guaranteed surface reconstruction algorithms that depended on sampling conditions on the input. These results required concepts from topology, provoking an interest in the subject. Topological problems also arise naturally in areas that do not deal directly with geometry. In robotics, researchers need to understand the connectivity of the *configuration space* of a robot for computing optimal trajectories that minimize resource consumption. In biology, the thermodynamics hypothesis states that proteins fold to their native states along such optimal trajectories. In sensor networks, determining coverage without localizing the sensors requires deriving global information from local connections. Once again the question of connectivity arises, steering us toward a topological understanding of the problem.

Like topology, computational topology is a large and diverse area. The aim of this chapter is not to be comprehensive, but to describe the fundamental concepts, methods, and structures that permeate the field. We begin with our objects of study, topological spaces and their combinatorial representations, in Section 3.2. We study these spaces through topological invariants, which we formalize in Section 3.3, classifying all surfaces, and introducing both combinatorial and algebraic invariants in the process. Sections 3.4 and 3.5 focus on homology and persistent homology, which are algebraic invariants that derive their popularity from their computability. Geometry and topology are intrinsically entangled, as revealed by Morse theory through additional structures in Section 3.6. For topological data analysis, we describe methods for deriving combinatorial structures that represent point sets in Section 3.7. We end the chapter with a brief discussion of geometric issues in Section 3.8.

3.2 Topological Spaces

The focus of this section is topological spaces: their definitions, finite representations, and data structures for efficient manipulation.

3.2.1 Topology

Metric spaces are endowed with a metric that defines open sets, neighborhoods, and continuity. Without a metric, we must prescribe enough structures to extend these concepts. Intuitively, a topological space is a set of points, each of which knows its neighbors. A **topology** on a set X is a subset $T \subseteq 2^X$ such that:

1. If $S_1, S_2 \in T$, then $S_1 \cap S_2 \in T$.
2. If $\{S_j \mid j \in J\} \subseteq T$, then $\cup_{j \in J} S_j \in T$.
3. $\emptyset, X \in T$.

A set $S \in T$ is an **open set** and its complement in X is **closed**. The pair $\mathbb{X} = (X, T)$ is a **topological space**. A set of points may be endowed with different topologies, but we will abuse notation by using $p \in \mathbb{X}$ for $p \in X$.

A function $f : \mathbb{X} \to \mathbb{Y}$ is **continuous**, if for every open set A in $\mathbb{Y}$, $f^{-1}(A)$ is open in $\mathbb{X}$. We call a continuous function a **map**. The **closure** $\overline{A}$ of A is the intersection of all closed sets containing A. The **interior** $\overset{\circ}{A}$ of A is the union of all open sets contained in A. The **boundary** ∂A of A is $\partial A = \overline{A} - \overset{\circ}{A}$.

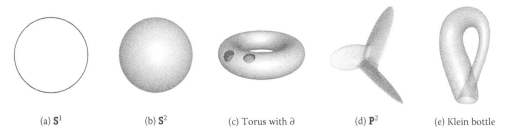

(a) $\mathbb{S}^1$ (b) $\mathbb{S}^2$ (c) Torus with ∂ (d) $\mathbb{P}^2$ (e) Klein bottle

FIGURE 3.1 Manifolds. (a) The only compact connected one-manifold is a circle $\mathbb{S}^1$. (b) The sphere is a two-manifold. (c) The surface of a donut, a torus, is also a two-manifold. This torus has two boundaries, so it is a manifold with boundary. Its boundary is two circles, a one-manifold. (d) A Boy's surface is a geometric immersion of the projective plane $\mathbb{P}^2$, a nonorientable two-manifold. (e) The Klein bottle is another nonorientable two-manifold.

A **neighborhood** of $x \in X$ is any $S \in T$ such that $x \in \mathring{S}$. A subset $A \subseteq X$ with **induced topology** $T_A = \{S \cap A \mid S \in T\}$ is a **subspace** of $\mathbb{X}$. A **homeomorphism** $f : \mathbb{X} \to \mathbb{Y}$ is a 1-1 onto function such that both f and f^{-1} are continuous. We say that $\mathbb{X}$ is **homeomorphic** to $\mathbb{Y}$, $\mathbb{X}$ and $\mathbb{Y}$ have the same **topological type**, and denote it as $\mathbb{X} \approx \mathbb{Y}$.

3.2.2 Manifolds

A topological space may be viewed as an abstraction of a metric space. Similarly, manifolds generalize the connectivity of d-dimensional Euclidean spaces $\mathbb{R}^d$ by being locally similar, but globally different. A **d-dimensional chart** at $p \in \mathbb{X}$ is a homeomorphism $\varphi : U \to \mathbb{R}^d$ onto an open subset of $\mathbb{R}^d$, where U is a neighborhood of p and *open* is defined using the metric (Apostol, 1969). A **d-dimensional manifold (d-manifold)** is a (separable Hausdorff) topological space $\mathbb{X}$ with a d-dimensional chart at every point $x \in \mathbb{X}$. We do not define the technical terms *separable* and *Hausdorff* as they mainly disallow pathological spaces. The circle or **1-sphere** $\mathbb{S}^1$ in Figure 3.1a is a 1-manifold as every point has a neighborhood homeomorphic to an open interval in $\mathbb{R}^1$. All neighborhoods on the **2-sphere** $\mathbb{S}^2$ in Figure 3.1b are homeomorphic to open disks, so $\mathbb{S}^2$ is a two-manifold, also called a **surface**. The **boundary** $\partial \mathbb{X}$ of a d-manifold $\mathbb{X}$ is the set of points in $\mathbb{X}$ with neighborhoods homeomorphic to $\mathbb{H}^d = \{x \in \mathbb{R}^d \mid x_1 \geq 0\}$. If the boundary is nonempty, we say $\mathbb{X}$ is a **manifold with boundary**. The boundary of a d-manifold with boundary is always a $(d-1)$-manifold without boundary. Figure 3.1c displays a **torus** with boundary, the boundary being two circles.

We may use a homeomorphism to place one manifold within another. An **embedding** $g : \mathbb{X} \to \mathbb{Y}$ is a homeomorphism onto its image $g(\mathbb{X})$. This image is called an **embedded submanifold**, such as the spaces in Figure 3.1a through c. We are mainly interested in *compact* manifolds. Again, this is a technical definition that requires the notion of a limit. For us, a compact manifold is closed and can be embedded in $\mathbb{R}^d$ so that it has a finite extent, i.e. it is bounded. An **immersion** $g : \mathbb{X} \to \mathbb{Y}$ of a compact manifold is a *local* embedding: an immersed compact space may self-intersect, such as the immersions in Figure 3.1d and e.

3.2.3 Simplicial Complexes

To compute information about a topological space using a computer, we need a finite representation of the space. In this section, we represent a topological space as a union of simple pieces, deriving a *combinatorial* description that is useful in practice. Intuitively, *cell complexes* are composed of Euclidean pieces glued together along seams, generalizing *polyhedra*. Due to their structural simplicity, simplicial complexes are currently a popular representation for topological spaces, so we describe their construction.

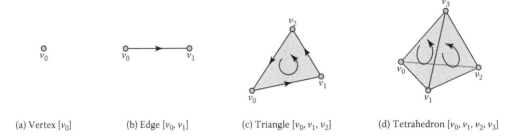

(a) Vertex $[v_0]$ (b) Edge $[v_0, v_1]$ (c) Triangle $[v_0, v_1, v_2]$ (d) Tetrahedron $[v_0, v_1, v_2, v_3]$

FIGURE 3.2 Oriented k-simplices, $0 \leq k \leq 3$. An oriented simplex induces orientation on its faces, as shown for the edges of the triangle and two faces of the tetrahedron.

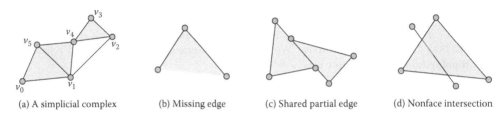

(a) A simplicial complex   (b) Missing edge   (c) Shared partial edge   (d) Nonface intersection

FIGURE 3.3 A simplicial complex (a) and three unions (b, c, d) that are not complexes. The triangle (b) does not have all its faces, and (c, d) have intersections not along shared faces.

A **k-simplex** σ is the convex hull of $k + 1$ affinely independent points $S = \{v_0, v_1, \ldots, v_k\}$, as shown in Figure 3.2. The points in S are the **vertices** of the simplex. A k-simplex σ is a k-dimensional subspace of $\mathbb{R}^d$, $\dim \sigma = k$. Since the defining points of a simplex are affinely independent, so is any subset of them. A simplex τ defined by a subset $T \subseteq S$ is a **face** of σ and has σ as a **coface**. For example, the triangle in Figure 3.2c is a coface of eight faces: itself, its three edges, three vertices, and the **(-1)-simplex** defined by the empty set, which is a face of every simplex. A **simplicial complex** K is a finite set of simplices such that every face of a simplex in K is also in K, and the nonempty intersection of any two simplices of K is a face of each of them. The **dimension** of K is the dimension of a simplex with the maximum dimension in K. The **vertices** of K are the zero-simplices in K. Figure 3.3 displays a simplicial complex and several badly formed unions. A simplex is **maximal**, if it has no proper coface in K. In the simplicial complex in the figure, the triangles and the edge $v_1 v_2$ are maximal.

The **underlying space** $|K|$ of a simplicial complex K is the topological space $|K| = \cup_{\sigma \in K} \sigma$, where we regard each simplex as a topological subspace. A **triangulation** of a topological space $\mathbb{X}$ is a simplicial complex K such that $|K| \approx \mathbb{X}$. We say $\mathbb{X}$ is **triangulable**, when K exists. Triangulations enable us to represent topological spaces compactly as simplicial complexes. For instance, the surface of the tetrahedron in Figure 3.2d is a triangulation of the sphere $\mathbb{S}^2$ in Figure 3.1b. Two simplicial complexes K and L are **isomorphic** iff $|K| \approx |L|$.

We may also define simplicial complexes without utilizing geometry, thereby revealing their combinatorial nature. An **abstract simplicial complex** is a set $\mathcal{S}$ of finite sets such that if $A \in \mathcal{S}$, so is every subset of A. We say $A \in \mathcal{S}$ is an **(abstract) k-simplex** of dimension k, if $|A| = k + 1$. A vertex is a 0-simplex, and the face and coface definitions follow as before. Given a (geometric) simplicial complex K with vertices V, let $\mathcal{S}$ be the collection of all subsets $\{v_0, v_1, \ldots, v_k\}$ of V such that the vertices $v_0, v_1, \ldots, v_k$ span a simplex of K. The collection $\mathcal{S}$ is the **vertex scheme** of K

and is an abstract simplicial complex. For example, the vertex scheme of the simplicial complex in Figure 3.3a is:

$$\left\{\begin{array}{l} \emptyset, \\ \{v_0\}, \{v_1\}, \{v_2\}, \{v_3\}, \{v_4\}, \{v_5\}, \\ \{v_0, v_1\}, \{v_0, v_5\}, \{v_1, v_2\}, \{v_1, v_4\}, \{v_1, v_5\}, \{v_2, v_3\}, \{v_2, v_4\}, \{v_3, v_4\}, \{v_4, v_5\}, \\ \{v_0, v_1, v_5\}, \{v_1, v_4, v_5\}, \{v_2, v_3, v_4\} \end{array}\right\} \quad (3.1)$$

Let S_1 and S_2 be abstract simplicial complexes with vertices V_1 and V_2, respectively. An **isomorphism** between S_1 and S_2 is a bijection $\varphi : V_1 \to V_2$, such that the sets in S_1 and S_2 are the same under the renaming of the vertices by φ and its inverse.

There is a strong relationship between the geometric and abstract definitions. Every abstract simplicial complex S is isomorphic to the vertex scheme of some simplicial complex K, which is its **geometric realization**. For example, Figure 3.3a is a geometric realization of vertex scheme (3.1). Two simplicial complexes are isomorphic iff their vertex schemes are isomorphic. As we shall see in Section 3.7, we usually compute simplicial complexes using geometric techniques, but discard the realization and focus on its topology as captured by the vertex scheme. As such, we refer to abstract simplicial complexes simply as simplicial complexes from now on.

A **subcomplex** is a subset $L \subseteq K$ that is also a simplicial complex. An important subcomplex is the **k-skeleton**, which consists of simplices in K of dimension less than or equal to k. The smallest subcomplex containing a subset $L \subseteq K$ is its **closure**, $\mathrm{Cl}\, L = \{\tau \in K \mid \tau \subseteq \sigma \in L\}$. The **star** of L contains all of the cofaces of L, $\mathrm{St}\, L = \{\sigma \in K \mid \sigma \supseteq \tau \in L\}$. The **link** of L is the boundary of its star, $\mathrm{Lk}\, L = \mathrm{Cl}\,\mathrm{St}\, L - \mathrm{St}\,(\mathrm{Cl}\, L - \{\emptyset\})$. Stars and links correspond to open sets and boundaries in topological spaces.

Suppose we fix an order on the set of vertices. An **orientation** of a k-simplex $\sigma \in K$, $\sigma = \{v_0, v_1, \ldots, v_k\}, v_i \in K$ is an equivalence class of orderings of the vertices of σ, where $(v_0, v_1, \ldots, v_k) \sim (v_{\tau(0)}, v_{\tau(1)}, \ldots, v_{\tau(k)})$ are equivalent orderings if the parity of the permutation τ is even. An **oriented simplex** is a simplex with an equivalence class of orderings, denoted as a sequence $[\sigma]$. We may show an orientation graphically using arrows, as in Figure 3.2. An oriented simplex **induces** orientations on its faces, where we drop the vertices not defining a face in the sequence to get the orientation. For example, triangle $[v_0, v_1, v_2]$ induces oriented edge $[v_0, v_1]$. Two k-simplices sharing a $(k-1)$-face τ are **consistently oriented**, if they induce different orientations on τ. A triangulable d-manifold is **orientable** if all d-simplices in any of its triangulations can be oriented consistently. Otherwise, the d-manifold is **nonorientable**.

3.2.4 Data Structures

There exist numerous data structures for storing cell complexes, especially simplicial complexes. In triangulations, all simplices are maximal and have the same dimension as the underlying manifold. For instance, all the maximal simplices of a triangulation of a two-manifold, with or without boundary, are triangles, so we may just store triangles and infer the existence of lower-dimensional simplices. This is the basic idea behind *triangle soup* file formats for representing surfaces in graphics, such as *OBJ* or *PLY*.

For topological computation, we require quick access to our neighbors. The key insight is to store both the complex and its dual at once, as demonstrated by the **quadedge** data structure for surfaces (Guibas and Stolfi, 1985). This data structure focuses on a directed edge e, from its origin $\mathrm{Org}(e)$ to its destination $\mathrm{Dest}(e)$. Each directed edge separates two regions to its left and right, as shown in Figure 3.4a. The data structure represents each edge in the triangulation with four edges: the original edge and edges $\mathrm{Rot}(e)$ directed from right to left, $\mathrm{Sym}(e)$ from $\mathrm{Dest}(e)$ to $\mathrm{Org}(e)$, and $\mathrm{Tor}(e)$ from left to right, as shown in Figure 3.4b. The edges e and $\mathrm{Sym}(e)$ are in **primal**

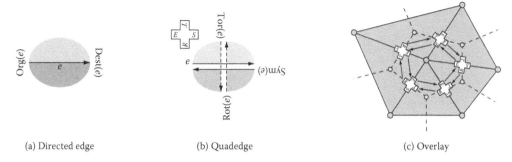

(a) Directed edge | (b) Quadedge | (c) Overlay

FIGURE 3.4 The quadedge data structure. (a) A directed edge e from Org(e) to Dest(e). (b) A quadedge is composed of four edges: e, Sym(e) from Dest(e) to Org(e), and dual edges Rot(e) and Tor(e) that go from right to left and in reverse, respectively. The edges are stored together in a single data structure. (c) A portion of a triangulation overlaid with dashed dual edges and quadedge data structure.

space and the edges Rot(e) and Tor(e) are in **dual** space. Intuitively, we get these edges by rotating counterclockwise by 90° repeatedly, moving between primal and dual spaces. In addition to storing geometric data, each edge also stores a pointer to the next counterclockwise edge with the same origin, returning it with operation Onext(e). Figure 3.4c shows a portion of a triangulation, its dual, and its quadedge data structure. The quadedge stores the four edges together in a single data structure, so all operations are $O(1)$. The operations form an *edge algebra* that allows us to walk quickly across the surface. For example, to get the next clockwise edge with the same origin, we can define Oprev $\equiv$ Rot $\circ$ Onext $\circ$ Rot, as can be checked readily on Figure 3.4c.

For arbitrary-dimensional simplicial complexes, we begin by placing a full ordering on the vertices. We then place all simplices in a hash table, using their vertices as unique keys. This allows for $O(1)$ access to faces, but not to cofaces, which require an inverse lookup table.

Computing orientability. Access to the dual structure enables easy computation of many topological properties, such as orientability. Suppose K is a cell complex whose underlying space is d-manifold. We begin by orienting one cell arbitrarily. We then use the dual structure to consistently orient the neighboring cells. We continue to spread the orientation until either all the cells are consistently oriented, or we arrive at a cell that is inconsistently oriented. In the former case, K is orientable, and in the latter, it is nonorientable. This algorithm takes time $O(n_d)$, where n_d is the number of d-cells.

3.3 Topological Invariants

In the last section, we focused on topological spaces and their combinatorial representations. In this section, we examine the properties of topologically equivalent spaces. The equivalence relation for topological spaces is the homeomorphism, which places spaces of the same topological type into the same class. We are interested in *intrinsic* properties of spaces within each class, that is, properties that are invariant under homeomorphism. Therefore, a fundamental problem in topology is the *homeomorphism problem*: Can we determine whether two objects are homeomorphic?

As we shall see, the homeomorphism problem is computationally intractable for manifolds of dimensions greater than three. Therefore, we often look for partial answers in the form of invariants. A **topological invariant** is a map f that assigns the same object to spaces of the same topological type, that is:

$$X \approx Y \implies f(X) = f(Y) \tag{3.2}$$

$$f(X) \neq f(Y) \implies X \not\approx Y \quad \text{(contrapositive)} \tag{3.3}$$

$$f(X) = f(Y) \implies \text{nothing (in general)} \tag{3.4}$$

Note that an invariant is only useful through the contrapositive. The **trivial** invariant assigns the same object to all spaces and is therefore useless. The **complete** invariant assigns different objects to nonhomeomorphic spaces, so the inverse of the implication (3.2) is true. Most invariants fall in between these two extremes. A topological invariant implies a classification that is coarser than, but respects, the topological type. In general, the more powerful an invariant, the harder it is to compute it, and as we relax the classification to be coarser, the computation becomes easier. In this section, we learn about several combinatorial and algebraic invariants.

3.3.1 Topological Type for Manifolds

For compact manifolds, the homeomorphism problem is well understood. In one dimension, there is only a single manifold, namely $\mathbb{S}^1$ in Figure 3.1a, so the homeomorphism problem is trivially solved. In two dimensions, the situation is more interesting, so we begin by describing an operation for connecting manifolds. The **connected sum** of two d-manifolds $\mathbb{M}_1$ and $\mathbb{M}_2$ is

$$\mathbb{M}_1 \# \mathbb{M}_2 = \mathbb{M}_1 - \mathbb{D}_1^d \bigcup_{\partial \mathbb{D}_1^d = \partial \mathbb{D}_2^d} \mathbb{M}_2 - \mathbb{D}_2^d, \tag{3.5}$$

where $\mathbb{D}_1^d$ and $\mathbb{D}_2^d$ are d-dimensional disks in $\mathbb{M}_1$ and $\mathbb{M}_2$, respectively. In other words, we cut out two disks and glue the manifolds together along the boundary of those disks using a homeomorphism, as shown in Figure 3.5 for two tori. The connected sum of g tori or g projective planes (Figure 3.1d) is a surface of **genus** g, and the former has g **handles**.

Using the connected sum, the homeomorphism problem is fully resolved for compact two-manifolds: Every triangulable compact surface is homeomorphic to a sphere, the connected sum of tori, or the connected sum of projective planes. Dehn and Heegaard (1907) give the first rigorous proof of this result. Rado (1925) completes their proof by showing that all compact surfaces are triangulable. A classic approach for this proof represents surfaces with **polygonal schema**, polygons whose edges are labeled by symbols, as shown in Figure 3.6 (Seifert and Threlfall, 2003). There are two edges for each symbol, and $\bar{a}$ represents an edge in the direction opposite to edge a. If we glue the edges according to direction, we get a compact surface. There is a **canonical form** of the polygonal schema for all surfaces. For the sphere, it is the special 2-gon in Figure 3.6a. For orientable manifolds

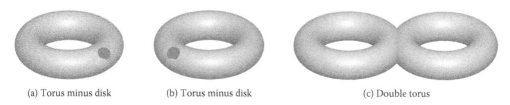

(a) Torus minus disk   (b) Torus minus disk   (c) Double torus

FIGURE 3.5 Connected sum. We cut out a disk from each two-manifold and glue the manifolds together along the boundary of those disks to get a larger manifold. Here, we sum two tori to get a surface with two handles.

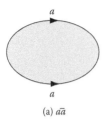

(a) $a\bar{a}$

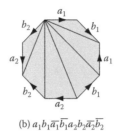

(b) $a_1 b_1 \bar{a_1} \bar{b_1} a_2 b_2 \bar{a_2} \bar{b_2}$

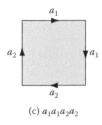

(c) $a_1 a_1 a_2 a_2$

FIGURE 3.6 Canonical forms for polygonal schema. The edges are glued according to the directions. (a) The sphere $\mathbb{S}^2$ has a special form. (b) A genus g orientable manifold is constructed of a $4g$-gon with the specified gluing pattern, so this is the double torus in Figure 3.5c. Adding diagonals, we triangulate the polygon to get a canonical triangulation for the surface. (c) A genus g nonorientable manifold is constructed out of a $2g$-gon with the specified gluing pattern. This 4-gon is the Klein bottle in Figure 3.1e, the connected sum of two projective planes in Figure 3.1d.

with genus g, it is a $4g$-gon with form $a_1 b_1 \bar{a_1} \bar{b_1} \ldots a_g b_g \bar{a_g} \bar{b_g}$, as shown for the genus two torus in Figure 3.6b. For nonorientable manifolds with genus g, it is a $2g$-gon with form $a_1 a_1 \ldots a_g a_g$, as shown for the genus two surface (Klein bottle) in Figure 3.6c. On the surface, the edges of the polygonal schema form loops called **canonical generators**. We can cut along these loops to flatten the surface into a topological disk, that is, the original polygon. We may also triangulate the polygon to get a minimal triangulation with a single vertex called the **canonical triangulation**, as shown in Figure 3.6b for the double torus.

In dimensions four and higher, Markov (1958) proves that the homeomorphism problem is undecidable. For the remaining dimension, $d = 3$, the problem is harder to resolve. Thurston (1982) conjectures that three-manifolds may be decomposed into pieces with uniform geometry in his *geometrization program*. Grigori Perelman completes Thurston's program in 2003 by eliminating geometric singularities that may arise from Ricci flow on the manifolds (Morgan and Tian, 2007).

Computing schema and generators. Brahana (1921) gives an algorithm for converting a polygonal schema into its canonical form. Based on this algorithm, Vegter and Yap (1990) present an algorithm for triangulated surfaces that runs in $O(n \log n)$ time, where n is the size of the complex. They also sketch an optimal algorithm for computing the canonical generators in $O(gn)$ time for a surface of genus g. For orientable surfaces, Lazarus et al. (2001) simplify this algorithm, describe a second optimal algorithm for computing canonical generators, and implement both algorithms. Canonical generators may be viewed as a one-vertex graph whose removal from the surface gives us a polygon. A **cut graph** generalizes this notion as any set of edges whose removal transforms the surface to a topological disk, a noncanonical schema. Dey and Schipper (1995) compute a cut graph for cell complexes via a breadth-first search of the dual graph.

3.3.2 The Euler Characteristic

Since the homeomorphism problem can be undecidable, we resort to invariants for classifying spaces. Let K be any cell complex and n_k be the number of k-dimensional cells in K. The **Euler characteristic** $\chi(K)$ is

$$\chi(K) = \sum_{k=0}^{\dim K} (-1)^k n_k. \tag{3.6}$$

The Euler characteristic is an integer invariant for $|K|$, so we get the same integer for different complexes with the same underlying space. For instance, the surface of the tetrahedron in Figure 3.2d is a triangulation of the sphere $\mathbb{S}^2$ and we have $\chi(\mathbb{S}^2) = 4 - 6 + 4 = 2$. A cube is a cell complex homeomorphic to $\mathbb{S}^2$, also giving us $\chi(\mathbb{S}^2) = 8 - 12 + 6 = 2$.

For compact surfaces, the Euler characteristic, along with orientability, provides us with a complete invariant. Recall the classification of surfaces from Section 3.3.1. Using small triangulations, it is easy to show $\chi(\text{Torus}) = 0$ and $\chi(\mathbb{P}^2) = 1$. Similarly, we can show that for compact surfaces $\mathbb{M}_1$ and $\mathbb{M}_2$, $\chi(\mathbb{M}_1 \# \mathbb{M}_2) = \chi(\mathbb{M}_1) + \chi(\mathbb{M}_2) - 2$. We may combine these results to compute the Euler characteristic of connected sums of manifolds. Let $\mathbb{M}_g$ be the connected sum of g tori and $\mathbb{N}_g$ be the connected sum of g projective planes. Then, we have

$$\chi(\mathbb{M}_g) = 2 - 2g, \tag{3.7}$$

$$\chi(\mathbb{N}_g) = 2 - g. \tag{3.8}$$

Computing topological types of 2-manifolds. We now have a two-phase algorithm for determining the topological type of an unknown triangulated compact surface $\mathbb{M}$: (1) Determine orientability and (2) compute $\chi(\mathbb{M})$. Using the algorithm in Section 3.2.4, we determine the orientability in $O(n_2)$, where n_2 is the number of triangles in $\mathbb{M}$. We then compute the Euler characteristic using Equation 3.6 in $O(n)$, where n is the size of $\mathbb{M}$. The characteristic gives us g, completing the classification of $\mathbb{M}$ in linear time.

3.3.3 Homotopy

A **homotopy** is a family of maps $f_t : \mathbb{X} \to \mathbb{Y}$, $t \in [0, 1]$, such that the associated map $F : \mathbb{X} \times [0, 1] \to \mathbb{Y}$ given by $F(x, t) = f_t(x)$ is continuous. Then, $f_0, f_1 : \mathbb{X} \to \mathbb{Y}$ are **homotopic** via the homotopy f_t, denoted $f_0 \simeq f_1$. A map $f : \mathbb{X} \to \mathbb{Y}$ is a **homotopy equivalence**, if there exists a map $g : \mathbb{Y} \to \mathbb{X}$, such that $f \circ g \simeq 1_{\mathbb{Y}}$ and $g \circ f \simeq 1_{\mathbb{X}}$, where $1_{\mathbb{X}}$ and $1_{\mathbb{Y}}$ are the identity maps on the respective spaces. Then, $\mathbb{X}$ and $\mathbb{Y}$ are **homotopy equivalent** and have the same **homotopy type**, denoted as $\mathbb{X} \simeq \mathbb{Y}$. The difference between the two spaces being *homeomorphic* versus being *homotopy equivalent* is in what the compositions of the two functions f and g are forced to be: In the former, the compositions must be *equivalent* to the identity maps; in the latter, they only need to be *homotopic* to them. A space with the homotopy type of a point is **contractible** or is **null-homotopic**. Homotopy is a topological invariant since $\mathbb{X} \approx \mathbb{Y} \Rightarrow \mathbb{X} \simeq \mathbb{Y}$. Markov's undecidability proof for the homeomorphism problem extends to show undecidability for homotopy equivalence for d-manifolds, $d \geq 4$ (Markov, 1958).

A **deformation retraction** of a space $\mathbb{X}$ onto a subspace $\mathbb{A}$ is a family of maps $f_t : \mathbb{X} \to \mathbb{X}$, $t \in [0, 1]$ such that f_0 is the identity map, $f_1(\mathbb{X}) = \mathbb{A}$, and $f_t|_{\mathbb{A}}$, the restriction of f_t to $\mathbb{A}$, is the identity map for all t. The family f_t should be continuous in the sense described above. Note that f_1 and the inclusion $i : \mathbb{A} \to \mathbb{X}$ give a homotopy equivalence between $\mathbb{X}$ and $\mathbb{A}$. For example, a cylinder is not homeomorphic to a circle, but homotopy equivalent to it, as we may continuously shrink a cylinder to a circle via a deformation retraction.

Computing homotopy for curves. Within computational geometry, the *homotopy problem* has been studied for curves on surfaces: Are two given curves homotopic? Suppose the curves are specified by n line segments in the plane, with n point obstacles, such as holes. Cabello et al. (2004) give an efficient $O(n \log n)$ time algorithm for simple paths, and an $O(n^{3/2} \log n)$ time algorithm for self-intersecting paths. For the general case, assume we have two closed curves of size k_1 and k_2 on a genus g two-manifold $\mathbb{M}$, represented by a triangulation of size n. Dey and Guha (1999) utilize combinatorial group theory to give an algorithm that decides if the paths are homotopic in $O(n + k_1 + k_2)$ time and space, provided $g \neq 2$ if $\mathbb{M}$ is orientable, or $g \neq 3, 4$ if $\mathbb{M}$ is nonorientable.

When one path is a point, the homotopy problem reduces to asking whether a closed curve is contractible. This problem is known also as the *contractibility problem*. While the above algorithms apply, the problem has been studied on its own. For a curve of size k on a genus g surface, Dey and Schipper (1995) compute a noncanonical polygonal schema to decide contractibility in suboptimal $O(n + k \log g)$ time and $O(n + k)$ space.

3.3.4 The Fundamental Group

The Euler characteristic is an invariant that describes a topological space via a single integer. *Algebraic topology* gives invariants that associate richer algebraic structures with topological spaces. The primary mechanism is via *functors*, which not only provide algebraic images of topological spaces, but also provide images of maps between the spaces. In the remainder of this section and the next two sections, we look at three functors as topological invariants.

The fundamental group captures the structure of homotopic loops within a space. A **path** in $\mathbb{X}$ is a continuous map $f : [0, 1] \to \mathbb{X}$. A **loop** is a path f with $f(0) = f(1)$, that is, a loop starts and ends at the same **basepoint**. The **trivial loop** never leaves its basepoint. The equivalence class of a path f under the equivalence relation of homotopy is $[f]$. A loop that is the boundary of a disk is a **boundary** and since it is contractible, it belongs to the class of the trivial loop. Otherwise, the loop is **nonbounding** and belongs to a different class. We see a boundary and two nonbounding loops on a torus in Figure 3.7. Given two paths $f, g : [0, 1] \to \mathbb{X}$ with $f(1) = g(0)$, the **product path** $f \cdot g$ is a path which traverses f and then g. The **fundamental group** $\pi_1(\mathbb{X}, x_0)$ of $\mathbb{X}$ and basepoint x_0 has the homotopy classes of loops in $\mathbb{X}$ based at x_0 as its elements, the equivalence class of the trivial loop as its identity element, and $[f][g] = [f \cdot g]$ as its binary operation. For a path-connected space $\mathbb{X}$, the basepoint is irrelevant, so we refer to the group as $\pi_1(\mathbb{X})$.

Clearly, any loop drawn on $\mathbb{S}^2$ is bounding, so $\pi_1(\mathbb{S}^2) \cong \{0\}$, where $\cong$ denotes group isomorphism and $\{0\}$ is the trivial group. For any $n \in \mathbb{Z}$, we my go around a circle n times, going clockwise for positive and counterclockwise for negative integers. Intuitively then, $\pi_1(\mathbb{S}^1) \cong \mathbb{Z}$. Similarly, the two nonbounding loops in Figure 3.7 generate the fundamental group of a torus, $\pi_1(\text{torus}) \cong \mathbb{Z} \times \mathbb{Z}$, where the surface of the torus turns the group Abelian.

The fundamental group was instrumental in two significant historical problems. Markov (1958) used the fundamental group to reduce the homeomorphism problem for manifolds to the *group isomorphism* problem, which was known to be undecidable (Adyan, 1955). Essentially, Markov's reduction is through the construction of a d-manifold, $d \geq 4$, whose fundamental group is a given group. In 1904, Poincaré conjectured that if the fundamental group of a three-manifold is trivial, then the manifold is homeomorphic to $\mathbb{S}^3$: $\pi_1(\mathbb{M}) \cong \{0\} \implies M \approx \mathbb{S}^3$. Milnor (2008) provides a detailed history of the **Poincaré Conjecture**, which was subsumed by Thurston's Geometrization program and completed by Perelman.

The fundamental group is the first in a series of **homotopy groups** $\pi_n(\mathbb{X})$ that extend the notion of loops to n-dimensional cycles. These invariant groups are very complicated and not directly computable from a cell complex. Moreover, for an n-dimensional space, only a finite number may be trivial, giving us an infinite description that is not computationally feasible.

FIGURE 3.7 Loops on a torus. The loop on the left is a boundary as it bounds a disk. The other two loops are nonbounding and together generate the fundamental group of a torus, $\pi_1(\text{torus}) \cong \mathbb{Z} \times \mathbb{Z}$.

Computation. For orientable surfaces, canonical generators or one-vertex cut graphs (Section 3.3.1) generate the fundamental group. While the fundamental group of specific spaces has been studied, there is no general algorithm for computing the fundamental group for higher-dimensional spaces.

3.4 Simplicial Homology

Homology is a topological invariant that is quite popular in computational topology as it is easily computable. Homology groups may be regarded as an algebraization of the first layer of geometry in cell complexes: how cells of dimension n attach to cells of dimension $n - 1$ (Hatcher, 2002). We define homology for simplicial complexes, but the theory extends to arbitrary topological spaces.

3.4.1 Definition

Suppose K is a simplicial complex. A **k-chain** is $\sum_i n_i[\sigma_i]$, where $n_i \in \mathbb{Z}$, $\sigma_i \in K$ is a k-simplex, and the brackets indicate orientation (Section 3.2.3). The **kth chain group $C_k(K)$** is the set of all chains, the free Abelian group on oriented k-simplices. For example, $[v_1, v_2] + 2[v_0, v_5] + [v_2, v_3] \in C_1(K)$ for the complex in Figure 3.3a. To reduce clutter, we will omit writing the complex below but emphasize that all groups are defined with respect to some complex. We relate chain groups in successive dimensions via the boundary operator. For k-simplex $\sigma = [v_0, v_1, \ldots, v_k] \in K$, we let

$$\partial_k \sigma = \sum_i (-1)^i [v_0, v_1, \ldots, \hat{v}_i, \ldots, v_n], \tag{3.9}$$

where $\hat{v}_i$ indicates that v_i is deleted from the sequence. The **boundary homomorphism $\partial_k : C_k \to C_{k-1}$** is the linear extension of the operator above, where we define $\partial_0 \equiv 0$. For instance,

$$\partial_1([v_1, v_2] + 2[v_0, v_5] + [v_2, v_3]) = \partial_1[v_1, v_2] + 2\partial_1[v_0, v_5] + \partial_1[v_2, v_3]$$
$$= (v_2 - v_1) + 2(v_5 - v_0) + (v_3 - v_2)$$
$$= -2v_0 - v_1 + v_3 + 2v_5,$$

where we have omitted brackets around vertices. A key fact, easily shown, is that $\partial_{k-1}\partial_k \equiv 0$ for all $k \geq 1$. The boundary operator connects the chain groups into a **chain complex C_***:

$$\ldots \to C_{k+1} \xrightarrow{\partial_{k+1}} C_k \xrightarrow{\partial_k} C_{k-1} \to \ldots \tag{3.10}$$

A k-chain with no boundary is a **k-cycle**. For example, $[v_1, v_2] + [v_2, v_3] + [v_3, v_4] - [v_1, v_4]$ is a 1-cycle as its boundary is empty. The k-cycles constitute the kernel of ∂_k, so they form a subgroup of C_k, the **kth cycle group Z_k**:

$$Z_k = \ker \partial_k = \{c \in C_k \mid \partial_k c = 0\}. \tag{3.11}$$

A k-chain that is the boundary of a $(k + 1)$-chain is a **k-boundary**. A k-boundary lies in the image of ∂_{k+1}. For example, $[v_0, v_1] + [v_1, v_4] + [v_4, v_5] + [v_5, v_0]$ is a one-boundary as it is the boundary of the two-chain $[v_0, v_1, v_5] + [v_1, v_4, v_5]$. The k-boundaries form another subgroup of C_k, the **kth boundary group B_k**:

$$B_k = \operatorname{im} \partial_{k+1} = \{c \in C_k \mid \exists d \in C_{k+1} : c = \partial_{k+1}d\}. \tag{3.12}$$

Cycles and boundaries are like loops and boundaries in the fundamental group, except the former may have multiple components and are not required to share basepoints. Since $\partial_{k-1}\partial_k \equiv 0$, the subgroups are nested, $B_k \subseteq Z_k \subseteq C_k$. The **$k$th homology group H_k** is:

$$H_k = Z_k/B_k = \ker \partial_k/\operatorname{im} \partial_{k+1}. \tag{3.13}$$

If $z_1, z_2 \in Z_k$ are in the same homology class, z_1 and z_2 are **homologous**, and denoted as $z_1 \sim z_2$. From group theory, we know that we can write $z_1 = z_2 + b$, where $b \in B_k$. For example, the one-cycles $z_1 = [v_1, v_2] + [v_2, v_3] + [v_3, v_4] + [v_4, v_1]$ and $z_2 = [v_1, v_2] + [v_2, v_4] + [v_4, v_1]$ are homologous in Figure 3.3a, as they both describe the hole in the complex. We have $z_1 = z_2 + b$, where $b = [v_2, v_3] + [v_3, v_4] + [v_4, v_2]$ is a one-boundary.

The homology groups are invariants for $|K|$ and homotopy equivalent spaces. Formally, $\mathbb{X} \simeq \mathbb{Y} \Rightarrow H_k(\mathbb{X}) \cong H_k(\mathbb{Y})$ for all k. The invariance of simplicial homology gave rise to the *Hauptvermutung* (principle conjecture) by Poincaré in 1904, which was shown to be false in dimensions higher than three (Ranicki, 1997). Homology's invariance was finally resolved through the axiomatization of homology and the general theory of *singular homology*.

3.4.2 Characterization

Since H_k is a finitely generated group, the standard structure theorem states that it decomposes uniquely into a direct sum

$$\bigoplus_{i=1}^{\beta_k} \mathbb{Z} \oplus \bigoplus_{j=1}^{m} \mathbb{Z}_{t_j}, \tag{3.14}$$

where $\beta_k, t_j \in \mathbb{Z}$, $t_j | t_{j+1}$, $\mathbb{Z}_{t_j} = \mathbb{Z}/t_j\mathbb{Z}$ (Dummit and Foote, 2003). This decomposition defines key invariants of the group. The left sum captures the free subgroup and its rank is the **kth Betti number** β_k of K. The right sum captures the torsion subgroup and the integers t_j are the **torsion coefficients** for the homology group. Table 3.1 lists the homology groups for basic two-manifolds.

For torsion-free spaces in three dimensions, the Betti numbers have intuitive meaning as a consequence of the *Alexander Duality*. β_0 counts the number of connected **components** of the space. β_1 is the dimension of any basis for the **tunnels**. β_2 counts the number of enclosed spaces or **voids**. For example, the torus is a connected component, has two tunnels that are delineated by the cycles in Figure 3.7, and encloses one void. Therefore, $\beta_0 = 1$, $\beta_1 = 2$, and $\beta_2 = 1$.

3.4.3 The Euler-Poincaré Formula

We may redefine our first invariant, the Euler characteristic from Section 3.3.2, in terms of the chain complex C_*. Recall that for a cell complex K, $\chi(K)$ is an alternating sum of n_k, the number of k-simplices in K. But since C_k is the free group on oriented k-simplices, its rank is precisely n_k. In other words, we have $\chi(K) = \chi(C_*(K)) = \sum_k (-1)^k \operatorname{rank}(C_k(K))$. We now denote the sequence of homology functors as H_*. Then, $H_*(C_*(K))$ is another chain complex:

$$\ldots \to H_{k+1} \xrightarrow{\partial_{k+1}} H_k \xrightarrow{\partial_k} H_{k-1} \to \ldots. \tag{3.15}$$

The maps between the homology groups are induced by the boundary operators: We map a homology class to the class that contains it. According to our new definition, the Euler characteristic of our chain

TABLE 3.1 Homology Groups of Basic Two-Manifolds

Two-manifold	H_0	H_1	H_2
Sphere	$\mathbb{Z}$	$\{0\}$	$\mathbb{Z}$
Torus	$\mathbb{Z}$	$\mathbb{Z} \times \mathbb{Z}$	$\mathbb{Z}$
Projective plane	$\mathbb{Z}$	$\mathbb{Z}_2$	$\{0\}$
Klein bottle	$\mathbb{Z}$	$\mathbb{Z} \times \mathbb{Z}_2$	$\{0\}$

is $\chi(H_*(C_*(K))) = \sum_k (-1)^k \operatorname{rank}(H_k) = \sum_k (-1)^k \beta_k$, the alternating sum of the Betti numbers. Surprisingly, the homology functor preserves the Euler characteristic of a chain complex, giving us the **Euler-Poincaré** formula:

$$\chi(K) = \sum_k (-1)^k n_k = \sum_k (-1)^k \beta_k, \tag{3.16}$$

where $n_k = |\{\sigma \in K \mid \dim \sigma = k\}|$ and $\beta_k = \operatorname{rank} H_k(K)$. For example, we know $\chi(\mathbb{S}^2) = 2$ from Section 3.3.2. From Table 3.1, we have $\beta_0 = 1$, $\beta_1 = 0$, and $\beta_2 = 1$ for $\mathbb{S}^2$, and $1 - 0 + 1 = 2$. The formula derives the invariance of the Euler characteristic from the invariance of homology. For a surface, it also allows us to compute one Betti number given the other two Betti numbers, as we may compute the Euler characteristic by counting simplices.

3.4.4 Computation

We may view any group as a $\mathbb{Z}$-module, where $\mathbb{Z}$ is the **ring of coefficients** for the chains. This view allows us to use alternate rings of coefficients, such as finite fields $\mathbb{Z}_p$ for a prime p, reals $\mathbb{R}$, or rationals $\mathbb{Q}$. Over a field F, a module becomes a vector space and is fully characterized by its dimension, the Betti number. This means that we get a full characterization for torsion-free spaces when computing over fields.

Computing connected components. For torsion-free spaces, we may use the **union-find** data structure to compute connected components or β_0 (Cormen et al., 2001, Chapter 21). This data structure maintains a collection of disjoint dynamic sets using three operations that we use below. For a vertex, we create a new set using MAKE-SET and increment β_0. For an edge, we determine if the endpoints of the edge are in the same component using two calls to FIND-SET. If they are not, we use UNION to unite the two sets and decrement β_0. Consequently, we compute β_0 in time $O(n_1 + n_0 \alpha(n_0))$, where n_k is the number of k-simplices, and α is the inverse of the Ackermann function which is less than the constant four for all practical values. Since the algorithm is incremental, it also yields Betti numbers for filtrations as defined in Section 3.5. Using duality, we may extend this algorithm to compute higher-dimensional Betti numbers for surfaces and three-dimensional complexes whose highest Betti numbers are known. Delfinado and Edelsbrunner (1995) use this approach to compute Betti numbers for subcomplexes of triangulations of $\mathbb{S}^3$.

Computing β_k over fields. Over a field F, our problem lies within linear algebra. Since $\partial_k \colon C_k \to C_{k-1}$ is a linear map, it has a matrix M_k with entries from F in terms of bases for its domain and codomain. For instance, we may use oriented simplices as respective bases for C_k and C_{k-1}. The cycle vector space $Z_k = \ker \partial_k$ is the null space of M_k and the boundary vector space $B_k = \operatorname{im} \partial_{k+1}$ is the range space of M_{k+1}. By Equation (3.13), $\beta_k = \dim H_k = \dim Z_k - \dim B_k$. We may compute these dimensions with two Gaussian eliminations in time $O(n^3)$, where n is the number of simplices in K (Uhlig, 2002). Here, we are assuming that the field operations take time $O(1)$, which is true for simple fields. The matrix M_k is very sparse for a cell complex and the running time is faster than cubic in practice.

Computing β_k over PIDs. The structure theorem also applies for rings of coefficients that are *principal ideal domains* (*PIDs*). For our purposes, a PID is simply a ring in which we may compute the greatest common divisor of a pair of elements. Starting with the matrix M_k, the **reduction algorithm** uses elementary column and row operations to derive alternate bases for the chain groups, relative to which the matrix for ∂_k has the diagonal **Smith normal form**:

$$
\begin{bmatrix}
b_1 & & & 0 & \\
& \ddots & & & 0 \\
0 & & b_{l_k} & & \\
\hline
& 0 & & & 0
\end{bmatrix}, \tag{3.17}
$$

where $b_j > 1$ and $b_j | b_{j+1}$. The torsion coefficients for H_{k-1} are now precisely those diagonal entries b_j that are greater than one. We may also compute the Betti numbers from the ranks of the matrices, as before.

Over $\mathbb{Z}$, neither the size of the matrix entries nor the number of operations in $\mathbb{Z}$ is polynomially bounded for the reduction algorithm. Consequently, there are sophisticated polynomial algorithms based on modular arithmetic (Storjohann, 1998), although reduction is preferred in practice (Dumas et al., 2003). It is important to reiterate that homology computation is easy over fields, but hard over $\mathbb{Z}$, as this distinction is a source of common misunderstanding in the literature.

Computing generators. In addition to the Betti numbers and torsion coefficients, we may be interested in actual descriptions of homology classes through representative cycles. To do so, we may modify Gaussian elimination and the reduction algorithm to keep track of all the elementary operations during elimination. This process is almost never used since applications for generators are motivated geometrically, but the computed generators are usually geometrically displeasing. We discuss geometric descriptions in Section 3.8.2.

3.5   Persistent Homology

Persistent homology is a recently developed invariant that captures the homological history of a space that is undergoing growth. We model this history for a space X as a **filtration**, a sequence of nested spaces $\emptyset = X_0 \subseteq X_1 \subseteq \dots \subseteq X_m = X$. A space with a filtration is a **filtered space**. We see a filtered simplicial complex in Figure 3.8. Note that in a filtered complex, the simplices are always added, but never removed, implying a *partial order* on the simplices. Filtrations arise naturally from many processes, such as the excursion sets of Morse functions over manifolds in Section 3.6 or the multiscale representations for point sets in Section 3.7.

Over fields, each space X_j has a kth homology group $H_k(X_j)$, a vector space whose dimension $\beta_k(X_j)$ counts the number of topological attributes in dimension k. Viewing a filtration as a growing space, we see that topological attributes appear and disappear. For instance, the filtered complex in Figure 3.8 develops a one-cycle at time 2 that is completely filled at time 5. If we could track an

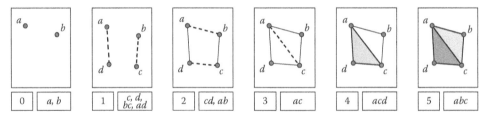

FIGURE 3.8 A simple filtration of a simplicial complex with newly added simplices highlighted and listed.

attribute through the filtration, we could talk about its *lifetime* within this growth history. This is the intuition behind the theory of *persistent homology*. Initially, Edelsbrunner et al. (2002) developed the concept for subcomplexes of $\mathbb{S}^3$ and $\mathbb{Z}_2$ coefficients. The treatment here follows later theoretical development that extended the concept to arbitrary spaces and field coefficients (Zomorodian and Carlsson, 2005).

3.5.1 Theory

A filtration yields a **directed space**

$$\emptyset = X_0 \overset{i}{\hookrightarrow} \cdots \overset{i}{\hookrightarrow} X_j \cdots \overset{i}{\hookrightarrow} X, \qquad (3.18)$$

where the maps i are the respective inclusions. Applying the H_k to both the spaces and the maps, we get another directed space

$$\emptyset = H_k(X_0) \xrightarrow{H_k(i)} \cdots \xrightarrow{H_k(i)} H_k(X_j) \xrightarrow{H_k(i)} \cdots \xrightarrow{H_k(i)} H_k(X) \qquad (3.19)$$

where $H_k(i)$ are the respective maps induced by inclusion. The **persistent homology** of the filtration is the structure of this directed space, and we would like to understand it by classifying and parameterizing it. Intuitively, we do so by building a single structure that contains all the complexes in the directed space. Suppose we are computing over the ring of coefficients R. Then, $R[t]$ is the **polynomial ring**, which we may grade with the **standard grading** Rt^n. We then form a graded module over $R[t]$, $\oplus_{i=0}^n H_k(i)$, where the R-module structure is simply the sum of the structures on the individual components, and where the action of t is given by $t \cdot (m_0, m_1, m_2, \ldots) = (0, \varphi_0(m_0), \varphi_1(m_1), \varphi_2(m_2), \ldots)$. and φ_i is the map induced by homology in the directed space. That is, t simply shifts elements of the module up in the gradation. In this manner, we encode the ordering of simplices within the coefficients from the polynomial ring. For instance, vertex a enters the filtration in Figure 3.8 at time 0, so $t \cdot a$ exists at time 1, $t^2 \cdot a$ at time 2, and so on.

When the ring of coefficients is a field F, $F[t]$ is a PID, so the standard structure theorem gives the unique decomposition of the $F[t]$-module:

$$\bigoplus_{i=1}^n \Sigma^{\alpha_i} F[t] \oplus \bigoplus_{j=1}^m \Sigma^{\gamma_j} F[t]/(t^{n_j}), \qquad (3.20)$$

where $(t^n) = t^n F[t]$ are the ideals of the graded field $F[t]$ and Σ^α denotes an α-shift upward in grading. Except for the shifts, the decomposition is essentially the same as the decomposition (3.14) for groups: the left sum describes free generators and the right sum the torsion subgroup.

Barcodes. Using decomposition (3.20), we may parameterize the isomorphism classes with a multiset of intervals:

1. Each left summand gives us $[\alpha_i, \infty)$, corresponding to a topological attribute that is created at time α_i and exists in the final structure.
2. Each right summand gives us $[\gamma_j, \gamma_j + n_j)$, corresponding to an attribute that is created at time γ_j, lives for time n_j, and is destroyed.

We refer to this multiset of intervals as a **barcode**. Like a homology group, a barcode is a homotopy invariant. For the filtration in Figure 3.8, the β_0-barcode is $\{[0, \infty), [0, 2)\}$ with the intervals describing the lifetimes of the components created by simplices a and b, respectively. The β_1-barcode is

$\{[2, 5), [3, 4)\}$ for the two one-cycles created by edges ab and ac, respectively, provided ab enters the complex after cd at time 2.

3.5.2 Algorithm

While persistence was initially defined for simplicial complexes, the theory extends to singular homology, and the algorithm extends to a broad class of filtered cell complexes, such as simplicial complexes, simplicial sets, Δ-complexes, and cubical complexes. In this section, we describe an algorithm that not only computes both persistence barcodes, but also descriptions of the representatives of persistent homology classes (Zomorodian and Carlsson, 2008). We present the algorithm for cell complexes over $\mathbb{Z}_2$ coefficients, but the same algorithm may be easily adapted to other complexes or fields.

Recall that a filtration implies a partial order on the cells. We begin by sorting the cells within each time snapshot by dimension, breaking other ties arbitrarily, and getting a *full order*. The algorithm, listed in pseudocode in the procedure PAIR-CELLS, takes this full order as input. Its output consists of persistence information: It partitions the cells into *creators* and *destroyers* of homology classes, and pairs the cells that are associated to the same class. Also, the procedure computes a generator for each homology class.

The procedure PAIR-CELLS is incremental, processing one cell at a time. Each cell σ stores its *partner*, its paired cell, as shown in Table 3.2 for the filtration in Figure 3.8. Once we have this pairing, we may simply read off the barcode from the filtration. For instance, since vertex b is paired with edge cd, we get interval $[0, 2)$ in the β_0-barcode. Each cell σ also stores its *cascade*, a k-chain, which is initially σ itself. The algorithm focuses the impact of σ's entry on the topology by determining whether $\partial\sigma$ is already a boundary in the complex using a call to the procedure ELIMINATE-BOUNDARIES. After the call, there are two possibilities on line 5:

1. If $\partial(cascade[\sigma]) = 0$, we are able to write $\partial\sigma$ as a sum of the boundary basis elements, so $\partial\sigma$ is already a $(k-1)$-boundary. But now, $cascade[\sigma]$ is a new k-cycle that σ completed. That is, σ creates a new homology cycle and is a *creator*.

2. If $\partial(cascade[\sigma]) \neq 0$, then $\partial\sigma$ becomes a boundary after we add σ, so σ destroys the homology class of its boundary and is a *destroyer*. On lines 6 through 8, we pair σ with the *youngest* cell τ in $\partial(cascade[\sigma])$, that is, the cell that has the most recently entered the filtration.

During each iteration, the algorithm maintains the following invariants: It identifies the ith cell as a creator or destroyer and computes its cascade; if σ is a creator, its cascade is a generator for the homology class it creates; otherwise, the boundary of its cascade is a generator for the boundary class.

TABLE 3.2 Data Structure after Running the Persistence Algorithm on the Filtration in Figure 3.8

σ	a	b	c	d	bc	ad	cd	ab	ac	acd	abc
partner $[\sigma]$		cd	bc	ad	c	d	b	abc	acd	ac	ab
cascade $[\sigma]$	a	b	c	d	bc	ad	cd	ab	ac	acd	abc
							ad	cd	cd		acd
							bc	ad	ad		
								bc			

Note: The simplices without partners, or with partners that come after them in the full order, are creators; the others are destroyers.

Pair-Cells (K)

```
1   for σ ← σ₁ to σₙ ∈ K
2       do partner[σ] ← ∅
3          cascade[σ] ← σ
4          Eliminate-Boundaries (σ)
5          if ∂(cascade[σ]) ≠ 0
6             then τ ← Youngest(∂(cascade[σ]))
7                  partner[σ] ← τ
8                  partner[τ] ← σ
```

The procedure Eliminate-Boundaries corresponds to the processing of one row (or column) in Gaussian elimination. We repeatedly look at the youngest cell τ in $\partial(cascade[\sigma])$. If it has no partner, we are done. If it has one, then the cycle that τ created was destroyed by its partner. We then add τ's partner's cascade to σ's cascade, which has the effect of adding a boundary to $\partial(cascade[\sigma])$. Since we only add boundaries, we do not change homology classes.

Eliminate-Boundaries(σ)

```
1   while ∂(cascade[σ]) ≠ 0
2       do τ ← Youngest(∂(cascade[σ]))
3          if partner[τ] = ∅
4             then return
5             else cascade[σ] ← cascade[σ] + cascade[partner[τ]]
```

Table 3.2 shows the stored attributes for our filtration in Figure 3.8 after the algorithm's completion. For example, tracing Eliminate-Boundaries (cd) through the iterations of the **while** loop, we get:

#	cascade[cd]	τ	partner[τ]
1	cd	d	ad
2	cd + ad	c	bc
3	cd + ad + bc	b	∅

Since $partner[b] = \emptyset$ at the time of bc's entry, we pair b with cd in Pair-Cells upon return from this procedure, as shown in the table. Table 3.2 contains all the persistence information we require. If $\partial(cascade[\sigma]) = 0$, σ is a creator and its cascade is a representative of the homology class it created. Otherwise, σ is a destroyer and its cascade is a chain whose boundary is a representative of the homology class σ destroyed.

3.6 Morse Theoretic Invariants

In the last three sections, we have looked at a number of combinatorial and algebraic invariants. In this section, we examine the deep relationship between topology and geometry. We begin with the following question: Can we define any smooth function on a manifold? Morse theory shows that the *topology* of the manifold regulates the *geometry* of the function on it, answering the question in the negative. However, this relationship also implies that we may study topology through geometry. Our study yields a number of combinatorial structures that capture the topology of the underlying manifold.

FIGURE 3.9 Selected tangent planes, indicated with disks, and normals on a torus.

3.6.1 Morse Theory

In this section, we generally assume we have a two-manifold $\mathbb{M}$ embedded in $\mathbb{R}^3$. We also assume the manifold is *smooth*, although we do not extend the notions formally due to lack of space, but depend on the reader's intuition Boothby (1986). Most concepts generalize to higher-dimensional manifolds with Riemannian metrics. Let p be a point on $\mathbb{M}$. A **tangent vector v_p to $\mathbb{R}^3$** consists of two points of $\mathbb{R}^3$: its **vector part** v and its **point of application** p. A tangent vector v_p to $\mathbb{R}^3$ at p is **tangent to $\mathbb{M}$ at** p if v is the velocity of some curve φ in $\mathbb{M}$. The set of all tangent vectors to $\mathbb{M}$ at p is the **tangent plane $T_p(\mathbb{M})$**, the best linear approximation to $\mathbb{M}$ at p. Figure 3.9 shows selected tangent planes and their respective normals on the torus.

Suppose we have a smooth map $h\colon \mathbb{M} \to \mathbb{R}$. The **derivative $v_p[h]$ of h with respect to v_p** is the common value of $(d/dt)(h \circ \gamma)(0)$, for all curves $\gamma \in \mathbb{M}$ with initial velocity v_p. The **differential dh_p** is a linear function on $T_p(\mathbb{M})$ such that $dh_p(v_p) = v_p[h]$, for all tangent vectors $v_p \in T_p(\mathbb{M})$. The differential converts vector fields into real-valued functions. A point p is **critical** for h if dh_p is the zero map. Otherwise, p is **regular**. Given local coordinates x, y on $\mathbb{M}$, the **Hessian** of h is

$$H(p) = \left[\begin{array}{cc} \frac{\partial^2 h}{\partial x^2}(p) & \frac{\partial^2 h}{\partial y \partial x}(p) \\ \frac{\partial^2 h}{\partial x \partial y}(p) & \frac{\partial^2 h}{\partial y^2}(p) \end{array} \right], \tag{3.21}$$

in terms of the basis $(\partial/\partial x(p), \partial/\partial y(p))$ for $T_p(\mathbb{M})$. A critical point p is **nondegenerate** if the Hessian is nonsingular at p, i.e., $\det H(p) \neq 0$. Otherwise, it is **degenerate**. A map h is a **Morse function** if all its critical points are nondegenerate. By the **Morse Lemma**, it is possible to choose local coordinates so that the neighborhood of a nondegenerate critical point takes the form $h(x, y) = \pm x^2 \pm y^2$. The **index** of a nondegenerate critical point is the number of minuses in this parameterization, or more generally, the number of negative eigenvalues of the Hessian. For a d-dimensional manifold, the index ranges from 0 to d. In two dimensions, a critical point of index 0, 1, or 2 is called a **minimum**, **saddle**, or **maximum**, respectively, as shown in Figure 3.10. The **monkey saddle** in the figure is the neighborhood of a degenerate critical point, a surface with three ridges and valleys. By the

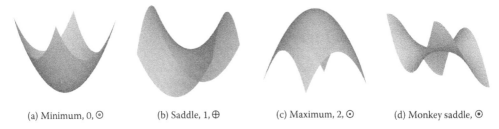

(a) Minimum, 0, ⊙   (b) Saddle, 1, ⊕   (c) Maximum, 2, ⊙   (d) Monkey saddle, ⊙

FIGURE 3.10 Neighborhoods of critical points, along with their indices, and mnemonic symbols. All except the Monkey saddle are nondegenerate.

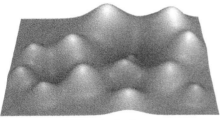
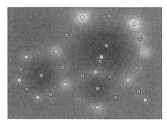

(a) Gaussian landscape   (b) Critical points

FIGURE 3.11 A Gaussian landscape. A Morse function defined on a two-manifold with boundary (a) has 3 minima, 15 saddles, and 13 peaks, as noted on the view from above (b). We get an additional minimum when we compactify the disk into a sphere.

Morse Lemma, it cannot be the neighborhood of a critical point of a Morse function. In higher dimensions, we still have maxima and minima, but more types of saddle points. In Figure 3.11, we define a Gaussian landscape on a two-manifold with boundary (a closed disk). The landscape has 3 lakes (minima), 15 passes (saddles), and 13 peaks (maxima), as shown in the orthographic view from above with our mnemonic symbols listed in Figure 3.10. To eliminate the boundary in our landscape, we place a point at $-\infty$ and identify the boundary to it, turning the disk into $\mathbb{S}^2$, and adding a minimum to our Morse function. This **one-point compactification** of the disk is done combinatorially in practice.

Relationship to homology. Given a manifold $\mathbb{M}$ and a Morse function $h: \mathbb{M} \to \mathbb{R}$, an **excursion set of h at height r** is $\mathbb{M}_r = \{m \in \mathbb{M} | h(m) \leq r\}$. We see three excursion sets for our landscape in Figure 3.12. Using this figure, we now consider the topology of the excursion sets $\mathbb{M}_r$ as we increase r from $-\infty$:

- At a minimum, we get a new component, as demonstrated in the leftmost excursion set. So, β_0 increases by one whenever r passes the height of a minimum point.
- At a saddle point, we have two possibilities. Two components may connect through a saddle point, as shown in the middle excursion set. Alternatively, the components may connect around the base of a maxima, forming a one-cycle, as shown in the rightmost excursion set. So, either β_0 increases or β_1 decreases by one at a saddle point.
- At a maximum, a one-cycle around that maximum is filled in, so β_1 is decremented.
- At regular points, there is no change to topology.

Therefore, minima, saddles, and maxima have the same impact on the Betti numbers as vertices, edges, and triangles. Given this relationship, we may extend the Euler-Poincaré formula (Equation 3.16):

$$\chi(\mathbb{M}) = \sum_k (-1)^k n_k = \sum_k (-1)^k \beta_k = \sum_k (-1)^k c_k, \qquad (3.22)$$

FIGURE 3.12 Excursion sets for the Gaussian landscape in Figure 3.11.

where n_k is the number of k-cells for any cellular complex with underlying space $\mathbb{M}$, $\beta_k = $ rank $H_k(\mathbb{M})$, and c_k is the number of Morse critical points with index k for any Morse function defined on $\mathbb{M}$. This beautiful equation relates combinatorial topology, algebraic topology, and Morse theory at once. For example, our compactified landscape is a sphere with $\chi(\mathbb{S}) = 2$. We also have $c_0 = 4$ (counting the minimum at $-\infty$), $c_1 = 15$, $c_2 = 13$, and $4 - 15 + 13 = 2$. That is, the topology of a manifold implies a relationship on the number of critical points of any function defined on that manifold.

Unfolding and cancellation. In practice, most of our functions are non-Morse, containing degenerate critical points, such as the function in Figure 3.13b. Fortunately, we may perturb any function infinitesimally to be Morse as Morse functions are dense in the space of smooth functions. For instance, simple calculus shows that we may perturb the function in Figure 3.13b to either a function with two nondegenerate critical points (Figure 3.13a) or zero critical points (Figure 3.13c). Going from Figure 3.13b to a is **unfolding** the degeneracy and its geometric study belongs to singularity theory (Bruce and Giblin, 1992). Going from Figure 3.13a through c describes the **cancellation** of a pair of critical points as motivated by excursion sets and Equation 3.22. Both concepts extend to higher-dimensional critical points.

The excursion sets of a Morse function form a filtration, as required by persistent homology in Section 3.5. The relationship between critical points and homology described in the last section implies a pairing of critical points. In one dimension, a minimum is paired with a maximum, as in Figure 3.13. In two dimensions, a minimum is paired with a saddle, and a saddle with a maximum. This pairing dictates an ordering for *combinatorial* cancellation of critical points and an identification of the significant critical points in the landscape. To remove the canceled critical points *geometrically* as in Figure 3.13c, we need additional structure in higher dimensions, as described in Section 3.6.3.

Nonsmooth manifolds. While Morse theory is defined only for smooth manifolds, most concepts are not intrinsically dependent upon smoothness. Indeed, the main theorems have been shown to have analogs within discrete settings. Banchoff (1970) extends critical point theory to polyhedral surfaces. Forman (1998) defines discrete Morse functions on cell complexes to create a discrete Morse theory.

3.6.2 Reeb Graph and Contour Tree

Given a manifold $\mathbb{M}$ and a Morse function $h \colon \mathbb{M} \to \mathbb{R}$, a **level set of h at height r** is the preimage $h^{-1}(r)$. Note that level sets are simply the boundaries of excursion sets. Level sets are called **isolines** and **isosurfaces** for two- and three-manifolds, respectively. A **contour** is a connected component of

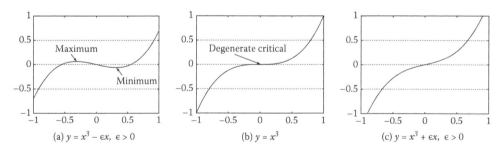

FIGURE 3.13 The function $y = x^3$ (b) has a single degenerate critical point at the origin and is not Morse. We may unfold the degenerate critical point into a maximum and a minimum (a), two nondegenerate critical points. We may also smooth it into a regular point (c). Either perturbation yields a Morse functions arbitrarily close to (b). The process of going from (a) to (c) is cancellation.

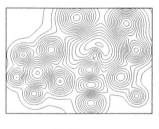
(a) Contours

(b) Contour tree

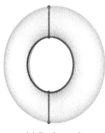
(c) Reeb graph

FIGURE 3.14 Topology of contours. (a) Contours from 20 level sets for the landscape in Figure 3.11. (b) The contour tree for the landscape, superimposed on the landscape. The bottom most critical point is the minimum at $-\infty$. (c) The Reeb graph for the torus with Morse function equivalent to its z coordinate. The function has one minimum, two saddles, and one maximum.

a level set. For two-manifolds, a contour is a loop and for three-manifolds, it is a void. In Figure 3.14a, we see the contours corresponding to 20 level sets of our landscape. As we change the level, contours appear, join, split, and disappear. To track the topology of the contours, we contract each contour to a point, obtaining the **Reeb graph** (Reeb, 1946). When the Reeb graph is a tree, it is known as the **contour tree** (Boyell and Ruston, 1963). In Figure 3.14b, we superimpose the contour tree of our landscape over it. In Figure 3.14c, we see the Reeb graph for a torus standing on its end, where the z coordinate is the Morse function.

Computation. We often have three-dimensional Morse functions, sampled on a regular voxelized grid, such as images from CT scans or MRI. We may linearly interpolate between the voxels for a piece-wise linear approximation of the underlying function. Lorensen and Cline (1987) first presented the *Marching cubes* algorithm for extracting isosurfaces using a table-based approach that exploited the symmetries of a cube. The algorithm has evolved considerably since then (Newman and Hong, 2006).

Carr et al. (2003) give a simple $O(n \log n + N\alpha(N))$ time algorithm for computing the contour tree of a discrete real-valued field interpolated over a simplicial complex with n vertices and N simplices. Not surprisingly, the algorithm first captures the joining and splitting of contours by using the union-find data structure, described in Section 3.4.4. Cole-McLaughlin et al. (2004) give an $O(n \log n)$ algorithm for computing Reeb graphs for triangulated two-manifolds, with or without boundary, where n is the number of edges in the triangulation.

3.6.3 Morse-Smale Complex

The critical points of a Morse function are locations on the two-manifold where the function is stationary. To extract more structure, we look at what happens on regular points. A **vector field** or **flow** is a function that assigns a vector $v_p \in T_p(\mathbb{M})$ to each point $p \in \mathbb{M}$. A particular vector field is the **gradient** ∇h which is defined by

$$\left\langle \frac{d\gamma}{dt} \cdot \nabla h \right\rangle = \frac{d(h \circ \gamma)}{dt}, \tag{3.23}$$

where γ is any curve passing through $p \in \mathbb{M}$, tangent to $v_p \in T_p(\mathbb{M})$. The gradient is related to the derivative, as $v_p[h] = v_p \cdot \nabla h(p)$. It is always possible to choose coordinates (x, y) so that the tangent vectors $\frac{\partial}{\partial x}(p)$, $\frac{\partial}{\partial y}(p)$ are orthonormal with respect to the chosen metric. For such coordinates, the gradient is given by the familiar formula $\nabla h = \left(\frac{\partial h}{\partial x}(p), \frac{\partial h}{\partial y}(p) \right)$. We integrate the gradient

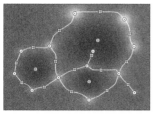

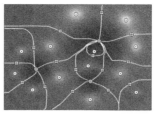

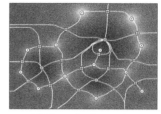

(a) Unstable 1-manifolds   (b) Stable 1-manifolds   (c) Morse-Smale complex

FIGURE 3.15 The Morse-Smale complex. The unstable manifolds (a) and stable manifolds (b) each decompose the underlying manifolds into a cell-complex. If the function is Morse-Smale, the stable and unstable manifolds intersect transversally to form the Morse-Smale complex (c).

to decompose $\mathbb{M}$ into regions of uniform flow. An **integral line** $\ell : \mathbb{R} \to \mathbb{M}$ is a maximal path whose tangent vectors agree with the gradient, that is, $\frac{\partial}{\partial s}\ell(s) = \nabla h(\ell(s))$ for all $s \in \mathbb{R}$. We call org $\ell = \lim_{s \to -\infty} \ell(s)$ the **origin** and dest $\ell = \lim_{s \to +\infty} \ell(s)$ the **destination** of the path ℓ. The integral lines are either disjoint or the same, cover $\mathbb{M}$, and their limits are critical points, which are integral lines by themselves. For a critical point p, the **unstable manifold** $U(p)$ and the **stable manifold** $S(p)$ are:

$$U(p) = \{p\} \cup \{y \in \mathbb{M} \mid y \in \text{im } \ell, \text{ org } \ell = p\}, \tag{3.24}$$

$$S(p) = \{p\} \cup \{y \in \mathbb{M} \mid y \in \text{im } \ell, \text{ dest } \ell = p\}. \tag{3.25}$$

If p has index i, its unstable manifold is an open $(2 - i)$-cell and its stable manifold is an open i-cell. For example, the minima on our landscape have open disks as unstable manifolds as shown in Figure 3.15a. In *geographic information systems*, a minimum's unstable manifold is called its **watershed** as all the rain in this region flows toward this minimum. The stable manifolds of a Morse function h, shown in Figure 3.15b, are the unstable manifolds of $-h$ as ∇ is linear. If the stable and unstable manifolds intersect only transversally, we say that our Morse function h is **Morse-Smale**. We then intersect the manifolds to obtain the **Morse-Smale complex**, as shown in Figure 3.15c. In two dimensions, the complex is a quadrangulation, where each quadrangle has one maximum, two saddle points, and one minimum as vertices, and uniform flow from its maximum to its minimum. In three dimensions, the complex is composed of hexahedra.

Cayley (1859) was the first to conceive of this structure in one of the earliest contributions to Morse theory. Independently, Maxwell (1870) gave full details of the complex as well as the relationship between the number of Morse critical points (Equation 3.22), but gave Cayley priority generously upon learning about the earlier contribution.

Computation.   Using Banchoff's critical point theory, Edelsbrunner et al. (2003b) extend Morse-Smale complexes to piece-wise linear two-manifolds, giving an algorithm for computing the complex as well as simplifying it through persistent homology. The complex has also been extended to three-manifolds (Edelsbrunner et al., 2003a), although its computation remains challenging (Gyulassy et al., 2007).

3.7   Structures for Point Sets

A principle problem within computational topology is recovering the topology of a finite point set. The assumption is that the point set, either acquired or simulated, is sampled from some underlying topological space, whose connectivity is lost during the sampling process. Also, one often assumes that the space is a manifold. In this section, we look at techniques for computing structures that topologically approximate the underlying space of a given point set.

3.7.1 A Cover and Its Nerve

We assume we are given a point set X embedded in $\mathbb{R}^d$, although the ambient space could also be any Riemannian manifold. Since the point set does not have any interesting topology by itself, we begin by approximating the underlying space by pieces of the embedding space. An **open cover** of X is $\mathcal{U} = \{U_i\}_{i \in I}$, $U_i \subseteq \mathbb{R}^d$, where I is an indexing set and $X \subseteq \cup_i U_i$. The **nerve** N of $\mathcal{U}$ is

1. $\emptyset \in N$, and
2. If $\cap_{j \in J} U_j \neq \emptyset$ for $J \subseteq I$, then $J \in N$.

Clearly, the Nerve is an abstract simplicial complex. Figure 3.16a displays a point set, an open cover of the points, and the nerve of the cover. We say $\mathcal{U}$ is a **good** cover if all U_i are contractible and so are all of their nonempty finite intersections. The cover in Figure 3.16a is not good, as the leftmost set is homotopy equivalent to a circle. According to Leray's classical *Nerve Lemma*, the nerve of a good cover is homotopy equivalent to the cover. Rotman (1988, Lemma 7.26) shows this result for homology groups and Bott and Tu (1982, Theorem 13.4) prove it for the fundamental group, but given some technical conditions, one gets a homotopy equivalence. This lemma is the basis of most methods for representing point sets. We search for good covers whose nerve will be our representation.

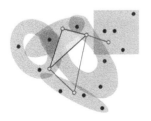

(a) Points, open cover, and nerve

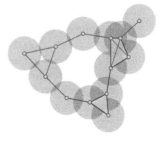

(b) Cover: Union of ε-balls; Nerve: Čech complex

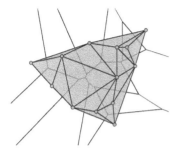

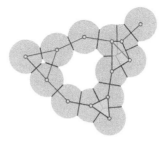

(c) Cover: Voronoi diagram; Nerve: Delaunay complex   (d) Cover: Restricted Voronoi diagram; Nerve: Alpha complex

FIGURE 3.16 Representing point sets. (a) Given a point set (black points), we represent it using the nerve of an open cover. (b) The Čech complex is the nerve of a union of ε-balls. (c) The Delaunay complex is the nerve of the Voronoi diagram. (d) The Alpha complex is the nerve of the restricted Voronoi cells.

3.7.2   Abstract Complexes

Our first cover is the union of ∈-balls centered around the input points, as shown in Figure 3.16b. The cover is clearly good. More generally, any cover consisting of geodesically convex neighborhoods will be good. The implicit assumption here is that each ball approximates the space *locally* whenever the sampling is dense enough. The nerve of this cover is called the **Čech complex**. The complex may have dimension higher than the original embedding space and be as large as the power set of the original point set in the worst case. In our example, the point set is in $\mathbb{R}^2$, but the complex contains a tetrahedron.

The **Vietoris-Rips complex** relaxes the Čech condition for simplex inclusion by allowing a simplex provided its vertices are pairwise within distance ∈ (Gromov, 1987). We first compute the 1-skeleton of the complex, a graph on the vertices. We then perform a series of **expansions** by dimension, where we add a simplex provided all its faces are in the complex. This expansion fills in the triangular hole in the Čech complex in Figure 3.16b, so the complex is not homotopy equivalent to the Čech complex.

3.7.3   Geometric Complexes

We next look at three methods that use techniques from *computational geometry* to compute complexes that are smaller than abstract complexes, but at the cost of intricate geometric computation that often does not extend to higher dimensions.

For $p \in X$, the **Voronoi cell** is the set of points in the ambient space closest to p, $V(p) = \{x \in \mathbb{R}^d \mid d(x,p) \leq d(x,y), \forall p \in X\}$. The **Voronoi diagram** decomposes $\mathbb{R}^d$ into Voronoi cells and is a good cover for $\mathbb{R}^d$ as all the cells are convex. The **Delaunay complex** is the nerve of the Voronoi diagram, as shown in Figure 3.16c (de Berg et al., 2000). The **restricted Voronoi diagram** is the intersection of union of ∈-balls and the Voronoi diagram, and the **alpha complex**, shown in Figure 3.16d, is the nerve of this cover (Edelsbrunner and Mücke, 1994). This complex is homotopy equivalent to the Čech complex, but is also embedded and has the same dimension as the ambient space. By construction, the alpha complex is always a subcomplex of the Delaunay complex, so we may compute the former by computing the latter. A deficiency of the methods that use ∈-balls is that they assume uniform sampling for the point set, which is rarely true in practice. Recently, Cazals et al. (2006) address this deficiency of the alpha complex by defining the *conformal alpha complex* which features a global scale parameter.

Giesen and John (2003) define a distance function based on the point set and define the **flow complex** to be the stable manifolds of this function (see Section 3.6.3 for definitions.) Like the alpha complex, the flow complex is homotopy equivalent to the Čech complex (Dey et al., 2003). de Silva and Carlsson (2004) define the **witness complex** using a relaxation of the Delaunay test. The complex is built on **landmarks**, a subset of the point sets. The key idea is that the nonlandmark points participate in the construction of the complex by acting as *witnesses* to the existence of simplices. While the witness complex utilizes additional geometry, it is not embedded like the alpha and flow complex, and only approximates the Čech complex. We usually build the one-skeleton of the witness complex and perform Vietoris-Rips expansions for higher-dimensional skeletons.

Computing the nearest neighbors.   An essential task in computing complexes is enumerating exact or approximate nearest neighbors. This is a well-studied problem with a rich array of results. Geometric solutions (Arya et al., 1998) are efficient and feasible in low dimensions (less than 20) as they have an exponential dependence on dimension, the *curse of dimensionality*. In higher dimensions, recent hashing techniques have resulted in practical algorithms (Andoni and Indyk, 2006). While lower bounds exist, they often depend on the model of computation (Chakrabarti and Regev, 2004).

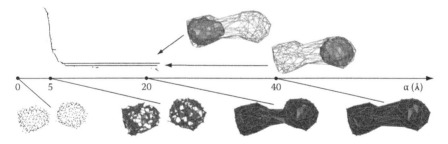

FIGURE 3.17 Application to biophysics. Below the α-axis are the point set (α = 0) and three complexes from its alpha complex filtration. Above is the persistence barcode: a multiset of 146 intervals. Most intervals are very short, but there are two long intervals corresponding to the two visualized voids. The point set contains the phosphate group coordinates of the unfused inner membranes of a phospholipid molecule.

3.7.4 Using Persistent Homology

All the complexes described in the last two sections describe one-parameter families of spaces. In each case, the parameter is a notion of scale or local feature size in the space. For instance, the Čech complexes require a radius ϵ for its cover. As we increase ϵ, we get filtered complexes as defined in Section 3.5.1. Therefore, we may apply persistent homology to capture the topology of the underlying point set. For instance, Kasson et al. (2007) use topological techniques for an application in *biophysics*. In Figure 3.17, we see four complexes from the alpha complex filtration of the point set at $\alpha = 0$, which describes the inner membrane of a bilayer vesicle, a primary cellular transport vehicle. We also see the β_2 persistence barcode above the axis, along with descriptions of the two significant voids that correspond to the two long intervals, both of which are computed with the persistence algorithm in Section 3.5.2. By analyzing more than 170,000 snapshots of fusion trajectories, the authors give a systematic topology-based method for measuring structural changes in membrane fusion.

3.8 Interactions with Geometry

As we noted in the introduction, computational topology was motivated initially by geometric problems that contained topological subproblems. Having discussed topological techniques, we end this chapter with a sample of geometric problems that involve topology.

3.8.1 Manifold Reconstruction

Given a finite set of (noisy) samples from a manifold, we would like to recover the original manifold. For 2-manifold, Dey (2007) surveys results from *computational geometry* that guarantee a reconstruction *homeomorphic* to the original surface, provided the sampling satisfies certain local geometric conditions. Recently, Boissonnat et al. (2007) use witness complexes to reconstruct manifolds in arbitrary dimensions. There is also a new focus on finding homotopy equivalent spaces. Niyogi et al. (2008) give an algorithm to learn the underlying manifold with high confidence, provided the data is drawn from a sampling probability distribution that has support on or near a submanifold of a Euclidean space.

3.8.2 Geometric Descriptions

Topological noise often creates significant problems for subsequent geometry processing of surfaces, such as mesh *decimation*, *smoothing*, and *parameterization* for *texture mapping* or *remeshing*. To simplify the surface topologically, we need descriptions of topological attributes that also take geometry into account. A deciding factor is how one *measures* the attributes geometrically, as the complexity of the resulting optimal geometric problem is often dependent on the nature of the underlying geometric measure.

Noncanonical polygonal schema.   Erickson and Har-Peled (2004) show that the problem of finding the *minimum* cut graph (Section 3.3.1) is NP-hard, when the geometric measure is either the total number of cut edges or their total lengths. They also describe a greedy algorithm that gives an $O(\log^2 g)$ approximation of the minimum cut graph in $O(g^2 n \log n)$ for a surface of genus g and combinatorial complexity n.

Homotopic paths in the plane.   For a single given path of size k_{in} in the plane with n point obstacles, Hershberger and Snoeyink (1994) give an optimal $O(nk_{in})$ time algorithm for finding the shortest homotopic path, minimizing either the Euclidean length or number of segments. Efrat et al. (2006) present an output-sensitive algorithm that runs in $O(n^{3/2} + k_{in} \log n + k_{out})$ time, where n is the number of obstacles, and k_{in} and k_{out} are the complexity of the input and output paths, respectively. Bespamyatnikh (2003) gives an $O(n \log^{1+\epsilon} n + k_{in} \log n + k_{out})$ time algorithm for simple paths and an $O(n^{2+\epsilon} + k \log^n)$ time algorithm for nonsimple paths, where $\epsilon > 0$ is an arbitrary small constant. The constants hidden in the notation depend on ϵ.

Generators for the fundamental group.   Colin de Verdière and Lazarus (2005) consider the special case of one-vertex cut graphs, which are sets of loops which are also a minimum presentation for the fundamental group. Given a set of loops, they give a polynomial-time algorithm that computes the shortest system of loops in the same homotopy class on an orientable surface, provided that the lengths of edges are uniform. For an orientable combinatorial surface with complexity n and a given basepoint, Erickson and Whittlesey (2005) show that a greedy algorithm can find the shortest set of loops that generate the fundamental group of the surface in $O(n \log n)$ using an application of Dijkstra's algorithm.

Generators for homology groups.   Erickson and Whittlesey (2005) also give a greedy algorithm that computes the shortest set of cycles that generate the first homology group over $\mathbb{Z}_2$ in time $O(n^2 \log n + n^2 g + n g^3)$, where g is the genus of the surface and n is the complexity of the surface. Zomorodian and Carlsson (2008) consider the problem of *localization*: finding local homology groups for arbitrary spaces and dimensions. Rather than using a particular measure, they include geometry in the input as a cover. They then construct a larger space whose persistent homology localizes the homology generators with respect to the given cover. Freedman and Chen (2008) define the size of a homology class to be the radius of the smallest geodesic ball within the ambient space. Given this measure, they use matroid theory to prove that an optimal basis may be computed by a greedy method in time $O(\beta^4 n^3 \log^2 n)$, where n is the size of the simplicial complex and β is the Betti number of the homology group.

3.9   Research Issues and Summary

Suppose we were given a million points in 100 dimensions and we wish to recover the topology of the space from which these points were sampled. Currently, none of our tools either scale to these many points or extend to this high a dimension. Yet, we are currently inundated with massive data sets from acquisition devices and computer simulations. Computational topology could provide powerful tools for understanding the structure of this data. However, we need both theoretical results as well as practical algorithms tailored to massive data sets for computational topology to become successful as a general method for data analysis.

3.10 Further Information

Surveys. The survey by Vegter (2004) contains results on knots, embeddings, and immersions not in this chapter. Edelsbrunner (2004) considers biological applications of computational topology. Joswig (2004) focuses on computation of invariants, including cohomology. Ghrist (2007) surveys recent results in persistent homology and its applications.

Books. Henle (1997) is an accessible introduction to topology. Hatcher (2002) is an excellent textbook on algebraic topology and is already in its 7th printing, although it is available freely on the web. Basener (2006) covers topology and its applications. Matsumoto (2002) is a monograph on Morse theory. Zomorodian (2005) expands on persistent homology and Morse-Smale complexes.

Software. PLEX (2006) is a library of MATLAB® routines that include construction of Čech, Vietoris-Rips and witness complexes, as well as computation of homology and persistent homology barcodes. CHomP (2008) computes homology of cubical complexes using reduction with tailored heuristics. CGAL (2008) is a computational geometry library that includes construction of alpha complexes. polymake's TOPAZ application supports computing simplicial homology and cohomology modules, as well as more sophisticated invariants, such as cup products (Gawrilow and Joswig, 2008). ANN (Mount and Arya, 2006) and E²LSH (Andoni, 2006) are software packages for finding approximate nearest neighbors in high dimensions.

Defining Terms

Betti number: In dimension k, the rank of the kth homology group, denoted β_k.

Contour: A connected component of a level set $h^{-1}(c)$ of a Morse function $h: \mathbb{M} \to \mathbb{R}$ defined on a manifold $\mathbb{M}$.

Critical point: A point on a manifold at which the differential of a Morse function is zero. Nondegenerate critical points on a two-manifold are minima, saddles, and maxima.

Euler characteristic: An integer invariant that relates combinatorial topology to algebraic topology and Morse theory: $\chi(\mathbb{M}) = \sum_k (-1)^k n_k = \sum_k (-1)^k \beta_k = \sum_k (-1)^k c_k$, where n_k is the number of k-cells for any cellular complex with underlying space $\mathbb{M}$, β_k is the kth Betti number, and c_k is the number of Morse critical points with index k for any Morse function define on $\mathbb{M}$.

Fundamental group: The group of homotopy classes of loops in a space.

Homeomorphism: A 1-1 onto continuous map whose inverse is continuous.

Homology group: An algebraic invariant $H_k = Z_k/B_k$, where Z_k is the group of k-cycles and B_k is the group of k-boundaries in the simplicial complex.

Homotopy: A family of maps $f_t : \mathbb{X} \to \mathbb{Y}$, $t \in [0, 1]$, such that the associated map $F : \mathbb{X} \times [0, 1] \to \mathbb{Y}$ given by $F(x, t) = f_t(x)$ is continuous.

Manifold: A topological space that is locally Euclidean.

Morse function: A function defined on a manifold whose critical points are nondegenerate.

Persistent homology: The homology of a growing space, which captures the lifetimes of topological attributes in a multiset of intervals called a barcode.

Reeb graph: A graph that captures the connectivity of contours. When it does not have cycles, it is called the contour tree.

Simplicial complex: A collection of finite sets such that if a set is in the complex, so are all its subsets. It may be visualized as a union of convex hulls of finitely independent points, glued together along shared faces.

Topological space: A point set with topology, a collection of sets that define the open neighborhoods of the points.

References

S. I. Adyan. The algorithmic unsolvability of problems concerning recognition of certain properties of groups. *Doklady Academy Nauk SSSR*, 103:533–535, 1955.

A. Andoni. E^2LSH: Nearest neighbors in high dimensional spaces, 2006. http://web.mit.edu/andoni/www/LSH/index.html.

A. Andoni and P. Indyk. Near-optimal hashing algorithms for near neighbor problem in high dimensions. In *Proceedings of the 47th Annual IEEE Symposium on Foundations of Computer Science*, pp. 459–468, Washington, D.C., 2006.

T. M. Apostol. *Calculus. Volume II: Multi-Variable Calculus and Linear Algebra, with Applications to Differential Equations and Probability*. John Wiley & Sons, Inc., New York, 2nd edition, 1969.

S. Arya, D. M. Mount, N. S. Netanyahu, R. Silverman, and A. Y. Wu. An optimal algorithm for approximate nearest neighbor searching in fixed dimensions. *Journal of the ACM*, 45(6):891–923, 1998.

T. F. Banchoff. Critical points and curvature for embedded polyhedral surfaces. *The American Mathematical Monthly*, 77(5):475–485, 1970.

W. F. Basener. *Topology and Its Applications*. John Wiley & Sons, Inc., Hoboken, NJ, 2006.

S. Bespamyatnikh. Computing homotopic shortest paths in the plane. *Journal of Algorithms*, 49:284–303, 2003.

J.-D. Boissonnat, L. J. Guibas, and S. Y. Oudot. Manifold reconstruction in arbitrary dimensions using witness complexes. In *Proceedings of 23rd ACM Symposium on Computational Geometry*, pp. 194–203, New York, 2007.

W. M. Boothby. *An Introduction to Differentiable Manifolds and Riemannian Geometry*. Academic Press, San Diego, CA, 2nd edition, 1986.

R. Bott and L. W. Tu. *Differential Forms in Algebraic Topology*. Springer-Verlag, New York, 1982.

R. L. Boyell and H. Ruston. Hybrid techniques for real-time radar simulation. In *Proceedings of the IEEE Fall Joint Computer Conference*, pp. 445–458, New York, 1963.

T. Brahana. Systems of circuits on 2-dimensional manifolds. *Annals of Mathematics*, 23:144–168, 1921.

J. W. Bruce and P. J. Giblin. *Curves and Singularities: A Geometrical Introduction to Singularity Theory*. Cambridge University Press, New York, 2nd edition, 1992.

S. Cabello, Y. Liu, A. Mantler, and J. Snoeyink. Testing homotopy for paths in the plane. *Discrete & Computational Geometry*, 31(1):61–81, 2004.

H. Carr, J. Snoeyink, and U. Axen. Computing contour trees in all dimensions. *Computational Geometry: Theory and Applications*, 24:75–94, 2003.

A. Cayley. On contours and slope lines. *Philosophical Magazine*, XVIII:264–268, 1859.

F. Cazals, J. Giesen, M. Pauly, and A. Zomorodian. The conformal alpha shape filtration. *The Visual Computer*, 22(8):531–540, 2006.

CGAL. Computational Geometry Algorithms Library, 2008. http://www.cgal.org.

A. Chakrabarti and O. Regev. An optimal randomised cell probe lower bound for approximate nearest neighbour searching. In *Proceedings of the 45th Annual IEEE Symposium on Foundations of Computer Science*, pp. 473–482, Washington, D.C., 2004.

CHomP. Computational Homology Project, 2008. http://chomp.rutgers.edu/.

K. Cole-McLaughlin, H. Edelsbrunner, J. Harer, V. Natarajan, and V. Pascucci. Loops in Reeb graphs of 2-manifolds. *Discrete & Computational Geometry*, 32:231–244, 2004.

É. Colin de Verdière and F. Lazarus. Optimal system of loops on an orientable surface. *Discrete & Computational Geometry*, 33(3):507–534, 2005.

T. H. Cormen, C. E. Leiserson, R. L. Rivest, and C. Stein. *Introduction to Algorithms*. The MIT Press, Cambridge, MA, 2001.

M. de Berg, M. van Kreveld, M. Overmars, and O. Schwarzkopf. *Computational Geometry: Algorithms and Applications*. Springer-Verlag, New York, 2nd edition, 2000.

V. de Silva and G. Carlsson. Topological estimation using witness complexes. In *Proceedings of Eurographics Symposium on Point-Based Graphics*, ETH Zurich, Switzerland, pp. 157–166, 2004.

M. Dehn and P. Heegaard. Analysis situs. *Enzyklopädie der Mathematischen Wissenschaften*, IIAB3: 153–220, 1907.

C. J. A. Delfinado and H. Edelsbrunner. An incremental algorithm for Betti numbers of simplicial complexes on the 3-sphere. *Computer Aided Geometric Design*, 12:771–784, 1995.

T. K. Dey. *Curve and Surface Reconstruction*, volume 23 of *Cambridge Monographs on Applied and Computational Mathematics*. Cambridge University Press, New York, 2007.

T. K. Dey and S. Guha. Transforming curves on surfaces. *Journal of Computer and System Sciences*, 58(2):297–325, 1999.

T. K. Dey and H. Schipper. A new technique to compute polygonal schema for 2-manifolds with application to null-homotopy detection. *Discrete & Computational Geometry*, 14(1):93–110, 1995.

T. K. Dey, J. Giesen, and M. John. Alpha-shapes and flow shapes are homotopy equivalent. In *Proc. ACM Symposium on Theory of Computing*, San Diego, CA, pp. 493–502, 2003.

J.-G. Dumas, F. Heckenbach, B. D. Saunders, and V. Welker. Computing simplicial homology based on efficient Smith normal form algorithms. In M. Joswig and N. Takayama, editors, *Algebra, Geometry, and Software Systems*, pp. 177–207, Springer-Verlag, Berlin, 2003.

D. Dummit and R. Foote. *Abstract Algebra*. John Wiley & Sons, Inc., New York, 3rd edition, 2003.

H. Edelsbrunner. Biological applications of computational topology. In J. E. Goodman and J. O'Rourke, editors, *Handbook of Discrete and Computational Geometry*, Chapter 32, pp. 1395–1412. Chapman & Hall/CRC, Boca Raton, FL, 2004.

H. Edelsbrunner and E. P. Mücke. Three-dimensional alpha shapes. *ACM Transactions on Graphics*, 13: 43–72, 1994.

H. Edelsbrunner, D. Letscher, and A. Zomorodian. Topological persistence and simplification. *Discrete & Computational Geometry*, 28:511–533, 2002.

H. Edelsbrunner, J. Harer, V. Natarajan, and V. Pascucci. Morse-Smale complexes for piecewise linear 3-manifolds. In *Proceedings of the 19th ACM Symposium on Computational Geometry*, pp. 361–370, New York, 2003a.

H. Edelsbrunner, J. Harer, and A. Zomorodian. Hierarchical Morse-Smale complexes for piecewise linear 2-manifolds. *Discrete & Computational Geometry*, 30:87–107, 2003b.

A. Efrat, S. G. Kobourov, and A. Lubiw. Computing homotopic shortest paths efficiently. *Computational Geometry: Theory and Applications*, 35(3):162–172, 2006.

J. Erickson and S. Har-Peled. Optimally cutting a surface into a disk. *Discrete & Computational Geometry*, 31(1):37–59, 2004.

J. Erickson and K. Whittlesey. Greedy optimal homotopy and homology generators. In *Proceedings of the 16th ACM-SIAM Symposium on Discrete Algorithms*, pp. 1038–1046, Philadelphia, PA, 2005.

R. Forman. Morse theory for cell complexes. *Advances in Mathematics*, 134(1):90–145, 1998.

D. Freedman and C. Chen. Measuring and localizing homology, 2008. arXiv:0705.3061v2.

E. Gawrilow and M. Joswig. `polymake`, 2008. http://www.math.tu-berlin.de/polymake/.

R. Ghrist. Barcodes: The persistent topology of data. *Bulletin of the American Mathematical Society (New Series)*, 45(1):61–75, 2008.

J. Giesen and M. John. The flow complex: A data structure for geometric modeling. In *Proceedings of the 14th Annual ACM-SIAM Symposium on Discrete Algorithms*, pp. 285–294, Philadelphia, PA, 2003.

M. Gromov. Hyperbolic groups. In S. Gersten, editor, *Essays in Group Theory*, pp. 75–263. Springer-Verlag, New York 1987.

L. J. Guibas and J. Stolfi. Primitives for the manipulation of general subdivisions and the computation of Voronoi diagrams. *ACM Transactions on Graphics*, 4:74–123, 1985.

A. Gyulassy, V. Natarajan, B. Hamann, and V. Pascucci. Efficient computation of Morse-Smale complexes for three-dimensional scalar functions. *IEEE Transactions on Visualization and Computer Graphics*, 13(6):1440–1447, 2007.

A. Hatcher. *Algebraic Topology*. Cambridge University Press, New York, 2002. http://www.math.cornell.edu/~hatcher/AT/ATpage.html.

M. Henle. *A Combinatorial Introduction to Topology*. Dover Publications, Inc., New York, 1997.

J. Hershberger and J. Snoeyink. Computing minimum length paths of a given homotopy class. *Computational Geometry: Theory and Applications*, 4(2):63–97, 1994.

M. Joswig. Computing invariants of simplicial manifolds, 2004. arXiv:math/0401176v1.

P. M. Kasson, A. Zomorodian, S. Park, N. Singhal, L. J. Guibas, and V. S. Pande. Persistent voids: A new structural metric for membrane fusion. *Bioinformatics*, 23(14):1753–1759, 2007.

F. Lazarus, G. Vegter, M. Pocchiola, and A. Verroust. Computing a canonical polygonal schema of an orientable triangulated surface. In *Proceedings of the 17th Annual ACM Symposium on Computational Geometry*, pp. 80–89, New York, 2001.

W. E. Lorensen and H. E. Cline. Marching cubes: A high resolution 3D surface construction algorithm. In *ACM SIGGRAPH Computer Graphics*, 21(4): 163–169, 1987.

A. A. Markov. Insolubility of the problem of homeomorphy. In *Proceedings of the International Congress of Mathematicians*, pp. 14–21, Edinburgh, U.K., 1958.

Y. Matsumoto. *An Introduction to Morse Theory*, volume 208 of *Iwanami Series in Modern Mathematics*. American Mathematical Society, Providence, RI, 2002.

J. C. Maxwell. On hills and dales. *The London, Edinburgh, and Dublin Philosophical Magazine and Journal of Science*, 40(269):421–425, December 1870.

J. Milnor. The Poincaré conjecture, 2008. http://www.claymath.org/millennium/Poincare_Conjecture/.

J. Morgan and G. Tian. *Ricci Flow and the Poincaré Conjecture*, volume 3 of *Clay Mathematics Monographs*. American Mathematical Society and Clay Mathematics Institute, Cambridge, MA, 2007.

D. M. Mount and S. Arya. ANN: A library for approximate nearest neighbor searching, 2006. http://www.cs.umd.edu/~mount/ANN/.

T. S. Newman and Y. Hong. A survey of the marching cubes algorithm. *Computers & Graphics*, 30(5): 854–879, 2006.

P. Niyogi, S. Smale, and S. Weinberger. Finding the homology of submanifolds with high confidence from random samples. *Discrete & Computational Geometry*, 39(1):419–441, 2008.

PLEX. Simplicial complexes in MATLAB, 2006. http://math.stanford.edu/comptop/programs/.

T. Rado. Über den begriff den Riemannschen fläschen. *Acta Szeged*, 2:101–121, 1925.

A. A. Ranicki, editor. *The Hauptvermutung Book*. Kluwer Academic Publishers, New York, 1997.

G. Reeb. Sur les points singuliers d'une forme de Pfaff complètement intégrable ou d'une fonction numérique. *Les Comptes rendus de l'Académie des sciences*, 222:847–849, 1946.

J. J. Rotman. *An Introduction to Algebraic Topology*. Springer-Verlag, New York, 1988.

H. Seifert and W. Threlfall. *Lehrbuch der Topologie*. AMS Chelsea Publishing, Providence, RI, 2003.

A. Storjohann. Computing Hermite and Smith normal forms of triangular integer matrices. *Linear Algebra and Its Applications*, 282(1–3):25–45, 1998.

W. Thurston. Three-dimensional manifolds, Kleinian groups and hyperbolic geometry. *Bulletin of the American Mathematical Society (New Series)*, 6(3):357–381, 1982.

F. Uhlig. *Transform Linear Algebra*. Prentice Hall, Upper Saddle River, NJ, 2002.

G. Vegter. Computational topology. In J. E. Goodman and J. O'Rourke, editors, *Handbook of Discrete and Computational Geometry*, Chapter 63, pp. 719–742. Chapman & Hall/CRC, Boca Raton, FL, 2004.

G. Vegter and C. K. Yap. Computational complexity of combinatorial surfaces. In *Proceedings of the sixth Annual ACM Symposium on Computational Geometry*, pp. 102–111, New York, 1990.

A. Zomorodian. *Topology for Computing*, volume 16 of *Cambridge Monographs on Applied and Computational Mathematics*. Cambridge University Press, New York, 2005.

A. Zomorodian and G. Carlsson. Computing persistent homology. *Discrete & Computational Geometry*, 33(2):249–274, 2005.

A. Zomorodian and G. Carlsson. Localized homology. *Computational Geometry: Theory and Applications*, 41:126–148, 2008.

4

Robot Algorithms

Konstantinos Tsianos
Rice University

Dan Halperin
Tel-Aviv University

Lydia Kavraki
Rice University

Jean-Claude Latombe
Stanford University

4.1   Introduction

People tend to have very different perceptions of what a robot is. For some people, a robot is an intelligent and sophisticated machine. A robot must have the capability to move autonomously and make decisions on how to accomplish its task without any human intervention. From this perspective, examples would include mobile robots used in space exploration, or looking a bit into the future, autonomous intelligent cars. Other people think of robots more in the context of industrial automation. In this context, a robot is typically a robotic arm used in assembly lines, for example, in car manufacturing. Those robots tend to perform the same repetitive motions in their highly controlled and predictable environment. Precision and accuracy are paramount for this kind of robots. Finally, it is not uncommon to treat a part or even a molecule as a robot. A major issue in production lines is how to process parts that arrive at random orientations and have to be oriented in a certain way before they are used for assembly. During the last few years, it has become a very important area of research to identify how big molecules—such as proteins—move, by adapting ideas that were initially developed for robots. This chapter covers some of the fundamental principles and presents several examples of algorithms that are currently used in robotics.

Robot algorithms differ in significant ways from traditional computer algorithms. The latter have full control over, and perfect access to the data they use, letting aside, for example, problems related to floating-point arithmetic. In contrast, robot algorithms eventually apply to physical objects in the real world, which they attempt to control despite the fact that these objects are subject to the independently and imperfectly modeled laws of nature. Data acquisition through sensing is also local and noisy. Robot algorithms hence raise controllability (or reachability) and observability (or recognizability) issues that are classical in control theory but not present in computer algorithms.

On the other hand, control theory often deals with well-defined processes in strictly confined environments. In contrast, robot tasks tend to be underspecified, which require addressing combinatorial issues ignored in control theory. For instance, to reach a goal among obstacles that are not represented in the input model but are sensed during execution, a robot must search "on the fly" for a collision-free path, a notoriously hard computational problem.

This blend of control and computational issues is perhaps the main characteristic of robot algorithms. It is presented at greater length in Section 4.2, along with other features of these algorithms. Section 4.3 then surveys specific areas of robotics (e.g., part manipulation, assembly sequencing, motion planning, and sensing) and presents algorithmic techniques that have been developed in those areas.

4.2 Underlying Principles

4.2.1 Robot Algorithms Control

The primary goal of a robot algorithm is to describe a procedure for controlling a subset of the real world—the workspace—in order to achieve a given goal, say, a spatial arrangement of several physical objects. The real world, which is subject to the laws of nature (such as gravity, inertia, friction), can be regarded as performing its own actions, for instance, applying forces. These actions are not arbitrary, and to some extent, they can be modeled, predicted, and controlled. Therefore, a robot algorithm should specify robot's operations whose combination with the (re-)actions of the real world will result in achieving the goal [8]. Note that the robot is itself an important object in the workspace; for example, it should not collide with obstacles. Therefore, the algorithm should also control the relation between the robot and the workspace. The robot's internal controller, which drives the actuators and preprocesses sensory data, defines the primitive operations that can be used to build robot algorithms.

The design of a robot algorithm requires identifying a set of relevant states of the workspace (one being the goal) and selecting operations that transform the workspace through a sequence of states so that the final state is the goal. But, due to various inaccuracies (one is in modeling physical laws), an operation may transform a state into one among several possible states. The algorithm can then use sensing to refine its knowledge during execution. In turn, because workspace sensing is imperfect, a state may not be directly recognizable, which means that no combination of sensors may be capable to return the state's identity. As a result, the three subproblems—choosing pertinent states, selecting operations to transit among these states toward the goal, and constructing state-recognition functions—are strongly interdependent and cannot be solved sequentially.

To illustrate part of the above discussion, consider the task of orienting a convex polygonal part P on a table using a robot arm equipped with a parallel-jaw gripper, to a desired goal orientation θ_g. This is a typical problem in industrial part feeding (Section 4.3.1.6). If an overhead vision system is available to measure P's orientation, we can use the following (simplified) algorithm:

ORIENT(P, θ_g)
1. Measure P's initial orientation θ_i
2. Move the gripper to the grasp position of P
3. Close the gripper
4. Rotate the gripper by $\theta_g - \theta_i$
5. Open the gripper
6. Move the gripper to a resting position

The states of interest are defined by the orientations of P, the position of the gripper relative to P, and whether the gripper holds P or does not. (Only the initial and goal orientations, θ_i and θ_g, are

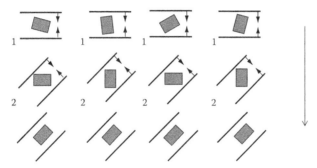

FIGURE 4.1 Orienting a convex polygonal part. Starting from four different initial positions, two squeezing operations can bring the part to the correct orientation. (From Goldberg, K.Y., *Algorithmica*, 10, 201, 1993. With permission.)

explicitly considered.) Step 1 acquires the initial state. Step 4 achieves the goal state. Steps 2 and 3 produce the intermediate states. Steps 5 and 6 achieve a second goal not mentioned above, that the robot be away from P at the end of the orientation operation.

A very different algorithm for this same part-orienting task consists of squeezing P several times between the gripper's jaws, at appropriately selected orientations of the gripper (see Figure 4.1). This algorithm, which requires no workspace sensing, is based on the following principle. Let P be at an arbitrary initial orientation. Any squeezing operation will achieve a new orientation that belongs to a set of $2n$ (n being the number of sides of P) possible orientations determined by the geometry of P and the orientation of the jaws. If P is released and squeezed again with another orientation of the gripper, the set of possible orientations of P can be reduced further. For any n-sided convex polygon P, there is a sequence of $2n - 1$ squeezes that achieves a single orientation of P (up to symmetries), for an arbitrary initial orientation of P [20].

The states considered by the second algorithm are individual orientations of P and sets of orientations. The state achieved by each squeeze is determined by the jaws' orientation and the previous state. Its prediction is based on understanding the simple mechanics of the operation. The fact that any convex polygon admits a finite sequence of squeezes ending at a unique orientation guarantees that any goal is reachable from any state. However, when the number of parameters that the robot can directly control is smaller than the number of parameters defining a state, the question of whether the goal state is reachable is more problematic (see Section 4.3.3).

State recognition can also be difficult. To illustrate, consider a mobile robot navigating in an office environment. Its controller uses dead-reckoning techniques to control the motion of the wheels. But these techniques yield cumulative errors in the robot's position with respect to a fixed coordinate system. For a better localization, the robot algorithm may sense environmental features (e.g., a wall, a door). However, because sensing is imperfect, a feature may be confused with a similar feature at a different place; this may occasionally cause a major localization mistake. Therefore, the robot algorithm must be designed such that enough environmental features will be sensed to make each successive state reliably recognizable.

To be more specific, consider the workspace of Figure 4.2. Obstacles are shown in bold lines. The robot is modeled as a point with perfect touch sensing. It can be commanded to move along any direction $\phi \in [0, 2\pi)$ in the plane, but imperfect control makes it move within a cone $\phi \pm \theta$, where the angle θ models directional uncertainty. The robot's goal is to move into G, the goal state, which is a subset of the wall W (for instance, G is an outlet for recharging batteries). The robot's initial location is not precisely known: it is anywhere in the disk I, the initial state. One candidate algorithm (illustrated in Figure 4.2a) first commands the robot to move perpendicularly to W until it touches it. Despite directionaluncertainty, the robot is guaranteed to eventually touch

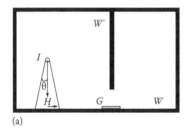 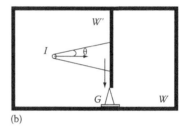

FIGURE 4.2 Mobile robot navigation with (a) unreliable goal recognition vs. (b) guaranteed goal recognition.

W, somewhere in the region denoted by H. From state H, it can slide along the wall (using touch to maintain contact) toward G. The robot is guaranteed to eventually reach G. But can it reliably recognize this achievement? The answer depends on the growth of the dead-reckoning localization error as the robot moves along W. Clearly, if this error grows by more than half the difference in the size of G and H, G is not reliably recognizable.

An alternative algorithm is possible, using the wall W' (Figure 4.2b). It commands the robot to first move toward W' until it touches it and then slide along it toward W. At the end of W', it continues heading toward W, but with directional uncertainty. The robot is nevertheless guaranteed to be in G when it senses that it has touched W.

4.2.2 Robot Algorithms Plan

Consider the following variant of the part-orienting task. Parts are successively fed with arbitrary orientations on a table by an independent machine. They have different and arbitrary convex polygonal shape, but whenever a part arrives, the feeding machine provides a geometric model of the part to the robot, along with its goal orientation. In the absence of a vision sensor, the multi-squeeze approach can still be used, but now the robot algorithm must include a planner to compute automatically the successive orientations of the gripper.

As another example, consider the pick-and-place task which requires a robot arm to transfer an object from one position to another. If the obstacles in the workspace are not known in advance, the robot algorithm needs sensing to localize them, as well as a planner to generate a collision-free path. If all obstacles cannot be sensed at once, the algorithm may have to interweave sensing, planning, and acting.

The point is that, for most tasks, the set of states that may have to be considered at the time of execution is too large to be explicitly anticipated. The robot algorithm must incorporate a planner. In first approximation, the planner can be seen as a separate algorithm that automatically generates a control algorithm for achieving a given task, e.g., [25]. The robot algorithm is the combination of the planning and the control algorithms. More generally, however, it is not sufficient to invoke the planner once, and planning and control are interleaved. The effect of planning is to dynamically change the control portion of the robot algorithm, by changing the set of states of the workspace that are explicitly considered.

Planning, which often requires exploring large search spaces, raises critical complexity issues. For example, finding a collision-free path for a three-dimensional linkage among polyhedral obstacles is PSPACE-hard [47], and the proof of this result provides strong evidence that any complete algorithm will require exponential time in the number of degrees of freedom. Planning the motion of a point robot among polyhedral obstacles, with bounded uncertainty in control and sensing, is NEXPTIME-hard [10].

The computational complexity of planning leads to looking for efficient solutions to restricted problems. For example, for part orienting, there exists a complete planning algorithm that computes a sequence of squeezes achieving a single orientation (up to symmetries) of a given convex polygonal part in quadratic time in the number of sides of the part [20]. Another way of dealing with complexity is to trade off completeness against time by accepting weaker variants of completeness. A complete planner is guaranteed to find a solution whenever one exists, and to notify that there exists none otherwise. A weaker variant is probabilistic completeness: if there exists a solution, the planner will find one only with high probability. This variant can be very useful if one can show that the probability of finding a solution (when one exists) tends rapidly toward 1 as the running time increases. In Section 4.3, we will present planning algorithms that embed similar approaches.

The complexity of a robot algorithm has also some interesting relations with the reachability and recognizability issues introduced in the previous subsection. We will mention several such relations in Section 4.3 (in particular, in Section 4.3.3).

The potentially high cost of planning and the fact that it may often have to be done online raise an additional issue. A robot algorithm must carefully allocate time between computations aimed at planning and computations aimed at controlling and sensing the workspace. If the workspace is changing,(say, under the influence of other agents), spending too much time on planning may result in obsolete control algorithms; on the other hand, not enough planning may yield irreversible failures [76]. The problem of allocating time between planning and control remains poorly understood, though several promising ideas have been proposed. For example, it has been suggested to develop planners that return a plan in whatever amount of time is allocated to them and can be called back later to incrementally improve the previous plan if more time is allocated to planning [7]. Deliberative techniques have been proposed to decide what amount of time should be given to planning and control and update this decision as more information is collected [43].

4.2.3   Robot Algorithms Reason about Geometry

Imagine a robot whose task is to maintain a botanic garden. To set and update its goal agenda, this robot needs knowledge in domains such as botany and fertilization. The algorithms using this knowledge can barely be considered parts of a robot algorithm. But, on the other hand, all robots, including gardener robots, accomplish tasks by eventually moving objects (including themselves) in the real world. Hence, at some point, all robots must reason about the geometry of their workspace. Actually, geometry is not enough, since objects have mass inducing gravitational and inertial forces, while contacts between objects generate frictional forces. All robots must, therefore, reason with classical mechanics. However, Newtonian concepts of mechanics translate into geometric constructs (e.g., forces are represented as vectors), so that most of the reasoning of a robot eventually involves dealing with geometry.

Computing with continuous geometric models raises discretization issues. Several planners, computing a robot's path, discretize the robot's free space in order to build a connectivity graph to which well-known search algorithms can be applied.

Consider, for example, the configuration space. A formal definition of this very important concept is given in Section 4.3.3, but for now just think of it as the set of all collision-free poses (configurations) of the robot. One discretization approach is to place a fine regular grid across configuration space and search that grid for a sequence of adjacent points in free space. The grid is just a computational tool and has no physical meaning. Its resolution is arbitrarily chosen despite its critical role in the computation: if it is too coarse, planning is likely to fail; if it is too fine, planning will take too much time. Instead, criticality-driven discretizations have been proposed, whose underlying principle is widely applicable. They consist of partitioning the continuous space of interest into cells, such that some pertinent property remains invariant over each cell and changes when the boundary separating two cells is crossed. The second part-orienting algorithm in Section 4.2.1 is based on such a discretization. The set of all possible orientations of the part is represented as the unit circle (the cyclic interval $[0, 2\pi)$). For a given orientation of the gripper, this circle can be partitioned into arcs such that, for all initial orientations of the part in the same arc, the part's final orientation will be the same after the gripper has closed its jaws. The final orientation is the invariant associated with the cell. From this decomposition, it is a relatively simple matter to plan a squeezing sequence.

Several such criticality-driven discretizations have been proposed for path planning, assembly sequence planning, motion planning with uncertainty, robot localization, object recognition, and so on, as will be described in Section 4.3. Several of them use ideas and tools originally developed in computational geometry, for instance, plane sweep, constructing arrangements, and constructing Voronoi diagrams.

Robot algorithms often require dealing with high-dimensional geometric spaces. Although criticality-based discretization methods apply to such spaces in theory (for instance, see [49]), their computational complexity is overwhelming. This has led the development of randomized techniques that efficiently approximate the topology and geometry of such spaces by random discretization. Such techniques have been particularly successful for building probabilistically complete planners (Section 4.3.3.4).

4.2.4   Robot Algorithms Have Physical Complexity

Just as the complexity of a computation characterizes the amount of time and memory this computation requires, we can define the physical complexity of a robot algorithm by the amount of physical resources it takes, for example, the number of "hands," the number of motions, or the number of beacons. Some resources, such as the number of motions, relate to the time spent for executing the algorithm. Others, such as the number of beacons, relate to the engineering cost induced by the algorithm. For example, one complexity measure of the multi-squeeze algorithm to orient a convex polygon (Section 4.2.1) is the maximal number of squeeze operations this algorithm performs. As another example, consider an assembly operation merging several parts into a subassembly. The number of subsets of parts moving relative to each other (each subset moving as a single rigid body) measures the number of hands necessary to hold the parts during the operation. The number of hands required by an assembly sequence is the maximal number of hands needed by an operation, over all operations in the sequence (Section 4.3.2). The number of fingers to safely grasp or fixture an object is another complexity measure (Section 4.3.1.1).

Though there is a strong conceptual analogy between computational and physical complexities, there are also major differences between the two notions. Physical complexity must be measured along many more dimensions than computational complexity. Moreover, while computational complexity typically measures an asymptotic trend, a tighter evaluation is usually needed for physical complexity since robot tasks involve relatively few objects.

One may also consider the inherent physical complexity of a task, a notion analogous to the inherent complexity of a computational problem. For example, to orient a convex polygon with n sides, $2n - 1$ squeezes may be needed in the worst case; no correct algorithm can perform better in all cases. By generating all feasible assembly sequences of a product, one could determine the number of hands needed for each sequence and return the smallest number. This number is a measure of the inherent complexity of assembling the product. No robot algorithm, to assemble this product, can require fewer hands.

Evaluating the inherent physical complexity of a task may lead to redefining the task, if it turns out to be too complex. For example, it has been shown that a product made of n parts may need up to n hands for its assembly (Section 4.3.1.1), thus requiring the delicate coordination of $n - 1$ motions. Perhaps, a product whose assembly requires several hands could be redesigned so that two hands are sufficient, as is the case for most industrial products. Indeed, designers strive to reduce physical complexity along various dimensions. For instance, many mass-produced devices are designed to be assembled with translations only, along very few directions (possibly a single one). The inherent physical complexity of a robot task is not a recent concept, but its formal application to task analysis is [54].

An interesting issue is how the computational and physical complexities of a robot algorithm relate to each other. For example, planning for mobile robot navigation with uncertainty is a provably hard computational problem (Section 4.3.5.6). On the other hand, burying wires in the ground or placing enough infrared beacons allows robots to navigate reliably at small computational cost. But isn't it too much? Perhaps, the intractability of motion planning with uncertainty can be eliminated with less costly engineering.

4.3 State of the Art and Best Practices

Robotics is a broad domain of research. In this subsection, we study a number of specific areas: part manipulation, assembly sequencing, and motion planning. For each area, we introduce problems and survey key algorithmic results.

Although we present the current research according to problem domain, there are several techniques that cross over many domains. One of the most frequently applied methods in robotics is the criticality-based discretization mentioned in Section 4.2.3. This technique allows us to discretize a continuous space without giving up the completeness or exactness of the solution. It is closely related to the study of arrangements in computational geometry [23]. When criticality-based discretization is done in a space representing all possible motions, it yields the so-called "nondirectional" data structures, which is another prevailing concept in robot algorithms and is exemplified in detail in Section 4.3.2.4.

Randomization is another important paradigm in robotics. Randomized techniques have made it possible to cope practically with robot motion planning with many degrees of freedom (Section 4.3.3.4). Also, randomized algorithms are often simpler than their deterministic counterparts and hence better candidates for efficient implementation. Randomization has recently been applied to solving problems in grasping as well as in many other areas that involve geometric reasoning.

Throughout this section, we interchangeably use the terms "body," "physical object," and "part" to designate a rigid physical object modeled as a compact manifold with boundary $B \subset \mathbb{R}^k$ ($k = 2$ or 3). B's boundary is also assumed piecewise-smooth.

4.3.1 Part Manipulation

Part manipulation is one of the most frequently performed operations in industrial robotics: parts are grasped from conveyor belts, they are oriented prior to feeding assembly workcells, and they are immobilized for machining operations.

4.3.1.1 Grasping

Part grasping has motivated various kinds of research, including the design of versatile mechanical hands, as well as simple, low-cost grippers. From an algorithmic point of view, the main goal is to compute "safe" grasps for an object whose model is given as input.

4.3.1.2 Force-Closure Grasp

Informally, a grasp specifies the positions of "fingers" on a body B. A more formal definition uses the notion of a wrench, a pair $[f, p \times f]$, where p denotes a point on the boundary ∂B of B, represented by its coordinate vector in a frame attached to B; f designates a force applied to B at p; and $\times$ is the vector cross-product. If f is a unit vector, the wrench is said to be a unit wrench. A finger is any tool that can apply a wrench.

A grasp of B is a set of unit wrenches $w_i = [f_i, p_i \times f_i]$, $i = 1, \ldots, p$, defined on B. For each w_i, if the contact is frictionless, f_i is normal to ∂B at p_i; otherwise, it can span a friction cone (Coulomb law of friction).

The notion of a safe grasp is captured by force closure. A force-closure grasp $\{w_i\}_{i=1,\ldots,p}$ on B is such that, for any arbitrary wrench w, there exists a set of real values $\{f_1, \ldots, f_p\}$ achieving $\sum_{i=1}^{p} f_i w_i = -w$. In other words, a force-closure grasp can resist any external wrench applied to B. If contacts are nonsticky, we require that $f_i \geq 0$, for all $i = 1, \ldots, p$, and the grasp is called positive. Here, we consider only positive grasps. A form-closure grasp is a positive force-closure grasp when all finger-body contacts are frictionless.

4.3.1.3 Size of a Form/Force-Closure Grasp

The following results characterize the physical complexity of achieving a safe grasp [40]:

- Bodies with rotational symmetry (e.g., discs in two-space, spheres and cylinders in three-space) admit no form-closure grasps.
- All other bodies admit a form-closure grasp with at most 4 fingers in two-space and 12 fingers in three-space.
- All polyhedral bodies have a form-closure grasp with seven fingers.
- With frictional finger-body contacts, all bodies admit a force-closure grasp that consists of three fingers in two-space and four fingers in three-space.

4.3.1.4 Testing Force Closure

A necessary and sufficient condition for a grasp $\{w_i\}_{i=1,\ldots,p}$ to achieve force closure in two-space (respectively, three-space) is that the finger wrenches w_i span a space F of dimension three (respectively, six) and that a strictly positive linear combination of them be zero. In other words, the origin of F (null wrench) should lie in the interior of the convex hull of the finger wrenches [40]. This condition provides an effective test for deciding in constant time whether a given grasp achieves force closure.

4.3.1.5 Computing Form/Force Closure Grasps

Most research has concentrated on computing grasps with two to four nonsticky fingers. Algorithms that compute a single force-closure grasp of a polygonal/polyhedral part in time linear in the part's complexity have been derived in [40].

Finding the maximal regions on a body where fingers can be positioned independently while achieving force closure makes it possible to accommodate errors in finger placement. Geometric algorithms for constructing such regions are proposed in [42] for grasping polygons with two fingers (with friction) and four fingers (without friction), and for grasping polyhedra with three fingers (with frictional contact capable of generating torques) and seven fingers (without friction). Grasping of curved obstacles is addressed in [45].

4.3.1.6 Fixturing

Most manufacturing operations require fixtures to hold parts. To avoid the custom design of fixtures for each part, modular reconfigurable fixtures are often used. A typical modular fixture consists of a workholding surface, usually a plane, that has a lattice of holes where locators, clamps, and edge fixtures can be placed. Locators are simple round pins, whereas clamps apply pressure on the part.

Contacts between fixture elements and parts are generally assumed to be frictionless. In modular fixturing, contact locations are restricted by the lattice of holes, and form closure cannot always be achieved. In particular, when three locators and one clamp are used on a workholding plane, there exist polygons of arbitrary size for which no form-closure fixture exists; but, if parts are restricted to be rectilinear with all edges longer than four lattice units, a form-closure fixture always exists [56].

When the fixturing kit consists of a latticed workholding plane, three locators, and one clamp, it is possible to find all possible placements of a given part on the workholding surface where form closure can be achieved, along with the corresponding positions of the locators and the clamp [9].

Algorithms for computing all placements of (frictionless) point fingers that put a polygonal part in form closure and all placements of point fingers that achieve "2nd-order immobility" [96] of a polygonal part are presented in [95]. Immobilizing hinged parts are discussed in [97].

4.3.1.7 Part Feeding

Part feeders account for a large fraction of the cost of a robotic assembly workcell. A typical feeder must bring parts at subsecond rates with high reliability. An example of a flexible feeder is given in [20] and described in Section 4.2.1.

Part feeding often relies on nonprehensile manipulation, which exploits task mechanics to achieve a goal state without grasping and frequently allows accomplishing complex feeding tasks with simple mechanisms [2]. Pushing is one form of nonprehensile manipulation [90]. Work on pushing originated in [38] where a simple rule is established to qualitatively determine the motion of a pushed object. This rule makes use of the position of the center of friction of the object on the supporting surface. Related results include a planning algorithm for a robot that tilts a tray with a planar part of known shape to orient it to a desired orientation [16], and an algorithm that computes the sequence of motions of a single articulated fence on a conveyor belt to achieve a goal orientation of an object [2]. A variety of interesting results on part feeding appear in the thesis [98].

4.3.2 Assembly Sequencing

Most mechanical products consist of multiple parts. The goal of assembly sequencing is to compute both an order in which parts can be assembled and the corresponding required movements of the parts. Assembly sequencing can be used during design to verify that the product will be easy to manufacture and service. An assembly sequence is also a robot algorithm at a high level of abstraction since parts are assumed free-flying, massless geometric objects.

4.3.2.1 Notion of an Assembly Sequence

An assembly A is a collection of bodies in some given relative placements. Subassemblies are separated if they are arbitrarily far apart from one another. An assembly operation is a motion that merges s separated subassemblies ($s \geq 2$) into a new subassembly, with each subassembly moving as a single body. No overlapping between bodies is allowed during the operation. The parameter s is called the number of hands of the operation. (Hence, a hand is seen here as a grasping or fixturing tool that can hold an arbitrary number of bodies in fixed relative placements.) Assembly partitioning is the reverse of an assembly operation.

An assembly sequence is a total ordering on assembly operations that merges the separated parts composing a new assembly into this assembly. The maximum, over all the operations in the sequence, of the number of hands of an operation is the number of hands of the sequence.

A monotone assembly sequence contains no operation that brings a body to an intermediate placement (relative to other bodies), before another operation transfers it to its final placement. Therefore, the bodies in every subassembly produced by such a sequence are in the same relative placements as in the complete assembly. Note that a product may admit no monotone assembly sequence for a given number of hands, while it may admit such sequences if more hands are allowed.

4.3.2.2 Number of Hands in Assembly

The number of hands needed for various families of assemblies is a measure of the inherent physical complexity of an assembly task (Section 4.2.4). It has been shown that an assembly of convex polygons in the plane has a two-handed assembly sequence of translations. In the worst case, s hands are necessary and sufficient for assemblies of s star-shaped polygons/polyhedra [41].

There exists an assembly of six tetrahedra without a two-handed assembly sequence of translations, but with a three-handed sequence of translations. Every assembly of five or fewer convex polyhedra admits a two-handed assembly sequence of translations. There exists an assembly of 30 convex polyhedra that cannot be assembled with two hands [51].

4.3.2.3 Complexity of Assembly Sequencing

When arbitrary sequences are allowed, assembly sequencing is PSPACE-hard. The problem remains PSPACE-hard even when the bodies are polygons, each with a constant maximal number of vertices [41]. When only two-handed monotone sequences are permitted and rigid motions are allowed, finding a partition of an assembly A into two subassemblies S and $A \setminus S$ is NP-complete. The problem remains NP-complete when both S and $A \setminus S$ are connected, and motions are restricted to translations [26]. These latter results were obtained by reducing in polynomial time any instance of the three-SAT problem to a mechanical assembly such that the partitioning of this assembly gives the solution of the three-SAT problem instance.

4.3.2.4 Monotone Two-Handed Assembly Sequencing

A popular approach to assembly sequencing is disassembly sequencing. A sequence that separates an assembly to its individual components is first generated and next reversed. Most existing assembly sequencers can only generate two-handed monotone sequences. Such a sequence is computed by partitioning the assembly and, recursively, the obtained subassemblies into two separated assemblies.

The nondirectional blocking graph (NDBG) is proposed in [54] to represent all the blocking relations in an assembly. It is a subdivision of the space of all allowable motions of separation into a finite number of cells such that within each cell the set of blocking relations between all pairs of parts remain fixed. Within each cell, this set is represented in the form of a directed graph, which is called the directional blocking graph (DBG). The NDBG is the collection of the DBGs over all the cells in the subdivision. The NDBG is one example of a data structure obtained by a criticality-driven discretization technique (Section 4.2.3).

We illustrate this approach for polyhedral assemblies when the allowable motions are infinite translations. The partitioning of an assembly consisting of polyhedral parts into two subassemblies is done as follows. For an ordered pair of parts P_i, P_j, the three-vector d is a blocking direction if translating P_i to infinity in direction d will cause P_i to collide with P_j. For each ordered pair of parts, the set of blocking directions is constructed on the unit sphere S^2 by drawing the boundary arcs of the union of the blocking directions (each arc is a portion of a great circle). The resulting collection of arcs partitions S^2 into maximal regions such that the blocking relation among the parts is the same for any direction inside such a region.

Next, the blocking graph is computed for one such maximal region. The algorithm then moves to an adjacent region and updates the DBG by the blocking relations that change at the boundary between the regions, and so on. After each time the construction of a DBG is completed, this graph is checked for strong connectivity in time linear in the number of its nodes and edges. The algorithm stops the first time it encounters a DBG that is not strongly connected, and it outputs the two subassemblies of the partitioning. The overall sequencing algorithm continues recursively with the resulting subassemblies. If all the DBGs that are produced during a partitioning step are strongly connected, the algorithm notifies that the assembly does not admit a two-handed monotone assembly sequence with infinite translations.

Polynomial time algorithms are proposed in [54] to compute and exploit NDBGs for restricted families of motions. In particular, the case of partitioning a polyhedral assembly by a single translation to infinity is analyzed in detail, and it is shown that partitioning an assembly of m polyhedra with a total of v vertices takes $O(m^2 v^4)$ time. Another case is where the separating motions are infinitesimal rigid motions. In such a case, partitioning the polyhedral assembly can be carried out efficiently and practically [93]. With the above algorithms, every feasible disassembly sequence can be generated in polynomial time.

4.3.3 Motion Planning

Motion planning is central to robotics, as motion planning algorithms can provide robots with the capability of deciding automatically which motions to execute to reach their goal. In its simplest form, the problem is known as path planning because the question is to find a collision free path from an initial to a final position. A more challenging and general version is captured by the term motion planning. The distinctive difference is that it is not enough to come up with a collision free path. In addition, the algorithm must compute the exact actions that the robot's actuators must perform to implement the computed path.

It is reasonable to expect that motion planning problems can present various difficulties depending on the type of robot at hand (e.g., planning for mobile robots, humanoids, reconfigurable robots, manipulators, etc.). Fortunately, all those differences can be abstracted with the use of the configuration space that is described below. In the rest of this section, several motion planning issues are discussed. For the case of just path planning, complete algorithms with a complexity analysis are presented together with more recent sampling-based approaches. Then the case of motion planning is considered under the prism of planning with differential constraints. The section ends with a discussion of several other motion planning variants that include planning in dynamic workspaces, planning with moving obstacles in the environment, multiple robots, movable objects, online planning, optimal planning, and dealing with uncertainties.

4.3.3.1 Configuration Space

The configuration space has been informally described in Section 4.2.3. At first sight, planning for a car on the highway looks very different from planning for an industrial robotic arm. It is possible though, to define a powerful abstraction that hides the robot's specific details and transforms the problem into finding a solution for a point robot that has to move from one position to another in some new space, called the configuration space.

A configuration of a robot $\mathcal{A}$ is any mathematical specification of the position and orientation of every body composing $\mathcal{A}$, relative to a fixed coordinate system. The configuration of a single body is called a placement or a pose.

The robot's configuration space is the set of all its configurations. Usually, it is a smooth manifold. We will always denote the configuration space of a robot by $\mathcal{C}$ and its dimension by m. Given a robot $\mathcal{A}$, we will let $\mathcal{A}(q)$ denote the subset of the workspace occupied by $\mathcal{A}$ at configuration q.

The number of degrees of freedom of a robot is the dimension m of its configuration space. We abbreviate "degree of freedom" by dof.

Given an obstacle B_i in the workspace, the subset $CB_i \subseteq \mathcal{C}$ such that, for any $q \in CB_i$, $\mathcal{A}(q)$ intersects B_i is called a C-obstacle. The union $CB = \cup_i CB_i$ plus the configurations that violate the mechanical limits of the robot's joints is called the C-obstacle region. The free space is the complement of the C-obstacle region in $\mathcal{C}$, that is, $\mathcal{C} \setminus CB$. In most practical cases, C-obstacles are represented as semialgebraic sets with piecewise smooth boundaries.

A robot's path is a continuous map $\tau : [0, 1] \rightarrow \mathcal{C}$. A free path is a path that entirely lies in free space. A semifree path lies in the closure of free space.

After the configuration space has been defined, solving a path problem for some robot becomes a question of computing a free or semifree path between two configurations, e.g., [35,36]. A complete planner is guaranteed to find a (semi)free path between two given configurations whenever such a path exists, and to report that no such path exists otherwise.

4.3.3.2 Complete Algorithms

Basic path planning for a three-dimensional linkage made of polyhedral links is PSPACE-hard [47]. The proof uses the robot's dofs to both encode the configuration of a polynomial space bounded Turing machine and design obstacles which force the robot's motions to simulate the computation of this machine. It provides strong evidence that any complete algorithm will require exponential time in the number of dofs. This result remains true in more specific cases, for instance, when the robot is a planar arm in which all joints are revolute. However, it no longer holds in some very simple settings; for instance, planning the path of a planar arm within an empty circle is in P. For a collection of complexity results on motion planning see [30].

Most complete algorithms first capture the connectivity of the free space into a graph either by partitioning the free space into a collection of cells (exact cell decomposition techniques) or by extracting a network of curves (roadmap techniques) [30]. General and specific complete planners have been proposed. The general ones apply to virtually any robot with an arbitrary number of dofs. The specific ones apply to a restricted family of robots usually having a fixed small number of dofs.

The general algorithm in [49] computes a cylindrical cell decomposition of the free space using the Collins method. It takes doubly exponential time in the number m of dofs of the robot. The roadmap algorithm in [10] computes a semifree path in time singly exponential in m. Both algorithms are polynomial in the number of polynomial constraints defining the free space and their maximal degree. Specific algorithms have been developed mainly for robots with 2 or 3 dofs. For a k-sided polygonal robot moving freely in a polygonal workspace, the algorithm in [24] takes $O((kn)^{2+\epsilon})$ time, where n is the total number of edges of the workspace, for any $\epsilon > 0$.

4.3.3.3 Heuristic Algorithms

Several heuristic techniques have been proposed to speedup path planning. Some of them work well in practice, but they usually offer no performance guarantee.

Heuristic algorithms often search a regular grid defined over configuration space and generate a path as a sequence of adjacent grid points. The search can be guided by a potential field, a function over the free space that has a global minimum at the goal configuration. This function may be constructed as the sum of an attractive and a repulsive field [28]. The attractive field has a single minimum at the goal and grows to infinity as the distance to the goal increases. The repulsive field is zero at all configurations where the distance between the robot and the obstacles is greater than some predefined value, and grows to infinity as the robot gets closer to an obstacle. Ideally, a robot could find its way to the goal by following a potential field that has only one global minimum at the goal (the potential field is then called a navigation function). Yet in general, the configuration space is usually a high dimensional and strangely shaped manifold, which makes it very hard to design potential fields that have no local minima where the robot may be trapped.

One may also construct grids at variable resolution. Hierarchical space decomposition techniques such as octrees and boxtrees have been used to that purpose [30].

4.3.3.4 Sampling-Based Algorithms

The complexity of path planning for robots with many dofs (more than 4 or 5) has led the development of computational schemes that trade off completeness against computational time. Those methods avoid computing an explicit representation of the free space. Instead, their focus is on producing a graph that approximately captures the connectivity of the free space. As the title of this section suggests, those planners compute this graph by sampling the configuration space. The general idea is that a large number of random configurations are sampled from the configuration space. Then the configurations that correspond to collisions are filtered out, and an attempt is made to connect pairs of collision free configurations to produce the edges of the connectivity graph.

One of the most popular such planners is the probabilistic roadmap (PRM) [4,27]. The original algorithm consists of a learning and a querying phase. In the learning phase, the roadmap is constructed. A sampling strategy is used to generate a large number of configuration samples. The strategy can be just uniform random sampling, although many sophisticated strategies such as bridge test, Gaussian, etc., have been proposed over the years [61,62]. Then, an attempt is made to connect every free sample to its neighboring free samples. For this process, a local planner is used that tries to interpolate between the two samples using a very simple strategy; for example, along a straight line in the configuration space. The intermediate configurations on the line must be checked for collisions [61,62,88] If no collisions are detected, the corresponding edge is added to the graph. After this phase is over, multiple planning queries can be solved using the same graph as a roadmap. The initial and goal configuration are connected to the roadmap by producing edges to one of their respective nearest neighbors on the roadmap. Then, planning is reduced to a graph search problem which can be solved very efficiently.

PRM was initially intended as a multiquery planner where the same roadmap can be reused. In many cases though, it is required to solve only a single query as fast as possible, and it is more important to explore the space toward some goal rather than trying to capture the connectivity of the whole space. Two very popular such planners are rapidly-exploring random trees (RRT) [68] and expansive-spaces trees (EST)* [69]. Both end up building a tree T, rooted at the initial configuration. EST proceeds by extending new edges out to the existing tree T. At each step, a node q_1 in the tree is selected according to a probability distribution inverse to the local sampling density. Then, a new random configuration q_2 is sampled in the neighborhood of q_1, and an attempt is made to create an edge between q_1 and q_2 using a local planner as before. Hopefully at some point, there will be some configuration node in the tree that can be connected to the goal. RRT has a slightly different strategy. At each iteration, a new random configuration q is produced, and then an attempt is made to connect q to its nearest neighbor in the tree. A popular way to make this process more effective is to select the goal itself as q with some small probability. Several extensions of the above algorithms can be found in [61,62].

There are a number of issues that can greatly affect the performance of sampling-based planners. The sampling strategy has already been discussed above and is very important to produce useful samples. Moreover, sampling-based planners make extensive use of collision checking primitives [63]. This is a functionality that has to be available to provide the answer to the question of whether a configuration is in collision with an obstacle, or not. Finally, most of those planners need some functionality for nearest neighbor's computations. This is a nontrivial task in general. The reason is that configuration spaces are usually strangely shaped manifolds where it is not always easy to define good distance metrics. This issue is discussed at the end of this chapter.

An interesting question is how many samples are required to solve a problem. In general, this is related to the completeness properties of those algorithms. It is understood that if after some time the produced roadmap or tree is unable to solve the problem, it could just be that the sampler was "unlucky" and did not produce enough good samples yet. In principle, sampling-based algorithms are not complete with respect to the definition given in previous sections. Instead, most of them can be proven to be probabilistically complete. This is a weaker notion of completeness that guarantees

---

* The term EST was not used in the original paper. It was introduced later [62].

that a solution will eventually be found if one exists. For this definition to be of practical importance, it is good to also demand that the convergence to the solution will be fast, that is, exponential in the number of samples. Nevertheless, attempts have been made to estimate the quality of the roadmap and the number of samples that will be required. For example, the results reported in [4] bound the number of samples generated by the algorithm in [27], under the assumption that the configuration space verifies some simple geometric property.

4.3.4 Motion Planning under Differential Constraints

Real robots are constrained by mechanics and the laws of physics. Moreover, it is commonly the case that a robot has more dofs than the number of actuators that can control those configuration space parameters (underactuation). This means that a planner needs to produce paths that respect those constraints and are thus implementable (feasible). Over the years, many terms have been used in the literature to describe these classes of problems. Constraints in a robot's motion that cannot be converted into constraints that involve no derivatives, such as nonintegrable velocity constraints in the configuration space, are usually referred to as nonholonomic. A more recent term is kinodynamic constraints. The latter describes second order constraints on both velocity and acceleration, and lately it has been used to describe problems that involve dynamics in general. A generic term that captures all these constraints is differential constraints [61].

Below nonholonomic constraints are examined in some more depth to better understand the controllability issues introduced in Section 4.2.1. This section ends with a description of the work on sampling-based planners that can be applied to problems with differential constraints.

4.3.4.1 Planning for Robots with Nonholonomic Constraints

The trajectories of a nonholonomic robot are constrained by $p \geq 1$ nonintegrable scalar equality constraints:

$$G(q(t), \dot{q}(t)) = \left(G^1(q(t), \dot{q}(t)), \ldots, G^p(q(t), \dot{q}(t))\right) = (0, \ldots, 0),$$

where $\dot{q}(t) \in T_{q(t)}(\mathcal{C})$ designates the velocity vector along the trajectory $q(t)$. At every q, the function $G_q = G(q,.)$ maps the tangent space* $T_q(\mathcal{C})$ into $\mathbb{R}^p$. If G_q is smooth and its Jacobian has full rank (two conditions that are often satisfied), the constraint $G_q(\dot{q}) = (0, \ldots, 0)$ constrains $\dot{q}$ to be in a linear subspace of $T_q(\mathcal{C})$ of dimension $m - p$. The nonholonomic robot may also be subject to scalar inequality constraints of the form $H^j(q, \dot{q}) > 0$. The subset of $T_q(\mathcal{C})$ that satisfies all the constraints on $\dot{q}$ is called the set $\Omega(q)$ of controls at q. A feasible path is a piecewise differentiable path whose tangent lies everywhere in the control set.

A car-like robot is a classical example of a nonholonomic robot. It is constrained by one equality constraint (the linear velocity must point along the car's axis so that the car is not allowed to move sideways). Limits on the steering angle impose two inequality constraints. Other nonholonomic robots include tractor-trailers, airplanes, and satellites.

A key question when dealing with a nonholonomic robot is: Despite the relatively small number of controls, can the robot span its configuration space? The study of this question requires introducing some controllability notions. Given an arbitrary subset $U \subset \mathcal{C}$, the configuration $q_1 \in U$ is said to be U-accessible from $q_0 \in U$ if there exists a piecewise constant control $\dot{q}(t)$ in the control set whose integral is a trajectory joining q_0 to q_1 that fully lies in U. Let $A_U(q_0)$ be the set of configurations U-accessible from q_0. The robot is said to be locally controllable at q_0 if for every neighborhood U of

---

* The tangent space $T_p(M)$ at a point p of a smooth manifold M is the vector space of all tangent vectors to curves contained in M and passing through p. It has the same dimension as M.

q_0, $A_U(q_0)$ is also a neighborhood of q_0. It is locally controllable if this is true for all $q_0 \in C$. Car-like robots and tractor-trailers that can go forward and backward are locally controllable [5].

Let X and Y be two smooth vector fields on C. The Lie bracket of X and Y, denoted by $[X, Y]$, is the smooth vector field on C defined by $[X, Y] = dY \cdot X - dX \cdot Y$, where dX and dY, respectively, denote the $m \times m$ matrices of the partial derivatives of the components of X and Y w.r.t. the configuration coordinates in a chart placed on C. The control Lie algebra associated with the control set Ω, denoted by $L(\Omega)$, is the space of all linear combinations of vector fields in Ω closed by the Lie bracket operation. The following result derives from the controllability rank condition theorem [5]:

A robot is locally controllable if, for every $q \in C$, $\Omega(q)$ is symmetric with respect to the origin of $T_q(C)$, and the set $\{X(q) \mid X(q) \in L(\Omega(q))\}$ has dimension m.

The minimal number of Lie brackets sufficient to express any vector in $L(\Omega)$ using vectors in Ω is called the degree of nonholonomy of the robot. The degree of nonholonomy of a car-like robot is two. Except at some singular configurations, the degree of nonholonomy of a tractor towing a chain of s trailers is $2 + s$. Intuitively, the higher the degree of nonholonomy the more complex (and the slower) the robot's maneuvers to perform some motions.

Let A be a locally controllable nonholonomic robot. A necessary and sufficient condition for the existence of a feasible free path of A between two given configurations is that they lie in the same connected component of the open free space. Indeed, local controllability guarantees that a possibly nonfeasible path can be decomposed into a finite number of subpaths, each short enough to be replaced by a feasible free subpath [31]. Hence, deciding if there exists a free path for a locally controllable nonholonomic robot has the same complexity as deciding if there exists a free path for the holonomic robot having the same geometry. Transforming a nonfeasible free path τ into a feasible one can be done by recursively decomposing τ into subpaths. The recursion halts at every subpath that can be replaced by a feasible free subpath. Specific substitution rules (e.g., Reeds and Shepp curves) have been defined for car-like robots [31]. The complexity of transforming a nonfeasible free path τ into a feasible one is of the form $O(\epsilon^d)$, where ϵ is the smallest clearance between the robot and the obstacles along τ, and d is the degree of nonholonomy of the robot. The algorithm in [5] directly constructs a nonholonomic path for a car-like or a tractor-trailer robot by searching a tree obtained by concatenating short feasible paths, starting at the robot's initial configuration. The planner is guaranteed to find a path if one exists, provided that the length of the short feasible paths is small enough. It can also find paths that minimize the number of cusps (changes of sign of the linear velocity).

Path planning for nonholonomic robots that are not locally controllable is much less understood. Research has almost exclusively focused on car-like robots that can only move forward. The algorithm in [17] decides whether there exists such a path between two configurations, but it runs in time exponential in obstacle complexity. The algorithm in [1] computes a path in polynomial time under the assumptions that all obstacles are convex, and their boundaries have a curvature radius greater than the minimum turning radius of the point (so-called moderate obstacles). Other polynomial algorithms [5] require some sort of discretization.

4.3.4.2 Planning for Robots with Differential Constraints

A robot's motion can generally be described by a set of nonlinear equations of motion $\dot{x} = f(x, u)$, together with some constraints $g(x, \dot{x}) \le 0$. x is the robot's state vector and u is a vector of control inputs. The robot can be abstracted into a point that moves in a state space. The state space is a superset of the aforementioned configuration space. For example, a car moving in a plane can be modeled using a five-dimensional state space where $x = (x_r, y_r, \theta, v, s)$. x_r and y_r provide the position of a reference point on the car, and together with orientation θ, they constitute the car's configuration. For the state description, we need the components for the linear velocity v and angle of the steering s. In this model, the car is controlled by two inputs: $u = (a, b)$. a is the car's linear acceleration and b is the car's steering velocity. Notice that this system can have bounds in its acceleration which is a second order constraint.

Sampling-based techniques are becoming increasingly popular for tackling problems with kinodynamic and in general differential constraints [61,62,72]. For this reason, this section will not extend beyond sampling-based planners. For planning under differential constraints, the planner is now called to search the state space described earlier. Since motion is constrained by nonlinear equations, collision checking is generalized to also check for invalid states (e.g., due to constraint violations). Moreover, the sampling process is modified. The planner typically produces a random set of controls that are applied at a given state for some time in an attempt to move the system toward a newly sampled state. For this process, the planner requires a function that can integrate the equations of motion forward in time. Notice that all the produced trajectories are feasible by construction. This is an advantage, since the planner avoids dealing with controllability issues. The disadvantage is that it becomes hard to drive the system to an exact goal state. To overcome this difficulty, it is sometimes easier to define a whole region of states that are acceptably close to the goal.

Adaptations of tree-based planners, such as RRT and EST, proved very successful in kinodynamic problems (for example, see [73,74]). In addition, many newer planners are specifically designed for problems with differential constraints [70,75] although they can apply to simpler problems as well. An interesting observation is that the planner only needs the integrator function as a black box. For this reason, it is possible to use a physical simulator to do the integration of motion [92]. In this way, it is possible to model more realistic effects and constraints such as inertia, gravity, and friction. Moreover, it becomes possible to do planning for systems for which the exact equations are not explicitly written.

4.3.5 Extensions to Motion Planning

Most of the topics discussed so far focused on difficulties presented by the nature of a robot itself. In practice, there are many interesting problems that extend beyond planning for a robot in a static known environment. In the following paragraphs, an attempt is made to cover the most important ones. For more detailed descriptions, the reader is referred to [30,60–62].

4.3.5.1 Dynamic Workspace

In the presence of moving obstacles, one can no longer plan a robot's motion as a mere geometric path. The path must be indexed by time and is then called a trajectory. The simplest scenario is when the trajectories of the moving obstacles are known in advance or can be accurately predicted. In that case, planning can be done in the configuration×time space ($\mathcal{C} \times [0, +\infty)$) of the robot. All workspace obstacles map to static forbidden regions in that space. A free trajectory is a free path in that space whose tangent at every point points positively along the time axis (or within a more restricted cone, if the robot's velocity modulus is bounded).

Computing a free trajectory for a rigid object in three-space among arbitrarily moving obstacles (with known trajectories) is PSPACE-hard if the robot's velocity is bounded, and NP-hard otherwise [48]. The problem remains NP-hard for a point robot moving with bounded velocity in the plane among convex polygonal obstacles translating at constant linear velocities [10]. A complete planning algorithm is given in [48] for a polygonal robot that translates in the plane among polygonal obstacles. The obstacles translate at fixed velocities. This algorithm takes time exponential in the number of moving obstacles and polynomial in the total number of edges of the robot and the obstacles.

In realistic scenarios, exact information about moving obstacle trajectories is not available. In those cases, a robot has to make conservative assumptions about where the obstacles will be in the future. For example, for problems on a plane, moving obstacles with bounded velocity can be anywhere within a disk whose radius grows in time. As long as the robot's path does not enter any of those disks at any moment in time, the robot's path will be collision free. This idea is described in [65] which also shows how to find time-optimal paths. Another approach that tries to address these issues is by employing an adaptive online replanning strategy [67,70]. In such approaches, the planner operates in a closed loop and is interleaved with sensing operations that try to keep track of the motions of the obstacles. A partial plan for some small time horizon is computed and implemented. Then using the latest available information, the planner is called again to compute a new plan. Such replanning techniques are also known as online planning and are discussed in a bit more detail in the relevant section below. For further reference, [94] is closely related to the topics discussed in this subsection.

4.3.5.2 Planning for Multiple Robots

The case of multiple robots can be trivially addressed by considering them as the components of a single robot, that is, by planning a path in the cross product of their configuration spaces. This product is called the composite configuration space of the robots, and the approach is referred to as centralized planning. This approach can be used to take advantage of the completeness properties that single robot algorithms have. Owing to the increased complexity, powerful planners need to be used (e.g., [78]).

One may try to reduce complexity by separately computing a path for each robot, before tuning the robots' velocities along their respective paths to avoid inter-robot collision (decoupled planning) [64]. This approach can be helped by assigning priorities to robots according to some measure of importance. Then, each robot has to respect the higher priorities robots when planning. Although inherently incomplete, decoupled planning may work well in some practical applications.

Planning for multiple robots can become really challenging if it has to be done distributedly, with each robot restricted to its own configuration space. This is a reasonable situation in practice where each robot may have its own limitations and goals and can interact only with the robots that are close to it. In [64], the robots move toward their independent goals. During the process, when robots come close, they form dynamic networks. The robots within a network solve a centralized motion planning problem to avoid collisions between each other using all available information within the network. Whenever robots get out of some communication range, networks are dissolved. For this kind of problems, planning can greatly benefit from utilizing the communication capabilities that robots typically have. In [76,77], a coordination framework for mulitrobot planning is described. In this work, each robot plans in its own state space and uses only local information obtained through communication with neighboring robots. This approach can deal with robots that have nontrivial kinodynamic constraints. Moreover, it is possible to guarantee collision avoidance between robots and static obstacles while the robots can move as a connected network. Coordination is achieved with a distributed message passing scheme.

4.3.5.3 Planning with Movable Objects

Many robot tasks require from the robot to interact with its environment by moving an object. Such objects are called movable objects and cannot move by themselves; they must be moved by a robot. These problems fall into the category of manipulation planning.

In [53], the robot A and the movable object M are both convex polygons in a polygonal workspace. The goal is to bring A and M to specified positions. A can only translate. To grasp M, A must have one of its edges that exactly coincides with an edge of M. While A grasps M, they move together as one rigid object. An exact cell decomposition algorithm is given that runs in $O(n^2)$ time after $O(n^3 \log^2 n)$ preprocessing, where n is the total number of edges in the workspace, the robot, and the movable object. An extension of this problem allowing an infinite set of grasps is solved by an exact cell decomposition algorithm in [3].

Heuristic algorithms have also been proposed. The planner in [29] first plans the path of the movable object M. During that phase, it only verifies that, for every configuration taken by M, there exists at least one collision-free configuration of the robot where it can hold M. In the second phase, the planner determines the points along the path of M where the robot must change grasps. It then computes the paths where the robot moves alone to (re)grasp M. The paths of the robot when it carries M are obtained through inverse kinematics. This planner is not complete, but it has solved complex tasks in practice.

Another scenario is that of a robot that needs to move inside a building where the doors are blocked by obstacles that have to be moved out of the way. Questions of interest are, how many obstacles must be moved, in which order must these obstacles be moved, and where should the robot put them so as not to block future motions. [66,91] presents a planner that can solve problems in class LP. An $LP_k \subseteq LP$ problem is one where a series of $k - 1$ obstacles must be moved before an obstacle can be reached and moved, so that a room becomes accessible through a blocked door.

4.3.5.4 Online Planning and Exploration

Online planning addresses the case where the workspace is initially unknown or partially unknown. As the robot moves, it acquires new partial information about the workspace through sensing. A motion plan is generated using the partial information that is available and updated as new information is acquired. With this planning strategy, a robot exhibits a "reactive or adaptive behavior" and can also move in the presence of moving obstacles or other robots.

In [67,70], the robot initially preprocesses the workspace and creates a roadmap as described in Section 4.3.3.4. Then the robot finds a path to its goal and starts following it, until an unexpected change, such as a moving obstacle, is sensed. Then, the robot quickly revises its path online, and chooses a new path on the roadmap that avoids regions that are invalidated by the obstacle. Another useful application where online planning is needed is that of exploration of unknown environments [89]. In that case, it is necessary that a robot plans online and revises its plan periodically after receiving new sensor information. Exploration can be done more efficiently when multiple robots are used. In [76], a team of robots explores an unknown environment. Each robot plans online and communicates with neighboring robots to coordinate their actions before computing a new plan. In this way, robots that have kinodynamic constraints manage to complete their task while avoiding all collisions between robots or workspace obstacles.

The main difficulty in online planning is that the planner has to run under a time budget and must generally be very fast in producing a new plan. Moreover, since the overall motion is a concatenation of small motions that use only partial information without a global view of the workspace, the quality of the paths can be bad, and it is very hard to establish path optimality.

A way of evaluating an online planner is competitive analysis. The competitive ratio of an online planner is the maximal ratio (over all possible workspaces) between the length of the path generated by the online algorithm and the length of the shortest path [44]. Competitive analysis is not restricted to path length and can be applied to other measures of performance as well.

4.3.5.5 Optimal Planning

There has been considerable research in computational geometry on finding shortest Euclidean paths [63] (Chapter 26), but minimal Euclidean length is usually not the most suitable criterion in robotics. Rather, one wishes to minimize execution time, which requires taking the robot's dynamics into account.

In optimal-time control planning, the input is a geometric free path τ parameterized by $s \in [0, L]$, the distance traveled from the starting configuration. The problem is to find the time parameterization $s(t)$ that minimizes travel time along τ, while satisfying actuator limits.

The dynamic equation of motion of a robot arm with m dofs can be written as $M(q)\ddot{q} + V(\dot{q}, q) + G(q) = \Gamma$, where q, $\dot{q}$, and $\ddot{q}$, respectively, denote the robot's configuration, velocity, and acceleration [12]. M is the $m \times m$ inertia matrix of the robot, V the m-vector (quadratic in $\dot{q}$) of centrifugal and Coriolis forces, and G the m-vector of gravity forces. Γ is the m-vector of the torques applied by the joint actuators.

Using the fact that the robot follows τ, this equation can be rewritten in the form: $m\ddot{s} + v\dot{s}^2 + g = \Gamma$, where m, v, and g are derived from M, V, and G, respectively. Minimum-time control planning becomes a two-point boundary value problem: find $s(t)$ that minimizes $t_f = \int_0^L ds/\dot{s}$, subject to $\Gamma = m\ddot{s} + v\dot{s}^2 + g$, $\Gamma_{\min} \leq \Gamma \leq \Gamma_{\max}$, $s(0) = 0$, $s(t_f) = L$, and $\dot{s}(0) = \dot{s}(L) = 0$. Numerical techniques solve this problem by finely discretizing the path τ.

To find a minimal-time trajectory, a common approach is to first plan a geometric free path and then iteratively deform this path to reduce travel time. Each iteration requires checking the new path for collision and recomputing the optimal-time control. No bound has been established on the running time of this approach or the goodness of its outcome. The problem is NP-hard for a point robot under Newtonian mechanics in three-space. The approximation algorithm in [13] computes a trajectory ϵ-close to optimal in time polynomial in both $1/\epsilon$ and the workspace complexity.

4.3.5.6 Dealing with Uncertainties in Motion and Sensing

In all of the topics discussed so far, it has been implicitly assumed that the robot has accurate knowledge about its own state at the present and possibly future times after executing some action. Unfortunately, this is not the case in practice. Real robots have a number of inherent mechanical imperfections that affect the result of an action and introduce motion errors. For example, the actuators introduce errors when executing a motion command, and there is usually discrepancy between how much the robot thinks it has moved as opposed to how much it has actually moved. The problem is more pronounced in mobile robots that have odometry errors produced by unaccounted for wheel slipage. The effect of such errors is that the robot no longer has the exact knowledge of its own state. Instead, it has to somehow infer its state from the available sensing information. To add to the overall problem, processing sensor information is computationally intensive and error-prone due to imperfect sensors. (Notice that reconstructing shapes of objects using simple sensors has attracted considerable attention [50].)

All these issues amount to the very fundamental problem of dealing with uncertainty in motion and sensing. A very simple example was already shown in Figure 4.2. A broad area of robotics algorithms focus on processing all the available information to reduce this uncertainty. There exist some algorithmic techniques for planning with uncertainty, e.g., [11,15] but maybe the most common framework for handling uncertainty is based on the theory of Bayesian estimation [60]. The robot is assumed to be at some state x_t which is unknown. The robot maintains a probability distribution $bel(x)$ called belief, that represents how likely it is that the true state is x. For Bayesian estimation, some system specific characteristics must be modeled. In particular, a state transition function must be given to describe how likely it is to move to some new state x_{t+1} when starting at some state x_t and performing an action/motion. Moreover, a measurement model must be available to describe how likely it is to take some specific measurement when being at some state. Given the belief at time t together with the latest action and sensor measurement, Bayesian estimation performs two steps in an attempt to compute the next belief $bel(x_{t+1})$. First, there is a prediction step where using $bel(x_t)$ and the state transition model, an estimate of $bel(x_{t+1})$ is computed. Then, this estimate is refined using the measurement model to validate the predictions against the latest sensor measurement. This technique is very general and powerful and has been successfully applied in many robotics estimation problems. Below, concrete examples of the most important estimation problems are briefly described. An extensive reference for such problems is [60].

4.3.5.6.1 Localization

One of the most fundamental problems in robotics is that of robot localization [22]. The goal is to maintain an accurate estimation of the robot's state at all times as the robot moves in a known environment for which a map is available. The easiest version is when the initial state is known with certainty. Then the problem is called position tracking, and the robot just needs to compensate for local uncertainty introduced every time it attempts to move, e.g., [33]. A very successful technique for such problem is the extended Kalman filter (EKF) [60,79]. A more challenging variant is called global localization [52,57,80]. This is the problem of estimating the robot's state when the initial location is also unknown. This kind of problems are usually solved using particle filters [59,60].

4.3.5.6.2 Mapping

The assumption that the robot's map is known does not always hold. In fact, the mapping problem tries to address exactly this issue. Here, the robot is assumed to have an accurate mechanism for tracking its position (e.g., using a GPS). Its target is to use sensor information to build an accurate map of the environment [55]. The map is usually described by the locations of all the features in the environment, called landmarks.

In a sense, this is the complementary problem to localization. In a mapping problem, the state vector describes the map, and $bel(x)$ is the distribution that describes how likely each version of a map is [82].

4.3.5.6.3 SLAM

One of the most exciting and challenging class of estimation problems is that of simultaneous localization and mapping or SLAM, where localization and mapping have to be tackled simultaneously. This is a hard problem since in most cases there is a demand to perform all the computations online as the robot moves. Moreover, this ends up being a "chicken and egg" problem where trusting a map would increase localization accuracy, while trusting localization can give better estimates about the position of the obstacles.

The typical difficulty for solving any SLAM problem is in data association. The current estimated map is represented by the locations of certain landmarks. As the robot moves, the sensors extract new sensed features, and the algorithm needs to decide whether a sensed feature represents a new landmark that has never been observed before or not. To solve this correspondence problem between the sensed features and the current list of observed landmarks, a standard way is to use maximum likelihood estimation techniques. Another problem that is also a benchmark for the effectiveness of SLAM algorithms is that of loop closure. As the robot is moving, it often happens that after spending a lot of time in some newly discovered part of the environment the robot returns to an area it has visited before, and closes a loop. The ability to detect that indeed a loop was closed can be a challenging problem especially for online SLAM algorithms. Finally, for SLAM problems, the state that needs to be estimated tends to be very high dimensional, and this leads to expensive computations for updating the state. There are a number of state-of-the-art algorithms for SLAM [58,81,83–85], and the area is an active research field.

4.4 Distance Computation

The efficient computation of (minimum) distances between bodies in two- and three-space is a crucial element of many algorithms in robotics [63].

Algorithms have been proposed to efficiently compute distances between two convex bodies. In [14], an algorithm is given which computes the distance between two convex polygons P and Q (together with the points that realize it) in $O(\log p + \log q)$ time, where p and q denote the number of vertices of P and Q, respectively. This time is optimal in the worst case. The algorithm is based on the observation that the minimal distance is realized between two vertices or between a vertex and an edge. It represents P and Q as sequences of vertices and edges and performs a binary search that eliminates half of the edges in at least one sequence at each step. For treatment of the analogous problem in three-dimensions see [100]. A widely tested numerical descent technique is described in [18] to compute the distance between two convex polyhedra; extensive experience indicates that it runs in approximately linear time in the total complexity of the polyhedra. This observation has recently been theoretically substantiated in [101].

Most robotics applications, however, involve many bodies. Typically, one must compute the minimum distance between two sets of bodies, one representing the robot and the other the obstacles. Each body can be quite complex, and the number of bodies forming the obstacles can be large. The cost of accurately computing the distance between every pair of bodies is often prohibitive. In that context, simple bounding volumes, such as parallelepipeds and spheres, have been extensively used to reduce computation time. They are often coupled with hierarchical decomposition techniques, such as octrees, boxtrees, or sphere trees (for an example, see [46]). These techniques make it possible to rapidly eliminate pairs of bodies that are too far apart to contribute the minimum distance.

When motion is involved, incremental distance computation has been suggested for tracking the closest points on a pair of convex polyhedra [34,86,87]. It takes advantage of the fact that the closest features (faces, edges, vertices) change infrequently as the polyhedra move along finely discretized paths.

4.5   Research Issues and Summary

In this chapter, we have introduced robot algorithms as abstract descriptions of processes consisting of motions and sensing operations in the physical space. Robot algorithms send commands to actuators and sensors in order to control a subset of the real world, the workspace, despite the fact that the workspace is subject to the imperfectly modeled laws of nature. Robot algorithms uniquely blend controllability, observability, computational complexity, and physical complexity issues, as described in Section 4.2. Research on robot algorithms is broad and touches many different areas. In Section 4.3, we have surveyed a number of selected areas in which research has been particularly active: part manipulation (grasping, fixturing, and feeding), assembly sequencing, motion planning (including basic path planning, nonholonomic planning, dynamic workspaces, multiple robots, and optimal-time planning), and sensing.

Many of the core issues reviewed in Section 4.2 have been barely addressed in currently existing algorithms. There is much more to understand in how controllability, observability, and complexity interact in robot tasks. The interaction between controllability and complexity has been studied to some extent for nonholonomic robots. The interaction between observability (or recognizability) and complexity has been considered in motion planning with uncertainty. But, in both cases, much more remains to be done.

Concerning the areas studied in Section 4.3, several specific problems remain open. We list a few below (by no means is this list exhaustive):

- Given a workspace W, find the optimal design of a robot arm that can reach everywhere in W without collision. The three-dimensional case is largely open. An extension of this problem is to design the layout of the workspace so that a certain task can be completed efficiently.
- Given the geometry of the parts to be manipulated, predict feeders' throughputs to evaluate alternative feeder designs. In relation to this problem, simulation algorithms have been used to predict the pose of a part dropped on a flat surface [39].

- In assembly planning, the complexity of an NDBG grows exponentially with the number of parameters that control the allowable motions. Are there situations where only a small portion of the full NDBG needs to be constructed?
- Develop efficient sampling techniques for searching the configuration space of robots with many degrees of freedom in the context of the scheme given in [4].
- Establish a nontrivial lower bound on the complexity of planning for a nonholonomic robot that is not locally controllable.

4.6 Further Information

For an introduction to robot arm kinematics, dynamics, and control, see [12]. More on robotic manipulation can be found in [90]. Robot motion planning and its variants are discussed in a number of books [30,61,62,71]. Research in all aspects of robotics is published in the *IEEE Transactions of Robotics and Automation* and the *International Journal of Robotics Research*, as well as in the proceedings of the *IEEE International Conference on Robotics and Automation* and the *International Symposium on Robotics Research* [19]. The *Workshop on Algorithmic Foundations of Robotics* (see, e.g., [21,32]) emphasizes algorithmic issues in robotics. Several computational geometry books contain sections on robotics or motion planning [6].

Defining Terms

Basic path planning problem: Compute a free or semifree path between two input configurations for a robot moving in a known and static workspace.

C-Obstacle: Given an obstacle B_i, the subset CB_i of the configuration space C such that, for any $q \in CB_i$, $A(q)$ intersects B_i. The union $CB = \cup_i CB_i$ plus the configurations that violate the mechanical limits of the robot's joints is called the C-obstacle region.

Complete motion planner: A planner guaranteed to find a (semi)free path between two given configurations whenever such a path exists, and to notify that no such path exists otherwise.

Configuration: Any mathematical specification of the position and orientation of every body composing a robot A, relative to a fixed coordinate system. The configuration of a single body is called a placement or a pose.

Configuration space: Set C of all configurations of a robot. For almost any robot, this set is a smooth manifold.

Differential constraint: A motion constraint that is both nonholonomic (nonintegrable) and kinodynamic (at least second order).

Free path: A path in free space.

Free space: The complement of the C-obstacle region in C, that is, $C \backslash CB$.

Kinodynamic constraint: A second order constraint in the configuration space, that is, for problems where both velocity and acceleration bounds are specified for the robot's motion.

Linkage: A collection of rigid objects, called links, in which some pairs of links are connected by joints (e.g., revolute and/or prismatic joints). Most industrial robot arms are serial linkages with actuated joints.

Nonholonomic constraint: Constraints in a robot's motion that cannot be converted into constraints that involve no derivatives. For example, nonintegrable velocity constraints in the configuration space.

Number of degrees of freedom: The dimension m of C.

Obstacle: The workspace W is often defined by a set of obstacles (bodies) B_i ($i = 1, \ldots, q$) such that $W = \mathbb{R}^k \backslash \bigcup_1^q B_i$.

Path: A continuous map $\tau : [0, 1] \to C$.

Probabilistically complete: An algorithm is probabilistically complete if given enough time, the probability of finding a solution goes to 1 when a solution exists.

Semifree path: A path in the closure of free space.

State space: The set of all robot's states. A superset of the configuration space that captures the robot's dynamics as well.

Trajectory: Path indexed by time.

Workspace: A subset of the two- or three-dimensional physical space modeled by $W \subset \mathbb{R}^k, k = 2, 3$.

Workspace complexity: The total number of features (vertices, edges, faces, etc.) on the boundary of the obstacles.

References

1. Agarwal, P.K., Raghavan, P., and Tamaki, H., Motion planning for a steering-constrained robot through moderate obstacles. *Proceedings of the 27th Annual ACM Symposium on Theory of Computing,* ACM Press, New York, pp. 343–352, 1995.
2. Akella, S., Huang, W., Lynch, K., and Mason, M.T., Planar manipulation on a conveyor with a one joint robot. In *Robotics Research,* Giralt, G. and Hirzinger, G., Eds., Springer-Verlag, Berlin, pp. 265–276, 1996.
3. Alami, R., Laumond, J.P., and Siméon, T., Two manipulation algorithms. In *Algorithmic Foundations of Robotics,* Goldberg, K.Y., Halperin, D., Latombe, J.C., and Wilson, R.H., Eds., AK Peters, Wellseley, MA, pp. 109–125, 1995.
4. Barraquand, J., Kavraki, L.E., Latombe, J.C., Li, T.Y., Motwani, R., and Raghavan, P., A Random sampling framework for path planning in large-dimensional configuration spaces. *Int. J. Robot. Res.,* 16(6), 759–774, 1997.
5. Barraquand, J. and Latombe, J.C., Nonholonomic multibody mobile robots: Controllability and motion planning in the presence of obstacles, *Algorithmica,* 10(2-3-4), 121–155, 1993.
6. de Berg, M., van Kreveld, M., Overmars, M., and Schwarzkopf, O., *Computational Geometry: Algorithms and Applications.* Springer-Verlag, New York, 2000.
7. Boddy M. and Dean T.L., Solving time-dependent planning problems. *Proceedings of the 11th International Joint Conference on Artificial Intelligence,* Detroit, MI, pp. 979–984, 1989.
8. Briggs, A.J., Efficient geometric algorithms for robot sensing and control. Report No. 95-1480, Department of Computer Science, Cornell University, Ithaca, NY, 1995.
9. Brost, R.C. and Goldberg, K.Y., Complete algorithm for designing planar fixtures using modular components. *IEEE Trans. Syst. Man Cyber.,* 12, 31–46, 1996.
10. Canny, J.F., *The Complexity of Robot Motion Planning.* MIT Press, Cambridge, MA, 1988.
11. Canny, J.F., On computability of fine motion plans, *Proceedings of the 1989 IEEE International Conference on Robotics and Automation,* Scottsdale, AZ, pp. 177–182, 1989.
12. Craig, J.J., *Introduction to Robotics: Mechanics and Control.* Addison-Wesley, Reading, MA, 2004.
13. Donald, B.R., Xavier, P., Canny, J.F., and Reif, J.H., Kinodynamic motion planning. *J. ACM,* 40, 1048–1066, 1993.
14. Edelsbrunner, H., Computing the extreme distances between two convex polygons. *J. Algorithms,* 6, 213–224, 1985.
15. Erdmann, M., Using backprojections for fine motion planning with uncertainty. *Int. J. Robot. Res.,* 5, 19–45, 1986.

16. Erdmann, M. and Mason, M.T., An exploration of sensorless manipulation. *IEEE Trans. Robot. Autom.*, 4(4), 369–379, 1988.
17. Fortune, S. and Wilfong, G.T., Planning constrained motions. *Proceedings of the 20th Annual Symposium on Theory of Computing*, ACM Press, New York, pp. 445–459, 1988.
18. Gilbert, E.G., Johnson, D.W., and Keerthi, S.S., A fast procedure for computing distance between complex objects in three-dimensional space. *IEEE Trans. Robot. Autom.*, 4, 193–203, 1988.
19. Giralt, G. and Hirzinger, G., Eds., *Robotics Research*, Springer-Verlag, Berlin, 1996.
20. Goldberg, K.Y., Orienting polygonal parts without sensors. *Algorithmica*, 10(2-3-4), 201–225, 1993.
21. Goldberg, K.Y., Halperin, D., Latombe, J.C., and Wilson, R.H., Eds., *Algorithmic Foundations of Robotics*, AK Peters, Wellesley, MA, 1995.
22. Guibas, L., Motwani, R., and Raghavan, P., The robot localization problem in two dimensions. *SIAM J. Comp.*, 26(4), 1121–1138, 1996.
23. Halperin, D. Arrangements. In *Handbook of Discrete and Computational Geometry*, Goodman, J.E. and O'Rourke, J., Eds., CRC Press, Boca Raton, FL, pp. 389–412, 2004.
24. Halperin, D. and Sharir, M., Near-quadratic algorithm for planning the motion of a polygon in a polygonal environment. *Discrete Comput. Geom.*, 16, 121–134, 2004.
25. Kant, K.G. and Zucker, S.W., Toward efficient trajectory planning: Path velocity decomposition. *Int. J. Robot. Res.*, 5, 72–89, 1986.
26. Kavraki, L.E. and Kolountzakis, M.N., Partitioning a planar assembly into two connected parts is NP-complete. *Inform. Process. Lett.*, 55, 159–165, 1995.
27. Kavraki, L.E., Švestka, P., Latombe, J.C., and Overmars, M., Probabilistic roadmaps for fast path planning in high dimensional configuration spaces. *IEEE Trans. Robot. Autom.*, 12, 566–580, 1996.
28. Khatib, O. Real-time obstacle avoidance for manipulators and mobile robots. *Int. J. Robot. Res.*, 5, 90–98, 1986.
29. Koga, Y., Kondo, K., Kuffner, J., and Latombe, J.C., Planning motions with intentions. *Proc. ACM SIGGRAPH'94*, ACM Press, Orlando, FL, pp. 395–408, 1994.
30. Latombe J.C., *Robot Motion Planning*, Kluwer Academic Publishers, Boston, MA, 1991.
31. Laumond, J.P., Jacobs, P., Taix, M., and Murray, R., A motion planner for nonholonomic mobile robots. *IEEE Trans. Robotic. Autom.*, 10, 577–593, 1994.
32. Laumond, J.P. and Overmars, M., Eds., *Algorithms for Robot Motion and Manipulation*, AK Peters, Wellesley, MA, 1997.
33. Lazanas, A. and Latombe, J.C., Landmark-based robot navigation. *Algorithmica*, 13, 472–501, 1995.
34. Lin, M.C. and Canny, J.F., A fast algorithm for incremental distance computation. *Proceedings of the 1991 IEEE International Conference on Robotics and Automation*, Sacramento, CA, pp. 1008–1014, 1991.
35. Lozano-Pérez, T., Spatial planning: A configuration space approach, *IEEE Trans. Comput.*, 32(2), 108–120, 1983.
36. Lozano-Pérez, T., Mason, M.T., and Taylor, R.H., Automatic synthesis of fine-motion strategies for robots, *Int. J. Robot. Res.*, 3(1), 3–24, 1984.
37. Lumelsky, V., A comparative study on the path length performance of maze-searching and robot motion planning algorithms. *IEEE Trans. Robot. Autom.*, 7, 57–66, 1991.
38. Mason, M.T., Mechanics and planning of manipulator pushing operations, *Int. J. Robot. Res.*, 5(3), 53–71, 1986.
39. Mirtich, B., Zhuang, Y., Goldberg, K., Craig, J.J., Zanutta, R., Carlisle, B., and Canny, J.F., Estimating pose statistics for robotic part feeders. *Proceedings of the 1996 IEEE International Conference on Robotics and Automation*, Minneapolis, MN, pp. 1140–1146, 1996.
40. Mishra B., Schwartz, J.T., and Sharir, M., On the existence and synthesis of multifinger positive grips, *Algorithmica*, 2, 541–558, 1987.
41. Natarajan, B.K., On planning assemblies. *Proceedings of the 4th Annual Symposium on Computational Geometry*, ACM Press, New York, pp. 299–308, 1988.

42. Nguyen, V.D., Constructing force-closure grasps. *Int. J. Robot. Res.*, 7, 3–16, 1988.
43. Nourbakhsh, I.R., Interleaving planning and execution. PhD thesis. Department of Computer Science, Stanford University, Stanford, CA, 1996.
44. Papadimitriou, C.H. and Yannakakis, M., Shortest paths without a map. *Theor. Comput. Sci.*, 84, 127–150, 1991.
45. Ponce, J., Sudsang, A., Sullivan, S., Faverjon, B., Boissonnat, J.D., and Merlet, J.P., Algorithms for computing force-closure grasps of polyhedral objects. In *Algorithmic Foundations of Robotics*, Golberg, K.Y., Halperin, D., Latombe, J.C., and Wilson, R.H., Eds., AK Peters, Wellesley, MA, pp. 167–184, 1995.
46. Quinlan, S., Efficient distance computation between non-convex objects. *Proceedings of the 1994 IEEE Conference Robotics and Automation*, San Diego, CA, pp. 3324–3329, 1994.
47. Reif J.H., Complexity of the Mover's problem and generalizations. *Proceedings of the 20th Annual Symposium on Foundations of Computer Science*, IEEE Computer Society, Washington, DC, pp. 421–427, 1979.
48. Reif, J.H. and Sharir, M., Motion planning in the presence of moving obstacles. *J. ACM*, 41(4), 764–790, 1994.
49. Schwartz, J.T. and Sharir, M., On the 'Piano Movers' problem: II. General techniques for computing topological properties of real algebraic manifolds, *Adv. Appl. Math.*, 4, 298–351, 1983.
50. Skiena, S.S., Geometric reconstruction problems. In *Handbook of Discrete and Computational Geometry*, Goodman, J.E. and O'Rourke, J., Eds., CRC Press, Boca Raton, FL, pp. 481–490, 2004.
51. Snoeyink, J. and Stolfi, J., Objects that cannot be taken apart with two hands. *Discrete Comput. Geom.*, 12, 367–384, 1994.
52. Talluri, R. and Aggarwal, J.K., Mobile robot self-location using model-image feature correspondence. *IEEE Trans. Robot. Autom.*, 12, 63–77, 1996.
53. Wilfong, G.T., Motion planning in the presence of movable objects. *Ann. Math. Artif. Intell.*, 3, 131–150, 1991.
54. Wilson, R.H. and Latombe, J.C., Reasoning about mechanical assembly. *Artif. Intell.*, 71, 371–396, 1995.
55. Zhang, Z. and Faugeras, O., A 3D world model builder with a mobile robot. *Int. J. Robot. Res.*, 11, 269–285, 1996.
56. Zhuang, Y., Goldberg, K.Y., and Wong, Y., On the existence of modular fixtures. *Proceedings of the IEEE International Conference on Robotics and Automation*, San Diego, CA, pp. 543–549, 1994.
57. Thrun, S., Fox, D., Burgard, W., and Dellaert, F., Robust Monte Carlo localization for mobile robots. *Artif. Intell.*, 128, 99–141, 2000.
58. Montemerlo, M., Thrun, S., Koller, D., and Wegbreit, B., FastSLAM: A factored solution to the simultaneous localization and mapping problem, *Proceedings of the AAAI National Conference on Artificial Intelligence*, Edmonton, AB, Canada, pp. 593–598, 2002.
59. Ferris, B., Hähnel, D., and Fox, D., Gaussian processes for signal strength-based location estimation. *Proceedings of Robotics Science and Systems*, Philadelphia, 2006.
60. Thrun, S., Burgard, W., and Fox, D., *Probabilistic Robotics*, MIT Press, Cambridge, MA, 2005.
61. LaValle, S.M., *Planning Algorithms*, Cambridge University Press, New York, 2006.
62. Choset, H., Lynch, K.M., Hutchinson, S., Kantor, G., Burgard, W., Kavraki, E.L., and Thrun, S., *Principles of Robot Motion*, MIT Press, Cambridge, MA, 2005.
63. Lin, M.C. and Manocha, D., Collision and proximity queries. In *Handbook of Discrete and Computational Geometry*, 2nd Ed., CRC Press, Boca Raton, FL, pp. 787–807, 2004.
64. Clark, C.M., Rock, S.M., and Latombe, J.-C., Motion planning for multiple mobile robots using dynamic networks, *Proc. IEEE Int. Conf. Robot. Autom.*, 3, pp. 4222–4227, 2003.
65. van den Berg, J.P. and Overmars, M.H., Planning the shortest safe path amidst unpredictably moving obstacles, *Proceedings of 6th International Workshop on Algorithmic Foundations of Robotics*, New York, 2006.

66. Nieuwenhuisen, D., Stappen, A.F. van der, and Overmars, M.H., An effective framework for path planning amidst movable obstacles, *Proceedings of International Workshop on the Algorithmic Foundations of Robotics*, New York, 2006.

67. van den Berg, J., Ferguson, D., and Kuffner, J., Anytime path planning and replanning in dynamic environments, *Proceedings of the IEEE International Conference on Robotics and Automation*, Orlando, FL, pp. 2366–2371, 2006.

68. Kuffner, J. and LaValle, S.M., RRT-connect: An efficient approach to single-query path planning, *Proceedings of the IEEE International Conference Robotics and Automation*, San Francisco, CA, pp. 995–1001, 2000.

69. Hsu, D., Latombe, J.C., and Motwani, R., Path planning in expansive configuration spaces. *Int. J. Comput. Geo. Appl.*, 9(4/5), 495–512, 1998.

70. Bekris, K.E. and Kavraki, L.K., Greedy but safe replanning under kinodynamic constraints, *Proc. IEEE Int. Conf. Robot. Autom.*, 704–710, 2007.

71. Laumond, J.P., *Robot Motion Planning and Control*, Springer-Verlag, Berlin, 1998.

72. Tsianos, K.I., Sucan, I.A., and Kavraki, L.E., Sampling-based robot motion planning: Towards realistic applications, *Comp. Sci. Rev.*, 1, 2–11, August 2007.

73. Kindel, R., Hsu, D., Latombe, J.C., and Rock, S., Kinodynamic motion planning amidst moving obstacles, *Proc. IEEE Int. Conf. Robot. Autom.*, 1, 537–543, 2000.

74. LaValle, S.M. and Kuffner, J., Randomized kinodynamic planning, *Int. J. Robot. Res.*, 20(5), 378–400, May 2001.

75. Ladd, A.M. and Kavraki, L.E. Motion planning in the presence of drift, underactuation and discrete system changes. In *Robotics: Science and Systems*, MIT, Boston, MA, June 2005.

76. Bekris, K.E., Tsianos, K.I., and Kavraki, L.E., Distributed and safe real-time planning for networks of second-order vehicles, *International Conference in Robot Communication and Coordination*, Athens, Greece, October 2007.

77. Bekris, K.E., Tsianos, K.I., and Kavraki, L.E., A decentralized planner that guarantees the safety of communicating vehicles with complex dynamics that replan online, International Conference on Intelligent Robots and Systems, Athens, Greece, October 2007.

78. Plaku, E., Bekris, K.E., Chen, B.Y., Ladd, A.M., and Kavraki, L.E., Sampling-based roadmap of trees for parallel motion planning, *IEEE Trans. Robot.*, 21(4), 597–608, August 2005.

79. Smith, R.C. and Cheeseman, P., On the representation and estimation of spatial uncertainty. *IJRR*, 5(4), 5668, 1987.

80. Kwok,C., Fox, D., and Meila, M., Real-time Particle Filters, *Proc. IEEE*, 92(2), 469–484, 2004.

81. Davison, A.J., Real-time simultaneous localisation and mapping with a single camera. *9th ICCV*, Vol. 2, IEEE Computer Society, Washington, DC, pp. 1403–1410, 2003.

82. Thrun, S., Robotic mapping: A survey. In *Exploring Artificial Intelligence in the New Millenium*, G. Lakemeyer and B. Nebel, Eds., Morgan Kaufmann Publishers, San Francisco, CA, 2002.

83. Dellaert, F. and Kaess, M. Square root SAM: Simultaneous localization and mapping via square root information smoothing, *Int. J. Robot. Res.*, 25(12), 1181–1203, 2006.

84. Durrant-Whyte, H. and Bailey, T. Simultaneous localization and mapping: Part I, *IEEE Robotics and Automation Magazine*, 13(2), 99–110, June 2006.

85. Bailey, T. and Durrant-Whyte, H., Simultaneous localization and mapping: Part II, *IEEE Robot. Autom. Mag.*, 13(2), 110–117, 2006.

86. Larsen, E., Gottschalk, S., Lin, M.C., and Manocha, D., Fast distance queries using rectangular swept sphere volumes, *IEEE International Conference on Robotics and Automation*, San Francisco, CA, vol. 4, pp. 3719–3726, 2000.

87. Zhang, L., Kim, Y.J., and Manocha, D., C-DIST: Efficient distance computation for rigid and articulated models in configuration space. *Proceedings of the ACM Symposium on Solid and Physical Modeling*, Beijing, China, pp. 159–169, 2007.

88. Schwarzer, F., Saha, M., and Latombe, J.C., Adaptive dynamic collision checking for single and multiple articulated robots in complex environments, *IEEE Trans. Robot.*, 21(3), 338–353, 2005.

89. Baker, C.R., Morris, A., Ferguson, D.I., Thayer, S., Wittaker, W., Omohundro, Z., Reverte, C.F., Hähnel, D., and Thrun, S., A Campaign in Autonomous Mine Mapping, *ICRA*, 2004–2009, 2004.

90. Mason, M.T., *Mechanics of Robotic Manipulation*, MIT Press, Cambridge, MA, 2001.

91. Stilman, M. and Kuffner, J., Planning among movable obstacles with artificial constraints. *Proc. Workshop on the Algorithmic Foundations of Robotics*, New York, July, pp. 119–135, 2006.

92. Ladd, A., Motion planning for physical simulation, PhD thesis, Rice University, Houstan, TX, 2007.

93. Guibas, L.J., Halperin, D., Hirukawa, H., Latombe, J.C., and Wilson, R.H., Polyhedral assembly partitioning using maximally covered cells in arrangements of convex polytopes, *Int. J. Comput. Geo. Appl.*, 8, 179–200, 1998.

94. Berg, J. van den, Path planning in dynamic environments, PhD thesis, Utrecht University, Utrecht, the Netherlands, 2007.

95. Stappen, A.F. van der, Wentink, C., and Overmars, M.H., Computing immobilizing grasps of polygonal parts, *Int. J. Robot. Res.*, 19(5), 467–479, 2000.

96. Rimon, E. and Burdick, J., Mobility of bodies in contact–I: A 2nd order mobility index for multiple-finger grasps, *IEEE Trans. Robot. Automat.*, 14(5), 2329–2335, 1998.

97. Cheong, J.-S., Stappen, A.F. van der, Goldberg, K., Overmars, M.H., and Rimon, E., Immobilizing hinged parts, *Int. J. Comput. Geo. Appl.*, 17, 45–69, 2007.

98. Berretty, R.-P.M., Geometric design of part feeders, PhD thesis, Utrecht University, Utrecht, the Netherlands, 2000.

99. Mitchell, J.S.B., Shortest paths and networks. In *Handbook of Discrete and Computational Geometry*, 2nd edition, Goodman, J.E. and O'Rourke, J., Eds., CRC Press, Boca Raton, FL, Chapter 27, pp. 607–642, 2004.

100. Dobkin, D.P. and Kirkpatrick, D.G., Determining the separation of preprocessed polyhedra—A unified approach, *ICALP*, 400–413, 1990.

101. Gartner, B. and Jaggi, M., Coresets for polytope distance, *Symp. Comput. Geo.*, 33–42, 2009.

5

Vision and Image Processing Algorithms

Concettina Guerra
Georgia Institute of Technology

5.1   Introduction

There is an abundance of algorithms developed in the field of image processing and computer vision ranging from simple algorithms that manipulate binary images based on local point operations to complex algorithms for the interpretation of the symbolic information extracted from the images.

Here we concentrate on algorithms for three central problems in image processing and computer vision that are representative of different types of algorithms developed in this area. The first problem, connectivity analysis, has been studied since the early days of binary images. It consists of separating the objects from the background by assigning different labels to the **connected components** of an image. The algorithms to identify connected components in binary images are rather straightforward: the first one is an iterative algorithm that performs several scans of the image and uses only local operations; the next two algorithms consist of only two scans of the image and use global information in the form of an equivalence table.

The second problem is that of grouping features extracted from the image (for instance, **edge points**) into parametric curves such as straight lines and circles. An algorithm to detect lines, based on the Hough transform, is described that maps the image data into a parameter space that is quantized into an accumulator array. The Hough transform is a robust technique since it is relatively insensitive to noise in the sensory data and to small gaps in a line.

The last problem is that of identifying and locating objects known a priori in an image, a problem that is generally called model-based object recognition. Our discussion will focus on the matching task, that is, finding correspondences between the image and model descriptions. This correspondence can be used to solve the localization problem, i.e., the determination of a geometrical transformation that maps the model into the observed object. Finding correspondences is difficult, due to the combinatorial nature of the problem and to noise and uncertainty in the data. We deal mainly with the case of planar objects undergoing rigid transformations in 3D space followed by scaled orthographic projections. We first assume that both the model and the image data are represented in terms of sets of feature points and describe two approaches (alignment and geometric hashing) to solve the point set matching problem. Then we consider model and image representations in terms of object **boundary** descriptions. We review dynamic programming algorithms for matching two sequences of boundary segments, which resemble the algorithm for the string editing problem. For multiresolution boundary representations, a tree dynamic programming algorithm is discussed. Extensions of the indexing techniques and of the algorithms based on the hypothesis-and-test paradigm to take advantage of the richer boundary descriptions are also described.

5.2 Connected Components

The connected component problem consists of assigning a label to each 1-pixel of a binary image so that two 1-pixels are assigned the same label if and only if they are connected. Two pixels are said to be connected if there is a path of 1-pixels adjacent along the horizontal and vertical directions that links them. (A different definition of connectivity includes the diagonal direction as well.) The set of connected pixels is called a connected component. Figure 5.1 shows a binary image and its connected components labeled with integers. A simple iterative algorithm to determine the connected components performs a sequence of scans over the image, propagating a label from each 1-pixel to its adjacent 1-pixels. The algorithm starts with an arbitrary labeling. More precisely, the algorithm works as follows.

CONNECTED COMPONENT ALGORITHM-Iterative
Step 1. (Initialization phase)
   Assign each 1-pixel a unique integer label.
Step 2. Scan the image top-down, left-to-right.
   Assign each 1-pixel the smallest between its own label and those
   of the adjacent pixels already examined in the scan sequence.
Step 3. Scan the image bottom-up, right-to-left.
   Like Step 2 above with a different scan order.
Alternate Step 2 and Step 3 until no changes occur.

0	1	1	1	1	0	0	0	0	0
0	1	1	0	0	0	0	0	0	0
0	1	0	0	0	0	0	0	0	1
0	0	0	0	0	0	0	0	0	1
0	1	1	0	0	1	0	0	1	1
0	1	0	0	0	1	0	0	0	1
0	1	0	0	0	1	0	0	0	1
0	0	0	1	1	1	0	0	0	1
0	0	0	0	1	1	1	0	0	0
0	0	0	0	0	0	0	0	0	0

(a)

0	1	1	1	1	0	0	0	0	0
0	1	1	0	0	0	0	0	0	0
0	1	0	0	0	0	0	0	0	2
0	0	0	0	0	0	0	0	0	2
0	3	3	0	0	4	0	0	2	2
0	3	0	0	0	4	0	0	0	2
0	3	0	0	0	4	0	0	0	2
0	0	0	4	4	4	0	0	0	2
0	0	0	0	4	4	4	0	0	0
0	0	0	0	0	0	0	0	0	0

(b)

FIGURE 5.1 (a) A 8×8 image and (b) its labeled connected components.

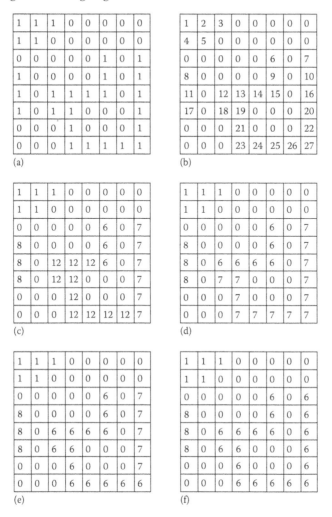

(a)

1	1	1	0	0	0	0	0
1	1	0	0	0	0	0	0
0	0	0	0	0	1	0	1
1	0	0	0	0	1	0	1
1	0	1	1	1	1	0	1
1	0	1	1	0	0	0	1
0	0	0	1	0	0	0	1
0	0	0	1	1	1	1	1

(b)

1	2	3	0	0	0	0	0
4	5	0	0	0	0	0	0
0	0	0	0	0	6	0	7
8	0	0	0	0	9	0	10
11	0	12	13	14	15	0	16
17	0	18	19	0	0	0	20
0	0	0	21	0	0	0	22
0	0	0	23	24	25	26	27

(c)

1	1	1	0	0	0	0	0
1	1	0	0	0	0	0	0
0	0	0	0	0	6	0	7
8	0	0	0	0	6	0	7
8	0	12	12	12	6	0	7
8	0	12	12	0	0	0	7
0	0	0	12	0	0	0	7
0	0	0	12	12	12	12	7

(d)

1	1	1	0	0	0	0	0
1	1	0	0	0	0	0	0
0	0	0	0	0	6	0	7
8	0	0	0	0	6	0	7
8	0	6	6	6	6	0	7
8	0	7	7	0	0	0	7
0	0	0	7	0	0	0	7
0	0	0	7	7	7	7	7

(e)

1	1	1	0	0	0	0	0
1	1	0	0	0	0	0	0
0	0	0	0	0	6	0	7
8	0	0	0	0	6	0	7
8	0	6	6	6	6	0	7
8	0	6	6	0	0	0	7
0	0	0	6	0	0	0	7
0	0	0	6	6	6	6	6

(f)

1	1	1	0	0	0	0	0
1	1	0	0	0	0	0	0
0	0	0	0	0	6	0	6
8	0	0	0	0	6	0	6
8	0	6	6	6	6	0	6
8	0	6	6	0	0	0	6
0	0	0	6	0	0	0	6
0	0	0	6	6	6	6	6

FIGURE 5.2 All the steps of the iterative algorithm for the input image (a).

Figure 5.2 shows a few steps of the algorithm (b–f) for the input image (a). It is clear that this algorithm is highly inefficient on conventional computers; however, it becomes attractive for implementation on parallel SIMD (single instruction multiple data stream) architectures where all the updates of the pixel's labels can be done concurrently. Another advantage of the iterative algorithm is that it does not require auxiliary memory, unlike the next two algorithms.

A common approach to finding the connected components is based on a two-pass algorithm. First the image is scanned in a top-bottom, left-to-right fashion and a label is assigned to each 1-pixel based on the value of adjacent 1-pixels already labeled. If there are no adjacent 1-pixels, a new label is assigned. Conflicting situations may arise in which a pixel can be given two different labels. Equivalent classes of labels are then constructed and stored to be used in the second pass of the algorithm to disambiguate such conflicts. During the second scan of the image, each label is replaced by the one selected as the representative of the corresponding equivalence class, for instance, the smallest one in the class if the labels are integers. The details follow.

CONNECTED COMPONENT ALGORITHM-Equivalence Table

Step 1. Scan the image top-down, left-to-right

    for each 1-pixel **do**

        if the upper and left adjacent pixels are all 0-pixels

            then assign a new label

        if the upper and left adjacent pixels have the same label

            then assign that label

        if only one adjacent pixel has a label

            then assign this label

        otherwise

            Assign the smaller label to the pixel.

            Enter the two equivalent labels into the equivalence table.

Step 2.

    Find equivalence classes.

    Choose a representative of each class

    (i.e., the smallest label).

Step 3. Scan the image top-down, left-to-right.

    Replace the label of each 1-pixel with its smallest equivalent one.

Step 2. can be done using the algorithm for UNION-FIND. The drawback of the above algorithm is that it may require a large number of memory locations to store the equivalent classes.

The next algorithm overcomes this problem by building a local equivalence table that takes into account only two consecutive rows. Thus, as the image is processed the equivalences are found and resolved locally. The algorithm works in two passes over the image.

CONNECTED COMPONENT ALGORITHM-Local Equivalence Table

Step 1. Scan the image top-down, left-to-right

    for each row *r* of the image **do**

        Initialize the local equivalence table for row *r*.

        for each 1-pixel of row *r* **do**

            if the upper and left adjacent pixels are all 0-pixels

                then assign a new label

            if the upper and left adjacent pixels have the same label

                then assign this label

            if only one adjacent pixel has a label

                then assign this label

            otherwise assign the smaller label to the pixel and

                enter the two equivalent labels into the local table.

        Find equivalence classes of the local table.

        Choose a representative of each class.

        for each 1-pixel of row *r* **do**

            replace its label with the smallest equivalent label.

Step 2. Scan the image bottom-up, right-to-left

    for each row *r* of the image **do**

        Initialize the local equivalence table for row *r*.

        for each 1-pixel of row *r* **do**

            for each adjacent pixel with a different label **do**

                enter the two equivalent labels into the local table.

Find equivalence classes of the local table.
Choose a representative of each class.
for each 1-pixel of row r **do**
    replace its label with the smallest equivalent label.

It has been shown by experiments that as the image size increases this algorithm works better than the classical algorithm on a virtual memory computer.

Modifications of the above algorithms involving different partitioning of the image into rectangular blocks are straightforward.

5.3 The Hough Transform

5.3.1 Line Detection

The Hough transform is a powerful technique for the detection of lines in an image [11,16]. Suppose we are given a set of image points—for instance, edge points or some other local feature points— and we want to determine subsets of them lying on straight lines. A straightforward but inefficient procedure to do that consists of determining for each pair of points the straight line through them and then counting the number of other points lying on it (or close to it).

The Hough transform formulates the problem as follows. Consider the following parametric representation of a line:

$$y = mx + b$$

where the parameters m, b are the slope and the intercept, respectively. Given a point (x_1, y_1), the equation $b = -x_1 m + y_1$, for varying values of b and m, represents the parameters of all possible lines through (x_1, y_1). This equation in the (b, m) plane is the equation of a straight line. Similarly, point (x_2, y_2) maps into the line $b = -x_2 m + y_2$ of the parameter space. These two lines intersect at a point (b', m') that gives the parameters of the line through the (x_1, y_1) and (x_2, y_2). More generally, points aligned in the (x, y) plane along the line with parameters (b', m') correspond to lines in the (b, m) plane intersecting at (b', m'). Figure 5.3 illustrates this property.

Thus the line detection problem is converted into the problem of finding intersections of sets of lines in the parameter space. We next see how to solve this latter problem efficiently in the presence of noise in the image data.

The parameter space is quantized into $h \times k$ cells that form an array A, called, accumulator array; each entry of the array or cell corresponds to a pair (b, m). In other words, the parameter b is discretized into h values $b_1, b_2, \ldots, b_h$, and m in k values $m_1, m_2, \ldots, m_k$, where b_h and m_k are the largest possible values for the two parameters. See Figure 5.4.

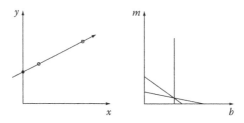

FIGURE 5.3 Mapping collinear image points into parameter space.

The line detection algorithm proceeds in two phases. Phase 1 constructs the accumulator array as follows.

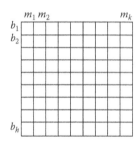

HOUGH ALGORITHM
    Initialize all entries of the accumulator array A to zero.
    for each image point (x, y) **do**
        for each m_i, $i = 1, k$ **do**
            $b \leftarrow -m_i x + y$
            round b to the nearest discretized value, say, b_j
            $A(b_j, m_i) \leftarrow A(b_j, m_i) + 1$.

FIGURE 5.4 The parameter space.

At the end of phase 1, the value t at $A(b, m)$ indicates that there are t image points along the line $y = mx + b$. Thus a maximum value in the array corresponds to the best choice of values for the line parameters describing the image data. Phase 2 of the algorithm determines such a peak in the accumulator array A or, more generally, the s largest peaks, for a given s.

Another way of looking at the Hough transform is as a "voting" process where the image points cast votes in the accumulator array.

The above parametric line representation has the drawback that the slope m approaches infinity for vertical lines, complicating the construction of the Hough table. Since vertical lines tend to occur frequently in real applications, this representation is rarely used. A better representation is

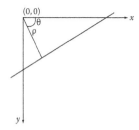

FIGURE 5.5 ρ, θ line representation.

$$\rho = x \cos \theta + y \sin \theta$$

where the two parameters ρ and θ denote the distance of the line from the origin and the angle of the normal line, respectively, as illustrated in Figure 5.5.

An important property of the Hough transform is its insensitivity to noise and to missing parts of lines.

We now analyze the time complexity of the Hough algorithm. Let n be the number of points (edges or other feature points) and $k \times m$ the size of the accumulator array. Phase 1 of the above algorithm requires $O(nm)$ operations. Phase 2, that is, finding the maximum in the array, requires $O(km)$ time for a total of $O(m \max(k, n))$ time.

5.3.2 Detection of Other Parametric Curves

The concept of the Hough transform can be extended in several ways. One extension is to other parametric curves with a reasonably small number of parameters; circles and ellipses are typical examples. Consider the case of circle detection. A circle can be represented by the equation

$$(x - a)^2 + (y - b)^2 = r^2$$

where the parameters a and b are the coordinates of the center and r is the radius. The transform maps the image points into a three-dimensional accumulator array indexed by discretized values of the center coordinates and of the radius. If it is known in advance that only circles with given radii may be present in an image, which is sometimes true in real applications, only a few two-dimensional (2D) subarrays need to be considered. If space is a concern this approach can be used only for few parametric curves, since the array dimensionality grows with the number of the curve parameters.

Another generalization is to arbitrary shapes represented by boundary points [2].

5.4 Model-Based Object Recognition

The problem of object recognition, central to computer vision, is the identification and localization of given objects in a scene. The recognition process generally involves two stages. The first is the extraction from sensed data of information relevant to the problem and the construction of a suitable symbolic representation. The second is that of matching the image and model representations to establish a correspondence between their elements. This correspondence can be used to solve the localization problem, that is the determination of a geometrical transformation that maps the model into the observed object.

Matching is a difficult task, for a number of reasons. First, a brute-force matching is equivalent to a combinatorial search with exponential worst-case complexity. Heuristics may be used to reduce the complexity by pruning the search space that must be explored. Approaches with polynomial time complexity have also been proposed; but the execution time and memory requirements remain high. Second, images do not present perfect data: noise and occlusion greatly complicate the task.

There have been many recognition methods and systems proposed to handle this complexity [7,10,13,19]. Surveys of recognition methods are [1,3,4,8,33]. The systems differ both in the choice of the representational models and of the matching algorithms used to relate the visual data to the model data. It is obvious that these two aspects influence each other and that the selection of a matching strategy heavily depends on which structures are used to represent data. Some methods use representational models that rely on few high-level features for fast interpretation of the image data. While this facilitates matching, the process of extracting high-level primitives from the raw data may be difficult and time consuming. Moreover, in the presence of occlusion the disappearance of few distinctive features may became fatal. Simpler features are easier to detect and tend to be dense, implying less sensitivity to noise and occlusion. However, low-level features do not have high discriminating power, thus affecting the time for matching.

There are a number of other issues that any recognition system needs to address:

1. *What class of objects is the system able to deal with?*
   The objects can be 2D or three-dimensional (3D). 3D objects may be smoothly curved or approximated by polyhedra. A restricted class of 3D objects includes flat objects, that is,objects with one dimension much smaller than the other two. Flat objects have received a lot of attention, because a projective transformation applied to this class of objects can be effectively approximated by an **affine transformation**, which is easier to handle. Another important distinction is whether the objects are rigid or deformable or composed of parts that are allowed to move with respect to one another.

2. *What type of geometric transformation is allowed?*
   There is a hierarchy of transformations that has been considered in computer vision: from Euclidean transformations (rotations, translations) to similarity transformations (rotations, translations, and scaling) to affine transformations (represented by a set of linear equations) to the more general projective transformations. To remove the effects of a transformation it is useful to represent the objects by means of features that are invariant under that class of transformations. As the class of transformations becomes more general, the invariant features become more complex and harder to extract. For example, a well-known simple invariant to Euclidean transformations is the distance between two points; however, distance is not invariant to similarity. For a projective transformation four points are needed to define an invariant (cross-ratio).

3. *Robustness*
   Is the recognition system able to deal with real data? Some available systems give good results in controlled environments, with good lighting conditions and with isolated objects. Noisy and cluttered images represent challenging domains for most systems.

This chapter reviews algorithms to match an image against stored object models. Topics that are not covered in this chapter include model representation and organization. The objects are assumed to be given either by means of sets of feature points or by their boundaries (2D or 3D). For an introduction to feature and boundary extraction, see the references listed at the end of this chapter.

The objects may have undergone an affine transformation; thus we do not consider here the more general class of projective transformations.

5.4.1 Matching Sets of Feature Points

In this section we consider recognition of rigid flat objects from an arbitrary viewpoint. Models in the database are represented by sets of feature points. We assume that an image has been preprocessed and feature points have been extracted. Such points might be edge points or corners or correspond to any other relevant feature. The following discussion is independent on the choice of the points and on the method used to acquire them. In fact, as we will see later, similar techniques can be applied when other more complex features, for instance line segments, are used instead of points.

In object recognition to obtain independence from external factors such as viewpoints, it is convenient to represent a shape by means of geometric invariants, that is, shape properties that do not change under a class of transformations. For a flat object a projective transformation can be approximated by an affine transformation. For the definition and properties of affine transformations see Section "Defining Terms."

5.4.1.1 Affine Matching

There are two main approaches proposed to match objects under affine transformations: alignment and geometric hashing. The first method, alignment, uses the hypothesis-and-test paradigm [17]. It computes an affine transformation based on an hypothesized correspondence between an object and model basis and then verifies the hypothesis by transforming the model to image coordinates and determining the fraction of model and image points brought into correspondence. This is taken as a measure of quality of the transformation. The above steps are repeated for all possible groups of three model and image points, since it is known that they uniquely determine an affine transformation.

The geometric hashing or, more generally, indexing methods [20,21], build at compile time a lookup table that encodes model information in a redundant way. At run time, hypotheses of associations between an observed object and the models can be retrieved from the table and then verified by additional processing. Thus much of the complexity of the task is moved to a preprocessing phase where the work is done on the models alone. Strategies based on indexing do not consider each model separately and are therefore more convenient than search-based techniques in applications involving large databases of models. Crucial to indexing is the choice of image properties to index the table; these can be groups of features or other geometric properties derived from such groups. In the following, we first concentrate on hash methods based on triplets of features points, then on methods based on segments of a boundary object decomposition.

5.4.1.2 Hypothesize-and-Test (Alignment)

Given an image I containing n feature points, alignment consists of the following steps.

ALIGNMENT
for each model M. Let m be the number of model points
    for each triple of model points **do**
        for each triple of image points **do**
            hypothesize that they are in correspondence and

compute the affine transformation based on this correspondence.
for each of the remaining $m - 3$ model points **do**
    apply that transformation.
Find correspondences between the transformed model points
    and the image points.
Measure the *quality* of the transformation
(based on the number of model points that are paired with image points.)

For a given model, these steps are repeated for all triples of model and image features. In the worst-case the number of hypotheses is $O(m^3 n^3)$. Thus the total time is $O(T_v m^3 n^3)$, where T_v is the time for the verification of the hypothesized mapping. Since a verification step is likely to take time polynomial in the number of model and image features, this approach takes overall polynomial time which represents a significant improvement over several exponential time approaches to matching proposed in the literature. In the following, a more detailed description of the verification phase is given. Once a transformation is found, it is used to superimpose the model and the object. A distance measure between the two sets of points must be defined so that recognition occurs when such a distance is below a given threshold. One definition of distance is just the number of model and image points that can be brought into correspondence. An approximation to this value can be computed by determining for any transformed model point if an image point can be found in a given small region around it (for instance a small square of pixels of fixed side). In this way an image point can be double-counted if it matches two different model points. Nevertheless, in most cases the above provides a good approximation at a reasonable computational cost.

The search for image points that match each transformed model point can be done sequentially over the image points in time $O(nm)$. Alternatively, it can be made more efficient with some preprocessing of the image points and the use of auxiliary data structures. A practical solution uses a lookup table indexed by the quantized values of the image points coordinates. An entry of the table contains the set of image points located in the small region represented by the cell in the quantized space. Each transformed model point is entered into the table to retrieve, in constant time, possible matching points. Other information besides location of image points can be used to index the table, for instance orientation, which is already available if the features points are computed through an edge operator.

The algorithm above iterates over all the k models in the database leading to a total time complexity $O(kT_v m^3 n^3)$.

5.4.1.3 Indexing

The second method, hashing or indexing, is a table lookup method. It consists of representing each model object by storing transformation-invariant information about it in a hash table. This table is compiled off-line. At recognition time, similar invariants are extracted from the sensory data I and used to index the table to find possible instances of the model. The indexing mechanism, referred to as geometric hashing for point set matching, is based on the following invariant. The coordinates of a point into a reference frame consisting of three noncollinear points are affine invariant. The algorithm consists of a preprocessing phase and a recognition phase.

INDEXING
Preprocessing phase
for each model M. Let m be the number of model points
    for each triple of noncollinear model points **do**
        form a basis (reference frame)
        for each of the $m - 3$ remaining model points **do**
            determine the point coordinates in that basis.

Use the triplet of coordinates (after a proper quantization)
as an index to an entry in the hash table, where the pair
(*M, basis*) is stored.

For k models the algorithm has an $O(km^4)$ time complexity. Notice that this process is carried out only once and off-line. At the end of the preprocessing stage an entry of the hash table contains a list of pairs: $(M_{i1}, basis_{i1}), (M_{i2}, basis_{i2}), \ldots, (M_{it}, basis_{it})$.

INDEXING

Recognition phase

Initialization

for each entry of the hash table **do**
    set a counter to 0.
Choose three noncollinear points of I as a basis.
    for each of the $n - 3$ remaining image points **do**
        determine its coordinates in that basis.
        Use the triplet of coordinates (after a proper quantization)
        as an index to an entry in the hash table and
        increment the corresponding counter.
Find the pair (*M, basis*) that achieved the maximum value
of the counter when summed over the hash table.

The last process can be seen as one of "voting," with a vote tallied by each image point to all pairs (*M, basis*) that appear at the corresponding entry in the table.

At the end of these steps after all votes have been cast, if the pair (*M, basis*) scores a large number of votes, then there is evidence that the model *M* is present in the image. If no pair achieves a high vote, then it might be that the selected basis in the image does not correspond to any basis in the model database, and therefore it is convenient to repeat the entire process with another selected basis in the image, until either a match is found or all bases of image points have been explored.

The time complexity of the recognition phase is $O(n)$ if a single selected image basis gives satisfactory results (high scores). Otherwise more bases need to be considered leading, in the worst case, to the time $O(n^4)$. In summary, the time complexity of recognition $T_{\text{recognition}}$ is bound by

$$O(n) \leq T_{\text{recognition}} \leq O\left(n^4\right)$$

There are a few issues that affect the performance of this approach. First, the choice of the basis. The three selected points of the basis should be far away to reduce the numerical error. Other issues are the sensitivity to quantization parameters and to noise. A precise analysis of this approach and a comparison with alignment under uncertainty of sensory data is beyond the scope of this chapter. The interested reader may refer to [15]. Although alignment is computationally more demanding, it is less sensitive to noise than hashing and able to deal with uncertainty under a bounded error noise model. On the other hand, indexing is especially convenient when large databases are involved, since it does not require to match each model separately and its time complexity is therefore not directly dependent on the size of the database.

One way to overcome the limitations of geometric hashing and make it less sensitive to noise is to use more complex global invariants to index the hash table at the expense of preprocessing time to extract such invariants. More complex invariants have a greater discriminating power and generate fewer false positive matches.

5.4.2 Matching Contours of Planar Shapes

It is natural to use contours as a representation for planar objects. Contour information can be in the form of a polygonal approximation, that is, a sequence of line segments of possibly varying lengths approximating the curve. A common boundary description is the eight-direction chain code [36] where all segments of the boundary decomposition have unit length and one of eight possible directions. Another boundary description is given in terms of concave/convex curve segments. Boundary representations are simple and compact, since a small number of segments suffices to accurately describe most shapes. Techniques to derive boundary descriptions are reviewed in the references listed at the end of this chapter.

Contours have been used in a variety of ways in object recognition to make the process faster and more accurate. First, the availability of contours allows computation of global invariant properties of a shape that can be used for fast removal of candidate objects when searching in the model database; only models found to have similar global properties need be considered for further processing. Global features that can be used include

- The number of segments of a convex/concave boundary decomposition
- For a close contour, a measure of the area over the length

A combination of the above features allows a reduction of the candidate objects. An object with extreme values of the global attributes is more easily recognized than an object with average attributes. Obviously there must be a sufficiently high tolerance in the removal process so that good choices are not eliminated.

In the following, we first describe techniques based on dynamic programing that have been developed for contour matching. Then we will see how the methodologies of indexing and alignment based on point sets can take advantage of the contour information in the verification phase for more reliable results. The same methodologies can be further modified so that contour segments become the basic elements of the matching, that is, they are used instead of the feature points to formulate hypotheses.

5.4.2.1 Dynamic Programming

A number of approaches use dynamic programing to match shape contours. A shape boundary is described by a string of symbols representing boundary segments. For instance, if the segments result from the decomposition of the boundary into convex/concave parts, there might be just three symbols (for convex, concave, and straight) or more if different degrees of convexity or concavity are considered. The matching problem becomes then a string matching problem [36]. We briefly review the string matching algorithm. Then we will discuss adaptations of the basic algorithm to shape recognition that take into account noise and distortion.

Let $A = a_0, \ldots, a_{n-1}$ and $B = b_0, \ldots, b_{m-1}$ be two strings of symbols. Three types of edit operations, namely, insertion, deletion, and change, are defined to transform A into B.

- Insertion: Insert a symbol a into a string, denoted as $\lambda \to a$ where λ is the null symbol
- Deletion: Delete a symbol from a string, denoted as $a \to \lambda$
- Change: Change one symbol into another, denoted as $a \to b$

A nonnegative real cost function $d(a \to b)$ is assigned to each edit operation $a \to b$. The cost of a sequence of edit operations that transforms A into B is given by the sum of the costs of the individual operations. The edit distance $D(A, B)$ is defined as the minimum of such total costs. Let $D(i, j)$ be the distance between the substrings $a_0, \ldots, a_i$ and $b_0, \ldots, b_j$. It is $D(n, m) = D(A, B)$. Let $D(0, 0) = 0$; then $D(i, j), 0 < i < n, 0 < j < m$ is given by

$$D(i,j) = \min \begin{cases} D(i-1,j) + d\,(a_i \rightarrow \lambda) \\ D(i,j-1) + d\,(\lambda \rightarrow b_j) \\ D(i-1,j-1) + d\,(a_i \rightarrow b_j) \end{cases} \qquad (5.1)$$

The matching problem can be seen as one of finding an optimal nondecreasing path in the 2D table $D(i,j)$ from the entry $(0,0)$ to the entry (n,m). If the elements of the table are computed horizontally from each row to the next one, then when computing $D(i,j)$ the values that are needed have already been computed. Since it takes constant time to compute $D(i,j)$, the overall time complexity is given by $O(nm)$. The space complexity is also quadratic.

5.4.2.2 Using Attributes

A number of variations of the above dynamic programing algorithm have been proposed to adapt it to the shape matching problem. First, attributes are associated to the symbols of the two strings. The choice of the attributes depends both on the type of transformations allowed for the shapes and on the types of boundary segments. Some of the commonly used attributes of concave/convex segments are the normalized length, and the degree of symmetry S_c. Let L_a be the length of segment a and L_A the total length of the segments of A. The normalized length is L_a/L_A. L_A is used as normalization factors to obtain scale invariance.

Let $f(l)$ be the curvature function along segment a.

$$S_a = \int_0^{L_a} \left(\int_0^s f(l)dl - 1/2 \int_0^{L_a} f(l)dl \right) ds$$

If $S_a = 0$ then the segment is symmetric, otherwise it is inclined to the left or to the right depending on whether S_a is positive or negative, respectively.

The attributes are used in the matching process to define the cost $d(i,j)$ of the edit operation that changes segment a_i into b_j. The cost $d(i,j)$ is defined as the weighted sum of the differences between the attributes of a_i and b_j.

Thus the cost function $d(i,j)$ is defined as

$$d(i,j) = w_1 |L_{a_i}/L_A - L_{b_j}/L_B| + w_2 \sigma(S_{a_i}, S_{b_j})$$

where $\sigma(S_a, S_b)$ is taken to be 0 if S_a and S_b are both positive or negative or both close to zero, otherwise is 1. In the above expression w_1 and w_2 are weights used to take into account the different magnitude of the two terms when summed over all edit operations.

Consider now a shape represented by a polygonal approximation. Typical choices for the attributes of a segment a are the normalized length and the angle θ_a that the segment forms with a reference axis. In this case, the cost function can be defined as

$$d(i,j) = w_1 |L_{a_i}/L_A - L_{b_j}/L_B| + w_2 \gamma(\theta_{a_i}, \theta_{b_j})$$

where, again, w_1 and w_2 are weights to make both terms to lie between 0 and 1.

5.4.2.3 Cyclic Matching

The above algorithm assumes that the correct starting points are known for the correspondences on both shapes. These could be, for instance, the topmost boundary segments when no rotation is present. A more reliable choice that is invariant to Euclidean transformations may be obtained by ranking the segments according to some criterion, for instance the normalized length or the curvature. The segments on the two curves with highest rank are chosen as the starting points for the dynamic programing algorithm. Alternatively, when the starting points are not available or cannot be computed reliably, cyclic shifts of one of the two strings are needed to try to match segments starting at any point in one of the boundaries. Let B be the shortest string, i.e., $m < n$. To solve the cyclic matching problem a table with n rows and $2m$ columns is built. Paths in the $D(i, j)$ table that correspond to optimal solutions can start at any column $1 \leq j \leq m$ of the first row and end at column $j + m$ of the last row. The dynamic programing algorithm is repeated for each new starting symbol of the shortest string. This brute-force approach solves the cyclic matching problem in time $O(nm^2)$. A more efficient $O(mn \log n)$ solution can be obtained by using divide-and-conquer [24].

5.4.2.4 Merge Operation

To reduce the effect of noise and distortion and improve matching accuracy, another variant of the above method suggests the use of a fourth operation, called merge [34], in addition to the insert, delete, and change operations. Merge allows change of any set of consecutive segments of one string into any set of consecutive segments in the other string. Let $a_{<k,i>}$ denote the sequence of k segments $a_{i-k+1} \cdots a_i$. The merge operation $a_{<k,i>} \rightarrow b_{<h,j>}$ attempts to change the combined segments from the first string into the combined segments of the second. For $k = h = 1$, the merge operation reduces to a change. It has now to be defined what are the attributes associated with the merged segments. First the length of $a_{<k,i>}$ is simply the normalized sum of the lengths of all segments in the sequence. As for the direction, assume $k = 2$. Then the angle $\theta_{a_{<2,i>}}$ can defined as follows:

$$
\begin{aligned}
\theta_{a_{2,i}} &= \theta_{a_{i-1}} + \left| \theta_{a_{i-1}} - \theta_{a_i} \right| \\
&\quad \text{if} \quad \theta_{a_{i-1}} < \theta_{a_i} \text{ and } \left| \theta_{a_{i-1}} - \theta_{a_i} \right| \leq 180 \\
&= \theta_{a_i} + \left| \theta_{a_{i-1}} - \theta_{a_i} \right| \\
&\quad \text{if} \quad \theta_{a_{i-1}} \geq \theta_{a_i} \text{ and } \left| \theta_{a_{i-1}} - \theta_{a_i} \right| \leq 180 \\
&= \theta_{a_{i-1}} + \left(360 - \left| \theta_{a_{i-1}} - \theta_{a_i} \right| \right) \\
&\quad \text{if} \quad \theta_{a_{i-1}} \geq \theta_{a_i} \text{ and } \left| \theta_{a_{i-1}} - \theta_{a_i} \right| > 180 \\
&= \theta_{a_i} + \left(360 - \left| \theta_{a_{i-1}} - \theta_{a_i} \right| \right) \\
&\quad \text{if} \quad \theta_{a_{i-1}} < \theta_{a_i} \text{ and } \left| \theta_{a_{i-1}} - \theta_{a_i} \right| > 180
\end{aligned}
$$

To include the merge operation in the above algorithm Equation 5.1 has to be replaced by

$$
D(i, j) = \min \begin{cases} D(i-1, j) + d\,(a_i \rightarrow \lambda) \\ D(i, j-1) + d\,(\lambda \rightarrow b_j) \\ D_{\text{merge}}(i, j) \end{cases}
$$

where

$$
D_{\text{merge}}(i, j) = \min_{1 \leq k \leq i, 1 \leq h \leq j} D(i-k, j-h) + d\left(a_{<k,i>}, b_{<h,j>} \right)
$$

When there is no limitation on the number of merged segments, the time complexity of the dynamic programming algorithm becomes $O(n^2 m^2)$.

In summary, the dynamic programming algorithm is able to deal with rotation, translation, and scaling through the combined use of a distance measure invariant to these transformations and of a cyclic mapping of the two boundaries. Furthermore, noise can be dealt with through the use of concatenated segments in both shapes.

5.4.2.5 Multiscale Tree Matching

Another approach to match contours of planar shapes uses multiscale data representations. Given a scene containing multiple objects, a full pyramid of images, taken at various levels of resolution, is first constructed (for instance, a Gaussian pyramid [6]). Then contours are extracted at all levels of resolution and each of them is decomposed into a sequence of segments (either curved or straight line segments) [25,26].

There are two main ways of matching multiscale objects. One approach, the coarse-to-fine matching, tries to find a good initial pair of corresponding segments at the coarsest level of resolution and to expand, at the finer levels, the hypothesized correspondences [28,29]. The coarse-to-fine approach has generally the disadvantage that mismatches at a coarse scale cause errors from which it is impossible to recover, since the algorithms usually proceed by subdividing the corresponding coarse elements into subelements. This limitation is heavy particularly for highly deformed shapes for which the matching is not very reliable at all levels. On the other hand, this method is fast, because it disregards large portions of the search space. Coarse-to-fine strategies have been successfully used in a variety of image processing and vision applications, including stereo matching, optical flow computation, etc.

Alternative approaches [9,35] try to overcome the limitations of a coarse-to-fine strategy by processing the multiresolution data starting from the finest level. The shape contours are represented at all scales by sequences of concave/convex segments. In [9], a planar object is modeled as a tree, in which a node corresponds to a multiscale boundary segment and an arc connects nodes at successive levels of resolution. The children of a given node describe the structural changes to that segment at a finer level of resolution. The problem of matching an object against a model is formulated as the one of determining the best mapping between nodes at all levels of the two corresponding trees, according to a given distance between segments [23,31]. The distance is chosen to be invariant under rotation, translation, and change of scale. The mapping has to satisfy the following constraint: for any path from a leaf node to the root of each of the two trees, there is exactly one node in the mapping. Thus, if a node is in the mapping none of its proper ancestors nor descendants is. Intuitively, such a constraint on the mapped nodes means that for any portion of the boundary at the finest resolution there is one matched segment at some resolution level that covers it.

This method applies to objects for which the entire boundaries are available, implying that it is able to deal with occlusion when this causes alterations in a boundary without breaking it into separate pieces.

We now describe a tree matching algorithm based on dynamic programming that has optimal $O(|T||T'|)$ time complexity, where $|T|$ and $|T'|$ are the number of nodes in the two trees. Let l be the number of leaves of T. $T[i]$ denotes the node of T whose position in the postorder traversal of T is i. Recall that in the postorder traversal the nodes are visited by recursively visiting the first subtree of the root, then the second subtree and so on, and finally visiting the root. In the postorder, $T[1], T[2], \ldots, T[i]$ is in general a forest. $anc(i)$ is the set of ancestor nodes of $T[i]$, including i itself. The postorder number of the father of node $T[i]$ is denoted by $p(i)$. $ll(i)$ and $rl(i)$ denote the postorder number of the leftmost leaf and the rightmost leaf, respectively, of the subtree rooted at $T[i]$; $ll(i) = rl(i) = i$ when $T[i]$ is a leaf node.

A measure of dissimilarity $d(i,j)$ between nodes i and j is defined as above as the weighted sum of the attributes of the segments. In addition $d(i,j)$ must contain a term that rewards matches at higher resolution levels.

The matching problem can be formulated as a minimization problem: find a set of pairs $M = \{(i_h, j_h)\}$ that satisfies the above constraint and minimizes the total cost function, that is,

$$F\left(T, T'\right) = Min_M \sum d\left(i_h, j_h\right)$$

The matching set M of two trees is shown in Figure 5.6. The nodes of the two trees are processed according to the left-to-right postorder numbering. Let $D(i,j)$ be the distance between the two forests $T[1], T[2], \ldots, T[i]$ and $T'[1], T'[2], \ldots, T'[j]$, that is the cost of the best mapping involving nodes

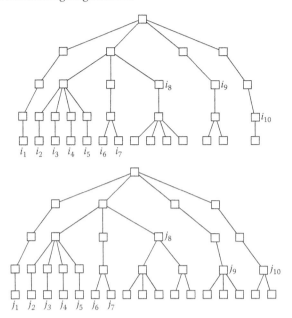

FIGURE 5.6 The matching set $M = \{(i_t, j_t)\}$ of two trees.

up to i and j and covering all leaves $1, \ldots, rl(i)$ and $1, \ldots, rl(j)$. $D(i,j) = F(T, T')$ when $i = |T|$ and $j = |T'|$. Let $D(0,0) = 0, D(0,j) = D(i,0) = \infty, i,j \neq 0$. $D(i,j)$ is given by the following relations:

- Case 1: if $i \neq p(i-1)$ and $j \neq p(j-1)$
  $D(i,j) = D(i-1, j-1) + d(i,j)$;
- Case 2: if $i \neq p(i-1)$ and $j = p(j-1)$
  $D(i,j) = \min\{D(i, j-1), D(i-1, ll(j)-1) + d(i,j)\}$;
- Case 3: if $i = p(i-1)$ and $j \neq p(j-1)$
  $D(i,j) = \min\{D(i-1, j), D(ll(i)-1, j-1) + d(i,j)\}$;
- Case 4: if $i = p(i-1)$ and $j = p(j-1)$
  $D(i,j) = \min\{D(i-1, j-1), D(i-1, j), D(i, j-1), D(ll(i)-1, ll(j)-1) + d(i,j)\}$

The above relations suggest the use of dynamic programing to solve the minimization problem. When computing $D(i,j)$, $1 \leq i \leq |T|$, $1 \leq j \leq |T'|$, all the values on the right side of the recurrence relation above have already been computed. Thus it takes constant time to extend the mapping to nodes i and j leading to a $O(|T||T'|)$ time algorithm. The space complexity is also quadratic.

5.4.2.6 Hypothesize-and-Test

A common scheme for matching uses three steps: a hypothesis generation step during which a few initial matchings are chosen and based on them a transformation is hypothesized, a prediction step that determines the matching of image and model features that are compatible with the initial hypothesis, and a verification step that evaluates all the resulting possible matchings. This scheme is quite general and can be applied to a variety of data representations. The alignment approach described in Section 5.4.2.1 is one such example. The contour information can be effectively used in the alignment method to increase the reliability of the match. As described in Section 5.4.2.5, hypotheses are generated based on small groups of features points brought into correspondence and a geometric transformation is derived for each hypothesis. Then a hierarchical verification process is applied [17]. A preliminary verification phase checks the percentage of model points that lie within reasonable error ranges of the corresponding image points. This step allows to rapidly eliminate false matches by considering only location and orientation of points (edge points). A more accurate and time-consuming verification procedure based on contour information is applied to the few surviving alignments. Each model segment is mapped through the hypothesized transformation into the image plane and all nearby image segments are determined. Then, for each pair of model and image segments, three cases are considered: (1) the segments are almost parallel and of approximately the same length; (2) the segments are parallel but one is much longer than the other; (3) the segments cross each other. The three cases contribute positive, neutral, or negative evidence to a match, respectively. The basic idea underlying this comparison strategy is that very unlikely two almost coincident segments can be the result of an accidental match. In summary, properties of segments such as length and parallelism can effectively contribute to the evaluation of the quality of a match at the expense of some additional processing.

The next application of the hypothesis-and-test paradigm to contour-based 2D shape descriptions generates hypotheses based on a limited number of corresponding segments. Since the choice of the initial pairing of segments strongly affects the performance of the process, the image segments are first ranked according to a given criterion, for instance length. Then segments in the image and the model are processed in decreasing ranking order of length. Additional constraints can be imposed before a hypothesis is generated. For each pair a and b of such segments, a hypothesis is generated if they are compatible. Compatibility is defined depending on the type of transformation. For similarity transformations a suitable definition can be: segments a and b are compatible (1) if the angle that a forms with its preceding segment along the boundary is close to the angle of b with its neighbor, and (2) assuming the scale factor is known, if the ratio of the lengths of the two segments is close to that value. The verification phase then tries to add more consistent pairings to the initial ones by using an appropriate definition of geometric consistency. The process stops as soon as a reasonably good match is found. We omit here the details of this last phase.

5.4.2.7 Indexing

The indexing methods, as described above, are based on the idea of representing an object by storing invariant information about groups of features in a hash table. Hypotheses of correspondences between an observed object and the models are retrieved from the table by indexing it with groups of features extracted from the object.

One approach to indexing uses information collected locally from the boundary shape in the form of the so-called super-segment [32]. A flat object can be approximated by a polygon. Since there exist many polygonal approximations of a boundary with different line fitting tolerances, several of them are used for the purpose of robustness. A super segment is a group of a fixed number of adjacent segments along the boundary, as in Figure 5.7. Supersegments are the basic elements of the indexing mechanism. The quantized angles between consecutive segments are used to encode each super-segment. Assume that there are n segments in a super segment, then the code is

$$(\alpha_1, \alpha_2, \ldots, \alpha_{n-1})$$

Other geometric features of a super segment can be used in the encoding to increase the ability of the system to distinguish between different super segments. One such feature is the eccentricity, that is the ratio of the length of the small and long axis of the ellipse representing the second moment of inertia. A quantized value of the eccentricity is added to the above list of angles.

FIGURE 5.7   A super-segment.

The preprocessing and matching phases of indexing are similar to those described in Section 5.4.1. Briefly, the super-segments of all models are encoded, and each code is used as a key for a table where the corresponding super segment is recorded as an entry. During the matching stage, all encoded super segments extracted from a scene provide indices to the hash table to generate the matching hypotheses.

A verification phase is needed to check the consistency of all multiple hypotheses generated by the matching phases. The hypotheses are first divided according to the models they belong to. For a model M_i, let $h_{i1}, \ldots, h_{it}$ be the hypotheses of associations between super-segments of the image with super-segments of M_i. To check that subsets of such hypotheses are consistent, the following heuristics has been proposed. If three hypotheses are found to be consistent, then the remaining that are found to be consistent with at least one of the three provide an instantiation of the model in the image. Consistency is defined on the basis of few geometric constraints such as the distance, the angle, and the direction. More precisely, the distances of corresponding super-segments must be in the same range. The distance between two super-segments is defined as the distance between the locations of the segments taken as the midpoints of the middle segments. Similarly, angles and directions of corresponding super-segments must be in the same range. The orientation of a super-segment is the vector of the predecessor and successor segments of the middle segment.

An alternative approach to indexing suggests the use of information spatially distributed over the object rather than at localized portions of the shape. Groups of points along the curved boundary are collected and indices are computed on the basis of geometrical relationships between the points and local properties at those points. These local measures can include location and tangent or higher order local information as local curve shape.

The difficulty associated with indexing methods arises from the large memory requirements for representing multidimensional tables. Another crucial point is the distribution of data over the table.

5.4.3 Matching Relational Descriptions of Shapes

5.4.3.1 Graph Matching

An object can be represented by a set of features and their relationships. This representation may take the form of a graph, where nodes correspond to features and arcs represent geometric and topological relationships between features. Similarly, a 3D object can have an object-centered representation consisting of a list of 3D primitives (surface patches, 3D edges, vertices, etc.) and relationships between primitives such as connections between surfaces, edges, etc. Recognition of an object becomes a graph isomorphism problem. Given a graph $G_1 = (V_1, E_1)$ corresponding to a model object and graph $G_2 = (V_2, E_2)$ corresponding to an observed object, the graph isomorphism can be formulated as follows. Find a one-to-one mapping between the vertices of the two graphs $f : V_1 \rightarrow V_2$ such that vertices of V_1 are adjacent iff and only if their corresponding vertices of V_2 are adjacent. In other words, for $u, v \in V_1$, it is $uv \in E_1$ if $f(u)f(v) \in E_2$. Above we considered only one relation for each graph, but the definition can easily extend to many relations. If an object is only partially visible in an image, a subgraph isomorphism can be used. Graph problems are covered in other chapters of this book. Here we sketch some commonly used heuristics to reduce the complexity of the algorithms for vision applications.

Some researchers [14] have approached the matching problem as one of search, using an interpretation tree. A node in the tree represents a pairing between an image feature and a model feature. Nodes at the first level of the tree represent all possible assignments of the first image feature. For each such assignment there a node at the second level for each pairing of the second image feature, and so on. A path from the root of the tree to a leaf represents a set of consistent pairings that is a solution to the correspondence problem from which a rigid transformation can be derived. Figure 5.8

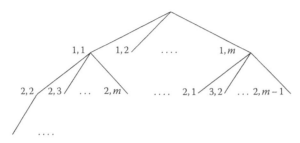

FIGURE 5.8 An interpretation tree with *n* levels and degree *m*.

shows the interpretation tree for *n* and *m* model and object features, respectively. The exponential complexity of the search can be reduced by pruning the interpretation tree using different types of constraints. Mostly geometric constraints involving angles and distances have been considered.

Another approach to cope with the complexity of a brute-force backtracking tree search is the introduction of forward checking and look-ahead functions [30] in the search. Given a partial solution, the idea of forward checking is to find a lower bound on the error of all complete solutions that include the given partial one so that the search can pruned at earlier stages.

5.5 Research Issues and Summary

Most of the work done on image connectivity analysis dates back at least 25 years, and this can be considered a solved problem. Some more recent work has been done on parallel algorithms for image connected component determination for a variety of interconnection networks, including meshes, hypercubes, etc.

The Hough transform is a popular method for line detection, but it is rarely used for more complex parametric curves, due to the high memory requirements. Parallel algorithms for the Hough transform have also been proposed.

Substantial effort has been devoted over the past years to the problem of model-based object recognition, and this is still an area of active research. A number of recognition systems have been designed and built that are successful for limited application domains. The general problem of recognition remains complex, even though polynomial time solutions to matching have been proposed. Practical approaches rely on heuristics based mainly on geometric constraints to reduce the complexity. An interesting issue not covered in this chapter is the effect of noise and spurious elements in the recognition problem. Some recent approaches explicitly take into account a noise error model and formulate the matching problem as one of finding a transformation of model and image data that is consistent with the error model.

Other approaches to match point sets under translation and rotation, developed in the field of computational geometry, are based on the computation of the **Hausdorff distance**.

The problem of finding correspondences is the main component of other important vision tasks, for instance of the stereo vision problem, which is the problem of deriving 3D information about an object from different views of the object.

It is also important to observe that some of the matching algorithms used in computer vision can be applied to other research areas, for instance molecular biology. The problem of matching the 3D structure of proteins differs from the point set matching problem mostly in the dimensionality of the data, few tens of feature points for an object and few hundreds of atoms in a protein.

5.6   Further Information

A good recent introduction to image processing is presented in *Machine Vision* by R. Jain.

Classical books are: *Computer Vision* by D. H. Ballard, C. M. Brown, and *Digital Picture Processing* by A. Rosenfeld and A. Kak.

A mathematically oriented presentation of the field of computer vision is in *Computer and Robot Vision* by R. M. Haralick and L. G. Shapiro.

Good survey papers on object recognition are [1,4,8,33]. Other papers presenting approaches related to the ones discussed here are [5,12,14,15,18,20,22,27].

Advances on research on computer vision and image processing are reported in the journals *IEEE Transactions on Pattern Analysis and Machine Intelligence, CVIP: Image Understanding, International Journal on Computer Vision, Pattern Recognition.* This list is by no means exhaustive.

The following is a list of some of the major conference proceedings that publish related work.

IEEE Computer Vision and Pattern Recognition Conference, IAPR International Conference on Pattern Recognition, International Conference on Computer Vision.

Defining Terms

Affine transformation: Affine transformations are a subgroup of projective transformations. When the depth of an object is small compared to its distance from the camera, the affine transformation effectively approximates a projective one. For a flat object there is a 2D affine transformation T between two different images of the same object:

$$T : \mathbf{x} \rightarrow A\mathbf{x} + \mathbf{t}$$

where A is 2×2 nonsingular matrix and $\mathbf{t}$ is a 2-vector.

An affine transformation maps parallel lines into parallel lines. Another well-known affine invariant is the following. Given three noncollinear points in the plane, they form a basis or reference frame in which the coordinates of any other point in the plane can be expressed. Such coordinates are affine invariant, meaning that they do not change when the points are transformed under an affine transformation. In other words, let $\mathbf{x}_{sm1}, \mathbf{x}_{sm2}, \mathbf{x}_{sm3}$ be the three noncollinear points, taken as a basis. Let $\mathbf{x}$ be another point and (α, β) its coordinates in the above basis, i.e.,

$$\mathbf{x} = \alpha\,(\mathbf{x}_{sm1} - \mathbf{x}_{sm3}) + \beta\,(\mathbf{x}_{sm2} - \mathbf{x}_{sm3}) + \mathbf{x}_{sm3}$$

The values α and β are invariant, that is,

$$T(\mathbf{x}) = \alpha\,(T\,(\mathbf{x}_{sm1}) - T\,(\mathbf{x}_{sm3})) + \beta\,(T\,(\mathbf{x}_{sm2}) - T\,(\mathbf{x}_{sm3})) + T\,(\mathbf{x}_{sm3})$$

The following fact is also known in affine geometry: Given three noncollinear points in the plane $\mathbf{x}_{sm1}, \mathbf{x}_{sm2}, \mathbf{x}_{sm3}$ and three corresponding points $\mathbf{x}'_{sm1}, \mathbf{x}'_{sm2}, \mathbf{x}'_{sm3}$ in the plane, there exists a unique affine transformation T such that $\mathbf{x}'_{sm1} = T(\mathbf{x}_{sm1})$ $\mathbf{x}'_{sm2} = T(\mathbf{x}_{sm2})$, and $\mathbf{x}'_{sm3} = T(\mathbf{x}_{sm3})$.

The affine transformation $T : \mathbf{x} \rightarrow A\mathbf{x} + \mathbf{t}$ can be determined from the set of corresponding points as follows. Let $\mathbf{m}_i, i = 1, 2, 3$ and $\mathbf{i}_i, i = 1, 2, 3$ be the set of corresponding model and image points, respectively, where

$$\mathbf{m}_i = \begin{pmatrix} m_{i,x} \\ m_{i,y} \end{pmatrix}$$

and

$$\mathbf{i}_i = \begin{pmatrix} i_{i,x} \\ i_{i,y} \end{pmatrix}$$

The six unknown parameters of the 2×2 matrix

$$A = \begin{pmatrix} a_{11} & a_{12} \\ a_{21} & a_{22} \end{pmatrix}$$

and of the vector

$$\mathbf{t} = \begin{pmatrix} t_1 \\ t_2 \end{pmatrix}$$

can be computed by solving the following system of equations

$$\begin{pmatrix} i_{1,x} & i_{1,y} \\ i_{2,x} & i_{2,y} \\ i_{3,x} & i_{3,y} \end{pmatrix} - \begin{pmatrix} m_{1,x} & m_{1,y} & 1 \\ m_{2,x} & m_{2,y} & 1 \\ m_{3,x} & m_{3,y} & 1 \end{pmatrix} \begin{pmatrix} a_{11} & a_{12} \\ a_{21} & a_{22} \\ t_1 & t_2 \end{pmatrix} = 0$$

Boundary: A closed curve that separates an image component from the background and/or other components.

Connected component: A set C of connected image points. Two points (i, j) and (h, k) are connected if there is a path from (i, j) to (h, k) consisting only of points of C

$$(i, j) = (i_0, j_0), (i_1, j_1), \ldots (i_t, j_t) = (h, k)$$

such that (i_s, j_s) is adjacent to $(i_{s+1}, j_{s+1}), 0 \leq s \leq t - 1$. Two definitions of adjacency are generally used, 4-adjacency and 8-adjacency. Point (i, j) has four 4-adjacent points, those with coordinates $(i-1, j), (i, j-1), (i, j+1),$ and $(i+1, j+1)$. The 8-adjacent points are, in addition to the four above, the points along the diagonals, namely: $(i - 1, j - 1), (i - 1, j + 1), (i + 1, j - 1),$ and $(i + 1, j + 1)$.

Digital line: The set of image cells that have a nonempty intersection with a straight-line.

Edge detector: A technique to detect intensity discontinuities. It yields points lying on the boundary between objects and the background. Some of most popular edge detectors are listed. Given an image I, the *gradient operator* is based on the computation of the first-order derivatives, $\delta I / \delta x$ and $\delta I / \delta y$. In a **digital image** there are different ways of approximating such derivatives. The first approach involves only four adjacent pixels.

$$\delta I / \delta x = I(x, y) - I(x - 1, y)$$
$$\delta I / \delta y = I(x, y) - I(x, y - 1)$$

Alternatively, using eight adjacent values:

$$\delta I / \delta x = [I(x + 1, y - 1) + 2I(x + 1, y) + I(x + 1, y + 1)]$$
$$- [I(x - 1, y - 1) + 2I(x - 1, y) + I(x - 1, y + 1)]$$

The computation of the above expressions can be obtained by applying the following *masks*, known as the *Sobel operators*, to all image points.

−1	−2	−1
0	0	0
1	2	1

−1	0	1
−2	0	2
−1	0	1

The *magnitude* of the gradient, denoted by $G(I(x, y))$, is

$$G(I(x, y)) = \left[(\delta I / \delta x)^2 + (\delta I / \delta y)^2 \right]^{1/2}$$

or simply

$$G(I(x, y)) = |\delta I/\delta x| + |\delta I/\delta y|$$

The Laplacian operator is a second-order derivative operator defined as

$$L[I(x, y)] = \delta^2 I/\delta x^2 + \delta^2 I/\delta y^2$$

or, in the digital version, as

$$L[I(x, y)] = L[I(x + 1, y) + I(x - 1, y) + I(x, y + 1) + I(x, y - 1)] - 4I(x, y)$$

The zero-crossing operator determines whether the digital Laplacian has a zero-crossing at a given pixel, that is whether the gradient edge detector has a relative maxima.

Edge point: An image point that is at the border between two image components, that is a point where there is an abrupt change in intensity.

Hausdorff distance: Let $A = \{a_1, a_2, \ldots, a_m\}$ and $B = \{b_1, b_2, \ldots, b_n\}$ two sets of points. The Hausdorff distance between A and B is defined as

$$H(A, B) = \max(h(A, B), h(B, A)),$$

where the one-way Hausdorff distance from A to B is

$$h(A, B) = \max_{a \in A} \min_{b \in B} d(a, b)$$

and $d(a, b)$ is the distance between two points a and b. The minimum Hausdorff distance is then defined as

$$D(A, B) = \min_{t \in E_2} H(t(A), B)$$

where E_2 is the group of planar motions and $t(A)$ is the transformed of A under motion t.

The corresponding Hausdorff decision problem for a given ϵ is deciding whether the minimum Hausdorff distance is bounded by ϵ. This last problem is generally solved as a problem of intersection of unions of disks in the transformation space.

Image, digital image: A 2D array I of regions or cells each with an assigned integer representing the intensity value or gray level of the cell. A binary image is an image with only two gray levels: 0 (white), 1 (black).

Polygonal approximation of a curve: A segmentation of a curve into piecewise linear segments that approximates it. One approach to determine it for an open curve is by considering the segment between its endpoints and recursively dividing a segment if the maximum distance between the segment and the curve points is above a given threshold. The segment is divided at the point of maximum distance.

References

1. Arman, F. and Aggarwal, J.K., Model-based object recognition in dense-range images—A review. *ACM Comput. Surv.*, 25(1), 6–43, 1993.
2. Ballard, D.H., Generalizing the Hough transform to detect arbitrary shapes. *Pattern Recogn.*, 3(2), 11–22, 1981.
3. Besl, P.J. and Jain, R.C., Three-dimensional object recognition. *Comput. Surv.*, 17(1), 75–145, 1985.
4. Binford, T.O., Survey of model-based image analysis systems. *Int. J. Robot. Res.*, 1(1), 18–64, 1982.
5. Bruel, T.M., Geometric aspects of visual object recognition. PhD dissertation, MIT, Cambridge, MA, 1992.

6. Burt, P.S., The pyramid as a structure for efficient computation. In *Multiresolution Image Processing and Analysis*, Ed., A. Rosenfeld, Germany, pp. 6–35, Springer-Verlag, Berlin, 1984.

7. Cass, T.A., Polynomial-time geometric matching for object recognition. *Proceedings of the European Conference on Computer Vision*, Graz, Austria, 1992.

8. Chin, R.T. and Dyer, C.R., Model-based recognition in robot vision. *ACM Comput. Surv.*, 18(1), 66–108, 1986.

9. Cantoni, V., Cinque, L., Guerra, C., Levialdi, S., and Lombardi, L., Recognizing 2D objects by a multiresolution approach. *12th International Conference on Pattern Recognition*, pp. 310–316, Israel, 1994.

10. Dasri, R., Costa, L., Geiger, D., and Jacobs, D., Determining the similarity of deformable shapes. *IEEE Workshop on Physics-Based Modeling in Computer Vision*, pp. 135–143, Cambridge, MA, 1995.

11. Duda, R.O. and Hart, P.E., Use of the Hough transformation to detect lines and curves in pictures. *Commun. ACM*, 15(1), 11–15, 1972.

12. Flynn, P.J. and Jain, A.K., 3D object recognition using invariant feature indexing of interpretation tables. *CVIP Image Underst.*, 55(2), 119–129, 1992.

13. Gorman, J.R., Mithcell, R., and Kuhl, F., Partial shape recognition using dynamic programming. *IEEE Trans. Pattern Anal. Mach. Intell.*, 10(2), 257–266, 1988.

14. Grimson, W.E.L., On the recognition of parameterized 2D objects. *Int. J. Comput. Vis.*, 3, 353–372, 1988.

15. Grimson, W.E., Huttenlocher, D.P., and Jacobs, D., A study of affine matching with bounded sensor error. *Inter. J. Comp. Vis.*, 13(1), 7–32, 1994.

16. Hough, P.V., Methods and means to recognize complex patterns. U.S. patent 3.069.654, 1962.

17. Huttenlocher, D.P. and Ullman, S., Recognizing solid objects by alignment with an image. *Inter. J. Comput. Vis.*, 5, 195–212, 1990.

18. Jacobs, D., Optimal matching of planar models in 3D scenes. *IEEE Conf. Comput. Vis. Pattern Recogn.*, 24, 269–274, 1991.

19. Lamdan, Y., Schwartz, J.T., and Wolfson, H.J., On the recognition of 3-D objects from 2-D images. *Proceedings of the IEEE International Conference on Robotics and Application*, IEEE Computer Soc., Los Alamitos, CA, 1988, pp. 1407–1413.

20. Lamdan, Y., Schwartz, J.T., and Wolfson, H.J., Affine invariant model-based object recognition. *IEEE Trans. Robot. Autom.*, 5, 578–589, 1990.

21. Lamdan, Y. and Wolfson, H., Geometric hashing: A general and efficient model-based recognition scheme. *Proceedings of the Second IEEE International Conference on Computer Vision*, pp. 238–249, Tampa, FL, 1988.

22. Lowe, D.G., Three-dimensional object recognition from single two-dimensional images. *Artif. Intell.*, 31, 355–395, 1987.

23. Lu, S.Y., A tree-to-tree distance and its application to cluster analysis. *IEEE Trans. Pattern Anal. Mach. Intell.*, 1(1), 219–224, 1971.

24. Maes, M., On a cyclic string-to-string correction problem. *Inf. Process. Lett.*, 35, 73–78, 1990.

25. Mokhtarian, F. and Mackworth, A.K., Scale-based descriptions and recognition of planar curves and two dimensional shapes. *IEEE Trans. Pattern Anal. Mach. Intell.*, 8(1), 34–43, 1986.

26. Mokhtarian, F., Silhouette-based isolated object recognition through curvature scale space. *IEEE Trans. Pattern Anal. Mach. Intell.*, 17(5), 539–544, 1995.

27. Pauwels, E.J., Moons, T., Van Gool, L.J., Kempeners, P., and Oosterlinck, A., Recognition of planar shapes under affine distortion. *Int. J. Comput. Vis.*, 14, 49–65, 1995.

28. Sakou, H., Yoda, H., and Ejiri, M., An algorithm for matching distorted waveforms using a scale-based description. *Proceedings of the IAPR Workshop on Computer Vision*, pp. 329–334, Tokyo, Japan, 1988.

29. Segen, J., Model learning and recognition of nonrigid objects. *Proceedings of the Conference on Computer Vision Pattern Recognition*, pp. 597–602, San Diego, CA, 1989.

30. Shapiro, L.G. and Haralick, R.M., Structural descriptions and inexact matching. *IEEE Trans. Pattern Anal. Mach. Intell.*, 3, 504–519, 1981.

31. Shasha, D. and Zhang, K., Fast algorithms for the unit cost editing distance between trees. *J. Algorithm*, 11, 581–621, 1990.

32. Stein, F. and Medioni, G., Structural hashing: Efficient three-dimensional object recognition. *Proceedings of the IEEE Conference Computer Vision and Pattern Recognition*, pp. 244–250, Maui, HI, 1991.

33. Suetens, P., Fua, P., and Hanson, A.J., Computational strategies for object recognition. *ACM Comput. Surv.*, 24(1), 6–61, 1992.

34. Tsai, W. and Yu, S., Attributed string matching with merging for shape recognition. *IEEE Trans. Pattern Anal. Mach. Intell.*, 7(4), 453–462, 1985.

35. Ueda, N. and Suzuki, S., Learning visual models from shape contours using multi-scale convex/concave structure matching. *IEEE Trans. Pattern Anal. Mach. Intell.*, 15(4), 337–352, 1993.

36. Wagner, R.A. and Fischer, M.J., The string-to-string correction problem. *J. Assoc. Comput. Mach.*, 21, 168–173, 1974.

6

Graph Drawing Algorithms

Peter Eades
University of Sydney

Carsten Gutwenger
Dortmund University of Technology

Seok-Hee Hong
University of Sydney

Petra Mutzel
Dortmund University of Technology

6.1 Introduction

Graphs are commonly used in computer science to model relational structures such as programs, databases, and data structures. For example:

- Petri nets are used extensively to model communications protocols or biological networks.
- Call graphs of programs are often used in CASE tools.
- Data flow graphs are used widely in Software Engineering; an example is shown in Figure 6.1a.
- Object oriented design techniques use a variety of graphs; one such example is the UML class diagram in Figure 6.1b.

One of the critical problems in using such models is that the graph must be drawn in a way that illuminates the information in the application. A good graph drawing gives a clear understanding of a structural model; a bad drawing is simply misleading. For example, a graph of a computer network is pictured in Figure 6.2a; this drawing is easy to follow. A different drawing of the same graph is shown in Figure 6.2b; this is much more difficult to follow.

A graph drawing algorithm takes a graph and produces a drawing of it. The graph drawing problem is to find graph drawing algorithms that produce good drawings. To make the problem more precise, we define a graph $G = (V, E)$ to consist of a set V of vertices and a set E of edges, that

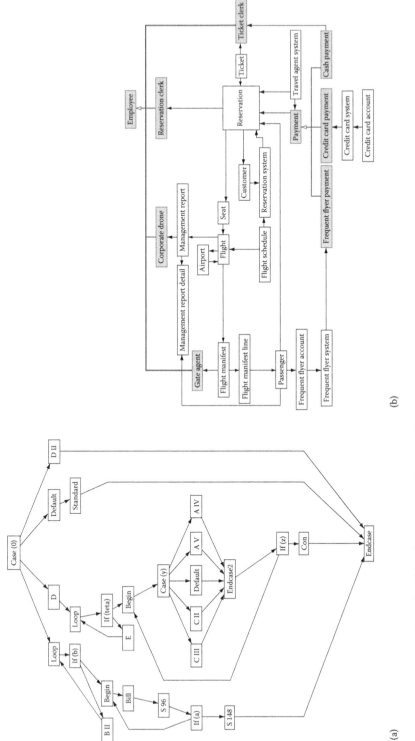

FIGURE 6.1 Two real-world graphs: (a) data flow diagram and (b) UML class diagram.

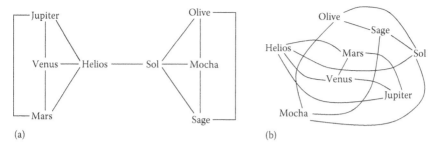

FIGURE 6.2 Two drawings of the same computer network.

is, unordered pairs of vertices. A drawing of G assigns a location (in two or three dimensions) to every vertex of G and a simple curve c_{uv} to every edge (u, v) of G such that the endpoints of c_{uv} are the locations of u and v. Notation and terminology of graph theory is given in [8].

The remainder of this chapter gives an overview of graph drawing (Section 6.2) and details algorithms for obtaining planar drawings (straight-line as well as orthogonal, see Sections 6.3.1 and 6.3.2) and symmetric drawings (Section 6.5). Since not every graph is planar, Section 6.4 describes methods for planarizing a nonplanar graph. A list of research problems is given in Section 6.6, and a review of the defining terms is given in Section 6.7.

6.2 Overview

In this section, we provide a brief overview of the graph drawing problem. Section 6.2.1 outlines the main conventions for drawing graphs, and Section 6.2.2 lists the most common "aesthetic criteria," or optimization goals, of graph drawing. A brief survey of some significant graph drawing algorithms is given in Section 6.2.3.

6.2.1 Drawing Conventions

Drawing conventions for graphs differ from one application area to another. Some of the common conventions are listed below.

- Many graph drawing methods output a grid drawing: the location of each vertex has integer coordinates. Some grid drawings appear in Figure 6.3.
- In a polyline drawing, the curve representing each edge is a polyline, that is, a chain of line segments. Polyline drawings are presented in Figures 6.3b and 6.1a. If each polyline is just a line segment, then the drawing is a straight-line drawing, as in Figure 6.3c.

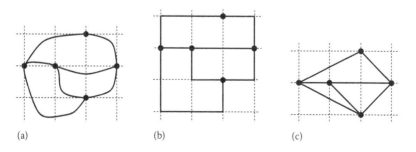

FIGURE 6.3 Examples of drawing conventions: (a) a grid drawing, (b) an orthogonal grid drawing, and (c) a straight-line drawing.

- In an orthogonal drawing, each edge is a polyline composed of straight line segments parallel to one of the coordinate axes. Orthogonal drawings are used in many application areas because horizontal and vertical line segments are easy to follow. Figures 6.3b and 6.1b are orthogonal drawings.

6.2.2 Aesthetic Criteria

The main requirement of a graph drawing method is that the output should be readable; that is, it should be easy to understand, easy to remember, and it should illuminate rather than obscure the application. Of course it is difficult to model readability precisely, since it varies from one application to another, and from one human to another; these variations mean that there are many graph-drawing problems. The problems can be classified roughly according to the specific optimization goals that they try to achieve. These goals are called aesthetic criteria; a list of some such criteria is given below.

- Minimization of the number of edge crossings is an aesthetic criterion, which is important in many application areas. The drawing in Figure 6.2a has no edge crossings, whereas the one in Figure 6.2b has ten.

  A graph that can be drawn in the plane with no edge crossings is called a planar graph. Methods for drawing planar graphs are described in Section 6.3.

  In general, visualization systems must deal with graphs that are not planar. To exploit the theory of planar graphs and methods for drawing planar graphs, it is necessary to planarize nonplanar graphs, that is, to transform them to planar graphs. A planarization technique is described in Section 6.4.

- Bends: In polyline drawings, it is easier to follow edges with fewer bends. Thus, many graph drawing methods aim to minimize, or at least bound, the number of edge bends. In the drawing in Figure 6.3b, there are six edge bends; the maximum number of bends on an edge is two. Algorithms for straight-line drawings (no bends at all) are given in Sections 6.3.1 and 6.5.

  Very few graphs have an orthogonal drawing with no bends, but there are a number of methods that aim to keep the number of bends in orthogonal drawings small. A method for creating two-dimensional planar orthogonal drawings of a planar graph (with a fixed embedding) with a minimum number of bends is described in Section 6.3.2.

- The vertex resolution of a drawing is the minimum distance between a pair of vertices. For a given screen size, we would like to have a drawing with maximum resolution.

  In most cases, the drawing is a grid drawing; this guarantees that the drawing has vertex resolution at least one. To ensure adequate vertex resolution, we try to keep the *size* of the grid drawing bounded. If a two dimensional grid drawing lies within an isothetic rectangle of width w and height h, then the vertex resolution for a unit square screen is at least $\max(1/w, 1/h)$. All the methods described in Section 6.3 give grid drawings with polynomially bounded size.

- Displaying symmetries in a drawing of a graph is an important criterion, as symmetry clearly reveals the hidden structure of the graph. For example, the drawing in Figure 6.2a displays one reflectional (or axial) symmetry. It clearly shows two isomorphic subgraphs with congruent drawings. Symmetric drawings are beautiful and easy to understand.

Maximizing the number of symmetries in a drawing is one of the desired optimization criteria in graph drawing. Formal models and algorithms for symmetric graph drawings are described in Section 6.5.

6.2.3 Drawing Methods

In Sections 6.3 through 6.5, we describe some graph drawing methods in detail. However, there are many additional approaches to graph drawing. Here, we briefly overview some of the more significant methods.

6.2.3.1 Force-Directed Methods

Force-directed methods draw a physical analogy between the layout problem and a system of forces defined on drawings of graphs. For example, vertices may be replaced with bodies that repel each other, and edges have been replaced with Hooke's law springs. In general, these methods have two parts:

The model: This is a "force system" defined by the vertices and edges of the graph. It is a physical model for a graph. The model may be defined in terms of "energy" rather than in terms of a system of forces; the force system is just the derivative of the energy system.

The algorithm: This is a technique for finding an equilibrium state of the force system, that is, a position for each vertex such that the total force on every vertex is zero. This state defines a drawing of the graph. If the model is stated as an energy system then the algorithm is usually stated as a technique for finding a configuration with locally minimal energy.

Force directed algorithms are easy to understand and easy to implement, and thus they have become quite popular [20,25]. Moreover, it is relatively easy to integrate various kinds of constraints. Recently, multilevel methods in combination with spatial data structures such as quad trees have been used to overcome the problems with large graphs. These methods are fairly successful with regular graphs (e.g., meshed-based), highly symmetric graphs, and "tree-like" graphs (see Figure 6.4). They work in both two and three dimensions. Comparisons of the many variations of the basic idea appear in [10,32].

6.2.3.2 Hierarchical Methods

Hierarchical methods are suitable for directed graphs, especially where the graph has very few directed cycles.

Suppose that $G = (V, E)$ is an acyclic directed graph. A layering of G is a partition of V into subsets $L_1, L_2, \ldots, L_h$, such that if $(u, v) \in E$ where $u \in L_i$ and $v \in L_j$, then $i > j$. An acyclic directed graph with a layering is a hierarchical graph.

Hierarchical methods convert a directed graph into a hierarchical graph, and draw the hierarchical graph such that layer L_i lies on the horizontal line $y = i$. A sample drawing is shown in Figure 6.1a. The aims of these methods include the following.

a. Represent the "flow" of the graph from top to bottom of the page. This implies that most arcs should point downward.

b. Ensure that the graph drawing fits the page; that is, it is not too high (not too many layers) and not too wide (not too many vertices on each layer).

c. Ensure that arcs are not too long. This implies that the y extent $|i - j|$ of an arc (u, v) with $u \in L_i$ and $v \in L_j$ should be minimized.

d. Reduce the number of edge crossings.

e. Ensure that arcs are as straight as possible.

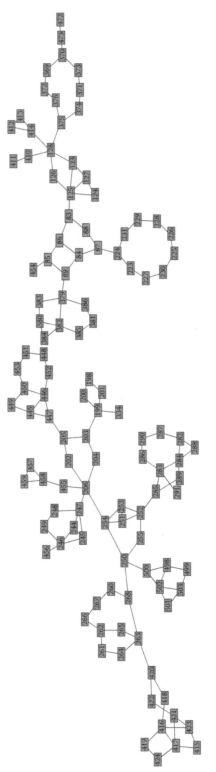

FIGURE 6.4 A force-directed drawing.

There are many hierarchical methods, but the following four steps are common to most.

1. Directed cycles are removed by temporarily reversing some of the arcs. The arcs that are reversed will appear pointing upward in the final drawing, and thus, the number of arcs reversed should be small to achieve (a) above.
2. The set of vertices is partitioned into layers. The partition aims to achieve (b) and (c) above.
3. Vertices within each layer are permuted so that the overall number of crossings is small.
4. Vertices are positioned in each layer so that edges that span more than one layer are as straight as possible.

Each step involves heuristics for NP-hard optimization problems. A detailed description of hierarchical methods appears in [22]. An empirical comparison of various hierarchical drawing methods appears in [7].

6.2.3.3 Tree Drawing Methods

Rooted trees are special hierarchical graphs, and hierarchical methods apply. We can assign vertices at depth k in the tree to layer $h - k$, where h is the maximum depth. The convention that layer i is drawn on the horizontal line $y = i$ helps the viewer to see the hierarchy represented by the tree. However, for trees, there are some simpler approaches; here, we outline one such method.

Note that the edge crossing problem is trivial for trees. The main challenge is to create a drawing with width small enough to fit the page. The Reingold–Tilford algorithm [48] is a simple heuristic method designed to reduce width. Suppose that T is an oriented binary rooted tree. We denote the left subtree by T_L and the right subtree by T_R. Draw subtrees T_L and T_R recursively on separate sheets of paper. Move the drawings of T_L and T_R toward each other as far as possible without the two drawings touching. Now, center the root of T between its children. The Reingold–Tilford method can be extended to nonbinary trees [50,12]. A typical tree drawing based on these ideas is shown in Figure 6.5.

6.2.3.4 3D Drawings

Affordable high quality 3D graphics in every PC has motivated a great deal of research in three-dimensional graph drawing over the last two decades. Three-dimensional graph drawings with a variety of aesthetics and edge representations have been extensively studied by the graph drawing community since early 1990. The proceedings of the annual graph drawing conferences document these developments.

Three-dimensional graph drawing research can be roughly divided into two categories: grid drawing and nongrid drawing. Grid drawings can be further divided into two categories based on edge representations: straight-line drawing and orthogonal drawing. Further, for each drawing

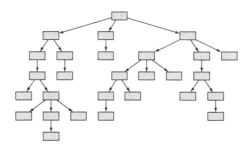

FIGURE 6.5 Tree drawn with an algorithm based on the ideas by Reingold–Tilford.

convention, one can define the following aesthetic (i.e., optimization) criteria: minimize volume for straight-line grid drawings and orthogonal grid drawings, minimize the number of bends in orthogonal grid drawings, or maximize symmetry in nongrid straight-line drawings.

Note that there is a recent survey on three-dimensional graph drawing [19], which covers the state of art survey on three dimensional graph drawing, in particular straight-line grid drawing and orthogonal grid drawing.

6.3   Planar Graph Drawing

A great deal of the long history of graph theory has been devoted to the study of representations of graphs in the plane. The concepts and results developed over centuries by graph theorists have proved very useful in recent visualization and VLSI layout applications. This section describes some techniques used in finding good drawings in the plane.

A representation of a planar graph without edge crossings in the plane is called a plane representation or a planar embedding. The graph is then called a plane graph. A planar embedding divides the plane into regions, which we call faces. The unbounded face is called the outer face. A planar embedding can be represented by a planar map, that is, a cyclic ordered list of the edges bounding each face. There are linear time algorithms for testing planarity of a given graph and, in the positive case, computing a planar map (see, e.g., [9]).

6.3.1   Straight Line Drawings

In this section, we consider the problem of constructing straight-line planar drawings of planar graphs. This problem predates the applications in information visualization and was considered by a number of mathematicians (e.g., [23]). The problem with these early approaches was that they offered poor resolution; that is, they placed vertices exponentially close together. The breakthrough came with the following theorem from [16].

THEOREM 6.1   *[16] Every n-vertex planar graph has a straight-line planar grid drawing which is contained within a rectangle of dimensions $O(n) \times O(n)$.*

The purpose of this section is to outline a constructive proof of Theorem 6.1. Roughly speaking, it proceeds as follows.

1. Dummy edges are added to the graph to ensure that it is a triangulated planar graph, that is, each face is a triangle.
2. The vertices are ordered in a certain way defined below.
3. The vertices are placed one at a time on the grid in the order defined in step 2.

The first step is not difficult. First we find a planar embedding, and then add edges to every face until it becomes a triangle. We present steps 2 and 3 in the following two subsections.

6.3.1.1   Computing the Ordering

Let $G = (V, E)$ be a triangulated plane graph with $n > 3$ vertices, with vertices u, v, and w on the outer face. Suppose that the vertices of V are ordered $v_1, v_2, \ldots, v_n$. Denote the subgraph induced by $v_1, v_2, \ldots, v_\ell$ by G_ℓ, and the outer face of G_ℓ by C_ℓ. The ordering $v_1 = u, v_2 = v, \ldots, v_n = w$ is a canonical ordering if for $3 \leq \ell \leq n - 1$, G_ℓ is two-connected, C_ℓ is a cycle containing the edge (v_1, v_2), $v_{\ell+1}$ is in the outer face of G_ℓ, and $v_{\ell+1}$ has at least two neighbors in G_ℓ, and the neighbors are consecutive on the path $C_\ell - (v_1, v_2)$.

LEMMA 6.1   [16] Every triangulated plane graph has a canonical ordering.

PROOF 6.1   The proof proceeds by reverse induction. The outer face of G is the triangle v_1, v_2, v_n. Since G is triangulated, the neighbors of v_n form a cycle, which is the boundary of the outer face of $G - \{v_n\} = G_{n-1}$. Thus, the lemma holds for $\ell = n - 1$.

Suppose that $i \leq n - 2$, and assume that $v_n, v_{n-1}, \ldots, v_{i+2}$ have been chosen so that G_{i+1} and C_{i+1} satisfy the requirements of the lemma. We need to choose a vertex w as v_{i+1} on C_{i+1} so that w is not incident with a chord of C_{i+1}. It is not difficult to show that such a vertex exists.   □

6.3.1.2   The Drawing Algorithm

The algorithm for drawing the graph begins by drawing vertices v_1, v_2, and v_3 at locations $(0, 0)$, $(2, 0)$, and $(1, 1)$, respectively. Then, the vertices $v_4, \ldots, v_n$ are placed one at a time, increasing in y coordinate. After v_k has been placed, we have a drawing of the subgraph G_k. Suppose that the outer face C_k of the drawing of G_k consists of the edge (v_1, v_2), and a path $P_k = (v_1 = w_1, w_2, \ldots, w_m = v_2)$; then, the drawing is constructed to satisfy the following three properties:

1. The vertices v_1 and v_2 are located at $(0, 0)$ and $(2k - 4, 0)$, respectively.
2. The path P_k increases monotonically in the x direction; that is, $x(w_1) < x(w_2) < \cdots < x(w_m)$.
3. For each $i \in \{1, 2, \ldots, m - 1\}$, the edge (w_i, w_{i+1}) has slope either $+1$ or -1.

Such a drawing is illustrated in Figure 6.6.

We proceed by induction to show how a drawing with these properties may be computed. The drawing of G_3 satisfies the three properties. Now, suppose that $k \geq 3$, and we have a drawing of G_k that satisfies the three properties. The canonical ordering of the vertices implies that the neighbors of v_{k+1} in G_k occur consecutively on the path P_k; suppose that they are $w_p, w_{p+1}, \ldots, w_q$. Note that the intersection of the line of slope $+1$ through w_p with the line of slope -1 through w_q is at a grid point $\mu(p, q)$ (since the Manhattan distance between two vertices is even). If we placed v_{k+1} at $\mu(p, q)$, then the resulting drawing would have no edge crossings, but perhaps the edge (w_p, v_{k+1}) would overlap with the edge (w_p, w_{p+1}), as in Figure 6.6. This can be repaired by moving all $w_{p+1}, w_{p+2}, \ldots, w_m$ one unit to the right. It is also possible that the edge (v_{k+1}, w_q) overlaps the edge (w_{q-1}, w_q), so we move all the vertices $w_q, w_{q+1}, \ldots, w_m$ a further unit to the right. This ensures that the newly added edges do not overlap with the edges of G_k. Now, we can place v_{k+1} safely at $\mu(p, q)$ as shown in Figure 6.7 (note that $\mu(p, q)$ is still a grid point). The three required properties clearly hold.

But there is a problem with merely moving the vertices on P_k: after the vertices $w_p, w_{p+1}, \ldots$ w_m are moved to the right, the drawing of G_k may have edge crossings. We must use a more comprehensive strategy to repair the drawing of G_k; we must move even more vertices. Roughly speaking, when moving vertex w_i to the right, we will also move the vertices to the right below w_i.

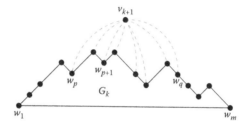

FIGURE 6.6 The subgraph G_k with v_{k+1} placed at $\mu(p, q)$.

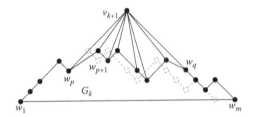

FIGURE 6.7 The subgraph G_k with v_{k+1} placed at $\mu(p, q)$ after moving the vertices $w_{p+1}, \ldots, w_{q-1}$ one unit and the vertices $w_q, \ldots, w_m$ two units to the right.

To this end, we define a sequence of sets $M_k(w_1), M_k(w_2), \ldots, M_k(w_m)$ below. For a given sequence $\alpha(w_1), \alpha(w_2), \ldots, \alpha(w_m)$ of nonnegative integers, the vertices of G_k are moved successively as follows: first, all vertices of $M_k(w_1)$ are moved by $\alpha(w_1)$, then all vertices of $M_k(w_2)$ are moved by $\alpha(w_2)$, and so on.

The sets M_k are defined recursively as follows: $M_3(v_1) = \{v_1, v_2, v_3\}$, $M_3(v_2) = \{v_2, v_3\}$, $M_3(v_3) = \{v_3\}$. To compute M_{k+1} from M_k, note that if v_{k+1} is adjacent to $w_p, w_{p+1}, \ldots, w_q$, then P_{k+1} is $(w_1, w_2, \ldots, w_p, v_{k+1}, w_q, \ldots, w_m)$. For each vertex u on this path, we must define $M_{k+1}(u)$. Roughly speaking, we add v_k to $M_k(w_i)$ if w_i is left of v_{k+1}, otherwise $M_k(w_i)$ is not altered. More precisely, for $1 \le i \le p$, we define $M_{k+1}(w_i)$ to be $M_k(w_i) \cup \{v_{k+1}\}$, and $M_{k+1}(v_{k+1})$ is $M_k(w_{p+1}) \cup \{v_{k+1}\}$. For $q \le j \le M$, we define $M_{k+1}(w_j)$ to be $M_k(w_j)$. It is not difficult to show that the sets satisfy the following three properties:

(a) $w_j \in M_k(w_i)$ if and only if $i \le j$.
(b) $M_k(w_m) \subset M_k(w_{m-1}) \subset \cdots \subset M_k(w_1)$.
(c) Suppose that $\alpha(w_1), \alpha(w_2), \ldots, \alpha(w_m)$ is a sequence of nonnegative integers and we apply algorithm *Move* to G_k; then the drawing of G_k remains planar.

These properties guarantee planarity.

6.3.1.3 Remarks

The proof of Theorem 6.1 constitutes an algorithm that can be implemented in linear time [15]. The area of the computed drawing is $2n - 4 \times n - 2$; this is asymptotically optimal, but the constants can be reduced to $(n - \Delta_0 - 1) \times (n - \Delta_0 - 1)$, where $0 \le \Delta_0 \le \lfloor (n - 1)/2 \rfloor$ is the number of cyclic faces of G with respect to its minimum realizer (see [51]).

A sample drawing appears in Figure 6.8, which also shows some of the problems with the method. If the input graph is relatively sparse, then the number of "dummy" edges that must be added can be significant. These dummy edges have considerable influence over the final shape of the drawing but do not occur in the final drawing; the result may appear strange (see Figure 6.8). Using a different ordering, the algorithm also works for triconnected and even for biconnected planar graphs [41,27]. This reduces the number of dummy edges considerably.

Although the drawings produced have relatively good vertex resolution, they are not very readable, because of two reasons: the angles between incident edges can get quite small, and the area is still too big in practice. These two significant problems can be overcome by allowing bends in the edges. Kant [41] modifies the algorithm described above to obtain a "mixed model algorithm;" this algorithm constructs a polyline drawing of a triconnected planar graph such that the size of the minimal angle is at least $\frac{1}{d}\pi$, where d is the maximal degree of a vertex. Figure 6.9 shows the same graph as in Figure 6.8 drawn with a modification of the mixed model algorithm for biconnected planar graphs (see [28]). The grid size of the drawing in Figure 6.8 is 38×19, whereas the size for the mixed model drawing is 10×8.

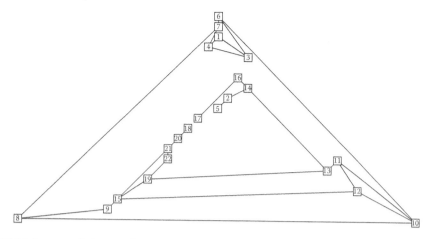

FIGURE 6.8 Sparse graph drawn with straight line edges.

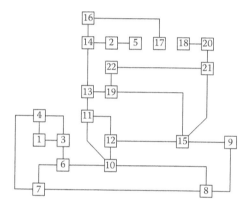

FIGURE 6.9 Sparse graph drawn with the mixed model algorithm.

6.3.2 Orthogonal Drawings

In polyline drawings, one aesthetic is to minimize the total number of bends. Computing an orthogonal drawing of a planar graph with the minimum number of bends is an NP-hard problem. However, for the restricted problem wherein a fixed planar embedding is part of the input, there is a polynomial time algorithm [49]. In this subsection, we will describe the transformation of this restricted bend minimization problem into a network flow problem.

6.3.2.1 Mathematical Preliminaries

Intuitively, the orthogonal representation of an orthogonal planar drawing of a graph describes its "shape;" it does not specify the length of the line-segments but determines the number of bends of each edge. The algorithm described in this section takes as input a planar embedding, represented as a planar map (that is a cyclic ordered list of the edges bounding each face). It produces as output an orthogonal representation with the minimum number of bends. From this, an actual drawing can be constructed via well-known compaction algorithms (see, e.g., [49,45] or Chapter 8).

Formally, an orthogonal representation is a function H assigning an ordered list $H(f)$ to each face f. Each element of $H(f)$ has the form (e, s, a), where e is an edge adjacent to f, s is a binary string, and $a \in \{90, 180, 270, 360\}$. If r is an element in $H(f)$, then e_r, s_r, and a_r denote the corresponding entries, that is, $r = (e_r, s_r, a_r)$. The list for an inner face f is ordered so that the edges e_r appear in clockwise order around f; thus, H consists of a planar map of the graph extended by the s- and a-entries. The kth bit in $s_r \in H(f)$ represents the kth bend while traversing the edge in face f in clockwise order: the entry is 0 if the bend produces a 90° angle (on the right hand side) and 1 otherwise. An edge without bend is represented by the zero-string ϵ. The number a_r represents the angle between the last line segment of edge e_r and the first line segment of edge e'_r, where $r' \in H(f)$ follows r in the clockwise order around f. For an outer face f_0, the element $r = (e_r, s_r, a_r)$ is defined similarly, just by replacing "clockwise order" by "counterclockwise order."

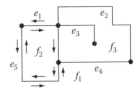

FIGURE 6.10   A graph and its orthogonal representation.

An orthogonal representation of Figure 6.10 is

$$H(f_1)=((e_1, \epsilon, 270), (e_5, 11, 90), (e_4, \epsilon, 270), (e_2, 1011, 90))$$
$$H(f_2)=((e_1, \epsilon, 90), (e_6, \epsilon, 180), (e_5, 00, 90))$$
$$H(f_3)=((e_2, 0010, 90), (e_4, \epsilon, 90), (e_6, \epsilon, 90), (e_3, 0, 360), (e_3, 1, 90))$$

Given an orthogonal representation H, the number of bends is obviously given by

$$B(H) = \frac{1}{2} \sum_{f} \sum_{r \in H(f)} |s_r|,$$

where $|s|$ is the length of string s. Geometrical observations lead to the following lemma that characterizes orthogonal representations.

LEMMA 6.2   The function H is an orthogonal representation of an orthogonal planar drawing of a graph if and only if the properties (P1) to (P4) below are satisfied.

(P1) There is a plane graph with maximal degree four, whose planar map corresponds to the e-entries in $H(f)$.

(P2) Suppose that $r = (e_r, s_r, a_r)$ and $r' = (e_{r'}, s_{r'}, a_{r'})$ are two elements in the orthogonal representation representing the same edge, that is, $e_r = e_{r'}$. Then, reversing the order of the elements in the string s_r and complementing each bit gives the string $s_{r'}$.

(P3) Suppose that ZEROS(s) denotes the number of 0's in string s, ONES(s), the number of 1's in string s, and

$$\text{ROT}(r) = \text{ZEROS}(s_r) - \text{ONES}(s_r) + (2 - a[r]/90).$$

Then, the faces described by H build rectilinear polygons, that is, polygons in which the lines are horizontal and vertical, if and only if

$$\sum_{r \in H(f)} \text{ROT}(r) = \begin{cases} +4 & \text{if } f \text{ is an inner face} \\ -4 & \text{otherwise.} \end{cases}$$

(P4) For all vertices v in the plane graph represented by H, the following holds: the sum of the angles between edges with end vertex v given in the a-entries of the corresponding elements is 360.

6.3.2.2 The Transformation into a Network Flow Problem

Suppose that $G = (V, E)$ is a plane graph and P is a planar map with outer face f_0, that is, for each face f of the embedding, $P(f)$ is a list of the edges on f in clockwise order (respectively counterclockwise order for f_0). Denote the set of faces of G defined by P by F.

We define a network $N(P) = (U, A, b, l, u, c)$ with vertex set U, arc set A, demand/supply b, lower bound l, capacity u, and cost c, as follows. The vertex set $U = U_V \cup U_F$ consists of the vertices in V and a vertex set U_F associated with the set of faces F in G. The vertices $i_v \in U_V$ supply $b(i_v) = 4$ units of flow. The vertices $i_f \in U_F$ have supply respectively demand of

$$b(i_f) = \begin{cases} -2|P(f)| + 4 & \text{if } f \text{ is inner face} \\ -2|P(f)| - 4 & \text{otherwise.} \end{cases}$$

The arc (multi-)set A consists of the two sets A_V and A_F, as follows:

- A_V contains the arcs (i_v, i_f), $v \in V, f \in F$, for all vertices v that are endpoints of edges in $P(f)$ (v is associated with an angle in face f). Formally, we define $E(v, f) := \{e \in P(f) \mid e = (u, v)\}$ and $A_V = \{(i_v, i_f) \mid e \in E(v, f)\}$. Each arc in A_V has lower bound $l(i_v, i_f) = 1$, upper bound $u(i_v, i_f) = 4$, and zero cost $c(i_v, i_f) = 0$. Intuitively, the flow in these arcs is associated with the angle at v in face f.

- A_F contains two arcs (i_f, i_g) and (i_g, i_f) for each edge $e = (f, g)$, where f and g are the faces adjacent to e. Arcs in A_F have lower bound $l(i_f, i_g) = 0$, infinite upper bound, and unit cost $c(i_f, i_g) = c(i_g, i_f) = 1$.

A graph and its network are shown in Figure 6.11. The value of the flow through the network is

$$B(U) = \sum_{v \in V} b(i_v) + \sum_{f \in F} b(i_f) = 4|V| + \sum_{f \neq f_0}(-2|P(f)| + 4) - 2|P(f_0)| - 4 = 4(|V| - |E| + |F| - 2) = 0.$$

The transformation defined above seems to be quite complex, but it can be intuitively interpreted as follows. Each unit of flow represents an angle of $90°$. The vertices distribute their 4 flow units among their adjacent faces through the arcs in A_V; this explains the lower and upper bounds of 1 and 4, respectively. Flow in these arcs does not introduce any bends; this explains the zero cost. In general, the flow arriving from vertices $i_v \in U_V$ at the face vertices $i_f \in U_F$ is not equal to the

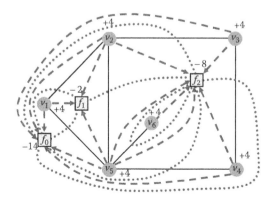

FIGURE 6.11 A graph and its network. The graph consists of vertices $v_1, \ldots, v_6$ and the solid edges. The nodes of the network are labeled with their supply and demand. The arc sets A_V and A_F are drawn dashed and dotted, respectively.

demand or supply of i_f. For adjusting the imbalances, flow will be sent through arcs in A_F. Each flow unit through such arcs will produce one bend. The flow conservation rule on vertices U_V forces the sum of the angles around the vertices to be $360°$ (see property (P4)). The conservation rule for vertices U_F forces the created faces to be rectilinear polygons (see (P3)). Most importantly, the cost of a given flow in the network $N(P)$ is equal to the number of bends in the constructed orthogonal representation.

LEMMA 6.3 [49] For any orthogonal representation H with planar map P, there exists an integer flow x in $N(P)$ with cost equal to the number of bends $b(H)$.

PROOF 6.3 The flow x in $N(P)$ can be constructed in the following way: With every arc (i_v, i_f) in A_V, we associate an element $r \in R(v, f)$, where $R(v, f) = \{r \in H(f) \mid e_r = (u, v)\}$. The flow value $x(i_v, i_f)$ arises from the a-entries in r: $x(i_v, i_f) = a_r/90$. For every arc (i_f, i_g) in A_F, we associate an element $r \in R(f, g)$, where $R(f, g) = \{r \in H(f) \mid e_r \in P(g)\}$. The flow is defined by the s-entries of r: $x(i_f, i_g) = \text{ZEROS}(s_r)$. The flow $x(i_f, i_g)$ represents the $90°$ angles within the region f along the edge represented by r separating f and g. The corresponding $270°$ angles in the face f are represented by their $90°$ angles in face g through $x(i_g, i_f)$.

We show that the function x is, indeed, a feasible flow with cost equal to the number of bends in H. Obviously, the lower and upper bounds of the arcs in A_V and A_F are satisfied by x. In order to show that the flow conservation rules at vertices in U are satisfied, we first consider the flow balance at a vertex $i_v \in U_V$ (using (P4)):

$$\sum_f x(i_v, i_f) - 0 = \sum_f \sum_{r \in R(v, f)} \frac{a_r}{90} = \sum_f \sum_{\{r \in H(f) \mid e_r = (u, v)\}} \frac{a_r}{90} = \sum_{e_r = (u, v)} \frac{a_r}{90} = 4 = b_{i_v}.$$

For vertices $i_f \in U_F$ (using (P2) and (P3)), we have

$$\sum_g x(i_f, i_g) - \sum_{v \in V} x(i_v, i_f) - \sum_{h \in F} x(i_h, i_f)$$

$$= -\sum_{r \in H(f)} \frac{a_r}{90} - \sum_{r \in H(f)} \text{ONES}(s_r) + \sum_{r \in H(f)} \text{ZEROS}(s_r)$$

$$= -2|P(f)| + \sum_{r \in H(f)} \text{ROT}(r) = -2|P(f)| \pm 4 = b_{i_f}.$$

The costs of the flow are equal to the number of bends in H:

$$B(H) = \frac{1}{2} \sum_f \sum_{r \in H(f)} |s_r| = \sum_{f \in F} \sum_{g \in F} x(i_f, i_g) = \sum_{a \in A_F} x_a = \sum_{a \in A} c_a x_a.$$

☐

LEMMA 6.4 [49] For any integer flow x in $N(P)$, there exists an orthogonal representation H with planar map P. The number of bends in H is equal to the cost of the flow x.

PROOF 6.4 We construct the orthogonal representation H as follows: the edge lists $H(f)$ are given by the planar map $P(f)$. Again, we associate an element $r \in R(v, f)$ with an arc (i_v, i_f) in A_V, and set

$a_r = 90 \cdot x(i_v, i_f)$. For the bit strings s, we associate with each $r \in R(f, g)$ a directed arc $(i_f, i_g) \in A_F$. Then, we determine the element $q \in H(g)$ with $q \neq r$ and $e_q = e_r$. We set

$$s_r = 0^{x(i_f, i_g)} 1^{x(i_g, i_f)} \quad \text{and} \quad s_q = 0^{x(i_g, i_f)} 1^{x(i_f, i_g)}.$$

We claim that the properties (P1) to (P4) are satisfied by the above defined e-, s- and a-entries. Obviously, (P1) and (P2) are satisfied by the construction. The sum of the angles between adjacent edges to v given by the a-entries is

$$\sum_{e_r = (u,v)} a_r = \sum_f x(i_v, i_f) \cdot 90 = 360,$$

thus showing (P4). Finally, it is not hard to show that the faces defined by H build rectilinear polygons (P3), since

$$\sum_{r \in H(f)} \text{ROT}(r) = 2|P(f)| + (-2|P(f)| \pm 4) = 0 \pm 4$$

The number of bends in the orthogonal representation is given by

$$B(H) = \frac{1}{2} \sum_f \sum_{r \in H(f)} |s_r| = \frac{1}{2} \sum_{(i_f, i_g) \in A_F} x(f, g) = \sum_{a \in A} c(a) x(a).$$

$\square$

In a network with integral costs and capacities, there always exists a min-cost flow with integer flow values. Hence we have the following theorem.

THEOREM 6.2 *[49] Let P be a planar map of a planar graph G with maximal degree 4. The minimal number of bends in an orthogonal planar drawing with planar map P is equal to the cost of a min-cost flow in N(P). The orthogonal representation H of any orthogonal planar drawing of G can be constructed from the integer-valued min-cost flow in N(P).*

The algorithm for constructing an orthogonal drawing is as follows. First, construct the network $N(P)$ in time $O(|U| + |A|)$. A min-cost flow on a planar graph with n vertices can be computed in time $O(n^2 \log n)$ (see Chapter 28 for the theory of network algorithms). In this special case, it is even possible to compute the min-cost flow in time $O(n^{\frac{7}{4}} \sqrt{\log n})$ [26]. The orthogonal representation can be constructed easily from the given flow. The length of the line segments can be generated by any compaction algorithm (see [49,45] or Chapter 8).

6.3.2.3 Extensions to Graphs with High Degree

The drawings produced by the bend minimization algorithm look very pleasant (see, e.g., Figures 6.13d and 6.17). Since the algorithm is restricted to planar graphs with maximal degree four, various models have been suggested for extending the network-flow method to planar graphs with high degree vertices. Tamassia et al. [4] have suggested the big nodes model where they decompose the high degree vertices into certain patterns of low degree vertices. Thereby, each edge is assigned to one of the new low degree vertices. This leads to layouts, in which the corresponding high degree nodes are drawn much bigger than the original ones (see Figure 6.12a). This can be overcome by placing vertices of original size into the centers of the big nodes. The quasiorthogonal drawing approach by Klau and Mutzel does not introduce such patterns, but simply integrates the high degree vertices

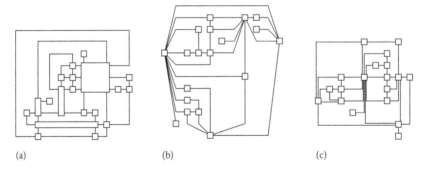

(a)   (b)   (c)

FIGURE 6.12 A graph with high degree vertices drawn with three different approaches: (a) big-nodes model, (b) quasiorthogonal drawing, and (c) Kandinsky layout.

into the network [44]. Also here, the high degree vertices expand their size, which is repaired again by a final replacement step (see Figure 6.12b).

Fößmeier and Kaufmann [24] extend the new network model so that it can deal with a coarse grid for the vertices and a fine grid for the edges (see Figure 6.12c). Unfortunately, the network model needs additional arcs and constraints leading to negative cycles. It is open if the min-cost flow in this Kandinsky model can be found in polynomial time. However, it can be formulated as an integer linear program, which can be solved to optimality for many practical instances using commercial ILP-solvers. There also exist approximation algorithms that theoretically guarantee a two-approximation of the optimal solution. Practical experiments have shown that the solutions found by these algorithms are very close to the optimal solutions [3].

6.4   Planarization

This section describes methods for transforming a nonplanar graph G into a planar graph. This transformation yields a planarized representation G' in which crossings are represented as additional nodes (called dummy nodes) with degree four. We then apply any planar drawing algorithm to G' and obtain a drawing of the original graph by simply replacing the dummy nodes in the layout of G' with edge crossings.

Planarization is a practically successful heuristic for finding a drawing with the minimum number of crossings [30]. The crossing minimization problem is NP-hard. For this problem, Buchheim et al. [11] presented an exact branch-and-cut algorithm that is able to solve small to medium size instances to optimality; cf. [13]. However, the algorithm is very complex in theory and practice so that it will not replace the planarization method in the future.

The planarization approach via planar subgraph was introduced by Batini et al. [5] as a general framework consisting of two steps: The first step computes a large planar subgraph, which serves as a starting point for the following step. The second step reinserts the remaining "nonplanar" edges, while trying to keep the number of crossings small. This is usually done by processing the edges to be inserted successively. Each edge insertion step needs to determine which edges will be crossed; replacing these crossings by dummy nodes with degree four results again in a planar graph that is taken as input for inserting the next edge. Finally, we obtain a planarized representation P of G. Figure 6.13 illustrates the application of this approach by an example, where just a single edge needs to be reinserted.

Finding a planar subgraph of maximum size is an NP-hard optimization problem, which can be solved using branch-and-cut algorithms, as has been shown by Jünger and Mutzel [39]. On the other

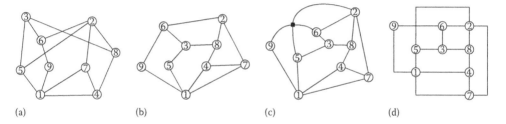

FIGURE 6.13 Example for the application of the planarization method: (a) input graph, (b) planar graph, (c) planarized representation, and (d) final drawing.

hand, we can resort to efficient heuristics, e.g., a maximal planar subgraph (i.e., a planar subgraph such that adding any further edge results in a nonplanar graph) can be computed in linear time as shown by Djidjev [18]. However, this algorithm is very complex and so far no implementation exists; a much easier approach is to use a planarity testing algorithm. The easiest way is to start with a spanning tree of the graph, and add the remaining edges successively; if adding an edge makes the graph nonplanar, we remove it again and proceed with the next edge.

6.4.1 Edge Insertion with Fixed Embedding

The standard technique for reinserting an edge into a planar graph works as follows. Let $e = (v, w)$ be the edge we want to insert into the planar graph P. We first compute a planar embedding Π of P, which fixes the order of edges around each vertex and thus the faces of the drawing. Finding a planar embedding works in linear time; see [9]. Next, we construct the geometric dual graph $\Pi^* = (V^*, E^*)$ of Π. Its vertices V^* are exactly the faces of Π, and it contains an edge (f, f') for each edge of P, where f and f' are the faces of Π separated by e.

Observe that crossing an edge corresponds to traversing an edge in Π^*; if we want to connect vertices v and w, we simply have to find a path in Π^* starting at a face adjacent to v and ending at a face adjacent to w. Therefore, we add two more vertices v^* and w^* (representing v and w, respectively) to Π^* and connect v^* with all faces adjacent to v, and analogously w^* with all faces adjacent to w, resulting in a graph $\Pi^*_{v,w}$. Then, connecting v and w with a minimum number of crossings in a drawing of Π corresponds to finding a shortest path from v^* to w^* in $\Pi^*_{v,w}$, which can be found in linear time by applying breadth-first search. Figure 6.14a shows an example for a geometric dual graph extended in this way, given by the squared vertices (plus vertex 2 and 5) and the dashed edges. The shortest path for inserting edge (2,5) is highlighted in (a), and the corresponding drawing is shown in (b).

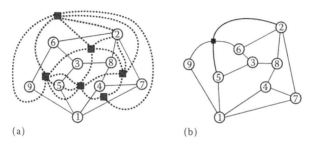

FIGURE 6.14 Inserting an edge $e = (2, 5)$ into a planar graph P with fixed embedding Π: (a) the (extended) geometric dual graph of Π (dashed edges) and (b) substituting the crossing with a dummy node.

6.4.2 Edge Insertion with Variable Embedding

The drawback of edge insertion with fixed planar embedding is that the quality of the solution heavily depends on the choice of the planar embedding. Fixing an unfavorable embedding may result in arbitrarily many unnecessary crossings, which could be avoided by choosing a different embedding. Edge insertion with variable embedding overcomes this weakness by also finding an "optimal" embedding. This requires that we are able to efficiently represent all possible planar embeddings of the graph; the challenging part here is that the possible number of a graph's planar embeddings can be exponential in its size. In the following section, we outline the necessary foundations and an efficient algorithm to solve the problems. For simplicity, we only consider biconnected graphs.

A compact and efficiently computable representation of all planar embeddings of a biconnected graph is given by decomposing the graph into its triconnected components, similar to the decomposition of a graph into its biconnected components. The triconnected components consist of three types of structures: serial, parallel, and triconnected components. All the components of a graph are related in a tree-like fashion. Figure 6.15 gives examples; the shaded regions represent subgraphs that share exactly two vertices (a separation pair) with the rest of the graph. The actual structure of a component—its skeleton—is obtained by shrinking the shaded parts to single, the so-called virtual edges; the part that was shrunk is also called the expansion graph of the virtual edge. The key observation is that we can enumerate all planar embeddings of a graph by permuting the components in parallel structures and mirroring triconnected structures; obviously, we cannot afford explicit enumeration in an algorithm. Typically, triconnected components are represented by a data structure called SPQR-tree introduced by Di Battista and Tamassia [17], which can be constructed in linear time and space; see [29].

The optimal edge insertion algorithm by Gutwenger et al. [31] proceeds as follows. Assume that we want to insert edge (v, w) into the planar and biconnected graph G. First, we compute the SPQR-tree of G and identify the (shortest) path from a component whose skeleton contains v to a component whose skeleton contains w. For each component on this path, we can compute the crossings contributed by that particular component individually. Starting with the component's skeleton, we replace all virtual edges whose expansion graph does not contain v or w (except as an endpoint of the virtual edge) by their expansion graphs; other virtual edges are split creating a representant for v or w, respectively. We observe that only the triconnected structures contribute required crossings (see Figure 6.16): For serial structures, this is trivial, and parallel structures can always be arranged such that the affected parts are neighbored. Triconnected structures can be handled analogously as in the fixed embedding case, since their skeletons themselves are triconnected and thus admit unique embeddings (up to mirroring), and it can be shown that it is permitted to choose arbitrary embeddings for the expansion graphs. Finally, the obtained crossings are linked together in the order given by the computed path in the SPQR-tree. The overall runtime of the algorithm is linear, thus even matching the algorithm for the fixed embedding case.

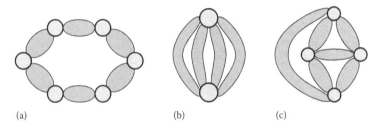

(a)   (b)   (c)

FIGURE 6.15 Triconnected components: (a) serial, (b) parallel, and (c) triconnected structures.

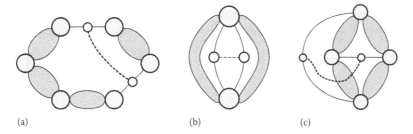

(a) (b) (c)

FIGURE 6.16 Edge insertion: serial (a) and parallel (b) components do not contribute any crossings; triconnected structures can be handled as the fixed embedding case.

6.4.3 Remarks

A graph drawn (by the software library OGDF [47]) using the planarization approach and network flow techniques (for drawing the planarized graph) is shown in Figure 6.17.

The question arises whether the insertion of an edge into a planar graph G using the optimal edge insertion algorithm will lead to a crossing minimum drawing. However, Gutwenger et al. [31] have presented a class of almost planar graphs for which optimal edge insertion does not lead to an approximation algorithm with a constant factor. An almost planar graph G is a graph containing an edge $e \in G$ such that $G - e$ is planar. On the other hand, Hliněný and Salazar [33] have shown that optimal edge insertion approximates the optimal number of crossings in almost planar graphs by a factor of Δ, where Δ denotes the maximum node degree in the given graph.

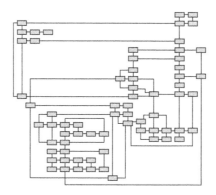

FIGURE 6.17 A graph drawn using the planarization approach and network flow techniques.

6.5 Symmetric Graph Drawing

Symmetry is a much admired property in visualization. Drawings of graphs in graph theory textbooks are often symmetric, because the symmetry clearly reveals the structure of the graph.

The aim of symmetric graph drawing is to take as input a graph G and construct a drawing of G that is as symmetric as possible. In general, constructing a symmetric graph drawing has two steps:

1. Detection of symmetries in an abstract graph
2. Display of the symmetries in a drawing of the graph

Note that the first step is much harder than the second step. As a result, most research in symmetric graph drawing has been devoted to the first step. In this survey, we also focus on symmetry detection algorithms.

We first start with a short review on symmetric graph drawing in two dimensions for a brief comparison.

6.5.1   Two-Dimensional Symmetric Graph Drawing

A symmetry of a two-dimensional figure is an isometry of the plane that fixes the figure. There are two types of two dimensional symmetries, *rotational* symmetry and *reflectional (or axial)* symmetry. Rotational symmetry is a rotation about a *point* and reflectional symmetry is a reflection with respect to an *axis*.

Symmetric graph drawing in two dimensions has been widely investigated by a number of authors for the last two decades. For details, see a recent survey on symmetric graph drawing in two dimensions (see, e.g., [21]). Here, we briefly summarize fundamental results.

The problem of determining whether a given graph can be drawn symmetrically is NP-complete in general [46]; however, exact algorithms are presented based on group-theoretic approaches [1]. Linear-time algorithms for constructing maximally symmetric drawings are available for restricted classes of graphs. A series of linear time algorithms developed for constructing maximally symmetric drawings of planar graphs, including trees, outerplanar graphs, triconnected planar graphs, biconnected planar graphs, one-connected planar graphs, and disconnected planar graphs (see, e.g., [46,36]).

6.5.2   Three-Dimensional Symmetric Graph Drawing

Symmetry in three dimensions is much richer than that in two dimensions. For example, a maximal symmetric drawing of the icosahedron graph in two dimensions shows six symmetries. However, the maximal symmetric drawing of the icosahedron graph in three dimensions shows 120 symmetries.

Recently, symmetric graph drawing in three dimensions has been investigated. The problem of drawing a given graph with the maximum number of symmetries in three dimensions is NP-hard [34]. Nevertheless, fast exact algorithms for general graphs based on group theory are available [1]. Also, there are linear time algorithms for constructing maximally symmetric drawings of restricted classes of graphs in three dimensions. Examples include trees, series parallel digraphs, and planar graphs [34,35,37].

6.5.3   Geometric Automorphisms and Three-Dimensional
Symmetry Groups

A symmetry of a drawing D of a graph $G = (V, E)$ induces a permutation of the vertices, which is an automorphism of the graph; this automorphism is displayed by the symmetry. We say that the automorphism is a *geometric* automorphism of a graph G [1]. The symmetry group of a drawing D of a graph G induces a subgroup of the automorphism group of G. The subgroup P is a *geometric automorphism group* if there is a drawing D of G, which displays every element of P as a symmetry of the drawing [1]. The size of a symmetry group is the number of elements of the group. The aim of symmetric graph drawing is to construct a drawing of a given graph with a maximum size geometric automorphism group.

The structure of three-dimensional symmetry groups can be quite complex. A complete list of all possible three dimensional symmetry groups can be found in [2]. However, all are variations on just three types: regular pyramid, regular prism, and Platonic solids, which are described below.

A *regular pyramid* is a pyramid with a regular g-gon as its base. It has only one g-fold rotation axis, called the *vertical rotation* axis, passing through the apex and the center of its base. It has g rotational symmetries, called *vertical rotations*, each of which is a rotation of $2\pi i/g$, $i = 0, 1, \ldots, g - 1$. Also, there are g *vertical reflections* in reflection planes, each containing the vertical rotation axis. In total, the regular pyramid has $2g$ symmetries.

A *regular prism* has a regular g-gon as its top and bottom face. There are $g + 1$ rotation axes, and they can be divided into two classes. The first one, called the *principal* axis or vertical rotation axis, is a g-fold rotation axis that passes through the centers of the two g-gon faces. The second class, called *secondary* axes or *horizontal rotation* axes, consists of g two-fold rotation axes that lie in a plane perpendicular to the principal axis. The number of rotational symmetries is $2g$. Also, there are g reflection planes, called vertical reflection planes, each containing the principal axis, and another reflection plane perpendicular to the principal axis, called *horizontal reflection*. Further, it has $g - 1$ *rotary reflections*, composition of a rotation and a reflection. In total, the regular prism has $4g$ symmetries.

Platonic solids have more axes of symmetry. The tetrahedron has four three-fold rotation axes and three two-fold rotation axes. It has 12 rotational symmetries and in total 24 symmetries. The octahedron has three four-fold rotation axes, four three-fold axes, and six two-fold rotation axes. It has 24 rotational symmetries and a full symmetry group of size 48. The icosahedron has six 5-fold rotation axes, ten 3-fold rotation axes, and fifteen 2-fold rotation axes. It has 60 rotational symmetries and a full symmetry group of size 120. Note that the cube and the octahedron are dual solids, and the dodecahedron and the icosahedron are dual.

Other symmetry groups are derived from those of the regular pyramid, the regular prism, and the Platonic solids. In fact, the variations have *less* symmetries than the three basic types. For example, a variation of regular pyramid is a pyramid with only vertical rotations, without the vertical reflections. For more details of three-dimensional symmetry groups and their relations, see [2].

6.5.4 Algorithm for Drawing Trees with Maximum Symmetries

Here, we sketch an algorithm for finding maximally symmetric drawings of trees in three dimensions [35].

Fortunately, maximally symmetric three-dimensional drawings of trees have only a few kinds of groups. First, note a trivial case: every tree has a planar drawing in the xy-plane; a reflection in this plane is a symmetry of the drawing. This kind of symmetry display, where every vertex is mapped to itself, is *inconsequential*.

LEMMA 6.5 [35] The maximum size three-dimensional symmetry group of a tree T is either inconsequential or one of the following three types:

1. Symmetry group of a regular pyramid
2. Symmetry group of a regular prism
3. Symmetry group of one of the Platonic solids

The symmetry finding algorithm for trees constructs all three possible symmetric configurations (i.e., pyramid, prism, and Platonic solids), and then chooses the configuration that has the maximum number of symmetries.

A vertex is *fixed* by a given three-dimensional symmetry if it is mapped onto itself by that symmetry. Note that the center of a tree T is fixed by every automorphism of T; thus, every symmetry of a drawing of T fixes the location of the center.

We first place the center of the input tree at the apex for the pyramid configuration, at the centroid for the prism and the Platonic solids configuration. Then, we place each subtree attached to the center to form a symmetric configuration.

The symmetry finding algorithm uses an auxiliary tree called the Isomorphism Class Tree. This tree is a fundamental structure describing the isomorphisms between subtrees in the input tree. Suppose that T is a tree rooted at r, and that $T_1, T_2, \ldots, T_k$ are the subtrees formed by deleting r. Then, the ICT tree I_T of T has a root node n_r corresponding to r. Each child of n_r in I_T corresponds to an isomorphism class of the subtrees $T_1, T_2, \ldots, T_k$. Information about each isomorphism class is stored at the node. The subtrees under these children are defined recursively. Using a tree isomorphism algorithm that runs in linear time, we can compute the ICT tree in linear time.

Thus, we can state the symmetry finding algorithm as follows.

1. Find the center of T and root T at the center.
2. Construct the Isomorphism Class Tree (ICT) of T.
3. Find three-dimensional symmetry groups of each type:
   a. Construct a pyramid configuration.
   b. Construct a prism configuration.
   c. Construct Platonic solids configuration.
4. Output the group of the configuration that has the maximum size.

There are some variations for each configuration, based on the number of subtrees fixed by the three dimensional symmetry group.

For the pyramid configuration, we construct a pyramid-type drawing by placing the center of the tree at the apex of a pyramid, some fixed subtrees about the rotation axis, and g isomorphic subtrees in the reflection planes that contain the side edges of the pyramid. The resulting drawing has the same symmetry group of the g-gon based pyramid.

The pyramid configuration can have up to two fixed subtrees, one on each side of the center on the rotation axis. For example, Figure 6.18a shows two fixed isomorphic subtrees, and Figure 6.18b shows two fixed nonisomorphic subtrees. Both drawings display four rotational symmetries and four reflectional symmetries. Note that if we do not fix the two subtrees on the rotation axes, the tree in Figure 6.18a can display only two rotational symmetries in three dimensions.

In order to find the three-dimensional pyramid symmetry group of maximum size, we need to consider each subtree as a possible candidate for a fixed subtree. Further, we need to compute the size of the rotational symmetry group of each subtree, as it affects the size of the rotational symmetry group of each configuration as a whole. For example, if the fixed subtrees in Figure 6.18b were

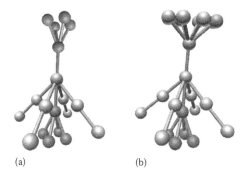

(a)   (b)

FIGURE 6.18 Pyramid configuration with two fixed subtrees.

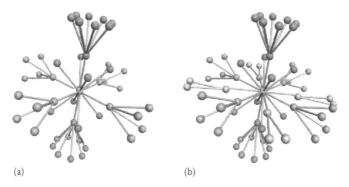

(a)   (b)

FIGURE 6.19 Prism configuration with fixed subtrees on the *xy*-plane and the *z*-axis.

redrawn to display only two rotational symmetries, then the size of the rotational symmetry group of the whole drawing would be merely two. The ICT can be used to find the fixed subtrees efficiently, and thus to compute a maximum size rotational group for the pyramid configuration in linear time.

The symmetries of a tree drawn in the prism configuration can be more complex than those of the pyramid configuration. The main reason is that a prism has two kinds of rotation axis: the first, called the principal axis, passes through the regular polygons at each end of the prism.

Suppose that the prism has a g-fold rotation about the principal axis. We assume that the principal axis is the z-axis. There are also g two-fold secondary axes in the xy-plane, each at an angle of $\pi i/g$ to the x-axis, for $i = 0, 1, \ldots, g - 1$. The prism has a rotational symmetry by an angle of π about each of these axes. The complexity of prism drawings of trees comes from the fact that each of these $g + 1$ rotation axes may fix subtrees. For example, see Figure 6.19. The drawing in Figure 6.19a shows two isomorphic fixed subtrees on the z-axis, and four isomorphic subtrees on the secondary axis. The drawing in Figure 6.19b shows two different types of four isomorphic subtrees on the secondary axis, in addition to the two isomorphic fixed subtrees on the z-axis.

Again, using the ICT, we can compute a maximum size rotational group for prism configuration in linear time.

The Platonic solids have many rotation axes. However, the symmetry groups of the Platonic solids are fixed, and we only need to test whether we can construct a three-dimensional drawing of a tree that has the same symmetry group as one of the Platonic solids. Using a method similar to the previous cases, we can test this in linear time.

For example, Figure 6.20 shows the fixed subtree on the four-fold axes of the cube configuration.

The main thrust of the algorithms for the pyramid, prism, and the Platonic solids is the use of the ICT to determine the fixed subtrees.

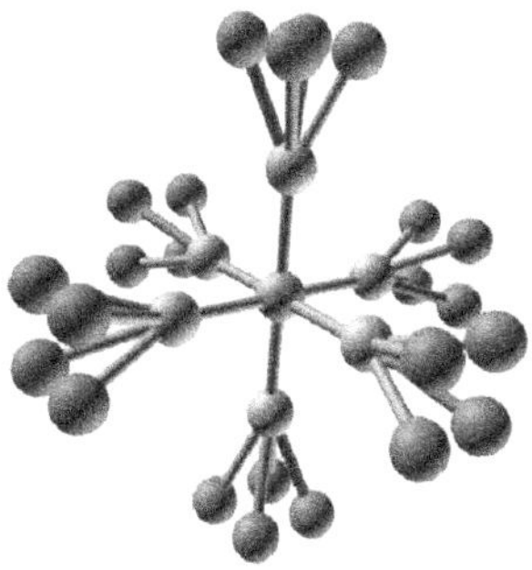

FIGURE 6.20 The cube configuration with fixed subtrees on four-fold axes.

Once we compute the maximum size three-dimensional symmetry group, one can use a linear time straight-line drawing algorithm that arranges subtrees in *wedges* and *cones* in [35].

The following theorem summarizes the main results of this section.

THEOREM 6.3 *[35] There is a linear time algorithm that constructs symmetric drawings of trees with maximum number of three-dimensional symmetries.*

6.6   Research Issues

The field of graph drawing is still developing and there are many fundamental questions to be resolved. Here, we list a few related to the above-mentioned drawing methods.

- The grid size for planar straight-line drawings is bounded below; there is an n-vertex plane graph G such that, for any straight-line grid drawing of G, each dimension of the grid is at least $\lfloor \frac{2(n-1)}{3} \rfloor$ even if the other dimension is allowed to be unbounded [16,14]. It is conjectured that this lower bound is almost achievable; more precisely, every planar graph with n vertices has a two-dimensional straight-line drawing on a grid of size $\lceil \frac{2n}{3} \rceil \times \lceil \frac{2n}{3} \rceil$. So far, the best known upper bound is $(n - \Delta_0 - 1) \times (n - \Delta_0 - 1)$, where $0 \leq \Delta_0 \leq \lfloor (n-1)/2 \rfloor$ is the number of cyclic faces of G with respect to its minimum realizer (see [51]).
- The network-flow approach by Tamassia for achieving a bend-minimum orthogonal planar drawing works only for plane graphs with maximum degree 4. So far there is no perfect model or method for extending the network-flow approach to plane graphs with maximum vertex degree > 4. A specific open research question is the problem of finding a polynomial time algorithm for computing bend-minimum drawings in the Kandinsky model for high-degree plane graphs.
- A lot of research has been done in the field of crossing minimization, but there are still open fundamental problems. Though there is a polynomial time algorithm to approximate the crossing number for almost planar graphs with bounded degree, it is unknown if there is such a polynomial time approximation (or even exact) algorithm for almost planar graphs in general. Furthermore, if we consider apex graphs, i.e., graphs containing a vertex whose removal leaves a planar graph, we do not even know if there is a polynomial time approximation algorithm in the case of bounded degree.
- Every triconnected planar graph G can be realized as the one-skeleton of a convex polytope P in $\mathbb{R}^3$ such that all automorphisms of G are induced by isometries of P (see, e.g., [21], Theorem by Mani). Is there a linear time algorithm for constructing a symmetric convex polytope of a triconnected planar graph?

6.7   Further Information

Books on graph drawing appeared in [6,42]. The proceedings of the annual international Graph Drawing Symposia are published in the Lecture Notes in Computer Science Series by Springer; see, for example, [43,38]. There exist various software packages for graph drawing (e.g., [40]). The *Graph Drawing E-Print Archive* (GDEA) is an electronic repository and archive for research materials on the topic of graph drawing (see, http://gdea.informatik.uni-koeln.de).

Defining Terms

Graph drawing: A geometric representation of a graph is a graph drawing. Usually, the representation is in either 2- or 3-dimensional Euclidean space, where a vertex v is represented by a point $p(v)$ and an edge (u, v) is represented by a simple curve whose endpoints are $p(u)$ and $p(v)$.

Edge crossing: Two edges cross in a graph drawing if their geometric representations intersect. The number of crossings in a graph drawing is the number of pairs of edges that cross.

Planar graph: A graph drawing in two-dimensional space is planar if it has no edge crossings. A graph is planar if it has a planar drawing.

Almost planar graph: A graph is almost planar if it contains an edge whose removal leaves a planar graph.

Planarization: Informally, *planarization* is the process of transforming a graph into a planar graph. More precisely, the transformation involves either removing edges (planarization by edge removal) or replacing pairs of nonincident edges by four-stars, as in Figure 6.21 (planarization by adding crossing vertices). In both cases, the aim of planarization is to make the number of operations (either removing edges or replacing pairs of nonincident edges by four-stars) as small as possible.

FIGURE 6.21 Planarization by crossing vertices.

Grid drawing: A grid drawing is a graph drawing in which each vertex is represented by a point with integer coordinates.

Straight-line drawing: A straight-line drawing is a graph drawing in which each edge is represented by a straight line segment.

Orthogonal drawing: An orthogonal drawing is a graph drawing in which each edge is represented by a polyline, each segment of which is parallel to a coordinate axis.

References

1. D. Abelson, S. Hong, and D. E. Taylor. Geometric automorphism groups of graphs. *Discrete Appl. Math.*, 155(17):2211–2226, 2007.
2. M. A. Armstrong. *Groups and Symmetry*. Springer-Verlag, New York, 1988.
3. W. Barth, P. Mutzel, and C. Yildiz. A new approximation algorithm for bend minimization in the kandinsky model. In D. Kaufmann and M. und Wagner, editors, *Graph Drawing*, volume 4372 of *Lecture Notes in Computer Science*, pp. 343–354. Springer-Verlag, Berlin, Germany, 2007.
4. C. Batini, E. Nardelli, and R. Tamassia. A layout algorithm for data flow diagrams. *IEEE Trans. Softw. Eng.*, SE-12(4):538–546, 1986.
5. C. Batini, M. Talamo, and R. Tamassia. Computer aided layout of entity-relationship diagrams. *J. Syst. Softw.*, 4:163–173, 1984.
6. G. Di Battista, P. Eades, R. Tamassia, and I. G. Tollis. *Graph Drawing*. Prentice Hall, Upper Saddle River, NJ, 1999.
7. G. Di Battista, A. Garg, G. Liotta, A. Parise, R. Tamassia, E. Tassinari, F. Vargiu, and L. Vismara. Drawing directed acyclic graphs: An experimental study. In S. North, editor, *Graph Drawing (Proceedings of the GD '96)*, volume 1190 of *Lecture Notes in Computer Science*, pp. 76–91. Springer-Verlag, Berlin, Germany, 1997.
8. J. A. Bondy and U. S. R. Murty. *Graph Theory with Applications*. Macmillan, London, U.K., 1976.
9. J. M. Boyer and W. Myrvold. On the cutting edge: Simplified o(n) planarity by edge addition. *J. Graph Algorithms Appl.*, 8(2):241–273, 2004.
10. F. J. Brandenburg, M. Himsolt, and C. Rohrer. An experimental comparison of force-directed and randomized graph drawing algorithms. In F. J. Brandenburg, editor, *Graph Drawing (Proceedings of the GD '95)*, volume 1027 of *Lecture Notes Computer Science*, pp. 76–87. Springer-Verlag, Berlin, Germany, 1996.
11. C. Buchheim, M. Chimani, D. Ebner, C. Gutwenger, M. Jünger, G. W. Klau, P. Mutzel, and R. Weiskircher. A branch-and-cut approach to the crossing number problem. *Discrete Optimization*, 5(2):373–388, 2008.
12. C. Buchheim, M. Jünger, and S. Leipert. Drawing rooted trees in linear time. *Softw. Pract. Exp.*, 36(6):651–665, 2006.

13. M. Chimani, C. Gutwenger, and P. Mutzel. Experiments on exact crossing minimization using column generation. In *Experimental Algorithms (Proceedings of the WEA 2006)*, volume 4007 of *Lecture Notes in Computer Science*, pp. 303–315. Springer-Verlag, Berlin, Germany, 2006.

14. M. Chrobak and S. Nakao. Minimum width grid drawings of planar graphs. In R. Tamassia and I. G. Tollis, editors, *Graph Drawing (Proceedings of the GD '94)*, volume 894 of *Lecture Notes in Computer Science*, pp. 104–110. Springer-Verlag, Berlin, Germany, 1995.

15. M. Chrobak and T. H. Payne. A linear time algorithm for drawing a planar graph on a grid. *Inf. Process. Lett.*, 54:241–246, 1995.

16. H. de Fraysseix, J. Pach, and R. Pollack. How to draw a planar graph on a grid. *Combinatorica*, 10(1):41–51, 1990.

17. G. Di Battista and R. Tamassia. On-line maintenance of triconnected components with SPQR-trees. *Algorithmica*, 15:302–318, 1996.

18. H. N. Djidjev. A linear algorithm for the maximal planar subgraph problem. In *Proceedings of the 4th International Workshop on Algorithms and Data Structures, Lecture Notes Computer Science*, pp. 369–380. Springer-Verlag, Berlin, Germany, 1995.

19. V. Dujmović and S. Whitesides. Three dimensional graph drawing. In R. Tamassia, editor, *Handbook of Graph Drawing and Visualization*. CRC Press, Boca Raton, FL, to appear.

20. P. Eades. A heuristic for graph drawing. *Congressus Numerantium*, 42:149–160, 1984.

21. P. Eades and S. Hong. Detection and display of symmetries. In R. Tamassia, editor, *Handbook of Graph Drawing and Visualisation*. CRC Press, Boca Raton, FL, to appear.

22. P. Eades and K. Sugiyama. How to draw a directed graph. *J. Inf. Process.*, 13:424–437, 1991.

23. I. Fary. On straight lines representation of planar graphs. *Acta Sci. Math. Szeged*, 11:229–233, 1948.

24. U. Fößmeier and M. Kaufmann. Drawing high degree graphs with low bend numbers. In F. J. Brandenburg, editor, *Graph Drawing (Proceedings of the GD '95)*, volume 1027 of *Lecture Notes in Computer Science*, pp. 254–266. Springer-Verlag, Berlin, Germany, 1996.

25. T. Fruchterman and E. Reingold. Graph drawing by force-directed placement. *Softw. Pract. Exp.*, 21(11):1129–1164, 1991.

26. A. Garg and R. Tamassia. A new minimum cost flow algorithm with applications to graph drawing. In S. North, editor, *Graph Drawing (Proceedings of the GD '96)*, volume 1190 of *Lecture Notes in Computer Science*, pp. 201–216. Springer-Verlag, Berlin, Germany, 1997.

27. C. Gutwenger and P. Mutzel. Grid embedding of biconnected planar graphs. Technical Report, Max-Planck-Institut Informatik, Saarbrücken, Germany, 1998.

28. C. Gutwenger and P. Mutzel. Planar polyline drawings with good angular resolution. In S. Whitesides, editor, *Graph Drawing*, volume 1547 of *Lecture Notes in Computer Science*, pp. 167–182. Springer-Verlag, Berlin, Germany, 1998.

29. C. Gutwenger and P. Mutzel. A linear time implementation of SPQR trees. In J. Marks, editor, *Graph Drawing (Proceedings of the GD 2000)*, volume 1984 of *Lecture Notes in Computer Science*, pp. 77–90. Springer-Verlag, Berlin, Germany, 2001.

30. C. Gutwenger and P. Mutzel. An experimental study of crossing minimization heuristics. In B. Liotta, editor, *Graph Drawing (Proceedings of the GD 2003)*, volume 2912 of *Lecture Notes in Computer Science*, pp. 13–24. Springer-Verlag, Berlin, Germany, 2004.

31. C. Gutwenger, P. Mutzel, and R. Weiskircher. Inserting an edge into a planar graph. *Algorithmica*, 41(4):289–308, 2005.

32. S. Hachul and M. Jünger. An experimental comparison of fast algorithms for drawing general large graphs. In P. Healy and N. S. Nikolov, editors, *Graph Drawing*, volume 3843 of *Lecture Notes in Computer Science*, pp. 235–250. Springer-Verlag, Berlin, Germany, 2006.

33. P. Hliněný and G. Salazar. On the crossing number of almost planar graphs. In M. Kaufmann and D. Wagner, editors, *Graph Drawing (Proceedings of the GD 2006)*, volume 4372 of *Lecture Notes in Computer Science*, pp. 162–173. Springer-Verlag, Berlin, Germany, 2007.

34. S. Hong. Drawing graphs symmetrically in three dimensions. In *Proceedings of the Graph Drawing 2001*, volume 2265 of *Lecture Notes in Computer Science*, pp. 189–204. Springer, Berlin, Germany, 2002.

35. S. Hong and P. Eades. Drawing trees symmetrically in three dimensions. *Algorithmica*, 36(2):153–178, 2003.

36. S. Hong and P. Eades. Drawing planar graphs symmetrically. II. Biconnected planar graphs. *Algorithmica*, 42(2):159–197, 2005.

37. S. Hong, P. Eades, and J. Hillman. Linkless symmetric drawings of series parallel digraphs. *Comput. Geom. Theor. Appl.*, 29(3):191–222, 2004.

38. S.-H. Hong, T. Nishizeki, and W. Quan, editors. *Graph Drawing 2007*, volume 4875 of *Lecture Notes in Computer Science*. Springer-Verlag, Berlin, Germany, 2008.

39. M. Jünger and P. Mutzel. Maximum planar subgraphs and nice embeddings: Practical layout tools. *Algorithmica, Special Issue on Graph Drawing*, 16(1):33–59, 1996.

40. M. Jünger and P. Mutzel. *Graph Drawing Software*. Springer-Verlag, Berlin, Germany, 2003.

41. G. Kant. Drawing planar graphs nicely using the *lmc*-ordering. In *Proceedings of the 33th Annual IEEE Symposium on Foundation of Computer Science*, pp. 101–110, Pittsburgh, PA, 1992.

42. M. Kaufmann and D. Wagner, editors. *Drawing Graphs: Methods and Models*, volume 2025 of *Lecture Notes in Computer Science Tutorial*. Springer-Verlag, Berlin, Germany, 2001.

43. M. Kaufmann and D. Wagner, editors. *Graph Drawing 2006*, volume 4372 of *Lecture Notes in Computer Science*. Springer-Verlag, Berlin, Germany, 2007.

44. G. Klau and P. Mutzel. Quasi-orthogonal drawing of planar graphs. Technical Report, Max-Planck-Institut Informatik, Saarbrücken, Germany, 1998.

45. G. W. Klau, K. Klein, and P. Mutzel. An experimental comparison of orthogonal compaction algorithms. In J. Marks, editor, *Graph Drawing (Proceedings of the GD 2000)*, volume 1984 of *Lecture Notes in Computer Science*, pp. 37–51. Springer-Verlag, Berlin, Germany, 2001.

46. J. Manning. Geometric symmetry in graphs. PhD thesis, Purdue University, West Lafayette, IN, 1990.

47. OGDF–Open Graph Drawing Framework, 2008. Open source software project, available via http://www.ogdf.net.

48. E. Reingold and J. Tilford. Tidier drawing of trees. *IEEE Trans. Softw. Eng.*, SE-7(2):223–228, 1981.

49. R. Tamassia. On embedding a graph in the grid with the minimum number of bends. *SIAM J. Comput.*, 16(3):421–444, 1987.

50. J. Q. Walker II. A node-positioning algorithm for general trees. *Softw. Pract. Exp.*, 20(7):685–705, 1990.

51. H. Zhang and X. He. Compact visibility representation and straight-line grid embedding of plane graphs. In F. Dehne, J.-R. Sack, and M. Smid, editors, *Algorithms and Data Structures (WADS 2003)*, volume 2748 of *Lecture Notes in Computer Science*, pp. 493–504. Springer-Verlag, Berlin, Germany, 2003.

7

Algorithmics in Intensity-Modulated Radiation Therapy

Danny Z. Chen
University of Notre Dame

Chao Wang
University of Notre Dame

Intensity-modulated radiation therapy (IMRT) is a modern cancer treatment technique aiming to deliver a prescribed conformal radiation dose to a target tumor while sparing the surrounding normal tissues and critical structures. In this chapter, we consider a set of combinatorial and geometric problems that arise in IMRT planning and delivery: (1) the static leaf sequencing problem, (2) the static leaf sequencing with error control problem, (3) the field splitting problem, (4) the dose simplification problem, (5) the dynamic leaf sequencing problem, and (6) the single-arc leaf sequencing problem. We discuss new efficient algorithms for these problems. The main ideas of the algorithms are to exploit the underlying geometric and combinatorial properties of the problems and transform them into graph problems such as shortest paths, optimal matchings, maximum flows, multicommodity demand flows, or linear programming problems. Some open problems and promising directions for future research are also given.

7.1 Introduction

The study of mathematical and geometric optimization problems that arise in the field of medicine is becoming a significant research area. Many medical applications call for effective and efficient algorithmic solutions for various discrete or continuous algorithmic problems. In this chapter, we will cover a number of combinatorial and geometric problems that arise in the research of intensity-modulated radiation therapy (IMRT).

IMRT is a modern cancer therapy technique that aims to deliver a highly conformal radiation dose to a target volume in 3-D (e.g., a tumor) while sparing the surrounding normal tissues and critical structures. Performing IMRT is based on the ability to accurately and efficiently deliver prescribed dose distributions of radiation, called intensity maps (IMs). An IM is specified by a set of nonnegative integers on a uniform 2-D grid (see Figure 7.1a). The value in each grid cell indicates the intensity level (in units) of radiation to be delivered to the body region corresponding to that IM cell. For real medical prescriptions, an IM cell is typically measured from 0.5 cm × 0.5 cm to 1 cm × 1 cm [29].

Currently, most IMRT delivery system consists of two major components: (1) a medical linear accelerator (LINAC), which generates photon beams used in IMRT, and (2) the multileaf collimator (MLC) [44,72,73]. An MLC is made of many pairs of tungsten alloy leaves of the same rectangular shape and size (see Figure 7.1b). The opposite leaves of a pair are aligned to each other. These leaves, which are controlled by a computer, can move left and right to form a rectilinear polygonal region, called an MLC-aperture. During IMRT treatment, the patient is positioned and secured to the treatment couch (see Figure 7.1c) and the LINAC delivers radiation beams to the tumor from various directions. The direction of the radiation beam is controlled by rotating the gantry to the desired angle. The MLC is mounted on the gantry and the cross section of the cylindrical radiation beam is shaped by an MLC-aperture to deliver a uniform dose to (a portion of) an IM [17,33,51–53,63,67,80,83].

Currently, there are three popular MLC systems that are used for clinical cancer treatment. The systems are the Elekta, the Siemens, and the Varian MLCs [44]. Depending on the actual MLC system in use, there are some differences among the required geometric shapes of the rectilinear polygonal regions that can be formed by the MLC. The details on the differences between these MLC systems will be discussed in Section 7.2.1.

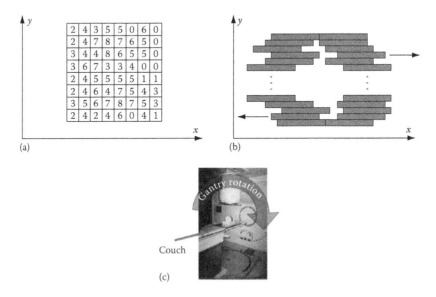

(a)

(b)

Couch

(c)

FIGURE 7.1 (a) An IM. (b) A schematic drawing of a multileaf collimator (MLC), where the shaded rectangles represent the MLC leaves which can move left and right (as indicated by the arrows) to form an MLC-aperture. (c) Patient couch and rotatable gantry.

Using a computer-controlled MLC, the IMs are usually delivered either statically, or dynamically. In the static approach [11–16,30,35,41,40,74,79], the MLC leaves do not move during irradiation, and reposition to form another beam shape only when the radiation is turned off. In the dynamic approach [28,29,31,45,69–71], the MLC leaves keep moving across an IM field while the radiation remains on. Note that MLC leaves are capable of changing their positions quickly compared to the total radiation time, and the actually delivered IMs can still remain to be of integer values in the dynamic approach. In both approaches, the gantry is fixed during irradiation. Arc-modulated radiation therapy (AMRT) is a newly emerging IMRT delivery technique [82]. AMRT differs from the traditional static and dynamic IMRT approach in that the gantry is rotating, typically along a single 3-D circular arc around the patient, during delivery. In the meantime, the MLC leaves also move quickly across the IM field to reproduce the prescribed IM.

A treatment plan for a given IM gives a precise description of how to control the MLC leaves to delivering the IM. For static IMRT, the treatment plan must also specify how to turn on/off the radiation beam source. Two key criteria are used to evaluate the quality of an IMRT treatment plan:

1. Delivery time (the efficiency): Minimizing the delivery time is important because it not only lowers the treatment costs but also increases the patient throughput. Short delivery time also reduces the recurrence of tumor cells. For static IMRT, the delivery time consists of two parts: (1) the beam-on time and (2) the machine setup time. Here, the beam-on time refers to the total amount of time when a patient is actually exposed to radiation, and the machine setup time refers to the time associated with turning on/off the radiation source and repositioning MLC leaves. Minimizing the beam-on time has been considered as an effective way to enhance the radiation efficiency of the machine (referred to as *MU* efficiency in the medical literature) as well as to reduce the patient's risk under radiation [10]. For dynamic IMRT or AMRT, since the radiation source is always on when delivering an IM, the treatment time equals to the beam-on time.

2. Delivery error (the accuracy): For various reasons, there is a discrepancy between the prescribed IM and actually delivered IM. We distinguish two types of delivery error: (1) approximation error and (2) tongue-and-groove error. The approximation error refers to the discrepancy of the prescribed IM and actually delivered IM that resides in the interior of the IM cells; and the tongue-and-groove error [68,79,81] refers to the discrepancy of the prescribed IM and actually delivered IM that appears on the boundary of the IM cells, which is due to the special geometric shape design, called tongue-and-groove design, of the MLC leaves [81] (Section 7.2.2 discusses more on its nature).

In this chapter, we consider the following problems that arise in the planning and delivery of IMRT: (1) the static leaf sequencing (SLS) problem, (2) the SLS with tongue-and-groove error control problem, (3) the field splitting problem, (4) the dose simplification problem, (5) the dynamic leaf sequencing problem (DLS), and (6) the single-arc leaf sequencing problem (SALS). The definitions of the above problems will be given in later sections.

The rest of this chapter is organized as follows. Section 7.2 discusses the MLC constraints and its tongue-and-groove feature. Sections 7.3 through 7.7 discuss the aforementioned problems and give efficient algorithms. Section 7.8 discusses some open problems and promising directions for future research.

7.2 Preliminaries

In this section, we first discuss the three popular MLC systems and their constraints, then characterize the tongue-and-groove error associated with IMRT treatments using such MLC systems.

7.2.1 Constraints of Multileaf Collimators

As mentioned in Section 7.1, there are three popular MLC systems currently used in clinical treatments [44]: the Elekta, Siemens, and Varian MLC systems. The mechanical structure of these MLCs, although is quite flexible, is not perfect in that it still precludes certain aperture shapes from being used [29,44,73] for treatment. In the following, we summarize the common constraints that appear in these MLC systems:

1. The minimum leaf separation constraint. This requires the distance between the opposite leaves of any MLC leaf pair (e.g., on the Elekta or Varian MLC) to be no smaller than a given value δ (e.g., $\delta = 1$ cm).

2. The interleaf motion constraint. On the Elekta or Siemens MLC, the tip of each MLC leaf is not allowed to surpass those of its neighboring leaves on the opposite leaf bank.

3. The maximum leaf spread constraint. This requires the maximum distance between the tip of leftmost left leaf and the tip of the rightmost right leaf of the MLC is no more than a threshold distance (e.g., 25 cm for Elekta MLCs). This constraint applies to all existing MLCs.

4. The maximum leaf motion speed constraint. This requires the MLC leaves cannot move faster than a threshold value (e.g., 3 cm/s for Varian MLCs). This constraint applies to all existing MLCs.

Figure 7.2 shows the constraints (1) and (2) of these MLC systems. The Elekta MLC is subject to both the minimum leaf separation and interleaf motion constraints. The Siemens MLC is subject to only the interleaf motion constraint. Hence a degenerate rectilinear y-monotone polygon can be formed by the Siemens MLC, by closing some leaf pairs (see Figure 7.2b). (A polygon P in the plane is called y-monotone if every line orthogonal to the y-axis intersects P at most twice. A y-monotone polygon is degenerate if its interior is not a single connected piece.) The Varian MLC is subject to the minimum leaf separation constraint, but allows interleaf motion. Thus, to "close" a leaf pair in the Varian system, we can move the leaf opening under the backup diaphragms (see Figure 7.2c). But, the Elekta MLC cannot "close" its leaf pairs in a similar manner due to its interleaf motion constraint.

Geometrically, on the Elekta, each MLC-aperture is a rectilinear y-monotone simple polygon whose minimum vertical "width" is $\geq$ the minimum separation value δ, while on the Siemens or Varian, an MLC-aperture can be a degenerate y-monotone polygon (i.e., with several connected components).

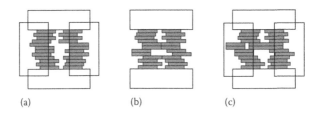

(a)   (b)   (c)

FIGURE 7.2 Illustrating the constraints of three different MLC systems. The shaded rectangles represent the MLC leaves, and the unshaded rectangles represent the backup diaphragms which form a bounding box of an MLC-aperture. (a) The Elekta MLC; (b) the Siemens MLC. (Notice that unlike the Elekta and Varian MLCs, the Siemens MLC only has a single pair of backup metal diaphragms.); and (c) the Varian MLC.

7.2.2 Tongue-and-Groove Design of the MLC Leaves

On most current MLCs, the sides of the leaves are designed to have a "tongue-and-groove" interlock feature (see Figure 7.3a). This design reduces the radiation leakage through the gap between two neighboring MLC leaves and minimizes the friction during leaf movement [68,73,79,81]. But, it also causes an unwanted underdose and leakage situation when an MLC leaf is used for blocking radiation (see Figure 7.3b and c). Geometrically, the underdose and leakage error caused by the tongue-and-groove feature associated with an MLC-aperture is a set of 3-D axis-parallel boxes $w \cdot l_i \cdot h$, where w is the (fixed) width of the tongue or groove side of an MLC leaf, l_i is the length of the portion of the ith leaf that is actually involved in blocking radiation, and $h = \alpha \cdot r$ is the amount of radiation leakage with α being the (fixed) leakage ratio and r being the amount of radiation delivered by that MLC-aperture. Figure 7.3b illustrates the height of the underdose and leakage error, and Figure 7.3c illustrates the width and length of the underdose and leakage error.

The tongue-or-groove error of an MLC-aperture is defined as the amount of underdose and leakage error occurred whenever the tongue side or groove side of an MLC leaf is used for blocking radiation. The tongue-or-groove error of an IMRT plan (i.e., a set of MLC-apertures) is the sum of the tongue-or-groove errors of all its MLC-apertures. The tongue-and-groove error occurs whenever the tongue side of an MLC leaf and the corresponding groove side of its neighboring leaf are both used for blocking radiation in any two different MLC-apertures of an IMRT plan (see Figure 7.4). Note that the tongue-or-groove error is defined on each individual MLC-aperture, while the tongue-and-groove error is defined by the relations between different MLC-apertures. Clearly, the tongue-and-groove error is a subset of the tongue-or-groove error. In medical physics, the tongue-and-groove error has received more attention than tongue-or-groove error because it

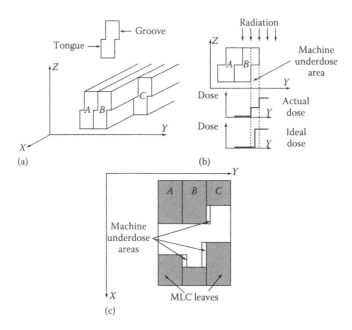

FIGURE 7.3 (a) Illustrating the tongue-and-groove interlock feature of the MLC in 3-D, where leaf B is used for blocking radiation. (b) When leaf B is used for blocking radiation, there is an underdose and leakage in the tongue or groove area. (c) The underdose and leakage areas of the tongue-and-groove feature on an MLC-aperture region.

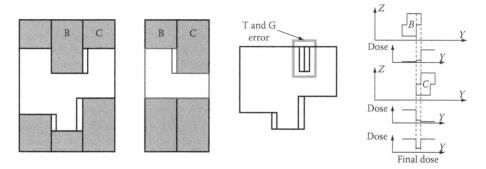

FIGURE 7.4 Illustrating the tongue-and-groove error. (a) and (b): Two MLC-apertures (the shaded rectangles represent MLC leaves). (c) When delivering the two MLC-apertures in (a) and (b) (one by one), the groove side of leaf B and tongue side of leaf C are both used for blocking radiation, causing a tongue-and-groove error in the area between the leaves B and C. (d) Illustrating a dose "dip" in the final dose distribution where a tongue-and-groove error occurs.

usually occurs in the middle of the delivered IMs, causing insufficient dose coverage to the tumor [79]. According to a recent study [30], tongue-and-groove error may cause a point dose error up to 10%, well beyond the allowed 3%–5% limit.

Chen et al. [24] introduced the notion of error-norm, which is closely related to the tongue-or-groove error. The error-norm $||A||_E$ of an IM $A = (a_{i,j})$ of size $m \times n$ is defined as

$$||A||_E = \sum_{i=1}^{m} \left(|a_{i,1}| + \sum_{j=1}^{n-1} |a_{i,j} - a_{i,j+1}| + |a_{i,n}| \right). \tag{7.1}$$

Chen et al. [23] proved that for any treatment plan S that exactly delivers IM A, the difference between the associated tongue-or-groove error T or $G(S)$ and the tongue-and-groove error T and $G(S)$ always equals to $||A||_E$. Thus minimizing tongue-or-groove error is equivalent to minimization tongue-and-groove error. Chen et al. [23] further showed that $||A||_E$ is the strict lower bound for T or $G(S)$, which implies that the tongue-and-groove error has a strict lower bound 0 for any input IM.

7.3 Static Leaf Sequencing

7.3.1 Problem Definition

The SLS problem arises in the static IMRT delivery. In this delivery approach, an IM is delivered as follows:

1. Form an MLC-aperture.
2. Turn on the beam source and deliver radiation to the area of the IM exposed by the MLC-aperture.
3. Turn off the beam source.
4. Reposition the MLC leaves to form another MLC-aperture, and repeat steps 2–4 until the entire IM is done.

In this setting, the boundary of each MLC-aperture does not intersect the interior of any IM cell, i.e., any IM cell is either completely inside or completely outside the polygonal region formed by the MLC-aperture. In delivering a beam shaped by an MLC-aperture, all the cells inside the region of the MLC-aperture receive the same integral amount of radiation dose (say, one unit), i.e., the numbers in all such cells are decreased by the same integer value. The IM is done when each cell has a value zero.

The SLS problem, in general, seeks a treatment plan S (i.e., a set of MLC-apertures together with their weights) for exactly delivering a given IM A such that the delivery time is minimized. Note that the weight of an MLC-aperture corresponds to the amount of radiation dose delivered to the area of the IM exposed by the MLC-aperture, and by "exactly delivering" we mean there is no approximation error but may have tongue-or-groove error. We will discuss how to minimize the tongue-or-groove error in static IMRT delivery in the next section.

Several variations of the SLS problem have been studied in the literature: (1) the SLS problem with minimum beam-on time (MBT) [3,7,10,13,46,54], (2) the SLS problem with minimum machine setup time [7,23,25,30,79], and (3) the SLS problem with MBT plus machine setup time [7,54]. Recall that for static IMRT, the delivery time equals the sum of beam-on time and machine setup time.

The SLS problem with MBT is polynomial time solvable [10]. Ahuja and Hamacher [3] formulated this problem as a minimum cost flow problem in a directed network and gave an optimal linear time algorithm for the case when the MLC is not subject to the minimum separation constraint and interleaf motion constraint, e.g., the Varian MLC. Kamath et al. [46] proposed a different approach by sweeping the IM field from left to right and generating MLC-apertures accordingly. They showed how to handle the minimum separation constraint and interleaf motion constraint during the sweeping and achieve MBT.

The SLS problem with minimum machine setup time is, in general, NP-hard. Burkard [18] proved that for IMs with at least two rows, the problem can be reduced from the subset sum problem [38]. Baatar et al. [7] further showed that NP-hardness holds even for input IMs with a single row, using a reduction from the 3-partition problem [38]. Chen et al. [22], independently, gave an NP-hardness proof of this problem based on a reduction from the 0-1 knapsack problem [38]. The SLS problem with MBT plus machine setup time is also NP-hard [10]. We are not aware of any efficient algorithms for the MLC problem with MBT plus machine setup time.

In this section, we will study the following 3-D SLS problem [79]: Given an IM, find a minimum set S of MLC-apertures (together with their weights) for delivering the IM (i.e., the size $|S|$ is minimized). Note that our goal is to minimize the number *of* MLC-apertures used for delivering an IM, which is equivalent to minimizing the machine setup time under the assumption that the setup time needed to go from one MLC-aperture shape to another shape is constant. While the MLC problem with MBT plus machine setup time seems more general, our formulation of the 3-D SLS problem also well captures the total delivery time. This is because the machine setup time for the MLC-apertures dominates the total delivery time [30,79], and algorithms that minimize the number of MLC-apertures used are desired to reduce the delivery time [79]. It should be pointed out that even though the SLS problem with the MBT is polynomial time solvable, the resulting plan can use a very large number of MLC-apertures [36], and thus a prolonged total delivery time.

A key special case of the SLS problem, called the basic 3-D SLS problem, is also of clinical value [11–16]: Given an IM, find a minimum set of MLC-apertures for the IM, such that each MLC-aperture has a unit height. Note that in the general SLS problem, the weight of each MLC-aperture can be any integer ≥ 1. Studying the basic case is important because the maximum heights of the majority of IMs used in current clinical treatments are around five, and an optimal solution for the basic case on such an IM is often very close to an optimal solution for the general case.

The rest of this section is organized as follows. In Section 7.3.2, we consider the special SLS case that the input IM has a single row. In Section 7.3.3, we consider the special SLS case that the input IM has only 0 or 1 in its cells. In Section 7.3.4, we study the 3-D SLS problem.

7.3.2   The 1-D SLS Problem

When the input IM is of a single row, the 3-D SLS problem reduces to the 1-D case. The input IM becomes a one-dimensional intensity profile and the output MLC-apertures become a set of leaf openings with weighted intensity levels (see Figure 7.5). Such openings are obtained by setting the left and right leaves of an MLC leaf pair to specific positions. The goal is to minimize the number of leaf openings.

The basic 1-D SLS problem, in which the output leaf openings are required to have unit weights, can be easily solved using a sweeping method proposed by Boyer and Yu [17]. The method first distinguishes the left and right ends along the intensity profile curve, and then uses a greedy approach to generate a delivery option, i.e., a valid pairing in which a left end is always paired with a right end on its right. An interesting question is how to generate all possible delivery options for a basic 1-D SLS instance. Yu [80] pointed out that for an IM row with N left ends and N right ends, there are in the worst case $N!$ delivery options. Webb [74] presented a general formula for determining the total number of delivery options for an arbitrary IM row, and showed that this number in general tends to be considerably smaller than $N!$. Luan et al. [56] gave an efficient algorithm for producing all possible delivery options that runs in optimal time, linearly proportional to the number of output delivery options. The key idea of the algorithm is to impose an order onto how the left and right ends should be paired: it recursively pairs the rightmost unpaired left end with an unpaired right end such that the pair does not define an illegal opening.

The general 1-D SLS problem, in which the output leaf openings are not required to have unit weights, is NP-hard [7,22]. It is quite easy to observe that a horizontal trapezoidal decomposition of the input intensity profile curve yields a 2-approximation solution for the general 1-D SLS problem (e.g., see Figure 7.5).

Bansal et al. [8] modeled the general 1-D SLS problem as a maximum partition problem of prefix positive zero sum vectors, and proposed several approximation algorithms with approximation ratio less than two. For the unimodal input intensity profile curves, i.e., of only one peak, they gave a $\frac{9}{7}$ approximation algorithm by reducing the general 1-D SLS problem to the set packing problem [43]. For arbitrary input intensity profile curves, they gave a $\frac{24}{13}$ approximation based on rounding the solution of a certain linear programming problem.

Independently, Chen et al. [21] studied the shape rectangularization (SR) problem of finding minimum set of rectangles to build a given functional curve, which turns out to be equivalent to the general 1-D SLS problem. They presented a $(\frac{3}{2} + \epsilon)$-approximation algorithm for the SR problem. They pointed out two interesting geometric observations: (1) For the SR problem, it is sufficient to consider only a special type of rectangle sets, called canonical rectangle sets, in which the left and right ends of the rectangles coincide with the left and right ends of the input intensity profiles; (2) An optimal rectangle set which is also canonical is isomorphic to a forest of weighted trees. Based on the above observations, they proposed a combinatorial optimization problem, called the primary block set (PBS) problem space [21], which is related to the SR problem in the sense that for any $\mu \geq 2$,

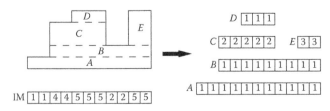

FIGURE 7.5   The leaf openings $A, B, C, D,$ and E for delivering the dose profile curve of an IM of a single row.

a μ-approximation PBS algorithm immediately implies a $(2 - \frac{1}{\mu})$-approximation SR algorithm. Further, they showed that the PBS problem can be reformulated, in polynomial time, as a multi-commodity demand flow (MDF) problem [19] on a path. Chekuri et al. [19] gave a polynomial time $(2 + \epsilon)$-approximation algorithm for the MDF problem on a path when the maximum demand d_{max} is less than or equal to the minimum capacity c_{min}. Chen et al. [21] extended Chekuri et al.'s result and gave a $(2 + \epsilon)$-approximation algorithm for the MDF problem on a path when $d_{max} \leq \lambda \cdot c_{min}$, where $\lambda > 0$ is any constant. This leads to a $(\frac{3}{2} + \epsilon)$-approximation algorithm for the SR problem when $M_f \leq \lambda \cdot m_f$, where m_f (or M_f) is the global positive minimum (or maximum) of the input intensity profile curve f.

7.3.3 The 2-D SLS Problem

When the input IM has only 0 or 1 in its cells, the 3-D SLS problem becomes a 2-D one (called the 2-D SLS problem), and an optimal set of MLC-apertures is just a minimum set of rectilinear y-monotone simple polygons. For MLCs with the minimum separation constraint, each rectilinear y-monotone simple polygon must have a minimum width $\geq \delta$ (the parameter δ represents the minimum separation constraint discussed in Section 7.2.1). As shown in Section 7.3.4, a good algorithm for partitioning a rectilinear polygonal region (possibly with holes) into a minimum set of rectilinear y-monotone simple polygons can become a key procedure for solving the general 3-D SLS problem.

Chen et al. [23] presented a unified approach for solving the minimum y-monotone partition problem on an n-vertex polygonal domain (possibly with holes) in various settings. This problem can be reduced to computing a maximum bipartite matching. But, due to the specific geometric setting, an explicit construction of the underlying bipartite graph for a maximum matching is too costly. They overcome this difficulty by reducing the matching problem to a maximum flow problem on a geometric graph. Since the matching information is implicitly stored in the flow graph, they use a special depth-first search to find an actual optimal matching. Thus, an $O(n^2)$ time algorithm is obtained for partitioning a polygonal domain (possibly with holes) into a minimum set of y-monotone parts, improving the previous $O(n^3)$ time algorithm [55] (which works only for the simple polygon case). The ideas can be extended to handling the minimum separation constraint by modifying the edge capacities of the flow graph.

7.3.4 The 3-D SLS Problem

In this subsection, we study the 3-D SLS problem, which seeks a minimum set S of MLC-apertures (together with their weights) for delivering the given IM.

7.3.4.1 The Basic 3-D SLS Problem

The basic 3-D SLS problem, in which all weights of the MLC-apertures are required to be 1, has been studied under different MLC constraints. For MLCs that are not subject to the minimum separation constraint nor the interleaf motion constraint, the problem can be easily reduced to the basic 1-D SLS problem (see Section 7.3.4), since each leaf pair is fully independent. For MLCs subjective to the minimum separation constraint but not the interleaf motion constraint, Kamath et al. [46] gave an algorithm that generates an optimal MLC-aperture set. The algorithm first generates an optimal MLC-aperture set without considering the minimum separation constraint, and then performs a sweeping of the IM field; during the sweeping, it repeatedly detects instances of violation and eliminates them accordingly.

For MLCs subjective to both the minimum separation and interleaf motion constraints, Chen et al. [25] modeled the corresponding 3-D SLS problem as an optimal terrain construction problem. To compute an optimal set of MLC-apertures for the IM, the algorithm builds a graph G: (1) Generate all distinct delivery options for each IM row, and let every vertex of G correspond to exactly one such delivery option; (2) for any two delivery options for two consecutive IM rows, put left-to-right directed edges whose weights are determined by the optimal bipartite matchings between the leaf openings of the two delivery options. The basic 3-D SLS problem is then solved by finding the shortest path in the graph G [25].

This algorithm, however, does not handle two crucial issues well. The first issue is on the worst-case time bound of the algorithm that has a multiplicative factor of $N!$. Recall that there can be $N!$ distinct delivery options in the worst case for every IM row (see Section 7.3.2). The second issue is on the optimality of the MLC-aperture set produced. Since the algorithm uses only the minimal delivery options for each IM row to compute a set of MLC-apertures (i.e., each such delivery option has the minimum number of leaf opening for building the corresponding IM row), the output may not be truly optimal. It is possible that a truly optimal MLC-aperture set for an IM need not use a minimal delivery option for each row; instead, some of the delivery options used may be defined by Steiner points [23].

Chen et al. [23] proposed a polynomial time basic 3-D SLS algorithm that can handle the above two issues. The key idea is based on new geometric observations which imply that it is sufficient to use only very few special delivery options, which they called canonical delivery options, for each IM row without sacrificing the optimality of the output MLC-aperture set. This produces guaranteed optimal quality solutions for the case when a constant number of Steiner points is used on each IM row.

7.3.4.2 The General 3-D SLS Problem

We now study the general 3-D **SLS** problem, in which the weights of the MLC-apertures can be arbitrary positive integers.

Several heuristic methods for the general 3-D SLS problem have been proposed [11–16,29,30, 35,41,74,79]. They all have several main steps: (1) choose the upper monotone boundary of an MLC-aperture (by using some simple criteria), (2) choose the lower monotone boundary, (3) check whether the two monotone boundaries enclose an MLC-aperture, and (4) output the MLC-aperture thus obtained. These steps are repeated until the entire IM is built. Depending on where the upper monotone boundary is placed, these algorithms are further classified as either the "sliding window" or "reducing level" methods. The sliding window methods always place the upper monotone boundary at the boundary of the planar projection of the remaining IM to be built, while the reducing level methods normally set the upper monotone boundary at the place with the maximum height of the remaining IM.

Chen et al. [23] presented a heuristic algorithm for the general 3-D SLS problem based on their solutions for the 2-D SLS problem (see Section 7.3.3) and basic 3-D SLS problem (see Section 7.3.4.1). To make use of the 2-D SLS algorithm and basic 3-D SLS algorithm, they partition the input IM into a "good" set of sub-IMs, such that each sub-IM can be handled optimally by one of these two algorithms (i.e., each resulting sub-IM must be of a uniform height, or the maximum height of each such sub-IM should be reasonably small, say, ≤ 5). The partition is carried out in a recursive fashion and for each recursive step, several partition schemes are tested and compared, and the algorithm finally chooses the best result using dynamic programming.

Luan et al. [57] gave two approximation general 3-D SLS algorithms for MLCs without the minimum separation constraint nor the interleaf motion constraint (e.g., the Varian MLC). The first is a $(\lceil \log h \rceil + 1)$-approximation where $h > 0$ is the largest entry in the IM. The second is a $2(\lceil \log D \rceil + 1)$-approximation algorithm where D is the maximum element of a set containing (1) all absolute differences between any two consecutive row entries over all rows, (2) the first entry of each row, (3) the last entry of each row, and (4) the value 1. The main ideas include (1)

2-approximation algorithm for the single-row case, and (2) an IM partition scheme that partitions the input IM into logarithmic number of sub-IMs with weights that are all powers of two. These two algorithms have a running time complexity of $O(mn \log h)$ and $O(mn \log D)$, respectively.

7.4 Static Leaf Sequencing with Tongue-and-Groove Error Control

In this section, we study the SLS problem with tongue-and-groove error control, i.e., we will seek to minimize both the delivery time and the tongue-and-groove error. As we will show later in this section, there is actually a trade-off between these two criteria. Thus this problem has been studied in two variations: (1) minimizing the delivery time subject to the constraint that the tongue-and-groove error is completely eliminated [24,49,65] and (2) minimizing the delivery time subject to the constraint that the tongue-and-groove error is no more than a given error bound [22].

Que et al. [65] gave a heuristic SLS algorithm that eliminates the tongue-and-groove error based on the "sliding-windows" method proposed by Bortfeld et al. [13]. The algorithm, however, does not guarantee optimal delivery time, either in terms of beam-on time or in terms of the number of MLC-apertures.

Kamath et al. [49] gave an SLS algorithm that minimizes beam-on time of the output plan subjective to the constraint that the tongue-and-groove error is completely eliminated. Their algorithm consists of a scanning scheme they proposed earlier [46] for solving the SLS problem with MBT (and without tongue-and-groove error control), and a modification scheme that rectifies that possible tongue-and-groove error violation.

Chen et al. [22] presented a graph modeling of the basic SLS problem (i.e., each MLC-aperture is of unit weight) with a tradeoff between the tongue-and-groove error and the number of MLC apertures, and an efficient algorithm for the problem on the Elekta MLC model, which has both the minimum separation and interleaf motion constraints. In their solution, the problem is formulated as a k-weight shortest path problem on a directed graph, in which each edge is defined by a minimum weight g-cardinality matching. Every such k-weight path specifies a set S of k MLC-apertures for delivering the given IM, and the cost of the path indicates the tongue-and-groove error of the set S of MLC-apertures. Chen et al. [22] also extended the above algorithm to other MLC models, such as Siemens and Varian, based on computing a minimum g-path cover of a directed acyclic graph (DAG). They used a partition scheme developed in paper [23] for handling the general case, i.e., each MLC-aperture can be of an arbitrary weight.

Chen et al. [24] presented an algorithm for SLS problem with minimum number of MLC-apertures subjective to the constraint that the tongue-and-groove error is completely eliminated. The main ideas of their algorithm include

- A novel IM partition scheme, called mountain reduction, for reducing a "tall" 3-D IM (mountain) to a small set of "low" sub-IM mountains that introduces no tongue-and-groove error. (In contrast, the partition scheme used in paper [22] may introduce tongue-and-groove error.) The key to their mountain reduction scheme is the profile-preserving mountain cutting (PPMC) problem, which seeks to cut an IM A into two IMs qB and C (i.e., $A = q \cdot B + C$, where $q \geq 2$ is a chosen integer), such that qB and C have the same profile as A. Two IMs M' and M'' are said to have the same *profile* if for all i, j, $M'_{i,j} \geq M'_{i,j+1} \Leftrightarrow M''_{i,j} \geq M''_{i,j+1}$ and $M'_{i,j} \leq M'_{i,j+1} \Leftrightarrow M''_{i,j} \leq M''_{i,j+1}$. They showed that the profile-preserving cutting introduces no tongue-and-groove error. They formulated this PPMC problem as a bottleneck shortest path (BSP) problem on a DAG G' of a pseudo-polynomial size. By exploiting interesting properties of this DAG, they compute the BSP by searching only a small (linear size) portion of G' and achieve an optimal linear time PPMC algorithm.

- An efficient graph-based algorithm for partitioning a sub-IM into a minimum number of MLC-apertures without tongue-and-groove error. They directly incorporated the zero tongue-and-groove error constraint with the previous graph-based algorithm proposed in paper [22], and by exploiting geometric properties, they show that the size of the graph can be significantly reduced and computing the weights of edges can be done in a much faster manner using a new matching algorithm.

7.5 Field Splitting

In this section, we study a few geometric partition problems, called field splitting, which arise in IMRT. Due to the maximum leaf spread constraint (see Section 7.2.1), an MLC cannot enclose an IM of a too large width. For example, on one of the most popular MLC systems called Varian, the maximum leaf spread constraint limits the maximum allowed field width to about 14.5 cm. Hence, this necessitates a large-width IM field to be split into two or more adjacent subfields, each of which can be delivered separately by the MLC subject to the maximum leaf spread constraint [32,42,75]. But, such IM splitting may result in a prolonged beam-on time and thus affect the treatment quality. The field splitting problem, roughly speaking, is to split an IM of a large width into several subfields whose widths are all no bigger than a threshold value, such that the total beam-on time of these subfields is minimized.

In this section, we will focus our discussion on the Varian MLCs, whose maximum spread threshold value (i.e., 14.5 cm) is the smallest among existing popular MLCs and where the field splitting is often needed for medium and large size tumor cases. Henceforth, we always assume the MLCs under discussion do not have the minimum separation constraint nor the interleaf motion constraint.

Engel [34] showed that for an IM M of size $m \times n$, when n is no larger than the maximum allowed field width w, the MBT of M is captured by the following formula

$$\text{MBT}(M) = \max_{i=1}^{m}\{M_{i,1} + \sum_{j=2}^{n} \max\{0, M_{i,j} - M_{i,j-1}\}\} \tag{7.2}$$

Engel also described a class of algorithms achieving this minimum value.

Geometrically, we distinguish three versions of the field splitting problem based on how an IM is split (see Figure 7.6c through e): (1) splitting using vertical lines; (2) splitting using y-monotone paths; (3) splitting with overlapping. Note that in versions (1) and (2), an IM cell belongs to exactly one subfield; but in version (3), a cell can belong to two adjacent subfields, with a nonnegative value in both subfields, and in the resulting sequence of subfields, each subfield is allowed to overlap only with the subfield immediately before or after it.

Kamath et al. [50] studied the field splitting using vertical lines (FSVL) problem with minimum total MBT, i.e., the sum of the MBTs of the resulting subfields. Their algorithm worked in a brute-force manner and split a size $m \times n$ IM using vertical lines into at most three subfields (i.e., $n \leq 3w$ for their algorithm, where w is the maximum allowed field width), and took $O(mn^2)$ time. Wu [77] formulated the FSVL problem that splits IMs of arbitrary widths into $k \geq 3$ subfields as a k-link shortest path problem in a directed acyclic graph. Each vertex in the graph represents a possible splitting line, and each edge represents the subfield enclosed by the two lines and is assigned a weight which equals the MBT for its associated subfield. With a carefully characterization of the intrinsic structures of the graph, an $O(mnw)$ time algorithm was achieved.

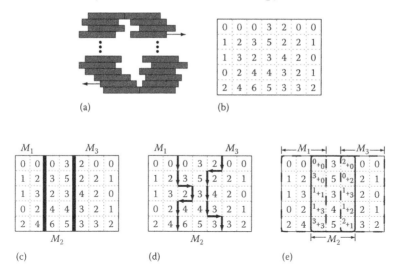

FIGURE 7.6 (a) An MLC. (b) An IM. (c)–(e) Examples of splitting an IM into three subfields, M_1, M_2, and M_3, using vertical lines, y-monotone paths, and with overlapping, respectively. The dark cells in (e) show the overlapping regions of the subfields; the prescribed dose in each dark cell is divided into two parts and allocated to two adjacent subfields.

Chen and Wang [27] studied the field splitting using y-monotone paths (FSMP) problem. Their key observation is that there are at most a number of $mw + 1$, instead of $O((w + 1)^m)$, candidates for the leftmost y-monotone path used in the splitting that need to be considered. They then showed that all these candidate paths can be enumerated efficiently by using a heap data structure and an interesting new method called MBT-sweeping [27]. Further, they exploited the geometric properties of the underlying field splitting problem to speed up the computation of the total MBT of the induced subfields. The resulting FSMP algorithm took polynomial time as long as the number of subfields is a constant.

The field splitting with overlapping (FSO) problem has also been studied. Kamath et al. [47] studied a special FSO case in which an IM is split into at most three overlapping subfields and the overlapping regions of the subfields are fixed; a greedy scanning algorithm is proposed that produces optimal solutions in $O(mn)$ time for this special case. In fact, it was pointed out in paper [27] that by considering all possible combinations of overlapping regions, the algorithm in [47] can be extended to solving the general FSO problem, i.e., the overlapping regions of the subfields are not fixed and the number d of resulting subfields is any integer ≥ 2, in $O(mnw^{d-2})$ time.

Chen and Wang [27] presented an $O(mn + mw^{d-2})$ time FSO algorithm, improving the time bound of Kamath et al.'s algorithm [47] by a factor of $\min\{w^{d-2}, n\}$. The algorithm hinges on a nontrivial integer linear programming (ILP) formulation of the field splitting with fixed overlapping (FSFO) problem. Basically, this is the FSO problem subject to the constraint that the sizes and positions of the d sought subfields are all fixed. They showed that the constraint matrix of the induced ILP problem is totally unimodular [64], and thus the ILP problem can be solved optimally by linear programming (LP). Further, they showed that the dual of this LP is a shortest path problem on a DAG, and the FSFO problem is solvable in totally $O(mn)$ time. The algorithm then reduced the original FSO problem to a set of $O(w^{d-2})$ FSFO problem instances. They pointed out that under the ILP formulation of the FSFO problem, with an $O(mn)$ time preprocess, each FSFO instance is solvable in only $O(m)$ time. This finally gave an $O(mn + mw^{d-2})$ time FSO algorithm.

7.6   Dose Simplification

In this section, we study the dose simplification problem that arises in IMRT. During the IMRT treatment planning process, the shapes, sizes, and relative positions of a tumor volume and other surrounding tissues are determined by 3-D image data, and an "ideal" radiation distribution is computed by a computer system. Without loss of generality, let the z-axis be the beam orientation. Then this "ideal" radiation distribution is a function defined on the xy-plane (geometrically, it is a 3-D functional surface above the xy-plane), which is usually of a complicated shape and not deliverable by the MLC. Thus, the "ideal" radiation distribution must be "simplified" to a discrete IM (see Section 7.1), i.e., a discrete IM approximates the "ideal" distribution under certain criteria. In some literature, an "ideal" radiation distribution is also referred to as a continuous IM.

The dose simplification problem has been studied under two major variants, depending on how the complexity of a (discrete) IM is defined:

- (Clustering) If the complexity of an IM is defined by the number of distinct intensity levels in the IM, then the dose simplification problem can be modeled as a constrained 1-D K-means clustering problem [9,20,61,78]: Given an "ideal" radiation distribution of size $m \times n$, the mn intensity values in the distribution need to be grouped into K clusters for the smallest possible number K, such that the maximum difference between any two intensity values in each cluster is no bigger than a given bandwidth parameter δ and the total sum of variances of the K clusters is minimized. The resulting clusters are then used to specify a discrete IM.

- (Shape Approximation) If the complexity of an IM is defined by the number of MLC-apertures required to deliver the IM, then the dose simplification problem can be modeled as a shape approximation problem: Given an "ideal" radiation distribution, find K MLC-apertures for the smallest possible number K, such that the approximation error between the "ideal" radiation distribution and the discrete IM (defined by the sum of the K MLC-apertures) is within a given error bound $\mathcal{E}$.

Several algorithms for the constrained 1-D K-means clustering problems have been given in medical literature and used in clinical IMRT planning systems [9,61,78]. These algorithms use heuristic numerical methods to determine the clusters iteratively [9,61,78], and can be trapped in a local minimal. To the best of our knowledge, no theoretical analysis has been given on their convergence speed.

Chen et al. [20] presented an efficient geometric algorithm for computing optimal solutions to the constrained 1-D K-means problem. They modeled the problem as a K-link shortest path problem on a weighted DAG. By exploiting the Monge property [1,2,62,76] of the DAG, the algorithm runs in $O\left(\min\left\{Kn, n2^{\sqrt{\log K \log \log n}}\right\}\right)$ time.

The shape approximation variant of the dose simplification problem appears to be much harder. Several papers [13,21,35,73] discussed a simple version of the problem: Given a 1-D dose profile functional curve f, find a minimum set of rectangles whose sum approximates f within a given error bound $\mathcal{E}$. Due to its relation with the SR problem (see Section 7.3.2), Chen et al. [21] called this special case the generalized shape rectangularization (GSR) problem and pointed out its NP-hardness. They also showed the relation between the GSR problem and the forest of block towers (FBT) problem: an optimal FBT solution immediately implies a 2-approximation GSR solution. Here, the FBT problem is a special GSR case in which the output rectangles are required to satisfy the inclusion–exclusion constraint [13,35], meaning that for any two rectangles, either their projected intervals on the x-axis do not intersect each other (except possibly at an endpoint), or one interval is fully contained in the other interval.

Some work has been done on the FBT problem in the medical field [13,35,73]. An approach based on Newton's method and calculus techniques was used in [35]; but, it works well only when the input curve f has few "peaks," and must handle exponentially many cases as the number of "peaks," of f increases. Chen et al. [21] presented an FBT algorithm that produces optimal solutions for arbitrary input curves. They made a set of geometric observations, which imply that only a finite set of rectangles needs to be considered as candidates for the sought rectangles, and a graph G on such rectangles can be built. They then modeled the FBT problem as a k-MST [4–6,37,39,59,60,66,84] problem in G. By exploiting the geometry of G, they gave an efficient dynamic programming-based FBT algorithm.

7.7 Leaf Sequencing for Dynamic IMRT and AMRT

In this section, we study the DLS problem and the SALS problem, i.e., the leaf sequencing problems for dynamic IMRT delivery and AMRT (see Section 7.1), respectively.

7.7.1 Dynamic Leaf Sequencing

Dynamic IMRT delivery is different from static delivery in that the radiation beam is always on and the leaf positions change with respect to time. The DLS problem seeks a set of trajectories of the left and right leaf tips of the MLC leaf pairs, i.e., the leaf tip position at any time, to deliver a given IM with minimum total delivery time. Unlike the SLS problem (see Section 7.3), we need to consider an additional MLC constraint, namely, the maximum leaf motion speed constraint (see Section 7.2.1), in the DLS problem. Note that the actual delivered radiation dose to an IM cell $A(i,j)$ in the IM A is proportional to the amount of time the cell is exposed to the radiation beam, (i.e., the total time the cell is inside the leaf opening formed by the ith MLC leaf pair) during the dynamic delivery.

DLS algorithms [13,28,45,58,70,69] were given for exactly delivering an input IM with a short delivery time, under the maximum leaf motion speed constraint. Spirou and Chui's algorithm [69] computes the MLC leaf trajectories for exactly delivering the input IM. Their algorithm scans all IM cells from left to right, and when entering a new cell, produces the corresponding leaf positions. They proved that the algorithm optimizes the delivery time under the assumption that on each IM row, the corresponding MLC leaf tips always move from the left boundary of the leftmost nonzero cell to the right boundary of the rightmost nonzero cell. Kamath et al. [48] presented a DLS algorithm that can handle the interleaf motion MLC constraint, which occurs in some MLC systems such as Elekta and Siemens. The algorithm interleaves a scanning procedure similar to Spirou and Chui's [69] with detecting and rectifying possible violations of the interleaf motion constraint. The algorithm makes the same assumption on the starting and ending leaf tip positions as the algorithm in paper [69].

Chen et al. [26] pointed out that the assumption on the starting and ending leaf tip positions made in papers [48,69] may sacrifice the optimality of the output DLS solution. For MLCs without the interleaf motion constraint, they modeled the DLS problem as the following coupled path planning (CPP) problem [26]. The problem is defined on a uniform grid R_g of size $n \times H$ for some integers n and H such that the length of each grid edge is one unit. A path on the plane is said to be xy-monotone if it is monotone with respect to both the x-axis and the y-axis. For an integer $c > 0$, an xy-monotone (rectilinear) path p along the edges of R_g is said to be c-steep if every vertical segment of p is of a length at least c (i.e., formed by c or more grid edges) and every horizontal segment has a unit length. The CPP problem is defined as follows: Given a nonnegative function f defined on the

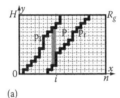

(a)   (b)

FIGURE 7.7 (a) Illustrating the CPP problem: p_l and p_r are two noncrossing c-steep paths with $c = 1$; the darkened area shows the vertical section on the ith column of the region enclosed by the two paths, whose length corresponds to the amount of actually delivered radiation dose. (b) The input (intensity) function f for the CPP problem, which is defined on $\{1, 2, \ldots, n\}$; the darkened area shows the value of f at the ith cell, which specifies the amount of prescribed dose at that cell.

integer set $\{1, 2, \ldots, n\}$ and positive integers c and Δ ($\Delta \leq H$), find two noncrossing c-steep paths on R_g, each starting at the bottom boundary and ending at the top boundary of R_g, such that the two paths, possibly with the bottom and top boundaries of R_g, enclose a (rectilinear) region P in R_g such that (1) for any column C_i of R_g, the vertical length of the intersection between C_i and P approximates the function value $f(i)$ (i.e., the value of f at i) within the given error bound Δ (see Figure 7.7a and b), and (2) the total sum of errors on P is minimized. Note that in the CPP problem, f corresponds to the intensity profile function specifying one row of an IM (see Figure 7.7b), and the two output paths specify the moving trajectories of the two MLC leaf tips, i.e., the leaf tip position (the x-coordinate) at any unit time (the y-coordinate). The maximum motion speed constraint of MLC is reflected in the c-steepness constraint on the paths. The CPP problem basically seeks to minimize the total approximation error of delivery within a given amount H units of delivery time.

Chen et al. [26] presented a novel approach based on interesting geometric observations for the CPP problem. The key idea is to formulate the problem as computing a shortest path in a weighted DAG of $O(nH\Delta)$ vertices and $O(nH^2\Delta^2)$ edges. They exploited a set of geometric properties, such as certain domination relations among the vertices, to speed up the shortest path computation, resulting in an $O(nH\Delta)$ time CPP algorithm. One unique feature of their algorithm is it computes a tradeoff between the delivery time and the approximation error.

7.7.2 Arc Modulated Radiation Therapy

Arc modulated radiation therapy (AMRT) is a newly patented IMRT delivery technique [82]. In an AMRT delivery, the beam source rotates along a single arc path in 3-D, and for every θ degrees (usually $\theta = 10$), a prescribed IM of size $m \times n$ is delivered toward the target regions. A key problem in AMRT delivery is the so-called SALS problem, which seeks to optimally convert a given set of K IMs (with $K = \frac{360}{\theta}$) into MLC leaf trajectories. The MLC is assumed to be free of the interleaf motion constraint, and thus each MLC leaf pair is treated independently.

Chen et al. [26] pointed out the close relation between the SALS problem and the constrained CPP problem, which is a special version the CPP problem (see Section 7.7.1) where the starting and ending points of the sought paths are given as part of the input. Based on a graph modeling, it was shown [26] that the SALS problem can be computed by solving the set of all constrained CPP problem instances. More specifically, for all possible combinations of starting and ending leaf pair positions, the optimal solutions are sought for the corresponding constrained CPP instances (in total, there are $O(n^4)$ problems instances). They proposed a nontrivial graph transformation scheme that allows a batch fashion computation of the instances. Further, the shortest path computation is accelerated by exploiting the Monge property of the transformed graphs. Consequently, an $O(n^4\Delta + n^2H\Delta^2)$ time algorithm was given for solving the set of all constrained CPP problem instances. The final SALS algorithm takes $O(Kmn^2\Delta(n^2 + H\Delta))$ time.

7.8 Concluding Remarks

As pointed out by Webb [73], two of the most important developments in improving the quality of radiation therapy are (1) the introduction of IMs in the treatment planning stage to approximate the geometric shape of the planning target volume, and (2) the usage of MLCs in the treatment delivery stage to geometrically shape the radiation beams. These developments are at the leading edge of the field and call for effective and efficient methods for treatment planning and delivery. In this chapter, we have discussed various geometric and combinatorial problems that arise in the treatment planning and the treatment delivery phases of IMRT.

There are many exciting and important computational problems in IMRT that are yet to be solved. We would like to discuss some of the open problems and research directions that are likely to receive considerable attention in the future:

- For the general 3-D SLS problem discussed in Section 7.3.4.2, the existing approximation algorithm by Luan et al. [57] has an approximation ratio related to the complexity of the input IM and applies to MLCs without the interleaf motion constraint and the minimum separation constraint. However, what about the MLCs (e.g., Elekta and Siemens MLCs) that have one or both of these constraints? And is it possible to achieve a polynomial time algorithm with better approximation ratio, say a constant?

- Current field splitting algorithms (see Section 7.5) all target at the Varian MLCs which are not subject to the interleaf motion constraint and seek to minimize the total beam-on time. Can one extend those field splitting algorithms to other types of MLCs? Also, is it possible to develop efficient field splitting algorithms that minimize the total number of MLC-apertures?

- The shape approximation variant of the dose simplification problem still remains to be intriguing. The FBT algorithm (see Section 7.6) presented is a weakly polynomial time algorithm. Is it possible to develop strongly polynomial time algorithms for the FBT problem? Also, can we develop approximation algorithms for the shape approximation variant of the dose simplification problem, which is known to be NP-hard?

- For the DLS problem, the CPP-based algorithm (see Section 7.7.1) produces truly optimal solutions for MLCs without the interleaf motion constraint. Is it possible to extend this algorithm so that it can handle MLCs with the interleaf motion constraint?

- Recent development of radiation therapy calls for the combination of IMRT with Image-guided radiation therapy (IGRT). IGRT is a process of using various imaging technologies to locate a tumor target prior to a radiation therapy treatment. This process is aimed to improve the treatment accuracy by eliminating the need for large target margins which have traditionally been used to compensate for errors in localization. Advanced imaging techniques using CT, MRI, and ultrasound are applied to accurately delineate treatment target. One key problem arising in IGRT is the so-called medical image registration problem, which aims to transform one set of medical image data into others (e.g., data of the same patient taken at different points in time). The main difficulty here is to cope with elastic deformations of the body parts imaged. Existing medical image registration algorithms generally require lots of human interaction in order to find a satisfactory transformation, and do not work well for deformable cases. We believe the underlying problem could be formulated by some special matching and flow problems in graphs.

Acknowledgment

This research was supported in part by NSF Grants CCF-0515203 and CCF-0916606, and NIH Grant 1R01-CA117997-01A2.

References

1. A. Aggarwal, M.M. Klawe, S. Moran, P. Shor, and R. Wilber. Geometric applications of a matrix-searching algorithm. *Algorithmica*, 2:195–208, 1987.
2. A. Aggarwal and J. Park. Notes on searching in multidimensional monotone arrays. In *Proceedings of the 29th Annual IEEE Symposium on Foundations of Computer Science*, IEEE Computer Society, Washington, DC, pp. 497–512, 1988.
3. R.K. Ahuja and H.W. Hamacher. A network flow algorithm to minimize beam-on time for unconstrained multileaf collimator problems in cancer radiation therapy. *Networks*, 45:36–41, 2005.
4. E.M. Arkin, J.S.B. Mitchell, and G. Narasimhan. Resource-constrained geometric network optimization. In *Proceedings of the 14th ACM Symposium on Computational Geometry*, pp. 307–316, ACM, New York, 1998.
5. S. Arora. Polynomial-time approximation schemes for euclidean TSP and other geometric problems. *J. ACM*, 45(5):753–782, 1998.
6. A. Arya and H. Ramesh. A 2.5 factor approximation algorithm for the *k*-MST problem. *Inf. Process. Lett.*, 65:117–118, 1998.
7. D. Baatar, M. Ehrgott, H.W. Hamacher, and G.J. Woeginger. Decomposition of integer matrices and multileaf collimator sequencing. *Discrete Appl. Math.*, 152:6–34, 2005.
8. N. Bansal, D. Coppersmith, and B. Schieber. Minimizing setup and beam-on times in radiation therapy. In *Proceedings of the 9th International Workshop on Approximation Algorithms for Combinatorial Optimization Problems*, LNCS, volume 4110, Springer-Verlag, Berlin, Germany, pp. 27–38, 2006.
9. W. Bär, M. Alber, and F. Nsslin. A variable fluence step clustering and segmentation algorithm for step and shoot IMRT. *Phys. Med. Biol.*, 46:1997–2007, 2001.
10. N. Boland, H.W. Hamacher, and F. Lenzen. Minimizing beam-on time in cancer radiation treatment using multileaf collimators. *Networks*, 43(4):226–240, 2004.
11. T.R. Bortfeld, A.L. Boyer, W. Schlegel, D.L. Kahler, and T.L. Waldron. Experimental verification of multileaf conformal radiotherapy. In A.R. Hounsell, J.M. Wilkinson, and P.C. Williams, editors, *Proceedings of the 11th International Conference on the Use of Computers in Radiation Therapy*, North Western Medical Physics Dept., Manchester, U.K., pp. 180–181, 1994.
12. T.R. Bortfeld, A.L. Boyer, W. Schlegel, D.L. Kahler, and T.L. Waldron. Realization and verification of three-dimensional conformal radiotherapy with modulated fields. *Int. J. Radiat. Oncol. Biol. Phys.*, 30:899–908, 1994.
13. T.R. Bortfeld, D.L. Kahler, T.J. Waldron, and A.L. Boyer. X-ray field compensation with multileaf collimators. *Int. J. Radiat. Oncol. Biol. Phys.*, 28:723–730, 1994.
14. T.R. Bortfeld, J. Stein, K. Preiser, and K. Hartwig. Intensity modulation for optimized conformal therapy. In *Proceedings Symposium Principles and Practice of 3-D Radiation Treatment Planning*, Munich, Germany, 1996.
15. A.L. Boyer. Use of MLC for intensity modulation. *Med. Phys.*, 21:1007, 1994.
16. A.L. Boyer, T.R. Bortfeld, D.L. Kahler, and T.J. Waldron. MLC modulation of x-ray beams in discrete steps. In A.R. Hounsell, J.M. Wilkinson, and P.C. Williams, editors, *Proceedings of the 11th International Conference on the Use of Computers in Radiation Therapy*, Stockport, U.K., pp. 178–179, 1994.

17. A.L. Boyer and C.X. Yu. Intensity-modulated radiation therapy using dynamic multileaf collimator. *Semin. Radiat. Oncol.*, 9(2):48–49, 1999.

18. R. Burkard. Open problem session. In *Oberwolfach Conference on Combinatorial Optimization*, Möhring, Berlin, Germany, November 24–29, 2002.

19. C. Chekuri, M. Mydlarz, and F.B. Shepherd. Multicommodity demand flow in a tree and packing integer problem. In *Proceedings of 30th International Colloquium on Automata, Languages and Programming*, LNCS, Springer-Verlag, Berlin, pp. 410–425, 2003.

20. D.Z. Chen, M.A. Healy, C. Wang, and B. Xu. Geometric algorithms for the constrained 1-D K-means clustering problems and IMRT applications. In *Proceedings of 1st International Frontiers of Algorithmics Workshop (FAW)*, LNCS, volume 4613, Springer-Verlag, Berlin, Germany, pp. 1–13, 2007.

21. D.Z. Chen, X.S. Hu, S. Luan, E. Misiolek, and C. Wang. Shape rectangularization problems in intensity-modulated radiation therapy. In *Proceedings of the 12th Annual International Symposium on Algorithms and Computation*, LNCS, volume 4288, Springer-Verlag, Berlin, Germany, pp. 701–711, 2006.

22. D.Z. Chen, X.S. Hu, S. Luan, S.A. Naqvi, C. Wang, and C.X. Yu. Generalized geometric approaches for leaf sequencing problems in radiation therapy. *Int. J. Comput. Geom. Appl.*, 16(2–3):175–204, 2006.

23. D.Z. Chen, X.S. Hu, S. Luan, C. Wang, and X. Wu. Geometric algorithms for static leaf sequencing problems in radiation therapy. *Int. J. Comput. Geom. Appl.*, 14(5):311–339, 2004.

24. D.Z. Chen, X.S. Hu, S. Luan, C. Wang, and X. Wu. Mountain reduction, block matching, and applications in intensity-modulated radiation therapy. *Int. J. Comput. Geom. Appl.*, 18(1–2):63–106, 2008.

25. D.Z. Chen, X.S. Hu, S. Luan, X. Wu, and C.X. Yu. Optimal terrain construction problems and applications in intensity-modulated radiation therapy. *Algorithmica*, 42:265–288, 2005.

26. D.Z. Chen, S. Luan, and C. Wang. Coupled path planning, region optimization, and applications in intensity-modulated radiation therapy. In *Proceedings of the 16th Annual European Symposium on Algorithms*, LNCS, Springer-Verlag, Berlin, pp. 271–283, 2008.

27. D.Z. Chen and C. Wang. Field splitting problems in intensity-modulated radiation therapy. In *Proceedings of the 12th Annual International Symposium on Algorithms and Computation*, LNCS, volume 4288, Springer-Verlag, Berlin, Germany, pp. 690–700, 2006.

28. D.J. Convery and M.E. Rosenbloom. The generation of intensity modulated fields for conformal radiotherapy by dynamic collimation. *Phys. Med. Biol.*, 37:1359–1374, 1992.

29. D.J. Convery and S. Webb. Generation of discrete beam-intensity modulation by dynamic multileaf collimation under minimum leaf separation constraints. *Phys. Med. Biol.*, 43:2521–2538, 1998.

30. J. Dai and Y. Zhu. Minimizing the number of segments in a delivery sequence for intensity-modulated radiation therapy with multileaf collimator. *Med. Phys.*, 28(10):2113–2120, 2001.

31. M.L.P. Dirkx, B.J.M. Heijmen, and J.P.C. van Santvoort. Trajectory calculation for dynamic multileaf collimation to realize optimized fluence profiles. *Phys. Med. Biol.*, 43:1171–1184, 1998.

32. N. Dogan, L.B. Leybovich, A. Sethi, and B. Emami. Automatic feathering of split fields for step-and-shoot intensity modulated radiation therapy. *Phys. Med. Biol.*, 48:1133–1140, 2003.

33. M.N. Du, C.X. Yu, J.W. Wong, M. Symons, D. Yan, R.C. Matter, and A. Martinez. A multi-leaf collimator prescription preparation system for conventional radiotherapy. *Int. J. Radiat. Oncol. Biol. Phys.*, 30(3):707–714, 1994.

34. K. Engel. A new algorithm for optimal multileaf collimator field segmentation. *Discrete Appl. Math.*, 152(1–3):35–51, 2005.

35. P.M. Evans, V.N. Hansen, and W. Swindell. The optimum intensities for multiple static collimator field compensation. *Med. Phys.*, 24(7):1147–1156, 1997.

36. M.C. Ferris, R.R. Meyer, and W. D'Souza. Radiation treatment planning: Mixed integer programming formulations and approaches. Optimization Technical Report 02-08, Computer Sciences Department, University of Wisconsin, Madison, WI, 2002.

37. M. Fischetti, H.W. Hamacher, K. Jørnsten, and F. Maffioli. Weighted *k*-cardinality trees: Complexity and polyhedral structure. *Networks*, 24:11–21, 1994.

38. M.R. Garey and D.S. Johnson. *Computers and Intractability: A Guide to the Theory of NP-Completeness*. Freeman, San Francisco, CA, 1979.

39. N. Garg. A 3-approximation for the minimum tree spanning *k* vertices. In *Proceedings of the 37th Annual IEEE Symposium on Foundations of Computer Science*, IEEE Computer Society, Washington, DC, pp. 302–309, 1996.

40. R.W. Hill, B.H. Curran, J.P. Strait, and M.P. Carol. Delivery of intensity modulated radiation therapy using computer controlled multileaf collimators with the CORVUS inverse treatment planning system. In *Proceedings of the 12th International Conference on the Use of Computers in Radiation Therapy*, Medical Physics Publishing Madison, WI, pp. 394–397, 1997.

41. T.W. Holmes, A.R. Bleier, M.P. Carol, B.H. Curran, A.A. Kania, R.J. Lalonde, L.S. Larson, and E.S. Sternick. The effect of MLC leakage on the calculation and delivery of intensity modulated radiation therapy. In *Proceedings of the 12th International Conference on the Use of Computers in Radiation Therapy*, Medical Physics Publishing, Madison, WI, pp. 346–349, 1997.

42. L. Hong, A. Kaled, C. Chui, T. Losasso, M. Hunt, S. Spirou, J. Yang, H. Amols, C. Ling, Z. Fuks, and S. Leibel. IMRT of large fields: Whole-abdomen irradiation. *Int. J. Radiat. Oncol. Biol. Phys.*, 54:278–289, 2002.

43. C.A.J. Hurkens and A. Schrijver. On the size of systems of sets every *t* of which have an SDR, with an application to the worst-case ratio of heuristics for packing problem. *SIAM J. Discrete Math.*, 2:68–72, 1989.

44. T.J. Jordan and P.C. Williams. The design and performance characteristics of a multileaf collimator. *Phys. Med. Biol.*, 39:231–251, 1994.

45. P. Kallman, B. Lind, and A. Brahme. Shaping of arbitrary dose distribution by dynamic multileaf collimation. *Phys. Med. Biol.*, 33:1291–1300, 1988.

46. S. Kamath, S. Sahni, J. Li, J. Palta, and S. Ranka. Leaf sequencing algorithms for segmented multileaf collimation. *Phys. Med. Biol.*, 48(3):307–324, 2003.

47. S. Kamath, S. Sahni, J. Li, J. Palta, and S. Ranka. A generalized field splitting algorithm for optimal IMRT delivery efficiency. *The 47th Annual Meeting and Technical Exhibition of the American Association of Physicists in Medicine (AAPM)*, Seattle, WA, 2005; *Med. Phys.*, 32(6):1890, 2005.

48. S. Kamath, S. Sahni, J. Palta, and S. Ranka. Algorithms for optimal sequencing of dynamic multileaf collimators. *Phys. Med. Biol.*, 49(1):33–54, 2004.

49. S. Kamath, S. Sahni, J. Palta, S. Ranka, and J. Li. Optimal leaf sequencing with elimination of tongue-and-groove underdosage. *Phys. Med. Biol.*, 49(3):N7–N19, 2004.

50. S. Kamath, S. Sahni, S. Ranka, J. Li, and J. Palta. Optimal field splitting for large intensity-modulated fields. *Med. Phys.*, 31(12):3314–3323, 2004.

51. H. Kobayashi, S. Sakuma, O. Kaii, and H. Yogo. Computer-assisted conformation radiotherapy with a variable thickness multi-leaf filter. *Int. J. Radiat. Oncol. Biol. Phys.*, 16(6):1631–1635, 1989.

52. D.D. Leavitt, F.A. Gibbs, M.P. Heilbrun, J.H. Moeller, and G.A. Takach. Dynamic field shaping to optimize stereotaxic radiosurgery. *Int. J. Radiat. Oncol. Biol. Phys.*, 21(5):1247–1255, 1991.

53. D.D. Leavitt, M. Martin, J.H. Moeller, and W.L. Lee. Dynamic wedge field techniques through computer-controlled collimator or motion and dose delivery. *Med. Phys.*, 17(1):87–91, 1990.

54. F. Lenzen. An integer programming approach to the multileaf collimator problem. Master's thesis, University of Kaiserslautern, Kaiserslautern, Germany, June 2000.

55. R. Liu and S. Ntafos. On decomposing polygons into uniformly monotone parts. *Inf. Process. Lett.*, 27:85–89, 1988.

56. S. Luan, D.Z. Chen, L. Zhang, X. Wu, and C.X. Yu. An optimal algorithm for computing configuration options of one-dimensional intensity modulated beams. *Phys. Med. Biol.*, 48(15):2321–2338, 2003.

57. S. Luan, J. Saia, and M. Young. Approximation algorithms for minimizing segments in radiation therapy. *Inf. Process. Lett.*, 101:239–244, 2007.

58. L. Ma, A. Boyer, L. Xing, and C.M. Ma. An optimized leaf-setting algorithm for beam intensity modulation using dynamic multileaf collimators. *Phys. Med. Biol.*, 43:1629–1643, 2004.

59. J.S.B. Mitchell. Guillotine subdivisions approximate polygonal subdivisions: A simple new method for the geometric *k*-MST problem. In *Proceedings of the 7th Annual ACM-SIAM Symposium on Discrete Algorithms*, Society of Industrial and Applied Mathematics, Philadelphia, PA, pp. 402–408, 1996.

60. J.S.B. Mitchell. Guillotine subdivisions approximate polygonal subdivisions: Part II—A simple polynomial-time approximation scheme for geometric TSP, *k*-MST, and related problems. *SIAM J. Comput.*, 28(4):1298–1309, 1999.

61. R. Mohan, H. Liu, I. Turesson, T. Mackie, and M. Parliament. Intensity-modulated radiotherapy: Current status and issues of interest. *Int. J. Radiat. Oncol. Biol. Phys.*, 53:1088–1089, 2002.

62. G. Monge. Déblai et Remblai. In *Mémories de l'Académie des Sciences*, Paris, France, 1781.

63. L.A. Nedzi, H.M. Kooy, E. Alexander, G.K. Svensson, and J.S. Loeffler. Dynamic field shaping for stereotaxic radiosurgery: A modeling study. *Int. J. Radiat. Oncol. Biol. Phys.*, 25(5):859–869, 1993.

64. G.L. Nemhauser and L.A. Wolsey. *Integer and Combinatorial Optimization*. John Wiley, New York, 1988.

65. W. Que, J. Kung, and J. Dai. "Tongue-and-groove" effect in intensity modulated radiotherapy with static multileaf collimator fields. *Phys. Med. Biol.*, 49:399–405, 2004.

66. R. Ravi, R. Sundaram, M.V. Marathe, D.J. Rosenkrantz, and S.S. Ravi. Spanning trees short and small. In *Proceedings of the 5th Annual ACM-SIAM Symposium on Discrete Algorithms*, Arlington, VA, pp. 546–555, 1994.

67. A. Schweikard, R.Z. Tombropoulos, and J.R. Adler. Robotic radiosurgery with beams of adaptable shapes. In *Proceedings of the 1st International Conference on Computer Vision, Virtual Reality and Robotics in Medicine*, LNCS, volume 905, Springer-Verlag, Berlin, Germany, pp. 138–149, 1995.

68. R.A.C. Siochi. Minimizing static intensity modulation delivery time using an intensity solid paradigm. *Int. J. Radiat. Oncol. Biol. Phys.*, 43(3):671–680, 1999.

69. S.V. Spirou and C.S. Chui. Generation of arbitrary intensity profiles by dynamic jaws or multileaf collimators. *Med. Phys.*, 21:1031–1041, 1994.

70. J. Stein, T. Bortfeld, B. Dorschel, and W. Schlegel. Dynamic X-ray compensation for conformal radiotherapy by means of multileaf collimations. *Radiother. Oncol.*, 32:163–173, 1994.

71. R. Svensson, P. Kallman, and A. Brahme. An analytical solution for the dynamic control of multileaf collimation. *Phys. Med. Biol.*, 39:37–61, 1994.

72. S. Webb. *The Physics of Three-Dimensional Radiation Therapy*. Institute of Physics Publishing, Bristol, Avon, U.K., 1993.

73. S. Webb. *The Physics of Conformal Radiotherapy—Advances in Technology*. Institute of Physics Publishing, Bristol, Avon, U.K., 1997.

74. S. Webb. Configuration options for intensity-modulated radiation therapy using multiple static fields shaped by a multileaf collimator. *Phys. Med. Biol.*, 43:241–260, 1998.

75. Q. Wu, M. Arnfield, S. Tong, Y. Wu, and R. Mohan. Dynamic splitting of large intensity-modulated fields. *Phys. Med. Biol.*, 45:1731–1740, 2000.

76. X. Wu. Optimal quantization by matrix searching. *J. Algorithms*, 12:663–673, 1991.

77. X. Wu. Efficient algorithms for intensity map splitting problems in radiation therapy. In *Proceedings of the 11th Annual International Computing and Combinatorics Conference*, LNCS, volume 3595, Springer-Verlag, Berlin, pp. 504–513, 2005.

78. Y. Wu, D. Yan, M.B. Sharpe, B. Miller, and J.W. Wong. Implementing multiple static field delivery for intensity modulated beams. *Med. Phys.*, 28:2188–2197, 2001.

79. P. Xia and L.J. Verhey. MLC leaf sequencing algorithm for intensity modulated beams with multiple static segments. *Med. Phys.*, 25:1424–1434, 1998.

80. C.X. Yu. Intensity-modulated arc therapy with dynamic multileaf collimation: An alternative to tomotherapy. *Phys. Med. Biol.*, 40:1435–1449, 1995.

81. C.X. Yu. Design Considerations of the sides of the multileaf collimator. *Phys. Med. Biol.*, 43(5): 1335–1342, 1998.

82. C.X. Yu, S. Luan, C. Wang, D.Z. Chen, and M. Earl. Single-arc dose painting: An efficient method of precision radiation therapy. Provisional patent, University of Maryland, College Park, MD, 2006.

83. C.X. Yu, D. Yan, M.N. Du, S. Zhou, and L.J. Verhey. Optimization of leaf positioning when shaping a radiation field with a multileaf collimator. *Phys. Med. Biol.*, 40(2):305–308, 1995.

84. A. Zelikovsky and D. Lozevanu. Minimal and bounded trees. In *Tezele Congresului XVIII Academilei Romano-Americane*, Kishinev, Moldova, pp. 25–26, 1993.

8

VLSI Layout Algorithms*

Andrea S. LaPaugh
Princeton University

One of the many application areas that has made effective use of algorithm design and analysis is computer-aided design (CAD) of digital circuits. Many aspects of circuit design yield to combinatorial models and the algorithmic techniques discussed in other chapters. In this chapter we focus on one area within the field of CAD: layout of very large scale integrated (VLSI) circuits, which is a particularly good example of the effective use of algorithm design and analysis. We will discuss specific problems in VLSI layout and how algorithmic techniques have been successfully applied. This chapter will not provide a broad survey of CAD techniques for layout, but will highlight a few problems that are important and have particularly nice algorithmic solutions. The reader may find more complete discussions of the field in the references discussed in Section 8.9 at the end of this chapter.

Integrated circuits are made by arranging active elements (usually transistors) on a planar substrate and interconnecting these elements with conducting wires that are also patterned on the planar substrate [41]. There may be several layers that can be used for wires, but there are restrictions on how wires in different layers can connect, as well as requirements of separations between wires and elements. Therefore, the layout of integrated circuits is usually modeled as a planar-embedding problem with several layers in the plane.

VLSI circuits contain millions of transistors. Therefore, it is not feasible to consider the positioning of each transistor separately. Transistors are organized into subcircuits called components; this may be done hierarchically, resulting in several levels of component definition between the individual

---

* Supported in part by the National Science Foundation, FAW award MIP-9023542.

transistors and the complete VLSI circuit. The layout problem for VLSI circuits becomes one of positioning components and their interconnecting wires on a plane, following the design rules, and optimizing some measure such as area or wire length. Within this basic problem structure are a multitude of variations rising from changes in design rules and flexibilities within components as to size, shape, and regions where wires may connect to components. Graph models are used heavily, both to model the components and interconnections themselves and to capture constraint between objects. Geometric aspects of the layout problem must also be modeled. Most layout problems are optimization problems and most are NP-complete. Therefore, heuristics are also employed heavily. In the following text, we present several of the best known and best understood problems in VLSI layout.

8.1   Background

We will consider a design style known as "general cell." In **general cell layout**, components vary in size and degree of functional complexity. Some components may be from a predesigned component library and have rigidly defined layouts (e.g., a register bank) and others may be full custom designs, in which the building blocks are individual transistors and wires.* Components may be defined hierarchically, so that the degree of flexibility in the layout of each component is quite variable. Other design styles, such as standard cell and gate array, are more constrained but share many of the same layout techniques.

The layout problem for a VLSI chip is often decomposed into two stages: placement of components and routing of wires. For this decomposition, the circuit is described as a set of components and a set of interconnections among those components. The components are first placed on the plane based on their size, shape, and interconnectivity. Paths for wires are then found to interconnect specified positions on the components. Thus, a placement problem is to position a set of components in a planar region; either the region is bounded, or a measure such as the total area of the region used is optimized. The area needed for the yet undetermined routing must be taken into account. A routing problem is, given a collection of sets of points in the plane, to interconnect each set of points (called a **net**) using paths from an allowable set of paths. The allowable set of paths captures all the constraints on wire routes. In routing problems, the width of wires is abstracted away by representing only the midline of each wire and ensuring enough room for the actual wires through the definition of the set of allowable paths.

This description of the decomposition of a layout problem into placement and routing is meant to be very general. To discuss specific problems and algorithms, we will use a more constrained model. In our model, components will be rectangles. Each component will contain a set of points along its boundary, the **terminals**. Sets of these terminals are the nets, which must be interconnected. A layout consists of a placement of the components in the plane and a set of paths in the plane that do not intersect the components except at terminals and interconnect the terminals as specified by the nets. The paths are composed of segments in various layers of the plane. Further constraints on the set of allowable paths define the routing style, and will be discussed for each style individually. The area of a layout will be the area of the minimum-area rectangle that contains the components and wire paths (see Figure 8.1).

While we still have a fairly general model, we have now restricted our component shapes to be rectangular, our terminals to be single points on component boundaries, and our routing paths to avoid components. (Components are thus assumed to be densely populated with circuitry.) While

---

* We will not discuss special algorithms for laying out transistors, used in tools called "cell generators" or "leaf-cell generators" for building components from low-level layout primitives. The reader is referred to [22] as a starting point for an investigation of this topic.

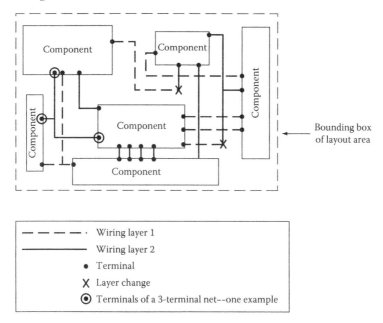

FIGURE 8.1 Example of a layout. This layout is rectilinear and has two layers for wiring.

these assumptions are common and allow us to illustrate important algorithmic results, there is quite a bit of work on nonrectangular components (e.g., [16,21]), more flexible terminals (see [37]), and "over-the-cell" routing (see [37]). Often, layouts are further constrained to be **rectilinear**. In rectilinear layouts, there is an underlying pair of orthogonal axes defining "horizontal" and "vertical" and the sides of the components are oriented parallel to these axes. The paths of wires are composed of horizontal and vertical segments. In our ensuing discussion, we too will often assume rectilinear layouts.

If a VLSI system is too large to fit on one chip, then it is first partitioned into chip-sized pieces. During partitioning, the goal is to create the fewest chips with the fewest connections between chips. Estimates are used for the amount of space needed by wires to interconnect the components on one chip. The underlying graph problem for this task is **graph partitioning**, which is discussed in the following text.

Closely related to the placement problem is the **floorplanning** problem. Floorplanning occurs before the designs of components in a general cell design have been completed. The resulting approximate layout is called a **floorplan**. Estimates are used for the size of each component, based on either the functionality of the component or a hierarchical decomposition of the component. Rough positions are determined for the components. These positions can influence the shape and terminal placement within each component as its layout is refined. For hierarchically defined components, one can work bottom up to get rough estimates of size, then top down to get rough estimates of position, then bottom up again to refine positions and sizes.

Once a layout is obtained for a VLSI circuit, either through the use of tools or by hand with a layout editor, there may still be room for improvement. **Compaction** refers to the process of modifying a given layout to remove extra spaces between features of the layout, spaces not required by design rules. Humans may introduce such space by virtue of the complexity of the layout task. Tools may place artificial restrictions on layout in order to have tractable models for the main problems of placement and routing. Compaction becomes a postprocessing step to make improvements too difficult to carry out during placement and routing.

8.2 Placement Techniques

Placement algorithms can be divided into two types: constructive initial placement algorithms and iterative improvement algorithms. A constructive initial placement algorithm has as input a set of components and a set of nets. The algorithm constructs a legal placement with the goal of optimizing some cost function for the layout. Common cost functions measure component area, estimated routing area, estimated total wire length, estimated maximum wire length of a net, or a combination of these. An iterative improvement algorithm has as input the set of components, set of nets, and an initial placement; it modifies the placement, usually repeatedly, to improve a cost function. The initial placement may be a random placement or may be the output of a constructive initial placement algorithm.

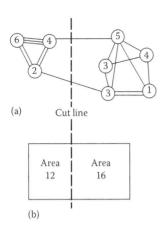

(a)   Cut line

(b)

FIGURE 8.2   Partitioning used in placement construction. (a) Partitioning the circuit: Each vertex represents a component; the area of the component is the number inside the vertex. Connections between components are represented by edges. Multiple edges between vertices indicate multiple nets connecting components. (b) Partitioning the layout rectangle proportionally to the partition of component area.

The iterative improvement of placements can proceed in a variety of ways. A set of allowable moves, that is, ways in which a placement can be modified, must be identified. These moves should be simple to carry out, including reevaluating the cost function. Typically, a component is rotated or a pair of components are exchanged. An iterative improvement algorithm repeatedly modifies the placement, evaluates the new placement, and decides whether the move should be kept as a starting point for further moves or as a tentative final placement. In the simplest of iterative improvement algorithms, a move is kept only if the placement cost function is improved by it. One more sophisticated paradigm for iterative improvement is simulated annealing (see Chapter 33 of *Algorithms and Theory of Computation Handbook, Second Edition: General Concepts and Techniques*). It has been applied successfully to the placement problem, for example, [36], and the floorplanning problem, for example, [4].

For general cell placement, one of the most widely used initial placement algorithms is a recursive partitioning method (see [24, p. 333]). In this method, a rectangular area for the layout is estimated based on component sizes and connectivity. The algorithm partitions the set of components into two roughly equal-sized subsets such that the number of interconnections between the two subsets is minimized and simultaneously partitions the layout area into two subrectangles of sizes equal to the sizes of the subsets (see Figure 8.2). This partitioning proceeds recursively on the two subsets and subrectangles until each component is in its own subset and the rectangle contains a region for each component. The recursive partitioning method is also heavily used in floorplanning [16].

The fundamental problem underlying placement by partitioning is **graph partitioning**.[*] Given a graph $G = (V, E)$, a vertex weight function $w : V \to N$, an edge cost function $c : E \to N$, and a balance factor $\beta \in [1/2, 1]$, the graph partitioning problem is to partition V into two subsets V_1 and V_2 such that

$$\sum_{v \in V_1} w(v) \le \beta \sum_{v \in V} w(v) \tag{8.1}$$

$$\sum_{v \in V_2} w(v) \le \beta \sum_{v \in V} w(v) \tag{8.2}$$

---

[*] Our definitions follow those in [24].

and the cost of the partition,

$$\sum_{e \in E \cap (V_1 \times V_2)} c(e)$$ (8.3)

is minimized. This problem is NP-complete (see [10, pp. 209,210]). Graph partitioning is a well-studied problem. The version we have defined is actually bipartitioning. Heuristics for this problem form the core of heuristic algorithms for more general versions of the partition problem where one partitions into more than two vertex sets. The hypergraph version of partitioning, in which each edge is a set of two or more vertices rather than simply a pair of vertices, is a more accurate version for placement, since nets may contain many terminals on many components. But heuristics for the hypergraph version again are based on techniques used for the graph bipartitioning problem.

Among many techniques for graph partitioning, two—the Kernighan–Lin algorithm [17], and simulated annealing—are best known. Both are techniques for iteratively improving a partition. The Kernighan–Lin approach involves exchanging pairs of vertices across the partition. It was improved in the context of layout problems by Fiduccia and Mattheyses [9], who moved a single vertex at a time. As applied to graph partitioning, simulated annealing also considers the exchange of vertices across the partition or the movement of a vertex from one side of the partition to the other. The methods of deciding which partitions are altered, which moves are tried, and when to stop the iteration process differentiate the two techniques.

Alternatives to iterative improvement have also received attention for circuit applications, especially as performance-related criteria have been added to the partitioning problem. Examples of these alternatives are spectral methods, based on eigenvectors of the Laplacian, and the use of network flow. The reader is referred to [1] or [32] for a more complete discussion of partitioning heuristics.

A second group of algorithms for constructing placements is based on agglomerative clustering. In this approach, components are selected one at a time and placed in the layout area according to their connectivity to components previously placed. Components are clustered so that highly connected sets of components are close to each other.

When the cost function for a layout involves estimating wire length, several methods can be used. The goal is to define a measure for each net that estimates the length of that net after it is routed. These estimated lengths can then be summed over all nets to get the estimate on the total wire length, or the maximum can be taken over all nets to get maximum net length. Two estimates are commonly used: (1) the half-perimeter of the smallest rectangle containing all terminals of a net and (2) the minimum Euclidean spanning tree of the net. Given a placement, the Euclidean spanning tree of a net is the spanning tree of a graph whose vertices are the terminals of the net, whose edges are all edges between vertices, and whose edge costs are the Euclidean distances in the given placement between the pair of terminals that are endpoints of the edges. Another often-used estimate is the minimum rectilinear spanning tree. This is because rectilinear layouts are common. For a rectilinear layout, the length of the shortest path between a pair of points, (x_a, y_a) and (x_b, y_b), is $|x_a - x_b| + |y_a - y_b|$ (assuming no obstacles). This distance, rather than the Euclidean distance, is used as the distance between terminals. (This is also called the L_1 or Manhattan metric.) A more accurate estimate of the wire length of a net would be the minimum Euclidean (or rectilinear) **Steiner tree** for the net. A Steiner tree for a set of points in the plane is a spanning tree for a superset of the points, that is, additional points may be introduced to decrease the length of the tree. Finding minimum Steiner trees is NP-hard (see [10]), while finding minimum spanning trees can be done in $O(|E| + |V| \log |V|)$ time for general graphs* and in $O(|V| \log |V|)$ time for Euclidean or rectilinear spanning trees (see Chapter 7 of *Algorithms and Theory of Computation Handbook, Second Edition:*

---

* Actually, using more sophisticated techniques, finding minimum spanning trees can be done in time almost linear in $|E|$.

General Concepts and Techniques and Chapter 1 of this book). The cost of a minimum spanning tree is an upper bound on the cost of a minimum Steiner tree. For rectilinear Steiner trees, the half-perimeter measure and two-thirds the cost of a minimum rectilinear spanning tree are lower bounds on the cost of a Steiner tree [14].

The minimum spanning tree is also useful for estimating routing area. The minimum spanning tree for each net is used as an approximation of the set of paths for the wires of the net. Congested areas of the layout can then be identified and space for routing allocated accordingly.

8.3 Compaction and the Single-Source Shortest Path Problem

Compaction can be done at various levels of design: an entire layout can be compacted at the level of transistors and wires; the layouts of individual components can be compacted; a layout of components can be compacted without changing the layout within components. To simplify our discussion, we will assume that layouts are rectilinear. For compaction, we model a layout as composed entirely of rectangles. These rectangles may represent the most basic geometric building blocks of the circuit: pieces of transistors and segments of wires, or may represent more complex objects such as complex components. We refer to these rectangles as the features of the layout. We distinguish two types of features: those that are of fixed size and shape and those that can be stretched or shrunk in one or both dimensions. For example, a wire segment may be able to stretch or shrink in one dimension, representing a lengthening or shortening of the wire, but be fixed in the other dimension, representing a wire of fixed width. We refer to the horizontal dimension of a feature as its width and the vertical dimension as its height.

Compaction is fundamentally a two-dimensional problem. However, two-dimensional compaction is very difficult. Algorithms based on branch and bound techniques for integer linear programming (see Chapter 31 of *Algorithms and Theory of Computation Handbook, Second Edition: General Concepts and Techniques*) have been developed, but none is efficient enough to use in practice (see [24], Chapter 6 of [34]). Therefore, the problem is commonly simplified by compacting in each dimension separately: first all features of a layout are pushed together horizontally as much as the design rules will allow, keeping their vertical positions fixed; then all features of a layout are pushed together vertically as much as the design rules will allow, keeping their horizontal positions fixed. The vertical compaction may in fact make possible more horizontal compaction, and so the process may be iterated. This method is not guaranteed to find a minimum area compaction, but, for each dimension, the compaction problem can be solved optimally. We are assuming that we start with a legal layout and that one-dimensional compaction cannot change order relationships in the compaction direction. That is, if two features are intersected by the same horizontal (vertical) line and one feature is to the left of (above) the other, then horizontal (vertical) compaction will maintain this property. The algorithm we present is based on the single-source shortest path algorithm (see Chapter 7 of *Algorithms and Theory of Computation Handbook, Second Edition: General Concepts and Techniques*). It is an excellent example of a widely used application of this graph algorithm.

The compaction approach we are presenting is called "constraint-based" compaction because it models constraints on and between features explicitly. We shall discuss the algorithm in terms of horizontal compaction, vertical compaction being analogous. We use a graph model in which vertices represent the horizontal positions of features; edges represent constraints between the positions of features. Constraints capture the layout design rules, relationships between features such as connectivity, and possibly other desirable constraints such as performance-related constraints. Design rules are of two types: feature-size rules and separation rules. Feature-size rules give exact sizes or minimum dimensions of features. For example, each type of wire has a minimum width; each transistor in a layout is of a fixed size. Separation rules require that certain features of a layout be at least a minimum distance apart to avoid electrical interaction or problems during fabrication. Connectivity constraints occur when a wire segment is allowed to connect to a component (or another wire segment) anywhere in a given interval along the component boundary. Performance requirements may dictate that certain elements are not too far apart. A detailed discussion of the issues in the extraction of constraints from a layout can be found in Chapter 6 of [34].

In the simplest case, we start with a legal layout and only consider feature-size rules and separation rules. We assume all wire segments connect at fixed positions on component boundaries and there are no performance constraints. Furthermore, we assume all features that have a variable width attach at their left and right edges to features with fixed width, for example, a wire segment stretched between two components. Then, we need only represent features with fixed width; we can use one variable for each feature. In this case, any constraints on the width of a variable-width feature are translated into constraints on the positions of the fixed-width features attached to either end (see Figure 8.3). We are left with only separation constraints, which are of the form

$$x_B \geq x_A + d_{min} \tag{8.4}$$

or equivalently

$$x_B - x_A \geq d_{min} \tag{8.5}$$

where B is a feature to the right of A, x_A is the horizontal position of A, x_B is the horizontal position of B, and the minimum separation between x_A and x_B is d_{min}. In our graph model, there is an edge from the vertex for feature A to the vertex for feature B with length d_{min}. We add a single extra source vertex to the graph and a 0-length edge from this source vertex to every other vertex in the graph. This source vertex represents the left edge of the layout. Then finding the longest path from this source vertex to every vertex in the graph will give the leftmost legal position of each feature—as if we had pushed each feature as far to the left as possible. Finding the longest path is converted to a single-source shortest path problem by negating all the lengths on edges. This is equivalent to rewriting the constraint as

$$x_A - x_B \leq -d_{min}. \tag{8.6}$$

From now on, we will write constraints in this form. Note that this graph is acyclic. Therefore, as explained in the following text, the single-source shortest path problem can be solved in time $O(n + |E|)$ by a **topological sort**, where n is the number of features and E is the set of edges in the constraint graph.

A topological sorting algorithm is an algorithm for visiting the vertices of a directed acyclic graph (DAG). The edges of a DAG induce a partial order on the vertices: $v < u$ if there is a (directed) path

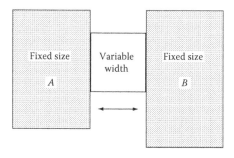

FIGURE 8.3 Generating separation constraints. The constraint on the separation between features A and B is the larger of the minimum separation between them and the minimum width of the flexible feature connecting them.

from v to u in the graph. A topological order is any total ordering of the vertices that is consistent with this partial order. A topological sorting algorithm visits the vertices in some topological order. In Chapter 7 of *Algorithms and Theory of Computation Handbook, Second Edition: General Concepts and Techniques*, a topological sorting algorithm based on depth-first search is presented. For our purposes, a modified breadth-first search approach is more convenient. In this modification, we visit a node as soon as all its immediate predecessors have been visited (rather than as soon as any single immediate predecessor has been visited). For a graph $G = (V, E)$ this algorithm can be expressed as follows:

TOPOLOGICAL SORT (G)

```
1   S ← all vertices with no incoming edges (sources)
2   U ← V
3   while S is not empty
4       do choose any vertex v from S
5           VISIT v
6           for each vertex u such that (v, u) ∈ E
7               do E ← E − {(v, u)}
8                   if u is now a source
9                       then S ← S ∪ {u}
10          U ← U − {v}
11          S ← S − {v}
12  if U is not empty
13      then error ▷ G is not acyclic
```

In our single-source shortest path problem, we start with only one source, s, the vertex representing the left edge of the layout. We compute the length of the shortest path from s to each vertex v, denoted as $\ell(v)$. We initialize $\ell(v)$ before line 3 to be 0 for $\ell(s)$ and ∞ for all other vertices. Then for each vertex v we select at line 4, and each edge (v, u) we delete at line 7 we update for all shortest paths that go through v by $\ell(u) \leftarrow \min\{\ell(u), \ell(v) + \text{length}(v, u)\}$. When the topological sort has completed, $\ell(v)$ will contain the length of the shortest path from s to v (unless G was not acyclic to begin with). The algorithm takes $O(|V| + |E|)$ time.

In our simplest case, all our constraints were minimum separation constraints. In the general case, we may have maximum separation constraints as well. These occur when connectivity constraints are used and also when performance constraints that limit the length of wires are used. Then we have constraints of the form

$$x_B \leq x_A + d_{max} \tag{8.7}$$

or equivalently

$$x_B - x_A \leq d_{max}. \tag{8.8}$$

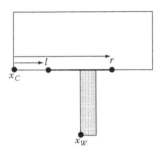

x_C

x_W

FIGURE 8.4   Modeling a connection that can be made along an interval of a component boundary.

Such a constraint is modeled as an edge from the vertex for feature B to the vertex for feature A with weight d_{max}. For example, to model a horizontal wire W that has an interval from l to r along which it can connect to a component C (see Figure 8.4), we use the pair of constraints

$$x_C - x_W \leq -l \tag{8.9}$$

and

$$x_W - x_C \leq r. \tag{8.10}$$

Once we allow both minimum and maximum separation constraints, we have a much more general linear constraint system. All constraints are still of the form

$$x - y \leq d, \tag{8.11}$$

but the resulting constraint graph need not be acyclic. (Note that equality constraints $y - x = d$ may be expressed as $y - x \leq d$ and $x - y \leq -d$. Equality constraints may also be handled in a preprocessing step that merges vertices that are related by equality constraints.) To solve the single-source shortest path algorithm, we now need the $O(|V||E|)$-time Bellman–Ford algorithm (see [6]). This algorithm only works if the graph contains no negative cycle. If the constraint graph is derived from a layout that satisfies all the constraints, this will be true, since a negative cycle represents an infeasible set of constraints. However, if the initial layout does not satisfy all the constraints, for example, the designer adds constraints for performance to a rough layout, then the graph may be cyclic. The Bellman–Ford algorithm can detect this condition, but since the set of constraints is infeasible, no layout can be produced.

If the constraint graph is derived from a layout that satisfies all the constraints, an observation by Maley [26] allows us to use Dijkstra's algorithm to compute the shortest paths more efficiently. To use Dijkstra's algorithm, the weights on all the edges must be positive (see Chapter 7 of *Algorithms and Theory of Computation Handbook, Second Edition: General Concepts and Techniques*). Maley observed that when an initial layout exists, the initial positions of the features can be used to convert all lengths to positive lengths as follows. Let p_A and p_B be initial positions of features A and B. The constraint graph is modified so that the length of an edge (v_A, v_B) from the vertex for A to the vertex for B becomes $\text{length}(v_A, v_B) + p_B - p_A$. Since the initial layout satisfies the constraint $x_A - x_B \leq \text{length}(v_A, v_B)$, we have $p_B - p_A \geq -\text{length}(v_A, v_B)$ and $p_B - p_A + \text{length}(v_A, v_B) \geq 0$. Maley shows that this transformation of the edge lengths preserves the shortest paths. Since all edge weights have been converted to positive lengths, Dijkstra's algorithm can be used, giving a running time of $O(|V| \log |V| + |E|)$ or $O(|E| \log |V|)$, depending on the implementation used.*

Even when the constraint graph is not acyclic and an initial feasible layout is not available, restrictions on the type or structure of constraints can be used to get faster algorithms. For example, Lengauer and Mehlhorn give an $O(|V| + |E|)$-time algorithm when the constraint graph has a special structure called a "chain DAG" that is found when the only constraints other than minimum separation constraints are those coming from flexible connections such as those modeled by Equations 8.9 and 8.10 (see [24, p. 614]). Liao and Wong [25] and Mata [28][†] present $O(|E_x| \times |E|)$-time algorithms, where E_x is the set of edges derived from constraints other than the minimum-separation constraints. These algorithms are based on the observation that $E - E_x$ is a DAG (as in our simple case earlier). Topological sort is used as a subroutine to solve the single-source shortest path problem with edges $E - E_x$. The solution to this problem may violate constraints represented by E_x. Therefore, after finding the shortest paths for $E - E_x$, positions are modified in an attempt to satisfy the other constraints (represented by E_x), and the single-source shortest path algorithm for $E - E_x$ is run again. This technique is iterated until it converges to a solution for the entire set of constraints or the set of constraints is shown to be infeasible, which is proven to be within $|E_x|$ iterations. If $|E_x|$ is small, this algorithm is more efficient than using Bellman–Ford.

The single-dimensional compaction that we have discussed ignores many practical issues. One major issue is the introduction of bends in wires. The fact that a straight wire segment connects two components may be an artifact of the layout, but it puts the components in lock-step during compaction. Adding a bend to the wire would allow the components to move independently, stretching the bend accordingly. Although the bend may require extra area, the overall area might improve through compaction with the components moving separately.

---

* The $O(|V| \log |V| + |E|)$ running time depends on using Fibonacci heaps for a priority queue. If the simpler binary heap is used, the running time is $O(|E| \log |V|)$. This comment also holds for finding minimum spanning trees using Prim's algorithm. See Chapter 7 of *Algorithms and Theory of Computation Handbook, Second Edition: General Concepts and Techniques* for a discussion on the running times of Dijkstra's algorithm and Prim's algorithm.

† The technique used by Mata is the same as that by Liao and Wong, but Mata has a different stopping condition for his search, which can lead to more efficient execution.

Another issue is allowing components to change their order from left to right or top to bottom. This change might allow for a smaller layout, but the compaction problem becomes much more difficult. In fact, a definition of one-dimensional compaction that allows for such exchanges is NP-complete (see [24, p. 587]). Practically speaking, such exchanges may cause problems for wire routing. The compaction problem we have presented requires that the topological relationships between the layout features remain unchanged while space is compressed.

8.4 Floorplan Sizing and Classic Divide and Conquer

The problem we will now consider, called **floorplan sizing**, is one encountered during certain styles of placement or floorplanning. With some reasonable assumptions about the form of the layout, the problem can be solved optimally by a polynomial-time algorithm that is an example of classic divide and conquer.

Floorplan sizing occurs when a floorplan is initially specified as a partitioning of a rectangular layout area, representing the chip, into subrectangles, representing components (see Figure 8.5a).

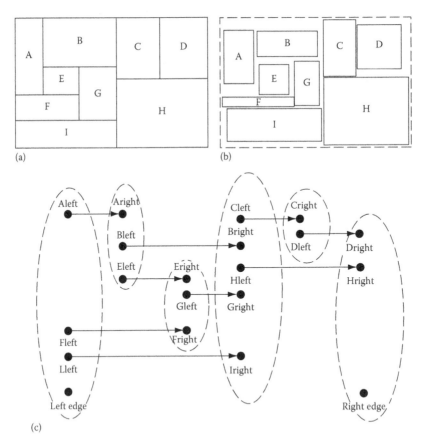

FIGURE 8.5 A floorplan and the derived layout. (a) A partition of a rectangle representing a floorplan. (b) A layout with actual components corresponding to the floorplan. (c) The horizontal constraint graph for the layout. Bidirectional 0-length edges are not shown. Instead, vertices constrained to have the same horizontal position due to chains of 0-length edges are shown in the same oval. They are treated as a single vertex.

Each subrectangle corresponds to some component. We assume that the rectangular partitioning is rectilinear. For this discussion, given a set of components C, by "a floorplan for C," we shall mean such a rectangular partition of a rectangle into $|C|$ subrectangles, and a one-to-one mapping from C to the subrectangles. This partition indicates the relative position of components, but the subrectangles are constrained only to have approximately the same area as the components (possibly with some bloating to account for routing), not to have the same aspect ratio as the components. When the actual components are placed in the locations, the layout will change (see Figure 8.5b). Furthermore, it may be possible to orient each component so that the longer dimension may be either horizontal or vertical, affecting the ultimate size of the layout. In fact, if the component layouts have not been completed, the components may be able to assume many shapes while satisfying an area bound that is represented by the corresponding subrectangle. We will formalize this through the use of a **shape function**.

Definition: A shape function for a component is a mapping $s : [w_{min}, \infty] \rightarrow [h_{min}, \infty]$ such that s is monotonically decreasing, where $[w_{min}, \infty]$ and $[h_{min}, \infty]$ are intervals of $\Re^+$.

The interpretation of $s(w)$ is that it is the minimum height (vertical dimension) of any rectangle of width w that contains a layout of the component. w_{min} is the minimum width of any rectangle that contains a layout and h_{min} is the minimum height. The monotonicity requirement represents the fact that if there is a layout for a component that fits in a $w \times s(w)$ rectangle, it certainly must fit in a $(w + d) \times s(w)$ rectangle for any $d \geq 0$; therefore, $s(w + d) \leq s(w)$. In this discussion we will restrict ourselves to piecewise linear shape functions.

Given an actual shape (width and height) for each component, determining the minimum width and minimum height of the rectangular layout area becomes a simple compaction problem as discussed earlier. Each dimension is done separately, and two constraint graphs are built. We will discuss the horizontal constraint graph; the vertical constraint graph is analogous. The reader should refer to Figure 8.5c. The horizontal constraint graph has a vertex for each vertical side of the layout rectangle and each vertical side of a subrectangle (representing a component) in the rectangular partition. There is a directed edge from each vertex representing the left side of a subrectangle to each vertex representing the right side of a subrectangle; this edge has a length that is the width of the corresponding component. There are also two directed edges (one in each direction) between the vertices representing any two overlapping sides of the layout rectangle or subrectangles; the length of these edges is 0. Thus, the vertex representing the left side of the layout rectangle has 0-length edges between it and the vertices representing the left sides of the leftmost subrectangles. Note that these constraints force two subrectangles that do not touch but are related through a chain of 0-length edges between one's left side and the other's right side to lie on opposite sides of a vertical line in the layout (e.g., components B and H in Figure 8.5a). This is an added restriction to the layout, but an important one for the correctness of the algorithm presented later for the sizing of **slicing floorplans**.

Given an actual width for each component and having constructed the horizontal constraint graph as described in the preceding paragraph, to determine the minimum width of the rectangular layout area, one simply finds the longest path from the vertex representing the left side of the layout rectangle to the vertex representing the right side of the layout rectangle. To simplify the problem to one in an acyclic graph, vertices connected by pairs of 0-length edges can be collapsed into a single vertex; only the longest edge between each pair of (collapsed) vertices is needed. Then topological sort can be used to find the longest path between the left side and the right side of the floorplan, as discussed in Section 8.3.

We now have the machinery to state the problem of interest:

Floorplan sizing: Given a set C of components, a piecewise linear shape function for each component, and a floorplan for C, find an assignment of specific shapes to the components so that the area of the layout is minimized.

Stockmeyer [39] showed that for general floorplans, the floorplan sizing problem is NP-complete. This holds even if the components are of fixed shape, but can be rotated 90°. In this case, the shape function of each component is a step function with at most two steps: $s(x) = d_2$ for $d_1 \leq x < d_2$ and $s(x) = d_1$ for $d_2 \leq x$, where the dimensions of the component are d_1 and d_2 with $d_1 \leq d_2$. However, for floorplans of a special form, called slicing floorplans, Stockmeyer gave a polynomial-time algorithm for the floorplan sizing problem when components are of fixed shape but can rotate. Otten [29] generalized this result to any piecewise-linear shape function. A slicing floorplan is one in which the partition of the layout rectangle can be constructed by a recursive cutting of a rectangular region into two subregions using either a vertical or horizontal line segment (see Figure 8.6). The rectangular regions that are not further par-

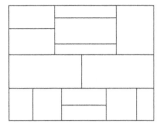

FIGURE 8.6   A slicing floorplan.

titioned are the subrectangles corresponding to components. The recursive partitioning method of constructing a placement discussed earlier produces a slicing floorplan.

A slicing floorplan can be represented by a binary tree. The root of the tree represents the entire layout rectangle and is labeled with the direction of the first cut. Other interior vertices represent rectangular subregions that are to be further subdivided, and the label on any such vertex is the direction of the cut used to subdivide it. The two children of any vertex are the rectangular regions resulting from cutting the rectangle represented by the vertex. The leaves of the binary tree represent the rectangles corresponding to components.

The algorithm for the sizing of a slicing floorplan uses the binary tree representation in a fundamental way. The key observation is that one only needs the shape functions of the two subregions represented by the children of a vertex to determine the shape function of a vertex. If the shape functions can be represented succinctly and combined efficiently for each vertex, the shape function of the root can be determined efficiently. We will present the combining step for shape functions that are step functions (i.e., piecewise constant) following the description in [24], since it illustrates the technique but is a straightforward calculation. Otten [29] shows how to combine piecewise linear slicing functions, but Lengauer [24] comments that arithmetic precision can become an issue in this case.

We shall represent a shape function that is a step function by a list of pairs $(w_i, s(w_i))$ for $0 \leq i \leq b_s$ and $w_0 = w_{\min}$. The interpretation is that for all x, $w_i \leq x < w_{i+1}$ (with $w_{b_s+1} = \infty$), $s(x) = s(w_i)$. Parameter b_s is the number of breaks in the step function. The representation of the function is linear in the number of breaks. (This represents a step function whose constant intervals are left closed and right open and is the most logical form of step function for shape functions. However, other forms of step functions also can be represented in size linear in the number of breaks.)

Given step functions, s_l and s_r, for the shapes of two children of a vertex, the shape function, s, for the vertex will also be a step function. When the direction of the cut is horizontal, the shape functions can simply be added, that is, $s(x) = s_l(x) + s_r(x)$. $w_{\min}$ for s is the maximum of $w_{\min,l}$ for s_l and $w_{\min,r}$ for s_r. Each subsequent break point for s_l or s_r is a break point for s, so that $b_s \leq b_{s_l} + b_{s_r}$. Combining the shape functions takes time $O(b_{s_l} + b_{s_r})$. When the direction is vertical, the step functions must first be inverted, the inverted functions combined, and then the combined function inverted back. The inversion of a step function s is straightforward and can be done in $O(b_s)$ time.

To compute the shape function for the root of a slicing floorplan, one simply does a postorder traversal of the binary tree (see Chapter 7 of *Algorithms and Theory of Computation Handbook, Second Edition: General Concepts and Techniques*), computing the shape function for each vertex from the shape functions for the children of the vertices. The number of breaks in the shape function for any vertex is no more than the number of breaks in the shape functions for the leaves of the subtree rooted at that vertex. Let b be the maximum number of breaks in any shape function of a component. Then, the running time of this algorithm for a slicing floorplan with n components is

$$T(n) \leq T(n_l) + T(n_r) + bn \tag{8.12}$$

$$\leq dbn, \tag{8.13}$$

where d is the depth of the tree (the length of the longest path from the root to a leaf). We have the following:

> Given an instance of the floorplan sizing problem that has a slicing floorplan and step functions as shape functions for the components, there is an $O(dbn)$-time algorithm to compute the shape function of the layout rectangle.

Given the shape function for the layout rectangle, the minimum area shape can be found by computing the area at each break in linear time in the number of breaks, which is at most $O(bn)$.

Shi [38] has presented a modification to this technique that improves the running time for imbalanced slicing trees. For general (nonslicing) floorplans, many heuristics have been proposed: for example, Pan and Liu [31] present a generalization of the slicing floorplan technique to general floorplans.

8.5 Routing Problems

We shall only discuss the most common routing model for general cell placement—the rectilinear **channel-routing** model. In this model, the layout is rectilinear. The regions of the layout that are not covered by components are partitioned into nonoverlapping rectangles, called **channels**. The allowed partitions are restricted so that each channel only has components touching its horizontal edges (a horizontal channel) or its vertical edges (a vertical channel). These edges will be referred to as the "top" and "bottom" of the channel, regardless of whether the channel is horizontal or vertical. The orthogonal edges, referred to as the "left edge" and "right edge" of the channel, can only touch another channel. These channels compose the area for the wire routing. There are several strategies for partitioning the routing area into channels, that is "defining" the channels, but most use maximal rectangles where possible (i.e., no two channels can be merged to form a larger channel). The layout area becomes a rectangle that is partitioned into subrectangles of two types: components and channels (see Figure 8.7).

Given a layout with channels defined, the routing problem can be decomposed into two subproblems: **global routing** and local or detailed routing. **Global routing** is the problem of choosing which channels will be used to make the interconnections for each net. Actual paths are not produced. By doing global routing first, one can determine the actual paths for wires in each channel separately. The problem of detailed routing is to determine these actual paths and is more commonly referred

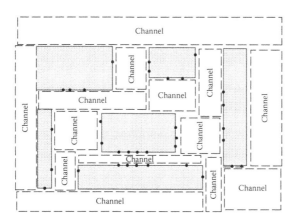

FIGURE 8.7 The decomposition of a layout into routing channels and components.

to as "channel routing." Of course, the segments of a wire path in each channel must join at the edges of the channel to produce an actual interconnection of terminals. To understand the approaches for handling this interfacing, we must have a more detailed definition of the channel-routing problem, which we give next.

The channel-routing problem is defined so that there are initial positional constraints in only one dimension. Recall that we define channels to abut components on only two parallel sides, the "top" and "bottom." This is so that the routes in the channel will only be constrained by terminal positions on two sides. The standard channel-routing problem has the following input: a rectangle (the channel) containing points (terminals) at fixed positions along its top and bottom edges, a set of nets that must be routed in the channel, and an assignment of each of the terminals on the channel to one of these nets. Moreover, two (possibly empty) subsets of nets are specified: one containing nets whose routing must interface with the left edge of the channel and one containing nets whose routing must interface with the right edge of the channel. The positions at which wires must intersect the left and right edges of the channel are not specified. The dimension of the channel from the left edge to the right edge is called the length of the channel; the dimension from the top edge to the bottom edge is the width of the channel (see Figure 8.8). Since there are no terminals on the left and right edges, the width is often taken to be variable, and the goal is to find routing paths achieving the connections specified by the nets and minimizing the width of the channel. In this case, the space needed for the wires determines the width of the channel. The length of the channel is more often viewed as fixed, although there are channel-routing models in which routes are allowed to cross outside the left and right edges of the channel.

We can now discuss the problem of interfacing routes at channel edges. There are two main approaches to handling the interfacing of channels. One is to define the channels so that all adjacent channels form $\top$s (no $+$s). Then, if the channel routing is done for the base of the $\top$ first, the positions of paths leaving the base of the $\top$ and entering the crosspiece of the $\top$ are fixed by the routing of the base and can be treated as terminal positions for the crosspiece of the $\top$. Using this approach places constraints on the routing order of the channels. We can model these constraints using a directed graph: there is a vertex, v_C, for each channel C and an edge from v_A to v_B if channel A and channel B abut with channel A as the base of the $\top$ and channel B as the crosspiece of the $\top$. The edge from v_A to v_B represents that channel A must be routed before channel B. This graph must be acyclic for the set of constraints on the order of routing to be feasible. If the graph is not acyclic, another method must be used to deal with some of the channel interfaces. Slicing floorplans are very good for this approach because if each slice is defined to be a channel, then the channel order constraint graph will be acyclic.

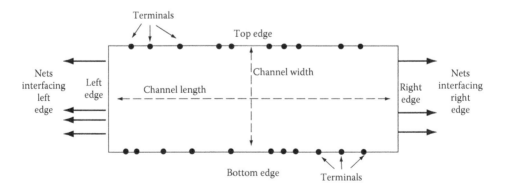

FIGURE 8.8 A channel. Nets interfacing at the left and right edges are not in a given order.

The second alternative for handling the interfaces of channels is to define a special kind of channel, called a **switch box**, which is constrained by terminal locations on all four sides. In this alternative, some or all of the standard channels abut switch boxes. Special algorithms are used to do the switch box routing, since, practically speaking, this is a more difficult problem than standard channel routing.

8.6 Global Routing

Since global routing need not produce actual paths for wires, the channels can be modeled by a graph, called the channel intersection graph. For each channel, project the terminals lying on the top and bottom of the channel onto the midline of the channel. Each channel can be divided into segments by the positions of these projected terminals. The channel intersection graph is an undirected graph with one vertex for each projected terminal and one vertex for each intersection of channels. There is one edge for each segment of a channel, which connects the pair of vertices representing the terminals and/or channel intersections bounding the segment. A length and a capacity can be assigned to each edge, representing, respectively, the length between the ends of the channel segment and the width of the channel. Different versions of the global routing problem use one or both of the length and capacity parameters.

Given the channel intersection graph, the problem of finding a global routing for a set of nets becomes the problem of finding a Steiner tree in the channel intersection graph for the terminals of each net. Earlier in this chapter, we defined Steiner trees for points in the plane. For a general graph $G = (V, E)$, a Steiner tree for a subset of vertices $U \subset V$ is a set of edges of G that form a tree in G and whose set of endpoints contains the set U. Various versions of the global routing problem are produced by the constraints and optimization criteria used. For example, one can simply ask for the minimum length Steiner tree for each net, or one can ask for a set of Steiner trees that does not violate capacity constraints and has total minimum length. For each edge, the capacity used by a set of Steiner trees is the number of Steiner trees containing that edge; this must be no greater than the capacity of the edge. Another choice is not to have constraints on the capacity of edges but to minimize the maximum capacity used for any edge. In general, any combination of constraints and cost functions on length and capacity can be used. However, regardless of the criteria, the global routing problem is invariably NP-complete. A more detailed discussion of variations of the problem and their complexity can be found in [24]. There, a number of sophisticated algorithms for Steiner tree problems are also discussed. Here we will only discuss two techniques based on basic graph algorithms: breadth-first search and Dijkstra's single-source shortest path algorithm.

The minimum Steiner tree problem is itself NP-complete (see pages 208–209 in [10]). Therefore, one approach to global routing is to avoid finding Steiner trees by breaking up each net into a collection of point-to-point connections. One way to do this is to find the minimum Euclidean or rectilinear spanning tree for the terminals belonging to each net (ignoring the channel structure), and use the edges of this tree to define the point-to-point connections. Then one can use Dijkstra's single-source shortest path algorithm on the channel intersection graph to find a shortest path for each point-to-point connection. Paths for connections of the same net that share edges can then be merged, yielding a Steiner tree. If there are no capacity constraints on edges, the quality of this solution is only limited by the quality of the approximation of a minimum Steiner tree by the chosen collection of point-to-point paths. If there are capacity constraints, then after solving each shortest path problem, one must remove from the channel intersection graph the edges whose used capacity already equals the edge capacity. In this case, the order in which nets and terminals within a net are routed is significant. Heuristics are used to choose this order. One can better approximate Steiner trees for the nets by, at each iteration for connections within one net, choosing a terminal not yet connected to any other terminals in the net as the source of Dijkstra's algorithm.

Since this algorithm computes the shortest path to every other vertex in the graph from the source, the shortest path, which connects to any other vertex in the channel intersection graph that is already on a path connecting terminals of the net can be used. Of course, there are variations on this idea.

For any graph, breadth-first search from a vertex v will find a shortest path from v to every other vertex in the graph when the length of each edge is 1. Breadth-first search takes time $O(|V| + |E|)$ compared to the best worst-case running time known for Dijkstra's algorithm: $O(|V| \log |V| + |E|)$ time. It is also very straightforward to implement. It is easy to incorporate heuristics that take into account the capacity of an edge already used and bias the search toward edges with little capacity used. Furthermore, breadth-first search can be started from several vertices simultaneously, so that all terminals of a net could be starting points of the search simultaneously. If it is adequate to view all channel segments as being of equal length, then the edge lengths can all be assigned value 1 and breadth-first search can be used. This might occur when the terminals are uniformly distributed and so divide channels into approximately equal-length segments. Alternatively, one can add new vertices to the channel intersection graph to further decompose the channel segments into unit-length segments. This can substantially increase $|V|$ and $|E|$ so that they are proportional to the dimensions of the underlying grid defining the unit of length rather than the number of channels and terminals in the problem. However, this allows one to use breadth-first search to compute shortest paths while modeling the actual lengths of channels. In fact, breadth-first search was developed by Lee [20]* for the routing of circuit boards in exactly this manner. He modeled the entire board by a grid graph, and modeled obstacles to routing paths as grid vertices that were missing from the graph. Each wire route was found by doing a breadth-first search in the grid graph.

8.7 Channel Routing

Channel routing is not one single problem, but rather a family of problems based on the allowable paths for wires in the channel. We will limit our discussion to grid-based routing. While both grid-free rectilinear and nonrectilinear routing techniques exist, the most basic techniques are grid-based. We assume that there is a grid underlying the channel, the sides of the channel lie on grid edges, terminals lie on grid points, and all routing paths must follow edges in the grid. For ease of discussion, we shall refer to channel directions as though the channel were oriented with its length running horizontally. The vertical segments of the grid that run from the top to the bottom of the channel are referred to as columns. The horizontal segments of the grid that run from the left edge to the right edge of the channel are referred to as tracks. We will consider channel-routing problems that allow the width of the channel to vary. Therefore, the number of columns, determining the channel length, will be fixed, but the number of tracks, determining the channel width, will be variable. The goal is to minimize the number of tracks used to route the channel.

The next distinction is based on how many routing layers are presumed. If there are ℓ routing layers, then there are ℓ overlaid copies of the grid, one for each layer. Routes that use the same edge on different layers do not interact and are considered disjoint. Routes change layer at grid points. The exact rules for how routes can change layers vary, but the most common is to view a route that goes from layer i to layer j ($j > i$) at a grid point as using layers $i + 1, \ldots, j - 1$ as well at the grid point.

---

* The first published description of breadth-first search was by E.F. Moore for finding a path in a maze. Lee developed the algorithm for routing in grid graphs under a variety of path costs. See the discussion on page 394 of [24].

One can separate the channel-routing problem into two subproblems: finding Steiner trees in the grid that achieve the interconnections, and finding a **layer assignment** for each edge in each Steiner tree so that the resulting set of routes is legal. One channel-routing model for which this separation is made is knock-knee routing (see Section 9.5 of [24]). Given any set of trees in the grid (not only those for knock-knee routes), a legal layer assignment can be determined efficiently for two layers. Maximal single-layer segments in each tree can be represented as vertices of a conflict graph, with segments that must be on different layers related by conflict edges. Then finding a legal two-layer assignment is equivalent to two-coloring the conflict graph. A more challenging problem is **via minimization**, which is to find a legal layer assignment that minimizes the number of grid points at which layer change occurs. This problem is also solvable in polynomial time as long as none of the Steiner trees contain a four-way split (a technical limitation of the algorithm). For more than two layers, both layer assignment and via minimization are NP-complete (see Section 9.5.3 of [24]).

We have chosen to discuss two routing models in detail: single-layer routing and two-layer **Manhattan routing**. Routing models that allow more than two layers, called multilayer models, are becoming more popular as technology is able to provide more layers for wires. However, many of the multilayer routing techniques are derived from single-layer or two-layer Manhattan techniques, so we focus this restricted discussion on those models. A more detailed review of channel-routing algorithms can be found in [19]. Neither model requires layer assignment. In single-layer routing there is no issue of layer assignment; in Manhattan routing, the model is such that the layer assignment is automatically derived from the paths. Therefore, for each model, our discussion need only address how to find a collection of Steiner trees that satisfy the restrictions of the routing model.

8.7.1 Manhattan Routing

Manhattan routing is the dominant two-layer routing model. It dates back to printed circuit boards [13]. It dominates because it finesses the issue of layer assignment by defining all vertical wire segments to be on one layer and all horizontal wire segments to be on the other layer. Therefore, a horizontal routing path and a vertical routing path can cross without interacting, but any path that bends at a grid point is on both layers at that point and no path for a disjoint net can use the same point. Thus, under the Manhattan model, the problem of routing can be stated completely in terms of finding a set of paths such that paths for distinct nets may cross but do not share edges or bend points.

Although Manhattan routing provides a simple model of legal routes, the resulting channel-routing problem is NP-complete. An important lower bound on the width of a channel is the **channel density**. The density at any vertical line cutting the channel (not necessarily a column) is the number of nets that have terminals both to the left and the right of the vertical line. The interpretation is that each net must have at least one wire crossing the vertical line, and thus a number of tracks equal to the density at the vertical line is necessary. For columns, nets that contain a terminal on the column are counted in the density unless the net contains exactly two terminals and these are at the top and bottom of the column. (Such a net can be routed with a vertical path the extent of the column.) The channel density is the maximum density of any vertical cut of the channel. In practice, the channel-routing problem is solved with heuristic algorithms that find routes giving a channel width within one or two of density, although such performance is not provably achievable.

If a channel-routing instance has no top terminal and bottom terminal on the same column, then a number of tracks equal to the channel density suffices and a route achieving this density can be solved in $O(m + n \log n)$ time, where n is the number of nets and m is the number of terminals. Under this restriction, the channel-routing problem becomes equivalent to the problem of **interval graph coloring**. The equivalence is based on the observation that if no column contains terminals at its top and bottom, then any terminal can connect to any track by a vertical

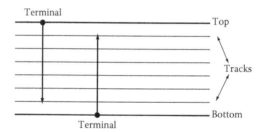

FIGURE 8.9 Connecting to tracks without conflicts. Each of the nets can reach any of the tracks with a vertical segment in the column containing the terminal.

segment that does not conflict with any other path (see Figure 8.9). Each net can use one horizontal segment that spans all the columns containing its terminals. The issue is to pack these horizontal segments into the minimum number of tracks so that no segments intersect. Equivalently, the goal is to color the segments (or intervals) with the minimum number of colors so that no two segments that intersect are of the same color. Hence, we have the relationship to interval graphs, which have a set of vertices that represent intervals on a line and edges between vertices of intersecting intervals.

A classic greedy algorithm can assign a set I of intervals to d tracks, where d is the **channel density**. Intervals are placed in tracks in order of their left endpoints; all tracks are filled with intervals from left to right concurrently (the actual position of each track does not matter). The set of interval endpoints is first sorted so that the combined order of left (starting) endpoints and right (ending) endpoints is known. At any point in the algorithm, there are tracks containing intervals that have not yet ended and tracks that are free to receive new intervals. A queue F is used to hold the free tracks; the function FREE inserts (enqueues) a track in F. The unprocessed endpoints are stored in a queue of points, P, sorted from left to right. The function DEQUEUE is the standard deletion operation for a queue (see Chapter 4 of *Algorithms and Theory of Computation Handbook, Second Edition: General Concepts and Techniques* or [6]). The algorithm starts with only one track and adds tracks as needed; variable t holds the number of tracks used.

INTERVAL-BY-INTERVAL ASSIGNMENT (I)

```
1    Sort the endpoints of the intervals in I and build queue P
2    t ← 1
3    FREE track 1
4    while P is not empty
5        do p ← DEQUEUE(P)
6            if p is the left endpoint of an interval i
7                then do   if F is empty
8                            then do t ← t + 1
9                                Put i in track t
10                           else do track ← DEQUEUE(F)
11                               Put i in track
12                else do  FREE the track containing the interval whose right endpoint is p
```

To see that this algorithm never uses more than a number of tracks equal to the density d of the channel, note that when t is increased at line 8 it is because the current t tracks all contain intervals that have not yet ended when the left endpoint p obtained in line 5 is considered. Therefore, when t is increased for the last time to value w, all tracks numbered less than w contain intervals that span the current p; hence $d \geq w$. Since no overlapping intervals are places in the same track, $w = d$. Therefore, the algorithm finds an optimal track assignment. Preprocessing to find the interval for each net takes time $O(m)$ and INTERVAL-BY-INTERVAL ASSIGNMENT has running time $O(|I| \log |I|) = O(n \log n)$, due to the initial sorting of the endpoints of I.

Once one allows terminals at both the top and the bottom of a column (except when all such pairs of terminals belong to the same net), one introduces a new set of constraints called vertical constraints. These constraints capture the fact that if net i has a terminal at the top of column c and net j has a terminal at the bottom of column c, then to connect these terminals to horizontal segments using vertical segments at column c, the horizontal segment for i must be above the horizontal segment for j. One can construct a vertical constraint graph that has a vertex v_i for each net i and a directed edge between v_i and v_j if there is a column that contains a terminal in i at the top and a terminal in j at the bottom. If one considers only routes that use at most one horizontal segment per net, then the constraint graph represents order constraints on the tracks used by the horizontal segments. If the vertical constraint graph is cyclic, then the routing cannot be done with one horizontal segment per net. If the vertical constraint graph is acyclic, it is NP-complete to determine if the routing can be achieved in a given number of tracks (see [24, p. 547]). Furthermore, even if an optimal or good routing using one horizontal segment per net is found, the number of tracks required is often substantially larger than what could be obtained using more horizontal segments. For these reasons, practical channel-routing algorithms allow the route for each net to traverse portions of several tracks. Each time a route changes from one track to another, it uses a section of a column; this is called a **jog** (see Figure 8.10).

Manhattan channel routing remains NP-complete even if unrestricted jogs are allowed (see [24, p. 541]). Therefore, the practical routing algorithms for this problem use heuristics. The earliest of these is by Deutsch [7]. Deutsch allows the route for a net to jog only in a column that contains a terminal of the net; he calls these jogs "doglegs" (see Figure 8.10). This approach effectively breaks up each net into two-point subnets, and one can then define a vertical constraint graph in which each vertex represents a subnet. Deutsch's algorithm is based on a modification of INTERVAL-BY-INTERVAL ASSIGNMENT called track-by-track assignment. Track-by-track assignment also fills tracks greedily from left to right but fills one track to completion before starting the next. Deutsch's basic algorithm does track-by-track assignment but does not assign an interval for a subnet to a track if the assignment would violate a vertical constraint. Embellishments on the basic algorithm try to improve its performance and minimize the number of doglegs. Others have also modified the approach

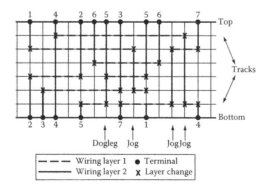

FIGURE 8.10 A channel routing in the two-layer Manhattan model showing jogs.

(see the discussion in Section 9.6.1.4 of [24]). The class of algorithms based on greedy assignment of sorted intervals is also known as "left-edge algorithms."

Manhattan channel routing is arguably the most widely used and detailed routing model, and many algorithms have been developed. The reader is referred to [37] or [19] for a survey of algorithms. In this chapter, we will discuss only one more algorithm—an algorithm that proceeds column-by-column in contrast to the track-by-track algorithm. The column-by-column algorithm was originally proposed by Rivest and Fiduccia [35] and was called by them a "greedy" router. This algorithm routes all nets simultaneously. It starts at the leftmost column of the channel and proceeds to the right, considering each column in order. As it proceeds left to right it creates, destroys, and continues horizontal segments for nets in the channel. Using this approach, it is easy to introduce a jog for a net in any column. At each column, the algorithm connects terminals at the top and the bottom to horizontal segments for their nets, starting new segments when necessary, and ending segments when justified. At each column, for each continuing net with terminals to the right, it may also create a jog to bring a horizontal segment of the route closer to the channel edge containing the next terminal to the right. Thus, the algorithm employs some "look ahead" in determining what track to use for each net at each column. The algorithm is actually a framework with many parameters that can be adjusted. It may create, for one net, multiple horizontal segments that cross the same column and may extend routes beyond the left and the right edges of the channel. It is a very flexible framework that allows many competing criteria to be considered and allows the interaction of nets more directly than strategies that route one net at a time. Many routing tools have adopted this approach.

8.7.2   Single-Layer Routing

From a practical perspective, single-layer channel routing is needed to route bus structures and plays a role in the routing of multilayer channels (e.g., [11]). For bus structures, additional performance constraints are often present (e.g., [30]). From a theoretical perspective, single-layer channel routing is of great significance because, even in its most general form, it can be solved optimally in polynomial time. There is a rich theory of single-layer detailed routing that has been developed not only for channel routing, but also for routing in more general regions (see [27]). The first algorithmic results for single-layer channel routing were by Tompa [40], who considered **river routing** problems. A river routing problem is a single-layer channel-routing problem in which each net contains exactly two terminals, one at the top edge of the channel and one at the bottom edge of the channel. The nets have terminals in the same order along the top and the bottom—a requirement if the problem is to be routable in one layer. Tompa considered unconstrained (versus rectilinear) wire paths and gave an $O(n^2)$-time algorithm for n nets to test routability and find the route that minimizes both the individual wire lengths and the total wire length when the width of the channel is fixed. This algorithm can be used as a subroutine within binary search to find the minimum-width channel in $O(n^2 \log n)$ time. Tompa also suggested how to modify his algorithm for the rectilinear case. Dolev et al. [8] built upon Tompa's theory for the rectilinear case and presented an $O(n)$-time algorithm to compute the minimum width of the channel and an $O(n^2)$-time algorithm to actually produce the rectilinear routing. The difference in running times comes from the fact that the routing may actually have n segments per net, and thus would take $O(n^2)$-time to generate (see Figure 8.11). In contrast, the testing for routability can be done by examining a set of constraints for the channel. The results were generalized to multiterminal nets by Greenberg and Maley [12], where the time to calculate the minimum width remains linear in the number of terminals.

We now present the theory that allows the width of a river-routing channel to be computed in linear time. Our presentation is an amalgam of the presentations in [8] and [23]. The heart of the

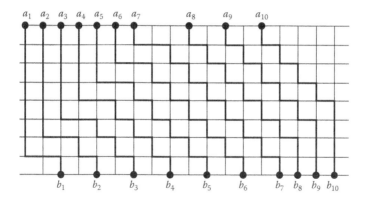

FIGURE 8.11 A river routing. The route for a single net may bend $O(n)$ times.

theory is the observation that for river routing, cut lines other than the vertical lines that define channel density also contribute to a lower bound for the width of the channel. This lower bound is then shown to be an upper bound as well. Indexing from left to right, let the ith net of a routing channel, $1 \leq i \leq n$, be denoted by the pair (a_i, b_i) where a_i is the horizontal position of its terminal on the top of the channel and b_i is the horizontal position of its terminal on the bottom of the channel. Consider a line from b_i to a_{i+k}, $k > 0$, cutting the channel. There are $k + 1$ nets that must cross this line. Measuring slopes from $0°$ to $180°$, if the line has slope $\geq 90°$, then $k + 1$ nets must cross the vertical ($90°$) line at b_i, and there must be $k + 1$ tracks. If the line has slope $< 90°$ and $> 45°$, then each vertical grid segment that crosses the line can be paired with a horizontal grid segment that must also cross the line and which cannot be used by a different net. Therefore, the line must cross $k + 1$ tracks. Finally, if the line has slope $\leq 45°$, then each horizontal grid segment that crosses the line can be paired with a vertical grid segment that must also cross the line and which cannot be used by a different net. Therefore, there must be $k + 1$ columns crossing the line, that is, $a_{i+k} - b_i \geq k$. Similarly, by considering a line from b_{i+k} to a_i, we conclude $k + 1$ tracks are necessary unless $b_{k+i} - a_i \geq k$. In the case of $k = 0$, a_i must equal b_i for no horizontal track to be required. Based on this observation, it can be proved that the minimum number of tracks t required by an instance of river routing is the least t such that for all $1 \leq i \leq n - t$

$$b_{i+t} - a_i \geq t \tag{8.14}$$

and

$$a_{i+t} - b_i \geq t. \tag{8.15}$$

To find the minimum such t in linear time, observe that $a_{i+k+1} \geq a_{i+k} + 1$ and $b_{i+k+1} \geq b_{i+k} + 1$. Therefore,

$$\text{if } b_{i+k} - a_i \geq k \text{ then } b_{i+k+1} - a_i \geq b_{i+k} + 1 - a_i \geq k + 1 \tag{8.16}$$

and

$$\text{if } a_{i+k} - b_i \geq k \text{ then } a_{i+k+1} - b_i \geq a_{i+k} + 1 - b_i \geq k + 1. \tag{8.17}$$

Therefore, we can start with $t = 0$ and search for violated constraints from $i = 1$ to $n - t$; each time a constraint of the form of (8.14) or (8.15) above is violated, we increase t by one and continue the search; t can be no larger than n. Let N denote the set of nets in a river routing channel, $|N| = n$. The following algorithm calculates the minimum number of tracks needed to route this channel.

RIVER-ROUTING WIDTH (N)

```
1   i ← 1
2   t ← 0
3   while i ≤ |N| − t
4       do if b_{i+t} − a_i ≥ t and a_{i+t} − b_i ≥ t
5           then do i ← i + 1
6           else do t ← t + 1
7   return t
```

The actual routing for a river-routing channel can be produced in a greedy fashion by routing one net at a time from left to right, and routing each net beginning with its left terminal. The route of any net travels vertically whenever it is not blocked by a previously routed net, travels horizontally right until it can travel vertically again or until it reaches the horizontal position of the right terminal of the net. This routing takes worst-case time $O(n^2)$, as the example in Figure 8.11 illustrates.

8.8   Research Issues and Summary

This chapter has given an overview of the design problems arising from the computer-aided layout of VLSI circuits and some of the algorithmic approaches used. The algorithms presented in this chapter draw upon the theoretical foundations discussed in other chapters. Graph models are predominant and are frequently used to capture constraints. Since many of the problems are NP-complete, heuristics are used. Research continues in this field, both to find better methods of solving these difficult problems—both in the efficiency and the quality of solution—and to model and solve new layout problems arising from the ever-changing technology of VLSI fabrication and packaging and the ever-increasing complexity of VLSI circuits. Layout techniques particular to increasingly popular technologies such as field-programmable gate arrays (FPGAs) have been and continue to be developed. System-on-chip (SoC) designs have brought new challenges due not only to their increased complexity but also to the use of a mixture of design styles for one chip.

A major theme of current research in VLSI layout is the consideration of circuit performance as well as layout area. As feature sizes continue to shrink, wire delay is becoming an increasingly large fraction of total circuit delay. Therefore, the delay introduced by routing is increasingly important. The consideration of performance has necessitated new techniques, not only for routing but also for partitioning and placement as well. In some cases, the techniques for area-based layout have been extended to consider delay. For example, the simulated annealing approach to placement has been modified to consider delay on critical paths. However, many researchers are developing new approaches for performance-based layout. For example, one finds the use of the techniques of linear programming, integer programming, and even specialized higher-order programming (see Chapters 30 and 31 of *Algorithms and Theory of Computation Handbook, Second Edition: General Concepts and Techniques*) in recent work on performance-driven layout. Clock-tree routing, to minimize clock delay and clock skew, is also receiving attention as an important component of performance-driven layout.

The reader is referred to the references given in "Further Information" for broader descriptions of the field and, in particular, for more thorough treatments of current research.

8.9   Further Information

This chapter has provided several examples of the successful application of the theory of combinatorial algorithms to problems in VLSI layout. It is by no means a survey of all the important problems and algorithms in the area. Several textbooks have been written on algorithms for VLSI layout, such as [24,34,37,42], and the reader is referred to these for more complete coverage of the area. Other review articles of interest are [33] and [5] on placement, [2] on geometric algorithms for layout, [18] on layout, [3] on relaxation techniques for layout problems, [19] on channel routing, [15] on layer assignment, and [1] and [32] on netlist partitioning. There are many conferences and workshops on CAD that contain papers presenting algorithms for layout. The *ACM/IEEE Design Automation Conference*, the *IEEE/ACM International Conference on Computer-Aided Design*, and the *ACM International Symposium on Physical Design* have traditionally been good sources of algorithms among these conferences. Major journals on the topic are *IEEE Transactions on Computer-Aided Design of Integrated Circuits and Systems*, *IEEE Transactions on VLSI Systems*, *ACM Transactions on Design Automation of Electronic Systems*, and *Integration, the VLSI Journal* published by North-Holland Publishing. Layout algorithms also appear in journals focusing on general algorithm development such as *SIAM Journal on Computing,* published by the Society of Industrial and Applied Mathematics, *Journal of Algorithms,* published by Elsevier, and *Algorithmica,* published by Springer-Verlag.

Defining Terms

Channel: A rectangular region for routing wires, with terminals lying on two opposite edges, called the "top" and the "bottom." The other two edges contain no terminals, but wires may cross these edges for nets that enter the channel from other channels. The routing area of a layout is decomposed into several channels.

Channel density: Orient a channel so that the top and the bottom are horizontal edges. Then the density at any vertical line cutting the channel is the number of nets that have terminals both to the left and the right of the vertical line. Nets with a terminal on the vertical line contribute to the density unless all of the terminals of the net lie on the vertical line. The channel density is the maximum density of any vertical cut of the channel.

Channel routing: The problem of determining the routes, that is, paths and layers, for wires in a routing channel.

Compaction: The process of modifying a given layout to remove extra space between features of the layout.

Floorplan: An approximate layout of a circuit that is made before the layouts of the components composing the circuit have been completely determined.

Floorplan sizing: Given a floorplan and a set of components, each with a shape function, finding an assignment of specific shapes to the components so that the area of the layout is minimized.

Floorplanning: Designing a floorplan.

General cell layout: A style of layout in which the components may be of arbitrary height and width and functional complexity.

Global routing: When a layout area is decomposed into channels, global routing is the problem of choosing which channels will be used to make the interconnections for each net.

Graph partitioning: Given a graph with weights on its vertices and costs on its edges, the problem of partitioning the vertices into some given number k of approximately equal-weight subsets such that the cost of the edges that connect vertices in different subsets is minimized.

Interval graph coloring: Given a finite set of n intervals $\{[l_i, r_i], 1 \le i \le n\}$, for integer l_i and r_i, color the intervals with a minimum number of colors so that no two intervals that intersect are of the same color. The graph representation is direct: each vertex represents an interval, and there is an edge between two vertices if the corresponding intervals intersect. Then coloring the interval graph corresponds to coloring the intervals.

Jog: In a rectilinear routing model, a vertical segment in a path that is generally running horizontally, or vice versa.

Layer assignment: Given a set of trees in the plane, each interconnecting the terminals of a net, an assignment of a routing layer to each segment of each tree so that the resulting wiring layout is legal under the routing model.

Manhattan routing: A popular rectilinear channel-routing model in which paths for disjoint nets can cross (a vertical segment crosses a horizontal segment) but cannot contain segments that overlap in the same direction at even a point.

Net: A set of terminals to be connected together.

Rectilinear: With respect to layouts, describes a layout for which there is an underlying pair of orthogonal axes defining "horizontal" and "vertical;" the features of the layout, such as the sides of the components and the segments of the paths of wires, are horizontal and vertical line segments.

River routing: A single-layer channel-routing problem in which each net contains exactly two terminals, one at the top edge of the channel and one at the bottom edge of the channel. The nets have terminals in the same order along the top and bottom—a requirement if the problem is to be routable in one layer.

Shape function: A function that gives the possible dimensions of the layout of a component with a flexible (or not yet completely determined) layout. For a shape function $s : [w_{min}, \infty] \rightarrow [h_{min}, \infty]$ with $[w_{min}, \infty]$ and $[h_{min}, \infty]$ subsets of $\Re^+$, $s(w)$ is the minimum height of any rectangle of width w that contains a layout of the component.

Slicing floorplan: A floorplan that can be obtained by the recursive bipartitioning of a rectangular layout area using vertical and horizontal line segments.

Steiner tree: Given a graph $G = (V, E)$ a Steiner tree for a subset of vertices U of V is a subset of edges of G that form a tree and contain among their endpoints all the vertices of U. The tree may contain other vertices than those in U. For a Euclidean Steiner tree, U is a set of points in the Euclidean plane, and the tree interconnecting U can contain arbitrary points and line segments in the plane.

Switch box: A rectangular routing region containing terminals to be connected on all four sides of the rectangle boundary and for which the entire interior of the rectangle can be used by wires (contains no obstacles).

Terminal: A position within a component where a wire attaches. Usually a terminal is a single point on the boundary of a component, but a terminal can be on the interior of a component and may consist of a set of points, any of which may be used for the connection. A typical set of points is an interval along the component boundary.

Topological sort: Given a directed, acyclic graph, a topological sort of the vertices of the graph is a total ordering of the vertices such that if vertex u comes before vertex v in the ordering, there is no directed path from v to u.

Via minimization: Given a set of trees in the plane, each interconnecting the terminals of a net, determining a layer assignment that minimizes the number of points (vias) at which a layer change occurs.

Acknowledgment

I would like to thank Ron Pinter for his help in improving this chapter.

References

1. Alpert, C.J. and Kahng, A.B., Recent directions in netlist partitioning: A survey, *Integration: The VLSI Journal.* 19, 1–81, 1995.
2. Asano, T., Sato, M., and Ohtsuki, T., Computational geometry algorithms. In *Layout Design and Verification,* Ohtsuki, T., Ed., pp. 295–347. North-Holland, Amsterdam, the Netherlands, 1986.
3. Chan, T.F., Cong, J., Shinnerl, J.R., Sze, K., Xie, M., and Zhang, Y., Multiscale optimization in VLSI physical design automation. In *Multiscale Optimization Methods and Applications,* Hager, W.W., Huang, S.-J., Pardalos, P.M., and Prokopyev, O.A., editors, pp. 1–68. Springer New York, 2006.
4. Chen, T.-C. and Chang, Y.-W., Modern floorplanning based on fast simulated annealing. In *Proceedings of the 2005 International Symposium on Physical Design,* pp. 104–112. ACM, New York, 2005.
5. Cong, J., Shinnerl, J.R., Xie, M., Kong, T., and Yuan, X., Large-scale circuit placement, *ACM Transactions on Design Automation of Electronic Systems.* 10(2), 389–430, 2005.
6. Cormen, T.H., Leiserson, C.E., Rivest, R.L., and Stein, C. *Introduction to Algorithms,* 2nd ed. MIT Press, Cambridge, MA, 2001.
7. Deutsch, D.N., A dogleg channel router. In *Proceedings of the 13th ACM/IEEE Design Automation Conference,* pp. 425–433. ACM, New York, 1976.
8. Dolev, D., Karplus, K., Siegel, A., Strong, A., and Ullman, J.D., Optimal wiring between rectangles. In *Proceedings of the 13th Annual ACM Symposium on Theory of Computing,* pp. 312–317. ACM, New York, 1981.
9. Fiduccia, C.M. and Mattheyses, R.M., A linear-time heuristic for improving network partitions. In *Proceedings of the 19th ACM/IEEE Design Automation Conference,* pp. 175–181. IEEE Press, Piscataway, NJ, 1982.
10. Garey, M.R. and Johnson, D.S., *Computers and Intractability: A Guide to the Theory of NP-Completeness.* W.H. Freeman, San Francisco, CA, 1979.
11. Greenberg, R.I., Ishii, A.T., and Sangiovanni-Vincentelli, A.L., MulCh: A multi-layer channel router using one, two and three layer partitions. In *IEEE International Conference on Computer-Aided Design,* pp. 88–91. IEEE Press, Piscataway, NJ, 1988.
12. Greenberg, R.I. and Maley, F.M., Minimum separation for single-layer channel routing. *Information Processing Letters.* 43, 201–205, 1992.
13. Hashimoto, A. and Stevens, J., Wire routing by optimizing channel assignment within large apertures. In *Proceedings of the 8th IEEE Design Automation Workshop,* pp. 155–169. IEEE Press, Piscataway, NJ, 1971.
14. Hwang, F.K., On Steiner minimal trees with rectilinear distance. *SIAM Journal on Applied Mathematics,* 30(1), 104–114, 1976.
15. Joy, D. and Ciesielski, M., Layer assignment for printed circuit boards and integrated circuits. *Proceedings of the IEEE.* 80(2), 311–331, 1992.
16. Kahng, A., Classical floorplanning harmful? In *Proceedings of the International Symposium on Physical Design,* pp. 207–213. ACM, New York, 2000.
17. Kernighan, W. and Lin, S., An efficient heuristic procedure for partitioning graphs. *Bell System Technical Journal,* 49, 291–307, 1970.
18. Kuh, E.S. and Ohtsuki, T., Recent advances in VLSI layout. *Proceedings of the IEEE.* 78(2), 237–263, 1990.
19. LaPaugh, A.S. and Pinter, R.Y., Channel routing for integrated circuits. In *Annual Review of Computer Science,* Vol. 4, J. Traub, Ed., pp. 307–363. Annual Reviews Inc., Palo Alto, CA, 1990.
20. Lee, C.Y., An algorithm for path connection and its applications. *IRE Transactions on Electronic Computers.* EC-10(3), 346–365, 1961.

21. Lee, T.-C., A bounded 2D contour searching algorithm for floorplan design with arbitrarily shaped rectilinear and soft modules. In *Proceedings of the 30th ACM/IEEE Design Automation Conference,* pp. 525–530. ACM, New York, 1993.

22. Lefebvre, M., Marple, D., and Sechen, C., The future of custom cell generation in physical synthesis. In *Proceedings of the 34th Design Automation Conference,* pp. 446-451. ACM, New York, 1997.

23. Leiserson, C.E. and Pinter, R.Y., Optimal placement for river routing. *SIAM Journal on Computing.* 12(3), 447–462, 1983.

24. Lengauer, T., *Combinatorial Algorithms for Integrated Circuit Layout.* John Wiley & Sons, West Sussex, England, 1990.

25. Liao, Y.Z. and Wong, C.K., An algorithm to compact a VLSI symbolic layout with mixed constraints. *IEEE Transactions on Computer-Aided Design.* CAD-2(2), 62–69, 1983.

26. Maley, F.M., An observation concerning constraint-based compaction. *Information Processing Letters.* 25(2), 119–122, 1987.

27. Maley, F.M., *Single-layer wire routing.* PhD thesis, Department of Electrical Engineering and Computer Science, MIT, Cambridge, MA, 1987.

28. Mata, J.M., Solving systems of linear equalities and inequalities efficiently. In *Congressus Numerantum,* vol 45, *Proceedings of the 15th Southeastern International Conference on Combinatorics,* Graph Theory and Computing, Utilitas Mathematica Pub. Inc., Winnipeg, Manitoba, Canada, 1984.

29. Otten, R.H.J.M., Efficient floorplan optimization. In *Proceedings of the International Conference on Computer Design: VLSI in Computers,* pp. 499–502. IEEE Press, Piscataway, NJ, 1983.

30. Ozdal, M.M. and Wong, M.D.F., A provably good algorithm for high performance bus routing. In *Proceedings of the IEEE/ACM International Conference on Computer-Aided Design.* pp. 830–837. IEEE Computer Society, Washington, DC, 2004.

31. Pan, P. and Liu, C.L., Area minimization for floorplans. *IEEE Transactions on Computer-Aided Design.* CAD-14(1), 123–132, 1995.

32. Papa, D.A. and Markov, I.L. Hypergraph partitioning and clustering. In *Approximation Algorithms and Metaheuristics,* Gonzalez, T., Ed.. CRC Press, Boca Raton, FL, 2006.

33. Preas, B.T. and Karger, P.G., Automatic placement: A review of current techniques. In *Proceedings of the 23rd ACM/IEEE Design Automation Conference,* pp. 622–629. IEEE Press, Piscataway, NJ, 1986.

34. Preas, B.T. and Lorenzetti, M.J., Ed., *Physical Design Automation of VLSI Systems.* Benjamins/Cummings, Menlo Park, CA, 1988.

35. Rivest, R.L. and Fiduccia, C.M., A "greedy" channel router. In *Proceedings of the 19th ACM/IEEE Design Automation Conference,* pp. 418–422. IEEE Press, Piscataway, NJ, 1982.

36. Sechen, C., Chip-planning, placement, and global routing of macro/custom cell integrated circuits using simulated annealing. In *Proceedings of the 25th ACM/IEEE Design Automation Conference,* pp. 73–80. IEEE Computer Society Press, Los Alamitos, CA, 1988.

37. Sherwani, N., *Algorithms for VLSI Physical Design Automation,* 3rd edn. Springer, New York, 1999.

38. Shi, W., An optimal algorithm for area minimization of slicing floorplans. In *IEEE/ACM International Conference on Computer-Aided Design,* pp. 480–484. IEEE Computer Society, Washington, DC, 1995.

39. Stockmeyer, L., Optimal orientations of cells in slicing floorplan designs. *Information and Control.* 57, 91–101, 1983.

40. Tompa, M., An optimal solution to a wire-routing problem. *Journal of Computer and System Sciences.* 23(2), 127–150, 1981.

41. Wolf, W.H., *Modern VLSI Design: System-on-Chip Design,* 3rd edn. Prentice-Hall, Englewood Cliffs, NJ, 2002.

42. Youssef, H. and Sait, S., *VLSI Physical Design Automation: Theory and Practice,* World Scientific, Singapore, 1999.

9

Cryptographic Foundations

Yvo Desmedt
University College London

9.1 Introduction

Cryptography studies methods to protect several aspects of data, in particular privacy and authenticity, against a malicious adversary who tries to break the security. In contrast with steganography, where the data and their existence are physically hidden, cryptography transforms the data mathematically, usually using a key. Cryptanalysis is the study of methods to break cryptosystems.

Cryptography has been studied for centuries [19,28], although initially it focused only on protecting privacy. Originally, it was used in the context of military and diplomatic communication. Nowadays, most of these historical cryptoschemes have no practical value since they have been cryptanalyzed, that is, broken. However, it should be noted that it has taken cryptanalysts (those researchers or technicians trying to break cryptosystems) more than 300 years to find a general method to solve polyalphabetic ciphers with repeating keywords [19] (see Section 9.2.4). This contrasts with popular modern cryptosystems, such as Data Encryption Standard (DES) and RSA (see Sections 9.5.1 and 10.3.2), that have only been around for a few decades, which brings us to modern cryptography now.

Modern cryptography differs from historical cryptography in many respects. First of all, mathematics plays a more important role than ever before. By means of probability theory, Shannon was able to prove that Vernam's one-time pad (see Section 9.4) is secure. Second, the rather new area of computational complexity has been used as a foundation for cryptography. Indeed, the concept of **public key**, which facilitates the use of cryptography (see Section 9.3.4), finds its origin there. Third, the widespread use of communication implies that cryptography is no longer a uniquely military topic. High-speed networks and computers are responsible for a world in which postal mail has almost been replaced by electronic communication in such applications as bank transactions, access to worldwide databases as in the World Wide Web, e-mail, etc. This also implies a whole new range of security needs that need to be addressed, for example: anonymity (see Chapter 18), authenticity (see Chapter 12), commitment and identification, law enforcement, nonrepudiation (see Chapter 12), revocation, secure distributed computation, timestamping, traceability, witnessing, etc.

To illustrate the concept, we will first describe some historical cryptosystems in Section 9.2 and explain how these can be broken. In Section 9.3, we will define cryptosystems.

9.2 Historical Cryptosystems

Here, we will discuss some historical cryptosystems to lay the foundation for describing how they can be broken. For a more complete survey of historical cryptosystems, the reader may refer to the literature.

9.2.1 The Caesar Cipher and Exhaustive Key Search

One of the oldest cryptosystems is Caesar cipher, often incorrectly cited as the first cryptosystem. Caesar replaced each symbol in the original text, which is now called **plaintext** or **cleartext**, by one that was three positions further in the alphabet, counted cyclically. The word "plaintext," for example, would become "sodlqwhaw" in this system. The result is called ciphertext. The problem with this scheme is that anyone who knows how the text is encoded can break it. To prevent this, a key is used.

To describe a more modern variant of the Caesar cipher, let n be the cardinality of the alphabet being used, which is 26 for the English alphabet, or 27 when the space symbol is included in the alphabet. (In many old cryptoschemes, the space symbol was dropped since it would facilitate breaking the code.) The first symbol of the plaintext is mapped into the number 0, the second into 1, etc. To **encrypt** with the Caesar cipher, one adds modulo n the key k to the symbol m, represented as an integer between 0 and $n - 1$. (Two integers, a and b, are equivalent modulo n, denoted as $a \equiv b \bmod n$, when a and b have the same nonnegative remainder when divided by n). The corresponding symbol, then, in the ciphertext is $c = m + k \bmod n$, where the equality indicates that $0 \leq c < n$. If a long enough message contains redundancy, as plain English does, then an **exhaustive search** of all possible keys will reveal the correct plaintext. **Decryptions** (the process that permits the person who knows the secret key to compute the plaintext from the ciphertext) with the wrong key will (likely) not produce an understandable text. Since it is feasible to test all possible keys, the keyspace in the Caesar cipher is too small.

9.2.2 Substitution Cipher and Ciphertext-Only Attack

We will now consider the substitution cipher. In plaintext, each symbol m is replaced by the symbol $E_k(m)$, specified by the key k. To allow unique decryption, the function E_k must be one-to-one. Moreover, if the same symbols are used in the ciphertext as in the plaintext, it must be a bijection. If the key can specify any such bijection, the cipher is called a simple substitution cipher. Obviously, for the English alphabet, there are $26! = 403291461126605635584000000$, roughly $4 * 10^{26}$, different keys. We will now discuss the security of the scheme, assuming that only the cryptanalyst knows the ciphertext and the fact that a substitution cipher was used. Such an attack is called a ciphertext-only attack. Note that an exhaustive key search would take too long on a modern computer. Indeed, a modern parallel computer can perform 10^{15} operations per second. For simplicity, assume that such a computer could perform 10^{15} symbol decryptions per second. One wonders then how long the ciphertext needs to be before one can be certain that the cryptanalyst has found a sufficiently correct key. This measure is called the unicity distance. Shannon's theory of secrecy [30] tells us that there are 28 symbols for an English text. An exhaustive key search would roughly take $3.6 * 10^5$ years before finding a sufficiently correct key. However, a much faster method for breaking a substitution cipher exists, which we will describe now.

In English, the letter "e" is the most frequently used. Furthermore, no other letter has a frequency of occurrence that comes close to that of "e." A cryptanalyst starts the procedure by counting how many times each letter appears in the ciphertext. When the ciphertext is long enough, the most frequent letter in the ciphertext corresponds to the letter "e" in the plaintext. The frequencies of the letters "T,O,A,N,I,R,S,H" are too similar to decide by which letter they have been substituted. Therefore, the cryptanalyst will use the frequency distribution of two or three consecutive letters, called a digram and a trigram. When the space symbols have been discounted, the most frequent digrams are: "th"; "e" as the first letter, decreasing in order as follows: "er,ed,es,en,ea"; and "e" as the second letter: "he,re." The digram "he" is also quite common. This permits the identification of the letter "h," and then the letter "t." The next step is to distinguish the vowels from the consonants. With the exception of the diagrams "ea,io," two vowels rarely follow one another. This allows one to identify the letter "n," since 4 out of 5 letters following "n" are vowels. Using similar properties of other digrams and trigrams, the full key is found. If mistakes are made, they are easily spotted, and one can recover using backtracking.

9.2.3 Ideal Ciphers and Known-Plaintext Attack

Redundancy in a language permits breaking a substitution cipher; however, one may question the security of a text if it is compressed first. Shannon proved that if all redundancy is removed by the source coder, a cryptanalyst using a ciphertext-only attack cannot find a unique plaintext solution. In fact, there are 26 meaningful plaintexts! Shannon called such systems ideal. However, one cannot conclude that such a system is secure. Indeed, if a cryptanalyst knows just one (not too short) plaintext and its corresponding ciphertext, finding the key and breaking all future ciphertexts encrypted with the same key is a straightforward procedure. Such an attack is known as a known-plaintext attack.

Other types of attacks are the chosen text attacks, which comprise chosen-plaintext and chosen-ciphertext attacks. In a chosen-plaintext attack, the cryptanalyst succeeds in having a plaintext of his choice being encrypted. In a chosen-ciphertext attack, it is a ciphertext of his choice that is decrypted. The cryptanalyst can, in this context, break the simple substitution cipher by having the string of all the different symbols in the alphabet encrypted or decrypted. Chosen-plaintext attacks in which the text is not excessively long are quite realistic in a commercial environment [12]. Indeed, company *A* could send a (encrypted) message about a potential collaboration to a local branch of company *B*. This company, after having decrypted the message, will most likely forward it to its headquarters, encrypting it with a key that the cryptanalyst in company *A* wants to break. In order to break it, it is sufficient to eavesdrop on the corresponding ciphertext. Note that a chosen-ciphertext attack is a little harder. It requires access to the output of the decryption device, for example, when the corresponding, and likely unreadable, text is discarded.

Although ideal ciphers are a nice information theoretical concept, their applicability is limited in an industrial context where standard letters, facilitating a known-plaintext attack, are often sent. Information theory is unable to deal with known-plaintext attack. Indeed, finding out whether a known-plaintext is difficult or not is a computational complexity issue. A similar note applies to the unicity distance. Many modern ciphers, such as Advanced Encryption Standards (AES) (see Section 9.5.3), have unicity distances shorter than that of the substitution cipher, but no method is known to break them in a very efficient way. Therefore, we will not discuss in further detail the results of ideal ciphers and unicity distance.

9.2.4 Other Historical Ciphers

Before finishing the discussion on historic cryptosystems, we will briefly mention the transposition cipher and the polyalphabetic ciphers with repeating keywords. In a transposition cipher, also known as a permutation cipher, the text is split into blocks of equal length, and the order of the letters in each block is mixed according to the key. Note that in a transposition cipher, the frequency of individual letters is not affected by encrypting the data, but the frequency of digrams is. It can be cryptanalyzed by trying to restore the distribution of digrams and trigrams. In most polyalphabetic ciphers, known as periodic substitution ciphers, the plaintext is also split into blocks of equal length, called the period d. One uses d substitution ciphers by encrypting the ith symbol ($1 \leq i \leq d$) in a block using the ith substitution cipher. The cryptanalysis is similar to the simple substitution cipher once the period d has been found. The Kasiski [20] method analyzes repetition in the ciphertext to find the exact period. Friedman [16] index of coincidence to find the period is beyond the scope of this introduction. Other types of polyalphabetic ciphers are running key ciphers, the Vernam's one-time pad (see Section 9.4), and rotor machines, such as Enigma.

Many modern cryptosystems are **polygram substitution** ciphers, which are a substitutions of many symbols at once. To make them practical, only a subset of keys is used. For example, DES (see Section 9.5.1) used in *Electronic Code Book* mode substitutes 64 bits at a time using a 56 bit key instead of a $\log_2(2^{64}!)$ bit key (which is longer than 2^{64} bits) if any polygram substitution is allowed. Substitution and transposition ciphers are examples of **block ciphers**, in which the plaintext and ciphertext are divided into strings of equal length, called blocks, and each block is encrypted one-at-a-time.

9.3 Definitions

As mentioned in Section 9.1, modern cryptography covers more than simply the protection of privacy, but to give all these definitions is beyond the scope of this chapter. We will focus on privacy and authenticity.

9.3.1 Privacy

DEFINITION 9.1   A cryptosystem used to protect privacy, also called an encryption scheme or system, consists of an encryption algorithm E and a decryption algorithm D. The input to E is a plaintext message $m \in M$ and a key k in the key space K. The algorithm might use randomness $r \in R$ as an extra input. The output of the encryption is called the ciphertext $c \in C$ and $c = E_k(m) = f_E(k, m, r)$.

The decryption algorithm (which may use randomness) has input a key $k' \in K'$ and a ciphertext $c \in C$ and outputs the plaintext m, so, $m = D_{k'}(E_k(m))$. To guarantee unique decryption, the following must be satisfied:

$$\text{for all } k \in K, \text{ for all } m, \text{ for all } m' \neq m, \text{ for all } r \text{ and } r' : f_E(k, m, r) \neq f_E(k, m', r').$$

9.3.1.1 Security

Clearly, in order to prevent any unauthorized person from decrypting the ciphertext, the decryption key k' must be secret. Indeed, revealing parts of it may help the cryptanalyst.

The types of attacks that a cryptanalyst can use have been informally described in Section 9.2. The most powerful attack is the adaptive chosen text attack in which the cryptanalyst employs several chosen texts. In each run, the cryptanalyst observes the output and adapts the next chosen text based on the previous ones and the output of previous attacks.

An encryption scheme is secure if, given public parameters and old plaintext-ciphertext pairs obtained using known-plaintexts and/or chosen text attacks, the possible new ciphertexts for messages m_0 and m_1 are indistinguishable. A sufficient condition is that they are indistinguishable from a random string of the same length uniformly chosen. We will discuss this further in the Sections 9.3.3 and 9.4.2.

9.3.2 Authenticity

While the terminology is rather standard for cryptosystem's protecting privacy, that for authenticity is not. Research and a better understanding of the topic have made it clear that using concepts as encryption and decryption makes no sense. This was done in the early stages when the concept was introduced.

DEFINITION 9.2 A cryptosystem used to protect authenticity, also called an authentication scheme or system, consists of an authenticator generation algorithm G and a verification algorithm V. The input to G is a message $m \in M$ and a key k' in the key space K'. The algorithm might use randomness $r \in R$ as an extra input. The output of the generation algorithm is the authenticated message (m, a) where m is the message, and a is the authenticator. In other words, $(m, a) = G_{k'}(m) = (m, f_G(k', m, r))$.

The inputs of the verification algorithm V (which may use randomness) are a key $k \in K$ and a string (m', a'). The message is accepted as authentic if $V(m', a', k)$ returns ACCEPT, else it is rejected. To guarantee that authentic messages are accepted, one needs that $V_k(G_{k'}(m))$ is (almost always) ACCEPT.

9.3.2.1 Security

It is clear that to prevent any unauthorized person from authenticating fraudulent messages, the key k' must be secret. Indeed, revealing parts of it may help the cryptanalyst.

The types of attacks that a cryptanalyst can use are similar to those that have been informally described in Section 9.2. The goal of the cryptanalyst has changed. It is to construct a new, not yet authenticated, message. The most powerful attack is the adaptive chosen text attack, in which the cryptanalyst employs several chosen messages which are given as input to G. The cryptanalyst observes the output of G in order to adapt the next chosen message based on previous ones and the output of previous attacks.

An authentication scheme is secure if, given public parameters and old message-authenticator pairs obtained using known message and/or chosen message attacks, the probability that any cryptanalyst can construct a a new pair (m', a') in which the verification algorithm V will accept as authentic is negligible. Chapter 12 discusses this in more details.

9.3.3 Levels of Security

Modern cryptography uses different models to define security. One distinguishes between: heuristic security, as secure as, proven secure, quantum secure, and unconditionally secure.

A cryptographic system or protocol is heuristically secure as long as no attack has been found. Many practical cryptosystems fall within this category.

One says that a cryptosystem or protocol is as secure as another if it can be proven that a new attack against one implies a new attack against the other and vice versa. A much stronger statement is that a system is proven secure.

To speak about proven security, one must first formally model what security is, which is not always obvious. A system or protocol is said to be proven secure relative to an assumption if one can prove that if the assumption is true, this implies that the formal security definition is satisfied for that system or protocol.

In all aforementioned cryptosystems, one usually assumes that the opponent, for example, the eavesdropper, has a bounded computer power. In the modern theoretical computer science model, this is usually expressed as having the running time of the opponent be bounded above by a polynomial in function of the security parameter, which often is the length of the secret key. Infeasible corresponds with a minimum running time bounded below by a superpolynomial in the length of the security parameter. Note that there is no need to use a polynomial versus superpolynomial model. Indeed, having a huge constant as a lower bound for a cryptanalytic effort would be perfectly satisfactory. For example, according to quantum physics, time is discrete. A huge constant could be the estimated number of time units which have elapsed since the (alleged) big bang.

A cryptosystem is unconditionally secure when the computer power of the opponent is unbounded, and it satisfies a formal definition of security. Although these systems are not based on mathematical or computational assumptions, usually these systems can only exist in the real world when true randomness can be extracted from the universe. In Section 9.4, we will discuss an unconditionally secure encryption scheme.

A special class of cryptosystems assumes the correctness of the laws of quantum physics. These cryptosystems are known as quantum cryptography [1].

9.3.4 Conventional Cryptography versus Public Key

We will now discuss whether k must remain secret and analyze the relationship between k' and k. If it is easy to compute k' from k, it is obvious that k must also remain secret. The key is unique to a sender-receiver pair. In this case, the cryptosystem is called a **conventional** or **symmetric cryptosystem**.

If, on the other hand, given k it is hard to compute k' and hard to compute a k'', which allows partial cryptanalysis, then the key k can be made public. The system, the concept of which was invented by Diffie and Hellman and independently by Merkle, is called a public key or **asymmetric cryptosystem**. This means that for privacy protection each receiver R will publish a personal k_R, and for authentication, the sender S makes k_S public. In the latter case, the obtained authenticator is called a **digital signature**, since anyone who knows the correct public key k_S can verify the correctness. The scheme is then called a signature scheme. Note that recently some have unified the formal definitions for authentication and digital signature using the terminology of "digital signature," which is unfortunately a poor choice of wording.

The public key k is considered a given input in the discussion on the security of encryption and authentication schemes (see Sections 9.3.1 and 9.3.2).

It was stated in the literature that digital signature schemes have the property that the sender cannot deny having sent the message. However, the sender can claim that the secret key was physically stolen. This would allow him to deny even having sent a message [29]. Such situations must be dealt with by an authority. Protocols have been presented in which the message is being deposited to a notary public or arbiter [25]. Schemes have been developed in which the arbiter does not need to know the message that was authenticated [9,10]. Another solution is digital timestamping [18] based on cryptography (the signer needs to alert an authority that his public key must have been stolen).

The original description of public key systems did not explain the importance of the authenticity of the public key [27]. Indeed, if it is not authentic, the one who created the fake public key can decrypt messages intended for the legitimate receiver, or can sign claiming to be the sender. So, the security is then lost. In practice, this problem is solved by using a certificate which is itself a digital signature(s) of the public key. It is provided by a known trusted entity(ies) who guarantee(s) that the public key of S is k_S.

We will now explain the need to use randomness in public key encryption systems. If no random input is used, the public key system is vulnerable to a partially known-plaintext attack. In particular, if the sender S, for example, a supervisor, uses a standard letter to send almost the same message to different receivers, for example, to inform them about their salary increase, the system does not guarantee privacy. Indeed, any of the receivers of these letters can exhaustively fill in the nonstandard part and encrypt the resulting letter using the public key of the receiver until the obtained ciphertext corresponds with the eavesdropped one! So, if no redundancy is used, the eavesdropper can always verify whether a particular message has been encrypted. It is easy to see that if no randomness is used, the resulting ciphertext will have a probability distribution which is 0 for all values, except for the deterministic encryption of the plaintext. Although these attacks are well known, very few practical public key systems take precautions against them. Goldwasser–Micali [17] called schemes avoiding this weakness as **probabilistic encryption schemes** and presented a first solution (see Section 10.7). It is secure against known-plaintext attack under a computational number theoretic assumption.

9.3.5 Practical Concerns

To be practical, the encryption, decryption, authentication, and verification algorithms must be efficient. In the modern theoretical computer science model, this is usually expressed as having a running time bounded by a polynomial in function of the length of the key and by stating that the length of the message is bounded by a polynomial in function of the length of the key.

It is clear that to be useful $M = \{0, 1\}^*$ (or have a polynomial length). However, this is often not the case. A mode or protocol is then needed to specify how the encryption and decryption algorithms (or the authentication and verification algorithms) are used on a longer text. For an example, see Section 9.5.4.

9.4 The One-Time Pad

The one-time pad (a conventional cryptosystem) and Shannon's analysis of its security are the most important discoveries in modern cryptography. We will first discuss the scheme, then give a formal definition of the security, prove it to be secure, and briefly discuss some of the applications of the one-time pad.

9.4.1 The Scheme

In Vernam's one-time pad, the key is (at least) as long as the message. Let the message be a string of symbols belonging to the alphabet (a finite set) S, for example, $\{0, 1\}$, on which a binary operation "$*$" is defined, for example, the exor (exclusive-or). We assume that $S(*)$ forms a group.

Before encrypting the message, the sender and receiver have obtained a secret key, a string, of which the symbols have been chosen uniformly random in the set S and independent. Let m_i, k_i, and c_i be the ith symbols of, respectively, the message, the key, and the ciphertext, each belonging to S. The encryption algorithm produces $c_i = m_i * k_i$ in S. To decrypt, the receiver computes $m_i = c_i * k_i^{-1}$. The key is used only once. This implies that if a new message needs to be encrypted, a new key is chosen, which explains the terminology: one-time pad.

It is trivial to verify that this is an encryption scheme. In the case $S(*) = Z_2(+) = \{0, 1\}(+)$, the integers modulo 2, the encryption algorithm, and the decryption algorithm are identical, and the operation corresponds with an exor (exclusive or). We will now define what privacy means.

9.4.2 Security

DEFINITION 9.3   Shannon defined an encryption system to be perfect when, for a cryptanalyst not knowing the secret key, the message m is independent of the ciphertext c, formally:

$$\text{prob}(\mathbf{m} = m \mid \mathbf{c} = E_k(m)) = \text{prob}(\mathbf{m} = m). \tag{9.1}$$

THEOREM 9.4   *The one-time pad is perfect.*

PROOF 9.4   Let the length of the message be l (expressed as the number of symbols). Then the message, key, and ciphertext belong to $S^l = S \times S \times \cdots \times S$. Since S^l is a group, it is sufficient to discuss the proof for the case $l = 1$. Let $c = E_k(m) = m * k$ in S. Now, if $k' = m^{-1} * c = k$, then $\text{prob}(\mathbf{c} = c \mid \mathbf{m} = m, \mathbf{k} = k') = 1$, else it is 0. Using this fact we obtain:

$$\text{prob}(\mathbf{m} = m, \mathbf{c} = c) = \sum_{k' \in S} \text{prob}(\mathbf{c} = c \mid \mathbf{m} = m, \mathbf{k} = k') \cdot \text{prob}(\mathbf{m} = m, \mathbf{k} = k')$$
$$= \text{prob}(\mathbf{m} = m, \mathbf{k} = k)$$
$$= \text{prob}(\mathbf{m} = m) \cdot \text{prob}(\mathbf{k} = k) \qquad (\mathbf{k} \text{ is independent of } \mathbf{m})$$
$$= \text{prob}(\mathbf{m} = m) \cdot \frac{1}{|S|} \qquad (\mathbf{k} \text{ is uniform})$$

Also, if $c = m' * k'$, then $\text{prob}(\mathbf{c} = E_k(m) \mid \mathbf{m} = m', \mathbf{k} = k') = 1$, else it is 0. This gives:

$$\text{prob}(\mathbf{c} = c) = \sum_{k',m'} \text{prob}(\mathbf{c} = c \mid \mathbf{m} = m', \mathbf{k} = k') \cdot \text{prob}(\mathbf{m} = m', \mathbf{k} = k')$$
$$= \sum_{\substack{k',m' \\ c=m'*k'}} \text{prob}(\mathbf{m} = m', \mathbf{k} = k')$$
$$= \sum_{m'} \text{prob}(\mathbf{m} = m') \cdot \text{prob}(\mathbf{k} = (m')^{-1} * c) \qquad (\mathbf{k} \text{ is independent of } \mathbf{m})$$
$$= \frac{1}{|S|} \cdot \sum_{m' \in S} \text{prob}(\mathbf{m} = m') \qquad (\mathbf{k} \text{ is uniform})$$
$$= \frac{1}{|S|},$$

implying that $\text{prob}(\mathbf{m} = m, \mathbf{c} = c) = \text{prob}(\mathbf{m} = m) \cdot \text{prob}(\mathbf{c} = c)$.   □

COROLLARY 9.5   In the one-time pad, the ciphertext has a uniform distribution.

This corollary corresponds with the more modern view on the definition of privacy. Note that it is easy to make variants of the one-time pad that are also perfect.

Shannon proved that the length of the key must be at least, what he called, the entropy (see Chapter 12) of the message. More recent work has demonstrated that the length of the key must be at least the length of the message [2,22].

9.4.3 Its Use

The use of the one-time pad to protect private communication is rather limited since a new secret key is needed for each message. However, many zero-knowledge interactive proofs use the principle that the product in a group of a uniformly chosen element with any element of the group, whatever distribution, gives a uniform element.

Also, the idea of stream cipher, in which the truly random tape is replaced by a pseudo-random tape finds its origin in the one-time pad.

9.5   Block Ciphers

As already observed in Section 9.2.4, many modern encryption schemes, in particular conventional encryption schemes, are (based on) polygram substitution ciphers. To obtain a practical scheme, the number of possible keys needs to be reduced from the maximum possible ones. To solve this problem, Shannon proposed the use of mixing transformation, in which the plaintext is iteratively transformed using substitution and transposition ciphers. Feistel adapted this idea. The outdated DES is a typical example of a Feistel scheme, which we describe in Section 9.5.1. The more modern scheme, called the AES, is given in Section 9.5.3. The official modes used for AES are discussed in Section 9.5.4.

9.5.1   The DES Algorithm

The messages belong to $\{0, 1\}^{64}$ and the key has 56 bits. To encrypt a message longer than 64 bits, a mode is used (see Section 9.5.4). Since the algorithm is outdated, we only discuss it briefly. A more complete description can be found in [8,33].

First, the detailed design criteria of the DES algorithm are classified, while the algorithm itself is public. The DES algorithm [33], as described by NBS (now called NIST), consists of three fundamental parts:

- The enciphering computation which follows a typical Feistel approach
- The calculation of $f(R, K)$
- The key schedule calculation

The Feistel part is described in Figure 9.1 and briefly described below.

In the enciphering computation, the input is first permuted by a fixed permutation *IP* from 64 bits into 64 bits. The result is split up into 32 left bits and 32 right bits, respectively. In Figure 9.1, this corresponds to L and R. Then a bitwise modulo 2 sum of the left part L_i and of $f(R_i, K_i)$ is carried out. After this transformation, the left and right 32 bit blocks are interchanged. From Figure 9.1, one can observe that the encryption operation continues iteratively for 16 steps or rounds. In the last round, no interchange of the finally obtained left and right parts is performed. The output is obtained by applying the inverse of the initial permutation *IP* to the result of the 16th round.

Owing to the symmetry, the decryption algorithm is identical to the encryption operation, except that one uses K_{16} as first subkey, K_{15} as second, etc.

9.5.2   Variants of DES

Since the introduction of the DES, several variants were proposed. The popularity of these variants is different from country to country. Some examples are FEAL, IDEA, and GOST. The security of these schemes varies. Note that DES is no longer secure due to fast exhaustive key search machines [15]. To avoid this weakness, double and triple encryption are used. Both use a 112 bit key. Double encryption DES is obtained by running $\text{DES}_{k_1} \circ \text{DES}_{k_2}$. Triple encryption uses DES as encryption and as decryption, denoted as DES^{-1} giving: $\text{DES}_{k_1} \circ \text{DES}_{k_2}^{-1} \circ \text{DES}_{k_1}$. However, the block length of the double and triple variants is too short for high security. So, AES is preferred.

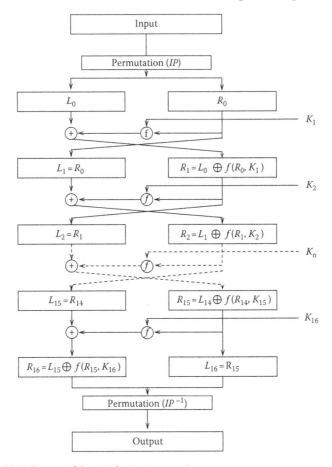

FIGURE 9.1 DES block diagram of the enciphering computation.

9.5.3 AES

The AES is a new encryption standard since November 2001. In contrast with the development of the DES, the development process of AES was open. AES was proposed by researchers from academia.

AES is a block cipher. The size of a plaintext/ciphertext block is 128 bits (the original scheme, called Rijndael allowed for larger block sizes, but these are not part of the standard). The key can have 128, 192, or 256 bits.

Figure 9.2 is a schematic representation of a typical round in the AES finalist. Depending on the key size block, the number of rounds varies, that is, 10, 12, or 14 rounds. The first and last round differ slightly from the other rounds.

A typical round in AES consists of:

Byte sub, which are fixed byte substitutions. In contrast with DES, only one S-box is used. It is a substitution, which was not the case in DES. The substitution is nonlinear. It consists of an inverse operation in a finite field $GF(2^8)$, followed by an affine (invertible) transformation over $GF(2)$.

Shift row, a permutation of the bytes.

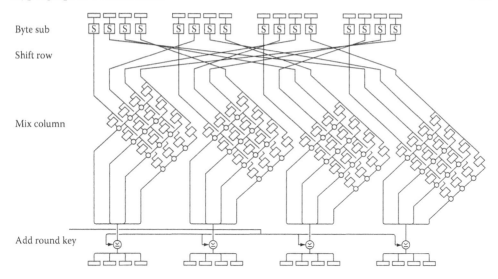

FIGURE 9.2 A typical one round of the AES finalist, as explained at the 23rd National Information Systems Security Conference. (From Daemen, J. and Rijmen, V., Rijndael, Presentation given at the 23rd National Information Systems Security Conference, October 16–19, 2000, http://csrc.nist.gov/encryption/aes/rijndael/misc/nissc2.pdf. With permission.)

Mix column, which are fixed linear combinations. Each linear combination over $GF(2^8)$ acts on 4 bytes and outputs 4 bytes.

Add round key, which performs an exor with the round key. The round key is obtained from the initial key by the key scheduling. The key scheduling uses linear combinations and the Byte Sub operation.

9.5.4 The Modes

Some of the modes that were used with the DES are outdated and have been replaced by new modes and/or updated modes. NIST recommends five confidentiality modes, one authentication mode, an authenticated encryption mode, and finally a high-throughput authenticated encryption mode. The AES Key Wrap mode is under preparation. Nonstandard modes have been developed and are, for example, used in Kerberos [3,21]. It is beyond the scope of this chapter to give all the details and properties for these modes; the reader can find them in several publications [7,13,14,26].

The five confidentiality modes are briefly surveyed. How these work for decryption is left as an exercise. When we refer to the Triple DES/AES encryption algorithm, we silently assume that each time a key is given as input.

The ECB mode, in encryption, works similarly to a substitution cipher. The plaintext is divided into full size blocks, m_i (64 bits for DES, Triple DES and 128 bits for AES). The ith input to DES/AES is the plaintext block m_i. The output corresponds to the ciphertext block c_i. This mode is not recommended. As we discussed in Section 9.2, most texts contain redundancy. If the same key is used for too long a time, most parts of the plaintext can be recovered.

In the CBC mode in encryption, the ciphertext block c_i is the 64/128 bit output of the DES/AES encryption algorithm. The input to the DES/AES is the bitwise modulo 2 sum (EXOR) of the 64/128 bit plaintext m_i and of the previous 64/128 bit ciphertext c_{i-1}. The initial vector used, IV, should be unpredictable.

In the CFB and OFB mode, the ciphertext block c_i and the plaintext block m_i are n bits, where $1 \leq n \leq 64/128$. In both modes, the ciphertext block $c_i = m_i \oplus \text{Select}_n(\text{DES}_k(d_i))$, where d_i is the input to DES (similarly for AES), $\oplus$ is the bitwise modulo 2 sum, and Select_n selects the n most significant bits.

The input d_i to the DES/AES in the CFB encryption mode is constructed from the $(64 - n)$ (or $128 - n$) least significant bits of d_{i-1} (the previous input), shifted to the left by n positions, and concatenated with c_{i-1}, the n bits of the ciphertext. The initial vector should be unpredictable.

In the OFB mode in encryption, DES/AES is used as a stream cipher (see Section 9.4.3). The output of DES/AES is used as a pseudo-random generator. The input d_i is the concatenation of the $(64 - n)$ (or $(128 - n)$) least significant bits of d_{i-1} (the previous input), shifted to the left by n positions, and the n most significant bits of $\text{DES}_k(d_{i-1})$ (similarly for AES). The initial vector must be unique to each plaintext.

The counter mode (CTR mode) in encryption is similar to the OFB mode, in the sense that both form stream ciphers. However, the input to the encryption algorithm is the output of a counter.

9.5.4.1 Authentication

While the CBC and CFB modes were originally considered for authentication with DES, for AES, a new mode exists. The original CBC-MAC should not be used since it is weak. The current recommendation is a variant of it, which was developed by academics. (Consult [14] for details.)

9.6 Research Issues and Summary

It is important to remark that cryptography does not solve all modern security problems. Cryptography, although rarely stated explicitly, assumes the existence of secure and reliable hard or software. Some research on secure distributed computation has allowed this assumption to be relaxed slightly. Also, in the communication context, modern cryptography only addresses part of a bigger problem. Spread spectrum techniques prevent jamming attacks, and reliable fault tolerant networks reduce the impact of the destruction of communication equipment.

Finally, it should be pointed out that cryptography is not sufficient to protect data. For example, it only eliminates the threat of eavesdropping while transmitted remotely. However, data can often be gathered at the source or at the destination using, for example, physical methods. These include theft (physical or virtual), the caption of electromagnetic radiation when data are displayed on a normal screen, etc.

This chapter introduced elementary principles of modern cryptography, including the concepts of conventional cryptosystems and public key systems, several different types of attacks, different levels of security schemes can have, one-time pad, and AES as an example of a block cipher.

9.7 Further Information

Since the introduction of public key, the research on cryptography has boomed. Readers interested in applied oriented cryptography should consult [23]. This book discusses block ciphers in great length. Those who prefer a textbook can consult [32].

The most known annual conferences on the topic of cryptography are Eurocrypt and Crypto, running since the early 1980s, of which the proceedings are published in Springer's *Lecture Notes in Computer Science*. Conferences and workshops running since the 1990s include Asiacrypt (which absorbed Auscrypt), FSE, and Public Key Cryptography (PKC). Some conferences focus on computer security issues as well as cryptography, for example, the ACM Conference on Computer and Communications Security. Several local and regional conferences are also organized, many with proceedings published in Springer's *Lecture Notes in Computer Science*, for example the IMA Conference on Cryptography and Coding in Britain. Some results on the topic have appeared in less specialized conferences such as FOCS, STOC, etc.

Articles on the topic appear in a wide variety of journals but unfortunately several years after the results have been presented at conferences. *The Journal of Cryptology* is dedicated to research on cryptography. *Design, Codes and Cryptography* is another journal. Some other specialized journals have a different focus, for example, Cryptologia is primarily on historic aspects of cryptography.

Defining Terms

Asymmetric cryptosystem: A cryptosystem in which given the key of one party (sender or receiver, depending from context), it is computationally difficult or, when using information theory, impossible to obtain the other parties secret key.

Block cipher: A family of cryptosystems in which the plaintext and ciphertext are divided into strings of equal length, called blocks, and each block is encrypted one-at-a-time.

Ciphertext: The result of an encryption operation.

Cleartext: The unencrypted, usually readable text.

Conventional cryptosystem: A cryptosystem in which the keys of all parties must remain secret.

Cryptanalysis: The study of methods to break cryptosystems.

Cryptanalyst: A person who (wants to) breaks cryptosystems.

Decryption: The operation that tranforms ciphertext into plaintext using a key.

Digital signature: The digital equivalent of a handwritten signature. A digital signature of a message is strongly message dependent and is generated by the sender using his/her secret key and a suitable public key cryptosystem.

Encrypt: The operation that tranforms plaintext into ciphertext.

Exhaustive search: A method to break a cryptosystem by trying all possible inputs, in particular all possible keys.

Plaintext: A synonym for cleartext, that is, the unencrypted text.

Polygram substitution cipher: A substitution cipher of many symbols at once.

Probabilistic encryption scheme: A public key system in which randomness is used, such that two encryptions of the same ciphertext give, very likely, different ciphertexts.

Public key: A key that is public, or a family of cryptosystems in which a key of one of the parties can be made public.

Symmetric cryptosystem: A system in which it is easy to find one party's key from the other party's key.

Acknowledgment

The first edition of this text was written while the author was at the University of Wisconsin – Milwaukee. Currently, he is BT Chair of Information Security and also funded by EPSRC EP/C538285/1.

References

1. C. H. Bennett and G. Brassard. An update on quantum cryptography. In *Advances in Cryptology. Proc. of Crypto 84 (Lecture Notes in Computer Science 196)*, pp. 475–480. Springer-Verlag, New York, 1985. Santa Barbara, CA, August 1984.
2. C. Blundo, A. De Santis, and U. Vaccaro. On secret sharing schemes. Technical report, Universita di Salerno, Fisciano (SA), Italy, 1995.
3. G. Brassard. *Modern Cryptology*, volume 325 of *Lecture Notes in Computer Science*. Springer-Verlag, New York, 1988.
4. J. Daemen and V. Rijmen. AES proposal: Rijndael. http://csrc.nist.gov/encryption/aes/rijndael/Rijndael.pdf.
5. J. Daemen and V. Rijmen. Rijndael. Presentation given at the 23rd National Information Systems Security Conference, October 16–19, 2000, http://csrc.nist.gov/encryption/aes/rijndael/misc/nissc2.pdf.
6. D. E. R. Denning. *Cryptography and Data Security*. Addison-Wesley, Reading, MA, 1982.
7. DES modes of operation. FIPS publication 81. *Federal Information Processing Standard*, National Bureau of Standards, U.S. Department of Commerce, Washington, DC, 1980.
8. Y. Desmedt. Cryptographic foundations. In M. Atallah, editor, *Handbook of Algorithms and Theory of Computation*, chapter 38. CRC, Boca Raton, FL, 1998.
9. Y. Desmedt and J. Seberry. Practical proven secure authentication with arbitration. In J. Seberry and Y. Zheng, editors, *Advances in Cryptology—Auscrypt '92, Proceedings (Lecture Notes in Computer Science 718)*, pp. 27–32. Springer-Verlag, New York, 1993. Gold Coast, Queensland, Australia, December, 1992.
10. Y. Desmedt and M. Yung. Arbitrated unconditionally secure authentication can be unconditionally protected against arbiter's attacks. In A. J. Menezes and S. A. Vanstone, editors, *Advances in Cryptology—Crypto '90, Proceedings (Lecture Notes in Computer Science 537)*, pp. 177–188. Springer-Verlag, New York, 1991. Santa Barbara, CA, August 11–15.
11. W. Diffie and M. E. Hellman. New directions in cryptography. *IEEE Trans. Inform. Theory*, IT–22(6), 644–654, November 1976.
12. W. Diffie and M. E. Hellman. Privacy and authentication: An introduction to cryptography. *Proc. IEEE*, 67, 397–427, March 1979.
13. Morris Dworkin. Recommendation for block cipher modes of operation: Methods and techniques. NIST Special Publication 800-38A, Gaithersburg, MD, 2001.
14. Morris Dworkin. Recommendation for block cipher modes of operation: The cmac mode for authentication. NIST Special Publication 800-38B, Gaithersburg, MD, 2005.
15. Electronic Frontier Foundation. *Cracking DES*. O'Reilly, Sebastol, CA, 1998.
16. W. F. Friedman. The index of coincidence and its applications in cryptography. Riverbank publication no. 22, Riverbank Labs, Geneva, IL, 1920.
17. S. Goldwasser and S. Micali. Probabilistic encryption. *J. Comput. Syst. Sci.*, 28(2), 270–299, April 1984.
18. S. Haber and W. S. Stornetta. How to time-stamp a digital document. *J. Cryptol.*, 3(2), pp. 99–111, 1991.
19. D. Kahn. *The Codebreakers*. MacMillan Publishing Co., New York, 1967.
20. F. W. Kasiski. *Die Geheimschriften und die Dechiffrir-kunst*. Mittler & Sohn, Berlin, 1863.
21. J. Kohl and B. C. Newmann. The Kerberos network authentication service. MIT Project Athena, Version 5.
22. J. L. Massey. Contemporary cryptology: An introduction. In G. J. Simmons, editor, *Contemporary Cryptology*, pp. 3–64. IEEE Press, New York, 1992.
23. A. Menezes, P. van Oorschot, and S. Vanstone. *Applied Cryptography*. CRC, Boca Raton, FL, 1996.
24. R. C. Merkle. Secure communications over insecure channels. *Commun. ACM*, 21, 294–299, 1978.

25. C. H. Meyer and S. M. Matyas. *Cryptography: A New Dimension in Computer Data Security*. John Wiley & Sons, New York, 1982.

26. Current modes, 2008 http://csrc.nist.gov/groups/ST/toolkit/BCM/current_modes.html

27. G. J. Popek and C. S. Kline. Encryption and secure computer networks. *ACM Comput. Surv.*, 11(4), pp. 335–356, December 1979.

28. L. Sacco. *Manuale di crittografia*. Rome, Italy, 2e edition riveduta e aumenta edition, 1936. Translated in English "Manual of Cryptography", Aegean Park Press, Laguna Hills, CA, 1977.

29. J. Saltzer. On digital signatures. *ACM Operating Syst. Rev.*, 12(2), 12–14, April 1978.

30. C. E. Shannon. Communication theory of secrecy systems. *Bell Syst. Tech. J.*, 28, 656–715, October 1949.

31. G. J. Simmons, editor. *Contemporary Cryptology*. IEEE Press, New York, 1992.

32. D. R. Stinson. *Cryptography: Theory and Practice*. CRC Press, Boca Raton, FL, 1995.

33. U.S. Department of Commerce, National Bureau of Standards. *Data Encryption Standard*, January 1977. *FIPS* PUB 46 (NBS Federal Information Processing Standards Publ.).

34. U.S. Department of Commerce, National Institute of Standards and Technology. *Advanced Encryption Standard*, *FIPS* PUB 197, November 2001.

10

Encryption Schemes

Yvo Desmedt
University College London

10.1 Introduction

Several conventional encryption schemes were discussed in Chapter 9. The concept of public key was introduced in Section 9.3.4. In this chapter, we will discuss some public key encryption systems based on number theory. First, we give the minimal number theory and algebraic background needed to understand these schemes from a mathematical viewpoint (see Section 10.2). Then, we present, in Section 10.3, the most popular schemes based on number theory. We explain in Section 10.4 the computational number theory required to understand why these schemes run in (expected) polynomial time. To avoid overburdening the reader with number theory, we will postpone the number theory needed to understand the computational aspect until Section 10.4. We briefly discuss security issues in Section 10.7.

10.2   Minimal Background

10.2.1   Algebra

For the reader who is not familiar with elementary algebra, we review the definition of a semigroup, monoid, group, etc.

DEFINITION 10.1   A set M with an operator "$*$," denoted as $M(*)$, is a semigroup if the following conditions are satisfied:

1. the operation is closed in M, i.e., $\forall a, b \in M : (a * b) \in M$,
2. the operation is associative in M, i.e., $\forall a, b, c \in M : (a * b) * c = a * (b * c)$,

When additionally

3. the operation has an identity element in M, i.e., $\exists e : \forall a \in M : a * e = e * a = a$,

we call $M(*)$ a monoid. When $M(*)$ is a monoid, we call the cardinality of M the order of M.

When using multiplicative notation ($*$), we will usually denote the identity element as 1 and when using additive notation ($+$) as 0.

DEFINITION 10.2   In a monoid $M(*)$, the element $a \in M$ has an inverse if an element denoted as a^{-1} exists such that:

$$a * a^{-1} = a^{-1} * a = e,$$

where e is the identity element in $M(*)$. Two elements $a, b \in M$ commute if $a * b = b * a$.

An element having an inverse is often called a unit. When working in a monoid, we define $a^{-n} = (a^{-1})^n$ and $a^0 = 1$.

DEFINITION 10.3   A monoid $G(*)$ is a group if each element in G has an inverse. It is an Abelian group if, additionally, any two elements of G commute.

If H is a monoid (group) and $H \subseteq G$, where G is also a monoid (group), then H is called a submonoid (subgroup).

DEFINITION 10.4   Let $M(*)$ be a monoid and $a \in M$. One defines

$$\langle a \rangle = \{a^k \mid k \in N\}$$

where N are the natural numbers, i.e., $\{0, 1, \ldots\}$. If $\langle a \rangle$ is a group, we call it a cyclic group and call the order of $\langle a \rangle$ the order of a, or ord(a).

Note that in modern cryptography, finite sets are more important than infinite sets. In this respect, the following result is interesting.

THEOREM 10.1   *Let $M(*)$ be finite monoid. If $a \in M$ has an inverse in $M(*)$, then $\langle a \rangle$ is an Abelian group. Also, ord(a) is the smallest positive integer m for which $a^m = 1$.*

PROOF   Since the set M is finite, positive integers k_1 and k_2 must exist, where $k_1 < k_2$, such that $a^{k_1} = a^{k_2}$. Now, since a is invertible, this means $a^{k_1-k_2} = 1$. This implies that $\langle a \rangle$ is an Abelian group. So, then $\langle a \rangle = \{a^k \mid 0 \le k < \mathrm{ord}(a)\}$.   □

We will now define what a ring is. Unfortunately, there are two different definitions for it in the literature, but we will use the one that is most relevant to modern cryptography.

DEFINITION 10.5   A set R with two operations $+$ and $*$ and two distinct elements 0 and 1 is a ring if:

1. $R(+)$ is an Abelian group with identity element 0.
2. $R(*)$ is a monoid with identity element 1.
3. R is distributive, i.e., $\forall a, b, c \in R$:

$$a * (b + c) = (a * b) + (a * c)$$
$$(a + b) * c = (a * c) + (b * c)$$

A ring R is commutative if it is commutative for the multiplication. A ring R in which $R^0 = R \setminus \{0\}$ is an Abelian group for the multiplication is called a field.

A subring is defined similarly as a subgroup.

10.2.2 Number Theory

10.2.2.1 Integers

We denote the set of integers as Z, and the positive integers as Z^+, i.e., $\{1, 2, \ldots\}$.

DEFINITION 10.6   If α is a real number, we call $\lfloor \alpha \rfloor$ the integer such that

$$\lfloor \alpha \rfloor \le \alpha < \lfloor \alpha \rfloor + 1. \tag{10.1}$$

THEOREM 10.2 $\forall a \in Z \, \forall b \in Z^+ \, \exists q, r \in Z : a = q \cdot b + r \text{ where } 0 \le r < b.$

PROOF   From 10.1, it follows that

$$0 \le \frac{a}{b} - \left\lfloor \frac{a}{b} \right\rfloor < 1.$$

Since $b > 0$, this gives

$$0 \le a - b \cdot \left\lfloor \frac{a}{b} \right\rfloor < b.$$

By taking $q = \lfloor a/b \rfloor$ and $r = a - b \cdot \lfloor a/b \rfloor$, we obtain the result.   □

The reader has probably recognized q and r as the quotient and non-negative remainder of the division of a by b. If the remainder is zero, then a is a multiple of b or b divides a, written as $b \mid a$. We also say that b is a factor of a. If b is different from 1 and a, then b is a nontrivial factor or a proper divisor of a.

One can categorize the positive integers based on the number of distinct positive divisors. The element 1 has one positive divisor. **Primes** have exactly two, and the other elements, called **composites**, have more than two. In cryptography, a prime number is usually denoted as p or q. Theorem 10.3 is easy to prove using induction and the following lemma.

LEMMA 10.1   When n is composite, the least positive nontrivial factor of n is a prime.

PROOF   Let $a, b, c \in Z^+$. It is easy to prove that if $b \mid a$ and $c \mid b$, then $c \mid a$. Also, if b is a nontrivial factor of a, then $1 < b < a$. Using contradiction, this implies the lemma.   □

THEOREM 10.3   *If $n \geq 2$, then n is the product of primes.*

10.2.2.2 Greatest Common Divisor

Using a definition other than the standard one and proving these to be equivalent will allow us to introduce several important results. First, we will define what an integral modulus is.

DEFINITION 10.7   A modulus is a subset of the integers which is closed under addition and subtraction. A modulus that contains only 0 is called the zero modulus.

THEOREM 10.4   *For each non-zero modulus, a positive integer d exists such that all the elements of the modulus are multiplies of d.*

PROOF   If such an element exists, it must clearly be the least positive integer. We will now prove its existence by contradiction. Suppose that a is not a multiple of d, then, using Theorem 10.2, we have $a = qd + r$, where q and r are integers such that $1 \leq r < d$. Clearly, r is then an element of the modulus and is smaller than d. We obtain a contradiction.   □

It is obvious that if a and b are integers, then $am + bn$, where m and n are any integers, forms a modulus. (For those familiar with ring theory, it implies that this modulus is an ideal.) We are now ready to define the greatest common divisor.

DEFINITION 10.8   When a and b are integers and are not zero, then the least positive integer d in the modulus $am + bn$ is the greatest common divisor of a and b, denoted as $\gcd(a, b)$ or (a, b). If $(a, b) = 1$, a and b are called co-prime.

COROLLARY 10.1   Theorem 10.4 implies:

$$\exists x, y \in Z : ax + by = \gcd(a, b) \tag{10.2}$$

$$\forall x, y \in Z : \gcd(a, b) \mid ax + by \tag{10.3}$$

$$\text{If } c \mid a \text{ and } c \mid b, \text{ then } c \mid \gcd(a, b). \tag{10.4}$$

PROOF   10.2 and 10.3 follow from Theorem 10.4 and 10.2 implies 10.4.   □

Due to 10.4, the definition of the greatest common divisor is equivalent to the traditional one.

10.2.2.3 Congruences

In Section 9.2.1, we defined what it means for two numbers to be equivalent modulo n. It is easy to see that this satisfies the definition of equivalence relation. The corresponding equivalence classes are called residue classes. When a is an integer, we let $\hat{a} = \{b \mid b \equiv a \bmod n\}$. When we work modulo n, we call $Z_n = \{\hat{0}, \hat{1}, \ldots, \widehat{n-1}\}$ the set of all residue classes. Other notations for Z_n, which we do not explain, are $Z/(n)$ and Z/nZ. The following theorem is trivial to prove.

THEOREM 10.5   *If $a \equiv b \pmod{n}$ and $c \equiv d \pmod{n}$, then $a + c \equiv b + d \pmod{n}$ and $a * c \equiv b * d \pmod{n}$.*

Due to this theorem, we can say that if $a \equiv b \bmod n$, that a and b are congruent modulo n. Now let A and B be two residue classes, and we define $A + B = \{a + b \mid a \in A \text{ and } b \in B\}$ and similarly $A * B$. Theorem 10.5 tells us that $A + B$ and $A * B$ are residue classes. This implies:

COROLLARY 10.2 Z_m *is a commutative ring.*

PROOF   Given that Z is a commutative ring, this follows easily from Theorem 10.5.   □

If one selects one representative out of each residue class, we call the resulting set of integers a complete residue system. It is easy to see that $\{0, 1, \ldots, m-1\}$ is a complete residue system. Adding (multiplying) two integers a and b in this complete residue system is easy by adding (multiplying) them as integers and taking the non-negative remainder after division by m. If c is the result, we write $c = a + b \bmod n$ ($c = a * b \bmod n$), as was done in Section 9.2.1. Using this addition (multiplication), one can view the ring Z_n as corresponding to the set $\{0, 1, \ldots, n-1\}$. We do not discuss this formally.

THEOREM 10.6   *Let $n \geq 2$. The element $a \in Z_n$ has an inverse modulo n if and only if $\gcd(a, n) = 1$.*

PROOF   If $\gcd(a, n) = 1$, then 10.2 implies that integers x and y exist such that $xa + yn = 1$, or $xa = 1 \bmod n$. So $x \equiv a^{-1}$.

Now if an inverse $a^{-1} \in Z_n$ exists, then $a \cdot a^{-1} \equiv 1 \bmod n$. Using the definition of congruence, this means that $n \mid (a \cdot a^{-1} - 1)$ or that $a \cdot a^{-1} - 1 = yn$, where y is an integer. From 10.3, this implies that $\gcd(a, n) \mid 1$.   □

COROLLARY 10.3 Z_n^*, *the set of elements in Z_n relatively prime to n is an Abelian group for the multiplication.*

COROLLARY 10.4   *When p is a prime, Z_p is a finite field.*

COROLLARY 10.5   *If $\gcd(a, n) = 1$, then the equation $ax \equiv b \bmod n$ has exactly one solution modulo n.*

PROOF   Assume that we had different solutions modulo n. Let us say x_1 and x_2. Then $ax_1 \equiv ax_2 \bmod n$, and since a has an inverse, we obtain that $x_1 \equiv x_2 \bmod n$. This is a contradiction.   □

10.2.2.4 Euler–Fermat's Theorem

The Euler–Fermat theorem is probably the most important theorem for understanding the RSA cryptosystem (see Section 10.3.2). We first give the following definition.

DEFINITION 10.9   The order of Z_n^* is denoted as $\phi(n)$. The function ϕ is called the Euler-totient function, or Euler function. If one selects one representative out of each residue class co-prime to n, we call the resulting set of integers a reduced residue system.

LEMMA 10.2   If $\{a_1, a_2, \ldots, a_{\phi(n)}\}$ is a reduced residue system, and $\gcd(k, n) = 1$, then $\{ka_1, ka_2, \ldots, ka_{\phi(n)}\}$ is a reduced residue system.

PROOF   First, $\gcd(ka_i, n) = 1$. Second, if $i \neq j$, then $ka_i \not\equiv ka_j \bmod n$, by contradiction.   □

This lemma implies the Euler–Fermat theorem.

THEOREM 10.7 $\forall b \in Z_n^* : b^{\phi(n)} \equiv 1 \bmod n$

PROOF   Let $\{a_1, a_2, \ldots, a_{\phi(n)}\}$ be a reduced residue system. Lemma 10.2 implies that

$$\prod_{h=1}^{\phi(n)} (ba_h) \equiv \prod_{h=1}^{\phi(n)} a_h \bmod n. \tag{10.5}$$

Since $\gcd(a_h, n) = 1$, a_h^{-1} exists, so:

$$\prod_{h=1}^{\phi(n)} \left(a_h^{-1} a_h b\right) \equiv \prod_{h=1}^{\phi(n)} a_h^{-1} a_h \bmod n, \text{ implying}$$

$$b^{\phi(n)} \equiv 1 \bmod n$$

□

COROLLARY 10.6   If $m \in Z_n^*$, then $m^{\phi(n)+a} \equiv m^a \bmod n$.

It is easy to see that when p is a prime, $\phi(p) = p - 1$. The next corollary is known as Fermat's little theorem.

COROLLARY 10.7   Let p be a prime. $\forall b \in Z_p : b^p \equiv b \bmod p$

10.3 Encryption Schemes

We will explain some encryption schemes from a mathematical viewpoint without addressing the algorithms needed. These are explained in Section 10.4.

10.3.1 Discrete Log

Given an element $b \in \langle a \rangle$, a cyclic group, we know that a k exists such that $b = a^k$ in this cyclic group. This k is called the discrete logarithm of b in the base a and often denoted as $k = \log_a(b)$. To have any value to cryptography, it must be hard to find this k. So, a proper group needs to be chosen, which we discuss later.

One of the first schemes that was based on discrete logarithm is a key distribution scheme. Here, we discuss the ElGamal encryption scheme [16].

10.3.1.1 Generating a Public Key

We assume that a finite group $\langle g \rangle (\cdot)$ has been chosen of a large enough order, and that q, a multiple of the order of the $\mathrm{ord}(g)$, is given (it is sufficient that not too large an upperbound on q is known). Note that q is not necessarily a prime. For simplicity, we assume that q is public. This information could be part of a person's public key. We also assume that the group operation $(\cdot)$ and the inverse of an element can be computed in (expected) polynomial time.

When Alice wants to generate her public key, she chooses a uniform random $a \in Z_q$ and computes $y_A = g^a$ in this group and makes y_A public.

10.3.1.2 ElGamal Encryption

To encrypt a message $m \in \langle g \rangle$ (otherwise a hybrid scheme is used, see Section 10.7), the sender finds the public key y_A of the receiver. The sender chooses* a uniformly random $k \in Z_q$ and sends as ciphertext $c = (c_1, c_2) = (g^k, m \cdot y_A^k)$ computed in $\langle g \rangle$.

10.3.1.3 ElGamal Decryption

To decrypt the ciphertext, the legitimate receiver knowing the secret key a computes $m' = c_2 \cdot (c_1^a)^{-1}$. It is easy to verify that $m' = m \cdot (g^a)^k \cdot (g^k)^{-a} = m$.

10.3.1.4 Suitable Group

As we already mentioned, to have any cryptographic value, the discrete logarithm must be hard. Unfortunately, there is no proof that the discrete logarithm is a hard problem. One can only state that so far no one has found an algorithm running in polynomial time for the general problem. For some groups, the problem is a little easier than in the general case. For some, the discrete logarithm is even easy. Indeed, for example, for the cyclic group $Z_n(+)$, the discrete logarithm corresponds with finding x such that $ax = b \bmod n$, which is easy as we will discuss in Sections 10.4.3 and 10.5.3.

Groups that are used in practice are the multiplicative group of a finite field (see also Section 10.5.9), or a subgroup of it and elliptic curve groups.

10.3.2 RSA

RSA, which is a heuristic cryptosystem, is an acronym for the inventors of the scheme, Rivest et al. [35]. It is basically an application of the Euler–Fermat theorem.

---

* It is possible that q is secret. For example, when the group $\langle g \rangle$ was selected by the one who constructed the public key. We then assume that a not too large upperbound on q is public and is used instead of q.

10.3.2.1 Generating a Public Key

To select her public key, Alice chooses two random primes p and q, large enough, and multiplies these to obtain $n = p \cdot q$. She chooses a random element $e \in Z^*_{\phi(n)}$ uniformly and computes $d = e^{-1} \bmod \phi(n)$. She publishes (n, e) as public key and keeps d as a secret key. Note that p, q and $\phi(n)$ need to remain secret.

10.3.2.2 Encryption

To encrypt a message, the sender finds the public key of the receiver, which we call (n, e). When $m \in Z_n$ (otherwise a hybrid scheme is used, see Section 10.7), the resulting ciphertext $c = m^e \bmod n$.

10.3.2.3 Decryption

To decrypt a ciphertext $c \in Z_n$, the legitimate receiver, let us say Alice, knowing her secret key d computes $m' = c^d \bmod n$.

We will now explain why the decryption works for the case $m \in Z^*_n$. This can easily be generalized for all $m \in Z_n$, using the Chinese Remainder theorem (see Theorem 10.22). Since $e \cdot d = 1 \bmod \phi(n)$, we have $e \cdot d = 1 + k\phi(n)$ for some integer k. So, due to Euler–Fermat's theorem (see Corollary 10.6), when $m \in Z^*_n$, we have that $c^d = m^{ed} = m^{k \cdot \phi(n)+1} = m \bmod n$. So $m' = m$.

10.3.2.4 Notes

RSA is also very popular as a signature scheme (see Chapter 12).

To speed up encryption, it has been suggested to choose $e = 3$, or a small e, or an e such that $w(e)$, the Hamming* weight of e, is small. Several attacks have been presented against such solutions. To avoid these, it seems best to choose e as described.

When m is chosen as a uniformly random element in Z_n, and p and q are large enough, no attack is known for finding m. It has been argued, without proof, (see Section 10.5.2 for more details) that this is as hard as factoring n.

10.4 Computational Number Theory: Part 1

When describing the RSA and the ElGamal public key systems, we silently assume many efficient algorithms. Indeed for the ElGamal system, we need efficient algorithms to

- Select an element of large enough order (when choosing the public key),
- Select a uniformly random element in Z_q (when generating a public key and when encrypting),
- Raise an element in a group to a power (when constructing the public key, when encrypting and decrypting),
- Multiply two elements in the group (when encrypting and decrypting),
- Compute an inverse, for example in Z^*_p (when decrypting), and
- Guarantee that $m \in \langle g \rangle$ (when encrypting).

For RSA we need algorithms to

- Select a random prime (when generating a public key),
- Multiply two large integers (when constructing a public key),

---

* The Hamming weight of a binary string is the number of ones in the string.

- Compute $\phi(n)$, or a multiple of it (when constructing a public key),
- Randomly select an element in $Z^*_{\phi(n)}$ (when constructing a public key),
- Compute an inverse (when constructing a public key), and
- Raise an element in a group to a power (when encrypting and decrypting).

We now discuss the necessary algorithms. To avoid repetition, we will proceed in a different order than listed above. If necessary, we will discuss on more number theory before giving the algorithm.

10.4.1 Multiplication and Modulo Multiplication

Discussing fast integer multiplication and modulo multiplication is a chapter in itself, and therefore, beyond the scope of this chapter. Although algorithms based on FFT are order wise very fast, the numbers used in modern cryptography are too small to compensate for the rather large constants in FFT based algorithms.

The algorithms used are the trivial ones learned in elementary school. However, usually the base is a power of 2, instead of 10.

10.4.2 Fast Exponentiation

Assume that we have a semigroup $M(*)$ in which an efficient algorithm exists for multiplying two elements. Since squaring an element might be faster [24] than multiplying, we allow for a separate algorithm square to square an element.

Input declaration: an element $a \in M$ and b a positive integer.
Output declaration: a^b in $M(*)$

function fastexpo(a, b)
begin
case

$b = 1$	then	fastexpo:= a
$b > 1$ and odd	then	fastexpo:= $a*$**fastexpo**$(a, b-1)$
b is even	then	fastexpo:= **square**(**fastexpo**$(a,b/2)$)

end

The above function uses, at maximum, $2|b|$ multiplications, where $|b|$ is the binary length of b.

10.4.3 Gcd and Modulo Inverses

Let $a \geq b \geq 0$. The algorithm to compute the greatest common divisor goes back to Euclid and is, therefore, called the Euclidean algorithm. It is based on the observation that $\gcd(a, b) = \gcd(a-b, b)$. This trick can be repeated until $0 \leq r = a - mb < b$, which exists (see Theorem 10.2). This gives the modern version of the Euclidean algorithm:

Input declaration: non-negative integers a, b where $a \geq b$
Output declaration: $\gcd(a, b)$

function gcd(a, b)
begin

$$\text{if } b = 0 \quad \text{then} \quad gcd := a$$
$$\text{else} \quad gcd := \mathbf{gcd}(b, a \bmod b)$$
end

Note: We assume that $a \bmod b$ returns the least non-negative remainder of a when divided by b. When $a = b = 0$, the function returns 0.

THEOREM 10.8 *The number of divisions required to compute the $gcd(a, b)$ when $0 < b < a$, is, at maximum, $1 + \lfloor \log_R a \rfloor$, where $R = (1 + \sqrt{5})/2$.*

PROOF We denote $r_{-1} = a$, $r_0 = b$, and r_i the remainder obtained in the ith division. Note that the $gcd(a, b)$ will be such a remainder. We define n such that $r_{n-1} = gcd(a, b)$, implying that $r_n = 0$. We call the ith quotient d_i. So, we have:

$$d_i = \left\lfloor \frac{r_{i-2}}{r_{i-1}} \right\rfloor \quad (1 \le i \le n) \tag{10.6}$$

$$r_{i-2} = d_i \cdot r_{i-1} + r_i \quad (1 \le i \le n) \tag{10.7}$$

$$r_{n-2} = d_n \cdot r_{n-1}.$$

Note that $r_i < r_{i-1}$ ($0 \le i \le n$), $r_{n-1} \ge 1 = f_2$, and $r_{n-2} \ge 2 = f_3$, where f_n is the nth Fibonacci number. Using this and the fact that $f_{n+1-i} = f_{n-i} + f_{n-i-1}$ and 10.7, it is easy to prove with inverse induction, starting from $i = n - 1$, that $r_i \ge f_{n+1-i}$. So, $a \ge f_{n+2}$ and $b \ge f_{n+1}$. Since $r_{n-2} \ge 2$, this implies the following. When the Euclidean algorithm takes exactly n divisions to compute $gcd(a, b)$, we have $a \ge f_{n+2}$ and $b \ge f_{n+1}$. So, if $b' < a' < f_{n+2}$, then there are, at maximum, $n - 1$ divisions when computing the greatest common divisor of a' and b' using the Euclidean algorithm.

Now since $R^2 = R + 1$, we have $R^{n-1} = R^{n-2} + R^{n-3}$. This implies, using induction, that $\forall n \ge 1 : R^{n-2} \le f_n \le R^{n-1}$. So, if $0 < b' < a'$ and $R^{n-2} < a' < R^n$, there are, at maximum, $n - 1$ divisions. This corresponds to saying that if $n - 2 < \log_R(a') < n$, or $n - 2 \le \lfloor \log_R(a') \rfloor < n$ there are, at maximum, $n - 1$ divisions.

An extended version of the Euclidean algorithm allows computing the inverse of b modulo a, where $0 < b < a$.

ALGORITHM 10.1
Input: positive integers a, b where $a > b > 0$
Output: $g = gcd(a, b)$, x and y, integers such that $ax + by = gcd(a, b)$ and $c = b^{-1} \bmod a$ if it exists (otherwise $c = 0$)
begin
$r_{-1} = a; r_0 = b; x_{-1} := 1; y_{-1} := 0; x_0 := 0; y_0 := 1; i := 0;$
while $r_i \ne 0$ **do**
    begin

$$
\begin{aligned}
i \quad &:= \quad i + 1; \\
d \quad &:= \quad \left\lfloor \frac{r_{i-2}}{r_{i-1}} \right\rfloor; \\
r_i \quad &:= \quad r_{i-2} - d * r_{i-1}; \\
x_i \quad &:= \quad x_{i-2} - d * x_{i-1}; \\
y_i \quad &:= \quad y_{i-2} - d * y_{i-1};
\end{aligned}
$$

    end

$g := r_{i-1}; x := x_{i-1}; y := y_{i-1};$
if $g = 1$ then

$$\text{if } i \bmod 2 = 1 \quad \text{then} \quad c := y$$
$$\text{else} \quad c := y + a$$

else $c := 0;$

end

Observe that only $d, r_i, r_{i-1}, x_i, x_{i-1}, y_i,$ and y_{i-1} are needed. So, a more careful implementation allows saving memory.

LEMMA 10.3   For $i \geq -1$, we have in Algorithm 10.1 that

$$ax_i + by_i = r_i. \tag{10.8}$$

PROOF   We use the definition of d_i in 10.6. It is easy to verify that 10.8 holds for $i = -1$ and $i = 0$. For larger values, we use induction and assume that 10.8 has been proven for $i - 1$ and $i - 2$. Observe that $x_i = x_{i-2} - d_i x_{i-1}$ and $y_i = y_{i-2} - d_i y_{i-1}$. So

$$ax_i + by_i = a(x_{i-2} - d_i x_{i-1}) + b(y_{i-2} - d_i y_{i-1}) = (ax_{i-2} + by_{i-2}) - d_i(ax_{i-1} + by_{i-1}) = r_{i-2} - d_i r_{i-1}$$

using the induction hypothesis. Since $r_i = r_{i-2} - d_i * r_{i-1}$, we have proven the claim. $\qquad\square$

THEOREM 10.9   *In Algorithm 10.1, $c = b^{-1} \bmod a$ if $b^{-1} \bmod a$ exists and $0 \leq c < a$.*

PROOF   We assume that $\gcd(a, b) = 1$. From the proof of Theorem 10.8 and Lemma 10.3, we have for the x and y returned by Algorithm 10.1 that $ax + by = \gcd(a, b)$. This implies that $c \equiv b^{-1} \bmod a$. So, we only need to prove that $0 \leq c < a$.

We let n be as in the proof of Theorem 10.8 and d_i be as in 10.6. We first claim that

$$(-1)^i y_i \geq 0 \tag{10.9}$$

when $-1 \leq i \leq n$. This is easy to verify for $i = -1$ and $i = 0$. We now assume that 10.9 is true for $i = k - 1$ and $i = k - 2$. Since $d_i > 0$ when $0 < b < a$ and using the recursive definition of y_i, we have $(-1)^k y_k = (-1)^k (y_{k-2} - d_k y_{k-1}) = (-1)^{k-2} y_{k-2} + (-1)^{k-1} d_k y_{k-1} \geq 0$ by the induction hypothesis.

Due to 10.9, we have $|y_i| = |y_{i-2}| + d_i * |y_{i-1}|$. Since $y_0 = 1$ and $d_i \geq 1$, we have

$$|y_1| \geq |y_0| \quad \text{and} \quad |y_i| > |y_{i-1}| \quad \text{for } 2 \leq i \leq n, \tag{10.10}$$

as is easy to verify using induction.

Finally, we claim that for all i ($0 \leq i \leq n$):

$$y_{i-1} r_i - y_i r_{i-1} = (-1)^{i+1} a \tag{10.11}$$

$$x_{i-1} r_i - x_i r_{i-1} = (-1)^i b \tag{10.12}$$

For $i = 0$, 10.11 is trivially satisfied. Assume that the equations are satisfied for $i = k - 1$. Noticing that $r_k = r_{k-2} - d_k r_{k-1}$ and $y_k = y_{k-2} - d_k y_{k-1}$, we obtain for $i = k$ that

$$y_{k-1} r_k - y_k r_{k-1} = y_{k-1}(r_{k-2} - d_k r_{k-1}) - (y_{k-2} - d_k y_{k-1}) r_{k-1}$$
$$= y_{k-1} r_{k-2} - y_{k-2} r_{k-1}$$
$$= -(y_{k-2} r_{k-1} - y_{k-1} r_{k-2}),$$

which, using the induction hypothesis, proves 10.11. Similarly, one can prove 10.12.

Since $r_n = 0$ and $r_{n-1} = \gcd(a, b) = 1$, 10.11 and 10.12 imply that $y_n = (-1)^n a$ and $x_n = (-1)^{n+1} b$, respectively. So if $n \geq 2$ then, using 10.10, $|y_{n-1}| < a$ and $|y_{n-1}| \neq 0$. If $n = 1$, then by the definition $x_1 = x_{-1} - d_1 * x_0 = 1$ and by 10.12 $x_1 = b$, so $b = 1$ and $y_{n-1} = y_0 = 1 < a$. Thus, $0 < |y_{n-1}| < a$. Using this and 10.9, we obtain that if n is odd $c = y_{n-1} < a$ and $c > 0$, else $-a < y_{n-1} < 0$, so $0 < c = a + y_{n-1} < a$.   □

10.4.4 Random Selection

We assume that the user has a binary random generator which outputs a string of independent bits with uniform distribution. Using this generator to output one bit is called a coin flip.

We will now describe how to select a natural number a with uniform distribution such that $0 \leq a \leq b$, where $2^{k-1} \leq b < 2^k$. One lets the generator output k bits. We view these bits as a binary representation of an integer x. If $x \leq b$, one outputs $a = x$, else one flips k new bits until $x \leq b$. The expected number of coin flips is bounded above by $2k$.

The case one requires that $b_1 \leq a \leq b_2$ is easy to reduce to the previous one. Indeed, choose a' uniformly such that $0 \leq a' \leq (b_2 - b_1)$ and add b_1 to the result.

Selecting an element in Z_n^* can be done by selecting an integer a such that $0 \leq a \leq n - 1$ and by repeating the procedure until $\gcd(a, n) = 1$. The expected number of coin flips is $O(\log \log(n) \cdot \log(n))$, which we do not prove.

Before we discuss how to select a prime, we will discuss more on number theory. This number theory will also be useful to explain the first probabilistic encryption scheme (see Section 10.7.1).

10.5 More Algebra and Number Theory

10.5.1 Primes and Factorization

LEMMA 10.4   If p is a prime and $p \mid a \cdot b$, then p divides a or b.

PROOF   If $p \nmid a$, then $\gcd(a, p) = 1$. So, due to Corollary 10.1, integers x, y exist such that $xa + yp = 1$, or $xab + ybp = b$. Since $p \mid ab$, this implies that $p \mid b$.   □

THEOREM 10.10   *Prime factorization of any integer n is unique, i.e., the primes $p_1, \ldots, p_k$ and the integers $a_i \geq 1$ such that*

$$n = \prod_{i=1}^{k} p_i^{a_i}$$

are unique.

PROOF   From Theorem 10.3, we know that n is a product of primes. Assume that

$$\prod_{i=1}^{k} p_i^{a_i} = \prod_{j=1}^{m} q_j^{b_j}, \tag{10.13}$$

where q_j are primes and $b_j \geq 1$. First, due to Lemma 10.4 and the fact that p_i and q_j are primes, a p_i is equal to some q_j and vice versa (by induction). So, $k = m$ and when renaming $p_i = q_i$. If $a_i > b_i$ and dividing both sides of 10.13 by $p_i^{b_i}$, we obtain a contradiction due to Lemma 10.4.   □

COROLLARY 10.8 The least common multiple of a and b, denoted as $\text{lcm}(a, b) = a \cdot b / \gcd(a, b)$.

PROOF The proof is left as an exercise. □

10.5.2 Euler-Totient Function

The algorithm used to compute $\phi(n)$ when constructing an RSA public key is trivially based on the following theorem. First, we define what, in number theory, is called a multiplicative function, and we also introduce the Möbius function.

DEFINITION 10.10 A real or a complex valued function defined on the positive integers is called an arithmetical or a number-theoretic function. An arithmetical function f is multiplicative if it is not the zero function, and if for all positive integers m and n where $\gcd(m, n) = 1$, we have $f(m \cdot n) = f(m) \cdot f(n)$. If the gcd restriction is removed, it is *completely multiplicative*.

DEFINITION 10.11 A square is an integer to the power 2. The Möbius function is defined as being

$$\mu(n) \quad = \quad \begin{cases} 0 & \text{if } n \text{ is divisible by a square different from 1,} \\ (-1)^k & \text{if } n = p_1 \cdot p_2 \cdots p_k, \text{ where } p_i \text{ are distinct primes.} \end{cases}$$

A number is squarefree if $\mu(n) \neq 0$.

LEMMA 10.5 If $n \geq 1$, we have

$$\sum_{d|n} \mu(d) = \left\lfloor \frac{1}{n} \right\rfloor = \begin{cases} 1 & \text{if } n = 1, \\ 0 & \text{if } n > 1. \end{cases} \tag{10.14}$$

PROOF When $n = 1$, $\mu(1) = 1$ since it is the product of 0 different primes. We now discuss the case that $n > 1$, and then accordingly to Theorem 10.10 $n = \prod_{i=1}^{k} p_i^{a_i}$, where p_i are distinct primes. In 10.14, only when d is squarefree, $\mu(d) \neq 0$, so

$$\sum_{d|n} \mu(d) = \mu(1) + \mu(p_1) + \cdots + \mu(p_k) + \mu(p_1 \cdot p_2) + \cdots + \mu(p_{k-1} \cdot p_k) + \cdots + \mu(p_1 \cdot p_2 \cdots p_k)$$

$$= 1 + \binom{k}{1} \cdot (-1) + \binom{k}{2} \cdot (-1)^2 + \cdots + \binom{k}{k}(-1)^k = (1 - 1)^k = 0$$

□

LEMMA 10.6 If $n \geq 1$ then

$$\phi(n) = n \cdot \sum_{d|n} \frac{\mu(d)}{d}.$$

PROOF From the definition of the Euler-totient function, we immediately have

$$\phi(n) = \sum_{k=1}^{n} \left\lfloor \frac{1}{\gcd(k, n)} \right\rfloor.$$

This can be rewritten, using Lemma 10.5, as

$$\phi(n) = \sum_{k=1}^{n} \sum_{d|\gcd(k,n)} \mu(d) = \sum_{k=1}^{n} \sum_{\substack{d|k \\ d|n}} \mu(d).$$

In the second sum, all d divide n. For such a fixed d, k must be a multiple of d, i.e., $k = md$. Now, $1 \le k \le n$, which is equivalent to the requirement that $1 \le m \le n/d$. Thus,

$$\phi(n) = \sum_{d|n} \sum_{m=1}^{\frac{n}{d}} \mu(d) = \sum_{d|n} \mu(d) \sum_{m=1}^{\frac{n}{d}} 1 = \sum_{d|n} \mu(d) \cdot \frac{n}{d}.$$

□

THEOREM 10.11 $\phi(n) = n \cdot \prod_{p|n} \left(1 - \frac{1}{p}\right)$, *where p are distinct primes.*

PROOF   When $n = 1$, no primes divide n, so the product is equal to 1. We will now consider the case where $n > 1$, so $n = \prod_{i=1}^{m} p_i^{a_i}$, where p_i are distinct primes and $a_i \ge 1$. Obviously,

$$\prod_{i=1}^{m} \left(1 - \frac{1}{p_i}\right) = 1 - \sum_{i=1}^{m} \frac{1}{p_i} + \sum_{\substack{i,j \\ i \ne j}} \frac{1}{p_i p_j} - \sum_{\substack{i,j,k \\ i \ne j \\ i \ne k \\ j \ne k}} \frac{1}{p_i p_j p_k} + \cdots + \frac{(-1)^m}{p_1 p_2 \cdots p_m}. \qquad (10.15)$$

Each term on the right hand side corresponds to a $\pm 1/d$, where d is a squarefree divisor of n. For those, the numerator is $\mu(d)$. For the other divisors of n, $\mu(d) = 0$, so the sum in 10.15 is $\sum_{d|n}(\mu(d)/d)$. Using Lemma 10.6, we have proven the theorem.   □

COROLLARY 10.9 $\phi(n)$ is multiplicative.

This corollary implies that, given the factorization of n, it is easy to compute $\phi(n)$. Assuming the correctness of the Extended Riemann Hypothesis and given $\phi(n)$, one can factor n in polynomial time, which we will not prove.

10.5.3 Linear Equations

THEOREM 10.12   *The equation*

$$ax = b \bmod n \qquad (10.16)$$

has no solution if $\gcd(a, n) \nmid b$, else it has $\gcd(a, n)$ solutions.

PROOF   If $ax = b \bmod n$, an integer y exists such that $ax + ny = b$. Statement 10.3 in Corollary 10.1 tells us that $\gcd(a, n)$ must divide b. From now on, we assume it does. We will prove that there are then $\gcd(a, n)$ solutions.

Corollary 10.5 tells us that if $\gcd(a, n) = 1$, the solution is unique. Suppose now that $d = \gcd(a, n)$ and $ax = b \bmod n$, then $d \mid n \mid (ax - b)$, which implies that

$$\frac{a}{d}x = \frac{b}{d} \bmod \frac{n}{d}.$$

This has a unique solution x_0 modulo n/d, since $\gcd(a/d, n/d) = 1$. Using the definition of congruence, this implies that x_0 is a solution of 10.16. Another implication is that all other solutions modulo n must have the form

$$x = x_0 + m \cdot \frac{n}{d} \quad \text{where } 0 \le m \le d - 1.$$

$\square$

10.5.4 Polynomials

The following theorem plays an important role in secret sharing (see Chapter 13).

THEOREM 10.13 *Let p be a prime and f a polynomial. The number of solutions (including repeating ones) of $f(x) = 0 \bmod p$ is, at maximum, the degree of f.*

PROOF The theorem is trivially satisfied when the number of solutions is zero. If a_1 is a solution, then $f(x) = (x - a_1)q_1(x) + r_1(x)$, where $r_1(x)$ is the remainder, i.e., has degree zero. Since $p \mid f(a_1)$ and $p \mid (a_1 - a_1)q_1(a_1)$, $r_1(x) = 0 \bmod p$. If a_1 is a solution of $q_1(x)$ obtain, $f(x) = (x - a_1)^2 q_1'(x) \bmod p$. In general, an $f_1(x)$ exists such that $f(x) = (x - a_1)^{h_1} f_1(x) \bmod p$ and $f_1(a_1) \not\equiv 0 \bmod p$. Clearly, the degree of $f_1(x)$ is $m - h_1$. Suppose that a_2 is another solution modulo p, then, $f(a_2) = (a_2 - a_1)^{h_1} f_1(a_2) = 0 \bmod p$. Since $a_2 \not\equiv a_1 \bmod p$, Lemma 10.4 implies $p \mid f_1(a_2)$, or, in other words, that $f_1(a_2) \bmod p$. So, using a similar argument as previously, we obtain that $f(x) = (x - a_1)^{h_1}(x - a_2)^{h_2} f_2(x)$, where a_1 and a_2 are solutions of $f_2(x) = 0$. This gives (using induction) the following:

$$f(x) = \left(\prod_{i=1}^{l} (x - a_i)^{h_i} \right) f_l(x),$$

where $f_l(x) = 0 \bmod p$ has no solutions. This immediately implies the theorem. $\square$

When p is composite, the theorem does not extend, as we will briefly discuss in Section 10.5.7.

10.5.5 Quadratic Residues

DEFINITION 10.12 The set of quadratic residue modulo n is denoted as $QR_n = \{x \mid \exists y \in Z_n^* : x \equiv y^2 \bmod n\}$. The set of quadratic nonresidues modulo n is $QNR_n = Z_n^* \setminus QR_n$.

THEOREM 10.14 *If p is an odd prime, then $|QR_p| = |QNR_p| = (p - 1)/2$.*

PROOF First $x^2 = a \bmod p$ has, at maximum, two solutions due to Theorem 10.13. Now, squaring all elements of Z_p^* and noticing that $x^2 \equiv (-x)^2 \bmod p$, we obtain the result. $\square$

DEFINITION 10.13   Let p be an odd prime. The Legendre symbol

$$\left(\frac{a}{p}\right) = (a \mid p) = \begin{cases} 1 & \text{if } a \in QR_p, \\ -1 & \text{if } a \in QNR_p, \\ 0 & \text{if } a \equiv 0 \bmod p. \end{cases}$$

We will now discuss Euler's criterion.

THEOREM 10.15   *Let p be an odd prime. $(a \mid p) = a^{(p-1)/2} \bmod p$.*

PROOF   The case $a \equiv 0 \bmod p$ is trivial, which we will now exclude. If $(a \mid p) = 1$, then an integer x exists such that $x^2 = a \bmod p$, so $a^{(p-1)/2} = x^{p-1} = 1 \bmod p$ due to Fermat's little theorem (Corollary 10.7). Since $|QR_p| = (p-1)/2$, the polynomial equation

$$a^{\frac{p-1}{2}} = 1 \bmod p \tag{10.17}$$

in a has at least $(p-1)/2$ solutions (see Theorem 10.14), and using Theorem 10.13, this implies exactly $(p-1)/2$ solutions. So, if $a^{(p-1)/2} = 1 \bmod p$, then $(a \mid p) = 1$.

Similarly, using Fermat's little theorem, the equation

$$a^{p-1} - 1 = 0 \bmod p \tag{10.18}$$

in a has exactly $p-1$ solutions. Now $a^{p-1}-1 = (a^{(p-1)/2}-1)(a^{(p-1)/2}+1)$. So, the $(p-1)/2$ solutions of 10.18 that are not solutions of 10.17 must be the solutions of the equation $(a^{(p-1)/2}+1) = 0 \bmod p$. Since $|QNR_p| = (p-1)/2$, and all solutions of 10.17 are quadratic residues, we have for all $a \in QNR_p$ that $(a^{(p-1)/2} + 1) = 0 \bmod p$. $\square$

COROLLARY 10.10   *If p is an odd prime, then $(a \mid p) \cdot (b \mid p) = (a \cdot b \mid p)$.*

COROLLARY 10.11   *If p is an odd prime, then $(-1 \mid p) = (-1)^{\frac{p-1}{2}}$.*

We now discuss Gauss' lemma.

THEOREM 10.16   *Let p be an odd prime, n an integer such that $p \nmid n$ and m the cardinality of the set*

$$A = \{a_i \mid a_i = k \cdot n \bmod p \text{ for } 1 \le k \le (p-1)/2 \text{ and } p/2 < a_i < p\}.$$

Then, $(n \mid p) = (-1)^m$.

PROOF   Let $B = \{b_k \mid b_k = k \cdot n \bmod p \text{ for } 1 \le k \le (p-1)/2\}$. We define $C = \{c_j\} = B \setminus A$, and let $|C| = l$. Observe that for all c_j and a_i we have that $p \ne a_i + c_j$, by contradiction. Indeed, otherwise $p \mid (a_i + c_j) = (k_i + k_j)n$, which is not possible. Therefore,

$$\left(\prod_{j=1}^{l} c_j\right) \cdot \left(\prod_{i=1}^{m} (p - a_i)\right) = \left(\frac{p-1}{2}\right)! \tag{10.19}$$

Now, trivially,

$$\left(\prod_{j=1}^{l} c_j\right) \cdot \left(\prod_{i=1}^{m} a_i\right) \equiv \prod_{k=1}^{\frac{p-1}{2}} (k \cdot n) = \left(\frac{p-1}{2}\right)! \cdot (n^{\frac{p-1}{2}}) \pmod{p}. \tag{10.20}$$

$$\equiv (-1)^m \cdot \left(\frac{p-1}{2}\right)! \pmod{p} \tag{10.21}$$

using 10.19 to obtain the last congruence. So, combining 10.20 and 10.21, we have $n^{(p-1)/2} \equiv (-1)^m \bmod p$, which gives the result using Euler's criterion. $\square$

COROLLARY 10.12   If p is an odd prime, then

$$\left(\frac{2}{p}\right) = \begin{cases} 1 & \text{if } p \equiv 1 \bmod 8 \text{ or } p \equiv 7 \bmod 8, \\ -1 & \text{if } p \equiv 3 \bmod 8 \text{ or } p \equiv 5 \bmod 8. \end{cases}$$

PROOF   It is easy to verify that when $n = 2$, the set $A = \{\lfloor (p+3)/4 \rfloor \cdot 2, \ldots, ((p-1)/2) \cdot 2\}$. So, $m = (p-1)/2 - \lfloor p/4 \rfloor$. Let $p = 8a + r$, where $r = 1, 3, 5,$ or 7. Then, m modulo 2 is respectively $0, 1, 1,$ and 0. $\square$

Using Gauss' lemma, one can prove the law of quadratic reciprocity. Since the proof is rather long (but not complicated), we refer the reader to the literature.

THEOREM 10.17   *If p and q are odd distinct primes, we have*

$$\left(\frac{p}{q}\right) \cdot \left(\frac{q}{p}\right) = (-1)^{\frac{(p-1)(q-1)}{4}}.$$

10.5.6 Jacobi Symbol

DEFINITION 10.14   Let $n = \prod_{i=1}^{h} p_i$, where p_i are (not necessarily distinct) primes and a an integer. The Jacobi symbol $(a \mid n) = \prod_{i=1}^{h}(a \mid p_i)$. The set $Z_n^{+1} = \{a \in Z_n \mid (a \mid n) = 1\}$.

Since $1 = p^0$, we have that $(a \mid 1) = 1$. Also, if n is a prime, it is obvious that the Jacobi symbol is the same as the Legendre symbol. We will discuss why there is no need to factor n to compute the Jacobi symbol. The following theorems are useful in this respect.

THEOREM 10.18   *The Jacobi symbol has the following properties:*

1. *If $a \equiv b \bmod n$, then $(a \mid n) = (b \mid n)$.*
2. *$(a \mid n) \cdot (a \mid m) = (a \mid n \cdot m)$.*
3. *$(a \mid n) \cdot (b \mid n) = (a \cdot b \mid n)$.*

PROOF   These follow immediately from the definition and Corollary 10.10. $\square$

THEOREM 10.19 *When n is an odd positive integer, we have $(-1 \mid n) = (-1)^{(n-1)/2}$.*

PROOF When a and b are odd, then trivially

$$(a-1)/2 + (b-1)/2 \equiv (ab-1)/2 \bmod 2. \tag{10.22}$$

This allows one to prove by induction that

$$\sum_{i=1}^{h} \frac{p_i - 1}{2} \equiv \frac{\left(\prod_{i=1}^{h} p_i\right) - 1}{2} \bmod 2. \tag{10.23}$$

The rest follows from Corollary 10.11. □

THEOREM 10.20 *If n is an odd positive integer, then*

$$\left(\frac{2}{n}\right) = \begin{cases} 1 & \text{if } n \equiv 1 \bmod 8 \text{ or } n \equiv 7 \bmod 8, \\ -1 & \text{if } n \equiv 3 \bmod 8 \text{ or } n \equiv 5 \bmod 8. \end{cases}$$

PROOF It is easy to verify that Corollary 10.12 can be rewritten as $(2 \mid p) = (-1)^{(p^2-1)/8}$. The rest of the proof is similar to that of the preceding theorem replacing 10.22 by

$$(a^2 - 1)/8 + (b^2 - 1)/8 \equiv (a^2 b^2 - 1)/8 \bmod 2,$$

where a and b are odd integers. Note that $k(k+1)$ is always even. □

THEOREM 10.21 *If m and n are odd positive integers and $\gcd(m, n) = 1$, then*

$$\left(\frac{m}{n}\right) \cdot \left(\frac{n}{m}\right) = (-1)^{\frac{(m-1)(n-1)}{4}}.$$

PROOF Let $m = \prod_{i=1}^{k} p_i$ and $n = \prod_{j=1}^{l} q_j$, where p_i and q_j are primes. From Theorem 10.18, we have

$$\left(\frac{m}{n}\right) \cdot \left(\frac{n}{m}\right) = \left(\prod_{i=1}^{k}\prod_{j=1}^{l}\left(\frac{p_i}{q_j}\right)\right) \cdot \left(\prod_{i=1}^{k}\prod_{j=1}^{l}\left(\frac{q_j}{p_i}\right)\right) = \left(\prod_{i=1}^{k}\prod_{j=1}^{l}\left(\frac{p_i}{q_j}\right) \cdot \left(\frac{q_j}{p_i}\right)\right)$$

$$= \prod_{i=1}^{k}\prod_{j=1}^{l}(-1)^{\frac{(p_i-1)(q_j-1)}{4}} = (-1)^{\frac{(m-1)(n-1)}{4}}$$

using Theorem 10.17 to obtain the second to the last equation and 10.23 to obtain the last. □

One could wonder whether $(a \mid n) = 1$ implies that $a \in QR_n$. Before disproving this, we discuss the Chinese Remainder Theorem.

10.5.7 Chinese Remainder Theorem

THEOREM 10.22 *If* $\gcd(n_1, n_2) = 1$, *then the system of equations*

$$x \equiv a_1 \bmod n_1 \tag{10.24}$$

$$x \equiv a_2 \bmod n_2 \tag{10.25}$$

has exactly one solution modulo $n_1 \cdot n_2$.

PROOF Due to 10.24 and the definition of modulo computation, x must have the form $x = a_1 + n_1 \cdot y$ for some y. Using 10.25, $a_1 + n_1 \cdot y \equiv a_2 \bmod n_2$, which has exactly one solution in y modulo n_2. This follows from Corollary 10.5 since $\gcd(n_1, n_2) = 1$. So, $n_1 y$ has only one solution modulo $n_1 \cdot n_2$, or $x = a_1 + n_1 \cdot y$ is unique modulo $n_1 \cdot n_2$. $\square$

As an application, we consider the equation $x^2 = a \bmod n$, where n is the product of two different primes, p and q. Since $p \mid n \mid (x^2 - a)$, a solution modulo n is also a solution modulo p (and q respectively). The Chinese Remainder Theorem tells us that it is sufficient to consider the solutions of $x^2 = a \bmod p$ and $x^2 = a \bmod q$, which we now discuss. If $(a \mid p) = 1$, then there are two solutions modulo p; when $(a \mid p) = -1$, there are no solutions modulo p; and finally when $(a \mid p) = 0$, there is one solution modulo p. So $a \in QR_n$ only if $(a \mid p) = 1$ and $(a \mid q) = 1$, implying that $(a \mid n) = 1$. However, the converse is obviously not true, so $(a \mid n) = 1$ does not necessarily imply that $a \in QR_n$.

10.5.8 Order of an Element

We describe some general properties of the order of an element.

LEMMA 10.7 Let $\langle \alpha \rangle (\cdot)$ be a cyclic group of order l, an integer. If $\alpha^m = 1$, then $l \mid m$.

PROOF From Theorem 10.1, we have that l is the smallest positive integer for which $\alpha^l = 1$. Assume that $l \nmid m$, then $m = ql + r$, where $0 < r < l$. So, $1 = \alpha^m = (\alpha^l)^q \cdot \alpha^r = 1 \cdot \alpha^r$. So, r is a smaller positive integer for which $\alpha^r = 1$. So, we have a contradiction. $\square$

LEMMA 10.8 Let $K(\cdot)$ be an Abelian group and $\alpha, \beta \in K$ with $k = \mathrm{ord}(\alpha)$ and $l = \mathrm{ord}(\beta)$. If $\gcd(k, l) = 1$, then $\mathrm{ord}(\alpha \cdot \beta) = \mathrm{ord}(\alpha) \cdot \mathrm{ord}(\beta)$.

PROOF Let us call m the order of $\alpha \cdot \beta$. Since $(\alpha\beta)^m = 1$, we have $\alpha^m = \beta^{-m}$. This implies that $\alpha^{lm} = \beta^{-lm} = (\beta^l)^{-m} = 1$ and $\beta^{-km} = (\alpha^k)^m = 1$. Using Lemma 10.7, we obtain that $k \mid lm$ and respectively $l \mid km$. Since $\gcd(k, l) = 1$, this implies that $k \mid m$ and $l \mid m$, and so, $kl \mid m$. Now, trivially, $(\alpha\beta)^{kl} = 1$. Using Lemma 10.7, this implies that kl must be the order of $\alpha\beta$. $\square$

THEOREM 10.23 Let $K(\cdot)$ be an Abelian group and let

$$m = \max_{\alpha \in K}(\mathrm{ord}(\alpha)).$$

We have that $\forall \beta \in K : \mathrm{ord}(\beta) \mid m$.

PROOF   We prove this by contradiction. When $\beta = 1$, the identity, the result is trivial. We now assume that $\beta \neq 1$. Let $d = \text{ord}(\beta)$ and suppose that $d \nmid m$. Then, a prime p and an integer $e \geq 1$ exit such that $p^e \mid d$, $p^e \nmid m$, but $p^{e-1} \mid m$. It is easy to verify that when $\text{ord}(\alpha) = m$ we have that $\text{ord}(\alpha^{p^{e-1}}) = m/(p^{e-1})$ and that $\text{ord}(\beta^{d/p^e}) = p^e$. Since $\gcd(m/(p^{e-1}), p^e) = 1$, Lemma 10.8 implies that $\text{ord}(\alpha^{p^{e-1}} \cdot \beta^{d/p^e}) = m \cdot p$, which is larger than m. So, we obtain a contradiction.   □

DEFINITION 10.15   The exponent, $\exp(K)$, of a finite group K is the smallest positive integer such that $\forall \beta \in K : \beta^{\exp(K)} = 1$.

COROLLARY 10.13   When K is a finite Abelian group, then $\exp(K) = \max_{\alpha \in K}(\text{ord}(\alpha))$.

The following theorem is used to prove that Z_p^*, where p is prime, is a cyclic group.

THEOREM 10.24   *Let K be a finite Abelian group. K is cyclic if and only if $\exp(K) = |K|$.*

PROOF   First, if $K = \langle \alpha \rangle$, then $|K| = \text{ord}(\alpha)$, so $\exp(K) = |K|$. Second, let K be a finite Abelian group such that $\exp(K) = |K|$. By Corollary 10.13, an element α exists such that $\exp(K) = \text{ord}(\alpha)$. Since $\exp(K) = |K|$, we have that $|K| = \text{ord}(\alpha) = |\langle \alpha \rangle|$, implying that K is cyclic.   □

10.5.9 Primitive Elements

Before proving that Z_p^* (p is a prime) is a cyclic group, we will define a primitive element.

DEFINITION 10.16   If the order of α in the group Z_n^* is $\phi(n)$, then we say that α is a primitive element of Z_n^*, or a primitive root $\bmod n$, or a generator of Z_n^*.

So, when Z_n^* has a primitive element, the group is cyclic.

THEOREM 10.25   *If p is prime, Z_p^* is cyclic.*

PROOF   Due to Fermat's little theorem (Corollary 10.7), we have that $\forall a \in Z_p^* : a^{|Z_p^*|} = 1$. So $\exp(Z_p^*) \leq |Z_p^*|$. By the definition of the exponent of a group, all elements of Z_p^* satisfy the equation $x^{\exp(Z_p^*)} - 1 = 0$. Now by Theorem 10.13, this equation has at most $\exp(Z_p^*)$ solutions in Z_p, therefore at most $\exp(Z_p^*)$ solutions in Z_p^*. So, $|Z_p^*| \leq \exp(Z_p^*)$. Since these results imply $|Z_p^*| = \exp(Z_p^*)$, using Theorem 10.24, the theorem is proven.   □

To prove that Z_{p^e} is cyclic when p is an odd prime, we use the following lemma.

LEMMA 10.9   When p is an odd prime, a primitive root $g \bmod p$ exists such that for all integers $e > 1$:

$$g^{\phi(p^{e-1})} \not\equiv 1 \bmod p^e. \tag{10.26}$$

PROOF   We will start with the case where $e = 2$. If g is a primitive root mod$p$ which satisfies 10.26, then there is nothing to prove. Else, when $g^{\phi(p^{e-1})} = 1 \bmod p^2$, we choose $g_0 = g + p$ (modulo p) as a primitive element. Now, using Theorem 10.11,

$$(g_0)^{\phi(p^{e-1})} \equiv (g+p)^{p-1} \equiv \sum_{i=0}^{p-1} \binom{p-1}{i} p^i g^{p-1-i} \equiv 1 + (p-1)pg^{p-2} + 0 \not\equiv 1 (\bmod p^2),$$

satisfying 10.26. So, we assume from now on that g satisfies 10.26 when $e = 2$.

The case $e > 2$ is proven by induction. We will assume that 10.26, up to $e \geq 2$, has been proven. Due to Euler–Fermat's theorem, $g^{\phi(p^{e-1})} \equiv 1 \bmod p^{e-1}$, or

$$g^{\phi(p^{e-1})} = 1 + lp^{e-1},$$

for an integer l. Due to our assumption, $p \nmid l$. Since $e \geq 2$, $\phi(p^e) = p\phi(p^{e-1})$. So,

$$g^{\phi(p^{(e-1)+1})} = (1+lp^{e-1})^p = \sum_{i=0}^{p} \binom{p}{i} l^i p^{i(e-1)} = 1 + plp^{e-1} + \frac{p(p-1)}{2}l^2 p^{2(e-1)} + rp^{3(e-1)}, \quad (10.27)$$

for some integer r. Since p is odd, we have that $2 \mid (p-1)$. This implies $p^{e+1} \mid p(p-1)p^{2(e-1)}/2$, since $e+1 \leq 2e-1$ when $e \geq 2$. Also, $e+1 \leq 3e-3$ when $e \geq 2$. So, modulo p^{e+1} Equation 10.27 becomes

$$g^{\phi(p^e)} \equiv 1 + lp^e \not\equiv 1 \bmod p^{e+1}.$$

$\square$

THEOREM 10.26   *When p is an odd prime and $e \geq 1$ is an integer, Z_{p^e} is cyclic.*

PROOF   The case $e = 1$ was proven in Theorem 10.25. When $e \geq 2$, we consider a g satisfying the conditions in Lemma 10.9 and call k the order of g modulo p^e. Since g is a primitive root modulo p, $p-1 \mid k$, so $k = m(p-1)$. From Euler–Fermat's theorem $k \mid \phi(p^e)$, it is implying that $m \mid p^{e-1}$, or $m = p^s$. So, $k = (p-1)p^s$. Now, $s = e-1$, otherwise we have a contradiction with Lemma 10.9. $\square$

DEFINITION 10.17   The Carmichael function $\lambda(n) = \exp(Z_n^*)$.

THEOREM 10.27   *We have that*

$$\lambda(2^k) = \begin{cases} 2^{k-1} & \text{if } k < 3 \\ 2^{k-2} & \text{if } k \geq 3 \end{cases}$$

PROOF   It is easy to verify that 1 and 3 are primitive roots modulo 2 and 4, respectively. When $k \geq 3$, we prove by induction that

$$\forall a \in Z_{2^k}^* : \quad a^{2^{k-2}} = 1 \bmod 2^k. \quad (10.28)$$

First, if $k = 3$, then $2^{k-2} = 2$. Now all a are odd, and $a^2 = (2l+1)^2 = 4l(l+1)+1$, for some l. Since $l(l+1)$ is even, $a^2 = 1 \bmod 8$. We will now assume that 10.28 is valid for k. This implies that for all odd integers $a^{2^{k-2}} = 1 + q2^k$, and squaring both sides gives $a^{2^{k-1}} = 1 + q2^{k+1} + q^2 2^{2k} \equiv 1 \bmod 2^{k+1}$. So, when $k \geq 3$ and $a \in Z_{2^k}$, then $\operatorname{ord}(a) \mid 2^{k-2}$. We now need to prove that an $a \in Z_{2^k}^*$ exists for which $\operatorname{ord}(a) = 2^{k-2}$. We take $a = 3$ and need to prove that $3^{2^{k-3}} \not\equiv 1 \bmod 2^k$. Using 10.28 instead of the Euler–Fermat theorem, the last part of the proof of Lemma 10.9 can easily be adapted to prove the claim, and this is left as an exercise.   □

The Chinese Remainder Theorem implies the following corollary.

COROLLARY 10.14   If $n = 2^{a_0} p_1^{a_1} \cdots p_k^{a_k}$, where p_i are different odd primes, then

$$\lambda(n) = \operatorname{lcm}(\lambda(2^{a_0}), \phi(p_1^{a_1}), \ldots, \phi(p_k^{a_k})).$$

COROLLARY 10.15   When n is the product of two different odd primes, as in RSA, Z_n^* is not cyclic.

10.5.10 Lagrange Theorem

The following theorem is well known in elementary algebra.

THEOREM 10.28   *The order of a subgroup H of a finite group G is a factor of the order of G.*

PROOF   Define the left coset of $x \in G$ relative to the subgroup H as $Hx = \{hx \mid h \in H\}$. First, any two cosets, let us say Hx and Hy, have the same cardinality. Indeed, the map mapping $a \in Hx$ to $ax^{-1}y$ is a bijection. Second, these cosets partition G. Obviously, a coset is not empty and any element $a \in G$ belongs to the coset Ha. Suppose now that b belongs to two different cosets Hx and Hy, then $b = h_1 x$, $h_1 \in H$ and $b = h_2 y$, where $h_2 \in H$. But then $y = h_2^{-1} h_1 x$, so $y \in Hx$. Then any element $z \in Hy$ will also belong to Hx, since $z = hy$ for some $h \in H$. So, we have a contradiction. Since each coset has the same cardinality and they form a partition, the result follows immediately.   □

10.6   Computational Number Theory: Part 2

10.6.1 Computing the Jacobi Symbol

Theorems 10.18–10.21 can easily be used to adapt the Euclidean algorithm to compute the Jacobi symbol.

Input declaration: integers a, n, where $0 \leq a < n$
Output declaration: $(a \mid n)$

function Jacobi(a, n)
begin

if $n = 1$
   then Jacobi := 1
   else if $a = 0$
         then Jacobi := 0
         else if $a = 1$

then Jacobi := 1
else if a is even
 then if $(n \equiv 3 \bmod 8)$ or $(n \equiv 5 \bmod 8)$
  then Jacobi := $-$**Jacobi**$(a/2 \bmod n, n)$
  else Jacobi := **Jacobi**$(a/2 \bmod n, n)$
 else if n is even
  then Jacobi := **Jacobi**$(a \bmod 2, 2) \cdot$ **Jacobi**$(a \bmod n/2, n/2)$
  else if $(a \equiv 3 \bmod 4)$ and $(n \equiv 3 \bmod 4)$
   then Jacobi := $-$**Jacobi**$(n \bmod a, a)$
   else Jacobi := **Jacobi**$(n \bmod a, a)$

end

Note that if $a \geq n$ then $(a \mid n) = (a \bmod n \mid n)$. The proof of correctness, the analysis of the algorithm, and a non-recursive version are left as exercises.

10.6.2   Selecting a Prime

The methods used to select a (uniform random) prime of a certain length consist in choosing a (uniformly random) positive integer of a certain length and then testing whether the number is a prime. There are several methods to test whether a number is a prime. Since the research on this topic is so extensive, it is worth a book in itself. We will, therefore, limit our discussion.

Primality testing belongs to **NP** ∩ **co** − **NP** as proven by Pratt [32]. In 2004, it was proven that the problem is in **P** [1], which was an unexpected result. The algorithm is unfortunately not very practical. In most applications in cryptography, it is sufficient to know that a number is likely a prime. We distinguish two types of primality tests, depending on who chooses the prime. We will first discuss the case where the user of the prime chooses it.

10.6.2.1   Fermat Pseudoprimes

We will start by discussing primality tests where the user chooses a uniformly random number of a certain length and tests for primality. In this case, a Fermat pseudoprime test is sufficient. The contrapositive of Fermat's little theorem (see Corollary 10.7) tells us that if a number is composite, a witness $a \not\equiv 0 \bmod n$ exists such that $a^{n-1} \not\equiv 1 \bmod n$, and then n is composite. A number n is called a Fermat pseudoprime to the base a if $a^{n-1} \not\equiv 1 \bmod n$. Although all primes are Fermat pseudoprimes, unfortunately, not all pseudoprimes are necessarily primes. The Carmichael numbers are composite numbers but have the property that for all $a \in Z_n^* : a^{n-1} = 1 \bmod n$. In other words, that $\lambda(n) \mid n - 1$. It is easy to verify that $n = 3 * 11 * 17$ satisfies this. Alford et al. [3] showed that there are infinitely many Carmichael numbers. We will use the following result.

THEOREM 10.29   *If n is a Carmichael number, then n is odd and squarefree.*

PROOF   2 is not a Carmichael number and if $n > 2$, then $2 \mid \lambda(n)$. Since n is a Carmichael number, $\lambda(n) \mid n - 1$, so n is odd. Suppose now that $p^2 \mid n$, where p is a prime. Since $p \neq 2$, using Corollary 10.14, we have that $p \mid \lambda(n)$. This implies that $p \mid n - 1$, since $\lambda(n) \mid n - 1$. Now, $p \mid n$ and $p \mid n - 1$, implying that $p \mid 1$, which is a contradiction.   □

In the case, where an odd number n is chosen uniformly random among those of a given length and one uses Fermat's primality test with $a = 2$, the probability that the obtained number is not a prime is sufficiently small for cryptographic applications. However, if the number n is not chosen by the user and is given by an outsider, Fermat's test makes no sense due to the existence of Carmichael numbers, which the outsider could choose. We will now discuss how to check whether a number given by an outsider is likely to be a prime.

10.6.2.2 Probabilistic Primality Tests

We will now describe the Solovay–Strassen algorithm in which the probability of receiving an incorrect statement that a composite n is prime can be made sufficiently small for each number n. The algorithm was also independently discovered by D. H. Lehmer.

We will assume that using a function call to **Random**, with input n, outputs a natural number a with a uniform distribution such that $1 \le a < n$.

> **Input declaration:** an odd integer $n > 2$.
> **Output declaration:** element of {likely-prime, composite}
>
> **function Solovay-Strassen**(n)
> **begin**
> $a := $ **Random**(n);   (so, $1 \le a \le n - 1$)
> if $\gcd(a, n) \ne 1$, then Solovay–Strassen:=composite
>   else if $(a \mid n) \not\equiv a^{\frac{n-1}{2}}$ mod n, then Solovay–Strassen:=composite
>   else Solovay–Strassen:=likely-prime
> **end**

To discuss how good this algorithm is, we first introduce the set

$$E(n) = \{a \in Z_n^* \mid (a \mid n) \equiv a^{\frac{n-1}{2}} \text{ mod } n\}.$$

LEMMA 10.10   Let $n \ge 3$ be an odd integer. We have that n is prime if and only if $E(n) = Z_n^*$.

PROOF   If n is an odd prime, the claim follows directly from Euler's criterion (Theorem 10.15).

We prove the converse using a contradiction. So, we assume that n is composite and $E(n) = Z_n^*$, which implies (by squaring) that $\forall a \in Z_n^* : a^{n-1} \equiv 1$ mod n. Thus, n is a Carmichael number, implying that n is squarefree (see Theorem 10.29). So, $n = pr$, where p is a prime and $\gcd(p, r) = 1$. Let b be a quadratic nonresidue modulo p and $a \equiv b$ mod p and $a \equiv 1$ mod r. Now, from Theorem 10.18, $(a \mid n) = (a \mid p)(a \mid r) = (b \mid p) = -1$. Thus, due to our assumption, $a^{(n-1)/2} \equiv -1$ mod n, implying that $a^{(n-1)/2} \equiv -1$ mod r, since $r \mid n \mid (a^{(n-1)/2} + 1)$. This contradicts with the choice of $a \equiv 1$ mod r.   □

THEOREM 10.30   *If n is an odd prime, then Solovay–Strassen(n) returns likely-prime. If n is an odd composite, then Solovay–Strassen(n) returns likely-prime with a probability less or equal to $1/2$.*

PROOF   The first part follows from Corollary 10.4 and Theorem 10.15. To prove the second part, we use Lemma 10.10. It implies that $E(n) \ne Z_n^*$. So, since $E(n)$ is a subgroup of Z_n^*, as is easy to verify $E(n)$, it is a proper subgroup. Applying Lagrange's theorem (see Theorem 10.28) implies that $|E(n)| \le |Z_n^*|/2 = \phi(n)/2 \le (n - 1)/2$, which implies the theorem.   □

It is easy to verify that the protocol runs in expected polynomial time. If, when running the Solovay–Strassen algorithm k times for a given integer n, it returns each time likely-prime, then the probability that n is composite is, at maximum, 2^{-k}.

It should be observed that the Miller–Rabin primality test algorithm has an even smaller probability of making a mistake.

10.6.3 Selecting an Element with a Larger Order

We will discuss how to select an element of large order in Z_p^*, where p a prime. On the basis of the Chinese Remainder Theorem, Pohlig–Hellman [30] proved that if $p - 1$ has only small prime factors, it is easy to compute the discrete logarithm in Z_p^*. Even if $p - 1$ has one large prime factor, uniformly picking a random element will not necessarily imply that it has (with high probability) a high order. Therefore, we will discuss how one can generate a primitive element when the prime factorization of $p - 1$ is given. The algorithm follows immediately from the following theorem.

THEOREM 10.31 *If p is a prime, then an integer a exists such that*

$$\text{for all primes } q \mid p - 1, \text{ we have } a^{\frac{p-1}{q}} \not\equiv 1 \bmod p. \tag{10.29}$$

Such an element a is a primitive root modulo p.

PROOF   A primitive element must satisfy 10.29 and exists due to Theorem 10.25. We now prove, using a contradiction, that an element satisfying 10.29 must be a primitive element. Let $l = \text{ord}(a)$ modulo p. First, Lemma 10.7 implies that $l \mid p - 1$. So, when $p - 1 = \prod_{i=1}^{m} q_i^{b_i}$, where q_i are different primes and $b_i \geq 1$, then $l = \prod_{i=1}^{m} q_i^{c_i}$, where $c_i \leq b_i$. Suppose $c_j < b_j$, then $a^{(p-1)/q_j} \equiv 1 \bmod p$, which is a contradiction. $\qquad\square$

As in the Solovay–Strassen algorithm, we will use a random generator.

> **Input declaration:** a prime p and the prime factorization of $p - 1 = \prod_{i=1}^{m} q_i^{b_i}$
> **Output declaration:** a primitive root modulo p
>
> **function generator**(p)
> **begin**
> repeat
>   generator:=**Random**(p);   (so, $1 \leq$ generator $\leq p - 1$)
>   until for all q_i (generator)$^{(p-1)/q_i} \neq 1 \bmod p$
> **end**

The expected running time follows from the following theorem.

THEOREM 10.32 Z_p^*, where p a prime, has $\phi(p - 1)$ different generators.

PROOF   Theorem 10.25 says that there is at least one primitive root. Assume that g is a primitive root modulo p and let $a \in Z_{p-1}^*$. Then an element $b = a^{-1} \bmod p - 1$ exists, so $ab = k(p-1)+1$. We claim that $h = g^a \bmod p$ is a primitive root. Indeed, $h^b = g^{ab} = g^{k(p-1)+1} = g \bmod p$. Therefore, any element $g^i = h^{bi} \bmod p$. So, we have proven that there are at least $\phi(p - 1)$ primitive roots.

Now if $d = \gcd(a, n) \neq 1$, then $h = g^a \bmod p$ cannot be a generator. Indeed, if h is the generator, then $h^x = 1 \bmod p$ implies that $x \equiv 0 \bmod p - 1$. Now, $h^x = g^{ax} = 1 \bmod p$, or $ax = 0 \bmod p - 1$, which has only $\gcd(a, p - 1)$ incongruent solutions modulo $p - 1$ (see Theorem 10.12). □

This algorithm can easily be adapted to construct an element of a given order. This is important in the DSS signature scheme. A variant of this algorithm allows one to prove that p is a prime.

10.6.4 Other Algorithms

We have only discussed computational algorithms needed to explain the encryption schemes explained in this chapter. However, so much research has been done in the area of computational number theory, that a series of books are necessary to give a decent, up-to-date description of the area. For example, we did not discuss an algorithm to compute square roots, nor did we discuss algorithms used to cryptanalyze cryptosystems, such as algorithms to factor integers and to compute discrete logarithms. It is the progress on these algorithms and their implementations that dictates the size of the numbers one needs to choose to obtain a decent security.

10.7 Security Issues

To truly estimate the security of public key systems, the area of computational complexity has to advance significantly. In its early stages of public key encryption, heuristic security (see Chapter 9) was used. Today, proven secure public key encryption schemes have been proposed. We briefly discuss the concepts of probabilistic encryption and semantic security, concepts that played major roles in the work on proven secure public key encryption. We then briefly survey the research on security against chosen text attacks. We briefly mention the work on proven secure hybrid encryption.

10.7.1 Probabilistic Encryption

10.7.1.1 Introduction

There is the following problem with the use of the first public key encryption schemes, as these were described originally. When resending the same message to a receiver, the corresponding ciphertext will be identical. This problem arises in particular when the message is short or when the uncertainty of the message (entropy) is low. When taking, for example, what is called today textbook RSA, the ciphertext $C = M^e \bmod n$ only depends on the message, and so suffers from this problem.

The idea of probabilistic encryption is also to use randomness to avoid this. The first such public key scheme was the Goldwasser–Micali probabilistic encryption scheme. Although it is not very practical, its historical contribution is too significant to be ignored.

10.7.1.2 Generating a Public Key in Goldwasser–Micali

When Alice wants to generate her public key, she first selects (randomly) two primes p and q of equal binary length. She computes $n = p \cdot q$. Then she repeats choosing a random element y until $(y \mid n) = 1$ and $y \in QNR_n$. Knowing the prime factorization of n, the generation of such a y will take expected polynomial time. She then publishes (n, y) as her public key.

Note that if $p \equiv q \equiv 3 \bmod 4$, $y = -1$ is such a number since $(-1 \mid p) = (-1 \mid q) = -1$, which is due to Corollary 10.11.

10.7.1.3 Encryption in Goldwasser–Micali

To encrypt a bit b, the sender uniformly chooses an $r \in Z_n^*$ and sends $c = y^b r^2 \bmod n$ as ciphertext.

10.7.1.4 Decryption in Goldwasser–Micali

To decrypt, Alice sets $b = 0$ if $(c \mid p) = 1$. If, however, $(c \mid p) = -1$, then $b = 1$. Due to Corollary 10.10, the decryption is correct.

10.7.2 Semantic Security

As discussed in Section 9.3.3, we must first model formally what security means. In the context of public key encryption, the weakest level of security usually studied is called semantic security. Basically, it requires that no polynomial time algorithm exists which can distinguish the encryptions of a string x from these of a string y of the same length (polynomial in the security parameter) substantially better than guessing. For a formal definition, see [18,19].

One can prove that the Goldwasser–Micali scheme is sematically secure if it is hard to decide whether a number $a \in Z_n^{+1}$ is a quadratic residue or not.

A practical problem with the Goldwasser–Micali scheme is the huge ciphertext expansion, i.e., the ciphertext is $|n|$ longer than the plaintext! Although no security proof was originally provided for the much more practical ElGamal encryption scheme (see Section 10.3.1), it is not too difficult to prove that it is semantically secure assuming the difficulty of the Decisional Diffie–Hellman problem. Although the problem finds its origin in the context of secure key exchange (or key agreement), it is one of the corner stones of making proven secure cryptosystems more practical. It assumes that no computer algorithm can distinguish in expected polynomial time between the choice of (g^a, g^b, g^{ab}) and (g^a, g^b, g^c), where a, b, and c are chosen uniformly random modulo $\mathrm{ord}(g)$, where, e.g., $\mathrm{ord}(g)$ is a prime.

10.7.3 Security against Chosen Ciphertexts

In several circumstances, semantic security alone does not guarantee an acceptable level of security against a sophisticated attack, as discussed in Section 9.2.3. Although proven secure public key encryption schemes were designed early on [28,34], one had to wait till 1998 for a rather practical scheme [14]. Although this scheme is substantially faster than earlier work, it is still more than twice as slow as ElGamal. For this reason, the scheme is not used in practice.

An approach dating back to the early days of public key encryption is the use of hybrid encryption. The basic idea is that a public key encryption scheme is only used to send a session key, chosen by the sender. A conventional encryption scheme is then used to send the actual message. The key of the conventional encryption is the session key sent using the public key encryption algorithm. Decryption is straightforward. Shoup [36] was the first to study how to use and adapt hybrid encryption in a way to achieve a proven secure approach (see also [15]). Their solution relies on underlying public key and conventional cryptosystems that are both secure against (adaptive) chosen ciphertext attacks. In 2004, it was shown that this condition can be relaxed, leading to faster schemes [25].

The above solutions to achieve proven security are based on the standard model, which is essentially using classical reduction proofs from computational complexity theory. An alternative approach is to use what has been called a random oracle. Recently, this approach has been found to be potentially flawed [12], although no practical schemes have been broken this way (so far). OAEP (Optimal Asymmetric Encryption Padding) [8] is a method to convert a deterministic scheme (such as RSA) to a probabilistic one (see also [17,37]). Its security relies on the more controversial random oracle.

10.8   Research Issues and Summary

This chapter surveyed some modern public key encryption schemes and the number theory and computational number theory required to understand these schemes. Public key schemes used in practice seem to be based on the discrete logarithm problem and the integer factorization problem. However, no proofs have been provided linking these actual assumptions to the actual cryptosystems. Proven secure encryption schemes (public key and hybrid) have come a long way in the last 25 years. Such schemes, today, are not widely used, although these are only a constant factor slower than the practical ones. The many years of research on this topic will only pay off when the speed ratio between heuristic schemes and proven secure ones (under the standard model) becomes close to 1. Although the adaptation of AES was a major success for academia (see Chapter 9), proven secure conventional encryption is still in its infancy.

The problem of making new public key encryption schemes that are secure and significantly faster than surveyed remains an important issue. Many researchers have addressed this and have usually been unsuccessful.

10.9   Further Information

The Further Information section in the previous chapter discusses information on public key encryption schemes.

The number theory required to understand modern public key systems can be found in many books on number theory such as [4,13,20,21,26] and the algebra in [9,10,22,23]. These books discuss detailed proofs of the law of quadratic reciprocity (Theorem 10.17).

More information on computational number theory, also called algorithmic number theory, can be found in the book by Bach and Shallit [6] (see also [2,7,11,29]). In the first volume, the authors have discussed number theoretical problems that can be solved in (expected) polynomial time. In the forthcoming second volume, they will discuss the number theoretical problems that are considered intractable today. For primality testing consult [3,27,33,38] and for factoring related work, e.g., [5,31]. Knuth's book on seminumerical algorithms [24] is a worthy reference book. ANTS (Algorithmic Number Theory Symposium) is a conference that takes place every two years. The proceedings of the conferences are published in the *Lecture Notes in Computer Science* (Springer-Verlag). Results on the topic have also appeared in a wide range of conferences and journals, such as FOCS and STOC.

Defining Terms

Composite: An integer with at least three different positive divisors.

Congruent: Two numbers a and b are congruent modulo c if they have the same non-negative remainder after division by c.

Cyclic group: A group that can be generated by one element, i.e., all elements are powers of that one element.

Discrete logarithm problem: Given α and β in a group, the discrete logarithm decision problem is to decide whether β is a power of α in that group. The discrete logarithm search problem is to find what this power is (if it exists).

Integer factorization problem: Is to find a nontrivial divisor of a given integer.

Order of an element: Is the smallest positive integer such that the element raised to this integer is the identity element.

Prime: An integer with exactly two different positive divisors, namely 1 and itself.

Primitive element: An element that generates all the elements of the group, in particular of the group of integers modulo a given prime.

Pseudoprime: An integer, not necessarily prime, that passes some test.

Quadratic nonresidue modulo n: Is an integer relatively prime to n, which is not congruent to a square modulo n.

Quadratic residue modulo n: Is an integer relatively prime to n, which is congruent to a square modulo n.

Random oracle: Is an assumption which states that some function (such as hash function) behaves as (is indistinguishable) from a true random function (a function chosen uniformly random).

Acknowledgment

The first edition of this text was written while the author was at the University of Wisconsin—Milwaukee. Currently, he is BT Chair of Information Security and also funded by EPSRC EP/C538285/1.

References

1. M. Agrawal, N. Kayal, and N. Saxena. Primes in P. *Annals of Mathematics*, 160(2), pp. 781–793, 2004.
2. A. V. Aho, J. E. Hopcroft, and J. D. Ullman. *The Design and Analysis of Computer Algorithms*. Addison-Wesley, Reading, MA, 1974.
3. W. R. Alford, A. Granville, and C. Pomerance. There are infinitely many Carmichael numbers. *Annals of Mathematics*, 140, 703–722, 1994.
4. T. M. Apostol. *Introduction to Analytic Number Theory*. Springer-Verlag, New York, 1976.
5. E. Bach. How to generate factored random numbers. *SIAM Journal on Computing*, 17(2), 179–193, April 1988.
6. E. Bach and J. Shallit. *Algorithmic Number Theory*, volume 1, Efficient Algorithms of *Foundation of Computing Series*. MIT Press, New York, 1996.
7. P. Beauchemin, G. Brassard, C. Crépeau, C. Goutier, and C. Pomerance. The generation of random numbers which are probably prime. *Journal of Cryptology*, 1(1), 53–64, 1988.
8. M. Bellare and P. Rogaway. Optimal asymmetric encryption—how to encrypt with RSA. In A. De Santis, editor, *Advances in Cryptology—Eurocrypt '94, Proceedings (Lecture Notes in Computer Science 950)*, pp. 341–358. Springer-Verlag, Berlin, 1995. Perugia, Italy, May 9–12.
9. E. R. Berlekamp. *Algebraic Coding Theory*. McGraw-Hill Book Company, New York, 1968.
10. E. R. Berlekamp. *Algebraic Coding Theory*. Aegen Park Press, Laguna Hills, CA, 1984.
11. G. Brassard and P. Bratley. *Algorithmics—Theory & Practice*. Prentice Hall, Englewood Cliffs, NJ, 1988.
12. R. Canetti, O. Goldreich, and S. Halevi. The random oracle methodology, revisited. *Journal of ACM*, 51(4), 557–594, 2004.
13. J. W. S. Cassels. *An Introduction to the Geometry of Numbers*. Springer-Verlag, New York, 1971.
14. R. Cramer and V. Shoup. A practical public key cryptosystem provably secure against advaptive chosen ciphertext attack. In H. Krawczyk, editor, *Advances in Cryptology—Crypto '98, Proceedings (Lecture Notes in Computer Science 1462)*, pp. 13–25. Springer-Verlag, Berlin, 1998. Santa Barbara, CA, August 23–27.
15. R. Cramer and V. Shoup. Design and analysis of practical public-key encryption schemes secure against adaptive chosen ciphertext attack. *SIAM Journal on Computing*, 33(1), 167–226, 2003.

16. T. ElGamal. A public key cryptosystem and a signature scheme based on discrete logarithms. *IEEE Transactions on Information Theory*, 31, 469–472, 1985.

17. E. Fujisaki, T. Okamoto, D. Pointcheval, and J. Stern. RSA-OAEP is secure under the RSA assumption. In Joe Kilian, editor, *Advances in Cryptology—CRYPTO 2001, (Lecture Notes in Computer Science 2139)*, pp. 260–274. Springer-Verlag, Berlin, 2001.

18. O. Goldreich. *The Foundations of Cryptography*. Cambridge University Press, Cambridge, U.K., 2001.

19. S. Goldwasser and S. Micali. Probabilistic encryption. *Journal of Computer and System Sciences*, 28(2), 270–299, April 1984.

20. G. Hardy and E. Wright. *An Introduction to the Theory of Numbers*. Oxford Science Publications, London, Great Britain, fifth edition, 1985.

21. L. K. Hua. *Introduction to Number Theory*. Springer-Verlag, New York, 1982.

22. T. Hungerford. *Algebra*. Springer-Verlag, New York, fifth edition, 1989.

23. N. Jacobson. *Basic Algebra I*. W. H. Freeman and Company, New York, 1985.

24. D. E. Knuth. *The Art of Computer Programming, Vol. 2, Seminumerical Algorithms*. Addison-Wesley, Reading, MA, third edition, 1997.

25. K. Kurosawa and Y. Desmedt. A new paradigm of hybrid encryption scheme. In M. Franklin, editor, *Advances in Cryptology—Crypto 2004, Proceedings (Lecture Notes in Computer Science 3152)*, pp. 426–442. Springer-Verlag, 2004. Santa Barbara, CA, August 15–19.

26. W. LeVeque. *Fundamentals of Number Theory*. Addison-Wesley, New York, 1977.

27. G. L. Miller. Riemann's hypothesis and tests for primality. *Journal of Computer and Systems Sciences*, 13, 300–317, 1976.

28. M. Naor and M. Yung. Public-key cryptosytems provably secure against chosen ciphertext attack. In *Proceedings of the Twenty Second Annual ACM Symposium on Theory of Computing, STOC*, pp. 427–437, May 14–16, 1990.

29. R. Peralta. A simple and fast probabilistic algorithm for computing square roots modulo a prime number. *IEEE Transactions on Information Theory*, IT-32(6), 846–847, 1986.

30. S. C. Pohlig and M. E. Hellman. An improved algorithm for computing logarithms over $GF(p)$ and its cryptographic significance. *IEEE Transactions on Information Theory*, IT-24(1), 106–110, January 1978.

31. C. Pomerance. The quadratic sieve factoring algorithm. In T. Beth, N. Cot, and I. Ingemarsson, editors, *Advances in Cryptology. Proceedings of Eurocrypt 84 (Lecture Notes in Computer Science 209)*, pp. 169–182. Springer-Verlag, Berlin, 1985. Paris, France, April 9–11, 1984.

32. V. R. Pratt. Every prime has a succinct certificate. *SIAM Journal on Computing*, 4(3), 214–220, 1975.

33. M. Rabin. Probabilistic algorithm for primality testing. *Journal of Number Theory*, 12, 128–138, 1980.

34. C. Rackoff and D. R. Simon. Non-interactive zero-knowledge proof of knowledge and chosen ciphertext attack. In J. Feigenbaum, editor, *Advances in Cryptology—Crypto '91, Proceedings (Lecture Notes in Computer Science 576)*, pp. 433–444. Springer-Verlag, 1992. Santa Barbara, CA, August 12–15.

35. R. L. Rivest, A. Shamir, and L. Adleman. A method for obtaining digital signatures and public key cryptosystems. *Communications of ACM*, 21, 294–299, April 1978.

36. V. Shoup. A composition theorem for universal one-way hash functions. In B. Preneel, editor, *Advances in Cryptology—Eurocrypt 2000, Proceedings (Lecture Notes in Computer Science 1807)*, pp. 445–452, Springer-Verlag, Berlin, 2000.

37. V. Shoup. OAEP reconsidered. In Joe Kilian, editor, *Advances in Cryptology—CRYPTO 2001 (Lecture Notes in Computer Science 2139)*, pp. 239–259. Springer-Verlag, Berlin, 2001.

38. R. Solovay and V. Strassen. A fast Monte-Carlo test for primality. *SIAM Journal on Computing*, 6(1), 84–85, erratum (1978), ibid, 7,118, 1977.

11

Cryptanalysis

Samuel S. Wagstaff, Jr.
Purdue University

11.1 Introduction

A cipher is a secret way of writing in which plaintext is enciphered into ciphertext under the control of a key. Those who know the key can easily decipher the ciphertext back into the plaintext. Cryptanalysis is the study of breaking ciphers, that is, finding the key or converting the ciphertext into the plaintext without knowing the key.

For a given cipher, key, and plaintext, let M, C, and K denote the plaintext, the corresponding ciphertext, and the key, respectively. If E_K and D_K represent the enciphering and deciphering functions when the key is K, then we may write $C = E_K(M)$ and $M = D_K(C)$. For all M and K, we must have $D_K(E_K(M)) = M$.

There are four kinds of cryptanalytic attacks. All four kinds assume that the forms of the enciphering and deciphering functions are known.

1. Ciphertext only: Given C, find K or at least M for which $C = E_K(M)$.

2. Known plaintext: Given M and the corresponding C, find K for which $C = E_K(M)$.

3. Chosen plaintext: The cryptanalyst may choose M. He is told C and must find K for which $C = E_K(M)$.

4. Chosen ciphertext: The cryptanalyst may choose C. He is told M and must find K for which $C = E_K(M)$.

The ciphertext only attack is hardest. To carry it out, one may exploit knowledge of the language of the plaintext, its redundancy, or its common words.

An obvious known plaintext attack is to try all possible keys K and stop when one is found with $C = E_K(M)$. This is feasible only when the number of possible keys is small.

The chosen plaintext and ciphertext attacks may be possible when the enemy can be tricked into enciphering or deciphering some text or after the capture of a cryptographic chip with an embedded, unreadable key.

The cryptanalyst has a priori information about the plaintext. For example, he knows that a string like "the" is more probable than the string "wqx." One goal of cryptanalysis is to modify the a priori probability distribution of possible plaintexts to make the correct plaintext more probable than the incorrect ones, although not necessarily certain. Shannon's [Sha49] information theory is used to formalize this process. It estimates, for example, the unicity distance of a cipher, which is the shortest length of the ciphertext needed to make the correct plaintext more probable than the incorrect ones.

11.2   Types of Ciphers

There is no general method of attack on all ciphers. There are many ad hoc methods which work on just one cipher or, at best, one type of cipher. We will list some of the kinds of ciphers and describe what methods one might use to break them. In the later sections, we will describe some of the more sophisticated methods of cryptanalysis.

Ciphers may be classified as transposition ciphers, substitution ciphers, or product ciphers.

Transposition ciphers rearrange the characters of the plaintext to form the ciphertext. For example, enciphering may consist of writing the plaintext into a matrix row-wise and reading out the ciphertext column-wise. More generally, one may split the plaintext into blocks of fixed length L and encipher by applying a given permutation, the key, to each block. It is easy to recognize transposition ciphers, because the frequency distribution of ciphertext letters is the same as the usual distribution of letters in the language of the plaintext. One may guess the period length L from the message length and the spacing between repeated strings. Once L is guessed, the permutation may be constructed by trial and error using the frequency distribution of pairs or triples of letters.

In a simple substitution cipher, the key is a fixed permutation of the alphabet. The ciphertext is formed from the plaintext by replacing each letter by its image under the permutation. These ciphers are broken by trial and error, comparing the frequency distribution of the ciphertext letters with that of the plaintext letters. For example, "e" is the most common letter in English. Therefore, the most frequent ciphertext letter is probably the image of "e." The frequency distribution of pairs or triples of letters helps, too. The pair "th" is common in English. Hence, a pair of letters with high frequency in the ciphertext might be the image of "th."

A homophonic substitution cipher uses a ciphertext alphabet larger than the plaintext alphabet, to confound the frequency analysis just described. The key is an assignment of each plaintext letter to a subset of the ciphertext alphabet, called the homophones of the plaintext letter. The homophones of different letters are disjoint sets, and the size of the homophone set is often proportional to the frequency of the corresponding letter in an ordinary plaintext. Thus, "e" would have the largest homophone set. Plaintext is enciphered by replacing each letter by a random element of its homophone set. As a result, a frequency analysis of single ciphertext letters will find that they have a uniform distribution, which is useless for cryptanalysis. Homophonic substitution ciphers may be broken by analysis of the frequency distribution of pairs and triples of letters.

Polyalphabetic substitution ciphers use multiple permutations (called alphabets) of the plaintext alphabet onto the ciphertext alphabet. The key is a description of these permutations. If the permutations repeat in a certain sequence, as in the Beaufort and Vigenère ciphers, one may break the cipher by first determining the number d of different alphabets and then solving d interleaved simple substitution ciphers. Either the Kasiski method or the index of coincidence may be used to find d.

The Kaski method looks for repeated ciphertext strings with the hope that they may be encipherments of the same plaintext word which occurs each time at the same place in the cycle of alphabets. For example, if the ciphertext "wqx" occurs twice, starting in positions i and j, then probably d divides $j - i$. When several repeated pairs are found, d may be the greatest common divisor of a subset of the differences in their starting points.

The index of coincidence (IC) (see Friedman [Fri76]) measures frequency variations of letters to estimate the size of d. Let $\{a_1, \ldots, a_n\}$ be the alphabet (plaintext or ciphertext). For $1 \leq i \leq n$, let F_i be the frequency of occurrence of a_i in a ciphertext of length N. The IC is given by the formula

$$\text{IC} = \binom{N}{2}^{-1} \sum_{i=1}^{n} \binom{F_i}{2}.$$

Then, IC is the probability that two letters chosen at random in the ciphertext are the same letter. One can estimate IC theoretically in terms of d, N, and the usual frequency distribution of letters in plaintext. For English, $n = 26$, and the expected value of IC is

$$\frac{1}{d}\left(\frac{N-d}{N-1}\right)(0.066) + \left(\frac{d-1}{d}\right)\left(\frac{N}{N-1}\right)\frac{1}{26}.$$

One estimates d by comparing IC, computed from the ciphertext, with its expected value for various d.

These two methods of finding d often complement each other in that the IC tells the approximate size of d, while the Kaski method gives a number that d divides. For example, the IC may suggest that d is 5, 6, or 7, and the Kaski method may predict that d probably divides 12. In this case, $d = 6$.

Rotor machines, like the German Enigma and Hagelin machines, are hardware devices which implement polyalphabetic substitution ciphers with very long periods d. They can be cryptanalyzed using group theory (see Barker [Bar77], Kahn [Kah67], and Konheim [Kon81]). The UNIX crypt(1) command uses a similar cipher. See Reeds and Weinberger [RW84] for its cryptanalysis.

A one-time pad is a polyalphabetic substitution cipher whose alphabets do not repeat. They are selected by a random sequence which is the key. The key is as long as the plaintext. If the key sequence is truly random and if it is used only once, then this cipher is unbreakable. However, if the key sequence is itself an ordinary text, as in a running-key cipher, the cipher can be broken by frequency analysis, beginning with the assumption that usually in each position both the plaintext letter and key letter are letters which occur with high frequency. Frequency analysis also works when a key is reused to encipher a second message. In this case, one assumes that in each position the plaintext letters from both messages are letters which occur with high frequency.

A polygram substitution cipher enciphers a block of several letters together to prevent frequency analysis. For example, the Playfair cipher enciphers pairs of plaintext letters into pairs of ciphertext letters. It may be broken by analyzing the frequency of pairs of ciphertext letters (see Hitt [Hit18] for details). The Hill cipher treats a block of plaintext or ciphertext letters as a vector and enciphers by matrix multiplication. If the vector dimension is two or three, then the cipher may be broken by frequency analysis of pairs or triples of letters. A known plaintext attack on a Hill cipher is an easy exercise in linear algebra. (Compare with linear feedback shift registers in the next section.)

The Pohlig–Hellman cipher and most public key ciphers are polygram substitution ciphers with a block length of several hundred characters. The domain and range of the substitution mapping are so large that the key must be a compact description of this function, such as exponentiation modulo a large number (see Sections 11.7 through 11.10 for cryptanalysis of such ciphers).

A product cipher is a cipher formed by composing several transposition and substitution ciphers. A classic example is the data encryption standard (DES), which alternates transposition and substitution ciphers in 16 rounds (see Diffie and Hellman [DH77] for general cryptanalysis of DES and

Section 11.5 for a powerful attack on DES and similar product ciphers). The advanced encryption standard (AES) is another product cipher discussed subsequently.

Ciphers are also classified as block or stream ciphers. All ciphers split long messages into blocks and encipher each block separately. Block sizes range from one bit to thousands of bits per block.

A block cipher enciphers each block with the same key.

A stream cipher has a sequence or stream of keys and enciphers each block with the next key. The key stream may be periodic, as in the Vigenère cipher or a linear feedback shift register, or not periodic, as in a one-time pad. Ciphertext is often formed in stream ciphers by exclusive-oring the plaintext with the key.

11.3   Linear Feedback Shift Registers

A linear feedback shift register is a device which generates a key stream for a stream cipher. It consists of an n-bit shift register and an XOR gate. Let the shift register hold the vector $R = (r_0, r_1, \ldots, r_{n-1})$. The inputs to the XOR gate are several bits selected (tapped) from fixed bit positions in the register. Let the 1 bits in the vector $T = (t_0, t_1, \ldots, t_{n-1})$ specify the tapped bit positions. Then the output of the XOR gate is the scalar product $TR = \sum_{i=0}^{n-1} t_i r_i \bmod 2$, and this bit is shifted into the shift register. Let $R' = (r'_0, r'_1, \ldots, r'_{n-1})$ be the contents of the register after the shift. Then $r'_i = r_{i-1}$ for $1 \leq i < n$ and $r'_0 = TR$. In other words, $R' \equiv HR \bmod 2$, where H is the $n \times n$ matrix with T as its first row, 1s just below the main diagonal and 0s elsewhere.

$$H = \begin{pmatrix} t_0 & t_1 & \cdots & t_{n-2} & t_{n-1} \\ 1 & 0 & \cdots & 0 & 0 \\ 0 & 1 & \cdots & 0 & 0 \\ \vdots & \vdots & \ddots & \vdots & \vdots \\ 0 & 0 & \cdots & 0 & 0 \\ 0 & 0 & \cdots & 1 & 0 \end{pmatrix}$$

The bit r_{n-1} is shifted out of the register and into the key stream. One can choose T so that the period of the bit stream is $2^n - 1$, which is the maximum possible period. If n is a few thousand, this length may make the cipher to appear secure, but the linearity makes it easy to break.

Suppose $2n$ consecutive key bits $k_0, \ldots, k_{2n-1}$ are known. Let X and Y be the $n \times n$ matrices

$$X = \begin{pmatrix} k_{n-1} & k_n & \cdots & k_{2n-2} \\ k_{n-2} & k_{n-1} & \cdots & k_{2n-3} \\ \vdots & \vdots & \ddots & \vdots \\ k_0 & k_1 & \cdots & k_{n-1} \end{pmatrix}$$

$$Y = \begin{pmatrix} k_n & k_{n+1} & \cdots & k_{2n-1} \\ k_{n-1} & k_n & \cdots & k_{2n-2} \\ \vdots & \vdots & \ddots & \vdots \\ k_1 & k_2 & \cdots & k_n \end{pmatrix}$$

It follows from $R' \equiv HR \bmod 2$ that $Y \equiv HX \bmod 2$, so H may be computed from $H \equiv YX^{-1} \bmod 2$. The inverse matrix $X^{-1} \bmod 2$ is easy to compute by Gaussian elimination or by the Berlekamp-Massey algorithm for n up to at least 10^5. The tap vector T is the first row of H, and the initial contents R of the shift register are $(k_{n-1}, \ldots, k_0)$.

See Ding et al. [DXS91] for more information about linear feedback shift registers and variations of them.

11.4 Meet in the Middle Attacks

One might think that a way to make block ciphers more secure is to use them twice with different keys. If the key length is n bits, a brute force known plaintext attack on the basic block cipher would take up to 2^n encryptions. It might appear that the double encryption $C = E_{K_2}(E_{K_1}(M))$, where K_1 and K_2 are independent n-bit keys, which would take up to 2^{2n} encryptions to find the two keys. The meet-in-the-middle attack is a known plaintext attack which trades time for memory and breaks the double cipher using only about 2^{n+1} encryptions and space to store 2^n blocks and keys. Let M_1 and M_2 be two known plaintexts, and let C_1 and C_2 be the corresponding ciphertexts. (A third pair may be needed.) For each possible key K, store the pair $(K, E_K(M_1))$ in a file. Sort the 2^n pairs by second component. For each possible key K, compute $D_K(C_1)$ and look for this value as the second component of a pair in the file. If it is found, the current key might be K_2, and the key in the pair found in the file might be K_1. Check whether $C_2 = E_{K_2}(E_{K_1}(M_2))$ to determine whether K_1 and K_2 are really the keys.

In the case of DES, $n = 56$, and the meet-in-the-middle attack requires enough memory to store 2^{56} plaintext-ciphertext pairs, more than is available now. But, some day, there may be enough memory to make the attack feasible.

11.5 Differential and Linear Cryptanalysis

Differential cryptanalysis is an attack on DES and some related ciphers, which is faster than exhaustive search. It requires about 2^{47} steps to break DES rather than the 2^{55} steps needed on an average to try keys until the correct one is found.

Note: Although there are 2^{56} keys for DES, one can test both a key K and its complement $\overline{K}$ with a single DES encipherment using the equivalence

$$C = \text{DES}_K(M) \iff \overline{C} = \text{DES}_{\overline{K}}(\overline{M}).$$

The adjective "differential" here refers to a difference modulo 2 or XOR. The basic idea of differential cryptanalysis is to form many input plaintext pairs M, M^* whose XOR $M \oplus M^*$ is constant and study the distribution of the output XORs $\text{DES}_K(M) \oplus \text{DES}_K(M^*)$. The XOR of two inputs to DES, or a part of DES, is called the input XOR, and the XOR of the outputs is called the output XOR.

DES (see Chapter 9) consists of permutations, XORs, and F functions. The F functions are built from an expansion (48 bits from 32 bits), S-boxes, and permutations. It is easy to see that permutations P, expansions E, and XORs satisfy these equations:

$$P(X) \oplus P(X^*) = P(X \oplus X^*),$$
$$E(X) \oplus E(X^*) = E(X \oplus X^*)$$

and

$$(X \oplus K) \oplus (X^* \oplus K) = X \oplus X^*.$$

These operations are linear (as functions on vector spaces over the field with two elements). In contrast, S-boxes are nonlinear. They do not preserve XORs:

$$S(X) \oplus S(X^*) \neq S(X \oplus X^*).$$

Indeed, if one tabulates the distribution of the four-bit output XOR of an S-box as a function of the six-bit input XOR, one obtains an irregular table. We show a portion of this table for the S-box S_1.

TABLE 11.1   XOR Distribution for S-box S_1

Input XOR	0H	1H	2H	3H	4H	5H	6H	7H	...	EH	FH
					Output XOR						
0H	64	0	0	0	0	0	0	0	...	0	0
1H	0	0	0	6	0	2	4	4	...	2	4
2H	0	0	0	8	0	4	4	4	...	4	2
3H	14	4	2	2	10	6	4	2	...	2	0
:	:	:	:	:	:	:	:	:	...	:	:
34H	0	8	16	6	2	0	0	12	...	0	6
:	:	:	:	:	:	:	:	:	...	:	:
3FH	4	8	4	2	4	0	2	4	...	2	2

The row and column labels are hexadecimal numbers. All counts are even numbers since a pair (X, X^*) is counted in an entry if and only if (X^*, X) is counted. Note also that if the input XOR is 0, then the output XOR is 0, too. Hence, the first row (0H) has the entry 64 in column 0H and 0 counts for the other 15 columns. The other rows have no entry greater than 16 and 20% to 30% 0 counts. The average entry size is $64/16 = 4$. The 64 rows for each S-box are all different.

Here is an overview of the differential cryptanalysis of DES. Encipher many plaintext pairs having a fixed XOR and save the ciphertext pairs. The input to the F function in the last round is $R_{15} = L_{16}$, which can be computed from the ciphertext by undoing the final permutation. Thus, we can find the input XOR and output XOR for each S-box in the last round. Consider one S-box. If we knew the output E of expansion and the input X to the S-box, we could compute 6 bits (K) of the key used here by $K = E \oplus X$, since the F function computes $X = E \oplus K$. For each of the 64 possible 6-bit blocks K of the key, count the number of pairs that result with the known output XOR using this key value in the last round. Compare this distribution with the rows of the XOR table for the S box. When enough pairs have been enciphered, the distribution will be similar to a unique row, and this will give six of the 48 key bits used in the last round. After this has been done for all eight S boxes, one knows 48 of the 56 key bits. The remaining 8 bits can be found quickly by trying the 256 possibilities.

The input XORs to the last round may be specified by using as plaintext XORs certain 64-bit patterns which, with a small positive probability, remain invariant under the first 15 rounds. Plaintext pairs which produce the desired input to the last round are called right pairs. Other plaintext pairs are called wrong pairs. In the statistical analysis, right pairs predict correct key bits while wrong pairs predict random key bits. When enough plaintext pairs have been analyzed, the correct key bits overcome the random bits by becoming the most frequently suggested values.

The above description gives the chosen plaintext attack on DES. A known plaintext attack may be performed as follows. Suppose we need m plaintext pairs to perform a chosen plaintext attack. Let $2^{32}\sqrt{2m}$ random known plaintexts be given together with their corresponding ciphertexts. Consider all $(2^{32}\sqrt{2m})^2/2 = 2^{64}m$ possible (unordered) pairs of known plaintext. Since the block size is 64 bits, there are only 2^{64} possible plaintext XOR values. Hence, there are about $2^{64}m/2^{64} = m$ pairs, giving each plaintext XOR value. Therefore, it is likely that there are m pairs with each of the plaintext XORs needed for a chosen plaintext attack.

Although we have described the attack for ECB mode, it works also for the other modes (CBC, CFB, and OFB), because it is easy to compute the real input and output of DES from the plaintext and ciphertext in these modes.

The linearity of parts of the DES algorithm mentioned earlier is the basis for Matsui's [Mat94] linear cryptanalysis. In this known plaintext attack on DES, one XORs together certain plaintext and ciphertext bits to form a linear approximation for the XOR of certain bits of the key. This approximation will be correct with some probability p. Irregularity in the definition of the S-boxes allows one to choose bit positions from the plaintext, ciphertext, and key for which the XOR approximation is valid with probability $p \neq 1/2$. This bias can be amplified by computing the XOR of the key bits for many different known plaintext-ciphertext pairs. When enough XORs of key bits have been guessed, one can compute the key bits by solving a system of linear equations over the field with 2 elements. Linear cryptanalysis of DES requires about 2^{43} known plaintext-ciphertext pairs to determine the 56-bit key.

Differential and linear cryptanalysis apply also to many substitution/permutation ciphers similar to DES. These include FEAL, LOKI, Lucifer, and REDOC-II. The same methods can break some hash functions such as Snefru and N-hash, by producing two messages with the same hash value.

See Biham and Shamir in [BS91] and [BS93] for more information about differential cryptanalysis. See Matsui [Mat94] for the details of linear cryptanalysis.

11.6 The Advanced Encryption Standard

In 2000, the US National Institute of Standards selected a cipher called Rijndael as the AES to replace DES. Like DES, AES performs bit operations (rather than arithmetic with large integers, say) and is very fast on a binary computer. Its design is public (unlike that of DES) and is based on the theory of finite fields. The specification of the algorithm may be found at the NIST web site [NIS].

AES is a block cipher and may be used in the same modes as DES. The block size and key length are independent and each may be 128, 192, or 256 bits. It has 10, 12, or 14 rounds, depending on the block size and key length. Differential and linear cryptanalysis cannot successfully attack AES because it has more nonlinearity than DES. If AES is reduced to four rounds, there is an attack called the "square attack" that can break this simplified cipher. With effort, the square attack can be extended to break AES with five or six rounds. For AES with at least seven rounds, no attack faster than exhaustive search of all possible keys is known to work.

Some ciphers, like DES and IDEA, have weak keys that are easy to identify in a chosen plaintext attack, perhaps because they produce repeated round subkeys. This weakness occurs for some ciphers whose nonlinear operations depend on the key value or whose key expansion is linear. As the fraction of weak keys is minuscule for most ciphers, it is unlikely that a randomly chosen key will be weak. However, they can be exploited in some one-way function applications, where the attacker can choose the key. The nonlinearity in AES resides in its fixed S-boxes and does not depend on the key. Also, the key expansion in AES is nonlinear. Therefore, AES has no weak keys and no restriction on key selection. See [NIS] for more information about the square attack and weak keys.

11.7 Knapsack Ciphers

We describe a variation of Shamir's attack on the Merkle–Hellman knapsack. A similar attack will work for almost all known knapsacks, including iterated ones. The only public key knapsack not yet broken is that of Chor and Rivest [CR85].

Let the public knapsack weights be the positive integers a_i for $1 \leq i \leq n$. An n-bit plaintext block $x_1, \ldots, x_n$ is enciphered as $E = \sum_{i=1}^{n} a_i x_i$.

Let the secret superincreasing knapsack weights be s_i for $1 \leq i \leq n$, where $\sum_{i=1}^{j} s_i < s_{j+1}$ for $1 \leq j < n$. Let the secret modulus M and multiplier W satisfy $1 < W < M$, $\gcd(W, M) = 1$ and $M > \sum_{i=1}^{n} s_i$. Then $a_i \equiv W^{-1} s_i \bmod M$ or $W a_i \equiv s_i \bmod M$ for $1 \leq i \leq n$.

To attack this cryptosystem, we rewrite the last congruence as $W a_i - M k_i = s_i$ for $1 \leq i \leq n$, where the k_i are unknown nonnegative integers. Divide by $M a_i$ to get

$$\frac{W}{M} - \frac{k_i}{a_i} = \frac{s_i}{M a_i}.$$

This equation shows that, at least for small i, the fraction k_i/a_i is a close approximation to the fraction W/M. The inequalities $\sum_{i=1}^{j} s_i < s_{j+1}$ and $\sum_{i=1}^{n} s_i < M$ imply $s_i \leq 2^{-n+i}M$. For almost all W, we have $a_i \geq M/n^2$. Hence,

$$\frac{W}{M} - \frac{k_i}{a_i} = \frac{s_i}{Ma_i} \leq \frac{2^{-n+i}}{a_i} \leq \frac{2^{-n+i}n^2}{M}.$$

We expect that all k_i and all a_i are about the same size as M, so that

$$\frac{k_i}{k_1} - \frac{a_i}{a_1} = O(2^{-n+i}n^2/M).$$

This says that the vector $(a_2/a_1, \ldots, a_n/a_1)$, constructed from the public weights, is a very close approximation to the vector $(k_2/k_1, \ldots, k_n/k_1)$, which involves numbers k_i from which one can compute the secret weights s_i. Given a vector like $(a_2/a_1, \ldots, a_n/a_1)$, one can find close approximations to it, like $(k_2/k_1, \ldots, k_n/k_1)$, by integer programming or the Lenstra–Lenstra–Lovasz theorem. After k_1 is found, M and W are easily determined from the fact that k_1/a_1 is a close approximation to W/M. Finally, Shamir showed that if W^*/M^* is any reasonably close approximation to W/M, then W^* and M^* can be used to decrypt every ciphertext.

See Brickell and Odlyzko [BO88] for more information about breaking knapsacks.

11.8 Cryptanalysis of RSA

Suppose an RSA cipher (see Rivest, Shamir, and Adleman [RSA78]) is used with public modulus $n = pq$, where $p < q$ are large secret primes, public enciphering exponent e, and secret deciphering exponent d.

Most attacks on RSA try to factor n. One could read the message without factoring n if the sender of the enciphered message packed only one letter per plaintext block. Then the number of different ciphertext blocks would be small (the alphabet size), and the ciphertext could be cryptanalyzed as a simple substitution cipher.

Here are four cases in which $n = pq$ might be easy to factor.

1. If p and q are too close, that is, if $q-p$ is not much larger than $n^{1/4}$, then n can be factored by Fermat's difference of squares method. It finds x and y so that $n = x^2 - y^2 = (x-y)(x+y)$, as follows. Let $m = \lceil \sqrt{n} \rceil$ and write $x = m$. If $z = x^2 - n$ is a square y^2, then $p = x - y$, $q = x + y$ and we are done. Otherwise, we add $2m + 1$ to z and 2 to m and then test whether $z = (x+1)^2 - n$ is a square, and so on. Most z can be eliminated quickly because squares must lie in certain congruence classes. For example, squares must be $\equiv 0, 1, 4,$ or 9 mod 16, which eliminates three out of four possible values for z. If $p = k\sqrt{n}$, where $0 < k < 1$, then this algorithm tests about $((1 - k)^2/(2k))\sqrt{n}$ zs to discover p. The difference of squares method normally is not at all used to factor large integers. Its only modern use is to ensure that no one forms an RSA modulus by multiplying two 100-digit primes having the same high order 45 digits, say. Riesel [Rie94] has a nice description of Fermat's difference of squares method.

2. If either $p - 1$ or $q - 1$ has a small ($<10^9$, say) largest prime factor, then n can be factored easily by Pollard's $p - 1$ factoring algorithm. The idea of this method is that if $p - 1$ divides Q, and p does not divide b, then p divides $b^Q - 1$. This assertion follows from Fermat's little theorem. Typically, Q is the product of all prime powers below some limit. The method is feasible when this limit is as large as 10^9. If $Q = \prod_{i=1}^{k} q_i$, when q_i are prime powers, then one computes

POLLARD (b, k, n)

1     $x \leftarrow b$
2   **for** $i \leftarrow 1$ **to** k
3   **do** $x \leftarrow x^{q_i} \bmod n$
4   **if** $(g \leftarrow \gcd(x - 1, n)) > 1$,
5   **then return** factor g of n

This method and the following one have analogues in which $p - 1$ and $q - 1$ are replaced by $p + 1$ and $q + 1$. See Riesel [Rie94] for more details of Pollard's $p - 1$ method. See Williams [Wil82] for details of the $p + 1$ method.

3. The number $g = \gcd(p - 1, q - 1)$ divides $n - 1$ and may be found by factoring $n - 1$, which may be much easier than factoring n. When g is small, as usually happens, it is not useful for finding p or q. But when g is large, it may help find p or q. In that case, for each large factor $g < \sqrt{n}$ of $n - 1$, one can seek a factorization $n = (ag + 1)(bg + 1)$.

4. If n is small, $n < 10^{150}$, say, then one can factor n by a general factoring algorithm like those described in the next section.

See Boneh [Bon99] for many other attacks on RSA.

11.9   Integer Factoring

We outline the quadratic sieve (QS) and the number field sieve (NFS), which are currently the two fastest known integer factoring algorithms that work for any composite number, even an RSA modulus.

Both of these methods factor the odd composite positive integer n by finding integers x, y so that $x^2 \equiv y^2 \bmod n$ but $x \not\equiv \pm y \bmod n$. The first congruence implies that n divides $(x - y)(x + y)$, while the second congruence implies that n does not divide $x - y$ or $x + y$. It follows that at least one prime factor of n divides $x - y$ and at least one prime factor of n does not divide $x - y$. Therefore, $\gcd(n, x - y)$ is a proper factor of n. (This analysis fails if n is a power of a prime, but an RSA modulus would never be a prime power because it would be easy to recognize.)

However, these two factoring methods ignore the condition $x \not\equiv \pm y \bmod n$ and seek many random solutions to $x^2 \equiv y^2 \bmod n$. If n is odd and not a prime power, at least half of all solutions with $y \not\equiv 0 \bmod n$ satisfy $x \not\equiv \pm y \bmod n$ and factor n. Actually, if n has k different prime factors, the probability of successfully factoring n is $1 - 2^{1-k}$, for each random congruence $x^2 \equiv y^2 \bmod n$ with $y \not\equiv 0 \bmod n$.

In QS, many congruences (called relations) of the form $z^2 \equiv q \bmod n$ are produced with q factored completely. One uses linear algebra over the field GF(2) with two elements to pair these primes and finds a subset of the relations in which the product of the qs is a square, y^2, say. Let x be the product of the zs in these relations. Then $x^2 \equiv y^2 \bmod n$, as desired.

The linear algebra is done as follows: Let $p_1, \ldots, p_r$ be all of the primes which divide any q. Then write the relations as:

$$z_i^2 \equiv q_i = \prod_{j=1}^{r} p_j^{e_{ij}} \bmod n, \quad \text{for } 1 \leq i \leq s,$$

where $s > r$. Note that $\prod_{j=1}^{r} p_j^{f_j}$ is the square of an integer if and only if every $f_i \equiv 0 \bmod 2$. Suppose $(t_1, \ldots, t_s)$ is not the zero vector in the s-dimensional vector space $GF(2)^s$ but does lie in the null space of the matrix $[e_{ij}]$, that is, $\sum_{i=1}^{s} e_{ij}t_i = 0$ in $GF(2)$. Then

$$\prod_{\substack{i=1 \\ t_i=1}}^{s} q_i = \prod_{j=1}^{r} \prod_{\substack{i=1 \\ t_i=1}}^{s} p_j^{e_{ij}}.$$

The exponent on p_j in this double product is $\sum_{i=1}^{s} e_{ij}t_i$, which is an even integer since the t vector is in the null space of the matrix. Therefore,

$$\prod_{\substack{i=1 \\ t_i=1}}^{s} q_i$$

is a square because each of its prime factors appears raised to an even power. We call this square y^2 and let

$$x = \prod_{\substack{i=1 \\ t_i=1}}^{s} z_i.$$

Then $x^2 \equiv y^2 \bmod n$.

In QS, the relations $z^2 \equiv q \bmod n$ with q factored completely are produced as follows: Small primes $p_1, \ldots, p_r$ for which n is a quadratic residue (square) are chosen as the factor base. If we choose many pairs a, b of integers with $a = c^2$ for some integer c, $b^2 \equiv n \bmod a$ and $|b| \le a/2$, then the quadratic polynomials

$$Q(t) = \frac{1}{a}[(at+b)^2 - n] = at^2 + 2bt + \frac{b^2 - n}{a}$$

will take integer values at every integer t. If a value of t is found for which the right hand side is factored completely, a relation $z^2 \equiv q \bmod n$ is produced, with $z \equiv (at+b)c^{-1} \bmod n$ and $q = Q(t) = \prod_{j=1}^{r} p_j^{f_j}$, as desired. No trial division by the primes in the factor base is necessary. A sieve quickly factors millions of $Q(t)$s. Let t_1 and t_2 be the two solutions of

$$(at+b)^2 \equiv n \bmod p_i \quad \text{in } 0 \le t_1, t_2 < p_i.$$

(A prime p is put in the factor base for QS only if this congruence has solutions, that is, only if n is a quadratic residue modulo p.) Then all solutions of $Q(t) \equiv 0 \bmod p_i$ are $t_1 + kp_i$ and $t_2 + kp_i$ for $k \in \mathbf{Z}$. In most implementations, $Q(t)$ is represented by one byte $Q[t]$ and $\log p_i$ is added to this byte to avoid division of $Q(t)$ by p_i, a slow operation. Let U denote the upper limit of the sieve interval, p denote the current prime p_i, and L denote the byte approximation to $\log p_i$. The two inner loops are

LOOP (p, L, t_1, U)
1      $t \leftarrow t_1$
2   **while** $t < U$
3   **do** $Q[t] \leftarrow Q[t] + L$
4               $t \leftarrow t + p$

and LOOP(p, L, t_2, U). After the sieve completes, one harvests the relations $z^2 \equiv q \bmod n$ from those t for which $Q[t]$ exceeds a threshold. Only at this point is $Q(t)$ formed and factored, the latter operation being done by trial division with the primes in the factor base.

The time QS that takes to factor n is about

$$\exp((1 + \epsilon(n))(\log n)^{1/2}(\log \log n)^{1/2}),$$

where $\epsilon(n) \to 0$ as $n \to \infty$.

We begin our outline of the NFS by describing the special number field sieve (SNFS), which factors numbers of the form $n = r^e - s$, where r and $|s|$ are small positive integers. It uses some algebraic number theory. Choose a small positive integer d, the degree of an extension field. Let k be the least positive integer for which $kd \geq e$. Let $t = sr^{kd-e}$. Let f be the polynomial $X^d - t$. Let $m = r^k$. Then $f(m) = r^{kd} - sr^{kd-e} = r^{kd-e}n$ is a multiple of n. The optimal degree for f is $((3 + \epsilon(e)) \log n/(2 \log \log n))^{1/3}$ as $e \to \infty$ uniformly for r, s in a finite set, where $\epsilon(e) \to 0$ as $e \to \infty$.

Let α be a zero of f. Let $K = \mathbf{Q}(\alpha)$. Assume f is irreducible, so the degree of K over $\mathbf{Q}$ is d. Let Q_n denote the ring of rational numbers with denominator coprime to n. The subring $Q_n[\alpha]$ of K consists of expressions $\sum_{i=0}^{d-1}(s_i/t_i)\alpha^i$ with $s_i, t_i \in \mathbf{Z}$ and $\gcd(n, t_i) = 1$. Define a ring homomorphism $\phi : Q_n[\alpha] \to \mathbf{Z}/n\mathbf{Z}$ by the formula $\phi(\alpha) = (m \bmod n)$. Thus, $\phi(a + b\alpha) \equiv a + bm \bmod n$.

For $0 < a \leq A$ and $-B \leq b \leq B$, SNFS uses a sieve (as in QS) to factor $a + bm$ and the norm of $a + b\alpha$ in $\mathbf{Z}$. The norm of $a + b\alpha$ is $(-b)^d f(-a/b)$. This polynomial of degree d has d roots modulo p which must be sieved, just like the two roots of $Q(t)$ in QS. A pair (a, b) is saved in a file if $\gcd(a, b) = 1$, and $a + bm$ and the norm of $a + b\alpha$ both have only small prime factors (ones in the factor base, say).

We use linear algebra to pair the prime factors just as in QS, except that now we must form squares on both sides of the congruences. The result is a nonempty set S of pairs (a, b) of coprime integers such that

$$\prod_{(a,b) \in S} (a + bm) \text{ is a square in } \mathbf{Z},$$

and

$$\prod_{(a,b) \in S} (a + b\alpha) \text{ is a square in } Q_n[\alpha].$$

Let the integer x be a square root of the first product. Let $\beta \in Q_n[\alpha]$ be a square root of the second product. We have $\phi(\beta^2) \equiv x^2 \bmod n$, since $\phi(a + b\alpha) \equiv a + bm \bmod n$. Let y be an integer for which $\phi(\beta) \equiv y \bmod n$. Then $x^2 \equiv y^2 \bmod n$, which will factor n with a probability of at least $1/2$.

In the general NFS, we must factor an arbitrary n for which no obvious polynomial is given. The key properties required for the polynomial are that it is irreducible, it has moderately small coefficients so that the norm of α is small, and we know a nontrivial root m modulo n of it. Research in good polynomial selection is still in progress. One simple choice is to let m be some integer near $n^{1/d}$ and write n in radix m as $n = c_d m^d + \cdots + c_0$, where $0 \leq c_i < m$. Then use the polynomial $f(X) = c_d X^d + \cdots + c_0$. With this choice, the time needed by NFS to factor n is

$$\exp(((64/9)^{1/3} + \epsilon(n))(\log n)^{1/3}(\log \log n)^{2/3}),$$

where $\epsilon(n) \to 0$ as $n \to \infty$. QS is faster than NFS for factoring numbers up to a certain size, and NFS is faster for larger numbers. This crossover size is between 100 and 120 decimal digits, depending on the implementation and the size of the coefficients of the NFS polynomial.

See Riesel [Rie94] for a description of the QS and NFS factoring algorithms. See also Lenstra and Lenstra [LL93] for more information about NFS.

11.10   Discrete Logarithms

The Diffie–Hellman key exchange, the ElGamal public key cryptosystem, the Pohlig–Hellman private key cryptosystem, and the digital signature algorithm could all be broken if we could compute discrete logarithms quickly, that is, if we could solve the equation $a^x = b$ in a large finite field. For simplicity, we assume that the finite field is either the integers modulo a prime p or the field with 2^n elements.

Consider first the exponential congruence $a^x \equiv b \bmod p$. By analogy to ordinary logarithms, we may write $x = \log_a b$ when p is understood from the context. These discrete logarithms enjoy many properties of ordinary logarithms, such as $\log_a bc = \log_a b + \log_a c$, except that the arithmetic with logarithms must be done modulo $p - 1$ because $a^{p-1} \equiv 1 \bmod p$. Neglecting powers of $\log p$, the congruence may be solved in $O(p)$ time and $O(1)$ space by raising a to successive powers modulo p and comparing each with b. It may also be solved in $O(1)$ time and $O(p)$ space by looking up x in a precomputed table of pairs $(x, a^x \bmod p)$ sorted by the second coordinate. Shanks' "giant step–baby step" algorithm solves the congruence in $O(\sqrt{p})$ time and $O(\sqrt{p})$ space as follows. Let $m = \lceil \sqrt{p-1} \rceil$. Compute and sort the m ordered pairs $(j, a^{mj} \bmod p)$, for j from 0 to $m - 1$, by the second coordinate. Compute and sort the m ordered pairs $(i, ba^{-i} \bmod p)$, for i from 0 to $m - 1$, by the second coordinate. Find a pair (j, y) in the first list and a pair (i, y) in the second list. This search will succeed because every integer between 0 and $p - 1$ can be written as a two-digit number ji in radix m. Finally, $x = (mj + i) \bmod (p - 1)$.

There are two probabilistic algorithms, the rho method and the lambda (or kangaroo) method, for solving $a^x \equiv b \bmod p$ which also have time complexity $O(\sqrt{p})$, but space complexity $O(1)$. See [Wag03] for details of these methods.

There are faster ways to solve $a^x \equiv b \bmod p$ using methods similar to the two integer factoring algorithms QS and NFS. Here is the analogue for QS. Choose a factor base of primes $p_1, \ldots, p_k$. Perform the following precomputation which depends on a and p but not on b. For many random values of x, try to factor $a^x \bmod p$ using the primes in the factor base. Save at least $k + 20$ of the factored residues:

$$a^{x_j} \equiv \prod_{i=1}^{k} p_i^{e_{ij}} \bmod p \quad \text{for } 1 \leq j \leq k + 20,$$

or equivalently

$$x_j \equiv \sum_{i=1}^{k} e_{ij} \log_a p_i \bmod (p - 1) \quad \text{for } 1 \leq j \leq k + 20.$$

When b is given, perform the following main computation to find $\log_a b$. Try many random values for s until one is found for which $ba^s \bmod p$ can be factored using only the primes in the factor base. Write it as

$$ba^s \equiv \prod_{i=1}^{k} p_i^{c_i} \bmod p$$

or

$$(\log_a b) + s \equiv \sum_{i=1}^{k} c_i \log_a p_i \bmod (p - 1).$$

Use linear algebra as in QS to solve the linear system of congruences modulo $p - 1$ for $\log_a b$. One can prove that the precomputation takes time

$$\exp((1 + \epsilon(p))\sqrt{\log p \log \log p}),$$

where $\epsilon(p) \to 0$ as $p \to \infty$, while the main computation takes time

$$\exp((0.5 + \epsilon(p))\sqrt{\log p \log \log p}),$$

where $\epsilon(p) \to 0$ as $p \to \infty$.

There is a similar algorithm for solving congruences of the form $a^x \equiv b \bmod p$ which is analogous to NFS and runs faster than the above for large p.

There is a method of Coppersmith [Cop84] for solving equations of the form $a^x = b$ in the field with 2^n elements, which is practical for n up to about 1000. Empirically, it is about as difficult to solve $a^x = b$ in the field with p^n elements as it is to factor a general number about as large as p^n. As these are the two problems which must be solved in order to break the RSA and the ElGamal cryptosystems, it is not clear which one of these systems is more secure for a fixed size of modulus.

11.11 Elliptic Curve Discrete Logarithms

All of the cryptographic algorithms whose security relies on the difficulty of solving the discrete logarithm problem $a^x = b$ in a large finite field have analogues in which the multiplicative group of the field is replaced by the additive group of points on an elliptic curve. The discrete logarithm problem in the field has this analogue on an elliptic curve E modulo a prime p: Given points P and Q on E, find an integer x so that $Q = xP$, where xP denotes the sum $P + P + \cdots + P$ of x copies of P. It is also given that such an integer x exists, perhaps because P generates E, that is, the curve consists of the points mP for all integers m.

The algorithms for solving $a^x \equiv b \bmod p$ with time or space complexity $O(p)$ given in the preceding section have analogues for solving the discrete logarithm problem $Q = xP$ on an elliptic curve E modulo p. By Hasse's theorem, the size of E is $O(p)$, so these algorithms also have time or space complexity $O(p)$. Similarly, there is an analogue of Shanks' "giant step–baby-step" algorithm which solves $Q = xP$ in time and space $O(\sqrt{p})$, ignoring small powers of $\log p$. Likewise, the rho and lambda algorithms have elliptic curve analogues with time complexity $O(\sqrt{p})$. In fact, these algorithms work in all groups, not just those mentioned here.

An important property of elliptic curves is that, for most curves, the analogues just mentioned are the fastest known ways to solve the elliptic curve discrete logarithm problem. The subexponential algorithms for solving $a^x \equiv b \bmod p$ which are similar to QS and NFS have no known analogues for general elliptic curves. Because of this hardness of the elliptic curve discrete logarithm problem, one may choose the modulus p of the curve much smaller than would be needed to make it hard to solve $a^x \equiv b \bmod p$. The modulus p of an elliptic curve only needs to be large enough so that it is not feasible for an attacker to perform $O(\sqrt{p})$ operations. According to NIST guidelines, the discrete logarithm problem for an elliptic curve modulo a 163-bit prime is about as hard as solving $a^x \equiv b$ modulo a 1024-bit prime. Since the arithmetic of multiplying numbers modulo p and adding points on an elliptic curve modulo p both have time complexity $O((\log p)^2)$, cryptographic algorithms using these operations will run faster when p is smaller. Thus, it is faster to use an elliptic curve in a cryptographic algorithm than to use the multiplicative group of integers modulo a prime large enough to have an equally hard discrete logarithm problem.

The MOV attack [MOV93] uses the Weil pairing to convert a discrete logarithm problem in an elliptic curve E modulo p into an equivalent discrete logarithm problem in the field with p^m elements, where the embedding degree m is a certain positive integer that depends on E. If m were small, subexponential methods would make it feasible to solve the equivalent discrete logarithm problem. Generally, m is a large positive integer, and the discrete logarithm problem in the field with p^m elements is infeasible. Balasubramanian and Koblitz [BK98] have shown that $m \geq 7$ for almost all elliptic curves. For such m, the discrete logarithm problem in the field with p^m elements is infeasible when p has at least 163 bits.

Weil descent converts an elliptic curve discrete logarithm problem to one in a hyperelliptic curve of high genus, where it can sometimes be solved. As with the MOV attack, this attack is infeasible for almost all elliptic curves. See Diem [Die04] for more details.

11.12   Research Issues and Summary

This chapter presents an overview of some of the many techniques of breaking ciphers. This subject has a long and rich history. We gave some historical perspective by mentioning some of the older ciphers whose cryptanalysis is well understood. Most of this article deals with the cryptanalysis of ciphers still being used today, such as DES, AES, RSA, Pohlig–Hellman, etc. We still do not know how to break these ciphers quickly, and research into methods for doing this remains intense today.

New cryptosystems are invented frequently. Most of them are broken quickly. For example, research continues on multistage knapsacks, some of which have not been broken yet, but probably will be soon. Some variations of differential cryptanalysis being studied now include higher-order differential cryptanalysis and a combination of linear and differential cryptanalysis. Some recent advances in integer factoring algorithms include the use of several primes outside the factor base in QS, the self-initializing QS, faster computation of the square root (β) in the NFS, better methods of polynomial selection for the general NFS, and the lattice sieve variation of the NFS. Most of these advances apply equally to computing discrete logarithms in a field with p elements. Several efforts are under way to accelerate the addition of points on an elliptic curve, a basic operation in cryptosystems that use elliptic curves. One approach uses an elliptic curve over the field with p^n elements, but whose coefficients lie in the field with p elements. Another approach uses hyperelliptic curves over a finite field. These methods accelerate cryptographic algorithms that use elliptic curves but could lead to failures in the algorithms. In both approaches the fields must have appropriate sizes and the coefficients must be chosen wisely to avoid the MOV and Weil descent attacks.

We have omitted discussion of many important ciphers in this short article. Even for the ciphers which were mentioned, the citations are far from complete. See Section 11.13 for general references.

11.13   Further Information

Research on cryptanalysis is published often in the journals *Journal of Cryptology, Cryptologia*, and *Computers and Security*. Such research appears occasionally in the journals *Algorithmica, AT&T Technical Journal, Communications of the ACM, Electronics Letters, Information Processing Letters, IBM Journal of Research and Development, IEEE Transactions on Information Theory, IEEE Spectrum, Mathematics of Computation*, and *Philips Journal of Research*.

Several annual or semi-annual conferences with published proceedings deal with cryptanalysis. CRYPTO has been held since 1981 and has published proceedings since 1982. See CRYPTO '82, CRYPTO '83, etc. EUROCRYPT has published its proceedings in 1982, 1984, 1985, 1987, 1988, etc. The conference AUSCRYPT is held in even numbered years since 1990, while ASIACRYPT is held in odd numbered years since 1991. ANTS, the Algorithmic Number Theory Symposium, began in 1994 and is held biannually.

Kahn [Kah67] gives a comprehensive history of cryptology up to 1967. Public key cryptography was launched by Diffie and Hellman [DH76]. Denning [Den83] and Konheim [Kon81] have lots of information about and examples of old ciphers and some information about recent ones. Biham and Shamir [BS93] is the basic reference for differential cryptanalysis of DES and related ciphers. Wagstaff [Wag03] emphasizes cryptanalysis of ciphers based on number theory.

Defining Terms

Breaking a cipher: Finding the key to the cipher by analysis of the ciphertext or, sometimes, both the plaintext and corresponding ciphertext.

Coprime: Relatively prime. Refers to integers having no common factor greater than 1.

Differential cryptanalysis: An attack on ciphers like DES which tries to find the key by examining how certain differences (XORs) of plaintext pairs affect differences in the corresponding ciphertext pairs.

Digital signature algorithm: A public-key algorithm for signing or authenticating messages. It was proposed by the National Institute of Standards and Technology in 1991.

Factor base: A fixed set of small primes, usually all suitable primes up to some limit, used in factoring auxiliary numbers in the quadratic and number field sieves.

Homophone: One of several ciphertext letters used to encipher a single plaintext letter.

Key stream: A sequence of bits, bytes, or longer strings used as keys to encipher successive blocks of plaintext.

Permutation or transposition cipher: A cipher which enciphers a block of plaintext by rearranging the bits or letters.

Quadratic residue modulo m: A number r which is coprime to m and for which there exists an x so that $x^2 \equiv r \bmod m$.

Sieve: A number theoretic algorithm in which, for each prime number p in a list, some operation is performed on every pth entry in an array.

Substitution cipher: A cipher which enciphers by replacing letters or blocks of plaintext by ciphertext letters or blocks under the control of a key.

XOR: Exclusive-or or sum of bits modulo 2.

References

[Bar77]   W. G. Barker. *Cryptanalysis of the Hagelin Cryptograph*. Aegean Park Press, Laguna Hill, CA, 1977.

[BK98]   R. Balasubramanian and N. Koblitz. The improbability that an elliptic curve has subexponential discrete logarithm problem under the Menezes-Okamoto-Vanstone algorithm. *J. Cryptol.*, 11 (2):141–145, 1998.

[BO88]   E. F. Brickell and A. M. Odlyzko. Cryptanalysis: A survey of recent results. *Proc. IEEE*, 76: 578–593, May, 1988.

[Bon99]   D. Boneh. Twenty years of attacks on the RSA cryptosystem. *Amer. Math. Soc. Notices*, 46:202–213, 1999.

[BS91]   E. Biham and A. Shamir. Differential cryptanalysis of DES-like cryptosystems. In A. J. Menezes and S. A. Vanstone, editors, *Advances in Cryptology–CRYPTO '90, Lecture Notes in Computer Science 537*, pages 2–21, Springer-Verlag, Berlin, 1991.

[BS93]   E. Biham and A. Shamir. *Differential Cryptanalysis of the Data Encryption Standard*. Springer-Verlag, New York, 1993.

[Cop84]   D. Coppersmith. Fast evaluation of logarithms in fields of characteristic two. *IEEE Trans. Inform. Theory*, 30:587–594, 1984.

[CR85]   B. Chor and R. L. Rivest. A knapsack type public key cryptosystem based on arithmetic in finite fields. In *Advances in Cryptology–CRYPTO '84*, pages 54–65, Springer-Verlag, Berlin, 1985.

[Den83]   D. E. Denning. *Cryptography and Data Security*. Addison-Wesley, Reading, MA, 1983.

[DH76]   W. Diffie and M. Hellman. New directions in cryptography. *IEEE Trans. Inform. Theory*, IT-22 (6):644–654, November, 1976.

[DH77]   W. Diffie and M. Hellman. Exhaustive cryptanalysis of the NBS data encryption standard. *Computer*, 10 (6):74–84, June, 1977.

[Die04]   C. Diem. The GHS attack in odd characteristic. *J. Ramanujan Math. Soc.*, 18:1–32, 2004.

[DXS91]   C. Ding, G. Xiao, and W. Shan. *The Stability Theory of Stream Ciphers, Lecture Notes in Computer Science 561*. Springer-Verlag, New York, 1991.

[Fri76]   W. F. Friedman. *Elements of Cryptanalysis*. Aegean Park Press, Laguna Hills, CA, 1976.

[Hit18]   Parker Hitt. *Manual for the Solution of Military Ciphers*, Second edition. Army Service Schools Press, Fort Leavenworth, KS, 1918.

[Kah67]   D. Kahn. *The Codebreakers*. Macmillan Co., New York, 1967.

[Kon81]   A. G. Konheim. *Cryptography, A Primer*. John Wiley & Sons, New York, 1981.

[LL93]   A. K. Lenstra and H. W. Lenstra, Jr. *The Development of the Number Field Sieve*. Springer-Verlag, New York, 1993.

[Mat94]   M. Matsui. Linear cryptanalysis method for DES cipher. In A. J. Menezes and S. A. Vanstone, editors, *Advances in Cryptology–EUROCRYPT '93*, pages 386–397, Springer-Verlag, Berlin, 1994.

[MOV93]   A. J. Menezes, T. Okamoto, and S. A. Vanstone. Reducing elliptic curve logarithms to logarithms in a finite field. *IEEE Trans. Inform. Theory*, 39 (5):1639–1646, 1993.

[NIS]   http://csrc.nist.gov/CryptoToolkit/aes/rijndael.

[Rie94]   H. Riesel. *Prime Numbers and Computer Methods for Factorization*, 2nd edition. Birkhäuser, Boston, MA, 1994.

[RSA78]   R. L. Rivest, A. Shamir, and L. Adleman. A method for obtaining digital signatures and public-key cryptosystems. *Commun. ACM*, 21 (2):120–126, February, 1978.

[RW84]   J. A. Reeds and P. J. Weinberger. File security and the UNIX crypt command. *AT&T Tech. J.*, 63:1673–1683, October, 1984.

[Sha49]   C. E. Shannon. Communication theory of secrecy systems. *Bell Syst. Tech. J.*, 28:656–715, 1949.

[Wag03]   S. S. Wagstaff, Jr. *Cryptanalysis of Number Theoretic Ciphers*. Chapman & Hall/CRC Press, Boca Raton, FL, 2003.

[Wil82]   H. C. Williams. A $p + 1$ method of factoring. *Math. Comp.*, 39:225–234, 1982.

12

Crypto Topics and Applications I

Jennifer Seberry
University of Wollongong

Chris Charnes
Institute of Applied Physics and CASED
Technical University Darmstadt

Josef Pieprzyk
Macquarie University

Rei Safavi-Naini
University of Calgary

12.1 Introduction

In this chapter, we discuss four related areas of cryptology, namely, authentication, hashing, message authentication codes (MACs), and digital signatures. These topics represent active and growing research topics in cryptology. Space limitations allow us to concentrate only on the essential aspects of each topic. The bibliography is intended to supplement our survey. We have selected those items which providean overview of the current state of knowledge in the above areas.

Authentication deals with the problem of providing assurance to a receiver that a communicated message originates from a particular transmitter, and that the received message has the same content as the transmitted message. A typical authentication scenario occurs in computer networks, where the identity of two communicating entities is established by means of authentication.

Hashing is concerned with the problem of providing a relatively short digest–fingerprint of a much longer message or electronic document. A hashing function must satisfy (at least) the critical requirement that the fingerprints of two distinct messages are distinct. Hashing functions have numerous applications in cryptology. They are often used as primitives to construct other cryptographic functions.

MACs are symmetric key primitives that provide message integrity against active spoofing by appending a cryptographic checksum to a message that is verifiable only by the intended recipient of the message. Message authentication is one of the most important ways of ensuring the integrity of information that is transferred by electronic means.

Digital signatures provide electronic equivalents of handwritten signatures. They preserve the essential features of handwritten signatures and can be used to sign electronic documents. Digital signatures can potentially be used in legal contexts.

12.2 Authentication

One of the main goals of a cryptographic system is to provide authentication, which simply means providing assurance about the content and origin of communicated messages.

Historically, cryptography began with secret writing, and this remained the main area of development until recently. With the rapid progress in data communication, the need for providing message integrity and authenticity has escalated to the extent that currently authentication is seen as an urgent goal of cryptographic systems.

Traditionally, it was assumed that a secrecy system provides authentication by the virtue of the secret key being only known to the intended communicants; this would prevent an adversary from constructing a fraudulent message. Simmons [60] argued that the two goals of cryptography are independent. He showed that a system that provides perfect secrecy might not provide any protection against authentication threats. Similarly, a system can provide perfect authentication without concealing messages.

In the rest of this chapter, we use the term communication system to encompass message transmission as well as storage. The system consists of a transmitter who wants to send a message, a receiver who is the intended recipient of the message, and an adversary who attempts to construct a fraudulent message with the aim of getting it accepted by the receiver unwittingly. In some cases, there is a fourth party, called the arbiter, whose basic role is to provide protection against cheating by the transmitter and/or the receiver. The communication is assumed to take place over a public channel, and hence the communicated messages can be seen by all the players. An authentication threat is an attempt by an adversary in the system to modify a communicated message or inject a fraudulent message into the channel. In a secrecy system, the attacker is passive, while in an authentication system, the adversary is active and not only observes the communicated messages and gathers information such as plaintext and ciphertext, but also actively interacts with the system to achieve his or her goal. This view of the system clearly explains Simmons' motivation for basing authentication systems on game theory.

The most important criteria that can be used to classify authentication systems are

- The relation between authenticity and secrecy
- The framework for the security analysis

The first criterion divides authentication systems into those that provide authentication with and without secrecy. The second criterion divides systems into systems with unconditional security, systems with computational security, and systems with provable security. Unconditional security implies that the adversary has unlimited computational power, while in computational security the resources of the adversary are bounded, and security relies on the fact that the required computation exceeds this bound. A provably secure system is in fact a subclass of computationally secure systems, and compromising the system is equivalent to solving a known difficult problem.

These two classifications are orthogonal and produce four subclasses. Below, we review the basic concepts of authentication theory, some known bounds and constructions in unconditional security, and then consider computational security.

12.2.1 Unconditional Security

A basic model introduced by Simmons [60] has remained the mainstay of most of the theoretical research on authentication systems. The model has the same structure as described in Section 12.2 but excludes the arbiter. To provide protection against an adversary, the transmitter and receiver use an authentication code (A-code). An A-code is a collection $\mathcal{E}$ of mappings called encoding rules (also called keys) from a set $\mathcal{S}$ of source states (also called transmitter states) into the set $\mathcal{M}$ of cryptogram (also called codewords). For A-codes without secrecy, also called Cartesian A-codes, a codeword uniquely determines a source state. That is, the set of codewords is partitioned into subsets each corresponding to a distinct source state. In a systematic Cartesian A-code, $\mathcal{M} = \mathcal{S} \times \mathcal{T}$ where $\mathcal{T}$ is a set of authentication tags and each codeword is of the form $s.t, s \in \mathcal{S}, t \in \mathcal{T}$ where '.' denotes concatenation. Let the cardinality of the set $\mathcal{S}$ of source states be denoted as k; that is, $|\mathcal{S}| = k$. Let $E = |\mathcal{E}|$ and $M = |\mathcal{M}|$. The encoding process adds key dependent redundancy to the message, so $k < M$. A key (or encoding rule) e determines a subset $\mathcal{M}_e \subset \mathcal{M}$ of codewords that are authentic under e.

The incidence matrix A of an A-code is a binary matrix of size $E \times M$ whose rows are labeled by encoding rules and columns by codewords, such that $A(e, m) = 1$ if m is a valid codeword under e, and $A(e, m) = 0$, otherwise.

An authentication matrix B of a Cartesian A-code is a matrix of size $E \times k$ whose rows are labeled by the encoding rules and columns by the source states, and $B(e, s) = t$ if t is the tag for the source state s under the encoding rule e.

To use the system, the transmitter and receiver must secretly share an encoding rule. The adversary does not know the encoding rule and uses an impersonation attack, which only requires a knowledge of the system, or a substitution attack in which the adversary waits to see a transmitted codeword, and then constructs a fraudulent codeword. The security of the system is measured in terms of the adversary's chance of success with the chosen attack. The adversary's chance of success in an impersonation attack is denoted by P_0, and in a substitution attack by P_1. The best chance an adversary has in succeeding in either of the above attacks is called the probability of deception, P_d.

An attack is said to be spoofing of order ℓ if the adversary has seen ℓ communicated codewords under a single key. The adversary's chance of success in this case is denoted by P_ℓ.

The chance of success can be defined using two different approaches. The first approach corresponds to an average case analysis of the system and can be described as the adversary's payoff in the game theory model. It has been used by a number of authors, including MacWilliams et al., [40], Fak [27], and Simmons [60]. The second is to consider the worst-case scenario. This approach is based on the relation between A-codes and error correcting codes (also called E-codes).

Using the game theory model, P_j is the value of a zero-sum game between communicants and the adversary. For impersonation

$$P_0 = \max_{m \in \mathcal{M}} (\text{payoff}(m)),$$

and for substitution

$$P_1 = \sum_{j=1}^{E} \sum_{m \in \mathcal{M}} P(m) \max_{m'} \text{payoff}(m, m'),$$

where $P(m)$ is the probability of a codeword m occurring in the channel, and payoff(m, m') is the adversary's payoff (best chance of success) when it substitutes an intercepted codeword m with a fraudulent one, m'.

12.2.2 Bounds on the Performance of the A-Code

The first types of bounds relate the main parameters of an A-code, that is, E, M, k, and hence are usually called combinatorial bounds. The most important combinatorial bound for A-codes with secrecy is

$$P_i \geq \frac{k-i}{M-i}, \quad i = 1, 2, \ldots$$

and for A-codes without secrecy is

$$P_i \geq k/M, \quad i = 1, 2, \ldots.$$

An A-code that satisfies these bounds with equality, that is, with $P_i = \frac{k-i}{M-i}$ for A-codes with secrecy and $P_i = k/M$ for Cartesian A-codes is said to provide perfect protection for spoofing of order i. The adversary's best strategy in spoofing of order i for such an A-code is to randomly select one of the remaining codewords.

A-codes that provide perfect protection for all orders of spoofing up to r are said to be r-fold secure. These codes can be characterized using combinatorial structures, such as orthogonal arrays and t-designs.

An orthogonal array $OA_\lambda(t, k, v)$ is an array with λv^t rows, each row of size k, from the elements of set X of v symbols, such that in any t columns of the array every t-tuple of elements of X occurs in exactly λ rows. Usually, t is referred to as the strength of the OA.

The following table gives an $OA_2(2, 5, 2)$ on the set $\{0, 1\}$:

0	0	0	0	0
1	1	0	0	0
0	0	0	1	1
1	1	0	1	1
1	0	1	0	1
0	1	1	0	1
1	0	1	1	0
0	1	1	1	0

A t-(v, k, λ) design is a collection of b subsets, each of size k of a set X of size v where every distinct subset of size t occurs exactly λ times.

The incidence matrix of a t-(v, k, λ) is a binary matrix, $A = (a_{ij})$, of size $b \times v$ such that $a_{ij} = 1$ if element j is in block i and 0 otherwise.

The following table gives a 3-$(8, 4, 1)$ design on the set $\{0, 1, 2, 3, 4, 5, 6, 7\}$:

7	0	1	3
7	1	2	4
7	2	3	5
7	3	4	6
7	4	5	0
7	5	6	1
7	6	0	2
2	4	5	6
3	5	6	0
4	6	0	1
5	0	1	2
6	1	2	3
0	2	3	4
1	3	4	5

with incidence matrix:

1	1	0	1	0	0	0	1
0	1	1	0	1	0	0	1
0	0	1	1	0	1	0	1
0	0	0	1	1	0	1	1
1	0	0	0	1	1	0	1
0	1	0	0	0	1	1	1
1	0	1	0	0	0	1	1
0	0	1	0	1	1	1	0
1	0	0	1	0	1	1	0
1	1	0	0	1	0	1	0
1	1	1	0	0	1	0	0
0	1	1	1	0	0	1	0
1	0	1	1	1	0	0	0
0	1	0	1	1	1	0	0

The main theorems relating A-codes with r-fold security and combinatorial structures are due to a number of authors, including Stinson [62], and Tombak and Safavi-Naini [70]. The following are the most general forms of these theorems.

Let the source be r-fold uniform. Then an A-code provides r-fold security against spoofing if and only if the incidence matrix of the code is the incidence matrix of a $(r + 1) - (M, k, \lambda)$ design.

In the above theorem, an r-fold uniform source is a source for which every string of r distinct outputs from the source has probability $\frac{1}{k(k-1)\cdots(k-r+1)}$.

Let $P_0 = P_1 = P_2 = \cdots = P_r = k/M$. Then the authentication matrix is a $OA(r + 1, k, \ell)$ where $\ell = M/k$.

The so-called information theoretic bounds characterize the adversary's chance of success using uncertainty measures (entropy). The first such bound for Cartesian A-codes, derived by MacWilliams et al. [40], is

$$P_1 \geq 2^{-\frac{H(M)}{2}}$$

where $H(M)$ is the entropy of the codeword space. The first general bound on P_0, due to Simmons [60], is

$$P_0 \geq 2^{-(H(E)-H(E|M))},$$

where $H(E)$ is the entropy of the key space and $H(E|M)$ is the conditional entropy of the key when a codeword is intercepted. Write $I(E; M)$ for the mutual information of E and M. Then,

$$P_0 \geq 2^{-I(E; M)}.$$

The above bound relates the adversary's best chance of success to the mutual information between the cryptogram space and the key space. A general form of this bound, proved independently by Rosenbaum [57] and Pei [44], is

$$P_\ell \geq 2^{-I(E; M^\ell)},$$

where $I(E; M^\ell)$ is the mutual information between a string of ℓ codewords and m^ℓ is the key.

Similar bounds for A-codes without secrecy were proved by MacWilliams et al. [40] and Walker [72].

A general bound on the probability of deception, P_{d_r}, derived by Rosenbaum [57], is

$$P_{d_r} \geq 2^{-\frac{H(E)}{r+1}}.$$

12.2.3 Other Types of Attack

Tombak and Safavi-Naini [69] consider other types of attacks, similar to those for secrecy systems. In a plaintext attack against A-codes with secrecy, the adversary not only knows the codeword but also knows the corresponding plaintext. In chosen content attack, the adversary wants to succeed with a codeword that has a prescribed plaintext. It is shown that by applying some transformation on the A-code it is possible to provide immunity against the above attacks.

A-codes with secrecy are generally more difficult to analyze than Cartesian A-codes. Moreover, the verification process for the former is not efficient. In the case of Cartesian A-codes, verification of a received codeword, $s.t$, amounts to recalculating the tag using the secret key and the source state s to obtain t' and comparing it to the received tag t. For an authentic codeword, we have $t = t'$. In the case of A-codes with secrecy, when a codeword m is received, the receiver must try all authentic codewords using his or her secret key, otherwise there must be an inverse algorithm that allows the receiver to find and verify the source state. The former process is costly, and the latter does not exist for a general A-code. For practical reasons, the majority of research has concentrated on Cartesian A-codes.

12.2.4 Efficiency

Authentication systems require secure transmission of key information prior to the communication and hence, similar to secrecy systems, it is desirable to have a small number of key bits.

Rees and Stinson [54] prove that for any (M, k, E) A-code that is onefold secure, $E \geq M$. For A-codes with secrecy that provide onefold security, Stinson [62] shows that

$$E \geq \frac{M^2 - M}{k^2 - k}.$$

Under similar conditions for Cartesian A-codes, Stinson [62] shows that

$$E \geq k(\ell - 1) + 1,$$

where $\ell = k/M$.

For A-codes with r-fold security, Stinson [64] shows that

$$E \geq \frac{M(M - 1) \cdots (M - r)}{k(k - 1) \cdots (k - r)}.$$

An A-code provides perfect authenticity of order r if $P_{d_r} = k/M$. In such codes, the probability of success of the spoofer does not improve with the interception of extra codewords.

The following bound is established by Tombak and Safavi-Naini [68] for codes of the above type:

$$E \geq \frac{M^{r+1}}{k^{r+1}}.$$

Their bound shows that the provision of perfect authenticity requires $\log M - \log k$ extra key bits on an average for every added order of spoofing. Hence, using the same key for authentication of more than one message is expensive.

A second measure of efficiency, often used for systematic Cartesian A-codes, is the size of the tag space for a fixed size of source space and probability of success in substitution. Stinson [65] shows that, for perfect protection against substitution, the size of key space grows linearly with the size of the source. Johansson et al. [35] show that if $P_1 > P_0$, A-codes with an exponential (in E) number of source states can be obtained.

12.2.5 A-Codes and E-Codes

An error correcting code provides protection against random channel error. The study of error-correcting codes was motivated by Shannon's Channel Capacity Theorem and has been a very active research area since the early 1950s. Error-correcting codes add redundancy to a message in such a way that a codeword corrupted by the channel noise can be detected and/or corrected. The main difference between an A-code and an E-code is that in the former redundancy depends on a secret key, while in the latter it only depends on the message being coded. There exists a duality between authentication codes and error correcting codes. In the words of Simmons [60], "... one (coding theory) is concerned with clustering the most likely alterations as closely about the original code as possible and the other (authentication theory) with spreading the optimal (to the opponent) alterations as uniformly as possible over $\mathcal{M}$".

The relation between E-codes and A-codes is explored in the work of Johansson et al. [35], who show that it is possible to construct E-codes from A-codes and vice-versa. Their work uses a worst case analysis approach in analyzing the security of A-codes. That is, in the case of substitution attack, they consider the best chance of success that an adversary has when it intercepts all possible codewords. This contrasts with the information theoretic (or game theory) approach in which the average success probability of the adversary over all possible intercepted codewords is calculated.

The work of Johansson et al. is especially useful as it allows the well-developed body of bounds and asymptotic results from the theory of error correcting codes to be employed in the context of authentication codes, to derive upper and lower bounds on the size of the source for A-codes with given E, T, and P_1.

12.2.6 Authentication with Arbiter

In the basic model of authentication discussed above, the adversary is an outsider, and we assume that the transmitter and the receiver are trustworthy. Moreover, because the key is shared by the transmitter and the receiver, the two players are cryptographically indistinguishable. In an attempt to model authentication systems in which the transmitter and the receiver are distinguished and to remove assumptions about the trustworthiness of the two, Simmons [60] introduced a fourth player called the arbiter. The transmitter and receiver have different keys, and the arbiter has access to all or part of the key information. The system has a key distribution phase during which keys satisfying certain conditions are chosen. After that there is a transmission phase during which the transmitter uses its key to produce a codeword and finally a dispute phase during which disputes are resolved with the aid of the arbiter. The arbiter in Simmons' model is active during the transmission phase and is assumed to be trustworthy. Yung and Desmedt [78] relax this assumption and consider a model in which the arbiter is only trusted to resolve disputes. Johansson [33] and Kurosawa [38] derive lower bounds on the probability of deception in such codes. Johansson [34] and Taylor [66] proposed constructions.

12.2.7 Shared Generation of Authenticators

Many applications require the power to generate an authentic message and/or to verify the authenticity of a message to be distributed among a number of players. An example of such a situation is multiple signatures in a bank account or court room. Desmedt and Frankel [24] introduced systems with shared generation of authenticators (SGA-systems), which have been studied by Safavi-Naini [58]. In such systems, there is a group P of transmitters created with a structure Γ that determines authorized subsets of P. Each player has a secret key which is used to generate a partial tag. The system has two phases. In the key distribution phase, a trusted authority generates keys for the transmitters and receivers and securely delivers them to them. In the communication phase, the trusted authority is not active. When an authorized group of transmitters wants to construct an authentic codeword, using its key, each group member generates a partial tag for the source state s which needs to be authenticated and sends it to a combiner. The combiner is a fixed algorithm with no secret input. It combines its inputs to produce a tag, t, to be appended to s. The receiver is able to authenticate this codeword using its secret key.

12.2.8 Multiple Authentication

As mentioned earlier, in the theory of A-codes, the possible attacks by the adversary are limited to impersonation and substitution. This means that the security of the system is only for one message communication; after that the key must be changed. To extend protection over more than one message transmission, there exist a number of alternatives. The most obvious ones use A-codes that provide perfect protection against spoofing of order ℓ. However, little is known about the construction of such codes, and it is preferable to use A-codes that provide protection against substitution for more that one message transmission. Vanroose et al. [77] suggest key strategies in which the communicants change their key after each transmitted codeword, using some pre-specified strategy. In this case, the key information shared by the communicants is the sequence of keys to be used for consecutive transmission slots. The resulting bounds on the probability of deception generalize the bounds given by the following authors: Pei [44], Rosenbaum [57], and Walker [72].

Another successful approach proposed by Wegman and Carter [75] uses a special class of hash functions together with a one time pad of random numbers. This construction is discussed in more detail in Section 12.5.

12.3 Computationally Secure Systems

The study of computationally secure A-systems is relatively informal, cf. Simmons [61]. The basic framework is similar to unconditionally secure systems. A simple computationally secure A-code can be obtained by considering $S = GF(2^{40})$ and $M = GF(2^{64})$. We use $\mathcal{E}$ to be the collection of DES [47] encryption functions and so $E = 2^{56}$. To construct the codeword corresponding to a source state s, using the key k, we append 24 zeros to s, and then use DES with key k to encrypt $s.0_{24}$, where 0_{24} is the string of 24 zeros.

It is easy to see that the above scheme is an A-code with secrecy. It allows the receiver to easily verify the authenticity of a received codeword by decrypting a received codeword and checking the existence of the string of zeros. If an adversary wants to impersonate the transmitter, its chance of success is 2^{-56}, which is the probability of guessing the correct key. For a substitution attack, the adversary waits to see a transmitted codeword. Then it uses all the keys to decrypt the codeword, and once a decryption of the right form (ending in 24 zeros) is obtained, a possible key is found. In general, there is more than one key with this property. On an average, there are $2^{56} \times 2^{40}/2^{64} = 2^{32}$ pairs (s, k) that produce the same cryptogram and hence the chance of guessing correctly is 2^{-32}. A better strategy for the adversary is to randomly choose a cryptogram and send it to the receiver. In this case, the chance of success is 2^{-24}, which is better than the previous case.

The security of computationally secure A-systems weakens rapidly as the adversary intercepts more cryptograms. Trying all possible keys on ℓ received cryptograms enables the adversary to uniquely identify the key, in this case the adversary succeeds with a probability of one.

Computationally secure A-systems without secrecy are obtained by appending an authenticator to the message which is verifiable by the intended receiver. The authenticator can be produced by a symmetric key algorithm or an asymmetric key algorithm. The former is the subject of the section on MAC, whilethe latter is discussed in the section on digital signatures.

12.4 Hashing

In many cryptographic applications, it is necessary to produce a relatively short fingerprint of a much longer message or electronic document. The fingerprint is also called a digest of the message. Ideally, a hash function should produce a unique digest of a fixed length for a message of an arbitrary length. Obviously, this is impossible as any hash function is, in fact, a many-to-one mapping. The properties required for secure hashing can be summarized as follows:

- Hashing should be a many-to-one function producing a digest that is a complex function of all bits of the message.
- A hash function should behave as a random function that creates a digest for a given message by randomly choosing an element from the whole digest space.
- For any pair of messages, it should be computationally difficult to find a collision; that is, distinct messages with the same digest.
- A hash function should be one-way; that is, it should be easy to compute the digest of a given message but difficult to determine the message corresponding to a given digest.

The main requirement of secure hashing is that it should be collision-free in the sense that finding two colliding messages is computationally intractable. This requirement must hold for long as well as short messages.

12.4.1 Strong and Weak Hash Functions

Hash functions can be broadly classified into two classes: *strong* one-way hash functions (also called collision-free hash functions) and weak one-way hash functions (also known as universal one-way hash functions). A strong one-way hash function is a function h satisfying the following conditions:

1. h can be applied to any message or document M of any size.
2. h produces a fixed size digest.
3. Given h and M, it is easy to compute the digest $h(M)$.
4. Given h, it is computationally infeasible to find two distinct messages M_1, M_2 that collide, that is, $h(M_1) = h(M_2)$.

On the other hand, a weak one-way hash function is a function that satisfies conditions 1, 2, 3 and the following:

5. Given h and a randomly chosen message M, it is computationally intractable to find another message M' that collides with M, that is, $h(M) = h(M')$.

Strong one-way hash functions are easier to use since there is no precondition on the selection of the messages. On the other hand, for weak one-way hash functions, there is no guarantee that a nonrandom selection of two messages is collision-free. This means that the space of easily found colliding messages must be small. Otherwise, a random selection of two messages would produce a collision with a nonnegligible probability.

12.4.2 Theoretic Constructions

Naor and Yung [43] introduced the concept of a universal one-way hash function (UOWHF) and suggested a construction based on a one-way permutation. In their construction, they employ the notion of a universal family of functions with collision accessibility property [75]. The above functions are defined as follows. Suppose $G = \{g \mid A \rightarrow B\}$ is a set of functions, G is strongly universal$_r$ if given any r distinct elements $a_1, \ldots, a_r \in A$, and any r elements $b_1, \ldots, b_r \in B$, there are $|G|/|B|^2$ functions which take a_1 to b_1, a_2 to b_2, etc. ($|G|$ and $|B|$ denote the cardinality of sets G and B, respectively.)

A strongly universal$_r$ family of functions G has the collision accessibility property if it is possible to generate in polynomial time of a function $g \in G$ that satisfies the equations

$$g(a_i) = b_i, \quad 1 \le i \le r.$$

Naor and Yung construct a family of UOWHFs by concatenating any one-way permutation with a family of strongly universal$_2$ hash functions having the collision accessibility property. In this construction, the one-way permutation provides the one-wayness of the UOWHF, and the strongly universal$_2$ family of hash functions provides the mapping to the small length output. When a function is chosen randomly and uniformly from the family, the output is distributed randomly and uniformly over the output space.

De Santis and Yung [23] construct hash functions from one-way functions with an almost-known preimage size. In other words, if an element of the domain is given, then with a polynomial uncertainty an estimate of the size of the preimage set is easily computable. A regular function is an example of such a function. (In a regular function, each image of an n-bit input has the same number of preimages of length n.)

Rompel [56] constructs a UOWHF from any one-way function. His construction is rather elaborate. Briefly, his idea is to transform any one-way function into a UOWHF through a sequence of complicated procedures. First, the one-way function is transformed into another one-way function such that for most elements of the domain except for a fraction, it is easy to find a collision. From this, another one-way function is constructed such that for most of the elements it is hard to find a collision. Subsequently, a length increasing one-way function is constructed for which it is hard to find collisions almost everywhere. Finally, this is turned into a UOWHF which compresses the input in a way that makes it difficult to find a collision.

12.4.3 Hashing Based on Block Ciphers

To minimize the design effort for cryptographically secure hash functions, the designers of hash functions tend to base their schemes on existing encryption algorithms. For example, sequential hashing can be obtained by dividing a given message into blocks and applying an encryption algorithm on the message blocks. The message block length must be the same as the block length of the encryption algorithm. If the message length is not a multiple of the block length, then the last block is usually padded with some redundant bits. To provide a randomizing element, an initial public vector is normally used. The proof of the security of such schemes relies on the collision freeness of the underlying encryption algorithm.

In the following, let E denote an arbitrary encryption algorithm. Let $E(K, M)$ denote the encryption of message M with key K using E; let IV denote the initial vector.

Rabin [50] showed that any private-key cryptosystem can be used for hashing. Rabin's scheme is the following. First the message is divided into blocks $M_1, M_2, \ldots$ of the same size as the block length of the encryption algorithm. In the case of DES, the message is divided into 64-bit blocks. To hash a message $M = (M_1, M_2, \ldots, M_t)$, the following computations are performed:

$$H_0 = IV$$
$$H_i = E(M_i, H_{i-1}) \quad i = 1, 2, \ldots, t$$
$$H(M) = H_t$$

where M_i is a message block, H_i are intermediate results of hashing, and $H(M)$ is the digest. Although Rabin's scheme is simple and elegant, it is susceptible to the so-called birthday attack when the size of the hash value is 64 bits.

Winternitz [76] proposes a scheme for the construction of a one-way hash function from any block cipher. In any good block cipher, given an input and an output, it should be difficult to determine the key, but from the key and the output, it should be easy to determine the input. The scheme uses an operation E^* defined by:

$$E^*(K \parallel M) = E(K, M) \oplus M.$$

On the basis of the above scheme, Davies [22] proposed the following hashing algorithm:

$$H_0 = IV$$
$$H_i = E(M_i, H_{i-1}) \oplus H_{i-1} \quad i = 1, 2, \ldots, t$$
$$H(M) = H_t.$$

If $E(K, M)$ is DES, then it may be vulnerable to attacks based on weak keys or a key-collision search. The meet-in-the-middle attack is avoided because $E(K, M)$ is a one-way function.

Merkle [42] proposed hashing schemes based on Winternitz's construction. These schemes use DES to produce digests of size ≈ 128 bits.

Their construction follows a general method for producing hash algorithms, called the meta method. This is same as the serial method of Damgård [21]. The description of the method is as follows. The message is first divided into blocks of 106 bits. Each 106-bit block M_i of data is concatenated with the 128-bit block H_{i-1}. The concatenation $X_i = M_i \parallel H_{i-1}$ contains 234 bits. Each block X_i is further divided into halves, X_{i1} and X_{i2}.

$$H_0 = IV$$
$$X_i = H_{i-1} \parallel M_i$$
$$H_i = E^*(00 \parallel \text{first 59 bits of } \{E^*(100 \parallel X_{1i})\} \parallel$$
$$\text{first 59 bits of } \{E^*(101 \parallel X_{2i})\}) \parallel$$
$$E^*(01 \parallel \text{first 59 bits of } \{E^*(110 \parallel X_{1i})\} \parallel$$
$$\text{first 59 bits of } \{E^*(111 \parallel X_{2i})\})$$
$$H(M) = H_t.$$

In this scheme, E^* is defined in the Winternitz construction. The strings 00, 01, 100, 101, 110, and 111 above are used to prevent the manipulation of weak keys.

12.4.4 Hashing Functions Based on Intractable Problems

Hashing functions can also be based on one-way functions such as exponentiation, squaring, knapsack (cf. Pieprzyk and Sadeghiyan [46]), and discrete logarithm. More recently, a group-theoretic construction using the SL_2 groups has been proposed by Tillich and Zémor [67].

A scheme based on RSA exponentiation as the underlying one-way function is defined by

$$H_0 = IV$$
$$H_i = (H_{i-1} \oplus M_i)^e \bmod N \quad i = 1, 2, \ldots, t$$
$$H(M) = H_t$$

where the modulus N and the exponent e are public. A correcting block attack can be used to compromise the scheme by appending or inserting a carefully selected last block message to achieve a desired hash value. To immunize the scheme against this attack, it is necessary to add redundancy to the message so that the last message block cannot be manipulated (cf. Davies and Price [22]). To ensure the security of RSA, N should be at least 512 bits in length, making the implementation of the above scheme very slow.

To improve the performance of the above scheme, the public exponent can be made small. For example, squaring can be used:

$$H_i = (H_{i-1} \oplus M_i)^2 \bmod N.$$

It is suggested that 64 bits of every message block be set to 0 to avoid a correcting block attack.

An algorithm for hashing based on squaring is proposed in Appendix D of the X.509 recommendations of the CCITT standards on secure message handling. The proposal stipulates that 256 bits of redundancy be distributed over every 256-bit message block by interleaving every four bits of the message with 1111, so that the total number of bits in each block becomes 512. The exponentiation algorithm, with exponent two, is then run on the modified message in CBC mode (cf. Pieprzyk and Sadeghiyan [46]). In this scheme, the four most significant bits of every byte in each block are set to 1. Coppersmith [19] shows how to construct colliding messages in this scheme.

Damgård [21] described a scheme based on squaring, which maps a block of n bits into a block of m bits. The scheme is defined by

$$H_0 = IV$$
$$H_i = \text{extract}(00111111\|H_{i-1}\|M_i)^2 \bmod N$$
$$H(M) = H_t.$$

In the above scheme, the role of extract is to extract m bits from the result of the squaring function. To obtain a secure scheme, m should be sufficiently large so as to thwart the birthday attack; this attack will be explained later. Moreover, extract should select those bits for which finding colliding inputs is difficult. One way to do this is to extract m bits uniformly. However, for practical reasons, it is better to bind them together in bytes. Another possibility is to extract every fourth byte. Daemen et al. [20] showed that this scheme can be broken.

Impagliazzo and Naor [32] propose a hashing function based on the knapsack problem. The description of the scheme is as follows. Choose at random numbers $a_1, \ldots, a_n$ in the interval $0, \ldots, N$, where n indicates the length of the message in bits, and $N = 2^\ell - 1$ where $\ell < n$. A binary message $M = M_1, M_2, \ldots, M_n$ is hashed as

$$H(M) = \sum_{i=1}^{n} a_i M_i \bmod 2^\ell.$$

Impagliazzo and Naor do not give any concrete parameters for the above scheme, but they have shown that it is theoretically sound.

Gibson [30] constructs hash functions whose security is conditional upon the difficulty of factoring certain numbers. The hash function is defined by

$$f(x) = a^x \pmod{n},$$

where $n = pq$, p and q are large primes, and a is a primitive element of the ring Z_n. In Gibson's hash function, n has to be sufficiently large to ensure the difficulty of factoring. This constraint makes the hash function considerably slower than the MD4 algorithm.

Tillich and Zémor [67] proposed a hashing scheme where the message digests are given by two-dimensional matrices with entries in the binary Galois fields GF(2^n) for $130 \le n \le 170$. The hashing functions are parameterized by the irreducible polynomials $P_n(X)$ of degree n over GF(2); their choice is left to the user. Their scheme has several provably secure properties: detection of local modification of text; and resistance to the birthday attack as well as other attacks. Hashing is fast as digests are produced by matrix multiplication in GF(2^n) which can be parallelized.

Messages (encoded as a binary strings) $x_1 x_2 \ldots$ of arbitrary length are mapped to products of a selected pair of generators $\{A, B\}$ of the group $SL(2, 2^n)$, as follows:

$$x_i = \begin{cases} A & \text{if } x_i = 0 \\ B & \text{if } x_i = 1. \end{cases}$$

The resulting product belongs to the (infinite) group $SL(2, GF(2)[X])$, where $GF(2)[X]$ is the ring of all polynomials over $GF(2)$. The product is then reduced modulo an irreducible polynomial of degree n (Euclidean algorithm), mapping it to an element of $SL(2, 2^n)$. The four n-bit entries of the reduced matrix give the $(3n + 1)$-bit message digest of $x_1 x_2 \ldots$.

Charnes and Pieprzyk [15] showed how to find irreducible polynomials which produce collisions for the SL_2 hash functions. Geiselmann [29] gave an algorithm for producing potential collisions for the $SL(2, 2^n)$ hashing scheme, which is independent of the choice of the irreducible polynomials. The complexity of his algorithm is that of the discrete logarithm problem in $GF(2^n)$ or $GF(2^{2n})$. However, no collisions in the specified range of the hash function have been found using this algorithm. Some pairs of rather long colliding strings are obtained for a toy example of $GF(2^{21})$.

12.4.5   Hashing Algorithms

Rivest [51] proposed a hashing algorithm called MD4. It is a software-oriented scheme that is especially designed to be fast on 32-bit machines. The algorithm produces a 128-bit output, so it is not computationally feasible to produce two messages having the same hash value. The scheme provides diffusion and confusion using three Boolean functions. The MD5 hashing algorithm is a strengthened version of MD4 [52]. MD4 has been broken by Dobbertin [25].

HAVAL stands for a one-way hashing algorithm with a variable length of output. It was designed at the University of Wollongong by Zheng et al. [79]. It compresses a message of an arbitrary length into a digest of either 128, 160, 192, 224, or 256 bits. The security level can be adjusted by selecting 3, 4, or 5 passes. The structure of HAVAL is based on MD4 and MD5. Unlike MD4 and MD5 whose basic operations are done using functions of three Boolean variables, HAVAL employs five Boolean functions of seven variables (each function serves a single pass). All functions used in HAVAL are highly nonlinear, 0-1 balanced, linearly inequivalent, mutually output-uncorrelated, and satisfy the strict avalanche criterion (SAC).

Charnes and Pieprzyk [14] proposed a modified version of HAVAL based on five Boolean functions of five variables. The resulting hashing algorithm is faster than the five pass, seven variable version of the original HAVAL algorithm. They use the same cryptographic criteria that were used to select the Boolean functions in the original scheme. Unlike the seven variable case, the choice of the Boolean functions is fairly restricted in the modified setting. Using the shortest algebraic normal form of the Boolean functions as one of the criteria (to maximize the speed of processing), it was shown in [14] that the Boolean functions are optimal. No attacks have been reported for the five variable version.

12.4.6   Attacks

The best method to evaluate a hashing scheme is to see what attacks an adversary can perform to find collisions. A good hashing algorithm produces a fixed length number which depends on all the bits of the message. It is generally assumed that the adversary knows the hashing algorithm. In a conservative approach, it is assumed that the adversary can perform an adaptive chosen message attack, where it may choose messages, ask for their digests, and try to compute colliding messages. There are several methods for using such pairs to attack a hashing scheme and calculate colliding messages. Some methods are quite general and can be applied against any hashing scheme; for example, the so-called birthday attack. Other methods are applicable only to specific hashing schemes. Some attacks can be used against a wide range of hash functions. For example, the so-called meet-in-the-middle attack is applicable to any scheme that uses some sort of block chaining in its structure. In another example, the so-called correcting block attack is applicable mainly to hash functions based on modular arithmetic.

12.4.6.1 Birthday Attack

The idea behind this attack originates from a famous problem from probability theory, called the birthday paradox. The paradox can be stated as follows. What is the minimum number of pupils in a classroom so the probability that at least two pupils have the same birthday which is greater than 0.5? The answer to this question is 23, which is much smaller than the value suggested by intuition. The justification for the paradox is as follows. Suppose that the pupils are entering the classroom one at a time, the probability that the birthday of the first pupil falls on a specific day of the year is equal to $\frac{1}{365}$. The probability that the birthday of the second pupil is not the same as the first one is equal to $1 - \frac{1}{365}$. If the birthdays of the first two pupils are different, the probability that the birthday of the third pupil is different from the first one and the second one is equal to $1 - \frac{2}{365}$. Consequently, the probability that t students have different birthdays is equal to $\left(1 - \frac{1}{365}\right)\left(1 - \frac{2}{365}\right)\dots\left(1 - \frac{t-1}{365}\right)$, and the probability that at least two of them have the same birthday is

$$P = 1 - \left(1 - \frac{1}{365}\right)\left(1 - \frac{2}{365}\right)\dots\left(1 - \frac{t-1}{365}\right).$$

It can be easily computed that for $t \geq 23$, this probability is greater than 0.5.

The birthday paradox can be employed for attacking hash functions. Suppose that the number of bits of the hash value is n, an adversary generates r_1 variations of a bogus message and r_2 variations of a genuine message. The probability of finding a bogus message and a genuine message that hash to the same digest is

$$P \approx 1 - e^{-\frac{r_1 r_2}{2^n}}$$

where $r_2 \gg 1$. When $r_1 = r_2 = 2^{\frac{n}{2}}$, the above probability is ≈ 0.63. Therefore, any hashing algorithm which produces digests of length around 64 bits is insecure, since the time complexity function for the birthday attack is $\approx 2^{32}$. It is usually recommended that the hash value should be around 128 bits to thwart the birthday attack.

This method of attack does not take advantage of the structural properties of the hash scheme or its algebraic weaknesses. It applies to any hash scheme. In addition, it is assumed that the hash scheme assigns a value to a message which is chosen with a uniform probability among all the possible hash values. Note that if the structure is weak or has certain algebraic properties, the digests do not have a uniform probability distribution. In such cases, it may be possible to find colliding messages with a better probability and fewer message-digest pairs.

12.4.6.2 Meet-in-the-Middle Attack

This is a variation of the birthday attack, but instead of comparing the digests, the intermediate results in the chain are compared. The attack can be made against schemes which employ some sort of block chaining in their structure. In contrast to birthday attack, a meet-in-the-middle attack enables an attacker to construct a bogus message with a desired digest. In this attack, the message is divided into two parts. The attacker generates r_1 variations on the first part of a bogus message. He starts from the initial value and goes forward to the intermediate stage. He also generates r_2 variations on the second part of the bogus message. He starts from the desired false digest and goes backward to the intermediate stage. The probability of a match in the intermediate stage is the same as the probability of success in the birthday attack.

12.4.6.3 Correcting-Block Attack

In a correcting block attack, a bogus message is concatenated with a block to produce a corrected digest of the desired value. This attack is often applied to the last block and is called correcting last block attack, although it can be applied to other blocks as well. Hash functions based on modular arithmetic are especially sensitive to the correcting last block attack. The introduction of redundancy into the message in these schemes makes finding a correcting block with the necessary redundancy difficult. However, it makes the scheme less efficient. The difficulty of finding a correcting block depends on the nature of the redundancy introduced.

12.4.6.4 Differential Analysis

Biham and Shamir [11] developed a method for attacking block ciphers, known as differential cryptanalysis. This is a general method for attacking cryptographic algorithms, including hashing schemes. Dobbertin [25] was the first to cryptanalyze the MD4 hashing algorithm, and he also made a preliminary cryptanalysis of MD5. In 2004 Wang et al. [74] announced they had found colliding pairs of messages for the hash functions MD4, MD5, HAVAL-128, and RIPEMD. More specifically, these differential attacks find two differing vectors with the same hash; they are not preimage attacks which attempt to find data set with a given hash. For example, the MD5 algorithm, which is used widely for password authentication systems (it became the most common Unix password hashing algorithm in the 1990s), is not affected by the attacks of [74]. In response to the attacks of [74], the Australian Government Defence Signals Directorate (DSD) issued a public bulletin in 2005 stating that: "DSD does not believe the recent results represent a practical threat to users of SHA-1 or MD5 for most purposes."

12.5   MAC

MAC provide message integrity and are one of the most important security primitives in current distributed information systems. A MAC is a symmetric key cryptographic primitive that consists of two algorithms. A MAC generation algorithm, $G = \{G_k : k = 1, \ldots, N\}$, takes an arbitrary message, s, from a given collection S of messages and produces a *tag*, $t = G_k(s)$, which is appended to the message to produce an authentic message, $m = (s.t)$. A MAC verification algorithm, $V = \{V_k(.) : k = 1, \ldots, N\}$, takes authenticated messages of the form $(s.t)$ and produces a true or false value, depending on whether the message is authentic. The security of a MAC depends on the best chance that an active spoofer has to successfully substitute a received message $(s.G_k(s))$ for a fraudulent one, $m' = (s', t)$, so that $V_k(m')$ produces a true result. In MAC systems, the communicants share a secret key and are therefore not distinguishable cryptographically.

The security of MACs can be studied from the point of view of unconditional or computational security.

Unconditionally secure MACs are equivalent to cartesian authentication codes. However, in MAC systems only multiple communications are of interest. In Section 12.2, A-codes that provide protection for multiple transmissions were discussed. In Section 12.5.1, we present a construction for a MAC that has been the basis of all the recent MAC constructions and have a number of important properties. It is provably secure; the number of key bits required is asymptotically minimal; and it has a fast implementation.

Computationally secure MACs have arisen from the needs of the banking community. They are also studied under other names, such as keyed hash functions and keying hash functions. In Section 12.5.3, we review the main properties and constructions of such MACs.

12.5.1 Unconditionally Secure MACs

When the adversary has unlimited computational resources, attacks against MAC systems and the analysis of security are similar to that of Cartesian A-codes. The adversary observes n codewords of the form $s_i.t_i$, $i = 1, \ldots, n$, in the channel and attempts to construct a fraudulent codeword $s.t$ which is accepted by the receiver. (This is the same as spoofing of order n in an A-code.) If the communicants want to limit the adversary's chance of success to p after n message transmissions, the number of authentication functions (number of keys) must be greater than a lower bound which depends on p. If the adversary's chance of success in spoofing of order i, $i = 1, \ldots, n$ is p_i, then at least $1/p_1 p_2 \ldots p_n$ keys are required; see Fak [27], Wegman and Carter [75] for a proof of this. For $p_i = p$, $i = 1, \ldots, n$, the required number of key bits is $-n \log_2 p$. That is, for every message, $-\log_2 p$ key bits are required. This is the absolute minimum for the required number of key bits.

Perfect protection is obtained when the adversary's best strategy is a random choice of a tag and appending it to the message; this strategy succeeds with probability $p = 2^{-k}$, if the size of the tag is k bits. In this case, the number of required key bits for every extra message is k.

Wegman and Carter [75] give a general construction for unconditionally secure MACs that can be used for providing protection for an arbitrary number of messages.

Their construction uses universal classes of hash functions. Traditionally, a hash function is used to achieve fast average performance over all inputs in varied applications, such as databases. By using a universal class of hash functions, it is possible to achieve provable average performance without restricting the input distribution.

Let $h : A \to B$ be a hash function mapping the elements of a set A to a set B. A strongly universal$_n$ class of hash function is a class of hash functions with the property that for n distinct elements $a_1, \ldots, a_n$ of A and n distinct elements $b_1, \ldots, b_n$ of B, exactly $|H|/(b^n)$ functions map a_i to b_i, for $i = 1, \ldots, n$. Strongly universal$_n$ hash functions give perfect protection for multiple messages as follows. The transmitter and the receiver use a publicly known class of strongly universal$_n$ hash functions, and a shared secret key determines a particular member of the class that they will use for their communication. Stinson [63] shows that a class of strongly universal$_2$ that maps a set of a elements to a set of b elements is equivalent to an orthogonal array $OA_\lambda(2, a, b)$ with $\lambda = |H|/b^2$. Similar results can be proved for strongly universal$_n$ classes of hash functions. Because of this equivalence, universal$_n$ hash functions are not a practically attractive solution. In particular, this proposal is limited by the constraints of constructing orthogonal arrays with arbitrary parameters. A good account of orthogonal arrays and other combinatorial structures can be found in the monograph by Beth et al. [10].

12.5.2 Wegman and Carter Construction

Wegman and Carter show that, instead of strongly universal$_n$, one can always use a strongly universal$_2$ family of hash functions together with a one time pad of random numbers. The system works as follows. Let B denote the set of tags consisting of the sequences of k bit strings. Let $\mathcal{H}$ denote a strongly universal$_2$ class of hash functions mapping S to B. Two communicants share a key that specifies a function $h \in \mathcal{H}$ together with a one-time pad containing k-bit random numbers. The tag for the jth message s_j is $h(s_j) \oplus r_j$, where r_j is the jth number on the pad. It can be proved that this system limits the adversary's chance of success to 2^{-k} as long as the pad is random and not used repeatedly. The system requires $nk + K$ bits of key, where K is the number of bits required to specify an element of $\mathcal{H}$, n is the number of messages to be authenticated, and k is the size of the tags.

This construction has a number of remarkable properties. First, for large n, the key requirement for the system approaches the theoretical minimum of k bits per message. This is because for large n the number of key bits is effectively determined by nk. Second, the construction of MAC with provable security for multiple communications is effectively reduced to the construction of a better studied primitive, that is, strongly universal$_2$ class of hash functions. Finally, by replacing the one-time pad with a pseudorandom sequence generator, unconditional security is replaced by computational security.

Wegman and Carter's important observation is the following. By not insisting on the minimum value for the probability of success in spoofing of order one, that is, allowing $p_1 > 1/k$, it is possible to reduce the number of functions and thus the required number of keys. This observation leads to the notion of almost strongly universal$_2$ class.

An ϵ-almost universal$_2$ (or ϵ-AU$_2$) class of hash functions has the following property. For any pair $x, y \in A, x \neq y$, the number of hash functions h with $h(x) = h(y)$ is at most equal to ϵ. The ϵ-almost strongly-universal$_2$ (or ϵ-ASU$_2$) hash functions have the added property that for any $x \in A, y \in B$, the number of functions with $h(x) = y$ is $|H|/|B|$. Using an ϵ-almost strongly universal$_2$ class of functions in the Wegman and Carter construction results in MAC systems for which the probability of success for an intruder is ϵ. Such MACs are called ϵ-otp-secure, see Krawczyk [37].

Stinson [65] gives several methods for combining hash functions of class AU$_2$ and ASU$_2$. The following theorem shows that an ϵ-ASU$_2$ class can be constructed from an ϵ-AU$_2$ class.

[65] Suppose H_1 is an ϵ_1-AU$_2$ class of hash functions from A_1 to B_1, and suppose H_2 is an ϵ_2-ASU$_2$ class of hash functions from B_1 to B_2, then there exists an ϵ-ASU$_2$ class H of hash functions from A_1 to B_2, where $\epsilon = \epsilon_1 + \epsilon_2$ and $|H| = |H_1| \times |H_2|$.

This theorem further reduces the construction of MACs with provable security to the construction of ϵ-AU$_2$ class of hash functions.

Several constructions for ϵ-ASU$_2$ hash functions are given by Stinson [65]. Johansson et al. [35] establish relationships between ASU$_2$ hash functions and error correcting codes. They use geometric error correcting codes to construct new classes of ϵ-ASU$_2$ hash function of smaller size. This reduces the key size.

Krawczyk [37] shows that in the Wegman–Carter construction, ϵ-ASU$_2$ hash functions can be replaced with a less demanding class of hash functions, called ϵ-otp-secure. The definition of this class differs from other classes of hash functions, in that it is directly related to MAC constructions and their security, in particular, to the Wegman–Carter construction.

Let $s \in S$ denote a message that is to be authenticated by a k bit tag $h(s) \oplus r$, constructed by Wegman and Carter's method. An adversary succeeds in a spoofing attack if he or she can find $s' \neq s, t' = h(s') \oplus r$, assuming that he or she knows H but does not know h and r. A class H of hash functions is ϵ-otp-secure if for any message no adversary succeeds in the above attack scenario with a probability greater than ϵ.

[37] A necessary and sufficient condition for a family H of hash functions to be ϵ-otp-secure is that

$$\forall a_1 \neq a_2, c, \ Pr_h \left(h\left(a_1 \right) \oplus h\left(a_2 \right) = c \right) \leq \epsilon.$$

The need for high speed MACs has increased with the progress in high speed data communication. A successful approach to the construction of such MACs uses hash function families in which the message is hashed by multiplying it by a binary matrix. Because hashing is achieved with exclusive-or operations, it can be efficiently implemented in software. An obvious candidate for such a class of hash functions, originally proposed by Wegman and Carter [13,75], is the set of linear transformations from A to B. It is shown that this forms an ϵ-AU$_2$ class of hash functions. However, the size of the key—the number of entries in the matrix—is too large, and too many operations are required for hashing. Later proposals by Krawczyk [37] and by Rogaway [55] are aimed at alleviating these problems and have a fast software implementation. The former uses Topelitz matrices [37], while the latter uses binary matrices with only three ones per column. In both cases, the resulting family is ϵ-AU$_2$.

The design of a complete MAC usually involves a number of hash functions which are combined by methods, similar to those proposed by Stinson [65]. The role of some of the hash functions is to produce high compression (small b), while others produce the desired spread and uniformity (see Rogaway [55]).

Reducing the key size of the hash function is especially important in practical applications, because the one-time pad is replaced by the output of a pseudorandom generator with a short key (of the order of 128 bits). Hence, it is desirable to have the key size of the hash function of similar order.

12.5.3 Computational Security

In the computationally secure approach, protection is achieved because excessive computation is required for a successful forgery. Although a hash value can be used as a checksum to detect random changes in the data, a secret key must be used to provide protection against active tampering. Methods for constructing MACs from hash functions have traditionally followed one of the following approaches: the so-called hash-then-encrypt and keying a hash function.

Hash-then-encrypt: To construct a MAC, for a message x with this method, the hash value of x is calculated, and the result is encrypted using an encryption algorithm. This is similar to signature generation, where a public key algorithm is replaced by a private key encryption function.

There are a number of drawbacks to this method. First, the overall scheme is slow. This is because the two primitives used in the construction, that is, the cryptographic hash functions and encryption functions are designed for other purposes and have extra security properties which are not strictly required in the construction. Although this construction can produce a secure MAC, the speed of the MAC is bounded by the speed of its constituent algorithms. For example, cryptographic hash functions are designed to be one-way. It is not clear whether this is a required property in the hash-then-encrypt construction, where the output of the hash function is encrypted and one-wayness is effectively obtained through the difficulty in finding the plaintext from the ciphertext.

Keying a Hash Function: In the second approach, a secret key is incorporated into a hashing algorithm. This operation is sometimes called keying a hash function (see Bellare et al. [4]). This method is attractive, because of the availability of hashing algorithms and their relative speed in software implementation; these algorithms are not subject to export restrictions.

Although this scheme can be implemented more efficiently in software than the previous scheme, the objection to the superfluous properties of the hash functions remains.

The keying method depends on the structure of the hash function. Tsudik [71] proposed three methods of incorporating the key into the data. In the secret prefix method, $G_k(s) = H(k \| s)$, while in the secret suffix, we have $G_k(s) = H(s \| k)$. Finally, the envelope method combines the previous two methods with $G_k(s) = H(k_1 \| s \| k_2)$ and $k = k_1 \| k_2$.

Instead of including the key into the data, the key information can be included into the hashing algorithm. In iterative hash functions such as MD5 and SHA, the key can be incorporated into the initial vector, compression function, or into the output transformation.

There have also been some attempts at defining and constructing secure keyed hash functions as independent primitives, namely by Berson et al. [9] and Bakhtiari et al. [1]. The former propose a set of criteria for secure keyed hash functions and give constructions using one-way hash functions. The latter argue that the suggested criteria for security are in most cases excessive; relaxing these allows constructions of more efficient secure keyed hash functions. Bakhtiari et al. also give a design of a keyed hash function from scratch. Their design is based mostly on intuitive principles and lacks a rigorous proof of security. A similar approach is taken in the design of MDx-MAC by Preneel and van Oorschot [48], which is a scheme for constructing a MAC from a MD5-type hash function.

12.5.3.1 Security Analysis of Computationally Secure MACs

The security analysis of computationally secure MACs has followed two different approaches. In the first approach, the security assessment is based on an analysis of some possible attacks. In the second approach, a security model is developed and used to examine the proposed MAC.

Security Analysis Through Attacks: Consider a MAC algorithm that produces MACs of length m using a k bit key. In general, an attack might result in a successful forgery, or in the recovery of the key. According to the classification given by Preneel and van Oorschot [49], a forgery in a MAC can be either existential—the opponent can construct a valid message and MAC without the knowledge of the key pair—or selective—the opponent can determine the MAC for a message of his choice. Protection against the former type of attack imposes more stringent conditions than the latter type of attack. A forgery is verifiable if the attacker can determine with a high probability whether the attack is successful. In a chosen text attack, the attacker is given the MACs for the messages of his own choice. In an adaptive attack, the attacker chooses text for which he can see the result of his previous request before forming his next request. In a key recovery attack, the aim of the attacker is to find the key. If the attacker is successful, he can perform selective forgery on any message of his choice, and the security of the system is totally compromised.

For an ideal MAC, any method to find the key is as expensive as an exhaustive search of $O(2^k)$ operations. If $m < k$, the attacker may randomly choose the MAC for a message with the probability of success equal to $1/2^m$. However, in this attack, the attacker cannot verify whether his attack has been successful.

The complexity of various attacks is discussed by several authors: Tsudik [71], Bakhtiari et al. [2], and Bellare et al. [5]. Preneel and van Oorschot [48,49] propose constructions resistant to such attacks. Some attacks can be applied to all MACs obtained using a specific construction method while other attacks are limited to particular instances of the method.

12.5.3.2 Formal Security Analysis

The main attempts at formalizing the security analysis of computationally secure MACs are due to Bellare et al. [4], and Bellare and Rogaway [6]. In both papers, an attack model is first carefully defined, and the security of a MAC with respect to that model is considered. Bellare et al. [4] use their model to prove the security of a generic construction on the basis of pseudorandom functions, while Bellare and Rogaway [6] use their model to prove the security of a generic construction on the basis of hash functions.

MAC from pseudorandom functions: The formal definition of security given by Bellare et al. [6] assumes that the adversary can ask the transmitter to construct tags for messages of his choice and also ask the receiver to verify chosen message and tag pairs. The number of these requests is limited, and a limited time t can be spent on the attack. Security of the MAC is expressed as an upper bound on the adversary's chance of succeeding in its best attack.

The construction proposed by Bellare et al. applies to any pseudorandom function. Their proposal, called XOR-MAC, basically breaks a message into blocks. For each block, the output of the pseudorandom function is calculated, and the outputs are finally XORed. Two schemes based on this approach are proposed: the randomized XOR scheme and the counter-based scheme.

The pseudorandom function used in the above construction can be an encryption function, such as DES, or a hash function, such as MD5. It is proved that the counter based scheme is more secure than the randomized scheme, and if DES is used, both schemes are more secure than CBC MAC. Some of the desirable features of this construction are parallelization and incrementality. The former means that message blocks can be fed into the pseudorandom function in parallel. The latter refers to the feature of calculating incrementally the value of the MAC for a message s' which differs from s in only a few blocks.

MAC from hash functions: The model used by Bellare and Rogaway [6] is similar to the above one. The adversary can obtain information by asking queries; however, in this case, queries are only addressed to the transmitter.

A family of functions $\{F_k\}$ is (ϵ, t, q, L)-secure MAC [5] if any adversary that is not given the key k, is limited to spend total time t, and sees the values of the function F_k computed on q messages $s_1, s_2, \ldots, s_q$ of its choice, each of length at most L, cannot find a message and tag pair $(s, t), s \neq s_i, i = 1, \ldots, q$, such that $t = F_k(s)$ with a probability better that ϵ.

Two general constructions for MAC from hash functions, the so-called NMAC (the nested construction) and HMAC (the hash-based MAC) are given and their security is formally proved in [4].

If the keyed compression function f is a (ϵ_f, t, q, L)-secure MAC and the keyed iterated hash function F is (ϵ_F, t, q, L)-weakly collision-resistant, then the NMAC is $(\epsilon_f + \epsilon_F, t, q, L)$-secure MAC.

Weak collision-resistance is a much weaker notion than the collision resistance of (unkeyed) hash functions, because the adversary does not know the secret key and finding collision is much more difficult in this case. More precisely, a family of keyed hash functions $\{F_k\}$ is (ϵ, t, q, L)-weakly collision-resistant if any adversary, that is not given the key k, is limited to spend total time t, and sees the values of the function F_k computed on q messages $m_1, m_2, \ldots, m_q$ of its choice, each of length at most L, cannot find messages m and m' for which $F_k(m) = F_k(m')$ with probability better that ϵ. With some extra assumptions, similar results are proved for the HMAC construction.

A related construction is the collisionful keyed hash function proposed by Gong [31]. In his construction, the collisions can be selected and the resulting function can be claimed to provide security against password guessing attacks. Bakhtiari et al. [3] explore the security of Gong's function and a key exchange protocol based on collisionful hash functions.

Two distinguishers for HMACs based on the hash functions HAVAL, MD4, MD5, SHA-0, and SHA-1 were described byKim et al. [36]. It was shown that these can be used to devise a forgery attack on the HMACs and NMACs based on these functions. (A distinguishing attack establishes that with some probability, the output of a stream cipher's keystream differs from a random bitstream.) Bellare [8] proves that HMAC is a pseudo randomfunction without the assumption that the underlying hash function is collisionresistant; see Section 12.4.6.

12.5.4 Applications

The main application of a MAC is to provide protection against active spoofing (see Carter and Wegman [13]). This is particularly important in open distributed systems such as the Internet. Other applications include secure password checking and software protection. MACs can be used to construct encryption functions and have been used in authentication protocols in place of encryption functions (cf. Bird et al. [12]). An important advantage of MAC functions is that they are not subject to export restrictions. Other applications of MAC functions are to protect software against viruses or to protect computer files against tampering. Integrity checking is an important service in a computer operating system which can be automated with software tools.

12.6 Digital Signatures

Digital signatures are meant to be the electronic equivalents of handwritten signatures. They should preserve the main features of handwritten signatures. Obviously, it is desirable that digital signatures be as legally binding as handwritten ones. There are three elements in every signature: the signer, the document, and the time of signing.

In most cases, the document already includes a timestamp. A digital signature must reflect both the content of the document and the identity of the signer. The signer is uniquely identified by its secret key. In particular, we require the signature to be

- Unique—a given signature reflects the document and can be generated by the signer only
- Unforgeable—it must be computationally intractable for an opponent to forge the signature
- Easy to generate by the signer and easy to verify by recipients
- Impossible to deny by the signer (nonrepudiation)

A digital signature differs from a handwritten signature such that it is not physically attached to the document on a piece of paper. Digital signatures have to be related both to the signer and the document by a cryptographic algorithm. Signatures can be verified by any potential recipient. Therefore, the verification algorithm must be public. Signature verification succeeds only when the signer and document match the signature.

There are two general classes of signature schemes:

- One-time signature schemes
- Multiple signature schemes

One-time signature schemes can be used to sign only one message. To sign a second message, the signature scheme has to be reinitialized; however, any signature can be verified repeatedly. Multiple signature schemes can be used to sign several messages without the necessity to reinitialize the signature scheme.

In practice, a signature scheme is required to provide a relatively short signature for a document of an arbitrary length. We sign the document by generating a signature for its digest. The hashing employed to produce the digest must be secure and collision free.

12.6.1 One-Time Signature Schemes

This class of signature schemes can be implemented using any one-way function. These schemes were first developed using private key cryptosystems. We follow the original notation. An encryption algorithm is used as a one-way function. To set up the signature scheme, the signer chooses a one-way function (encryption algorithm). The signer selects an index k (secret key) randomly and uniformly from the set of keys, K. The index determines an instance of the one-way function, that is, $E_k : \Sigma^n \to \Sigma^n$, where $\Sigma = \{0, 1\}$; it is known only by the signer. Note that n has to be large enough to avoid birthday attacks.

12.6.1.1 Lamport Scheme

Lamport's scheme [39] generates signatures for n-bit messages. To sign a message, the signer first chooses randomly n key pairs:

$$(K_{10}, K_{11}), (K_{20}, K_{21}), \ldots, (K_{n_0}, K_{n_1}). \tag{12.1}$$

The pairs of keys are kept secret and are known to the signer only. Next, the signer creates two sequences, S and R:

$$S = \{(S_{10}, S_{11}), (S_{20}, S_{21}), \ldots, (S_{n_0}, S_{n_1})\},$$
$$R = \{(R_{10}, R_{11}), (R_{20}, R_{21}), \ldots, (R_{n_0}, R_{n_1})\}. \tag{12.2}$$

The elements of S are selected randomly and the elements of R are cryptograms of S so

$$R_{ij} = E_{K_{ij}}\left(S_{ij}\right) \quad \text{for } i = 1, \ldots, n \text{ and } j = 0, 1, \tag{12.3}$$

where E_K is the encryption function of the selected symmetric cryptosystem. S and R are stored in a read-only public register; they are known by the receivers.

The signature of an n-bit message $M = (m_1, \ldots, m_n)$, $m_i \in \{0, 1\}$ for $i = 1, \ldots, n$, is a sequence of cryptographic keys,

$$S(M) = \{K_{1i_1}, K_{2i_2}, \ldots, K_{ni_n}\} \tag{12.4}$$

where $i_j = 0$ if $m_j = 0$; otherwise $i_j = 1$, $j = 1, \ldots, n$. A receiver validates the signature $S(M)$ by verifying whether suitable pairs of S and R match each other for known keys.

12.6.1.2 Rabin Scheme

In Rabin's scheme [50], a signer begins the construction of the signature by generating $2r$ keys at random:

$$K_1, K_2, \ldots, K_{2r}. \tag{12.5}$$

The parameter r is determined by the security requirements. The K_i are secret and known only to the signer. Next, the signer creates two sequences which are needed by the recipients to verify the signature. The first sequence,

$$S = \{S_1, S_2, \ldots, S_{2r}\}$$

comprises binary blocks chosen at random by the signer. The second,

$$R = \{R_1, R_2, \ldots, R_{2r}\},$$

is created using the sequence S, $R_i = E_{K_i}(S_i)$ for $i = 1, \ldots, 2r$. S and R are stored in a read-only public register.

The signature is generated using the following steps. The message to be signed M is enciphered under keys $K_1, \ldots, K_{2r}$. The cryptograms

$$E_{K_1}(M), \ldots, E_{K_{2r}}(M) \tag{12.6}$$

form the signature $S(M)$. The pair $(S(M), M)$ is sent to the receivers.

To verify the signature, a receiver selects randomly a $2r$-bit sequence σ of r-ones and r-zeros. A copy of σ is forwarded to the signer. Using σ, the signer forms an r-element subset of the keys with the property that K_i belongs to the subset if and only if the ith element of the $2r$-bit sequence is '1'; $i = 1, \ldots, 2r$. The subset of keys is then communicated to the receiver. To verify the key subset, the receiver generates and compares r suitable cryptograms of S with the originals kept in the public register.

12.6.1.3 Matyas–Meyer Scheme

Matyas and Meyer [41] propose a signature scheme based on the DES algorithm. However, any one-way function can be used in this scheme.

The signer first selects a random matrix $U = [u_{i,j}]$ $i = 1, \ldots, 30, j = 1, \ldots, 31$ and $u_{i,j} \in \Sigma^n$. Using U, a 31×31 matrix $KEY = [k_{i,j}]$ is constructed for $k_{i,j} \in \Sigma^n$. The first row of KEY matrix is chosen at random, the other rows are

$$k_{i+1,j} = E_{k_{i,j}}\left(u_{i,j}\right)$$

for $i = 1, \ldots, 30$ and $j = 1, \ldots, 31$. Finally, the signer installs the matrix U and the vector $(k_{31,1}, \ldots, k_{31,31})$ (the last row of KEY) in a public registry. To sign a message $m \in \Sigma^n$, the cryptograms

$$c_i = E_{k_{31,i}}(m) \quad \text{for } i = 1, \ldots, 31$$

are computed. The cryptograms are considered as integers and ordered according to their values so $c_{i_1} < c_{i_2} < \cdots < c_{i_{31}}$. The signature of m is the sequence of keys

$$SG_k(m) = \left(k_{i_1,1}, k_{i_2,2}, \ldots, k_{i_{31},31} \right).$$

The verifier takes the message m, recreates the cryptograms c_i, and orders them in increasing order. Next, the signature-keys are put into the "empty" matrix KEY in the entries indicated by the ordered sequence of c_i's. The verifier then repeats the signer's steps and computes all keys below the keys of the signature. The signature is accepted if the last row of KEY is identical to the row stored in the registry.

12.6.2 Signature Schemes Based on Public-Key Cryptosystems

12.6.2.1 RSA Signature Scheme

First, a signer sets up the RSA system [53] with the modulus $N = p \times q$, where the two primes p and q are fixed. Next a random decryption key $d \in Z_N$ is chosen; the encryption key e is

$$e \times d \equiv 1 \ (\text{mod } (p - 1, q - 1)).$$

The signer publishes both the modulus N and the decryption key d.

Given a message $M \in Z_N$, the signature generated by the signer is

$$S \equiv M^e \ (\text{mod } N).$$

Since the decryption key is public, anyone can verify whether

$$M \equiv S^d \ (\text{mod } N).$$

The signature is considered to be valid if this congruence is satisfied. RSA signatures are subject to various attacks which exploit the commutativity of exponentiation.

12.6.2.2 ElGamal Signature Scheme

The signature scheme due to ElGamal [26] is based on the discrete logarithm problem. The signer chooses a finite field GF(p) for a sufficiently large prime p. A primitive element $g \in$ GF(p) and a random integer $r \in$ GF(p) are fixed. Next, the signer computes

$$K \equiv g^r \ (\text{mod } p)$$

and announces K, g, and p. To sign a message $M \in$ GF(p), the signer selects a random integer $R \in$ GF(p) such that gcd $(R, p - 1) = 1$ and calculates

$$X \equiv g^R \ (\text{mod } p).$$

Using this data, following congruence is solved

$$M \equiv r \times X + R \times Y \ (\text{mod } p - 1)$$

for Y using Euclid's algorithm. The signature of M is the triple (M, X, Y). Note that r and R are kept secret by the signer. The recipient of the signed message forms

$$A \equiv K^X X^Y \ (\text{mod } p)$$

and accepts the message M as authentic if $A \equiv g^M \ (\text{mod } p)$. It is worth to note that knowledge of the pair (X, Y) does not reveal the message M. As a matter of fact, there are many pairs matching the message—for every pair (r, R), there is a pair (X, Y).

Since discrete exponent systems can be based on any cyclic group, the ElGamal signature scheme can be extended to this setting.

12.6.3 Special Signatures

Sometimes, additional conditions are imposed on digital signatures. Blind signatures are useful in situations where the message to be signed should not be revealed to the signer. Unlike typical digital signatures, the undeniable versions require the participation of the signer to verify the signature. Fail-stop signatures are used whenever there is a need for protection against a very powerful adversary. As these signatures require interactions among the parties involved, the signatures are sometimes called signature protocols.

12.6.3.1 Blind Signatures

The concept of blind signatures was introduced by Chaum [17]. They are applicable to situations where the holder of a message M needs to get M signed by a signer (which could be a public registry) without revealing the message. This can be done with the following steps:

- The holder of the message first encrypts it.
- The holder sends a cryptogram of the message to the signer.
- The signer generates the signature for the cryptogram and sends it back to the holder.
- The holder decodes the encryption and obtains the signature of the message.

This scheme works if the encryption and signature operations commute, for example, the RSA scheme can be used to implement blind signatures.

Assume that the signer has set up a RSA signature scheme with modulus N and public decryption key d, the holder of the message M selects at random an integer $k \in Z_N$ and computes the cryptogram

$$C \equiv M \times k^d \ (\text{mod } N).$$

The cryptogram C is now sent to the signer who computes the blind signature

$$S_C \equiv \left(M \times k^d \right)^e \ (\text{mod } N).$$

The blind signature S_C is returned to the holder who computes the signature for M as follows:

$$S_M \equiv S_C \times k^{-1} \equiv M^e \ (\text{mod } N).$$

It is not necessary to have special signature schemes to generate blind signatures. It is enough for the holder of the message to use a secure hash function $h()$. To get a (blind) signature from the signer, the holder first compresses the message M using $h()$. The digest $D = h(M)$ is sent to the signer. After signing the digest, the signature $SG_k(D)$ is communicated to the holder who attaches the message M to the signature $SG_k(D)$. Note that the signer cannot recover the message M from the digest, since $h()$ is a one-way hash function. Also the holder cannot cheat by attaching a "false" message unless collisions for the hash function can be found.

12.6.3.2 Undeniable Signatures

The concept of undeniable signatures is due to Chaum and van Antwerpen [18]. The characteristic feature of undeniable signatures is that signatures cannot be verified without the cooperation of the signer. Assume we have selected a large prime p and a primitive element $g \in GF(p)$, both p and g are public. The signer randomly selects its secret $k \in GF(p)$ and announces $g^k \pmod p$. For a message M, the signer creates the signature

$$S \equiv M^k \pmod p.$$

Verification needs the cooperation of the verifier and signer and proceeds as follows.

- The verifier selects two random numbers $a, b \in GF(p)$ and sends $C \equiv S^a (g^k)^b \pmod p$ to the signer.
- The signer computes k^{-1} such that $k \times k^{-1} \equiv 1 \pmod{p-1}$ and returns $d = C^{k^{-1}} \equiv M^a \times g^b \pmod p$ to the verifier.
- The verifier accepts or rejects the signature as genuine depending on whether $d \equiv M^a \times g^b \pmod p$.

There are two possible ways in which a verification can fail. Either the signer has tried to disavow a genuine signature or the signature is indeed false. The first possibility is prevented by incorporating a disavowal protocol. The protocol requires two runs for verification. In the first run, the verifier randomly selects two integers $a_1, b_1 \in GF(p)$ and sends $C_1 \equiv S^{a_1} (g^k)^{b_1} \pmod p$ to the signer. The signer returns $d_1 = C_1^{k^{-1}}$ to the verifier. The verifier checks whether

$$d_1 \neq M^{a_1} \times g^{b_1} \pmod p.$$

If the congruence is not satisfied, the verifier repeats the process using a different pair $a_2, b_2 \in GF(p)$. The verifier concludes that S is a forgery if and only if

$$\left(d_1 g^{-b_1} \right)^{a_2} \equiv \left(d_2 g^{-b_2} \right)^{a_1} \pmod p;$$

otherwise, the signer is cheating.

12.6.3.3 Fail-Stop Signatures

The concept of fail-stop signatures was introduced by Pfitzmann and Waidner [45]. Fail-stop signatures protect signatures against a powerful adversary. As usual, the signature is produced by a signer who holds a particular secret key. There are, however, many other keys which can be used to produce the same signature and which thus works with the original public key. There is a high probability that the key chosen by the adversary differs from the key held by the signer. Fail-stop signatures provide signing and verification algorithms as well as an algorithm to detect forgery.

Let k be a secret key known to the signer only and K be the public key. The signature on a message M is denoted as $s = SG_k(M)$. A fail-stop signature must satisfy the following conditions:

- An opponent with unlimited computational power can forge a signature with a negligible probability. More precisely, an opponent who knows the pair $(s = SG_k(M), M)$ and the signer's public key K can create a collection of all keys $\mathcal{K}_{s,M}$ such that $k^* \in \mathcal{K}_{s,M}$ if and only if $s = SG_{k^*}(M) = SG_k(M)$. The size of $\mathcal{K}_{s,M}$ has to increase exponentially as a function of the security parameter n. Without knowing the secret k, the opponent can only randomly choose an element from $K_{s,M}$. Let this element be k^*. If the opponent signs another message $M^* \neq M$, it is a requirement that $s^* = SG_{k^*}(M^*) \neq SG_k(M^*)$ with a probability close to one.

- There is a polynomial-time algorithm that produces a proof of forgery as output, when given the following inputs: a secret key k, a public key K, a message M, a valid signature s, and a forged signature s^*.

- A signer with polynomially bounded computing power cannot construct a valid signature that it can later deny by proving it to be a forgery.

Clearly, after the signer has provided a proof of forgery, the scheme is compromised and is no longer used. This is why it is called "fail-stop."

12.7   Research Issues and Summary

In this chapter, we discussed authentication, hashing, MACs, and digital signatures. We have presented the fundamental ideas underlying each topic and indicated the current research developments in these topics. This is reflected in our list of references.

We will now summarize the topics covered in this chapter.

Authentication deals with the problem of providing assurance to a receiver that a communicated message originates from a particular transmitter, and that the received message has the same content as the transmitted message. A typical and widely used application of authentication occurs in computer networks. Here the problem is to provide a protocol to establish the identity of two parties wishing to communicate or make transactions via the network. Motivated by such applications, the theory of authentication codes has developed into a mature area of research, drawing from several areas of mathematics.

Hashing algorithms provide a relatively short digest of a much longer input. Hashing must satisfy the critical requirement that the digests of two distinct messages are distinct. Hashing functions constructed from block encryption ciphers are an important type of hashing function. They have numerous applications in cryptology. Algebraic methods have also been proposed as a source of good hashing functions. These offer some provable security properties.

MACs are symmetric key primitives providing message integrity against active spoofing, by appending a cryptographic checksum to a message that is verifiable only by the intended recipient of the message. Message authentication is one of the most important ways of ensuring the integrity of information communicated by electronic means. This is especially relevant in the rapidly developing sphere of electronic commerce.

Digital signatures are the electronic equivalents of handwritten signatures. They are designed so as to preserve the essential features of handwritten signatures. They can be used to sign electronic documents and have potential applications in legal contexts.

12.8   Further Information

Current research in cryptology is represented in the proceedings of the conferences CRYPTO, EUROCRYPT, ASIACRYPT, and AUSCRYPT. More specialized conferences deal with topics such as hashing, fast software encryption, and network security. The proceedings are published by Springer in their LNCS series. *The Journal of Cryptology, IEEE Proceedings on Information Theory, Information Processing Letters, Designs Codes and Cryptography,* and other journals publish extended versions of the articles that were presented in the above-mentioned conferences. The International Association for Cryptologic Research ePrint Server (http://eprint.iacr.org) is a source of electronically distributed recent papers which are immediately accessible. The papers are placed on the server by the authors and undergo almost no refereeing, other than a superficial check for relevance to cryptology.

Defining Terms

Authentication: One of the main two goals of cryptography (the other is secrecy). An authentication system ensures that messages transmitted over a communication channel are authentic.

Cryptology: The art/science of design and analysis of cryptographic systems.

Digital signatures: An asymmetric cryptographic primitive that is the digital counterpart of a signature and links a document to a unique person.

Distinguishing attack: A concept derived from formal models of security. A distinguisher is a statistical measure which allows an adversary to distinguish between the output of a particular cipher and the output of a random source with a nonnegligible probability.

Encryption algorithm: Transforms an input text by "mixing" it with a randomly chosen bit string—the key—to produce the cipher text. In a symmetric encryption algorithm, the plain text can be recovered by applying the key to the cipher text.

Hashing: Hashing is accomplished by applying a function to an arbitrary length message to create a digest/hash value, which is usually of fixed length.

Key: An input provided by the user of a cryptographic system. This piece of information is kept secret and is the source of security in a cryptographic system. Sometimes a part of key information is made public, in which case the secret part is the source of security.

Message authentication codes: A symmetric cryptographic primitive that is used for providing authenticity.

Plain text, cipher text: The cipher text is the "scrambled" version of an original source—the plain text. It is assumed that the scrambled text, produced by an encryption algorithm, can be inspected by persons not having the key and not reveal the content of the source.

References

1. Bakhtiari, S., Safavi-Naini, R., and Pieprzyk, J., Keyed hash functions, *Cryptography: Policy and Algorithms Conference, LNCS*, Vol. 1029, pp. 201–214, Springer-Verlag, Berlin, 1995.
2. Bakhtiari, S., Safavi-Naini, R., and Pieprzyk, J., Practical and secure message authentication, *Proceedings of the Second Annual Workshop on Selected Areas in Cryptography (SAC'95)*, pp. 55–68, Ottawa, Canada, May 1995.
3. Bakhtiari, S., Safavi-Naini, R., and Pieprzyk, J., Password-based authenticated key exchange using collisionful hash functions, *Proceedings of the Australasian Conference on Information Security and Privacy, LNCS*, Vol. 1172, pp. 299–310, Springer-Verlag, Berlin, 1996.
4. Bellare, M., Canetti, R., and Krawczyk, H., Keying hash functions for message authentication, *Proceedings of the Crypto'96, LNCS*, Vol. 110, pp. 1–15, Springer-Verlag, Berlin, 1996.
5. Bellare, M., Kilian, J., and Rogaway, P., The security of cipher block chaining, *Proceedings of the Crypto'94, LNCS*, Vol. 839, pp. 348–358, Springer-Verlag, Berlin, 1994.
6. Bellare, M. and Rogaway, P., The exact security of digital signatures—how to sign with RSA and Rabin. *Proceedings of the Eurocrypt'96, LNCS*, Vol. 1070, pp. 399–416, Springer-Verlag, Berlin, May 1996.
7. Bellare, M. and Kohno, T., Hash functions balance and its impact on birthday attacks, *Proceedings of the Eurocrypt'04, LNCS*, Vol. 3027, Springer-Verlag, Berlin, 2004.

8. Bellare, M., New proofs for NMAC and HMAC: Security without collision resistance, *Proceedings of the Crypto'06, LNCS*, Vol. 4117, pp. 602–619, Springer-Verlag, Berlin, 2006.

9. Berson, T.A., Gong, L., and Lomas, T.M.A., Secure, keyed, and collisionful hash functions, TR SRI-CSL-94-08, *SRI International*, Dec. 1993, Revised version (Sept. 2, 1994).

10. Beth, T., Jungnickel, D., and Lenz, H., *Design Theory*, 2nd Edition, Cambridge University Press, Cambridge, U.K., 1999.

11. Biham, E. and Shamir, A., Differential cryptanalysis of DES-like Cryptosystems, *Journal of Cryptology*, 4, 3–72, 1991.

12. Bird, R., Gopal, I., Herzberg, A., Janson, P., Kutten, S., Molva, R., and Yung, M., The KryptoKnight family of light-weight protocols for authentication and key distribution, *IEEE/ACM Transactions on Networking*, 3, 31–41, 1995.

13. Carter, J.L. and Wegman, M.N., Universal class of hash functions, *Journal of Computer and System Sciences*, 18(2), 143–154, 1979.

14. Charnes, C. and Pieprzyk, J., Linear nonequivalence versus nonlinearity, *Proceedings of the Auscrypt'92, LNCS*, Vol. 718, pp. 156–164, Springer-Verlag, Berlin, 1993.

15. Charnes, C. and Pieprzyk, J., Attacking the SL_2 Hashing scheme, *Proceedings of the Asiacrypt'94, LNCS*, Vol. 917, pp. 322–330, Springer-Verlag, Berlin, 1995.

16. Charnes, C., O'Connor, L., Pieprzyk, J., Safavi-Naini, R., and Zheng, Y., Comments on GOST encryption algorithm, *Proceedings of the Eurocrypt'94, LNCS*, Vol. 950, pp. 433–438, Springer-Verlag, Berlin, 1995.

17. Chaum, D., Blind signatures for untraceable payments, *Proceedings of the Crypto 82*, 199–203, Plenum Press, New York, 1983.

18. Chaum, D. and Van Antwerpen, H., Undeniable signatures, *Proceedings of the Crypto 89, LNCS*, Vol. 435, pp. 212–217, Springer-Verlag, Berlin, 1990.

19. Coppersmith, D., Analysis of ISO/CCITT Document X.509 Annex D. Internal Memo, IBM T. J. Watson Center, June 11, 1989.

20. Daemen, J., Govaerts, R., and Vandewalle, J., A framework for the design of one-way hash functions including cryptanalysis of Damgard one-way function based on a cellular automaton, *Proceedings of the Asiacrypt'91, LNCS*, Vol. 739, pp. 82–97, Springer-Verlag, Berlin, 1993.

21. Damgård, I., A design principle for hash functions, *Proceedings of the Crypto'89, LNCS*, Vol. 435, pp. 416–427, Springer-Verlag, Berlin, 1990.

22. Davies, D.W. and Price, W.L., The application of digital signatures based on public-key cryptosystems, *Proceedings of the Fifth International Computer Communications Conference*, pp. 525–530, Atlanta, FL, Oct. 1980.

23. De Santis, A. and Yung, M., On the design of provably-secure cryptographic hash functions, *Proceedings of the Eurocrypt'90, LNCS*, Vol. 473, pp. 377–397, Springer-Verlag, Berlin, 1990.

24. Desmedt, Y. and Frankel, Y., Shared generation of authenticators and signatures, *Proceedings of the Crypto'91, LNCS*, Vol. 576, pp. 457–469, Springer-Verlag, Berlin, 1992.

25. Dobbertin, H., Cryptanalysis of MD4, *Proceedings of the Fast Software Encryption Workshop, LNCS*, Vol. 1039, pp. 71–82, Springer-Verlag, Berlin, 1996.

26. ElGamal, T., A public key cryptosystem and a signature scheme based on discrete logarithms, *IEEE Transactions on Information Theory*, 31, 469–472, 1985.

27. Fak, V., Repeated use of codes which detect deception, *IEEE Transactions on Information Theory*, 25(2), 233–234, Mar. 1979.

28. Mironv, I. and Zhang, L., Applications of SAT solvers to cryptanalysis of hash functions, *Proceedings of the Theory and Applications of Satisfiability Testing*, SAT, 102–115, 2006. ePrinthttp://eprint.iacr.org/2006/254.

29. Geiselmann, W., A note on the hash function of Tillich and Zémor, *Cryptography and Coding, LNCS*, Vol. 1025, pp. 257–263, Springer-Verlag, Berlin, 1995.

30. Gibson, J.K., Discrete logarithm hash function that is collision free and one way, *IEE Proceedings-E,* 138(6), 407–427, 1991.
31. Gong, L., Collisionful keyed hash functions with selectable collisions, *Information Processing Letters,* 55, 167–170, 1995.
32. Impagliazzo, R. and Naor, M., Efficient cryptographic schemes as provably secure as subset sum, *Proceedings of the 30th IEEE Symposium on Foundations of Computer Science,* pp. 236–241, IEEE Computer Society, Washington, DC, 1989.
33. Johansson, T., Lower bound on the probability of deception in authentication with arbitration, *IEEE Transaction on Information Theory,* 40, 1573–1585, 1994.
34. Johansson, T., Authentication codes for nontrusting parties obtained from rank metric codes, *Designs, Codes and Cryptography,* 6, 205–218, 1995.
35. Johansson, T., Kabatianskii, G., and Smeets, B., On the relation between *A*-codes and codes correcting independent errors, *Proceedings of the Eurocrypt'93, LNCS,* Vol. 765, pp. 1–11, Springer-Verlag, Berlin, 1994.
36. Kim, J., Biryukov, A., Preneel, B., and Hong, S., On the security of HMAC and NMAC based on HAVAL, MD4, MD5, SHA-0 and SHA-1, *Proceedings of the Security in Communication Networks 20006, LNCS,* Vol. 4116, Springer-Verlag, Berlin, 2006. ePrinthttp://eprint.iacr.org/2006/187.
37. Krawczyk, H., LFSR-based hashing and authentication, *Proceedings of the Crypto'94, LNCS,* Vol. 839, pp. 129–139, Springer-Verlag, Berlin, 1994.
38. Kurosawa, K., New bounds on authentication code with arbitration, *Proceedings of the Crypto'94, LNCS,* Vol. 839, pp. 140–149, Springer-Verlag, Berlin, 1994.
39. Lamport, L., Constructing digital signatures from a one-way function, TR CSL-98, *SRI International,* Oct. 1979.
40. MacWilliams, F.J., Gilbert, E.N., and Sloane, N.J.A., Codes which detect deception, *Bell System Technical Journal,* 53(3), 405–424, 1974.
41. Matyas, S.M. and Meyer, C.H., Electronic signature for data encryption standard, *IBM Technical Disclosure Bulletin,* 24(5), 2335–2336, 1981.
42. Merkle, R.C., One way hash functions and DES, *Proceedings of the Crypto'89, LNCS,* Vol. 435, pp. 428–446, Springer-Verlag, Berlin, 1990.
43. Naor, M. and Yung, M., Universal one-way hash functions and their Cryptographic applications, *Proceedings of the 21st ACM Symposium on Theory of Computing,* pp. 33–43, Seattle, WA, 1989.
44. Pei, D., Information theoretic bounds for authentication codes and PBIB, *Presented at the Rump session of Asiacrypt'91,*
45. Pfitzmann, B. and Waidner, M., Fail-stop signatures and their applications, *Proceedings of the Securicom'91,* pp. 338–350, Paris, France, 1991.
46. Pieprzyk, J. and Sadeghiyan, B., *Design of Hashing Algorithms, LNCS,* Vol. 756, Springer-Verlag, New York, 1993.
47. Piper, F. and Beker, H., *Cipher Systems,* Northwood Books, London, 1982.
48. Preneel, B. and van Oorschot, P.C., MDx-MAC and building fast MACs from hash functions, *Proceedings of the Eurocrypt'96, LNCS,* Vol. 1070, pp. 1–14, Springer-Verlag, Berlin, 1996.
49. Preneel, B. and van Oorschot, P.C., On the security of two MAC Algorithms, *Proceedings of the Eurocrypt'96, LNCS,* Vol. 1070, pp. 19–32, Springer-Verlag, Berlin, 1996.
50. Rabin, M.O., Digitalized signatures. In: *Foundations of Secure Computation,* R.A. Demillo et al., (eds.), pp. 155–168, Academic Press, New York, 1978.
51. Rivest, R.L., The MD4 message digest algorithm, *Proceedings of the Crypto'90, LNCS,* Vol. 537, pp. 303–311, Springer-Verlag, Berlin, 1991.
52. Rivest, R.L., *RFC 1321: The MD5 Message-Digest Algorithm,* Internet Activities Board, Apr. 1992.
53. Rivest, R.L., Shamir, A., and Adleman, L.M., A method for obtaining digital signatures and public-key cryptosystems, *Communications of the ACM,* 21(2), 120–126, 1978.

54. Rees, R.S. and Stinson, D.R., Combinatorial characterization of authentication codes II, *Designs Codes and Cryptography*, 7, 239–259, 1996.

55. Rogaway, P., Bucket hashing and its application to fast message authentication, *Proceedings of the Crypto'95, LNCS*, Vol. 963, pp. 30–42, Springer-Verlag, Berlin, 1995.

56. Rompel, J., One-way functions are necessary and sufficient for secure signatures, *Proceedings of the 22nd ACM Symposium on Theory of Computing*, pp. 387–394, Baltimore, MD, 1990.

57. Rosenbaum, V., A lower bound on authentication after having observed a sequence of messages, *Journal of Cryptology*, 6(3), 135–156, 1993.

58. Safavi-Naini, R., Three systems for shared generation of authenticators, *Proceedings of the Cocoon'96, LNCS*, Vol. 1090, pp. 401–411, Springer-Verlag, Berlin, 1996.

59. Safavi-Naini. R. and Charnes, C., MRD hashing, *Designs Codes and Cryptography*, 37, 229–242, 2005.

60. Simmons, G.J., A game theory model of digital message authentication, *Congressus Numerantium*, 34, 413–424, 1982.

61. Simmons, G.J., A survey of information authentication, *Contemporary Cryptology: The Science of Information Integrity*, pp. 379–419, IEEE Press, 1992.

62. Stinson, D.R., Combinatorial characterisation of authentication codes, *Proceedings of the Crypto'91, LNCS*, Vol. 576, pp. 62–73, Springer-Verlag, Berlin, 1991.

63. Stinson, D.R., Combinatorial techniques for universal hashing, *Journal of Computer and System Sciences*, 48, 337–346, 1994.

64. Stinson, D.R., The combinatorics of authentication and secrecy codes, *Journal of Cryptology*, 2(1), 23–49, 1990.

65. Stinson, D.R., Universal hashing and authentication codes, *Designs, Codes and Cryptography*, 4, 369–380, 1994.

66. Taylor, R., Near optimal unconditionally secure authentication, *Proceedings of the Eurocrypt'94, LNCS*, Vol. 950, pp. 245–255, Springer-Verlag, Berlin, 1995.

67. Tillich, J.-P. and Zémor, G., Hashing with SL_2, *Proceedings of the Crypto'94, LNCS*, Vol. 839, pp. 40–49, Springer-Verlag, Berlin 1994.

68. Tombak, L. and Safavi-Naini, R., Authentication codes that are r-fold secure against spoofing, *Proceedings of the 2nd ACM Conference on Computer and Communication Security*, pp. 166–169, ACM, New York, 1994.

69. Tombak, L. and Safavi-Naini, R., Authentication codes in plaintext and content-chosen attacks, *Designs, Codes and Cryptography*, 6, 83–99, 1995.

70. Tombak, L. and Safavi-Naini, R., Combinatorial characterization of A-codes with r-fold security, *Proceedings of the Asiacrypt'94, LNCS*, Vol. 917, pp. 211–223, Springer-Verlag, Berlin, 1995.

71. Tsudik, G., Message authentication with one-way hash functions, *IEEE Infocom'92*, 2055–2059, Florence, Italy, May 1992.

72. Walker, M., Information theoretic bounds for authentication schemes, *Journal of Cryptology*, 2(3), 133–138, 1990.

73. Wagnera D. and Goldberg, I., Proofs of security for the Unix password hashing algorithm, *Proceedings of the Asiacrytp'2000, LNCS*, Vol. 1976, pp. 560–572, Springer-Verlag, Berlin, 2000.

74. Wang, X., Feng, D., Lai, X., and Yu, H., Collisions for hash functions MD4, MD5, HAVAL-128, and RIPEMD, rump session Crypto'04. ePrinthttp://eprint.iacr.org/2004/199.

75. Wegman, M.N. and Carter, J.L., New hash functions and their use in authentication and set equality, *Journal of Computer and System Sciences*, 22, 265–279, 1981.

76. Winternitz, R.S., Producing a one-way hash function from DES, *Proceedings of the Crypto'83*, pp. 203–207, Plenum Press, New York, 1984.

77. Vanroose, P., Smeets, B., and Wan, Z.-X., On the construction of authentication codes with secrecy and codes withstanding spoofing attacks of order $L \geq 2$, *Proceedings of the Eurocrypt'90, LNCS*, Vol. 473, pp. 306–312, Springer-Verlag, Berlin, 1990.

78. Yung, M. and Desmedt, Y., Arbitrated unconditionally secure authentication can be unconditionally protected against arbiter's attack, *Proceedings of the Crypto'90, LNCS*, Vol. 537, pp. 177–188, Springer-Verlag, Berlin, 1990.

79. Zheng, Y., Pieprzyk, J., and Seberry, J., HAVAL—a one-way hashing algorithm with variable length ofoutput, *Proceedings of the Auscrypt'92, LNCS*, Vol. 718, pp. 83–104, Springer-Verlag, Berlin, 1993.

13

Crypto Topics and Applications II

Jennifer Seberry
University of Wollongong

Chris Charnes
Institute of Applied Physics and CASED Technical University Darmstadt

Josef Pieprzyk
Macquarie University

Rei Safavi-Naini
University of Calgary

13.1 Introduction

In this chapter we continue the exposition of crypto topics that was begun in the previous chapter. This chapter covers secret sharing, threshold cryptography, signature schemes, and finally quantum key distribution and quantum cryptography. As in the previous chapter, we have focused only on the essentials of each topic. We have selected in the bibliography a list of representative items, which can be consulted for further details.

First we give a synopsis of the topics that are discussed in this chapter.

Secret sharing is concerned with the problem of how to distribute a secret among a group of participating individuals, or entities, so that only predesignated collections of individuals are able to recreate the secret by collectively combining the parts of the secret that were allocated to them. There are numerous applications of secret-sharing schemes in practice. One example of secret sharing occurs in banking. For instance, the combination to a vault may be distributed in such a way that only specified collections of employees can open the vault by pooling their portions of the combination. In this way the authority to initiate an action, e.g., the opening of a bank vault, is divided for the purposes of providing security and for added functionality, such as auditing, if required.

Threshold cryptography is a relatively recently studied area of cryptography. It deals with situations where the authority to initiate or perform cryptographic operations is distributed among a group of individuals. Many of the standard operations of single-user cryptography have counterparts in threshold cryptography.

Signature schemes deal with the problem of generating and verifying (electronic) signatures for documents. A subclass of signature schemes is concerned with the shared-generation and the shared-verification of signatures, where a collaborating group of individuals are required to perform these actions.

A new paradigm of security has recently been introduced into cryptography with the emergence of the ideas of quantum key distribution and quantum cryptography. While classical cryptography employs various mathematical techniques to restrict eavesdroppers from learning the contents of encrypted messages, in quantum cryptography the information is protected by the laws of physics.

13.2 Secret Sharing

13.2.1 Introduction

Secret sharing is concerned with the problem of distributing a secret among a group of participating individuals, or entities, so that only predesignated collections of individuals are able to recreate the secret by collectively combining their shares of the secret.

The earliest and the most widely studied type of secret-sharing schemes are called (t, n)-threshold schemes. In these schemes the access structure—a precise specification of the participants authorized to recreate the secret, comprises all possible t-element subsets of an n-element set.

The problem of realizing, i.e., implementing, secret-sharing schemes for threshold structures was solved independently by Blakley [17] and Shamir [92] in 1979. Shamir's solution is based on the property of polynomial interpolation in finite fields; Blakley formulated and solved the problem in terms of finite geometries.

In a (t, n)-threshold scheme, each of the n participants holds some shares (also called shadows) of the secret. The parameter $t \leq n$ is the threshold value. A fundamental property of a (t, n)-threshold scheme is that the secret can only be recreated if at least t shareholders combine their shares, but less than t shareholders cannot recreate the secret. The fact that the key can be recovered from any t of the shares makes threshold schemes very useful in key management. Threshold schemes can tolerate the invalidation of up to $n - t$ shares—the secret can still be recreated from the remaining intact shares.

Secret-sharing schemes are also used to control the authority to perform critical actions. For example, a bank vault can be opened only if say, any two out of three trusted employees of the bank agree to do so by combining their partial knowledge of the vault combination. In this case, even if any one of the three employees is not present at any given time the vault can still be opened, but no single employee has sufficient information about the combination to open the vault.

Secret-sharing schemes that do not reveal any information about the shared secret to unauthorized individuals are called perfect. This notion will be formally defined in Section 13.2.2. In this survey, we discuss both perfect and nonperfect schemes, as the latter schemes are proving to be useful in various secret-sharing applications.

Besides the (t, n)-threshold structures, more general access structures are encountered in the theory of secret sharing. These will be considered in Section 13.2.8. General access structures apply to situations where the trust status of the participants is not uniform. For example, in the bank scenario described earlier, it could be considered more secure to authorize either the bank manager or any two out of three senior employees to open the vault.

Since Blakley's and Shamir's papers have appeared, the study of secret sharing has developed into an active area of research in cryptography. A fundamental problem of the theory and practice of secret sharing is the issue of how to implement secret-sharing schemes for arbitrary access structures. We shall discuss later some of the solutions to this problem. Simmons [96] gives numerous examples of practical situations, which require secret-sharing schemes. He also gives a detailed account of the geometric approach to secret sharing. Stinson's [102] survey is broader and more condensed.

Simmons [95] discusses secret-sharing schemes with extended capabilities. He argues that there are realistic applications in which schemes with extended capabilities are required. In our discussion, we will assume that there is a key distribution center (KDC) that is trusted unconditionally.

13.2.2 Models of Secret Sharing

A common model of secret sharing has two phases. In the initialization phase, a trusted entity—the dealer—distributes shares of a secret to the participants via secure means. In the reconstruction phase, the authorized participants submit their shares to a combiner, which reconstructs the secret on their behalf. It is assumed that the combiner is an algorithm, which only performs the task of reconstructing the secret. We denote the sets of all possible secrets and shares by $\mathcal{K}$ and $\mathcal{S}$ respectively; the set of participants in a scheme is denoted by $\mathcal{P}$. Secret-sharing schemes can be modeled by the information theory concept of entropy (cf. [52]). This approach was initiated by Karnin et al. [63] and developed further by Capocelli et al. [28].

A secret-sharing scheme is a collection of two algorithms. The first (the dealer) is a probabilistic mapping:

$$\mathcal{D} : \mathcal{K} \to \mathcal{S}_1 \times \mathcal{S}_2 \times \cdots \times \mathcal{S}_n$$

where $\mathcal{S}_i \subset \mathcal{S}$ ($i = 1, 2, \ldots, n$) and $\mathcal{S}_i$ is a subset of shares which is used to generate a share for the participant $P_i \in \mathcal{P}$. The second (the combiner) is a function:

$$\mathcal{C} : \mathcal{S}_{i_1} \times \mathcal{S}_{i_2} \times \cdots \times \mathcal{S}_{i_t} \to \mathcal{K}$$

such that if the corresponding subset of participants $\{P_{i_1}, P_{i_2}, \ldots, P_{i_t}\}$ belongs to the access structure Γ, it produces the secret $K \in \mathcal{K}$, i.e.,

$$H\left(K \mid P_{i_1}, P_{i_2}, \ldots, P_{i_t}\right) = 0. \tag{13.1}$$

The combiner fails to recompute the secret if the subset of participants does not belong to the access structure Γ, i.e.,

$$H\left(K \mid S_l\right) \geq 0 \tag{13.2}$$

for $S_l = \{s_{i_1}, s_{i_2}, \ldots, s_{i_j}\}$ and $S_l \notin \Gamma$.

In Equation 13.1, $H(K \mid P_{i_1}, P_{i_2}, \ldots, P_{i_t})$ is calculated with respect to the shares of the participants. A secret-sharing scheme is called perfect if $H(K \mid S_l) = H(K)$ for any unauthorized subset of participants, i.e., not belonging to an access structure Γ (cf. Section 13.2.8).

The following result is proved by Karnin et al. [63]. A necessary condition for a perfect threshold scheme is that for each share s_i, the inequality $H(s_i) \geq H(K)$ holds.

Most of the secret-sharing schemes that we discuss satisfy this inequality, but we will also consider in Section 13.2.12 schemes that do not satisfy this inequality; these are called nonperfect schemes.

13.2.2.1 The Matrix Model

A matrix representation of perfect secret-sharing schemes was introduced by Brickell and Stinson [26]. The matrix model is often used in theoretical investigations of secret sharing.

In this model, a perfect secret-sharing scheme is formulated as a matrix M that is known by all the participants P in the scheme. The $|P| + 1$ columns of M are indexed as follows. The first column corresponds to the dealer D, the remaining columns are indexed by the remaining participants in P. Each row of M contains one of the possible keys K that is to be shared in column D, and the shares of K are located in the remaining columns. When the dealer wants share K, a row r that has K in the D-column is chosen uniformly and randomly. The dealer distributes the shares of K to each participant using the matrix M, i.e., participant P_j receives the entry $M_{r,j}$ as his share.

The general requirements of a perfect secret scheme translate into the following combinatorial conditions in the matrix model; see Stinson [102], and Blundo et al. [20]. Suppose that Γ is an access structure:

1. If $B \in \Gamma$ and $M(r, P) = M(r', P)$ for all $P \in B$, then $M(r, D) = M(r', D)$.
2. If $B \notin \Gamma$, then for every possible assignment f of shares to the participants in B, say $f = (f_P : P \in B)$, a nonnegative integer $\lambda(f, B)$ exists such that

$$\left| \{ r : M(r, P) = f_P \; \forall P \in B, \; M(r, D) = K \} \right| = \lambda(f, B)$$

   is independent of the value of K.

13.2.2.2 Information Rate

The information rate of secret-sharing schemes was studied by Brickell and Stinson [26]. It is a measure of the amount of information that the participants need to keep secret in a secret-sharing scheme. The information rate of a participant P_i in a secret-sharing scheme with $|S_i|$ shares is

$$\rho_i = \frac{\log_2 |K|}{\log_2 |S_i|}.$$

The information rate of the scheme, denoted ρ, is defined to be the minimum of the ρ_i.

A proof of the fact that $\rho \leq 1$ is given by Stinson [102]. This result motivates the definition of ideal secret-sharing schemes.

A perfect secret-sharing scheme is called ideal if $\rho = 1$; that is, if the size of each participants share, measured in the number of bits, equals the size of the secret.

We now define another measure used to quantify the comparison between secret-sharing schemes (cf. Section 13.2.8).

$\rho^*(\Gamma)$ is the maximum value of ρ for any perfect secret-sharing scheme realizing the access structure Γ.

For any access structure it is desirable to implement a secret-sharing scheme with the information rate close to 1. This minimizes the amount of information that needs to be kept secret by the participants, which means that there is a greater chance of the scheme remaining secure. For example, a (t, n)-threshold scheme implemented by Shamir's method is ideal, but when the scheme is modified to prevent cheating as proposed by Tompa and Woll [107], it is no longer ideal (cf. Section 13.2.7).

13.2.3 Some Known Schemes

We now describe several well-known threshold secret-sharing schemes.

13.2.3.1 Blakley's Scheme

Blakley [17] implements threshold schemes using projective spaces over finite fields $GF(q)$. A projective space $PG(t, q)$ is obtained from the corresponding $(t + 1)$-dimensional vector space

$V(t + 1, q)$ by omitting the zero vector of $V(t + 1, q)$, and identifying two vectors v and v', which satisfy the relation $v = \lambda v'$ for some nonzero λ in $GF(q)$. This equivalence relation partitions the vectors of $V(t + 1, q)$ into equivalence classes, i.e., the lines through the origin of $V(t + 1, q)$. These are the points of $PG(t, q)$, and there are $(q^t - 1)/(q - 1)$ such points. Similarly, each k-dimensional subspace of $V(t + 1, q)$ corresponds to a $(k - 1)$-dimensional subspace of $PG(t, q)$. And every point of $PG(t, q)$ lies on $(q^t - 1)/(q - 1)$ $(t - 1)$-dimensional subspaces, which are called the hyperplanes of $PG(t, q)$.

To realize a (t, n)-threshold scheme, the secret is represented by a point p chosen randomly from $PG(t, q)$; each point p belongs to $(q^t - 1)/(q - 1)$ hyperplanes. The shares of the secret are the n hyperplanes, which are randomly selected and distributed to the participants. If q is sufficiently large and n is not too large, then the probability that any t of the hyperplanes intersect in some point other than p is close to zero; cf. Blakley [17]. Thus in general the secret can be recovered from any t of the n shares. The secret cannot be recovered from the knowledge of less than t hyperplanes, as these will intersect only in some subspace containing p. This scheme is not perfect, since a coalition of unauthorized insider participants has a greater chance of guessing the secret than an unauthorized group of outsider participants.

Blakley's geometric solution to the secret-sharing problem has grown into an active area of research. We will cover some of these developments in this survey.

13.2.3.2 Simmons' Scheme

Simmons formulates secret-sharing schemes in terms of affine spaces instead of projective spaces. The reasons for using affine spaces instead of projective spaces are explained by Simmons [96]. (There is a correspondence between projective spaces and affine spaces, cf. Beth et al. [14].) Briefly, an affine space $AG(n, q)$ consists of points—the vectors of $V(n, q)$, and a hierarchy of l-dimensional subspaces for $l \leq n$ and their cosets. These correspond to the equivalence classes in projective geometry mentioned earlier, and are called the flats of $AG(n, q)$. The equivalence classes of lines, planes, etc., of $AG(t, q)$ are the 1-dimensional, 2-dimensional, etc., flats. A hyperplane is a flat of codimension one. To realize a (t, n)-threshold scheme in $AG(t, q)$, the secret is represented by a point p chosen randomly from $AG(t, q)$, which lies on a publicly known line V_d (lines have q points). A hyperplane V_i of the indicator variety is selected so that V_i intersects V_d in a single point p. The shares of the secret are the subsets of points of V_i. An authorized subset of participants, which spans V_i, enables the reconstruction of the secret. If an unauthorized subset of participants attempts to reconstruct the secret, their shares will only span a flat that intersects V_d in the empty set. Thus they gain no information about the secret. The precise amount of information gained by the unauthorized participants about the secret can be expressed in terms of the defining parameters of $AG(n, q)$. These schemes are perfect. Simmons [96] gives a detailed explanation of the implementation of secret-sharing schemes using projective and affine spaces.

13.2.3.3 Shamir's Scheme

Shamir's [92] scheme realizes (t, n)-access structures using polynomial interpolation over finite fields. In his scheme, the secrets S belong to a prime power finite field $GF(q)$, which satisfies $q \geq n + 1$. In the initialization phase, the dealer $\mathcal{D}$ chooses n distinct nonzero elements $\{x_1, \ldots, x_n\}$ from $GF(q)$ and allocates these to participants $\{P_1, \ldots, P_n\}$. This correspondence is publicly known, and has undesirable side effects if any of the participants are dishonest; see Section 13.2.7. However for now, we will assume that all the participants obey faithfully the protocol for reconstructing the secret.

Fix a random element of $GF(q)$ as the secret K. The shares of K are created using the following protocol:

1. $\mathcal{D}$ chooses $a_1, a_2, \ldots, a_{t-1}$ from $GF(q)$ randomly, uniformly, and independently.
2. Let $a(x)$ be a polynomial of degree at most $t - 1$, defined as $a(x) = K + a_1 x + a_2 x^2 + \cdots + a_{k-1} x^{k-1}$.
3. The shares of the secret key are $y_i = a(x_i)$, for $1 \leq i \leq n$.

With the aforementioned data, if any t out of the n participants $\{x_{i_1}, \ldots, x_{i_t}\}$ combine their shares $\{y_{i_1}, \ldots, y_{i_t}\}$, then by Lagrangian interpolation, there is a unique polynomial of degree at most $(t-1)$ passing through the points $\{(x_{i_1}, y_{i_1}), \ldots, (x_{i_t}, y_{i_t})\}$. So the combined shares of the t participants can be used to recreate the polynomial $a(x)$, and hence the secret, which is $K = a(0)$.

The relation between the secret and the shares is given by Lagrange's interpolation formula:

$$K = \sum_{j=1}^{t} y_{i_j} b_j, \qquad (13.3)$$

where the b_j are defined as

$$b_j = \prod_{\substack{1 \leq k \leq t, \\ k \neq j}} \frac{x_{i_k}}{x_{i_k} - x_{i_j}}.$$

Shamir's scheme is computationally efficient in terms of the computational effort required to create the shares and to recover the secret. Also the share size is optimal in an information theoretic sense, cf. Definition 13.2.2.

The reconstruction phase in Shamir's scheme can be also considered as a system of linear equations, which are defined by the shares K_i. If t shares are submitted to the combiner, the system of linear equations

$$y_{i_j} = K + a_1 x_{i_j} + a_2 x_{i_j}^2 + \cdots + a_{k-1} x_{i_j}^{k-1}, \quad j = 1, \ldots, t$$

can be solved for the unknowns $K, a_1, a_2, \ldots, a_{t-1}$ because the determinant of this system of equations is a nonsingular Vandermonde determinant (the $\{x_1, \ldots, x_n\}$ are pair-wise distinct). However, if $t - 1$ participants try to reconstruct the secret, they face the problem of solving $t - 1$ linear equations in t unknowns. This system of equations has one degree of freedom. Consequently, $t - 1$ participants do not obtain any information about the secret, as K was selected uniformly and randomly from $GF(q)$. Shamir's system is perfect.

13.2.3.4 A (t,t) Threshold Scheme

Karnin et al. [63] describe a secret-sharing scheme which realizes (t, t)-access structures. The interest in such schemes is that they can be used as the basis for other cryptographic constructions.

In their scheme, the set of secrets S is the ring of residue classes Z_m, where m is any integer. (In applications m is large.) The secret K is shared using the following algorithm:

1. $\mathcal{D}$ secretly chooses randomly, uniformly, and independently $t-1$ elements $y_1, y_2, \ldots, y_{t-1}$ from Z_m; the y_t are defined as

$$y_t = K - \sum_{i=1}^{t-1} y_i \bmod m.$$

2. Participant P_i for $1 \leq i \leq t$ receives the share y_i from $\mathcal{D}$.

The above system is perfect as the following argument shows. The set of shares of $l < t$ participants attempting to reconstruct the secret either contains the share $y_t = K - \sum_{i=1}^{t-1} y_i \bmod m$, or not. In both cases the (unauthorized) participants lack the necessary information to determine K. Shamir's scheme with $t = n$ provides an alternative construction of (t, t)-threshold schemes, using the fields $GF(q)$ instead of Z_m.

13.2.4 Threshold Schemes and Discrete Logarithms

The discrete logarithm has been widely employed in the literature to transform threshold schemes into conditionally secure schemes with extra properties. This idea is exploited in the papers by Benaloh [4], Beth [15], Charnes et al. [32], Charnes and Pieprzyk [33], Lin and Harn [69], Langford [66], and Hwang and Chang [58].

It is a consequence of the linearity of Equation 13.3 that Shamir's scheme can be modified to obtain schemes having enhanced properties, such as disenrollment capability, in which shares from one or more participants can be made incapable of forming an updated secret. (The formal analysis of schemes with this property was given by Blakley et al. [18].) Let $a(x)$ be a polynomial and let $a(i)$ be the shares in Shamir's scheme. In the modified threshold scheme proposed in [32], $g^{a(0)}$ is the secret and the shares are $s_i = g^{c_i}$, $c_i = a(i)$. A generator g of the cyclic group of the field $GF(2^n)$ is chosen so that $2^n - 1$ is a Mersenne prime.

The modified (t, n)-threshold schemes can disenrol participants whose shares have been compromised either through loss or theft, and still maintain the original threshold level. In the event that some of the original shares are compromised, the KDC can issue using a public authenticated channel a new generator g' of the cyclic group of $GF(2^n)$. The shareholders can calculate their new shares s_i' from the initial secret data according to

$$s_i' = g'^{c_i}.$$

Hwang and Chang [58] used a similar setting to obtain dynamic threshold schemes.

Threshold schemes with disenrollment capability, without the assumption of the intractability of the discrete logarithm problem, can be based on families of threshold schemes. The properties of these schemes are studied in a paper by Charnes et al. [31]; here we provide the basic definition.

A threshold scheme family (TSF) is defined by an $(m \times n)$ matrix of shares $[s_{i,j}]$, such that

1. Any row $(s_{i,1}, s_{i,2}, \ldots, s_{i,n})$ represents an instance of $TS_{r_i}(t_i, n)$, where $i = 1, \ldots, m$.
2. Any column $(s_{1,j}, s_{2,j}, \ldots, s_{m,j})$ represents an instance of $TS_{c_j}(t_j, m)$, where $j = 1, \ldots, n$.

A family of threshold schemes in which all rows and all columns are ideal schemes is called an ideal *threshold scheme family*, or ITS family for short. In these schemes it is possible to alter dynamically the threshold values by moving from one level of the matrix to another.

Lin and Harn [69], and Langford [66] use the discrete logarithm to transform Shamir's scheme into a conditionally secure scheme which does not require a trusted KDC. A similar approach is used by Langford [66] to obtain a threshold signature scheme. Beth [15] describes a protocol for verifiable secret sharing for general access structures based on geometric schemes. The discrete logarithm problem is used to encode the secret and the shares so that they can be publicly announced for verification purposes.

It should be noted that the definition of disenrollment used in [32] is not the same as that of Blakley et al. [18]. Blakley et al. established a lower bound on the number of bits required to encode the shares in schemes that can disenrol participants. Their bound shows that this number grows linearly with the number of participants that the scheme can disenrol. They also present two geometric (t, n)-threshold schemes, which meet this bound.

It is interesting to note that Benaloh [4] used the discrete logarithm to transform Shamir's scheme, but for a very different purpose. One of the properties of the discrete logarithm is that the sum of the discrete logarithms of the shares of a secret is equal to the discrete logarithm of the product of the shares of the secret. This property has an application in secret-ballot elections (cf. Benaloh [4]) where, in contrast with schemes earlier mentioned the discrete logarithm problem is required to be tractable.

The homomorphic property introduced by Benaloh [4] has prompted the question whether similar schemes can be set up in noncommutative groups—other than the additive and the multiplicative groups of finite fields. It is an open problem to find useful applications of homomorphic schemes in abelian groups.

13.2.5 Error Correcting Codes and Secret Sharing

McEliece and Sarwate [75] observed that Shamir's scheme is closely related to Reed–Solomon codes [76]. The advantage of this formulation is that the error correcting capabilities of the Reed–Solomon codes can be translated into desirable secret-sharing properties.

Let $(\alpha_0, \alpha_1, \ldots, \alpha_{q-1})$ be a fixed list of the nonzero elements of a finite field $GF(q)$ containing q elements. In a Reed–Solomon code, an information word $\mathbf{a} = (a_0, a_1, \ldots, a_{k-1})$, $a_i \in GF(q)$, is encoded into the codeword $\mathbf{D} = (D_1, D_2, \ldots, D_{r-1})$, where $D_i = \sum_{j=0}^{k-1} a_j \alpha_i^j$. In this formulation, the secret is $a_0 = -\sum_{i=1}^{r-1} D_i$ and the shares distributed to the participants are the D_i.

In the above formulation of threshold schemes, algorithms, such as the errors-and-erasures decoding algorithm, can be used to correct t out of s shares where $s - 2t \geq k$ in a (k, n)-threshold scheme, if for some reason these shares were corrupted. The algorithm will also locate which invalid shares D_i were submitted, either as a result of deliberate tampering or as a result of storage degradation.

Karnin et al. [63] realize threshold schemes using linear codes. Massey [71] introduced the concept of minimal codewords, and proved that the access structure of a secret-sharing scheme based on a $[n, k]$ linear code is determined by the minimal codewords of the dual code. To realize a (t, n)-threshold scheme, a linear $[n + 1, t; q]$ code $\mathcal{C}$ over $GF(q)$ is selected. If G is the generator matrix of $\mathcal{C}$ and $s \in GF(q)$ is the secret, then the information vector $\mathbf{s} = (s_0, s_1, \ldots, s_{t-1})$ is any vector satisfying $s = \mathbf{s} \cdot \mathbf{g}^T$, where $\mathbf{g}^T$ is the first column vector of G. The codeword corresponding to $\mathbf{s}$ is $sG = (t_0, t_1, \ldots, t_n)$. Each participant in the scheme receives t_i as its share and t_0 is the secret. To recover the secret, first the linear dependency between $\mathbf{g}$ and the other column vectors in the (public) generator matrix G is determined. If $\mathbf{g} = \sum x_j \mathbf{g}_j$ is the linear relation, the secret is given by $\sum x_j t_{i_j}$, where $\{t_{i_1}, t_{i_2}, \ldots, t_{i_t}\}$ is a set of t shares.

Renvall and Ding [84] consider the access structures of secret-sharing schemes based on linear codes as used by McEliece and Sarwate, and Karnin et al. They determine the access structures that arise from $[n + 1, k, n - k + 2]$ MDS codes—codes which achieve the singleton bound [76]. Bertilsson and Ingemarsson [13] use linear block codes to realize secret-sharing schemes for general access structures. Their algorithm takes a description of an access structure by a monotone Boolean formula Γ, and outputs the generator matrix of a linear code that realizes Γ.

13.2.6 Combinatorial Structures and Secret Sharing

There are various connections between combinatorial structures and secret sharing, cf. [14]. Stinson and Vanstone [105], and Schellenberg and Stinson [88] study threshold schemes based on combinatorial designs. Stinson [102] uses balanced incomplete blocks designs to obtain general bounds on the information rate ρ^* of schemes with access structure based on graphs (cf. Section 13.2.10).

Street [100] surveys defining sets for t-designs and critical sets for Latin squares, with the view of applying these concepts to multilevel secret-sharing schemes, in which a hierarchical structure can be imposed on the shares. To illustrate these methods, we give an example of a $(2, 3)$-threshold scheme based on a small Latin square, cf. Chaudhry and Seberry [36]. For an example of a scheme with a hierarchical share structure, cf. Street [100].

Let $(i, j; k)$ denote that the value k is in the position (i, j) of the Latin square:

$$L = \begin{pmatrix} 1 & 2 & 3 \\ 2 & 3 & 1 \\ 3 & 1 & 2 \end{pmatrix}.$$

The shares of the secret, which is L, are $S = \{(2, 1; 1), (3, 2; 1), (1, 3; 3)\}$.

More recently, critical sets in Room squares have been used to realize multilevel secret-sharing schemes, cf. Chaudhry and Seberry [36]. Some other approaches to multilevel schemes are considered in the papers by Beutelspacher [16] and Cooper et al. [38]. The schemes based on Latin and Room squares are examples of nonperfect schemes, which will be discussed in Section 13.2.12.

13.2.7 The Problem of Cheaters

So far we have assumed that the participants in a secret-sharing scheme are honest and obey the reconstruction protocol. However, there are conceivable situations where a dishonest clique of participants (assuming an honest KDC) may attempt to defraud the honest participants by altering the shares they were issued.

In the McEliece and Sarwate formulation of Shamir's scheme, invalid shares can be identified. Schemes with this capability are said to have the cheater identification property. A weaker capability ascertains that invalid shares were submitted in the reconstruction phase without necessarily locating the source of these shares; this is called cheater detection.

Tompa and Woll [107] show that the public knowledge of the abscissae in Shamir's scheme allows a clique of dishonest participants to modify their shares resulting in the recreation of an invalid secret K'. Suppose that participants $i_1, i_2, \ldots, i_t$ agree to pool their shares in order to recreate the secret. A dishonest participant, say i_1, can determine a polynomial $\Delta(x)$ of degree at most $t - 1$ from $\Delta(0) = -1$ and $\Delta(i_2) = \Delta(i_3) = \cdots = \Delta(i_t) = 0$ using Lagrangian interpolation. Instead of the share originally issued by the dealer, the cheater submits the modified share $a(i_1) + \Delta(i_1)$. Lagrangian interpolation of points using the modified share will result in the polynomial $a(x) + \Delta(x)$ being recreated, instead of the intended polynomial $a(x)$. Now the constant term is $a(0) + \Delta(0) = K - 1$, a legal but incorrect secret. The honest participants believe that the secret is $K - 1$, while the cheater will privately recover the correct secret since $K = (K - 1) + 1$.

To prevent this type of cheating, Tompa and Woll define the shares in their modified scheme to be: $(x_1, d_1), (x_2, d_2), \ldots, (x_n, d_n)$. The dealer chooses randomly and uniformly a permutation $(x_1, x_2, \ldots, x_n)$ of n distinct elements from $\{1, 2, \ldots, q - 1\}$, and $d_i = a(x_i)$. The modified scheme resists the aforementioned attack for up to $t - 1$ cheaters. The expected running time of the scheme is polynomial in $k, n, \log s$, and $\log(1/\epsilon)$, where ϵ is a designated security parameter of the scheme and the secret k is chosen from $\{0, 1, \ldots, s - 1\}$. The overhead is that the participants need to keep secure two shares instead of the usual single share.

Using the error correcting code approach to secret sharing [75], Ogata and Kurosawa [81] formulated the problem of cheaters in secret-sharing schemes by introducing d_{cheat}. This is more appropriate than the minimum Hamming distance d_{min} of the related error correcting code in situations where only the correct secret needs to be recovered (as opposed to identifying the cheaters in the scheme). For general access structures Γ (see the following text), Ogata and Kurosawa prove that $d_{min} \le d_{cheat} = n - \max_{B \notin \Gamma}|B|$. Secret-sharing schemes for which the two bounds coincide are called maximally distance separable, or (MDS) schemes.

Brickell and Stinson [25] modified Blakley's geometric (t, n)-scheme and obtained a scheme in which cheaters can be detected and identified. Blakley et al. [18] proved that this scheme can disenrol participants, cf. Section 13.2.4. To set up the scheme, the dealer performs certain computations, such as checking that the shadows are in general position. It is an open problem whether these computations can be done efficiently as the numbers of participants increase.

The problem of secret sharing without the usual assumptions about the honesty of the participants, or even the KDC, has been considered in the literature. For example, in verifiable secret sharing it is not assumed that the dealer is honest. This problem is studied by Chor et al. [37]. The problem is how to convince the participants in a (t, n)-threshold scheme that every subset of t shares of a share set $\{s_1, s_2, \ldots, s_n\}$ defines the same secret. This is called t-consistency. In Shamir's scheme, t-consistency is equivalent to the condition that interpolation on the points $(1, s_1), (2, s_2), \ldots, (n, s_n)$ yields a polynomial of degree at most $t - 1$. As application of homomorphic schemes, Benaloh [4] gives an interactive proof that Shamir's scheme is t-consistent.

13.2.8 General Access Structures

A complete discussion of secret sharing requires the notion of a general access structure.

Ito et al. [61] describe a method to realize secret-sharing schemes for general access structures. They observe that for most applications of secret sharing it suffices to consider monotone access structures (MAS) defined as follows.

Given a set $\mathcal{P}$ of n participants ($|\mathcal{P}| = n$), an MAS on $\mathcal{P}$ is a family of subsets $\mathcal{A} \subseteq 2^{\mathcal{P}}$ such that

$$\mathcal{A} \subseteq \mathcal{A}' \subseteq \mathcal{P} \Rightarrow \mathcal{A}' \in \mathcal{A} \tag{13.4}$$

The intersection $\mathcal{A}_1 \cap \mathcal{A}_2$ and the union $\mathcal{A}_1 \cup \mathcal{A}_2$ of two MASs is a MAS. If $\mathcal{A}$ is a MAS, then $2^{\mathcal{P}} \setminus \mathcal{A} = \bar{\mathcal{A}}$ is not a MAS. Any MAS can be expressed equivalently by a monotone Boolean function. Conversely, any Boolean expression without negations represents a MAS.

In view of the above observations, we consider the minimal authorized subsets of an access structure $\mathcal{A}$ on $\mathcal{P}$. A set $B \in \mathcal{A}$ is minimal authorized, if for each proper subset A of B, it is the case that $A \notin \mathcal{A}$. The set of minimal authorized subsets of $\mathcal{A}$ is called the basis. An access structure $\mathcal{A}$ is the unique closure of the basis, i.e., all subsets of $\mathcal{P}$ that are supersets of the basis elements.

Some examples of inequivalent access structures on four participants are given by the following monotone formulae: $\Gamma_1 = P_1P_2P_3 + P_1P_2P_4 + P_1P_3P_4 + P_2P_3P_4$ — a $(3, 4)$-threshold scheme; $\Gamma_2 = P_1P_2 + P_3P_4$; $\Gamma_4 = P_1P_2 + P_2P_3 + P_3P_4$. In these formulae, the P_i's represent the participants in the scheme (sometimes the literals A, B, etc., are used). The authorized subsets in the access structure are specified precisely by these formulae. For example, Γ_2 stipulates that either P_1 AND P_2 OR P_3 AND P_4 are the authorized subsets. (It is known that no threshold scheme can realize the access structure defined by Γ_2. For a proof, cf. Benaloh and Leichter [7].)

The inequivalent access structures on three and four participants, and the information rates of secret-sharing schemes realizing these structures are given by Simmons et al. [97] and Stinson [102]. The information rates of all inequivalent access structures on five participants are given in [73]. It should be remarked that a practical examination of access structures is probably limited to five participants, since the number of equivalence classes of monotone Boolean formulae becomes too great to consider for more than five participants. However, Martin and Jackson [73] provide inductive methods using which the information rates of an access structure Γ is related to the information rates of smaller access structures that are "embedded" in Γ.

Secret-sharing schemes for nonmonotone access structures have also been investigated, cf. Simmons [96].

13.2.9 Realizing General Access Structures

Ito et al. [61] were the first to show how to realize secret-sharing schemes for general access structures. Benaloh and Leichter [7] simplified the method of Ito et al.

They show that any monotone access structure can be recognized by a monotone Boolean circuit. In a monotone circuit, each variable corresponds to an element of $\mathcal{P}$. The circuit outputs a true value only when the set of variables which take a true value correspond to an authorized subset of $\mathcal{P}$, i.e., belongs to the access structure. Monotone circuits are described by Boolean formulae which involve only AND and OR operators. Using Benaloh and Leichter's method, one can realize any access structure as a composite of subsecrets. The subsecrets are shared across AND gates by (t, t)-threshold schemes for appropriate t, and all the inputs to the OR gates have the same value.

Simmons et al. [97] showed how cumulative arrays, first studied by Ito et al. [61], can be used to realize geometric secret-sharing schemes for general access structures.

A cumulative array $C_{\mathcal{A}} = (\mathcal{S}, f)_{\mathcal{A}}$ for the access structure $\mathcal{A}$ is a pair comprising of the share set $\mathcal{S} = \{s_1, s_2, \ldots\}$, and the dealer function $f : \mathcal{P} \to 2^{\mathcal{S}}$ that assigns subset of shares to each participant.

As an example, consider the following access structure:

$$\mathcal{A} = \text{closure } \{\{P_1, P_2\}, \{P_2, P_3\}, \{P_3, P_4\}, \{P_1, P_4\}\},$$

where $\mathcal{P} = \{P_1, P_2, P_3, P_4\}$. Let $\mathcal{S} = \{s_1, s_2\}$. A cumulative array for this access structure is $f(P_1) = s_1$, $f(P_2) = s_2, f(P_3) = s_1$, and $f(P_4) = s_2$.

Perfect geometric secret-sharing schemes are obtained from cumulative arrays as follows. Choose a projective space $V_i = PG(m - 1, q)$, where m is the number of columns in the cumulative array. In V_i, let $\{s_i, \ldots, s_m, K\}$ be $m + 1$ points, such that no m points lie on a hyperplane of V_i — the points are in general position. A domain variety V_d is chosen so that $V_i \cap V_d = \{K\}$. The set of shares in the geometric scheme is $\{s_i, \ldots, s_m\}$ and K is the secret. The shares are distributed using the cumulative array: participant P_i receives share s_j if and only if the (i, j) entry of the array is one. Note that it could be difficult to verify the general position hypothesis for large m, cf. Brickell and Stinson [25]. Jackson and Martin [62] show that any geometric secret-sharing scheme realizing an access structure is "contained" in the cumulative array, which realizes the access structure.

For any access structure $\mathcal{A}$ on the set $\mathcal{P}$, there is a unique minimal cumulative array. Thus to implement geometric secret-sharing schemes with the minimal number of shares, we need to consider only minimal cumulative arrays. It remains only to have a means by which the minimal cumulative array can be calculated given an arbitrary monotone Boolean function Γ. Such a method was first given by Simmons et al. [97]. It relies on minimizing the Boolean expression, which results when the AND and OR operators in Γ are exchanged.

An alternative method for calculating minimal cumulative arrays is described by Charnes and Pieprzyk [34]. This method has the advantage that the complete truth table of Γ is not required for some Γ, thereby avoiding an exponential time computation. For general Boolean expressions, as the number of variables increases the time complexity of the aforementioned method and that of [97] are the same.

To describe the method of [34], we require the following.

The representative matrix M_Γ of a monotone Boolean function $\Gamma(P_1, P_2, \ldots, P_n)$, expressed as a disjunctive sum of r products of n variables, is an $n \times r$ matrix with rows indexed by the P_i and columns by the product terms of the P_i [34]. The (i, j)-entry is one if P_i occurs in the jth product, and is zero otherwise. For example, if $\Gamma = P_1 P_2 + P_2 P_3 + P_3 P_4$, then M_Γ is the following matrix:

	$P_1 P_2$	$P_2 P_3$	$P_3 P_4$
P_1	1	0	0
P_2	1	1	0
P_3	0	1	1
P_4	0	0	1

Suppose that $\Gamma(P_1, P_2, \ldots, P_n)$ is a monotone formula expressed in minimal disjunctive form, i.e., a disjunctive sum of products of the P_i and no product term is contained in any other representative set. Let M_Γ be its representative matrix.

[34] A subset $\{P_l, P_m, \ldots\}$ of the variables of $\Gamma(P_1, P_2, \ldots, P_n)$ is a relation *set* if $P_l P_m \ldots$ is represented in M_Γ by the all ones vector.

In the previous representative matrix $\{P_1, P_3\}, \{P_2, P_3\}$, and $\{P_2, P_4\}$ are the minimal representative sets, i.e., not contained in any other representative set. The Boolean formula derived from these sets is $P_1 P_3 + P_2 P_3 + P_2 P_4$.

Let $\Gamma(P_1, P_2, \ldots, P_n)$ be a monotone formula and M_Γ its representative matrix [34]. Let $\mathcal{R}$ be the collection of minimal relation sets of M_Γ. Then the representative matrix whose rows are indexed by the variables P_i and columns by product terms derived from $\mathcal{R}$ is the minimal cumulative array for $\mathcal{A}$.

Thus, using the above theorem the matrix

	$P_1 P_3$	$P_2 P_3$	$P_2 P_4$
P_1	1	0	0
P_2	0	1	1
P_3	1	1	0
P_4	0	0	1

is the minimal cumulative array for $\Gamma = P_1 P_2 + P_2 P_3 + P_3 P_4$. To realize Γ as a geometric scheme, we require a projective space $V_i = PG(2, q)$. The secret $K \in V_d$, and the shares $\{s_1, s_2, s_3\}$ are points chosen in general position in V_i. The cumulative array specifies the distribution of the shares: P_1 receives share $\{s_1\}$; P_2 receives shares $\{s_2, s_3\}$; P_3 receives shares $\{s_1, s_2\}$; P_4 receives share $\{s_3\}$. It can be easily verified that only the authorized subsets of participants can recreate the secret, e.g., the combined shares of P_1 and P_2 span V_i, hence these participants can recover the secret as $V_i \cap V_d = \{K\}$. But unauthorized participants, e.g., P_1 and P_3, cannot recover the secret.

An algorithm for calculating cumulative arrays, based on Theorem 13.2.9, is given in [34]. This algorithm is efficient for those Γ for which M_Γ contains columns with many zeros. Thus in the previous example, the combinations $\{P_1, P_2\}, \{P_1, P_4\}$, and $\{P_3, P_4\}$ cannot produce relation sets and can be ignored. Further computational savings are obtained if the Boolean formula has symmetries; i.e., permutations of the participants that do not change Γ.

13.2.10   Ideal and Other Schemes

Brickell [23] gives a vector space construction for realizing ideal secret-sharing schemes for certain types of access structures, Γ. Let ϕ be a function

$$\phi : \mathcal{P} \cup \{\mathcal{D}\} \rightarrow GF(q)^d$$

with the property that $\phi(D)$ can be expressed as a linear combination of the vectors in $< \phi(P_i) :$ $P_i \in B >$ if and only if B is an authorized subset, i.e., $B \in \Gamma$. Then, for any such ϕ, the distribution rules (cf. Section 13.2.2.1) are for any vector $\mathbf{a} = (a_1, \ldots, s_d)$ in $GF(q)^d$, a distribution rule is given by the inner product of $\mathbf{a}$ and $\phi(x)$ for every $x \in \mathcal{P} \cup \{\mathcal{D}\}$. Under the previous conditions, the collection of distribution rules is an ideal secret-sharing scheme for Γ. A proof of this result can be found in a paper by Stinson [102].

Shamir's (t, n)-threshold scheme is an instance of the vector space construction. Access structures $\Gamma(G)$, whose basis is the edge set of certain undirected graphs, can also be realized as ideal schemes by this construction. In particular the access structure $\Gamma(G)$, where $G = (V, E)$ is a complete multigraph, can be realized as an ideal scheme. A proof of this is given by Stinson [102].

A relation between ideal secret sharing schemes and matroids was established by Brickell and Davenport [24]. The matroid theory counterpart of a minimal linearly dependent set of vectors in a vector space is called a circuit. A coordinatizable matroid is one that can be mapped into a vector space over a field in a way that preserves linear independence. Brickell and Davenport [24] prove the following theorem about coordinatizable matroids.

Suppose the connected matroid $\mathcal{M} = (X, \mathcal{I})$ is coordinatizable over a finite field [24]. Let $x \in X$ and let $\mathcal{P} = X \backslash \{x\}$. Then there exists an ideal scheme for the connected access structure having basis $\Gamma_0 = \{C \backslash \{x\} : x \in C \in \mathcal{C}\}$, where $\mathcal{C}$ denotes the set of circuits of $\mathcal{M}$.

Not all access structures can be realized as ideal secret-sharing schemes. This was first established by Benaloh and Leichter [7]. They proved that the access structure on four participants specified by the monotone formula $\Gamma = P_1 P_2 + P_2 P_3 + P_3 P_4$ cannot be realized by an ideal scheme. The relation between the size of the shares and the secret for Γ was made precise by Capocelli et al. [28]. They proved the following information theoretic bound.

For the access structure $\Gamma = $ closure $\{\{P_1, P_2\}, \{P_2, P_3\}, \{P_3, P_4\}\}$ on four participants $\{P_1, P_2, P_3, P_4\}$, the inequality $H(P_2) + H(P_3) \geq 3H(K)$ holds for any secret-sharing scheme realizing Γ.

From this theorem, it follows that the information rate ρ of any secret-sharing scheme realizing Γ satisfies the bound $\rho \leq \frac{2}{3}$. Bounds are also derived by Capocelli et al. [28] for the maximum information rate ρ^* of access structures $\Gamma(G)$, where the graph G is a path P_n $(n \geq 3)$; a cycle C_n, $n \geq 6$, for n even and $n \geq 5$, for n odd; or any tree T_n.

13.2.11 Realizing Schemes Efficiently

In view of bounds on the information rates of secret-sharing schemes, it is natural to ask whether there exist schemes whose information rates equal the known bounds. For example, for $\Gamma = P_1 P_2 + P_2 P_3 + P_3 P_4$ one is interested in the realizations of Γ with $\rho = \frac{2}{3}$.

Stinson [102] used a general method, called decomposition construction, to build larger schemes starting from smaller ideal schemes. In this method, the basis Γ_0 of an access structure is decomposed into smaller access structures, as $\Gamma_0 = \cup \Gamma_k$, where the Γ_k are the basis of the constituent access structures which can be realized as ideal schemes. From such decompositions of access structures, Stinson [102] derived a lower bound $\rho^*(\Gamma) \geq \ell/R$, where ℓ and R are two quantities defined in terms of the ideal decomposition of Γ_0. The decomposition construction and its precursor, the graph decomposition construction (cf. Blundo et al. [20]), can be formulated as linear programming problems in order to derive the best possible information rates that are obtainable using these constructions.

Other ways of realizing schemes with optimal or close to optimal information rates are considered by Charnes and Pieprzyk [35]. Their method combines multiple copies of cumulative arrays using the notion of composite shares—combinations of the ordinary shares in cumulative arrays. This procedure is stated as an algorithm that outputs a cumulative array with the best information rate. It is not clear how efficient this algorithm is as the number of participants increases. However, the optimal information rates for access structures on four participants given by Stinson [102] can be attained by combining cumulative arrays.

13.2.12 Nonperfect Schemes

It is known that in nonperfect secret sharing schemes the size of the shares is less that the size of the secret, i.e., $H(s_i) < H(K)$. Because of this inequality, a nonperfect scheme can be used to disperse a computer file to n sites, in such a way that the file can be recovered from its images that are held at any t of the sites for $t \leq n$. Moreover, this can be done so that the size of the images is less than the size of the original file resulting in an obvious saving of disk space. Making backups of computer files with this method provides insurance against the loss or the destruction of valuable data. For details, cf. Karnin et al. [63].

A formal analysis of nonperfect secret sharing schemes is given by Ogata et al. [80]. Their analysis characterizes, using information theory, secret-sharing schemes in which the participants not belonging to an access structure do gain some information about the secret. This possibility is precluded in perfect secret-sharing schemes.

Ogata et al. [80] define a nonperfect scheme in using a triple of access sets $(\Gamma_1, \Gamma_2, \Gamma_3)$, which partition the set of all subsets of the participants $\mathcal{P}$. Γ_1 is the family of access subsets, Γ_2 is the family of semiaccess subsets; and Γ_3 is the family of nonaccess subsets. The participants belonging to the semiaccess subsets are able to obtain some, but not complete information about the secret. The participants belonging to the nonaccess subsets gain no information about the secret.

The ramp schemes of Blakley and Meadows [19] are examples of nonperfect schemes where the access structure consists of semiaccess subsets. Another way of viewing ramp schemes is that the collective uncertainty about a secret gradually decreases as more participants join the collective.

Ogata et al. [80] prove a lower bound on the size of the shares in nonperfect schemes. They also characterize nonperfect schemes for which the size of the shares is $|K|/2$.

Ogata and Kurosawa [79] establish a general lower bound for the sizes of shares in nonperfect schemes. They show that there is an access hierarchy for which the size of the shares is strictly larger than this bound. It is in general a difficult problem to realize nonperfect secret-sharing schemes with the optimum share size, as in the case of perfect schemes.

13.3 Threshold Cryptography

There are circumstances in cryptography where an action requires to be executed by a group of people. For example, to transfer money from a bank a manager and a clerk need to concur. A bank vault can be opened only if three high-ranking bank employees cooperate. A ballistic missile can be launched only if two officers authorize the action.

Democratic groups usually exhibit a flat relational structure where every member has equal rights. On the other hand, in hierarchical groups, the privileges of group members depend on their position in the hierarchy. A member on the level $i - 1$ inherits all the privileges from the level i, as well as additional privileges specific to its position.

Unlike single-user cryptography, threshold or society-oriented cryptography allows groups to perform cryptographic operations, such as encryption, decryption, and signature. A trivial implementation of group-oriented cryptography can be achieved by concatenating secret-sharing schemes and a single user cryptosystem. This arrangement is usually unacceptable as the cooperating subgroup must first recover the cryptographic key. Having access to the key can compromise the system, as its use is not confined to the requested operation. Ideally, the cooperating participants should perform their private computations in one go. Their partial results are then sent to a so-called combiner who calculates the final result. Note that at no point is the secret key exposed.

A group-oriented cryptosystem is usually set up by a dealer who is a trusted authority. The dealer generates all the parameters, distributes elements via secure channels if the elements are secret, or broadcasts the parameters if they need not be protected. After setting up a group cryptosystem, the dealer is no longer required, as all the necessary information has been deposited with the participants of the group cryptosystem.

If some participants want to cooperate to perform a cryptographic operation, they use a combiner to perform the final computations on behalf of the group. The final result is always correct if the participants belong to the access structure and follow the steps of the algorithm. The combiner fails if the participants do not belong to the access structure, or if the participants do not follow the algorithm (that is, they cheat). The combiner need not be trusted; it suffices to assume that it will perform some computations reliably but not necessarily all.

The access structure is the collection of all subsets of participants authorized to perform an action. An example is a (t, n)-threshold scheme, where any t out of n participants are authorized subsets $t \leq n$.

Threshold cryptography provides tools for groups to perform the following tasks:

- Threshold encryption—a group generates a valid cryptogram which can later be decrypted by a single receiver.
- Threshold decryption—a single sender generates a valid cryptogram that can be decrypted by a group.
- Threshold authentication—a group of senders agrees to co-authenticate the message so that the receiver can decide whether the message is authentic or not.
- Threshold signature (multisignature)—a group signs a message that is later validated by a single verifier.
- Threshold pseudorandom generation.

13.3.1   Threshold Encryption

Public-key cryptography can be used as a basis for simple group encryption. Assume that a receiver wants to have a communication channel from a group of n participants $\mathcal{P} = \{P_1, \ldots, P_n\}$. Further suppose that the receiver can decrypt a cryptogram only if all participants cooperate, i.e., a (n, n)-threshold encryption system. Group encryption works as follows.

Assume that the group and the receiver agree to use the RSA cryptosystem with the modulus $N = pq$. The receiver first computes a pair of keys: one for encryption e and the other for decryption d, where $e \times d \equiv 1 \bmod (p - 1)(q - 1)$. Both keys are secret. The factors p and q are known by the receiver only. The encryption key is communicated to the dealer (via a secure channel). The dealer selects $n - 1$ shares e_i of the encryption key at random from the interval $[0, e/n]$. The last share is

$$e_n = e - \sum_{i=1}^{n-1} e_i.$$

Each share e_i is communicated to participant P_i via a secure channel $(i = 1, \ldots, n)$.

Now if the group wants to send a message m to the receiver, each participant P_i prepares a partial cryptogram $c_i \equiv m^{e_i} \bmod N$ $(i = 1, \ldots, n)$. After collecting n partial cryptograms, the receiver can recover the message $m \equiv (\prod_{i=1}^{n} c_i)^d \bmod N$. Note that the receiver also plays the role of a combiner. Moreover, the participants need not reconstruct the secret encryption key e and at no stage of decryption is the encryption key revealed—this is a characteristic feature of threshold cryptography.

Many existing secret-key algorithms, such as the DES [78], the LOKI [27], the FEAL [93], or the Russian GOST [99], are not homomorphic. These algorithms cannot be used for threshold encryption. The homomorphic property is necessary in order to generate shares of the key so that partial cryptograms can be combined into a cryptogram for the correct message, cf. [4].

Threshold encryption has not received a great deal of attention, perhaps because of its limited practical significance.

13.3.2 Threshold Decryption

Hwang [57] proposes a cryptosystem for group decryption based on the discrete logarithm problem. In his system, it is assumed that the sender knows the participants of the group. The sender encrypts the message using a predetermined (either private or public key) cryptosystem with a secret key known to the sender only. The sender then distributes the secret key among the group of intended receivers using Shamir's (t, n)-threshold scheme. Any t cooperating participants can recover the decryption key and decrypt the cryptogram. In Hwang's scheme, key distribution is based on the Diffie–Hellman [45] protocol. Thus the security of his scheme is equivalent to the security of the discrete logarithm problem. However, the main problem with the above solution is that the key can be recovered by a straightforward application of secret sharing. This violates the fundamental requirement that the decryption key must never be revealed to the group (or the combiner).

We consider now an implementation of a scheme for (t, n)-threshold decryption. The group decryption used here is based on the ElGamal public-key cryptosystem [47] and is described by Desmedt and Frankel [41].

The system is set up by the dealer $\mathcal{D}$ who first chooses a Galois field $GF(q)$ such that $q - 1$ is a Mersenne prime and $q = 2^\ell$. Further $\mathcal{D}$ selects a primitive element $g \in GF(q)$ and a nonzero random integer $s \in GF(q)$. The dealer computes $y = g^s \bmod q$ and publishes the triple (g, q, y) as the public parameters of the system. The dealer then uses Shamir's (t, n)-threshold scheme to distribute the secret s among the n shareholders in such a way that for any subset $\mathcal{B}$ of t participants, the secret $s = \sum_{P_i \in \mathcal{B}} s_i \bmod (q - 1)$ (all calculations are performed in $GF(q)$).

Suppose that user A wants to send a message $m \in GF(q)$ to the group. A first chooses at random an integer $k \in GF(q)$ and computes the cryptogram $c = (g^k, m y^k)$ for the message m.

Assume that $\mathcal{B}$ is an authorized subset, so it contains at least t participants. The first stage of decryption is executed separately by each participant $P_i \in \mathcal{B}$. P_i takes the first part of the cryptogram and computes $(g^k)^{s_i} \bmod q$. The result is sent to the combiner, who computes $y^k = g^{ks} = \prod_{i \in \mathcal{B}} g^{ks_i}$, and decrypts (using the multiplicative inverse y^{-k}) the cryptogram

$$m \equiv m y^k \times y^{-k} \bmod p.$$

Group decryption can also be based on a combination of the RSA cryptosystem [85] and Shamir's threshold scheme. The scheme described by Desmedt and Frankel [42] works as follows. The dealer D computes the modulus $N = pq$, where p, q are strong primes, i.e., $p = 2p' + 1$ and $q = 2q' + 1$ (p' and q' are large and distinct primes). The dealer selects at random an integer e such that e and $\lambda(N)$ are coprime ($\lambda(N)$ is the least common multiple of two integers $p - 1$ and $q - 1$, so $\lambda(N) = 2p'q'$). Next D publishes e and N as the public parameters of the system, but keeps p, q, and d secret (d satisfies the congruence $ed = 1 \bmod \lambda(N)$). It is clear that computing d is easy for the dealer who knows $\lambda(N)$, but is difficult—equivalent to the factoring of N—to someone who does not know $\lambda(N)$. The dealer then uses Shamir's scheme to distribute the secret $s = d - 1$ amongst n participants. The shares are denoted as s_i and any t cooperating participants (the set $\mathcal{B}$) can retrieve the secret. We have,

$$s = \sum_{i \in \mathcal{B}} s_i \bmod \lambda(N).$$

Group decryption of the cryptogram $c \equiv m^e \bmod N$ starts from individual computations. Each $P_i \in \mathcal{B}$ calculates a partial cryptogram $c^{s_i} \bmod N$. All the partial cryptograms are sent to the combiner who recovers the message

$$m = \prod_{i \in \mathcal{B}} c^{s_i} \times c \equiv c^{(\sum_{i \in \mathcal{B}} s_i + 1)} \equiv c^d \equiv \left(m^e\right)^d \bmod N.$$

Again the secret $s = d - 1$ is never exposed during the decryption.

Ghodosi et al. [53] proposed a solution to the problem of group decryption which does not require a dealer. It uses the RSA cryptosystem and Shamir's threshold scheme. The system works under the assumption that all participants from $\mathcal{P} = \{P_1, \ldots, P_n\}$ have their entries in a public registry (white pages). The registry provides the public parameters of a given participant. A participant P_i has N_i, e_i as its RSA entry in the registry, and this entry cannot be modified by an unauthorized person.

The sender first selects the group $\mathcal{P} = \{P_1, \ldots, P_n\}$. For the message m $(0 < m < \prod_{i=1}^n N_i)$, the sender computes

$$m_i \equiv m \bmod N_i$$

for $i = 1, \ldots, n$. Next the sender selects at random a polynomial $f(x)$ of degree at most t over $GF(p)$, where $p < \min_i N_i$. Let

$$f(x) = a_0 + a_1 x + \cdots + a_{t-1} x^{t-1}.$$

The sender computes $c_i = f(x_i)$ for public x_i, $k = f(0)$, $c_i^{e_i} \bmod N_i$, and $m_i^k \bmod N_i$ $(i = 1, \ldots, n)$. Finally, the sender merges $c_i^{e_i} \bmod N_i$ into C_1 and $m_i^k \bmod N_i$ into C_2 using the Chinese Remainder Theorem. The sender broadcasts the tuple (N, p, t, C_1, C_2) as the cryptogram.

The participants check whether they are the intended recipients, for instance, by finding the gcd (N_i, N). Note that the sender can give the list of all participants instead of the modulus N. A participant P_i first recovers the pair $(c_i^{e_i} \bmod N_i)$ and $(m_i^k \bmod N_i)$ from C_1 and C_2, respectively. Using its secret key d_i, the participant retrieves c_i. The c_i are now broadcast so that each participant can reconstruct $f(x)$ and find $k = f(0)$. Note that none of the participants can cheat as it can be readily verified whether c_i' satisfies the congruence

$$c_i'^{e_i} \equiv C_1 \bmod N_i.$$

Knowing k, each participant finds the message $m_i \equiv C_2^{k^{-1}} \bmod N_i$. Although k is public, only participant P_i can find $k^{-1} \times k \bmod (p_i - 1)(q_i - 1)$ from his knowledge of the factorization of $N_i = p_i q_i$. Lastly, all the partial messages are communicated to the combiner who recovers the message m by the Chinese Remainder Theorem.

13.4 Signature Schemes

A signature scheme consists of two algorithms: signature generation and signature verification. Each of these algorithms can be collaboratively performed. A shared-generation scheme allows a group of signers to collaboratively sign a document. In a signature scheme with shared verification, the signature verification requires the collaboration of a group. We examine the two types of systems and note that the two can be combined if necessary.

13.4.1 Shared Generation Schemes

In these schemes, a signer group P of n participants has a public/private key pair. The private part is shared among members of the group such that each member has part of the private key that is not known to anyone else. The signature scheme is usually based on one of the well-known signature schemes, such as ElGamal, Schnorr, RSA, and Fiat–Shamir.

The group is created with an access structure that determines the authorized groups of signers. A special case of shared-generation schemes is the multisignature scheme, in which the collaboration of all members in P is necessary. Most systems proposed for shared generation are of the multisignature type, or its generalization, (t, n)-threshold signature. In the latter type of signature, each subgroup $p, p \subset P$ of size t can generate the signature.

A shared-generation scheme can be sequential or simultaneous. In a sequential scheme, each member of the group signs the message and forwards it to the next group member. In some schemes, after the first signer the message is not readable and all subsequent signers must blindly sign the message. In a simultaneous scheme, each group member forms a partial signature that is sent to a combiner who forms the final signature.

There are a number of issues that differentiate shared-generation systems:

1. Mutually trusted party: a system may need a mutually trusted party who is usually active during the key generation phase; it chooses the group secret key and generates secrets for all group members. In systems without a trusted party, each signer produces his secret key and participates in a protocol with other signers to generate the group public key.

2. The security of most signatures schemes is based on the intractability of one of the following problems: discrete logarithm or integer factorization. Shared-generation schemes based on ElGamal and Schnorr signature schemes use the former, while those based on RSA and Fiat–Shamir use the latter.

3. Using many/few interactions for producing signature. The amount of interaction between the signers and the trusted third party varies in different schemes.

There are properties—some essential and some desirable—that a shared-generation scheme must satisfy. The essential properties are as follows:

A1 Signature generation must require the collaboration of all members of the authorized group and no signer in the group should be able to deny his signature. Verification must be possible by any outsider.

A2 An unauthorized group should not be able to forge the signature of an authorized group. It should not also be possible for an authorized group to forge the signature of another authorized group.

A3 No secret information should be derivable from the released group and partial signatures.

The desirable properties are as follows:

1. Each signer must have the same power and be able to see the message that he is signing.

2. The order of signing in a sequential scheme should not be fixed.

3. The size of the multisignature should be comparable to, preferably the same as, the size of the individual signature.

For a (t, n) threshold signature scheme, (A1) and (A2) reduce to

B1 From any t partial signature the group signature should be easily derivable.

B2 Knowledge of $t - 1$ or fewer partial signatures should not reduce the chance of forgery of an unauthorized group.

13.4.2 Constructions

The earliest proposals for shared-generation schemes are by Itakura and Nakamura [60] and by Boyd [21]. Boyd's scheme is a (n, n)-threshold group signature based on RSA, in which if $n > 2$ most participants must blindly sign the message.

13.4.2.1 Threshold RSA Signature

Desmedt and Frankel [42] construct a simultaneous threshold (t, n) RSA signature that requires a trusted third party to generate and distribute the group public key and the secret keys of the signers.

Their scheme works as follows. In the initialization stage, a trusted KDC (dealer) selects at random a polynomial of degree $t - 1$: $f(x) = a_0 + a_1 x + a_2 x^2 + \cdots + a_{t-1} x^{t-1}$. The group secret key k is fixed as $a_0 = f(0)$. The dealer gives $y_i = f(x_i)$ to participant P_i, for each i, via a secure channel. The computations are performed in $Z_{\lambda(N)}$, where $\lambda = 2p'q'$ and $p = 2p' + 1, q = 2q' + 1$. To sign a message m ($0 \leq m < N$), each participant $P_i \in B$ calculates their partial signature $s_i = m^{k_i} \bmod N$ and transmits the result to the combiner. The combiner computes the signature S of the message m according to the following equation:

$$S = m \times \prod_{P_i \in B} s_i = m \times \prod_{\substack{P_i \in B \\ i=1}}^{t} m^{k_i} = m \times m^{d-1} = m^d \bmod N.$$

The signature verification is similar to the conventional RSA signature scheme.

13.4.2.2 Threshold Signature Based on Discrete Logarithm

Ohta and Okamoto [82] propose a sequential multisignature scheme based on the Fiat–Shamir signature scheme. In their scheme, the order of signing is not restricted but the scheme requires a trusted center for key generation.

A variation of group signature is undeniable group signature, in which verification requires the collaboration of signers. The signature scheme has a "commitment phase" during which t group members work together to sign a message, and a "verification phase" during which all signers work together to prove the validity of the signature to an outsider. Harn and Yang [56] propose two (t, n)-threshold schemes with $t = 1$ and $t = n$. Their schemes do not require a trusted third party and the algorithm is based on the discrete logarithm problem.

Harn [55] proposes three simultaneous multisignature schemes based on the difficulty of discrete logarithm. Two of these schemes do not require a trusted third party. We briefly review one of the schemes. We use the notation of Harn [55].

Let KDC denote the key distribution center. The KDC selects

1. p, a large prime, in the range $2^{511} \leq p \leq 2^{512}$.
2. q, a prime divisor of $p - 1$.
3. $\{a_i, i = 0, \ldots, t - 1\}$ and $f(x) = a_0 + a_1 x + \cdots + a_{t-1} x^{t-1} \pmod q$ where $0 < a_i < q$.
4. α, where $\alpha = h^{(p-1)/q} \pmod p > 1$. α is a generator with order q in GF(p); p, q and α are made public.

The KDC computes the group public key $y = \alpha^{f(0)} \bmod p$, where $f(0)$ is the group secret key. The KDC also computes public keys for all group members as

$$y_i = \alpha^{f(x_i)} \pmod p, \quad \text{for } i = 1, 2, \ldots, n$$

where $f(x_i) \bmod q$ is the share of participant i from the group secret key. (Note that, since α is a generator with order q in GF(p), $\alpha^r \bmod p = \alpha^{r \bmod q}$, for any nonnegative integer r.)

In order to generate the group signature on a message m, each participant of a group B ($|B| \geq t$) randomly selects an integer, $k_i \in [1, q - 1]$, and computes a public value, $r_i = \alpha^{k_i} \bmod p$ and broadcasts r_i to all members in B. Knowing all the r_i ($i \in B$), each member of the group B computes

$$r = \prod_{i \in B} r_i \pmod p.$$

Participant i computes his partial signature as

$$s_i = m' \times f(x_i) \times \left(\prod_{\substack{i,j \in B \\ i \neq j}} \frac{x_j}{(x_i - x_j)} \right) - k_i \times r \,(\mathrm{mod}\ p)$$

where $H(m) = m'$ (H is a one-way and collision free hash function) and transmits (r_i, s_i) to a designated combiner.

Once the combiner receives the partial signature (r_i, s_i), it is verified using

$$y_i^{(m')}{}^{\left(\prod_{\substack{i,j \in B \\ i \neq j}} \frac{x_j}{(x_i - x_j)} \right)} = \alpha^{s_i} r_i^r \,(\mathrm{mod}\ p).$$

If all the partial signatures are verified, then the combiner calculates the group signature (r, s) on message m, where $s = \sum_{i \in B} s_i \,(\mathrm{mod}\ q)$.

An outsider who receives the signature (r, s) on the message m can verify the validity of the signature using the check $y^{m'} = \alpha^s r^r \,(\mathrm{mod}\ p)$. This check works because

$$f(0) = \sum_{i \in B} f(x_i) \prod_{\substack{i,j \in B \\ i \neq j}} \frac{x_j}{(x_i - x_j)} \,(\mathrm{mod}\ q)$$

and thus,

$$y^{m'} = \left(\alpha^{f(0)} \right)^{m'} = \alpha^{\left(\sum_{i \in B} f(x_i) \prod_{\substack{i,j \in B \\ i \neq j}} \frac{x_j}{(x_i - x_j)} \right)^{m'}}$$

$$= \prod_{i \in B} y_i^{(m')}{}^{\left(\prod_{\substack{i,j \in B \\ i \neq j}} \frac{x_j}{(x_i - x_j)} \right)} = \prod_{i \in B} \left(r_i^r a_i^s \right) = r^r \alpha^s.$$

An interesting security problem in these schemes, as discussed by Desmedt and Frankel [42] and by Harn [55], is that if more than t signers collaborate they can find the secrets of the system with a high probability, and thus identify the rest of the shareholders. Possible solutions to this problem in the case of discrete logarithm-based schemes can be found in a paper by Li et al. [68].

A concept related to threshold signature is t-resilient digital signatures. In these schemes, n members of a group can collaboratively sign a message even if there are t dishonest members. Moreover no subset of t dishonest members can forge a signature.

Desmedt [40] shows that a t-resilient signature scheme with no trusted center can be constructed for any signature scheme using a general multiparty protocol. Cerecedo et al. [29] present efficient protocols for the shared generation of signatures based on the discrete logarithm problem, Schnorr's scheme, and variants of the ElGamal scheme. Their protocols are based on an efficient multiparty protocol for the shared computation of products and they do not need a trusted party. Park and Kurosawa [83] discuss a (t, n)-threshold scheme based on the discrete logarithm, more precisely a version of digital signature standard (DSS), which does not require multiplication and only uses linear combination for the combination of shares.

Chang and Liou [30] and Langford [66] propose other signature schemes based on the discrete logarithm problem.

13.4.3   Shared Verification of Signatures

Signature schemes with shared verification are not commonly found in the literature.

De Soete et al. [98] propose a system for the shared verification of signatures. But their system is not really a signature scheme in the sense that it does not produce a signature for every message. Each user has a secret that enables her/him to verify herself/himself to others. It requires at least two verifiers for the secret to be verified.

Laih and Yen [65] argue that in some cases it might be necessary to sign a message such that only a specified group of participants can verify the signed message. The main requirements of such schemes are

1. A can sign any message M for any specified group B.
2. Only the specified group can validate the signature of A. No other group, except B, can validate the signature of A on M.
3. B should not be able to forge A's signature on M for another user C, even if B and C conspire.
4. No one should be able to forge A's signature on another message M'.
5. If A disavows his signature, it must be possible for a third party to resolve the dispute between A and B.

The scheme proposed by Laih and Yen [65] is based on Harn's scheme, which is an efficient ElGamal type shared-generation scheme. In the proposed scheme a group of signers can create a digital shared-generation scheme for a specified group, who can collectively check the validity of the signature. The secret key of the users is chosen by the users themselves and each group has a public key for signature generation or verification. Since the private key of the verifiers is not known, dispute settlement by a third party requires an extra protocol between the third party and the verifiers.

13.5   Quantum Key Distribution—Quantum Cryptography

While classical cryptography employs various mathematical techniques to restrict eavesdroppers from learning the contents of encrypted messages, in quantum mechanics the information is protected by the laws of physics. In classical cryptography absolute security of information cannot be guaranteed. However on the quantum level there is a law called the Heisenberg uncertainty principle. This states that even the most refined measurement on a quantum object cannot reveal everything about the state of the object before the measurement. This is because the object may be altered by simply taking the measurement. The Heisenberg uncertainty principle and quantum entanglement can be exploited in a system of secure communication, often referred to as "quantum cryptography" [11]. Quantum cryptography provides the means for two parties to exchange a enciphering key over a private channel with complete security of communication.

There are least two main types of quantum cryptosystems for the key distribution:

- *BB-protocol*: Cryptosystems with encoding based on two nonsimultaneously measurable observables proposed by Wiesner [110] and Bennett and Brassard [10]
- *EPR-type*: Cryptosystems with encoding built upon quantum entanglement and the Bell Theorem proposed by Ekert [48]

A typical Quantum key distribution (QKD) protocol is comprised of the following stages:

1. Random number generation by Alice
2. Quantum communication
3. Sifting

4. Reconciliation
5. Estimation of Eve's partial information gain
6. Privacy amplification
7. Authentication of public messages
8. Key confirmation

For detailed explanations of these terms we refer the reader to [3], here we will give an informal explanation of some of these steps using the following simple example.

A *BB* quantum cryptosystem includes a transmitter, Alice, and a receiver, Bob. Alice may use the transmitter to send photons in one of the four polarizations: 0°, 45°, 90°, or 135°. Bob at the other end uses the receiver to measure the polarization. According to the laws of quantum mechanics, Bob's receiver can distinguish between rectilinear polarizations (0 and 90; basis I), or it can quickly be reconfigured to discriminate between diagonal polarizations (45 and 135; basis II); it can never, however, distinguish both types. The key distribution requires several steps.

Alice sends photons with one of the four polarizations which are chosen at random. For each incoming photon, Bob chooses at random the type of measurement: either the rectilinear type or the diagonal type. Bob records the results of the measurements but keeps them secret. Subsequently Bob publicly announces the type of measurement (but not the results) and Alice tells the receiver which measurements were correct—the measurement is done in the same basis as the preparation. Alice and Bob (the sender and the receiver) keep all cases in which Bob's measurements were of the correct type. These cases are then translated into bits (1's and 0's) and thereby become the key. An eavesdropper (Eve) is bound to introduce errors to this transmission because she does not know in advance the type of polarization of each photon and quantum mechanics does not allow her to acquire the sharp values of two nonsimultaneously measurable observables (here the rectilinear and diagonal polarizations).

The two legitimate users of the quantum channel, Alice and Bob, test for eavesdropping by revealing a random subset of the key bits and checking (in public) the error rate. Although they cannot prevent eavesdropping, they will never be fooled by Eve because any effort to tap the channel, however subtle and sophisticated, will be detected. Whenever they are not satisfied with the security of the channel they can try to set up the key distribution again. Privacy amplification is used to distill a secret key between Alice and Bob from these interactions, cf. Bennett et al. [12].

In our example, we will consider two types of measurement: we consider \ and / to be one type (diagonal) and | and — to be the other (rectilinear).

1. Alice's polarization	|	\	—	|	/	—	—	—	—	\	/
2. Bits Alice sent	0	0	1	0	0	1	1	1	0	1	1
3. Bob's polarization	|	\	|	\	|	|	/	|	—	\	—
4. Bits Bob registered	0	0	1	1	0	1	0	1	0	1	1
5. Alice states publicly whether Bob's polarization was correct or not	Yes	Yes	Yes	No	No	Yes	No	Yes	Yes	Yes	No
6. Alice's remaining bits	0	0	1	*x*	*x*	1	*x*	1	0	1	*x*
7. Bob's remaining bits *x* means discard	0	0	1	*x*	*x*	1	*x*	1	0	1	*x*
8. Alice and Bob compare bits	0	0	1	*x*	*x*	1	*x*	1	0	1	*x*
	0	0	1	*x*	*x*	1	*x*	1	0	1	*x*
chosen at random	OK							OK		OK	
9. Alice's remaining bits	*x*	0	1	*x*	*x*	1	*x*	*x*	0	*x*	*x*
10. Bob's remaining bits	*x*	0	1	*x*	*x*	1	*x*	*x*	0	*x*	*x*
11. The key		0	1			1			0		

The basic idea of cryptosystems of *EPR*-type is as follows. A sequence of correlated particle pairs is generated, with one member of each pair being detected by Alice and the other by Bob (e.g., a pair prepared in a Bell state, whose polarizations are measured by Alice and Bob). To eavesdrop on this communication, Eve would have to detect a particle to read the signal, and retransmit it in order to conceal her presence. However, the act of detection of one particle of a pair destroys its quantum correlation with the other. Thus Alice and Bob can easily verify whether this has been done, without revealing the results of their own measurements, by communication over an open channel.

In the BB-protocol, Eve does not gain any information about the key material by passively monitoring Alice and Bob's public channel communications; however, it is essential that these messages are authenticated. Otherwise Eve could perform a man-in-the-middle attack in which he/she masquerades as Bob to Alice and as Alice to Bob, while forming separate keys with each. This scenario can be avoided if Alice and Bob append an authentication tag to their public messages, which is computed using a keyed hash function (cf. Wegman & Carter Construction, Section 12.5.2). Upon receiving a message, Alice and Bob can each verify that the received tag value matches the value computed from the message using the keyed hash function. This requires that Alice and Bob share a short initial key, which needs to be kept secret only for the duration of the transfer of Alice's photons to Bob. The QKD procedure produces large quantities of shared long-term secret bits, a few of which can be used to authenticate the next QKD session.

In the original proposal of Bennett and Brassard, the shared keys produced by quantum key distribution would be used directly for encryption as a one-time pad (encryption-mode QKD). With present day technology (c 2009), it is more practical to use QKD for the transfer or the generation of conventional symmetric cryptographic keys.

13.5.1 Practicalities—Quantum Cryptography

The field of quantum cryptography was pioneered by Wiesner [110]. Around 1970 at Columbia University in New York he showed how quantum effects could in theory be used to produce "quantum bank notes" that are immune to counterfeiting. The first feasible cryptosystems were proposed between 1982 and 1984 by the American physicist Charles H. Bennett of IBM's Thomas J. Watson Research Center in Yorktown Heights in the United States and by the Canadian expert in cryptography, Gilles Brassard from the University of Montreal [10]. In 1991 Artur Ekert [48] developed the idea along slightly different lines. This system is based on another aspect of quantum theory called entanglement.

By the early 1980s, the theory of secure quantum key distribution based on both the Heisenberg uncertainty principle and the quantum entanglement had evolved into a proposal for a testable system, though this was by no means easy to set up experimentally. The first apparatus, constructed by Bennett, Brassard and colleagues in 1989 at IBM's research center, was capable of transmitting a secret key over a distance of approximately 30 cm [9]. The practical use of this technology was established when the first demonstrations over optical fiber were implemented in the early 1990s.

Since then, other researchers have looked at systems based on correlations of another quantum property of light called phase. Phase is a measure of how far a photon has gone in its cycle of vibration. Information about the key is encoded in this property of phase instead of polarization. This has the advantage that with current technology, phase is easier to handle over a long distance. Since 1991 John Rarity and Paul Tapster of Britain's [49,106,108] Defense Research Agency in Malvern have been developing a system to increase the transmission distance. They have designed and tested an optical system good enough to transmit photons that stay correlated in phase over several hundred meters. With improvements in the technology of optical fibers and semiconductor photo detectors, which allowed better transmission and detection and thereby reducing background errors, the maximum transmission distance achieved by the mid-1990s was approximately 10 km.

In theory, cryptosystems based on entanglement should also allow quantum keys to be stored by storing photons without performing any measurements. At present, however, photons cannot be kept correlated longer than a small fraction of a second, so they are not a good medium for information storage. But a fraction of a second is long enough for a photon to cover a long distance, so photons are suitable for sending information and for key distribution.

Since the mid-1990s there have been important advances in quantum cryptography, such as

- Increase in transmission distance to about 150 km
- Implementations of practical systems (from the optical table to a 19″ rack)
- Development of commercial products

Current research in QKD focuses on implementations, which use weak laser pulses, single-photon sources, entangled pairs, and continuous variables. A comprehensive evaluation of the transmission distances that can be achieved by these technologies and their relative merits can be found in the Quantum Information and Technology Roadmapping Project [3]. The review papers [54,74,101] provide details of the current experimental implementations of QKD.

In 2006, IdQuantique (Switzerland) and Senetas (Australia) released commercially a hybrid quantum/classical encryption system. This system uses quantum key distribution to securely exchange encryption keys, which are used by a high speed classical network for encryption at speeds up to 10 Gbps. It was used in October 2007 to secure the transmission of federal election results in the State of Geneva. The system features AES 256 bit encryption, BB84 and SARG [87] protocols for QKD, HMAC-SHA-1 authentication for the classical link, and Wegmann & Carter for the QKD link. The cyryptographic keys are refreshed at the rate of 1 key/min up to four encryption appliances, and transmitted by a quantum link channel of length up to 80 km on single mode dark fiber.

13.5.2   Shor's Quantum Factoring Algorithm

Mathematicians have tried hard to solve the key distribution problem, and in the 1970s a clever mathematical discovery called "public-key" systems gave an elegant solution. Public-key cryptosystems avoid the key distribution problem, but unfortunately their security depends on unproven mathematical assumptions, such as the difficulty of factoring large integers (RSA, the most popular public key cryptosystem, gets its security from the difficulty of factoring large numbers). An enemy who knows your public key can in principle calculate your private key because the two keys are mathematically related; however, the difficulty of computing the private key from the respective public key is exactly that of factoring big integers.

Difficulty of factoring grows rapidly with the size, i.e., number of digits, of the number we want to factor. To see this, take a number N with ℓ decimal digits ($N \approx 10^{\ell}$) and try to factor it by dividing it by 2, 3,..., $\sqrt{N}$ and checking the remainder. In the worst case, approximately $\sqrt{N} \approx 10^{\ell/2}$ divisions may be needed to solve the problem—an exponential increase as a function of ℓ. Now imagine computer capable of performing 10^{10} divisions per second. The computer can then factor any number N, using the trial division method, in about $\sqrt{N}/10^{10}$ s. Take a 100-digit number N, so that $N \approx 10^{100}$. The computer will factor this number in about 10^{40} s which is much longer than 10^{17} s—the estimated age of the universe!

It seems that factoring big numbers will remain beyond the capabilities of any realistic computing devices and unless mathematicians or computer scientists come up with an efficient factoring algorithm public-key cryptosystems will remain secure. Or will they? As it turns out we know that this is not the case; the classical, purely mathematical, theory of computation is not complete simply because it does not describe all physically possible computations. In particular, it does not describe computations that can be performed by quantum devices. Indeed, recent work in quantum computation shows that a quantum computer can factor much faster than any classical computer.

Quantum computers can compute faster because they can accept as input not just one number, but a coherent superposition of many different numbers, and subsequently perform a computation (a sequence of unitary operations) on all of these numbers simultaneously. This can be viewed as a massive parallel computation, but instead of having many processors working in parallel we have only one quantum processor performing a computation that affects all the components of the state vector. To see how it works let us describe Shor's factoring using a quantum computer composed of two quantum registers.

Consider two quantum registers, each register composed of a certain number of two-state quantum systems, which we call "qubits" (quantum bits). We take the first register and place it in a quantum superposition of all the possible integer numbers it can contain. This can be done by starting with all qubits in the $|0\rangle$ states and applying a simple unitary transformation to each qubit that creates a superposition of $|0\rangle$ and $|1\rangle$ states:

$$|0\rangle \longrightarrow \frac{1}{\sqrt{2}}(|0\rangle + |1\rangle). \tag{13.5}$$

Imagine a two-qubit register, for example. After this procedure the register will be in a superposition of all four numbers it can contain

$$\frac{1}{\sqrt{2}}(|0\rangle + |1\rangle)\frac{1}{\sqrt{2}}(|0\rangle + |1\rangle) = \frac{1}{2}(|00\rangle + |01\rangle + |10\rangle + |11\rangle) \tag{13.6}$$

where $|00\rangle$ is binary for 0, $|01\rangle$ is binary for 1, $|10\rangle$ is binary for 2, and finally $|11\rangle$ which is binary for 3.

Then we perform an arithmetical operation that takes advantage of quantum parallelism by computing the function $F_N(x)$ for each number x in the superposition. The values of $F_N(x)$ are placed in the second register so that after the computation the two registers become entangled:

$$\sum_x |x\rangle |F_N(x)\rangle \tag{13.7}$$

(we have ignored the normalization constant). Now we perform a measurement on the second register. The measurement yields a randomly selected value $F_N(k)$ for some k. The state of the first register immediately after the measurement, due to the periodicity of $F_N(k)$, is a coherent superposition of all states $|x\rangle$ such that $x = k,\ k+r,\ k+2r, \ldots$, i.e., all x for which $F_N(x) = F_N(k)$. The value k is randomly selected by the measurement: therefore, the state of the first register is subsequently transformed via a unitary operation that sets any k to 0 (i.e., $|k\rangle$ becomes $|0\rangle$ plus a phase factor) and modifies the period from r to a multiple of $1/r$. This operation is known as the quantum Fourier transform. The first register is then ready for the final measurement and yields an integer, which is the best whole approximation of a multiple of $1/r$. From this result r, and subsequently factors of N, can be easily calculated (see the following text). The execution time of the quantum factoring algorithm can be estimated to grow as a quadratic function of ℓ, and numbers 100 decimal digits long can be factored in a fraction of a second!

When the first quantum factoring devices are built, the security of public-key cryptosystems will vanish. The mathematical solution to the key distribution problem is shattered by the power of quantum computation. Does it leave us without any means to protect our privacy? Fortunately quantum mechanics after destroying classical ciphers comes to rescue our privacy and offers its own solution to the key distribution problem.

The main reference for this brief account of quantum cryptanalysis is Shor's paper [94]. Ekert and Jozsa [50] give a comprehensive exposition of Shor's algorithm for factoring on a quantum computer, which includes some relevant background in number theory, computational complexity theory, and quantum computation as well as remarks about possible experimental realizations.

13.5.2.1 Factoring

Quantum factoring of an integer N is based on calculating the period of the function $F_N(x) = a^x \pmod{N}$. We form the increasing powers of a until they start to repeat with a period we denote r. Once r is known, the factors of N can be found using the Euclidean algorithm to find the greatest common divisor of $a^{r/2} \pm 1$ and N.

Suppose we want to factor 91. Let us take a number at random, say 28, and raise it to the powers $2, 3, \ldots$. After 12 iterations we find the number 28 repeated and so we use $r = 12$. Hence we want to find $\gcd(91, 28^6 \pm 1)$, which we find to be 1 and 13, respectively. From here we can factorize 91. Classically, calculating r is as difficult as trying to factor N and the execution time is exponential in the number of digits in N. Quantum computers can find r in time, which grows only as a quadratic function of the number of digits of N.

13.5.3 Practicalities—Quantum Computation

The idea of a quantum computer is simple, however, its realization is not. Quantum computers require a coherent, controlled evolution for a period of time, which is necessary to complete the computation. Many view this requirement as an insurmountable experimental problem; however, others believe that technological progress will sooner or later make such devices feasible. In an ordinary, classical computer, all the bits have a definite state at a given instant in time, say 0 1 1 0 0 0 1 0 1 0 . . .. However, in a quantum computer the state of the bits is described by a pure quantum state of the form

$$\Psi = a|0\,1\,1\,0\,0\,0\,1\,0\,1\ldots\rangle + b|1\,1\,1\,0\,0\,0\,0\,0\,1\ldots\rangle. \tag{13.8}$$

The coefficients $a, b, \ldots$ are complex numbers and the probability that the computer is in the state 0 1 1 0 0 0 1 0 1 . . . is $|a|^2$, that it is in the state 1 1 1 0 0 0 0 0 1 . . . is $|b|^2$, and so on. However, describing the state of the computer by the quantum state of 13.8 does not merely imply the ordinary uncertainties that we describe using probabilities. For instance, the phases of the complex coefficients $a, b, \ldots$ have genuine significance: they can describe interference between different states of the computer, which is very useful for quantum computation. The quantum state declares that the computer exists in all of its computational basis states simultaneously so long as that state is not measured; when we do choose to measure it, a particular computational basis state will be observed with the prescribed probability.

Since quantum computers have the potential to perform certain computational tasks more efficiently than their classical counterparts, there has been an intense international research effort aimed at realizing these devices. In early research on the synthesis of quantum logical circuits, DiVincenzo [46] showed how to construct the quantum analogue of the one-bit NOT, or inverter gate, with spectroscopic techniques that have been well known in physics for over 50 years. He established that the three-qubit AND operation can be performed using three XOR gates and four single-qubit rotations.

There are currently several promising approaches to the realization of a quantum computer: nuclear magnetic resonance (NMR), trapped ions, neutral atoms, cavity QED, optical, solid state, and superconducting devices. The potentials of these technologies are evaluated by the DiVincenzo criteria in the Quantum Information and Technology Roadmapping Project [2], which also gives medium- and long-term objectives for the realization of a quantum computer.

A prominent example of this technology is the Cirac Zoller proposal for a scalable quantum computer based on a string of trapped ions whose electronic states represent the quantum bits of information. In this scheme, quantum logical gates involving any subset of ions are realized by coupling the ions through their collective quantized motion. In 2003, the CNOT quantum gate according to the Cirac Zoller proposal was experimentally demonstrated by Schmidt-Kaler et al. [89]

In 2001, Vandersypen et al. [109] realized experimentally Shor's algorithm using liquid-state NMR techniques. In 2007, a linear optics realization of Shor's algorithm was demonstrated by Lanyon et al. [67]. Both teams chose the simplest instance of this algorithm, i.e., factorization of "$N = 15$". The linear optics experiment represents an essential step toward the full realization of Shor's algorithm and scalable linear optics quantum computation.

13.6 Further Information

As in the previous chapter we mention the conferences CRYPTO, EUROCRYPT, ASIACRYPT, AUSCRYPT, and conferences dealing with security such as ACISP. Information can also be found in Journals such as the *Communications of the ACM*. Quantum cryptography is also covered in the physics literature, e.g., *Europhysics Letters, Physical Review Letters,* and QIPC Strategic Report: Quantum Information Processing and Communication in Europe http://qist.ect.it/Reports/reports.htm

Acknowledgments

We thank Julian Fay of Senetas Australia, Gregoire Ribordy of id Quantique Switzerland and Professor Gernot Alber of IAP TUD Darmstadt for providing comments and suggestions which have greatly improved our exposition.

References

1. Van Assche, G., Cardinal, J., and Cerf, N.J., Reconciliation of a quantum-distributed gaussian key, *IEEE Transactions on Information Theory*, 50, 2004, arXiv:quant-ph/0107030.
2. ARDA, A quantum information and technology roadmapping project, Part 1: Quantum computation, LA-OR-04-1778, 2004, http://qist.lanl.gov/qcomp_map/shtml.
3. ARDA, A quantum information and technology roadmapping project, Part 2: Quantum cryptography, LA-OR-04-4085, 2004, http://qist.lanl.gov/qcrypt_map/shtml.
4. Benaloh, J.C., Secret sharing homomorphisms: Keeping shares of a secret secret. *Proceedings of Crypto'86, LNCS* Vol. 263, pp. 251–260, Springer-Verlag, Berlin, 1987.
5. Bellare, M. and Rogaway, P., Robust computational secret sharing and a unified account of classical secret sharing goals, http://eprint.iacr.org/2006/449.
6. Belenkiy, M. Disjunctive multi-level secret sharing, http://eprint.iacr.org/2008/018.
7. Benaloh, J. and Leichter, J., Generalized secret sharing and monotone functions, *Proceedings of Crypto'88, LNCS*, Vol. 403, pp. 27–35, Springer-Verlag, Berlin, 1989.
8. Bennett, C.H., Quantum cryptography using any two nonorthogonal states, *Physical Review Letters,* 68, 3121–3124, 1992.
9. Bennett, C.H., Bessette, F., Brassard, G., Salvail, L., and Smolin, J., Experimental quantum cryptography, *Journal of Cryptology,* 5, 3–28, 1992.
10. Bennett, C.H. and Brassard, G., Quantum cryptography: Public-key distribution and coin tossing, *Proceedings of the IEEE International Conference on Computers, Systems and Signal Processing,* IEEE, pp. 175–179, New York, 1984.
11. Bennett, C.H., Brassard, G., and Ekert, A.K., Quantum cryptography, *Scientific American,* pp. 50–57, Oct. 1992.

12. Bennett, C.H., Brassard, G., and Robert, J.-M., Privacy amplification by public discussion, *SIAM Journal on Computing*, 17(2), 210–229, 1988.

13. Bertilson, M. and Ingemarsson, I., A construction of practical secret sharing schemes using linear block codes, *Proceedings of AUSCRYPT'92, LNCS*, Vol. 718, pp. 67–79, Springer-Verlag, Berlin, 1993.

14. Beth, T., Jungnickel, D., and Lenz, H., *Design Theory, 2nd edition*, Cambridge University Press, Cambridge, U.K., 1999.

15. Beth, T., Multifeature security through homomorphic encryption, *Proceedings of Asiacrypt'94, LNCS*, Vol. 917, pp. 3–17, Springer-Verlag, Berlin, 1993.

16. Beutelspacher, A., Enciphered geometry: Some applications of geometry to cryptography, *Discrete Applied Mathematics*, 37, 59–68, 1988.

17. Blakley, G.R., Safeguarding cryptographic keys, *Proceedings of N.C.C., AFIPS Conference Proceedings*, Vol. 48, pp. 313–317, Montvale, NJ, 1979.

18. Blakley, B., Blakley, G.R., Chan, A.H., and Massey, J.L., Threshold schemes with disenrollment, *Proceedings of Crypto'92, LNCS*, Vol. 740, pp. 546–554, Springer-Verlag, Berlin, 1992.

19. Blakley, G.R. and Meadows, C., Security of ramp schemes, *Proceedings of CRYPTO'84, LNCS*, Vol. 196, pp. 242–268, Springer-Verlag, Berlin, 1985.

20. Blundo, C., De Santis, A., Stinson, D.R., and Vaccaro, V., Graph decompositions and secret sharing schemes, *Journal of Cryptology*, 8(1), 39–64, 1995.

21. Boyd, C., Digital Multisignatures, In *Cryptography and Coding*, Beker, H. and Piper, F., Eds., pp. 241–246. Clarendon Press, Gloucestershire, GL, 1989.

22. Brassard, G., *Modern Cryptology: A Tutorial*, Springer, Berlin, 1988.

23. Brickell, E.F., Some ideal secret sharing schemes, *Journal of Combinatorial Mathematics and Combinatorial Computing*, 6, 105–113, 1989.

24. Brickell, E.F. and Davenport, D.M., On the classification of ideal secret sharing schemes, *Journal of Cryptology*, 4, 123–134, 1991.

25. Brickell, E.F. and Stinson, D.R., The detection of cheaters in threshold schemes, *Proceedings of Crypto'88, LNCS*, Vol. 403, pp. 564–577, Springer-Verlag, Berlin, 1990.

26. Brickell, E.F. and Stinson, D.R., Some improved bounds on the information rate of perfect secret sharing schemes, *Journal of Cryptology*, 5, 153–166, 1992.

27. Brown, L., Kwan, M., Pieprzyk, J., and Seberry, J., Improving resistance to differential cryptanalysis and the redesign of LOKI, in *Advances in Cryptology—Proceedings of ASIACRYPT '91*, Imai, R.R.H. and Matsumoto, T., Eds., Vol. 739 of *Lecture Notes in Computer Science*, pp. 36–50, Springer-Verlag, Berlin, 1993.

28. Capocelli, R., DeSantis, A., Gargano, L., and Vaccaro, V., On the size of shares for secret sharing schemes, *Proceedings of CRYPTO '91, LNCS*, Vol. 576, pp. 101–113, Springer-Verlag, Berlin, 1992.

29. Cerecedo, M., Matsumoto, T., and Imai, H., Efficient and secure multiparty generation of digital signatures based on discrete logarithms, *IEICE Transactions on Fundamentals* of *Electronics, Communications, and Computer Science*, E76-A(4), 531–545, Apr. 1993.

30. Chang, C.-C. and Liou, F.-Y., A digital multisignature scheme based upon the digital signature scheme of a modified ElGamal public key cryptosystem, *Journal of Information Science and Engineering*, 10, 423–432, 1994.

31. Charnes, C., Pieprzyk, J., and Safavi-Naini, R., Families of threshold schemes, *Proceedings of 1994 IEEE International Symposium on Information Theory*, Trondheim, Norway, 1994.

32. Charnes, C., Pieprzyk, J., and Safavi-Naini, R., Conditionally secure secret sharing schemes with disenrollment capability, *2nd ACM Conference on Computer and Communications Security*, Nov. 2–4, Fairfax, VA, pp. 89–95, ACM 1994.

33. Charnes, C. and Pieprzyk, J., Disenrollment capability of conditionally secure secret sharing schemes, *Proceedings of International Symposium on Information Theory and Its Applications (ISITA'94)*, Nov. 20–25, 1994, pp. 225–227, Sydney, Australia, IEA, NCP 94/9 1994.

34. Charnes, C. and Pieprzyk, J., Cumulative arrays and generalised Shamir secret sharing schemes, *17th Australasian Computer Science Conference, Australian Computer Science Communications,* 16(1), 519–528, 1994.

35. Charnes, C. and Pieprzyk, J., Generalised cumulative arrays and their application to secret sharing schemes, *18th Australasian Computer Science Conference, Australian Computer Science Communications,* 17(1), 61–65, 1995.

36. Chaudhry, G. and Seberry, J., Secret sharing schemes based on Room squares, *Combinatorics, Complexity and Logic, DMTCS'96,* pp. 158–167, Springer-Verlag, Berlin, 1996.

37. Chor, B., Goldwasser, S., Micali, S., and Awerbuch, B., Verifiable secret sharing and achieving simultaneity in the presence of faults, *Proceedings of 26th Annual IEEE Symposium on Foundations of Computer Science,* pp. 383–395, IEEE, Portland 1985.

38. Cooper, J., Donovan, D., and Seberry, J., Secret sharing schemes arising from Latin squares, *Bulletin of the ICA,* 12, 33–43, 1994.

39. Damgard, I. and Thorbek, R., Linear integer secret sharing and distributed exponentiation, http://eprint.iacr.org/2006/044.

40. Desmedt, Y., Society and group oriented cryptography: A new concept, In *Advances in Cryptology—Proceedings of CRYPTO '87,* Pomerance, C., Ed., LNCS, Vol. 293, pp. 120–127, Springer-Verlag, Berlin, 1988.

41. Desmedt, Y. and Frankel, Y., Threshold cryptosystems, in *Advances in Cryptology—Proceedings of CRYPTO '89,* Brassard, G., Ed., Vol. 435 of *Lecture Notes in Computer Science,* pp. 307–315, Springer-Verlag, Berlin, 1990.

42. Desmedt, Y. and Frankel, Y., Shared generation of authenticators and signatures, in *Advances in Cryptology—Proceedings of CRYPTO'91,* Feigenbaum, J., Ed., LNCS, Vol. 576, pp. 457–469, Springer-Verlag, Berlin, 1992.

43. Deutsch, D., Quantum theory, the Church-Turing principle and the universal quantum computer, *Proceedings of the Royal Society of London, Series A,* 400, 96–117, 1985.

44. Deutsch, D., Quantum computational networks, *Proceedings of the Royal Society of London, Series A,* 425, 73–90, 1989.

45. Diffie, W. and Hellman, M., New directions in cryptography, *IEEE Transactions on Information Theory,* IT-22, 644–654, Nov. 1976.

46. DiVincenzo, D.P., Quantum computation, *Science,* 270, 255–261, 1995.

47. ElGamal, T., A public key cryptosystem and a signature scheme based on discrete logarithms, *IEEE Transactions on Information Theory,* IT-31, 469–472, Jul. 1985.

48. Ekert, A.K., Quantum cryptography based on Bell's theorem, *Physical Review Letters,* 67, 661–663, 1991.

49. Ekert, A.K., Rarity, J.G., Tapster, P.R., and Palma, G.M., Practical quantum cryptography based on two-photon interferometry, *Physical Review Letters,* 69, 1293–1295, 1992.

50. Ekert, A.K., and Josza, R., Shor's algorithm for factoring numbers, *Review of Modern Physics,* 68(30), 733–753, 1996.

51. Feldman, J., Malkin, J., Servedio, R.A., and Stein, C., On the capacity of secure network coding, *Proceedings of the 42nd Annual Allerton Conference on Communication, Control and Computing,* Urbana, IL, 2004.

52. Gallagher, R.G., *Information Theory and Reliable Communications,* John Wiley & Sons, New York, 1968.

53. Ghodosi, H., Pieprzyk, J., and Safavi-Naini, R., Dynamic threshold cryptosystems, *Proceedings of PRAGOCRYPT'96,* CTU Publishing House, Prague, Part 1, pp. 370–379, 1996.

54. Gisin, N., Ribordy, G., Titel, W., and Zbinden, H., Quantum cryptography, *Reviews of Modern Physics,* 74(1), 145–195, 2002.

55. Harn, L., Group-oriented (t, n) threshold digital signature scheme and digital multisignature, *IEEE Proceedings—Computers and Digital Techniques,* 141(5), 307–313, Sep. 1994.

56. Harn, L. and Yang, S., Group-oriented undeniable signature schemes without the assistance of a mutually trusted party, In *Advances in Cryptology—Proceedings of AUSCRYPT '92,* Seberry, J. and Zheng, Y., Eds., *LNCS,* Vol. 718, pp. 133–142, Springer-Verlag, Berlin, 1993.

57. Hwang, T., Cryptosystem for group oriented cryptography, *Proceedings of Eurocrypt'90, LNCS,* Vol. 473, pp. 353–360, Springer-Verlag, Berlin, 1990.

58. Hwang, S.-J. and Chang, C.-C., A dynamic secret sharing scheme with cheater detection, *Proceedings of ACISP'96, LNCS,* Vol. 1172, NSW, Australia, pp. 48–55, 1996.

59. Imai, H., Mueller-Quade, J., Tuyls, P., Nascimento, A., Tuyls, P., and Winter, A., An information theoretical model for quantum secret sharing schemes, *Quantum Information and Computation,* 5(1), 69–80, 2005.

60. Itakura, K. and Nakamura, K., A public-key cryptosystem suitable for digital multisignature, *NEC Research & Development,* 71, Oct. 1983.

61. Ito, M., Saito, A., and Nishizeki, T., Secret sharing scheme realising general access structure, *Proceedings of Globecom'87,* 99–102, Tokyo, 1987.

62. Jackson, W.-A. and Martin, K.M., Cumulative arrays and geometric secret sharing schemes, *Proceedings of Auscrypt'92, LNCS,* Vol. 718, pp. 49–55, Springer-Verlag, Berlin, 1993.

63. Karnin, E.D., Greene, J.W., and Hellman, M.E., On secret sharing systems, *IEEE Transactions on Information Theory,* IT-29(1) 35–41, 1983.

64. Krawczyk, H., Secret sharing made short, *Proceedings of Crypto'93, LNCS,* Vol. 773, pp. 136–146, Springer-Verlag, Berlin, 1994.

65. Laih, C.-S. and Yen, S.-M., Multi-signature for specified group of verifiers, *Journal of Information Science and Engineering,* 12(1), 143–152, Mar. 1996.

66. Langford, S.K., Threshold DSS signatures without a trusted party, *Proceedings of Crypto'95, LNCS,* Vol. 963, pp. 397–409, Springer-Verlag, Berlin, 1996.

67. Lanyon, B.P. et al., Experimental demonstration of Shor's algorithm with quantum entanglement, *Physical Review Letters,* 99, 250505, 2007. arXiv:quant-ph/07051398.

68. Li, C.-M., Hwang, T., and Lee, N.-Y., Threshold-multisignature schemes where suspected forgery implies traceability of adversarial shareholders, In *Advances in Cryptology—Proceedings of EURO-CRYPT '94,* De Santis, A., Ed., *LNCS* Vol. 950, pp. 194–204, Springer-Verlag, Berlin, 1995.

69. Lin, H.-Y. and Harn, L., A generalized secret sharing scheme with cheater detection. *Proceedings of Asiacrypt'91, LNCS,* Vol. 739, pp. 149–158, Springer-Verlag, Berlin, 1993.

70. Lo, H. and Lütkenhaus, N. Quantum cryptography: From theory to practice, arXiv:quant-ph/0702202. A generalized secret sharing scheme with cheater detection.

71. Massey, J.L., Minimal codewords and secret sharing, *Proceedings of 6th Joint Swedish-Russian Workshop in Information Theory,* Molle, Sweden, pp. 276–279, 1993.

72. Massey, J.L., Provable security–myth or reality?, *ISPEC,* Hangzhou, China, 2006.

73. Martin, K. and Jackson, W.-A., Perfect secret sharing schemes on five participants, *Designs Codes and Cryptography,* 9, 267–286, 1996.

74. Maurer, W., Helwig, W., and Silberhorn, C., Recent developments in quantum key distribution: Theory and practice, *Annale of Physics (Leipzig),* No. 2–3, 2008. arXiv:quant-ph/0712.0517.

75. McEliece, R.J. and Sarwate, D.V., On sharing secrets and Reed-Solomon codes, *Communications of the ACM,* 24(9), 683–584, 1981.

76. MacWilliams and Sloane, N.J.A., *Theory of Codes,* North Holland, Amsterdam, the Netherlands 1977.

77. Muller, A., Breguet, J., and Gisin, N., Experimental demonstration of quantum cryptography using polarised photons in optical fibre over more than 1 km, *Europhysics Letters,* 23, 383–388, 1993.

78. National Bureau of Standards, Federal Information Processing Standard (FIPS), *Data Encryption Standard,* 46 edn., U.S. Department of Commerce, Washington, DC, Jan. 1977.

79. Ogata, W. and Kurosawa, K., Lower bound on the size of shares of nonperfect secret sharing schemes, *Proceedings of Asiacrypt'94, LNCS,* Vol. 917, pp. 33–41, Springer-Verlag, Berlin, 1995.

80. Ogata, W., Kurosawa, K., and Tsujii, S., Nonperfect secret sharing schemes, *Proceedings of Auscrypt'92, LNCS*, Vol. 718, pp. 56–66, Springer-Verlag, Berlin, 1993.

81. Ogata, W. and Kurosawa, K., MDS secret sharing scheme secure against cheaters, *IEEE Transactions on Information Theory*, 46(3), 1078–1081, 2000.

82. Ohta, K. and Okamoto, T., A digital multisignature scheme based on the Fiat-Shamir scheme, In *Advances in Cryptology—Proceedings of ASIACRYPT '91*, Rivest, R.L., Imai, H., and Matsumoto, T., Eds., Vol. 739 of *Lecture Notes in Computer Science*, pp. 139–148, Springer-Verlag, Berlin, 1993.

83. Park, C. and Kurosawa, K., New ElGamal type threshold digital signature scheme, *IEICE Transactions on Fundamentals of Electronics, Communications and Computer Sciences*, E79(1), 86–93, Jan. 1996.

84. Renvall, A. and Ding, C., The access structure of some secret sharing schemes, *Proceedings of ACISP'96, LNCS*, Vol. 1172, pp. 67–86, Springer-Verlag, Berlin, 1996.

85. Rivest, R., Shamir, A., and Adleman, L., A method for obtaining digital signatures and public-key cryptosystems, *Communications of the ACM*, Vol. 21, pp. 120–126, Feb. 1978.

86. Rivest, R., Shamir, A., and Adleman, L., On digital signatures and public-key cryptosystems, MIT Laboratory for Computer Science, Technical Report, MIT/LCS/TR-212, Jan. 1979.

87. Scarani, V., Acin, A., Ribordy, V., and Gisin, V., *Physical Review Letters*, 92, 057901, 2004.

88. Schellenberg, P.J. and Stinson, D.R., Threshold schemes from combinatorial designs, *Journal of Combinatorial Mathematics and Combinatorial Computing*, 5, 143–160, 1989.

89. Schmidt-Kaler, F. et al., Realization of the Cirac-Zoller controlled-NOT quantum gate, *Nature*, 422, 408–411, 2003.

90. Schneier, B., *Applied Cryptography: Protocols, Algorithms, and Source Code in C*, John Wiley & Sons, New York, 1994.

91. Pieprzyk, J., Hardjono, T., and Seberry, J., *Fundamentals of Computer Security*, Springer-Verlag, Berlin, 2003.

92. Shamir, A., How to share a secret, *Communications of the ACM*, 22(11), 612–613, 1979.

93. Shimizu, A. and Miyaguchi, S., Fast data encipherment algorithm FEAL, *Advances in Cryptology—Proceedings of EUROCRYPT '87*, Chaum, D. and Price, W., Eds., Vol. 304 of *Lecture Notes in Computer Science*, pp. 267–278, Springer-Verlag, Berlin, 1987.

94. Shor, P.W., Algorithms for quantum computation: Discrete log and factoring, *Proceedings of the 35th Annual Symposium on the Foundations of Computer Science*, pp. 124–134, Goldwasser, S., Ed., IEEE Computer Society Press, Los Alamitos, CA, 1994.

95. Simmons, G.J., How to (really) share a secret, *Proceedings of Crypto'88, LNCS*, Vol. 403, pp. 390–448, Springer-Verlag, Berlin, 1989.

96. Simmons, G.J., An introduction to shared secret and/or shared control schemes and their application, In *Contemporary Cryptology—The Science of Information Integrity*, pp. 441–497, Simmons, G.J., Ed., IEEE Press, New York, 1992.

97. Simmons, G.J., Jackson, W.-A., and Martin, K., The geometry of shared secret schemes, *Bulletin of the ICA*, 1, 71–88, 1991.

98. De Soete, M., Quisquater, J.-J., and Vedder, K., A signature with shared verification scheme, In Brassard, J., Ed., *Advances in Cryptology—Proceedings of CRYPTO '89*, Vol. 435 of *Lecture Notes in Computer Science*, pp. 253–262. Springer-Verlag, Berlin, 1990.

99. NSBS. Processing Information Systems. Cryptographic Protection. Cryptographic Algorithm, GOST 28147-89, C1–C26, National Soviet Bureau of Standards, 1989.

100. Street, A.P., Defining sets for *t*-designs and critical sets for Latin squares, *New Zealand Journal of Mathematics*, 21, 133–144, 1992.

101. Scarani, V. at al., A framework for practical quantum cryptography, arXiv:quant-ph/0802.4155.

102. Stinson, D.R., An explication of secret sharing schemes, *Designs, Codes and Cryptography*, 2, 357–390, 1992.

103. Stinson, D.R., Bibliography on secret sharing schemes, http://www.cacr.math.uwaterloo.ca/dstinson/pubs.html.

104. Stinson, D.R., _Cryptography Theory and Practice,_ 3rd Edition, CRC Press, Boca Raton, FL, 2005.
105. Stinson, D.R. and Vanstone, S.A., A combinatorial approach to threshold schemes, _SIAM Journal of Discrete Mathematics,_ 1, 230–236, 1988.
106. Tapster, P.R., Rarity, J.G., and Owens, P.C.M., Violation of Bell's inequality over 4 km of optical fibre, _Physical Review Letters,_ 73, 1923–1926, 1994.
107. Tompa, M. and Woll, H., How to share a secret with cheaters, _Journal of Cryptology,_ 1, 133–138, 1988.
108. Townsend, P.D., Rarity, J.G., and Tapster, P.R., Enhanced single photon fringe visibility in a 10 km-long prototype quantum cryptography channel, _Electronic Letters,_ 29, 1291–1293, 1993.
109. Vandersypen, L.K.M. et al., Experimental realization of Shor's quantum factoring algorithm using nuclear magnetic resonance, _Nature,_ 414, 883-887, 2001.
110. Wiesner, S., Conjugate coding, _SIGACT News,_ 15, 78–88, 1983. (Original manuscript written circa 1970.)

14

Secure Multiparty Computation

Keith B. Frikken
Miami University

14.1 Introduction

A plethora of mutually beneficial information collaboration opportunities exist when organizations and individuals share information. Some examples of such collaborations include (1) hospitals sharing information about patients with a rare disease in order to develop better treatments; (2) suppliers and retailers sharing their holding costs, inventory on hand, and other supply-chain information in order to reduce their costs; and (3) finding interorganization purchase patterns by sharing customer records to perform collaborative data mining. Clearly, there are many potential benefits, both monetary and societal, for such collaborations, however, one significant drawback is concerns about confidentiality and privacy that are associated with sharing information. Returning to the examples above, sensitivity concerns include (1) patients' information is private and there may be laws preventing such sharing of information (e.g., in the United States, HIPAA restricts the sharing of medical information); (2) the supplier and retailer may be concerned that their cost information could be used against them in a future interaction; and (3) sharing customers' information may violate the customers' privacy. Thus, while significant benefit may result from such collaborations, these collaborations may not occur due to privacy and confidentiality concerns.

    Secure protocols (also called privacy-preserving or confidentiality-preserving protocols) are cryptographic techniques that obtain the result of information collaboration while preserving privacy and confidentiality. More specifically, secure protocols are cryptographic techniques for computing functions (i.e., collaboration outcomes) over distributed inputs while revealing only the result of the function. Thus, secure protocols are a way to have the best of both worlds—we can obtain the benefit of collaboration without the drawback of revealing too much information. While this may seem like an intractable goal for many problems, there have been many general results which state, under various assumptions, that any function computable in polynomial time can also be computed securely with polynomial communication and polynomial computation. The problem of computing any function in a secure manner is called secure multiparty computation (SMC).

This chapter provides a description of techniques for building secure protocols from an applied standpoint. That is, this chapter is not an exhaustive description of SMC techniques, but rather this chapter describes a few key building blocks that facilitate the creation of several secure protocols. For the reader that is interested in a formal description of cryptographic primitives and SMC we refer you to [15,16]. The rest of this chapter is organized as follows: in Section 14.2 secure protocols are defined more formally. Section 14.3 presents a survey of known results for secure protocols. Section 14.4 describes practicality problems for the general SMC results along with methods that help overcome these problems. In Sections 14.5 through 14.7 several specific techniques for two-party secure protocols are described, and in Section 14.8 we summarize this chapter.

14.2   What Is a "Secure" Protocol?

Before describing techniques for "secure" protocols, one must define what is meant by "secure." Before formally defining this notion, we provide some intuition into the definition. One trivial way to securely compute any function is for all parties to send their input to a fully trusted third party (TTP), and this TTP then computes the desired result and sends the output of the function to each party. Obviously, it is unlikely that such a TTP exists, but this protocol achieves the best possible outcome—that is, the only information revealed about inputs is the output of the protocol and inferences that can be made from this output along with one or more inputs.* Thus a protocol is considered "secure" if it is no worse than the above-described TTP protocol. More specifically, a secure protocol is a protocol that reveals only the result of the function and inferences that can be deduced from this output with one or more input values.

The following is a simple example of a secure protocol that helps clarify the above intuition. Suppose N people are sitting around a table lamenting that they are vastly underpaid; to help justify their complaints, this group has a genuine interest in finding the mean salary of everyone at the table, but due to privacy concerns, they do not want to reveal individual salaries to the group. Now if there are only two people at the table, then when given the result (i.e., the average salary) and one person's salary, one can deduce the other person's salary. Thus for $N = 2$ a secure protocol is as simple as having both parties reveal their salaries to each other, because one party's value can be deduced from the result and the other input. However, when $N \geq 3$, such exact inferences cannot be made, so a more complicated protocol is needed.

One way to compute the average salary (for $N \geq 3$) is as follows. The first person chooses a random number R from a range of values much larger than the range of salaries. This person adds his salary to R, writes the result on a slip of paper, and then passes this piece of paper to the next person at the table. When this person receives the slip of paper, he adds his salary to the value on the paper, writes down the new sum on a different piece of paper, destroys the original piece of paper, and passes the new paper to the person next to him. The paper passing continues until everyone has incorporated their salary into the sum. After everyone has had a turn, the final piece of paper is passed back to the first person. The first person then subtracts R from the sum to learn the sum of everyone's salary. He then divides this sum by the number of people to learn the average salary and then he reveals the average salary to everyone at the table.

---

* In a multiple party protocol it is possible that participants will collude and share inputs with each other.

Let us examine the properties of this average salary protocol. By adding a large random value R to the first person's salary, this person's salary is hidden with high probability (furthermore the probability is controllable by the range from which R is chosen). More generally, assuming that there is no collusion between members, then all that the ith person learns from the protocol is the sum of the first $i-1$ salaries and R, and since R is chosen from a large range, the sum of the first $i-1$ salaries is hidden with high probability. Of course, there are some problems with this protocol: (1) the protocol is not collusion-resistant—for example, if the 2nd and 4th parties collaborate together, then they could learn the 3rd party's salary; (2) the first person can learn the actual result and announce a different result; and (3) any participant can modify the value that is being passed around the group to increase or reduce the average.

Clearly, there are different types of adversaries that must be considered when creating secure protocols. There are two basic types of adversaries that are considered in the SMC literature. The first adversary type is called an honest-but-curious (HBC) adversary model (also called semi-honest or passive adversaries). In this adversary model, the adversary will faithfully follow the specified protocol, but after the protocol has completed it will attempt to learn additional information about other inputs. While this model is unrealistic in many regards, if an efficient protocol cannot be developed for this simplified model, then there is little hope that a protocol can be developed for a more realistic adversary model. The second adversary model is a more realistic adversary model, and it is often called the malicious or active adversary model. A malicious adversary will deviate arbitrarily* from the protocol in order to gain some advantage over the other participants. The advantages that an adversary can try to obtain include (1) to learn additional information about other inputs; (2) to modify the result of the protocol; (3) to abort the protocol prematurely (perhaps after the adversary learns a partial result it aborts the protocol to prevent others from learning the results).

14.2.1 Definition of Two-Party HBC Secure Protocols

We are now ready to formalize the above informal definition for two parties in the HBC adversary model. Recall that a two-party protocol is secure if everything that can be computed from the protocol can be deduced from the output and one of the input values. To make this more concrete suppose Alice and Bob are engaging in a protocol to compute a function f. This protocol involves one or more message exchanges between Alice and Bob; we will call these messages the transcript of the protocol. It must be shown that the transcript sent from Alice and Bob (and vica versa) does not reveal too much information. More specifically, the transcript sent from Alice to Bob should be computable from the result of the protocol and Bob's inputs, otherwise the protocol would reveal more information to Bob than the results (a similar statement must be made about the transcript from Bob to Alice). In the remainder of this section we formalize this definition. The reader that is already familiar with the cryptographic literature should feel free to skip the remainder of this section, and we refer the reader that is interested in more details about these definitions to [15,16].

As any cryptographic protocol will fail with some probability (one can always guess the other party's inputs), it is necessary to define an acceptable probability threshold for which a protocol is deemed secure. Clearly, a fixed constant probability is undesirable since an adversary can simply repeat the attack several times in order to increase the probability. The standard technique used in the cryptographic literature is the notion of a *negligible* probability. More specifically, a function $\mu(k)$ is negligible in k if for any polynomial p, for large enough k, $\mu(k) < \frac{1}{p(k)}$. A protocol is deemed secure if the adversary's success probability is negligible in a security parameter. One advantage of

---

* Adversaries are usually modeled as probabilistic-polynomial time adversaries, and this is the only bound placed on malicious adversaries.

using a negligible probability is that even if the adversary repeats the attack a polynomial number of times, the success probability will still be negligible.

As mentioned above, to show that a protocol is secure, it must be shown that the communication transcript from the protocol can be simulated when given the results and one party's input. One way to do this is to require that the communication transcript be exactly generatable from the output of the protocol, but this is very limiting. However, if the adversary is computationally-bounded, then it is enough to generate a transcript that is so close to the real transcript that the adversary cannot distinguish between the real and simulated transcripts. In the cryptographic literature, this notion of "close enough" is called computational indistinguishability. More formally, two random probability ensembles $\{X_n\}_{n \in N}$ and $\{Y_n\}_{n \in N}$ are *computationally indistinguishable* if for any probabilistic polynomial time algorithm D the following value is negligible in n:

$$|Pr[D(X_n, 1^n) = 1] - Pr[D(Y_n, 1^n) = 1]|$$

It is now possible to define a secure two-party protocol in the HBC model. Suppose that Alice and Bob have respective inputs x_A and x_B, and that at the end of the protocol Alice should learn $f_A(x_A, x_B)$ and Bob should learn $f_B(x_A, x_B)$. Note that in many cases f_A and f_B will be the same function, but to make the definition as general as possible, it incorporates the possibility that Alice and Bob will learn different results. Now given a protocol Π, Alice's view of the protocol (i.e., all messages she receives from Bob) is denoted by $\{view_A^\Pi(x_A, x_B)\}_{x_A, x_B \in \{0,1\}^*}$, and similarly Bob's view is denoted by $\{view_B^\Pi(x_A, x_B)\}_{x_A, x_B \in \{0,1\}^*}$. The protocol Π is said to be secure in the HBC model if there exist probabilistic polynomial time simulation algorithms S_A and S_B, such that (1) $\{S_A(x_A, f_A(x_A, x_B))\}_{x_A, x_B \in \{0,1\}^*}$ and $\{view_A^\Pi(x_A, x_B)\}_{x_A, x_B \in \{0,1\}^*}$ are computationally indistinguishable and (2) $\{S_B(x_B, f_B(x_A, x_B))\}_{x_A, x_B \in \{0,1\}^*}$ and $\{view_B^\Pi(x_A, x_B)\}_{x_A, x_B \in \{0,1\}^*}$ are computationally indistinguishable. To see how this definition matches the intuition from before, this definition states that an efficient algorithm exists that can simulate all of the messages sent by Bob (resp. Alice) when given only Alice's (resp. Bob's) input and output. Thus anything learned from the protocol must also be learnable from the result alone, and therefore the protocol reveals only the result and inferences derived from this result.

14.2.2 Beyond Two-Party and HBC Adversaries

The definition in the previous section can be extended to the malicious adversary model, but as this definition is quite cumbersome we opt for an informal discussion (we refer the interested reader to [16] for more details). Reconsider the TTP setting for computing a function securely, while this protocol is unrealistic it does provide an "ideal" goal for an SMC protocol. That is, a protocol will be secure if for any attack in the real protocol, there is another attack (with similar adversary complexity) in the ideal protocol with the TTP. Consider a malicious adversary in the TTP setting, this adversary can do the following actions: (1) refuse to participate in the protocol; (2) abort the protocol prematurely; and (3) modify its inputs. A protocol is secure in the malicious model if for any PPT adversary in the real protocol there is a malicious adversary in the TTP setting that can achieve the same* results. One limitation of the two-party secure computation is fairness of the protocols, that is, one party will learn its result before the other party and then can stop participating in the protocol to prevent the other party from obtaining its output. Unfortunately, the impossibility results in Byzantine agreement [11] imply that fairness is not achievable in two-party computation (e.g., a strict majority of the parties must be honest).

---

* That is, computationally indistinguishable.

14.3 Survey of General SMC Results

The first protocol for computing any two-party function securely in the semi-honest model was proposed in [30], and was later improved in [31] (in the latter we refer to this scheme as Yao's scheme), to require only a constant number of communication rounds. The basic idea of this approach is to build a logical circuit for the function in question, and then to use a secure protocol to blindly evaluate the circuit (for details on how this is accomplished, see Section 14.5). While the original scheme did not have a detailed security proof, the scheme was proven secure in [21]. Furthermore, Yao's scheme has been implemented in the FAIRPLAY system [22], and it was shown that this scheme is practical for some problems. Several protocols have extended Yao's system to make it secure in the malicious model [22,24,29]).

With regards to multiparty computation, it was shown in [17] that as long as a majority of the participants are honest, it is possible to securely compute any function. In [23] a scheme was given to securely compute any function in such a model in a constant number of rounds. There have been several extensions to more sophisticated adversary models, including (but not limited to) adaptive adversaries that corrupt participants after the protocol has began [5] and secure protocols against computationally unbounded adversaries [2,6].

14.4 Methods for Practical Secure Protocols

In this section, we consider techniques for making secure protocols practical. Since the general results of SMC state that there is a secure protocol for any number of participants, a first approach is to have all participants engage in a secure protocol as depicted in Figure 14.1. Typically, such schemes require a portion (e.g., a strict majority) of the participants to be honest. These protocols are robust in that after the initial rounds of such protocols, even when some of the participants abort the protocol (intentionally or unintentionally) the rest of the participants can still compute the result. While this natural protocol architecture does allow for secure computation there are several disadvantages including (1) all participants must be online at the same time which may be difficult in some environments and (2) for large numbers of participants these protocols are very expensive. In summary, this architecture works very well for a small number of participants (especially for two people), but it does not scale well.

To further emphasize the problems with this initial architecture consider creating a secure protocol for an auction system, where several bidders are bidding on an item held by a seller. Clearly, bidders will want to protect their bid values. In this architecture all of the bidders and the seller would have to agree upon a fixed time to engage in a protocol to compute the results. In many auctions this is impractical, because the identities of the bidders is unknown until they make a bid, and furthermore the number of bidders could make this protocol impractical.

Thus, for many problems having all participants engage in a protocol together is unlikely to be practical. One way to mitigate this problem is for many of the participants to delegate their portion of the protocol to another party. Of course, participants would not want to reveal their values to their representatives, but if the inputs could be split among different representatives so that no small group of the representatives can learn the values, then this may be acceptable. In the next section we describe various techniques for splitting values and then in Section 14.4.2 we formalize this representative-based architecture.

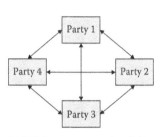

FIGURE 14.1 Example of first architecture.

14.4.1 Splitting Values

The following are three techniques for splitting a value x among n participants.

Sharing: In sharing approaches all n participants are required to recover the value. One example, of this approach is that party i has a value x_i such that $x = x_1 \oplus x_2 \oplus \cdots \oplus x_n$ where $\oplus$ is XOR. To split x in such a manner, $n - 1$ random values are chosen for $x_1, \ldots, x_{n-1}$ and x_n is set to $x \oplus x_1 \oplus \cdots \oplus x_{n-1}$. Another similar sharing approach is to store x as the sum of n values modulo some value. That is, the shares of a value x are $x_1, \ldots, x_n$ where $x = x_1 + x_2 + \cdots + x_n \bmod N$ for some value N.

When given values in a modularly additive split format, then one can perform certain arithmetic operations on the split values. For example, if two values x and y are split among a group of participants then without communicating this group can compute $x + y$ (modulo the splitting modulus) by having each member add up their shares of x and y. Also, a group of participants can multiply their shared value by a constant by each multiplying their individual values by the constant.

Threshold sharing: Another approach to sharing values is to split the values in such a way that t participants are required to recover the value $(t < n)$. A classic technique for achieving such sharing is Shamir's secret sharing [28]. In this approach a $t - 1$ degree polynomial P is chosen such that $P(0) = x$. Furthermore, participant i is given $(i, P(i))$. Now if any t parties share their values, they can interpolate their values to recover P and thus learn $P(0)$ (i.e., x). However, $t - 1$ or less participants do not have enough information to reconstruct P and thus x is still protected. This is often referred to as a (t, n) threshold sharing scheme.

When given values split using such a sharing approach, it is possible to perform some mathematical operations on the values without having the participants communicate. Suppose that values x and y are split with a (t, n) sharing scheme and where each participant's points have the same x-coordinate. Then it is possible to compute $x + y$ in a threshold (t, n) simply by having each participant add the y-coordinates of their points. Furthermore, it is also possible to compute $x * y$ in a $(2t, n)$ shared manner by having each participant multiply the y-coordinates of their points. Finally, it is trivial to multiply a (t, n) shared value by a constant and obtain a (t, n) shared value by having each participant multiply their y-coordinate by the constant.

Encoding/value: Suppose that $x \in \{0, 1\}$ and that $n = 2$. In this case a specialized form of splitting is achieved by having one party learn two large random values v_0 and v_1 and having the other party learn v_i. Thus the first party knows the semantics of the encoding, but does not know the actual value. On the other hand, the second party knows the encoded value but does not know what it means. To split a value in this form the party owning x chooses v_0 and v_1 and sends them to party one and then sends v_x to party two. Of course, this can be extended to ℓ values in a larger range in a natural way. Furthermore, this idea can be extended to multiple parties by splitting the encodings among the parties (using either sharing or threshold sharing) and giving all parties the value.

14.4.2 Representative-Based Architectures for SMC

An architecture (formalized in [9]) that increases scalability is to separate the participants into three, not necessarily mutually exclusive, groups: input servers, computation servers, and output servers. The input servers split their inputs (using the techniques described in the previous section) among the computation servers so that no small group of computation servers can recover the inputs. The computation servers then engage in a secure protocol to compute the results in a split fashion. And finally, the split results are sent to the output servers who then learn the results. This representative-based architecture is depicted in Figure 14.2. This architecture mitigates many of the disadvantages of the original protocol architecture described earlier. That is, the input servers do not all need to be online at the same time because they can submit their values to the computation servers and then go offline. Furthermore, the number of computation servers is typically much smaller than the number

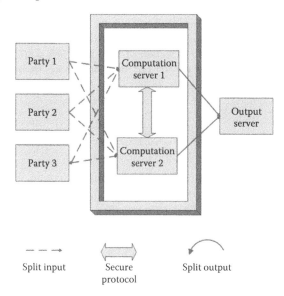

FIGURE 14.2 Example of representative-based architecture.

of input servers, and thus this approach is more scalable. The downside of this representative-based approach is that this usually requires some level of trust in an outside party (i.e., a party that is neither an input nor an output server).

In [25] a version of this representative-based architecture was used to provide a practical privacy-preserving auction. In this scheme a third party called the auctioneer is utilized. The sole purpose of the auctioneer is to help the seller compute the winning bid. The bidders split their bids between the auctioneer and the seller so that neither individual learns information about the bids. Once the auctioneer and the seller have received all bids, the seller and the auctioneer engage in a secure protocol to compute the winning bidder along with the winning bid in a split fashion. This information is then revealed to the seller and to the bidders. The only trust assumption is that the bidders trust the auctioneer not to collude with the seller, and this level of trust is more reasonable than the fully trusted third party approach, and may be applicable in many practical situations.

Since the focus of this chapter is practical SMC techniques, the focus will be on two-party SMC techniques (as these are more likely to be practical). Thus for problems with more than two parties we will assume a representative-based architecture with two computational servers.

14.5 Logical Circuit Based Approaches

In this section we discuss a technique for computing any two-party function in a secure manner. The main focus is on the HBC model, but extensions to the malicious model are also discussed. The principal idea is that we will represent the function f as a logical circuit C_f. In [31], a technique was described that securely evaluates a logical circuit with communication and computation proportional to the number of gates in the circuit and with a constant number of rounds; in [21] this technique was proven secure. Now, any function computable in polynomial time can be computed with a polynomially sized logical circuit, and so these two things imply that any function computable in polynomial-time can be computed securely with polynomial communication and computation and a constant number of rounds. In the remainder of this section we first describe the cryptographic primitives needed for Yao's construction in Section 14.5.1. Then Yao's scrambled circuit evaluation technique for the HBC adversary model is described in Section 14.5.2. In Section 14.5.3 extensions to the malicious model are described. Finally, in Section 14.5.4 applications of Yao's approach are described.

14.5.1   Cryptographic Primitives

One building block that is used is a form of symmetric key encryption. The encryption scheme for Yao's scrambled circuit evaluation requires specific properties: elusive range (it is difficult to choose a value that is a valid ciphertext for a particular key k) and efficiently verifiable range (given a ciphertext and a key it is possible to determine if the ciphertext is a possible encryption with the key k). These properties are possible for cryptosystems, and we refer the reader to [21] for more information about these properties. In the remainder of the paper we use the notation $Enc(M, k)$ to denote the encryption of the message M with the key k.

The other core building block needed for Yao's scrambled circuit evaluation is 1-out-of-2 chosen oblivious transfer; oblivious transfer (OT) was introduced in [27]. In the original OT protocols the sender would have a message and the receiver would obtain the value with probability one-half. In [10], a variation was introduced where the sender has two messages and the receiver obtains message one with probability one-half and receives message two with probability one-half; furthermore, the sender would not know which message the sender obtained. While these two versions of OT seem different, they were proved to be equivalent in [7]. The version that will be used in the remainder of the chapter is chosen 1-out-of-k OT (also called all or nothing disclosure of secrets) and was introduced in [3]. In this version of OT the sender has k messages and the receiver obtains a specific message that the receiver gets to choose. We will denote this protocol as $OT_1^k((M_1, M_2, \ldots, M_k), i)$ where the receiver learns M_i and the sender learns nothing.

The following is a protocol described in [1] for chosen 1-out-of-2 OT. Note that there are many other protocols for OT in the literature, but as OT is not the focus of this chapter only one such protocol is described (see Figure 14.3).

The above protocol for OT requires a single round (once the setup has been done), and it requires $O(1)$ modular exponentiations and other computations. We now give a sketch of security argument for the above protocol. First, an honest receiver will obtain M_b, because $\beta_b^r = (g^{s_b})^r = (g^r)^{s_b} = \gamma_b$. However, the receiver cannot obtain M_{1-b}, because this would imply that the receiver also learns

---

Input: The sender inputs two messages M_0 and M_1, and the receiver inputs $b \in \{0, 1\}$.
Output: The receiver obtains M_b.

1. The sender chooses a large prime p, a generator g of Z_p^*, and a value C in Z_p^* where the receiver does not know the discrete log of C. The values p, g, and C are sent to the receiver. Note that this step needs to be done only once for several OT protocols.

2. The receiver chooses a random value r computes two values α_0 and α_1 where $\alpha_b = g^r$ and $\alpha_{1-b} = \frac{C}{g^r}$. The receiver sends α_0 to the sender.

3. The sender computes $\alpha_1 = \frac{C}{\alpha_0}$. It then chooses two random values s_0 and s_1, and computes: $\beta_i = g^{s_i}$, $\gamma_i = \alpha_i^{s_i}$, and $\delta_i = M_i \oplus \gamma_i$ for $i \in \{0, 1\}$. It then sends β_0, β_1, δ_0, and δ_1 to the receiver.

4. The receiver computes $\beta_b^r = (g^{s_b})^r = (g^r)^{s_b} = \gamma_b$. The receiver then computes $\delta_b \oplus \gamma_b = M_b$.

FIGURE 14.3 Oblivious transfer protocol.

γ_{1-b}, which is $\left(\frac{C}{g^r}\right)^{s_{1-b}}$, which is $g^{ms_{1-b}}$ for some value m. The receiver does not know this value, because otherwise the receiver would know the discrete logarithm of C. Thus the receiver knows g^m and $g^{s_{1-b}}$, and needs to calculate $g^{ms_{1-b}}$ (which is the Diffie–Hellman problem). The sender is unaware of which item the receiver is choosing because $\frac{C}{g^r}$ and g^r are indistinguishable for a randomly chosen (and unknown) value r. For a more detailed argument as to why this scheme is secure see [1].

14.5.2 Simulating Circuits with Yao's Method

A high-level overview of Yao's approach is as follows: one party is labeled the circuit generator and the other party is labeled the circuit evaluator. For each wire in the circuit the generator creates two encodings (one for 0 and one for 1), and the evaluator will learn the encoding that corresponds to the actual value of each wire without knowing what the encoding corresponds to. A crucial piece of this protocol is a gate encoding that allows the evaluator to obtain the gate's output wire's encoding when given the gate's input wires' encodings. Finally, to learn the final result the evaluator is given the mapping between the circuit's output wire encodings and their values. In what follows we describe the details of this process.

Setup: Assume that Alice is playing the role of the generator and that Bob is the evaluator. Suppose the circuit in question consists of wires $w_1, \ldots, w_n$ that are partitioned into four mutually exclusive sets A (Alice's inputs), B (Bob's inputs), I (intermediate wires), and O (output wires). Furthermore, suppose these wires are connected to a set of binary gates $G_1, \ldots, G_m$ and that each wire is either in $A \cup B$ or is an output from some gate. We denote the actual value of the wire w_i by v_i, and we denote the binary function that corresponds to gate G_i ad g_i, e.g., $g_i : \{0, 1\} \times \{0, 1\} \to \{0, 1\}$. Finally, the input wires to G_i are denoted by x_i and y_i, and the output wire is denoted by z_i. That is, $v_{z_i} = g(v_{x_i}, v_{y_i})$.

Circuit generation: For each wire w_i Alice randomly chooses two encryption keys $w_i[0]$ and $w_i[1]$, and she also chooses a random bit $\lambda_i \in \{0, 1\}$. One of the goals of the protocol is for Bob to learn $(v_i \oplus \lambda_i) || w_i[v_i \oplus \lambda_i]$ for each wire in the circuit. Note that the λ value hides from Bob the actual value of a wire.

Furthermore, for each gate G_i, the generator creates four messages, denoted by $G_i[0, 0]$, $G_i[0, 1]$, $G_i[1, 0]$, and $G_i[1, 1]$, where $G_i[a, b]$ is $Enc(s_i[a, b] || w_{z_i}[s_i[a, b]], w_{x_i}[a] \oplus w_{y_i}[b])^*$ where $s_i[a, b] = g_i(a \oplus \lambda_{x_i}, b \oplus \lambda_{y_i}) \oplus \lambda_{z_i}$.

Circuit evaluation: The circuit generator and the evaluator do the following:

1. For each $a \in A$, Alice sends $(v_a \oplus \lambda_a) || w_a[v_a \oplus \lambda_a]$ to Bob.
2. For each $b \in B$, Alice and Bob engage in $OT_1^2(\{\lambda_b || w_b[\lambda_b], \bar{\lambda}_b || w_b[\bar{\lambda}_b]\}; v_b)$ where Alice plays the role of the sender.
3. Alice sends to Bob $G_i[0, 0]$, $G_i[0, 1]$, $G_i[1, 0]$, and $G_i[1, 1]$ for every gate.
4. For each $o \in O$, Alice sends λ_o to Bob.

After Bob receives the above information, he evaluates the circuit. First, for each wire $i \in A \cup B$, Bob knows $(v_i \oplus \lambda_i) || w_i[v_i \oplus \lambda_i]$ (from Steps 1 and 2 in the protocol above). Once Bob has learned the encodings for wires x_j and y_j, he computes $Dec(G_j[v_{x_j} \oplus \lambda_{v_{x_j}}, v_{y_j} \oplus \lambda_{v_{y_j}}], w_{x_j}[v_{x_j} \oplus \lambda_{v_{x_j}}] \oplus w_{y_j}[v_{y_j} \oplus \lambda_{v_{y_j}}])$

---

* Note that this notation indicates that $s_i[a, b] || w_{z_i}[s_i[a, b]]$ is encrypted with the key $w_{x_i}[a] \oplus w_{y_i}[b]$, and thus to be able to decrypt this value one would need both $w_{x_i}[a]$ and $w_{y_i}[b]$.

to obtain the encoding of the output wire of G_j (i.e., wire z_j). That is, this value is $(v_{z_j} \oplus \lambda_{v_{z_j}})||w_{z_j}[v_{z_j} \oplus \lambda_{v_{z_j}}]$. Once Bob has these values for all wires, he can compute v_o for all wires in O, since he knows λ_o (from Step 4 of the protocol above).

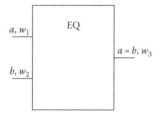

FIGURE 14.4　Example circuit.

To clarify the details of the above protocol, consider the following example. Suppose we have a small circuit where Alice has a single input bit a and Bob has a single input b. Furthermore, the output of the circuit is the predicate $a = b$. Note that this can be viewed as a circuit with a single equality gate (i.e., a not XOR gate). Furthermore, we will label the circuit as in Figure 14.4. To make this a concrete example, let us suppose that $a = 0$ and $b = 1$.

For each wire w_i Alice will choose encoding values $w_i[0]$ and $w_i[1]$ and λ_i. Let us suppose that Alice chooses $\lambda_1 = 0$, $\lambda_2 = 1$, and $\lambda_3 = 1$. Alice will send to Bob the values $(a \oplus \lambda_1)||w_1[a \oplus \lambda_1]$, that is she sends $0||w_1[0]$ to Bob. Alice and Bob will also engage in a 1-out-of-2 OT to reveal $0||w_2[0]$ (this corresponds to Bob's input wire).

For the gate information, Alice will calculate the following four messages:

- $s_1[0,0] = EQ(\lambda_1, \lambda_2) \oplus \lambda_3 = 1$
- $s_1[0,1] = EQ(\lambda_1, \bar{\lambda_2}) \oplus \lambda_3 = 0$
- $s_1[1,0] = EQ(\bar{\lambda_1}, \lambda_2) \oplus \lambda_3 = 0$
- $s_1[1,1] = EQ(\bar{\lambda_1}, \bar{\lambda_2}) \oplus \lambda_3 = 1$

Thus the gate information that Alice uses will be as follows:

- $G_1[0,0] = Enc(1||w_3[1], w_1[0] \oplus w_2[0])$
- $G_1[0,1] = Enc(0||w_3[0], w_1[0] \oplus w_2[1])$
- $G_1[1,0] = Enc(0||w_3[0], w_1[1] \oplus w_2[0])$
- $G_1[1,1] = Enc(1||w_3[1], w_1[1] \oplus w_2[1])$

Once Bob has this gate information, and he has the values $0||w_1[0]$ and $0||w_2[0]$ the only gate message he can decrypt is $G_1[0,0]$, and he thus receives $1||w_3[1]$. If this was to be used as an intermediate wire, he would not know the actual value of wire w_3, because it depends on λ_3 which he does not know. However, if this is an output wire, then Alice will reveal λ_3 and he will learn that the result is actually 0.

14.5.3　What about Malicious Behavior

Consider a malicious adversary in Yao's protocol. If the malicious player is the evaluator, then there is little that this player can do other than change his inputs. However, if the generator is malicious then this adversary can create any circuit with the same topology as the desired function. One mechanism that has been suggested to mitigate this attack is a cut-and-choose approach [22,24,25,29]. In this approach the generator creates several versions of the circuit and sends them all to the evaluator. The evaluator then requests that the generator "open" (i.e., give them all wire keys) a subset of the circuits. The evaluator then verifies that this subset of circuits was created properly.

In the simplest instantiation of this scheme the generator creates N circuits, and the evaluator verifies $N-1$ of the circuits. In this case, if a malicious generator wants to evaluate a faulty circuit, the malicious generator's chances of not being caught is $\frac{1}{N}$. In a more complicated approach the evaluator verifies $\frac{N}{2}$ circuits and computes the result for the other $\frac{N}{2}$ circuits. If the evaluated circuits' outputs differ then the evaluator sets the result to the most frequent output. In this case a cheating adversary is successful with a probability that is exponentially small in N. Many of the details on how this cut-and-choose model is implemented have been omitted, but we refer the reader to [22,24,25,29] for more details.

14.5.4 Using Circuits

To use the scrambled circuit evaluation approach one needs to construct a circuit for the problem at hand. One approach that was used in FAIRPLAY [22] was to build a compiler that converted a programming language into a circuit. However, depending on the needs of the application one may decide to construct the circuits by hand. Some common circuits are described below.

- To test two n-bit values, $a_1, \ldots, a_n$ and $b_1, \ldots, b_n$ for equality one can use the following circuit: $\bigwedge_{i=1}^{n} \overline{(a_i \oplus b_i)}$. Note that this circuit requires $O(n)$ gates.
- To test if an n-bit value $a_1, \ldots, a_n$ is greater than another n-bit value $b_1, \ldots, b_n$ one can use the following circuit: $\bigvee_{i=1}^{n} (a_i \wedge \bar{b}_i \wedge \bigwedge_{j=1}^{i-1} \overline{((a_j \oplus b_j)}))$. Note that this circuit has $O(n)$ gates.

One difficulty with using this approach is the difficulty with constructing circuit for some problems. Furthermore, circuits are inefficient for some problems such as indirect access into a list of values, that is if there is a list of items $v_1, \ldots, v_n$ given and an index $i \in [1, n]$ is computed by the circuit, then it requires $O(n)$ computation to obtain v_i in the circuit.

14.6 Computing on Encrypted Data

Because of the results in the previous section, one may wonder if all secure protocol problems have been solved. While the above techniques do imply that any polynomially computable function can be securely evaluated in the semi-honest model with polynomial communication, it is believed that for many problems the general solutions may not be practical. However, in some situations an efficient domain-specific protocol exists for some problems [18]. In this section we focus on one class of problems that have solutions based on arithmetic expressions.

14.6.1 Homomorphic Encryption

A homomorphic encryption scheme allows computation using encrypted values; this is useful because it facilitates some protocols based on arithmetic that are more efficient than their circuit counterparts. The idea behind these types of protocols is that the values can be encrypted and then someone can compute the encryption of a result without knowing the values. Specific homomorphic encryption schemes are described in [8,26]. In [26] the arithmetic is done mod n where n is an RSA modulus, and in [8], the arithmetic is done mod n^k where n is an RSA modulus and k is an integer. We now formally describe the properties of a homomorphic encryption scheme.

1. *Public key:* The systems are public-key encryption schemes in that anyone with the public parameters of the scheme can encrypt, but only those with the private parameters can decrypt.
2. *Semantically secure:* We require that the scheme be semantically-secure as defined in [19]. That is, given the following game between a probabilistic polynomial time (PPT) adversary A and a challenger C:

   a. C creates a public–private key pair (E, D) for the encryption system and sends E to A.

   b. A generates two messages M_0 and M_1 and sends them back to C.

c. C picks a random bit $b \in \{0, 1\}$ and sends $E(M_b)$ to A.

d. A outputs a guess $b' \in \{0, 1\}$.

We say that A wins the game if $b' = b$. We define the advantage of A for a security paramater k to be $Adv_A(k) = Pr[(b = b') - (1/2)]$. We say that the scheme is semantically secure if $Adv_A(k)$ is negligible in k.

3. *Additive:* Given the $E(x)$ and $E(y)$ and the public parameters of the scheme, one can compute $E(x + y)$ by calculating $E(x) * E(y)$.

4. *Multiplication:* A natural extension of the previous property is that when given $E(x)$ and c, one can compute $E(xc)$ by calculating $E(x)^c$.

5. *Re-encryption:* When given an encryption $E(x)$, one can compute another encryption of x, simply by multiplying the original encryption by $E(0)$.

14.6.2   Scalar Product

It should not come as a surprise that homomorphic encryption facilitates efficient protocols that compute some arithmetic expression. One example of such a protocol is the calculation of the scalar product of two vectors. That is given $\vec{A} = \; <a_1, \ldots, a_n>$ and $\vec{B} = \; <b_1, \ldots, b_n>$, the goal is to compute $\vec{A} \circ \vec{B} = \sum_{i=1}^{n} (a_i * b_i)$. One protocol for secure scalar product was introduced in [14], and Figure 14.5 is such a protocol.

14.6.3   Polynomial Operations

Another application of homomorphic encryption is the ability to compute various polynomial operations. This has been useful for set operations (which are described in detail in the next section), including [12,13,20]. To encrypt a polynomial with homomorphic encryption, each coefficient of the polynomial is encrypted with the encryption scheme. That is, the encryption of $P(x) = p_n x^n + \cdots + p_1 x + p_0$ is $E(p_n), \ldots, E(p_1), E(p_0)$, and we denote this value by $E_{poly}(P)$. Given an encrypted polynomial it is possible to perform some polynomial operations, including

1. *Polynomial evaluation:* Given $E_{poly}(P)$ and z it is possible to compute $E(P(z))$. If $P(x) = p_n x^n + \cdots + p_1 x + p_0$, then $E(P(z)) = E(p_n z^n + \cdots + p_1 z + p_0) = E(p_n z^n) * \cdots * E(p_1 z) * E(p_0) = E(p_n)^{z^n} * \cdots * E(p_1)^z * E(p_0)$ which can be computed with knowledge of $E_{poly}(P)$ and z.

2. *Polynomial addition:* Given $E_{poly}(P)$ and $E_{poly}(Q)$ it is possible to compute $E_{poly}(P + Q)$. Assume that $P(x) = p_n x^n + \cdots + p_1 x + p_0$, and $Q(x) = q_n x^n + \cdots + q_1 x + q_0$, if P and Q have different degrees, then one of them can be padded with 0's to make the degrees the same. Now $P + Q(x) = (p_n + q_n) x^n + \cdots + (p_1 + q_1) x + (p_0 + q_0)$, thus $E_{poly}(P + Q)$

Input: Alice has a vector $\vec{A} = \; <a_1, \ldots, a_n>$ and Bob has a vector $\vec{B} = \; <b_1, \ldots, b_n>$.
Output: Alice learns $\vec{A} \circ \vec{B}$.

1. *Setup:* Alice creates parameters for a semantically secure additively homomorphic encryption scheme and sends Bob the public parameters; we denote encryption by E. Note that this setup phase has to happen only once for multiple executions of the protocol.

2. Alice sends to Bob the values $E(a_1), \ldots, E(a_n)$.

3. Bob calculates the following value: $E(a_1)^{b_1} * \cdots * E(a_n)^{b_n} * E(0)$. It is straightforward to verify that this value is $E(\vec{A} \circ \vec{B})$, and the multiplication by $E(0)$ is to re-randomize the encryption. Bob sends this value to Alice.

4. Alice decrypts the result and learns $\vec{A} \circ \vec{B}$.

FIGURE 14.5 Two-party HBC protocol for scalar product.

is $E(p_n + q_n), \ldots, E(p_1 + q_1), E(p_0 + q_0)$ which is just $E(p_n) * E(q_n), \ldots, E(p_1) * E(q_1),$ $E(p_0) * E(q_0)$.

3. *Polynomial multiplication:* Given $E_{poly}(P)$ and Q it is possible to compute $E_{poly}(P * Q)$. Assume that $P(x) = p_n x^n + \cdots + p_1 x + p_0$, and $Q(x) = q_m x^m + \cdots + q_1 x + q_0$. Now, $P * Q(x) = s_{n+m} x^{n+m} + \cdots + s_1 x + s_0$ where $s_i = \sum_{j=0}^{i} (p_j * q_{i-j})$.* Furthermore, $E(s_i) = \prod_{j=0}^{i} E(p_j)^{q^{i-j}}$ which can be computed from knowledge of $E_{poly}(P)$ and Q.

4. *Polynomial differentiation:* Given $E_{poly}(P)$, it is possible to compute $E_{poly}(P')$ where P' is the first derivative of P. Note that other derivatives can be computed by repeating this process. Assume that $P(x) = p_n x^n + \cdots + p_1 x + p_0$, then $P'(x) = n p_n x^{n-1} + \cdots + p_1$, and so the new coefficients can be computed with the knowledge of $E_{poly}(P)$.

14.6.4 Set Operations

The ability to perform polynomial operations is useful for computing set operations (set union, set intersection, etc.) [12,13,20]. Set operations are useful for other privacy-preserving computations (such as privacy-preserving data mining). The basic idea behind many of these protocols is to represent the sets as polynomials and then perform set operations on these sets by doing polynomial operations.

To represent a set $S = \{s_1, \ldots, s_m\}$, the polynomial $P_S(x) = (x - s_1) \cdots (x - s_m)$ is used. Note that $P_S(z) = 0$ if and only if $z \in S.$[†] Using the polynomial operations from the previous section, we can do some types of set operations.

1. *Polynomial evaluation:* Given $E_{poly}(P_S)$, one can compute $E(P_S(z))$, and this will be an encryption of 0 iff $z \in S$. Thus this is a method for detecting if an element is in a set.

2. *Polynomial addition:* The polynomial $P_S + P_T$ will be 0 at all values in $S \cap T$. Thus this is useful for computing set intersection.

3. *Polynomial multiplication:* The polynomial $P_S * P_T$ will be the polynomial that represents the multiset union of S and T.

4. *Polynomial derivation:* The polynomial P_S' will be 0 for all items that are in S two or more times, and thus this is useful for eliminating duplicates in a multiset.

The above building blocks can be combined together to form protocols for set intersection and set union. Figure 14.6 is a simplified version of the protocol in [12] for set intersection, and Figure 14.7 describes a simplified protocol for secure two-party set union that was introduced in [13].

14.7 Composing Secure Protocols

One may wonder how to put secure building blocks together to form a secure protocol. However, secure protocols are not always composable. For example, suppose Alice has a point in Cartesian space and that Bob has a point in Cartesian space. Furthermore, suppose that a threshold distance T is known to both Alice and Bob. Furthermore, suppose that the goal of the protocol is for Alice to learn if the distance between her and Bob's points is smaller than T. One way of doing this is to use a secure protocol to compute the distance between the two points and to reveal this value to Alice. Then Alice would compute the result from this value and T. Clearly, the resulting protocol would not be secure, because it reveals the distance between the two points. On a positive note, it

---

* Note that $p_\ell = 0$ for $\ell > n$ and $q_\ell = 0$ for $\ell > m$.
† It is worth noting that if we are doing modular arithmetic over a large base that the more correct statement is: $P_S(z) = 0$ if $z \in S$ and if $z \notin S$, then $P_S(z) \neq 0$ with high probability.

Input: Alice has a set $A = \{a_1, \ldots, a_n\}$ and Bob has a set $B = \{b_1, \ldots, b_m\}$.
Output: Alice learns $A \cap B$.
Steps:

1. Alice creates a key pair for a semantically secure homomorphic encryption scheme and sends the public parameters to Bob. We denote encryption with this scheme by E and decryption by D.

2. Alice computes a polynomial $P(x) = (x - a_1) * \cdots * (x - a_n) = c_n x^n + c_{n-1} x^{n-1} + \cdots + c_1 x + c_0$. She then sends $E(P) = E(c_n), \ldots, E(c_1), E(c_0)$ to Bob.

3. For each item $b_i \in B$, Bob computes $E(P(b_i) * r_i + b_i)$ for a randomly chosen value r_i and sends all of these values to Alice. Note that $E(P(b_i) * r_i + b_i)$ is $E(b_i)$ if $b_i \in A$ and is a random value otherwise.

4. Alice decrypts all of the values that she receives from Bob and outputs all decrypted values that are in her set.

FIGURE 14.6 Two-party HBC protocol for set intersection.

Input: Alice has a set $A = \{a_1, \ldots, a_n\}$ and Bob has a set $B = \{b_1, \ldots, b_m\}$.
Output: Alice learns $A \cup B$.
Steps:

1. Alice creates a key pair for a semantically secure homomorphic encryption scheme and sends the public parameters to Bob. We denote encryption with this scheme by E and decryption by D.

2. Alice computes a polynomial $P(x) = (x - a_1) * \cdots * (x - a_n) = c_n x^n + c_{n-1} x^{n-1} + \cdots + c_1 x + c_0$. She then sends $E(P) = E(c_n), \ldots, E(c_1), E(c_0)$ to Bob.

3. For each item $b_i \in B$, Bob computes a tuple $(E(P(b_i) * r_i) \; ; \; E(P(b_i) * r_i * b_i))$ for a randomly chosen value r_i and sends all of these values to Alice. Note that if $b_i \in A$ then this tuple will be $(0 \; ; \; 0)$ and if $b_i \notin A$ then this tuple will be $(E(R) \; ; \; E(R * b_i))$ for some random value R. In the latter case, b_i is recoverable from the decryption of the tuple by multiplying the second value the inverse of the first value.

4. Alice decrypts all of the tuple that she receives from Bob. If a tuple is $(0 \; ; \; 0)$ then she does nothing and otherwise she recovers the value by multiplying the second value the inverse of the first value. She outputs all recovered values along with her own set.

FIGURE 14.7 Two-party HBC protocol for set union.

was shown in [4] that protocols can be composed in certain circumstances. More specifically, if the protocol can be proven secure when each secure building block is replaced with the ideal protocol for the building block (i.e., with the trusted third party solution), then the protocol that uses the secure protocols (instead of the ideal protocols) is also secure.

The following is a contrived example of a secure protocol composition, suppose we are trying to calculate the intersection of two sets A and B that are known respectively to Alice and Bob. Suppose we have two building blocks: (1) CARDINALITY$(A; B)$—which outputs the $|A \cap B|$, and (2) SETINT$(A, |A \cap B|; B)$ that outputs $A \cap B$. In this case one way to compute this value would be to first run CARDINALITY(A, B) to obtain $|A \cap B|$, and then use SETINT$(A, |A \cap B|; B)$ to calculate the result. If the individual protocols SETINT and CARDINALTITY are secure then this composition is also secure, because revealing the cardinality as an intermediate input is not a privacy violation because it is also revealed by the final result.

14.8 Summary

In summary, secure protocols provide a way to compute a function over distributed inputs, without revealing anything other than the result of the protocol. Of course, unless the function is independent of the inputs, the result will reveal some information about the inputs into the protocol, but a secure protocol should not reveal anything about the inputs other than what can be deduced from the output. In some cases, the output may be too revealing, i.e., the output of the function reveals too much information. However, when the result of a specific information collaboration is not too revealing, secure protocols are a promising technique, because they allow the result of the collaboration to be computed while preserving the privacy of the inputs. In fact, secure protocols have been created for many application domains, including auctions, data mining, set operations, benchmarking, and privacy-preserving surveys.

References

1. M. Bellare and S. Micali. Non-interactive oblivious transfer and applications. In *CRYPTO '89: Proceedings on Advances in Cryptology*, pages 547–557, Springer-Verlag, New York, 1989.
2. M. Ben-Or, S. Goldwasser, and A. Wigderson. Completeness theorems for non-cryptographic fault-tolerant distributed computation. In *Proceedings of the Twentieth Annual ACM Symposium on Theory of Computing*, pages 1–10, ACM Press, New York, 1988.
3. G. Brassard, C. Crepeau, and J. Robert. All-or-nothing disclosure of secrets. In *CRYPTO*, pages 234–238, Springer-Verlag, London, U.K., 1986.
4. R. Canetti. Security and composition of multiparty cryptographic protocols. *Journal of Cryptology*, 13(1):143–202, 2000.
5. R. Canetti, U. Feige, O. Goldreich, and M. Naor. Adaptively secure multi-party computation. In *Proceedings of the Twenty-Eighth Annual ACM Symposium on Theory of Computing*, pages 639–648, ACM Press, New York, 1996.
6. D. Chaum, C. Crépeau, and I. Damgård. Multiparty unconditionally secure protocols. In *Proceedings of the Twentieth Annual ACM symposium on Theory of Computing*, pages 11–19, ACM Press, New York, 1988.
7. C. Crepeau. Equivalence between two flavours of oblivious transfers. In *CRYPTO*, pages 350–354, Springer-Verlag, London, U.K., 1987.
8. I. Damgård and M. Jurik. A generalisation, a simplification and some applications of paillier's probabilistic public-key system. In *PKC '01: Proceedings of the 4th International Workshop on Practice and Theory in Public Key Cryptography*, LNCS 1992, pages 119–136, Springer-Verlag, Berlin, Germany, 2001.
9. I. Damgård and Y. Ishai. Constant-round multiparty computation using a black-box pseudorandom generator. In *Advances in Cryptology (CRYPT0 05)*, LNCS 3621, pages 378–394, Springer-Verlag, Berlin, Germany, 2005.
10. S. Even, O. Goldreich, and A. Lempel. A randomized protocol for signing contracts. *Communications of the ACM*, 28(6):637–647, 1985.
11. P. Feldman and S. Micali. Optimal algorithms for byzantine agreement. In *Proceedings of the Twentieth Annual ACM Symposium on Theory of Computing*, pages 148–161, ACM Press, New York, 1988.
12. M. Freedman, K. Nissim, and B. Pinkas. Efficient private matching and set intersection. In *Proceedings of Advances in Cryptology—EUROCRYPT '04*, LNCS, 3027, pages 1–19, Springer-Verlag, Berlin, Germany, 2004.
13. K. Frikken. Privacy-preserving set union. In *ACNS 2007*, LNCS, 2007, pages 237–252, Springer-Verlag, Berlin, Germany, 2007.
14. B. Goethals, S. Laur, H. Lipmaa, and T. Mielikainen. On private scalar product computation for privacy-preserving data mining. In *The 7th Annual International Conference on Information Security and Cryptology (ICISC 2004)*, pages 104–120, Seoul, Korea, 2004.
15. O. Goldreich. *Foundations of Cryptography: Volume I Basic Tools*. Cambridge University Press, Cambridge, U.K., 2001.
16. O. Goldreich. *Foundations of Cryptography: Volume II Basic Application*. Cambridge University Press, Cambridge, U.K., 2004.

17. O. Goldreich, S. Micali, and A. Wigderson. How to play any mental game. In *Proceedings of the Nineteenth Annual ACM Conference on Theory of Computing*, pages 218–229, ACM Press, New York, 1987.

18. S. Goldwasser. Multi party computations: Past and present. In *Proceedings of the Sixteenth Annual ACM Symposium on Principles of Distributed Computing*, pages 1–6, ACM Press, New York, 1997.

19. S. Goldwasser and S. Micali. Probabilistic encryption. *Journal of Computer and System Sciences*, 28(2):270–299, 1984.

20. L. Kissner and D. Song. Privacy-preserving set operations. In *Proceedings of Advances in Cryptology–CRYPTO '05*, LNCS, 3621, 2005. Full version appears at http://www.cs.cmu.edu/leak/.

21. Y. Lindell and B. Pinkas. A proof of yao's protocol for secure two-party computation. Cryptology ePrint Archive, Report 2004/175, 2004. http://eprint.iacr.org/.

22. D. Malkhi, N. Nisan, B. Pinkas, and Y. Sella. Fairplay—A secure two-party computation system. In *Proceedings of Usenix Security*, pages 287–302, San Diego, CA, 2004.

23. S. Micali and P. Rogaway. Secure computation (abstract). In *CRYPTO '91: Proceedings of the 11th Annual International Cryptology Conference on Advances in Cryptology*, pages 392–404, Springer-Verlag, Berlin, Germany, 1992.

24. P. Mohassel and M. Franklin. Efficiency tradeoffs for malicious two-party computation. In *Public Key Cryptography Conference (PKC)*, LNCS, 3958, pages 458–473, Springer-Verlag, Berlin, Germany, 2006.

25. M. Naor, B. Pinkas, and Re. Sumner. Privacy preserving auctions and mechanism design. In *EC '99: Proceedings of the 1st ACM Conference on Electronic Commerce*, pages 129–139, ACM Press, New York, 1999.

26. P. Paillier. Public-key cryptosystems based on composite degree residuosity classes. In *Advances in Cryptology: EUROCRYPT '99*, LNCS, 1592, pages 223–238, Springer-Verlag, Berlin, Germany, 1999.

27. M. Rabin. How to exchange secrets with oblivious transfer. Technical Report TR-81, Aiken Computation Lab, Harvard University, Cambridge, MA, 1981.

28. A. Shamir. How to share a secret. *Communications of the ACM*, 22(11):612–613, 1979.

29. D. Woodruff. Revisiting the efficiency of malicious two-party computation. In Naor, M. ed., *Eurocrypt 2007*, LNCS, vol. 4515, Springer, Heidelberg, pages 79–96, 2007.

30. A.C. Yao. Protocols for secure computations. In *Proceedings of the 23th IEEE Symposium on Foundations of Computer Science*, pages 160–164, IEEE Computer Society Press, Washington, DC, 1982.

31. A.C. Yao. How to generate and exchange secrets. In *Proceedings of the 27th Annual IEEE Symposium on Foundations of Computer Science*, pages 162–167, IEEE Computer Society Press, Washington, DC, 1986.

15

Voting Schemes

Berry Schoenmakers
Technical University of Eindhoven

15.1 Introduction

Electronic voting is probably the single most controversial application in the field of information security. Almost any area in information security, from computer security and cryptographic issues to human psychology and legal issues, is brought together when designing secure electronic voting systems. Not surprisingly, a wide variety of approaches have been used and a multitude of voting systems have been proposed.

This chapter is about cryptographic schemes for electronic voting, or voting schemes for short, whose main task is to facilitate a secure and private way of casting and counting votes.* The first wave of interest in voting schemes started in the early 1980s with the publication of Chaum's paper on anonymous communication [8]. Subsequently, approaches to voting were discovered, where the emphasis was on concepts and feasibility, much like other work in cryptography in that decade. A second wave started in the 1990s with the emergence of the World Wide Web, which created a huge interest in remote voting. Also, the emphasis shifted to efficiency concerns resulting in quite

---

* The game-theoretic aspect of what choices to offer voters to ensure as much as possible that the most favored candidates actually win is outside the scope of this chapter. Social choice theory, with results such as Arrow's impossibility theorem in the 1950s (generalizing voting paradoxes such as Condorcet's paradox from the late eighteenth century), deals with these issues. Ideally, voting schemes are sufficiently versatile to support any electoral system such as plurality voting, approval voting, single transfer voting, etc., but often voting schemes are biased toward a particular electoral system.

practical schemes. And, currently, we are experiencing a third wave because of the "Florida 2000" U.S. election fiasco, which renewed interest in voting from polling stations leading to voting schemes that combine cryptographic and physical aspects.

Voting schemes are intended to form the cryptographic core of electronic voting systems. The general goal of these schemes is to eliminate as many security problems as possible, thereby limiting the number and the extent of the residual assumptions needed to ensure the security of the overall voting system.

For example, proper encryption and authentication of votes ensure that no illegal modifications of existing votes and insertions of extra votes are possible, whereas appropriate redundancy techniques from distributed computing (realizing Byzantine fault tolerance; see also Chapter 29 of *Algorithm and Theory of Computation Handbook, Second Edition: General Concepts and Techniques*) ensure that no votes can be deleted from the system. However, these basic techniques which are commonly used to implement the main goals in information security, namely, confidentiality, integrity, and availability (CIA) are not sufficient for the special combinations of security properties found in electronic voting. For example, a strict level of ballot secrecy and full verifiability of the election result can only be achieved by using special forms of encryption and authentication rather than conventional ones. The strongest voting schemes thus employ advanced cryptographic techniques, such as threshold cryptography, homomorphic encryption, blind signatures, and zero-knowledge proofs, thereby eliminating the need for trusting individual (or small groups of) entities. This goes beyond what is common in the voting machine industry, where the design, the implementation, and the operation of voting machines are essentially split between a very limited number of key players (often, involving just one company, at best two or three companies). The challenge is to find the sweet spot with the best trade-off between cryptography and other security measures.

15.2 Problem Description

The main reason why voting schemes pose a challenge is the desire to achieve some form of ballot secrecy, i.e., hiding people's votes. Secret ballot elections are important to facilitate free choice such that voters need not be concerned or afraid of the consequences of making a particular choice. Even stronger, voters should not be able to willingly exchange their votes for money or other favors. Of course, by choosing a particular candidate voters also expect some favorable results in the long run, but rich candidates should not be able to buy poor people's votes by alleviating their immediate needs.

Ballot secrecy must be maintained while ensuring at the same time that only eligible voters are able to cast votes (one (wo)man, one vote). The difficulty is to authenticate the voters, or rather to authenticate the secret ballots, while not creating any links between voters and their votes. In traditional elections, voters are present in person and use some physical token to cast their vote, e.g., by using a small ball of a particular color (a colored pea or bean, called a "ballotta" in Italian) or marking a paper ballot form. Before counting the votes, these tokens are shuffled to remove any link between voters and votes (advances in forensic analysis may also necessitate cleaning of the tokens, or even require voters to wear gloves).

For remote voting, the physical presence of the voter is not required. Still, voters need to be able to cast their votes in an authenticated but secret manner. Postal voting is a traditional way to facilitate remote voting. The voter sends its voted ballot sealed inside two envelopes, where the outer envelope is signed (and linked to the voter) and the inner envelope is anonymous. Each outer envelope will be checked and if okay the inner envelope is put in a ballot box. After shuffling the ballot box, the inner envelopes are opened and the votes are counted.

In general, the essence of the voting problem can be captured by formulating it as a problem in secure multiparty computation. Each voter V_i provides its vote v_i, $1 \leq i \leq n$, as private input, and the object is to compute the election result $f(v_1, \ldots, v_n)$ correctly. A simple example is a referendum, where $v_i \in \{0, 1\}$ either represents a yes-vote ($v_i = 1$) or a no-vote ($v_i = 0$), and the election result is the sum of the votes $v_1 + v_2 + \cdots + v_n$. Here, it is understood that the list of eligible voters $V_1, \ldots, V_n$ is known (and has been compiled as intended—voter registration is beyond the scope of voting schemes).

Voting schemes solving the voting problem should thus enable the correct computation of $f(v_1, \ldots, v_n)$, while hiding any further information on the votes $v_1, \ldots, v_n$, and should do nothing else. Another way to state this is that we want to achieve the effect of an ideal election, which could be run using a single (completely) trusted party as follows: each voter identifies itself to the trusted party, and then tells the trusted party its vote in private; once all votes have been cast, the trusted party announces the election result. A voting scheme must emulate such an ideal election, though without relying on a single trusted party. Instead, multiple parties (or entities) will be involved in the protocols constituting a voting scheme, thereby reducing the level of trust put into each party individually.

This succinct formulation encompasses many security properties. For example, by stating that only the election result should be revealed, it means that no intermediate results or results per county or precinct should be revealed. Similarly, as only the trusted party learns the votes, voters cannot convince anyone else of how they voted. Also, there is the premise that the trusted party can actually identify voters upon accepting their votes. As identification of voters cannot be solved by cryptographic means only, we have to assume some external mechanism. For instance, a common assumption is that each (eligible) voter holds a key pair for authentication purposes, where the public key is registered as part of a public key infrastructure (PKI). Voter authentication may also involve biometrics to ensure that someone is taking part in person in the voting protocol.

Voting schemes consist of several protocols involving entities, such as election officials, voters, talliers, and observers (or scrutinizers). Protocols for voting and tallying cover the main tasks of a voting scheme, while additional protocols for setting up various parts of the system and possibly for various verification tasks need to be provided as well. As explained earlier, the goal for a voting scheme is to emulate the trusted party as close as possible, under reasonable assumptions. We will use a basic measure of resilience to indicate how well a voting scheme emulates the trusted party. A scheme will be called **k-resilient**, if at least k parties need to be corrupted in order to compromise the scheme. Similarly, a scheme is called **k-resilient** w.r.t. a security property, if at least k parties need to be corrupted to compromise the security property.

The ideal scheme outlined earlier is thus 1-resilient, but this is often unacceptable. Generally, a $(k + 1)$-resilient system is better than a k-resilient system. In some cases, ∞-resilience is possible, for security properties that hold unconditionally. However, k-resilience (w.r.t. a security property) is not an absolute measure. Another important factor is the total number of parties running the system: if more parties are involved, it is generally easier to find a collusion of corrupted parties. For example, a 1-resilient system in which any of 10 involved parties can be corrupted is worse than a 1-resilient system in which any of five involved parties can be corrupted. Thus, from a security point of view, introducing additional parties to a voting system must be done with care.

In this chapter, we focus on the voting protocol and the closely connected tally protocol. We will discuss a variety of ways to design the voting schemes of sufficient resilience. Important aspects are the assumptions needed and the building blocks used (the mere use of which may imply further assumptions). A full security analysis of a voting scheme is often elusive, because determining the security properties and their resilience, including a complete list of assumptions, is a complex task.

15.3 Building Blocks

In this section we recapitulate several (cryptographic) primitives that are typically used in the design of advanced voting schemes. Our description of each primitive may be deceptively simple, in no way reflecting the potential complexity or the subtlety of implementations for the primitive. However, application of these primitives may incur considerable costs, and similar primitives need not be interchangeable.

15.3.1 Communication Primitives

Voting schemes are formulated in terms of a communication model that describes what type of channels are available between which pairs or subsets of the entities involved. The communication model hides the implementation details of these channels, which are often far from trivial and may also come with a substantial cost. Voting schemes sharing the same communication model can be compared meaningfully, but when different types of channels are used the implementation details of the channels may be required for a useful comparison.

A **bulletin board**, or **public authenticated broadcast channel**, lets a sender broadcast a message such that everybody sees the same message. Each sender has a designated section on the bulletin board implying that senders are authenticated. In principle, a bulletin board requires a complex protocol involving techniques, such as Byzantine agreement (see Chapter 29 of *Algorithm and Theory of Computation Handbook, Second Edition: General Concepts and Techniques*).

The basic goal of an **anonymous channel** [8] is to hide the identity of the sender, whereas the receiver is not necessarily anonymous. A sender is able to transmit messages to an intended receiver, and possibly allows for an acknowledgement by the receiver as well. In particular, an anonymous channel must withstand traffic analysis, where an adversary monitors the activity throughout an entire network to see who is communicating with whom.

A **private channel** allows for information-theoretically private communication between two nodes, protecting against eavesdroppers, and possibly providing end-to-end authentication as well. Private channels are typically used as building blocks in protocols for unconditionally secure multiparty computation (e.g., see [3]). Private channels are often not realistic to assume, as keys for one-time pad encryption (see Chapter 9 of this book) must have been set up previously, requiring an untappable channel (see following text).

A **secure channel** is similar to a private channel, but usually provides computational security only (for encryption and/or authentication). Secure channels provide a means for abstracting away everything required to ensure a protected link between two nodes. Commonly, secure channels are implemented by combining some form of key exchange resulting in session keys used for symmetric encryption and authentication. Secure sockets layer (SSL) is a practical example.

Finally, **untappable channels** have been introduced [5] to accommodate totally unobservable communication preventing the adversary from capturing any of the information transmitted. The analogy here is that one may observe two persons whispering to each other, but one has no way of listening in. An untappable channel is meant to be implemented physically, by some form of out-of-band communication.

Combinations of these channels are possible as well, but care must be applied to see what is sensible, and to see what it takes for a good implementation. For example, an anonymous channel is already hard to implement, let alone an untappable anonymous channel. Furthermore, one can refine these notions by considering directed links, which results in notions like one-way untappable channels.

15.3.2 Authentication Primitives

Voter authentication is a basic issue, which must be addressed by any voting scheme. An important distinction is whether the authentication mechanism is external or internal to the voting scheme. An external authentication mechanism provides the link with the list of eligible (or registered) voters, hence forms an integral part of any voting scheme. Depending on the voting scheme, further internal authentication mechanisms can be used, typically providing some form of anonymous signatures as described in the following.

A wide range of voter authentication mechanisms is used, generally using a mixture of physical and digital techniques. External authentication of voters may be as simple as matching voters' names against a list (electoral roll) by an election official at a polling station. Similarly, internal authentication may be done implicitly by having successfully authenticated voters line up for a voting machine, or by handing out a ticket to them, which must be handed to the person at the voting machine. Some basic internet voting schemes use sealed envelopes containing unique authentication codes, which are sent by postal mail to the addresses of all registered voters. Here, external authentication relies on the sealed envelopes indeed reaching the intended voters, whereas the codes are used for internal authentication, possibly without the printer keeping track of who gets which code—to provide some level of anonymity.

Cryptographic techniques are used to increase the resilience of voting schemes. External authentication of voters can be based on any conventional symmetric technique (incl. password-based authentication) or asymmetric technique (digital signatures, incl. one-time signatures or PKI-based signatures), where the main security concern is protection against impersonation attacks; see Chapter 12 of this book for more information.

For internal authentication, more advanced authentication primitives may be used, which we will collectively call **anonymous signatures**. Ordinary digital signatures are verified against the public key of the signer revealing the identity of the signer. In many applications, however, there is a need for digital signatures providing the authentication of messages without revealing the identity of the signer. Anonymous signature schemes aim at various levels of privacy for the signer, examples of which with relevance to electronic voting, are blind signatures [9] (see Chapter 12 of this book), group signatures [11], restrictive blind signatures [6], ring signatures [38], and list signatures [7]. In these schemes, all signatures are verified with respect to the public key of a designated party, called the issuer, which uses the corresponding private key to assist in the generation of signatures. The key property is that the issuer does not learn the contents of the messages thus signed nor is able to recognize the actual signatures thus produced.

The proper use of anonymous signatures usually requires the availability of anonymous channels. Moreover, special care must be taken, e.g., to prevent that additional blind signatures can be issued, which are then used to cast additional votes or to replace votes cast by legitimate voters (which may go unnoticed as long as the total number of votes does not exceed the total number of voters). Use of threshold cryptography to enforce that (blind) signatures are issued jointly by a number of parties (rather than a single issuer) and this helps to achieve an acceptable level of resilience; see Chapter 13 of this book for some more details.

15.3.3 Encryption Primitives

Basic symmetric and asymmetric cryptosystems as well as some more advanced and related techniques have been covered extensively in previous chapters. Voting schemes use a large variety of such techniques to let voters encrypt their votes in some secure way. Here, it should be stressed that "vote encryption" is meant in a broad sense. Apart from encryption using a cryptosystem, we also use it to cover any application of secret-sharing and/or commitment schemes. For instance, vote encryption may also be done by having a voter distributing shares of its vote to a number of parties: this way, the voter performs the role of the dealer in a secret-sharing scheme, and decryption is done by reconstructing the secret (vote) from the shares. Similarly, vote encryption may actually be done by publishing a commitment to the vote, in which case decryption is done by later opening the commitment.

Throughout this chapter, we will use $[\![v]\!]$ as an abstract notation for any such encryption of a vote v. A basic feature of a vote encryption $[\![v]\!]$, which is suppressed from the notation, is the fact that $[\![v]\!]$ is obtained in a probabilistic way. This ensures semantic security, which means that given a vote encryption $[\![v]\!]$ one cannot get any information on vote v. In particular, if $[\![v]\!]$ and $[\![v']\!]$ are produced by two different voters, one cannot even decide if $v = v'$ holds, or not.

Two particular features for encryption schemes used in voting are homomorphic encryption and threshold decryption. Informally, **homomorphic encryption** satisfies the property that the product of two encryptions is an encryption of the sum of the corresponding messages:

$$[\![x]\!][\![y]\!] = [\![x + y]\!],$$

for messages x and y. For a basic example, let $\langle g \rangle = \{1, g, g^2, \dots, g^{q-1}\}$ be a cyclic group, written multiplicatively, of large prime order q. Hence, g is a generator of order q, $g^q = 1$. Let $h \in \langle g \rangle$ be a public key. Then, an additively homomorphic ElGamal encryption [16] of a message $x \in \mathbb{Z}_q$ is computed as

$$[\![x]\!] = (A, B) = (g^r, h^r g^x),$$

where r is chosen uniformly at random in $\mathbb{Z}_q$. The homomorphic property holds if we define the product of two encryptions (A, B) and (C, D) in the obvious way as (AC, BD), and we define $+$ as addition modulo q. Given the private key $\alpha = \log_g h$, encryption (A, B) is decrypted as

$$x = \log_g (b/a^\alpha).$$

Compared to standard ElGamal encryption, recovering message x from an additively homomorphic ElGamal encryption $[\![x]\!]$ thus involves the computation of a discrete logarithm in $\langle g \rangle$, which is in general hard. However, if x is known to be from a limited set of values, one may still recover x efficiently (e.g., if $0 \le x < u$, one may find x by a straightforward $O(u)$ linear search, or by Pollard's $O(\sqrt{u})$ lambda method [36]).

A strongly related example is Pedersen's commitment scheme [35] in which commitments are computed as

$$[\![x]\!] = h^r g^x,$$

where r is chosen uniformly at random in $\mathbb{Z}_q$. Hence, Pedersen commitments are just elements of $\langle g \rangle$, and the homomorphic property clearly holds. Strictly speaking, $[\![x]\!]$ is not an encryption in the sense that it completely determines x (knowing g and h). The idea is that by publishing the value of $[\![x]\!]$, one is committed to the value of x meaning that one can at a later stage open $[\![x]\!]$ in only one way by revealing the pair (r, x) (assuming that it is infeasible for the committer to compute the discrete log $\log_g h$): indeed, knowledge of any $(r, x) \ne (r', x')$ with $h^r g^x = h^{r'} g^{x'}$ implies that one knows $\log_g h = (x' - x)/(r - r')$ as well. Also, for a given value of $[\![x]\!] = h^r g^x$ and an hypothesized value x' for the committed message there exists exactly one $r' \in \mathbb{Z}_q$ such that $[\![x]\!] = h^{r'} g^{x'}$, which implies that even with unlimited computational power one is not able to extract any information on x from $[\![x]\!]$. The value of x is thus said to be unconditionally (or information-theoretically) hidden.

Threshold decryption is another feature, which is particularly useful when constructing voting schemes. Briefly, for **threshold decryption** the private key of a public key cryptosystem is generated in such a way that each of ℓ parties obtains a share of the private key such that any t parties can cooperate to jointly decrypt a given encryption. The critical security property is that the private key is never revealed to anyone as part of the decryption protocol (nor as part of the key generation protocol). See Chapter 13 of this book for more details on threshold cryptography, where it is shown how threshold decryption can be realized, starting from Shamir's threshold secret-sharing scheme.

Voting schemes rely on threshold decryption to prevent that a single entity or single master key suffices to decrypt votes. As shown in [34], there is also an efficient way to generate the key pair for discrete log-based cryptosystems, such as ElGamal.

15.3.4 Verification Primitives

The category of verification primitives provide some further functionality that is not covered by the primitives discussed so far. A **zero-knowledge proof** is the main example of a verification primitive, which allows a prover to convince a skeptical verifier of the validity of a given statement without revealing any information beyond the truth of the statement. The theory of zero-knowledge proofs tells us that any so-called NP-statement can actually be proved in zero-knowledge (see, e.g., [21]). We will see some concrete zero-knowledge proofs when discussing particular voting schemes later on; these proofs are generally noninteractive, which means that the proof is produced by the prover on its own, and may then be verified independently by multiple verifiers (much like a digitally signed message can be verified multiple times, by different verifiers).

A **designated verifier proof** is a special type of noninteractive zero-knowledge proof, which is convincing only to a single verifier, specified by the proof [26]. The idea is to refer to a particular verifier V by using its public key pk_V, say, and to provide a noninteractive zero-knowledge proof for a statement of the form "I know the private key sk_V for public key pk_V or statement Φ holds." This proof will not be convincing to anyone else than verifier V, because only V is sure that the prover did not know sk_V, hence statement Φ must actually hold. Note that this approach works only if the verifier properly protects its private key sk_V. Designated verifier proofs are typically used to achieve some level of receipt-freeness in voting schemes (see Section 15.6).

15.4 Classification of Voting Schemes

The design and the analysis of electronic voting systems is a truly complex endeavor, which is approached in many different ways, revolving around many types of voting schemes. The literature on voting schemes is vast, and consists mostly of rather informal conference papers. Finding a useful, pragmatic way to group the known and yet unknown voting schemes is therefore a challenging task.

In the following text, we will identify a few characteristics, which are all formulated in terms of building blocks used and other properties, but without reference to which security properties hold (hence classification can be done by inspecting the description of the voting system rather than performing a security analysis). Based on these characteristics we present a very brief taxonomy distinguishing two major types of voting schemes.

15.4.1 Characteristics

A basic characteristic is whether the voting protocol requires interaction between voters. Almost every voting protocol avoids interaction such that voters may cast their votes independently and in parallel to each other. However, see, e.g., [28] for a voting protocol in which a cryptographic counter is updated in turn by each of the voters. Interaction may also be limited, e.g., to all voters belonging to a given precinct, say. In this chapter, we focus on voting protocols without interaction between voters, which is the common case.

The main technical characteristics we use are derived from the contents and the properties of the voted ballot. The voted ballot (ballot, for short) consists of the information recorded by the voting servers (or machines), upon completion of the voting protocol by a voter. As such, the voted ballot is the interface between the voting stage and the tallying stage: it comprises all the information that is created during the voting protocol as a function of the voters' votes, and which is needed for successful tallying after the election closes.

A major characteristic is whether the voted ballot is anonymous vs. identifiable. Normally, an anonymous voted ballot is submitted via an anonymous channel preventing any link with the voter's identity. For identifiable voted ballots, there is no need for an anonymous channel. A subtle point is that even if voted ballots are anonymous, there may be further information available from the voting stage (possibly, in combination with information from other stages) that helps in identifying voters. For instance, the IP address from which a vote is cast may help in identifying a voter, but an IP address is in general not part of the voted ballot.

Another important characteristic is whether the voted ballot contains the vote in the clear (derivable from public information only), or in encrypted form. If the vote is in the clear, then intermediate election results will be available, which is usually not allowed. Encryption of votes in some form is always needed for the voted ballots to prevent that intermediate election results can be computed. Thus, to count the votes based on the voted ballots at least one secret key should be required and preferably shared between several parties.

As a final characteristic, we look at how votes (or, voted ballots) are authenticated, and in particular if authentication is done in an identifiable way or an anonymous way. Proper authentication should ensure that each voter casts at most one vote.

15.4.2 A Brief Taxonomy

For the purpose of this chapter, we will just consider two characteristics: whether voted ballots are anonymous or not, and how voted ballots are authenticated. As for the other characteristics we will simply assume that the voting protocol requires no interaction between voters, and further that voted ballots are not in the clear, hence encrypted (or committed) in some way.

The main question is how authentication and encryption of voted ballots are accomplished. Given that a voted ballot contains the vote in encrypted form, we consider three basic possibilities: (1) outer authentication (a signed encrypted vote), (2) inner authentication (an encrypted signed vote), or (3) both outer and inner authentications (a signed encrypted signed vote). Outer authentication can be checked as part of the voting protocol (and also later), and invalid voted ballots can be discarded immediately. Inner authentication can only be checked during the tallying stage (and also later) after the decryption of the votes.

The main distinction is anonymous vs. identifiable voted ballots.

15.4.2.1 Type I: Anonymous Voted Ballots

An anonymous voted ballot, as stored upon completion of the voting protocol, contains no information that helps in identifying the voter. Clearly, an anonymous channel should be used for the delivery of such voted ballots to prevent that the voting server or anyone monitoring communications may correlate voted ballots with voters or their computers.

The voted ballots should be authenticated to ensure that each eligible voter is able to cast at most one valid vote. Anonymous signatures are used for this purpose, both for outer authentication and inner authentication.

15.4.2.2 Type II: Identifiable Voted Ballots

For identifiable voted ballots, the voted ballot may be delivered in a nonanonymous way, e.g., posted to a bulletin board. For outer authentication one can use ordinary digital signatures (possible relying on a PKI). For inner authentication, if present, anonymous signatures need to be used, however, as this signature will typically be revealed in conjunction with the vote (thus the case of inner authentication only does not occur as this corresponds to an anonymous voted ballot).

15.4.3 Examples

Well-known examples of type I schemes are [17] and follow-up papers, such as [25,32,33]. In these examples, voters need to get a blind signature (preferably, a threshold blind signature, where the power of the issuer is shared among a number of parties) prior to the election, which is then used to authenticate an encrypted vote (preferably, a threshold encryption, such that the power to decrypt is shared among a number of parties). An anonymous channel is used to sent such an anonymously authenticated encrypted votes to the voting server, which is supposed to publish these data objects on a bulletin board. Once the election closes, the votes may be counted by threshold decrypting each of the voted ballots with valid authentication.

Well-known examples of Type II schemes are the homomorphic schemes of [13] and follow-up work such as [2,4,15,16,39,41], and the mix-based schemes, such as [19,31,40]. In these schemes, the voter may directly post a digitally signed encrypted vote to a bulletin board. We will consider these schemes in more detail in the next sections, emphasizing the universal (public) verifiability property of these schemes.

However, type II schemes are not necessarily universally verifiable. This depends on the details of the vote encryptions and the tally protocols. A limited level of verifiability may, for instance, be achieved by letting each voter encrypt its vote v together with a random bit string R of sufficient length resulting in an encrypted vote of the form $[\![v, R]\!]$. (This idea was already used in an early voting protocol by Merritt [30].) The encrypted votes are mixed before decryption takes place, however, without providing a zero-knowledge proof of validity (unlike in verifiable mixing, see the following text). The tallied votes are published together with the random strings so that voters can check their votes. A major drawback of this approach is that all voters are required to actively check their votes, which may be unrealistic. Moreover, it is not clear how to resolve disputes and in particular how to maintain ballot secrecy when resolving disputes: for example, how should a voter efficiently prove (in zero-knowledge) that its pair (v, R) is not present among the millions of tallied votes?

15.5 Verifiable Elections

We will now focus on two main approaches for achieving universally verifiable elections. We will show how electronic elections can be made universally (or, publicly) verifiable in much the same way as digital signatures are. Namely, to verify a digital signature all that one needs are (a) the public key of the signer, (b) a message, and (c) a purported signature. By applying a given predicate, represented as a formula or algorithm, to these three data values, it is decided unequivocally whether the signature is valid or not (w.r.t. the given message and public key). Moreover, the verification can be repeated by anyone who is interested and as many times as deemed necessary, always giving the same result.

Similarly, to verify an election one needs (a) the joint public key of the talliers, (b) the encrypted votes, and (c) the election result. By applying a given predicate to these data values, it can now be decided unequivocally whether the election result is valid or not. Admittedly, the predicate for electronic elections is a bit more involved than for digital signatures, but the underlying principle is the same: the predicate evaluates to true (or, accept) if and only if the election result corresponds exactly to the votes, which are given in an encrypted form only.

The digital signature analogy also works when considering the use of the private keys. To produce the election result, the talliers need to use their (shared) private key similar to a signer using its private key to produce a digital signature. Namely, once tallying is completed, resulting in a valid election result, there is no need to repeat this process ever. As such, the purpose of a recount is lost! However, as with digital signatures, the verification of the election result is universal meaning that it can be performed by anyone and repeated as many times as desired.

The difficulty with verifiable elections is of course that ballot secrecy must be maintained as well—at all times. Since we are considering a type II election, it is known which voter produced which encrypted vote (because the encrypted votes are digitally signed). The hard cryptographic problem that needs to be solved is thus formulated as follows. Suppose a threshold homomorphic cryptosystem $[\![\cdot]\!]$ (cf. Section 15.3) has been set up between a plurality of parties (talliers) $P_1, \ldots, P_\ell$. We wish to design a protocol TALLY satisfying a specification of the following form:

Protocol TALLY
Input: $[\![v_1]\!], \ldots, [\![v_n]\!]$ + Validity Proofs
Output: $f(v_1, \ldots, v_n)$ + Validity Proof

Here, each $[\![v_i]\!]$ denotes a probabilistically encrypted vote v_i. The election result is defined as some function f of these votes. A common example is $f(v_1, \ldots, v_n) = \sum_{i=1}^n v_i$ for yes/no votes, where $v_i = 1$ means "yes" and $v_i = 0$ means "no". But, function f may potentially hide even more information on the votes, e.g., $f(v_1, \ldots, v_n) \equiv \sum_{i=1}^n v_i > n/2$, indicating whether there is a strict majority or not (but revealing nothing else, like how small or large the majority is). A possibility at the other extreme is $f(v_1, \ldots, v_n) = (v_1, \ldots, v_n)$, where all votes are simply revealed in the same order as in the input.

Each encrypted vote $[\![v_i]\!]$ is required to be accompanied by a validity proof, which is commonly used to show that v_i is within a given range. In addition, the validity proof must provide a cryptographic link with the voter's identity, which we will denote by VID_i. By this way one prevents encrypted votes getting duplicated by other voters (without actually knowing the contents of the duplicated votes).

We will consider two types of tallying in more detail: homomorphic tallying and mix-based tallying. Both approaches are built using a threshold homomorphic cryptosystem, but in different ways. The challenge in designing the protocols is to find efficient ones, and in particular, to find efficient ways of rendering universally verifiable validity proofs.

15.5.1 Homomorphic Tallying

The basic premise of homomorphic tallying is that individual votes can somehow be added in a meaningful way. Hence, the election result takes the following form:

$$f(v_1, \ldots, v_n) = \sum_{i=1}^n v_i.$$

The addition of votes may be defined in various ways. The case of yes/no votes corresponds to integer addition of votes $v_i \in \{0, 1\}$. But, for example, elections based on approval voting in a race with m candidates also fit this pattern: votes $v_i \in \{0, 1\}^m$ are binary m-tuples, which are added componentwise. This allows each voter to indicate whether they approve of each candidate or not, and candidates with the highest numbers of vote (approvals) win. As a further example, consider voting on a scale, where voters express how strong they (dis)favor a given proposition by using votes $v_i \in \{-2, -1, 0, 1, 2\} \simeq \{--, -, \circ, +, ++\}$. Using integer addition to compute $\sum_{i=1}^n v_i$, one can determine how much, on average, the electorate (dis)favors the proposition.

A protocol for homomorphic tallying can in general be described as follows: one simply multiplies together all encrypted votes v_i which—by the homomorphic property of the cryptosystem— results in an encryption of the sum of the votes. Protocol TALLY thus consists of the following steps:

1. Check the validity proof of every $[\![v_i]\!]$, and discard all invalid ones.
2. Multiply all the valid encryptions: $C = \prod_i [\![v_i]\!] = [\![\sum_{i=1}^n v_i]\!]$.
3. Jointly decrypt C to obtain $T = \sum_i v_i$ plus a validity proof.

The validity proof in the final step shows that T is indeed the plaintext contained in C implying that $T = \sum_i v_i$. The validity proofs for the given encryptions $[\![v_i]\!]$ prevent malicious voters from entering multiple votes. For instance, in a binary (yes/no) election, voters are supposed to enter an encryption $[\![0]\!]$ or $[\![1]\!]$, but not encryptions like $[\![2]\!]$ (twice "yes") or $[\![-7]\!]$ (seven times "no"). As long as $0 \le T \le n$, the presence of such illegal encryptions may remain unnoticed. A validity proof for $[\![v_i]\!]$ shows that $v_i \in \{0, 1\}$ without revealing any further information on v_i.

To verify the election result T, anyone interested executes the following steps:

1. Check that encryption C is equal to the product of all valid encryptions $[\![v_i]\!]$.
2. Check the validity proof showing that T is the threshold-decryption of C.

The validity proofs are checked against the public key of the threshold homomorphic cryptosystem $[\![\cdot]\!]$.

As a concrete example, we consider the simplest case of the voting scheme of [16], namely, the case of yes/no votes in combination with additively homomorphic ElGamal encryption. The knowledge of the public key h, say, of the ElGamal threshold homomorphic cryptosystem, as set up by the talliers, suffices for encryption as well as validation of the votes. As explained in Section 15.3, the encryption of a vote $v \in \{0, 1\}$ takes the following form:

$$[\![v]\!] = (A, B) = (g^r, h^r g^v) \tag{15.1}$$

for a uniformly random value $r \in \mathbb{Z}_q$.

Since such an encryption $[\![v]\!]$ can easily be formed for any value $v \in \mathbb{Z}_q$, a validity proof is required to show that either $v = 0$ or $v = 1$ without revealing which of the two cases holds. In other words, the proof must be zero-knowledge. Technically, the validity proof for $[\![v]\!]$ amounts to showing the existence of a $v \in \{0, 1\}$ and an $r \in \mathbb{Z}_q$ such that $A = g^r$ and $B = h^r g^v$, which can be reformulated by stating that either $\log_g A = \log_h B$ holds (if $v = 0$) or $\log_g A = \log_h(B/g)$ holds (if $v = 1$).

Let $\mathcal{H}$ denote a suitable cryptographic hash function. Applying the theory of [14] to the Chaum-Pedersen protocol for proving the equality of discrete logarithms [12], the noninteractive zero-knowledge validity proof is a 4-tuple (d_0, r_0, d_1, r_1) computed as follows:

$$
\begin{aligned}
& d_{1-v} \in_R \mathbb{Z}_q \\
& r_{1-v} \in_R \mathbb{Z}_q, \\
& d_v = \mathcal{H}(a_0, b_0, a_1, b_1, A, B, \text{VID}_i) - d_{1-v} \pmod{q}, \\
& r_v = w - r d_v \pmod{q},
\end{aligned}
$$

where

$$
\begin{aligned}
& a_{1-v} = g^{r_{1-v}} A^{d_{1-v}}, \quad b_{1-v} = h^{r_{1-v}}(B/g^{1-v})^{d_{1-v}}, \\
& a_v = g^w, \quad b_v = h^w, \quad \text{for } w \in_R \mathbb{Z}_q.
\end{aligned}
$$

A proof (d_0, r_0, d_1, r_1) is valid for encryption (A, B) and voter identity VID_i w.r.t. public key h if and only if

$$d_0 + d_1 = \mathcal{H}(g^{r_0} A^{d_0}, h^{r_0} B^{d_0}, g^{r_1} A^{d_1}, h^{r_1}(B/g)^{d_1}, A, B, \text{VID}_i) \pmod{q}. \tag{15.2}$$

Note that the order in which a_0, b_0, a_1, b_1 are computed when generating the proof is determined by the value of v but the verification of the proof is done independently of v.

In the random oracle model (which means that $\mathcal{H}$ is viewed as an ideal hash function, selected uniformly at random from all functions of the appropriate type), it can be proved that these validity proofs release no information on the vote v. The voter's identity VID_i (a bit string unique to the voter) is included as one of the inputs to $\mathcal{H}$ in order to bind the proof to a particular voter. This prevents voters from duplicating other voters' votes (copying both the encryption and the validity proof); even though voters would not know the value of the votes they are duplicating, the basic requirement of independence would be violated. For the same reason, no partial election results should be available during an election (and it is usually not even allowed to announce the results of exit polls during election day).

So, in principle, the talliers need to perform a single joint decryption only. For more advanced elections, such as approval voting and voting on a scale (see previous text), vote encryptions and the validity proofs get a bit more complicated; e.g., approval voting for a race with m candidates can be done by running m instances of a yes/no election (one per candidate) in parallel.

15.5.2 Mix-Based Tallying

In mix-based tallying the basic idea is to mimic the use of a ballot box in traditional paper-based elections. Once the election is closed, or each time a ballot is inserted into the ballot box, the ballot box is shaken so as to remove any dependency on the order in which the ballots were actually inserted during the election.

Translated to our setting, the election result takes the following form:

$$f(v_1, \ldots, v_n) = (v_{\pi(1)}, \ldots, v_{\pi(n)})$$

for a permutation π. To ensure ballot secrecy, permutation π will be generated uniformly at random between the talliers by means of the following TALLY protocol:

1. Check the validity proof of every $[\![v_i]\!]$, and discard all invalid ones.
2. The talliers take turns in verifiably mixing the list of encrypted votes resulting in a final list of permuted votes $[\![v'_1]\!], \ldots, [\![v'_n]\!]$.
3. Jointly decrypt every $[\![v'_i]\!]$ to obtain v'_i plus a validity proof.

Clearly, if at least one of the talliers is honest, the final list will indeed be randomly permuted. The validity proof in the first step is simply a proof of plaintext knowledge proving that the contents of the encryption was known to the voter upon casting the vote. This prevents vote duplication, which may be particularly harmful when mix-based tallying is used: by duplicating the encrypted vote of voter X, say, once or several times, and by checking for duplicates in the output of the TALLY protocol one may learn what X voted.

The final step is the same as for homomorphic tallying except that decryption must be done for each encrypted vote separately. To verify the election result one performs these steps:

1. Check that the list of valid encryptions $[\![v_i]\!]$ is computed correctly.
2. Check the validity proof for each mix performed by a tallier.
3. Check the validity proof for each decrypted permuted vote.

As a concrete example, we consider verifiable mixing of homomorphic ElGamal encryptions. On input of a list of encryptions $C = (c_1, \ldots, c_n) = ([\![m_1]\!], \ldots, [\![m_n]\!])$, one sets

$$d_i = [\![0]\!] \, c_{\pi(i)} \tag{15.3}$$

$i = 1, \ldots, n$, for a random permutation π. The output is the list of encryptions $D = (d_1, \ldots, d_n)$. Here, each occurrence of $[\![0]\!] = (g^{r_i}, h^{r_i})$ denotes a probabilistic encryption of 0 using its own random value $r_i \in \mathbb{Z}_q$, and due to the homomorphic property, $[\![0]\!] c_{\pi(i)}$ and $c_{\pi(i)}$ both contain plaintext $m_{\pi(i)}$. Therefore, $[\![0]\!] c_{\pi(i)}$ is called a **random reencryption** of $c_{\pi(i)}$. Combined with randomly permuting the encryptions it is ensured that one cannot tell which entry of the output list corresponds to which entry of the input list—of course, relying on the semantic security of the ElGamal cryptosystem. Since encryptions are never decrypted (only reencrypted) access to the private decryption key is not required for mixing.

To make mixing verifiable, it must be proved (in zero-knowledge) that the lists C and D represent the same multiset of plaintexts. Efficient zero-knowledge proofs for this task are quite advanced. To give some intuition we first describe a simple but rather inefficient way ([40], based on similarity with zero-knowledge proofs for graph isomorphism). The prover generates an additional list E computed in the same way as D was computed from C, using a fresh permutation and randomness. Then a challenge bit is given, which indicates whether the prover should show how E corresponds to C or to D (by revealing which permutation and which randomness transforms list E into the requested one). It is not difficult to see that by this way no information is leaked on how D exactly corresponds to C. However, if D is not a valid mix of C, then E can correspond to C or to D but not to both at the same time! So, the probability of being caught is at least $1/2$. By repeating this protocol k times, cheating will remain undetected with probability at most 2^{-k}.

The computational complexity of this protocol is rather high, namely $O(kn)$, modular exponentiations, for n voters and security parameter k. With more advanced techniques, however, it is possible to achieve a computational complexity of $O(n)$ modular exponentiations.

Neff [31] starts by stating that the polynomial $c(x) = \prod_{i=1}^{n}(x - m_i)$ corresponding to the input list C should be identical to the polynomial $d(x) = \prod_{i=1}^{n}(x - m_{\pi(i)})$ corresponding to the output list D. A key observation is that these polynomials are defined over the message space Z_q, where q is a very large prime (e.g., $q \approx 2^{256}$), so to prove that $c(x) = d(x)$ it suffices to evaluate the polynomial $c(x) - d(x)$ at a random value $x' \in_R Z_q$: if $c(x) \neq d(x)$, the probability that $c(x') - d(x') = 0$ is bounded by n/q, which is negligibly small (as the number of roots of $c(x) - d(x)$ is at most its degree n). Neff uses this property in the design of an efficient verifiable mix requiring $O(n)$ modular exponentiations only.

Furukawa and Sako [19] approach the problem in terms of permutation matrices. The relation (Equation 15.3) between the input and output encryptions can be written as

$$d_i = [\![0]\!] \prod_{j=1}^{n} c_j^{A_{ij}}$$

for an $n \times n$ permutation matrix A. The key property for their zero-knowledge proof is that A is a permutation matrix if and only if for all $1 \leq i, j, k \leq n$

$$\sum_{h=1}^{n} A_{hi}A_{hj} = \delta_{ij}, \quad \sum_{h=1}^{n} A_{hi}A_{hj}A_{h,k} = \delta_{ij}\delta_{jk},$$

where δ_{ij} denotes the Kronecker delta ($\delta_{ij} = 1$ if $i = j$, and $\delta_{ij} = 0$ otherwise). The entries of A are numbers modulo q, for a prime q, hence A is defined over a finite field. Furukawa and Sako show how to prove that A satisfies these conditions, without leaking any further information on A, and without requiring more than $O(n)$ time (counting modular exponentiations).

Follow-up papers, such as by Groth [22] and by Furukawa [18], improve upon Neff's protocol and Furukawa-Sako's protocol, respectively. For instance, [18] uses the observation that if $q \neq 1$ (mod 3), then A is a permutation matrix if and only if $\sum_{h=1}^{n} A_{hi}A_{hj}A_{h,k} = \delta_{ij}\delta_{jk}$, which leads to some improvements under a mild restriction on q. All these verifiable mixes use $O(n)$ modular exponentiations and $O(nk)$ bits communication, where the hidden constant is reasonably small (say, around 5, not counting the modular exponentiations for the random reencryptions).

The mixes described so far are known as reencryption mixes. So-called decryption mixes have also been considered in the literature. In a decryption mix (e.g., see [18]), the mixing stage and the decryption stage are combined such that each tallier needs to be active only once: each tallier uses its share of the private key to partially decrypt the elements on the input list and at the same time mixes the list as before. The zero-knowledge proofs for decryption mixes are slightly more complicated than the proofs for reencryption mixes, but decryption mixes may lead to an overall performance gain. Reencryption mixes, however, are more generally applicable, and in the context of universally verifiable elections are more flexible, as the roles of mixers and talliers can be decoupled.

For a basic decryption mix, one assumes that each mix will actually contribute to the decryption. In other words, one uses (t, ℓ)-threshold decryption with $t = \ell$. To tolerate faulty or corrupt mixes, one can instead use (t, ℓ)-threshold decryption with $t < \ell$. However, the set of t mixes going to perform the decryption must be decided upon before the mixing starts; and one needs to backup and let all mixes do some extra work if any mix drops out.

15.5.3 Other Tallying Methods

If applicable, homomorphic tallying is very simple and efficient. However, the complexity of the validity proof for each vote $[\![v_i]\!]$ may become the bottleneck for increasingly complex votes (e.g., single transfer voting [STV] where each vote v_i is an ordered list of candidates may lead to prohibitively expensive validity proofs). For mixed-based tallying, the complexity is hardly sensitive to the range of possible values for votes v_i; here, the bottleneck is that each of the talliers must operate in turn handling $O(n)$ encryptions.

By using techniques from secure multiparty computation, however, other methods for tallying are conceivable, which are potentially more efficient or flexible. For example, an alternative to mixed-based tallying is sorting-based tallying:

$$f(v_1, \ldots, v_n) = (w_1, \ldots, w_n),$$

where $w_1 \leq w_2 \leq \cdots \leq w_n$ and $\{v_1, \ldots, v_n\} = \{w_1, \ldots, w_n\}$ (as a multiset). Hence, the votes are output in sorted order destroying any link with how the votes were entered. A protocol along these lines uses secure comparators, which put two encrypted input values in sorted order (e.g., on input $[\![a]\!]$, $[\![b]\!]$, output $[\![0]\!][\![\min(a, b)]\!]$, $[\![0]\!][\![\max(a, b)]\!]$). Each secure comparator may be implemented as a joint protocol between the talliers, involving one or more joint decryptions (see, e.g., [20]). Sorting-based tallying is then done by arranging these secure comparators in a sorting network. Practical sorting networks such as odd-even mergesort and bitonic sort, are of depth $O(\log^2 n)$ with each layer consisting of $O(n)$ comparators where the hidden constants are small (see, e.g., [29, Section 5.3.4]).

15.6 Receipt-Freeness and Incoercibility

So far, we have considered the problem of maintaining ballot secrecy during the tallying stage, where it must be prevented that anyone is able to find out who voted what. This protects against large scale fraud, where a small number of people directly involved in running the election (insiders such as, election officials, talliers, auditors, equipment manufacturers) compromise ballot secrecy of many, potentially targeted individual voters. The severity of such a large scale, centralized violation of ballot secrecy is exacerbated by the fact that it can remain completely unnoticed, without leaving any traces in the operational systems. For example, if vote encryption depends on a single (nonshared) master key, a copy of the master key suffices to read the contents of any encrypted vote. The existence of such copies of the master key cannot be excluded—in a verifiable way—and in fact some copies may be required as backups.

So, no one should be able to find out what voters voted, except the voters themselves! A tautology, but nevertheless a serious issue, because many remote voting schemes actually allow voters to tell others what they voted for in a verifiable way. This leads to problems known as voter coercion and vote trading, where voters either unwillingly or willingly give up the rights to their vote.

Whereas centralized violations of ballot secrecy by insiders, as mentioned earlier, may remain completely unnoticed by the voters, it is clear that voters will be completely aware of any (attempts to) voter coercion or vote trading. This means that large scale coercion is risky as the chances of detection are high: any of the affected voters involved may leak to the press what happened. And, e.g., internet sites facilitating vote trading will be hard to hide as well.

A technical aspect is whether a voter is able to convince the coercer or vote buyer of having voted in a particular way, i.e., whether a voter is able to produce a receipt proving how it voted. A voting scheme is said to be receipt-free if voters cannot produce such receipts. Note that the ideal voting scheme mentioned in Section 15.2 was indeed receipt-free as the voters tell their vote to the trusted party in private. In the following we will see to what extent receipt-freeness can be achieved for cryptographic voting schemes. The property of incoercibility is even stronger than receipt-freeness, and is mostly beyond the scope of cryptographic techniques: e.g., a coercer may force a voter not to vote at all.

15.6.1 Coercion in Paper-Based Elections

For postal voting, it is hard to rule out that others watch someone casting a vote, willingly or unwillingly. This leads to problems such as family voting where the ballots are marked by the dominating family members. However, even when people vote in person from a voting booth at a polling station, coercion is still possible as voters may produce receipts in subtle ways—of course, these days people should be asked to hand in their mobile phones, spy cams, etc., before entering the (Faraday cage like) voting booth, and this should be checked using similar equipments as for airport security.

Two well-known attacks illustrating the problem of coercion in paper-based elections are as follows. Chain voting is a simple attack, which makes the essential use of the fact that spare ballot forms are (and should be) not readily available. To mount a chain voting attack, the coercer must first get hold of a ballot form. Next, the coercer marks this ballot form in the desired way, and hands the premarked ballot form to a voter, upon entering the polling station, and the voter is kindly asked to return with an empty ballot form. The empty ballot form serves a receipt convincing the coercer that the voter put the premarked ballot form in the ballot box and may be used to repeat the attack.

A so-called Italian attack is applicable when ballot forms allow voters to cast their votes each in a unique way. Sometimes this may be done by marking the paper ballot in a particular way using recognizable marks. A more subtle way, which also works for electronic voting, applies when voters are supposed to rank the candidates (by listing them in the preferred order on the ballot form), such as in single transfer voting (STV) elections. In that case, the coercer will demand the intended winner to be put at the top followed by the remaining candidates in an order unique for the targeted voter. Later, the coercer will check if the requested ballot appears on the list of voted ballots.

15.6.2 Receipt-Freeness for Encrypted Votes

The secrecy of a probabilistic public key encryption actually depends on two factors: the secrecy of the private decryption key held by the receiver and the secrecy of the randomness used by the sender. Hence the necessity of secure (pseudo)random generators both during key generation and encryption. But what if the sender willingly discloses the randomness?

For instance, if ElGamal encryption is used to encrypt a vote v as $[\![v]\!] = (A, B) = (g^r, h^r g^v)$, then revealing the random value r to a coercer proves that the vote contained in $[\![v]\!]$ is equal to v. Indeed, the coercer first checks that $A = g^r$ holds, and if so, checks that $b = h^r g^v$ holds as well. In other words, the value of r serves as a receipt. Clearly, this is true for probabilistic public key cryptosystems in general: the randomness used to compute an encryption can be used as a receipt.

Obviously, for revealing r to the coercer the voter needs to have access to this value. If a software voting client is used, running on a platform controlled by the voter, such as a PC or a smart phone, it will be quite easy to extract the value of r. In fact, since the voting protocol is known (and preferably based on an openly available specification), the voter (or the coercer, for that matter) may program its own voting client with the additional feature of copying the value of r to the output. Hence, a way to make the voting protocol receipt-free is to execute it on a platform beyond the control of the voter, such as a smart card or a direct-recording electronic (DRE) voting machine, with the immediate drawback that the voter is not ensured anymore that its vote is recorded as intended and the additional problem that the voter must trust the smart card or the voting machine with its vote.

A more refined approach is to use what it is known as a randomizer, an idea reminiscent of the use of the wallet-with-observer paradigm due to Chaum (see Chapter 44 of first edition). A randomizer, which may be implemented as a smart card or as a part attached to a voting machine, runs a protocol with a voter to jointly produce an encrypted vote together with a validity proof. To this end, the voter and the randomizer communicate with each other over an untappable channel.

The randomizer will generate part of the randomness used to encrypt the vote and to generate the validity proof. The basic idea is to let the randomizer produce a random reencryption of the voter's encryption, and modify the validity proof (as generated by the voter) accordingly. So, the voter starts out by sending a homomorphic encryption $[\![v]\!]$ of its vote v to the randomizer over the untappable channel. The randomizer will then reply with a reencryption of the form $[\![0]\!][\![v]\!]$ accompanied by a designated verifier proof for the validity of the reencryption. This proof is only meaningful to the voter as it is bound to the public key of the voter. The tricky part is how to divert the validity proof of the voter for $[\![v]\!]$ to a validity proof for $[\![0]\!][\![v]\!]$. This is done by first jointly generating a proof for $[\![v]\!]$ and then adjusting it to a proof for $[\![0]\!][\![v]\!]$.

As an example we apply Hirt's technique [23, Chapter 5 of *Algorithm and Theory of Computation Handbook, Second Edition: General Concepts and Techniques*] to the yes/no voting protocol described in the previous section. The diverted proof looks as follows. Let $[\![v]\!] = (A, B) = (g^r, h^r g^v)$ denote the voter's encryption of vote v. The randomizer reencrypts this to $(A'', B'') = (g^{r'}, h^{r'})(A, B)$, where $r' \in_R Z_q$. As before, the voter will contribute a proof (d_0, r_0, d_1, r_1) for (A, B) satisfying (Equation 15.2), except that this time the challenge hash is applied to appropriately blinded versions of (a_0, b_0, a_1, b_1). This results in a proof $(d_0'', r_0'', d_1'', r_1'')$ for the validity of encryption (A'', B'') satisfying (Equation 15.2). The contribution of the randomizer to the proof consists of $d_0', d_1', r_0', r_1' \in Z_q$ subject to $d_0' + d_1' = 0$, such that

$$d_0'' = d_0 + d_0', \quad d_1'' = d_1 + d_1'$$

$$r_0'' = r_0 + r_0' - r' d_0'', \quad r_1'' = r_1 + r_1' - r' d_1''.$$

Note that the randomizer does not learn what vote v is. And, the voter does not know how to open (A'', B'') as the value of r' remains hidden to the voter. And, even if the coercer forces the voter to compute (A, B) and the proof (d_0, r_0, d_1, r_1) in a particular way, it is still possible to simulate the protocol transcript where the voter actually uses a different vote.

A randomizer may be implemented as a tamper-resistant smart card. The voter's software client will communicate with the smart card privately. Voters are required to use the randomizer, which can be ensured by a digital signature of the randomizer on the reencrypted vote. The smart cards should be beyond the reach of the coercers, and one needs to assume that the smart cards use good randomness. If the voter has a choice between several randomizers, all of these must be noncoerced, otherwise the coercer will force the voter to use a coerced one.

The use of randomizers can be seen as an add-on solution. One may decide per election, and also per category of voters, whether and how many randomizers are required. As large scale vote trading or coercion may be hard to hide anyway, one needs to weigh the use of cryptographic measures, such as randomizers, against other measures, such as severe punishments.

15.7 Untrusted Voting Clients

A basic assumption in the voting schemes described earlier is that each voter has a PC, a mobile phone, or any other computer device at its disposal, which can be used as a voting client. The voting protocol is assumed to be implemented on such a device, and, given an open specification of the voting protocol, voters may in principle build their own implementations (or, pick one from many vendors who comply with the open specification).

In many elections, such as traditional elections from polling stations, there is no possibility for voters to use their own computer devices and software. Instead, one needs to trust equipment provided at a polling station, e.g., running software that cannot be monitored. This raises the question whether, e.g., verifiable elections are still possible in such a setting.

Similarly, for paper-based elections verifiability is limited: election observers must continuously monitor all proceedings and trace exactly what happens to each of the paper ballots. There is no reliable way for checking whether one's voted ballot is actually included in the final tally. Also, using numbered ballot forms and giving each voter a copy of the marked ballot form may violate ballot secrecy and is clearly not receipt-free. More advanced techniques are needed to ensure in a verifiable way that each and every legitimately cast paper ballot—and no other ones—will be tabulated.

Much recent work has thus been focused on addressing the issue of ensuring that votes are cast as intended, complementing the work on verifiable tally protocols, which ensures that votes are counted as cast. Work in this direction was initiated by Chaum (see, e.g., [10] and references therein) and by Neff (see, e.g., [1] and references therein, incl. "Practical High Certainty Intent Verification for Encrypted Votes" by Neff, 2004).

To illustrate the ideas behind these new approaches, we briefly describe the common idea behind a range of proposals by Randell and Ryan (see, e.g., [37] and references therein). Consider an election in which voters must pick one candidate from a list of candidates. The idea is that each paper ballot is perforated vertically with the left-hand side containing the names of the candidates in a random order and the right-hand side containing an empty box for each candidate. To cast a vote, the voter will (1) mark the box of the candidate of its choice, (2) tear the ballot form into two parts, (3) destroy the left-hand half, and (4) drop the right-hand half into a ballot box—possibly after receiving a copy of this part. Since the candidates were listed in a random order on the left-hand half, the marked right-hand half does not reveal which candidate was chosen.

The ballot forms are numbered, though, in such a way that each right-hand half can be linked to an encryption of information indicating how the marked right-hand half should be interpreted. Cryptographically, the technique is similar to the use of masked ballots as introduced in [39], and used in [16] as well for yes/no votes. Similar to Equation 15.1, one precomputes an encryption of the form:

$$[\![b]\!] = (g^r, h^r g^b), \tag{15.4}$$

for a uniformly random value $r \in \mathbb{Z}_q$, and a uniformly random bit $b \in_R \{0, 1\}$. Depending on the value of b, the ballot form will list the options in the order no-yes or yes-no. Together with the position of the mark on the marked ballot, encryption $[\![b]\!]$ is then publicly transformed into an encryption $[\![v]\!]$ of the intended vote v. The encrypted votes $[\![v]\!]$ may then be tallied in a verifiable way, as described earlier.

The production of the ballot forms and the associated encryptions forms a critical part of these voting systems. Encryptions, such as (Equation 15.4), must be produced in a private yet secure way such that they correspond to the printed ballot forms, but without revealing the contents of these encryptions. Thus, this problem is best viewed as a (nontrivial) problem in secure multiparty computation as well. The challenge is to achieve efficient solutions for these problems.

These new approaches are still under scrutiny and several issues need to be addressed. For instance, even though a receipt (e.g., a copy of the marked right-half) does not reveal the vote, such a receipt nevertheless opens a possibility for coercion because of the so-called randomization attacks due to Schoenmakers who noted this issue first for the voting scheme of [24] (see also [27]). In a randomization attack, a coercer will simply specify for each voter to mark a particular box on the right-hand half—regardless of which candidate this box corresponds to. The coercer can easily check whether a voter complied to the specified rule (e.g., "mark the box at the bottom of the form"). Since the candidates are listed in a random order, different on each ballot form, the coercer thus forces a voter to effectively cast a random vote, which may have a deciding influence on an election.

15.8 Further Information

As described in the introduction, research in voting schemes developed in three waves, starting in the 1980s, 1990s, and 2000s, respectively. The first wave started from within the emerging cryptographic community, and many research papers on electronic voting can thus be found in the cryptographic literature, which is described extensively in Chapters 9–13 and 17, of this book and Chapter 44 of first edition. Information on the practical deployment of electronic voting schemes can be found on the Internet since the mid-1990s when the first experiments started. Interest in both theoretical and practical aspects of electronic voting has been growing ever since, and especially since "Florida 2000" there is also a lot of political and societal interest as well.

Research in electronic voting is shaping up witnessed by a flurry of workshops devoted to the topic. Researchers from various backgrounds now meet on a regular basis. Examples of workshops are those under the flag of the IAVOSS (see www.iavoss.org) including WOTE and other workshops, the EVT workshops colocated with the USENIX Security Symposium, and the EVOTE workshops organized by e-voting.cc.

Traditionally, many countries and organizations work on their own voting systems and implementations, and the business has been limited to a relatively small number of companies. The market is opening up though. A major impetus to this process will be the development of open standards capturing the main voting schemes as described in this chapter. In particular, when focusing on type II schemes with identifiable voted ballots, a lot of flexibility regarding the network delivery of the voted ballots (e.g., by email, by https, over a cable TV network, . . .) and regarding the outer authentication of the voted ballots will be possible, whereas the encryption of the votes will be prescribed by the standards. Implementations of the key generation, voting, tally, and verification protocols adhering to these standards may then be offered by a multitude of suppliers, incl. noncommercial ones.

Even though electronic voting will remain controversial for years to come—and remote voting even more so—it's the author's opinion that inevitably remote electronic voting will start to be used for major elections in the coming decades. With the advance of communication and computation technologies, instant large scale elections in which everyone votes online within a time frame of one hour (or one minute?) will be possible. And, how else would one vote in virtual communities, or in games like Second Life? Future generations may have an innate distrust of paper-based elections, once paper becomes something archaic, used only by magicians or other tricksters.

Defining Terms

Anonymous channel: A communication channel that hides the identity of the sender, and possibly allows for an acknowledgment by the receiver as well.

Blind signature: A signature scheme where the signer cannot recognize the message-signature pairs resulting from the signing protocol.

Bulletin Board: A publicly readable broadcast channel, possibly with authenticated write operations.

Computational security: If a security property holds subject to a computational intractability assumption.

Designated verifier proof: A noninteractive zero-knowledge proof bound to a particular public key that only convinces the holder of the corresponding private key.

Homomorphic encryption: A public key cryptosystem in which the product of two encryptions is equal to an encryption of the sum of the corresponding plaintexts.

Information theoretic security: If a security property holds regardless of the attacker's computational power.

Private channel: A communication channel protected against eavesdropping, and possibly providing end-to-end authentication as well, achieving information-theoretic security.

Secure channel: A communication channel protected against eavesdropping, and possibly providing end-to-end authentication as well, achieving computational security.

Threshold decryption: A public key cryptosystem in which decryption requires the participation of a certain number of parties, each holding a share of the private key.

Untappable channel: A totally unobservable (out-of-band) communication channel.

Validity proof: A zero-knowledge proof showing the validity of an input, or step performed in a protocol.

Zero-knowledge proof: A proof for a statement without giving away why it is true.

References

1. B. Adida and C.A. Neff. Ballot casting assurance. In *Proceedings of the USENIX/Accurate Electronic Voting Technology (EVT) Workshop 2006*, Berkeley, CA, 2006.
2. O. Baudron, P.-A. Fouque, D. Pointcheval, J. Stern, and G. Poupard. Practical multi-candidate election system. In *Proceedings of the 20th ACM Symposium on Principles of Distributed Computing (PODC '01)*, pp. 274–283, New York, 2001. A.C.M.
3. M. Ben-Or, S. Goldwasser, and A. Wigderson. Completeness theorems for non-cryptographic fault-tolerant distributed computation. In *Proceedings of the 20th Symposium on Theory of Computing (STOC '88)*, pp. 1–10, New York, 1988. A.C.M.
4. J. Benaloh. Verifiable secret-ballot elections. PhD thesis, Yale University, Department of Computer Science Department, New Haven, CT, September 1987.
5. J. Benaloh and D. Tuinstra. Receipt-free secret-ballot elections. In *Proceedings of the 26th Symposium on Theory of Computing (STOC '94)*, pp. 544–553, New York, 1994. A.C.M.
6. S. Brands. Untraceable off-line cash in wallet with observers. In *Advances in Cryptology—CRYPTO '93*, volume 773 of *Lecture Notes in Computer Science*, pp. 302–318, Berlin, Germany, 1994. Springer-Verlag.
7. S. Canard, B. Schoenmakers, M. Stam, and J. Traoré. List signature schemes. *Discrete Applied Mathematics*, 154:189–201, 2006. Special issue on Coding and Cryptology.
8. D. Chaum. Untraceable electronic mail, return addresses, and digital pseudonyms. *Communications of the ACM*, 24(2):84–88, 1981.
9. D. Chaum. Blind signatures for untraceable payments. In D. Chaum, R.L. Rivest, and A.T. Sherman, editors, *Advances in Cryptology—CRYPTO '82*, pp. 199–203, New York, 1983. Plenum Press.
10. D. Chaum. Secret ballot receipts: True voter-verifiable elections. *IEEE Security & Privacy*, 2(1):38–47, Jan.–Feb. 2004.
11. D. Chaum and E. van Heyst. Group signatures. In *Advances in Cryptology—EUROCRYPT '91*, volume 547 of *Lecture Notes in Computer Science*, pp. 257–265, Berlin, Germany, 1991. Springer-Verlag.
12. D. Chaum and T.P. Pedersen. Wallet databases with observers. In *Advances in Cryptology—CRYPTO '92*, volume 740 of *Lecture Notes in Computer Science*, pp. 89–105, Berlin, Germany, 1993. Springer-Verlag.

13. J. Cohen and M. Fischer. A robust and verifiable cryptographically secure election scheme. In *Proceedings of the 26th IEEE Symposium on Foundations of Computer Science (FOCS '85)*, pp. 372–382, Portland, OR, 1985. IEEE Computer Society.

14. R. Cramer, I. Damgård, and B. Schoenmakers. Proofs of partial knowledge and simplified design of witness hiding protocols. In *Advances in Cryptology—CRYPTO '94*, volume 839 of *Lecture Notes in Computer Science*, pp. 174–187, Berlin, Germany, 1994. Springer-Verlag.

15. R. Cramer, M. Franklin, B. Schoenmakers, and M. Yung. Multi-authority secret ballot elections with linear work. In *Advances in Cryptology—EUROCRYPT '96*, volume 1070 of *Lecture Notes in Computer Science*, pp. 72–83, Berlin, Germany, 1996. Springer-Verlag.

16. R. Cramer, R. Gennaro, and B. Schoenmakers. A secure and optimally efficient multi-authority election scheme. In *Advances in Cryptology—EUROCRYPT '97*, volume 1233 of *Lecture Notes in Computer Science*, pp. 103–118, Berlin, Germany, 1997. Springer-Verlag.

17. A. Fujioka, T. Okamoto, and K. Ohta. A practical secret voting scheme for large scale elections. In *Advances in Cryptology—AUSCRYPT '92*, volume 718 of *Lecture Notes in Computer Science*, pp. 244–251, Berlin, Germany, 1992. Springer-Verlag.

18. J. Furukawa. Efficient, verifiable shuffle decryption and its requirement of unlinkability. In *Public Key Cryptography—PKC '04*, volume 2947 of *Lecture Notes in Computer Science*, pp. 319–332, Berlin, Germany, 2004. Springer-Verlag.

19. J. Furukawa and K. Sako. An efficient scheme for proving a shuffle. In *Advances in Cryptology—CRYPTO '01*, volume 2139 of *Lecture Notes in Computer Science*, pp. 368–387, Berlin, Germany, 2001. Springer-Verlag.

20. J. Garay, B. Schoenmakers, and J. Villegas. Practical and secure solutions for integer comparison. In *Public Key Cryptography—PKC '07*, volume 4450 of *Lecture Notes in Computer Science*, pp. 330–342, Berlin, Germany, 2007. Springer-Verlag.

21. O. Goldreich. *Foundations of Cryptography—Basic Tools*. Cambridge University Press, Cambridge, U.K., 2001.

22. J. Groth. A verifable secret shuffle of homomorphic encryptions. In *Public Key Cryptography—PKC '03*, volume 2567 of *Lecture Notes in Computer Science*, pp. 145–160, Berlin, Germany, 2003. Springer-Verlag.

23. M. Hirt. Multi-party computation: Efficient protocols, general adversaries, and voting. PhD thesis, ETH ZURICH, Constance, Germany, 2001.

24. M. Hirt and K. Sako. Efficient receipt-free voting based on homomorphic encryption. In *Advances in Cryptology—EUROCRYPT '00*, volume 1807 of *Lecture Notes in Computer Science*, pp. 539–556, Berlin, Germany, 2000. Springer-Verlag.

25. P. Horster, M. Michels, and H. Petersen. Blind multisignature schemes and their relevance for electronic voting. In *Proceedings of the 11th Annual Computer Security Applications Conference*, pp. 149–155. New Orleans, LA, 1995. IEEE Computer Society.

26. M. Jakobsson, K. Sako, and R. Impagliazzo. Designated verifier proofs and their applications. In *Advances in Cryptology—EUROCRYPT '96*, volume 1070 of *Lecture Notes in Computer Science*, pp. 143–154, Berlin, Germany, 1996. Springer-Verlag.

27. A. Juels, D. Catalano, and M. Jakobsson. Coercion resistant electronic elections. In *Proceedings of the 2005 ACM Workshop on Privacy in the Electronic Society (WPES 2005)*, pp. 61–70, Alexandria, VA, 2005. A.C.M.

28. J. Katz, S. Myers, and R. Ostrovsky. Cryptographic counters and applications to electronic voting. In *Advances in Cryptology—EUROCRYPT '01*, volume 2045 of *Lecture Notes in Computer Science*, pp. 78–92, Berlin, Germany, 2001. Springer-Verlag.

29. D.E. Knuth. *The Art of Computer Programming (volume 3: Sorting and Searching)*, 2nd edn. Addison Wesley, Reading, MA, 1998.

30. M. Merritt. Cryptographic protocols. PhD thesis, Georgia Institute of Technology, Atlanta, GA, February 1983.

31. C.A. Neff. A verifiable secret shuffle and its application to e-voting. In *Proceedings of the 8th ACM Conference on Computer and Communications Security 2001*, pp. 116–125, New York, 2001. ACM Press.

32. M. Ohkubo, F. Miura, M. Abe, A. Fujioka, and T. Okamoto. An improvement on a practical secret voting scheme. In M. Mambo and Y. Zheng, editors, *ISW '99*, volume 1729 of *Lecture Notes in Computer Science*, pp. 225–234, Berlin, Germany, 1999. Springer-Verlag.

33. T. Okamoto. Receipt-free electronic voting schemes for large scale elections. In *Security Protocols. 5th International Workshop Proceedings*, pp. 25–35, Berlin, Germany, 1997. Springer-Verlag.

34. T. Pedersen. A threshold cryptosystem without a trusted party. In *Advances in Cryptology— EUROCRYPT '91*, volume 547 of *Lecture Notes in Computer Science*, pp. 522–526, Berlin, Germany, 1991. Springer-Verlag.

35. T.P. Pedersen. Distributed provers and verifiable secret sharing based on the discrete logarithm problem. PhD thesis, Aarhus University, Computer Science Department, Aarhus, Denmark, March 1992.

36. J.M. Pollard. Monte Carlo methods for index computation (mod p). *Mathematics of Computation*, 32(143):918–924, 1978.

37. B. Randell and P.Y.A. Ryan. Voting technologies and trust. *IEEE Security & Privacy*, 4(5):50–56, Sept.–Oct. 2006.

38. R. Rivest, A. Shamir, and Y. Tauman. How to leak a secret. In *Advances in Cryptology—ASIACRYPT '01*, volume 2248 of *Lecture Notes in Computer Science*, pp. 552–565, Berlin, Germany, 2001. Springer-Verlag.

39. K. Sako and J. Kilian. Secure voting using partially compatible homomorphisms. In *Advances in Cryptology—CRYPTO '94*, volume 839 of *Lecture Notes in Computer Science*, pp. 411–424, Berlin, Germany, 1994. Springer-Verlag.

40. K. Sako and J. Kilian. Receipt-free mix-type voting scheme—A practical solution to the implementation of a voting booth. In *Advances in Cryptology—EUROCRYPT '95*, volume 921 of *Lecture Notes in Computer Science*, pp. 393–403, Berlin, Germany, 1995. Springer-Verlag.

41. B. Schoenmakers. A simple publicly verifiable secret sharing scheme and its application to electronic voting. In *Advances in Cryptology—CRYPTO '99*, volume 1666 of *Lecture Notes in Computer Science*, pp. 148–164, Berlin, Germany, 1999. Springer-Verlag.

16

Auction Protocols

Vincent Conitzer
Duke University

The word "auction" generally refers to a mechanism for allocating one or more resources to one or more parties (or **bidders**). Generally, once the allocation is determined, some amount of money changes hands; the precise monetary transfers are determined by the auction process. While in some auction protocols, such as the English auction, bidders repeatedly increase their bids in an attempt to outbid each other, this is not an essential component of an auction. There are many other auction protocols, and we will study some of them in this chapter.

Auctions have traditionally been studied mostly by economists. In recent years, computer scientists have also become interested in auctions, for a variety of reasons. Auctions can be useful for allocating various computing resources across users. In artificial intelligence, they can be used to allocate resources and tasks across multiple artificially intelligent "agents." Auctions are also important in electronic commerce: there are of course several well-known auction Web sites, but additionally, search engines use auctions to sell advertising space on their results pages. Finally, increased computing power and improved algorithms have made new types of auctions possible—most notably **combinatorial** auctions, in which multiple items are for sale in the same auction, and bidders can bid on bundles of items.

We begin this chapter by studying single-item auctions. Even though most computer scientists are perhaps more interested in combinatorial auctions, single-item auctions allow us to more easily introduce certain concepts that are also of key importance in combinatorial auctions.

16.1   Standard Single-Item Auction Protocols

In this section, we review some basic protocols for auctioning a single item. The reader is encouraged to think about which of these protocols are similar to each other; we will discuss relationships among them shortly.

English. The English auction is the most familiar protocol to most people. In an English auction, every bidder is allowed to place a bid higher than the current highest bid. If at some point, no bidder wishes to place a higher bid, then the bidder with the current highest bid wins the item, and pays her bid.

Dutch. The Dutch auction proceeds in the opposite direction from the English auction. In a Dutch auction, an initial price is set that is very high, after which the price is gradually decreased. At any moment, any bidder can claim the item. She then wins the item and has to pay the current price.

Japanese. In a Japanese auction, the initial price is zero; the price is then gradually increased. A bidder can leave the room when the price becomes too high for her. Once there is only one bidder remaining, that bidder wins the item, and pays the price at which the last other bidder left the room.

First-price sealed-bid. In a first-price sealed-bid auction, each bidder communicates a bid privately to the auctioneer—say, in a sealed envelope. The auctioneer then opens all the envelopes; the bidder with the highest bid wins the item, and pays the bid that she placed.

Second-price sealed-bid (also known as **Vickrey**). The second-price sealed-bid auction proceeds exactly as the first-price sealed-bid auction, except the highest bidder (who still wins the item) now pays the second-highest bid, instead of her own.

Let us consider which of these auction protocols are similar to each other. Perhaps the most obvious similarity is between the English and the Japanese auctions. For both, there is a price that is rising, and the last remaining bidder wins. There is a distinction, however: in an English auction, two bidders may be bidding each other up, while a third bidder quietly sits by, even though she remains interested in the item. In this case, the first two bidders are unaware that they have another competitor. In a Japanese auction, this situation cannot occur.

The Japanese auction and the second-price sealed-bid auction are also closely related (and the English auction is related to the second-price sealed-bid auction in a similar way). Suppose, for a second, that each bidder in a Japanese auction decides at the beginning of the auction on the price at which she will leave the room. Of course, other strategies are possible: a bidder may base how long she stays in the room on which other bidders are still left. However, *if* the bidders follow the former kind of strategy, then the bidder who, at the beginning, chose the highest price will end up winning, and she will end up paying the second-highest price selected by a bidder—similarly to the second-price sealed-bid auction.

The Dutch auction and the first-price sealed-bid auction are even more closely related. Similarly to the Japanese auction, in a Dutch auction, a bidder may decide at the beginning on the price at which she will claim the item (if this price is reached). In fact, unlike in the Japanese auction, there is little else that a bidder can do in terms of strategizing. In a Japanese auction, a bidder can let her bidding strategy depend on who else is left; but in a Dutch auction, there is nothing to condition her strategy on, since the only event that can happen is that someone else claims the item—but at that point the auction is over and it no longer matters what anyone does. Now, the bidder who chooses the highest price will win, and pay that price—similarly to the first-price sealed-bid auction. Because of this argument, the Dutch and first-price sealed-bid auctions are usually considered strategically equivalent.

We will see some other single-item auctions later in this chapter. For most of this chapter, we will focus on sealed-bid auctions. As we have seen, for each of the auctions studied so far, there is a roughly equivalent sealed-bid auction; in a sense, this is true for *any* auction, as we will see shortly. Nevertheless, there are reasons to use English, Japanese, and Dutch auctions (more generally, **ascending** and **descending** auctions). One reason is that they allow bidders to postpone certain decisions until later. For example, if a bidder in a Dutch auction is deciding whether to claim the item at $60 or $50, she may as well wait until the price drops to, say, $70, before she starts to think about what she will do. If another bidder claims the item before that, the former bidder will have saved herself some unnecessary agonizing. This is related to **preference elicitation**, which we will discuss toward the end of this chapter.

16.2 Valuations and Utilities

How a bidder should bid in an auction depends on how much the item for sale is worth to her. In this chapter, we will assume that each bidder can determine how much the item is worth to her, and that events in the auction will not change her assessment. That is, each bidder i has an unchanging **valuation** v_i for the item, which she can determine at the beginning of the auction. This assumption is not always realistic. For example, if a bidder sees that other bidders are bidding aggressively on an item, this may be evidence to her that, before the auction, those bidders inspected the item in person and found it to be of good quality. This evidence may improve the first bidder's perception of—and hence, valuation for—the item. Settings such as these, where some bidders have private information that would affect the valuation of other bidders for the item, are known as **interdependent valuations** settings. Most research assumes away the possibility of interdependent valuations, and we will do so in this chapter.

In general, each bidder has a **utility** for each outcome of the auction, and acts to maximize her expected utility. We will assume that a bidder's utility for winning the item and having to pay π_i is $u_i = v_i - \pi_i$, and her utility for not winning the item and having to pay π_i is $u_i = -\pi_i$. (Generally, losing bidders will not be made to pay anything; however, later in this chapter, we will see an auction that makes payments to losing bidders, in which case π_i is negative.) Thus, we are assuming that a bidder's utility function decomposes into separate valuation and payment components, and that utility is linear in money. This assumption is known as the **quasilinear preferences** assumption. It, too, is not always realistic, for the following reasons. In general, one's utility may be strictly concave in money (that is, one may have **decreasing marginal utility** for money: the utility of having another (say) dollar may decrease as one accumulates more money), since at some point one runs out of uses for money. Also, in general, the effect of money on utility may depend on whether one has won the item: for example, if a bidder wins a pair of skis in an auction, she needs money to travel on a skiing vacation (and hence has high marginal utility for money), whereas if she does not win, she has less use for additional money (and hence has low marginal utility for money). Nevertheless, the quasilinearity assumption is usually made, and we will do so in this chapter.

Another assumption that is implicit in the above text is that a bidder who does not win the item does not care about which other bidder wins the item, and that bidders do not care about how much other bidders pay. This assumption is known as the **no externalities** assumption. Once again, this assumption is not always realistic: a bidder may prefer to see the item end up with a friend rather than with an enemy, or she may prefer to see the other bidders run out of money so that they will not compete in future auctions. Again, we will not go into detail on this in this chapter.

16.3 Strategic Bidding

It does not always make sense for a bidder to simply bid her true valuation. For example, if bidder i bids her true valuation v_i in a first-price sealed-bid auction, then even if she wins, her utility will be $v_i - \pi_i = v_i - v_i = 0$. Hence, she should bid lower than her true valuation to have any chance of obtaining positive utility. But how much lower should she bid? Intuitively, this should depend on her beliefs about the other bidders' valuations. If she expects to be the only bidder who is seriously interested in obtaining the item, she can place a low bid and still probably win; whereas if she expects there to be many other competitive bidders, she should bid closer to her true valuation to have a decent chance of winning. However, it is not obvious how to calculate her optimal bid precisely, even given a probability distribution over the others' valuations. This is because she cannot expect the other bidders to bid their true valuations, either: they also need to bid below their true valuations to have a chance of obtaining positive utility. And precisely how much lower is optimal for them to bid depends, in turn, on how the first bidder bids.

16.3.1 Solving the First-Price Sealed-Bid Auction

To resolve this circularity, we need to turn to **game theory**, which studies settings in which each bidder's (or, more generally, **agent's**) optimal course of action depends on the actions of the other bidders. To apply game theory to the first-price sealed-bid auction, we first need to introduce the concept of a **strategy**. In a sealed-bid auction, a strategy for bidder i is a function $s_i : \mathbb{R}^{\geq 0} \to \mathbb{R}^{\geq 0}$, where $s_i(v_i)$ is the bid that i will place if her true valuation is v_i. That is, for *every* valuation that the bidder may have, the strategy specifies what she should bid. This may appear somewhat excessive: if the bidder already knows her true valuation, why should she have to specify what she would have bid if her valuation had been different? The reason that we need to think about this is that the *other* bidders do not know bidder i's valuation, and *they* need to think about what i would do for each valuation. In turn, bidder i needs to think about what the other bidders will do, and hence also needs to think about what they think she will do.

Let us suppose that for each bidder i, there is a (commonly known) prior probability distribution p_i over her valuation v_i; moreover, let us assume that the valuations are drawn independently. Hence, each bidder i knows her own valuation v_i exactly, but for every other bidder $j \neq i$, i's probability distribution over j's valuation is p_j. Now, *if* bidder i knows the strategies of the other bidders, then for every bid that she might place, she can evaluate her expected utility; and of course she should choose one that maximizes her expected utility. As is typically done in game theory, we will look for an **equilibrium**, which prescribes a strategy for every bidder such that, for every bidder, for every possible valuation for that bidder, her strategy will prescribe a bid that maximizes her expected utility, given the other strategies. Formally, a **Bayes–Nash equilibrium** consists of a strategy $s_i : \mathbb{R}^{\geq 0} \to \mathbb{R}^{\geq 0}$ for every bidder such that, for every bidder i, for every $v_i \in \mathbb{R}^{\geq 0}$, and for every alternative bid $\hat{v}_i \in \mathbb{R}^{\geq 0}$:

$$\int_{v_{-i}} \left(\prod_{j \neq i} p_j(v_j) \right) u_i(v_i, s_{-i}(v_{-i}), s_i(v_i)) dv_{-i} \geq \int_{v_{-i}} \left(\prod_{j \neq i} p_j(v_j) \right) u_i(v_i, s_{-i}(v_{-i}), \hat{v}_i) dv_{-i}$$

Let us dissect this complicated inequality. First, the notation $-i$ is shorthand for "the bidders other than i," so that v_{-i} is shorthand for $v_1, \ldots, v_{i-1}, v_{i+1}, \ldots, v_n$. The notation $\hat{v}_i$ is generally used for bidder i's bid, not necessarily equal to her true valuation v_i. $u_i(v_i, \hat{v}_{-i}, \hat{v}_i)$ is the utility that bidder i obtains if her true valuation is v_i, but she bid $\hat{v}_i$, and the other bidders bid $\hat{v}_{-i}$. In the first-price sealed-bid auction, $u_i(v_i, \hat{v}_{-i}, \hat{v}_i) = v_i - \hat{v}_i$ if $\hat{v}_i$ is higher than all the bids in $\hat{v}_{-i}$, and it is 0 otherwise. Now we can see that the inequality says that, if the bidders other than i follow their strategies, then i's expected utility for bidding $s_i(v_i)$ should be at least equal to her expected utility for bidding any other $\hat{v}_i$—that is, she should not be able to do better by not following the strategy s_i, given that the other bidders are indeed following their strategies s_{-i}.

We will now give an example of an equilibrium. Suppose that each p_i is a uniform distribution over $[0, 1]$. We will show that the strategies defined by $s_i(v_i) = v_i(n - 1)/n$ (where n is the number of bidders) constitute an equilibrium. (We will not get into detail here on how one might have actually *derived* these strategies, but there are techniques for doing so.) Suppose that all other bidders $(-i)$ indeed follow these strategies. Then, the expected utility for i of bidding $\hat{v}_i \leq (n-1)/n$ is $(\hat{v}_i n/(n-1))^{n-1}(v_i - \hat{v}_i)$, because the probability that a given other bidder bids less than $\hat{v}_i$ is $\hat{v}_i n/(n-1)$. (There is no reason to bid more than $(n-1)/n$, because no other bidder will bid more than $(n-1)/n$.) Using simple calculus, one can check that this expression is maximized by setting $\hat{v}_i$ equal to $v_i(n-1)/n$—exactly as the strategy prescribes! This proves that these strategies indeed constitute an equilibrium.

16.3.2   Solving the Second-Price Sealed-Bid Auction

Now, let us turn to the second-price sealed-bid auction. As it turns out, the analysis needed to solve this auction is not nearly as complicated. In fact, in the second-price sealed-bid auction, it is always optimal for a bidder to bid her true valuation, regardless of the other bids! That is, the strategy $s_i(v_i) = v_i$ is a **dominant strategy**. While this may come as a surprise at first, it is not so difficult to see why it is true. Suppose, for a second, that bidder i can actually see the others' bids before placing her own bid. Let us consider the value $\hat{v}_{max}$, the highest bid among the other bidders. Bidder i effectively has only two choices: to bid higher than $\hat{v}_{max}$, and obtain utility $v_i - \hat{v}_{max}$; or to bid lower than $\hat{v}_{max}$, and obtain utility 0. Clearly, she should do the former if and only if $v_i > \hat{v}_{max}$. But this is exactly what would happen if she just bid her true valuation—for which she does not even need to know the others' bids! Hence, by bidding her true valuation, she performs as well as she could have performed even if she had known the others' bids. This implies that bidding truthfully is also a Bayes–Nash equilibrium, although the result is much stronger than that. For example, in the first-price auction, if one knows the bids of the other bidders, then certainly one might be better off bidding differently from the equilibrium that we derived for that auction (which we derived under the assumption that bidders do not know each other's valuations). Hence, the strategies in the first-price auction equilibrium are not dominant strategies. Mechanisms in which revealing one's true valuation is a dominant strategy (such as the second-price sealed-bid auction) are called (**dominant-strategies**) **incentive compatible**, **strategy-proof**, or simply **truthful**.

16.4   Revenue Equivalence

Now that we have analyzed how bidders should bid in these two auctions, let us ask the following question: which one obtains more revenue for the seller, in expectation? The answer is not immediately obvious: naïvely, one might say that the first-price auction should result in more revenue, since after all it charges the highest bid rather than the second-highest; but then again, in equilibrium, the bids are lower in the first-price auction. Which of these two effects is stronger?

For the case of independent uniform priors over $[0, 1]$, we can compute the expected revenues using the equilibrium strategies from above text. For the first-price auction, the probability that all bids are below a given value b is $(bn/(n-1))^n$, which is also the probability that the revenue will be below b. That is, this expression gives the cumulative density function of the revenue of the first-price auction, and using it one can compute the expected revenue to be $(n-1)/(n+1)$. For the second-price auction, the probability that there is at most one bid higher than b is $b^n + nb^{n-1}(1-b)$, which is also the probability that the revenue will be below b. That is, this expression gives the cumulative density function of the revenue of the second-price auction, and using it one can compute the expected revenue to be $(n-1)/(n+1)$—the same as that for the first-price auction! This is no accident: it is a special case of the following result, which is known as the **revenue equivalence theorem** (Myerson, 1981; Riley and Samuelson, 1981).

THEOREM 16.1   *Suppose that the bidders' valuations are independent and identically distributed over a continuous interval [L, H], and that there are no "gaps" in this distribution. Then, any two auction mechanisms that*

1. *in equilibrium always allocate the item to the bidder with the highest valuation, and*
2. *give a bidder with valuation L an expected utility of 0,*

will result in the same expected revenue for the seller.

(There are more general versions of this result.) In the next section, we will see single-item auction mechanisms that result in different expected revenues, because they violate one of the two conditions in the theorem. From this point on, we will study only truthful mechanisms. This is justified by a result known as the **revelation principle** (Gibbard, 1973; Green and Laffont, 1977; Myerson, 1979, 1981), which states (roughly) that, if bidders bid strategically, then for every mechanism that is not a truthful mechanism, there is a truthful mechanism that performs equally well.

16.5   Auctions with Different Revenues

Suppose that we have a prior distribution over each bidder's valuation, and that we wish to design a mechanism that maximizes expected revenue. It is easy to see that running, say, a second-price sealed-bid auction is not always optimal. For example, suppose there is only one bidder. The second-price sealed-bid auction will never collect any revenue in this case, because there is no second bidder. However, we can also make a **take-it-or-leave-it offer** to the one bidder: the bidder will obtain the item if and only if it is worth more than some fixed value k to her, in which case she will pay k; otherwise, the seller keeps the item. This will generate a revenue of k at least some of the time. While this may not seem like an auction, setting a **reserve price** of k in a second-price sealed-bid auction will have the same effect. (In such an auction, if only one bid is above the reserve price, then that bid pays the reserve price.)

In general, the auction that maximizes expected revenue is known as the **Myerson auction** (Myerson, 1981), and it proceeds as follows. For each bidder i, compute her **virtual valuation** $\psi_i(\hat{v}_i)$ as a function of her bid, as follows:

$$\psi_i(\hat{v}_i) = \hat{v}_i - (1 - F_i(\hat{v}_i))/f_i(\hat{v}_i)$$

Here, F_i is the cumulative density function of i's valuation, and f_i is its derivative, the probability density function. The bidder with the highest virtual valuation wins, unless this bidder has a virtual valuation below 0, in which case nobody wins. The price that the winning bidder pays is the lowest value that she could have bid while still winning. For example, if each bidder's valuation is drawn from the uniform distribution over [0, 1], then the Myerson auction becomes a second-price sealed-bid auction with a reserve price of 1/2. This is because

$$\psi_i(1/2) = 1/2 - (1 - F_i(1/2))/f_i(1/2) = 1/2 - (1 - 1/2)/1 = 0$$

It does not always make sense to try to maximize expected revenue. In some settings, our main goal is to allocate the item efficiently (that is, to the bidder that values it most), and payments are merely a necessary nuisance in achieving this goal. For example, suppose that several parties jointly own an item, and they wish to run an auction amongst themselves to decide on a single owner. What should happen to the revenue of this auction? It seems to make sense to redistribute it back to the bidders themselves, but doing so effectively changes the auction mechanism. For example, suppose that the bidders run a second-price sealed-bid auction for the item, and then redistribute the revenue of this auction equally (each bidder receives $1/n$ of the revenue). Unlike the second-price sealed-bid auction without redistribution, this auction is actually *not* truthful: the second-highest bidder now has an incentive to increase her bid to drive up the price that the highest bidder pays, because the second-highest bidder will receive a fraction of this price.

Fortunately, it turns out that we can redistribute at least some of the revenue while maintaining truthfulness. One auction mechanism that achieves this is the following [independently invented on at least three different occasions (Bailey, 1997; Porter et al., 2004; Cavallo, 2006)]. Let us define v^2_{-i} to be the second-highest bid among bidders *other than* bidder i. For the top two bidders, this is the third-highest bid overall (v^3); for the remaining $n - 2$ bidders, it is the second-highest bid (v^2). We run the second-price sealed-bid auction, and we redistribute v^2_{-i}/n to bidder i. This redistribution payment does not affect bidders' incentives in bidding, because no bidder can affect her own redistribution payment. Hence, the auction remains truthful. Additionally, the total redistributed is $2v^3/n + (n - 2)v^2/n \leq v^2$—so the total redistributed is no more than is collected from the second-price sealed-bid auction. A total of $2(v^2 - v^3)/n$ is not redistributed. This money must be given to someone else (but not someone whom the bidders care about, since that might affect their incentives), or, say, burned. It is impossible to achieve efficient allocation without ever wasting any money, but it is possible to waste even less money (either on average or in the worst case): this is achieved by also letting bidder i's redistribution payment depend on $v^3_{-i}, v^4_{-i}, \ldots, v^{n-1}_{-i}$ (Guo and Conitzer, 2007; Moulin, 2007).

It is interesting to note that the revenue equivalence theorem from above text does not apply to Myerson's auction because the first condition is not satisfied; it does not apply to the redistribution mechanisms because the second condition is not satisfied.

16.6 Complementarity and Substitutability

Now that we have studied single-item auctions, let us consider settings where multiple items are for sale. One possibility is to sell each item in a separate single-item auction; these auctions can be held simultaneously (**parallel** auctions) or back-to-back (**sequential** auctions). If, for each item, each bidder's valuation for that item does not depend on which other items she wins, then the individual auctions are entirely separate events, and we can apply the techniques that we have studied up to this point. However, this is not always a realistic assumption. For example, if the items for sale are a plane ticket to Las Vegas and a hotel reservation in Las Vegas, then it may be that the bidder's valuation for the plane ticket alone is 200, her valuation for the hotel reservation alone is 100, but her valuation for both together is 500. The package of both items is worth more than the sum of its parts, that is, the items are **complementary**.

Now suppose that the plane ticket is auctioned first, and the hotel reservation second (both in second-price sealed-bid auctions). How much should the bidder bid in the first auction? If she bids 200 for the ticket, she may lose to a bidder bidding 201, only to later find out she could have won the hotel reservation for 101, so that she regrets not bidding higher in the first auction. However, if she bids 400 for the ticket, she may win it at a price of 399, only to later find out that the hotel reservation sells for 1000, so that she regrets winning the ticket. It is not clear what the bidder should do—she no longer has a dominant strategy. Moreover, the resulting allocation of items can be inefficient.

Another possibility is that the package of items is worth *less* than the sum of its parts, in which case the items are said to be **substitutable**. For example, if reservations for two different hotels are for sale, a bidder may value each individual reservation at 100, but the package of both reservations at 150. In sequential auctions, substitutability can cause problems similar to those caused by complementarity. Both can also cause similar problems in parallel auctions.

Instead of making the bidders agonize over the prices at which items in future or parallel auctions are likely to sell, an alternative is to let each bidder report *all* her valuations, one for each subset of the items, and decide on the allocation of items to bidders based on that information. This is what is done in a combinatorial auction, and it circumvents the problems that parallel and sequential auctions run into when there are complementarities and substitutabilities.

16.7 Combinatorial Auctions

In a **combinatorial auction**, a set I of multiple items is (simultaneously) for sale, and bidders can bid on any **bundle** (that is, subset) of items. If we again make the assumptions that bidders' valuations for the items do not change based on other bidders' private information, utilities are quasilinear, and there are no externalities, then each bidder i has a privately held valuation function $v_i : 2^I \rightarrow \mathbb{R}^{\geq 0}$, where $v_i(S)$ is i's valuation for bundle $S \subseteq I$; and the utility of bidder i when she wins bundle S and pays π_i is $v_i(S) - \pi_i$. Generally it is assumed that $v_i(\emptyset) = 0$, and additionally that for $S \subseteq S'$, $v_i(S) \leq v_i(S')$. (The latter assumption is often called **free disposal**: receiving additional items can never decrease a bidder's valuation, because at worst the additional items can simply be discarded.) We will start by looking at sealed-bid combinatorial auctions. An immediate problem with this approach is that in general, each bidder must reveal $2^m - 1$ real numbers (where $m = |I|$), one for each nonempty bundle. Once m gets to be somewhat large, this becomes impractical. However, there is usually some structure in the bidders' valuation functions, so that they can be represented more concisely.

One very restrictive, but commonly studied assumption about this structure is that bidders are **single-minded**. A bidder i is single-minded if there exists some bundle S_i and some real number v_i such that $v_i(S) = v_i$ if $S_i \subseteq S$, and $v_i(S) = 0$ otherwise. That is, there is a single bundle of items that the bidder wishes to obtain; she will simply discard any additional items, and if she fails to obtain even one item within her desired bundle, her valuation drops to 0. If bidders are single-minded, then a bid can be represented simply as an ordered pair $(S_i, v_i) \in 2^I \times \mathbb{R}^{\geq 0}$.

A single-minded bid cannot represent even fairly straightforward valuation functions, such as **additive** valuation functions. (A valuation function v_i is additive if for all $S \subseteq I$, $v_i(S) = \sum_{s \in S} v_i(\{s\})$. That is, there are no complementarities or substitutabilities.) We would like to give bidders some more flexibility, by providing them with a richer **bidding language** in which to describe their valuation function. One such bidding language is the **OR language**, which effectively allows a bidder to submit multiple single-minded bids. Formally, a bid in the OR language takes the form (S_1, v_1) OR (S_2, v_2) OR $\ldots$ OR (S_k, v_k). Such a bid is interpreted as follows: for any subset $T \subseteq \{1, \ldots, k\}$ with the property that for any $j_1, j_2 \in T, j_1 \neq j_2$, we have $S_{j_1} \cap S_{j_2} = \emptyset$, $v_i(\bigcup_{j \in T} S_j) = \sum_{j \in T} v_j$. That is, the auctioneer can accept any subcollection of the single-minded bids within the OR-bid, as long as there is no overlap between the accepted S_j. (To be precise, the last $=$ symbol should really be a $\geq$ symbol, because it is possible that the same set $\bigcup_{j \in T} S_j$ (or a subset thereof) can be written as a union of disjoint S_j in a different way that results in a greater sum of v_j. The bidder's valuation for the subset is the maximum value that can be obtained in this way. Hence, equality holds only if there is no better way to write the bundle as a union of disjoint S_j. For example, given the bid $(\{A\}, 1)$ OR $(\{A, B\}, 3)$ OR $(\{B, C\}, 3)$ OR $(\{C\}, 2)$, we have $v_i(\{A, B, C\}) = v_i(\{A, B\}) + v_i(\{C\}) = 5 > 4 = v_i(\{A\}) + v_i(\{B, C\})$.) The OR language allows for representing additive valuations, simply by OR-ing together singleton bundles.

Nevertheless, there are valuation functions that the OR language cannot capture. For example, suppose there are two items, a and b, and that bidder i's valuation function is given by $v_i(\{a\}) = 1, v_i(\{b\}) = 1, v_i(\{a, b\}) = 1$. That is, she wants either item, and having both items is no more useful to her than having a single one. This function cannot be represented in the OR language: the bid would have to contain the terms $(\{a\}, 1)$ and $(\{b\}, 1)$, but this would already imply that $v_i(\{a, b\}) \geq 2$. A language that *can* capture this valuation function is the **XOR language**. The difference between the OR and XOR languages is that the auctioneer can accept at most *one* of the single-minded bids that are XORed together, even if they do not overlap. For example, the above valuation function is easily expressed as $(\{a\}, 1)$ XOR $(\{b\}, 1)$: this bid implies a valuation of only 1 for the bundle $\{a, b\}$, since it is not possible to accept both single-minded bids in the bid. Using XORs, we can in fact represent *any* valuation function, by using a single-minded bid for every possible bundle and XOR-ing them all together. Of course, this is not a very concise representation. Unfortunately, even representing additive valuation functions can require exponentially long bids if we use only XORs. But it is also possible to use ORs and XORs simultaneously, to get the best of both worlds. For example, the bid $((\{a\}, 1)$ XOR $(\{b\}, 1))$ OR $(\{c\}, 2)$ indicates a value of $1 + 2 = 3$ for the bundle $\{a, b, c\}$. There are other bidding languages that are not based on ORs and XORs, but we will not discuss them in this chapter.

16.8   The Winner Determination Problem

Now that we have considered how to bid in a (sealed-bid) combinatorial auction, we must consider how to determine who wins what. Of course, we cannot award a single item to two different bidders. But this still leaves plenty of options. One natural approach is to maximize **efficiency**, the total value generated. That is, if S_i is the bundle of items that we award to bidder i, then we should maximize $\sum_{i=1}^{n} \hat{v}_i(S_i)$ (under the constraint that $S_i \cap S_j = \emptyset$ for all $i \neq j$). This optimization problem is known as the **winner determination problem** (**WDP**).

As it turns out, the WDP is computationally hard. Even if we consider the special case where each bidder submits a single-minded bid, the problem turns out to be equivalent to WEIGHTED-SET-PACKING, which is NP-hard (Rothkopf et al., 1998) and inapproximable (Sandholm, 2002). On the other hand, the problem does not become any harder if we allow bidders to use ORs, since a bidder using ORs is effectively submitting multiple single-minded bids. In fact, even if we allow XORs, the problem in a sense becomes no harder: this is because, for the purpose of solving the WDP, we can transform an instance with XORs into one with only ORs using the following trick (Fujishima et al., 1999; Nisan, 2000). Given a bid of the form (S_1, v_1) XOR (S_2, v_2), we create a new "dummy" item, d, and replace the bid by $(S_1 \cup \{d\}, v_1)$ OR $(S_2 \cup \{d\}, v_2)$. Even though there is an OR between these two bids, they cannot both be accepted, since they have an item in common; moreover, since no other bids mention this item, everything else remains unaffected. Because of this, in the next two subsections, we will focus on algorithms for the single-minded case only. We will consider optimal algorithms for both the general case and some special cases. We will postpone the discussion of approximation algorithms until later, because we will require a particular kind of approximation algorithm in this context.

16.8.1   General-Purpose Winner Determination Algorithms

Given single-minded bids $\{(S_i, \hat{v}_i)\}$, one straightforward way to solve the winner determination problem is to solve the following integer program, which uses a binary variable b_i to indicate whether bid i is accepted:

maximize $\sum_i b_i \hat{v}_i$
subject to
for each $s \in I, (\sum_{i:s \in S_i} b_i) \leq 1$
for each $i, b_i \in \{0, 1\}$

The main constraint of this integer program ensures that each item is awarded at most once. Software packages such as CPLEX can be used to solve such an integer program; since the WDP is NP-hard, it should come as no surprise that solving integer programs is also NP-hard.

One interesting aside is that in some settings, bids can be accepted **partially**. For example, if we have three bids, $(\{a, b\}, 2)$, $(\{a, c\}, 2)$, and $(\{b, c\}, 2)$, it may be possible to accept *half* of each bid, awarding half of a and half of b to the first bidder for a value of 1, half of a and half of c to the second bidder for a value of 1, and half of b and half of c to the third bidder for a value of 1. This gives us a total value of 3 (we note that if we cannot accept bids partially, then we can obtain only 2). If accepting bids partially is possible, then we can solve the WDP by modifying the above program slightly. We make the b_i continuous variables, replacing the last constraint by

for each $i, 0 \leq b_i \leq 1$

At this point, the program has become a linear program, and linear programs can be solved in polynomial time (Khachiyan, 1979). However, in the remainder of this chapter we will assume that it is not possible to accept bids partially.

An alternative approach to solving the general WDP is to write a search algorithm based on techniques from artificial intelligence; for an overview of work along this line, see Sandholm (2006). It should be noted that such algorithms are in many ways similar to algorithms for solving integer programs.

Yet another option is to use the following dynamic programming approach (Rothkopf et al., 1998). For any subset $S \subseteq I$, let $w(S)$ be the maximum total value that can be obtained using only items in S (that is, if we threw away the items in $I - S$). Let $B(S)$ be the collection of all proper subsets $S' \subset S$ such that there is at least one bid on exactly S'. Then, we have

$$w(S) = \max\{\max_i \hat{v}_i(S), \max_{S' \in B(S)} w(S') + w(S - S')\}$$

Since the occurrences of w on the right-hand side involve subsets smaller than S, we can use dynamic programming to compute $w(S)$ for every subset, starting with the small ones and working our way up to I—and $w(I)$ gives the maximum value that can be obtained overall. This algorithm runs in $O(n3^m)$ time (where m is the number of items). It is straightforward to extend the dynamic program to keep track not only of the values that can be obtained, but also of the bids that need to be accepted to obtain these values.

16.8.2 Special-Purpose Winner Determination Algorithms

The WDP is NP-hard in general. Nevertheless, if the bids have some structure, then the WDP is sometimes solvable in polynomial time. For example, suppose that there are only (single-minded) bids on *pairs* of items (Rothkopf et al., 1998). Certainly, if multiple bids bid on the same pair of items, it never makes sense to accept a bid that is not the highest. Hence, we know the value of pairing any two given items together for sale (namely, the highest bid for that pair), and the only decision left is which items to pair together. This is a MAXIMUM-WEIGHTED-MATCHING problem, which can be solved in polynomial time. (This can easily be extended to also allow for bids on individual items—for example, by adding dummy items.) Unfortunately, if we allow for bids on sets of three items, the problem becomes NP-hard (by reduction from EXACT-COVER-BY-3-SETS).

As another example, suppose that the items are arranged as the vertices of a graph, and that every bid is on a bundle of items that constitutes a *connected component* in the graph. This is always possible by adding an edge between every pair of items (that is, making the graph a complete graph), but we will be interested in restricted classes of graphs. In particular, if the graph is a tree or a cycle, or, more generally, has **bounded treewidth**, then the WDP can be solved in polynomial time using dynamic programming (Sandholm and Suri, 2003; Conitzer et al., 2004). To use a result like this, we can either collect the bids first and then find a graph with which they are consistent, or we can specify the graph beforehand and require all bids to be consistent with this graph. A later result generalizes this even further by considering **hypertree decompositions** (Gottlob and Greco, 2007).

There are various other works on structure that bids may have that makes the WDP easier (Tennenholtz, 2000; Penn and Tennenholtz, 2000; Sandholm and Suri, 2003). Even if the bidders' valuations are not likely to have the required structure exactly, one possibility is to force the bidders to only use bids with this structure. This comes at the loss of some economic efficiency, because bidders can no longer express their exact valuations; nevertheless, it is generally better than reverting to single-item auctions.

16.9 The Generalized Vickrey Auction

So far, we have not yet considered how much a winning bidder should pay in a combinatorial auction. We could simply make such a bidder pay her bid (that is, her reported valuation for the bundle she won), resulting in a first-price sealed-bid combinatorial auction. As in the case of a first-price sealed-bid single-item auction, no bidder would ever bid her true valuation function, since this would guarantee that even if she wins something, she will have a utility of 0. We recall that an auction is truthful if revealing one's true valuation function is always optimal, regardless of the others' bids. Can we create a truthful combinatorial auction? A natural approach is to try to generalize the (truthful) second-price sealed-bid (aka. Vickrey) auction to the combinatorial setting. There are multiple ways in which one can generalize the Vickrey auction. For example, one can charge a winning bidder the highest other bid that was placed on exactly the same bundle. To see that this is not a good idea, consider the following example. Suppose that the bids are $(\{a\}, 1)$, $(\{b\}, 1)$, and $(\{a, b\}, 5)$, all from different bidders. The third bidder would win both items, and this bidder would pay 0, because nobody else bid on the bundle $\{a, b\}$. This intuitively feels wrong, since there was demand for the items from the other bidders. If the third bidder had bid $(\{a\}, 5)$ instead, she would have had to pay 1; thus, by bidding for *more* items, the bidder actually pays *less*! This also shows that this particular generalization is not truthful: if the third bidder is in fact interested only in a, she is still better off bidding on both items, and just throwing b away.

Fortunately, there is another generalization that does work. Let $\hat{V}$ be the total value of the accepted bids. Let $\hat{V}_{-i}$ be the total value that would have resulted if i had never entered the auction. (Computing this requires solving the winner determination problem again, this time without i.) Then, if S_i is the bundle that bidder i wins (possibly the empty bundle), she must pay $\hat{V}_{-i} - (\hat{V} - \hat{v}_i(S_i))$. This expression is the difference between how much the other bidders would have valued the allocation that would have resulted if i had never been present, and how much they value the current allocation. (For this reason, it is sometimes said that i pays the **externality** that she imposes on the other bidders.) Let us consider the above example with bids $(\{a\}, 1)$, $(\{b\}, 1)$, and $(\{a, b\}, 5)$. If the third bidder had not been present, the first two bids would have been accepted, so $V_{-3} = 2$. Hence, bidder 3 pays $2 - (5 - 5) = 2$. Let us make the example slightly richer, by adding bids $(\{c\}, 2)$ and $(\{a, c\}, 5)$, again from different bidders. Now, the third and fourth bidders win, for a total value of $5 + 2 = 7$. Without the third bidder, the second and fifth bidders would have won, for a total value of $1 + 5 = 6$. Hence, the third bidder must pay $6 - (7 - 5) = 4$. Without the fourth bidder, again, the second and fifth bidders would have won, for a total value of 6. Hence, the fourth bidder must pay $6 - (7 - 2) = 1$.

This way of computing payments is usually called the **Generalized Vickrey Auction (GVA)**. It is sometimes also called the **Clarke** mechanism, or the **VCG** mechanism [for Vickrey, Clarke, and Groves (Vickrey, 1961; Clarke, 1971; Groves, 1973)]. (Clarke and VCG refer to generalizations beyond auctions.) The GVA has several nice properties. For one, it is truthful. To see this, we note that bidder i's eventual utility is $v_i(S_i) - \pi_i = v_i(S_i) - (\hat{V}_{-i} - (\hat{V} - \hat{v}_i(S_i))) = v_i(S_i) + (\hat{V} - \hat{v}_i(S_i)) - \hat{V}_{-i}$. It is impossible for i to affect $\hat{V}_{-i}$; therefore, i can focus on maximizing $v_i(S_i) + (\hat{V} - \hat{v}_i(S_i)) = v_i(S_i) + \sum_{j \neq i} \hat{v}_j(S_j)$. There is only one way in which this expression depends on i's bid $\hat{v}_i$: her bid affects the chosen allocation $S_1, \ldots, S_n$. Now, if i had complete control over the chosen allocation (which she does not, but let us suppose for a second that she does), then she would choose the S_j to maximize $v_i(S_i) + \sum_{j \neq i} \hat{v}_j(S_j)$. The winner determination algorithm, on the other hand, chooses the S_j to maximize $\sum_j \hat{v}_j(S_j) = \hat{v}_i(S_i) + \sum_{j \neq i} \hat{v}_j(S_j)$. The only difference between the two expressions is that the first uses v_i, and the second uses $\hat{v}_i$. But then, if bidder i truthfully reports $\hat{v}_i = v_i$, the two expressions will be the same, and the winner determination algorithm will choose exactly the allocation that maximizes i's utility! Hence, the GVA is truthful.

A very observant reader may have noticed that in the proof of truthfulness, the only property of the payment term $\hat{V}_{-i}$ that we used is that i cannot affect it with her bid. Therefore, if truthfulness is all that we care about, we can replace the term $\hat{V}_{-i}$ in the payment expression with any other term that does not depend on i's bid. The mechanisms that can be obtained in this way are the **Groves** mechanisms (the G in VCG). The GVA mechanism, however, does have some additional nice properties that not all Groves mechanisms have. For one, it satisfies **voluntary participation** (aka. **individual rationality**): a bidder never receives negative utility as a result of participating in the auction, as long as she bids truthfully. This is because $\hat{V} \geq \hat{V}_{-i}$ (if this were false, it would mean that we had chosen a suboptimal allocation), and hence $v_i(S_i) - \pi_i = v_i(S_i) + (\hat{V} - \hat{v}_i(S_i)) - \hat{V}_{-i} \geq v_i(S_i) - \hat{v}_i(S_i)$; and the last expression is zero if i reports truthfully. A final nice property of the GVA is the **nondeficit** (aka. **weak budget balance**) property: at least as much money is collected from the bidders as is given to them. In fact, no money is ever given to a bidder. This is because $\hat{V}_{-i} \geq \hat{V} - \hat{v}_i(S_i)$ ($\hat{V} - \hat{v}_i(S_i)$ is the value of an allocation that is feasible even when i is not present), and hence $\pi_i = \hat{V}_{-i} - (\hat{V} - \hat{v}_i(S_i)) \geq 0$.

16.10 Collusion and False-Name Bidding

The GVA is truthful, so it is not possible for an individual bidder to benefit from misreporting her valuation function. However, if multiple bidders simultaneously misreport, that is, they **collude**, then it is possible that all of them benefit from this. To some extent, this problem occurs even in a single-item Vickrey auction. For example, suppose that there are three bidders with valuations 1, 3, and 4. If the third bidder can convince the second bidder not to place any bid, then the third bidder has to pay only 1 instead of 3. The third bidder may even pay the second bidder 1 for staying out, so that they each increase their utility by 1 from this. [In general, the colluders need some protocol for colluding (Graham and Marshall, 1987; Leyton-Brown et al., 2000, 2002), but we will not concern ourselves with that here.]

Still, there is a limit on what colluders can achieve in a single-item Vickrey auction. For example, they can never win the item at a price lower than the highest bid by a noncolluder; nor can they reduce the seller's revenue below what the seller would have made if the colluders had not participated. It turns out that in a combinatorial auction, not even these properties are true. For example, consider a GVA with only two items, a and b. Suppose two bidders have each placed a bid $(\{a, b\}, 1)$. If these are the only bidders, then one of them will win, and pay 1. However, let us now suppose that there are two additional bidders (the colluders): one of them bids $(\{a\}, 2)$ and the other bids $(\{b\}, 2)$. The colluders then win. How much does each colluder pay? If we remove one of the colluders, then the other colluder still wins—that is, the remainder of the allocation does not change. Thus, each colluder (individually) imposes no externality on the other bidders, and hence pays 0. The colluders benefit from this (assuming they have some value for the items they receive), and the auction's revenue has actually decreased as a result of the additional bids. More detail on collusion in the GVA can be found, for example, in work by Ausubel and Milgrom (2006) and Conitzer and Sandholm (2006).

The same example can also be used to demonstrate a different vulnerability of the GVA. Suppose that the auction is run in an open, anonymous environment such as the Internet. In such an auction, it is usually possible for a single bidder to participate in the auction under multiple identifiers (for example, e-mail addresses). Thus, given two other bids $(\{a, b\}, 1)$, a single bidder can place a bid $(\{a\}, 2)$ under one identifier, and a bid $(\{b\}, 2)$ under another identifier. As a result, the "false-name" bidder will win both items, and, as before, the price charged to each bid is 0. Hence, bidders sometimes have an incentive to bid under multiple identifiers; that is, the GVA is not **false-name-proof** (Yokoo et al., 2001, 2004).

It should be emphasized that this manipulation cannot be performed simply by submitting multiple bids under a single identifier [for example by bidding $(\{a\}, 2)$ OR $(\{b\}, 2)$]. In this case, to compute the bidder's GVA payment, we would remove the bidder's OR-bid in its entirety, so that the resulting payment would be 1. In the example where the bidder uses two different identifiers, there would be no problem if we could tell that the two identifiers correspond to the same real bidder, because in that case we would remove both identifiers' bids simultaneously to compute the bidder's GVA payment. Unfortunately, it is generally not possible to tell which identifiers were created by the same bidder.

One may wonder whether we can address some of these problems by using a combinatorial auction mechanism other than the GVA. It has been shown that *any* mechanism that avoids these issues (revenue nonmonotonicity, collusion, and false-name bidding) must have some unnatural properties (Rastegari et al., 2007). Still, several false-name-proof combinatorial auction mechanisms have been designed (Yokoo et al., 2001; Yokoo, 2003; Yokoo et al., 2004). It is also possible to make the use of multiple identifiers suboptimal by **verifying** the identities of some of the bidders (Conitzer, 2007).

16.11 Computationally Efficient Truthful Combinatorial Auctions

Another problem with the GVA is that it requires us to solve the winner determination problem to optimality. In fact, to compute the GVA payments, we need to solve up to n additional WDP instances: for each winning bidder, we need to solve the problem again with that bidder omitted. [It should be noted that at least in some settings, some of the computational work can be reused across the different instances (Hershberger and Suri, 2001).] An obvious idea is to not solve the WDP to optimality, but rather to use an approximation algorithm that returns a solution that is close to optimal. However, this effectively changes the mechanism, and there is no reason to think that desirable properties such as truthfulness and voluntary participation will continue to hold (Nisan and Ronen, 2001).

Let us consider the special case of single-minded bidders. One natural approximation algorithm is the following. Sort the bids (S_i, v_i) by $v_i / |S_i|$, the per-item value of the bid, in descending order. Then, consider the bids in this order, and accept any bid that can still be accepted. For example, suppose the bids are, in sorted order, $(\{a\}, 11)$, $(\{b, c\}, 20)$, $(\{a, d\}, 18)$, $(\{a, c\}, 16)$, $(\{c\}, 7)$, $(\{d\}, 6)$. The algorithm will accept the first two bids; the next three bids can then no longer be accepted, because one of their items has already been allocated; finally, the last bid is accepted. The total value of this allocation is $11 + 20 + 6 = 37$ (which is less than the $20 + 18 = 38$ that could have been obtained by accepting just the second and third bids). Now, if we wish to calculate the first bidder's GVA payment using this approximation algorithm, we must remove this bid, and run the algorithm again. After removing the first bid, the algorithm will actually accept the bids $(\{b, c\}, 20)$ and $(\{a, d\}, 18)$, the optimal solution. Hence, the first bidder's (approximated) GVA payment is $38 - (37 - 11) = 12$. This is more than the bidder's valuation! It follows that this approximation of the GVA mechanism does not satisfy voluntary participation (and hence it is also not truthful, because the bidder would have been better off bidding 0).

However, we *can* use this approximation algorithm for the WDP to obtain a truthful mechanism that satisfies voluntary participation: we just need to compute the payments somewhat differently (Lehmann et al., 2002). For each winning bid, consider the first bid in the sorted list that was forced out by this bid. The ratio $v_i/|S_i|$ for that bid is what the winning bid must pay per item. For example, in the above instance, the first bid forced out by the winning bid ($\{a\}, 11$) is ($\{a, d\}, 18$). Hence, the bid ($\{a\}, 11$) pays $18/|\{a, b\}| = 9$ per item—and since it wins only one item (a), that means it pays 9. As for the winning bid ($\{b, c\}, 20$), the first bid forced out by it is ($\{c\}, 7$) (it is *not* ($\{a, c\}, 16$), because this bid was already forced out by ($\{a\}, 11$)), so the bid pays $7/|\{c\}| = 7$ per item, and since it wins two items, it pays 14. The final winning bid, ($\{d\}, 6$), forces no other bids out and hence pays 0. Given that bidders are single-minded, this mechanism satisfies voluntary participation and truthfulness (each winning bidder pays the lowest amount that they could have bid while still winning).

The above approximation algorithm for the WDP has a worst-case approximation ratio of m, the number of items. If we sort the bids by $v_i/\sqrt{|S_i|}$ instead, then the approximation ratio is improved to $\sqrt{m}$.

There is a significant body of work on computationally efficient truthful combinatorial auctions: see, for example, Nisan and Ronen (2000); Mu'alem and Nisan (2002); Bartal et al. (2003); Archer et al. (2003); Dobzinski et al. (2006); Bikhchandani et al. (2006); Dobzinski and Nisan (2007a,b).

16.12 Iterative Combinatorial Auctions and Preference Elicitation

In a perfect world, every bidder would have a valuation function that can be concisely expressed in the bidding language of the combinatorial auction. Unfortunately, in reality, this is often not the case. This does not mean that bidders usually submit extremely long bids (such as an XOR of $2^m - 1$ bundles): this is too impractical, not only because it requires the communication of an exponential amount of information, but also because *determining* one's valuation for a given bundle is generally a nontrivial task. Instead, bidders bid on a few bundles on which they think their bids will be competitive. But they may not know exactly on which bundles they would be competitive, and if they do not bid on the right bundles, this results in decreased economic welfare.

A potential remedy for this is to, during the auction, give the bidders feedback on how they are doing in the auction. This means that we must abandon the sealed-bid format, and instead consider **iterative** auction mechanisms. We have already seen some examples of iterative auctions in the single-item context, namely the English, Japanese, and Dutch auctions. For example, in the English auction, a bidder knows if she is currently winning, and if she is not, she can choose to raise her bid. The English and the Japanese auctions are **ascending** auctions.

It turns out that we can also create ascending combinatorial auctions. As an example, let us study the **iBundle** ascending combinatorial auction (Parkes and Ungar, 2000). This auction maintains, for each bidder i, for each bundle S, a current price $p_i(S)$. In each round of the auction, the bidder is supposed to choose the bundle(s) most attractive to her at her current prices, that is, the set $\arg\max_S v_i(S) - p_i(S)$, and bid $p_i(S)$ on these bundles. The exception is if $v_i(S) - p_i(S) < 0$ for every bundle S, in which case the bidder is supposed to drop out. If a bidder follows this strategy, she is said to **bid straightforwardly**. At the end of the round, the winner determination problem is solved with the submitted bids. Then, for every active bidder i that is not winning anything, for every bundle S that she bid on, the price $p_i(S)$ is increased by some predetermined amount ϵ. Eventually, there will be a round where every active bidder wins something, and at this point the auction terminates with the current allocation and payments. This auction is known to have some nice properties: for example, if the bidders' valuations satisfy a condition known as **buyer submodularity**, then straightforward bidding is an **ex-post equilibrium**, and the GVA outcome results (Ausubel and Milgrom, 2002). Informally, buyer submodularity means that the more bidders are already present, the less having additional bidders adds to the final allocation value. Strategies are said to be in ex-post equilibrium if for each bidder i, it is optimal to follow the strategy, assuming that the other bidders follow their strategies (but regardless of what the other bidders' valuations are). Numerous other iterative combinatorial auctions have been proposed [for an overview, see (Parkes, 2006)], and inherent limitations of this approach have also been studied (Blumrosen and Nisan, 2005).

More generally, the auctioneer can sequentially ask the bidders various **queries** about their valuation functions, until the auctioneer has enough information to determine the final outcome. If the final outcome is (always) the GVA outcome, then responding truthfully to the auctioneer is an ex-post equilibrium. This flexible query-based approach is generally referred to as **preference elicitation** (Conen and Sandholm, 2001). Common queries include **value queries** ("What is your valuation for bundle S?") and **demand queries** ("Given these prices, which bundle(s) do you prefer?"). Demand queries can use either **item prices**, where the price for a bundle is the sum of the prices of the individual items, or **bundle prices**, where each bundle has a separate price (as we saw in the iBundle auction above). For some restricted classes of valuation functions, it has been shown that a polynomial number of queries suffices to learn a bidder's valuation function completely (if we assume that the valuation function lies in that class). For example, a valuation function can be elicited using a number of (bundle-price) demand queries that is polynomial in the length of the function's XOR-representation (Lahaie and Parkes, 2004), though in general an exponential number of queries is required if only item prices (and value queries) are allowed (Blum et al., 2004). Various other results have been obtained on classes of valuation functions that can(not) be elicited using a polynomial number of queries (Zinkevich et al., 2003; Santi et al., 2004; Blumrosen and Nisan, 2005; Conitzer et al., 2005; Lahaie et al., 2005).

Without any restrictions on the valuation functions, negative results are known: for example, solving the winner determination problem in general requires an exponential amount of communication, regardless of what types of query are used (Nisan and Segal, 2005).

16.13 Additional Topics

In this final section, we mention a few additional topics and provide some references for the interested reader.

There are several important variants of (combinatorial) auctions, including (**combinatorial**) **reverse auctions** and (**combinatorial**) **exchanges**. In a reverse auction, the auctioneer seeks to *buy* one or more items, and the bidders submit bids indicating how much they need to be compensated to provide the items. In an exchange, bidders can act as buyers as well as sellers. While these variants display some significant similarities to regular (forward) auctions, there are also important differences (Sandholm et al., 2002).

Quite a few researchers have tried to generalize Myerson's expected-revenue maximizing auction to combinatorial auctions; this turns out to be surprisingly difficult (Avery and Hendershott, 2000; Armstrong, 2000; Conitzer and Sandholm, 2004; Likhodedov and Sandholm, 2004, 2005). A different take on this problem is to design **competitive** auctions, which obtain a revenue that is within a factor of the revenue that can be obtained with a single sale price (Goldberg et al., 2006).

A final important direction is the design of **online auctions**. "Online" here does not refer to Internet auctions; rather, it refers to settings in which the bidders arrive at and depart from the auction over time, and allocation decisions must be made before all the bidders have arrived. (More detail can be found in, for example, Lavi and Nisan, 2000; Awerbuch et al., 2003; Blum et al., 2003; Friedman and Parkes, 2003; Kleinberg and Leighton, 2003; Hajiaghayi et al., 2004, 2005; Bredin and Parkes, 2005; Blum et al., 2006; Babaioff et al., 2007; Hajiaghayi et al., 2007; Parkes and Duong, 2007.)

References

A. Archer, C.H. Papadimitriou, K. Talwar, and E. Tardos. An approximate truthful mechanism for combinatorial auctions with single parameter agents. In *Proceedings of the Annual ACM-SIAM Symposium on Discrete Algorithms (SODA)*, Baltimore, MD, 2003.

M. Armstrong. Optimal multi-object auctions. *Review of Economic Studies*, 67:455–481, 2000.

L. Ausubel and P. Milgrom. Ascending auctions with package bidding. *Frontiers of Theoretical Economics*, 1, 2002(1), Article 1.

L.M. Ausubel and P. Milgrom. The lovely but lonely Vickrey auction. In P. Cramton, Y. Shoham, and R. Steinberg, editors, *Combinatorial Auctions*, Chapter 1. MIT Press, Cambridge, MA, 2006.

C. Avery and T. Hendershott. Bundling and optimal auctions of multiple products. *Review of Economic Studies*, 67:483–497, 2000.

B. Awerbuch, Y. Azar, and A. Meyerson. Reducing truth-telling online mechanisms to online optimization. In *Proceedings of the Annual Symposium on Theory of Computing (STOC)*, San Diego, CA, June 9–11, 2003.

M. Babaioff, N. Immorlica, and R. Kleinberg. Matroids, secretary problems, and online mechanisms. In *Proceedings of the Annual ACM-SIAM Symposium on Discrete Algorithms (SODA)*, pp. 434–443, New Orleans, LA, January 7–9, 2007.

M.J. Bailey. The demand revealing process: To distribute the surplus. *Public Choice*, 91:107–126, 1997.

Y. Bartal, R. Gonen, and N. Nisan. Incentive compatible multi-unit combinatorial auctions. In *Theoretical Aspects of Rationality and Knowledge (TARK)*, Bloomington, IN, 2003.

S. Bikhchandani, S. Chatterji, R. Lavi, A. Mu'alem, N. Nisan, and A. Sen. Weak monotonicity characterizes deterministic dominant strategy implementation. *Econometrica*, 74(4):1109–1132, 2006.

A. Blum, V. Kumar, A. Rudra, and F. Wu. Online learning in online auctions. In *Proceedings of the Annual ACM-SIAM Symposium on Discrete Algorithms (SODA)*, pp. 202–204, Baltimore, MD, January 12–14, 2003.

A. Blum, J. Jackson, T. Sandholm, and M. Zinkevich. Preference elicitation and query learning. *Journal of Machine Learning Research*, 5:649–667, 2004.

A. Blum, T. Sandholm, and M. Zinkevich. Online algorithms for market clearing. *Journal of the ACM*, 53(5):845–879, 2006.

L. Blumrosen and N. Nisan. On the computational power of iterative auctions. In *Proceedings of the ACM Conference on Electronic Commerce (EC)*, pp. 29–43, Vancouver, BC, Canada, 2005.

J. Bredin and D. Parkes. Models for truthful online double auctions. In *Proceedings of the 21st Annual Conference on Uncertainty in Artificial Intelligence (UAI)*, pp. 50–59, Edinburgh, U.K. 2005.

R. Cavallo. Optimal decision-making with minimal waste: Strategyproof redistribution of VCG payments. In *International Conference on Autonomous Agents and Multi-Agent Systems (AAMAS)*, pp. 882–889, Hakodate, Japan, 2006.

E.H. Clarke. Multipart pricing of public goods. *Public Choice*, 11:17–33, 1971.

W. Conen and T. Sandholm. Preference elicitation in combinatorial auctions: Extended abstract. In *Proceedings of the ACM Conference on Electronic Commerce (EC)*, pp. 256–259, Tampa, FL October 2001.

V. Conitzer and T. Sandholm. Self-interested automated mechanism design and implications for optimal combinatorial auctions. In *Proceedings of the ACM Conference on Electronic Commerce (EC)*, pp. 132–141, New York, 2004.

V. Conitzer and T. Sandholm. Failures of the VCG mechanism in combinatorial auctions and exchanges. In *International Conference on Autonomous Agents and Multi-Agent Systems (AAMAS)*, pp. 521–528, Hakodate, Japan, 2006.

V. Conitzer, J. Derryberry, and T. Sandholm. Combinatorial auctions with structured item graphs. In *Proceedings of the National Conference on Artificial Intelligence (AAAI)*, pp. 212–218, San Jose, CA, 2004.

V. Conitzer, T. Sandholm, and P. Santi. Combinatorial auctions with k-wise dependent valuations. In *Proceedings of the National Conference on Artificial Intelligence (AAAI)*, pp. 248–254, Pittsburgh, PA, 2005.

V. Conitzer. Limited verification of identities to induce false-name-proofness. In *Theoretical Aspects of Rationality and Knowledge (TARK)*, pp. 102–111, Brussels, Belgium, 2007.

S. Dobzinski and N. Nisan. Limitations of vcg-based mechanisms. In *Proceedings of the Annual Symposium on Theory of Computing (STOC)*, San Diego, CA, June 11–13, pp. 338–344, 2007.

S. Dobzinski and N. Nisan. Mechanisms for multi-unit auctions. In *Proceedings of the ACM Conference on Electronic Commerce (EC)*, pp. 346–351, San Diego, CA, USA, 2007.

S. Dobzinski, N. Nisan, and M. Schapira. Truthful randomized mechanisms for combinatorial auctions. In *Proceedings of the Annual Symposium on Theory of Computing (STOC)*, Seattle, WA, pp. 644–652, 2006.

E. Friedman and D. Parkes. Pricing WiFi at Starbucks – Issues in online mechanism design. In *Proceedings of the ACM Conference on Electronic Commerce (EC)*, pp. 240–241, San Diego, CA, 2003.

Y. Fujishima, K. Leyton-Brown, and Y. Shoham. Taming the computational complexity of combinatorial auctions: Optimal and approximate approaches. In *Proceedings of the Sixteenth International Joint Conference on Artificial Intelligence (IJCAI)*, pp. 548–553, Stockholm, Sweden, August 1999.

A. Gibbard. Manipulation of voting schemes: A general result. *Econometrica*, 41:587–602, 1973.

A. Goldberg, J. Hartline, A. Karlin, M. Saks, and A. Wright. Competitive auctions. *Games and Economic Behavior*, 55:242–269, 2006.

G. Gottlob and G. Greco. On the complexity of combinatorial auctions: Structured item graphs and hypertree decompositions. In *Proceedings of the ACM Conference on Electronic Commerce (EC)*, pp. 152–161, San Diego, CA, 2007.

D.A. Graham and R.C. Marshall. Collusive bidder behavior at single-object second-price and English auctions. *Journal of Political Economy*, 95(6):1217–1239, 1987.

J. Green and J.-J. Laffont. Characterization of satisfactory mechanisms for the revelation of preferences for public goods. *Econometrica*, 45:427–438, 1977.

T. Groves. Incentives in teams. *Econometrica*, 41:617–631, 1973.

M. Guo and V. Conitzer. Worst-case optimal redistribution of VCG payments. In *Proceedings of the ACM Conference on Electronic Commerce (EC)*, pp. 30–39, San Diego, CA, 2007.

M.T. Hajiaghayi, R. Kleinberg, and D.C. Parkes. Adaptive limited-supply online auctions. In *Proceedings of the ACM Conference on Electronic Commerce (EC)*, pp. 71–80, New York, 2004.

M.T. Hajiaghayi, R. Kleinberg, M. Mahdian, and D.C. Parkes. Online auctions with re-usable goods. In *Proceedings of the ACM Conference on Electronic Commerce (EC)*, pp. 165–174, Vancouver, BC, Canada, 2005.

M.T. Hajiaghayi, R. Kleinberg, and T. Sandholm. Automated online mechanism design and prophet inequalities. In *Proceedings of the National Conference on Artificial Intelligence (AAAI)*, Vancouver, BC, Canada, 2007.

J. Hershberger and S. Suri. Vickrey prices and shortest paths: What is an edge worth? In *Proceedings of the Annual Symposium on Foundations of Computer Science (FOCS)*, Las Vegas, NV, October 14–17, 2001.

L. Khachiyan. A polynomial algorithm in linear programming. *Soviet Mathematics Doklady*, 20:191–194, 1979.

R. Kleinberg and T. Leighton. The value of knowing a demand curve: Bounds on regret for on-line posted-price auctions. In *Proceedings of the Annual Symposium on Foundations of Computer Science (FOCS)*, pp. 594–605, Cambridge, MA, October 11–14, 2003.

S. Lahaie and D. Parkes. Applying learning algorithms to preference elicitation. In *Proceedings of the ACM Conference on Electronic Commerce (EC)*, pp. 180–188, New York, 2004.

S. Lahaie, F. Constantin, and D.C. Parkes. More on the power of demand queries in combinatorial auctions: Learning atomic languages and handling incentives. In *Proceedings of the Nineteenth*

International Joint Conference on Artificial Intelligence (IJCAI), pp. 959–964, Edinburgh, U.K., July 30–August 5, 2005.

R. Lavi and N. Nisan. Competitive analysis of incentive compatible on-line auctions. In *Proceedings of the ACM Conference on Electronic Commerce (EC)*, pp. 233–241, Minneapolis, MN, 2000.

D. Lehmann, L.I. O'Callaghan, and Y. Shoham. Truth revelation in rapid, approximately efficient combinatorial auctions. *Journal of the ACM*, 49(5):577–602, 2002.

K. Leyton-Brown, Y. Shoham, and M. Tennenholtz. Bidding clubs: Institutionalized collusion in auctions. In *Proceedings of the ACM Conference on Electronic Commerce (EC)*, Minneapolis, MN, 2000.

K. Leyton-Brown, Y. Shoham, and M. Tennenholtz. Bidding clubs in first-price auctions. In *Proceedings of the National Conference on Artificial Intelligence (AAAI)*, pp. 373–378, Edmonton, Canada, 2002.

A. Likhodedov and T. Sandholm. Methods for boosting revenue in combinatorial auctions. In *Proceedings of the National Conference on Artificial Intelligence (AAAI)*, pp. 232–237, San Jose, CA, 2004.

A. Likhodedov and T. Sandholm. Approximating revenue-maximizing combinatorial auctions. In *Proceedings of the National Conference on Artificial Intelligence (AAAI)*, Pittsburgh, PA, 2005.

H. Moulin. Efficient, strategy-proof and almost budget-balanced assignment, March 2007. Working Paper.

A. Mu'alem and N. Nisan. Truthful approximate mechanisms for restricted combinatorial auctions. In *Proceedings of the National Conference on Artificial Intelligence (AAAI)*, pp. 379–384, Edmonton, Canada, July 2002.

R. Myerson. Incentive compatibility and the bargaining problem. *Econometrica*, 41(1):61–73, 1979.

R. Myerson. Optimal auction design. *Mathematics of Operations Research*, 6:58–73, 1981.

N. Nisan and A. Ronen. Computationally feasible VCG mechanisms. In *Proceedings of the ACM Conference on Electronic Commerce (EC)*, pp. 242–252, Minneapolis, MN, 2000.

N. Nisan and A. Ronen. Algorithmic mechanism design. *Games and Economic Behavior*, 35:166–196, 2001.

N. Nisan and I. Segal. The communication requirements of efficient allocations and supporting prices. *Journal of Economic Theory*, 129:192–224, 2005.

N. Nisan. Bidding and allocation in combinatorial auctions. In *Proceedings of the ACM Conference on Electronic Commerce (EC)*, pp. 1–12, Minneapolis, MN, 2000.

D.C. Parkes and Q. Duong. An ironing-based approach to adaptive online mechanism design in single-valued domains. In *Proceedings of the National Conference on Artificial Intelligence (AAAI)*, Vancouver, BC, Canada, 2007.

D. Parkes and L. Ungar. Iterative combinatorial auctions: Theory and practice. In *Proceedings of the National Conference on Artificial Intelligence (AAAI)*, pp. 74–81, Austin, TX, August 2000.

D. Parkes. Iterative combinatorial auctions. In P. Cramton, Y. Shoham, and R. Steinberg, editors, *Combinatorial Auctions*, Chapter 3. MIT Press, Cambridge, MA, 2006.

M. Penn and M. Tennenholtz. Constrained multi-object auctions and *b*-matching. *Information Processing Letters*, 75(1–2):29–34, July 2000.

R. Porter, Y. Shoham, and M. Tennenholtz. Fair imposition. *Journal of Economic Theory*, 118:209–228, 2004.

B. Rastegari, A. Condon, and K. Leyton-Brown. Revenue monotonicity in combinatorial auctions. In *Proceedings of the National Conference on Artificial Intelligence (AAAI)*, Vancouver, BC, Canada, 2007.

J.G. Riley and W.F. Samuelson. Optimal auctions. *American Economic Review*, 71:381–392, June 1981.

M. Rothkopf, A. Pekeč, and R. Harstad. Computationally manageable combinatorial auctions. *Management Science*, 44(8):1131–1147, 1998.

T. Sandholm and S. Suri. BOB: Improved winner determination in combinatorial auctions and generalizations. *Artificial Intelligence*, 145:33–58, 2003.

T. Sandholm, S. Suri, A. Gilpin, and D. Levine. Winner determination in combinatorial auction generalizations. In *International Conference on Autonomous Agents and Multi-Agent Systems (AAMAS)*, pp. 69–76, Bologna, Italy, July 2002.

T. Sandholm. Algorithm for optimal winner determination in combinatorial auctions. *Artificial Intelligence*, 135:1–54, January 2002.

T. Sandholm. Optimal winner determination algorithms. In P. Cramton, Y. Shoham, and R. Steinberg, editors, *Combinatorial Auctions*, pp. 337–368. MIT Press, Cambridge, MA, 2006.

P. Santi, V. Conitzer, and T. Sandholm. Towards a characterization of polynomial preference elicitation with value queries in combinatorial auctions. In *Conference on Learning Theory (COLT)*, pp. 1–16, Banff, AB, Canada, 2004.

M. Tennenholtz. Some tractable combinatorial auctions. In *Proceedings of the National Conference on Artificial Intelligence (AAAI)*, Austin, TX, August 2000.

W. Vickrey. Counterspeculation, auctions, and competitive sealed tenders. *Journal of Finance*, 16:8–37, 1961.

M. Yokoo, Y. Sakurai, and S. Matsubara. Robust combinatorial auction protocol against false-name bids. *Artificial Intelligence*, 130(2):167–181, 2001.

M. Yokoo, Y. Sakurai, and S. Matsubara. The effect of false-name bids in combinatorial auctions: New fraud in Internet auctions. *Games and Economic Behavior*, 46(1):174–188, 2004.

M. Yokoo. The characterization of strategy/false-name proof combinatorial auction protocols: Price-oriented, rationing-free protocol. In *Proceedings of the Eighteenth International Joint Conference on Artificial Intelligence (IJCAI)*, pp. 733–742, Acapulco, Mexico, 2003.

M. Zinkevich, A. Blum, and T. Sandholm. On polynomial-time preference elicitation with value queries. In *Proceedings of the ACM Conference on Electronic Commerce (EC)*, pp. 176–185, San Diego, CA, 2003.

17

Pseudorandom Sequences and Stream Ciphers

Andrew Klapper
University of Kentucky

17.1 Introduction

This chapter concerns the generation of **pseudorandom sequences** and the role of these sequences in **stream ciphers.** Pseudorandom sequences are also used in various probabilistic algorithms that arise in cryptography. In this latter context, however, the role of pseudorandomness is essentially the same as in other probabilistic algorithms and we leave this aspect of pseudorandomness to other chapters.

The only completely secure cryptosystem is the **one-time pad**. In this private key system the message alphabet is the integers modulo an integer m. The key or keystream is a sequence of symbols from the message alphabet generated uniformly independently at random. The key is known to both the sender and receiver. To encrypt a message, the key is added to the message symbol by symbol modulo m. From the point of view of Shannon's **information theory** this system is unconditionally secure. That is, an adversary who knows any subset of the symbols of the key (or, equivalently, any set of known plaintext/ciphertext pairs) can determine any other symbol of the message with probability no better than guessing. Such a system is further advantageous because encryption is as fast as the method of generating the random symbols. The drawback is that sharing the key between the sender and the receiver is no easier than sharing the message, since the key is as large as the message. Thus the one-time pad is only practical when the sender and receiver have access to a secure channel at some point (when they can share the key) and plan to use an insecure channel to share a message at some later time. Such situations are rare.

A stream cipher uses a pseudorandom sequence as the keystream in place of a truly random one. By pseudorandom we mean that the key is apparently random by various criteria that are related to the effectiveness of the system. Since the actual encryption is simple and well understood, the issues surrounding the effectiveness of a stream cipher have to do mainly with the generation of keys. The symbols of the key must be difficult to determine from a subset of the symbols and the key must be efficiently generated. There is widespread belief in the cryptographic community that stream ciphers are less secure than either well-designed block ciphers or public key systems. Thus they are considered practical only if they achieve much higher speed. We stress, however, that the relative security of types of cryptosystems is largely a matter of belief, often depending on unproved complexity theoretic assumptions.

Stream ciphers can be implemented either in hardware, which has the advantage of speed of execution, or in software, which has the advantages of speed of implementation and portability. The choice often affects the choice of message alphabet. A binary alphabet is typically used in hardware implementations. An alphabet of size 256 is often used in software implementations. In much of this chapter we treat the case of stream ciphers based on binary sequences, since this case has been more extensively studied. In many instances the analysis is considerably simpler with binary sequences. Also, hardware implementations are essentially finite state devices. Their output sequences are thus eventually periodic and this is often an assumption about the sequences used in stream ciphers.

As with most cryptography there are two aspects to the study of stream ciphers: their design and their cryptanalysis. Of course there is a relationship. Stream ciphers must be designed to resist any general cryptanalytic attacks and the cryptanalysis of a system depends on its design.

17.1.1   Classification and Modes of Stream Ciphers

In the greatest generality a stream cipher is a parametrized nonterminating finite automaton with output. More precisely, it consists of

1. A state space $\mathcal{A}$, a key space $\mathcal{K}$, and a message alphabet $\mathcal{M}$
2. A state change function $F : \mathcal{K} \times \mathcal{A} \times \mathcal{M} \to \mathcal{A}$
3. An output function $g : \mathcal{K} \times \mathcal{A} \times \mathcal{M} \to \mathcal{M}$

For a given message sequence $x_0, x_1, \ldots \in \mathcal{M}$, initial state $\alpha_0 \in \mathcal{A}$, and key $k \in \mathcal{K}$, the stream cipher generates a state sequence and output (i.e., ciphertext) sequence by

$$\alpha_{i+1} = F(k, \alpha_i, x_i)$$
$$y_i = g(k, \alpha_i, x_i).$$

Most commonly
$$g(k, \alpha, x) = x + h(k, \alpha) \bmod m,$$

where $\mathcal{M}$ is the integers modulo an integer m. In this case the sequence $z_i = h(k, \alpha_i)$ is called the keystream and the pair (F, h) is called a **keystream generator**.

Keystream generators may be classified as synchronous or self-synchronizing. In a self-synchronizing generator the state depends on the previous few output symbols,

$$\alpha_{i+1} = F(y_i, \ldots, y_{i-r+1}).$$

In effect the state consists of the previous r output symbols. The advantage of such a system is that if symbols are lost, then the receiver can resynchronize after r output symbols have been received. The disadvantage is that distorted symbols cause error propagation for r symbols.

In a synchronous generator the keystream is independent of the message sequence. That is, $F : \mathcal{K} \times \mathcal{A} \to \mathcal{A}$. Synchronous generators are unable to recover from lost symbols without resetting to an initial state. However, an external mechanism can be added to solve the synchronization problem. Distorted symbols result in no error propagation. Nonetheless, in most applications using noisy channels it is likely that an underlying error correction mechanism will be used. Synchronous generators are the most commonly used stream ciphers.

Synchronous generators are often restricted to one of two special cases. In counter mode, $\alpha_{i+1} = F(\alpha_i)$ and $z_i = h(k, \alpha_i)$. An example of a counter mode generator is the cyclotomic generator, described below. In output feedback mode, $\alpha_{i+1} = F(k, \alpha_i)$ and $z_i = h(\alpha_i)$. Most of the feedback register based generators described in this chapter run in output feedback mode.

17.2 Underlying Principles

The mathematical tools used to study cryptographically strong pseudorandom sequences are algebra, information theory, and probability theory. In this section various formalizations of the notion of randomness are described. We also describe the structure and analysis of **linear feedback shift registers**. These are fundamental building blocks in the design of keystream generators. Of particular importance in this analysis is the theory of finite fields [22,23].

17.2.1 Randomness

There are several different notions of randomness that have been applied to sequences. These include information theoretic randomness, statistical randomness, unpredictability (a complexity theoretic notion), and Kolmogorov complexity. The first three have the greatest relevance for cryptography, and we describe them in more detail next.

17.2.1.1 Information Theoretic Randomness

The mathematical analysis of cryptographic systems was initiated by Shannon's invention of information theory [37]. Shannon's approach was to analyze the information that can be derived about a system by an adversary who has unlimited computational resources. The analysis is probabilistic. The cryptanalyst has a known ciphertext and knows the probability distribution of plaintexts, keys, and ciphertexts. The a priori distribution on the plaintexts arises by assuming they lie in some restricted class of strings such as a natural language. By conditioning this distribution on the known ciphertext, a new distribution on plaintext is determined. The cryptanalyst is successful if one plaintext has conditional probability close to one.

Let X be a discrete random variable whose distribution is given by $p_i = Pr(X = i)$. The **entropy** of X is defined by

$$H(X) = - \sum_{i, p_i \neq 0} p_i \log (p_i).$$

The entropy is a measure of uncertainty about X. Shannon gave a set of properties that the uncertainty intuitively should have and proved that the so-defined entropy function is the unique function that has them. For any X, $H(X) \leq \log(n)$, with equality if and only if $p_i = 1/n$ for all i. The notion of entropy also plays a role in compression. It is a lower bound on the average number of bits required to encode a finite set. An introduction to information theory and its role in coding theory and cryptography can be found in the book by Welsh [38].

If X and Y are two random variables, the notion of entropy extends naturally to the joint entropy $H(X, Y)$, the conditional entropy $H(X|Y)$ (the uncertainty about X if Y is known), and the mutual information

$$I(X; Y) = H(X) - H(X|Y)$$

between X and Y (the information common to X and Y). If $I(X; Y) = 0$, then X and Y are independent, so Y reveals nothing about X. Conversely, if $I(X; Y) = H(X)$, i.e., is maximal, then $H(X|Y) = 0$ so there is no uncertainty about X when Y is known.

Now consider the case of a cryptosystem. Suppose that $X^n = x_1, x_2, \ldots, x_n$ is an n-bit plaintext, K is a key, and $Y^n = y_1, y_2, \ldots, y_n$ is the corresponding n-bit ciphertext. In general,

$$H(K|Y^n) = H(X^n|Y^n) + H(K|X^n, Y^n).$$

If $I(X^n; Y^n) = 0$, then the ciphertext reveals nothing of the plaintext and the system is said to have perfect secrecy. In this case

$$H(K) \geq H(K|Y^n) \geq H(X^n|Y^n) = H(X^n).$$

Thus the uncertainty about the key must be at least as great as the uncertainty about the plaintext. It follows that the average length of the minimal length encoding of the key must be at least as great as that of the plaintext. That is, to achieve perfect secrecy the key must be as long as the plaintext. This is a fundamental limit on the power of private key encryption.

At the other extreme, if $I(X^n; Y^n)$ is asymptotic to $H(X^n)$, then for long enough plaintexts knowledge of the ciphertext essentially determines the plaintext and the system can be broken (although it still may require an exhaustive search to find the plaintext).

In the intermediate cases $0 < I(X^n; Y^n) < rH(X^n)$ for some positive constant $r < 1$. For arbitrarily long messages there is some uncertainty about the plaintext. It follows that

$$H(K|Y^n) > (1 - r)H(X^n)$$

is bounded away from zero. Such a cryptosystem is called ideally secure. This means that the key can never be determined from the ciphertext with perfect certainty. However, the uncertainty may be small enough that significant portions of the plaintext are revealed.

Shannon also defined a measure of how much ciphertext is necessary to determine the key. The unicity distance n_u is the minimum length n such that $H(K|Y^n) \sim 0$. Under reasonable assumptions, we have

$$n_u = \frac{H(K)}{1 - h},$$

where h is the information rate of the plaintext (so $1 - h$ is the redundancy).

17.2.1.2 Statistical Randomness

A sequence is statistically random if various statistical properties are close to those of a truly random sequence. The failure of these properties to hold does not necessarily lead directly to a cryptanalytic attack. However, such failure is reason for concern that an attack may be found in the future. Following are some of the randomness criteria that have been considered for periodic binary sequences. Let $\mathbf{a} = a_0, a_1, a_2, \ldots$ denote such a sequence. Its period, denoted $\rho(\mathbf{a})$, is the least positive integer n such that $a_i = a_{i+n}$ for every i. The period of a keystream must be large.

17.2.1.2.1 Balance
A sequence is balanced if the number of occurrences of 0 equals the number of occurrences of 1 in each period. In the extreme, a highly unbalanced sequence allows a cryptanalyst to read most of a message either directly or by complementing all bits. In less extreme cases, unbalanced sequences have been shown to leak essential information [18]. Of course if $\rho(\mathbf{a})$ is odd, then $\mathbf{a}$ cannot be perfectly balanced.

17.2.1.2.2 Subsequence Distribution
If r is any integer, we can count the number of occurrences of each binary r-tuple $b_0, \ldots, b_{r-1}$ in a period of $\mathbf{a}$ (more precisely, we count the number of indices i such that $0 \leq i \leq n - 1$, and

$a_i = b_0, a_1 = b_{i+1}, \ldots a_{i_r-1} = b_{r-1})$. It is desirable that the distribution of occurrences be close to uniform. A sequence of period 2^r such that every binary r-tuple occurs exactly once in a period is known as a de Bruijn sequence.

17.2.1.2.3 Autocorrelations

The autocorrelation with shift τ of $\mathbf{a}$ is defined as

$$A_{sma}(\tau) = \sum_{i=0}^{\rho(sma)-1} (-1)^{a_i + a_{i+\tau}},$$

or, equivalently, the number of bits in one period of $\mathbf{a}$ and a τ shift of $\mathbf{a}$ that are equal minus the number of bits that are not equal. If $\mathbf{a}$ is independent of its τ shift, then the autocorrelation is zero, while if for each i, a_i determines $a_{i+\tau}$, then the autocorrelation is plus or minus $\rho(\mathbf{a})$. Thus the autocorrelation measures the extent to which a sequence is independent of its shifts. The failure of the autocorrelation to be close to zero can sometimes be used to derive essential information about a cryptosystem.

17.2.1.2.4 Run Property

A run in a sequence is a maximal subsequence consisting of only zeros or only ones. A binary sequence can be thought of as a series of runs of varying lengths. A sequence has the run property if the distribution of runs is close to what would be expected of a truly random sequence. If a run starts at position i, then it has length r with probability 2^{-r}.

17.2.1.2.5 Nonlinearity

Generally, linearity can be exploited in constructing cryptanalytic attacks. Thus, loosely speaking, the sequence should not exhibit linear structure. This statement can be interpreted in various ways. For example, the sequence defines a function from its set of indices to $\{0, 1\}$. Whenever we interpret the index set as an algebraic structure (say, modular integers or a finite field), this function should be highly nonlinear. Various approaches have been taken to defining the degree to which such a function is nonlinear.

A Boolean function f on n bits satisfies the strict avalanche criterion (or SAC) if $f(\bar{x}) + f(\bar{x} + \bar{y})$ is a balanced function of $\bar{x}$ for every $\bar{y}$ with $||\bar{y}|| = 1$. Here $||\bar{y}||$ denotes the number of ones in $\bar{y}$, known as its Hamming weight. That is, changing any single bit of the input to f gives a function that is uncorrelated with f. The function f satisfies $SAC(k)$ if holding any k bits constant results in a function that satisfies SAC. Preneel et al. gave conditions under which various $SAC(k)$ hold [28]. More generally, f satisfies the propagation criterion of degree k if $f(\bar{x}) + f(\bar{x} + \bar{y})$ is a balanced function of $\bar{x}$ for every $\bar{y}$ with $1 \le ||\bar{y}|| \le k$.

The Walsh transform of f is defined as

$$\hat{F}(\bar{w}) = \sum_{\bar{x}} (-1)^{f(\bar{x}) + \bar{w} \cdot \bar{x}}.$$

Parseval's theorem says that

$$\sum_{\bar{w}} \left(\hat{F}(\bar{w}) \right)^2 = 2^{2n}.$$

A function is bent if each Walsh transform achieves exactly the average value implied by Parseval's theorem,

$$\left| \hat{F}(\bar{w}) \right| = 2^{n/2}$$

for every w [30]. That is, the maximum correlation of f with a linear function is as small as possible.

The notions of **correlation immunity** and the degree of algebraic nonlinearity are described in subsequent sections.

17.2.1.3 Unpredictability

In a very different attempt to define a notion of randomness suitable for cryptography, Yao [39] and Blum and Micali [3] considered what would make it effectively impossible for an adversary to determine the next bit of a sequence if a prefix were known. The adversary in their model is assumed to have limited resources, so their definition of unpredictability is complexity theoretic.

In Blum and Micali's model a generator G inputs a security parameter n and a random number $0 \leq i < 2^n$ and outputs a pseudorandom bit sequence **a**. Such a generator is a **cryptographically strong pseudorandom bit generator** or CSPRB generator if the following hold.

1. The bits a_j are easy to generate. That is, it should take time polynomial in n to output the jth bit.
2. The bits are unpredictable. Given G, n, and $a_0, \ldots, a_{j-1}$, but not i, it should be computationally infeasible to predict a_j with probability significantly greater than $1/2$.

More precisely, suppose the output from G has length $p(n)$ if n is the security parameter. A predicting family is a polynomial (in n) size family of circuits

$$C = \left\{ C_n^j : j < p(n) \right\}$$

such that each C_n^j has j inputs and one output. If the input is $a_0, \ldots, a_{j-1}$, and the output is d, let $P_{j,i}$ be the probability that $d = a_j$. Then G passes the **next bit test** C if for every polynomial $q(n)$, every large enough n, every $j < p(n)$, and every $i < 2^n$,

$$P_{j,i} < \frac{1}{2} + \frac{1}{q(n)}.$$

A generator is perfect if it passes all polynomial size next bit tests. A different, but equivalent, formulation was given by Yao. In Section 17.3.3, we describe Blum and Micali's construction of a generator that passes the next bit test assuming the hardness of the discrete log problem.

17.2.2 Feedback Shift Registers

Many devices that are proposed for generating cryptographically strong pseudorandom sequences are based on linear feedback shift registers or LFSRs for short [14]. An LFSR of length r has an r-bit state vector that is updated by shifting by one position and filling the vacated position by a linear function of the previous state. More specifically, if we let the state be $\bar{a} = (a_0, \ldots, a_{r-1})$, $a_i \in \{0, 1\}$, then there is a linear feedback function

$$f(\bar{a}) = \left(\sum_{i=0}^{r-1} c_i a_i \right) \bmod 2,$$

where each c_i is a bit. The state is updated by replacing $\bar{a}$ with $(a_1, \ldots, a_{r-1}, f(\bar{a}))$. A diagram of an LFSR is given in Figure 17.1.

In hardware the state change can be thought of as tapping the cells whose corresponding c_is equal 1 and adding the values modulo 2. Thus LFSRs are extremely fast, especially when implemented in hardware and when the number of tapped cells is small. They can be designed to generate sequences of large period—up to $2^r - 1$. A sequence of period $2^r - 1$ output by an LFSR of length r is called

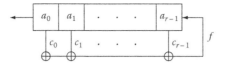

FIGURE 17.1 Linear feedback shift register.

an *m*-**sequence**. These sequences have many of the desirable statistical properties mentioned in Section 17.2.1.2. They are optimally balanced for odd period sequences. Every subsequence of length $t \leq r$ occurs 2^{r-t} times, except the all-zero subsequence which occurs $2^{r-t} - 1$ times. The shifted autocorrelations all equal -1. The distribution of subsequences of each length up to r is nearly uniform and the distribution on runs is nearly perfect.

There is a useful algebraic theory of LFSRs. To describe this, some algebraic background is needed. If $q = 2^r$, then there is a unique finite field with q elements, called $GF(q)$. An element α of $GF(q)$ is called primitive if every nonzero element of $GF(q)$ is a power of α. (Compare to Section 10.5 of Chapter 10 of this book). Primitive elements exist in every $GF(q)$. The minimal degree polynomial with coefficients in $GF(2) = \mathbf{Z}/2\mathbf{Z} = \{0, 1\}$ for which α is a root is called a primitive polynomial. We also need the trace function, defined by

$$Tr_1^r(x) = x + x^2 + x^4 + \cdots + x^{2^{r-1}}.$$

The trace function maps $GF(q)$ to $GF(2)$, is nonzero, and is $GF(2)$-linear, thinking of $GF(q)$ as a vector space over $GF(2)$. In fact, every such linear function is of the form $x \mapsto Tr_1^r(\gamma x)$ for some $\gamma \in GF(q)$. These notions generalize to a setting where the base field $GF(2)$ is replaced by an arbitrary finite field of arbitrary characteristic. This makes it possible to extend the following analysis of LFSR sequences to LFSRs whose entries lie in an arbitrary finite field.

We can associate to any eventually periodic sequence **a** the generating function

$$g(x) = \sum_{i=0}^{\infty} a_i x^i.$$

We can associate to any LFSR with feedback function $\sum_{i=0}^{r-1} c_i a_i$ the connection polynomial

$$q(x) = \sum_{i=1}^{r} c_{r-i} x^i - 1.$$

These are defined over $GF(2)$. Then $g(x)$ is a rational function and, when represented as a quotient of two relatively prime polynomials, the denominator is the connection polynomial of the smallest LFSR that outputs **a**. An LFSR sequence is an m-sequence if and only if the connection polynomial is primitive. The period of **a** is the least n such that $g(x)$ divides $x^n - 1$. Also, it can be shown that if the connection polynomial of an LFSR is irreducible, then there are elements $\gamma, \alpha \in GF(2^r)$ such that

$$a_i = Tr_1^r\left(\gamma \alpha^i\right).$$

The sequence **a** is an m-sequence if and only if α is primitive. It is these algebraic structures that make it possible to analyze many of the statistical properties of m-sequences mentioned above.

Despite their nice statistical properties m-sequences are cryptologically weak. This is due to the linearity of the feedback function and the associated algebraic structures.

If **a** is any sequence, then the size of the smallest LFSR that outputs **a** is called the **linear span** or linear complexity of **a**. We denote this quantity by $\lambda(\mathbf{a})$. There is an algorithm, due to Berlekamp and Massey [24] which, given $2\lambda(\mathbf{a})$ bits of a sequence, outputs a description of a minimal length

Berlekamp–Massey (**a**)

```
1      input aᵢs until the first nonzero a_{k−1} is found
```

1   input a_is until the first nonzero a_{k-1} is found
2   $g \leftarrow a_{k-1} \cdot x^{k-1}$
3   $p_{k-1} \leftarrow 0$
4   $q_{k-1} \leftarrow 1$
5   $m \leftarrow k - 1$
6   $p_k \leftarrow x^{k-1}$
7   $q_k \leftarrow x^{k-1} + 1$
8   **while** there are more bits
9   **do**   input a new bit a_k
10            $g \leftarrow g + a_k x^k$
11   **if** $g \cdot q_k \equiv p_k \left(\mathrm{mod}\, x^{k+1} \right)$
12   **then** $q_{k+1} \leftarrow q_k$
13                        $p_{k+1} \leftarrow p_k$
14   **else** $q_{k+1} \leftarrow q_k + x^{k-m} q_m$
15                        $p_{k+1} \leftarrow p_k + x^{k-m} p_m$
16   **if** $\deg (q_{k+1}) > \deg (q_k)$
17   **then** $m \leftarrow k$
18                $k = k + 1$
19   **return** q_k, p_k

FIGURE 17.2 The Berlekamp–Massey algorithm.

LFSR that generates **a**. This algorithm is given in Figure 17.2. At the kth stage the best rational representation of the generating function modulo x^k is found. Details of the proof that the algorithm achieves this and converges in $2\lambda(\mathbf{a})$ steps were given by Massey [24]. Furthermore, the Berlekamp–Massey algorithm can be made efficient by computing the product in step 11 from previous values in linear time. Thus if the algorithm examines T bits of a sequence, then it runs in time $\mathcal{O}(T^2)$. If the period of a sequence is exponentially larger than its linear span, then this time is quite small. This is the case for m-sequences. Thus LFSRs are unsuitable for generating sequences for stream ciphers. Moreover, sequences generated by other means must have large linear span or they will still be susceptible to the Berlekamp–Massey algorithm.

A major goal of research on stream ciphers has been to design efficient keystream generators whose output has large linear span. Despite the cryptographic weakness of LFSRs, their speed, simplicity, and ability to be analyzed make them important building blocks for stream ciphers. They are useful as well in other areas such as spread spectrum systems, radar systems, and Monte Carlo simulation. Many variations on LFSRs have been proposed that gain cryptographic strength by introducing some nonlinearity. Several of these are described in Section 17.3.

It has long been known that a small change in a sequence can result in an enormous change in the linear span. For example, the all 0 sequence has linear span 0, but if we change every nth position to a 1, then the resulting sequence has linear span n (any register of length less than n that outputs this sequence must reach an all 0 state, and will output all 0s from then on, which is a contradiction). Suppose **a** is any sequence of period n and **b** is another sequence with the same period that differs from **a** in a small number k of bits per period. If a cryptanalyst is given a prefix $\bar{u} = a_0, \ldots, a_{m-1}$ of **a**, she can run the Berlekamp–Massey algorithm for every m-tuple that differs from $\bar{u}$ in at most k positions. Among the resulting generators will be one that outputs **b**, although it may be problematic deciding which generator to select. This led Ding et al. [10] to define the sphere complexity. Let $O(\mathbf{a}, k)$ be the set of sequences of period n that differ from **a** in at least 1 and at most k positions. Then the sphere complexity of **a** is defined as

$$SC_k(\mathbf{a}) = \min_{smb \in O(sma, k)} \lambda(\mathbf{b}).$$

Although less important than the linear span, it is desirable that the sphere complexity of a sequence be large.

LFSRs can be generalized to feedback registers whose entries are elements of an arbitrary fixed finite field $GF(q)$. The coefficients a_i are arbitrary elements of $GF(q)$. Such a register outputs a sequence of elements of $GF(q)$. Essentially everything we have said about LFSRs applies in this more general setting, although some slight modification is necessary in steps 14 and 15 of the Berlekemp–Massey algorithm.

17.3 State of the Art

The importance of cryptanalysis is threefold. First, it reveals weaknesses in existing systems so that users know what to avoid. Second, it provides a limited means of certification: users are more confident in a system that has withstood attack for a reasonable time. This is important in an area where formal proofs of security seem unlikely and systems often intentionally defy formal analysis. Third, the study of cryptanalysis often reveals general principles for the design of future cryptosystems. In addition to the Berlekamp–Massey algorithm there have been several general methods of cryptanalysis of stream ciphers. In Section 17.3.1 we discuss **correlation attacks** [12,26,36] and 2-adic rational approximation [20].

The ultimate goal of research on stream ciphers is to provide fast methods of securely transmitting data. Methods that have been proposed in recent years include feedback shift register based methods such as **nonlinear filter generators** [15]; **nonlinear combiners** [31]; clock-controlled shift registers [13]; shrinking generators [7]; and cyclotomic generators [9]. There have also been several recent proposals of generators that are not based on shift registers, and are suitable for software implementations. These include RC4 and SEAL [29]. In Section 17.3.2 we survey these approaches.

In contrast to public key cryptosystems, the theoretical foundations of stream ciphers are generally weak or only weakly connected to practice. In Section 17.3.3 we discuss complexity theoretic models for stream cipher security [3,39]; and models for security against broad generalizations of the Berlekamp–Massey algorithm [19].

17.3.1 Cryptanalysis

In this section we describe general methods of cryptanalyzing stream ciphers.

17.3.1.1 Correlation Attacks

Consider a situation in which one or more feedback registers are made to interact to produce an output sequence. Suppose a cryptanalyst knows the structure of the feedback registers and how they interact but not the start states. This is a typical arrangement when keystream generators are implemented in hardware. The goal of the cryptanalyst then is to determine the initial states of the registers from a known segment of key stream. If the generator is constructed so that the state is large enough, then an exhaustive search is infeasible. The idea behind a correlation attack is to find statistical correlations between keystream bits and state bits. These correlations can then be used to improve searches for the initial states. Typically, exhaustive search is performed on the start state of one of the underlying registers until the correlations match the prediction.

This general framework has been used to attack combination generators, nonlinear filter generators, and various clock-controlled shift registers. These attacks are described in greater detail in the sections below on the specific generators.

17.3.1.2 2-Adic Rational Approximation

The method of 2-adic rational approximation is based on a class of feedback registers invented by Goresky and Klapper [20] that is analogous to LFSRs. These registers, called **feedback with carry shift registers** or FCSRs, are based on algebra over the integers and 2-adic numbers just as LFSRs are based on algebra over polynomials and power series. An FCSR of length r has an r-bit state vector $\bar{a} = (a_0, \ldots, a_{r-1})$ plus an integer memory m. The state is updated similarly to that of an LFSR, but the addition is performed as integers rather than modulo 2. The low bit is fed back to the register and the high bits are retained in m as a carry to the next state change. More precisely, there is a linear feedback function

$$f(\bar{a}, m) = m + \sum_{i=0}^{r-1} c_i a_i,$$

where each c_i is a bit. The state is updated by replacing $\bar{a}$ with

$$\left(a_1, \ldots, a_{r-1}, f(\bar{a}, m) \bmod 2\right)$$

and replacing m by

$$\left\lfloor f(\bar{a}, m)/2 \right\rfloor.$$

A diagram of an FCSR is given in Figure 17.3.

FCSRs are very simple and fast devices that can output sequences that are exponentially larger than the size of the register. There are many analogies between LFSRs and FCSRs. Associated with the output sequence **a** is the 2-adic number

$$\alpha = \sum_{i=0}^{\infty} a_i 2^i.$$

The algebra of 2-adic numbers is like that of power series, but addition and multiplication are performed with carry to higher terms instead of modulo 2. Associated with the FCSR is the connection number

$$q = \sum_{i=1}^{r} c_{r-i} x^i - 1.$$

An FCSR sequence is always eventually periodic. The associated 2-adic number α is rational, and q is the denominator of a rational representation of α. Thus the cryptanalytic problem of finding the smallest FCSR that outputs a given sequence is equivalent to the problem of finding the minimal rational representation for a 2-adic number. This problem was solved by de Weger's algorithm. An adaptive version appears in [20], where the notion of 2-adic span (the minimal number of bits of storage used by an FCSR that outputs **a**) is defined. As with linear span, in order for a sequence to be cryptographically strong it must have large 2-adic span. It was further shown that sequences generated by summation combiners (see Section 17.3.2.1) have relatively low 2-adic span. Various

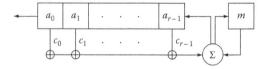

FIGURE 17.3 Feedback with carry shift register.

generalizations of FCSRs have been proposed by exploiting known generalizations of the algebra of 2-adic numbers.

Despite this cryptanalysis, maximal period FCSR sequences (ones for which 2 is a primitive root modulo the connection number) have many desirable statistical properties. Their periods are exponentially larger than the sizes of their generators, they are balanced, their arithmetic correlations vanish, and they are nearly de Bruijn sequences. Thus they are potential substitutes for m-sequences as building blocks for keystream generators.

17.3.2 Keystream Generation

In this section we survey various methods of generating sequences for stream ciphers. Much of the research in this area has concentrated on generating sequences with large linear span and immunity to correlation attacks.

17.3.2.1 Linear Congruential Generators

Linear congruential generators are often suggested as pseudorandom generators because they are simple, have large period, and are readily available on computer systems. However, they are highly linear and hence are cryptographically weak.

A linear congruential generator is determined by an integer modulus, m, and a pair of integers, u and v, with $0 < u < m$ and $0 \le v < m$. A sequence of integers $a_0, a_1, a_2, \ldots$ is generated by choosing a_0 arbitrarily and computing

$$a_n = u a_{n-1} + v \bmod m.$$

If v is relatively prime to m and u has maximal order modulo m (that is, $u^i = 1 \bmod m$ only if $\phi(m)$ divides i, where ϕ is Euler's function), then this sequence has maximal period $\phi(m)$.

Several attacks on linear congruential generators have appeared in the literature. The one due to Boyar assumes that u, v, and m are unknown and the cryptanalyst does not know the $\log(\log(m))$ low order bits of each a_n [4]. Lagarias and Reeds have described an attack on a generalization of the linear congruential generator in which the linear function is replaced by an arbitrary polynomial.

17.3.2.2 Nonlinear Combiners

A nonlinear combiner takes the outputs from a set of k LFSRs and combines them with a nonlinear function $h : GF(2)^k \to GF(2)$. We denote by a^j the output from the jth LFSR. Then the output $\mathbf{b}$ of the nonlinear combiner is the sequence whose ith bit is $b_i = h(a_i^1, \ldots, a_i^k)$. A diagram of a nonlinear combiner with $k = 2$ is given in Figure 17.4.

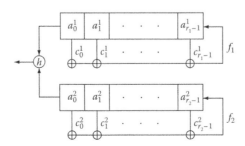

FIGURE 17.4 Nonlinear combiner.

The Geffe generator is a special case using three LFSRs. The output from the third register is used to select between the first two. That is, $h(a^1, a^2, a^3) = a^3 a^1 \oplus (\neg a^3) a^2$. The period of the Geffe generator is $n_1 n_2 n_3$ and the linear span is

$$\lambda\left(\mathbf{a}^1\right) \lambda\left(\mathbf{a}^3\right) + \lambda\left(\mathbf{a}^2\right)\left(1 + \lambda\left(\mathbf{a}^3\right)\right).$$

However, state information is leaked since

$$\text{Prob}\left(h\left(a^1, a^2, a^3\right) = a^1\right) = \text{Prob}\left(h\left(a^1, a^2, a^3\right) = a^2\right) = \frac{3}{4}.$$

Another special case is the *threshold generator*. This generator combines k LFSR sequences by outputting a 1 if and only if the majority of the outputs are 1. That is,

$$h\left(a^1, \ldots, a^k\right) = \begin{cases} 1 & \text{if } \displaystyle\sum_{i=1}^{k} a^i > \frac{k}{2} \\ 0 & \text{otherwise.} \end{cases}$$

In case $k = 3$ the period is $n_1 n_2 n_3$ and the linear span is

$$\lambda\left(\mathbf{a}^1\right) \lambda\left(\mathbf{a}^2\right) + \lambda\left(\mathbf{a}^2\right) \lambda\left(\mathbf{a}^3\right) + \lambda\left(\mathbf{a}^1\right) \lambda\left(\mathbf{a}^3\right).$$

Once again, however, there is a positive correlation between the output sequence **b** and each sequence $\mathbf{a}^i$. Specifically, the mutual information $I(\mathbf{b}; \mathbf{a}^i)$ is 0.189 bits.

In the general case it is known that the period is bounded by

$$\rho(\mathbf{b}) \leq \text{lcm}\left(\rho\left(\mathbf{a}^1\right), \ldots, \rho\left(\mathbf{a}^k\right)\right)$$

and the linear span is bounded by

$$\lambda(\mathbf{b}) \leq h^*\left(\lambda\left(\mathbf{a}^1\right), \ldots, \lambda\left(\mathbf{a}^k\right)\right),$$

where h^* is h thought of as a polynomial over the integers [32]. Further, Key [17] showed that these inequalities become equalities when the $\mathbf{a}^j$ are m-sequences with relatively prime periods. These results generalize to sequences over arbitrary finite fields.

Nonlinear combiners can be further generalized by allowing the combining function to retain a small amount of memory, say m bits. Such a combiner is specified by a pair of functions,

$$h : GF(2)^k \times GF(2)^m \rightarrow GF(2)$$

and

$$u : GF(2)^k \times GF(2)^m \rightarrow GF(2)^m.$$

Thus if the state of the memory is c, then the combiner outputs $b_i = h(a_i^1, \ldots, a_i^k, c)$ and updates the memory by $c = u(a_i^1, \ldots, a_i^k, c)$. Of particular interest is the summation combiner which adds its input sequence with carry, using the extra memory bits to save the carry. That is,

$$h\left(a^1, \ldots, a^k, c\right) = c + \sum_{j} a^j \bmod 2$$

and

$$u\left(a^1, \ldots, a^k, c\right) = \left\lfloor \left(c + \sum_{j} a^j\right) \Big/ 2 \right\rfloor,$$

where we treat c as an integer. Rueppel [31] showed that if the input sequences to a summation combiner are m-sequences with relatively prime periods $\rho_1, \ldots, \rho_k$, then the output of the sequence has period $\prod_i \rho_i$ and linear span close to its period. Summation combiners are susceptible to a 2-adic rational approximation attack.

Combiner generators are vulnerable to correlation attacks. This was first observed by Siegenthaler. The cryptanalyst is assumed to know the combining function h and the individual LFSRs, but not the initial states of the LFSRs. This is equivalent to saying that the cryptanalyst knows the individual LFSR sequences but not the phase shifts that gave rise to the keystream. The idea of Siegenthaler's correlation attack is to pick one of the input sequences to h and compute the correlation of each phase shift with the known bits of the keystream until one is found that matches the predicted correlation. The time needed to determine the initial states of all the LFSRs is proportional to the sum of their periods times the number of known keystream bits. This is much smaller than the time needed to do an exhaustive search of all states, which is proportional to at least the products of the periods of the sequences. An attack such as this that attacks a piece of a generator at a time is sometimes called a divide-and-conquer attack.

To compute statistics it is assumed that each input to h is a sequence of independent uniformly distributed binary random variables. Let $\mathbf{b}$ be the output sequence and suppose $b_0, \ldots, b_{n-1}$ are known. For each j let p_j be the probability that the jth input to h equals the output. If p_j is close to $1/2$, then n must be enormous. On the other hand suppose p_j is far from $1/2$. For each phase shift τ the correlation

$$C_{smaj,smb}(\tau) \overset{def}{=} \frac{1}{n} \sum_{i=0}^{n-1} (-1)^{a_{i+\tau}^j} (-1)^{b_i}$$

is computed. For some T the phase shifts that give the T best correlations are chosen as candidates. (Best means closest to $2p_j - 1$, the a priori expected correlation.) If n and T are large enough, then the probability that the correct phase shift is in the candidate set is large. For example, suppose the length of the jth LFSR is 41, $p_j = 0.75$, and $n = 300$. Then the probability that the correct phase shift is among the best 1000 candidates is 0.98.

In order to build a combiner that resists this correlation attack, it is necessary that all the probabilities p_j be close to $1/2$. However, even if this is the case it may be possible to find a correlation between the output from h and a small subset of its inputs. In this case a similar divide-and-conquer attack can be used. The goal is then to find a combining function h that has no such correlations. This gives rise to the notion of **correlation immunity**, which was defined by Siegenthaler [34].

DEFINITION 17.1 A function $h(x_1, x_2, \ldots, x_k) : GF(2)^k \rightarrow GF(2)$ is mth order correlation immune if the random variable given by any m-tuple of x_js is statistically independent of $h(x_1, x_2, \ldots, x_k)$.

This condition is equivalent to the condition that, for every choice of binary vector $\bar{w} = (w_1, \ldots, w_k)$ with $||\bar{w}|| \leq m$, $h(x_1, x_2, \ldots, x_k)$ is statistically independent of the inner product $\bar{w} \cdot \bar{x}$. It can also be shown that h is mth order correlation immune if and only if $\hat{H}(\bar{w}) = 0$ whenever $1 \leq ||\bar{w}|| \leq m$ (where $\hat{H}(\bar{w})$ is the Walsh transform of h).

Siegenthaler showed that there is a trade-off between the order of correlation immunity and the attainable nonlinearity. The nonlinear order of h is the maximum number of variables in any monomial in the algebraic normal form of h. If h is mth-order correlation immune and $1 \leq m < k$, then its nonlinear order is at most $k - m$. This is lowered to $k - m - 1$ if h is balanced and $m \neq k - 1$, and this is tight. Since both measures cannot be large, memoryless nonlinear combiners are cryptographically weak.

Combiners with memory were invented in part to remedy this. For example, a combiner with output function

$$h(\bar{x}, \bar{y}) = \sum_{i=1}^{k} x_i + g(\bar{y}) \bmod 2$$

is $(n-1)$st-order immune for any nonzero g and any state change function u. Thus g and u can be chosen to satisfy any nonlinearity conditions. Meier and Staffelbach [26] observed that the correlation immunity of such combiners arises because the linear functions applied to input bits in computing the correlations fail to take earlier bits into account. However, if at stage i one considers all bits

$$\left\{ a_\ell^j : 0 \le \ell \le i \text{ and } 1 \le j \le k \right\},$$

then there must be correlations between the ith output bit b_i and linear functions of the form

$$\sum_{\ell=0}^{i} \sum_{j=1}^{k} w_{\ell,j} a_\ell^j.$$

For the summation combiner and for general combiners with a single bit of memory Meier and Staffelbach found explicit linear functions for which these correlations are large. This analysis was generalized to combiners with arbitrary amounts of memory by Golić [12], who showed that if the size of the memory is sufficiently large, then the correlations between inputs and outputs can be kept small.

17.3.2.3 Nonlinear Filter Generators

A nonlinear filter generator applies a nonlinear function h to the state of a linear feedback shift register to produce the output sequence $\mathbf{b}$. Thus if the register has length r, then h is a function from $GF(2)^r$ to $GF(2)$. For speed and ease of implementation it is desirable that h have few terms when expressed as a polynomial. If n is the period of the underlying LFSR sequence, then it is possible to generate any sequence of period dividing n by a nonlinear filter generator. However, when the function h is expressed as a polynomial, it may have as many terms as the period of the sequence. Key [17] showed that if h has degree d then

$$\lambda(\mathbf{b}) \le \sum_{i=1}^{d} \binom{r}{i}.$$

This result was generalized by Chan et al. [6] to registers with nonlinear feedback functions. It follows from Key's result that h must have high degree. Lower bounds are the real need for cryptographic purposes but such results have been rare. Kumar and Scholtz [21] showed that if h is a bent function and 4 divides r, then

$$\lambda(\mathbf{b}) \ge 2^{r/4} \binom{r/2}{r/4} \sim 2 \frac{2^{r/2}}{\sqrt{\pi r}}.$$

Weaker lower bounds for more general classes of filter generators have been shown by Rueppel [31]. It is unknown how to construct filter generators with maximal linear span where h has few terms.

Chan and Games considered the following generalization. Let $\mathbf{a}$ be an m-sequence over a finite field $GF(q)$. Let $h : GF(q) \to GF(2)$. Let $b_i = h(a_i)$. The sequence $\mathbf{b}$ is called a geometric sequence.

If q is even, then h can be represented algebraically and results of Herlestam [16] and Brynielsson can be used to show that if $h(x) = \sum_{i=0}^{q-1} c_i x^i$, then

$$\lambda(\mathbf{b}) = \sum_{c_i \neq 0} \lambda(\mathbf{a})^{||i||}$$

$$\leq q^{\log_2(\lambda(sma)+1)}.$$

To be cryptographically strong, a register of this size would need a linear span close to $q^{\lambda(\mathbf{a})}$. Chan and Games showed that if q is odd, then $\lambda(\mathbf{b})$ can be made as large as $q^{\lambda(sma)-1}$. Furthermore, geometric sequences have optimal autocorrelations (and in some cases low cross-correlations). However, Klapper later showed that if one considers these sequences as sequences over $GF(q)$ (that simply happen to have only two values), then the linear span is low and, by exploiting the imbalance of the sequences, the parameter q can be found with a probabilistic attack [18].

Filter generators are vulnerable to correlation attacks. This was first observed by Siegenthaler [35]. The easiest way to see this is to consider a filter generator whose underlying LFSR has length r as a nonlinear combiner with r input sequences. The LFSRs generating the input sequences are identical, but the initial state of the ith register is the second state of the $(i-1)$st register. Then the same correlation attack that was used on a nonlinear combiner can be used. In fact the attack is now faster since all the underlying registers have the same structure. It is only necessary to cycle through the set of initial states once looking for correlations.

17.3.2.4 Clock-Controlled Generators

A quite different way to introduce nonlinearity in a generator is to irregularly clock certain parts of the generator. A survey of these clock-controlled generators as of 1989 was given by Gollman and Chambers [13]. In a simple case, two LFSRs are used: L_1 and L_2 of lengths n_1 and n_2, respectively. We are also given a function

$$f : \{0, 1\}^{n_1} \to \mathbf{Z}.$$

At each step we clock L_1 once and extract $f(s_i)$ where $s_i = (s_{i,1}, \ldots, s_{i,n_1})$ is the ith state of L_1. Register L_2 then changes state (i.e., is clocked) $f(s_i)$ times and the bit produced by the last state change is taken as the ith output of the clock-controlled generator (if $\mathbf{b}$ is the output from L_2, this is $c_i = b_{\sigma(i)}$, where $\sigma(i) = \sum_{i=0}^{i} f(s_i)$). The case when $f(s_i) = s_{i,1}$ is called the stop-and-go generator and is weak, since each change in the output reveals that a 1 has been generated by L_1. Also, there is a large correlation between consecutive output bits. The strength is improved by taking $f(s_i) = 1 + s_{i,1}$, giving rise to the step-once-twice generator.

For general clock-controlled generators, if ρ_i is the period of L_1 and T is the sum of the f values of the states of L_1 in one period, then the period of the output sequence $\mathbf{c}$ is

$$\rho(\mathbf{c}) = \rho_1 \rho_2 / \gcd(T, \rho_2).$$

Assuming $\gcd(T, \rho_2) = 1$, the period is maximal. We make this assumption for the remainder of this discussion. The linear span of $\mathbf{c}$ is upper bounded by $\lambda(\mathbf{c}) \leq n_2 \rho_1$, hence is large but not maximal. If L_1 generates an m-sequence, then the stop-and-go generator achieves the upper bound. If L_1 and L_2 generate the same m-sequence, then both the stop-and-go and step-once-twice generators achieve the maximum linear span [1].

Several authors have described correlation attacks on clock-controlled shift registers. For example, Golić showed that in some circumstances the feedback polynomial and the initial state of the clocked register can be determined [11].

A variation on the stop-and-go generator, called the alternating step generator, was proposed by Günther. Two stop-and-go generators are used, sharing the same control (L_1) register. The outputs

are then added modulo 2. Let ρ_2 and ρ_2' be the periods of the L_2 registers, which have lengths n_2 and n_2'. Assume that the control register
produces a de Bruijn sequence of period 2^m and the connection polynomials for the L_2 registers are irreducible. Then the output period
is $2^m \rho_1 \rho_2$. The output linear span satisfies

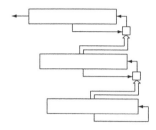

$$(n_2 + n_2') \, 2^{m-1} < \lambda(\mathbf{c}) \le (n_2 + n_2') \, 2^m.$$

Günther also gave conditions under which the distribution of subsequences and the autocorrelations are close to ideal. Unfortunately,
the alternating step generator is vulnerable to a divide-and-conquer
attack against the control register.

FIGURE 17.5 Cascaded clock
controlled shift register of
height $= 3$.

Clock-controlled shift registers can be extended by using a cascade of registers, each output
sequence clocking the next register. The structure may be modified so that the output from stage
$k - 1$ both clocks stage i and is added to the output of stage k modulo 2. A diagram of a cascaded
clock controlled shift register of height 3 is given in Figure 17.5.

If 0,1 clocking is used (that is, a cascaded stop-and-go generator), each LFSR has maximal period
and length n, and all LFSRs have distinct primitive connection polynomials, then the output period
is $(2^n - 1)^k$ and the linear span is $n(2^n - 1)^{k-1}$. For arbitrary s, t clocking suppose the LFSRs have
irreducible degree d connection polynomials and period p with $p^2 \nmid (2^{p-1} - 1)$. Then the output
period is p^n and the output linear span is at least $d(p^n - 1)/(p - 1)$. The distribution of subsequences
of any fixed length approaches the uniform distribution as k approaches infinity.

Correlation attacks may be used to recover state information about the last stage of a cascaded
clock controlled shift register. Another attack based on iteratively reconstructing the states of the
registers is possible in some circumstances [5]. The cryptanalyst is assumed to know the individual
shift register sequences but not the correct phase shifts. She attempts to determine the phase shift
of each stage in turn, starting with the last stage. In some cases if she guesses the phase of a given
stage and reverses the register one step at a time, then the reconstructed output from the preceding
stage eventually locks into the correct phase. This allows the reconstruction to be repeated at the
preceding stage.

With a self-clocking generator a single LFSR is used to clock itself. The case when the clocking
function satisfies $f(s_i) = d$ if $s_{i,1} = 0$ and $f(s_i) = k$ if $s_{i,1} = 1$ is called the $[d, k]$ self-decimation
generator, to which we restrict our attention. We also assume the underlying LFSR has maximal
period $2^n - 1$. The state graph of a $[d, k]$ self-decimation generator may not be purely cyclic. If $d \ne k$,
then the (eventual) period is at most $(3/4)(2^n - 1)$. In case $\gcd(d, 2^n - 1) = 1$ and $2d \equiv k \mod 2^n - 1$
or $2^{n-1} d \equiv k \mod 2^n - 1$ the period is exactly $(2/3)(2^n - 1)$. The distribution of short subsequences
is close to uniform. There is evidence that the linear span is at least 2^{n-1} but this has not been proved.
The drawback to the $[d, k]$-self decimation generator is that each output bit reveals a state bit and
reveals which state bit is revealed by the next output bit. One way of avoiding this is to use different
state bits for output and clocking control.

17.3.2.5 The Shrinking Generator

A somewhat different type of clocking occurs in the shrinking generator [7]. Again, we start with
a pair of LFSRs, L_1 and L_2, with lengths n_1 and n_2 and output sequences $\mathbf{a}$ and $\mathbf{b}$. At each stage, if
$a_i = 1$, then b_i is output. Otherwise no bit is output. Thus the output $\mathbf{c}$ is a shrunken version of
$\mathbf{b}$: $c_j = b_{i_j}$ if i_j is the position of the jth 1 in $\mathbf{a}$. If $\mathbf{a}$ and $\mathbf{b}$ are m-sequences and $\rho(\mathbf{a})$ and $\rho(\mathbf{b})$ are
relatively prime, then $\mathbf{c}$ has period

$$\rho(\mathbf{c}) = \rho(\mathbf{b}) 2^{n_1 - 1} = \left(2^{n_2} - 1\right) 2^{n_1 - 1}.$$

The linear span of **c** satisfies

$$n_1 2^{n_2-2} < \lambda(\mathbf{c}) \le n_1 2^{n_2-1}.$$

It can also be shown that the distribution on fixed length subsequences in **c** is close to uniform. One drawback is the irregularity of the output. A string of zeros in **a** leads to a delay in generating the next bit. Buffering can be used to alleviate this problem. A variation called the self-shrinking generator, where a single register clocks itself, has also been considered.

17.3.2.6 Cyclotomic Generator

Ding [9] proved a lower bound on the linear span of a sequence based only on its period. If n is an integer with prime factorization

$$n = \prod_{i=1}^{t} p_i^{e_i},$$

q is a prime power relatively prime to n, and **a** is a periodic sequence of period n over $GF(q)$, then

$$\lambda(\mathbf{a}) \ge \max\left\{\mathrm{ord}_{p_i}(q) : 1 \le i \le t\right\}.$$

The sphere complexity is similarly bounded. If $k < \min\{||\mathbf{a}||, n - ||\mathbf{a}||\}$, where $||\mathbf{a}||$ is the Hamming weight of a single period of **a**, then

$$SC_k(\mathbf{a}) \ge \max\left\{\mathrm{ord}_{p_i}(q) : 1 \le i \le t\right\}.$$

In particular, if the period n is prime and q is a primitive element modulo n, then the linear span is at least $n - 1$, as is the sphere complexity for $k < \min\{||\mathbf{a}||, n - ||\mathbf{a}||\}$. When n is prime, various conditions guarantee that 2 is a primitive element modulo n. For example, (1) $n4t \pm 1$, t an odd prime; or (2) $n = t_1 t_2 \pm 1$, each t_i an odd prime and $2^{t_i} \not\equiv -1 \bmod n$.

Sequences satisfying the hypotheses of the above results can be produced by a class of generators called cyclotomic generators whose analysis is based on the theory of cyclotomic numbers. These generators use a base register that counts by ones modulo n. A function h is then applied to the value of the counter to produce an output bit. The simplest case is the cyclotomic generator of order $2k$, for which

$$h(i) = \left(i^{(n-1)/2k} \bmod n\right) \bmod 2.$$

Here $(x \bmod n) \bmod 2$ means reduce x modulo n to a residue in the range 0 to $n - 1$, then take the parity. It can further be shown that the autocorrelations of these sequences are ideal. Several generalizations of this generator have also been considered by Ding [9].

17.3.2.7 RC4

RC4 is a byte-oriented stream cipher designed by Rivest for RSA Data Security, Inc. in 1987. It is intended for use in software. Its details were unpublished and proprietary until 1994, when they were anonymously leaked on the sci.crypt newsgroup. The state consists of two nonnegative integer variables x and y, each less than 256, and an array P of 256 bytes. P must contain a permutation of $\{0, 1, \ldots, 255\}$. A sequence of bytes is generated by the pseudocode in Figure 17.6. All addition is modulo 256.

RC4(**a**)

```
1    while more bytes are needed
2        x ← x + 1
3        y ← y + P[x]
4        swap P[x] and P[y]
5        output P[P[x] + P[y]]
```

FIGURE 17.6 The RC4 stream cipher.

The key is an array of bytes, $K[0], \ldots, K[m-1]$ for some m. This is used to initialize the state as follows: The variables x and y are set to 0 and the array P is set to $(0, 1, \ldots, 255)$. Then the RC4 algorithm in Figure 17.6 is iterated 256 times, with line 3 replaced by

$$y \leftarrow y + P[x] + K[x \bmod m],$$

and with line 5 deleted. Finally, x and y are reset to 0. Very little is publicly known about the security of RC4. RSA Data Security, Inc. claims the algorithm is immune to linear and differential cryptanalysis [8], has no small cycles, and is highly nonlinear.

17.3.2.8 SEAL

Software-optimized encryption algorithm (SEAL) was proposed in 1993 by Rogaway and Copper-smith [29]. Since it is quite new, little is known about its cryptanalysis. SEAL is designed for use in software. It depends on a 32-bit architecture and uses eight 32-bit registers and about 3 kB of cache. It is claimed to have a data rate ten to thirty times faster than DES, the most popular block cipher.

A basic design principle of SEAL is to generate a large table in a preprocessing stage. This stage is relatively slow but is intended to take place concurrently with the (generally slow) key exchange at the beginning of a session. In such a setting the preprocessing of SEAL incurs little extra cost. The table is generated by a complex mix of bit-wise ands, bit-wise ors, bit-wise exclusive-ors, bit-wise complements, concatenations, shifts, and additions modulo 2^{32}.

The input to SEAL is a 160 bit string a and a message size L. For each a, a function SEAL$_a$ is determined from the set of positive integers to the set of infinite binary strings. Encryption is performed by computing the bit-wise exclusive or of the nth message and the first L bits of SEAL$_a(n)$. In practice only the first $128 \lceil L/128 \rceil$ bits of SEAL$_a(n)$ are generated. Again, the computation of SEAL$_a(n)$ is a complex mix of bit-wise ands, bit-wise ors, bit-wise exclusive-ors, bit-wise complements, con-catenations, shifts, and additions modulo 2^{32}, plus table lookups using the tables generated in the preprocessing stage.

The assumed strength of SEAL depends on the fact that, if a is chosen uniformly at random, then SEAL$_a(n)$ is computationally indistinguishable from a random L-bit function of n.

17.3.3 Universal Security

In this section we consider the existence of universally secure generators from several theoretical viewpoints. We start with complexity theoretic considerations.

17.3.3.1 Blum–Micali Discrete Log Generator

Blum and Micali designed a generator whose security is based on the assumed hardness of the discrete log problem [3]. Suppose p is a prime number and g is a generator of Z_p^*, the multiplicative group of integers modulo p. If $x \in Z_p^*$, then the discrete logarithm of x with respect to g is the unique integer k such that $x = g^k \bmod p$. The discrete log problem is to find k given p, g, and x. This problem is widely believed to be computationally infeasible. If x is a quadratic residue modulo p, then its discrete log is of the form $2t$ with $0 \le t < (p-1)/2$. Its square roots are g^t and $g^{t+(p-1)/2}$. The latter is called the principle square root. The discrete log problem is polynomial time reducible to the problem of determining whether an element is a principle square root. Define the predicate

$$\psi_p(x) = \begin{cases} 1 & \text{if } x < (p-1)/2, \\ 0 & \text{otherwise.} \end{cases}$$

The state of a Blum–Micali generator is an element $x \in Z_p^*$. At each stage the generator outputs $\psi(x)$ and changes state to $g^x \bmod p$. It can be shown that if a certain assumption about the infeasibility of solving the discrete log problem with polynomial size circuits is true, then the Blum–Micali generator is perfect in the sense of Section 17.2.1.3 in Section 17.2 when the initial state (seed) is chosen randomly.

17.3.3.2 Other Perfect Generators

The Blum–Micali generator is actually an example of a general construction that can be based on any one-way permutation f (in the case of the discrete log generator, $f(x) = g^x$) and binary predicate B on the domain of the predicate (in the case of the discrete log generator, $B(x) = \psi_p(x)$). There is a notion of an unpredictable predicate that guarantees that the resulting generator is perfect. The goal is to design a predicate whose unpredictability is equivalent to some problem that is infeasible. Such a predicate is called hard core for f. In reality we have no proof of infeasibility for the problems that arise this way, so belief in the security of such generators depends on belief in the infeasibility of the underlying problem (e.g., the discrete log problem).

One such construction is based on the infeasibility of inverting RSA ciphertexts. If $n = pq$ is a product of two primes and e is relatively prime to $(p - 1)(q - 1)$, then the domain is Z_n^*. The permutation is $f(x) = x^e$ and the hard core predicate is the least significant bit. It can be shown that if RSA ciphertexts cannot be inverted in expected polynomial time, then the RSA generator is perfect.

Another such construction is based on the infeasibility of computing square roots modulo a product of two primes that are congruent to 3 modulo 4 [2]. This problem is known to be equivalent to factoring the modulus. Here the domain is the set of quadratic residues modulo an RSA modulus (i.e., a product of two large primes), the permutation is squaring, and the hard core predicate is the least significant bit. It can be shown that if it is impossible to compute modular square roots with polynomial size circuits on a polynomial fraction of the moduli n, correctly on all but a polynomial fraction of Z_n^*, then the quadratic residue generator is perfect.

There have been several recent results that show that other bits (in addition to the least significant bit) are hard core, thus expanding the suite of perfect generators. For example, Näslund showed that any bit in a random linear function modulo a random prime is hard core for any one-way function.

17.3.3.3 Provable Security

Several attacks on stream ciphers synthesize a generator given a prefix of the keystream. The synthesized generator belongs to a specific class of generators $\mathcal{F}$. If enough bits are available, the attack should produce the smallest generator (in the sense of the size of the state space) in $\mathcal{F}$ that outputs the given sequence. The Berlekamp–Massey and the 2-adic rational approximation algorithms are examples of such attacks.

Klapper [19] studied the question of whether there exists a family of generators that resists all such attacks. For a family $\mathcal{F}$ and a sequence **a** the size of the smallest generator is denoted $\lambda_{\mathcal{F}}(\mathbf{a})$. Such an attack A is called effective if it runs in polynomial time and is successful whenever the number of bits available is at least a fixed polynomial in $\lambda_{\mathcal{F}}(\mathbf{a})$. The existence of an effective $\mathcal{F}$-synthesizing algorithm implies that $\lambda_{\mathcal{F}}(\mathbf{a})$ is a measure of security for stream ciphers, analogous to the linear span. By a diagonalization argument it can be shown that there exists a family of efficiently generated sequences S such that, for any effective algorithm synthesizing generators in a family $\mathcal{F}$, $\lambda_{\mathcal{F}}(\mathbf{a})$ grows superpolynomially in the log of the period of $\mathbf{a} \in S$. Various generalizations of this result are possible for weaker notions of security.

A quite different approach to provable security was taken by Maurer [25]. He considered stream ciphers based on the availability of a public global source of randomness. Maurer described such a randomized stream cipher that is perfectly secure with high probability. The proof of this result is based on Shannon's information theory. The sender and receiver need only share a short key, while the cryptanalyst must examine an infeasible number of random bits in order to attack the system. The drawback to this system is the difficulty in making a source of a large number of truly random bits publicly available.

17.4 Research Issues and Summary

In this article we provide a survey of techniques related to pseudorandom sequence generation and its role in cryptography. Stream ciphers, the resulting systems, are generally much faster than other cipher systems. Thus they are useful when high-speed secure communications is needed. Proving their security, however, is more difficult than that of many public key systems. Techniques for generating sequences that have cryptographically strong properties are presented, as well as techniques for cryptanalyzing sequences. A number of measures of randomness are also described.

A variety of sequence generators are presented. Many of these are based on modifications of linear feedback shift registers to eliminate vulnerability to Berlekamp–Massey type attacks. In most cases the best that can be said is that the generators resist certain known attacks and have various desirable randomness properties. In some cases the security of a generator is described in complexity theoretic (and in particular asymptotic) terms. Unfortunately such generators tend to be slow.

A few of the cryptanalytic techniques presented are general and can be tried on any stream cipher. Such techniques give rise to general measures of security. More often, cryptanalytic techniques are specific to generators or classes of generators. Nonetheless, they can sometimes be adapted to other classes of generators. Attempting the cryptanalysis of a generator is important for its certification as a secure system.

One focus of current research is the development of new techniques of keystream generation. There are no known provably secure efficient keystream generators, so their need persists. When each new technique of cryptanalysis is developed, new generators are needed that resist the new attack, as well as all old attacks. The difficulty faced by designers of keystream generators is that the presence of enough structure to prove that a generator resists known attacks and has other desirable properties often leads to the development of new cryptanalytic techniques to which the generator is vulnerable. The ideal would be to design a type of keystream generator that resists all computationally feasible cryptanalytic attacks, is efficient, and has an efficient description. This is a goal for security defined in various models.

Simultaneously, considerable research is focused on developing new cryptanalytic techniques. The challenge is to find exploitable structure in systems that are often designed to have little clear structure or to combine several types of apparently incompatible structure. The search for cryptanalytic tools may at first seem perverse and counter productive. However, when a system has weaknesses it is important that potential users be aware of this. These weaknesses may be known already to hostile cryptanalysts who are unlikely to advertise their knowledge.

17.5 Further Information

Research on pseudorandom sequences and stream ciphers is extensively published in the *IEEE Transactions on Information Theory* and the *Journal of Cryptology*.

Many major results appear initially in the proceedings of the annual conferences Crypto, held in Santa Barbara, California, Eurocrypt, held in various countries in Europe, and Asiacrypt, held in various countries in Asia and Australia. The proceedings of these conferences appear in the Springer-Verlag *Lecture Notes in Computer Science* series. Crypto and Eurocrypt are held under the auspices of the IACR, the International Association for Cryptologic Research.

Many papers have also appeared at specialty workshops such as "Fast Software Encryption," whose proceedings also appear in the Springer-Verlag *Lecture Notes in Computer Science* series.

Information about the IACR and about many conferences on cryptography can be found at the IACR web-site, http://www.iacr.org/~iacr/.

The chapter by Rueppel in Simmon's *Contemporary Cryptography* [36] provides a more detailed summary of the state of the art as of 1990.

Menezes, van Oorschot, and Vanstone's *CRC Handbook of Applied Cryptograpy* [27] provides a thorough summary as of 1996.

A thorough treatment of the basic analysis of linear feedback shift registers can be found in Golomb's classic *Shift Register Sequences* [14].

Many aspects of shift registers, stream ciphers, and linear span are treated in Rueppel's *Analysis and Design of Stream Ciphers* [31].

Many of the more practical aspects of stream ciphers, as well as a thorough bibliography, can be found in Schneier's *Applied Cryptography* [33].

Good sources for the mathematical foundations of the study of pseudorandom sequences are Lidl and Niederreiter's *Finite Fields* [22] and McEliece's *Finite Fields for Computer Scientists and Engineers* [23].

Defining Terms

Autocorrelation: The number of positions where a periodic binary sequence agrees with a shift of itself minus the number of places where it disagrees.

Balance: The number of zeros minus the number of ones in a single period of a periodic binary sequence.

Clock-controlled generator: A keystream generator in which one LFSR is used to determine which output symbols of a second LFSR are used as the final output.

Correlation attack: A cryptanalytic attack on a keystream generator based on improving key search by exploiting correlations between output bits and initial state or seed bits.

Correlation immunity: A measure of resistance of a nonlinear combiner to correlation attacks.

Cryptographically strong pseudorandom bit generator (CSPRB): An efficient family of keystream generators such that it is infeasible to predict bits with probability significantly greater than one half.

Entropy: A formal measure of the uncertainty in a random variable.

Feedback with carry shift register (FCSR): A feedback register similar to an LFSR, but where the addition in the feedback function is performed with carry, the carry being retained for the next stage in extra memory.

Information theory: The mathematical study of the information content of random variables.

Keystream generator: A device or algorithm for generating a pseudorandom sequence.

Linear feedback shift register (LFSR): A device for generating infinite periodic sequences. The state updates by shifting its state vector one position and generating a new state symbol as a linear function of the old state vector. The output is the symbol shifted out of the register.

Linear span: Also called the linear complexity, the length of the shortest linear feedback shift register that outputs a given sequence.

m-Sequence: A maximal period linear feedback shift register sequence.

Next bit test: A family of circuits used to predict the next bit of a sequence given a prefix of the sequence.

Nonlinear combiner: A keystream generator in which the outputs of several LFSRs are combined by a nonlinear function. The combining function may have a small amount of memory.

Nonlinear filter generator: A linear feedback shift register modified so the output is computed as a nonlinear function of the state.

One-time pad: A stream cipher in which the key bits are chosen independently at random.

Pseudorandom sequence: An infinite periodic sequence that behaves like a truly random sequence with respect to various measures of randomness.

Stream cipher: A private key cryptosystem in which the key is added to the plaintext symbol by symbol modulo the size of the plaintext alphabet.

Strict avalanche condition (SAC): A nonlinearity condition. A function satisfies SAC if changing any single bit results in a function uncorrelated with f.

References

1. Beth, T. and Piper, F., The stop-and-go generator. In *Advances in Cryptology—Eurocrypt '84*, T. Beth, N. Cot, and I. Ingemarsson (Eds.), *Lecture Notes in Computer Science*, Vol. 209, 88–92, Springer-Verlag, Berlin, Germany, 1985.
2. Blum, L., Blum, M., and Shub, M., A simple unpredictable pseudo-random number generator. *SIAM J. Comput.*, 15, 364–383, 1986.
3. Blum, M. and Micali, S., How to generate cryptographically strong sequences of pseudorandom bits. *SIAM J. Comput.*, 13, 850–864, 1984.
4. Boyar, J., Inferring sequences produced by a linear congruential generator missing low-order bits. *J. Cryptol.*, 1, 177–184, 1989.
5. Chambers, W. and Gollmann, D., Lock-in effect in cascades of clock-controlled shift-registers. In *Advances in Cryptology—Eurocrypt '88*, G. Günther (Ed.), *Lecture Notes in Computer Science*, Vol. 330, 331–344, Springer Verlag, Berlin, Germany, 1988.
6. Chan, A., Goresky, M., and Klapper, A., On the linear complexity of feedback registers. *IEEE Trans. Inform. Theory*, 36, 640–645, 1990.
7. Coppersmith, D., Krawczyk, H., and Mansour, Y., The shrinking generator. In *Advances in Cryptology—Crypto '93*, D. Stinson (Ed.), *Lecture Notes in Computer Science*, Vol. 773, 22–39, Springer-Verlag, New York, 1994.
8. Ding, C., The differential cryptanalysis and design of natural stream ciphers. In *Fast Software Encryption: Proceedings of 1993 Cambridge Security Workshop*, R. Anderson (Ed.), *Lecture Notes in Computer Science*, Vol. 809, 101–120, Springer-Verlag, Berlin, Germany, 1994.
9. Ding, C., Binary cyclotomic generators. In *Fast Software Encryption: Proceedings of 1994 Leuven Security Workshop*, B. Perneel (Ed.), *Lecture Notes in Computer Science*, Vol. 1008, 29–60, Springer-Verlag, Berlin, Germany, 1995.
10. Ding, C., Xiao, G., and Shan, W., *The Stability Theory of Stream Ciphers, Lecture Notes in Computer Science*, Vol. 561, Springer-Verlag, Berlin, Germany, 1991.
11. Golić, J., Fast correlation attacks on irregularly clocked shift registers. In *Advances in Cryptology—Eurocrypt '95*, J. Quisqater (Ed.), *Lecture Notes in Computer Science*, Vol. 921, 248–262, Springer-Verlag, Berlin, Germany, 1995.
12. Golić, J., Correlation properties of a general binary combiner with memory. *J. Cryptol.*, 9, 111–126, 1996.
13. Gollman, D. and Chambers, W., Clock-controlled shift registers: A review. *IEEE J. Sel. Area. Commun.*, 7, 525–533, 1989.
14. Golomb, S., *Shift Register Sequences*, Aegean Park Press, Laguna Hills, CA, 1982.
15. Groth, E., Generation of binary sequences with controllable complexity. *IEEE Trans. Inform. Theory*, IT-17, 288–296, 1971.

16. Herlestam, T., On function of linear shift register sequences. In *Advances in Cryptology—Eurocrypt '85*, E. Pichler (Ed.), *Lecture Notes in Computer Science*, Vol. 219, 119–129, Springer-Verlag, Berlin, Germany, 1985.

17. Key, E., An analysis of the structure and complexity of nonlinear binary sequence generators. *IEEE Trans. Inform. Theory*, IT-22, 732–736, 1976.

18. Klapper, A., The vulnerability of geometric sequences based on fields of odd characteristic. *J. Cryptol*, 7, 33–51, 1994.

19. Klapper, A., On the existence of secure feedback registers. In *Advances in Cryptology—Eurocrypt '96*, U. Maurer (Ed.), *Lecture Notes in Computer Science*, Vol. 1070, 256–267, Springer-Verlag, Berlin, Germany, 1996.

20. Klapper, A. and Goresky, M., Feedback shift registers, 2-adic span, and combiners with memory. *J. Cryptol.*, 10, 111–147, 1997.

21. Kumar, V. and Scholtz, R., Bounds on the linear span of bent sequences. *IEEE Trans. Inform. Theory*, IT-29, 854–862, 1983.

22. Lidl, R. and Niederreiter, H., *Finite Fields: Encyclopedia of Mathematics*, Vol. 20, Cambridge University Press, Cambridge, MA, 1983.

23. McEliece, R., *Finite Fields for Computer Scientists and Engineers*, Kluwer Academic Publishers, Boston, MA, 1987.

24. Massey, J., Shift register sequences and BCH decoding. *IEEE Trans. Inform. Theory*, IT-15, 122–127, 1969.

25. Maurer, U., A provably-secure strongly-randomized cipher. In *Advances in Cryptology—Crypto '90*, S. Vanstone (Ed.), *Lecture Notes in Computer Science*, Vol. 473, 361–373, Springer-Verlag, New York, 1991.

26. Meier, W. and Staffelbach, O., Correlation properties of combiners with memory in stream ciphers, *J. Cryptol.*, 5, 67–86, 1992.

27. Menezes, A., van Oorschot, P., and Vanstone, S., *CRC Handbook of Applied Cryptography*, CRC Press, Boca Raton, FL, 1996.

28. Preneel, B., Van Leekwijk, W., Van Linden, L., Govaerts, R., and Vandewalle, J., Boolean functions satisfying higher order propagation criteria. In *Advances in Cryptology—Eurocrypt '90*, I. Damgård (Ed.), *Lecture Notes in Computer Science*, Vol. 473, 161–173, Springer-Verlag, Berlin, Germany, 1990.

29. Rogaway, P. and Coppersmith, D., A software optimized encryption algorithm. In *Fast Software Encryption: Proceedings of 1993 Cambridge Security Workshop*, R. Anderson (Ed.), *Lecture Notes in Computer Science*, Vol. 809, 56–63, Springer-Verlag, Berlin, Germany, 1993.

30. Rothaus, O., On bent functions. *J. Combin. Theory (A)*, 20, 300–305, 1976.

31. Rueppel, R., *Analysis and Design of Stream Ciphers*, Springer-Verlag, New York, 1986.

32. Rueppel, R. and Staffelbach, O., Products of sequences with maximum linear complexity. *IEEE Trans. Inform. Theory*, IT-33, 124–131, 1987.

33. Schneier, B., *Applied Cryptography*, John Wiley & Sons, New York, 1996.

34. Siegenthaler, T., Correlation-immunity of nonlinear combining functions for cryptographic applications. *IEEE Trans. Inform. Theory*, IT-30, 776–780, 1984.

35. Siegenthaler, T., Cryptanalyst's representation of nonlinearly filtered ml-sequences. In *Advances in Cryptology—Eurocrypt'85*, E. Pichler (Ed.), *Lecture Notes in Computer Science*, Vol. 219, 103–110, Springer-Verlag, Berlin, Germany, 1986.

36. Simmons, G., Ed., *Contemporary Cryptography*, IEEE Press, New York, 1992.

37. Shannon, C., Communication theory of secrecy systems. *Bell Syst. Tech. J.*, 28, 656–715, 1949.

38. Welsh, D., *Codes and Cryptography*, Clarendon Press, Oxford, U.K., 1988.

39. Yao, A., Theory and applications of trapdoor functions. In *Proceedings, 23rd IEEE Symposium on Foundations of Computer Science, 1982*, 80–91, IEEE Computer Society Press, Los Alamitos, CA, 1982.

18

Theory of Privacy and Anonymity

Valentina Ciriani
Università degli Studi di Milano

Sabrina De Capitani di Vimercati
Università degli Studi di Milano

Sara Foresti
Università degli Studi di Milano

Pierangela Samarati
Università degli Studi di Milano

18.1 Introduction

The increased power and interconnectivity of computer systems and the advances in memory sizes, disk storage capacity, and networking bandwidth allow data to be collected, stored, and analyzed in ways that were impossible in the past due to the restricted access to the data and the expensive processing (in both time and resources) of them. Huge data collections can be analyzed by powerful techniques (e.g., data mining techniques [12]) and sophisticated algorithms thus making possible *linking attacks* combining information available through different sources to infer information that was not intended for disclosure. For instance, by linking deidentified medical records (i.e., records where the explicit identifiers such as the social security numbers (SSN) have been removed) with other publicly available data or by looking at unique characteristics found in the released medical data, a data observer will most certainly be able to reduce the uncertainty about the identities of the users to whom the medical records refer, or—worse—to determine them exactly. This identity disclosure often implies leakage of sensitive information, for example, allowing data observers to infer the illness of patients. The need for *privacy* is therefore becoming an issue that most people are concerned about. Although there are many attempts to create a unified and simple definition of privacy, privacy by its own nature is a multifaceted concept that may encompass several meaning, depending on different contexts.

In this chapter, we focus our attention on the technological aspect of privacy within today's global network infrastructure, where users interact with remote information sources for retrieving data or for using online services. In such a context, privacy involves the following three different but related concepts.

- *Privacy of the user*. It corresponds to the problem of protecting the relationship of users with a particular action and outcome. Since we focus on network infrastructures, we are interested in protecting the relationship between a user and the messages the user sends, which can be, for example, queries for retrieving some information or requests for using a particular online service.

- *Privacy of the communication*. It corresponds to the classical problem of protecting the confidentiality of personal information when transmitted over the network (or with other forms of communication), and to the problem of protecting the privacy of a request to a service provider hiding the content of the request from every party, as well as from the service provider.

- *Privacy of the information*. It involves privacy policies as well as technologies for ensuring proper data protection. A basic requirement of a privacy policy is to establish both the responsibilities of the data holder with respect to data use and dissemination, and the rights of the user to whom the information refers to regulate such use and dissemination. Each user should be able to control further disclosure, view data collected about the user and, possibly, make or require corrections.

We now discuss these three aspects more in details.

18.1.1   Privacy of the User

Privacy of the user concerns protecting the identities of the parties that communicate through a network to avoid possible attacks that have the main purpose to trace who is communicating with whom, or who is interacting with which server or searching for which data. This problem can be solved by providing techniques and protocols that guarantee an *anonymous communication* among different parties. In particular, anonymous communication is about the protection of the relationship between senders and the messages they send, called *sender anonymity*, between recipients and the messages they receive, called *recipient anonymity*, or both, meaning that anonymous senders send messages to anonymous recipients. In the literature, there are a number of approaches providing anonymous communication.

Mix networks, first proposed by Chaum [8], are a way of achieving anonymity on communication networks. Intuitively, a mix is a special node in a network that relays messages in such a way that an outside observer cannot link an outgoing message with an incoming message. Several mix nodes can also be chained to relay a message anonymously. Since their introduction, a large amount of research on mixnets has been performed and different mixnet topologies have been studied and compared [36].

Onion routing [38] is another solution for ensuring anonymous connections. The main goal of onion routing is to guarantee that malicious users cannot determine the contents of messages flowing from a sender to a recipient, and that they cannot verify whether the sender and the recipient are communicating with each other. Onion routing also provides sender anonymity while preserving the ability of the recipient to send a reply to the sender. With onion routing, a network includes a set of *onion routers*, which work as the ordinary routers, combined with mixing properties. When a sender wants to communicate anonymously with a particular recipient, an anonymous connection has to be set up. The sender therefore connects to a particular onion router that prepares an *onion*. An onion is a layered data structure including information about the route of the anonymous connection. In particular, the onion router randomly selects other onion routers and generates a message for each of such a router, providing it with symmetric keys for decrypting messages, and telling it which the next hop (an onion router or the final recipient) in the path will be. Note that anonymity is only provided from the first to the last onion router. The connections from the sender to the first onion router, and from the last onion router to the recipient are not anonymous.

Tor [17] allows users to communicate anonymously on the Internet. Tor is primarily used for anonymous TCP-based applications. Basically, messages are encrypted and then sent through a randomly chosen path of different servers in such a way that an observer cannot discover the source and the destination of the message.

Crowds [34] provides anonymity based on the definition of groups of users, called *crowds*, who collectively perform requests. In this way, each request is equally likely to originate from any user in the crowd. Each user is represented by a local *jondo* stored on the user's machine that receives a request from the user and removes from such a request all identifying information. A jondo acts as a Web proxy that can forward both the user's and other users' requests to the end server, or to another jondo. In this way, since a jondo cannot tell whether or not a request has been initiated by the previous jondo (or the one before it, and so on), users maintain their anonymity within the crowd. The communications along a path of jondos is encrypted and such a path remains the same for the whole session, meaning that requests and replies follow the same path.

18.1.2 Privacy of the Communication

Privacy of the communication is related to two aspects: (1) protection of the confidentiality of personal information sent through a network; and (2) protection of the content of requests to prevent, for example, user profiling. The confidentiality of the personal information transmitted over the network can be ensured by adopting protocols (e.g., SSL) that encrypt it. The need for protecting also the request content arises because there are many real-world scenarios where the request content could be misused by service providers. For instance, consider a medical database that contains information about known illnesses, symptoms, treatment options, specialists, and treatment costs. Suppose that a user submits a query to the medical database to collect information about a specific illness. Any other user, including the database administrator, who is able to observe such a query can then infer that the requestor, or a near relative/friend, might suffer from that specific illness.

The problem of protecting the request content is known as Private Information Retrieval (PIR) problem [10]. There are many different ways for formally defining the PIR problem. The first formulation assumes that a database can be modeled as an N-bit string and a user is interested in *privately retrieving* the ith bit of the N bits stored at the server, meaning that the server does not know which is the bit the user is interested in. Starting from this formulation, many results and advancements have been obtained. In particular, depending on the assumptions about the computational power of the service provider, PIR proposals can be classified into two main classes: *theoretical PIR* and *computational PIR*. Theoretical PIR does not make any assumption about the computational power of the service provider and the privacy of the request must then be provided independently from the computational power of an attacker. Computational PIR assumes that a request is private if the service provider must solve a computationally intractable problem to break it. A naive solution to the theoretical PIR problem consists in completely downloading the database at the client side and performing the research on the local copy of the data. Such a solution however has a high communication cost, especially when the database contains a huge data collection. In [10] the authors prove that if there is only one copy of the database stored on one service provider, there is no solution to the theoretical PIR problem that is better than the solution corresponding to the download of the whole database. By contrast, if there are m copies of the data been stored at different noncommunicating servers, the theoretical PIR problem can be solved by submitting m independent requests (each single request does not provide any information about the bit on which the user is interested in) to the corresponding service providers. The m results are then combined by the user to obtain the answer to their request [5,10,23]. The computational PIR problem is typically solved by encrypting the request submitted to the service provider [7,9,26]. By exploiting properties of the encryption function, the server computes the encrypted result of the request, which can be decrypted only by the requestor.

Note that the PIR problem can also be seen as a Secure Multiparty Computation (SMC) problem [19,41]. The main goal of SMC is to allow a set of parties to compute the value of a function f on private inputs provided by the parties without revealing to them the private inputs of the other parties. The SMC problem is usually solved by modeling function f through a Boolean circuit, where gates correspond to protocols that require the collaboration among parties, since they know the input and share knowledge on the output. Other examples of SMC problems are represented by privacy-preserving statistical analysis [18] and privacy-preserving data mining [4,30]. The SMC problem will be discussed in more details in Chapter 14.

18.1.3 Privacy of the Information

Privacy of the information is related to the definition of privacy policies expressing and combining different protection requirements, as well as the development of technologies for ensuring proper data protection. An important aspect of the data protection issue relates to the protection of the identities of the users to whom the data refer, meaning that the *anonymity* of the users must be guaranteed. Anonymity does not imply that no information at all is released, but requires that information released be nonidentifiable. More precisely, given a collection of personal information about a user, the privacy of the user is protected if the value of the user's personal information is kept private. Whenever a subset of the personal information can be used to identify the user, the anonymity of the user depends on keeping such a subset private. For instance, a recent study [22] shows that in 2000, 63% of the U.S. population was uniquely identifiable by gender, ZIP code, and full date of birth. This means that while these individual pieces of information do not identify a user and therefore the user is anonymous according to the usual meaning of the term, the combined knowledge of gender, ZIP code, and full date of birth may result in the unique identification of a user. Note also that anonymity is always related to the identification of a user rather than the specification of that user. For instance, a user can be univocally identified through her SSN but in the absence of an information source that associates that SSN with a specific identity, the user is still anonymous. Ensuring proper anonymity protection then requires the investigation of the following different issues [13].

- *Identity disclosure protection.* Identity disclosure occurs whenever it is possible to reidentify a user, called *respondent*, from the released data. Techniques for limiting the possibility of reidentifying respondents should therefore be adopted.
- *Attribute disclosure protection.* Identity disclosure protection alone does not guarantee privacy of sensitive information because all the respondents in a group could have the same sensitive information. To overcome this issue, mechanisms that protect sensitive information about respondents should be adopted.
- *Inference channel protection.* Given the possibly enormous amount of data to be considered, and the possible interrelationships between data, it is important that the security specifications and enforcement mechanisms provide automatic support for complex security requirements, such as those due to inference and data association channels [15].

Protection of anonymity is a key aspect in many different contexts, where the identities of the users (or organizations, associations, and so on) to whom the data refer have been removed, encrypted, or coded. The identity information removed or encoded to produce anonymous data includes names, telephone numbers, and SSNs. Although apparently anonymous, the deidentified data may contain other identifying information that uniquely or almost uniquely distinguishes the user [14,20,21], such as the gender, ZIP code, and full date of birth mentioned above. By linking such an information to publicly available databases (e.g., many countries provide access to public records to their citizens thus making available a broad range of personal information) associating them to the user's identity, a data recipient can determine to which user each piece of released information belongs, or restrict the user's uncertainty to a specific subset of users. To avoid this, specific data protection norms apply to data collected for a given purpose that state that such data can be further elaborated for historical, statistical, or scientific purposes, provided that appropriate safeguards are applied. In general, the "appropriate safeguards" depend on the method in which the data are released. In the past, data were principally released in the form of *macrodata*, which are statistics on users or organizations usually presented as two-dimensional tables, and through *statistical databases*, which are databases whose users may retrieve only aggregate statistics [1]. Macrodata protection techniques are based on the *selective obfuscation of sensitive cells* [13]. Techniques for protecting statistical databases follow two main approaches [16]. The first approach restricts the statistical queries that can be made (e.g., queries that identify a small/large number of tuples) or the data that can be published. The second approach provides protection by returning to the user a modified result. The modification can be enforced directly on the stored data or at run time (i.e., when computing the result to be returned to the user). Many situations require today that the specific *microdata*, that is, data actually stored in the database and not an aggregation of them, are released. The advantage of releasing microdata instead of specific precomputed statistics is an increased flexibility and availability of information for the users. At the same time, however, microdata, releasing more specific information, are subject to a greater risk of privacy breaches. Several techniques have been proposed in the literature to protect the disclosure of information that should be kept private [11,13]. In this chapter, we focus on *k*-anonymity [35], an approach that, compared to the other commonly used approaches (e.g., sampling, swapping values, and adding noise to the data while maintaining some overall statistical properties of the resulting table), has the great advantage of protecting respondents' identities while releasing truthful information.

In the remainder of this chapter, we describe the *k*-anonymity concept (Section 18.2) and illustrate some proposals for its enforcement (Sections 18.3 and 18.4). Note that *k*-anonymity protects identity disclosure, while remaining exposed to attribute disclosure [35]. We will see then how some researchers have just started proposing extensions to *k*-anonymity to protect also attribute disclosure [31] (Section 18.5).

18.2 *k*-Anonymity: The Problem

Although *k*-anonymity is a concept that applies to any kind of data, for simplicity its formulation considers data represented by a *relational table*. Formally, let $\mathcal{A}$ be a set of attributes, $\mathcal{D}$ be a set of domains, and *dom*: $\mathcal{A} \rightarrow \mathcal{D}$ be a function that associates with each attribute $A \in \mathcal{A}$ a domain $D = dom(A) \in \mathcal{D}$, containing the set of values that A can assume. A *tuple t* over a set $\{A_1, \ldots, A_p\}$ of attributes is a function that associates with each attribute A_i a value $v \in dom(A_i)$, $i = 1, \ldots, p$.

DEFINITION 18.1   (**Relational table**) Let $\mathcal{A}$ be a set of attributes, $\mathcal{D}$ be a set of domains, and *dom* : $\mathcal{A} \rightarrow \mathcal{D}$ be a function associating each attribute with its domain. A *relational table T* over a finite set $\{A_1, \ldots, A_p\} \subseteq \mathcal{A}$ of attributes, denoted $T(A_1, \ldots, A_p)$ is a set of tuples over the set $\{A_1, \ldots, A_p\}$ of attributes.

Notation *dom*(A, T) denotes the domain of attribute A in T, $|T|$ denotes the number of tuples in T, and $t[A]$ represents the value v associated with attribute A in t. Similarly, $t[A_1, \ldots, A_k]$ denotes the subtuple of t containing the values of attributes $\{A_1, \ldots, A_k\}$. By extending this notation, $T[A_1, \ldots, A_k]$ represents the subtuples of T containing the values of attributes $\{A_1, \ldots, A_k\}$, that is, the projection of T over $\{A_1, \ldots, A_k\}$, keeping duplicates.

In the remainder of this chapter, the relational table storing the data to be protected is called *private table*, denoted PT. Each tuple t in PT reports data referred to a specific *respondent* (usually

	SSN	Name	ZIP	MaritalStatus	Sex	Disease
1			22030	married	F	hypertension
2			22030	married	F	hypertension
3			22030	single	M	obesity
4			22032	single	M	HIV
5			22032	single	M	obesity
6			22032	divorced	F	hypertension
7			22045	divorced	M	obesity
8			22047	widow	M	HIV
9			22047	widow	M	HIV
10			22047	single	F	obesity

FIGURE 18.1 An example of private table PT.

an individual). For instance, Figure 18.1 illustrates an example of private table over attributes SSN, Name, Zip, MaritalStatus, Sex, and Disease containing 10 tuples. Here, the identity information, that is, attributes SSN and Name, have been removed. The attributes (columns) of PT can be classified as follows.

- *Identifiers.* Attributes that uniquely identify a respondent. For instance, attribute SSN uniquely identifies the person with which it is associated.
- *Quasi-identifiers.* Attributes that, in combination, can be linked with external information, reducing the uncertainty over the identities of all or some of the respondents to whom information in PT refers. For instance, the set of attributes ZIP, MaritalStatus, and Sex may represent a quasi-identifier that can be exploited for linking PT with an external information source that associates ZIP, MaritalStatus, and Sex with Name and Address.
- *Confidential attributes.* Attributes that contain sensitive information. For instance, attribute Disease can be considered sensitive.
- *Nonconfidential attributes.* Attributes that do not fall into any of the categories above, that is, they do not identify respondents, cannot be exploited for linking, and do not contain sensitive information. For instance, attribute FavoriteColor is a nonconfidential attribute.

Although all the explicit identifiers (e.g., SSN) are removed from a private table PT, that is public or semipublic released, the privacy of the respondents is at risk, since quasi-identifiers can be used to link each tuple in PT to a limited number of respondents. The main goal of k-anonymity is therefore to protect the released data against possible reidentification of the respondents to whom the data refer.

The k-anonymity requirement is formally stated by the following definition [35].

DEFINITION 18.2　(k-anonymity requirement) Each release of data must be such that every combination of values of quasi-identifiers can be indistinctly matched to at least k respondents.

The k-anonymity requirement implicitly assumes that the data owner knows how many respondents each released tuple matches. This information can be known precisely only by explicitly linking the released data with externally available data. Since it is not realistic to assume that the data owner knows the data in external tables, the definition of k-anonymity translates the k-anonymity requirement in terms of the released data themselves. k-anonymity requires each respondent to be indistinguishable with respect to at least other $k - 1$ respondents in the released table, as stated by the following definition.

DEFINITION 18.3 (*k*-**anonymity**) Let $T(A_1, \ldots, A_p)$ be a table, and `QI` be a quasi-identifier associated with it. *T* is said to satisfy *k-anonymity* with respect to `QI` iff each sequence of values in *T*[`QI`] appears at least with *k* occurrences in *T*[`QI`].

The definition of *k*-anonymity represents a sufficient (but not a necessary) condition for the *k*-anonymity requirement. As a matter of fact, if a subset of `QI` appears in a publicly available table, the combination of the released *k*-anonymous table with the public table never allows an adversary to associate each released tuple with less than *k* respondents. For instance, the private table in Figure 18.1 is 1-anonymous w.r.t. `QI` = {`ZIP, MaritalStatus, Sex`}, since there are unique combinations of values of the quasi-identifying attributes (e.g., ⟨22030, single, M⟩). However, the same table is 2-anonymous with respect to `QI` = {`MaritalStatus`}. To correctly enforce *k*-anonymity, it is then necessary to identify the quasi-identifiers. The identification of quasi-identifiers depends on which data are known to a potential adversary. Since different adversaries may have different knowledge, and it seems highly improbable to exactly know which data are available to each adversary, the original *k*-anonymity proposal [35] assumes to define a unique quasi-identifier for each private table, including all attributes that are possibly externally available.

18.2.1 Generalization and Suppression

Although different data disclosure protection techniques have been developed (e.g., scrambling, swapping values, and adding noise), *k*-anonymity is typically enforced by combining two nonperturbative protecting techniques, called *generalization* and *suppression*, which have the advantage of preserving the truthfulness of the information, in contrast to other techniques that compromise the correctness of the information [13].

18.2.1.1 Generalization

Generalization consists in substituting the values in a column (or cell) of `PT`[`QI`] with more general values. Each attribute *A* in `PT` is initially associated with a *ground domain* $D = dom(A, \text{PT})$. The generalization maps each value $v \in D$ to a more general value $v' \in D'$, where D' is a *generalized domain* for *D*. Given a ground domain *D*, we distinguish two classes of generalizations depending on how the generalized domain *D'* is defined.

- *Hierarchy-based generalization.* This technique is based on the definition of a generalization hierarchy for each attribute in `QI`, where the most general value is at the root of the hierarchy and the leaves correspond to the most specific values (i.e., to the values in the ground domain for the attribute). A hierarchy-based generalization maps the values represented by the leaf vertices with one of their ancestor vertices at a higher level (see Section 18.3). As an example, attribute `ZIP` can be generalized by suppressing, at each generalization step, the rightmost digit.
- *Recoding-based generalization.* This technique is based on the *recoding into intervals* protection method [13] and typically assumes a total order relationship among values in the considered domain. The ground domain of each attribute in `QI` is partitioned into possibly disjoint intervals and each interval is associated with a label (e.g., the extreme interval values). A recoding-based generalization maps the values in the ground domain with the intervals they belong to (see Section 18.4).

It is interesting to note that one of the main differences between hierarchy-based and recoding-based generalizations is that while hierarchies need to be predefined, intervals for the recoding are usually computed at runtime during the generalization process. Accordingly, the *k*-anonymity algorithms adopting a hierarchy-based generalization will receive as an input the generalization hierarchies associated with attributes in QI, while the *k*-anonymity algorithms adopting a recoding-based generalization will have the additional task of defining the intervals.

18.2.1.2 Suppression

Suppression consists in removing from the private table a cell, a column, a tuple, or a set thereof. The intuition behind the introduction of suppression is that the combined use of generalization and suppression can reduce the amount of generalization necessary to satisfy the *k*-anonymity constraint. As a matter of fact, when a limited number of outliers (i.e., tuples with less than *k* occurrences) would force a great amount of generalization, their suppression allows the release of a less general (more precise) table. For instance, the suppression of tuples 3, 6, 7, and 10 of the private table in Figure 18.1 satisfies 2-anonymity with respect to QI = {ZIP, MaritalStatus, Sex} without any generalization.

18.2.2 Classification of *k*-Anonymity Techniques

Generalization and suppression can be applied at different granularity levels, which correspond to different approaches and solutions to the *k*-anonymity problem, and introduce a taxonomy of *k*-anonymity solutions [11]. Such a taxonomy is orthogonal to the type of generalization (i.e., hierarchy- or recoding-based) adopted.

Generalization can be applied at the level of:

- *Attribute (AG)*: Generalization is performed at the level of column; a generalization step generalizes all the values in the column.
- *Cell (CG)*: Generalization is performed on individual cells; as a result a generalized table may contain, for a specific column, values at different generalization levels.
  Generalizing at the cell level has the advantage of allowing the release of more specific values (since generalization can be confined to specific cells rather than hitting whole columns). However, besides a higher complexity of the problem, a possible drawback in the application of generalization at the cell level is the complication arising from the management of values at different generalization levels within the same column.

Suppression can be applied at the level of

- *Tuple (TS)*: Suppression is performed at the level of row; a suppression operation removes a whole tuple.
- *Attribute (AS)*: Suppression is performed at the level of column; a suppression operation obscures all the values of a column.
- *Cell (CS)*: Suppression is performed at the level of single cells; as a result a *k*-anonymized table may wipe out only certain cells of a given tuple/attribute.

Figure 18.2 summarizes the different combinations of generalization and suppression at all possible granularity levels (including combinations where one of the two techniques is not adopted). It is interesting to note that the application of generalization and suppression at the same granularity level is equivalent to the application of generalization only (**AG_≡AG_AS** and **CG_≡CG_CS**), since suppression can be modeled as a generalization of all domain values to a unique value. Combinations **CG_TS** (cell generalization, tuple suppression) and **CG_AS** (cell generalization, attribute suppression) are not applicable since the application of generalization at the cell level implies the application of suppression at that level too.

Generalization	Suppression			
	Tuple	Attribute	Cell	None
Attribute	AG_TS	AG_AS $\equiv$ AG_	AG_CS	AG_ $\equiv$ AG_AS
Cell	CG_TS not applicable	CG_AS not applicable	CG_CS $\equiv$ CG_	CG_ $\equiv$ CG_CS
None	_TS	_AS	_CS	– not interesting

FIGURE 18.2 Classification of *k*-anonymity techniques.

18.3 *k*-Anonymity with Hierarchy-Based Generalization

The problem of *k*-anonymizing a private table by exploiting generalization and suppression has been widely studied and a number of approaches have been proposed. In this section we focus on those solutions that adopt a hierarchy-based generalization (together with suppression) to achieve *k*-anonymity. Before discussing such solutions, we first formally define how generalization can be performed by adopting a predefined hierarchy and we then discuss the problem complexity.

For each ground domain $D \in \mathcal{D}$, we assume the existence of a set of generalized domains. The set of all ground and generalized domains is denoted Dom. The relationship between a ground domain and domains generalization of it is formally defined as follows.

DEFINITION 18.4 (**Domain generalization relationship**) Let Dom be a set of ground and generalized domains. A *domain generalization relationship*, denoted $\leq_D$, is a partial order relation on Dom that satisfies the following conditions:

C1: $\forall D_i, D_j, D_z \in$ Dom: $D_i \leq_D D_j, D_i \leq_D D_z \Rightarrow D_j \leq_D D_z \vee D_z \leq_D D_j$
C2: all maximal elements of Dom are singleton

Condition C1 states that, for each domain D_i, the set of its generalized domains is totally ordered and each D_i has at most *one* direct generalized domain D_j. This condition ensures determinism in the generalization process. Condition C2 ensures that all values in each domain can always be generalized to a single value. The definition of the domain generalization relationship implies the existence, for each domain $D \in$ Dom, of a totally ordered hierarchy, called *domain generalization hierarchy* and denoted DGH$_D$. Each DGH$_D$ can be graphically represented as a chain of vertices, where the top element corresponds to the singleton generalized domain, and the bottom element corresponds to D.

Analogously, a *value generalization relationship*, denoted $\leq_V$, can also be defined that associates with each value $v_i \in D_i$ a unique value $v_j \in D_j$, where D_j is the direct generalization of D_i. The definition of the value generalization relationship implies the existence, for each domain $D \in$ Dom, of a partially ordered hierarchy, called *value generalization hierarchy* and denoted VGH$_D$. Each VGH$_D$ can be graphically represented as a tree, where the root element corresponds to the unique value in the top domain in DGH$_D$, and the leaves correspond to the values in D. Figure 18.3 shows an example of domain and value generalization hierarchies for attributes ZIP, Sex, and MaritalStatus. Attribute ZIP, with domain $Z_0 = \{22030, 22032, 22045, 22047\}$, is generalized by suppressing, at each step, the rightmost digit. Attribute Sex, with domain $S_0 = \{M, F\}$, is generalized to the not_released value (domain S_1). Attribute MaritalStatus, with domain $M_0 = \{$single, married, divorced, widow$\}$, is first generalized to the been_married and never_married values (domain M_1), and then to the not_released value (domain M_2).

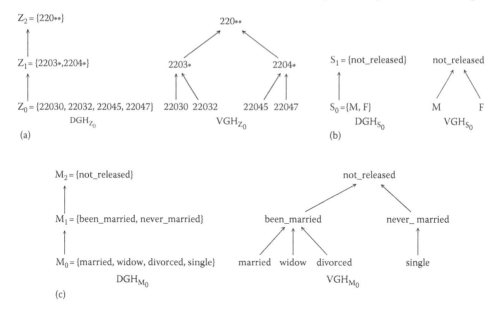

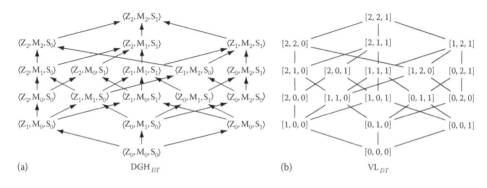

FIGURE 18.3 Examples of domain and value generalization hierarchies: (a) ZIP. (b) Sex. (c) MaritalStatus.

Since most k-anonymity approaches work on sets of attributes, the definition of domain gen-eralization hierarchy is extended to tuples of domains. A domain tuple $DT = \langle D_1, \ldots, D_n \rangle$ is an ordered set of domains composed through the Cartesian product to impose coordinate-wise order among domains. Since each domain $D_i \in DT$ is characterized by a total order domain generalization hierarchy DGH_{D_i}, domain tuple DT is characterized by a domain generalization hierarchy DGH_{DT} defined as $\mathrm{DGH}_{DT} = \mathrm{DGH}_{D_1} \times \cdots \times \mathrm{DGH}_{D_n}$. DGH_{DT} is a lattice where the bottom element is DT and the top element is the tuple composed of all top elements in DGH_{D_i}, $i = 1, \ldots, n$. Each path from the bottom to the top element in DGH_{DT} is called *generalization strategy* and represents a possi-ble strategy for generalizing quasi-identifier $QI = \{A_1, \ldots, A_n\}$, where $dom(A_i) = D_i$, $i = 1, \ldots, n$. Figure 18.4 illustrates the domain generalization hierarchy built on the domain tuple $\langle Z_0, M_0, S_0 \rangle$ and obtained through the combination of the domain generalization hierarchies DGH_{Z_0}, DGH_{M_0}, and DGH_{S_0} illustrated in Figure 18.3.

FIGURE 18.4 Domain generalization hierarchy (a) and distance vector hierarchy (b) for $DT = \langle Z_0, M_0, S_0 \rangle$.

Given a private table PT and its quasi-identifier QI, the application of generalization and suppression produces a *generalized table T* containing less information (more general values and less tuples) than PT, formally defined as follows.

DEFINITION 18.5 (Generalized table) Let $T_i(A_1,\ldots,A_n)$ and $T_j(A_1,\ldots,A_n)$ be two tables defined on the same set of attributes. Table T_j is said to be a *generalization (with tuple suppression)* of table T_i, denoted $T_i \preceq T_j$, iff:

1. $|T_j| \leq |T_i|$.
2. $\forall A \in \{A_1, \ldots, A_n\}$: $dom(A, T_i) \leq_D dom(A, T_j)$.
3. It is possible to define an injective mapping associating each tuple $t_j \in T_j$ with a tuple $t_i \in T_i$, such that $t_i[A] \leq_V t_j[A]$, for all $A \in \{A_1, \ldots, A_n\}$.

Given a private table PT and its quasi-identifier QI, there may exist different generalized tables that satisfy k-anonymity. Among all possible generalized tables, we are interested in a table that minimizes information loss, meaning that is k-anonymous and does not remove, through generalization and suppression, more information than necessary. As an example, consider the private table in Figure 18.1 with QI = {ZIP, MaritalStatus, Sex} and suppose that $k = 3$. The generalized table in Figure 18.5a corresponding to the top element $\langle Z_2, M_2, S_1\rangle$ in DGH$_{DT}$, which is composed of 10 identical tuples (all the cells in each column have been generalized to the same value), is 3-anonymous but it removes more information than necessary. In fact, the generalized table corresponding to $\langle Z_2, M_1, S_0\rangle$ in Figure 18.5b is 3-anonymous and it is more specific than the table corresponding to $\langle Z_2, M_2, S_1\rangle$, since it contains more specific values for MaritalStatus and Sex attributes.

The goal of k-anonymity is to compute a generalized k-anonymous table, while maintaining as much information as possible. This concept is captured by the definition of k-*minimal generalization*. The formal definition of k-minimal generalization requires the introduction of the concept of *distance vector*.

DEFINITION 18.6 (Distance vector) Let $T_i(A_1, \ldots, A_n)$ and $T_j(A_1, \ldots, A_n)$ be two tables such that $T_i \preceq T_j$. The distance vector of T_j from T_i is the vector $DV_{i,j} = [d_1, \ldots, d_n]$, where d_z, $z = 1, \ldots, n$, is the length of the *unique* path between $dom(A_z, T_i)$ and $dom(A_z, T_j)$ in the domain generalization hierarchy DGH$_{D_z}$.

The dominance relationship $\leq$ on integers is then extended on the set of distance vectors as follows. Given two distance vectors $DV = [d_1, \ldots, d_n]$ and $DV' = [d'_1, \ldots, d'_n]$, $DV \leq DV'$ iff $d_i \leq d'_i$,

ZIP	MaritalStatus	Sex
220 * *	not_released	not_released
220 * *	not_released	not_released
220 * *	not_released	not_released
220 * *	not_released	not_released
220 * *	not_released	not_released
220 * *	not_released	not_released
220 * *	not_released	not_released
220 * *	not_released	not_released
220 * *	not_released	not_released
220 * *	not_released	not_released

(a) $\langle Z_2, M_2, S_1\rangle$

ZIP	MaritalStatus	Sex
220 * *	been_married	F
220 * *	been_married	F
220 * *	never_married	M
220 * *	never_married	M
220 * *	never_married	M
220 * *	been_married	F
220 * *	been_married	M
220 * *	been_married	M
220 * *	been_married	M

(b) $\langle Z_2, M_1, S_0\rangle$

ZIP	MaritalStatus	Sex
2203*	been_married	F
2203*	been_married	F
2203*	never_married	M
2203*	never_married	M
2203*	never_married	M
2203*	been_married	F
2204*	been_married	M
2204*	been_married	M
2204*	been_married	M

(c) $\langle Z_1, M_1, S_0\rangle$

FIGURE 18.5 Examples of generalized tables.

$i = 1, \ldots, n$. For each $\mathtt{PT[QI]}$ defined on domain tuple DT, we can therefore define a partially ordered *hierarchy of distance vectors*, denoted VL_{DT}, containing the distance vectors of all the generalized tables of $\mathtt{PT[QI]}$. Each VL_{DT} can be graphically represented as an isomorphic lattice to DGH_{DT}. The height of a distance vector DV in VL_{DT}, denoted **height**(DV, VL_{DT}), is equal to the sum of the elements in DV. The height of VL_{DT} corresponds to the height of its top element. As an example, Figure 18.4b illustrates the VL_{DT} lattice defined on $DT = \langle \mathrm{Z}_0, \mathrm{M}_0, \mathrm{S}_0 \rangle$ and therefore isomorphic to the DGH in Figure 18.4a.

Note that given an element in DGH_{DT} and the corresponding generalized table, Definition 18.5 allows any amount of suppression to achieve k-anonymity. However, we are interested in a table obtained by suppressing the minimum number of tuples necessary to achieve k-anonymity at a given level of generalization.

It is worth noting that, given a hierarchy of distance vectors VL_{DT}, the generalized tables corresponding to the distance vectors at the same height and ensuring minimality in suppression may suppress a different number of tuples. Since the joint use of generalization and suppression permits to maintain as much information as possible in the released table, the question is whether it is better to generalize, loosing data precision, or to suppress, loosing completeness. The compromise proposed by Samarati [35] consists in establishing a threshold $\mathtt{MaxSup}$ to the maximum number of tuples that can be suppressed; within this threshold, generalization decides minimality. Given this threshold on the number of tuples that can be suppressed, a *k-minimal generalization* is defined as follows.

DEFINITION 18.7   (*k*-Minimal generalization) Let T_i and T_j be two tables such that $T_i \preceq T_j$, and let $\mathtt{MaxSup}$ be the specified threshold of acceptable suppression. T_j is said to be a *k-minimal generalization* of table T_i iff:

1. T_j satisfies k-anonymity enforcing minimal required suppression, that is, T_j satisfies k-anonymity and $\forall T_z : T_i \preceq T_z, DV_{i,z} = DV_{i,j}, T_z$ satisfies k-anonymity $\Rightarrow |T_j| \geq |T_z|$
2. $||T_i| - |T_j|| \leq \mathtt{MaxSup}$
3. $\forall T_z : T_i \preceq T_z$ and T_z satisfies conditions 1 and 2 $\Rightarrow \neg(DV_{i,z} < DV_{i,j})$.

This definition states that a generalization T_j of a table T_i is k-minimal if it satisfies k-anonymity (condition 1), it does not suppress more tuples than $\mathtt{MaxSup}$ (condition 2), and there does not exist another generalization of T_i satisfying these conditions and characterized by a distance vector lower than that of T_i (condition 3). As an example, consider the private table in Figure 18.1 with QI = $\{\mathtt{ZIP}, \mathtt{MaritalStatus}, \mathtt{Sex}\}$ and suppose that $k = 3$ and $\mathtt{MaxSup} = 2$. The generalized table in Figure 18.5c is a k-minimal generalization for PT. As a matter of fact, it is 3-anonymous (condition 1), it suppresses less than 2 tuples (condition 2), and the generalized tables characterized by distance vectors lower than $[1, 1, 0]$ do not satisfy these two conditions: the generalized table corresponding to $[1, 0, 0]$ requires to suppress at least 7 tuples; and the generalized tables corresponding to $[0, 1, 0]$ and $[0, 0, 0]$ require to suppress all the tuples.

Given a private table PT, there may exist more than one k-minimal generalization since DGH_{DT} is a lattice and two solutions may be noncomparable. Furthermore, the definition of k-minimal generalization only captures the concept that the least amount of generalization and the minimal required suppression to achieve k-anonymity is applied. Different *preference criteria* can be applied in choosing a preferred minimal generalization, among which [35]:

- *minimum absolute distance* prefers the generalization(s) with the smallest absolute distance, that is, with the smallest total number of generalization steps (regardless of the hierarchies on which they have been taken);

- *minimum relative distance* prefers the generalization(s) with the smallest relative distance, that is, that minimizes the total number of relative steps (a step is made relative by dividing it over the height of the domain hierarchy to which it refers),
- *maximum distribution* prefers the generalization(s) with the greatest number of distinct tuples,
- *minimum suppression* prefers the generalization(s) that suppresses less tuples, that is, the one with the greatest cardinality.

18.3.1 Problem Complexity

All the models with hierarchies investigated in the literature (**AG_TS**, **AG_**, **CG_**, and **_CS**), as well as **_AS**, are NP-hard. The complexity results of all these models with hierarchies derive from the NP-hardness of **_CS** and **_AS**. To formally prove the NP-hardness of such problems, and without loss of generality, each table T can be seen as a matrix of m rows (tuples) and n columns (attributes). Each row x_i, $i = 1, \ldots, m$, is an n-dimensional string defined on an alphabet Σ, where each $c \in \Sigma$ corresponds to an attribute value. For instance, the projection over QI of the private table in Figure 18.1 is a matrix with 10 rows and each row is a string of three characters of the alphabet $\Sigma = \{22030, 22032, 22045, 22047, \text{married}, \text{single}, \text{divorced}, \text{widow}, \text{F,M}\}$.

The NP-hardness of **_AS** has been proved for $|\Sigma| \geq 2$ [32], while the NP-hardness of the **_CS** problem for $|\Sigma| \geq 3$ has been proved with a reduction from the "Edge partition into triangles" problem [3], which is an improvement on the NP-hardness proved for **_AS** when the size of alphabet Σ is equal to the number n of attributes. NP-hardness of **_CS** and **_AS** clearly implies NP-hardness of **CG_** and **AG_**, respectively. This implication holds since suppression can be considered as a special case of generalization, where all hierarchies have height of 1. Note also that NP-hardness of **AG_** implies NP-hardness of **AG_TS**, where, as in the existing proposals, tuple suppression is regulated with the specification of a maximum number of tuples that can be suppressed. Nevertheless, the computational complexity of the general **AG_TS** model (where the number of tuples that can be suppressed is no more a constant value given as input, but it should be minimized as part of the solution) is still an open issue. Note that the decisional versions of **_AS**, **_CS**, **AG_**, **AG_TS**, and **CG_** are obviously in NP [3].

We now give the proof of NP-hardness for **_CS** and we briefly describe the result for **_AS**.

THEOREM 18.1   *The _CS 3-anonymity problem is NP-hard even for a ternary alphabet.*

PROOF   The proof [3] is a reduction from the NP-hard problem of "Edge partition into triangles" [25], which can be formulated as follows: *given a graph $G = (V, E)$ with $|E| = 3m$ for some integer m, can the edges of G be partitioned into m triangles with disjoint edges?*

To facilitate the description of the reduction from "Edge partition into triangles," the proof will first describe the simpler reduction from "Edge partition into triangles and 4-stars" and then it will describe a reduction from "Edge partition into triangles." The overall proof therefore consists in two main steps. First, we show that, given a graph $G = (V, E)$ with $|E| = 3m$, we can construct a table T such that an optimal 3-anonymous solution for T costs at most $9m$ (i.e., T is obtained by suppressing at most $9m$ cells from PT) if and only if G can be partitioned into a collection of m disjoint triangles and 4-stars (a 4-star is a graph with 3 edges and 4 vertices, having one degree 3 and the others degree 1). Second, we show how to define a table T' from G such that an optimal 3-anonymous solution for T' costs at most $9m \lceil \log_2(3m + 1) \rceil$ if and only if G can be partitioned into a collection of m disjoint triangles. In the following, we use character $*$ to denote suppressed cells.

Edge partition into triangles and 4-stars. Given a graph $G = (V, E)$ with $|E| = 3m$ and $|V| = n$, we can construct a table T that contains $3m$ rows (one for each edge) and n attributes (one for each

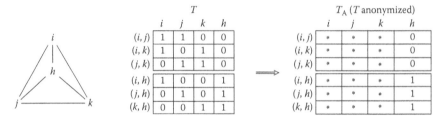

FIGURE 18.6 An example of construction of T from a graph, as described in the proof of Theorem 18.1.

vertex). The row corresponding to edge (i, j) has 1 in positions corresponding to attributes i and j; 0 otherwise.

Suppose that G can be partitioned into a collection of m disjoint triangles and 4-stars. Consider first a triangle with vertices i, j, and k. By suppressing the cells in the rows corresponding to edges (i, j), (i, k), and (j, k) and columns i, j, and k, we get 3 identical rows containing 3 *s each and 0 anywhere else. Consider now a 4-star with vertices i, j, k, and h and edges (i, h), (j, h), and (k, h). By suppressing the cells in the rows corresponding to edges (i, h), (j, h), and (k, h) and columns i, j, and k, we get 3 identical rows containing 3 *s each, with a single 1 (corresponding to attribute h) and 0 anywhere else. Since for each triangle and for each 4-star we obtain three identical generalized tuples by suppressing 9 cells, table T is 3-anonymous of cost $9m$. As a simple example, consider the tables in Figure 18.6, where the first three rows correspond to a triangle and the last three rows to a 4-star. The anonymized table on the right side shows the suppressed cells.

Suppose to have a 3-anonymous table T_A of cost at most $9m$ that is a generalization of T. We want to show that the graph G corresponding to T can be partitioned into a collection of m disjoint triangles and 4-stars. Since G is a simple graph and therefore there are no edges with the same end vertices, any 3 rows in the corresponding table T are distinct and differ in at least 3 positions; otherwise there would be multiple edges. This implies that each modified row in the anonymized table T_A contains at least 3 *s, as explained above, and the overall number of *s in T_A is at least $9m$. From the previous hypothesis, we have that the cost of T_A is exactly $9m$, and each row of T_A contains 3 *s. Therefore T_A contains only clusters of 3 identical rows. In fact, any tuple in the anonymized table T_A belonging to a group of more than 3 equal tuples would contain at least 4 *s. The corresponding graph contains then only edges composing either triangles or 4-stars. In fact, while each modified row in a triangle has 3 *s and 0 elsewhere (see Figure 18.6), each modified tuple in a 4-star contains 3 *s, a single 1, and 0 elsewhere (see Figure 18.6). No other configuration is possible. The overall solution corresponds to a partition of edges into triangles and 4-stars.

We now show the reduction from the "Edge partition into triangles" problem.

Edge partition into triangles. To prove that _CS 3-anonymity is reduced from "Edge partition into triangles," the informal idea is to make 4-stars costing more *s than triangles. A slightly different construction of the table from a graph $G = (V, E)$ with $|E| = 3m$ and $|V| = n$ is then required. Let $t = \lceil \log_2(3m + 1) \rceil$. We define a new table T' where each row has t blocks of n columns. Consider an arbitrary ordering of the edges in E and express the rank of an edge $e = (v_1, v_2)$ in binary notation $b_1 b_2 \ldots b_t$. In the tuple corresponding to edge e, each block has 0 in all columns but the columns corresponding to the vertices v_1 and v_2, for which two configurations are possible: $conf_0$ has a 1 in column v_1 and a 2 in column v_2, and $conf_1$ has a 2 in column v_1 and a 1 in column v_2. The lth block in the row corresponding to e has configuration $conf_{b_l}$. For instance, with respect to the graph shown in Figure 18.6, the edges are ranked from 1 (001 in binary notation) to 6 (110 in binary notation). Figure 18.7 illustrates table T' corresponding to the graph. We now show that the cost of the optimal 3-anonymity solution of T' is at most $9mt$ if and only if E can be partitioned into m disjoint triangles.

	i	j	k	h	i	j	k	h	i	j	k	h
001 (i,j)	1	2	0	0	1	2	0	0	2	1	0	0
010 (i,k)	1	0	2	0	2	0	1	0	1	0	2	0
011 (j,k)	0	1	2	0	0	2	1	0	0	2	1	0
100 (i,h)	2	0	0	1	1	0	0	2	1	0	0	2
101 (j,h)	0	2	0	1	0	1	0	2	0	2	0	1
110 (k,h)	0	0	2	1	0	0	2	1	0	0	1	2

T' $\implies$ T'_A (T' anonymized)

	i	j	k	h	i	j	k	h	i	j	k	h
(i,j)	*	*	*	0	*	*	*	0	*	*	*	0
(i,k)	*	*	*	0	*	*	*	0	*	*	*	0
(j,k)	*	*	*	0	*	*	*	0	*	*	*	0
(i,h)	*	*	*	1	*	*	*	*	*	*	*	*
(j,h)	*	*	*	1	*	*	*	*	*	*	*	*
(k,h)	*	*	*	1	*	*	*	*	*	*	*	*

FIGURE 18.7 An example of construction of T' from the graph in Figure 18.6, as described in the proof of Theorem 18.1

Suppose that E can be partitioned into m disjoint triangles. As previously discussed, every triangle in such a partition corresponds to a cluster with $3t$ *s in each tuple. Thus, we get a 3-anonymity solution of cost $9mt$.

Suppose to have a 3-anonymity solution of cost at most $9mt$. Again, any 3 tuples differ in at least $3t$ columns and the cost of any solution is at least $9mt$. Hence, the solution cost is exactly $9mt$ and each modified row has exactly $3t$ *s. Thus, each cluster has exactly 3 equal rows. We now show, by contradiction, that the corresponding edges form a triangle and cannot be a 4-star. Suppose, on the contrary, that the 3 rows form a 4-star. Let h be the common vertex and consider the integer $\{1, 2\}$ assigned by each of the 3 edges to h in $conf_0$. Two of the three edges must have assigned the same digit to h. However, since these two edges differ in rank, there must exist at least one block where they have a different configuration (and therefore, a different digit in the column corresponding to h). Thus, the rows corresponding to the 3 edges contain an additional $*$ corresponding to column h in addition to the $3t$ *s corresponding to the remaining 3 columns (e.g., see the last 3 tuples of Figure 18.7). This contradicts the fact that each row has exactly $3t$ *s. Therefore, E can be partitioned into m disjoint triangles.   □

The corresponding theorem for _AS has been proved with a proof similar to the one that was proposed for _CS [32]. The result is the following.

THEOREM 18.2   *The _AS problem for $k > 2$ is NP-hard for any alphabet Σ such that $|\Sigma| \geq 2$.*

The proof is a reduction from the "k-dimensional perfect matching" problem: *given a k-hypergraph $G = (V, E)$ is there a subset S of G with $|V|/k$ hyperedges such that each vertex of G is contained in exactly one hyperedge of S?*

By orchestrating Theorems 18.1 and 18.2 and the observations given in this section, we have the following result.

COROLLARY 18.1   Problems _AS, _CS, AG_, AG_TS, and CG_ are NP-hard for $k > 2$, and for any alphabet Σ such that $|\Sigma| > 2$.

18.3.2 Algorithms for k-Anonymity

Since the hierarchy-based generalization has been proposed first, the solutions based on such a type of hierarchy have been studied in more depth than the ones based on the recoding-based generalization. Here, we present two exact algorithms, both belonging to the **AG_TS** class, and both aimed at finding a k-minimal solution for a given PT [28,35]. We also briefly describe the most important approximation algorithms proposed for k-anonymizing a private table.

Besides the two exact algorithms described in the following, Sweeney [37] proposed an exact algorithm (**AG_TS** class) that exhaustively examines all potential generalizations (with suppression) for identifying a minimal one satisfying the k-anonymity requirement. This approach is however impractical for large datasets.

18.3.2.1 Samarati's Algorithm

The first algorithm introduced for k-anonymizing a private table PT was proposed along with the definition of k-anonymity [35]. This algorithm follows the minimum absolute distance criterion (see Section 18.3) to determine a k-minimal generalization of PT.

Given a private table PT such that $|PT| \geq k$,* the algorithm restricts its analysis to $PT[QI]$, where $QI = \{A_1, \ldots, A_n\}$ is defined on the domain tuple $DT = \langle D_1, \ldots, D_n \rangle$ (i.e., $dom(A_i, PT) = D_i$, $i = 1, \ldots, n$).

Given a domain generalization hierarchy DGH_{DT}, each path from the bottom to the top element in DGH_{DT} represents a generalization strategy that is characterized by a locally minimal generalization defined as follows.

DEFINITION 18.8 (**Locally minimal generalization**) Let DGH_{DT} be a domain generalization hierarchy, k be the anonymity threshold, and MaxSup be the suppression threshold. Any path in DGH_{DT} is characterized by a *locally minimal generalization*, which is the lowest generalization in the path that satisfies k-anonymity and suppresses a number of tuples lower than MaxSup.

A locally minimal generalization represents the k-anonymous generalization that maintains the most information with respect to a given generalization strategy.

Example 18.1

Consider the table in Figure 18.1 and the domain generalization and distance vector hierarchies in Figure 18.4. If $k = 3$ and $MaxSup = 2$, along path $\langle Z_0, M_0, S_0 \rangle \rightarrow \langle Z_1, M_0, S_0 \rangle \rightarrow \langle Z_2, M_0, S_0 \rangle \rightarrow \langle Z_2, M_1, S_0 \rangle \rightarrow \langle Z_2, M_2, S_0 \rangle \rightarrow \langle Z_2, M_2, S_1 \rangle$, the generalized table corresponding to $\langle Z_2, M_1, S_0 \rangle$ (see Figure 18.5b) is a locally minimal generalization.

The definition of locally minimal generalization can be exploited to compute k-minimal generalizations, on the basis of the following theorem.

THEOREM 18.3 [35] *Let* $T_i(A_1, \ldots A_n) = PI[QI]$ *be a table to be generalized and let* $DT = \langle D_1, \ldots, D_n \rangle$ $(D_z = dom(A_z, T_i), z = 1, \ldots, n)$ *be the domain tuple of its attributes. Every k-minimal generalization of T_i is a locally minimal generalization for some strategy in DGH_{DT}.*

Note that a locally minimal generalization may not correspond to a k-minimal generalization, since a generalization may be locally minimal along one path but not along another one. With respect to Example 18.1, the locally minimal generalization $\langle Z_2, M_1, S_0 \rangle$ is not a k-minimal generalization, since the generalized table corresponding to $\langle Z_1, M_1, S_0 \rangle$ (see Figure 18.5c) is k-anonymous, suppresses a number of tuples lower than MaxSup and contains more specific values for the ZIP attribute. Exploiting Theorem 18.3, a naive method to compute a k-minimal generalization consists in finding all locally minimal generalizations by visiting all the paths in DGH_{DT}, starting from the bottom and stopping at the first generalization that both satisfies k-anonymity and suppresses a number of tuples

---

* Note that if $1 \leq |PT| < k$, a k-anonymous version of PT does not exist.

lower than MaxSup. By discarding the locally minimal generalizations that are dominated by other locally minimal generalizations, only k-minimal generalizations of PT are maintained. However, since the number of paths in DGH$_{DT}$ may be very high, such a naive strategy is not viable. The key idea exploited by Algorithm 18.1 in Figure 18.8 to reduce the computational time is that the number of tuples that need to be suppressed to satisfy k-anonymity decreases while going up in a path.

THEOREM 18.4 *[35] Let T_i=PT[QI] be a table to be generalized and T_j and T_z be two of its generalizations (i.e., $T_i \preceq T_j$ and $T_i \preceq T_z$) enforcing minimal required suppression. Then, $DV_{i,j} < DV_{i,z} \Rightarrow |T_j| \leq |T_z|$.*

From Theorem 18.4, it follows that given two generalized tables T_j and T_z of PT such that $DV_{i,j} < DV_{i,z}$, if T_j is k-anonymous and suppresses a number of tuples lower than MaxSup, also T_z is k-anonymous and suppresses a number of tuples lower than MaxSup. Furthermore, if there is no solution that guarantees k-anonymity suppressing a number of tuples lower than MaxSup at height h, there cannot exist a solution at height $h' < h$ that guarantees it. Such a consequence is exploited by the Algorithm 18.1 to compute a k-minimal generalization for PT[QI], by performing a binary search on the heights of distance vectors.

The algorithm takes as input the projection T over quasi-identifying attributes of the private table PT, the anonymity constraint k, the suppression threshold MaxSup, and the domain generalization hierarchies for the attributes composing the quasi-identifier. During the initialization phase, the algorithm first builds VL$_{DT}$ (i.e., the distance vector lattice based on the domain generalization hierarchies) and a matrix VT, with a row for each distinct tuple t in T with less than k occurrences and a column for each distinct tuple in T. Each entry VT[x, y] of the matrix contains the distance vector $[d_1, \ldots, d_n]$ between tuples $x = \langle v_1 \ldots v_n \rangle$ and $y = \langle v'_1 \ldots v'_n \rangle$, where $d_i, i = 1, \ldots, n$, is the distance from the domain of v_i and v'_i to the domain to which they generalize to the same value. The binary search phase visits the lattice VL$_{DT}$ as follows. Variable *high* is first initialized to the height of VL$_{DT}$ and all the generalized tables corresponding to distance vectors at height $\lfloor \frac{high}{2} \rfloor$ are first evaluated. If there is at least one generalized table satisfying k-anonymity and suppressing a number of tuples lower than MaxSup, the algorithm evaluates the generalized tables at height $\lfloor \frac{high}{4} \rfloor$, otherwise those at height $\lfloor \frac{3high}{4} \rfloor$, and so on. The algorithm stops when it reaches the lowest height in VL$_{DT}$ where there is at least a generalized table satisfying k-anonymity and the MaxSup constraint.

To check whether or not at a given height h there is a table that satisfies k-anonymity and the MaxSup suppression threshold, for each distance vector *vec* at height h the algorithm calls function **Satisfy**. Function **Satisfy** computes the number of tuples *req_sup* that need to be suppressed for achieving k-anonymity in table T_{vec} corresponding to the *vec* distance vector. For each row x in VT, **Satisfy** computes the sum c of the occurrences of tuples y (column of the matrix) such that VT[x, y]$\leq$*vec*. As a matter of fact, all tuples such that VT[x, y]$\leq$*vec* will be generalized to the same value in T_{vec} as x. Therefore, if $c < k$, c is added to *req_sup* since the considered tuples will be outliers for T_{vec} and will be therefore suppressed to satisfy k-anonymity. If *req_sup* is lower than MaxSup, T_{vec} satisfies k-anonymity and the MaxSup threshold and **Satisfy** returns *true*; otherwise it returns *false*.

Example 18.2

Consider the private table in Figure 18.1 with QI = {ZIP, MaritalStatus, Sex} and the corresponding hierarchies in Figure 18.4. Also, suppose that $k = 3$, and MaxSup = 2. The algorithm builds matrix VT illustrated in Figure 18.9, which is composed of seven rows (all quasi-identifying values appear less than 3 times in PT) and 7 columns since there are 7 distinct values for QI in PT ($t_1 = t_2$, $t_4 = t_5$, and $t_8 = t_9$).

Algorithm 18.1 (Samarati's Algorithm)

INPUT
T = PT[QI]: private table
k: anonymity requirement
MaxSup: suppression threshold
$\forall A \in$ QI, DGH$_A$: domain generalization hierarchies

OUTPUT
sol: distance vector corresponding to the k-minimal generalization of PT[QI]

MAIN
Let VL$_{DT}$ be the distance vector hierarchy
/* matrix initialization phase */
Outlier:= ∅ /* set of outlier values */
order(T) /* order the tuples in the table */
V:= ∅ /* set of distinct tuples */
counter[t_1] := 1 /* first tuple */
for i:=2 ... $|T|$ **do**
    if $t_i \neq t_{(i-1)}$ **then** /* if the i-th is different from the $(i-1)$th tuple */
    V:= $V \cup \{t_i\}$
    counter[t_i] := 1
    if *counter*[$t_{(i-1)}$]< k **then** *Outlier*:= *Outlier* $\cup \{t_{(i-1)}\}$
    else *counter*[t_i] := *counter*[t_i] + 1
for j:=1 ... $|V|$ **do**
    for i:=1 ... $|Outlier|$ **do**
        VT[i,j] := **distance**(v_j,*outlier*$_i$)
/* binary search phase */
low:= 0
high:= **height**(⊤,VL$_{DT}$)
sol:= ⊤
while *low*<*high* **do**
    h:= $\lfloor \frac{low+high}{2} \rfloor$
    Vectors:= {*vec*|**height**(*vec*,VL$_{DT}$)=h}
    reach_k:= *false*
    while *Vectors*≠ ∅ ∧ *reach_k*≠*true* **do**
        let *vec* be a vector in *Vectors*
        Vectors := *Vectors* − {*vec*}
        if **Satisfy**(T,*vec*,VT,$|Outlier|$,$|V|$) **then**
            sol:= *vec*
            reach_k := *true*
    if *reach_k*=*true* **then** *high* := h
    else *low* := h+1
return(*sol*)

SATISFY(T,*vec*,VT,*rows*,*columns*)
req_sup := 0
for i:=1 ... *rows* **do**
    c := 0
    for j:=1 ... *columns* **do**
        if VT[i,j]≤*vec* **then**
            c := c + *counter*[v_j]
    if c<k **then** *req_sup* := *req_sup* + c
if *req_sup*<MaxSup **then return**(*true*)
else return(*false*)

FIGURE 18.8 Algorithm that computes a k-minimal generalization.

		1	2	3	4	5	6	7
		t_1/t_2	t_3	t_4/t_5	t_6	t_7	t_8/t_9	t_{10}
1	t_1/t_2	[0,0,0]	[0,2,1]	[1,2,1]	[1,1,0]	[2,1,1]	[2,1,1]	[2,2,0]
2	t_3	[0,2,1]	[0,0,0]	[1,0,0]	[1,2,1]	[2,2,0]	[2,2,0]	[2,0,1]
3	t_4/t_5	[1,2,1]	[1,0,0]	[0,0,0]	[0,2,1]	[2,2,0]	[2,2,0]	[2,0,1]
4	t_6	[1,1,0]	[1,2,1]	[0,2,1]	[0,0,0]	[2,0,1]	[2,1,1]	[2,2,0]
5	t_7	[2,1,1]	[2,2,0]	[2,2,0]	[2,0,1]	[0,0,0]	[1,1,0]	[1,2,1]
6	t_8/t_9	[2,1,1]	[2,2,0]	[2,2,0]	[2,1,1]	[1,1,0]	[0,0,0]	[0,2,1]
7	t_{10}	[2,2,0]	[2,0,1]	[2,0,1]	[2,2,0]	[1,2,1]	[0,2,1]	[0,0,0]

FIGURE 18.9 An example of the VT matrix computed with respect to the table in Figure 18.1.

By exploiting the hierarchies in Figure 18.4, the algorithm computes the distance vectors between pairs of tuples. For instance, VT[1,4] = [1, 1, 0], which is the distance vector of the generalized table at which $t_1 = t_2 = \langle 22030, \text{married}, F \rangle$ and $t_6 = \langle 22032, \text{divorced}, F \rangle$ are generalized to the same value $\langle 2203*, \text{been\_married}, F \rangle$.

Once VT has been computed, the algorithm starts with the binary search phase and evaluates first the generalizations at height $\lfloor 5/2 \rfloor = 2$, that is, the tables corresponding to distance vectors $[2, 0, 0], [1, 1, 0], [1, 0, 1], [0, 1, 1]$, and $[0, 2, 0]$. The table corresponding to $[1, 1, 0]$ satisfies 3-anonymity suppressing only one tuple. It is easy to see that for each row but the last one of the VT matrix in Figure 18.9, there are 3 tuples that are generalized to the same tuple. The algorithm then evaluates tables at height $\lfloor 5/4 \rfloor = 1$, that is, the tables corresponding to distance vectors $[1, 0, 0], [0, 1, 0]$, and $[0, 0, 1]$. Since none of these tables is 3-anonymous and suppresses a number of tuples lower than MaxSup, the solution returned by the algorithm is $[1, 1, 0]$, which corresponds to $\langle Z_1, M_1, S_0 \rangle$.

The time complexity of the algorithm in Figure 18.8 is exponential in the number (n) of attributes in QI.

First, the algorithm builds the domain generalization hierarchy for QI (VL_{DT}) from the given domain generalization hierarchies of the individual attributes in QI. This step requires a complexity of $O(|DGH|)$ in time. Second, the algorithm sorts the tuples in the table with a standard sorting algorithm in polynomial time. The third step initializes the matrix VT, and it requires a time proportional to its dimension ($O(m^2 \cdot n)$). The last step computes a minimal generalization with a binary search in the lattice that verifies the anonymity of the generalized tables corresponding to each node of the lattice considered. Note that, even if this is a binary search, the lattice contains, in the middle level, a huge number of possible generalizations. For instance, even if we assume that QI contains only attributes with domain generalization hierarchies of height 2, the number of possible generalizations in the middle level of the corresponding DGH is $\binom{n}{n/2} \in O(2^{n/2})$. Therefore, in the worst case, the time complexity of the binary search phase is upperbounded by $|DGH| \cdot (n \cdot m)$, where $|DGH| = n \cdot \prod_{i=1}^{n} h_i$ with h_i the height of the domain generalization hierarchy of the single attribute A_i in QI. The overall time complexity of the algorithm is then $O((n \cdot m) \cdot (m + |DGH|))$, where $|DGH| \in O(n \cdot (h_{max})^n)$ and $h_{max} = max\{h_1, \ldots, h_n\}$.

18.3.2.2 Incognito

The *Incognito* algorithm [28] has been proposed for computing a k-minimal generalization by using a domain generalization hierarchy DGH_{DT} on the domain tuple of the quasi-identifier, where the vertices in DGH_{DT} are only the vertices that correspond to k-anonymous generalized tables.

Since the computational cost of the k-anonymization algorithms is mainly due to the verification of the k-anonymity condition on a great number of generalizations of the private table PT, Incognito reduces the number of tables for which this verification is necessary by exploiting the observation that if a table T with quasi-identifier QI is k-anonymous, T is k-anonymous with respect to any quasi-identifiers $Q \subset QI$.

DEFINITION 18.9 (Subset property) Let $T(A_1, \ldots, A_n)$ be a table and $QI = \{A_1, \ldots, A_m\}$ be its quasi-identifier. If T is k-anonymous with respect to QI, T is k-anonymous with respect to any $Q \subseteq QI$.

This subset property represents a necessary (not sufficient) condition for k-anonymity. As a matter of fact, a table T can be k-anonymous with respect to QI only if T is k-anonymous with respect to any subset of QI. Incognito then excludes a priori the generalizations in DGH_{DT} that cannot be k-anonymous with respect to QI. To this purpose, Incognito iteratively builds all the domain generalization hierarchies on subsets of QI, excluding at each step the vertices representing generalizations that do not satisfy k-anonymity (see Figure 18.10).

At the first iteration, for each $A_i \in QI$, Incognito builds DGH_{D_i} with $D_i = dom(A_i, PT)$. For each DGH_{D_i}, Incognito then follows a bottom-up breadth-first search on the domain generalization hierarchy. If a table T_v corresponding to vertex v in DGH_{D_i} satisfies k-anonymity, Incognito marks true

---

Algorithm 18.2 (Incognito Algorithm)

INPUT
$T = PT[QI]$: private table where $QI = \{A_1, \ldots, A_n\}$ and $DT = \langle dom(A_1, PT), \ldots, dom(A_n, PT) \rangle$
k: anonymity requirement
$\forall D \in DT$, DGH_D: domain generalization hierarchies where for all $v \in DGH_D$, **mark**(v) is set false

OUTPUT
DGH_D: restricted version of DGH_{DT}

MAIN
for $i=1 \ldots n$ **do**
$\quad \mathcal{D}^i := \{D \mid D \subseteq DT \wedge |D| = i\}$ /* all subsets of i elements in QI */
$\quad$**for each** $D \in \mathcal{D}^i$ **do**
$\quad\quad$**if** $i \neq 1$ **then** $DGH_D :=$ **Compose**(D)
$\quad\quad$**for** $h=0 \ldots$ **height**($\top$) **do** /* $\top$ represents the root node in DGH_D */
$\quad\quad\quad$/* **height**(v) denotes the length of the path from the bottom element in DGH_D to v */
$\quad\quad\quad \mathcal{V}^h := \{v \mid v \in DGH_D \wedge$ **height**(v)$=h \wedge$ **mark**(v)$=false\}$
$\quad\quad\quad$**for each** $v \in \mathcal{V}^h$ **do**
$\quad\quad\quad\quad$let T_v be the generalized table corresponding to vertex v
$\quad\quad\quad\quad$**if** **Satisfy**(T_v, k) **then**
$\quad\quad\quad\quad\quad$**for each** $v' \mid v' \in DGH_D \wedge v \leq v'$ **do** **mark**(v'):$=true$
return(DGH_D)

COMPOSE(D)
$DGH_D := (V, E)$
$i := 1$
for each $D_i \in \{D' \subset D : |D'| = |D| - 1\}$ **do**
$\quad V_i := \{v \mid v \in DGH_{D_i} \wedge$ **mark**(v)$=true\}$
$\quad i := i + 1$
$V := V_1 \times \ldots \times V_i$
for each $v \in V$ **do** **mark**(v) := $false$
$E := \{(v_i, v_j) \mid v_i, v_j \in V, v_i \leq v_j, \not\exists v_z \in V, v_i \leq v_z \wedge v_z \leq v_j\}$
return(DGH_D)

FIGURE 18.10 Algorithm that computes reduced generalization hierarchies.

v and all vertices v', such that $v \leq v'$, without explicitly computing the generalizations corresponding to v' to check if they are k-anonymous.*

At the second iteration, for each subset $\{A_i, A_j\} \subseteq \text{QI}$, Incognito builds $\text{DGH}_{\langle D_i, D_j \rangle}$ with $D_i = dom(A_i, \text{PT})$ and $D_j = dom(A_j, \text{PT})$. The domain generalization hierarchy $\text{DGH}_{\langle D_i, D_j \rangle}$ is built by combining (through function **Compose**) DGH_{D_i} and DGH_{D_j}. For the subset property the **Compose** function removes from $\text{DGH}_{\langle D_i, D_j \rangle}$ all those vertices that represent domain tuples containing generalized domains for D_i (or D_j) marked false in DGH_{D_i} (or DGH_{D_j}).

At iteration i, the algorithm builds the domain generalization hierarchies on subsets of QI composed of i attributes, using the vertices marked true in the domain generalization hierarchies computed at iteration $i - 1$. It terminates at iteration $i = |\text{QI}|$, when it evaluates the domain generalization hierarchy for the whole quasi-identifier.

Example 18.3

Consider the private table in Figure 18.1 with QI = {ZIP, MaritalStatus, Sex} and assume that $k = 3$ and MaxSup $= 2$. Figure 18.11 illustrates, on the left-hand side, the complete domain generalization hierarchies for all the subsets of QI, and on the right-hand side, the subhierarchies computed by Incognito at each iteration.

Iteration 1

- $\text{DGH}_{\langle Z_0 \rangle}$. Vertices $\langle Z_0 \rangle$, $\langle Z_1 \rangle$, and $\langle Z_2 \rangle$ are marked true, since table $T_{\langle Z_0 \rangle}$ satisfies 3-anonymity by suppressing a number of tuples lower than MaxSup.
- $\text{DGH}_{\langle M_0 \rangle}$. Vertex $\langle M_0 \rangle$ is marked false, since to satisfy 3-anonymity, in table $T_{\langle M_0 \rangle}$ we need to suppress more than 3 tuples. Vertex $\langle M_1 \rangle$ and vertex $\langle M_2 \rangle$ are marked true since table $T_{\langle M_1 \rangle}$ satisfies 3-anonymity by suppressing a number of tuples lower than MaxSup.
- $\text{DGH}_{\langle S_0 \rangle}$. Vertices $\langle S_0 \rangle$ and $\langle S_1 \rangle$ are marked true, since table $T_{\langle S_0 \rangle}$ satisfies 3-anonymity by suppressing a number of tuples lower than MaxSup.

Iteration 2

- $\text{DGH}_{\langle Z_0, M_0 \rangle}$. Since $\langle M_0 \rangle$ has been marked false in the previous iteration, this hierarchy does not include vertices $\langle Z_0, M_0 \rangle$, $\langle Z_1, M_0 \rangle$, and $\langle Z_2, M_0 \rangle$. Vertex $\langle Z_0, M_1 \rangle$ is marked false, since $T_{\langle Z_0, M_1 \rangle}$ satisfies 3-anonymity only if more than 3 tuples are suppressed. Vertices $\langle Z_0, M_2 \rangle$, $\langle Z_1, M_2 \rangle$, $\langle Z_2, M_2 \rangle$, $\langle Z_1, M_1 \rangle$, and $\langle Z_2, M_1 \rangle$ are instead marked true, since tables $T_{\langle Z_0, M_2 \rangle}$ and $T_{\langle Z_1, M_1 \rangle}$ satisfy 3-anonymity by suppressing a number of tuples lower than MaxSup.
- $\text{DGH}_{\langle Z_0, S_0 \rangle}$. Vertex $\langle Z_0, S_0 \rangle$ is marked false since $T_{\langle Z_0, S_0 \rangle}$ satisfies 3-anonymity only if more than MaxSup tuples re suppressed. Vertices $\langle Z_0, S_1 \rangle$, $\langle Z_1, S_1 \rangle$, $\langle Z_2, S_1 \rangle$, $\langle Z_1, S_0 \rangle$, and $\langle Z_2, S_0 \rangle$ are marked true, since tables $T_{\langle Z_0, S_1 \rangle}$ and $T_{\langle Z_1, S_0 \rangle}$ satisfy 3-anonymity by suppressing a number of tuples lower than MaxSup.
- $\text{DGH}_{\langle M_0, S_0 \rangle}$. Since $\langle M_0 \rangle$ has been marked false in the previous iteration, this hierarchy does not include vertices $\langle M_0, S_0 \rangle$ and $\langle M_0, S_1 \rangle$. All the other vertices in the hierarchy are marked true, since table $T_{\langle M_1, S_0 \rangle}$ satisfies 3-anonymity by suppressing a number of tuples lower than MaxSup.

Iteration 3

- $\text{DGH}_{\langle Z_0, M_0, S_0 \rangle}$. Since $\text{DGH}_{\langle Z_0, M_0 \rangle}$ does not contain vertices $\langle Z_0, M_0 \rangle$, $\langle Z_1, M_0 \rangle$, and $\langle Z_2, M_0 \rangle$, and vertex $\langle Z_0, M_1 \rangle$ has been marked false, this hierarchy does not contain vertices

---

* Incognito applies the monotonicity property of k-anonymity stating that if a table T is k-anonymous all its generalized tables are k-anonymous.

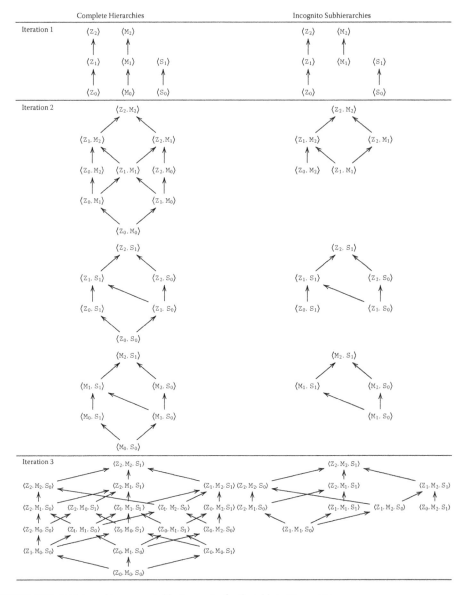

FIGURE 18.11 Subhierarchies computed by Incognito for the table in Figure 18.1.

$\langle Z_0, M_0, S_0 \rangle$, $\langle Z_1, M_0, S_0 \rangle$, $\langle Z_2, M_0, S_0 \rangle$, $\langle Z_0, M_0, S_1 \rangle$, $\langle Z_1, M_0, S_1 \rangle$, $\langle Z_2, M_0, S_1 \rangle$, $\langle Z_0, M_1, S_0 \rangle$, and $\langle Z_0, M_1, S_1 \rangle$. Analogously, since vertex $\langle Z_0, S_0 \rangle$ has been marked false in $\mathrm{DGH}_{\langle Z_0, S_0 \rangle}$, this hierarchy does not contain vertex $\langle Z_0, M_2, S_0 \rangle$. Vertices $\langle Z_1, M_1, S_0 \rangle$, $\langle Z_1, M_1, S_1 \rangle$, $\langle Z_1, M_2, S_0 \rangle$, $\langle Z_1, M_2, S_1 \rangle$, $\langle Z_2, M_1, S_0 \rangle$, $\langle Z_2, M_1, S_1 \rangle$, $\langle Z_2, M_2, S_0 \rangle$, and $\langle Z_2, M_2, S_1 \rangle$ are marked true, since table $T_{\langle Z_1, M_1, S_0 \rangle}$ satisfies 3-anonymity by suppressing a number of tuples lower than **MaxSup**. Analogously, vertex $\langle Z_0, M_2, S_1 \rangle$ is marked true, since table $T_{\langle Z_0, M_2, S_1 \rangle}$ satisfies 3-anonymity by suppressing a number of tuples lower than **MaxSup**.

Incognito iteratively builds the domain hierarchies of increasing dimension until the hierarchy DGH$_{DT}$ containing the entire QI is built. While in most cases some hierarchies are not built entirely, since some vertices are removed at iteration i due to the results obtained at iteration $i-1$, in the worst case all the hierarchies are completely built. Let Dom be the set of all ground and generalized domains for the quasi-identifier QI, n be the number of attributes composing QI, and m be the number of tuples in PT. We can note that each subset of Dom appears *exactly* in one of the domain generalization hierarchies on subsets of QI, therefore the cost of all the hierarchies computed in the worst case by Incognito is upperbounded by $(n \cdot 2^{\text{Dom}})$, where n is the upperbound of the cost of each single vertex (since each vertex has at most one component for each attribute in QI) and 2^{Dom} is the number of possible vertices. Since Incognito, in the worst case, tests if table PT, generalized according to each subset of Dom, is k-anonymous and suppresses a number of tuples lower than MaxSup, its complexity is in $O((n \cdot m) \cdot n \cdot 2^{\text{Dom}})$.

18.3.2.3 Approximation Algorithms

Exact algorithms for solving the k-anonymity problem for **AG_TS** and **AG_** are, due to the complexity of the problem, exponential in the size of the quasi-identifier. The exact algorithms for models **_CS** and **CG_** can be much more expensive, since the computational time can be exponential in the number of cells in the table.

The first approximation algorithm for **_CS** was proposed by Meyerson and Williams [32] and guarantees a $O(k \log(k))$-approximation. The best-known polynomial approximation algorithm for **_CS** guarantees a $O(k)$-approximate solution [2]. Such an $O(k)$-approximate algorithm constructs a complete weighted graph from the original private table PT. Each vertex in the graph corresponds to a tuple in PT, and the edges are weighted with the number of different attribute values between the two tuples represented by extreme vertices. The algorithm then constructs, starting from the graph, a forest composed of trees containing at least k vertices each, which represent the groups of tuples of at least k elements that are generalized to the same value for k-anonymization. Some cells in the vertices are suppressed to guarantee that all the tuples in the same tree have the same quasi-identifier value (i.e., to achieve k-anonymity). The cost of a vertex is evaluated as the number of cells suppressed, and the cost of a tree is the sum of the costs of its vertices. The cost of the final solution is equal to the sum of the costs of its trees. In constructing the forest, the algorithm limits the maximum number of vertices in a tree to be $3k-3$. Partitions with more than $3k-3$ elements are decomposed, without increasing the total solution cost. With the construction of trees with no more than $3k-3$ vertices, the authors prove that their solution is a $O(k)$-approximation.

An approximation algorithm for **CG_** is described in [3] as a direct extension of the approximation algorithm for **_CS** [2]. For taking into account the generalization hierarchies, each edge has a weight that is computed as follows. Given two tuples t_i and t_j and an attribute A, the generalization cost $h_{t_i,t_j}(A)$ associated with A is the lowest level of the value generalization hierarchy VGH$_{dom(A,\text{PT})}$ such that tuples t_i and t_j have the same generalized value for A. The weight $w(e)$ of the edge $e = (t_i, t_j)$ is therefore $w(e) = \Sigma_A h_{t_i,t_j}(A)/l_A$, where l_A is the number of levels in VGH$_{dom(A,\text{PT})}$. The solution of this algorithm is guaranteed to be a $O(k)$-approximation.

Recently, Park and Shim [33] described an algorithm for **_CS** with $O(\log k)$-approximation ratio, but the overall time complexity is $O((nm)^{2k})$.

Besides algorithms that compute k-anonymized tables for any value of k, ad hoc algorithms for specific values of k have also been proposed. For instance, to find better results for Boolean attributes, in the cases where $k=2$ or $k=3$, an ad hoc approach has been provided [3]. The algorithm for $k=2$ exploits the minimum-weight [1, 2]-factor built on the graph constructed for the 2-anonymity. The [1, 2]-factor for a graph G is a spanning subgraph of G built using only vertices with no more than 2 outgoing edges. Such a subgraph is a vertex-disjoint collection of edges and pairs of adjacent vertices and can be computed in polynomial time. Each component in the subgraph is treated as a cluster, and a 2-anonymized table is obtained by suppressing each cell for which the vectors in the cluster differ in value. This procedure is a 1.5-approximation algorithm. The approximation algorithm for $k = 3$ is similar and guarantees a 2-approximation solution.

18.4   *k*-Anonymity with Recoding-Based Generalization

The algorithms described in the previous section assume that the generalizations follow a predefined hierarchy. However, also a recoding-based generalization can be adopted for k-anonymity [6,24,27] that exploits an order relationship among values in the attribute domains. The algorithms adopting a recoding-based generalization are therefore not forced to follow a predefined strategy for generalizing values.

DEFINITION 18.10   (Recoding function) Let $D = \{v_1, \ldots, v_x\}$ be a domain on which a total order relationship is defined. A *recoding* function $\rho : D \rightarrow 2^{|D|}$ for D partitions the domain in a set of (possibly disjoint) intervals $I_1, \ldots, I_y$ $(y \leq x)$, such that $\bigcup_{i=1}^{y} I_i = D$ and $\forall v_i \in I_a, \forall v_j \in I_b$, with $a \neq b, v_i \leq v_j$ iff $a < b$.

As an example, consider attribute MaritalStatus and the VGH_{M_0} in Figure 18.3. A possible order among the values in M_0 is {married, widow, divorced, single}. A recoding function can partition M_0 into two intervals $I_1 = $ [married, widow] and $I_2 = $ [divorced, single].

Recoding-based generalization associates with each interval I_j a value v_j representing the values of the ground domain belonging to I_j. Any value in PT that belongs to I_j is then generalized to v_j.

The main advantage of a recoding-based generalization strategy is that it is possible to analyze solutions that would not be investigated adopting a hierarchy-based generalization, and this might improve the quality of the final solution. However, since recoding-based generalization strategies are based on the total order among attribute values, they are influenced by such an ordering, which may be difficult to define when the attribute domain does not have a characterizing ordering relationship. Furthermore, when the domain is categorical,* it might be difficult to represent the values obtained by the generalization process, since the set of values that need to be indistinguishable in the released table might not be similar. In some cases, the only way to describe an interval of values is the direct enumeration of its elements. For instance, the two subsets [married, widow] and [divorced, single] are described by enumerating their elements, while the set [married, widow, divorced] can be described as been_married.

18.4.1   Algorithms for *k*-Anonymity

We now describe two k-anonymization algorithms adopting recoding-based generalization: k-Optimize and Mondrian multidimensional k-anonymity.

18.4.1.1   *k*-Optimize

Bayardo and Agrawal [6] propose an **AG_TS** algorithm called *k-Optimize*, based on a recoding-based generalization defined on the quasi-identifying attributes. k-Optimize determines a solution minimizing a predefined cost function that measures the loss of information due to generalization and suppression.

---

* An attribute is categorical if it can assume a limited and specified set of values on which arithmetic operations cannot be defined. For instance, attribute MaritalStatus is categorical.

k-Optimize assumes the existence of a total order relationship among the attributes composing QI, and a total order relationship on each of the domains of the quasi-identifying attributes. It uses these total order relationships to associate an integer value, called *index*, with each interval in any domain of the quasi-identifier attributes. Note that, at initialization time, any value in any domain of QI represents an interval itself. The index assignment reflects the total order relationship among quasi-identifying attributes and within their domains.

Example 18.4

Consider the private table in Figure 18.1 with QI = {ZIP, MaritalStatus, Sex} and suppose that ZIP precedes MaritalStatus that, in turn, precedes Sex. Suppose also that the values within the domain of each of these attributes follow the same order as the leaves in the hierarchies in Figure 18.3. Figure 18.12 illustrates the index values assigned assuming that each value represents an interval. As it is visible from the figure, due to the order among attributes, the index values associated with the intervals of the ZIP domain are lower than the index values associated with the intervals of the domains of MaritalStatus and Sex. Furthermore, within each domain the assignment of the index values follow the total order among intervals.

Given a set of indexes $\mathcal{I}$, a generalization is represented by the union of individual index values. As an example, with reference to Figure 18.12, the union of index values 1 and 2 means that the ZIP values 22030 and 22032 have been generalized to the same indistinguishable value. Since only contiguous index values can be unioned for generalization purposes, their union is represented by the least index value. This implies that given a set $\mathcal{I}$ of index values, if an index value I does not appear in $\mathcal{I}$, the index I has been generalized to the nearest value appearing in $\mathcal{I}$ and lower than I. For instance, with respect to the set of indexes in Figure 18.12, the set $\mathcal{I} = \{1, 3, 5, 8, 9, 10\}$ implicitly indicates that index 2 has been generalized to the same value as 1, 4 to the same value as 3, and 6 and 7 to the same value as 5. Since the least value in any attribute domain will certainly appear in any generalization, it can be omitted in the representation of $\mathcal{I}$, assuming that it always implicitly belongs to the generalized domain. Consequently, the set $\mathcal{I} = \{1, 3, 5, 8, 9, 10\}$ can be represented as $\mathcal{I} = \{3, 8, 10\}$. This set of index values represents the following interval values: $\{[1,2],[3,4]\}$, corresponding to $\{[22030,22032],[22045,22047]\}$; $\{[5,6,7],[8]\}$, corresponding to $\{[married,widow,divorced],[single]\}$; and $\{[9],[10]\}$, corresponding to $\{[M],[F]\}$. Note that the empty set $\{\,\}$ represents the most general anonymization. For instance, with reference to Example 18.4, $\{\,\}$ corresponds to the generalizations $\{1\}$ for attribute ZIP, $\{5\}$ for attribute MaritalStatus, and $\{9\}$ for attribute Sex, which in turn correspond to the generalized values ZIP: $\{[22030, 22032, 22045, 22047]\}$; MaritalStatus: $\{[married,widow,divorced,single]\}$; and Sex: $\{[M,F]\}$.

Each generalized table that can be obtained from PT adopting a recoding-based generalization can be represented as a subset of the indexes $\mathcal{I}$ associated with the original domains of quasi-identifying attributes. Each of these solutions is associated with a cost, reflecting the information loss caused by the chosen generalization and by the tuples suppressed to achieve *k*-anonymity.

To compute the optimal generalized table, *k*-Optimize builds a *set enumeration tree* over the set $\mathcal{I}$ of index values, where each node corresponds to a generalized table of the original private table PT. The root node of the tree is the empty set. The children of a node *node* enumerate the sets that can be formed by appending a single element of $\mathcal{I}$ to *node*, with the restriction that this single element

ZIP				MaritalStatus				Sex	
⟨[22030]	[22032]	[22045]	[22047] ⟩	⟨[married]	[widow]	[divorced]	[single]⟩	⟨[M]	[F]⟩
1	2	3	4	5	6	7	8	9	10

FIGURE 18.12 Index assignment to attributes ZIP, MaritalStatus, and Sex.

Algorithm 18.3 (*k*-Optimize Algorithm)

INPUT
$\mathcal{I}$: set of index values corresponding to the original domains of PT[QI]
k: anonymity requirement

OUTPUT
$\mathcal{I}'$: set of index values representing the *k*-anonymous generalized table

MAIN
root := { }
best.cost := ∞
best.sol := *null*
Optimize(*root*,*best*)

OPTIMIZE(*node*,*best*)
if Satisfy(*node*) **then**
   cost := **Cost**(*node*)
   if *cost*≤*best.cost* **then**
     best.cost := *cost*
     best.sol := *node*
   for each $i \in \{idx | idx \in \mathcal{I} \wedge (\forall j \in node, idx \geq j \vee node = \emptyset)\}$ **do**
     child := *node* ∪ {*i*}
     lb := **LowerBound**(*child*)
     if *lb*≤*best.cost* **then**
       best := **Optimize**(*child*,*best*)
     else
       Prune(*root*,*node*) /* prune nodes having *node* as a subset */
return(*best*)

FIGURE 18.13 *k*-Optimize algorithm adopting a preorder traversal strategy.

must follow every element already in *node*, according to the given total order. Figure 18.14 illustrates a set enumeration tree over $\mathcal{I} = \{2, 3, 4\}$ (e.g., this tree could represent the set enumeration tree for attribute ZIP). The consideration of a tree guarantees the existence of a unique path between the root and each node. The visit of the set enumeration tree using a standard traversal strategy is equivalent to the evaluation of each possible solution to the *k*-anonymity problem.

Algorithm *k*-Optimize (see Figure 18.13) visits the set enumeration tree built on the basis of the given private table PT, keeping track of the best generalization (variable *best*) found at each step. For each visited node *node* in the tree that satisfies the *k*-anonymity constraint (function **Satisfy**), it computes (function **Cost**) the cost *cost* of the generalization strategy represented by *node*. If the cost of the solution associated with *node* is *lower* than *best.cost*, that is, the lowest cost found during the visit at that point, *node* becomes the new locally best solution and its cost becomes the comparison term for the following visited nodes. Since the number of nodes in the tree is $2^{|\mathcal{I}|}$, this approach is not practical and the authors have therefore proposed heuristics and pruning strategies to reduce the computational costs of the *k*-Optimize algorithm [6]. The idea is that if none of the nodes in the subtree rooted at a child *child* of *node* can have a cost lower than the locally optimal solution computed before visiting *child*, *k*-Optimize prunes the subtree rooted at *child* without visiting the solutions it contains. To this purpose, for each visited node *node*, *k*-Optimize computes the lower bound of the cost function for the subtree rooted at each of its children *child* (function **LowerBound**), that is, the lowest possible cost that a solution in the subtree can have. The lower bound for a subtree *child* is computed by noting that the solutions in the subtree rooted at *child* need to suppress a greater

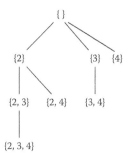

FIGURE 18.14 An example of set enumeration tree over set $\mathcal{I} = \{2, 3, 4\}$ of indexes.

number of tuples than *node* to guarantee *k*-anonymity, since they represent more specific solutions. If this lower bound is higher than the locally optimal solution, the subtree rooted at *child* is pruned. Note that when a subtree is pruned also additional nodes can be removed from the tree (procedure **Prune**). For instance, consider the set enumeration tree in Figure 18.14 and suppose that node {2, 4} is pruned. This means that all the solutions that contain index values 2 and 4 are not optimal, therefore also node {2, 3, 4} can be pruned.

k-Optimize can always compute the best solution in the space of the generalization strategies. Since the algorithm tries to improve the solution at each visited node evaluating the corresponding generalization strategy, it is possible to set a maximum computational time, and obtain a good, but nonoptimal, solution.

18.4.1.2 Mondrian Multidimensional Algorithm

LeFevre et al. [27] propose a **CG_** algorithm for anonymizing a private table PT based on *multidimensional global recoding*, by extending the single-dimensional recoding-based generalization to the multidimensional case.

On the basis of the order of values defined for all the domains of attributes composing QI, a multidimensional generalization function defines a set of *multidimensional regions*. These regions correspond to the intervals defined in the single-dimensional scenario. Let $QI = \{A_1, \ldots, A_n\}$ be a quasi-identifier and $\mathcal{D} = \{D_1, \ldots, D_n\}$ be the set of ground domains of the attributes $A_1, \ldots, A_n$. Each domain in $\mathcal{D}$ is a dimension for the multidimensional space, where PT[QI] can be represented as a set of points: each tuple $t \in$ PT[QI] represents a point in the multidimensional space defined by $\mathcal{D}$ and its coordinates are the values assumed by the quasi-identifying attributes in t.

A multidimensional region within such a space is represented by two tuples $p = (p_1, \ldots, p_n)$ and $v = (v_1, \ldots, v_n)$ such that $p_i \leq v_i, i = 1, \ldots, n$. A tuple $t = (t_1, \ldots, t_n)$ belongs to the region represented by $\langle p, v \rangle$ if $p_i \leq t_i \leq v_i, i = 1, \ldots, n$. A multidimensional region can therefore be seen as an *n*-dimensional rectangular, whose edges are parallel to the axis. As an example, consider the private table in Figure 18.1 and suppose that $QI = \{\texttt{MaritalStatus}, \texttt{Sex}\}$. Figure 18.15a provides a graphical representation of the multidimensional space determined by $\mathcal{D} = \{M_0, S_0\}$, where the domain of attribute Sex is represented on the *x*-axis and the domain of attribute MaritalStatus

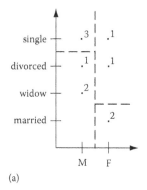

MaritalStatus	Sex
married	F
married	F
single	M
single	M
single	M
(widow, divorced, single)	F
(married, widow, divorced)	M
(married, widow, divorced)	M
(married, widow, divorced)	M
(widow, divorced, single)	F

(a) (b)

FIGURE 18.15 An example of strict multidimensional partitioning (a) and corresponding generalized table (b).

is represented on the y-axis. Regions are delimited by lines parallel to the axis and points are associated with the number of occurrences of the corresponding quasi-identifier values in PT[QI]. A *strict multidimensional partitioning* is a set of multidimensional regions that cover the whole space defined on $\mathcal{D}$.

DEFINITION 18.11   (**Strict multidimensional partitioning**) Let QI be a quasi-identifier and $\mathcal{D} = \{D_1, \ldots, D_n\}$ be the set of domains for the attributes composing it. A *strict multidimensional partitioning* is a set of *nonoverlapping* multidimensional regions covering the space defined by $\mathcal{D}$.

A strict multidimensional partitioning of the space defined by $\mathcal{D}$ represents the generalized table T obtained by making all the tuples belonging to the same multidimensional region indistinguishable, that is, by generalizing the tuples in the same region to the same generalized tuple. For instance, Figure 18.15b represents a possible generalized table corresponding to the regions in Figure 18.15a, where the generalized values of attributes MaritalStatus and Sex are obtained by listing all values that belong to a given region.

A strict multidimensional partitioning for a private table PT is k-anonymous if any multidimensional region in the space defined by $\mathcal{D}$ contains either zero or at least k tuples of PT. The problem of computing the *optimal* (i.e., the one that minimizes information loss) strict k-anonymous multidimensional partitioning is NP-hard, since its decisional version can be reduced from the well-known partition problem [27]. It is important to note here that this result is not implied by the NP-hardness of the hierarchy-based k-anonymity problem demonstrated in [32], since the two problems are different and cannot be reduced one to the other.

Given the set of points along with the number of their occurrences induced by private table PT on the multidimensional space defined by $\mathcal{D}$, a *multidimensional cut* for such a set of points is an axis-parallel cut that produces two disjoint sets of points. A multidimensional cut perpendicular to dimension D_i, corresponding to attribute A_i in QI, is performed at a given value v of D_i. Any tuple t such that $t[A_i] \leq v$ will belong to one partition, while any tuple t such that $t[A_i] > v$ will belong to the other one. A multidimensional cut is *allowable* with respect to a given k-anonymity constraint if and only if the resulting regions contain a set of points representing at least k tuples of the original table (i.e., the sum of the occurrences of the points in the region obtained is at least k). The recursive application of allowable multidimensional cuts on the original set of points representing PT generates a multidimensional strict partitioning.

DEFINITION 18.12   (**Minimal strict multidimensional partitioning**) A strict multidimensional partitioning, composed of regions $R_1, \ldots, R_r$, is *minimal* if there does not exist an allowable multidimensional cut for any of the sets of points in its regions.

For instance, for $k = 2$, the strict multidimensional partitioning in Figure 18.15a is minimal, since none of its regions can be further split obtaining two regions containing at least 2 tuples each.

The maximum number of points contained in any region R_i in a minimal strict multidimensional partitioning is $2n(k-1) + o$, where $n = |QI|$ is the number of dimensions (i.e., attributes composing the quasi-identifier), k is the k-anonymity constraint, and o is the maximum number of occurrences of a quasi-identifier value in PT (i.e., the maximum number of copies of the same point in the multidimensional space), as formally proved in [27].

The algorithm proposed in [27], and illustrated in Figure 18.16, is a greedy solution to the minimal multidimensional k-anonymity problem. The algorithm receives in input the set *partition* of points together with their occurrences that represent a private table PT, and the k value. The algorithm cuts the given n-dimensional space in two nonoverlapping spaces and is recursively applied to each

Algorithm 18.4 (Mondrian Algorithm)

INPUT
partition=$\{p|p = \langle p_1, \ldots, p_n \rangle \wedge p_i \in D_i, i = 1, \ldots, n \rangle\}$: set of points that represent private table PT[QI]
k: anonymity requirement

OUTPUT
T: k-anonymized table for PT[QI]

MAIN
Anonymize(*partition*)

ANONYMIZE(*partition*)
if(no allowable multidimensional cut for *partition*) **then**
    return Generalize(*partition*) /* generalize all the tuples in partition to the same value */
else
    dim := **ChooseDimension**(*partition*) /* choose the dimension for the cut */
    fs := **frequency_set**(*partition, dim*) /* frequency of the values in the domain of *dim* */
    split_val := **find_median**(*fs*) /* find the split point as the median for *dim* wrt *fs* */
    lhs := $\{p \in partition | p_{dim} \leq split\_val\}$ /* first partition created */
    rhs := $\{p \in partition | p_{dim} > split\_val\}$ /* second partition created */
    return Anonymize(*rhs*)$\cup$**Anonymize**(*lhs*) /* recursive call */

FIGURE 18.16 Mondrian multidimensional k-anonymity algorithm.

resulting space, until the minimal strict multidimensional partitioning is reached. At each recursive call, the algorithm chooses the dimension with the widest (normalized) range of values. It then splits the region by performing a strict multidimensional cut perpendicular to the chosen dimension, adopting as a split value the median of *partition* projected on the chosen dimension. If the cut is allowable, the algorithm is recursively called on each of the two regions obtained; otherwise the cut is not performed. As an example, consider the private table in Figure 18.1 and suppose that QI={MaritalStatus, Sex}. Figure 18.17 represents an example of step-by-step execution of the algorithm in Figure 18.16, where $k = 2$. The algorithm first chooses the Sex dimension and performs a cut at value "M," obtaining the regions in Figure 18.17b. The algorithm is then recursively called on the region characterized by Sex = M. This call causes a cut on attribute MaritalStatus at value "single." The two resulting regions (Figure 18.17c) cannot be further partitioned and, therefore the algorithm executes the recursive call on the region characterized by Sex = F. This call causes a cut on attribute MaritalStatus at value "widow." The two resulting regions cannot be further partitioned. The result obtained by the algorithm is then represented in Figure 18.17d.

This algorithm is greedy, since it chooses the dimension to exploit for the next cut on the basis of the local properties of the multidimensional region. It represents a $O(n)$-approximation

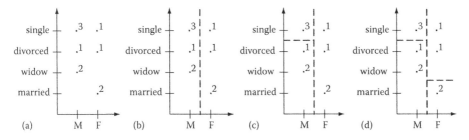

FIGURE 18.17 Spatial representation (a) and possible partitioning (b–d) of the table in Figure 18.1.

algorithm for the *k*-anonymity problem (with recoding-based generalization and adopting strict multidimensional cuts), supposing o/k to be a constant ratio. The time complexity for the given algorithm is $O((n \cdot m) \cdot log(n \cdot m))$, when $(n \cdot m)$ is the number of cells composing the original private table.

Besides strict multidimensional partitioning, also *relaxed* multidimensional partitioning can be adopted for the *k*-anonymity problem. The main difference of relaxed multidimensional partitioning with respect to strict multidimensional partitioning is that the regions covering the multidimensional space defined by $\mathcal{D}$ can be potentially overlapping. When adopting relaxed multidimensional partitioning, each tuple in PT is generalized following the values of one of the regions to which the point corresponding to the tuple belongs. Therefore, duplicate values in PT can be generalized to different values in the released table. The Mondrian algorithm proposed can be adapted to operate also with relaxed multidimensional partitioning. In this case, it computes a 2-approximate solution of the problem.

18.5   Attribute Disclosure Protection

Even if *k*-anonymity represents a solution to the problem of identity disclosure, since it protects respondents from attacks aimed at reducing the uncertainty about their identities, it does not protect from attribute disclosure. As a matter of fact, attribute disclosure is possible even on tables protected against identity disclosure.

Machanavajjhala, Gehrke, and Kifer define two possible attacks to *k*-anonymous tables: *homogeneity attack* (already noted in [35]) and *background knowledge attack* [31]. Consider a *k*-anonymized table, where there is a sensitive attribute and suppose that all tuples with a specific value for the quasi-identifier have the same sensitive attribute value. Under these homogeneity assumptions, if an attacker knows the quasi-identifier value of a respondent and knows that this respondent belongs to the population represented in the table, the attacker can infer which is the value for the sensitive attribute for the known respondent, since all the tuples with the given quasi-identifier value have the same value for the sensitive attribute. For instance, consider the 3-anonymous table in Figure 18.18, where Disease is a sensitive attribute. If Alice knows that her friend Hellen is a married female living in an area with ZIP code 22030, she can infer that Hellen suffers from hypertension, since all tuples with ⟨2203*,been_married,F⟩ as a quasi-identifier value are characterized by Disease = "hypertension."

The background knowledge attack is instead based on a priori knowledge of the attacker of some additional external information. For instance, with reference to the 3-anonymous table in Figure 18.18, suppose that Alice knows that Bob is a single man living in 22032 area. Alice can

ZIP	MaritalStatus	Sex	Disease
2203*	been_married	F	hypertension
2203*	been_married	F	hypertension
2203*	never_married	M	obesity
2203*	never_married	M	HIV
2203*	never_married	M	obesity
2203*	been_married	F	hypertension
2204*	been_married	M	obesity
2204*	been_married	M	HIV
2204*	been_married	M	HIV

FIGURE 18.18 An example of 3-anonymous table.

then infer that Bob suffers of obesity or HIV. Suppose now that Alice knows that Bob is thin. Alice can infer with probability equal to 1 that Bob suffers of HIV.

To prevent homogeneity and background knowledge attacks, Machanavajjhala, Gehrke, and Kifer introduce the notion of ℓ-*diversity* [31].

DEFINITION 18.13 (ℓ-**Diversity principle**) Let $T(A_1, \ldots, A_n, S)$ be a table, $QI = \{A_i, \ldots, A_n\}$ be its quasi-identifier, ℓ be a user-defined threshold, and S be a sensitive attribute. A set of tuples in T having the same value for QI, called q-block, is said to be ℓ-*diverse* if it contains at least ℓ different values for S. T is said to be ℓ-*diverse* if all its q-blocks are ℓ-*diverse*.

If a k-anonymous table is ℓ-diverse, the homogeneity attack is no more applicable, since each q-block has at least ℓ (≥ 2) distinct sensitive attribute values. Analogously, the background knowledge attack becomes more complicate as ℓ increases, because the attacker needs more knowledge to individuate a unique value associable with a predefined respondent. For instance, if $\ell = 2$, only two of the three q-blocks in the table in Figure 18.18 are ℓ-diverse, while the third is not ℓ-diverse ($\ell = 1$).

Machanavajjhala et al. [31] prove that ℓ-diversity satisfies the monotonicity property with respect to DGH$_{DT}$. Monotonicity means that if T_i guarantees ℓ-diversity, any T_j, such that $T_i \preceq T_j$ satisfies ℓ-diversity. This implies that any algorithm originally thought for k-anonymity can also be used to achieve ℓ-diversity, by simply checking the ℓ-diversity property every time a table is tested for k-anonymity.

Note that the original definition of ℓ-diversity considers the presence of one sensitive attribute only and cannot be simply extended to sets of attributes, since ℓ-diversity may be violated even if each sensitive attribute separately satisfies it. If the private table contains more than one sensitive attribute, ℓ-diversity can be achieved by ensuring that the given table is ℓ-diverse with respect to each sensitive attribute S_i separately, running the algorithm considering, as a quasi-identifier, QI unioned with all sensitive attributes but S_i.

After the introduction of ℓ-diversity, the problem of attribute disclosure has received much attention, and different proposals have been studied [39,40]. p-sensitive k-anonymity [39] is another approach very similar to ℓ-diversity that considers the case of table PT with more than one sensitive attribute.

(α, k)-anonymity [40] takes a different approach to the attribute disclosure problem. It supposes that not all the values in the domain of a sensitive attribute are equally sensitive. For instance, the obesity value for attribute Disease is not sensitive, while the HIV value is sensitive. To the aim of protecting the association of quasi-identifying values with sensitive values, (α, k)-anonymity imposes a different constraint to k-anonymous tables. If $dom(S, \text{PT})$ has only one sensitive value s, every q-block must have a relative frequency of s not greater than α, meaning that the number of occurrences of s in the block divided by the cardinality of the block cannot be greater than α.

DEFINITION 18.14 ((α,k)-**Anonymity**) Let $T_i(A_1, \ldots, A_n, S)$ and $T_j(A_1, \ldots, A_n, S)$ be two tables with $QI = \{A_1, \ldots, A_n\}$ such that $T_i[QI] \preceq T_j[QI]$, S be a sensitive attribute, s be the only sensitive value in $dom(S, T_i)$, and $0 < \alpha < 1$ be a user-defined threshold. T_j is α-deassociated with respect to QI and s if the relative frequency of s in every q-block is not greater than α. T_j is said to be (α, k)-*anonymous* if it satisfies both the k-anonymity and the α-deassociation constraints.

For instance, consider the 3-anonymous table in Figure 18.18, and suppose that the unique sensitive value for Disease is HIV. Suppose also that $\alpha = 0.4$ and $k = 3$. The table in Figure 18.18 is not (0.4, 3)-anonymous because the last q-block does not satisfy the 0.4-deassociation constraint, since the sensitive value appears twice in a block of 3 tuples with a relative frequency of 0.67. The other two q-blocks are 0.4-deassociated.

The (α, k)-anonymization problem is NP-hard since, considering a binary alphabet, it can be reduced from the edge partition into the 4-cliques problem [40]. Since also (α, k)-anonymity is a monotonic property with respect to DGH_{DT}, the classical k-anonymity algorithms can be adopted to achieve it. However, a local recoding strategy that adopts a top-down approach (i.e., starting from a completely generalized table, it selectively specializes the table without violating (α, k)-anonymity constraint) can produce better results [40].

Although ℓ-diversity protects data against attribute disclosure, this protection leaves space to attacks based on the distribution of values inside q-blocks. Li et al. [29] show that ℓ-diversity suffers from two attacks: the *skewness attack* and the *similarity attack*. As an example of skewness attack, consider the 2-diverse table in Figure 18.19 and the binary sensitive attribute `Diabetes`, which is characterized by a skewed distribution. The table has a q-block having 2 out of 3 tuples with a positive value for `Diabetes` and only one tuple with a negative value for it. Even if the table satisfies 2-diversity, it is possible to infer that respondents in the given q-block have 67% probability of contracting diabetes, compared to the 30% of the whole population. Furthermore, if the values for the sensitive attribute S in a q-block are distinct but semantically similar, an external recipient can however infer important information on it by applying a similarity attack. For instance, with reference to the table in Figure 18.19, it is possible to infer that single males living in the 2203* area have the cholesterol value between 250 and 260, since all the tuples in this q-block have these values for the considered sensitive attribute.

To counteract these attacks, Li et al. [29] introduce the t-closeness requirement, which is a stronger requirement than ℓ-diversity.

DEFINITION 18.15 (**t-Closeness principle**) Let $T_i(A_1, \ldots, A_n, S)$ and $T_j(A_1, \ldots, A_n, S)$ be two tables such that $T_i[A_1, \ldots, A_n] \preceq T_j[A_1, \ldots, A_n]$, S be a sensitive attribute, and t be a user-defined threshold. A q-block in T_j satisfies t-closeness if the distance between the distribution of S in this q-block and the distribution of S in T_i is lower than t. T_j satisfies t-closeness if all its q-blocks satisfy t-closeness.

By imposing that the distribution of sensitive values in the released table must be similar to the distribution in the private table, t-closeness helps in preventing both skewness and similarity attacks. As a matter of fact, all the sets of tuples with the same `QI` value in the released table have approximately the same sensitive value distribution as the whole original population. t-closeness is a difficult property to achieve since t-closeness requires to measure the distance between two distributions of values, either numerical or categorical. In [29], the authors propose to adopt the Earth Mover's Distance (EMD) measure. The advantage of this measure is that, as demonstrated in the paper, it can be easily integrated with the Incognito algorithm due to its generalization and

ZIP	MaritalStatus	Sex	Diabetes	Cholesterol
2203*	been_married	F	N	230
2203*	been_married	F	N	220
2203*	never_married	M	Y	250
2203*	never_married	M	Y	260
2203*	never_married	M	N	250
2203*	been_married	F	N	275
2204*	been_married	M	N	285
2204*	been_married	M	Y	210
2204*	been_married	M	N	190

FIGURE 18.19 An example of 3-anonymous and 2-diverse table.

subset properties, which imply monotonicity with respect to both the number of attributes and the generalization level chosen.

18.6 Conclusions

The management of privacy in today's global infrastructure is a complex issue, since it requires the combined application of solutions coming from technology (technical measures), legislation (law and public policy), ethics, and organizational/individual policies and practices. We focused on the technological aspect of privacy, which involves three different but related dimensions: privacy of the user, privacy of the communication, and privacy of the information (data protection). The chapter discussed the data protection dimension of privacy, which is more generally related to the collection, management, use, and protection of personal information. The data protection aspect requires the investigation of different problems including the problem of protecting the identities (anonymity) of the users to whom the data refer. This problem is becoming more and more difficult because of the increased information availability and ease of access as well as the increased computational power provided by today's technology. Many techniques have been developed for protecting data released publicly or semipublicly from improper disclosure.

In this chapter, we presented a specific technique that has been receiving considerable attention recently, and that is captured by the notion of *k-anonymity*. We discussed the concept of *k*-anonymity and described two classes of algorithms for its enforcement, which differ by the generalization technique (i.e., hierarchy-based or recoding-based) adopted. We also discussed recent proposals for attribute disclosure protection, which are aimed at extending the *k*-anonymity algorithms with additional properties.

Acknowledgments

This work was supported in part by the EU, within the Seventh Framework Programme (FP7/2007–2013) under grant agreement n° 216483, and by the Italian MIUR, within PRIN 2006, under project 2006099978.

References

1. N. Adam and J. Wortman. Security-control methods for statistical databases: A comparative study. *ACM Computing Surveys*, 21(4):515–556, 1989.
2. G. Aggarwal, T. Feder, K. Kenthapadi, R. Motwani, R. Panigrahy, D. Thomas, and A. Zhu. Anonymizing tables. In *Proceedings of the 10th International Conference on Database Theory (ICDT'05)*, Edinburgh, Scotland, January 2005.
3. G. Aggarwal, T. Feder, K. Kenthapadi, R. Motwani, R. Panigrahy, D. Thomas, and A. Zhu. Approximation algorithms for *k*-anonymity. *Journal of Privacy Technology*, November 2005.
4. R. Agrawal and R. Srikant. Privacy-preserving data mining. In *Proceedings of the ACM SIGMOD Conference on Management of Data*, Dallas, TX, May 2000.
5. A. Ambainis. Upper bound on communication complexity of private information retrieval. In *Proceedings of the 24th International Colloquium on Automata, Languages and Programming*, Bologna, Italy, July 1997.
6. R. J. Bayardo and R. Agrawal. Data privacy through optimal *k*-anonymization. In *Proceedings of the 21st International Conference on Data Engineering (ICDE'05)*, Tokyo, Japan, April 2005.
7. C. Cachin, S. Micali, and M. Stadler. Computationally private information retrieval with polylogarithmic communication. In *Proceedings of EUROCRYPT'99*, Prague, Czech Republic, May 1999.

8. D. Chaum. Untraceable electronic mail, return addresses, and digital pseudonyms. *Communications of the ACM*, 24(2):84–88, February 1981.

9. B. Chor and N. Gilboa. Computationally private information retrieval (extended abstract). In *Proceedings of the 29th Annual ACM Symposium on Theory of Computing*, El Paso, TX, May 1997.

10. B. Chor, E. Kushilevitz, O. Goldreich, and M. Sudan. Private information retrieval. *Journal of ACM*, 45(6):965–981, April 1998.

11. V. Ciriani, S. De Capitani di Vimercati, S. Foresti, and P. Samarati. *k*-anonymity. In T. Yu and S. Jajodia, editors, *Security in Decentralized Data Management*. Springer, Berlin, Heidelberg, 2007.

12. V. Ciriani, S. De Capitani di Vimercati, S. Foresti, and P. Samarati. *k*-anonymous data mining: A survey. In C. C. Aggarwal and P. S. Yu, editors, *Privacy-Preserving Data Mining: Models and Algorithms*. Springer-Verlag, New York, 2007.

13. V. Ciriani, S. De Capitani di Vimercati, S. Foresti, and P. Samarati. Microdata protection. In T. Yu and S. Jajodia, editors, *Security in Decentralized Data Management*. Springer, Berlin, Heidelberg, 2007.

14. L. Cox. A constructive procedure for unbiased controlled rounding. *Journal of the American Statistical Association*, 82(398):520–524, 1987.

15. S. Dawson, S. De Capitani di Vimercati, P. Lincoln, and P. Samarati. Maximizing sharing of protected information. *Journal of Computer and System Sciences*, 64(3):496–541, May 2002.

16. D. Denning. Inference controls. In *Cryptography and Data Security*. Addison-Wesley Publishing Company, Reading, MA, 1982.

17. R. Dingledine, N. Mathewson, and P. Syverson. Tor: The second-generation onion router. In *Proceedings of the 12th USENIX Security Symposium*, San Diego, CA, August 2004.

18. W. Du and M. Atallah. Privacy-preserving cooperative statistical analysis. In *Proceedings of the 17th Annual Computer Security Applications Conference*, New Orleans, LA, December 2001.

19. W. Du and M. Atallah. Secure multi-party computation problems and their applications: A review and open problems. In *Proceedings of the 2001 Workshop on New Security Paradigms*, Cloudcroft, New Mexico, September 2001.

20. G. Duncan, S. Keller-McNulty, and S. Stokes. Disclosure risk vs. data utility: The R-U confidentiality map. Technical report, Los Alamos National Laboratory, Los Alamos, NM, 2001. LA-UR-01-6428.

21. Federal Committee on Statistical Methodology. Statistical policy working paper 22, Washington, DC, May 1994. Report on Statistical Disclosure Limitation Methodology.

22. P. Golle. Revisiting the uniqueness of simple demographics in the us population. In *Proceedings of the Workshop on Privacy in the Electronic Society*, Alexandria, VA, October 2006.

23. Y. Ishai and E. Kushilevitz. Improved upper bounds on information-theoretic private information retrieval (extended abstract). In *Proceedings of the 31st annual ACM Symposium on Theory of Computing*, Atlanta, GA, May 1999.

24. V. S. Iyengar. Transforming data to satisfy privacy constraints. In *Proceedings of the 8th ACM SIGKDD Internationale Conference on Knowledge Discovery and Data Mining*, Edmonton, Alberta, Canada, 2002.

25. V. Kann. Maximum bounded h-matching is max snp-complete. *Information Processing Letters*, 49:309–318, 1994.

26. E. Kushilevitz and R. Ostrovsky. Replication is not needed: Single database, computationally-private information retrieval. In *Proceedings of the 38th Annual Symposium on Foundations of Computer Science*, Miami Beach, FL, October 1997.

27. K. LeFevre, D. DeWitt., and R. Ramakrishnan. Mondrian multidimensional *k*-anonymity. In *Proceedings of the International Conference on Data Engineering (ICDE'06)*, Atlanta, GA, April 2006.

28. K. LeFevre, D. J. DeWitt, and R. Ramakrishnan. Incognito: Efficient full-domain *k*-anonymity. In *Proceedings of ACM SIGMOD Conference on Management of Data*, Baltimore, MD, June 2005.

29. N. Li, T. Li, and S. Venkatasubramanian. *t*-closeness: Privacy beyond *k*-anonymity and ℓ-diversity. In *Proceedings of the 23nd International Conference on Data Engineering*, Istanbul, Turkey, April 2007.

30. Y. Lindell and B. Pinkas. Privacy preserving data mining. *Journal of Cryptology*, 15(3):177–206, June 2002.

31. A. Machanavajjhala, J. Gehrke, and D. Kifer. ℓ-density: Privacy beyond k-anonymity. In *Proceedings of the International Conference on Data Engineering (ICDE'06)*, Atlanta, GA, April 2006.

32. A. Meyerson and R. Williams. On the complexity of optimal k-anonymity. In *Proceedings of the 23rd ACM-SIGMOD-SIGACT-SIGART Symposium on the Principles of Database Systems*, Paris, France, June 2004.

33. H. Park and K. Shim. Approximate algorithms for k-anonymity. In *Proceedings of the ACM SIGMOD International Conference on Management of Data*, Beijing, China, June 2007.

34. M. Reiter and A. Rubin. Anonymous Web transactions with crowds. *Communications of the ACM*, 42(2):32–48, February 1999.

35. P. Samarati. Protecting respondents' identities in microdata release. *IEEE Transactions on Knowledge and Data Engineering*, 13(6):1010–1027, November 2001.

36. K. Sampigethaya and R. Poovendran. A survey on mix networks and their secure applications. *Proceedings of the IEEE*, 94(12):2142–2181, December 2006.

37. L. Sweeney. Achieving k-anonymity privacy protection using generalization and suppression. *International Journal on Uncertainty, Fuzziness and Knowledge-based Systems*, 10(5):571–588, 2002.

38. P. Syverson, D. Goldschlag, and M. Reed. Anonymous connections and onion routing. In *Proceedings of the 18th Annual Symposium on Security and Privacy*, Oakland, CA, May 1997.

39. T. M. Truta and B. Vinay. Privacy protection: p-sensitive k-anonymity property. In *Proceedings of the 22nd International Conference on Data Engineering Workshop (ICDEW'06)*, Atlanta, GA, April 2006.

40. R. C.-W. Wong, J. Li, A. W.-C. Fu, and K. Wang. (α,k)-anonymity: An enhanced k-anonymity model for privacy preserving data publishing. In *Proceedings of the 12th ACM SIGKDD International Conference on Knowledge Discovery and Data Mining*, Philadelphia, PA, August 2006.

41. A. Yao. Protocols for secure computations. In *Proceedings of the 23rd Annual IEEE Symposium on Foundations of Computer Science*, Chicago, IL, November 1982.

19

Database Theory: Query Languages

Nicole Schweikardt
Goethe-Universität Frankfurt am Main

Thomas Schwentick
Technische Universität Dortmund

Luc Segoufin
INRIA and ENS Cachan

This chapter gives an introduction to the theoretical foundations of query languages for relational databases. It thus addresses a significant part of database theory. Special emphasis is put on the expressive power of query languages and the computational complexity of their associated evaluation and static analysis problems.

19.1 Introduction

Most personal or industrial data is simply stored in files and accessed via simple programs. This approach works well for small applications but is not generic and does not scale. New applications require new software, and classical software can hardly cope with huge data sets.

Database management systems (DBMS) have been built to provide a generic solution to this issue. Usually, a DBMS enforces a clear distinction between how the data is stored on disk and how it is accessed via queries. When querying a database, one should not be concerned about how and where the data is physically stored, but only with the logical structure of the data. This concept is called the data independence principle.

Moreover a DBMS comes with an optimization engine containing evaluation heuristics, index structures, and data statistics that greatly improve the performance of the system and induce high scalability. Typical DBMS now handle gigabytes of data easily.

The roots of database theory lie in the work of Codd on the relational model, identifying the relational calculus and the relational algebra. Several other models were later proposed, e.g., the object oriented model and, more recently, the semistructured model of XML. The relational model is now the most widely used by DBMS. For this reason, and for lack of space, we mainly consider the relational model in this chapter. Only in the last section we shall briefly discuss other major data models.

Theoretical database research has covered areas such as the design of query languages, schema design, query evaluation (clustering, indexing, optimization heuristics, etc.), storage and transaction management, and concurrency control to name just a few. In this chapter we only address the theory of query languages, which forms the core of database theory. For gaining an in-depth knowledge of database theory we refer the reader to the textbooks [2,3,83], to the survey articles that regularly appear in the *Database Principles Column* of *SIGMOD Record*, and to the *Proceedings of the ACM Symposium on Principles of Database Systems (PODS)* and the *International Conference on Database Theory (ICDT)*. For theoretical results on schema design and integrity constraints we further refer to Kanellakis' handbook chapter [69].

The intention of this chapter is not to give a complete historical account. Sometimes we favor more recent presentations over the original papers. Full references can usually be found in the cited work.

The present chapter is organized as follows: In Section 19.2 we introduce the relational model and its basic definitions. In the same section, we also describe the aspects of query languages we are mostly interested in—expressive power, evaluation complexity, and complexity of static analysis. In Section 19.3 we discuss the most basic relational query language, conjunctive queries. In Section 19.4 we describe the relational algebra and the relational calculus whose expressive power is usually considered a yardstick for relational query languages. We study restricted query languages that can be evaluated more efficiently and allow for automatic static analysis in Section 19.5. In contrast, in Section 19.6 we cover more expressive query languages which support recursion and counting. In Section 19.7 we conclude with a brief survey on query languages for some other data models.

19.2   General Notions

In this section we define the general concepts used in database theory. We present the relational model and the notions of query and query language. We also introduce the key properties of query languages relevant for this article: their expressive power, how they are evaluated, their complexity, and optimization and static analysis that can be performed on them.

19.2.1   The Relational Model

In the early years of databases, when it became clear that file systems are not an adequate solution for storing and processing large amounts of interrelated data, several database models were proposed, including the hierarchical model and the network model (see, e.g., [102]).

One central idea at the time was that querying a database should not depend on how and where data is actually stored. This is known as the data independence principle.

One of the biggest breakthroughs in computer science came when Codd introduced the relational model in 1970 [25]. In this chapter we will focus on the relational model, as it is still dominating the databases industry. Furthermore, most classical results in database theory have been obtained for the relational model.

Operas	Composer	Piece
	Puccini	Turandot
	Wagner	Tristan
	Britten	Peter Grimes
	Puccini	Tosca
	Monteverdi	Orfeo
	Bizet	Carmen

People	Artist	Type	Birthday
	Puccini	Composer	12/22
	Nilsson	Soprano	05/17
	Pavarotti	Tenor	10/12
	Bizet	Composer	10/25
	Callas	Soprano	12/2

Cast	Theatre	Piece	Artist
	Scala	Turandot	Pavarotti
	Bayreuth	Tristan	Nilsson
	Covent Garden	Peter Grimes	Vickers
	Met	Tosca	Callas
	Scala	Turandot	Raisa

FIGURE 19.1 An example database with three relations.

In a nutshell, the basic idea of relational databases is to store data in tables or, seen from a more mathematical point of view, in relations. Figure 19.1 displays our simple running example of a relational database with three relations containing information on operas. Each table header gives some structural information on the relation, called the relation schema. Formally, a relation schema R consists of the relation name (Operas in the first table), the number of columns, its arity arity(R), and names for each column, the attributes. The actual content of a relation is given by the set of rows of the table, the tuples of the relation. Each tuple has one entry for each attribute of the relation. We assume that each entry comes from a fixed, infinite domain of potential database elements. Elements in this domain are often called constants or data values.

A database schema σ is simply a finite set of relation schemas where no two relations have the same name. For the purpose of this article, we ignore the fact that a database schema usually also includes a set of integrity constraints like key or foreign key constraints. We refer to [3,69] for a discussion of the theoretical issues raised by integrity constraints.

Finally, a database instance (or, for short, database) D over a database schema σ has one (finite) relation R^D of arity arity(R) for each relation schema R of σ. If the database instance is clear from the context, we often write R instead of R^D. Each element in R^D is called a tuple *of R* in D. The set of values occurring in a database instance D, its active domain, is denoted by adom(D).

19.2.2 Queries

Of course, the main purpose of storing data in a database is to be able to query it. As an example, someone might be interested in knowing all the operas written by Puccini. The result of this query on our example database in Figure 19.1 consists of Turandot and Tosca. More precisely, in the terminology of relational databases, it consists of the relation with the two unary tuples having Turandot and Tosca in their Piece column, respectively.

In general, a query q is just a mapping which takes a database instance D and maps it to a relation $q(D)$ of fixed arity. As we aim at evaluating queries by computers, we further require that this mapping be computable.

Some subtleties are associated with the term computable query, which we discuss next. First of all, the notion of computability is usually defined for functions mapping strings to strings. Thus, to fit this definition, we have to represent each database instance and each query result as a string. A complication arises from the fact that the tuples of a relation are unordered. —This is actually where the correspondence between tables and relations fails: when we represent a relation by a table, we have to put the rows into some order. But this order is not considered as fixed for a particular relation. Thus, all row permutations of a table represent the same relation. — Coming back to the computability issue: there are at least as many strings representing a relation as there are permutations of its tuples. However, the result of the query should not depend on the particular order chosen in the encoding of a relation. More precisely, if one encoding is a permutation of the other, then the output for the one should be a corresponding permutation of the output for the other. Similarly, if the data values of a database are changed consistently, the values in the result should change accordingly. A query fulfilling all these requirements is called generic, and by computable query we actually mean computable generic query.*

The following list of queries on our database example from Figure 19.1 will be used for illustrating the concepts introduced later.

1. List the artists performing in an opera written by Puccini.
2. List the theaters playing Puccini or Bizet.
3. Is there an artist performing in at least two theaters?
4. Is any theater showing a piece whose composer's birthday is 12/22?
5. Is any artist performing in an opera whose composer's birthday is the same as the artist's birthday?
6. List all artists who never performed in an opera written by Bizet.
7. List all artists who have performed in Bayreuth but neither at the Scala nor at the Met.
8. Is the total number of operas in the database even?

Apart from differences in the complexities of the queries, one can already observe a difference between queries with a yes/no answer, like queries (3) and (4) above, and queries that produce a set of tuples like queries (1) and (2). We refer to the former kind as Boolean queries.†

19.2.3 Query Languages

So far we have stated our example queries in natural language. This is (to date) not suitable for processing queries with a computer. Thus, there is a need for query languages that allow users to pose queries in a semantically unambiguous way. It is important to remark that one wants to avoid the use of general programming languages for querying databases for various reasons: they usually require more effort, they are error-prone, and they are not conducive to query optimization.

Ideally, a query language allows users to formulate their queries in a simple and intuitive way, without having any special proficiency in the technicalities of the database besides knowledge of the (relevant part of the) database schema. In particular, the user should not need to specify how the query is processed but only which properties the result should have. Query languages of this kind are called declarative.

---

* The first formal definition of a database query was given in [18]. A detailed treatment of the genericity issue and the question how constants can be handled is found in [3,63].

† Technically, Boolean queries can be seen as 0-ary queries, where the answer "no" corresponds to the empty result set and the answer "yes" corresponds to the result set containing the empty tuple.

The evaluation of a query is usually done in several stages:

1. A compilation transforms it into an algebra expression (see Section 19.3.5).
2. Using heuristic rules, this expression is rewritten into one that promises a more efficient evaluation.
3. From the latter expression, different query evaluation plans are constructed (e.g., taking into account different access paths for the data), and one of them is chosen based on statistical information on the actual content of the current database.
4. This evaluation plan is executed using efficient algorithms for each single operation.

Several important issues arise.

Expressive power: What can and what cannot be expressed in the query language at hand?

Complexity of evaluation: How complex is it to actually evaluate the queries expressible in the query language?

Complexity of static analysis: How difficult is it to analyze and optimize queries to ensure a good evaluation performance?

Of course, SQL is the lingua franca for relational databases. Nevertheless, in this article we will concentrate on languages (e.g., the relational algebra and the relational calculus) which are better suited for theoretical investigations of the above-mentioned questions. Actually, these languages were developed first and SQL can be conceived as a practical syntax for the relational calculus. The compilation of SQL into the algebra is based on the fundamental result that the calculus can be translated into the algebra.

19.2.4 Expressive Power

The expressive power of a query language is the set of queries it can express. This is an important measure for comparing different query languages. For instance, it can tell whether some features are redundant or not. Understanding the expressive power of a query language is a challenging task. Showing that a query is not expressible amounts to proving a lower bound, and lower bounds are often difficult to get.

Nevertheless, the close relationship of relational query languages with mathematical logic allows to apply methods from that field to gain insight in the expressive abilities and limitations of query languages. Indeed, there has been a strong mutual interaction with finite model theory [81].

19.2.5 Evaluation and Its Complexity

It is not surprising that there is a trade-off between the expressive power of a query language and the computational resources needed to evaluate a query stated in this language. This evaluation complexity can be studied from different angles, depending on the scenario at hand. We will quickly describe some of these different aspects next.

The first distinction is between Boolean (yes/no) queries and queries with a relation as output. In the latter case one might ask any of the following two questions:

a. What effort is needed to compute the full query result?
b. Given a tuple, what effort is needed to determine whether it is contained in the query result?

Many of the complexity investigations concentrate on decision problems, thus they mostly deal with Boolean queries or with question (b) above. Nevertheless, many results can be easily transferred from Boolean queries to general queries. In fact, most query languages Q have the property that an algorithm for efficiently evaluating Boolean Q-queries can be used to construct an algorithm that efficiently evaluates arbitrary (non-Boolean) Q-queries: Given a database D and a query q whose result is a relation of arity r, a naive approach is to successively consider each possible result tuple $\bar{t}$, evaluate the Boolean query *"Does $\bar{t}$ belong to $q(D)$?"*, and output $\bar{t}$ if the answer is "yes." Then, however, the delay between outputting two consecutive tuples in $q(D)$ might be rather long, as a large number of candidate tuples (and according Boolean queries) might have to be processed prior to finding the next tuple that belongs to the query result. This delay can be avoided if, prior to checking whether tuple $\bar{t} = (t_1, \ldots, t_r)$ belongs to $q(D)$, the algorithm first checks whether $q(D)$ contains any tuple whose first component is t_1—and if the answer is "no"—tuples with first component t_1 will not be further considered. This way of exploring the space of potential result tuples leads to an algorithm for computing $q(D)$ such that the delay between outputting any two consecutive tuples in $q(D)$ requires to process only a number of Boolean queries that is polynomial in the size of the database and the query q.* In this way, an efficient algorithm for evaluating Boolean Q-queries leads to an efficient algorithm for evaluating arbitrary Q-queries (provided that Q satisfies some mild closure properties ensuring, e.g., that if $q \in Q$ and t_1 is a data value, then the query "Does $q(D)$ contain a tuple whose first component is t_1?" is also expressible in Q). Bearing this in mind, Boolean queries will be in the focus of our exposition.

We refer to the algorithmic problem of evaluating a Boolean query in a query language Q by Eval(Q).

The second distinction has to do with the durability of queries and databases. Sometimes the same query is posed millions of times against an ever changing database. It may therefore be "compiled" once and forever, and it is reasonable to spend quite some effort on optimizing the query evaluation plan. In this scenario, it makes sense to consider the query as a fixed entity and to express the complexity in terms of the size of the database only. This is called the data complexity of a query. A further interest in data complexity stems from the observation that very often the database is by magnitudes larger than the query.

In other scenarios, the database never changes, but lots of different queries are posed against it. Then, one is interested in measuring the cost in terms of the size of the query, this is called query complexity. In the most general scenario of combined complexity, the database changes and many different queries are asked, and therefore the complexity is measured in the size of both, the query and the database. For most query languages, the data complexity is considerably lower than the combined complexity, whereas the query complexity usually is the same as the combined complexity. We therefore will restrict attention to data complexity and combined complexity.

We express complexities in terms of standard complexity classes like PTime, NP, ExpTime, LogSpace, and PSpace. We will also mention some parallel complexity classes like AC^0 (the class of all problems solvable by uniform constant depth, polynomial size circuits with *not*, *and*, and *or* gates of unbounded fan-in), the class TC^0 (the analog of AC^0 where also *threshold* gates are available), and LogCFL (the class of all problems that are logspace-reducible to a context-free language); for precise definitions we refer the reader to [109]. Recall that

$$AC^0 \subset TC^0 \subseteq \text{LogSpace} \subseteq \text{LogCFL} \subseteq \text{PTime} \subseteq \text{NP} \subseteq \text{PSpace} \subseteq \text{ExpTime}.$$

As mentioned before, it is usually a fair assumption that databases are big while the queries are small. Thus, algorithms which are bad in terms of the size of the query but perform well in terms of the database size are often considered as feasible. A systematic way of studying phenomena of this kind is provided by the framework of parameterized complexity. We will present some results

---

* A systematic study of the so-called polynomial (or even constant) delay algorithms has been initiated recently, see [8,9,29,37,51].

in this vein and refer to the (somewhat feasible) class FPT of fixed parameter tractable problems and the (presumably infeasible) classes W[1], W[P], and AW[*], for which the following inclusions hold: FPT $\subseteq$ W[1] $\subseteq$ W[P] and W[1] $\subseteq$ AW[*]. For the precise parametric notions we refer to the survey [52] and the book [40].

19.2.6 Static Analysis and Its Complexity

In the context of data complexity we already mentioned queries that are evaluated many times and therefore deserve to be optimized toward fast evaluation. Even queries that are evaluated once on very large data deserve optimizations. Query optimizers use cost models to decide what optimizations are worth doing. There are usually several ways of compiling a query into an evaluation plan and, even more, there are already several ways of expressing a query in the query language at hand. This often corresponds to several equivalent characterizations of the same query and sometimes induces radically different evaluation procedures after compilation. But which one should the system use? This task is attacked during query optimization.

Many optimization tasks rely on three simple questions: (1) Does query q ever produce a nonempty result? (2) Does query q_1 always produce the same result as query q_2? (3) Does query q_1 always produce a subset of the results of query q_2? We refer to the first question as the satisfiability problem (or, nonemptiness problem) to the second as the equivalence problem, and to the third as the containment problem. By Sat(Q) we denote the algorithmic problem of deciding for a given query in language Q whether there is at least one database on which the query has a nonempty result. We write Equiv(Q) and Cont(Q) for the problems of deciding whether for given queries q_1, q_2 from Q and every database D, $q_1(D) = q_2(D)$, respectively, $q_1(D) \subseteq q_2(D)$. We abbreviate the former by $q_1 \equiv q_2$ and the latter by $q_1 \subseteq q_2$.

Of course $q_1 \equiv q_2$ iff $q_1 \subseteq q_2$ and $q_2 \subseteq q_1$. Furthermore, q_1 is satisfiable iff it is not equivalent to the query that always produces the empty result set. On the other hand, the equivalence problem reduces to the containment problem and, if the query language is closed under complementation and conjunction, the containment problem reduces to the emptiness problem.

The whole area of reasoning about semantic properties of queries is called static analysis. In terms of automatic static analysis, one is interested in finding out whether static analysis for a given query language is decidable at all and, if so, what its exact complexity is.

19.2.7 Conclusion

The expressive power, the complexity of evaluation, and static analysis are correlated properties of a query language. More expressive power usually increases the complexity of query evaluation and static analysis. But even if two query languages have the same expressive power, they may vastly differ in terms of the complexity of static analysis and query evaluation. The reason for this is that even if every query of one language can be translated into an equivalent query of the other language, the translation may turn a short query into a huge one. Thus, apart from expressive power and complexity, the succinctness of queries is also a natural measure for comparing query languages. Although well-investigated for languages used in specification and automated verification, a systematic study of the succinctness of database query languages has started only recently (see, e.g., [33,57]). Somewhat related is Papadimitriou's work on the complexity of knowledge representation, where he compares the succinctness of various representation formalisms (see e.g., [45,87]).

There is a trade-off between the expressive power of a query language and the complexity of query evaluation. As an example, if the data complexity of a query language is in PTime then this query language cannot express any NP-complete property unless PTime $=$ NP. Succinctness considerations can help to investigate the relationship between different languages with respect to combined complexity.

In the end, designers usually try to get the best expressive power with the least complexity for the application area at hand.

19.3 The Simplest Language: Conjunctive Queries

We start with a simple query language, the conjunctive queries, whose expressive power is subsumed by most of the query languages we will consider and which is already able to express many common "every day queries." In particular, it corresponds to the very basic features of SQL.

After introducing the rule-based conjunctive queries, we study their expressive power and complexity. We then turn to a restriction, the *acyclic* conjunctive queries, for which query evaluation and static analysis have considerably lower complexity. Finally, we present a different mechanism, the SPJR algebra, for defining conjunctive queries.

19.3.1 Conjunctive Queries: Definition

We recall query (1) from Section 19.2.2, asking for all artists performing in an opera written by Puccini. In SQL this can be simply expressed by

> SELECT Artist
> FROM CAST, OPERAS
> WHERE CAST.Piece = OPERAS.Piece AND Composer = Puccini

This query can be expressed more concisely in the following rule-based way:

$$\text{PARTISTS}(x) \; : - \; \text{CAST}(z, y, x), \; \text{OPERAS}(\text{Puccini}, y). \tag{19.1}$$

This expression can be interpreted as a "tuple producing facility" in the following way: for each assignment of values to the variables x, y, and z, for which (z, y, x) is a tuple in the CAST relation and (Puccini, y) is a tuple in the OPERAS relation, (x) is a tuple in the result relation PARTISTS.

As an example, the assignment $x \mapsto$ Pavarotti, $y \mapsto$ Turandot, $z \mapsto$ Scala fulfills all requirements and produces the tuple (Pavarotti).

In general, a conjunctive query consists of a single *rule* of the form

$$Q(\bar{x}) : - R_1(\bar{t}_1), \ldots, R_\ell(\bar{t}_\ell),$$

where Q is the name of the relation that is defined by the query. Its arity is the arity of the tuple $\bar{x}$ (the PARTISTS query, for example, has arity one). The atom to the left of the symbol $: -$ is called the head of the query, whereas the expression to the right of the symbol $: -$ is called the body of the query. The tuples $\bar{x}$ and $\bar{t}_1, \ldots, \bar{t}_\ell$ consist of variables and/or constants. The atoms $R_i(\bar{t}_i)$ are such that R_i is the name of a relation occurring in the database schema and whose arity coincides with the length of the tuple $\bar{t}_i$. The $\bar{t}_i$ are not necessarily disjoint and altogether have to contain all variables of $\bar{x}$. The above query's result $Q(D)$ over a database D is obtained as follows: for each possible assignment α of values to the variables present in $\bar{x}, \bar{t}_1, \ldots, \bar{t}_\ell$ such that, for each $i \in \{1, \ldots, \ell\}$, the resulting tuple $\alpha(\bar{t}_i)$ belongs to the database relation R_i^D, the tuple $\alpha(\bar{x})$ is in $Q(D)$.

The set of all conjunctive queries is denoted by CQ. The reader should note that the name "conjunctive queries" is really justified: a variable assignment has to fulfill a conjunction of conditions to produce an output tuple.

19.3.2 Limitations of the Expressive Power of CQ

Given two databases D_1 and D_2 over the same schema, we write $D_1 \subseteq D_2$ if $R^{D_1} \subseteq R^{D_2}$, for each relation name R. A query q is said to be monotone if $D_1 \subseteq D_2$ implies $q(D_1) \subseteq q(D_2)$. For example, query (1) is monotone while query (6), asking for artists who never performed in an opera written by Bizet, is not: by adding the tuple (Scala, Carmen, Pavarotti) to the CAST relation, Pavarotti would no longer be in the result of that query.

It is not hard to obtain the following (see, e.g., [3] for a proof):

THEOREM 19.1 *CQ can define only monotone queries.*

As a consequence, query (6) is not expressible in CQ. Even though monotonicity is a useful tool for testing nonexpressiveness in CQ, it does not yield a complete characterization of CQ. In fact, there are monotone queries that cannot be expressed in CQ. An example is query (2) asking about theaters playing Puccini or Bizet.

19.3.3 Complexity of CQ

Next we discuss the complexity of evaluation and static analysis for conjunctive queries.

Evaluation. The naive way of evaluating a conjunctive query is by trying out each possible variable assignment, resulting in about $|\text{adom}(D)|^k$ steps if k is the overall number of variables in the query. The complexity of this method is exponential in the number of variables and thus in the size of the query. If the query is considered fixed, however, it is polynomial in the size of the database. Using the terminology introduced in Section 19.2.5, the following complexity results hold.

THEOREM 19.2

a. *The data complexity of* Eval(CQ) *is in* AC^0 *[65] and thus, in particular, in* LogSpace.
b. *The combined complexity of* Eval(CQ) *is NP-complete [20].*
c. *The parameterized complexity of* Eval(CQ) *(with the size of the query as parameter) is* W[1]*-complete [88].*

Statement (c) basically says that, as in the above naive algorithm, an exponent that grows with increasing k can never be avoided. Specifically, it says that if the (widely believed) complexity theoretic conjecture "W[1] $\neq$ FPT" is true, then there is no algorithm that solves Eval(CQ) in time $f(k) \cdot |\text{adom}(D)|^c$, where f is an arbitrary computable function, k is the size of the input query, D is the input database, and c is an arbitrary integer. It is not difficult to see that statement (c) is equivalent to the following statement: if W[1] $\neq$ FPT, then there does not exist a pair $(A_{\text{opt}}, A_{\text{eval}})$ of algorithms such that A_{opt} is an algorithm of arbitrary complexity that optimizes an input conjunctive query q, and A_{eval} is an algorithm that takes as input the optimized query and a database D and computes the query's result $q(D)$ in time polynomial in the size of D (see [52] for details).

Statement (b) seems to indicate that the evaluation of conjunctive queries against a relational database is intractable, clearly contradicting everyday experience. Some explanations for this counter intuitive phenomenon will be given in Section 19.4.3. The proof of statement (b) is by a straightforward reduction of the NP-complete clique problem for undirected graphs to the query evaluation problem for conjunctive queries: an input instance (G, k) for the clique problem is simply mapped to a database representing the graph G and a conjunctive query asking whether G contains a clique of size k.

Static Analysis. It is not difficult to see that every conjunctive query $q := Q(\bar{x}) : -R_1(\bar{t}_1), \cdots , R_\ell(\bar{t}_\ell)$ is satisfiable. In fact, there is a canonical database D_q witnessing this: to construct D_q just put, for every i, the tuple $\bar{t}_i$ into relation R_i. Then, $Q(D_q)$ at least contains the tuple $\bar{x}$. For example, the canonical database for the example query (19.1) has the tuple (z, y, x) in CAST and (Puccini, x) in

OPERAS, hereby viewing x, y, and z as ordinary data values. Thus, the problem Sat(CQ) is trivially solvable, since every conjunctive query is satisfiable.

As a matter of fact, a very similar approach also works for Cont(CQ).

THEOREM 19.3   *[20] Let $q_1(\bar{x})$ and $q_2(\bar{x})$ be two conjunctive queries with the same free variables $\bar{x}$. Then $q_1 \subseteq q_2$ if and only if $\bar{x} \in q_2(D_{q_1})$.*

This theorem is very often stated in terms of homomorphisms: Given two databases D_1 and D_2 we say that $h : \mathrm{adom}(D_1) \rightarrow \mathrm{adom}(D_2)$ is a homomorphism from D_1 to D_2 if, for each relation R, $\bar{a} \in R^{D_1}$ implies $h(\bar{a}) \in R^{D_2}$. The homomorphism theorem [20] then states that, with the notation of Theorem 19.3, $q_1 \subseteq q_2$ if and only if there is a homomorphism from D_{q_2} to D_{q_1} fixing $\bar{x}$. The following is a corollary of Theorem 19.3 and Theorem 19.2b.

THEOREM 19.4   *[20] The containment problem* Cont(CQ) *and the equivalence problem* Equiv(CQ) *are NP-complete.*

Furthermore, conjunctive queries can be minimized in the following sense: There is an algorithm which takes as input a conjunctive query q and outputs an equivalent conjunctive query q' such that the number of atoms in the body of q' is as small as possible (cf., e.g., the textbooks [3,102] for details).

19.3.4   Acyclic Conjunctive Queries

We have seen that the combined complexity and the parameterized complexity of evaluating conjunctive queries are NP-complete and W[1]-complete, respectively, and thus the worst-case complexity can, to the best of our knowledge, be expected to be exponential in the size of the query. Nevertheless, for practical purposes and if the query structure is "simple," there are smarter evaluation algorithms than the naive "test all possible variable assignments" approach mentioned above.

To illustrate this, let us consider the query (4) asking whether any theatre plays a piece whose composer's birthday is 12/22. We can express this by the Boolean conjunctive query

$$\text{BANSWER}() \;:-\; \text{CAST}(z, y, x),\; \text{OPERAS}(x', y),\; \text{PEOPLE}(x', \text{Composer}, 12/22).$$

One possible evaluation plan for this query is as follows. First, combine CAST and OPERAS to obtain an intermediate relation R in the following way: combine each tuple in CAST with all tuples in OPERAS that have the same entry in the Piece column (due to the two occurrences of y). For our example database, the resulting relation R is shown in Figure 19.2. Afterward, let S be the relation (also shown in Figure 19.2) that consists of all tuples $\bar{t}$ from R for which there is a tuple in PEOPLE with entries Composer and 12/22 in the Type and Birthday column and whose entry in the Artist column coincides with $\bar{t}$'s entry in the Composer column. Finally, the answer returned by the BANSWER query is "yes" if and only if the relation S is nonempty.

The evaluation order of this strategy can be depicted as a tree as shown in Figure 19.3a. Note that in this tree, the relation atoms of the query occur at the leaves, and the inner nodes correspond to intermediate results. From the root, the final result can be obtained by dropping some (maybe all or none) of the columns.

Even though it does not hurt in this small example, it could be annoying that the arity of the intermediate relation R is larger than the arities of the input relations. Furthermore, R is basically the cartesian product of two relations, joined in their Piece column. For a larger query this might result in a wide table which on a "real life" database could grow very large.

R	Theatre	Piece	Artist	Composer
	Scala	Turandot	Pavarotti	Puccini
	Bayreuth	Tristan	Nilsson	Wagner
	Covent Garden	Peter Grimes	Vickers	Britten
	Met	Tosca	Callas	Puccini
	Scala	Turandot	Raisa	Puccini

S	Theatre	Piece	Artist	Composer
	Scala	Turandot	Pavarotti	Puccini
	Met	Tosca	Callas	Puccini
	Scala	Turandot	Raisa	Puccini

FIGURE 19.2 Intermediate results for the BANSWER query.

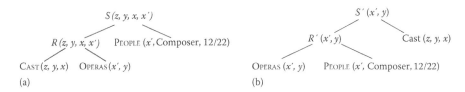

FIGURE 19.3 Two different evaluation trees for the PSoprano query. (a) An evaluation tree. (b) A more efficient evaluation tree.

Nevertheless, a closer look at the query shows that these problems can be avoided when using a different evaluation plan, depicted in Figure 19.3b. Even though at first sight it looks similar to the original plan, it has an important extra property: all tuples that enter the intermediate relation R' have to be tuples from relation OPERAS, and the same holds true for relation S'. In particular, the intermediate relations do not grow in arity, and their content is never obtained by an expensive product operation but rather by a kind of filter operation selecting tuples from one relation, controlled by the other relation. One could say that each intermediate relation is guarded by one of the input relations. Evaluation trees with this property are sometimes called join trees, and queries for which such a join tree exists are called acyclic. We denote the class of acyclic conjunctive queries by ACQ. For example, query (4) is acyclic whereas query (5) is not.

If a join tree is given for a Boolean query q, the query can be evaluated by processing the join tree bottom-up. It is easy to see that each intermediate step can be performed in polynomial time. As it is possible to test in polynomial time (even in LogSpace [47,91]) whether a join tree exists, and to actually compute one if this is the case, it follows that Eval(ACQ) is in PTime [111].

Note that the intuition given above is accurate only for Boolean queries. But analogous notions of join trees and acyclic queries also exist for non-Boolean queries (see [3] for details) that can be evaluated efficiently by processing a join tree in a bottom-up phase followed by a top-down phase and, possibly, another bottom-up phase (see [39,111] for an explanation of the bottom-up and top-down phases). The class of acyclic queries has been characterized in various ways (cf. [3]); the term "acyclic" is due to a characterization referring to acyclic hypergraphs [12].

A precise complexity analysis yields the following result, showing that query evaluation and static analysis of acyclic queries not only belong to PTime but can even be efficiently parallelized.

THEOREM 19.5 *[47]*

a. *The combined complexity of* Eval(ACQ) *is LogCFL-complete.*
b. *The containment problem* Cont(ACQ) *is LogCFL-complete.*

In this sense acyclic queries behave nicely, and it is natural to wonder whether there are further classes of well-behaved queries. Indeed, there are several extensions and variations of the notion of acyclicity.

One line of variations is based on the treewidth of query graphs (see [39,47] for the basic concept and references to the literature). For a conjunctive query q, the graph of q is the graph whose vertices are the variables of q and which has an (undirected) edge between two vertices whenever the corresponding variables occur in the same atom in the body of q. Given a class C of graphs, let CQ[C] be the set of conjunctive queries q such that the graph of q is in C. For instance, if A is the set of acyclic graphs, then CQ[A] is a subset of ACQ. In order to generalize this we let, for each number k, CQ$_k$ be the class of conjunctive queries whose graph has treewidth at most k. It is known that, for any number k, an analogue of Theorem 19.5 holds (a PTime upper bound was obtained already in [23]), and static analysis is also tractable for conjunctive queries of bounded treewidth.

THEOREM 19.6   *[47]*

    a. *For any number $k \geq 1$, the combined complexity of* Eval(CQ$_k$) *is* LogCFL-*complete.*
    b. *For any number $k \geq 1$,* Cont(CQ$_k$) *is* LogCFL-*complete.*

When the schema is fixed, relative to a plausible complexity theoretic assumption, even a precise characterization of the tractable graph-based classes of conjunctive queries is known.

THEOREM 19.7   *[58] Assume that* W[1] $\neq$ FPT *and let C be a recursively enumerable class of graphs. Then the combined complexity of* Eval(CQ[C]) *is in* PTime *if and only if C has bounded treewidth.*

Even more, an extension of the above result providing a complete characterization of all tractable classes of conjunctive queries over a fixed schema has been obtained in [53].

When (the arity of) the schema is not fixed, a larger class of tractable queries can be found by considering the hypergraph of a query instead of the graph. The hypergraph of query q has one hyperedge per query atom A (in the body of q) which contains all nodes corresponding to variables in A. In [48], the concepts of hypertree decompositions and hypertreewidth were introduced. It was shown that the acyclic conjunctive queries are precisely the queries whose hypergraph has a hypertree decomposition of hypertree-width 1, and that for each number $k \geq 1$, the statement of Theorem 19.5a can be generalized from ACQ to the class of all conjunctive queries that have a hypertree decomposition of hypertreewidth at most k. Various other classes of tractable queries based on hypergraph decompositions have been proposed in the literature; we refer to [94] for a survey.

19.3.5 Relational View

So far we considered only SQL and rule-based conjunctive queries as query languages. They are both declarative in the sense that they specify answers by their properties rather than by operations used to construct them. We already talked about evaluation plans in Section 19.3.4. A concise way to express evaluation plans is offered by the relational algebra. It consists of a few simple operators for manipulating relations. For the following it is useful to think of a relation as a table where each row (column) of the table corresponds to a tuple (attribute) of the relation.

The operators of the relational algebra that are needed for expressing conjunctive queries are (1) extraction of rows of a table (selection), (2) extraction of columns of a table (projection), (3) gluing together two tables along some columns (join), and (4) renaming of columns. For example, $\sigma_{Composer="Puccini"}(OPERAS)$ is a selection which extracts all rows where the composer (i.e., the first attribute) is "Puccini". The expression $\pi_{Theatre,Piece}(CAST)$ extracts the first two columns of the CAST relation. Note that the resulting table only has four rows. The expression $CAST \bowtie OPERAS$ is a join which combines each row of the CAST table with each row of the OPERAS table, provided they have the same value, in each column with the same name, i.e., provided they match on the Piece attribute. Thus, the resulting table is the relation R depicted in Figure 19.2; it has four columns and five rows. Query (1) can thus be expressed by $\pi_{Artist}(\sigma_{Composer='Puccini'}(CAST \bowtie OPERAS))$, whereas the query asking for all sopranos who perform in an opera written by Puccini is expressed by

$$\pi_{Artist}\left(\sigma_{Composer="Puccini"}(CAST \bowtie OPERAS) \bowtie \sigma_{Type="Soprano"}(PEOPLE)\right).$$

The algebra consisting of the four operators of selection, projection, join, and renaming is called the SPJR algebra. Note that the SPJR algebra can express unsatisfiable queries, e.g., $\sigma_{Composer="Puccini"}$ $(\sigma_{Composer="Verdi"}(OPERAS))$ will never return any opera. There is a straightforward polynomial time algorithm to test whether an SPJR query is satisfiable.

However, the ability to express unsatisfiable queries is the only difference between SPJR and CQ with respect to expressive power. It is not difficult to see the following (cf. [3] for details).

THEOREM 19.8 *CQ and satisfiable SPJR have the same expressive power. Moreover, queries can be translated from either language to the other in polynomial time.*

Recall that basic SQL queries are formed using the following syntax:

```
select attributes
from relations
where test.
```

Conjunctive queries are essentially those which can be expressed by SQL queries where the *test* part contains only conjunctions of equality tests.

19.4 The Gold Standard: Codd-Complete Languages

We recall from Section 19.3.2 that conjunctive queries can only express monotone queries, i.e., queries which never produce smaller result sets if something is added to the database. Of course, there are interesting nonmonotone queries, e.g., query (6) from Section 19.2.2. Furthermore, simple disjunctions like query (2) are not expressible in CQ either. In this section we consider extensions of CQ that are capable of expressing such queries: Codd-equivalent languages such as the relational algebra, the relational calculus, and the nonrecursive rule-based queries. After introducing these query languages, we study their expressive power and their complexity.

19.4.1 Codd-Equivalent Query Languages: Definition

The rule-based queries of Section 19.3.1 can be extended by (1) allowing queries to consist of more than one rule, (2) allowing relations defined by one or more rules to be used in the body of other rules, and (3) allowing negated atoms in the body of rules. A query is thus a finite set of *rules* of the form

$$Q(\bar{x}) :- A_1(\bar{t}_1), \dots, A_\ell(\bar{t}_\ell)$$

where the atoms $A_i(\bar{t}_i)$ are either of the form $S(\bar{t}_i)$ or of the form $\neg S(\bar{t}_i)$, where S is either a relation name of the database schema or the name of a relation symbol used in the head of one of the rules of the query.*

Nevertheless, recursion is not allowed. More formally, the following directed graph is not allowed to contain a directed cycle: the graph's nodes are the relation names, and there is an edge from R_1 to R_2 if R_1 appears in the body of a rule having R_2 in its head.

As an example, the following rules

$$
\begin{aligned}
\text{STARS}(x) \quad &:- \quad \text{CAST}(\text{Scala}, y, x) \\
\text{STARS}(x) \quad &:- \quad \text{CAST}(\text{Met}, y, x) \\
\text{RESULT}(x) \quad &:- \quad \text{CAST}(\text{Bayreuth}, y, x), \ \neg\text{STARS}(x)
\end{aligned}
$$

describe query (7) from Section 19.2.2, selecting all artists who have performed in Bayreuth but neither at the Scala nor at the Met. We call such queries nonrecursive rule-based queries.

The same expressive power can be obtained in two other ways:

1. By adding the relational union and difference operators to the SPJR algebra one gets the full relational algebra.

2. The relational calculus consists of the logical formulas of the predicate calculus (i.e., first-order logic FO) which use the relations of the database schema plus equality as relation symbols, and do not use any function symbols. We will sometimes briefly write FO to denote the relational calculus.

Query (7) from Section 19.2.2, for example, can be expressed by the relational algebra expression

$$
\pi_{\text{Artist}}(\sigma_{\text{Theatre}=\text{"Bayreuth"}}(\text{CAST})) - \pi_{\text{Artist}}(\sigma_{\text{Theatre}=\text{"Met"}}(\text{CAST}) \cup \sigma_{\text{Theatre}=\text{"Scala"}}(\text{CAST})))
$$

and by the relational calculus formula $\varphi_{\text{RESULT}}(x) :=$

$$
\left(\exists y \ \text{CAST}(\text{Bayreuth}, y, x) \ \wedge \ \forall y \ \neg\big(\text{CAST}(\text{Scala}, y, x) \vee \text{CAST}(\text{Met}, y, x)\big)\right)
$$

When fixing the precise semantics of the relational calculus, some care needs to be taken to decide over which domain variables should range. One possible solution is to let the quantifiers range only over elements in the active domain of the underlying database; another way is to let them range over the entire domain of potential data values but to restrict the syntax in a "safe" way to avoid infinite query results and to ensure efficient evaluation. A similar problem occurs in the context of negations in nonrecursive rule-based queries. Again, to avoid infinite query results, one can either restrict attention to the active domain of the underlying database or impose the syntactic restriction that in every rule each variable has to occur in at least one positive atom of the rule's body. It turns out that both variants have the same expressive power. A discussion of these issues can be found in [3]. In the rest of this chapter we will assume that all quantifiers range over the active domain.

The widely used query language SQL combines and extends features of both, relational calculus and algebra.

Codd's following theorem summarizes one of the most fundamental results in database theory.

THEOREM 19.9 *[26] Relational calculus, relational algebra, and nonrecursive rule-based queries have the same expressive power.*

Furthermore, translations are effective, and the translation from relational calculus to relational algebra and vice versa can be done in polynomial time. In particular, it is always possible to compile

---

* Recall from Section 19.3.1 that the *body* of a rule consists of the atoms to the right of the symbol : − and the head of a rule is the atom to the left of the symbol : −.

a query expressed in the relational calculus into an expression of the relational algebra. The latter can then be evaluated efficiently.

Query languages with (at least) the expressive power of the relational algebra or the relational calculus are called Codd-complete.* Languages which have exactly the same expressive power as the relational algebra are called Codd-equivalent.

It should be noted that conjunctive queries (cf. Section 19.3) correspond exactly to formulas of the relational calculus that use only conjunction and existential quantification, i.e., to formulas of the form $\exists \bar{x} \left(R_1(\bar{x}_1) \wedge \cdots \wedge R_\ell(\bar{x}_\ell) \right)$.

19.4.2 Limitations of the Expressive Power

Even though Codd-equivalent query languages like the relational algebra can express many everyday queries against a relational database, they still cannot express "everything." Theorem 19.9 is a key to understanding the precise expressive power of such languages: it relates them to the relational calculus, therefore allowing logic-based methods to prove that certain queries cannot be expressed.

In a nutshell, Codd-equivalent languages cannot count and cannot express recursion. For example, they can neither ask "Which artist performed in more operas than any other artist?" nor "Is the total number of operas in the database even?" Furthermore, in a database that consists of a parent–child relation, they cannot ask whether A is an ancestor of B.

All these statements can be directly proved by Ehrenfeucht–Fraïssé games, as is explained, e.g., in [81]. Very often, however, the impossibility to express a certain query can also be concluded in a simpler way, either by using 0-1 laws or by using locality. We present both notions next.

0-1 laws. Let q be a Boolean query and let σ be its schema, i.e., the set of relation names it mentions. For the moment, we consider only databases with schema σ whose active domain is an initial segment $\{1, \ldots, n\}$ of the natural numbers. We are interested in the ratio of databases of size n on which q yields the answer "yes" compared to all databases of size n, when n approaches infinity. More precisely, we denote by $\mu_n(q)$ the number of database instances over σ with active domain $\{1, \ldots, n\}$ on which q evaluates to "yes," divided by the number of all databases with schema σ and active domain $\{1, \ldots, n\}$. A Boolean query q is almost surely true (respectively, almost surely false) if the limit $\mu(q) := \lim_{n \to \infty} \mu_n(q)$ exists and is 1 (respectively, 0).

For instance, it is not hard to see that for query (8) from Section 19.2.2, i.e., for the query EVENOPERAS asking whether the number of operas in the database is even, $\mu(\text{EVENOPERAS}) = 1/2$.

A query language $\mathcal{Q}$ is said to have the 0-1 law if every Boolean query of $\mathcal{Q}$ that does not mention any constants from the domain of potential data values is almost surely true or almost surely false.

THEOREM 19.10 *[38,44] Codd-equivalent query languages have the 0-1 law.*

A simple consequence of this is that EVENOPERAS cannot be expressed by Codd-equivalent query languages. Also, many other counting queries q either have no limit $\mu(q)$ or a limit different from 0 and 1 and thus are not expressible by Codd-equivalent languages. Note, however, that counting "up to a constant threshold" is possible with Codd-complete languages; for example query (3), asking whether there is an artist starring in at least two theaters, can be expressed in the relational calculus via $\exists x \exists y_1 \exists z_1 \exists y_2 \exists z_2 \left(\text{CAST}(z_1, y_1, x) \wedge \text{CAST}(z_2, y_2, x) \wedge \neg z_1 = z_2 \right)$.

Locality. Unfortunately, the 0-1 law is not a very natural tool in the presence of non-Boolean queries. However, locality arguments can often be used to extend inexpressibility results to non-Boolean queries. In a nutshell, a query language is called local if it cannot express queries that depend on

---

* Codd himself called these query languages relationally complete [26].

an unbounded number of tuples "connecting" one data item with another one (for example, the query ANCESTOR(x, y) which asks whether person x is an ancestor of person y in a parent–child database).

To be more precise, the Gaifman-graph G_D of a database D is the undirected graph whose vertices are the elements of the active domain of the database, and there is an edge between two vertices whenever they appear in a tuple of a relation of D. The distance between two elements of D is their distance in G_D. For any number k, the k-neighborhood $N_k^D(\bar{a})$ of a tuple $\bar{a}$ of elements is the sub-database induced by all elements of D that have distance at most k from some element of $\bar{a}$. For example, the k-neighborhood of a person in the parent-child database consists of all persons to which he or she is related by at most k parent–child tuples and all tuples of the database containing only such persons.

A query q is called Gaifman-local if there exists a number k such that for any database D and any tuples $\bar{a}$ and $\bar{b}$ of D (of the right arity for q), if whenever $N_k^D(\bar{a})$ is isomorphic to $N_k^D(\bar{b})$ then $\bar{a} \in q(D)$ iff $\bar{b} \in q(D)$.

Obviously, there is no such k for the ANCESTOR query, i.e., the ANCESTOR query is *not* Gaifman-local. The following theorem therefore tells us that Codd-equivalent languages cannot express this query.

THEOREM 19.11   *[41,62] Codd-equivalent languages can only express Gaifman-local queries.*

Theorem 19.11 can be considered a formalization of the intuitive statement that Codd-equivalent languages lack recursion.

There are other notions of locality like Hanf-locality and bounded number of degrees property that hold for Codd-equivalent languages and some of their extensions. In particular, Hanf-locality can be used to prove that certain Boolean queries respecting the 0-1 law (e.g., connectivity of a graph) cannot be expressed by Codd-equivalent languages. We refer the interested reader to [80,81] and the references therein.

19.4.3   Complexity

Evaluation. Each operator of the relational algebra can be evaluated in a straightforward way. For example, the naive processing of a join operation $R_1 \bowtie R_2$ roughly requires $|R_1| \cdot |R_2|$ steps since each tuple in R_1 is combined with all the tuples in R_2. In general, the evaluation of a query which involves intermediate results of arity k can be evaluated on a database D in time $O(|\mathrm{adom}(D)|^k)$. Likewise, formulas of the relational calculus can be evaluated by essentially turning each quantifier into a FOR-loop, resulting in a similar complexity.

Parts (b) and (c) of the following theorem show that, in the worst case, this upper bound cannot be significantly improved unless some widely believed complexity theoretic assumptions fail. Recall that we denote the relational calculus by FO because it is based on first-order logic formulas. Although formulated for FO, the following theorem also holds for the relational algebra.

THEOREM 19.12

  a. *The data complexity of* Eval(FO) *is in* AC^0 *[65] and thus, in particular, in* LogSpace.
  b. *The combined complexity of* Eval(FO) *is* PSpace-complete *[99,106].*
  c. *The parameterized complexity of* Eval(FO) *(with the size of the query as parameter) is* AW[*]-complete *[36].*

The results above show that evaluating Codd-equivalent queries in a scenario where the query is not fixed is rather difficult in general. Just as in the case of Theorem 19.2 for conjunctive queries, this seems to contradict the empirical experience that SQL queries can usually be evaluated reasonably fast. The explanation for this has several facets.

First of all, Theorems 19.2 and 19.12 talk about *worst cases*. The queries constructed in the proofs are of a very complicated structure that usually does not occur in practice. This observation is the starting point for many investigations on how the structure of queries influences the evaluation complexity. We will come back to this issue in Section 19.5.

A second aspect is that in practice, queries are rather small whereas databases are large. Thus, data complexity seems to be a better measure than combined complexity—and Theorem 19.12a tells us that the data complexity is very low.

Furthermore, the structure of the database can have an impact on the complexity, and usually databases are not "arbitrarily complicated." In fact, the parameterized complexity of Eval(FO) gets feasible when attention is restricted to certain classes of database instances. For example, in [32] it is shown that over classes of databases that locally exclude a minor (see [32] for the definition), the parameterized complexity of Eval(FO) is fixed parameter tractable, i.e., belongs to the complexity class FPT. This result subsumes most of the previously known fixed parameter tractability results for query evaluation; see [54] for a survey. Nevertheless, restricting the structure of databases does not help to improve the evaluation complexity in the setting of Theorem 19.12b since the combined complexity of Eval(FO) is PSpace-complete already on databases with only two data items.

Finally, the massive amounts of data handled by database systems usually reside in external memory. When processing such data, the input/output communication between fast internal memory and slower external memory is a major performance bottleneck: during the time required for a single random access to external memory, a huge number of computation steps could be performed on data present in internal memory. Indeed, modern database technology uses clever heuristics to minimize the costs caused by accesses to external memory (cf., e.g., [90,108]). Classical complexity classes such as PTime and NP, however, measure complexity only by counting the total number of computation steps, but do not take into account the existence of different storage media. In recent years, machine models that distinguish between external memory and internal memory have also been proposed and studied (for an overview, see the surveys [96,108] and the references therein).

Static analysis. In general, static analysis for Codd-complete languages is impossible.

THEOREM 19.13   *[101] The satisfiability problem* Sat(FO) *is undecidable.*

As a consequence, one immediately obtains that also the equivalence problem Equiv(FO) and the containment problem Cont(FO) are undecidable. The next section presents a couple of restrictions of FO for which static analysis is decidable.

19.5   Toward Lower Complexity: Restricted Query Languages

In this section we revisit the idea that queries in practice, even if they go beyond conjunctive queries, are not arbitrarily complicated. We consider restrictions of the relational calculus FO (and other Codd-equivalent languages) which are not Codd-complete, but for which static analysis is decidable and the combined complexity of query evaluation is considerably lower than that of full FO.

First, we concentrate on simple extensions of conjunctive queries by adding either union or inequalities. Afterward, we consider a restriction of the relational calculus in which the number of variables is bounded. Finally, we consider a variant of the relational algebra in which the use of joins is restricted. We will see that in the latter case the resulting query language corresponds to a variant of the relational calculus in which the use of quantifiers is restricted, and also to a variant of the nonrecursive rule-based queries in which single rules are based on acyclic conjunctive queries.

19.5.1 Extensions of Conjunctive Queries

There are several simple ways to extend the rule-based approach of CQ in order to gain more expressive power. For instance one could allow other kinds of atoms in the body of a rule, typically atoms of the form $x \neq y$. One could also consider defining a query using several rules instead of just a single rule. We denote by CQ($\neq$) the extension of CQ allowing inequality atoms in the body of a rule. The class UCQ of unions of conjunctive queries is the extension of CQ allowing a finite number of rules in the definition of a query. We use UCQ($\neq$) to denote the combination of the two extensions. For instance, query (3) of Section 19.2.2 can be expressed in CQ($\neq$), and query (2) can be expressed in UCQ, while none of them is expressible in CQ.

It is easy to see that UCQ, CQ($\neq$), and UCQ($\neq$) have exactly the same data complexity and combined complexity as CQ. The satisfiability problem Sat(CQ($\neq$)) (and therefore also Sat(UCQ($\neq$))) can be decided in polynomial time. The containment problem for the three languages is still decidable, but slightly more difficult than that of CQ. In the theorem below, Π_2^p refers to the second level of the polynomial time hierarchy (recall that (NP $\cup$ co-NP) $\subseteq \Pi_2^p \subseteq$ PSpace).

THEOREM 19.14

  a. *If $\mathcal{Q}$ is one of the query languages CQ($\neq$), UCQ, UCQ($\neq$), then*

  i. *The data complexity of Eval($\mathcal{Q}$) is in AC^0 and thus, in particular, in LogSpace.*

  ii. *The combined complexity of Eval($\mathcal{Q}$) is NP-complete.*

  b. *The containment problem for CQ($\neq$), UCQ($\neq$) is Π_2^p-complete [93,105]. The containment problem for UCQ is NP-complete.*

For more information on extensions of conjunctive queries see [73,93,105] and the references therein.

19.5.2 Bounding the Number of Variables

A natural way of restricting the relational calculus is to bound the number of variables used in formulas. For each number $k \geq 1$ let FO^k be the restriction of FO to formulas that use at most k variables. Note that variables may be re-quantified inside a formula.

It turns out that bounding the number of variables to k improves the evaluation complexity for every k but enables static analysis only for $k = 2$.

THEOREM 19.15

  a. *For each $k \geq 2$, the combined complexity of Eval(FO^k) is PTime-complete [107].*
  b. *The satisfiability problem Sat(FO^2) is decidable [84].*
     As a consequence, Equiv(FO^2) and Cont(FO^2) are also decidable.
  c. *For each $k \geq 3$, the satisfiability problem Sat(FO^k) is undecidable* [14,59,68].*
     As a consequence, also Equiv(FO^k) and Cont(FO^k) are undecidable.

---

* Undecidability of (not necessarily finite) satisfiability of FO^3 follows from [68]. That finite satisfiability is undecidable as well follows from [59], see [14] for a discussion.

In the same way as full FO, the k-variable fragment FO^k also has a rule-based counterpart, the so-called NRSD-programs of strict treewidth at most $(k-1)$. This is a restriction of the nonrecursive rule-based queries where the query graph of every single rule has a strict tree decomposition of width at most $(k-1)$; see [39] for details.

A survey on FO^k and related finite variable logics can be found in [49].

19.5.3 The Semijoin Algebra

When looking at the relational algebra one notices that, in terms of complexity of query evaluation, the most troublesome operation is join. The arity of $R_1 \bowtie R_2$ is usually larger than that of R_1 and R_2, and the size of $R_1 \bowtie R_2$ can be as large as the product of $|R_1|$ and $|R_2|$. An interesting restriction of the join operator $\bowtie$ is the semijoin operator $\ltimes$. Given two relations R and S, $R \ltimes S$ consists of all tuples of R that can be joined with some tuple in S in the sense of the operator $\bowtie$. In particular, $R \ltimes S$ has the same attribute names as R, and the result of $R \ltimes S$ always is a subset of R.

The semijoin algebra SA (cf. [76,77]) is the variant of relational algebra where the join operator $\bowtie$ is replaced by the semijoin operator $\ltimes$. Strictly speaking, the semijoin algebra is defined in a slightly different framework where attributes do not have names and are addressed via column numbers instead. The reader might think of the semijoin algebra as being equipped with operators for selection, projection, renaming, union, difference, semijoin, and an additional operator with which columns of a relation can be duplicated (see [78] for details).

It is not difficult to see that for every database D and every semijoin algebra query q, the result $q(D)$ consists of so-called stored tuples, i.e., tuples that are obtained from tuples in D by projecting to and, possibly, duplicating some attributes. In particular, for each fixed semijoin algebra query q, the output size of q is at most linear in its input size, i.e., the number of tuples in the query result $q(D)$ is at most linear in the number n of tuples in the input database D. Of course, this also holds for all subexpressions of q. A remarkable result from [77] shows that the reverse is also true: Any relational algebra query all of whose subexpressions compute relations of size $O(n)$ is in fact expressible in the semijoin algebra. Furthermore, every query not expressible in the semijoin algebra has a subexpression that may produce results of size $\Omega(n^2)$.

The complexity of query evaluation of the semijoin algebra is much lower than that of the relational algebra, and static analysis is decidable.

THEOREM 19.16 *[76]*

a. Eval(SA) *can be solved in time* $O(k \cdot n)$, *where k denotes the size of the query and n denotes the size of the input database.*
   In particular, the combined complexity of Eval(SA) *is in* PTime.

b. *The satisfiability problem* Sat(SA) *is* ExpTime-*complete.*
   As a consequence, Equiv(SA) *and* Cont(SA) *are also* ExpTime-*complete.*

The semijoin algebra cannot express all conjunctive queries, but at least it can express all acyclic conjunctive queries whose result is a set of stored tuples. Moreover, it can also express many natural nonconjunctive queries, e.g., query (7): the relational algebra formulation of this query given in Section 19.4.1 in fact belongs to the semijoin algebra.

In the same way as the relational algebra, the semijoin algebra also has a logical and a rule-based counterpart.

The logical counterpart is the guarded fragment of FO, denoted by GF (see [6]). GF is a fragment of FO where first-order quantifications have to be guarded by atomic formulas. More precisely, GF is defined as follows: All atomic formulas belong to GF. If φ and ψ belong to GF then also $\neg\varphi$,

$(\varphi \wedge \psi)$ and $(\varphi \vee \psi)$ belong to GF. If α is an atomic formula and φ is a GF-formula whose free variables belong to the variables of α, then for every tuple $\bar{x}$ of variables, the formulas $\exists \bar{x} (\alpha \wedge \varphi)$ and $\forall \bar{x} (\alpha \rightarrow \varphi)$ belong to GF. A GF-formula is called *strictly guarded* if it either has no free variable or it is of one of the forms $(\alpha \wedge \varphi)$ and $\exists \bar{x} (\alpha \wedge \varphi)$, where α is an atomic formula and φ is a GF-formula whose free variables belong to the variables of α. We say that a GF-formula is guarded by stored tuples if it is a disjunction of strictly guarded GF-formulas.

The rule-based counterpart of SA is based on the restriction of the non-recursive rule-based queries where every single rule is a variant of an acyclic conjunctive query in which also negated atoms may occur. The resulting queries are called strictly acyclic NRSD-programs; for the precise definition we refer to [39]. A slightly different rule-based characterization is proposed in [46], the so-called recursion-free Datalog LITE.

The following theorem summarizes the relation between SA, GF, and their rule-based counterparts.

THEOREM 19.17

    a. SA *has the same expressive power as the class of* GF-*formulas that are guarded by stored tuples [76]*.

    b. *The class of* GF-*formulas that are guarded by stored tuples has the same expressive power as the strictly acyclic NRSD-programs [39]*.

    c. *Recursion-free Datalog LITE can express the same Boolean queries as sentences in the guarded fragment GF [46]*.

The translations between SA, GF, and the rule-based languages that are provided by the proof of the Theorem 19.17 are effective. The translations from GF to SA, to strictly acyclic NRSD-programs, and to recursion-free Datalog LITE, respectively, can be done in polynomial time. The translations in the opposite directions are a bit more involved.[*]

19.6 Toward More Expressiveness: Recursion and Counting

As pointed out in Section 19.4.2, Codd-equivalent query languages neither support recursion nor counting. Having these limitations in mind, it is natural to extend Codd-equivalent languages by recursion or counting capabilities while maintaining as much of the desirable algorithmic properties as possible. The query language SQL, in particular, contains constructs for counting and, starting with the SQL:1999 standard, also for expressing queries involving recursion. Corresponding extensions of the relational algebra and relational calculus have been investigated in [80].

In this section, we consider rule-based query languages with recursion and logics that are enhanced by fixpoint operators and counting. Furthermore, we discuss the possibility of finding a query language that precisely expresses the polynomial-time computable queries. Finally, we briefly consider more expressive languages that are capable of defining also queries of higher complexity.

19.6.1 Rule-Based Queries with Recursion

The query language datalog is a rule-based language that allows more than one rule, recursion, but no negation. A simple example is the following program which defines in a parent–child database a relation $\text{ANCESTOR}(x, y)$ consisting of all pairs (a, b) for which a is an ancestor of b:

---

    [*] The translations from [39,46,76] induce an exponential blow-up in terms of the size of the queries. It remains open if more efficient translations exist.

$$\text{ANCESTOR}(x, y) \quad : - \quad \text{PARENT}(x, y)$$
$$\text{ANCESTOR}(x, y) \quad : - \quad \text{PARENT}(x, z), \text{ANCESTOR}(z, y).$$

More formally, a datalog program over a database schema σ consists of finitely many finite rules of the form $Q(\bar{x}) : -R_1(\bar{t}_1), \ldots, R_\ell(\bar{t}_\ell)$ where the relation symbol Q does not belong to σ, and each variable in $\bar{x}$ occurs in at least one of the tuples $\bar{t}_1, \ldots, \bar{t}_\ell$. Relation symbols occurring in the head of some rule are called intensional relation symbols, whereas the symbols in σ are called extensional relation symbols. The body of each rule of a datalog program consists of atoms with intensional or extensional relation symbols. The schema σ_P of a datalog program P consists of the symbols in σ and the intensional relation symbols of P.

Given a database D of schema σ, a datalog program P is evaluated as follows: start with empty intensional relations and proceed step-wise, by adding tuples that satisfy a rule of the program, until nothing changes. Formally, the semantics of a datalog program P can be defined in various equivalent ways. One possibility is to associate with P an operator T_P, the so-called immediate consequence operator, which maps a database E of schema σ_P to a database $T_P(E)$ of the same schema: Extensional relations R remain unchanged, i.e., $R^{T_P(E)} = R^E$. For each intensional relation symbol Q, the relation $Q^{T_P(E)}$ is obtained as follows: take all rules of P whose head contains Q, view each of these rules as a conjunctive query, and let $Q^{T_P(E)}$ be the union of the results of these conjunctive queries when applied to database E.

The result of P when applied to a database D of schema σ is the database $P(D)$ of schema σ_P obtained as follows: let D^0 be the extension of D to schema σ_P where all intensional relations are empty, and repeatedly apply the operator T_P to obtain databases $D^1 := T_P(D^0)$, $D^2 := T_P(D^1)$, $D^3 := T_P(D^2)$, etc. that is, D^i is obtained by starting with D^0 and applying the operator T_P for i times. As T_P is a monotone operator*, the sequence of the D^i is increasing, i.e., $D^0 \subseteq D^1 \subseteq D^2 \subseteq \cdots$. Since the active domain of each D^i consists of constants occurring in P and of elements from the active domain of the original (finite) database D, a fixpoint will be reached eventually, i.e., there exists a number j such that $D^j = D^{j+1}$. The result of P on D is defined as $P(D) := D^j$. It is not difficult to see that j is of size polynomial in the size of D and that $P(D)$ can be computed in time polynomial in the size of D (with the exponent depending on the particular datalog program P). Furthermore, $P(D)$ is actually the *least* fixpoint of T_P that contains D.

A datalog query is a datalog program together with a designated intensional relation symbol which specifies the relation defined by the query. In the following we write "Datalog" to denote the class of all datalog queries.

Concerning limitations of the expressive power it should be noted that, similarly as CQ (see Theorem 19.1), Datalog can define only monotone queries (this immediately follows from the fact that T_P is a monotone operator). Thus, the expressive power of Datalog is incomparable to the expressive power of Codd-equivalent query languages: On the one hand, there are simple nonmonotone queries, e.g., query (6) from Section 19.2.2, that are expressible in the relational calculus FO but not in Datalog. On the other hand, there are recursive queries, e.g., the ANCESTOR query, that can be expressed in Datalog but not in FO.

There are many results on algorithmic properties of Datalog. We only mention the main results here; for a survey the reader is referred to [31]. In terms of query evaluation, the following holds.

THEOREM 19.18

a. *The data complexity of* Eval(Datalog) *is PTime-complete (implicit in [64,106]).*

b. *The combined complexity of* Eval(Datalog) *is ExpTime-complete (implicit in [106]).*

---

* In the sense that for all databases E and E' of schema σ_P, $E \subseteq E'$ implies that $T_P(E) \subseteq T_P(E')$.

Concerning the worst-case complexity of Eval(Datalog), it is known (even without relying on any complexity theoretic assumption) that the exponential dependence on the size of the input query cannot be avoided: there exists a sequence of (Boolean) Datalog queries q_k of size polynomial in k, such that $q_k(D)$ can be computed in time $|\text{adom}(D)|^k$ but not in time $|\text{adom}(D)|^{k-1}$. The proof even holds for a suitable fixed database schema σ; the arity of the intensional relation symbols of q_k, however, has to grow with increasing k. When restricting attention to relations of a fixed arity, the parameterized complexity of Eval(Datalog) is known to be W[1]-complete, i.e., the same as for conjunctive queries (cf. Theorem 19.2c). Details can be found in [88].

In terms of static complexity the following holds.

THEOREM 19.19

  a. *The satisfiability problem* Sat(Datalog) *is decidable.*

  b. *The equivalence problem* Equiv(Datalog) *and the containment problem* Cont(Datalog) *are undecidable [98].*

The proof of (a) is easy (see [3] for details). Concerning equivalence and containment, it should be noted that they are undecidable for datalog queries. Their "uniform" variants for datalog programs, asking whether all intensional relations defined by one program are equivalent to, respectively, included in the corresponding intensional relations of another datalog program, are decidable [92]. Furthermore, query containment becomes decidable if one of the two involved queries is nonrecursive [22,99].

Apart from equivalence and containment, another key problem relevant for static analysis of Datalog is the *boundedness problem*, asking for a given datalog query q whether or not the recursion depth necessary for evaluating q is bounded by a number that only depends on the query but not on the input database. This problem is closely related to the problem whether a given datalog query is expressible in FO, i.e., without using recursion. It is known that the boundedness problem is undecidable [42].

In the literature, various restrictions of Datalog have been identified for which the problems of boundedness, equivalence, and containment are decidable (see, e.g., [22,28] and the references therein). Also variants of Datalog that can be evaluated in linear time w.r.t. data complexity have been considered, e.g., Datalog LITE [46].

19.6.2 Relational Calculus with Fixpoints

In Section 19.6.1 we have seen how to extend conjunctive queries with a recursion mechanism via fixpoints. It is natural to consider a similar extension for the relational calculus.

Similarly to the immediate consequence operator T_P from Section 19.6.1, a formula of the relational calculus FO can define an operator on relations as follows: Let R be a relation symbol that is not present in the given database schema σ, and let k be the arity of R. Let $\varphi(\bar{x})$ be a formula with k free variables over the extended schema $\sigma \cup \{R\}$. Then, on any database D of schema σ, the formula $\varphi(\bar{x})$ defines an operator φ^D between k-ary relations over the active domain of D. Given a k-ary relation $\hat{R}$, $\varphi^D(\hat{R})$ is the result of the query $\varphi(\bar{x})$ when applied to the extension of D in which the relation symbol R is interpreted by the relation $\hat{R}$.

Analogous to the iterated application of the operator T_P in Section 19.6.1, we can now consider the iterated application of the operator φ^D, yielding a sequence of relations $R^0 := \emptyset$, $R^1 := \varphi^D(R^0)$, $R^2 := \varphi^D(R^1), \dots$, i.e., R^i is obtained by starting with the empty relation and applying the operator φ^D for i times. Note that, unlike with datalog, there exist FO-formulas φ (for example, the formula $\varphi(\bar{x}) := \neg R(\bar{x})$) for which the sequence $R^0, R^1, R^2, \dots$ is not increasing and does not reach a fixpoint.

There are several natural ways of ensuring that only those operators are considered for which the sequence $R^0, R^1, R^2, \ldots$ is increasing and eventually reaches a fixpoint. In the following, we present two of them: monotone operators and inflationary operators.

Monotone operators and the logic LFP. It is easy to see that if φ^D is a monotone operator, then $R^0 \subseteq R^1 \subseteq R^2 \subseteq \cdots$, and a fixpoint will be reached eventually, i.e., there exists a j such that $R^j = R^{j+1}$. Furthermore, this fixpoint is in fact the *least* fixpoint* of the operator φ^D. So it would be natural to semantically restrict attention to those formulas φ for which the operator φ^D is monotone for all databases D. Unfortunately, this kind of monotonicity of a formula is undecidable (see, e.g., the textbook [81]). Of course it makes little sense to define a query language based on an undecidable property of formulas, because then it is not even decidable if a given string belongs to the query language or not.

Fortunately, there is a second option which enforces monotonicity at the syntactic (rather than the semantic) level: restrict attention to those formulas φ in which R occurs only positively, i.e., within the scope of an even number of negations. It is straightforward to see that this implies that on every database D the operator φ^D is monotone. The opposite is of course not true but it turns out that any fixpoint obtained by a monotone formula can also be obtained by a positive formula [61]. Hence the syntactic restriction is harmless in terms of expressive power of the corresponding fixpoint logics.

This syntactic restriction leads to the least fixpoint logic (LFP), which extends first-order logic FO by the following rule: If $\varphi(\bar{x})$ is an LFP-formula in which R occurs only positively, then $[\text{lfp}_{R,\bar{x}}\varphi](\bar{x})$ is also an LFP-formula. The semantics of this formula is as follows: The result of query $[\text{lfp}_{R,\bar{x}}\varphi](\bar{x})$ on a database D is the limit of the sequence $R^0 := \emptyset$, $R^1 := \varphi^D(R^0)$, $R^2 := \varphi^D(R^1)$, $R^3 := \varphi^D(R^2)$ etc., and thus the least fixpoint of φ^D.

For example, if $\varphi(x, y)$ is the formula $\text{PARENT}(x, y) \vee \exists z(\text{PARENT}(x, z) \wedge R(z, y))$, then $[\text{lfp}_{R,xy}\varphi](x, y)$ is an LFP-formula that defines the ancestor relation for parent–child databases.

Inflationary operators and the logic IFP. Another way of ensuring that the considered sequence of relations $R^0, R^1, R^2, \ldots$ is increasing is to use instead of φ^D the operator I_φ^D which maps a relation $\hat{R}$ to the relation $I_\varphi^D(\hat{R}) := \hat{R} \cup \varphi^D(\hat{R})$. By definition, the sequence based on this operator, i.e., the sequence $R^0 := \emptyset$, $R^1 := I_\varphi^D(R^0)$, $R^2 := I_\varphi^D(R^1)$, $\ldots$ is increasing and thus eventually reaches a fixpoint (no matter what $\varphi(\bar{x})$ looks like). This fixpoint is called the inflationary fixpoint of φ. The inflationary fixpoint logic (IFP) extends first-order logic FO by the following rule: If $\varphi(\bar{x})$ is a IFP-formula, then $[\text{ifp}_{R,\bar{x}}\varphi](\bar{x})$ is also an IFP-formula. The semantics of this formula is as follows: The result of query $[\text{ifp}_{R,\bar{x}}\varphi](\bar{x})$ on a database D is the fixpoint reached by the sequence $R^0 := \emptyset$, $R^1 := I_\varphi^D(R^0)$, $R^2 := I_\varphi^D(R^1)$, $R^3 := I_\varphi^D(R^2)$, etc.

Complexity and expressive power of LFP and IFP. From the definition of the logics it is not difficult to see that the data complexity of evaluating queries definable in LFP or IFP belongs to PTime. In fact, query evaluation of these languages has the following complexity.

THEOREM 19.20

 a. *The data complexity of* Eval(IFP) *and* Eval(LFP) *is PTime-complete [64,106].*
 b. *The combined complexity of* Eval(IFP) *and* Eval(LFP) *is ExpTime-complete [106].*

Since LFP and IFP are extensions of FO, static analysis of these languages is impossible in general (cf. Theorem 19.13). Concerning the expressive power of LFP and IFP, the following is known.

---

* That is, $R^j \subseteq S$, for every relation S with $\varphi^D(S) = S$.

THEOREM 19.21 *[61] IFP can express exactly the same queries as LFP.*

Note that in the definition of the logics LFP and IFP, nesting of fixpoints is explicitly allowed. Restricting attention to formulas where just a single fixpoint operator may be applied to a first-order formula does not change the expressive power, since nested fixpoints can always be simulated by a single fixpoint (of potentially higher arity) [64].

It is also possible to get the same expressive power using an extension of Datalog with negation. Even though the definition of the semantics of this extension is not an obvious issue (see [3]), it turns out that with the so-called well-founded semantics it has the same expressive power as IFP and LFP [43].

Even though recursion adds a lot of expressive power to the relational calculus, it does not help to count. Indeed, the following holds.

THEOREM 19.22 *[13] LFP and IFP have the 0-1 law.*

Consequently, e.g., the EVENOPERAS query, i.e., query (8) from Section 19.2.2, is not definable in LFP or IFP.

For more details on fixpoint logics the reader is referred to the textbooks [3,81].

19.6.3 Relational Calculus with Counting

SQL has several numerical features and counting features which are actually used in practice much more often than the recursion mechanisms. In Section 19.4.2 we have seen, however, that the relational calculus basically cannot count: it even cannot express the query EVENOPERAS, asking whether the number of operas in the database is even.

The simplest way to extend the relational calculus with counting facilities is to explicitly include them in the syntax: We consider the extension FO+C of FO with counting quantifiers. FO+C is a two-sorted logic with the second sort being the natural numbers. The formula

$$\exists i \, \left(\exists j \, (i{=}j{+}j) \; \wedge \; \exists ! i \, y \, (\exists x \, \text{OPERAS}(x, y)) \right)$$

is an example of a FO+C-formula which expresses the EVENOPERAS query. Specifically, the formula states that there is a number i which is even (since there exists an integer j with $i = j + j$) such that the number of names y of pieces listed in the OPERAS-relation is exactly i. This formula combines the three kinds of quantifiers allowed in FO+C: Apart from the usual quantifiers ranging over elements of the active domain, it uses quantifiers of the form $\exists i$ that range over natural numbers (to be precise, the variables ranging over natural numbers only take values which are at most the size of the active domain of the underlying database). Furthermore, if variable i is interpreted by a natural number n, then a formula of the form $\exists ! i \, y \, \varphi(y)$ expresses that there are exactly n distinct elements a (in the active domain of the underlying database) for which $\varphi(a)$ holds. Apart from these quantifiers, FO+C also contains arithmetic predicates such as the linear order $<$, the addition $+$, and the multiplication $\times$.

Since FO+C contains FO, static analysis is impossible in general (cf. Theorem 19.13). The complexity of query evaluation for FO+C is not much higher than that of the relational calculus.

THEOREM 19.23

   a. *The data complexity of* Eval(FO+C) *is in* TC^0 *[11] and thus, in particular, in LogSpace.*
   b. *The combined complexity of* Eval(FO+C) *is PSpace-complete.*

Concerning the proof of (b), the inclusion in PSpace is straightforward, and the PSpace-hardness immediately follows from Theorem 19.12b.

By definition, FO+C extends the relational calculus with facilities for counting and for doing arithmetic. However, FO+C fails to express simple recursive queries, e.g., the ANCESTOR query. Actually, it is still Gaifman-local [79]. For a more detailed overview of FO+C and related logics, we refer to [79,95] and the references therein.

19.6.4 The Quest for PTime

As we have seen in Sections 19.6.2 and 19.6.3, recursion or counting can be added to FO while keeping the data complexity of query evaluation in polynomial time. Often, a query is considered tractable iff it can be evaluated in polynomial time w.r.t. data complexity. It would, of course, be desirable to have a query language that is capable of expressing exactly the tractable queries, i.e., exactly those queries that can be evaluated in polynomial time w.r.t. data complexity. The quest for such a query language is often stated as "Is there a language capturing PTime?"

We have seen that both FO+C and IFP fail to capture all of PTime: for example, FO+C cannot express the ANCESTOR query and IFP cannot express the EVENOPERAS query, but both queries can easily be evaluated in time polynomial in the size of the underlying database. It thus is natural to wonder whether the combination of IFP and FO+C (as it accounts for counting and recursion at the same time) does capture PTime. Let IFP+C be the language extending FO with both, inflationary fixpoints and counting. IFP+C has several nice properties, e.g., it can express the EVENOPERAS query and the ANCESTOR query, and its data complexity of query evaluation belongs to PTime. However it fails to capture all PTime-computable queries.

THEOREM 19.24 *[16] There exists a query that can be evaluated in polynomial time (w.r.t. data complexity) but that is not expressible in* IFP+C.

Nevertheless, when restricting attention to particular classes of databases, IFP+C and, in some cases, even IFP do capture all of PTime. The following theorem summarizes some results in this vein. Here, an ordered database is a database which contains a relation that is a linear order on the database's active domain.

THEOREM 19.25

a. IFP *captures* PTime *on the class of all ordered databases [64,106].*

b. IFP+C *captures* PTime *on all classes of databases of bounded treewidth [56] and on the class of databases corresponding to planar graphs [50].*

Note that (a) implies that IFP and IFP+C have the same expressive power on the class of ordered databases. Specifically, they express exactly those queries that can be evaluated in polynomial time data complexity.

Several generalizations of Theorem 19.25 (to larger classes of databases) and 19.24 (to certain restricted classes of databases) are known; see, e.g., [30,34].

The question of whether there is a query language or logic capturing PTime originated in the work of Chandra and Harel [18,19] and is considered one of the main challenging open problems in database theory and finite model theory. The logic must be reasonable in a sense defined by Gurevich [60] (in particular it must have an effective syntax). More details on the "quest for PTime" can be found in [55,86] and the references given therein.

19.6.5 Beyond PTime

All the query languages considered so far in this chapter are capable of expressing only queries that can be evaluated in polynomial time w.r.t. data complexity. Sometimes, however, there is a need for query languages that can define also more complex queries.

A simple way of extending the expressive power of LFP or IFP is to drop the requirement that the sequence $R^0, R^1, R^2, \ldots$ is monotone: Similarly to the definition of LFP, we consider the sequence obtained by iterated application of the operator φ^D, starting with $R^0 := \emptyset$. Now, however, φ may be an arbitrary formula such that the sequence $R^0, R^1, R^2, \ldots$ is not necessarily increasing. Thus, depending on the particular formula φ and the database D at hand, the sequence may or may not reach a fixpoint. If no fixpoint is reached, the partial fixpoint is, by convention, defined to be the empty relation $\emptyset$. This leads to the partial fixpoint logic PFP which extends first-order logic FO by the following rule: If $\varphi(\bar{x})$ is a PFP-formula, then $[\mathrm{pfp}_{R,\bar{x}}\varphi](\bar{x})$ is also a PFP-formula. The semantics of this formula is as follows: The result of query $[\mathrm{pfp}_{R,\bar{x}}\varphi](\bar{x})$ on a database D is the fixpoint of the sequence $R^0 := \emptyset$, $R^1 := \varphi^D(R^0)$, $R^2 := \varphi^D(R^1)$, $R^3 := \varphi^D(R^2)$, etc., if such a fixpoint exists; otherwise, it is defined to be the empty relation $\emptyset$.

Notice that all the intermediate relations R^i are of size polynomial in the size of the database. Furthermore, convergence (respectively, nonconvergence) to a fixpoint can be detected in PSpace. Altogether, this leads to an algorithm for evaluating a PFP-query in space polynomial in the size of the underlying database (where the exponent depends on the particular PFP-query).

Some more results on PFP are summarized in the following theorem.

THEOREM 19.26

   a. *The data complexity of* Eval(PFP) *is* PSpace-*complete* [106].
   b. PFP *has the 0-1 law* [74].
   c. PFP *captures* PSpace *on the class of all ordered databases* [106].
   d. IFP *has the same expressive power as* PFP *if, and only if,* PTime $=$ PSpace [5].

It is straightforward to see that all IFP queries can be expressed in PFP, i.e., PFP has at least the expressive power of IFP. Part (d) of Theorem 19.26 tells us that showing that the expressive power of PFP is strictly larger than that of IFP (on the class of arbitrary databases) is no easier than showing the (widely believed but, up to date, unproven) complexity theoretic assumption that PTime $\neq$ PSpace. On the other hand, part (b) implies that the EVENOPERAS query, i.e., query (8) from Section 19.2.2, cannot be expressed in PFP. This query, however, can easily be evaluated in polynomial time. Thus, on the class of arbitrary databases, PFP is not capable of expressing all tractable queries, let alone, all PSpace computable queries. Thus, PFP does not capture PSpace on the class of all databases. When restricting attention to the class of all ordered databases, however, Theorem 19.26c tells us that PFP precisely captures PSpace.

To conclude this section let us briefly consider a different (and widely used) way of extending the expressive power of query languages: Most database systems allow embedding of SQL queries into more powerful programming languages like C++ and java. This gives the user the capabilities to express arbitrary queries. The price to pay, however, is twofold: First of all, no optimization is performed except for, possibly, the SQL parts of the query. Furthermore, embedding a query language in a programming language implies that it is possible to specify queries that are not computable, i.e., queries for which there exists no evaluation algorithm that, on any input database D, stops after a finite number of steps and outputs the query result.

Several formalizations of the approach of embedding a query language into a programming language have been considered in the literature. We refer to Chapter 17 of [3] for a discussion on such and other highly expressive languages.

19.7 Toward More Flexibility: Query Languages for Other Data Models

Throughout the previous sections we restricted attention to the relational data model. In the present section we conclude with a brief discussion of some other major data models.

Set semantics vs. bag semantics. The relational model is based on the so-called set semantics: no relation can have two (or more) identical tuples. During the evaluation of a query, however, it is often the case that identical tuples are generated, for instance when projecting a relation on one of its attributes. Eliminating duplicates is a costly operation as it often requires to sort the tuples in order to identify duplicates. Moreover, in some applications, counting the numbers of duplicates may be desirable, e.g., for the query "How many shows are performed in the MET each year?". Therefore, in most relational database systems duplicates are not eliminated unless explicitly requested by the query (this is the role of the SELECT DISTINCT construct of SQL).

Instead of the set semantics (considered throughout the previous sections of this chapter), the relational model can also be equipped with a bag semantics. With bag semantics, the number of occurrences of a tuple in the output of a CQ query corresponds to the number of variable assignments satisfying the query.

With respect to query evaluation, the change from set semantics to bag semantics does not cause a considerable change in complexity (at least not for the "worst-case" analysis), since the complexity of intermediate sorting steps usually is dominated by the complexity of evaluating joins.

However, going from set semantics to bag semantics does change the complexity of static analysis, since the containment problem and the equivalence problem must take into account the number of occurrences of tuples in the query result. For example, two queries are equivalent iff they produce the same tuples with the same multiplicities. This makes a dramatic difference. Indeed, considering the class CQ of conjunctive queries under bag semantics, it is known that the equivalence problem Equiv(CQ) is as hard as testing graph isomorphism (and hence it is in NP) [21]. For the containment problem under bag semantics, however, it is still an open question whether Cont(CQ) is decidable.

For the class UCQ of unions of conjunctive queries and for the extension CQ($\neq$) of conjunctive queries where also inequalities are allowed as atoms, the containment problems Cont(UCQ) and Cont(CQ($\neq$)) become undecidable [66,67]. Recall from Theorem 19.14 that under set semantics these two problems are decidable. Concerning, however, the equivalence problem under bag semantics, it is known that Equiv(CQ($\neq$)) is decidable and belongs to PSpace [27].

The nested relational model. A further limitation of the relational model is given by the so-called first normal form which restricts attribute values to being atomic. In this sense, relations in the relational model are flat. The nested relational model (or complex value model) contains nested-type constructors which allow to build nested relations from atomic types by using tuple constructors and set constructors. Apart from suitable generalizations of the operators of the (flat) relational algebra, it has operators for nesting and unnesting relations. An equivalence between the logical calculus and the algebra can be established just as in the flat case [1,89]. Interestingly, the nested operators do not add any expressive power to the relational algebra for flat queries on flat databases [89]. This result was generalized in [110] for nonflat queries. On the other hand, adding a powerset operator to the nested relational algebra strongly extends its expressive power, e.g., it allows to express the ANCESTOR query [1]. Apart from the references mentioned above, we refer to [3,15] for more detailed information.

Object-oriented databases. Object-oriented databases further extend the nested relational model toward the object oriented paradigm. Each object, or entity, is given a unique identifier (OID). In the relational model this would correspond to adding an extra "ID" attribute to each relation and requiring that this attribute be a key. The model also allows OID as a possible attribute for an object. Query languages can be derived from those for nested relations, cf., for example, the language OQL [24], which was implemented in the O_2 database system [10]. A logical foundation for the object-oriented model was presented in [71]. For further information we refer to [3,4,17,35,104].

Constraint databases. Another feature of the relational model is that all relations are finite. There are many applications where this restriction can be seen as a limitation, for instance in spatial databases where a relation may correspond to a set of points in the plane. The constraint database framework is an elegant way of extending the relational model beyond finite relations. The idea is to manipulate implicit presentations of relations instead of explicit presentations that list all tuples. This is done by first specifying a logical framework, for instance first-order logic on the field of reals. Then, each (possibly infinite) relation is represented by a formula of the logical framework. In the case of first-order logic over the real field, this defines precisely the semialgebraic relations. Then, depending of the logical framework and on the query language at hand, query evaluation can or cannot be performed effectively. Typically, if the logical framework is first-order logic and admits quantifier elimination, as in the real field, then first-order queries can be evaluated (actually in PTime data complexity for the case of the real field). Since the seminal paper on constraint databases [70], this area has generated a lot of theoretical work concerning various logical frameworks, and several prototypes have been developed. The main application is in spatiotemporal databases. For a short introduction to constraint databases we refer to [103]; a comprehensive survey is provided in the book [75].

The semistructured data model. With the Internet, data is geographically distributed, and vast amounts of data are frequently exchanged between several database (and nondatabase) systems. The rigid structure of a relational database, enforced by its schema, makes it difficult to use the relational data model for transferring data from one site to another. To this end a new, more flexible model has been introduced: the semistructured model (see [2] for a comprehensive overview). The general idea is that data is now "self-describing," i.e., its structure is part of the data. The most widespread implementation of the semistructured model is the Extensible Markup Language (XML), specified by the World Wide Web Consortium. An XML document is a well-formed nested sequence of opening and closing *tags* in between which text and data values can be found. The tag structure can be seen as a labeled tree; it provides the "structure" of the document. A crucial difference with the relational setting is that data is now queried also by its position in the (possibly deeply nested) document tree. Hence, XML query languages must combine navigation in the document tree with other, more classical, database functionalities such as joins. Many query languages for XML have been proposed. They are designed for differing purposes, e.g., navigating to (or extracting) positions in a document, transforming documents into new documents, and posing Boolean queries about documents. The most important XML query languages, XPath, XSLT, and XQuery are maintained by the World Wide Web Consortium and are still under development, see http://www.w3.org/XML/ for up-to-date information.

The theoretical foundations build upon concepts from automata theory and mathematical logic, among them tree automata, monadic second-order logic, and logics with navigating features similar to those in temporal logic. For details see [72,82,97] and the references therein.

The data stream model. The data stream model considers data that is not stored but, instead, arrives in multiple, continuous, rapid, and time-varying streams. Typical application areas for which data stream processing is relevant are, e.g., IP network traffic analysis, mining text message streams, or processing data generated by sensor networks. In all these application areas it is not feasible to simply load the arriving data into a traditional relational DBMS and query it there. Instead, the streaming data has to be processed on-the-fly by using only a limited amount of memory. Instead of the precise query answers provided by traditional DBMS, queries against data streams are often evaluated by randomized algorithms that produce approximate solutions. Systems-oriented overviews of query languages for data streams and general purpose data stream mangagement systems (DSMS) can be found in [7,100]. A survey of machine models and lower bounds for stream-based query processing is given in [96]. For an introduction to efficient algorithms for data stream processing see [85].

Many other data models have been considered in the literature, each with its own query languages, and it seems that there will always be new data models and database applications. The area of query languages will thus be evolving in the foreseeable future, and there remain a lot of challenges for research.

Acknowledgments

We would like to thank Victor Vianu, Leonid Libkin, Martin Grohe, Moshe Vardi, and an anonymous referee for helpful remarks on an earlier version of this chapter.

References

1. S. Abiteboul and C. Beeri. The power of languages for the manipulation of complex values. *VLDB Journal*, 4(4):727–794, 1995.
2. S. Abiteboul, P. Buneman, and D. Suciu. *Data on the Web: From Relations to Semistructured Data and XML*. Morgan Kaufmann, San Francisco, CA, 2000.
3. S. Abiteboul, R. Hull, and V. Vianu. *Foundations of Databases*. Addison-Wesley, Boston, MA, 1995.
4. S. Abiteboul and P. C. Kanellakis. Object identity as a query language primitive. *Journal of the ACM*, 45(5):798–842, 1998.
5. S. Abiteboul and V. Vianu. Computing with first-order logic. *Journal of Computer and System Sciences*, 50(2):309–335, 1995.
6. H. Andréka, I. Németi, and J. van Benthem. Modal languages and bounded fragments of predicate logic. *Journal of Philosophical Logic*, 27(3):217–274, 1998.
7. B. Babcock, S. Babu, M. Datar, R. Motwani, and J. Widom. Models and issues in data stream systems. In *Proceedings of the 21st ACM Symposium on Principles of Database Systems (PODS)*, pages 1–16, ACM, New York, 2002.
8. G. Bagan. MSO queries on tree decomposable structures are computable with linear delay. In *Proceedings of the EACSL International Conference on Computer Science Logic (CSL)*, volume 4207 of *Lecture Notes in Computer Science*, pages 167–181, Springer-Verlag, Berlin, Germany, 2006.
9. G. Bagan, A. Durand, and E. Grandjean. On acyclic conjunctive queries and constant delay enumeration. In *Proceedings of the EACSL International Conference on Computer Science Logic (CSL)*, volume 4646 of *Lecture Notes in Computer Science*, pages 208–222, Springer-Verlag, Berlin, Germany, 2007.
10. F. Bancilhon, C. Delobel, and P. C. Kanellakis, editors. *Building an Object-Oriented Database System. The Story of O2*. Morgan Kaufmann, San Francisco, CA, 1992.
11. D. A. Mix Barrington, N. Immerman, and H. Straubing. On Uniformity within NC^1. *Journal of Computer and System Sciences*, 41(3):274–306, 1990.
12. C. Beeri, R. Fagin, D. Maier, and M. Yannakakis. On the desirability of acyclic database schemes. *Journal of the ACM*, 30(3):479–513, 1983.
13. A. Blass, Y. Gurevich, and D. Kozen. A zero-one law for logic with a fixed-point operator. *Information and Control*, 67(1–3):70–90, 1985.
14. E. Börger, E. Grädel, and Y. Gurevich. *The Classical Decision Problem*. Perspectives in mathematical logic. Springer-Verlag, Berlin, Germany, 1997.
15. P. Buneman, S. A. Naqvi, V. Tannen, and L. Wong. Principles of programming with complex objects and collection types. *Theoretical Computer Science*, 149(1):3–48, 1995.
16. J.-y. Cai, M. Fürer, and N. Immerman. An optimal lower bound on the number of variables for graph identifications. *Combinatorica*, 12(4):389–410, 1992.

17. R. G. G. Cattell, editor. *The Object Database Standard: ODMG-93*. Morgan Kaufmann, San Mateo, CA, 1994.

18. A. K. Chandra and D. Harel. Computable queries for relational data bases. *Journal of Computer and System Sciences*, 21(2):156–178, 1980.

19. A. K. Chandra and D. Harel. Structure and complexity of relational queries. *Journal of Computer and System Sciences*, 25(1):99–128, 1982.

20. A. K. Chandra and P. M. Merlin. Optimal implementation of conjunctive queries in relational data bases. In *Proceedings of the 9th Annual ACM SIGACT Symposium on the Theory of Computing (STOC)*, pages 77–90, ACM, New York, 1977.

21. S. Chaudhuri and M. Y. Vardi. Optimization of *real* conjunctive queries. In *Proceedings of the 12th ACM Symposium on Principles of Database Systems (PODS)*, pages 59–70, ACM, New York, 1993.

22. S. Chaudhuri and M. Y. Vardi. On the equivalence of recursive and nonrecursive datalog programs. *Journal of Computer and System Sciences*, 54(1):61–78, 1997.

23. C. Chekuri and A. Rajaraman. Conjunctive query containment revisited. *Theoretical Computer Science*, 239(2):211–229, 2000.

24. S. Cluet. Designing OQL: Allowing objects to be queried. *Information Systems*, 23(5):279–305, 1998.

25. E. F. Codd. A relational model of data for large shared data banks. *Communication of the ACM*, 13(6):377–387, 1970.

26. E. F. Codd. Relational completeness of data base sublanguages. In R. Rustin, editor, Database systems, pages 65–98, *Prentice Hall and IBM Research Report RJ 987*, San Jose, CA, 1972.

27. S. Cohen, W. Nutt, and Y. Sagiv. Deciding equivalences among conjunctive aggregate queries. *Journal of the ACM*, 54(2), 2007.

28. S. S. Cosmadakis, H. Gaifman, P. C. Kanellakis, and M. Y. Vardi. Decidable optimization problems for database logic programs (preliminary report). In *Proceedings of the 12th Annual symposium on the Theory of Computing (STOC)*, pages 477–490, ACM, New York, 1988.

29. B. Courcelle. Linear delay enumeration and monadic second-order logic. *Discrete Applied Mathematics*, 157(12):2675–2700, 2009.

30. E. Dahlhaus and J. A. Makowsky. Query languages for hierarchic databases. *Information and Computation*, 101(1):1–32, 1992.

31. E. Dantsin, T. Eiter, G. Gottlob, and A. Voronkov. Complexity and expressive power of logic programming. *ACM Computing Surveys*, 33(3):374–425, 2001.

32. A. Dawar, M. Grohe, and S. Kreutzer. Locally excluding a minor. In *Proceedings of the 22nd Annual Symposium on Logic in Computer Science (LICS)*, pages 270–279, IEEE Computer Society, Washington, DC, 2007.

33. A. Dawar, M. Grohe, S. Kreutzer, and N. Schweikardt. Model theory makes formulas large. In *Proceedings of the International Conference on Algorithms, Languages and Programming (ICALP)*, volume 4596 of *Lecture Notes in Computer Science*, pages 913–924, Springer-Verlag, Berlin, Germany, 2007.

34. A. Dawar and D. Richerby. The power of counting logics on restricted classes of finite structures. In *Proceedings of the 21st international Workshop and the 16th Annual Conference of the EACSL on Computer Science Logic (CSL)*, volume 4646 of *Lecture Notes in Computer Science*, pages 84–98, Springer-Verlag, Berlin, Germany, 2007.

35. J. Van den Bussche, D. Van Gucht, M. Andries, and M. Gyssens. On the completeness of object-creating database transformation languages. *Journal of the ACM*, 44(2):272–319, 1997.

36. R. G. Downey, M. R. Fellows, and U. Taylor. The parameterized complexity of relational database queries and an improved characterization of W[1]. In *Combinatorics, Complexity, and Logic, volume 39 of Proceedings of DMTCS*, pages 194–213, Springer-Verlag, Berlin, Germany, 1996.

37. A. Durand and E. Grandjean. First-order queries on structures of bounded degree are computable with constant delay. *ACM Transactions on Computational Logic*, 8(4), 2007.

38. R. Fagin. Probabilities on finite models. *Journal of Symbolic Logic*, 41(1):50–58, 1976.

39. J. Flum, M. Frick, and M. Grohe. Query evaluation via tree-decompositions. *Journal of the ACM*, 49(6):716–752, 2002.

40. J. Flum and M. Grohe. *Parameterized Complexity Theory*. Springer-Verlag, Berlin, Germany, 2006.

41. H. Gaifman. On local and non-local properties. In J. Stern, editor, *Proceedings of the Herbrand Symposium, Logic Colloquium '81*, pages 105–135, North Holland, Amsterdam, the Netherlands, 1982.

42. H. Gaifman, H. G. Mairson, Y. Sagiv, and M. Y. Vardi. Undecidable optimization problems for database logic programs. *Journal of the ACM*, 40(3):683–713, 1993.

43. A. Van Gelder. The alternating fixpoint of logic programs with negation. In *Proceedings of the 8th ACM Symposium on Principles of Database Systems (PODS)*, pages 1–10, ACM, New York, 1989.

44. Y. V. Glebskii, D. I. Kogan, M. A. Liogon'kii, and V. A. Talanov. Range and degree of realizability of formulas in the restricted predicate calculus. *Kibernetika*, 2:17–28, 1969.

45. G. Gogic, H. A. Kautz, C. H. Papadimitriou, and B. Selman. The comparative linguistics of knowledge representation. In *International Joint Conference on Artificial Intelligence (IJCAI)*, pages 862–869, Montreal, QC, 1995.

46. G. Gottlob, E. Grädel, and H. Veith. Datalog LITE: A deductive query language with linear time model checking. *ACM Transactions on Computational Logic*, 3(1):42–79, 2002.

47. G. Gottlob, N. Leone, and F. Scarcello. The complexity of acyclic conjunctive queries. *Journal of the ACM*, 48(3):431–498, 2001.

48. G. Gottlob, N. Leone, and F. Scarcello. Hypertree decompositions and tractable queries. *Journal of Computer and System Sciences*, 64(3):579–627, 2002.

49. M. Grohe. Finite variable logics in descriptive complexity theory. *Bulletin of Symbolic Logic*, 4:345–398, 1998.

50. M. Grohe. Fixed-point logics on planar graphs. In *Proceedings of the IEEE Conference on Logic in Computer Science (LICS)*, pages 6–15, IEEE Computer Society, Washington, DC, 1998.

51. M. Grohe. Generalized model-checking problems for first-order logic. In *Proceedings of the 18th Annual Symposium on Theoretical Aspects of Computer Science (STACS'01)*, volume 2010 of *Lecture Notes in Computer Science*, pages 12–26, Springer-Verlag, Berlin, Germany, 2001.

52. M. Grohe. The parameterized complexity of database queries. In *Proceedings of the 20th ACM Symposium on Principles of Database Systems (PODS)*, pages 2–92, ACM, New York, 2001.

53. M. Grohe. The complexity of homomorphism and constraint satisfaction problems seen from the other side. *Journal of the ACM*, 54(1), 2007.

54. M. Grohe. Logic, graphs, and algorithms. In G. Flum, E. Grädel, and T. Wilke, editors, *Logic and Automata: History and Perspectives*, pages 357–422, Amsterdam University Press, Amsterdam, the Netherlands, 2007.

55. M. Grohe. The quest for a logic capturing PTIME. In *Proceedings of the 23rd Annual IEEE Symposium on Logic in Computer Science (LICS)*, pages 267–271, IEEE Computer Society, Washington, DC, 2008.

56. M. Grohe and J. Mariño. Definability and descriptive complexity on databases of bounded tree-width. In *Proceedings of the 7th international conference on Database Theory (ICDT)*, volume 1540 of *Lecture Notes in Computer Science*, pages 70–82, Springer-Verlag, Berlin, Germany, 1999.

57. M. Grohe and N. Schweikardt. Comparing the succinctness of monadic query languages over finite trees. *RAIRO - Theoretical Informatics and Applications*, 38:343–373, 2004.

58. M. Grohe, T. Schwentick, and L. Segoufin. When is the evaluation of conjunctive queries tractable? In *Proceedings of the 33rd annual ACM Symposium on the Theory of Computing (STOC)*, pages 657–666, ACM, New York, 2001.

59. Y. Gurevich and I. Koryakov. Remarks on Berger's paper on the domino problem. *Siberian Mathematical Journal*, 13:319–321, 1999.

60. Y. Gurevich. Logic and the challenge of computer science. In E. Boerger, editor, *Current Trends in Theoretical Computer Science*, pages 1–57, Computer Science Press, New York, 1988.

61. Y. Gurevich and S. Shelah. Fixed-point extensions of first order logic. *Annals of Pure and Applied Logic*, 32:265–280, 1986.

62. L. Hella, L. Libkin, and J. Nurmonen. Notions of locality and their logical characterizations over finite models. *Journal of Symbolic Logic*, 64:1751–1773, 1999.

63. R. Hull. Relative information capacity of simple relational database schemata. In *Proceedings of the 3rd ACM Symposium on Principles of Database Systems (PODS)*, pages 97–109, ACM, New York, 1984.

64. N. Immerman. Relational queries computable in polynomial time. *Information and Control*, 68(1–3): 86–104, 1986.

65. N. Immerman. Expressibility and parallel complexity. *SIAM Journal on Computing*, 18(3):625–638, 1989.

66. Y. E. Ioannidis and R. Ramakrishnan. Containment of conjunctive queries: Beyond relations as sets. *ACM Transactions on Database Systems*, 20(3):288–324, 1995.

67. T. S. Jayram, P. G. Kolaitis, and E. Vee. The containment problem for REAL conjunctive queries with inequalities. In *Proceedings of the 21st ACM Symposium on Principles of Database Systems (PODS)*, pages 80–89, ACM, New York, 2006.

68. A. Kahr, E. Moore, and H. Wang. Entscheidungsproblem reduced to the ∀∃∀ case. *Proceedings of the National Academy of Sciences USA*, 48:365–377, 1962.

69. P. C. Kanellakis. Elements of relational database theory. In J. van Leeuwen, editor, *Handbook of Theoretical Computer Science*, volume B, Chapter 16, pages 1073–1155, Elsevier Science Publishers, St. Louis, MO, 1990.

70. P. C. Kanellakis, G. M. Kuper, and P. Z. Revesz. Constraint query languages. In *Proceedings of the 9th ACM Symposium on Principles of Database Systems (PODS)*, pages 299–313, ACM, New York, 1990.

71. M. Kifer, G. Lausen, and J. Wu. Logical foundations of object-oriented and frame-based languages. *Journal of the ACM*, 42(4):741–843, 1995.

72. N. Klarlund, T. Schwentick, and D. Suciu. XML: Model, schemas, types, logics, and queries. In J. Chomicki, R. van der Meyden, and G. Saake, editors, *Logics for Emerging Applications of Databases*, pages 1–41, Springer-Verlag, Berlin, Germany, 2003.

73. P. G. Kolaitis, D. L. Martin, and M. N. Thakur. On the complexity of the containment problem for conjunctive queries with built-in predicates. In *Proceedings of the 17th ACM Symposium on Principles of Database Systems (PODS)*, pages 197–204, ACM, New York, 1998.

74. P. G. Kolaitis and M. Y. Vardi. Infinitary logics and 0-1 laws. *Information and Computation*, 98(2):258–294, 1992.

75. G. M. Kuper, L. Libkin, and J. Paredaens, editors. *Constraint Databases*. Springer-Verlag, Berlin, Germany, 2000.

76. D. Leinders, M. Marx, J. Tyszkiewicz, and J. Van den Bussche. The semijoin algebra and the guarded fragment. *Journal of Logic, Language and Information*, 14(3):331–343, 2005.

77. D. Leinders and J. Van den Bussche. On the complexity of division and set joins in the relational algebra. *Journal of Computer and System Sciences*, 73(4):538–549, 2007.

78. D. Leinders and J. Van den Bussche. Repetitions and permutations of columns in the semijoin algebra. *RAIRO – Theoretical Informatics and Applications*, 43(2):179–187, 2009.

79. L. Libkin. Logics with counting and local properties. *ACM Transactions on Computational Logic*, 1(1):33–59, 2000.

80. L. Libkin. Expressive power of SQL. *Theoretical Computer Science*, 296:379–404, 2003.

81. L. Libkin. *Elements of Finite Model Theory*. Springer-Verlag, Berlin, Germany, 2004.

82. L. Libkin. Logics for unranked trees: An overview. *Logical Methods in Computer Science*, 2(3): 2006.

83. D. Maier. *The Theory of Relational Databases*. Computer Science Press, Rockville, MD, 1983.

84. M. Mortimer. On languages with two variables. *Zeitschrift für mathematische Logik und Grundlagen der Mathematik*, 21(8):135–140, 1975.

85. S. Muthukrishnan. Data streams: Algorithms and applications. *Foundations and Trends in Theoretical Computer Science*, 1(2), 2005.

86. A. Nash, J. B. Remmel, and V. Vianu. PTIME queries revisited. In *Proceedings of the 10th International Conference on Database Theory (ICDT)*, volume 3363 of *Lecture Notes in Computer Science*, pages 274–288, Springer-Verlag, Berlin, Germany, 2005.

87. C. H. Papadimitriou. The complexity of knowledge representation. In *IEEE Conference on Computational Complexity*, pages 244–248, Washington, DC, 1996.

88. C. H. Papadimitriou and M. Yannakakis. On the complexity of database queries. *Journal of Computer and System Sciences*, 58(3):407–427, 1999.

89. J. Paredaens and D. Van Gucht. Possibilities and limitations of using flat operators in nested algebra expressions. In *Proceedings of the 7th ACM Symposium on Principles of Database Systems (PODS)*, pages 29–38, ACM, New York, 1988.

90. R. Ramakrishnan and J. Gehrke. *Database Management Systems*. McGraw-Hill, New York, 2002.

91. O. Reingold. Undirected ST-connectivity in log-space. In *Proceedings of the 37th Annual ACM Symposium on the Theory of Computing (STOC)*, pages 376–385, ACM, New York, 2005.

92. Y. Sagiv. Optimizing datalog programs. In J. Minker, editor, *Foundations of Deductive Databases and Logic Programming*, pages 659–698, Morgan Kaufmann, San Francisco, CA, 1988.

93. Y. Sagiv and M. Yannakakis. Equivalences among relational expressions with the union and difference operators. *Journal of the ACM*, 27(4):633–655, 1980.

94. F. Scarcello. Query answering exploiting structural properties. *SIGMOD Record*, 34(3):91–99, 2005.

95. N. Schweikardt. Arithmetic, first-order logic, and counting quantifiers. *ACM Transactions on Computational Logic*, 6(3):634–671, 2005.

96. N. Schweikardt. Machine models and lower bounds for query processing. In *Proceedings of the 26th ACM Symposium on Principles of Database Systems (PODS)*, pages 41–52, ACM, New York, 2007.

97. T. Schwentick. Automata for XML—A survey. *Journal of Computer and System Sciences*, 73(3):289–315, 2007.

98. O. Shmueli. Decidability and expressiveness Aspects of logic queries. In *Proceedings of the 6th ACM Symposium on Principles of Database Systems (PODS)*, pages 237–249, ACM, New York, 1987.

99. L. J. Stockmeyer. The complexity of decision problems in automata and logic. PhD thesis, MIT, Cambridge, MA, 1974.

100. M. Stonebraker, U. Çetintemel, and S. B. Zdonik. The 8 requirements of real-time stream processing. *SIGMOD Record*, 34(4):42–47, 2005.

101. B. A. Trakhtenbrot. The impossibility of an algorithm for the decision problem for finite models. *Doklady Academii Nauk SSSR*, 70:569–572, 1950. In Russian; translated into English in *American Mathematical Society Translations*, Series 2: 23, 1963.

102. J. D. Ullman. *Principles of database and knowledge-base systems, Volume I*. Computer Science Press, Inc., New York, 1988.

103. J. Van den Bussche. Constraint databases: A tutorial introduction. *SIGMOD Record*, 29(3):44–51, 2000.

104. J. Van den Bussche, D. Van Gucht, M. Andries, and M. Gyssens. On the completeness of object-creating database transformation languages. *Journal of the ACM*, 44(2):272–319, 1997.

105. R. van der Meyden. The complexity of querying indefinite data about linearly ordered domains. *Journal of Computer and System Sciences*, 54(1):113–135, 1997.

106. M. Y. Vardi. The complexity of relational query languages (extended abstract). In *Proceedings of the 14th Annual Symposium on the Theory of Computing (STOC)*, pages 137–146, ACM, New York, 1982.

107. M. Y. Vardi. On the complexity of bounded-variable queries. In *Proceedings of the 14th ACM Symposium on Principles of Database Systems (PODS)*, pages 266–276, ACM, New York, 1995.

108. J. S. Vitter. External memory algorithms and data structures: Dealing with massive data. *ACM Computing Surveys*, 33:209–271, 2001.

109. H. Vollmer. *Introduction to Circuit Complexity*. Springer-Verlag, Berlin, Germany, 1999.

110. L. Wong. Normal forms and conservative extension properties for query languages over collection types. *Journal of Computer and System Sciences*, 52(3):495–505, 1996.

111. M. Yannakakis. Algorithms for acyclic database schemes. In *Proceedings of the 7th International Conference on Very Large Data Bases (VLDB)*, pages 82–94, IEEE Press, Pislatauray, NJ, 1981.

20

Scheduling Algorithms

David Karger
Massachusetts Institute of Technology

Cliff Stein
Dartmouth College

Joel Wein
Polytechnic Institute of New York University

20.1 Introduction

Scheduling theory is concerned with the optimal allocation of scarce resources to activities over time. The practice of this field dates to the first time two humans contended for a shared resource and developed a plan to share it without bloodshed. The theory of the design of algorithms for scheduling is younger, but still has a significant history—the earliest papers in the field were published more than forty years ago.

Scheduling problems arise in a variety of settings, as is illustrated by the following examples:

Example 20.1

Consider the central processing unit of a computer that must process a sequence of jobs that arrive over time. In what order should the jobs be processed in order to minimize, on average, the time that a job is in the system from arrival to completion?

Example 20.2

Consider a team of five astronauts preparing for the reentry of their space shuttle into the atmosphere. There is a set of tasks that must be accomplished by the team before reentry. Each task must be carried out by exactly one astronaut, and certain tasks cannot be started until other tasks are completed. Which tasks should be performed by which astronaut, and in which order, to ensure that the entire set of tasks is accomplished as quickly as possible?

Example 20.3

Consider a factory that produces different sorts of widgets. Each widget must first be processed by machine 1, then machine 2, and then machine 3, but different widgets require different amounts of processing time on different machines. The factory has orders for batches of widgets; each order has a date by which it must be completed. In what order should the machines work on different widgets in order to insure that the factory completes as many orders as possible on time?

More generally, scheduling problems involve jobs that must scheduled on machines subject to certain constraints to optimize some objective function. The goal is to specify a schedule that specifies when and on which machine each job is to be executed.

Researchers have studied literally thousands of scheduling problems, and it would be impossible even to enumerate all known variants in the space of this chapter. Our goal is more modest. We wish to make the reader familiar with an assortment of algorithmic techniques that have proved useful for solving a large variety of scheduling problems. We will demonstrate these techniques by drawing from a collection of "basic problems" that model important issues arising in many scheduling problems, while at the same time remaining simple enough to permit elegant and useful analysis. These basic problems have received much attention, and their centrality was reinforced by two influential surveys [13,29]. All three examples above fit into the basic problem framework.

In this survey we focus exclusively on algorithms that provably run, in the worst case, in time polynomial in the size of the input. If the algorithm always gives an optimum solution, we call it an exact algorithm. Many of the problems that we consider, however, are $\mathcal{N}P$-hard, and it thus seems unlikely that polynomial-time algorithms exist to solve them. In these cases we will be interested in approximation algorithms; we define a ρ-approximation algorithm to be an algorithm that runs in polynomial time and delivers a solution of value at most ρ times the optimum.

The rest of this chapter is organized as follows. We complete this introduction by laying out a standard framework covering the basic scheduling problems and a notation for describing them. We then explore various techniques that can be used to solve them. In Section 20.2 we present a collection of heuristics that use some simple rule to assign a priority to each job and then schedule the jobs in priority order. These heuristics are useful both for solving certain problems optimally in polynomial time, and for giving simple but high-quality approximations for certain $\mathcal{N}P$-hard scheduling problems. Many scheduling problems require a more complex approach than a simple priority rule; in Section 20.3 we study algorithms that are more sophisticated in their greedy choices. In Section 20.4 we discuss the application of some basic tools of combinatorial optimization, such as network optimization and linear programming, to the design of scheduling algorithms. We then turn exclusively to $\mathcal{N}P$-hard problems. In Section 20.5 we introduce the notion of a relaxation of a problem, and show how to use relaxations to design approximation algorithms. Finally, in Section 20.6 we discuss enumeration and scaling techniques by which certain other $\mathcal{N}P$-hard scheduling problems can be approximated arbitrarily closely in polynomial time.

20.1.1 The Framework of Basic Problems

A scheduling problem is defined by three separate elements: the **machine environment**, the optimality criterion, and a set of side constraints and characteristics. We first discuss the simplest machine environment, and use that to introduce a variety of **optimality criteria** and side constraints. We then introduce and discuss more complex machine environments.

20.1.1.1 The One-Machine Environment

In all of our scheduling problems we begin with a set $\mathcal{J}$ of n jobs, numbered $1, \ldots, n$. In the one-machine environment we have one machine that can process at most one job at a time. Each job j has a processing requirement p_j; namely, it requires processing for a total of p_j units of time on the machine. If each job must be processed in an uninterrupted fashion, we have a nonpreemptive scheduling environment, whereas if a job may be processed for a period of time, interrupted and continued at a later point in time, we have a preemptive environment. A schedule S for the set $\mathcal{J}$ specifies, for each job j, which p_j units of time the machine uses to process job j. Given a schedule S, we denote the completion time of job j in schedule S by C_j^S.

The goal of a scheduling algorithm is to produce a "good" schedule, but the definition of "good" will vary depending on the application. In Example 20.2 above, the goal is to process the entire batch of jobs as quickly as possible, or, in other words, to minimize the completion time of the last job finished in the schedule. In Example 20.1 we care less about the completion time of the last job in the batch as long as, on average, the jobs receive good service. Therefore, given a set of jobs and a machine environment, we must specify an optimality criterion; the goal of a scheduling algorithm will be to construct a schedule that optimizes this criterion. The two optimality criteria discussed in our examples are among the most basic optimality criteria: the average completion time of a schedule and its makespan. We define the makespan $C_{\max}^S = \max_j C_j^S$ of a schedule S to be the maximum completion time of any job in S, and the average completion of schedule S to be $\frac{1}{n} \sum_{j=1}^n C_j^S$. Note that optimizing the average completion time is equivalent to optimizing the sum of completion times $\sum_{j=1}^n C_j^S$.

We next turn to side constraints and characteristics that modify the one-machine environment. A number of side constraints and characteristics are possible; for example, we must specify whether or not preemption is allowed. Two other possible constraints model the arrival of jobs over time or the possibility of logical dependence between jobs. In a scheduling environment with release date constraints, we associate with each job j a release date r_j; job j is only available for processing at time r_j or later. In a scheduling environment with precedence constraints we are given a partial order $\prec$ on the set $\mathcal{J}$ of jobs; if $j' \prec j$ then we may not begin processing job j until job j' is completed.

Although we are early in our discussion of scheduling models, we already have enough information to define a number of problems. We refer to various scheduling problems in the now-standard notation defined by [13]. A problem is denoted by $\alpha|\beta|\gamma$, where (1) α denotes the machine environment, (2) β denotes various side constraints and characteristics, and (3) γ denotes an optimality criterion.

For the one-machine environment α is 1. For the optimality criteria we have introduced so far, γ is either $\sum C_j$ or $C_{\max}$. At this point in our discussion, β is a subset of r_j, *prec*, and *pmtn*, where these denote respectively the presence of (nontrivial) release date constraints, precedence constraints, and the ability to schedule preemptively. Any of the side constraints not explicitly listed are assumed not to be present—e.g., we default to a nonpreemptive model unless *pmtn* is given in the side constraints. As an illustration, $1||\sum C_j$ denotes the problem of nonpreemptively scheduling independent jobs on one machine so as to minimize their average completion time, while $1|r_j|\sum C_j$ denotes the variant of the problem in which jobs have release dates. As another example, $1|r_j, pmtn, prec|C_{\max}$ denotes the problem of preemptively scheduling jobs with release dates and precedence constraints on one machine so as to minimize their makespan. Note that Example 20.1, given above, can be modeled by $1|r_j|\sum C_j$, or, if preemption is allowed, by $1|r_j, pmtn|\sum C_j$.

Two other possible elements of a scheduling application might lead to different objective functions in the one-machine environment. It is possible that not all jobs are of equal importance, and thus, when measuring average service provided to a job, one might wish to weight the average so as to give more importance to certain jobs. We model this by assigning a weight $w_j > 0$ to each job j, and generalize the $\sum C_j$ criterion to the average weighted completion time of a schedule, $\frac{1}{n}\sum_{j=1}^{n} w_j C_j$. In the scheduling notation this optimality criterion is denoted by $\sum w_j C_j$.

It is also possible that each job j may have an associated due date d_j by which it should be completed. This gives rise to two different optimality criteria. Given a schedule S, we define $L_j = C_j^S - d_j$ to be the lateness of job j, and we will be interested in constructing a schedule that minimizes $L_{max} = \max_{j=1}^{n} L_j$, the maximum lateness of any job in the schedule. Alternatively, we concern ourselves with constructing a schedule that maximizes the number of jobs that complete by their due dates. To capture this, given a schedule S we define $U_j = 0$ if $C_j^S \le d_j$ and $U_j = 1$ otherwise; we can thus describe our optimality criterion as the minimization of $\sum U_j$, or more generally, $\sum w_j U_j$. As illustrations, $1|r_j|L_{max}$ denotes the problem of nonpreemptively scheduling, on one machine, jobs with release dates and due dates so as to minimize the maximum lateness of any job, and $1|prec|\sum w_j U_j$ denotes the problem of nonpreemptively scheduling precedence-constrained jobs on one machine so as to minimize the total (summed) weight of the late jobs. Deadlines are not listed in the side constraints since they are implicit in the objective function.

Finally, we will consider one scheduling problem that deals with a more general optimality criterion. For each job j, we let $f_j(t)$ be any function that is nondecreasing with the completion time of the job, and, with respect to a schedule S, define $f_{max} = \max_{j=1}^{n} f_j(C_j^S)$. The specific problem that we will consider (in Section 20.3) is $1|prec|f_{max}$—the scheduling of precedence-constrained jobs on one machine so as to minimize the maximum value of $f_j(C_j)$ over all $j \in \mathcal{J}$.

20.1.1.2 More Complex Machine Environments: Parallel Machines and the Shop

Having introduced all of the optimality criteria, side characteristics and conditions that we will use in this survey, we now discuss more complex machine environments.

We first discuss parallel machine environments. In these environments we are given m machines. A job j with processing requirement p_j can be processed on any one of the machines, or, if preemption is allowed, started on one machine, and when preempted potentially continued on another machine. A machine can process at most one job at a time and a job can be processed by at most one machine at a time.

In the identical parallel machine environment the machines are identical, and job j requires p_j units of processing time when processed on any machine. In the uniformly related machines environment each machine i has a speed $s_i > 0$, and thus job j, if processed entirely on machine i, would take a total of p_j/s_i time to process. In the unrelated parallel machines environment we model machines that have different capabilities and thus their relative performance on a job is unrelated. In other words, the speed of machine i on job j, s_{ij}, depends on both the machine and the job; job j requires p_j/s_{ij} processing time on machine i. We define $p_{ij} = p_j/s_{ij}$.

In the shop environment, which primarily models various sorts of production environments, we again have m machines. In this setting a job j is made up of operations, with each operation requiring processing on a specific one of the m machines. Different operations may take different amounts of time (possibly 0). In the open shop environment, the operations of a job can be processed in any order, as long as no two operations are processed on different machines simultaneously. In the job shop environment, there is a total order on the operations of a job, and one operation cannot be started until its predecessor in the total order is completed. A special case of the job shop is the flow shop, in which the order of the operations is the same—each job requires processing on the same machines and in the same order, but different jobs may require different amounts of processing on the same machine. Typically in the flow shop and open shop environment, each job is processed exactly once on each machine.

In the scheduling notation, the identical, uniformly related and unrelated machine environments are denoted respectively by P, Q, and R. The open, flow and job shop environments are denoted by O, F, and J. When the environment has a fixed number of machines the number is included in the environment specification; so, for example, P2 denotes the environment with two identical parallel machines. Note that Example 20.2 can be modeled by $P5|prec|C_{max}$, and Example 20.3 can be modeled by $F3|r_j|\sum U_j$.

20.2 Priority Rules

The most obvious approach to solving a scheduling problem is a greedy one: whenever a machine becomes available, assign some job to it. A more sophisticated variant of this approach is to give each job a priority derived from the particular optimality criterion, and then, whenever a machine becomes available, assign the available job of highest priority to it. In this section we discuss such scheduling strategies for one machine, parallel machine and shop problems. In all of our algorithms, the priority of a job can be determined without reference to other jobs. This typically gives a simple scheduling algorithm that runs in $O(n \log n)$ time—the bottleneck being the time needed to sort the jobs by priority. We also discuss the limitations of these approaches, giving examples where they do not perform well.

20.2.1 One Machine

We first focus on algorithms for single-machine problems in which we give each job a priority, sort by priorities, and schedule in this order. To establish the correctness of such algorithms, it is often possible to apply an interchange argument. Suppose that there is an optimal schedule with jobs processed in nonpriority order. It follows that some adjacent pair of jobs in the schedule has inverted priorities. We show that if we swap these two jobs, the scheduling objective function is improved, thus contradicting the claim that the original schedule was optimal.

20.2.1.1 Average Weighted Completion Time: $1\|\sum w_j C_j$

In perhaps the simplest scheduling problem, our objective is to minimize the sum of completion times $\sum C_j$. Intuitively, it makes sense to schedule the largest job at the end of the schedule to ensure that it does not contribute to the delay on any other job. We formalize this by defining the shortest processing time (SPT) algorithm: order the jobs by nondecreasing processing time (breaking ties arbitrarily) and schedule in that order.

THEOREM 20.1 SPT *is an exact algorithm for* $1\|\sum C_j$.

PROOF To establish the optimality of the schedule constructed by SPT we use an interchange argument. Suppose for the purpose of contradiction that the jobs in the optimal schedule are not scheduled in nondecreasing order of completion time. Then there is some pair of jobs j and k such that j immediately precedes k in the schedule but $p_j > p_k$.

Suppose we exchange jobs j and k. All jobs other than j and k still start, and thus complete, at the same time as they did before the swap. All that changes is the completion times of jobs j and k. Suppose that originally job j started at time t and ended at time $t + p_j$, so that job k started at time $t + p_j$ and finished at time $t + p_j + p_k$. It follows that the original contribution of these two jobs to the sum of completion times, namely $(t + p_j) + (t + p_j + p_k) = 2t + 2p_j + p_k$, is replaced by their new contribution of $2t + 2p_k + p_j$. This gives a net decrease of $p_j - p_k$ in $\sum C_j$, which is positive if $p_j > p_k$, implying that our original ordering was not optimal—a contradiction.

This algorithm and its proof of optimality generalize to the optimization of average weighted completion time, $1\|\sum w_j C_j$. Intuitively, we would like to schedule as much weight as possible with each unit of processing time. This suggests scheduling jobs in nonincreasing order of w_j/p_j; the optimality of this rule can be established by a simple generalization of the previous interchange argument.

THEOREM 20.2   [39] *Scheduling jobs in nonincreasing order of w_j/p_j gives an optimal schedule for $1\|\sum w_j C_j$.*

20.2.1.2 Maximum Lateness: $1\|L_{\max}$

A simple greedy algorithm also solves $1\|L_{\max}$, in which we seek to minimize the maximum job lateness. A natural strategy is to schedule the job that is closest to being late, which suggests the earliest duedate (EDD) algorithm: order the jobs by nondecreasing due dates (breaking ties arbitrarily) and schedule in that order.

THEOREM 20.3   [23] EDD *is an exact algorithm for $1\|L_{\max}$.*

PROOF   We again use an interchange argument to prove that the schedule constructed by EDD is optimal. Assume without loss of generality that all due dates are distinct, and number the jobs so that $d_1 < d_2 < \cdots < d_n$. Among all optimal schedules, we consider the one with the fewest inversions, where an inversion is a pair of jobs j, k such that $j < k$ but k is scheduled before j. Suppose the given optimal schedule S is not the EDD schedule. Then there is a pair of jobs j and k such that $d_j < d_k$ but k immediately precedes j in the schedule.

Suppose we exchange jobs j and k. This does not change the completion time or lateness of any job other than j and k. We claim that we can only decrease $\max(L_j, L_k)$, so we do not increase the maximum lateness. Furthermore, since $j < k$, swapping jobs j and k decreases the number of inversions in the schedule. It follows that the new schedule has the same or better lateness than the original one but fewer inversions, a contradiction.

To prove the claim, note that in schedule S $C_j^S > C_k^S$ but $d_j < d_k$. It follows that $\max(L_j^S, L_k^S) = C_j^S - d_j$. Under the exchange, job j's completion time, and thus lateness, decreases. Job k's completion time rises to C_j^S, but this gives it a lateness of $C_j^S - d_k < C_j^S - d_j$. Thus, the maximum of the two latenesses has decreased.

20.2.1.3 Preemption and Release Dates

We now consider the more complex one-machine environment in which jobs may arrive over time, as modeled by the introduction of release dates. The greedy heuristics of Sections 20.2.1.1 and 20.2.1.2 are not immediately applicable, since jobs of high priority might be released relatively late and thus not be available for processing before jobs of lower priority. The most natural idea to cope with this complication is to always process the available (released) job of highest priority.

In a preemptive setting, this would mean, upon the release of a job of higher priority, preempting the currently running job and switching to the "better" job. We will show that this idea in fact yields optimal scheduling algorithms.

We thus define the shortest remaining processing time (SRPT) algorithm: at each point in time, schedule the job with SRPT, preempting when jobs of shorter processing time are released. We also generalize EDD : upon the release of jobs with earlier dues dates than the job currently being processed, preempt the current job and process the job with EDD.

THEOREM 20.4 [2,22] SRPT *is an exact algorithm for* $1|r_j, pmtn|\sum C_j$, *and* EDD *is an exact algorithm for* $1|r_j, pmtn|L_{max}$

PROOF As before, we argue by contradiction, using a similar greedy exchange argument. However, instead of exchanging entire jobs, we exchange pieces of jobs, which is now allowed in our preemptive environment.

We focus on $1|r_j, pmtn|\sum C_j$. Consider a schedule in which available job j with the SRPT is not being processed at time t, and instead available job k is being processed. Let p'_j and p'_k denote the remaining processing times for jobs j and k after time t, so $p'_j < p'_k$. In total, $p'_j + p'_k$ time is spent on jobs j and k after time t. We now perform an exchange. Take the first p'_j units of time that were devoted to either of jobs j and k after time t, and use them instead to process job j to completion. Then, take the remaining p'_k units of time that were spent processing jobs j and k, and use them to schedule job j. This exchange preserves feasibility since both jobs were released by time t.

In the new schedule, all jobs other than j and k have the same completion times as before. Job k finishes when job j originally finished. But job j, which needed $p'_j < p'_k$ additional work, finishes before job k originally finished. Thus we have reduced $C_j + C_k$ without increasing any other completion time, meaning we have reduced $\sum C_j$, a contradiction.

The argument that EDD solved $1|r_j, pmtn|L_{max}$ goes much the same way. If at time t, job j with the earliest remaining due date is not being processed and job k with a later due date is, we reallocate the time spent processing job k to job j. This makes job j finish earlier, and makes job k finish when job j did originally. This cannot increase objective function value.

By considering how SRPT and EDD function if all jobs are available at time 0, we conclude that on one machine, in the absence of release dates, the ability to preempt jobs does not yield schedules with improved $\sum C_j$ or L_{max} optimality criteria. This is not the case when jobs have release dates; intuitively, a problem such as $1|r_j|\sum C_j$ seems more difficult, as one cannot simply preempt the current job for a newly-arrived better one, but rather must decide whether to start a worse job or wait for the better one. This intuition about the additional difficulty of this setting is justified— $1|r_j|\sum C_j$ and $1|r_j|L_{max}$ are in fact $\mathcal{N}P$-complete problems. We discuss approximation algorithms for these problems in Sections 20.5.2 and 20.6.3.

We also note that these ideas have their limitations, and do not generalize to the $\sum w_j C_j$ criterion– $1|r_j, pmtn|\sum w_j C_j$ is $\mathcal{N}P$-hard. Finally, we note that SRPT and EDD are on-line algorithms—their decisions about which job to schedule currently do not require any information about which jobs are to be released in the future. See [38] for a comprehensive survey of online scheduling.

20.2.2   The Two-Machine Flow Shop

We now consider a more complex machine environment in which we want to minimize the makespan in a flow shop. In general, this problem is $\mathcal{N}P$-hard, even in the case of three machines. However, in the special case of the two-machine flow shop $F2\|C_{max}$, a priority-based ordering approach due to Johnson [24] yields an exact algorithm. We denote the operations of job j on the first and second machines as a pair (a_j, b_j). Intuitively, we want to get jobs done on the first machine as quickly as possible so as to minimize idleness on the second machine due to waiting for jobs from the first machine. This suggests using an SPT rule on the first machine. On the other hand, it would be useful to process the jobs with large b_j as early as possible on the second machine, while machine 1 is still running, so they will not create a large tail of processing on machine 2 after machine 1 is finished. This suggests some kind of longest processing time first (LPT) rule for machine 2.

We now formalize this intuition. We partition our jobs into two sets. A is the set of jobs j for which $a_j \leq b_j$, while B is the set for which $a_j > b_j$. We construct a schedule by first ordering all the jobs in A by nondecreasing a_j value, and then all the jobs in B by nonincreasing b_j values. We process jobs in this order on both machines. This is called Johnson's rule.

It may be surprising that we do not reorder jobs to process them on the second machine. It turns out that for two-machine flow shops, such reordering is not necessary. A schedule in which all jobs are processed in the same order is called a permutation schedule.

LEMMA 20.1   An instance of F2$\|C_{\max}$ always has an optimal schedule that is a permutation schedule.

Note that for three or more machines there is not necessarily an optimal permutation schedule.

PROOF   Consider any optimal schedule, and number the jobs according to the time at which they complete on machine 1. Suppose that job k immediately precedes job j in the order in which jobs are completed on machine 2, but $j < k$. Let t be the time at which job k is started on machine 2. It follows that job k has completed on machine 1 by time t. Numbering $j < k$ means that j is processed earlier than k on machine 1, so it follows that job j also has completed on machine 1 by time t. Therefore, we can swap the order of jobs j and k on machine 2, and still have a legal schedule (since no other job's start time changes) with the same makespan. We can continue performing such swaps until there are none left to be done, implying that jobs on machine 2 are processed in the same order as those on machine 1.

Having limited our search for optimal schedules to permutation schedules, we present a clever argument given by Lawler et al. [29] to establish the optimality of the permutation schedule specified by Johnson's rule.

Renumber the jobs according to the ordering given by Johnson's rule. Notice that in a permutation schedule for F2$\|C_{\max}$, there must be a job k that is started on machine 2 immediately after its completion on machine 1; for example, the job that starts immediately after the last idle time on machine 2. The makespan of the schedule is thus determined by the processing times of k jobs on machine 1 and $n - k + 1$ jobs on machine 2, which is just a sum of $n + 1$ processing times. If we reduce all the a_i and b_i by the same value p, then every sum of $n + 1$ processing times decreases by $(n + 1)p$, so the makespan of every permutation schedule is reduced by $(n + 1)p$.

Now note that if a job has $a_i = 0$ it is scheduled first in some optimal permutation schedule, since it delays no jobs on machine 1 and only "buys time" for jobs that are processed later than it on machine 2. Similarly, if a job has $b_i = 0$, it is scheduled last in some optimal schedule.

Therefore, we can construct an optimal permutation schedule by repeatedly finding the minimum operation size amongst all the a_j and b_j values of the unscheduled jobs, subtracting that value from all of the operation sizes, and then scheduling the job with the new zero processing time according to the above rules. Now observe that the schedule constructed is exactly the schedule that orders the jobs by Johnson's rule. We have therefore proved the following.

THEOREM 20.5 [24] *Johnson's rule yields an optimal schedule for* $F2\|C_{max}$.

20.2.3 Parallel Machines

We now turn to the case of parallel machines. In the move to parallel machines, many problems that are easily solvable on one machine become $\mathcal{NP}$-hard; the focus therefore tends to be on approximation algorithms. In some cases, the simple priority-based rules we used for one machine generalize well. That is, we assign a priority to every job, and, whenever a machine becomes available, it starts processing the job that has the highest remaining priority. The schedules created by such algorithms, which immediately give work to any machine that becomes idle, will be referred to as busy schedules.

In this section, we also introduce a new method of analysis. Instead of arguing correctness based on interchange arguments, we give lower bounds on the quality of the optimal schedule. We then show that our algorithm produces a schedule whose quality is within some factor of the lower bound, thus demonstrating a fortiori that it is within this factor of the optimal schedule. This is a general technique for approximation, and it has the pleasing feature that we are able to guarantee that we are within a certain factor of the optimal value, without knowing what that optimal value is. Sometimes we can show that our greedy algorithm achieves the lower bound, thus demonstrating that the algorithm is actually optimal.

In this section, we devote most of our attention to the problem of minimizing the makespan (schedule length) on m parallel machines, and study the behavior of the greedy algorithm for the problem. We remark that for the average-completion-time problem $P\|\sum C_j$, the greedy SPT algorithm also turns out to yield an optimal schedule. We discuss this further in Section 20.4.1.

As was mentioned in Section 20.2.1, $P\|C_{max}$ is trivial when $m = 1$, as any schedule with no idle time will be optimal. Once we have more than one machine, things become more complicated. With preemption, it is possible to greedily construct an optimal schedule in polynomial time. In the nonpreemptive setting, however, it is unlikely that there is a polynomial time exact algorithm, since the problem is $\mathcal{NP}$-complete via a simple reduction from the $\mathcal{NP}$-complete partition problem [7]. We will thus focus on finding an approximately optimal solution. First, we will show that any busy schedule gives a 2-approximation. We will then see how this can be improved with a slightly smarter algorithm, the LPT algorithm, which is a 4/3-approximation algorithm. In Section 20.6 we will show that a more complicated algorithm can guarantee an even better quality of approximation.

Our analyses of these algorithms are all based on comparing their performance to certain lower bounds on the quality of the optimal schedule; their performance compared to the optimum can only be better. Our algorithms will make use of two simple lower bounds on the makespan C^*_{max} of the optimal schedule:

$$C^*_{max} \geq \sum_{j=1}^{n} p_j/m \tag{20.1}$$

$$C^*_{max} \geq p_j \quad \text{for all jobs } j. \tag{20.2}$$

The first lower bound says that the schedule is at least as long as the average machine load, and the second says that the schedule is at least as long as the size of any job. To demonstrate the power of these lower bounds, we begin with the preemptive problem, $P|pmtn|C_{max}$. In this case, we show how to find a schedule that matches the maximum of the two lower bounds given above. We then use the lower bounds to establish performance guarantees for approximation algorithms for the nonpreemptive case.

20.2.3.1 Minimizing C_{max} with Preemptions

We give a simple algorithm, called McNaughton's wrap-around rule [32], that creates an optimal schedule for P|$pmtn$|C_{max} with at most $m - 1$ preemptions. This algorithm is different from many scheduling algorithms in that it creates the schedule machine by machine, rather than over time.

Observing that the lower bounds Equations 20.1 and 20.2 still apply to preemptive schedules, we will give a schedule of length $D = \max\{\sum_j p_j/m, \max_j p_j\}$. We order the jobs arbitrarily. Then we begin placing jobs on the machines, in order, filling machine i up until time D before starting machine $i + 1$. Thus, a job of length p_j may be split, assigned to the last t units of time of machine i and the first $p_j - t$ units of time on machine $i + 1$, for some t. It is now easy to verify that since there are no more than mD units to be processed, every job is scheduled, and because $D - t \geq p_j - t$ for any t, a job is scheduled on at most one machine at any time. Thus we have created an optimal preemptive schedule.

THEOREM 20.6   [32] *McNaughton's wrap-around rule gives an optimal schedule for* P|$pmtn$|C_{max}.

20.2.3.2 List Scheduling for P‖C_{max}

In contrast to P|$pmtn$|C_{max}, P‖C_{max} is $\mathcal{NP}$-hard. We consider the performance of the list scheduling (LS) algorithm, which is a generic greedy algorithm: whenever a machine becomes available, process any unprocessed job.

THEOREM 20.7   [11] LS *is a 2-approximation algorithm for* P‖C_{max}.

PROOF   Let j' be the last job to finish in the schedule constructed by LS and let $s_{j'}$ be the time that j' begins processing. C_{max} is therefore $s_{j'} + p_{j'}$. All machines must be busy up to time $s_{j'}$, since otherwise job j' could have been started earlier. The maximum amount of time that all machines can be busy is $\sum_{j=1}^{n} p_j/m$, and so we obtain that

$$C_{max} \leq s_{j'} + p_{j'}$$
$$\leq \sum_{j=1}^{n} p_j + p_{j'}$$
$$\leq C_{max}^* + C_{max}^* = 2C_{max}^*.$$

The last inequality comes from lower bounds Equations 20.1 and 20.2 above.

This algorithm can easily be implemented in $O(n + m)$ time. By a similar analysis, the algorithm guarantees an approximation of the same quality even if the jobs have release dates [14].

20.2.3.3 Longest Processing Time First for P‖C_{max}

It is useful to think of the analysis of LS in the following manner. Every job starts being processed before time $\sum_{j=1}^{n} p_j$, and hence the schedule length is no more than $\sum_{j=1}^{n} p_j$ plus the length of the longest job that is running at time $\sum_{j=1}^{n} p_j$.

This motivates the natural idea that it is good to run the longer jobs early in the schedule and the shorter jobs later. This is formalized in the LPT rule: sort jobs in nonincreasing order of processing time and list schedule in that order.

THEOREM 20.8   [12] LPT *is a 4/3-approximation algorithm for* $P\|C_{max}$.

PROOF   We start by simplifying the problem. Suppose that j', the last job to finish in our schedule, is not the last job to start. Remove all jobs that start after time $s_{j'}$. This does not affect the makespan of our schedule, since these jobs must have run on other machines. Furthermore, it can only decrease the optimal makespan for the modified instance. Thus, if we prove an approximation bound for this new instance, it applies a fortiori to our original instance.

We can therefore assume that the last job to finish is the last to start, namely the smallest job. In this case, by the analysis of Theorem 20.7 above, LPT returns a schedule of length no more than $C^*_{max} + p_{min}$. We now consider two cases:

Case 1: $p_{min} \leq C^*_{max}/3$. In this case $C^*_{max} + p_{min} \leq C^*_{max} + (1/3)C^*_{max} \leq (4/3)C^*_{max}$.

Case 2: $p_{min} > C^*_{max}/3$. In this case, all jobs have $p_j > C^*_{max}/3$, and hence in the optimal schedule there are at most 2 jobs per machine. Number the jobs in order of nonincreasing p_j. If $n \leq m$, then the optimal schedule trivially puts one job on each machine. We thus consider the remaining case with $m < n \leq 2m$. In this case, we claim that, for each $j = 1, \ldots, m$ the optimal schedule pairs job j with job $2m + 1 - j$ if $2m + 1 - j \leq n$ and places job j by itself otherwise. This can be shown to be optimal via a simple interchange argument. We finish the proof by observing that this is exactly the schedule that LPT would construct.

This algorithm needs to sort the jobs, and can be implemented in $O(m + n \log n)$ time. If we are willing to spend substantially more time, we can obtain a $(1 + \epsilon)$-approximation algorithm for any fixed $\epsilon > 0$; see Section 20.6.

20.2.3.4   List Scheduling for $P|prec|C_{max}$

Even when our input contains precedence constraints, LS is still a 2-approximation algorithm. Given a precedence relation *prec*, we say that a job is available at time t if all its predecessors have completed processing by time t. Recall that in LS, whenever a machine becomes idle, any available job is scheduled. Before giving the algorithm, we give one additional lower bound that is relevant to scheduling with precedence constraints. Let $j_{i_1}, j_{i_2}, \ldots, j_{i_k}$ be any set of jobs such that $j_{i_1} \prec j_{i_2} \prec \cdots \prec j_{i_k}$, then

$$C^*_{max} \geq \sum_{\ell=1}^{k} p_{i_\ell}. \tag{20.3}$$

In other words, the total processing time of any chain of jobs is a lower bound on the makespan.

THEOREM 20.9   [11] LS *is a 2-approximation algorithm for* $P|prec|C_{max}$.

PROOF   Let j_1 be the last job to finish. Define j_2 to be the latest-finishing predecessor of j_1, and inductively define $j_{\ell+1}$ to be the latest-finishing predecessor of j_ℓ, continuing until reaching j_k, a job with no predecessors. Let $C = \{j_1, \ldots, j_k\}$. We partition time into two sets, A, the points in time when a job in C is running, and B, the remaining time. Observe that during all times in B, all machines must be busy, for if they were not, there would be a job from C that had all its predecessors completed and would be ready to run. Hence, $C_{max} \leq |A| + |B| \leq \sum_{j \in C} p_j + \sum_{j=1}^{n} p_j \leq 2C^*_{max}$, where the last inequality follows by applying lower bounds Equations 20.1 and 20.3. Note that $|A|$ is the total length of intervals in A.

For the case when all processing times are exactly one, P|$prec$|C_{max} is solvable in polynomial time if there are only two machines [27], and is $\mathcal{N}P$-complete if there are an arbitrary number of machines [40]. The complexity of the problem in the case when there are a fixed constant number of machines is one of the more famous open problems in scheduling.

20.2.3.5 List Scheduling for O‖C_{max}

LS can also be applied to O‖C_{max}. Recall that in this problem, each job must be processed for disjoint intervals of time on several different machines. By an analysis similar to that used for P‖C_{max}, we will show that any algorithm that constructs a busy schedule for O‖C_{max} is a 2-approximation algorithm. Let P_{max} be the maximum total processing time, summed over all machines, for any one job, and let Π_{max} be the maximum total processing time, summed over all jobs, of any one machine. Clearly, both P_{max} and Π_{max} are lower bounds on the makespan of the optimal schedule. We show that any busy schedule has makespan at most $P_{max} + \Pi_{max}$.

To see this, consider the machine M' that finishes processing last, and consider the last job j' to finish on machine M'. At any time during the schedule, either M' is processing a job or job j' is being processed (if neither of these is true, then LS would require that j' be running on M', a contradiction). However, the total length of time during which j' undergoes processing is at most P_{max}. During all the remaining time in the schedule, machine M' must be busy. But machine M' is busy for at most Π_{max} time units. Thus the total length of the schedule is at most $P_{max} + \Pi_{max}$, as claimed. Since $P_{max} + \Pi_{max} \leq C^*_{max} + C^*_{max} = 2C^*_{max}$, we obtain

THEOREM 20.10 **(Racsmány, see [3])** *LS is a 2-approximation algorithm for* O‖C_{max}.

20.2.4 Limitations of Priority Rules

For many problems, simple scheduling rules do not yield good schedules, and thus given a scheduling problem, the algorithm designer should be careful about applying one of these rules without justification. In particular, for many problems, particularly those with precedence constraints and release dates, the optimal schedule has unforced idle time. That is, if one is constructing the schedule over time, there may be a time t when there is an idle machine m and an available job j, but scheduling job j on machine m at time t will yield a suboptimal schedule.

Consider the problem Q‖C_{max} and recall that for P‖C_{max}, LS is a 2-approximation algorithm. Consider a two-job two-machine instance in which $s_1 = 1, s_2 = x, p_1 = 1, p_2 = 1$, and $x > 2$. Then LS, SPT, and LPT all schedule one job on machine 1 and one on machine 2, and the makespan is thus 1. However, the schedule that places both jobs on machine 2 has makespan $2/x < 1$. By making x arbitrarily large, we see that none of these simple algorithms, which all have approximation ratio at least $x/2$, have bounded approximation ratios.

For this problem there is actually a simple heuristic that comes within a factor of 2 of optimal, but for some problems, such as Q|$prec$|C_{max} and R‖C_{max}, there is no simple algorithm known that comes anywhere close to optimal. We also note that even though LS is a 2-approximation for O‖C_{max}, for F‖C_{max} busy schedules can be of makespan $\Omega(m)$ times optimal [10].

20.3 Sophisticated Greedy Approaches

As we have just argued, for many problems, the priority algorithms that consider jobs in isolation, as in Section 20.2, are not sufficient. In this section, we consider algorithms that do more than sort jobs by some priority measure—they take other jobs into account when making a decision about where to schedule a job. The algorithms we study here are "incremental" in nature: they start with an empty solution and grow it, one job at a time, until the optimal solution is revealed. At each step the decision about which job to add to the growing solution is made greedily, but is based on the current context of jobs which have already been scheduled. We present two examples which are classic examples of the dynamic programming paradigm, and several others that are more specialized.

All the algorithms share an analysis based on the idea of optimal substructure. Namely, if we consider the optimal solution to a problem, we can often argue that its "subparts" (e.g., prefixes of the optimal schedule) are optimal solutions to "subproblems" (e.g., the problem of scheduling the set of jobs in that prefix). This lets us argue that as our algorithms build their solution incrementally, they are building optimal solutions to bigger and bigger subproblems of the original problem, until they reach an optimal solution to the entire problem.

20.3.1 An Incremental Greedy Algorithm for $1\|f_{max}$

The first problem we consider is $1\|f_{max}$, which was defined in Section 20.1. In this problem, each job has some nondecreasing penalty function on its completion time C_j, and the goal is to find a schedule minimizing the maximum $f_j(C_j)$. As one example, $1\|L_{max}$ is captured by setting $f_j(t) = t - d_j$.

A greedy strategy still applies, when suitably modified. It is convenient, instead of talking about scheduling the "most penalizing" (e.g., EDD) job first, to talk about scheduling the "least penalizing" (e.g., latest due date) job last. Let $p(\mathcal{J}) = \sum_{j \in \mathcal{J}} p_j$ be the total processing time of the entire set of jobs. Note that some job must complete at time $p(\mathcal{J})$. We find the job j that minimizes $f_j(p(\mathcal{J}))$, and schedule this job last. We then (recursively) schedule all the remaining jobs before j so as to minimize their maximum penalty. We call this algorithm Least-Cost-Last.

Observe the difference between this and our previous scheduling rules. In this new scheme, we cannot determine the best job to schedule second-to-last until we know which job is scheduled last (we need to know the processing time of the last job in order to know the processing time of the recursive subproblem). Thus, instead of a simple $O(n \log n)$-time sorting algorithm based on absolute priorities, we are faced with an algorithm that inspects k jobs in order to identify the job to be scheduled kth, giving a total running time of $O(n + (n - 1) + \cdots + 1) = O(n^2)$.

This change in algorithm is matched by a change in analysis. Since the notion of which job is worst can change as the schedule is constructed, there is no obvious fixed priority to which we can apply a local exchange argument. Instead, as with $P|pmtn|C_{max}$ in Section 20.2.3.1, we show that our algorithm's greedy decisions are in agreement with a provable lower bound on the quality of the optimal schedule. Our algorithm produces a schedule that matches the lower bound and must therefore be optimal.

Let $f_{max}^*(S)$ denote the optimal value of the objective function if we are only scheduling the jobs in S. Consider the following two facts about f_{max}^*:

$$f_{max}^*(\mathcal{J}) \geq \min_{j \in N} f_j(p(\mathcal{J}))$$

$$f_{max}^*(\mathcal{J}) \geq f_{max}^*(\mathcal{J} - \{j\})$$

The first of these statements follows from the fact that some job must be scheduled last. The second follows from the fact that if we have an optimal schedule for $\mathcal{J}$ and remove a job from the schedule, then we do not increase the completion time of any job. Therefore, since the f_j are increasing functions, we do not increase any penalty.

We use these inequalities to prove by induction that our schedule is optimal. According to our scheduling rule, we schedule last the job j minimizing $f_j(p(\mathcal{J}))$. By induction, this gives us a schedule with objective $\max\{f_j(p(\mathcal{J})), f_{max}^*(\mathcal{J} - \{j\})\}$. But since each of these quantities is (by the equations above) a lower bound on the optimal $f_{max}^*(\mathcal{J})$, we see that in fact we obtain a schedule whose value is a lower bound on $f_{max}^*(\mathcal{J})$, and thus must in fact equal $f_{max}^*(\mathcal{J})$.

20.3.1.1 Extension to 1|*prec*|f_{max}

Our argument from Section 20.3.1 continues to apply even if we introduce precedence constraints. In the 1|*prec*|f_{max} problem, a partial order on jobs is given, and we must build a schedule that does not start a job until all jobs preceding it in the partial order have completed. Our above algorithm applies essentially unchanged to this case. Note that the last job in the schedule must be a job with no successors. We therefore build an optimal schedule by scheduling last the job j that, among jobs with no successors, minimizes $f_j(P(\mathcal{J}))$. We then recursively schedule all other jobs before it. The proof of optimality goes exactly as before, using the fact that if L is the set of all jobs without successors, then

$$f^*_{max}(\mathcal{J}) \geq \min_{j \in L} f_j(P(\mathcal{J}))$$

This is the same as the first equation above, except that the minimum is taken only over jobs without successors. The remainder of the proof proceeds unchanged.

THEOREM 20.11 [26] `Least-Cost-Last` *is an exact algorithm for* 1|*prec*|f_{max}.

It should also be noted that, once again, the fact that our algorithm is greedy makes preemption a moot point. One job needs to finish last, and it immediately follows that we can do no better than executing all of that job last. Thus, our greedy algorithm continues to be optimal.

20.3.1.2 An Alternative Approach

Moore [33] gave a different approach to 1‖f_{max} that may be faster in some cases. His scheme is based on a reduction to the maximum lateness problem and its solution by the EDD rule. To see how an algorithm for L_{max} can be applied to 1‖f_{max}, suppose we want to know whether there is a schedule with $f_{max} \leq B$. We can decide this as follows. Give each job j a deadline d_j equal to the maximum t for which $f_j(t) \leq B$. It is easy to see that a schedule has $f_{max} \leq B$ precisely when every job finishes by its specified deadline, i.e., $L_{max} \leq 0$. Thus, we have converted the feasibility problem for f_{max} into an instance of the lateness problem. The optimization problem may therefore be solved by a binary search for the correct value of B.

20.3.2 Dynamic Programming for 1‖ $\sum w_j U_j$

We now consider 1‖$\sum w_j U_j$ problem, in which the goal is to minimize the total weight of late jobs. This problem is weakly $\mathcal{NP}$-complete. That is, although it is $\mathcal{NP}$-complete, for integral weights it is possible to solve the problem exactly in $O(n \sum w_j)$ time, which is polynomial if the w_j are bounded by a polynomial. The necessary algorithm is a classical dynamic program that builds the solution out of solutions to smaller problems (a detailed introduction to dynamic programming can be found in many algorithms textbooks, see, for example [6]). This $O(n \sum w_j)$ dynamic programming algorithm has several consequences. First, it immediately yields an $O(n^2)$-time algorithm for 1‖$\sum U_j$ problem–just take all weights to be 1. Furthermore, we will show in Section 20.6 that this algorithm can be used to derive a fully polynomial approximation scheme for the general problem that finds a schedule with $\sum w_j U_j$ within $(1 + \epsilon)$ of the optimum in time polynomial in $1/\epsilon$ and n.

The first observation to make is that under this objective, a schedule partitions the jobs into two types: those completed by their due dates, and those not completed. Clearly, we might as well process all the jobs that meet their due date before processing any that do not. Furthermore, the processing order of these jobs might as well be determined using the EDD rule from Section 20.2.1.2 when all jobs can be completed by their due date (implying nonpositive maximum lateness), EDD, which minimizes maximum lateness, will clearly find a schedule that does so.

It is therefore convenient to discuss feasible subsets of jobs that can all be scheduled together to complete by their due dates. The question of finding a minimum weight set of late jobs can then be equivalently restated as finding a maximum weight feasible subset of the jobs.

To solve this problem, we aim to solve a harder one: namely, to identify the fastest-completing maximum weight feasible subset. We do so via dynamic programming. Order the jobs according to increasing due date. Let T_{wj} denote the minimum completion time of a weight w-or-greater feasible subset of $1, \ldots, j$, or ∞ if there is no such subset. Note that $T_{0j} = 0$, while $T_{w0} = \infty$ for all $w > 0$. We now give a dynamic program to compute the other values T_{wj}. Consider the fastest completing weight w-or-greater feasible subset S of $\{1, \ldots, j+1\}$. Either $j + 1 \in S$ or it is not. If $j + 1 \notin S$, then $S \subseteq \{1, \ldots, j\}$ and is then clearly the fastest completing weight w-or-greater subset of $\{1, \ldots, j\}$, so S completes in time T_{wj}. If $j + 1 \in S$, then since we can schedule feasible subsets using EDD , $j + 1$ can be scheduled last. The jobs preceding it have weight at least $w - w_{j+1}$, and clearly form the minimum-completion-time subset of $1, \ldots, j$ with this weight. Thus, the completion time of this feasible set is $T_{w-w_{j+1},j} + p_{j+1}$. It follows that

$$T_{w,j+1} = \begin{cases} \min \left(T_{wj}, T_{w-w_{j+1}} + p_{j+1} \right) & \text{if } T_{w-w_{j+1},j} + p_j \le d_{j+1} \\ T_{wj} & \text{otherwise} \end{cases}$$

Now observe that there is clearly no feasible subset of weight exceeding $\sum w_j$, so we can stop our dynamic program once we reach this value of w. This takes $O(n \sum w_j)$ time. Once we have all the values T_{wj}, we can find the maximum weight feasible subset by identifying the largest value of w for which some T_{wj} (and thus T_{wn}) is finite.

This gives a standard $O(n \sum w_j)$ time dynamic program for computing T_{wn} for every relevant value w; the maximum w for which T_{wn} is finite is the maximum total weight of jobs that can be completed by their due date.

THEOREM 20.12   [30] *Dynamic programming yields an $O(n \sum w_j)$-time algorithm for exactly solving $1 \| \sum w_j U_j$.*

We remark that a similar dynamic program can be used to solve the problem in time $O(n \sum p_j)$, which is effective when the processing times are polynomially bounded integers. We also note that a quite simple greedy algorithm due to Moore [33] can solve the unweighted $1 \| \sum U_j$ problem in $O(n \log n)$ time.

20.3.3   Dynamic Programming for $P \| C_{\max}$

For a second example of the applicability of dynamic programming, we return to the $\mathcal{N}P$-hard problem $P \| C_{\max}$, and focus on a special case that is solvable in polynomial time—the case in which the number of different job processing times is bounded by a constant. While this special case might appear to be somewhat contrived, in Section 20.6 we will show that it can form the core of a polynomial approximation scheme for $P \| C_{\max}$.

LEMMA 20.2   [18] *Given an instance of $P \| C_{\max}$ in which the p_j take on at most s distinct values, there exists an algorithm which finds an optimal solution in time $n^{O(s)}$.*

PROOF   Assume for now that we are given a target schedule length T. We again use dynamic programming. Let the different processing times be $z_1, \ldots, z_s$. The key observation is that the set of jobs on a machine can be described by an s-dimensional vector $V = (v_1, \ldots, v_s)$, where v_k is the number of jobs of length z_k. There are at most n^s such vectors since each entry has value at most n. Let $\mathcal{V}$ be the set of all such vectors whose total processing time (that is, $\sum v_i z_i$) is less than T. In the optimal schedule, every machine is assigned a set of jobs corresponding to a vector from this set. We now define $M(x_1, \ldots, x_s)$ to be the minimum number of machines needed to schedule a job set consisting of x_i jobs of size z_i, for $i = 1, \ldots, s$. We observe that

$$M(x_1, \ldots, x_s) = 1 + \min_{v \in \mathcal{V}} M(x_1 - v_1, \ldots, x_s - v_s).$$

The minimization is over all possible vectors that could be processed by the "first" machine counted by the quantity 1, and the recursive expression denotes the best way to process the remaining work. Thus we need to compute a table with n^s entries, where each entry depends on $O(n^s)$ other entries, and therefore the computation takes time $O(n^{2s})$.

It remains to handle the assumption that we know T. The easiest way to do this is to perform a binary search on all possible values of T. A slightly more sophisticated approach is to search only over the $O(n^s)$ makespans of vectors describing sets of jobs, as one of these clearly determines the makespan of the solution.

20.4 Matching and Linear Programming

Networks and linear programs are central themes in combinatorial optimization, and are useful tools in the solution of many problems. Therefore, it is not surprising that these techniques can be applied profitably to scheduling problems as well. In this section, we discuss applications of bipartite matching and linear programming to the exact solution of certain scheduling problems; in Section 20.5 we will revisit both techniques in the design of approximation algorithms for $\mathcal{NP}$-hard problems.

20.4.1 Applications of Matching

Given a bipartite graph on two sets of vertices A and B and an edge set $E \subseteq A \times B$, a matching M is a subset of the edges, such that each vertex A and B is an endpoint of at most one edge of M. A natural matching that is useful in scheduling problems is one that matches jobs to machines; the matching constraints force each job to be scheduled on at most one machine, and each machine to be processing at most one job. If A has no more vertices than B, we call a matching perfect if every vertex of A is in some matching edge. It is also possible to assign weights to the edges, and define the weight of a matching to be the sum of the weights of the matching edges. The key fact that we use in this section is that minimum-weight perfect matchings can be computed in polynomial time (see, e.g., [1]).

20.4.1.1 Matching to Schedule Positions for R$\| \sum C_j$

In this section we give a polynomial-time algorithm for R$\| \sum C_j$ that matches jobs to positions in the schedule on each machine. For any schedule, let κ_{ik} be the kth-from-last job to run on machine i, and let ℓ_i be the number of jobs that run on machine i. By observing that the completion time of a job is equal to the sum of the processing times of the jobs that run before it, we have that

$$\sum_j C_j = \sum_{i=1}^{m} \sum_{k=1}^{\ell_i} C_{\kappa_{ik}} = \sum_{i=1}^{m} \sum_{k=1}^{\ell_i} \sum_{x=k}^{\ell_i} p_{i,\kappa_{xi}} = \sum_{i=1}^{m} \sum_{k=1}^{\ell_i} k p_{i,\kappa_{ki}}. \tag{20.4}$$

From this, we see that the kth from last job to run on a machine contributes exactly k times its processing time to the sum of completion times. Based on this observation, Horn [21] and

Bruno et al. [4] proposed formulating $R\|\sum C_j$ problem as a minimum-weight bipartite matching problem. We define a bipartite graph $G = (V, E)$ with $V = A \cup B$ as follows. A will contain n vertices v_j, one for each of the n jobs $j = 1, \ldots, n$. B will contain nm nodes w_{ik}, where vertex w_{ik} represents the kth-from-last position on machine i, for $i = 1, \ldots, m$ and $k = 1, \ldots, n$. We include in E an edge (v_j, w_{ik}) between every node in A and every node in B. Using Equation 20.4 we define the weights on the edges from A to B as follows: edge (v_j, w_{ik}) is assigned weight kp_{ij}.

We now argue that a minimum-weight perfect matching in this graph corresponds to an optimal schedule. First, note that for each valid schedule there is a perfect matching in G. Not every perfect matching in G corresponds to a schedule, since a job might be assigned to the kth from last position while less than k jobs are assigned to that machine; however, such a perfect matching is clearly not of minimal weight—a better matching can be obtained by pushing the $k' < k$ jobs assigned to that machine into the k' from last slots. Therefore, a schedule of minimum total completion time corresponds to a minimum-weight perfect matching in the bipartite graph.

THEOREM 20.13   [4,21] *There is a polynomial-time algorithm for* $R\|\sum C_j$.

In the special case of parallel identical machines, it remains true that the kth-from-last job to run on a machine contributes exactly k times its processing time to the sum of completion times. Since in this case the processing time of each job is the same on any machine, the algorithm is clear: schedule the m largest jobs last on each machine, schedule the next m largest jobs next to last, etc. The schedule constructed is exactly that constructed by the SPT algorithm.

COROLLARY 20.1   [5] SPT *is an exact algorithm for* $P\|\sum C_j$.

20.4.1.2   Matching Jobs to Machines: $O|pmtn|C_{\max}$

For our second example of the utility of matching, we give an algorithm for $O|pmtn|C_{\max}$ due to Gonzalez and Sahni [9]. This algorithm will not find just one matching, but rather a sequence of matchings, each of which will correspond to a partial schedule, and then concatenate all of these partial schedules together. Recall from our discussion of $O\|C_{\max}$ in Section 20.2 that two lower bounds on the makespan of a nonpreemptive schedule are the maximum machine load $\Pi_{\max}$ and the maximum job size $P_{\max}$. Both of these remain lower bounds when preemption is allowed. In the nonpreemptive setting, a simple greedy algorithm gives a schedule with makespan bounded by $P_{\max} + \Pi_{\max}$. We now show that when preemption is allowed, matching can be used to achieve a makespan equal to $\max(P_{\max}, \Pi_{\max})$.

The intuition behind the algorithm is the following. Consider the schedule at any point in time. At this time, each machine is processing at most one job. In other words, the schedule at each point in time defines a matching between jobs and machines. We aim to find a matching that forms a part of the optimal schedule, and process jobs according to it for some time. Our goal is that processing the matched jobs on their matched machines for some amount of time t, and adjusting $P_{\max}$ and $\Pi_{\max}$ to reflect the decreased remaining processing requirements, should reduce $\max(P_{\max}, \Pi_{\max})$ by t. It follows that if we repeat this process for a total amount of time equal to $\max(P_{\max}, \Pi_{\max})$, we will reduce $\max(P_{\max}, \Pi_{\max})$ to 0, implying that there is no work remaining in the system.

What properties should our matching of jobs to machines have? Recall that our goal is to reduce our lower bound. Call a job tight if its total processing cost is $P_{\max}$. Call a machine tight if its total load is $\Pi_{\max}$. Clearly, it is necessary that every tight job undergo processing in our matching, since otherwise we will fail to subtract t from $P_{\max}$. Similarly, it is necessary that every tight machine be in the matching in order to ensure that we reduce $\Pi_{\max}$ by t. Lastly, we can only execute the matching for t time if every job–machine pair in the matching actually requires t units of processing. In other words, we are seeking a matching in which every tight machine and job is matched, and each matching edge requires positive processing time. Such a matching is referred to as a decrementing set. That it always exists is a nontrivial fact (about stochastic matrices) whose proof is beyond the scope of this survey; we refer the reader to Lawler and Labetoulle's presentation of this algorithm [28].

To find a decrementing set, we construct a (bipartite) graph with a node representing each job and machine, and include an edge between machine node i and job node j if job j requires a nonzero amount of processing on machine i. In this graph we require a matching that matches each tight machine or job node; this can easily be found with a variant of traditional matching algorithms. Note that we must include the nontight nodes in the matching problem, since tight nodes can be matched to them.

Once we have found our decrementing set via matching, we have machines execute the jobs matched to them until one of the matched jobs completes its work on its machine, or until a new job or machine becomes tight (this can happen because some jobs and machines are not being processed in the matching). Whenever this happens, we find a new decrementing set. For simplicity, we assume that $P_{max} = \Pi_{max}$; this can easily be arranged by adding dummy operations, which can only make our task harder. Since our decrementing set includes every tight job and machine, it follows that executing for time t will reduce both P_{max} and Π_{max} by t. It follows that after $P_{max} = \Pi_{max}$ time, both quantities will be reduced to 0. Clearly this means that we are done in time equal to the lower bound.

One might worry that the number of decrementing set calculations we must perform could be nonpolynomially bounded, making our approximation algorithm too slow. But this turns out not to happen. We only compute a new decrementing set when a job or machine finishes or when a new job or machine becomes tight. Each job–processor pair can finish only once, meaning that this occurs only nm times during our schedule. Also, each job or machine stays tight forever after it becomes tight; thus, new tight jobs and machines only occur $n + m$ times. Thus, constructing our schedule of optimal length requires only $mn + m + n$ matching computations.

THEOREM 20.14   [9] *There is a polynomial-time algorithm for* $O|pmtn|C_{max}$, *that finds an (optimal) schedule of makespan* $\max(P_{max}, \Pi_{max})$.

20.4.2 Linear Programming

We now discuss the application of linear programming to the design of scheduling algorithms. A linear program is given by a vector of variables $x = (x_1, \ldots, x_n)$, a set of m linear constraints of the form $a_{i1}x_1 + a_{i2}x_2 + \cdots + a_{in}x_n \leq b_i, 1 \leq i \leq m$, and a cost vector $c = (c_1, \ldots, c_n)$; the goal is to find an x that satisfies these constraints and that minimizes $cx = c_1x_1 + c_2x_2 + \cdots + c_nx_n$. Alternatively but equivalently, some of the inequality constraints might be given as equalities, and/or we may have no objective function and desire simply to find a feasible solution to the set of constraints. Many optimization problems can be formulated as linear programs, and thus solved efficiently, since a linear program can be solved in polynomial time [25].

In this section we consider $R|pmtn|C_{max}$. To model this problem as a linear program, we use nm variables $x_{ij}, 1 \leq i \leq m, 1 \leq j \leq n$. Variable x_{ij} denotes the fraction of job j that is processed on machine i; for example, we would interpret a linear-programming solution with $x_{1j} = x_{2j} = x_{3j} = \frac{1}{3}$ as assigning $\frac{1}{3}$ of job j to machine 1, $\frac{1}{3}$ to machine 2 and $\frac{1}{3}$ to machine 3.

We now consider what sorts of linear constraints on the x_{ij} are necessary to ensure that they describe a valid solution to an instance of $R|pmtn|C_{max}$. Clearly the fraction of a job assigned to any machine must be nonnegative, so we will create nm constraints

$$x_{ij} \geq 0.$$

In any schedule, we must fully process each job. We capture this requirement with the n constraints:

$$\sum_{i=1}^{m} x_{ij} = 1, \quad 1 \leq j \leq n.$$

Note that, along with the previous constraints, these constraints imply that $x_{ij} \leq 1 \; \forall \, i, j$.

Our objective, of course, is to minimize the makespan D of the schedule. Recall that the amount of processing that job j would require, if run entirely on machine i, is p_{ij}. Therefore, for a set of fractional assignments x_{ij}, we can determine the amount of time machine i will work: it is just $\sum x_{ij} p_{ij}$, which must be at most D. We model this with the m constraints

$$\sum_{j=1}^{n} p_{ij} x_{ij} \leq D, \quad \text{for } i = 1, \ldots, m.$$

Finally, we must ensure that no job is processed for more than D time; we model this with the n constraints

$$\sum_{i=1}^{m} x_{ij} p_{ij} \leq D, \quad 1 \leq j \leq n.$$

To summarize, we formulate the problem as the following linear program:

$$\min \; D \tag{20.5}$$

$$\sum_{i=1}^{m} x_{ij} = 1, \quad \text{for } j = 1, \ldots, n, \tag{20.6}$$

$$\sum_{j=1}^{n} p_{ij} x_{ij} \leq D, \quad \text{for } i = 1, \ldots, m, \tag{20.7}$$

$$\sum_{i=1}^{n} p_{ij} x_{ij} \leq D, \quad \text{for } j = 1, \ldots, n, \tag{20.8}$$

$$x_{ij} \geq 0, \quad \text{for } i = 1, \ldots, m, \; j = 1, \ldots, n. \tag{20.9}$$

It is clear that any feasible schedule for our problem yields an assignment of values to the x_{ij} that satisfies the constraints of our above linear program. However, it is not completely clear that solving the linear program yields a solution to the scheduling problem; this linear program does not specify the ordering of the jobs on a specific machine, but simply assigns the jobs to machines while constraining the maximum load on any machine. It thus fails to explicitly require that a job not be processed simultaneously on more than one machine.

Interestingly enough, we can resolve this difficulty with an application of open shop scheduling. We define an open shop problem by creating an operation o_{ij} for each positive variable x_{ij}, and define the size of o_{ij} to be $x_{ij} p_{ij}$. We then find an optimal preemptive schedule for this instance, using the matching-based algorithm discussed in Section 20.4 We know that both the maximum machine load and maximum job size of this open shop instance are bounded above by D, and therefore the makespan of the resulting open shop schedule is at most D. If we now reinterpret the operations of each job in the open shop schedule as fragments of the original job in the unrelated machines instance, we see that we have given a preemptive schedule of length D in which no two fragments of a job are scheduled simultaneously.

We thus have established the following.

THEOREM 20.15 [28] *There is an exact algorithm for* $R|pmtn|C_{\max}$.

We will see further applications of linear programming to the development of approximation algorithms for $\mathcal{N}P$-hard scheduling problems in Section 20.5.

20.5 Using Relaxations to Design Approximation Algorithms

We now turn exclusively to the design of approximation algorithms for $\mathcal{N}P$-hard scheduling problems. Recall that a ρ-approximation algorithm is one that is guaranteed to find a solution with value within a multiplicative factor of ρ of the optimum. Many of the approximation algorithms in this area are based on a relaxation of the $\mathcal{N}P$-hard problem. A relaxation of a problem is a version of the problem with some of the requirements or constraints removed ("relaxed"). For example, we might consider $1|r_j, pmtn|\sum C_j$ to be a relaxation of $1|r_j|\sum C_j$ in which the "no preemption" constraint has been relaxed. A second example of a relaxation might be a version of the problem in which we relax the constraint that a machine processes at most one job at a time; a solution to this relaxation may have several jobs scheduled at one time on the same machine.

A solution to the original problem is a solution to the relaxation, but a solution to the relaxation is not necessarily a solution to the original problem. This is clearly illustrated by our nonpreemptive/preemptive example—a nonpreemptive schedule is a legal solution to a preemptive problem, although perhaps not an optimal one, but the converse is not true. It follows that in the case of a minimization problem, the value of the optimal solution to the relaxation is a not-necessarily-tight lower bound on the optimal solution to the original problem.

An idea that has proven quite useful is to define first a relaxation of the problem which can be solved in polynomial time, and then to give an algorithm to convert the relaxation's solution into a valid solution to the original problem, with some degradation in the quality of solution. The key to making this work well is to find a relaxation that preserves enough of the structure of the original problem to make the optimal relaxed solution "similar" to the original optimum, so that the relaxed solution does not degrade too much when converted to a valid solution.

In this section we discuss two sorts of relaxations of scheduling problems and their use in the design of approximation algorithms, namely the preemptive version of a nonpreemptive problem and a linear-programming relaxation of a problem.

There are generally two different ways to infer a valid schedule from the relaxed solution: one is to infer an assignment of jobs to machines while the other is to infer a job ordering. We give examples of both methods.

Before going any further, we introduce the notion of a relaxed decision procedure (RDP), which we will use both in Section 20.5.1 and later in Section 20.6. A ρ-RDP for a minimization problem accepts as input a target value T, and returns either **no,** asserting that no solution of value $\leq T$ exists, or returns a solution of value at most ρT. A polynomial-time ρ-RDP can easily be converted into a ρ-approximation algorithm for the problem via binary search for the optimum T; see [18,19] for more details. This simple idea is quite useful, as it essentially lets us assume that we know the value T of an optimal solution to a problem. (Note that this is a different use of the word relax than the term "relaxation.")

20.5.1 Rounding a Fractional Assignment to Machines: R∥C_{max}

In this section we give a 2-RDP for R∥C_{max}. Recall the linear program that we used in giving an algorithm for R|$pmtn$|C_{max}. If, instead of the constraints $x_{ij} \geq 0$, we could constrain the x_{ij} to be 0 or 1, the solution would constitute a valid nonpreemptive schedule. Furthermore, note that these integer constraints combined with the constraints 20.7 make the constraints 20.8 unnecessary (if a job is assigned integrally to a machine, constraint 20.7 ensures that is a fast enough machine, thus satisfying constraint 20.8 for that job). In other words, the following formulation has a feasible solution if and only if there is a nonpreemptive schedule of makespan D.

$$\sum_{i=1}^{m} x_{ij} = 1, \quad \text{for } j = 1, \ldots, n, \tag{20.10}$$

$$\sum_{j=1}^{n} p_{ij} x_{ij} \leq D, \quad \text{for } i = 1, \ldots, m, \tag{20.11}$$

$$x_{ij} \in \{0, 1\}, \quad \text{for } i = 1, \ldots, m, \, j = 1, \ldots, n. \tag{20.12}$$

This is an example of an integer linear program, in which the variables are constrained to be integers. Unfortunately, in contrast to linear programming, finding a solution to an integer linear program is $\mathcal{N}P$-complete. However, this integer programming formulation will still be useful. A very common method for obtaining a relaxation of an optimization problem is to formulate it as an integer linear program, and then to relax the integrality constraints. One obtains a fractional solution and then rounds the fractions to integers in a fashion that (hopefully) does not degrade the solution too dramatically.

In our setting, we relax the constraints 20.12 to $x_{ij} \geq 0$. We will also add an additional set of constraints that will ensure that the fractional solutions to this linear program have enough structure to be useful for approximation. Specifically, we disallow any part of a job j being processed on a machine i on which it could not complete in D time in a nonpreemptive schedule. Specifically, we include the following constraints:

$$x_{ij} = 0, \quad \text{if } p_{ij} \geq D. \tag{20.13}$$

(In fact, instead of adding constraints, we can simply remove such variables from the linear program.) As argued above, this constraint is actually implicit in the integer program given by the constraints 20.10 through 20.12, but was no longer guaranteed when we relaxed the integer constraints. Our new constraints can be seen as a replacement for the constraints 20.8 that we did not need in the integer formulation. Note also that these new constraints are only linear constraints when D is fixed. This is why we use an RDP instead of taking the more obvious approach of writing a linear program to minimize D.

To recap, constraints 20.10, 20.11, and 20.13 along with $x_{ij} \geq 0$ constitute a linear-programming relaxation of R∥C_{max}. Our RDP attempts solve this relaxation, obtaining a solution $\bar{x}_{ij}, 1 \leq i \leq m$, $1 \leq j \leq n$. If there is no feasible solution, our RDP can output **no**—nonpreemptive schedule has makespan D or less. If the linear program is feasible, we will give a way to derive an integral assignment of jobs to machines from the fractional solution. Our job is made much easier by the fact, which we cite from the theory of linear programming, that we can find a so-called basic solution of this linear program that has at most $n + m$ positive variables. Since these $n + m$ positive variables must be distributed amongst n jobs, there are at most m jobs that are assigned in a fractional fashion to more than one machine.

We may now state our rounding procedure. For each (machine, job) pair (i, j) such that $\bar{x}_{ij} = 1$, we assign job j to machine i. We call the schedule of these jobs S_1. For the remaining at most m jobs, we simply construct a matching of the jobs to machines such that each job is matched to a machine it is already partially assigned to. We schedule each job on the machine to which it is matched, and call the schedule of these jobs S_2.

We defer momentarily the question of whether such a matching exists, and analyze the makespan of the resulting schedule, which is at most the sum of the makespans of S_1 and S_2. Since the x_{ij} form a feasible solution to the relaxed linear program, the makespan of S_1 is at most D. Since S_2 schedules at most one job per machine, and assigns j to i only if $x_{ij} > 0$, meaning $p_{ij} < D$, the makespan of S_2 is at most D (this argument is the reason we had to add constraint 20.13 to our linear program). Thus the overall schedule has length at most $2D$.

The argument that a matching always exists is somewhat complex and can only be sketched here. We create a graph G in which there is one node for each machine and one for each job, and an edge between each machine node i and job node j if $x_{ij} > 0$. We are again helped by the theory of linear programming, as the linear program we solved is a generalized assignment problem. As a result, for any basic solution, the structure of G is a forest of trees and 1-trees, which are trees with one edge added; for further details see [1]. We need not consider jobs that are already integrally assigned, so for every pair (i, j) such that $x_{ij} = 1$, we remove from G the nodes representing machine i, job j and their mutual edge (note that the constraints imply that this machine and job is not connected to any other machine or job). In the forest that remains, the only leaves are machine nodes, since every remaining job node represents a job that is fractionally assigned by the linear program and thus has an edge to at least two machines.

It is now straightforward to find a matching in G. We first consider the 1-trees, and in particular consider the unique cycle in each 1-tree. The nodes in these cycles alternate between machine nodes and job nodes, with an equal number of each. We arbitrarily choose an orientation of the cycle and assign each job to the machine that follows it in the oriented cycle. We then remove all of the matched nodes from G. What remains is a forest of trees; furthermore, it is possible that for each of these trees we have created at most one new leaf that is a job node. We then root each of the trees in the forest, either at its leaf job node, or, if it does not have one, at an arbitrary vertex. Finally, we assign each job node to one of its children machine nodes in the rooted tree. Each machine node has at most one parent, and thus is assigned at most one job. We have thus successfully matched all job nodes to machine nodes, as we required.

Thus, there exists a 2-RDP for $R\|C_{max}$, and we have the following theorem.

THEOREM 20.16   [31] *There is a 2-approximation algorithm for $R\|C_{max}$.*

20.5.2   Inferring an Ordering from a Preemptive Schedule for $1|r_j|\sum C_j$

In this section and the next we discuss techniques for inferring an ordering of jobs from a relaxation. In this section we consider the problem $1|r_j|\sum C_j$. Recall, as mentioned in Section 20.2.1, that this problem is $\mathcal{N}P$-hard. However, we can find a good relaxation by the simple expedient of allowing preemption. Specifically, we use $1|r_j, pmtn|\sum C_j$ as a relaxation of $1|r_j|\sum C_j$. We have seen that $1|r_j, pmtn|\sum C_j$ can be solved without linear programming, simply by using the SRPT rule. We will make use of this relaxation by extracting from it the order of completion of the jobs in the optimal preemptive schedule, and create a nonpreemptive schedule with the same order of completion.

Our algorithm, which we call Convert-Preempt-Schedule, is as follows. We first obtain an optimal preemptive schedule P for the instance in question. We then order the jobs in their order of completion in P; assume by renumbering that $C_1^P \leq \cdots \leq C_n^P$. We schedule the jobs nonpreemptively in the same order. If at some point the next job in the order has not been released, we wait idly until its release date and then schedule it.

THEOREM 20.17 [35] `Convert-Preempt-Schedule` *is a 2-approximation algorithm for* $1|r_j|\sum C_j$.

PROOF The nonpreemptive schedule N constructed by `Convert-Preempt-Schedule` can be understood as follows. For each job j, consider the point of completion of the last piece of j scheduled in P, insert p_j extra units of time into the schedule at the completion point of j in P (delaying by an additional p_j time the part of the schedule after C_j^P) and then schedule j nonpreemptively in the newly inserted block of length p_j. Then, remove from the schedule all of the time originally allocated to processing job j. Finally, cut out any idle time in the resulting schedule that can be removed without changing the scheduled order of the jobs or violating a release date constraint. The result is exactly the schedule computed by `Convert-Preempt-Schedule`.

Note that the completion of job j is only delayed by insertion of blocks for jobs that finish earlier in P and hence

$$C_j^N \leq C_j^P + \sum_{k \leq j} p_k.$$

However, $\sum_{k \leq j} p_k \leq C_j^P$, since all of these jobs completed before j in P, and therefore

$$\sum_{j=1}^{n} C_j^N \leq 2 \sum_{j=1}^{n} C_j^P.$$

The theorem now follows from the fact that the total completion time of the optimal preemptive schedule is a lower bound on the total completion time of the optimal nonpreemptive schedule.

20.5.3 An Ordering from a Linear-Programming Relaxation for $1|r_j, prec|\sum w_j C_j$

In this section we generalize the techniques of Section 20.5.2, applying them not to a preemptive schedule but instead to a linear-programming relaxation of $1|r_j, prec|\sum w_j C_j$.

20.5.3.1 The Relaxation

We begin by describing the linear-programming relaxation of our problem. Unlike our previous relaxation, this one does not arise from relaxing the integrality constraints of an integer linear program. Rather, we give several classes of inequalities that would be satisfied by feasible solutions to $1|r_j, prec|\sum w_j C_j$. These constraints are necessary but not sufficient to describe a valid solution to the problem.

The linear-programming formulation that we considered for $R\|C_{max}$ assigned jobs to machines but captured no information about the ordering of jobs on a machine. For $1|r_j, prec|\sum w_j C_j$ the ordering of jobs on a machine is a critical element of a high-quality solution, so we seek a formulation that can model this. We do this by making time explicit in the formulation: we will have n variables C_j, one for each of the n jobs; C_j will represent the completion time of job j in a schedule.

Consider the following formulation in these C_j variables, solutions to which correspond to optimal solutions of $1|r_j, prec|\sum w_j C_j$.

$$\text{minimize} \sum_{j=1}^{n} w_j C_j \tag{20.14}$$

subject to

$$C_j \geq r_j + p_j, \quad j = 1, \ldots, n, \tag{20.15}$$

$$C_k \geq C_j + p_k, \quad \text{for each pair } j, k \text{ such that } j \prec k, \tag{20.16}$$

$$C_k \geq C_j + p_k \quad \text{or} \quad C_j \geq C_k + p_j, \quad \text{for each pair } j, k. \tag{20.17}$$

Unfortunately, the last set of constraints are not linear constraints. Instead, we use a class of valid inequalities, introduced by Wolsey [41] and Queyranne [37]. Recall that we denote the entire set of jobs $\{1, \ldots, n\}$ as $\mathcal{J}$, and, for any subset $S \subseteq \mathcal{J}$, we define $p(S) = \sum_{j \in S} p_j$ and $p^2(S) = \sum_{j \in S} p_j^2$. We claim that for any feasible one-machine schedule (independent of constraints and objective)

$$\sum_{j \in S} p_j C_j \geq \frac{1}{2} \left(p^2(S) + p(S)^2 \right), \quad \text{for each } S \subseteq \mathcal{J}. \tag{20.18}$$

We show that these inequalities are satisfied by the completion times of any valid schedule for one machine and thus in particular by the completion times of a valid schedule for $1 | r_j, prec | \sum w_j C_j$.

LEMMA 20.3 [37,41] Let $C_1, \ldots, C_n$ be the completion times of jobs in any feasible schedule on one machine. Then the C_j must satisfy the inequalities

$$\sum_{j \in S} p_j C_j \geq \frac{1}{2} \left(p(S)^2 + p^2(S) \right) \quad \text{for each } S \subseteq \mathcal{J}. \tag{20.19}$$

PROOF We assume that the jobs are indexed so that $C_1 \leq \cdots \leq C_n$. Consider first the case $S = \{1, \ldots, n\}$. Clearly for any job j, $C_j \geq \sum_{k=1}^{j} p_k$. Multiplying by p_j and summing over all j, we obtain

$$\sum_{j=1}^{n} p_j C_j \geq \sum_{j=1}^{n} p_j \sum_{k=1}^{j} p_k = \frac{1}{2} \left(p^2(S) + p(S)^2 \right).$$

Thus Equation 20.19 holds for $S = \{1, \ldots, n\}$. The general case follows from the fact that for any other set of jobs S, the jobs in S are feasibly scheduled by the schedule of $\{1, \ldots, n\}$—just ignore the other jobs. So we may view S as our entire set of jobs and apply the previous argument.

In the special case of $1 || \sum w_j C_j$ the constraints 20.19 give an exact characterization of the problem [37,41]; specifically, any set of C_j that satisfy these constraints must describe the completion times of a feasible schedule, and thus these linear constraints effectively replace the disjunctive constraints 20.17. When we extend the formulation to include constraints 20.15 and 20.16, we no longer have an exact formulation, but rather a linear-programming relaxation of $1 | r_j, prec | \sum w_j C_j$.

We note that this formulation has an exponential number of constraints; however, it can be solved in polynomial time by the use of the ellipsoid algorithm for linear programming [37,41]. We also note that in the special case in which we just have release dates, a slightly strengthened version can (remarkably) be solved optimally in $O(n \log n)$ time [8].

20.5.3.2 Constructing a Schedule from a Solution to the Relaxation

We now show that a solution to this relaxation can be converted efficiently to an approximately optimal schedule. For simplicity, we ignore release dates and consider only $1 | prec | \sum w_j C_j$. Our approximation algorithm, which we call Schedule-by-$\bar{C}_j$, is simple to state. We first solve the linear-programming relaxation given by Equations 20.14, 20.16 and 20.18 and call the solution $\bar{C}_1, \ldots, \bar{C}_n$; we renumber the jobs so that $\bar{C}_1 \leq \bar{C}_2 \leq \cdots \leq \bar{C}_n$. We then schedule the jobs in the order $1, \ldots, n$. Since there are no release dates there is no idle time. Note that this ordering of the jobs respects the precedence constraints, because if the $\bar{C}_j$ satisfy Equation 20.14 then $j \prec k$ implies that $\bar{C}_j < \bar{C}_k$.

To analyze $\texttt{Schedule-by-}\bar{C}_j$, we begin by understanding why it is not an optimal algorithm. Unfortunately, $\bar{C}_1 \leq \cdots \leq \bar{C}_n$ being a feasible solution to Equation 20.18 does not guarantee that, in the schedule in which job j is designated to complete at time $\bar{C}_j$ (thus defining its start time), at most one job is scheduled at any point in time. More formally, the intervals $(\bar{C}_j - p_j, \bar{C}_j], j = 1, \ldots, n$, are not constrained to be disjoint. If $\bar{C}_1 \leq \cdots \leq \bar{C}_n$ actually corresponded to a valid schedule, then $\bar{C}_j$ would be no less than $\sum_{k=1}^{j} p_k$ for all j. We will see that, although the formulation does not guarantee this property, it does yield a relaxation of it, which is sufficient for the purposes of approximation.

THEOREM 20.18   [17] $\texttt{Schedule-by-}\bar{C}_j$ *is a 2-approximation algorithm for* $1|prec|\sum w_j C_j$.

PROOF   Since $\bar{C}_j$ optimized a relaxation, we know that $\sum w_j \bar{C}_j$ is a lower bound on the true optimum. It therefore suffices to show that our algorithm gets within a factor of 2 of this lower bound. So we let $\tilde{C}_1, \ldots, \tilde{C}_n$ denote the completion times in the schedule found by $\texttt{Schedule-by-}\bar{C}_j$, and show that $\sum w_j \tilde{C}_j \leq 2 \sum w_j \bar{C}_j$.

Since the jobs have been renumbered so that $\bar{C}_1 \leq \cdots \leq \bar{C}_n$, taking $S = \{1, \ldots, j\}$ gives

$$\tilde{C}_j = p(S).$$

We now show that $\bar{C}_j \geq \frac{1}{2} p(S)$. (Again, if the $\bar{C}_j$ were feasible completion times in an actual schedule, we would have $\bar{C}_j \geq p(S)$. This relaxed version of the property is the key to the approximation.)

We use inequality Equation 20.18 for $S = \{1, 2, \ldots, j\}$.

$$\sum_{k=1}^{j} p_k \bar{C}_k \geq \frac{1}{2} \left(p^2(S) + p(S)^2 \right) \geq \frac{1}{2} p(S)^2. \tag{20.20}$$

Since $\bar{C}_k \leq \bar{C}_j$, for each $k = 1, \ldots, j$, we have

$$\bar{C}_j p(S) = \bar{C}_j \sum_{k=1}^{j} p_k \geq \sum_{k=1}^{j} \bar{C}_k p_k \geq \frac{1}{2} p(S)^2.$$

Dividing by $p(S)$, we obtain that $\bar{C}_j$ is at least $\frac{1}{2} p(S)$. We therefore see that $\tilde{C}_j \leq 2\bar{C}_j$ and the result follows.

20.6   Polynomial Approximation Schemes Using Enumeration and Rounding

For certain $\mathcal{N}P$-hard scheduling problems there is a limit to our ability to approximate them in polynomial time; for example, Lenstra, Shmoys, and Tardos proved that there is no ρ-approximation algorithm, with $\rho < 3/2$, for $R\|C_{\max}$ unless $\mathcal{P} = \mathcal{N}P$ [31]. For certain problems, however, we can approximate their optimal solutions arbitrarily closely in polynomial time. In this section we present three polynomial time approximation schemes (PTAS); that is, polynomial-time algorithms that, for any constant $\rho > 1$, deliver a solution whose objective value is at most ρ times optimal. The running time will depend on ρ—the smaller ρ is, the slower the algorithm will be.

We will present two approaches to the design of such algorithms. The first approach is based on rounding processing times or weights to small integers so that we can apply pseudopolynomial-time algorithms such as that for $1\|\sum w_j U_j$. A second approach is based on identifying the "important" jobs—those that have the greatest impact on the solution—and processing them separately. In one version, illustrated for $P\|C_{max}$, we round the large jobs so that there are only a constant number of large job sizes, schedule them using dynamic programming, and then schedule the small jobs arbitrarily. In a second version, illustrated for $1|r_j|L_{max}$, we enumerate all possible schedules for the large jobs, and then fill in the small jobs around them.

20.6.1 From Pseudopolynomial to PTAS: $1\|\sum w_j U_j$

In Section 20.3, we gave an $O(n \sum w_j)$ time algorithm for $1\|\sum w_j U_j$. Since this gives an algorithm that runs in polynomial time when the weights are polynomial in n, a natural idea is to try to reduce any instance to such a special case. We will scale the weights so that the optimal solution is bounded by a polynomial in n; this will allow us to apply our dynamic programming algorithm to weights of polynomial size.

Assume for now that we know W^*, the value of $\sum w_j U_j$ in the optimal schedule. Multiply every weight by $n/(\epsilon W^*)$; now the optimal $\sum w_j U_j$ becomes n/ϵ. Clearly, a schedule with $\sum w_j U_j$ within a multiplicative $(1 + \epsilon)$-factor of optimum under these weights is also within a multiplicative $(1 + \epsilon)$-factor of optimum under the original weights. Thus, it suffices to find a schedule with $\sum w_j U_j$ at most $(1 + \epsilon)n/\epsilon = n/\epsilon + n$ under the new weights.

To do so, increase the weight of every job to the next larger integer. This increases the weight of each job by at most 1 and thus, for any schedule, increases $\sum w_j U_j$ by at most n. Under these new weights, $\sum w_j U_j$ for the original optimal schedule is now at most $n/\epsilon + n$, so the optimal schedule under these integral weights has $\sum w_j U_j \leq n/\epsilon + n$. Since all weights are integers, we can apply the dynamic programming algorithm of Section 20.3 to find an optimal schedule for the rounded instance. Since we only rounded up, the same schedule under the original weights can only have a smaller $\sum w_j U_j$. Thus, we find a schedule with weight at most $n/\epsilon + n$ in the (scaled) original weights, i.e., a $(1 + \epsilon)$ times optimum schedule.

The running time of our dynamic program is proportional to n times the sum of the (new) weights. This might be a problem, since the weights can be arbitrarily large. However, any job with new weight exceeding $n/\epsilon + n$ must be scheduled before its deadline. We therefore identify all such jobs, and modify our dynamic program: T_{wj} becomes the minimum time needed to complete all these jobs that must complete by their deadlines plus other jobs from $1, \ldots, j$ of total weight w. The dynamic programming argument goes through unchanged, but now we consider only jobs of weight $O(n/\epsilon)$. It follows that the largest value of w that we consider is $O(n^2/\epsilon)$, which means that the total running time is $O(n^3/\epsilon)$.

It remains to deal with our assumption that we know W^*. One approach is to use the RDP scheme that performs a binary search on W^*. Of course, we do not expect to arrive at W^* exactly, but note that an estimate will suffice. If we test a value W' with $W^*/\alpha \leq W' \leq W^*$, the analysis above will go through with the running time increased by a factor of α. So we can wait for the RDP binary search to bring us within (say) a constant factor of W^* and then solve the problem.

Of course, if the weights w are extremely large, our binary search could go through many iterations before finding a good value of W'. An elegant trick lets us avoid this problem. We will solve the following problem: find a schedule that minimizes the weight of the maximum-weight late job. The value of this schedule, W', is clearly a lower bound on W^*, as all schedules that minimize $\sum w_j U_j$ must have a late job of weight at least W'. Further, W^* is at most nW', since the schedule returned must have at most n late jobs each of which has weight at most W'. Hence our value W' is within a factor of n of optimal. Thus $O(\log n)$ binary search steps suffice to bring us within a constant factor of W^*.

To compute W', we formulate a $1||f_{max}$ problem. For each job j,

$$f_j(C_j) = \begin{cases} w_j & \text{if } C_j > d_j \\ 0 & \text{if } C_j \le d_j. \end{cases}$$

This will compute the schedule that minimizes the weight of the maximum weight late job. By the results in Section 20.2.1, we know we can compute this exactly in polynomial time.

THEOREM 20.19   *There exists a $O(n^3 (\log n)/\epsilon)$-time $(1+\epsilon)$-approximation algorithm for $1||\sum w_j U_j$.*

20.6.2   Rounding and Dynamic Programming for $P||C_{max}$

We now return to the problem of $P||C_{max}$. Recall that in Lemma 20.2 we gave a polynomial-time algorithm for a special case in which there are a constant number of different job sizes. For the general case, we will focus mainly on the big jobs. We will round and scale these jobs so that there is at most a constant number of sizes of big jobs, and apply the dynamic programming algorithm of Section 20.3 to these rounded jobs. We then finish up by scheduling the small jobs greedily. By the definition of big and small, the overall contribution of the small jobs to the makespan will be negligible.

We will give a $(1 + \epsilon)$-RDP for this problem that can be transformed as before into a $(1 + \epsilon)$-approximation algorithm. We therefore assume that we have a target optimum schedule length T, We also assume for the rest of this section that $\epsilon T, \epsilon^2 T, \epsilon^{-1}$ and ϵ^{-2} are integers. The proofs can easily be modified to handle the case of arbitrary rational numbers.

We first show how to handle the large jobs.

LEMMA 20.4   [18] Let I be an instance of $P||C_{max}$, let T be a target schedule length, and $\epsilon > 0$. Assume that all $p_j \ge \epsilon T$. Then, for this case, there is a $(1 + \epsilon)$-RDP for $P||C_{max}$.

PROOF   We assume $T \ge \max_j p_j$, since otherwise we immediately know that the problem is infeasible. Form instance I' from I with processing times p'_j by rounding each p_j down to an integer multiple of $\epsilon^2 T$. This creates an instance in which

1. $0 \le p_j - p'_j \le \epsilon^2 T$
2. There are at most $\frac{T}{\epsilon^2 T} = \frac{1}{\epsilon^2}$ different job sizes
3. In any feasible schedule, each machine has at most $\frac{T}{\epsilon T} = \frac{1}{\epsilon}$ jobs

Thus, we can apply Lemma 20.2 to instance I' and obtain an optimal solution to this scheduling problem; let its makespan be D. If $D > T$, then we know that there is no schedule of length $\le T$ for I, since job sizes in I' are no greater than those in I. In this case we can answer "no schedule of length $\le T$ exists." If $D \le T$, then we will answer "there exists a schedule of length $\le (1+\epsilon)T$." We now show that this answer will be correct. We simply take our schedule for I' and replace the rounded jobs with the original jobs from I. By 1 and 3 in the above list, we add at most $\epsilon^2 T$ to the processing time of each job, and since there are at most $\frac{1}{\epsilon}$ jobs per machine, we add at most ϵT to the processing time per machine. Thus we can create a schedule with makespan at most $T + \epsilon T = (1 + \epsilon)T$.

We now give the complete algorithm. The idea will be to remove the "small" jobs, use Lemma 20.4 to schedule the remaining jobs, and then add the small jobs back greedily. Given input I_0, target schedule length T, and $\rho = 1 + \epsilon > 1$, we execute the following algorithm.

Let R be the set of jobs with $p_j \le \epsilon T$. Let $I = I_0 - R$
Apply Lemma 20.4 to I, T, and ρ.
If this algorithm returns no,
($\dagger$)   then output "no schedule of length $\le T$ exists."
        else
                for each job j in R
                        if there is a machine i with load $\le T$,
                                then add job j to machine i
($*$)   else return "no schedule of length $\le T$ exists"
                        return "yes, a schedule of length $\le \rho T$ exists"

THEOREM 20.20   [18] *The algorithm above is a ρ-RDP for* P$\|C_{\max}$.

PROOF   If the algorithm outputs "yes, a schedule of length $\le \rho T$ exists," then it has constructed such a schedule, and is clearly correct. If the algorithm outputs "no schedule of length $\le T$ exists" online ($\dagger$), then it is because no schedule of length T exists for instance I. But instance I is a subset of the original jobs and so if no schedule exists for I, then no schedule exists for I_0, and the output is correct. If the algorithm outputs "no schedule of length $\le T$ exists" online ($*$), then at this point in the algorithm, every machine must have more than T units of processing on it. Thus, we have that $\sum_j p_j > mT$, which means that no schedule of length $\le T$ exists.

The running time is dominated by the dynamic programming in Lemma 20.2. It is polynomial in n, but the exponent is a polynomial in $1/\epsilon$. While for ρ very close to 1, the running time is prohibitively large, for larger, fixed values of ρ, a modified algorithm yields good schedules with near-linear running times; see [18] for details.

20.6.3   Exhaustive Enumeration for $1|r_j|L_{\max}$

We now turn to the problem of minimizing the maximum lateness in the presence of release dates. Recall from Section 20.2 that without release dates EDD is an exact algorithm for this problem. Once we add release dates the problem becomes $\mathcal{NP}$-hard. As we think about approximation algorithms, we come upon an immediate obstacle, namely that the objective function can be 0 or even negative, and hence a solution of value at most $\rho C_{\max}^*$ is clearly impossible. In order to get around this, we must guarantee that the value of the objective function is positive. One simple way to do so is to decrease all the d_j's uniformly by some value δ. This decreases the objective value by exactly δ and does not change the structure of the optimal solution. In particular, if we pick δ large enough so that all the d_j's are negative, we are guaranteed that $L_{\max}$ is positive.

Forcing d_j to be negative is somewhat artificial and so we do not concentrate on this interpretation (note that by taking δ arbitrarily large, we can make any algorithm into an arbitrarily good approximation algorithm). We instead use an equivalent but natural delivery time formulation which, in addition to modeling a number of applications, is a key subroutine in computational approaches to shop scheduling problems [29]. In this formulation, each job, in addition to having a release date r_j and a processing time p_j, has a delivery time q_j. A delivery time is an amount of time that must elapse between the completion time of a job on a machine and when it is truly considered finished. Our objective is now to minimize $\max_j\{C_j + q_j\}$. To see the connection to our original problem, note that by setting $q_j = -d_j$ (recall that we made all d_j negative, so all q_j are positive), the delivery-time problem is equivalent to minimizing maximum lateness, and in fact we will overload L_j and define it as $C_j + q_j$.

20.6.3.1 Jackson's Rule Is a 2-Approximation Algorithm

In the delivery-time model, EDD translates to Longest Delivery Time First. This is often referred to as Jackson's rule [23]. Let $L^*_{\max}$ be the optimum maximum lateness. The following two lower bounds for this problem are the easily derived analogs of Equations 20.1 and 20.2:

$$L^*_{\max} \geq \sum_j p_j, \tag{20.21}$$

$$L^*_{\max} \geq r_j + p_j + q_j \quad \text{for all } j. \tag{20.22}$$

LEMMA 20.5 Jackson's Rule is a 2-approximation algorithm for the delivery time version of $1|r_j|L_{\max}$.

PROOF Let j' be a job for which $L_{j'} = L_{\max}$. Since Jackson's rule creates a schedule with no unforced idle time, we know that there is no idle time between time $r_{j'}$ and $C_{j'}$. Let J' be the set of jobs that run between $r_{j'}$ and $C_{j'}$. Then

$$L_{j'} = C_{j'} + q_{j'} \tag{20.23}$$

$$= r_{j'} + \sum_{j \in J'} p_j + q_{j'} \tag{20.24}$$

$$\leq (r_{j'} + q_{j'}) + \sum_j p_j \tag{20.25}$$

$$= 2L^*_{\max}, \tag{20.26}$$

where the last line follows by applying the two lower bounds (20.21) and (20.22).

20.6.3.2 A PTAS Using Enumeration

The presentation of this section follows that of Hall [16]. The original approximation scheme for this problem is due to Hall and Shmoys [15].

To obtain an algorithm with an improved performance guarantee, we need to look more carefully at when Jackson's rule can go wrong. Let s_j be the starting time of job j, let $r_{\min}(S) = \min_{j \in S} r_j$, let $q_{\min}(S) = \min_{j \in S} q_j$, and recall that $p(S) = \sum_{j \in S} p_j$. Then clearly, for any $S \subseteq J$

$$L^*_{\max} \geq r_{\min}(S) + p(S) + q_{\min}(S). \tag{20.27}$$

Now consider a job j' for which $L_{j'} = L_{\max}$. Let t_i be the latest time before $s_{j'}$ at which the machine is idle, and let a be the job that runs immediately after this idle time. Let S be the set of jobs that run between s_a and $C_{j'}$. We call S a critical section. Because of the idle time immediately before s_a, we know that for all $j \in S$, $r_j \geq r_a$. In other words we have a set of jobs, all of which were released after time r_a, and which end with the job that achieves $L_{\max}$. Now if for all $j \in S$ $q_j \geq q_{j'}$, then we claim that $L_{j'} = L^*_{\max}$. This follows from the fact that

$$L_{\max} = L_{j'} = r_a + p(S) + q_{j'} = r_{\min}(S) + p(S) + q_{\min}(S),$$

and that the right-hand side, by Equation 17.27, is also a lower bound on L^*_{max}. So, as long as, in a critical section, the job with the shortest delivery time is last, we have an optimal schedule. Thus, if Jackson's rule is not optimal, there must be a job b in the critical section which has $q_b < q_{j'}$. We call the latest-scheduled job in the critical section with $q_b < q_{j'}$ an interference job. The following lemma shows the relationship between the interference job and its effect on L_{max}.

LEMMA 20.6   Let b be an interference job in a schedule created by Jackson's rule. Then $L_{max} < L^*_{max} + p_b$.

Thus, if interference jobs have small processing times, Jackson's rule does very well. To make sure that this is the case, we will handle the large jobs separately, to ensure that they are not interference jobs, and then use Jackson's rule on the remaining jobs.

Let us assume for now that we know the optimal schedule for instance I. Let s^*_j be the starting time of job j in the optimal schedule, and let $\delta > 0$ be a parameter to be chosen later. Partition the jobs into small jobs $S = \{j : p_j < \delta\}$ and big jobs $B = \{j : p_j \geq \delta\}$. We create instance $\tilde{I}$ as follows: if $j \in S$, then $\tilde{r}_j = r_j$, $\tilde{p}_j = p_j$, and $\tilde{q}_j = q_j$, otherwise, $\tilde{r}_j = s^*_j$, $\tilde{p}_j = p_j$, and $\tilde{q}_j = L^*_{max}(I) - p_j - s^*_j$. Instance $\tilde{I}$ is no easier than instance I, since we have not decreased any release dates or delivery times. Yet, the optimal schedule for I remains an optimal schedule for $\tilde{I}$, by construction. In $\tilde{I}$ we have given the large jobs a release date equal to their optimal starting time, and a delivery time that is equal to the schedule length minus their completion time, and hence have constrained the large jobs to run exactly when they would run in the optimal schedule for instance I. Thus, in an optimal schedule for $\tilde{I}$, the big jobs run at exactly time $\tilde{r}_j$ and have $L_j = \tilde{r}_j + \tilde{p}_j + \tilde{q}_j = L^*_{max}$.

Now we claim that if we run Jackson's rule on $\tilde{I}$, the big jobs will not be interference jobs.

LEMMA 20.7   If we run Jackson's rule on $\tilde{I}$, no job $b \in B$ will be an interference job.

PROOF   Assume that some job $b \in B$ is an interference job. As above, define the critical section, and jobs a and j'. Since b is an interference job, we know that $\tilde{q}_{j'} > \tilde{q}_b$ and $\tilde{r}_{j'} > \tilde{r}_b$. We also know that $\tilde{r}_b = s^*_b$, and so j' must run after b in the optimal schedule for I. Applying Equation 20.27 to the set consisting of jobs b and j', we get that

$$L^*_{max} \geq \tilde{r}_b + \tilde{p}_b + \tilde{p}_{j'} + \tilde{q}_{j'} \geq r_b + p_b + p_{j'} + \tilde{q}_b = L^*_{max} + p_{j'},$$

which is a contradiction.

So if we run Jackson's rule on $\tilde{I}$, we get a schedule whose length is at most $L^*_{max}(I) + \delta$. Choosing $\delta = \epsilon \sum_j p_j$, and recalling that $L^*_{max} \geq \sum_j p_j$, we get a schedule of length at most $(1 + \epsilon)L^*_{max}$. Further, there can be at most $\sum_j p_j / (\epsilon \sum_j p_j) = 1/\epsilon$ big jobs. The only problem is that we don't know $\tilde{I}$.

We now argue that it is not necessary to know $\tilde{I}$. First, observe that the set of big jobs is purely a function of the input, and ϵ. Now, if we knew the starting times of the big jobs in the optimal schedule for I, we would know $\tilde{I}$, and could run Jackson's rule on the job in S, inserting the big jobs at the appropriate time. This implies a numbering of the big jobs, i.e., each big job j_i is, for some k, the kth job in the schedule for $\tilde{I}$. Thus, we really only need to know k, and not the starting time for job j_i. Thus we just enumerate all possible numberings for the big jobs. There are $n^{1/\epsilon}$ such numberings. Given a numbering, we can run Jackson's rule on the small jobs, and insert the big jobs at the appropriate places in $O(n \log n)$ time, and thus we get an algorithm that in $O(n^{1+1/\epsilon} \log n)$ time finds a schedule with $L_{max} \leq (1 + \epsilon)L^*_{max}$.

20.7 Research Issues and Summary

In this chapter we have surveyed some of the basic techniques for deterministic scheduling. Scheduling is an old and therefore mature field, but important opportunities for research contributions remain. In addition to some of the outstanding open questions (see the survey by Lawler et al. [29]) it is our feeling that the most meaningful research contributions will be either new and innovative techniques for attacking old problems or new problem definitions that model more realistic applications.

There are other schools of approach to the design of algorithms for scheduling, such as those relying on techniques from artificial intelligence or from computational optimization. It will be quite valuable to forge stronger connections between these different approaches to solving scheduling problems.

20.8 Further Information

We conclude by reminding the reader what this chapter is not. In no way is this chapter a comprehensive survey of even the most basic and classical results in scheduling theory, and it is certainly not an up-to-date survey on the field. It also essentially entirely ignores "nontraditional" models, and does not touch on stochastic scheduling or on any of the other approaches to scheduling and resource allocation. The reader interested in a comprehensive survey of the field should consult the textbook by Pinedo [34] and the survey by Lawler et al. [29]. These sources provide pointers to a number of other references. In addition, we also recommend an annotated bibliography by Hoogeveen et al. that contains information on recent results in scheduling theory [20], the surveys by Queyranne and Schulz on polyhedral formulations [36], by Hall on approximation algorithms [16], and by Sgall on online scheduling [38]. Research on deterministic scheduling theory is published in many journals; for example see *Mathematics of Operations Research, Operations Research, SIAM Journal on Computing,* and *Journal of the ACM.*

Defining Terms

n: Number of jobs.

m: Number of machines.

p_j: Processing time of job j.

C_j^S: Completion time of job j in schedule S.

w_j: Weight of job j.

r_j: Release date of job j; job j is unavailable for processing before time r_j.

d_j: Due date of job j.

L_j: $= C_j - d_j$ the lateness of job j.

U_j: 1 is job j is scheduled by d_j and 0 otherwise.

$\alpha|\beta|\gamma$: Denotes scheduling problem with machine environment α, optimality criterion γ, and side characteristics and constraints denoted by β.

Machine environments:

1: One machine.

P: Parallel identical machines.

Q: Parallel machines of different speeds.

R: Parallel unrelated machines.

    O: Open shop.

    F: Flow shop.

    J: Job shop.

Possible characteristics and constraints:

    pmtn: Job preemption allowed.

      r_j: Jobs have nontrivial release dates.

    prec: Jobs are precedence-constrained.

Optimality criteria:

    $\sum C_j$: Average (sum) of completion times.

    $\sum w_j C_j$: Weighted average (sum) of completion times.

    C_{max}: Makespan (schedule length).

    L_{max}: Maximum lateness over all jobs.

    $\sum U_j$: Number of on-time jobs.

    $\sum w_j U_j$: Weighted number of on-time jobs.

Acknowledgments

We are grateful to Jan Karel Lenstra and David Shmoys for helpful comments.

References

1. Ahuja, R.K., Magnanti, T.L., and Orlin, J.B., *Network Flows: Theory, Algorithms, and Applications.* Prentice Hall, Englewood Cliffs, NJ, 1993.
2. Baker, K.R., *Introduction to Sequencing and Scheduling.* John Wiley & Sons, New York, 1974.
3. Bárány, I. and Fiala, T., Többgépes ütemezési problémák közel optimális megoldása. *Szigma–Mat.–Közgazdasági Folyóirat,* 15, 177–191, 1982.
4. Bruno, J.L., Coffman, E.G., and Sethi, R., Scheduling independent tasks to reduce mean finishing time. *Communications of the ACM,* 17, 382–387, 1974.
5. Conway, R.W., Maxwell, W.L., and Miller, L.W., *Theory of Scheduling.* Addison-Wesley, Reading, MA, 1967.
6. Cormen, T.H., Leiserson, C.E., and Rivest, R.L., *Introduction to Algorithms.* MIT Press/McGraw-Hill, Cambridge, MA, 1990.
7. Garey, M.R. and Johnson, D.S., *Computers and Intractability: A Guide to the Theory of NP-Completeness.* W.H. Freeman and Company, New York, 1979.
8. Goemans, M., A supermodular relaxation for scheduling with release dates. In *Proceedings of the Fifth Conference on Integer Programming and Combinatorial Optimization,* 288–300, June 1996. Published as *Lecture Notes in Computer Science,* Vol. 1084, Springer-Verlag, Berlin, Germany, 1996.
9. Gonzalez, T. and Sahni, S., Open shop scheduling to minimize finish time. *Journal of the ACM,* 23, 665–679, 1976.
10. Gonzalez, T. and Sahni, S., Flowshop and jobshop schedules: Complexity and approximation. *Operations Research,* 26, 36–52, 1978.
11. Graham, R.L., Bounds for certain multiprocessor anomalies. *Bell System Technical Journal,* 45, 1563–1581, 1966.

12. Graham, R.L., Bounds on multiprocessing anomalies. *SIAM Journal of Applied Mathematics,* 17, 263–269, 1969.
13. Graham, R.L., Lawler, E.L., Lenstra, J.K., and Rinnooy Kan, A.H.G., Optimization and approximation in deterministic sequencing and scheduling: A survey. *Annals of Discrete Mathematics,* 5, 287–326, 1979.
14. Gusfield, D., Bounds for naive multiple machine scheduling with release times and deadlines. *Journal of Algorithms,* 5, 1–6, 1984.
15. Hall, L. and Shmoys, D.B., Approximation schemes for constrained scheduling problems. In *Proceedings of the 30th Annual Symposium on Foundations of Computer Science,* 134–141. IEEE, Research Triangle Park, NC, October 1989.
16. Hall, L.A., *Approximation Algorithms for NP-Hard Problems,* Chapter 1. Hochbaum, D. (Ed.), PWS Publishing, Boston, MA, 1997.
17. Hall, L.A., Schulz, A.S., Shmoys, D.B., and Wein, J., Scheduling to minimize average completion time: Off-line and on-line approximation algorithms. *Mathematics of Operations Research,* 22, 513–544, August 1997.
18. Hochbaum, D.S. and Shmoys, D.B., Using dual approximation algorithms for scheduling problems: Theoretical and practical results. *Journal of the ACM,* 34, 144–162, 1987.
19. Hochbaum, D., Ed., *Approximation Algorithms.* PWS, Boston, MA, 1997.
20. Hoogeveen, J.A., Lenstra, J.K., and van de Velde, S.L., Sequencing and scheduling. In *Annotated Bibliographies in Combinatorial Optimization.* Interscience series in discrete mathematics and optimization, pp. 181–197, Dell'Amico, M., Maffioli, F., and Martello, S. (Eds.), John Wiley & Sons, Chichester, U.K., 1997.
21. Horn, W., Minimizing average flow time with parallel machines. *Operations Research,* 21, 846–847, 1973.
22. Horn, W., Some simple scheduling algorithms. *Naval Research Logistics Quarterly,* 21, 177–185, 1974.
23. Jackson, J.R., Scheduling a production line to minimize maximum tardiness. Management Science Research Project Research Report 43, University of California, Los Angeles, CA, 1955.
24. Johnson, S.M., Optimal two- and three-stage production schedules with setup times included. *Naval Research Logistics Quarterly,* 1(1), 61–68, 1954.
25. Khachiyan, L.G., A polynomial algorithm in linear programming (in Russian). *Doklady Adademiia Nauk SSSR,* 224, 1093–1096, 1979.
26. Lawler, E.L., Optimal sequencing of a single machine subject to precedence constraints. *Management Science,* 19, 544–546, 1973.
27. Lawler, E.L., *Combinatorial Optimization: Networks and Matroids.* Holt, Rinehart and Winston, New York, 1976.
28. Lawler, E.L. and Labetoulle, J., On preemptive scheduling of unrelated parallel processors by linear programming. *Journal of the ACM,* 25, 612–619, 1978.
29. Lawler, E.L., Lenstra, J.K., Rinnooy Kan, A.H.G., and Shmoys, D.B., Sequencing and scheduling: Algorithms and complexity. In *Handbooks in Operations Research and Management Science,* Graves, S.C., Rinnooy Kan, A.H.G., and Zipkin, P.H. (Eds.), Vol. 4., *Logistics of Production and Inventory,* North-Holland, Amsterdam, the Netherlands, 445–522, 1993.
30. Lawler, E.L. and Moore, J.M., A functional equation and its application to resource allocation and sequencing problems. *Management Science,* 16, 77–84, 1969.
31. Lenstra, J.K., Shmoys, D.B., and Tardos, É., Approximation algorithms for scheduling unrelated parallel machines. *Mathematical Programming,* 46, 259–271, 1990.
32. McNaughton, R., Scheduling with deadlines and loss functions. *Management Science,* 6, 1–12, 1959.
33. Moore, J.M., An *n*-job, one machine sequencing algorithm for minimizing the number of late jobs. *Management Science,* 15, 102–109, 1968.
34. Pinedo, M., *Scheduling: Theory, Algorithms and Systems.* Prentice Hall, Englewood Cliffs, NJ, 1995.

35. Phillips, C., Stein, C., and Wein, J., Scheduling jobs that arrive over time. *Mathematical Programming*, B82(1–2), 199–224, 1998.

36. Queyranne, M. and Schulz, A.S., Polyhedral approaches to machine scheduling. Technical Report 474/1995, Technical University of Berlin, Berlin, Germany, 1994.

37. Queyranne, M., Structure of a simple scheduling polyhedron. *Mathematical Programming*, 58, 263–285, 1993.

38. Sgall, J., On-line scheduling—A survey. In *On-Line Algorithms, Lecture Notes in Computer Science*. Fiat, A. and Woeginger, G. (Eds.), Springer-Verlag, Berlin, Germany, 1997. To appear.

39. Smith, W.E., Various optimizers for single-stage production. *Naval Research Logistics Quarterly*, 3, 59–66, 1956.

40. Ullman, J.D., NP-complete scheduling problems. *Journal of Computer and System Sciences*, 10, 384–393, 1975.

41. Wolsey, L.A., Mixed integer programming formulations for production planning and scheduling problems. Invited talk at the 12th International Symposium on Mathematical Programming, MIT Press, Cambridge, MA, 1985.

21

Computational Game Theory: An Introduction

Paul G. Spirakis
Research Academic Computer Technology Institute and Patras University

Panagiota N. Panagopoulou
Research Academic Computer Technology Institute and Patras University

21.1 Introduction

Game theory was founded by von Neumann and Morgenstern and can be defined as the study of mathematical models of interactive decision making. Game theory provides general mathematical techniques for analyzing situations in which two or more individuals, called players, make decisions that will influence one another's welfare. The dominant aspect of game theory is the belief that each player is rational, in the sense that he/she makes decisions consistently in pursuit of his/her own objectives, and this rationality is common knowledge.

This chapter is an introduction to the basic concepts and the advances of a new field, that of computational (or algorithmic) game theory. This field is a fertile interaction between the very deep and older field of game theory on the one hand, and of the younger field of algorithms and complexity on the other.

In the seminal paper of (Papadimitriou, 2001), the interaction between game theory and theoretical computer science was considered as a potential "object of the next great modeling adventure" of the field of computer scientists. In that work, it was pointed out that a fusion of algorithmic ideas and game theoretic concepts might yield to the most appropriate mathematical tools and insights for understanding the socio-economic complexity of the Internet. Furthermore, some game theoretic concepts (e.g., Nash equilibria) are related to fundamental computational issues, such as finding solutions that are guaranteed to exist via nonconstructive methods, or approximating them. On the other hand, game theoretic concepts have been used by theoretical computer scientists as a means of studying the computational complexity of algorithms: proving lower bounds can be seen as a game between the algorithm designer and an adversary.

In this chapter we deal with some basic topics of computational game theory. Namely, we study the computational complexity of Nash equilibria (which constitute the most important solution concept in game theory) and review the related algorithms proposed in the literature. Then, given the apparent difficulty of computing exact Nash equilibria, we study the efficient computation of approximate notions of Nash equilibria. Next we deal with several computational issues related to the class of congestion games, which model the selfish behavior of individuals when competing on the usage of a common set of resources. Finally, we study the price of anarchy (in the context of congestion games), which is defined as a measure of the performance degradation due to the lack of coordination among the involved players.

As this new field of computational game theory evolves, important research results have already appeared in several directions that are not touched here, because of lack of space and because of our selection of what are the most basic topics. For example, mechanism design, cooperative games, Bayesian games and their computational issues are not considered in this chapter. We hope that the interested reader will be motivated by this chapter to also look into those new research lines.

21.2   Underlying Principles

21.2.1   Strategic Form Games

In the language of game theory, a game refers to any situation in which two or more decision makers interact. In this chapter we focus on finite games in strategic form:

DEFINITION 21.1   A finite strategic form game is any Γ of the form

$$\Gamma = \langle N, (C_i)_{i \in N}, (u_i)_{i \in N} \rangle, \tag{21.1}$$

where
  N is a finite nonempty set, and, for each $i \in N$
  C_i is a finite nonempty set
  u_i is a function from $C = \times_{j \in N} C_j$ into the set of real numbers $\mathbb{R}$

In the aforementioned definition, N is the set of players in the game Γ. For each player $i \in N$, C_i is the set of actions available to player i. When the strategic form game Γ is played, each player i must choose one of the actions in the set C_i. For each combination of actions $c = (c_j)_{j \in N} \in C$ (one for each player), the number $u_i(c)$ represents the payoff that player i would get in this game if c were the combination of actions implemented by the players. When we study a strategic form game, we assume that all the players choose their actions simultaneously.

Given any strategic form game $\Gamma = \langle N, (C_i)_{i \in N}, (u_i)_{i \in N} \rangle$, a (mixed) strategy σ_i for any player $i \in N$ is a probability distribution over his/her set of actions C_i. Therefore, for each $c_i \in C_i$, the number $\sigma_i(c_i)$ represents the probability that player i would choose c_i. A pure strategy for player i is a strategy that poses probability 1 to a specific action in C_i and zero to the rest; we slightly abuse notation and denote c_i the pure strategy that poses probability 1 to action $c_i \in C_i$. Let $\Delta(C_i)$ denote the set of all possible strategies of player i, that is the set of all possible distributions on C_i.

The support of strategy $\sigma_i \in \Delta(\sigma_i)$ is the subset of actions of player i that are assigned strictly positive probability:

$$\text{support}(\sigma_i) = \{c_i \in C_i \mid \sigma_i(c_i) > 0\}. \tag{21.2}$$

A strategy profile $\sigma = (\sigma_i)_{i \in N}$ is a combination of strategies, one for each player, so the set of all strategy profiles is $\times_{j \in N} \Delta(C_j)$.

For any strategy profile σ, let $u_i(\sigma)$ denote the expected payoff that player i would get when the players independently choose their actions according to σ. That is,

$$u_i(\sigma) = \sum_{c \in C} \left(\prod_{j \in N} \sigma_j(c_j) \right) u_i(c) \quad \forall i \in N. \tag{21.3}$$

For any strategy profile σ, any player $i \in N$, and any $\tau_i \in \Delta(C_i)$, we denote (τ_i, σ_{-i}) the strategy profile in which the ith component is τ_i while all the other components are as in σ. Thus

$$u_i(\tau_i, \sigma_{-i}) = \sum_{c \in C} \left(\prod_{j \in N \setminus \{i\}} \sigma_j(c_j) \right) \tau_i(c_i) u_i(c) \quad \forall i \in N. \tag{21.4}$$

A game $\Gamma = \langle N, (C_i)_{i \in N}, (u_i)_{i \in N} \rangle$ is symmetric if all the players share the same action set, and the payoffs for playing a particular strategy depend only on the other strategies employed, not on who is playing them. Formally:

DEFINITION 21.2   A finite game in strategic form $\Gamma = \langle N, (C_i)_{i \in N}, (u_i)_{i \in N} \rangle$ is symmetric if $C_i = C_j = \hat{C}$ for all $i, j \in N$, and $u_i(c_i, c_{-i}) = u_j(c_j, c_{-j})$ for $c_i = c_j \in \hat{C}$ and $c_{-i} = c_{-j} \in \hat{C}^{|N|-1}$, for all $i, j \in N$.

21.2.2   Nash Equilibria

The most important solution concept in a strategic game is the Nash equilibrium (Nash, 1951). A Nash equilibrium is a strategy profile $\hat{\sigma}$ that corresponds to a steady state, in the sense that no player has a reason to change his/her strategy if everyone else adheres to $\hat{\sigma}$.

DEFINITION 21.3   Given a strategic form game $\Gamma = \langle N, (C_i)_{i \in N}, (u_i)_{i \in N} \rangle$, a strategy profile $\hat{\sigma}$ is a Nash equilibrium of Γ if

$$u_i(\hat{\sigma}) \geq u_i(\sigma_i, \hat{\sigma}_{-i}) \quad \forall i \in N, \forall \sigma_i \in \Delta(C_i), \tag{21.5}$$

or equivalently

$$u_i(\hat{\sigma}) \geq u_i(c_i, \hat{\sigma}_{-i}) \quad \forall i \in N, \forall c_i \in C_i, \tag{21.6}$$

or equivalently

$$\hat{\sigma}_i(c_i) > 0 \implies c_i \in \arg\max_{d_i \in C_i} u_i(d_i, \hat{\sigma}_{-i}) \quad \forall i \in N, \forall c_i \in C_i. \tag{21.7}$$

Thus, in a Nash equilibrium $\hat{\sigma}$, $\hat{\sigma}_i$ is a *best reply* to $\hat{\sigma}_{-i}$ for all $i \in N$, in the sense that it maximizes the expected payoff received by player i given the strategies $\hat{\sigma}_{-i}$ chosen by the rest of the players.

We can now state the general existence theorem of (Nash, 1951):

THEOREM 21.1   *Given any finite game in strategic form* $\Gamma = \langle N, (C_i)_{i \in N}, (u_i)_{i \in N} \rangle$, *there exists at least one Nash equilibrium* $\hat{\sigma} \in \times_{i \in N} \Delta(C_i)$.

A Nash equilibrium is pure if all players' strategies are pure.

21.2.3 Bimatrix Games

Bimatrix games are two-player games such that the payoff functions can be described by two real matrices R and C. In particular, an $n \times m$ bimatrix game $\Gamma = \langle R, C \rangle$ is a two-player strategic game, where R and C are $n \times m$ matrices. The n rows of R and C correspond to the actions of player 1 (the row player) and the m columns of R and C correspond to the actions of player 2 (the column player). Denote $[n]$ the set of actions available to the row player, that is, $[n] = \{1, 2, \ldots, n\}$. Similarly, let $[m] = \{1, 2, \ldots, m\}$ be the action set of the column player. For any matrix A, denote a_{ij} the element in the ith row and jth column of A. Then the payoff that the players receive in the pure strategy profile $(i, j) \in [n] \times [m]$ are r_{ij} for the row player and c_{ij} for the column player.

A strategy $\mathbf{x} = (x_1 \ \ldots \ x_n)^{\mathrm{T}}$ for the row player is a probability distribution on rows, written as an $n \times 1$ vector of probabilities. We denote $\mathbb{P}^n$ the set of all probability vectors in $\mathbb{R}^n$, that is,

$$\mathbb{P}^n = \left\{ \mathbf{x} \in \mathbb{R}^n \mid \sum_{i=1}^{n} x_i = 1 \quad \text{and} \quad x_i \geq 0 \ \forall i \in [n] \right\}. \tag{21.8}$$

Similarly, a strategy for the column player is any $\mathbf{y} \in \mathbb{P}^m$. The expected payoff that the players receive in the strategy profile $(\mathbf{x}, \mathbf{y})$ are $\mathbf{x}^{\mathrm{T}} R \mathbf{y}$ for the row player and $\mathbf{x}^{\mathrm{T}} C \mathbf{y}$ for the column player.

In the following, we denote by $\mathbf{e}_i$ the probability vector in which the ith component is 1 and all other components are 0. The size of $\mathbf{e}_i$ will be clear from the context.

Using the aforementioned notation, we restate the definition of a Nash equilibrium for the case of a bimatrix game:

DEFINITION 21.4   A strategy profile $(\mathbf{x}, \mathbf{y})$ is a Nash equilibrium for the $n \times m$ bimatrix game $\Gamma = \langle R, C \rangle$ if

1. For all $i \in [n]$, $\mathbf{e}_i^{\mathrm{T}} R \mathbf{y} \leq \mathbf{x}^{\mathrm{T}} R \mathbf{y}$ and
2. For all $j \in [m]$, $\mathbf{x}^{\mathrm{T}} C \mathbf{e}_j \leq \mathbf{x}^{\mathrm{T}} C \mathbf{y}$.

While in a Nash equilibrium no player can increase his/her payoff by unilaterally changing his/her strategy, in an ϵ-Nash equilibrium no player can increase his/her payoff by more than ϵ by unilaterally changing his/her strategy:

DEFINITION 21.5   For any $\epsilon \geq 0$ a strategy profile $(\mathbf{x}, \mathbf{y})$ is an ϵ-Nash equilibrium for the $n \times m$ bimatrix game $\Gamma = \langle R, C \rangle$ if

1. For all $i \in [n]$, $\mathbf{e}_i^{\mathrm{T}} R \mathbf{y} \leq \mathbf{x}^{\mathrm{T}} R \mathbf{y} + \epsilon$ and
2. For all $j \in [m]$, $\mathbf{x}^{\mathrm{T}} C \mathbf{e}_j \leq \mathbf{x}^{\mathrm{T}} C \mathbf{y} + \epsilon$.

A stronger notion of approximate Nash equilibrium is the ϵ-well supported Nash equilibrium (or well-supported ϵ-approximate Nash equilibrium). This is an ϵ-Nash equilibrium with the additional property that each player plays only approximately best-response pure strategies with nonzero probability:

DEFINITION 21.6   (ϵ-well-supported Nash equilibrium) For any $\epsilon \geq 0$ a strategy profile $(\mathbf{x}, \mathbf{y})$ is an ϵ-well-supported Nash equilibrium for the $n \times m$ bimatrix game $\Gamma = \langle R, C \rangle$ if

1. For all $i \in [n]$,

$$x_i > 0 \implies \mathbf{e}_i^{\mathrm{T}} R \mathbf{y} \geq \mathbf{e}_k^{\mathrm{T}} R \mathbf{y} - \epsilon \quad \forall k \in [n]$$

2. For $j \in [m]$,
$$y_j > 0 \implies \mathbf{x}^T C\mathbf{e}_j \geq \mathbf{x}^T C\mathbf{e}_k - \epsilon \quad \forall k \in [m].$$

Note that both notions of approximate equilibria are defined with respect to an additive error term ϵ. Although (exact) Nash equilibria are known not to be affected by any positive scaling of the payoff matrices, it is important to mention that approximate notions of Nash equilibria are indeed affected. Therefore, the commonly used assumption in the literature when referring to approximate Nash equilibria is that the bimatrix game is positively normalized, that is, both matrices entries lie in $[0, 1]$, and this assumption is adopted here. This is mainly done for the sake of comparison of the results on approximate equilibria. Therefore, when referring to ϵ-Nash equilibria and ϵ-well-supported Nash equilibria, ϵ lies in $[0, 1]$ and we seek for some ϵ as close to 0 as possible.

21.3 Computational Complexity of Nash Equilibria

21.3.1 The Complexity Class PPAD

The complexity class PPAD (standing for Polynomial Parity Argument in a Directed graph), introduced in (Papadimitriou, 1994), is described in terms of an algorithm, which implicitly defines a finite, though exponentially large in the size of the input, directed graph consisting of lines and cycles and having a standard source node. We are asking for any node, other than the standard source, with indegree + outdegree = 1. The existence of such a node is established via the graph-theoretic parity argument in a directed graph, which states that a directed graph G, each node of which has indegree at most 1 and outdegree at most 1, is a disjoint union of paths and cycles. Specifically, the number of endpoints (i.e., sources and sinks) must be even. Thus, given one endpoint, there must be another. A problem in PPAD is essentially a search problem for which the solution sought is guaranteed to exist by an inefficiently constructive proof.

PPAD is defined by its complete problem ANOTHER END OF DIRECTED LINES (or AEL in short) (Papadimitriou, 1994): The input instance of the problem implicitly defines a directed graph G of an exponential number of vertices (in the size of the input), each with at most one incoming edge and at most one outgoing edge. A starting vertex $\mathbf{0}$ with indegree 0 and outdegree 1 is given, and the output is a sink node, or a source other than $\mathbf{0}$. However, the graph is not directly given as the input, since then the problem would be solvable in polynomial time. Instead, the graph is generated by a given boolean circuit of polynomial size (e.g., $\log |G|$).

Formally, AEL is defined as follows. The input instance is a pair $(C, 0^n)$ where C is a boolean circuit with n input bits. For each $x \in \{0, 1\}^n$, let $C(x)$ denote the output of C when the input is x. A directed graph $G = (V, E)$ with $V = \{0, 1\}^n$ can be generated by C as follows:

1. For each vertex $v \in V$, $C(v)$ is an ordered pair (u, w). Each of u and w is either a vertex in V or nil.

2. Edge $(u, v) \in E$ if and only if v is the second component of $C(u)$ and u is the first component of $C(v)$.

C must satisfy $C(0^n) = (\text{nil}, u \neq \text{nil})$ and the first component of $C(u)$ is 0^n. These conditions on C imply that 0^n is a source of G, and we need to find either a source or a sink in G other than 0^n.

A search problem is in PPAD if it can be reduced in polynomial time to AEL. Moreover, it is PPAD-complete if there is a polynomial time reduction from AEL to it.

21.3.2 Complexity of Nash Equilibrium

Despite the certain existence of Nash equilibria in finite games, the computational complexity of finding a Nash equilibrium used to be a wide open problem for several years. Call r-NASH the problem of computing a Nash equilibrium in a game with r players. (Daskalakis et al., 2006a) showed that r-NASH is PPAD-complete for all $r \geq 4$. The proof used a reduction from a new three-dimensional discrete fixed point problem named three-dimensional Brouwer (that was proved to be PPAD-complete) to a particular type of game. Two concurrent and independent works, (Chen and Deng, 2005) and (Daskalakis and Papadimitriou, 2005), showed that three-Nash is PPAD-complete as well; the former used a direct reduction from 4-Nash while the latter used a variant of the proof in (Daskalakis et al., 2006a).

Finally, in (Chen and Deng, 2006) it was shown that two-Nash is PPAD-complete via a reduction from two-dimensional Brouwer, which is defined as follows:

two-dimensional Brouwer: The input is a pair $(C, 0^n)$ where C is a circuit with $3n$ input bits and 6 output bits $\Delta_1^+, \Delta_1^-, \Delta_2^+, \Delta_2^-, \Delta_3^+, \Delta_3^-$. It specifies a function ϕ of a very special form. Let

$$A^n = \{\mathbf{r} \in \mathbb{Z}^3 \mid 0 \leq r_i \leq 2^n - 1, \quad i = 1, 2, 3\} \tag{21.9}$$

and

$$B^n = \{\mathbf{r} \in A^n \mid \exists i : r_i = 0 \text{ or } r_i = 2^n - 1\}. \tag{21.10}$$

For each $\mathbf{r} \in A^n$, we define a cubelet in $[0, 1]^3$ as

$$\{\mathbf{q} \in \mathbb{R}^3 \mid r_i 2^{-n} \leq q_i \leq (r_i + 1)2^{-n}, \quad i = 1, 2, 3\} \tag{21.11}$$

and let $\mathbf{c_r}$ denote its center point. Function ϕ is defined on the set of 2^{3n} centers. For every center $\mathbf{c_r}$ (where $\mathbf{r} \in A^n$), $\phi(\mathbf{c_r}) \in \{\mathbf{e}^1, \mathbf{e}^2, \mathbf{e}^3, \mathbf{e}^4\} \subset \mathbb{Z}^3$, and $\phi(\mathbf{c_r})$ is specified by the output bits of $C(\mathbf{r})$ as follows:

- If $\Delta_1^+ = 1$ and the other five bits are 0, then $\phi(\mathbf{c_r}) = \mathbf{e}^1 = (1, 0, 0)$
- If $\Delta_2^+ = 1$ and the other five bits are 0, then $\phi(\mathbf{c_r}) = \mathbf{e}^2 = (0, 1, 0)$
- If $\Delta_3^+ = 1$ and the other five bits are 0, then $\phi(\mathbf{c_r}) = \mathbf{e}^3 = (0, 0, 1)$
- If $\Delta_1^- = \Delta_2^- = \Delta_3^- = 1$ and the other three bits are 0, then $\phi(\mathbf{c_r}) = \mathbf{e}^4 = (-1, -1, -1)$.

For all $\mathbf{r} \in A^n$, the six output bits of $C(\mathbf{r})$ are guaranteed to fall into one of the four cases aforementioned. C also satisfies the following boundary condition: For each $\mathbf{r} \in B^n$, if there exists $1 \leq i \leq 3$ such that $r_i = 0$, letting ℓ be the largest index such that $r_\ell = 0$, then $\phi(\mathbf{c_r}) = \mathbf{e}^\ell$; otherwise, $\phi(\mathbf{c_r}) = \mathbf{e}^4$. A vertex of the cubelet is called panchromatic if, among the eight cubelets adjacent to it, there are four that have all four vectors $\mathbf{e}^1, \mathbf{e}^2, \mathbf{e}^3$, and $\mathbf{e}^4$. The output of three-dimensional Brouwer is a panchromatic vertex of ϕ.

Now let $U = (C, 0^n)$ be an input instance of 3-dimensional Brouwer and let m be the smallest integer such that $K = 2^m > (|C| + n)^2$. (Chen and Deng, 2006) show how to construct a two-player game G^U in which both players have $2K$ strategies and prove that every ϵ-Nash equilibrium $(\mathbf{x}, \mathbf{y})$ of G^U must satisfy a set of constraints. These constraints imply that, given any ϵ-Nash equilibrium of game G^U where $\epsilon = 2^{-(m+4n)}$, a panchromatic vertex of function ϕ can be computed in polynomial time. This implies that 3-dimensional Brouwer reduces to 2-Nash.

21.4   Algorithms for Computing Nash Equilibria

21.4.1   *n*-Person Games

We will now describe a general procedure for finding Nash equilibria of any finite strategic form game $\Gamma = \langle N, (C_i)_{i \in N}, (u_i)_{i \in N} \rangle$. Although there are infinitely many strategy profiles, there is only a finite number of subsets of C that can be supports of Nash equilibria, so we can find all Nash equilibria of Γ by exhaustively searching over all possible supports.

For each player $i \in N$, let D_i be some nonempty subset of C_i, representing the set of actions of i that are assigned positive probability. If there exists a Nash equilibrium σ with support $\times_{i \in N} D_i$, then there must exist numbers $w_i \in \mathbb{R}$ for all $i \in N$ such that the following equations are satisfied:

$$\sum_{c_{-i} \in C_{-i}} \left(\prod_{j \in N \setminus \{i\}} \sigma_j(c_j) \right) u_i(d_i, c_{-i}) = w_i \quad \forall i \in N, \forall d_i \in D_i \tag{21.12}$$

$$\sigma_i(e_i) = 0 \quad \forall i \in N, \forall e_i \in C_i \setminus D_i \tag{21.13}$$

$$\sum_{c_i \in D_i} \sigma_i(c_i) = 1 \quad \forall i \in N. \tag{21.14}$$

Condition 21.12 asserts that each player gets the same payoff w_i if he/she chooses any of his/her actions that are assigned positive probability. Conditions 21.13 and 21.14 follow from the assumption that σ is a strategy profile with support $\times_{i \in N} D_i$.

These three conditions give us $\sum_{i \in N}(|C_i| + 1)$ equations in the same number of unknowns (namely, the payoffs w_i and the probabilities $\sigma_i(c_i)$ for each $i \in N, c_i \in C_i$). For games with more than two players, Equation 21.12 becomes nonlinear in σ; however, we can still solve the system of Equations 21.12 through 21.14.

Note however that the solutions of Equations 21.12 through 21.14 do not necessarily correspond to Nash equilibria of Γ since there are three difficulties that may arise. First, no solution may exist. Second, a solution might not correspond to a strategy profile σ if $\sigma_i(d_i) < 0$ for some $i \in N, d_i \in D_i$. So we must require

$$\sigma_i(d_i) \geq 0 \quad \forall i \in N, \forall d_i \in D_i. \tag{21.15}$$

Third, a solution may fail to be an equilibrium if there exists a strategy $e_i \in C_i \setminus D_i$ for some player $i \in N$ that would give player i a better payoff against σ_{-i}. So we must require

$$w_i \geq u_i(e_i, \sigma_{-i}) \quad \forall i \in N, \forall e_i \in C_i \setminus D_i. \tag{21.16}$$

Any solution $((\sigma_i)_{i \in N}, (w_i)_{i \in N})$ to Equations 21.12 through 21.14 that also satisfies Equations 21.15 and 21.16 is a Nash equilibrium of Γ. Furthermore, if there is no solution that satisfies Equations 21.12 through 21.16 then there is no Nash equilibrium with support $\times_{i \in N} D_i$. Nash's existence theorem guarantees that, if we exhaustively search over all possible $\times_{i \in N} |C_i|$ supports, we will find at least one support $\times_{i \in N} D_i$ for which Equations 21.12 through 21.16 are satisfied.

21.4.2 Two Person Games

Consider an $n \times m$ bimatrix game $\Gamma = \langle R, C \rangle$. Let $N = [n]$ and $M = [m]$ denote the action set of the row and column player respectively. Given a fixed strategy $\mathbf{y} \in \mathbb{P}^m$ for the column player, a best-response of the row player to $\mathbf{y}$ is a probability vector $\mathbf{x} \in \mathbb{P}^n$ that maximizes the expression $\mathbf{x}^T R \mathbf{y}$. Therefore, $\mathbf{x}$ is a solution to the linear program

$$\begin{aligned} \text{maximize} \quad & \mathbf{x}^T R \mathbf{y} \\ \text{subject to} \quad & \sum_{i \in N} x_i = 1 \\ & x_i \geq 0 \quad \forall i \in N. \end{aligned} \tag{21.17}$$

The dual of the linear program (Equation 21.17) is

$$\begin{aligned} \text{minimize} \quad & u \\ \text{subject to} \quad & u \geq (R\mathbf{y})_i \quad \forall i \in N. \end{aligned} \tag{21.18}$$

By the strong duality theorem of linear programming, Equations 21.17 and 21.18 have the same optimal value. Let

$$E = [1\ 1\ \ldots\ 1] \in \mathbb{R}^{1 \times N} \tag{21.19}$$

$$e = 1. \tag{21.20}$$

Then, a feasible solution $\mathbf{x}$ is optimal if and only if there is a dual solution u fulfilling $u \geq (Ry)_i$ for all $i \in N$ and $\mathbf{x}^T Ry = u$, that is $\mathbf{x}^T Ry = \mathbf{x}^T E^T u$, or equivalently

$$\mathbf{x}^T (E^T u - Ry) = 0. \tag{21.21}$$

Since $\mathbf{x}$ and $E^T u - Ry$ are nonnegative, Equation 21.21 states that the have to be complementary in the sense that they cannot both have positive components in the same position. Therefore, for each positive component of $\mathbf{x}$, the respective component of $E^T u - Ry$ is zero and u is the maximum of the components in Ry. Thus, any pure strategy $i \in N$ of the row player is a best response to $\mathbf{y}$ if and only if the ith component of $E^T u - Ry$ is zero.

Similarly, given a fixed strategy $\mathbf{x} \in \mathbb{P}^n$ for the row player, a best-response of the column player to $\mathbf{x}$ is a probability vector $\mathbf{y} \in \mathbb{P}^m$ that maximizes the expression $\mathbf{x}^T C\mathbf{y}$. Therefore, $\mathbf{y}$ is a solution to the Linear Program

$$\begin{array}{ll} \text{maximize} & \mathbf{x}^T C\mathbf{y} \\ \text{subject to} & \sum_{j \in M} y_j = 1 \\ & y_j \geq 0 \quad \forall j \in M. \end{array} \tag{21.22}$$

The dual of the linear program (Equation 21.22) is

$$\begin{array}{ll} \text{minimize} & v \\ \text{subject to} & v \geq (C^T \mathbf{x})_j \quad \forall j \in M. \end{array} \tag{21.23}$$

Now let

$$F = [1\ 1\ \ldots\ 1] \in \mathbb{R}^{1 \times M} \tag{21.24}$$

$$f = 1. \tag{21.25}$$

Here, a primal-dual pair $(\mathbf{y}, v)$ of feasible solutions is optimal if and only if

$$\mathbf{y}^T (F^T v - C^T \mathbf{x}) = 0. \tag{21.26}$$

The earlier conditions for both players yield:

THEOREM 21.2　*The bimatrix game $\Gamma = \langle R, C \rangle$ has the Nash equilibrium $(\mathbf{x}, \mathbf{y})$ if and only if for suitable $u, v \in \mathbb{R}$*

$$E\mathbf{x} = e \tag{21.27}$$

$$F\mathbf{y} = f \tag{21.28}$$

$$(E^T u - Ry)_i \geq 0 \quad \forall i \in N \tag{21.29}$$

$$(F^T v - C^T \mathbf{x})_j \geq 0 \quad \forall j \in M \tag{21.30}$$

$$x_i \geq 0 \quad \forall i \in N \tag{21.31}$$

$$y_j \geq 0 \quad \forall j \in M \tag{21.32}$$

and Equations 21.21 and 21.26 hold.

The conditions in the aforementioned theorem define a special case of a linear complementarity problem (LCP) (von Stengel, 2002). The most important method for finding a solution of the LCP defined by the theorem is the Lemke–Howson algorithm, which we will describe later in this section.

21.4.2.1 Zero-Sum Games

Consider the case where $\Gamma = \langle R, C \rangle$ is a zero-sum game, that is $C = -R$. If the row player chooses strategy $\mathbf{x}$, he/she can be sure of winning only $\min_\mathbf{y} \mathbf{x}^T R \mathbf{y}$. Thus, the optimal choice for the row player is given by $\max_\mathbf{x} \min_\mathbf{y} \mathbf{x}^T R \mathbf{y}$. Similarly, the optimal choice for the column player is given by $\max_\mathbf{y} \min_\mathbf{x} \mathbf{x}^T C \mathbf{y} = \min_\mathbf{y} \max_\mathbf{x} \mathbf{x}^T R \mathbf{y}$.

Then, the problem of computing the column player's optimal strategy can be expressed as

$$
\begin{aligned}
&\text{minimize} && u \\
&\text{subject to} && u - (R\mathbf{y})_i \geq 0 \quad \forall i \in N
\end{aligned}
$$

$$
\sum_{j \in M} y_j = 1 \tag{21.33}
$$

$$
y_j \geq 0 \quad \forall j \in M. \tag{21.34}
$$

The dual of this linear program has the form

$$
\begin{aligned}
&\text{maximize} && v \\
&\text{subject to} && v - (R^T\mathbf{x})_j \leq 0 \quad \forall j \in M
\end{aligned}
$$

$$
\sum_{i \in N} x_i = 1 \tag{21.35}
$$

$$
x_i \geq 0 \quad \forall i \in N. \tag{21.36}
$$

It is easy to verify that the aforementioned linear program describes the problem of finding an optimal strategy for the row player. Therefore, the problem of computing a Nash equilibrium for a zero-sum game is solvable in polynomial time. Moreover, by strong LP duality, $\max_\mathbf{x} \min_\mathbf{y} \mathbf{x}^T R \mathbf{y} = \min_\mathbf{y} \max_\mathbf{x} \mathbf{x}^T R \mathbf{y}$.

21.4.2.2 The Lemke–Howson Algorithm

In this section we present the classical algorithm by (Lemke and Howson, 1964) that computes a Nash equilibrium of a nondegenerate bimatrix game $\Gamma = \langle R, C \rangle$.

DEFINITION 21.7 A bimatrix game $\Gamma = \langle R, C \rangle$ is called nondegenerate if the number of pure best responses to a strategy never exceeds the size of its support.

It is useful to assume that the actions sets of the players are disjoint. In particular, we assume that the action set of the row player is $N = \{1, \ldots, n\}$ and the action set of the column player is $M = \{n+1, \ldots, n+m\}$. For each $i \in N$ and $j \in M$, define

$$
X(i) = \{\mathbf{x} \in \mathbb{P}^n \mid x_i = 0\} \tag{21.37}
$$

$$
X(j) = \left\{\mathbf{x} \in \mathbb{P}^n \mid \sum_{i=1}^n c_{ij} x_i \geq \sum_{i=1}^n c_{ik} x_i \quad \forall k \in M\right\} \tag{21.38}
$$

$$
Y(j) = \{\mathbf{y} \in \mathbb{P}^m \mid y_j = 0\} \tag{21.39}
$$

$$Y(i) = \left\{ \mathbf{y} \in \mathbb{P}^m \mid \sum_{j=1}^{m} r_{ij} y_j \geq \sum_{j=1}^{n} r_{kj} y_j \quad \forall k \in N \right\}. \tag{21.40}$$

Any strategy $\mathbf{x} \in \mathbb{P}^n$ and $\mathbf{y} \in \mathbb{P}^m$ is labelled with certain elements of $N \cup M$ as follows: for each $k \in N \cup M$, $\mathbf{x}$ has label k if $x \in X(k)$ and $\mathbf{y}$ has label k if $\mathbf{y} \in Y(k)$. So the labels of a strategy $\mathbf{x}$ ($\mathbf{y}$) of the row (column) player denote the actions of the row (column) player that are assigned zero probability and the actions of the column (row) player that are best responses to $\mathbf{x}$ ($\mathbf{y}$). Since, in any Nash equilibrium of the game, any action of a player is either a best response to the opponent's strategy or is played with probability zero, it follows that a strategy profile $(\mathbf{x}, \mathbf{y})$ is a Nash equilibrium if and only if $\mathbf{x}$ and $\mathbf{y}$ are completely labelled, that is, the union of the labels of $\mathbf{x}$ and $\mathbf{y}$ is the set $N \cup M$.

THEOREM 21.3   *A strategy profile $(\mathbf{x}, \mathbf{y})$ is a Nash equilibrium of the bimatrix game $\Gamma = \langle R, C \rangle$ if and only if, for all $k \in N \cup M$, $\mathbf{x} \in X(k)$ or $\mathbf{y} \in Y(k)$ (or both).*

     Now, since the game is nondegenerate, only finitely many strategies of the row player have n labels and only finitely many strategies of the column player have m labels. Using these finitely many strategies we define two graphs G_1 and G_2 as follows. Let G_1 be the graph whose nodes are the strategies $\mathbf{x} \in \mathbb{P}^n$ of the row player with exactly n labels, with an additional node $\mathbf{0} \in \mathbb{R}^n$ that has all labels in N. There is an edge between any two nodes $\mathbf{x}, \mathbf{x}'$ of G_1 if and only if $\mathbf{x}$ and $\mathbf{x}'$ differ in exactly one label. Similarly, let G_2 be the graph whose nodes are the strategies $\mathbf{y} \in \mathbb{P}^m$ of the column player with exactly m labels, with an additional node $\mathbf{0} \in \mathbb{R}^m$ that has all labels in M. There is an edge between any two nodes $\mathbf{y}, \mathbf{y}'$ of G_2 if and only if $\mathbf{y}$ and $\mathbf{y}'$ differ in exactly one label.

     Now define the product graph $G_1 \times G_2$ as the graph whose nodes are all pairs $(\mathbf{x}, \mathbf{y})$ such that $\mathbf{x}$ is a node of G_1 and $\mathbf{y}$ is a node of G_2. There is an edge between any two nodes $(\mathbf{x}, \mathbf{y})$ and $(\mathbf{x}', \mathbf{y}')$ if and only if (1) $\mathbf{x} = \mathbf{x}'$ is a node of G_1 and $(\mathbf{y}, \mathbf{y}')$ is an edge of G_2 or (2) $\mathbf{y} = \mathbf{y}'$ is a node of G_2 and $(\mathbf{x}, \mathbf{x}')$ is an edge of G_1.

     For any $k \in N \cup M$, call a node $(\mathbf{x}, \mathbf{y})$ of $G_1 \times G_2$ k-almost completely labelled if any label $\ell \in N \cup M \setminus \{k\}$ is a label of $\mathbf{x}$ or a label of $\mathbf{y}$. Since two adjacent nodes $\mathbf{x}$ and $\mathbf{x}'$ in G_1 have exactly $n - 1$ common labels, the edge $((\mathbf{x}, \mathbf{y}), (\mathbf{x}', \mathbf{y}))$ is also called k-almost completely labelled if $\mathbf{y}$ has the remaining m labels except k. Similarly, since two adjacent nodes $\mathbf{y}$ and $\mathbf{y}'$ in G_2 have exactly $m - 1$ common labels, the edge $((\mathbf{x}, \mathbf{y}), (\mathbf{x}, \mathbf{y}'))$ is called k-almost completely labelled if $\mathbf{x}$ has the remaining n labels except k.

     A Nash equilibrium $(\mathbf{x}, \mathbf{y})$ is a node in $G_1 \times G_2$ adjacent to exactly one node $(\mathbf{x}', \mathbf{y}')$ that is k-almost completely labelled. In particular, if k is a label of $\mathbf{x}$, then $\mathbf{x}$ is joined to the node $\mathbf{x}'$ in G_1 sharing the remaining $n - 1$ labels, and $\mathbf{y} = \mathbf{y}'$. If k is a label of $\mathbf{y}$, then $\mathbf{y}$ is joined to the node $\mathbf{y}'$ in G_2 sharing the remaining $m - 1$ labels, and $\mathbf{x} = \mathbf{x}'$. Moreover, a k-almost completely labelled node $(\mathbf{x}, \mathbf{y})$ that is completely labelled has exactly 2 neighbors in $G_1 \times G_2$. These are obtained by dropping the unique duplicate label that $\mathbf{x}$ and $\mathbf{y}$ have in common, joining to an adjacent node either in G_1 and keeping $\mathbf{y}$ fixed, or in G_2 and keeping $\mathbf{x}$ fixed. This defines a unique k-almost completely labelled path in $G_1 \times G_2$ connecting the completely labelled artificial equilibrium $(\mathbf{0}, \mathbf{0})$ to an actual equilibrium of Γ. The Lemke–Howson algorithm starts from $(\mathbf{0}, \mathbf{0})$, follows the path where label k is missing, and terminates at a Nash equilibrium of Γ.

THEOREM 21.4   (Lemke and Howson, 1964; Shapley, 1974) *Let $\Gamma = \langle R, C \rangle$ be a nondegenerate bimatrix game and $k \in N \cup M$. Then the set of k-almost completely labelled nodes and edges in $G_1 \times G_2$ consists of disjoint paths and cycles. The endpoints of the paths are the equilibria of the game and the artificial equilibrium $(\mathbf{0}, \mathbf{0})$. The number of Nash equilibria of Γ is odd.*

Savani and von Stengel (2004) proved that the Lemke–Howson algorithm may require an exponential number of steps for a specific class of inputs. Moreover, note that the Lemke–Howson algorithm can be extended to degenarate bimatrix games as well, by "lexicographically perturbation" (von Stengel, 2002).

21.5 Approximate Equilibria

21.5.1 A Nonpolynomial Algorithm

In Althöfer (1994) it is showed that, for any $n \times 1$ probability vector $\mathbf{p}$ there exists an $n \times 1$ probability vector $\hat{\mathbf{p}}$ with logarithmic support in n, so that for a fixed matrix C, $\max_j \left| \mathbf{p}^T C \mathbf{e}_j - \hat{\mathbf{p}}^T C \mathbf{e}_j \right| \leq \epsilon$, for any constant $\epsilon > 0$. Using this fact, Lipton et al. (2003) proved that, for any bimatrix game and for any constant $\epsilon > 0$, there exists an ϵ-Nash equilibrium with only logarithmic support (in the number n of available pure strategies). The proof is based on the probabilistic method and yields a nonpolynomial algorithm for computing an ϵ-Nash equilibrium for *any* $\epsilon > 0$.

We will now formally state and prove the aforementioned result. First we give the definition of a *k*-uniform strategy: Given an $n \times m$ bimatrix game $\Gamma = \langle R, C \rangle$, a *k*-uniform strategy for the row (column) player is the uniform distribution on a multiset S of actions of the row (column) player, with $|S| = k$.

THEOREM 21.5 (Lipton et al., 2003) *For any Nash equilibrium $(\hat{\mathbf{x}}, \hat{\mathbf{y}})$ of an $n \times n$ bimatrix game $\Gamma = \langle R, C \rangle$ and for every $\epsilon > 0$, there exists, for every $k \geq \frac{12 \ln n}{\epsilon^2}$, a pair of k-uniform strategies $\mathbf{x}$, $\mathbf{y}$, such that:*

1. $(\mathbf{x}, \mathbf{y})$ is an ϵ-Nash equilibrium,
2. $|\mathbf{x}^T R \mathbf{y} - \hat{\mathbf{x}}^T R \hat{\mathbf{y}}| < \epsilon$,
3. $|\mathbf{x}^T C \mathbf{y} - \hat{\mathbf{x}}^T C \hat{\mathbf{y}}| < \epsilon$.

PROOF For the given $\epsilon > 0$, fix some $k \geq \frac{12 \ln n}{\epsilon^2}$. Form a multiset A by sampling k times from the set of actions of the row player, independently at random and according to the distribution $\hat{\mathbf{x}}$, and a multiset B by sampling k times from the set of actions of the column player, independently at random and according to the distribution $\hat{\mathbf{y}}$. Let $\mathbf{x}$ be the strategy for the row player that assigns probability $\frac{1}{k}$ to each member of A and 0 each action not in A. Similarly, let $\mathbf{y}$ be the strategy for the row player that assigns probability $\frac{1}{k}$ to each member of B and 0 each action not in B. Note that, if an action occurs α times in the multiset, then it is assigned probability $\frac{\alpha}{k}$.

Define the following events:

$$\phi_1 = \left\{ |\mathbf{x}^T R \mathbf{y} - \hat{\mathbf{x}}^T R \hat{\mathbf{y}}| < \epsilon/2 \right\} \tag{21.41}$$

$$\pi_{1,i} = \left\{ \mathbf{e}_i^T R \mathbf{y} < \mathbf{x}^T R \mathbf{y} + \epsilon \right\} \quad \forall i \in \{1, \dots, n\} \tag{21.42}$$

$$\phi_2 = \left\{ |\mathbf{x}^T C \mathbf{y} - \hat{\mathbf{x}}^T C \hat{\mathbf{y}}| < \epsilon/2 \right\} \tag{21.43}$$

$$\pi_{2,j} = \left\{ \mathbf{x}^T C \mathbf{e}_j < \mathbf{x}^T C \mathbf{y} + \epsilon \right\} \quad \forall j \in \{1, \dots, n\} \tag{21.44}$$

$$\text{GOOD} = \phi_1 \cap \phi_2 \bigcap_{i=1}^{n} \pi_{1,i} \bigcap_{j=1}^{n} \pi_{2,j}. \tag{21.45}$$

We wish to show that $\Pr\{\text{GOOD}\} > 0$, which would imply that there exists some choice of the multisets A and B such that $\mathbf{x}$ and $\mathbf{y}$ satisfy the three conditions in the statement of the theorem.

In order to bound the probabilities of the events $\pi_{1,i}$'s and $\pi_{2,j}$'s we introduce the following events:

$$\psi_{1,i} = \left\{ \mathbf{e}_i^T R\mathbf{y} < \mathbf{e}_i^T R\hat{\mathbf{y}} + \epsilon/2 \right\} \quad \forall i \in \{1,\dots,n\} \tag{21.46}$$

$$\psi_{2,j} = \left\{ \mathbf{x}^T C\mathbf{e}_j < \hat{\mathbf{x}}^T C\mathbf{e}_j + \epsilon/2 \right\} \quad \forall i \in \{1,\dots,n\}. \tag{21.47}$$

Now assume that

$$|\mathbf{x}^T R\mathbf{y} - \hat{\mathbf{x}}^T R\hat{\mathbf{y}}| < \epsilon/2 \tag{21.48}$$

or equivalently

$$-\epsilon/2 < \mathbf{x}^T R\mathbf{y} - \hat{\mathbf{x}}^T R\hat{\mathbf{y}} < \epsilon/2. \tag{21.49}$$

Furthermore, assume that

$$\mathbf{e}_i^T R\mathbf{y} < \mathbf{e}_i^T R\hat{\mathbf{y}} + \epsilon/2 \tag{21.50}$$

for some $i \in \{1,\dots,n\}$. Then,

$$\mathbf{e}_i^T R\mathbf{y} < \mathbf{e}_i^T R\hat{\mathbf{y}} + \epsilon/2 \tag{21.51}$$

$$\leq \hat{\mathbf{x}}^T R\hat{\mathbf{y}} + \epsilon/2 \quad \text{(since } (\hat{\mathbf{x}}, \hat{\mathbf{y}}) \text{ is a Nash equilibrium)} \tag{21.52}$$

$$< \mathbf{x}^T R\mathbf{y} + \epsilon \quad \text{(due to assumption 21.49).} \tag{21.53}$$

Therefore,

$$\psi_{1,i} \cap \phi_1 \subseteq \pi_{1,i} \quad \forall i \in \{1,\dots,n\} \tag{21.54}$$

and similarly it can be shown that

$$\psi_{2,j} \cap \phi_2 \subseteq \pi_{2,j} \quad \forall j \in \{1,\dots,n\}. \tag{21.55}$$

The expression $\mathbf{e}_i^T R\mathbf{y}$ is essentially a sum of k independent random variables each of expected value $\mathbf{e}_i^T R\hat{\mathbf{y}}$. Each such random variable takes values in the interval $[0,1]$. Therefore, we can apply a standard tail inequality (Hoeffding, 1963) in order to bound the probability of $\psi_{1,i}^c$ (i.e., the complement of event $\psi_{1,i}$) occurring, and thus we get:

$$\Pr\{\psi_{1,i}^c\} \leq \exp\left(-\frac{k\epsilon^2}{2}\right). \tag{21.56}$$

Similarly,

$$\Pr\{\psi_{2,j}^c\} \leq \exp\left(-\frac{k\epsilon^2}{2}\right). \tag{21.57}$$

In order to bound the probabilities of the events ϕ_1^c and ϕ_2^c we define the following events:

$$\phi_{1a} = \left\{ |\mathbf{x}^T R\hat{\mathbf{y}} - \hat{\mathbf{x}}^T R\hat{\mathbf{y}}| < \epsilon/4 \right\} \tag{21.58}$$

$$\phi_{1b} = \left\{ |\mathbf{x}^T R\mathbf{y} - \mathbf{x}^T R\hat{\mathbf{y}}| < \epsilon/4 \right\} \tag{21.59}$$

$$\phi_{2a} = \left\{ |\hat{\mathbf{x}}^T C\mathbf{y} - \hat{\mathbf{x}}^T C\hat{\mathbf{y}}| < \epsilon/4 \right\} \tag{21.60}$$

$$\phi_{2b} = \left\{ |\mathbf{x}^T C\mathbf{y} - \hat{\mathbf{x}}^T C\mathbf{y}| < \epsilon/4 \right\}. \tag{21.61}$$

We can easily see that $\phi_{1a} \cap \phi_{1b} \subseteq \phi_1$ and $\phi_{2a} \cap \phi_{2b} \subseteq \phi_2$. The expression $\mathbf{x}^T R \hat{\mathbf{y}}$ is a sum of k independent random variables, each of expected value $\hat{\mathbf{x}}^T R \hat{\mathbf{y}}$ and each taking values in the interval $[0, 1]$. Therefore, we can apply the Hoeffding bound again and get:

$$\Pr\{\phi_{1a}^c\} \le 2 \exp\left(-\frac{k\epsilon^2}{8}\right) \tag{21.62}$$

and, using similar arguments,

$$\Pr\{\phi_{1b}^c\} \le 2 \exp\left(-\frac{k\epsilon^2}{8}\right). \tag{21.63}$$

Therefore,

$$\Pr\{\phi_1^c\} \le 4 \exp\left(-\frac{k\epsilon^2}{8}\right). \tag{21.64}$$

Similar reasoning yields that

$$\Pr\{\phi_2^c\} \le 4 \exp\left(-\frac{k\epsilon^2}{8}\right). \tag{21.65}$$

Now

$$\Pr\{\text{GOOD}^c\} \le \Pr\{\phi_1^c\} + \Pr\{\phi_2^c\} + \sum_{i=1}^{n} \Pr\{\pi_{1,i}^c\} + \sum_{j=1}^{n} \Pr\{\pi_{2,j}^c\} \tag{21.66}$$

$$\le \Pr\{\phi_1^c\} + \Pr\{\phi_2^c\} + \sum_{i=1}^{n} \left(\Pr\{\psi_{1,i}^c\} + \Pr\{\phi_1^c\}\right)$$

$$+ \sum_{j=1}^{n} \left(\Pr\{\psi_{2,j}^c\} + \Pr\{\phi_2^c\}\right) \tag{21.67}$$

$$\le 8n \exp\left(-\frac{k\epsilon^2}{8}\right) + 8 \exp\left(-\frac{k\epsilon^2}{8}\right) + 2n \exp\left(-\frac{k\epsilon^2}{2}\right). \tag{21.68}$$

Now, since $k \ge 12 \ln n / \epsilon^2$,

$$\Pr\{\text{GOOD}^c\} \le 8n \exp\left(-\frac{3 \ln n}{2}\right) + 8 \exp\left(-\frac{3 \ln n}{2}\right)$$

$$+ 2n \exp\left(-6 \ln n\right) \tag{21.69}$$

$$= 8n \cdot n^{-\frac{3}{2}} + 8 \cdot n^{-\frac{3}{2}} + 2n \cdot n^{-6} \tag{21.70}$$

$$< 1, \tag{21.71}$$

and therefore, $\Pr\{\text{GOOD}\} > 0$ as needed.

COROLLARY 21.1 (Lipton et al., 2003) For an $n \times n$ bimatrix game $\Gamma = \langle R, C \rangle$, there exists an algorithm for computing all k-uniform ϵ-Nash equilibria, for any $\epsilon > 0$.

PROOF   For the given $\epsilon > 0$, fix $k = \frac{12 \ln n}{\epsilon^2}$. By exhaustive search, we can find all possible pairs of multisets A and B. For each such pair, we can check in polynomial time if the pair of k-uniform strategies is an ϵ-Nash equilibrium. By Theorem 21.5, at least one pair of multiset exists such that the corresponding k-uniform strategies are ϵ-Nash equilibrium strategies. The running time of the algorithm is nonpolynomial, since there are

$$\binom{n+k-1}{k}^2 \leq \left(\frac{(n+k-1)e}{k}\right)^{2k} = O\left(n^{\frac{\ln n}{\epsilon^2}}\right)$$

possible pairs of multisets to check.

As pointed out in Althöfer (1994), no algorithm that examines supports smaller than about $\ln n$ can achieve an approximation better than $\frac{1}{4}$. Moreover, (Chen et al., 2006) proved the following:

THEOREM 21.6 (Chen et al., 2006) *The problem of computing a $\frac{1}{n^{\Theta(1)}}$-Nash equilibrium of a $n \times n$ bimatrix game is PPAD-complete.*

Theorem 21.6 asserts that, unless PPAD $\subseteq$ P, there exists no fully polynomial time approximation scheme for computing equilibria in bimatrix games. However, this does not rule out the existence of a polynomial approximation scheme for computing an ϵ-Nash equilibrium when ϵ is an absolute constant, or even when $\epsilon = \Theta\left(\frac{1}{\text{poly}(\ln n)}\right)$. Furthermore, as observed in (Chen et al., 2006), if the problem of finding an ϵ-Nash equilibrium were PPAD-complete when ϵ is an absolute constant, then, due to Theorem 21.5, all PPAD problems would be solved in $O(n^{\ln n})$ time, which is unlikely to be the case.

Two concurrent and independent works (Daskalakis et al., 2006b; Kontogiannis et al., 2006) were the first to make progress in providing ϵ-Nash equilibria for bimatrix games and some constant $0 < \epsilon < 1$. In particular, the work of (Kontogiannis et al., 2006) proposes a simple linear-time algorithm for computing a $\frac{3}{4}$-Nash equilibrium for any bimatrix game:

THEOREM 21.7 (Kontogiannis et al., 2006) *Consider any $n \times m$ bimatrix game $\Gamma = \langle R, C \rangle$ and let $r_{i_1,j_1} = \max_{i,j} r_{ij}$ and $c_{i_2,j_2} = \max_{i,j} c_{ij}$. Then the pair of strategies $(\hat{x}, \hat{y})$ where $\hat{x}_{i_1} = \hat{x}_{i_2} = \hat{y}_{j_1} = \hat{y}_{j_2} = \frac{1}{2}$ is a $\frac{3}{4}$-Nash equilibrium for Γ.*

The aforementioned technique can be extended so as to obtain a parameterized, stronger approximation:

THEOREM 21.8 (Kontogiannis et al., 2006) *Consider a $n \times m$ bimatrix game $\Gamma = \langle A, B \rangle$. Let λ_1^* (λ_2^*) be the minimum, among all Nash equilibria of Γ, expected payoff for the row (column) player and let $\lambda = \max\{\lambda_1^*, \lambda_2^*\}$. Then, there exists a $\frac{2+\lambda}{4}$-Nash equilibrium that can be computed in time polynomial in n and m.*

The work of (Daskalakis et al., 2006b) provides a simple algorithm for computing a $\frac{1}{2}$-Nash equilibrium: Pick an arbitrary row for the row player, say row i. Let $j = \arg\max_{j'} c_{ij'}$. Let $k = \arg\max_{k'} r_{k'j}$. Thus, j is a best-response column for the column player to the row i, and k is a best-response row for the row player to the column j. Let $\hat{x} = \frac{1}{2}e_i + \frac{1}{2}e_k$ and $\hat{y} = e_j$, that is, the row player plays row i or row k with probability $\frac{1}{2}$ each, while the column player plays column j with probability 1. Then:

THEOREM 21.9 (Daskalakis et al., 2006b) *The strategy profile $(\hat{x}, \hat{y})$ is a $\frac{1}{2}$-Nash equilibrium.*

Later, there was a series of results improving the constant for polynomial time constructions of approximate Nash equilibria. A polynomial construction (based on Linear Programming) of a 0.38-Nash equilibrium is presented in (Daskalakis et al., 2007), and consequently Bosse et al. (2007) proposed a 0.36392-Nash equilibrium based on the solvability of zero sum bimatrix games. Finally, Tsaknakis and Spirakis (2007) proposed a new methodology for determining approximate Nash equilibria of bimatrix games and based on that, they provided a polynomial time algorithm for computing 0.3393-approximate Nash equilibria. To our knowledge this is currently the best result for approximate Nash equilibria in bimatrix games.

For the more demanding notion of well-supported approximate Nash equilibrium, Daskalakis et al. (2006b) propose an algorithm, which, under a quite interesting and plausible graph theoretic conjecture, constructs in polynomial time a $\frac{5}{6}$-well-supported Nash equilibrium. However, the status of this conjecture is still unknown. In Daskalakis et al. (2006b), it is also shown how to transform a $[0,1]$-bimatrix game to a win–lose bimatrix game (i.e., a bimatrix game where each matrix entry is either 0 or 1) of the same size, so that each ϵ-well-supported Nash equilibrium of the resulting game is $\frac{1+\epsilon}{2}$-well-supported Nash equilibrium of the original game.

The work of Kontogiannis and Spirakis (2007) provides a polynomial algorithm that computes a $\frac{1}{2}$-well-supported Nash equilibrium for arbitrary win–lose games. The idea behind this algorithm is to split evenly the divergence from a zero-sum game between the two players and then solve this zero-sum game in polynomial time (using its direct connection to Linear Programming). The computed Nash equilibrium of the zero-sum game considered is indeed proved to be also a $\frac{1}{2}$-well-supported Nash equilibrium for the initial win–lose game. Therefore:

THEOREM 21.10 (Kontogiannis and Spirakis, 2007) *For any win–lose bimatrix game, there is a polynomial time constructible profile that is a $\frac{1}{2}$-well-supported Nash equilibrium of the game.*

In the same work, Kontogiannis and Spirakis (2007) parameterize the aforementioned methodology in order to apply it to arbitrary bimatrix games. This new technique leads to a weaker ϕ-well-supported Nash equilibrium for win–lose games, where $\phi = \frac{\sqrt{5}-1}{2}$ is the golden ratio. Nevertheless, this parameterized technique extends nicely to a technique for arbitrary bimatrix games, which assures a $\left(\frac{\sqrt{11}}{2} - 1\right)$-well-supported Nash equilibrium in polynomial time:

THEOREM 21.11 (Kontogiannis and Spirakis, 2007) *For any bimatrix game, a 0.658-well-supported Nash equilibrium is constructible in polynomial time.*

21.6 Potential Games and Congestion Games

21.6.1 Potential Games

Potential games, defined in (Monderer and Shapley, 1996), are games with the property that the incentive of all players to unilaterally deviate from a pure strategy profile can be expressed in one global function, the potential function.

Fix an arbitrary game in strategic form $\Gamma = \langle N, (C_i)_{i \in N}, (u_i)_{i \in N} \rangle$ and some vector $\mathbf{b} = (b_1, \ldots, b_{|N|}) \in \mathbb{R}_{>0}^{|N|}$. A function $P : C \to \mathbb{R}$ is called

- An ordinal potential for Γ if, $\forall c \in C, \forall i \in N, \forall a \in C_i$,

$$P(a, c_{-i}) - P(c) > 0 \iff u_i(a, c_{-i}) - u_i(c) > 0. \tag{21.72}$$

- A **b**-potential for Γ if, $\forall c \in C, \forall i \in N, \forall a \in C_i$,

$$P(a, c_{-i}) - P(c) = b_i \cdot (u_i(a, c_{-i}) - u_i(c)). \tag{21.73}$$

- An exact potential for Γ if it is a **b**-potential for Γ where $b_i = 1$ for all $i \in N$.

It is straightforward to see that the existence of an ordinal, exact, or **b**-potential function P for a finite game Γ guarantees the existence of pure Nash equilibria in Γ: each local optimum of P corresponds to a pure Nash equilibrium of Γ and vice versa. Thus, the problem of finding pure Nash equilibria of a potential game Γ is equivalent to finding the local optima for the optimization problem with state space the pure strategy space C of the game and objective the potential function of the game.

Furthermore, the existence of a potential function P for a game $\Gamma = \langle N, (C_i)_{i \in N}, (u_i)_{i \in N} \rangle$ implies a straightforward algorithm for constructing a pure Nash equilibrium of Γ: The algorithm starts from an arbitrary strategy profile $c \in C$ and, at each step, one single player performs a selfish step, that is, switches to a pure strategy that strictly improves his/her payoff. Since the payoff of the player increases, P increases as well. When no move is possible, that is, when a pure strategy profile $\hat{c}$ is reached from which no player has an incentive to unilaterally deviate, then $\hat{c}$ is a pure Nash equilibrium and a local optimum of P. This procedure however does not imply that the computation of a pure Nash equilibrium can be done in polynomial time, since the improvements in the potential can be very small and too many.

21.6.2 Congestion Games

Rosenthal (1973) introduced a class of games, called congestion games, in which each player chooses a particular subset of resources out of a family of allowable subsets for his/her (his/her action set), constructed from a basic set of primary resources for all the players. The delay associated with each primary resource is a non-decreasing function of the number of players who choose it, and the total delay received by each player is the sum of the delays associated with the primary resources he/she chooses. Each game in this class possesses at least one Nash equilibrium in pure strategies, which follows from the existence of an exact potential.

A congestion model $\langle N, E, (\Pi_i)_{i \in N}, (d_e)_{e \in E} \rangle$ is defined as follows. N denotes the set of players $\{1, \ldots, n\}$. E denotes a finite set of resources. For $i \in N$ let Π_i be the set of strategies of player i, where each $\varpi_i \in \Pi_i$ is a nonempty subset of resources. For $e \in E$ let $d_e : \{1, \ldots, n\} \to \mathbb{R}$ denote the delay function, where $d_e(k)$ denotes the cost (e.g., delay) to each user of resource e, if there are exactly k players using e.

The congestion game associated with this congestion model is the game in strategic form $\langle N, (\Pi_i)_{i \in N}, (u_i)_{i \in N} \rangle$, where the payoff functions u_i are defined as follows: Let $\Pi \equiv \times_{i \in N} \Pi_i$. For all $\varpi = (\varpi_1, \ldots, \varpi_n) \in \Pi$ and for every $e \in E$ let $\sigma_e(\varpi)$ be the number of users of resource e according to the configuration ϖ: $\sigma_e(\varpi) = |\{i \in N : e \in \varpi_i\}|$. Define $u_i : \Pi \to \mathbb{R}$ by $u_i(\varpi) = -\sum_{e \in \varpi_i} d_e(\sigma_e(\varpi))$.

In a network congestion game the families of subsets Π_i are represented implicitly as paths in a network. We are given a directed network $G = (V, E)$ with the edges playing the role of resources, a pair of nodes $(s_i, t_i) \in V \times V$ for each player i and the delay function d_e for each $e \in E$. The strategy set of player i is the set of all paths from s_i to t_i. If all origin–destination pairs (s_i, t_i) of the players coincide with a unique pair (s, t) we have a single-commodity network congestion game and then all users share the same strategy set, hence the game is symmetric.

In a weighted congestion model we allow the users to have different demands, and thus affect the resource delay functions in a different way, depending on their own weights. A weighted congestion model $\langle N, (w_i)_{i \in N}, E, (\Pi_i)_{i \in N}, (d_e)_{e \in E} \rangle$ is defined as follows. N denotes the set of players $\{1, 2, \ldots, n\}$, w_i denotes the demand of player i and E denotes a finite set of resources. For $i \in N$ let Π_i be the set of strategies of player i, where each $\varpi_i \in \Pi_i$ is a nonempty subset of resources. For each resource

$e \in E$ let $d_e(\cdot)$ be the delay per user that requests its service, as a function of the total usage of this resource by all the users.

The weighted congestion game associated with this congestion model is the game in strategic form $\langle (w_i)_{i \in N}, (\Pi_i)_{i \in N}, (u_i)_{i \in N} \rangle$, where the payoff functions u_i are defined as follows. For any configuration $\varpi \in \Pi$ and for all $e \in E$, let $\Lambda_e(\varpi) = \{i \in N : e \in \varpi_i\}$ be the set of players using resource e according to ϖ. The cost $\lambda^i(\varpi)$ of user i for adopting strategy $\varpi_i \in \Pi_i$ in a given configuration ϖ is equal to the cumulative delay $\lambda_{\varpi_i}(\varpi)$ on the resources that belong to ϖ_i:

$$\lambda^i(\varpi) = \lambda_{\varpi_i}(\varpi) = \sum_{e \in \varpi_i} d_e(\theta_e(\varpi)) \tag{21.74}$$

where, for all $e \in E$, $\theta_e(\varpi) \equiv \sum_{i \in \Lambda_e(\varpi)} w_i$ is the load on resource e with respect to the configuration ϖ. The payoff function for player i is then $u_i(\varpi) = -\lambda^i(\varpi)$. A configuration $\varpi \in \Pi$ is a pure Nash equilibrium if and only if, for all $i \in N$,

$$\lambda_{\varpi_i}(\varpi) \leq \lambda_{\pi_i}(\pi_i, \varpi_{-i}) \quad \forall \pi_i \in \Pi_i \tag{21.75}$$

where (π_i, ϖ_{-i}) is the same configuration as ϖ except for user i that has now been assigned to path π_i.

In a weighted network congestion game the strategy sets Π_i are represented implicitly as $s_i - t_i$ paths in a directed network $G = (V, E)$.

Since the payoff functions u_i of a congestion game can be implicitly computed by the resource delay functions d_e, in the following we will denote a general (weighted or unweighted) congestion game by $\langle N, E, (\Pi_i)_{i \in N}, (w_i)_{i \in N}, (d_e)_{e \in E} \rangle$. We drop $(w_i)_{i \in N}$ from this notation when referring to an unweighted congestion game.

The following theorem (Rosenthal, 1973; Monderer and Shapley, 1996) proves the strong connection of unweighted congestion games with the exact potential games.

THEOREM 21.12 (Rosenthal, 1973; Monderer and Shapley, 1996) *Every (unweighted) congestion game is an exact potential game.*

PROOF Fix an arbitrary (unweighted) congestion game $\Gamma = \langle N, E, (\Pi_i)_{i \in N}, (d_e)_{e \in E} \rangle$. For any pure strategy profile $\varpi \in \Pi$, the function

$$\Phi(\varpi) = \sum_{e \in \cup_{i \in N} \varpi_i} \sum_{k=1}^{\sigma_e(\varpi)} d_e(k) \tag{21.76}$$

(introduced in (Rosenthal, 1973)) is an exact potential function for Γ.   □

The converse of Theorem 21.12 does not hold in general; however, Monderer and Shapley (1996) proved that every (finite) exact potential game Γ is isomorphic to an unweighted congestion game.

21.6.2.1 The Complexity Class PLS

A crucial class of problems containing the family of weighted congestion games is PLS (standing for polynomial local search) (Johnson et al., 1988). This is the subclass of total functions in NP that are guaranteed to have a solution because of the fact that "every finite directed acyclic graph has a sink."

A local search problem Π has a set of instances D_Π which are strings. To each instance $x \in D_\Pi$, there corresponds a set $S_\Pi(x)$ of solutions, and a standard solution $s_0 \in S_\Pi(x)$. Each solution $s \in S_\Pi(x)$ has a cost $f_\Pi(s, x)$ and a neighborhood $N_\Pi(s, x)$. The search problem is, given an instance $x \in D_\Pi$, to find a locally optimal solution $s^* \in S_\Pi(x)$. That is, if the objective is to minimize the cost function, then we are seeking for some $s^* \in \arg\min_{s \in N_\Pi(s^*, x)} \{f_\Pi(s, x)\}$. Similarly, if the objective is to maximize the cost function, then we are seeking for some $s^* \in \arg\max_{s \in N_\Pi(s^*, x)} \{f_\Pi(s, x)\}$.

DEFINITION 21.8 A local search problem Π is in PLS if its instances D_Π and solutions $S_\Pi(x)$ for all $x \in D_\Pi$ are binary strings, there is a polynomial p such that the length of the solutions $S_\Pi(x)$ is bounded by $p(|x|)$, and there are three polynomial-time algorithms A_Π, B_Π, C_Π with the following properties:

1. Given a binary string x, A_Π determines whether $x \in D_\Pi$, and if so, returns some initial solution $s_0 \in S_\Pi(x)$.
2. Given an instance $x \in D_\Pi$ and a string s, B_Π determines whether $s \in S_\Pi(x)$, and if so, computes the value $f_\Pi(s, x)$ of the cost function at s.
3. Given an instance $x \in D_\Pi$ and a solution $s \in S_\Pi(x)$, C_Π determines whether s is a local optimum of $f_\Pi(\cdot, x)$ in its neighborhood $N_\Pi(s, x)$, and if not, returns a neighbor $s' \in N_\Pi(s, x)$ having a better value (i.e., $f_\Pi(s', x) < f_\Pi(s, x)$ for a minimization problem and $f_\Pi(s', x) > f_\Pi(s, x)$ for a maximization problem).

21.6.2.2 Complexity of Pure Nash Equilibria in Congestion Games

Fabrikant et al. (2004) characterized the complexity of computing pure Nash equilibria in congestion games. They showed that, for unweighted single commodity network congestion games, a pure Nash equilibrium can be constructed in polynomial time. On the other hand, they showed that, even for symmetric congestion games it is PLS-complete to compute a pure Nash equilibrium.

THEOREM 21.13 (Fabrikant et al., 2004) *There is a polynomial time algorithm for finding a pure Nash equilibrium in symmetric network congestion games.*

PROOF The algorithm is a reduction to min-cost flow. Given the single-source single-destination network $G = (V, E)$ and the delay functions d_e, we replace each edge e with n parallel edges between the same nodes, each with capacity 1, and with costs $d_e(1), \ldots, d_e(n)$. It is straightforward to see that any min-cost flow in the resulting network (which can be computed in polynomial time) is integral and it corresponds to a pure strategy profile of the congestion game that minimizes its exact potential function, thus, it corresponds to a pure Nash equilibrium.

Note that the aforementioned theorem holds only for symmetric network congestion games, and not for all symmetric congestion games.

THEOREM 21.14 (Fabrikant et al., 2004) *It is PLS-complete to find a pure Nash equilibrium in unweighted congestion games of the following sorts:*

1. *General congestion games.*
2. *Symmetric congestion games.*
3. *Multi-commodity network congestion games.*

PROOF   In order to prove (1) we shall use the following problems:

NOTALLEQUAL3SAT: Given a set N of binary variables and a collection C of clauses such that $\forall c \in C, |c| \leq 3$, is there an assignment of values to the variables so that no clause has all its literals assigned the same value?

POSNAE3FLIP: Given an instance (N, C) of NOTALLEQUAL3SAT with positive literals only and a weight function $w : C \rightarrow \mathbb{R}$, find an assignment of values to the variables such that the total weight of the unsatisfied clauses and the totally satisfied (i.e., with all their literals set to 1) cannot be decreased by a unilateral flip of the value of any variable.

POSNAE3FLIP is known to be PLS-complete. Given an instance of POSNAE3FLIP, we construct a congestion game as follows. The set of players is exactly the set of variables. For each three-clause c of weight w_c we have two resources e_c and e'_c, with delay 0 if there are two or fewer players, and w_c otherwise. Player v has two actions: one action contains all resources e_c for clauses that contain v, and one action that contains all resources e'_c for the same clauses. Smaller clauses are implemented similarly. Clearly, a flip of a variable corresponds to the change in the pure strategy of the corresponding player. The changes in the total weight due to a flip equal the changes in the cumulative delay over all the resources. Thus, any pure Nash equilibrium of the congestion game is a local optimum (and therefore a solution) of the POSNAE3FLIP problem, and vice versa.

The proof of (2) is by reduction of the nonsymmetric case to the symmetric case. Given an unweighted congestion game $\Gamma = \langle N, E, (\Pi_i)_{i \in N}, (d_e)_{e \in E} \rangle$, we construct a symmetric congestion game $\hat{\Gamma} = \langle N, E, \hat{\Pi}, (\hat{d}_e)_{e \in E} \rangle$ as follows. First we add to the set of resources n distinct resources: $\hat{E} = E \cup \{e_i\}_{i \in N}$. The delays of these resources are $\hat{d}_{e_i}(k) = M$ for some sufficiently large constant M if $k \geq 2$, and 0 if $k = 1$. The old resources maintain the same delay functions. Each player has the same action set $\hat{\Pi} = \cup_{i \in N} \{\pi_i \cup \{e_i\} : \pi_i \in \Pi_i\}$. By setting the constant M sufficiently large, in any pure Nash equilibrium of Γ each of the distinct resources is used by exactly one player (these resources act as if they have capacity 1 and there are only n of them). Therefore, in any pure Nash equilibrium, each one of the n players chooses a different subset of resources. So, for any pure Nash equilibrium in $\hat{\Gamma}$ we can easily get a pure Nash equilibrium in Γ by simply dropping the unique new resource used by each of the players. This is done by identifying the "anonymous" players of $\hat{\Gamma}$ according to the unique resource they use, and match them with the corresponding players of Γ.

The proof of (3) is rather complicated and will not be presented here; we refer the interested reader to (Fabrikant et al., 2004).

Next, we deal with the existence and the tractability of pure Nash equilibria in weighted network congestion games. In Fotakis et al. (2005) it was shown that it is not always the case that a pure Nash equilibrium exists:

THEOREM 21.15   (Fotakis et al., 2005) *There exist instances of weighted single-commodity network congestion games with resource delays being either linear or 2-wise linear function of the loads, for which there is no pure Nash equilibrium.*

Furthermore, it was shown that there may exist no exact potential function even for the simplest case of weighted congestion games:

THEOREM 21.16   (Fotakis et al., 2005) *There exist weighted single-commodity network congestion games which are not exact potential games, even when the resource delays are identical to their loads.*

This fact, however, does not rule out the possibility that there exists some non-exact potential for a weighted network congestion game. Actually, Fotakis et al. (2005) proved the existence of a **b**-potential function for any weighted multi-commodity network congestion game with linear resource delays:

THEOREM 21.17   (Fotakis et al., 2005) *For any weighted multi-commodity network congestion game with linear resource delays (i.e., $d_e(x) = a_e x + c_e$, $e \in E$, $a_e, c_e \geq 0$), at least one pure Nash equilibrium exists and can be computed in pseudo-polynomial time.*

PROOF   The result follows from the existence of the **b**-potential function

$$\Phi(\omega) = \sum_{e \in E} \left(a_e \theta_e^2(\omega) + c_e \theta_e(\omega) + \sum_{i:\omega_i = e} (a_e w_i^2 + c_e w_i) \right) \tag{21.77}$$

where $b_i = \frac{1}{2w_i}$ for all $i \in N$.

21.7 The Price of Anarchy

In a gamelike situation, the price of anarchy (or coordination ratio), introduced in (Koutsoupias and Papadimitriou, 1999), is a measure of the performance degradation due to the selfish behavior of the involved players (and therefore due to the lack of coordination among the players). In general, it is defined with respect to a global social cost function, capturing the performance of a strategy profile. Then the price of anarchy equals the ratio of the worst social cost over all Nash equilibria and the optimum value of the social cost.

21.7.1 The Price of Anarchy in Finite Games

In the following, we will focus on the characterization of the price of anarchy in the context of network congestion games. The social cost in this setting can be defined in several ways, such as the maximum congestion over all edges of the network, the average congestion of the edges, the sum of the costs over all players, and so on. In the following, we define the social cost as the maximum cost paid over all agents.

Formally, consider a weighted network congestion game $\Gamma = \langle N, E, (\Pi_i)_{i \in N}, (w_i)_{i \in N}, (d_e)_{e \in E} \rangle$ and let $G = (V, E)$ be the corresponding network. Let $\mathbf{p} = (p_i^j)_{i \in N, j \in \Pi_i}$ be an arbitrary strategy profile, that is, p_i^j is the probability that player $i \in N$ chooses action $j \in \Pi_i$. Then the social cost in this congestion game is

$$SC(\mathbf{p}) = \sum_{\omega \in \Pi} \Pr(\mathbf{p}, \omega) \cdot \max_{i \in N} \{ \lambda_{\omega_i}(\omega) \} \tag{21.78}$$

where $\Pr(\mathbf{p}, \omega) = \prod_{i \in N} p_i^{\omega_i}$ is the probability of pure strategy profile ω occurring, with respect to the strategy profile $\mathbf{p}$.

The social optimum of this game is defined as

$$OPT = \min_{\omega \in \Pi} \left\{ \max_{i \in N} \{ \lambda_{\omega_i}(\omega) \} \right\}. \tag{21.79}$$

The price of anarchy is then defined as the ratio of the social cost of the worst Nash equilibrium and *OPT*:

$$R = \max_{\mathbf{p} \text{ is a NE}} \frac{SC(\mathbf{p})}{OPT}. \tag{21.80}$$

In the seminal paper (Koutsoupias and Papadimitriou, 1999) it was proved that the price of anarchy for the special case of a network consisting of two parallel links is $R = 3/2$, while for m parallel links, $R = \Omega\left(\frac{\log m}{\log \log m}\right)$ and $R = O\left(\sqrt{m \log m}\right)$. For m identical parallel links, Mavronicolas and Spirakis (2001) proved that $R = \Theta\left(\frac{\log m}{\log \log m}\right)$ for the fully mixed Nash equilibrium, that is, for the Nash equilibrium where each player poses nonzero probability to each link. For the case of m identical parallel links it was shown in (Koutsoupias et al., 2003) that $R = \Theta\left(\frac{\log m}{\log \log m}\right)$. In Czumaj and Vöecking (2002), it was finally shown that $R = \Theta\left(\frac{\log m}{\log \log m}\right)$ for the general case of nonidentical parallel links (and players of varying demands).

As pointed out in (Fotakis et al., 2005), there exist weighted network congestion games with resource delays linear to their loads for which the price of anarchy is unbounded. This result holds even for the simple case of an ℓ-layered network, that is, a single-commodity network (V, E) where every source–destination path has length exactly ℓ and each node lies on a directed source–destination path. However, if resource delays are identical to their loads, the following holds:

THEOREM 21.18 (Fotakis et al., 2005) *The price of anarchy of any single-commodity network congestion game with resource delays identical to their loads, is at most*

$$8e\left(\frac{m}{\log \log m} + 1\right),$$

where $m = |E|$ is the number of available resources, that is, the number of edges of the network.

Christodoulou and Koutsoupias (2005) considered the price of anarchy of pure Nash equilibria in congestion games with linear latency functions. For asymmetric games, they showed that the price of anarchy when the social cost is defined as the maximum congestion over all resources is $\Theta(\sqrt{n})$ where n is the number of players. For all other cases of symmetric or asymmetric games and for both maximum and average social cost, the price of anarchy was shown to be $5/2$, and these results were extended to latency functions that are polynomials of bounded degree.

Gairing et al. (2006) studied an interesting variant of the price of anarchy in weighted congestion games on parallel links, where the social cost is the expectation of the sum, over all links, of latency costs; each latency cost is modeled as a certain polynomial function evaluated at the delay incurred by all users choosing the link.

The main argument for using worst-case demands in the definition of the price of anarchy is that the distribution of the players' demands is not known, and the worst-case distribution prevailing in this definition is the one in which the worst-case demand occurs with probability one. Mavronicolas et al. (2005) introduced the notion of *Diffuse Price of Anarchy* in order to remove this assumption while avoiding to assume full knowledge about the distribution of demands. Roughly speaking, the diffuse price of anarchy is the worst-case, over all allowed probability distributions, of the expectation (according to each specific probability distribution) of the ratio of social cost over optimum in the worst-case Nash equilibrium.

21.7.2　The Price of Anarchy in Selfish Wardrop Flows

In the following we study the price of anarchy for a different model of a network congestion game: We are given a network with distinguished source–destination pairs of nodes, a rate of traffic (flow demand) between each pair, and a latency function for each edge specifying the time needed to traverse the edge given the congestion. The objective is to route traffic such that the total latency is minimized.

21.7.2.1　Flows and Game Theory

Consider a directed network $G = (V, E)$ with vertex set V, edge set E, and k source–destination pairs of vertices $\{s_1, t_1\}, \ldots, \{s_k, t_k\}$. Let $[k] = \{1, \ldots, k\}$. Denote the set of (simple) $s_i - t_i$ paths by $\mathcal{P}_i$, and define $\mathcal{P} = \cup_{i=1}^{k} \mathcal{P}_i$. A *flow* is a function $f : \mathcal{P} \to \mathbb{R}_{\geq 0}$. In the following, and for the sake of simplicity, we use the notation f_P instead of $f(P)$ to refer to the flow of path $P \in \mathcal{P}$.

For a fixed flow f define $f_e = \sum_{P \in \mathcal{P}: e \in P} f_P$. Associated to each pair $\{s_i, t_i\}$ $(i \in [k])$ is a finite and positive flow demand r_i, which captures the amount of flow needed to be routed between source s_i and destination t_i. A flow f is said to be feasible if, for all $i \in \{1, \ldots, k\}$, $\sum_{P \in \mathcal{P}_i} f_P = r_i$. For each edge $e \in E$, we are given a latency function $\ell_e : \mathbb{R}_{\geq 0} \to \mathbb{R}_{\geq 0}$. We assume that, for all $e \in E$, function $\ell_e()$ is nondecreasing, continuously differentiable, and $\ell_e(x)x$ is convex. A network routing instance (or simply an instance) is the triple $(G, (r_i)_{i \in [k]}, (\ell_e)_{e \in E})$.

The path latency $\ell_P(f)$ of path $P \in \mathcal{P}$ with respect to flow f is the sum of the latencies of the edges in P, that is, $\ell_P(f) = \sum_{e \in P} \ell_e(f_e)$. The *cost* $C(f)$ of a flow f is defined as the total latency incurred by f, that is,

$$C(f) = \sum_{P \in \mathcal{P}} \ell_P(f) f_P \tag{21.81}$$

$$= \sum_{P \in \mathcal{P}} \left(\sum_{e \in P} \ell_e(f_e) \right) f_P \tag{21.82}$$

$$= \sum_{e \in E} \left(\sum_{P \in \mathcal{P}: e \in P} f_P \right) \ell_e(f_e) \tag{21.83}$$

$$= \sum_{e \in E} \ell_e(f_e) f_e. \tag{21.84}$$

DEFINITION 21.9　A flow f feasible for instance $(G, (r_i)_{i \in [k]}, (\ell_e)_{e \in E})$ is at Nash equilibrium if, for all $i \in [k]$, for all $P_1, P_2 \in \mathcal{P}_i$ such that $P_1 \neq P_2$, and for all $\delta \in [0, f_{P_1}]$, we have $\ell_{P_1}(f) \leq \ell_{P_2}(\tilde{f})$, where

$$\tilde{f} = \begin{cases} f_P - \delta & \text{if } P = P_1 \\ f_P + \delta & \text{if } P = P_2 \\ f_P & \text{otherwise} \end{cases} \tag{21.85}$$

LEMMA 21.1　(Wardrop's principle) A flow f feasible for instance $(G, (r_i)_{i \in [k]}, (\ell_e)_{e \in E})$ is at Nash equilibrium if and only if, for all $i \in [k]$ and for all $P_1, P_2 \in \mathcal{P}_i$ with $f_{P_1} > 0$, $\ell_{P_1}(f) \leq \ell_{P_2}(f)$.

In particular, if f is at Nash equilibrium, then all $s_i - t_i$ paths to which f assigns some positive amount of flow have equal latency, say $L_i(f)$. We can thus express the cost $C(f)$ of a flow f at Nash equilibrium as follows:

$$C(f) = \sum_{P \in \mathcal{P}} \ell_P(f) f_P \tag{21.86}$$

$$= \sum_{i=1}^{k} \sum_{P \in \mathcal{P}_i} \ell_P(f) f_P \tag{21.87}$$

$$= \sum_{i=1}^{k} \sum_{P \in \mathcal{P}_i} L_i(f) f_P \tag{21.88}$$

$$= \sum_{i=1}^{k} L_i(f) \sum_{P \in \mathcal{P}_i} f_P \tag{21.89}$$

$$= \sum_{i=1}^{k} L_i(f) r_i. \tag{21.90}$$

21.7.2.2 Optimal Flows

An optimal flow is a feasible flow that minimizes total latency. Since the cost of a flow f can be expressed as $C(f) = \sum_{e \in E} \ell_e(f) f_e$, the problem of finding the optimal flow is a special case of the following nonlinear program:

Program NLP:
minimize $\sum_{e \in E} c_e(f_e)$
subject to $\sum_{P \in \mathcal{P}_i} f_P = r_i \qquad \forall i \in [k]$
$f_e = \sum_{P \in \mathcal{P}: e \in P} f_P \qquad \forall e \in E$
$f_P \geq 0 \qquad \forall P \in \mathcal{P}$

where, in our problem, $c_e(f_e) = \ell_e(f_e) f_e$.

For simplicity we have given a formulation with an exponential number of variables (since the number of paths is in general exponential in the number of edges), but it is not difficult to give an equivalent compact formulation (with decision variables only on edges and explicit conservation constraints) that requires only polynomially many variables and constraints.

Now, since function $\ell_e(f_e) f_e$ is convex, its local optima and global optima coincide. Denote $c_e'(x) = \frac{d}{dx} c_e(x)$ the derivative of c_e and let $c_P'(f) = \sum_{e \in P} c_e'(f_e)$. Then:

LEMMA 21.2 (Beckmann et al., 1956; Dafermos and Sparrow, 1969) A flow f is optimal for a convex program of the form NLP if and only if, for every $i \in [k]$ and $P_1, P_2 \in \mathcal{P}_i$ with $f_{P_1} > 0$, $c_{P_1}'(f) \leq c_{P_2}'(f)$.

Lemmata 21.1 and 21.2 yield:

COROLLARY 21.2 (Beckmann et al., 1956) Let $(G, (r_i)_{i \in [k]}, (\ell_e)_{e \in E})$ be an instance in which $\ell_e(x) x$ is convex for all $e \in E$, and let $\ell_e^*(f_e) = (\ell_e(f_e) f_e)' = \ell_e(f_e) + \ell_e'(f_e) f_e$. Then a flow f feasible for $(G, (r_i)_{i \in [k]}, (\ell_e)_{e \in E})$ is optimal if and only if it is at Nash equilibrium for instance $(G, (r_i)_{i \in [k]}, (\ell_e^*)_{e \in E})$.

The following lemma, originally proved by Beckmann et al. (1956) and later reproved by Dafermos and Sparrow (1969), guarantees the existence of flows at Nash equilibrium:

LEMMA 21.3   An instance $(G, (r_i)_{i \in [k]}, (\ell_e)_{e \in E})$ with continuous, nondecreasing latency functions admits a feasible flow at Nash equilibrium. Moreover, if $f, \tilde{f}$ are flows at Nash equilibrium, then $C(f) = C(\tilde{f})$.

21.7.2.3 The Price of Anarchy

DEFINITION 21.10   For an instance $(G, (r_i)_{i \in [k]}, (\ell_e)_{e \in E})$ with an optimal flow f^* and a flow at Nash equilibrium f, the *price of anarchy* R is

$$R = \frac{C(f)}{C(f^*)}. \tag{21.91}$$

Roughgarden (2002) proved that for general, continuous, nondecreasing latency functions, it is actually the class of allowable latency functions and not the specific topology of a network that determines the price of anarchy. In the same setting, Roughdarden and Tardos (2002) showed that the cost of a flow at Nash equilibrium is at most equal to the cost of an optimal flow for double flow demands.

Roughdarden and Tardos (2002) focused on the case of linear latency functions, that is, $\ell_e(x) = a_e x + b_e$ for all $e \in E$. By giving specialized versions of Lemmas 21.1 and 21.2 for this setting, they gave an upper bound of $4/3$ to the price of anarchy (as well as examples for which this bound is tight).

Defining Terms

Bimatrix game: A bimatrix game is a two-player finite strategic form game, such that the players' payoffs can be described by two real matrices R and C, where each row corresponds to an action of the first player and each column corresponds to an action of the second player. The payoffs when the first player chooses his/her ith action and the second player chooses his/her jth action are given by the (i, j)th element of matrix R for the first player, and by the (i, j)th element of matrix C for the second player.

Finite strategic form game: A finite strategic form game is defined by a nonempty finite set of players, and, for each player, a nonempty finite set of actions and a payoff function mapping each combination of actions (one for each player) to a real number.

Mixed strategy: A mixed strategy for a player is any probability distribution on his/her set of actions.

Nash equilibrium: A strategy profile is a Nash equilibrium if no player can increase his/her payoff by unilaterally deviating from the profile.

ε-Nash equilibrium: A strategy profile is an ε-Nash equilibrium if no player can increase his/her payoff by more than ε by unilaterally deviating from the profile.

Price of anarchy: In a gamelike situation, the price of anarchy (or coordination ratio) is a measure of the performance degradation due to the selfish behavior of the involved players (and therefore due to the lack of coordination among the players). In general, it is defined with respect to a global social cost function, capturing the performance of a strategy profile. Then the price of anarchy equals the ratio of the worst social cost over all Nash equilibria and the optimum value of the social cost.

Pure strategy: A pure strategy for a player is a mixed strategy that poses probability 1 to exactly one of his/her actions, and 0 to all the rest.

Strategy profile: A strategy profile is a combination of (pure or mixed) strategies, one for each player.

Support: The support of a strategy of some player is the subset of actions that are assigned nonzero probability.

ϵ-**Well-supported Nash equilibrium:** A strategy profile is an ϵ-well-supported Nash equilibrium if it is an ϵ-Nash equilibrium with the additional property that each player plays with nonzero probability only those actions that guarantee him/her a payoff no less than the payoff that the specific strategy profile gives him/her, in addition to ϵ.

Acknowledgments

This work was partially supported by the Future and Emerging Technologies Unit of EC (IST priority - 6th FP) under contract No. FP6-021235-2 (project ARRIVAL), and by the General Secretariat for Research and Technology of the Greek Ministry of Development within the programme PENED 2003. We wish to thank Alexis Kaporis for his contribution to Section 21.7.2.

References

I. Althöfer. On sparse approximations to randomized strategies and convex combinations. *Linear Algebra and Applications*, 199:339–355, 1994.

M. Beckmann, C. B. McGuire, and C. B. Winsten. *Studies in the Economics of Transportation*. Yale University Press, New Haven, CT, 1956.

H. Bosse, J. Byrka, and E. Markakis. New algorithms for approximate Nash equilibria in bimatrix games. In *Proceedings of the 3rd Workshop on Internet and Network Economics (WINE'07)*, Volume 4858 of *LNCS*, pages 17–29, Springer-Verlag, Berlin, Germany, 2007.

X. Chen and X. Deng. 3-Nash is PPAD-complete. In *Electronic Colloquium on Computational Complexity (ECCC)*, Report TR05-134, 2005.

X. Chen and X. Deng. Settling the complexity of 2-player Nash-equilibrium. In *Proceedings of the 47th Annual IEEE Symposium on Foundations of Computer Science (FOCS'06)*, pages 261–272, IEEE Computer Society, Washington, DC, 2006.

X. Chen, X. Deng, and S. Teng. Computing Nash equilibria: Approximation and smoothed complexity. In *Electronic Colloquium on Computational Complexity (ECCC)*, Report TR06-023, 2006.

G. Christodoulou and E. Koutsoupias. The price of anarchy of finite congestion games. In *Proceedings of the 37th Annual ACM Symposium on Theory of Computing (STOC '05)*, pages 67–73, ACM, New York, 2005.

A. Czumaj and B. Vöecking. Tight bounds for worst-case equilibria. In *Proceedings of the 13th ACM-SIAM Symposium on Discrete Algorithms (SODA '02)*, pages 413–420, ACM, New York, 2002.

S. C. Dafermos and F. T. Sparrow. The traffic assignment problem for a general network. *Journal of Research of the National Bureau of Standards, Series B*, 73B(2):91–118, 1969.

C. Daskalakis and C. Papadimitriou. Three-player games are hard. In *Electronic Colloquium on Computational Complexity (ECCC)*, Report TR05-139, 2005.

C. Daskalakis, P. Goldberg, and C. Papadimitriou. The complexity of computing a Nash equilibrium. In *Proceedings of the 38th Annual ACM Symposium on Theory of Computing (STOC'06)*, pages 71–78, ACM, New York, 2006a.

C. Daskalakis, A. Mehta, and C. Papadimitriou. A note on approximate Nash equilibria. In *Proceedings of the 2nd Workshop on Internet and Network Economics (WINE'06)*, Volume 4286 of *LNCS*, pages 297–306, Springer-Verlag, Berlin, Germany, 2006b.

C. Daskalakis, A. Mehta, and C. Papadimitriou. Progress in approximate Nash equilibrium. In *Proceedings of the 8th ACM Conference on Electronic Commerce (EC'07)*, pages 355–358, ACM, New York, 2007.

A. Fabrikant, C. H. Papadimitriou, and K. Talwar. The complexity of pure Nash equilibria. In *Proceedings of the 36th ACM Symposium on Theory of Computing*, pages 281–290, ACM, New York, 2004.

D. Fotakis, S. Kontogiannis, and P. Spirakis. Selfish unsplittable flows. *Theoretical Computer Science*, 348(2–3):129–366, 2005.

M. Gairing, T. Lücking, M. Mavronicolas, and B. Monien. The price of anarchy for polynomial social cost. *Theoretical Computer Science*, 369(1–3):116–135, 2006.

W. Hoeffding. Probability inequalities for sums of bounded random variables. *Journal of the American Statistical Association*, 58:13–30, 1963.

D. Johnson, C. Papadimitriou, and M. Yannakakis. How easy is local search? *Journal of Computer and System Sciences*, 37:79–100, 1988.

S. Kontogiannis and P. G. Spirakis. Efficient algorithms for constant well supported approximate equilibria in bimatrix games. In *Proceedings of the 34th International Colloquium on Automata, Languages and Programming (ICALP'07, Track A: Algorithms and Complexity)*, pages 595–606, Springer, Wroclaw, Poland, 2007.

S. Kontogiannis, P. N. Panagopoulou, and P. G. Spirakis. Polynomial algorithms for approximating Nash equilibria of bimatrix games. In *Proceedings of the 2nd Workshop on Internet and Network Economics (WINE'06)*, Volume 4286 of *LNCS*, pages 286–296, Springer-Verlag, Berlin, Germany, 2006.

E. Koutsoupias and C. Papadimitriou. Worst case equilibria. In *16th Annual Symposium on Theoretical Aspects of Computer Science (STACS '99)*, pages 404–413, Springer, Trier, Germany, 1999.

E. Koutsoupias, M. Mavronicolas, and P. Spirakis. Approximate equilibria and ball fusion. *ACM Transactions on Computer Systems*, 36:683–693, 2003.

C. E. Lemke and J. T. Howson. Equilibrium points of bimatrix games. *Journal of the Society for Industrial and Applied Mathematics*, 12: 413–423, 1964.

R. J. Lipton, E. Markakis, and A. Mehta. Playing large games using simple strategies. In *Proceedings of the 4th ACM Conference on Electronic Commerce (EC'03)*, pages 36–41, ACM, New York, 2003.

M. Mavronicolas and P. Spirakis. The price of selfish routing. In *Proceedings of the 33rd Annual ACM Symposium on Theory of Computing (STOC '01)*, pages 510–519, ACM, New York, 2001.

M. Mavronicolas, P. Panagopoulou, and P. Spirakis. Cost sharing mechanism for fair pricing of resource usage. *Algorithmica*, 52:19–43, 2008.

D. Monderer and L. Shapley. Potential games. *Games and Economic Behavior*, 14:124–143, 1996.

J. Nash. Noncooperative games. *Annals of Mathematics*, 54:289–295, 1951.

C. H. Papadimitriou. On the complexity of the parity argument and other inefficient proofs of existence. *Journal of Computer and System Sciences*, 48:498–532, 1994.

C. H. Papadimitriou. Algorithms, games, and the internet. In *Proceedings of the 32nd Annual ACM Symposium on Theory of Computing (STOC '01)*, pages 749–753, ACM, New York, 2001.

R. W. Rosenthal. A class of games possessing pure-strategy Nash equilibria. *International Journal of Game Theory*, 2:65–67, 1973.

T. Roughdarden and E. Tardos. How bad is selfish routing? *Journal of the ACM*, 49(2):236–259, 2002.

T. Roughgarden. The price of anarchy is independent of the network topology. In *Proceedings of the 34th Annual ACM Symposium on Theory of Computing (STOC '02)*, pages 428–437, ACM, New York, 2002.

R. Savani and B. von Stengel. Exponentially many steps for finding a Nash equilibrium in a bimatrix game. In *Proceedings of the 45th Annual IEEE Symposium on Foundations of Computer Science (FOCS'04)*, pages 258–267, IEEE Computer Society, Washington, DC, 2004.

L. S. Shapley. A note on the Lemke-Howson algorithm. *Mathematical Programming Study 1: Pivoting and Extensions*, pages 175–189, Springer, Berlin, Heidelberg, 1974.

H. Tsaknakis and P. Spirakis. An optimization approach for approximate Nash equilibria. In *Proceedings of the 3rd Workshop on Internet and Network Economics (WINE'07)*, Volume 4858 of *LNCS*, pages 42–56, Springer-Verlag, Berlin, Germany, 2007.

B. von Stengel. Computing equilibria for two-person games. In R.J. Aumann and S. Hart, editors, *Handbook of Game Theory with Economic Applications*, Volume 3, pages 1723–1759, Elsevier, Amsterdam, the Netherlands, 2002.

22

Artificial Intelligence Search Algorithms

Richard E. Korf
University of California, Los Angeles

22.1 Introduction

Search is a universal problem-solving mechanism in artificial intelligence (AI). In AI problems, the sequence of steps required for the solution of a problem is not known a priori, but often must be determined by a trial-and-error exploration of alternatives. The problems that have been addressed by AI search algorithms fall into three general classes: single-agent pathfinding problems, game playing, and constraint-satisfaction problems.

Classic examples in the AI literature of single-agent pathfinding problems are the sliding-tile puzzles, including the 3 × 3 eight puzzle (see Figure 22.1) and its larger relatives the 4 × 4 fifteen

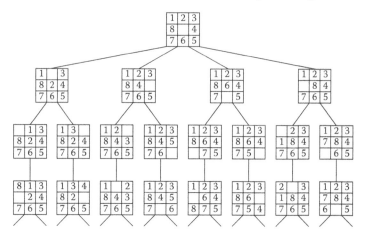

FIGURE 22.1 Eight puzzle search tree fragment.

puzzle and 5 × 5 twenty-four puzzle. The eight puzzle consists of a 3 × 3 square frame containing eight numbered square tiles and an empty position called the blank. The legal operators are to slide any tile that is horizontally or vertically adjacent to the blank into the blank position. The problem is to rearrange the tiles from some random initial configuration into a particular desired goal configuration. The sliding-tile puzzles are common testbeds for research in AI search algorithms because they are very simple to represent and manipulate, yet finding optimal solutions to the $N \times N$ generalization of the sliding-tile puzzles is NP-complete [38]. Another well-known example of a single-agent pathfinding problem is Rubik's cube. Real-world examples include theorem proving, the traveling salesman problem, and vehicle navigation. In each case, the task is to find a sequence of operations that maps an initial state to a goal state.

A second class of search problems include games, such as chess, checkers, Othello, backgammon, bridge, poker, and Go. The third category is constraint-satisfaction problems, such as the N-queens problem or Sudoku. The task in the N-queens problem is to place N queens on an $N \times N$ chessboard, such that no two queens are on the same row, column, or diagonal. The Sudoku task is to fill each empty cell in a 9 × 9 matrix with a digit from zero through nine, such that each row, column, and nine 3 × 3 submatrices contain all the digits zero through nine. Real-world examples of constraint-satisfaction problems are ubiquitous, including boolean satisfiability, planning, and scheduling applications.

We begin by describing the problem-space model on which search algorithms are based. Brute-force searches are then considered including breadth-first, uniform-cost, depth-first, depth-first iterative-deepening, bidirectional, frontier, and disk-based search algorithms. Next, we introduce heuristic evaluation functions, including pattern databases, and heuristic search algorithms including pure heuristic search, the A* algorithm, iterative-deepening-A*, depth-first branch-and-bound, the heuristic path algorithm, and recursive best-first search. We then consider single-agent algorithms that interleave search and execution, including minimin lookahead search, real-time-A*, and learning-real-time-A*. Next, we consider game playing, including minimax search, alpha-beta pruning, quiescence, iterative deepening, transposition tables, special-purpose hardware, multiplayer games, and imperfect and hidden information. We then examine constraint-satisfaction algorithms, such as chronological backtracking, intelligent backtracking techniques, constraint recording, and local search algorithms. Finally, we consider open research problems in this area. The performance of these algorithms, in terms of the costs of the solutions they generate, the amount of time the algorithms take to execute, and the amount of computer memory they require are of central concern throughout. Since search is a universal problem-solving method, what limits its applicability is the efficiency with which it can be performed.

22.2 Problem Space Model

A problem space is the environment in which a search takes place [32]. A problem space consists of a set of states of the problem, and a set of operators that change the state. For example, in the eight puzzle, the states are the different possible permutations of the tiles, and the operators slide a tile into the blank position. A problem instance is a problem space together with an initial state and a goal state. In the case of the eight puzzle, the initial state would be whatever initial permutation the puzzle starts out in, and the goal state is a particular desired permutation. The problem-solving task is to find a sequence of operators that map the initial state to a goal state. In the eight puzzle the goal state is given explicitly. In other problems, such as the N-Queens Problem, the goal state is not given explicitly, but rather implicitly specified by certain properties that must be satisfied by any goal state.

A problem-space graph is often used to represent a problem space. The states of the space are represented by nodes of the graph, and the operators by edges between nodes. Edges may be undirected or directed, depending on whether their corresponding operators are reversible or not. The task in a single-agent path-finding problem is to find a path in the graph from the initial node to a goal node. Figure 22.1 shows a small part of the eight puzzle problem-space graph.

Although most problem spaces correspond to graphs with more than one path between a pair of nodes, for simplicity they are often represented as trees, where the initial state is the root of the tree. The cost of this simplification is that any state that can be reached by two different paths will be represented by duplicate nodes in the tree, increasing the size of the tree. The benefit of a tree is that the absence of multiple paths to the same state greatly simplifies many search algorithms.

One feature that distinguishes AI search algorithms from other graph-searching algorithms is the size of the graphs involved. For example, the entire chess graph is estimated to contain over 10^{40} nodes. Even a simple problem like the Twenty-four puzzle contains almost 10^{25} nodes. As a result, the problem-space graphs of AI problems are never represented explicitly by listing each state, but rather are implicitly represented by specifying an initial state and a set of operators to generate new states from existing states. Furthermore, the size of an AI problem is rarely expressed as the number of nodes in its problem-space graph. Rather, the two parameters of a search tree that determine the efficiency of various search algorithms are its branching factor and its solution depth. The branching factor is the average number of children of a given node. For example, in the eight puzzle the average branching factor is $\sqrt{3}$, or about 1.732. The solution depth of a problem instance is the length of a shortest path from the initial state to a goal state, or the length of a shortest sequence of operators that solves the problem. For example, if the goal were in the bottom row of Figure 22.1, the depth of the problem instance represented by the initial state at the root would be three moves.

22.3 Brute-Force Search

The most general search algorithms are brute-force searches, since they do not require any domain-specific knowledge. All that is required for a brute-force search is a state description, a set of legal operators, an initial state, and a description of the goal state. The most important brute-force techniques are breadth-first, uniform-cost, depth-first, depth-first iterative-deepening, bidirectional, and frontier search. In the descriptions of the algorithms in the following text, to generate a node means to create the data structure corresponding to that node, whereas to expand a node means to generate all the children of that node.

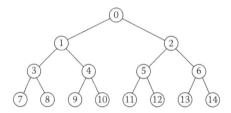

FIGURE 22.2 Order of node generation for breadth-first search.

22.3.1 Breadth-First Search

Breadth-first search expands nodes in order of their depth from the root, generating one level of the tree at a time until a solution is found (see Figure 22.2). It is most easily implemented by maintaining a first-in first-out queue of nodes, initially containing just the root, and always removing the node at the head of the queue, expanding it, and adding its children to the tail of the queue.

Since it never generates a node in the tree until all the nodes at shallower levels have been generated, breadth-first search always finds a shortest path to a goal. Since each node can be generated in constant time, the amount of time used by breadth-first search is proportional to the number of nodes generated, which is a function of the branching factor b and the solution depth d. Since the number of nodes in a uniform tree at level d is b^d, the total number of nodes generated in the worst case is $b + b^2 + b^3 + \cdots + b^d$, which is $O(b^d)$, the asymptotic time complexity of breadth-first search.

The main drawback of breadth-first search is its memory requirement. Since each level of the tree must be stored in order to generate the next level, and the amount of memory is proportional to the number of nodes stored, the space complexity of breadth-first search is also $O(b^d)$. As a result, breadth-first search is severely space-bound in practice, and will exhaust the memory available on typical computers in a matter of minutes.

22.3.2 Uniform-Cost Search

If all edges do not have the same cost, then breadth-first search generalizes to uniform-cost search. Instead of expanding nodes in order of their depth from the root, uniform-cost search expands nodes in order of their cost from the root. At each step, the next node n to be expanded is one whose cost $g(n)$ is lowest, where $g(n)$ is the sum of the edge costs from the root to node n. The nodes are stored in a priority queue. This algorithm is similar to Dijkstra's single-source shortest-path algorithm [7]. The main difference is that uniform-cost search runs until a goal node is chosen for expansion, while Dijkstra's algorithm runs until every node in a finite graph is chosen for expansion.

Whenever a node is chosen for expansion by uniform-cost search, a lowest-cost path to that node has been found. The worst-case time complexity of uniform-cost search is $O(b^{c/e})$, where c is the cost of an optimal solution, and e is the minimum edge cost. Unfortunately, it also suffers the same memory limitation as breadth-first search.

22.3.3 Depth-First Search

Depth-first search removes the space limitation of breadth-first and uniform-cost search by always generating next a child of the deepest unexpanded node (see Figure 22.3). Both algorithms can be implemented using a list of unexpanded nodes, with the difference that breadth-first search manages the list as a first-in first-out queue, whereas depth-first search treats the list as a last-in first-out stack. More commonly, depth-first search is implemented recursively, with the recursion stack taking the place of an explicit node stack.

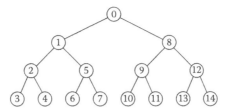

FIGURE 22.3 Order of node generation for depth-first search.

The advantage of depth-first search is that its space requirement is only linear in the maximum search depth, as opposed to exponential for breadth-first search. The reason is that the algorithm only needs to store a stack of nodes on the path from the root to the current node. The time complexity of a depth-first search to depth d is $O(b^d)$, since it generates the same set of nodes as breadth-first search, but simply in a different order. Thus, as a practical matter, depth-first search is time-limited rather than space-limited.

The primary disadvantage of depth-first search is that it may not terminate on an infinite tree, but simply go down the left-most path forever. For example, even though there are a finite number of states of the eight puzzle, the tree fragment shown in Figure 22.1 can be infinitely extended down any path, generating an infinite number of duplicate nodes representing the same states. The usual solution to this problem is to impose a cutoff depth on the search. Although the ideal cutoff is the solution depth d, this value is rarely known in advance of actually solving the problem. If the chosen cutoff depth is less than d, the algorithm will fail to find a solution, whereas if the cutoff depth is greater than d, a large price is paid in execution time, and the first solution found may not be an optimal one.

22.3.4 Depth-First Iterative-Deepening

Depth-first iterative-deepening (DFID) combines the best features of breadth-first and depth-first search [18,46]. DFID first performs a depth-first search to depth one, then starts over, executing a complete depth-first search to depth two, and continues to run depth-first searches to successively greater depths, until a solution is found (see Figure 22.4).

Since it never generates a node until all shallower nodes have been generated, the first solution found by DFID is guaranteed to be via a shortest path. Furthermore, since at any given point it is executing a depth-first search, saving only a stack of nodes, and the algorithm terminates when it finds a solution at depth d, the space complexity of DFID is only $O(d)$.

Although DFID spends extra time in the iterations before the one that finds a solution, this extra work is usually insignificant. To see this, note that the number of nodes at depth d is b^d, and each of these nodes are generated once, during the final iteration. The number of nodes at depth $d-1$ is b^{d-1},

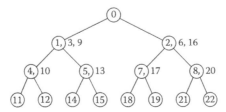

FIGURE 22.4 Order of node generation for depth-first iterative-deepening search.

and each of these are generated twice, once during the final iteration, and once during the penultimate iteration. In general, the number of nodes generated by DFID is $b^d + 2b^{d-1} + 3b^{d-2} + \cdots + db$. This is asymptotically $O(b^d)$ if b is greater than one, since for large values of d the lower order terms are insignificant. In other words, most of the work goes into the final iteration, and the cost of the previous iterations is relatively small. The ratio of the number of nodes generated by DFID to those generated by breadth-first search on a tree is approximately $b/(b-1)$. In fact, DFID is asymptotically optimal in terms of time and space among all brute-force shortest-path algorithms on a tree [18].

If the edge costs differ from one another, then one can run an iterative deepening version of uniform-cost search, where the depth cutoff is replaced by a cutoff on the $g(n)$ cost of a node. At the end of each iteration, the threshold for the next iteration is set to the minimum cost of all nodes generated but not expanded on the previous iteration.

On a graph with multiple paths to the same node, however, breadth-first search may be much more efficient than depth-first or depth-first iterative-deepening search. The reason is that a breadth-first search can easily detect all duplicate nodes, whereas a depth-first search can only check for duplicates along the current search path. Thus, the complexity of breadth-first search grows only as the number of states at a given depth, while the complexity of depth-first search depends on the number of paths of a given length. For example, in a square grid, the number of nodes within a radius r of the origin is $O(r^2)$, whereas the number of paths of length r is $O(3^r)$, since there are three children of every node, not counting its parent. Thus, in a graph with a large number of very short cycles, breadth-first search is preferable to depth-first search, if sufficient memory is available. For two approaches to the problem of pruning duplicate nodes in depth-first search, see [8,47].

22.3.5 Bidirectional Search

Bidirectional search is a brute-force algorithm that requires an explicit goal state instead of simply a test for a goal condition [36]. The main idea is to simultaneously search forward from the initial state, and backward from the goal state, until the two search frontiers meet. The path from the initial state is then concatenated with the inverse of the path from the goal state to form the complete solution path.

Bidirectional search still guarantees optimal solutions. Assuming that the comparisons for identifying a common state between the two frontiers can be done in constant time per node, by hashing for example, the time complexity of bidirectional search is $O(b^{d/2})$ since each search need only proceed to half the solution depth. Since at least one of the searches must be breadth-first in order to find a common state, the space complexity of bidirectional search is also $O(b^{d/2})$. As a result, bidirectional search is space bound in practice.

22.3.6 Frontier Search

Best-first searches, such as breadth-first, uniform-cost, and bidirectional search, store both a closed list of expanded nodes, and an open list of nodes that have been generated but not yet expanded. Another approach to the memory limitation of these algorithms, called frontier search, is to store only the open list, and delete nodes once they are expanded [25]. To keep from regenerating the parent of a node, frontier search stores with each node a bit for each operator that would generate an expanded node. This reduces the space complexity of a breadth-first search, for example, from the size of the problem space to the width of the problem space, or the maximum number of nodes at any given depth. For example, the size of a two dimensional grip graph is quadratic in the radius, while the width of the graph is only linear. Alternatively, the immediate parents of those nodes on the open list can be stored [49].

The main drawback of frontier search is that once the search completes, the solution path is not available, since the expanded nodes have been deleted. One way to reconstruct the solution path is to perform a bidirectional frontier search, from both the initial state and the goal state. Once the two search frontiers meet, then we have a middle node approximately half-way along an optimal solution path. Then, we can use divide and conquer frontier search to recursively generate a path from the initial state to the middle state, and from the middle state to the goal state. Frontier search can also be used to reduce the memory of heuristic best-first searches described in the following text.

22.3.7 Disk-Based Search Algorithms

Another approach to the memory limitation of graph-search algorithms is to store search nodes on magnetic disk instead of main memory. Currently, two terabyte disks can be bought for about $300, which is almost three orders of magnitude cheaper than semiconductor memory. The drawback of disk storage, however, is that the latency to access a single byte can be as much as ten milliseconds, which is about five orders of magnitude slower than semiconductor memory. The main reason to store generated nodes is to detect duplicate nodes, which is usually accomplished by a hash table in memory. A hash table cannot be directly implemented on magnetic disk, however, because of the long latencies for random access.

An alternative technique is called delayed duplicate detection. Rather than checking for duplicate nodes every time a node is generated, duplicate detection is delayed, for example, until the end of a breadth-first search iteration. At that point the queue of nodes on disk can be sorted by their state representation, using only sequential access algorithms, bringing duplicate nodes to adjacent locations. Then, a single sequential pass over the sorted queue can merge duplicate nodes in preparation for the next iteration of the search. This technique can also be combined with frontier search to further reduce the storage needed. See [27] for more details.

Another disk-based search algorithm is called structured duplicate detection [53].

22.3.8 Combinatorial Explosion

The problem with all brute-force search algorithms is that their time complexities often grow exponentially with problem size. This is called combinatorial explosion, and as a result, the size of problems that can be solved with these techniques is quite limited. For example, while the eight puzzle, with about 10^5 states, is easily solved by brute-force search, and the fifteen puzzle, with over 10^{13} states, can be searched exhaustively using brute-force frontier search, storing nodes on disk [26], the 5×5 twenty-four puzzle, with 10^{25} states, is completely beyond the reach of brute-force search.

22.4 Heuristic Search

In order to solve larger problems, domain-specific knowledge must be added to improve search efficiency. In AI, heuristic search has a general meaning, and a more specialized technical meaning. In a general sense, the term heuristic is used for any advice that is often effective, but is not guaranteed to work in every case. Within the heuristic search literature, however, the term heuristic usually refers to the special case of a heuristic evaluation function.

22.4.1 Heuristic Evaluation Functions

In a single-agent path-finding problem, a heuristic evaluation function estimates the cost of an optimal path between a pair of states. For a fixed goal state, a heuristic evaluation $h(n)$ is a function of a node n, which estimates the distance from node n to the goal state. For example, the Euclidean or airline distance is an estimate of the highway distance between a pair of locations. A common heuristic function for the sliding-tile puzzles is called the Manhattan distance. It is computed by counting the number of moves along the grid that each tile is displaced from its goal position, and summing these values over all tiles.

The key properties of a heuristic evaluation function are that it estimate actual cost, and that it is efficiently computable. For example, the Euclidean distance between a pair of points can be computed in constant time. The Manhattan distance between a pair of states can be computed in time proportional to the number of tiles.

Most heuristic functions are derived by generating a simplified version of the problem to be solved, then using the cost of an optimal solution to the simplified problem as a heuristic evaluation function for the original problem. For example, if we remove the constraint that we must drive on the roads, the cost of an optimal solution to the resulting helicopter navigation problem is the Euclidean distance. In the sliding-tile puzzles, if we remove the constraint that a tile can only be slid into the blank position, then any tile can be moved to any adjacent position at any time. The optimal number of moves required to solve this simplified version of the problem is the Manhattan distance.

Since they are derived from relaxations of the original problem, such heuristics are also lower bounds on the costs of optimal solutions to the original problem, a property referred to as admissibility. For example, the Euclidean distance is a lower bound on road distance between two points, since the shortest path between a pair of points is a straight line. Similarly, the Manhattan distance is a lower bound on the actual number of moves necessary to solve an instance of a sliding-tile puzzle, since every tile must move at least as many times as its Manhattan distance to its goal position, and each move moves only one tile.

22.4.2 Pattern Database Heuristics

Pattern databases are heuristics that are precomputed and stored in lookup tables for use during search [1]. For example, consider a table with a separate entry for each possible configuration of a subset of the tiles in a sliding tile puzzle, called the pattern tiles. Each entry stores the minimum number of moves required to move the pattern tiles from the corresponding configuration to their goal configuration. Such a value is a lower bound on the number of moves needed to reach the goal from any state where the pattern tiles are in the given configuration, and can be used as an admissible heuristic during search.

To construct such a table, we perform a breadth-first search backward from the goal state, where a state is defined by the configuration of the pattern tiles. As each such configuration is reached for the first time, we store in the pattern database the number of moves needed to reach that configuration. This table is computed once before the search, and can be used for multiple searches that share the same goal state. The number of tiles in a pattern is limited by the amount of memory available to store an entry for each possible configuration of pattern tiles.

If when computing a pattern database we only count moves of the pattern tiles, then we can add the values from different pattern databases based on disjoint sets of tiles, without overestimating actual cost [24]. Thus, we can partition all the tiles in a problem into disjoint sets of pattern tiles, and sum the values from the different databases to get a more accurate but still admissible heuristic function.

A number of algorithms make use of heuristic evaluations, including pure heuristic search, the A* algorithm, iterative-deepening-A*, depth-first branch-and-bound, the heuristic path algorithm, recursive best-first search, and real-time-A*. In addition, heuristic evaluations can be employed in bidirectional search, and are used in two-player games as well.

22.4.3 Pure Heuristic Search

The simplest of these algorithms, pure heuristic search, expands nodes in order of their heuristic values $h(n)$ [9]. As a best-first search, it maintains a closed list of those nodes that have been expanded, and an open list of those nodes that have been generated but not yet expanded. The algorithm begins with just the initial node on the open list. At each step, a node on the open list with the minimum $h(n)$ value is expanded, generating all of its children, and is placed on the closed list. The heuristic function is applied to each of the children, and they are placed on the open list, sorted by their heuristic values. The algorithm continues until a goal node is chosen for expansion.

In a graph with cycles, multiple paths will be found to the same node, and the first path found may not be the shortest. When a shorter path is found to an open node, the shorter path is saved and the longer one discarded. When a shorter path to a closed node is found, the node is moved to open, and the shorter path is associated with it. The main drawback of pure heuristic search is that since it ignores the cost of the path so far to node n, it does not find optimal solutions.

22.4.4 A* Algorithm

The A* algorithm [14] combines features of uniform-cost search and pure heuristic search to efficiently compute optimal solutions. A* is a best-first search in which the cost associated with a node is $f(n) = g(n) + h(n)$, where $g(n)$ is the cost of the path from the initial state to node n, and $h(n)$ is the heuristic estimate of the cost of a path from node n to a goal. At each point a node with lowest f value is chosen for expansion. Ties among nodes of equal f value are broken in favor of nodes with lower h values. The algorithm terminates when a goal node is chosen for expansion.

A* finds an optimal path to a goal if the heuristic function $h(n)$ is admissible, meaning it never overestimates actual cost. For example, since airline distance never overestimates actual highway distance, and Manhattan distance never overestimates actual moves in the sliding-tile puzzles, A* using these evaluation functions will find optimal solutions to these problems. In addition, A* makes the most efficient use of a given heuristic function in the following sense: among all shortest-path algorithms using a given heuristic function $h(n)$, A* expands the fewest number of nodes [5].

The main drawback of A*, and indeed of any best-first search, is its memory requirement. Since the open and closed lists are stored in memory, A* is severely space-limited in practice, and will exhaust the available memory on current machines in minutes. For example, while it can be run successfully on the eight puzzle, it cannot solve most random instances of the fifteen puzzle before exhausting memory. Frontier-A* only stores the open list, which ameliorates the memory limitation, but does not eliminate it. An additional drawback of A* is that managing the open and closed lists is complex, and takes time.

22.4.5 Iterative-Deepening-A*

Just as depth-first iterative-deepening solved the space problem of breadth-first search, iterative-deepening-A* (IDA*) eliminates the memory constraint of A*, without sacrificing solution optimality [18]. Each iteration of the algorithm is a depth-first search that keeps track of the cost, $f(n) = g(n) + h(n)$, of each node generated. As soon as a node is generated whose cost exceeds a threshold for that iteration, it is deleted, and the search backtracks before continuing along a different path. The cost threshold is initialized to the heuristic estimate of the initial state, and is increased in each successive iteration to the lowest cost of all the nodes that were generated but not expanded during the previous iteration. The algorithm terminates when a goal state is reached whose cost does not exceed the current threshold.

Since IDA* performs a series of depth-first searches, its memory requirement is linear in the maximum search depth. In addition, if the heuristic function is admissible, IDA* finds an optimal solution. Finally, by an argument similar to that presented for depth-first iterative-deepening, IDA* expands the same number of nodes, asymptotically, as A* on a tree, provided that the number of nodes grows exponentially with solution cost. These facts, together with the optimality of A*, imply that IDA* is asymptotically optimal in time and space over all heuristic search algorithms that find optimal solutions on a tree. The additional benefits of IDA* are that it is much easier to implement, and often runs faster than A*, since it does not incur the overhead of managing the open and closed lists. Using appropriate admissible heuristic functions, IDA* can optimally solve random instances of the fifteen puzzle [18], the twenty-four puzzle [24], and Rubik's cube [22].

22.4.6 Depth-First Branch-and-Bound

For many problems, the maximum search depth is known in advance, or the search tree is finite. For example, the traveling salesman problem (TSP) is to visit each of a given set of cities and return to the starting city in a tour of shortest total distance. The most natural problem space for this problem consists of a tree where the root node represents the starting city, the nodes at level one represent all the cities that could be visited first, the nodes at level two represent all the cites that could be visited second, and so on. In this tree, the maximum depth is the number of cities, and all candidate solutions occur at this depth. In such a space, a simple depth-first search guarantees finding an optimal solution using space that is only linear in the depth.

The idea of depth-first branch-and-bound (DFBnB) is to make this search more efficient by keeping track of the best solution found so far. Since the cost of a partial tour is the sum of the costs of the edges traveled so far, whenever a partial tour is found whose cost equals or exceeds the cost of the best complete tour found so far, the branch representing the partial tour can be pruned, since all its descendents must have equal or greater cost. Whenever a lower-cost complete tour is found, the cost of the best tour is updated to this lower cost. In addition, an admissible heuristic function, such as the cost of the minimum spanning tree of the remaining unvisited cities, can be added to the cost so far of a partial tour to increase the amount of pruning. Finally, by ordering the children of a given node from smallest to largest estimated total cost, a low-cost solution can be found more quickly, further improving the pruning efficiency.

Interestingly, IDA* and DFBnB exhibit complementary behavior. Both are guaranteed to return an optimal solution using only linear space, assuming that their cost functions are admissible. In IDA*, the cost threshold is always a lower bound on the optimal solution cost, and increases in each iteration until it reaches the optimal cost. In DFBnB, the cost of the best solution found so far is always an upper bound on the optimal solution cost, and decreases until it reaches the optimal cost. While IDA* never expands any nodes whose cost exceeds the optimal cost, its overhead consists of expanding some nodes more than once. While DFBnB never expands any node more than once, its overhead consists of expanding some nodes whose cost exceed the optimal cost. For problems whose search trees are of bounded depth, or for which it is easy to construct a good solution, such as the TSP, DFBnB is usually the algorithm of choice for finding an optimal solution. For problems with infinite search trees, or for which it is difficult to construct a low-cost solution, such as the sliding-tile puzzles or Rubik's cube, IDA* is usually the best choice. Both IDA* and DFBnB suffer the same limitation of all linear-space search algorithms, however, which is that they can generate exponentially more nodes on a graph with many short cycles, relative to a best-first search.

22.4.7 Complexity of Finding Optimal Solutions

The time complexity of a heuristic search algorithm depends on the accuracy of the heuristic function. For example, if the heuristic evaluation function is an exact estimator, then A* runs in linear time, expanding only those nodes on an optimal solution path. Conversely, with a heuristic that returns zero everywhere, A* becomes brute-force uniform-cost search, with exponential complexity. In general, the effect of a good heuristic function is to reduce the effective depth of search required [23].

22.4.8 Heuristic Path Algorithm

Since the complexity of finding optimal solutions to these problems is still exponential, even with a good heuristic function, in order to solve significantly larger problems, the optimality requirement must be relaxed. An early approach to this problem was the heuristic path algorithm (HPA) [35]. HPA is a best-first search, where the cost of a node n is computed as $f(n) = g(n) + w * h(n)$. Varying w produces a range of algorithms from uniform-cost search ($w = 0$), through A* ($w = 1$), to pure heuristic search ($w = \infty$). Increasing w beyond 1 generally decreases the amount of computation, while increasing the cost of the solution generated. This trade-off is often quite favorable, with small increases in solution cost yielding huge savings in computation [21]. Furthermore, it can be shown that the solutions found by this algorithm are guaranteed to be no more than a factor of w greater than optimal [2], but often are significantly better.

Breadth-first search, uniform-cost search, pure heuristic search, A*, and the heuristic path algorithm are all special cases of best-first search. In each step of a best-first search, the node that is best according to some cost function is chosen for expansion. These best-first algorithms differ only in their cost functions: the depth of node n for breadth-first search, $g(n)$ for uniform-cost search, $h(n)$ for pure heuristic search, $g(n) + h(n)$ for A*, and $g(n) + w * h(n)$ for the heuristic path algorithm.

22.4.9 Recursive Best-First Search

The memory limitation of the heuristic path algorithm can be overcome simply by replacing the best-first search with IDA* using the same weighted evaluation function. However, with $w \geq 1$, IDA* is no longer a best-first search, since the total cost of a child can be less than that of its parent, and thus nodes are not necessarily expanded in best-first order. An alternative algorithm is recursive best-first search (RBFS) [21]. RBFS simulates a best-first search in space that is linear in the maximum search depth, regardless of the cost function used. Even with an admissible cost function, RBFS generates fewer nodes than IDA*, and is generally superior to IDA*, except for a small increase in the cost per node generation.

It works by maintaining on the recursion stack the complete path to the current node being expanded, as well as all siblings of nodes on that path, along with the cost of the best node in the subtree explored below each sibling. Whenever the cost of the current node exceeds that of some other node in the previously expanded portion of the tree, the algorithm backs up to their closest common ancestor, and continues the search down the new path. In effect, the algorithm maintains a separate threshold for each subtree diverging from the current search path. See [21] for full details on RBFS.

22.5 Interleaving Search and Execution

In the discussion above, it is assumed that a complete solution can be computed, even before the first step of the solution is executed. This is in contrast to the situation in two-player games, discussed below, where because of computational limits and uncertainty due to the opponent's moves, search and execution are interleaved, with each search determining only the next move to be made. This paradigm is also applicable to single-agent problems. In the case of autonomous vehicle navigation, for example, information is limited by the horizon of the vehicle's sensors, and it must physically move to acquire more information. Thus, only a few moves can be computed at a time, and those moves must be executed before computing the next. Below we consider algorithms designed for this scenario.

22.5.1 Minimin Search

Minimin search determines individual single-agent moves in constant time per move [19]. The algorithm searches forward from the current state to a fixed depth determined by the informational or computational resources available. At the search horizon, the A* evaluation function $f(n) = g(n) + h(n)$ is applied to the frontier nodes. Since all decisions are made by a single agent, the value of an interior node is the minimum of the frontier values in the subtree below that node. A single move is then made to a neighbor of the current state with the minimum value.

Most heuristic functions obey the triangle inequality characteristic of distance measures. As a result, $f(n) = g(n) + h(n)$ is guaranteed to be monotonically nondecreasing along a path. Furthermore, since minimin search has a fixed depth limit, we can apply depth-first branch-and-bound to prune the search tree. The performance improvement due to branch-and-bound is quite dramatic, in some cases extending the achievable search horizon by a factor of 5 relative to brute-force minimin search on sliding-tile puzzles [19].

Minimin search with branch-and-bound is an algorithm for evaluating the immediate neighbors of the current node. As such, it is run until a best child is identified, at which point the chosen move is executed in the real world. We can view the static evaluation function combined with lookahead search as simply a more accurate, but computationally more expensive, heuristic function. In fact, it provides an entire spectrum of heuristic functions trading off accuracy for cost, depending on the search horizon.

22.5.2 Real-Time-A*

Simply repeating minimin search for each move ignores information from previous searches and results in infinite loops. In addition, since actions are committed based on limited information, often the best move may be to undo the previous move. The principle of rationality is that backtracking should occur when the estimated cost of continuing the current path exceeds the cost of going back to a previous state, plus the estimated cost of reaching the goal from that state. Real-time-A* (RTA*) implements this policy in constant time per move on a tree [19].

For each move, the $f(n) = g(n) + h(n)$ value of each neighbor of the current state is computed, where $g(n)$ is now the cost of the edge from the current state to the neighbor, instead of from the initial state. The problem solver moves to a neighbor with the minimum $f(n)$ value, and stores in the previous state the best $f(n)$ value among the remaining neighbors. This represents the $h(n)$ value of the previous state from the perspective of the new current state. This is repeated until a goal is reached. To determine the $h(n)$ value of a previously visited state, the stored value is used, while for a new state the heuristic evaluator is called. Note that the heuristic evaluator may employ minimin lookahead search with branch-and-bound as well.

In a finite problem space in which there exists a path to a goal from every state, RTA* is guaranteed to find a solution, regardless of the heuristic evaluation function [19]. Furthermore, on a tree, RTA* makes locally optimal decisions based on the information available at the time of the decision.

22.5.3 Learning-Real-Time-A*

If a problem is to be solved repeatedly with the same goal state but different initial states, we would like our algorithm to improve its performance over time. Learning-real-time-A* (LRTA*) is such an algorithm. It behaves almost identically to RTA*, except that instead of storing the second-best f

value of a node as its new heuristic value, it stores the best value instead. Once one problem instance is solved, the stored heuristic values are saved and become the initial values for the next problem instance. While LRTA* is less efficient than RTA* for solving a single problem instance, if it starts with admissible initial heuristic values, over repeated trials its heuristic values eventually converge to their exact values, at which point the algorithm returns optimal solutions.

22.6 Game Playing

The second major application of heuristic search algorithms in AI is game playing. One of the original challenges of AI, which in fact predates the term artificial intelligence, was to build a program that could play chess at the level of the best human players [48], a goal finally achieved in 1997 [18]. The earliest work in this area focussed on two-player perfect information games, such as chess and checkers, but recently the field has broadened considerably.

22.6.1 Minimax Search

The standard algorithm for two-player perfect-information games is minimax search with heuristic static evaluation [44]. The simplest version of the algorithm searches forward to a fixed depth in the game tree, limited by the amount of time available per move. At this search horizon, a heuristic evaluation function is applied to the frontier nodes. In this case, the heuristic is a function that takes a board position and returns a number that indicates how favorable that position is for one player relative to the other. For example, a very simple heuristic evaluator for chess would count the total number of pieces on the board for one player, weighted by their relative strength, and subtract the weighted sum of the opponent's pieces. Thus, large positive values would correspond to strong positions for one player, called MAX, whereas negative values of large magnitude would represent advantageous situations for the opponent, called MIN.

Given the heuristic evaluations of the frontier nodes, values for the interior nodes in the tree are recursively computed according to the minimax rule. The value of a node where it is MAX's turn to move is the maximum of the values of its children, while the value of a node where MIN is to move is the minimum of the values of its children. Thus, at alternate levels of the tree, the minimum or the maximum values of the children are backed up. This continues until the values of the immediate children of the current position are computed, at which point one move is made to the child with the maximum or minimum value, depending on whose turn it is to move.

22.6.2 Alpha-Beta Pruning

One of the most elegant of all AI search algorithms is alpha-beta pruning. It was invented in the late 1950s, and a thorough treatment of the algorithm can be found in [28]. The idea, similar to branch-and-bound, is that the minimax value of the root of a game tree can be determined without examining all the nodes at the search frontier.

Figure 22.5 shows some examples of alpha-beta pruning. Only the pictured nodes are generated by the algorithm, with the heavy black lines indicating pruning. At the square nodes MAX is to move, while at the circular nodes it is MIN's turn. The search proceeds depth-first to minimize the memory required, and only evaluates a frontier node when necessary. First, nodes e and f are statically evaluated at 4 and 5, respectively, and their minimum value, 4, is backed up to their parent node d. Node h is then evaluated at 3, and hence the value of its parent node g must be less than or equal to 3, since it is the minimum of 3 and the unknown value of its right child. The value of node c must be 4 then, because it is the maximum of 4 and a value that is less than or equal to 3. Since

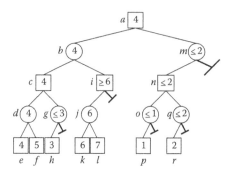

FIGURE 22.5 Alpha-beta pruning example.

we have determined the minimax value of node c, we do not need to evaluate or even generate any more children of node g.

Similarly, after statically evaluating nodes k and l at 6 and 7, respectively, the backed up value of their parent node j is 6, the minimum of these values. This tells us that the minimax value of node i must be greater than or equal to 6, since it is the maximum of 6 and the unknown value of its right child. Since the value of node b is the minimum of 4 and a value that is greater than or equal to 6, it must be 4, and hence we prune the remaining children of node i.

The right half of the tree shows an example of deep pruning. After evaluating the left half of the tree, we know that the value of the root node a is greater than or equal to 4, the minimax value of node b. Once node p is evaluated at 1, the value of its parent node o must be less than or equal to 1. Since the value of the root is greater than or equal to 4, the value of node o cannot propagate to the root, and hence we need not generate any more children of node o. A similar situation exists after the evaluation of node r as 2. At that point, the value of node o is less than or equal to 1, and the value of node q is less than or equal to 2; hence, the value of node n, which is the maximum of the values of nodes o and q, must be less than or equal to 2. Furthermore, since the value of node m is the minimum of the value of node n and that of its brothers, and node n has a value less than or equal to 2, the value of node m must also be less than or equal to 2. This causes the remaining children of node m to be pruned, since the value of the root node a is greater than or equal to 4. Thus, we compute the minimax value of the root of the tree to be 4 by generating only seven leaf nodes in this case.

Since alpha-beta pruning performs a minimax search while pruning much of the tree, its effect is to allow a deeper search in the same amount of time. This raises the question of how much does alpha-beta improve performance? The efficiency of alpha-beta pruning depends upon the order in which nodes are encountered at the search frontier. For any set of frontier node values, there exists some ordering of the values such that alpha-beta will not perform any cutoffs at all. In that case, all frontier nodes must be evaluated, and the time complexity is $O(b^d)$.

On the other hand, there is an optimal or perfect ordering in which every possible cutoff is realized. In that case, the asymptotic time complexity is reduced from $O(b^d)$ to $O(b^{d/2})$. Another way of viewing the perfect ordering case is that for the same amount of computation, one can search twice as deep with alpha-beta pruning as without. Since the search tree grows exponentially with depth, doubling the search horizon is a dramatic improvement.

In between worst-possible ordering and perfect ordering is random ordering, which is the average case. Under random ordering of the frontier nodes, alpha-beta pruning reduces the asymptotic time complexity to approximately $O(b^{3d/4})$ [33]. This means that one can search $4/3$ times as deep with alpha-beta than with simple minimax, yielding a 33% improvement in search depth.

In practice, however, the time complexity of alpha-beta is closer to the best case of $O(b^{d/2})$ due to node ordering. The idea of node ordering is that instead of generating the children of a node in an arbitrary order, we can order the tree based on static evaluations of the interior nodes. In particular, the children of MAX nodes are expanded in decreasing order of their static values, while the children of MIN nodes are expanded in increasing order of their static values.

22.6.3   Other Techniques

A wide variety of additional techniques are employed in modern game-playing programs, including quiescent search, iterative deepening, transposition tables, opening books, end-game databases, and special-purpose hardware. We consider each of these in turn.

22.6.3.1   Quiescence

The idea of quiescence is that the static evaluator should not be applied to positions whose values are unstable, such as those occurring in the middle of a piece trade. In those positions, a small secondary search is conducted until the static evaluation becomes more stable. In games such as chess or checkers, this can be achieved by not statically evaluating any position that allows capture moves, but exploring capture moves one level deeper.

22.6.3.2   Iterative Deepening

Iterative deepening is used to solve the problem of where to set the search horizon [45], and in fact predated its use as a memory-saving device in single-agent search. In a tournament game, there is a limited amount of time allowed for moves. Unfortunately, it is very difficult to accurately predict how long it will take to perform an alpha-beta search to a given depth. The solution is to perform a series of searches to successively greater depths. When time runs out, the move recommended by the last completed search is made.

22.6.3.3   Transposition Tables

The search graphs of most games contain multiple paths to the same node, often reached by making the same moves in a different order, referred to as a transposition of the moves. Since alpha-beta is a depth-first search, it does not detect such duplicate nodes. A transposition table is a table of previously encountered game states, together with their backed-up minimax values. Whenever a new state is generated, if it is stored in the transposition table, its stored value is used instead of researching the tree below the node. Transposition tables can also be used in single-agent depth-first searches to detect some duplicate nodes.

22.6.3.4   Opening Books

Just like people, programs often learn and store the best first few moves of a game, since many games start from a common initial state. Such an opening book can be hand-coded by a human game expert, or automatically learned from deep off-line searches.

22.6.3.5   End-Game Databases

In many games, such as checkers, the number of positions near the end of the game is small relative to the mid-game. This allows the precomputation and storage of end-game databases, which contain the true value, such as win, lose, or draw, of a large number of positions near the end of the game. For example, the Chinook checkers program stores on disk a database of all positions with 10 or fewer pieces on the board, along with their exact values [42]. When a position in the database is encountered during search, the stored value is used, rather than searching the position any further.

22.6.3.6 Special-Purpose Hardware

While the basic algorithms are described above, much of the performance advances in computer chess in previous decades came from faster hardware. The faster the machine, the deeper it can search in the time available, and the better it plays. Despite the rapidly advancing speed of general-purpose computers, the best machines in the 1980s and 1990s included special-purpose hardware designed and built only to play chess. For example, IBM's DeepBlue could evaluate about 200 million chess positions per second [17].

22.6.4 Significant Milestones

The most significant milestones in this area include defeating the human world champion at chess, and solving the game of checkers.

In May 1997, IBM's DeepBlue defeated Gary Kasparov, the world champion, in a six-game exhibition match, achieving a long-anticipated goal in artificial intelligence [18]. Currently, the best chess machines are general-purpose computers that rely entirely on software for their performance, and play comparably to the best humans.

In April 2007, Jonathan Schaeffer's Chinook group at the University of Alberta announced that they had solved the game of checkers [43], proving that when both sides play perfectly, checkers is a draw. This is by far the most complex game solved to date, using two dozen computers running for about 18 years. The combination of its opening book, its mid-game search, and 10-piece end-game databases allows Chinook to play perfect checkers.

22.6.5 Multiplayer Games, Imperfect and Hidden Information

Minimax search with static evaluation and alpha-beta pruning is most appropriate for two-player games with perfect information and alternating moves among the players. This paradigm extends in a straightforward way to more than two players, but alpha-beta becomes much less effective [20].

Games with chance elements, such as the roll of the dice in backgammon, for example, tend to limit search algorithms because of the need to search over all possible chance outcomes. In addition to chance, card games have information that is available to some players but hidden from others, such as cards in different hands in Bridge. Poker is a very difficult challenge in this area, combining all of the above complexities, as well as active deception and the need to model the opponents.

One technique that has been effective in handling hidden information is Monte-Carlo sampling [12]. Given a decision to be made, such as the play of a card in bridge, we can randomly generate a set of hands that is consistent with the information known about those hands at the current point in time. Given this particular set of hands, we then use perfect-information techniques, such as alpha-beta minimax, to determine the optimal play in this case. We then repeat the experiment, generating another random set of hands, consistent with the information available, and compute the optimal move in that case. After generating about 100 different random hands, we then play the card that was most often the optimal card to play over all the randomly generated hands.

22.7 Constraint-Satisfaction Problems

In addition to single-agent path-finding problems and game playing, the third major application of heuristic search is constraint-satisfaction problems. The N-queens problem and the popular Sudoku puzzle are classic examples. Other examples include graph coloring, boolean satisfiability, and scheduling problems.

Constraint-satisfaction problems are usually modelled as follows: There is a set of variables, a set of values for each variable, and a set of constraints on the values that the variables can be assigned.

A unary constraint on a variable specifies a subset of all possible values that can be assigned to that variable. A binary constraint between two variables specifies which possible combinations of assignments to the pair of variables is legal. For example, in the N-queens problem, the variables would represent individual queens, and the values would be their positions on the board. The constraints are binary constraints on each pair of queens that prohibit them from occupying positions on the same row, column, or diagonal.

22.7.1 Chronological Backtracking

The brute-force approach to constraint satisfaction is called chronological backtracking. One selects an order for the variables, and an order for the values, and starts assigning values to the variables one at a time. Each assignment is made so that all constraints involving any of the instantiated variables are satisfied. The reason for this is that once a constraint is violated, no assignment to the remaining variables can possibly satisfy that constraint. Once a variable is reached which has no remaining legal assignments, then the last variable that was assigned is reassigned to its next legal value. The algorithm continues until either one or all complete, consistent assignments are found, resulting in success, or all possible assignments are shown to violate some constraint, resulting in failure. Figure 22.6 shows the tree generated by chronological backtracking to find all solutions to the four-queens problem. The tree is searched depth-first to minimize memory requirements.

22.7.2 Intelligent Backtracking

One can improve the performance of chronological backtracking using a number of techniques, such as variable ordering, value ordering, backjumping, and forward checking.

The order in which variables are instantiated can have a large effect on the size of a search tree. The idea of variable ordering is to order the variables from most constrained to least constrained [10,37].

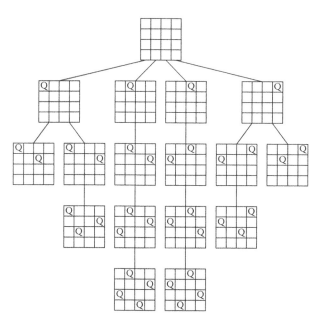

FIGURE 22.6 Tree generated to solve the four queens problem.

For example, if any variable has only a single value remaining that is consistent with the previously instantiated variables, it should be assigned that value immediately. In general, the variables should be instantiated in increasing order of the size of their remaining domains. This can either be done statically at the beginning of the search, or dynamically, reordering the remaining variables each time a variable is assigned a new value.

The order in which the values of a given variable are chosen determines the order in which the tree is searched. Since it does not effect the size of the tree, it makes no difference if all solutions are to be found. If only a single solution is required, however, value ordering can decrease the time required to find a solution. Thus, one should order the values from least constraining to most constraining, in order to minimize the time required to find a first solution [6,13].

The idea of backjumping is that when an impasse is reached, instead of simply undoing the last decision made, the decision that actually caused the failure should be modified [11]. For example, consider a three-variable problem where the variables are instantiated in the order x, y, z. Assume that values have been chosen for both x and y, but that all possible values for z conflict with the value chosen for x. In chronological backtracking, the value chosen for y would be changed, and then all the possible values for z would be tested again, to no avail. A better strategy in this case is to go back to the source of the failure, and change the value of x, before trying different values for y and z.

When a variable is assigned a value, forward checking checks each remaining uninstantiated variable to make sure that there is at least one assignment for each of them that is consistent with the current assignments. If not, the original variable is reassigned to its next value. Forward checking subsumes backjumping.

22.7.3 Constraint Recording

In a constraint-satisfaction problem, some constraints are explicitly specified, and others are implied by explicit constraints. Implicit constraints may be discovered either during a backtracking search, or in advance in a preprocessing phase. The idea of constraint recording is that once these implicit constraints are discovered, they should be saved explicitly so that they do not have to be rediscovered.

A simple example of constraint recording in a preprocessing phase is called arc consistency [10, 29,31]. For each pair of variables x and y that are related by a binary constraint, we remove from the domain of x any values that do not have at least one corresponding consistent assignment to y, and vice versa. In general, several iterations may be required to achieve complete arc consistency. Path consistency is a generalization of arc consistency where instead of considering pairs of variables, we examine triples of constrained variables. The effect of performing arc or path consistency before backtracking is that the resulting search space can be dramatically reduced. In some cases, this preprocessing of the constraints can eliminate the need for search entirely.

22.7.4 Local Search Algorithms

Backtracking searches a space of consistent partial assignments to variables, in the sense that all constraints among instantiated variables are satisfied, and looks for a complete consistent assignment to the variables, or in other words a solution. An alternative approach is to search a space of inconsistent but complete assignments to the variables, until a consistent complete assignment is found. This approach is known as heuristic repair [30] or more generally stochastic local search. For example, in the N-queens problem, this algorithm places all N queens on the board at the same time, and then moves queens one at a time until a solution is found. The natural heuristic, called min-conflicts, moves a queen that is in conflict with the most other queens, to a position where it conflicts with the fewest other queens. What is surprising about this simple strategy is how well it performs, relative to backtracking. While backtracking techniques can solve on the order of hundred-queen problems, heuristic repair can solve million-queen problems, often with only about 50 individual queen moves!

This strategy has been extensively explored in the context of boolean satisfiability. The problem of boolean satisfiability starts with a formula in propositional logic in conjunctive normal form. This is an AND of a set of clauses, each of which is an OR of a set of literals, each of which is either an atomic proposition or a negated atomic proposition. For example, the boolean formula $(a + b) \cdot (a' + b')$ is in conjunctive normal form if $+$ represents OR, $\cdot$ represents AND, and $'$ represents negation. The problem is to determine if there exists some assignment to the variables such that the entire formula evaluates to true, or in other words, all the clauses are satisfied. For example, the above formula is satisfiable, since assigning a to true and b to false will satisfy both clauses. If each clause has three literals, called 3-SAT, the problem is NP-complete.

The greedy satisfiability (GSAT) algorithm [40] starts with a random assignment of boolean values to the variables of a satisfiability problem, and flips the assignment of a variable that results in the largest net increase in the number of satisfied clauses. It continues until either a solution is found, or until a specified number of flips are performed, at which point it starts over with a different random initial assignment. WalkSAT [41] adds another random element to this algorithm. With a certain probability, it flips a variable that GSAT would flip, and with one minus that probability, it chooses a random unsatisfied clause, and flips a randomly chosen variable in that clause.

The main drawback of these local-search approaches is that they are not complete, in that they are not guaranteed to find a solution in a finite amount of time, even if one exists. If there is no solution, these algorithms will run forever, whereas complete backtracking algorithms will eventually discover that a problem is not solvable.

The best complete algorithms for boolean satisfiability are based on the DPLL algorithm [3]. These algorithms assign values to variables one at a time, and propagate the consequences of assignments using unit propagation, backtracking when a partial assignment cannot be extended any further. Both complete algorithms and stochastic local search are areas of active research for boolean satisfiability. Currently, local search algorithms seem to perform best on difficult random problems that are satisfiable, while the complete algorithms tend to perform better on more structured problems.

While constraint-satisfaction problems appear somewhat different from single-agent path-finding problems and two-player games, there is a strong similarity among the algorithms employed. For example, backtracking can be viewed as a form of branch-and-bound, where a node is pruned when any constraint is violated. Similarly, heuristic repair can be viewed as a heuristic search where the evaluation function is the total number of constraints that are violated, and the goal is to find a state with zero constraint violations.

22.8 Research Issues and Summary

22.8.1 Research Issues

A primary research problem in this area is the development of faster algorithms. All the above algorithms are limited by efficiency either in the size of problems that they can solve optimally, or in the quality of the decisions they can make, or solutions they can find within practical time limits. Thus, there is a continual demand for faster algorithms.

One approach to faster algorithms is parallel search. Most search algorithms have a tremendous amount of potential parallelism, since the basic step of node generation and evaluation is often performed billions or trillions of times. As a result, many such algorithms are readily parallelized with nearly linear speedups. The algorithms that are difficult to parallelize are branch-and-bound algorithms, such as alpha-beta pruning, because the results of searching one part of the tree determine whether another part of the tree needs to be examined at all.

Since the performance of a search algorithm depends critically on the quality of the heuristic evaluation function, another important research area is the automatic generation of such functions. This was pioneered in the area of two-player games by Arthur Samuel's landmark checkers program that learned to improve its evaluation function through repeated play [39]. The development of pattern databases, described above, is another approach to automatic heuristic construction.

While the first games to be addressed with AI techniques were two-player perfect information games such as chess and checkers, the field has branched out to consider games with random elements, such as backgammon, hidden information, such as card games, and multiple players. Recently the game of poker has seen a great deal of activity, both from search-based and game-theoretic approaches. The game of Go continues to be very challenging for computers, both because of its large branching factor, and the difficulty of evaluating nonterminal positions.

Another important area in computer game-playing is developing automated opponents for video games. The electronic entertainment industry is larger than the motion-picture industry, and most of these games have agents that must be computer controlled. In addition to low-level issues such as path planning for multiple agents, these games require high-level strategy as well.

22.8.2 Summary

We have described search algorithms for three different classes of problems. In single-agent path-finding problems, the task is to find a sequence of operators that map an initial state to a desired goal state. Much of the work in this area has focussed on finding optimal solutions to such problems, often making use of admissible heuristic functions to speed up the search without sacrificing solution optimality. In the area of game playing, finding optimal solutions is infeasible, and research has focussed on algorithms for making the best move decisions possible given a limited amount of computing time. This approach has also been applied to single-agent problems as well. In constraint-satisfaction problems, the task is to find a state that satisfies a set of constraints. While all three of these types of problems are different, the same set of ideas, such as brute-force search and heuristic evaluation functions, can be applied to all three.

22.9 Further Information

The classic reference in this area is [34]. A number of papers were collected in an edited volume devoted to search [16]. The constraint-satisfaction area is covered by [4], and stochastic local search is covered in [15]. Most new research in this area initially appears in the Proceedings of the Association for the Advancement of on Artificial Intelligence (AAAI) or the Proceedings of the International Joint Conference on Artificial Intelligence (IJCAI). The Preeminent journals in this area include *Artificial Intelligence* (AIJ), and the *Journal of Artificial Intelligence Research* (JAIR).

Defining Terms

Admissible: A heuristic is said to be admissible if it never overestimates actual distance from a given state to a goal. An algorithm is said to be admissible if it always finds an optimal solution to a problem if one exists.

Branching factor: The average number of children of a node in a problem-space graph.

Constraint-satisfaction problem: A problem where the task is to identify a state that satisfies a set of constraints.

Depth: The length of a shortest path from the initial state to a goal state.

Heuristic evaluation function: A function from a state to a number. In a single-agent problem, it estimates the distance from the state to a goal. In a game, it estimates the merit of the position with respect to one player.

Node expansion: Generating all the children of a given state.

Node generation: Creating the data structure that corresponds to a problem state.

Operator: An action that maps one state into another state, such as a twist of Rubik's cube.

Problem instance: A problem space together with an initial state of the problem and a desired set of goal states.

Problem space: A theoretical construct in which a search takes place, consisting of a set of states and a set of operators.

Problem-space graph: A graphical representation of a problem space, where states are represented by nodes, and operators are represented by edges.

Search: A trial-and-error exploration of alternative solutions to a problem.

Search tree: A problem-space graph with a unique path to each node.

Single-agent path-finding problem: A problem where the task is to find a sequence of operators that map an initial state to a goal state.

State: A configuration of a problem, such as the arrangement of the parts of a Rubik's cube at a given point in time.

Acknowledgment

This work was supported in part by NSF Grant IIS-0713178.

References

1. Culberson, J. and J. Schaeffer, Pattern databases, *Computational Intelligence*, 14(3), 318–334, 1998.
2. Davis, H.W., A. Bramanti-Gregor, and J. Wang, The advantages of using depth and breadth components in heuristic search, in *Methodologies for Intelligent Systems 3*, Z.W. Ras and L. Saitta (Eds.), North-Holland, Amsterdam, the Netherlands, 1989, pp. 19–28.
3. Davis, M., G. Logemann, and D. Loveland, A machine program for theorem proving, *Journal of the Association for Computing Machinery*, 5(7), 394–397, 1962.
4. Dechter, R., *Constraint Processing*, Morgan-Kaufmann, San Francisco, CA, 2003.
5. Dechter, R. and J. Pearl, Generalized best-first search strategies and the optimality of A*, *Journal of the Association for Computing Machinery*, 32(3), 505–536, July 1985.
6. Dechter, R. and J. Pearl, Network-based heuristics for constraint-satisfaction problems, *Artificial Intelligence*, 34(1), 1–38, 1987.
7. Dijkstra, E.W., A note on two problems in connexion with graphs, *Numerische Mathematik*, 1, 269–271, 1959.
8. Dillenburg, J.F. and P.C. Nelson, Improving the efficiency of depth-first search by cycle elimination, *Information Processing Letters*, 45(1), 5–10, 1993.
9. Doran, J.E. and D. Michie, Experiments with the graph traverser program, *Proceedings of the Royal Society A*, 294, 235–259, 1966.
10. Freuder, E.C., A sufficient condition for backtrack-free search, *Journal of the Association for Computing Machinery*, 29(1), 24–32, 1982.

11. Gaschnig, J., Performance measurement and analysis of certain search algorithms, PhD thesis. Department of Computer Science, Carnegie-Mellon University, Pittsburgh, PA, 1979.

12. Ginsberg, M.L., GIB: Imperfect information in a computationally challenging game, *Journal of Artificial Intelligence Research*, 14, 303–358, 2001.

13. Haralick, R.M. and G.L. Elliott, Increasing tree search efficiency for constraint satisfaction problems, *Artificial Intelligence*, 14, 263–313, 1980.

14. Hart, P.E., N.J. Nilsson, and B. Raphael, A formal basis for the heuristic determination of minimum cost paths, *IEEE Transactions on Systems Science and Cybernetics*, 4(2), 100–107, 1968.

15. Hoos, H.H. and T. Stutzle, *Stochastic Local Search: Foundations and Applications*, Morgan-Kaufmann, San Francisco, CA, 2004.

16. Kanal, L. and V. Kumar (Eds.), *Search in Artificial Intelligence*, Springer-Verlag, New York, 1988.

17. Campbell, M., A.J. Hoane, and F. Hsu, Deep blue, *Artificial Intelligence*, 134(1–2), 57–83, 2002.

18. Korf, R.E., Depth-first iterative-deepening: An optimal admissible tree search, *Artificial Intelligence*, 27(1), 97–109, 1985.

19. Korf, R.E., Real-time heuristic search, *Artificial Intelligence*, 42(2–3), 189–211, 1990.

20. Korf, R.E., Multi-player alpha-beta pruning, *Artificial Intelligence*, 48(1), 99–111, 1991.

21. Korf, R.E., Linear-space best-first search, *Artificial Intelligence*, 62(1), 41–78, 1993.

22. Korf, R.E., Finding optimal solutions to Rubik's cube using pattern databases, *Proceedings of the Fourteenth National Conference on Artificial Intelligence (AAAI-97)*, Providence, RI, July 1997, pp. 700–705.

23. Korf, R.E., M. Reid, and S. Edelkamp, Time complexity of Iterative-Deepening-A*, *Artificial Intelligence*, 129(1–2), 199–218, 2001.

24. Korf, R.E. and A. Felner, Disjoint pattern database heuristics, *Artificial Intelligence*, 134(1–2), 9–22, 2002.

25. Korf, R.E, W. Zhang, I. Thayer, and H. Hohwald, Frontier search, *Journal of the Association for Computing Machinery*, 52(5), 715–748, 2005.

26. Korf, R.E. and P. Schultze, Large-scale, parallel breadth-first search, *Proceedings of the National Conference on Artificial Intelligence (AAAI-05)*, Pittsburgh, PA, July 2005, pp. 1380–1385.

27. Korf, R.E., Linear-time disk-based implicit graph search, *Journal of the ACM*, 55(6), 26-1–26-40, 2008.

28. Knuth, D.E. and R.E. Moore, An analysis of alpha-beta pruning, *Artificial Intelligence*, 6(4), 293–326, 1975.

29. Mackworth, A.K., Consistency in networks of relations. *Artificial Intelligence*, 8(1), 99–118, 1977.

30. Minton, S., M.D. Johnston, A.B. Philips, and P. Laird, Minimizing conflicts: A heuristic repair method for constraint satisfaction and scheduling problems, *Artificial Intelligence*, 58(1–3), 161–205, 1992.

31. Montanari, U., Networks of constraints: Fundamental properties and applications to picture processing, *Information Science*, 7, 95–132, 1974.

32. Newell, A. and H.A. Simon, *Human Problem Solving*, Prentice-Hall, Englewood Cliffs, NJ, 1972.

33. Pearl, J., The solution for the branching factor of the alpha-beta pruning algorithm and its optimality, *Communications of the Association of Computing Machinery*, 25(8), 559–564, 1982.

34. Pearl, J., *Heuristics*, Addison-Wesley, Reading, MA, 1984.

35. Pohl, I., Heuristic search viewed as path finding in a graph, *Artificial Intelligence*, 1, 193–204, 1970.

36. Pohl, I., Bi-directional search, in *Machine Intelligence 6*, B. Meltzer and D. Michie (Eds.), American Elsevier, New York, 1971, pp. 127–140.

37. Purdom, P.W., Search rearrangement backtracking and polynomial average time, *Artificial Intelligence*, 21(1–2), 117–133, 1983.

38. Ratner, D. and M. Warmuth, Finding a shortest solution for the NxN extension of the 15-puzzle is intractable, *Proceedings of the Fifth National Conference on Artificial Intelligence (AAAI-86)*, Philadelphia, PA, 1986, pp. 168–172.

39. Samuel, A.L., Some studies in machine learning using the game of checkers, in *Computers and Thought*, E. Feigenbaum and J. Feldman (Eds.), McGraw-Hill, New York, 1963, pp. 71–105.
40. Selman, B., H. Levesque, and D. Mitchell, A new method for solving hard satisfiability problems, *Proceedings of the Tenth National Conference on Artificial Intelligence (AAAI-92)*, San Jose, CA, July 1992, pp. 440–446.
41. Selman, B., H. Kautz, and B. Cohen, Noise strategies for improving local search, *Proceedings of the Twelfth National Conference on Artificial Intelligence (AAAI-94)*, Seattle, WA, August 1994. pp. 337–343.
42. Schaeffer, J., Y. Bjornsson, N. Burch, R. Lake, P. Lu, and S. Sutphen, Building the checkers 10-piece endgame databases, in H.J. van den Herik, Iida, H., and Heinz, E.A., editors, *Advances in Computer Games 10*, Kluwer, Boston, MA, 2003, pp. 193–210.
43. Schaeffer, J., N. Burch, Y. Bjornsson, A. Kishimoto, M. Mueller, R. Lake, P. Lu, and S. Sutphen, Checkers is solved, *Science*, 317, 1518–1522, 2007.
44. Shannon, C.E., Programming a computer for playing chess, *Philosophical Magazine*, 41, 256–275, 1950.
45. Slate, D.J. and L.R. Atkin, CHESS 4.5 - The Northwestern University chess program, in *Chess Skill in Man and Machine*, P.W. Frey (Ed.), Springer-Verlag, New York, 1977, pp. 82–118.
46. Stickel, M.E. and W.M. Tyson, An analysis of consecutively bounded depth-first search with applications in automated deduction, *Proceedings of the International Joint Conference on Artificial Intelligence (IJCAI-85)*, Los Angeles, CA, August 1985, pp. 1073–1075.
47. Taylor, L. and R.E. Korf, Pruning duplicate nodes in depth-first search, *Proceedings of the National Conference on Artificial Intelligence (AAAI-93)*, Washington, DC, July 1993, pp. 756–761.
48. Turing, A.M., Computing machinery and intelligence, *Mind*, 59, 433–460, 1950. Also in *Computers and Thought*, E. Feigenbaum and J. Feldman (Eds.), McGraw-Hill, New York, 1963.
49. Zhou, R. and E.A. Hansen, Sparse-memory graph search, *Proceedings of the 18th International Joint Conference on Artificial Intelligence (IJCAI-03)*, Acapulco, Mexico, August 2003, pp. 1259–1266.
50. Zhou, R. and E.A. Hansen, Structured duplicate detection in external-memory graph search, *Proceedings of the 19th National Conference on Artificial Intelligence (AAAI-04)*, San Jose, CA, July 2004, pp. 683–688.

23

Algorithmic Aspects of Natural Language Processing

Mark-Jan Nederhof
University of St Andrews

Giorgio Satta
University of Padua

23.1 Introduction

Examples of natural languages are Chinese, English, and Italian. They are called natural as they evolved in a more or less natural way without too many deliberate considerations. This sets them apart from **formal languages**, among which are programming languages, which are designed to allow easy processing by computer algorithms. Typically, programs in programming languages, such as C or Java, can be processed (compiled) in close to linear time in their length.

One particular feature that most programming languages have in common, and that allows for their fast processing, is the absence of **ambiguity**. That is, only one structure, called a **parse** or **parse tree**, can be assigned to any program, and this parse can have only one meaning. Furthermore, the design of many programming languages is such that the single parse can be found deterministically, which means that every parsing step contributes a fragment of the resulting parse. As parses have a size linear in the length of the input, this explains why parsing is possible in linear time. Subsequent processing of the parse, for example in order to compile to machine code, is also commonly possible in close to linear time.

Natural languages are quite different in this respect. Like programs in a programming language, sentences in a natural language can be assigned parses, but often the sentences are ambiguous and allow more than one parse. Even for a single parse, there may be ambiguity in the meanings of words or expressions. The existence of ambiguity in natural language is witnessed by frequent misunderstandings in daily life, but it is also an essential feature of poetry and puns.

The field of **natural language processing** (NLP) studies algorithms, tools, and techniques for the automatic processing of natural languages. A related if not synonymous term is computational linguistics, which stresses that the field can be seen as a subfield of **linguistics**, which is the study of (natural) language.

Language can be investigated from different perspectives. At the lowest level, phonetics and phonology study the sounds that spoken language consists of and the rules that govern them. Morphology is the study of the internal structure of words. **Syntax** is the study of how words are combined to form sentences. The meaning and the use of language are studied by semantics and pragmatics, and the structure of human communication is studied by discourse.

The problem of parsing, with which we started our exposition, concerns the syntax of language. Although interesting algorithms exist for the other levels of language as well, this chapter will concentrate on algorithms related to syntax, as these form the most mature and well-understood part of NLP.

Next to parsing, many other tasks studied in NLP are relevant to syntax. One such task is **grammar induction**, which involves finding the syntactic structure of a language on the basis of examples. Another task is **machine translation**, which includes transforming syntactic structure from one language to that of another. A selection of algorithms that concerns these tasks will be discussed further in the remainder of this chapter.

23.2 Example

We illustrate the problem of ambiguity by an example. Familiarity with context-free grammars (CFGs) (see Chapter 20 of *Algorithms and Theory of Computation Handbook: General Concepts and Techniques*) is assumed.

The sentence:

(1) Our company is training workers

has one obvious meaning to most readers. However, in principle, at least two other meanings exist. This becomes clear if we replace some words by other words of the same type:

(2) Our problem is training workers
(3) Our company hires training workers

The three readings of (1) correspond to the three parses in Figure 23.1 assuming the CFG in the following. This grammar is intended for illustrative purposes and represents a rough approximation of a tiny part of English at best. We let S stand for "sentence," NP for "noun phrase," N for "noun,"

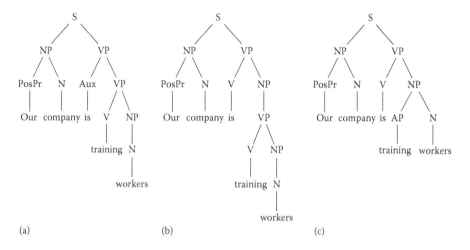

(a)   (b)   (c)

FIGURE 23.1 Three parses for "our company is training workers."

PosPr for "possessive pronoun," VP for "verb phrase," Aux for "auxiliary verb," V for "verb," and AP for "adjective phrase."

$$
\begin{array}{rcl}
S & \to & NP\ VP \\
NP & \to & N \\
NP & \to & PosPr\ N \\
NP & \to & AP\ N \\
NP & \to & VP \\
N & \to & company \\
N & \to & workers \\
PosPr & \to & our \\
VP & \to & Aux\ VP \\
VP & \to & V\ NP \\
Aux & \to & is \\
V & \to & training \\
V & \to & is \\
AP & \to & training
\end{array}
$$

An important observation is that the number of parses can be exponential in the length of the sentence. An easy way to illustrate this is to extend the above example by conjunction, to allow sentences of the following form:

(4) Robin's company is training workers and Sandy's company is training workers and [...]

Such a sentence has at least 3^k parses, where k is the number of times the ambiguous construction from (1) is repeated.

The exponential behavior entails that in practice it is not feasible to enumerate all parses of a sentence. Therefore, one often represents the set of all parses by a structure commonly referred to as a parse forest, also called more explicitly a shared-packed parse forest. The term "parse forest" has become popular since the release of Tomita's book (1986), but the underlying concept existed long before that as we see in the next section.

Two observations underlie parse forests. The first is that alternative subparses of a substring with the same nonterminal at the root can be "packed" together, and be treated as one with regard to larger subparses of larger substrings. In Figure 23.1, this pertains to the two subparses for "training workers" with NP at the root in parses (b) and (c). The second observation is that an identical subparse can be "shared" among several larger subparses. An example in Figure 23.1 is the subparse of "training workers" with root labeled VP, which is the same in parses (a) and (b).

A graphical representation of a parse forest is given in Figure 23.2. Packing is represented by a rectangle enclosing two or more nodes. We see sharing where a node has more than one parent.

In the following section, we discuss the construction of parse forests. Thereafter, we also address the problem of selecting the intended parse tree within a parse forest.

23.3 Context-Free Parsing by Intersection

The underlying principle of parse forests was already discovered by Bar-Hillel et al. (1964), who found a constructive proof that the intersection of a context-free language and a regular language is again a context-free language. The input to this construction is a CFG and a finite automaton. Instead of a graph representation as in Figure 23.2, the output is itself a CFG, which is called the intersection grammar. In this representation, sharing amounts to multiple occurrences of a nonterminal in right-hand sides, and packing amounts to multiple occurrences of a nonterminal in left-hand sides.

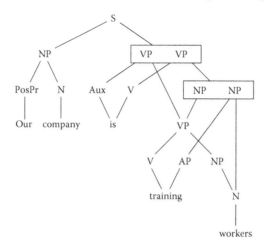

FIGURE 23.2 The three parses in a parse forest.

$[0, S, 5]$	$\rightarrow$	$[0, NP, 2] [2, VP, 5]$	
$[0, NP, 2]$	$\rightarrow$	$[0, PosPr, 1] [1, N, 2]$	
$[0, PosPr, 1]$	$\rightarrow$	$[0, our, 1]$	
$[1, N, 2]$	$\rightarrow$	$[1, company, 2]$	
$[2, VP, 5]$	$\rightarrow$	$[2, Aux, 3] [3, VP, 5]$	
$[2, VP, 5]$	$\rightarrow$	$[2, V, 3] [3, NP, 5]$	
$[2, Aux, 3]$	$\rightarrow$	$[2, is, 3]$	
$[2, V, 3]$	$\rightarrow$	$[2, is, 3]$	
$[3, NP, 5]$	$\rightarrow$	$[3, VP, 5]$	
$[3, NP, 5]$	$\rightarrow$	$[3, AP, 4] [4, N, 5]$	
$[3, VP, 5]$	$\rightarrow$	$[3, V, 4] [4, NP, 5]$	
$[3, V, 4]$	$\rightarrow$	$[3, training, 4]$	
$[3, AP, 4]$	$\rightarrow$	$[3, training, 4]$	
$[4, NP, 5]$	$\rightarrow$	$[4, N, 5]$	
$[4, N, 5]$	$\rightarrow$	$[4, workers, 5]$	

$[0, our, 1]$	$\rightarrow$	our
$[1, company, 2]$	$\rightarrow$	company
$[2, is, 3]$	$\rightarrow$	is
$[3, training, 4]$	$\rightarrow$	training
$[4, workers, 5]$	$\rightarrow$	workers

FIGURE 23.3 Parse forest in the representation as intersection grammar.

We simplify the discussion here by assuming a special type of finite automaton that recognizes just one string, say $w = a_1 \cdots a_n$. The states of the automaton represent the positions between adjacent symbols. In addition, the initial state 0 and the final state n represent the positions at the left and right ends of w. By this simplification, the nonterminals of the intersection grammar are triples $[i, X, j]$, where i and j are input positions, and X is a terminal or nonterminal from the input grammar. For each subparse with root labeled $[i, X, j]$ $(0 \leq i \leq j \leq n)$ in the intersection grammar, there is a corresponding subparse of $a_{i+1} \cdots a_j$ with root labeled X in the original grammar. Triples of this form, as well as related objects to be discussed in the sequel, are also referred to as (parse) items. The start symbol of the intersection grammar is $[0, S, n]$, where S is the start symbol of the input grammar.

As an example, Figure 23.3 shows the parse forest from Figure 23.2 in the representation as intersection grammar.

Let $\mathcal{G}$ be the input CFG and w the input string of length $|w| = n$. An easy way to construct the intersection grammar $\mathcal{G}_\cap$ is by the following algorithm:

INTERSECTION$(\mathcal{G}, w)$ {let $a_1 \cdots a_n = w$}

1: $\mathcal{G}_\cap \leftarrow$ CFG with start symbol $[0, S, n]$ and empty set of rules
2: **for all** rules $A \rightarrow X_1 \cdots X_m$ from $\mathcal{G}$ **do**

3:   **for all** sequences of positions $i_0, \ldots, i_m$ $(0 \leq i_0 \leq \cdots \leq i_m \leq n)$ **do**

4:     add the rule $[i_0, A, i_m] \rightarrow [i_0, X_1, i_1] \cdots [i_{m-1}, X_m, i_m]$ to $\mathcal{G}_\cap$

5: **for all** i $(1 \leq i \leq n)$ **do**

6:     add the rule $[i - 1, a_i, i] \rightarrow a_i$ to $\mathcal{G}_\cap$

7: **return** $\mathcal{G}_\cap$

In general, the above algorithm produces more rules than needed. Formally, we say a nonterminal in a CFG is generating if at least one terminal string can be derived from that nonterminal. We say a nonterminal is reachable if a string containing that nonterminal can be derived from the start symbol. A nonterminal is called useless, if it is nongenerating or unreachable or both. A grammar is called reduced, if it does not contain any rules with useless nonterminals.

In the running example, $[0, VP, 2]$ is nongenerating, as "our company" is not a verb phrase. Therefore, the presence of a rule $[0, VP, 2] \rightarrow [0, Aux, 1] [1, VP, 2]$ makes the intersection grammar nonreduced.

The set of generating nonterminals can be computed by the following algorithm. It is applicable to any CFG, not just to those with nonterminals of the form $[i, A, j]$. We let Σ denote the terminal alphabet, and Σ^* the set of strings of terminal symbols including the empty string ε.

GENERATING$(\mathcal{G})$

1: $OLDGEN \leftarrow \emptyset$

2: $GEN \leftarrow \{A \mid A \rightarrow v, v \in \Sigma^*\}$

3: **while** $GEN \neq OLDGEN$ **do**

4:   $OLDGEN \leftarrow GEN$

5:   $GEN \leftarrow \{A \mid A \rightarrow \alpha, \alpha \in (\Sigma \cup OLDGEN)^*\}$

6: **return** GEN

This algorithm can be implemented to run in linear time in the number of nonterminal occurrences in $\mathcal{G}$ in the following way. For each rule, we maintain a counter containing the number of nonterminals in the right-hand side that has not yet been added to GEN. If a nonterminal A is newly added to GEN, the counters of rules in which A occurs in the right-hand side are decremented. If a counter of any rule becomes 0, the left-hand side nonterminal is added to GEN unless it has already been added. By this implementation, at most one step is needed for each nonterminal occurrence in the grammar.

The problem of finding the set of reachable nonterminals can also be solved in linear time, by simple reduction to graph-reachability. We conclude that a grammar can be reduced in linear time by computing the generating and reachable nonterminals, and then eliminating all rules in which nonterminals that are not generating or not reachable occur.

In practice, it is preferable to combine the construction of an intersection grammar with its reduction, which diminishes the space requirements of intermediate results. The procedure starts by computing a set of generating nonterminals. The difference with the earlier procedure is that this phase now precedes the explicit construction of rules of the intersection grammar.

The algorithm does, however, store and return items of the form $[i, A \rightarrow \alpha \bullet \beta, j]$, where $A \rightarrow \alpha\beta$ is a rule from $\mathcal{G}$ and $\alpha \neq \varepsilon$. Informally, the dot divides the right-hand side into a (nonempty) prefix and a suffix, and the input positions i and j delimit a substring that can be derived from the prefix. More precisely, if we assume $\alpha = X_1 \cdots X_m$, then an item of this form represents that $[i_0, X_1, i_1]$, $\ldots, [i_{m-1}, X_m, i_m]$ are all generating nonterminals for some choice of $i = i_0, i_1, \ldots, i_{m-1}, i_m = j$. An item $[i, A \rightarrow \alpha \bullet \beta, j]$ can thereby be seen as a partial result toward establishing that $[i, A, j']$ may be a generating nonterminal, for some $j' \geq j$. Such items together with the familiar items of the form $[i, A, j]$ are gathered in a single set GEN.

The algorithm maintains an agenda with newly obtained items that are still to be put in GEN. New items are also combined with existing items to derive yet more items, until the agenda is empty.

GENERATINGINTERSECTION$(\mathcal{G}, w)$ {let $a_1 \cdots a_n = w$}

1: $GEN \leftarrow \emptyset$
2: $AGENDA \leftarrow \{[i-1, a_i, i] \mid 1 \le i \le n\} \cup \{[i, A, i] \mid A \to \varepsilon, 0 \le i \le n\}$
3: **while** $AGENDA \ne \emptyset$ **do**
4:    remove some $ITEM$ from $AGENDA$
5:    **if** $ITEM \notin GEN$ **then**
6:      $GEN \leftarrow GEN \cup \{ITEM\}$
7:      **if** $ITEM = [j, X, k]$ **then**
8:        **for all** $[i, A \to \alpha \bullet X\beta, j] \in GEN$ **do**
9:          $AGENDA \leftarrow AGENDA \cup \{[i, A \to \alpha X \bullet \beta, k]\}$
10:        **for all** $A \to X\beta$ **do**
11:          $AGENDA \leftarrow AGENDA \cup \{[j, A \to X \bullet \beta, k]\}$
12:      **if** $ITEM = [i, A \to \alpha \bullet X\beta, j]$ **then**
13:        **for all** $[j, X, k] \in GEN$ **do**
14:          $AGENDA \leftarrow AGENDA \cup \{[i, A \to \alpha X \bullet \beta, k]\}$
15:      **if** $ITEM = [i, A \to \alpha \bullet, j]$ **then**
16:        $GEN \leftarrow GEN \cup \{[i, A, j]\}$
17: **return** GEN

The rules of the reduced intersection grammar are now constructed in a top–down manner, which ensures that only reachable nonterminals are considered. The process is initiated by a call INTERSECTIONFILTERED$(0, S, n)$, where S is the start symbol of the input grammar and $n = |w|$. A set $DONE$, which is initially $\emptyset$, is maintained to prevent that rules would be constructed more than once by repeated calls of INTERSECTIONFILTERED with the same arguments. All new rules are composed of nonterminals that were found to be generating earlier. This guarantees that the intersection grammar is reduced.

INTERSECTIONFILTERED(i, X, j)

1: **if** $[i, X, j] \notin DONE$ **then**
2:    $DONE \leftarrow DONE \cup \{[i, X, j]\}$
3:    **if** X is terminal a **then**
4:      add $[i, a, j] \to a$ to $\mathcal{G}_\cap$
5:    **else** {X is nonterminal A}
6:      **if** $i = j$ and $A \to \varepsilon$ **then**
7:        add $[i, A, j] \to \varepsilon$ to $\mathcal{G}_\cap$
8:      **for all** $A \to X_1 \cdots X_m$ $(m > 0)$ and sequences
          $[i_0, A \to X_1 \cdots X_{m-1} X_m \bullet, i_m], [i_{m-1}, X_m, i_m],$
          $[i_0, A \to X_1 \cdots X_{m-1} \bullet X_m, i_{m-1}], [i_{m-2}, X_{m-1}, i_{m-1}],$
          $\ldots,$
          $[i_0, A \to X_1 \bullet \cdots X_{m-1} X_m, i_1], [i_0, X_1, i_1] \in GEN,$
       where $i_0 = i$ and $i_m = j$ **do**
9:        add $[i_0, A, i_m] \to [i_0, X_1, i_1] \cdots [i_{m-1}, X_m, i_m]$ to $\mathcal{G}_\cap$
10:        **for all** k $(1 \le k \le m)$ **do**
11:          INTERSECTIONFILTERED(i_{k-1}, X_k, i_k)

If r is the length of the longest right-hand side of a rule from the input grammar $\mathcal{G}$, then the size $|\mathcal{G}_\cap|$ of $\mathcal{G}_\cap$ is $\mathcal{O}(|\mathcal{G}| \cdot n^{r+1})$. The size of a CFG is defined as the sum of the number of nonterminal and terminal occurrences in its rules. Also the time and space complexities of the intersection process itself are given by $\mathcal{O}(|\mathcal{G}| \cdot n^{r+1})$, irrespective of whether reduction is integrated into the intersection process, or done afterward.

$$\frac{}{[i-1, a_i, i]} \quad \{1 \leq i \leq n \quad \text{(a)} \qquad\qquad \frac{[j, X, k]}{[j, A \to X \bullet \beta, k]} \quad \{A \to X\beta \quad \text{(d)}$$

$$\frac{}{[i, A, i]} \quad \begin{cases} A \to \varepsilon \\ 0 \leq i \leq n \end{cases} \quad \text{(b)} \qquad\qquad \frac{[i, A \to \alpha \bullet X\beta, j]}{[j, X, k]} \\ \overline{[i, A \to \alpha X \bullet \beta, k]} \quad \text{(e)}$$

$$\frac{[i, A \to \alpha \bullet, j]}{[i, A, j]} \quad \text{(c)}$$

FIGURE 23.4 The algorithm GENERATINGINTERSECTION as deduction system.

If the input grammar is in binary form (that is $r = 2$), then the complexity is $\mathcal{O}(|\mathcal{G}| \cdot n^3)$. This is also the time and space complexities of the best practical recognition and parsing algorithms for CFGs. (See, for example, the CYK algorithm discussed in Chapter 20 of *Algorithms and Theory of Computation Handbook: General Concepts and Techniques*). Grammars that are not in binary form can be easily brought in binary form by a construction that is linear in the size of the input grammar.

It is also possible to integrate binarization into the construction of the intersection grammar by using items of the form $[i, A \to \alpha \bullet \beta, j]$, $\alpha \neq \varepsilon$, as nonterminals. The intersection grammar then contains, among others, rules of the form $[i, A \to \alpha X \bullet \beta, k] \to [i, A \to \alpha \bullet X\beta, j]\ [j, X, k]$ and $[i, A \to X \bullet \beta, j] \to [i, X, j]$.

An important observation is that GENERATINGINTERSECTION has time complexity $\mathcal{O}(|\mathcal{G}| \cdot n^3)$ but its space complexity is only $\mathcal{O}(|\mathcal{G}| \cdot n^2)$. If the objective is recognition rather than parsing, then it suffices to check whether $[0, S, n]$ is in *GEN* and the application of INTERSECTIONFILTERED is not needed.

The extraction of a single and arbitrary parse from *GEN* has time complexity $\mathcal{O}(|\mathcal{G}| \cdot n^2)$. This can be explained as follows. There are $\mathcal{O}(|\mathcal{G}| \cdot n)$ items of the form $[i_0, A \to X_1 \cdots X_k \bullet \cdots X_m, i_k]$, $k \geq 2$, that are consistent with a given parse of the input string. Such items are considered while we are extracting the parse from *GEN* in a top–down manner, much as in INTERSECTIONFILTERED. For each, we may need to check at most $\mathcal{O}(n)$ values i_{k-1} between i_0 and i_k to determine whether $[i_0, A \to X_1 \cdots \bullet X_k \cdots X_m, i_{k-1}]$, $[i_{k-1}, X_k, i_k] \in GEN$.

One may extend GENERATINGINTERSECTION such that it stores a list of the relevant i_{k-1} with items $[i_0, A \to X_1 \cdots X_k \bullet \cdots X_m, i_k]$. This allows a single parse to be extracted in $\mathcal{O}(|\mathcal{G}| \cdot n)$ time, but the storage costs then grow to $\mathcal{O}(|\mathcal{G}| \cdot n^3)$.

Algorithms, such as GENERATINGINTERSECTION, are sometimes presented in the form of a deduction system. A deduction system consists of a collection of inference rules, each consisting of a list of antecedents, which stand for items that we have already derived, and, below a horizontal line, the consequent, which stands for an item that we derive from the antecedents. At the right of an inference rule, we may also write a number of side conditions, which need to be fulfilled for the inference rule to be applicable. The side conditions here refer to rules from the grammar and the input string.

A deduction system has a natural interpretation as dynamic programming algorithm, exemplified by our code for GENERATINGINTERSECTION, as realization of the deduction system in Figure 23.4. An easy way to determine the time complexity of the dynamic programming algorithm is to look at the number of possible instantiations of each inference rule. The inference rule in Figure 23.4e is the most expensive as it involves three input positions and one position within a grammar rule. The number of applications is therefore $\mathcal{O}(|\mathcal{G}| \cdot n^3)$, which confirms our earlier observations about the time complexity of GENERATINGINTERSECTION.

23.4 Lexicalization

A central issue in modeling the syntax of natural languages is the extreme sensitivity to the choice of lexical elements, that is, the words of the language. Let us return to the Example in Section 23.2.

The appropriate parse of "our company is training workers" is that of Figure 23.1a. However, if we replace the word "company" with the word "problem," the appropriate parse becomes the structure reflected by Figure 23.1b instead. A CFG of the kind presented in Section 23.2 cannot model such effects, as it treats "company" and "problem" uniformly as nouns (strings derived from N).

A solution is to incorporate a terminal symbol (lexical element) in each nonterminal. This terminal is such that it plays an important role in the syntactic and semantic contents of the derived string. The model we consider here is called bilexical context-free grammar (2-LCFG). In various guises, this model is used extensively in natural language parsing. It allows us to write rules of the form S[training] → NP[company] VP[training], which expresses that a noun phrase whose main element is "company" can combine with a verb phrase whose main element is "training." By the same token, one might exclude a rule of the form S[training] → NP[problem] VP[training], since, typically, a problem cannot be the subject of training.

More precisely, a 2-LCFG is a CFG with nonterminal symbols of the form $A[a]$, where a is a terminal symbol and A is a symbol drawn from a small set of so-called delexicalized nonterminals, which we denote as V_D. Every rule in a 2-LCFG has one of the following forms: $A[b] \rightarrow B[b] \, C[c]$, $A[c] \rightarrow B[b] \, C[c]$, or $A[a] \rightarrow a$. Note that in binary rules (those with two members in the right-hand side), the terminal symbol associated with the left-hand side nonterminal is always inherited from one of the nonterminals in the right-hand side.

For technical reasons, we assume the existence of a dummy terminal \$ to the right of each sentence, and nowhere else. A dummy nonterminal with the same name allows terminal \$ to be derived by $\$[\$] \rightarrow \$$. The start symbol of a 2-LCFG is $S[\$]$ and there are rules expanding it to $S[a] \, \$[\$]$, where a is typically the main verb of a sentence to be derived. Figure 23.5 presents a possible parse tree for the running example, assuming a 2-LCFG.

Let Σ denote the set of terminal symbols. In the worst case, a 2-LCFG can have $\Theta(|V_D|^3 \cdot |\Sigma|^2)$ binary rules. Whereas V_D is typically small, the set Σ can grow very large in practical applications. When parsing with a 2-LCFG, it is therefore preferable to restrict the grammar to those rules that contain lexical elements actually occurring within the input sentence w. Based on the time complexity of general context-free parsing, as discussed in the previous sections, we then obtain a time complexity of $\mathcal{O}(|V_D|^3 \cdot |w|^5)$, under the assumption that $|w| < |\Sigma|$, which always holds in practical applications. In terms of sentence length, this is much worse than the time complexity of unlexicalized parsing.

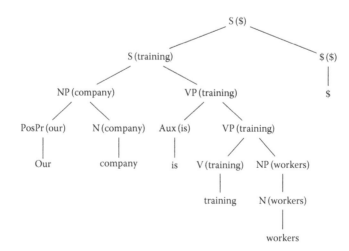

FIGURE 23.5 A parse of "our company is training workers," assuming a 2-LCFG.

$$\frac{}{[h-1,A,h,h]}\left\{\begin{array}{l}A[a_h]\rightarrow a_h\\ 1\le h\le n+1\end{array}\right.\quad\text{(a)}\qquad\qquad \frac{[i,A,h,j]}{[i,A,h,-]}\quad\text{(d)}$$

$$\frac{\begin{array}{c}[-,B,h,j]\\ [j,C,h',k]\end{array}}{[B,h,A,j,k]}\left\{A[a_h]\rightarrow B[a_h]\;C[a_{h'}]\right.\quad\text{(b)}\qquad\qquad \frac{[i,A,h,j]}{[-,A,h,j]}\quad\text{(e)}$$

$$\frac{\begin{array}{c}[i,B,h',j]\\ [j,C,h,-]\end{array}}{[i,j,A,C,h]}\left\{A[a_h]\rightarrow B[a_{h'}]\;C[a_h]\right.\quad\text{(c)}\qquad\qquad \frac{\begin{array}{c}[i,B,h,j]\\ [B,h,A,j,k]\end{array}}{[i,A,h,k]}\quad\text{(f)}$$

$$\frac{\begin{array}{c}[j,C,h,k]\\ [i,j,A,C,h]\end{array}}{[i,A,h,k]}\quad\text{(g)}$$

FIGURE 23.6 Deduction system for bilexical recognition. We assume $w = a_1 \cdots a_n$, $a_{n+1} = \$$.

FIGURE 23.7 Illustration of the use of inference rules (e), (b), and (f) of bilexical recognition.

The time complexity can be reduced by a factor of $|w|$ by a recognition algorithm that was specifically designed for lexicalized grammars. It uses items of the form $[i, A, h, j]$. In terms of the items used in Section 23.3, this has the same meaning as $[i, A[a_h], j]$, which now includes a lexicalized nonterminal $A[a_h]$. A new feature is that some steps of the algorithm require temporarily ignoring either i or j, which is indicated by substituting one or the other by a hyphen.

We also need items of the form $[B, h, A, j, k]$ to indicate that $[-, B, h, j]$ and $[j, C, h', k]$ were derived, for some C and h' such that $A[a_h] \rightarrow B[a_h]\; C[a_{h'}]$ is a rule. This represents an intermediate step in establishing $[i, A, h, k]$, temporarily ignoring the left boundary i of the substring derived from the left child $B[a_h]$, and forgetting the right child $C[a_{h'}]$. In a following step, i is reconstituted by access to an original item $[i, B, h, j]$.

The items of the form $[i, j, A, C, h]$ have a symmetrical meaning, that is, they indicate that $[i, B, h', j]$ and $[j, C, h, -]$ were derived, for some B and h' such that $A[a_h] \rightarrow B[a_{h'}]\; C[a_h]$ is a rule.

The algorithm is given in Figure 23.6 as deduction system, and an illustration of the use of an important combination of inference rules is given in Figure 23.7. The deduction of $[i, A, h, k]$ from $[i, B, h, j]$ and $[j, C', h, k]$ via a grammar rule $A[a_h] \rightarrow B[a_h]\; C[a_{h'}]$ is very similar to a step of the CYK algorithm for general CFGs. The current algorithm does the same in three different steps represented by inference rules (e), (b), and (f). Each of these involves no more than four input positions and three delexicalized nonterminals. This corresponds to $\mathcal{O}(|V_D|^3 \cdot n^4)$ applications of each inference rule, which is also the total time complexity of the algorithm.

23.5 Probabilistic Parsing

In natural language systems, parsing is commonly one stage of processing among several others. The effectiveness of the stages that follow parsing generally relies on having obtained a small set of

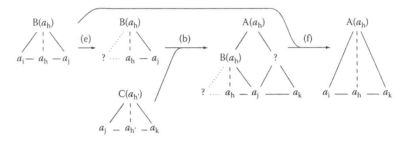

preferred parses, ideally only one from among the full set of parses, represented as a parse forest or intersection grammar. This is called (syntactic) disambiguation. There are roughly two ways to achieve this. First, some kind of filter may be applied to the full set of parses, to reject all but a few. This filter may look at the meanings of words and phrases, for example, and may be based on linguistic knowledge that is very different in character from the grammar that was used for parsing.

A second approach is to augment the parsing process so that probabilities are attached to parses and subparses. The higher the probability of a parse or subparse, the more confident we are that it is correct. This is called probabilistic parsing. The simplest form of probabilistic parsing relies on an assignment of probabilities to individual rules from a CFG. These probabilities are then multiplied upon a combination of rules to form parses.

As an example, consider the following probabilistic CFG, which extends the grammar from the running example with the probabilities between parentheses. As before, the example is meant to illustrate technical concepts, but has no linguistic pretences.

S	$\rightarrow$	NP VP	(1)
NP	$\rightarrow$	N	(0.4)
NP	$\rightarrow$	PosPr N	(0.3)
NP	$\rightarrow$	AP N	(0.2)
NP	$\rightarrow$	VP	(0.1)
N	$\rightarrow$	company	(0.6)
N	$\rightarrow$	workers	(0.4)
PosPr	$\rightarrow$	our	(1)
VP	$\rightarrow$	Aux VP	(0.3)
VP	$\rightarrow$	V NP	(0.7)
Aux	$\rightarrow$	is	(1)
V	$\rightarrow$	training	(0.9)
V	$\rightarrow$	is	(0.1)
AP	$\rightarrow$	training	(1)

In the aforementioned grammar, the probabilities of rules with a given nonterminal in the left-hand side always sum to 1. A probabilistic grammar for which this holds is called proper. The condition of properness is strongly related to the fact that CFGs are a generative formalism, that is, a grammar defines a set of objects by offering a number of operations to step-wise turn (partial) objects into larger objects. In the case of probabilistic CFGs that are proper, the range of applicable operations at each step of a left-most derivation forms a probability distribution.

With the aforementioned grammar, the most probable parse of "our company is training workers" is that of Figure 23.1a. Its probability is the product of the probabilities of all applied rules (or more precisely, rule occurrences). This is $1 \cdot 0.3 \cdot 1 \cdot 0.6 \cdot 0.3 \cdot 1 \cdot 0.7 \cdot 0.9 \cdot 0.4 \cdot 0.4 = 0.0054432$.

With a probabilistic input grammar $\mathcal{G}$, the intersection grammar $\mathcal{G}_\cap$ constructed for a given input string w is also probabilistic. A rule of the form $[i_0, A, i_m] \rightarrow [i_0, X_1, i_1] \cdots [i_{m-1}, X_m, i_m]$ is assigned the same probability as the rule $A \rightarrow X_1 \cdots X_m$ in the input grammar $\mathcal{G}$. The rules of the form $[i - 1, a_i, i] \rightarrow a_i$ are assigned the probability 1.

The intersection grammar is not proper, in general, but it has the attractive feature that parses of w in $\mathcal{G}$ have the same probability as the corresponding parses in $\mathcal{G}_\cap$. This means that the problem of finding the most probable parse of w in $\mathcal{G}$ can be translated to the problem of finding the most probable parse in $\mathcal{G}_\cap$. The latter problem is simpler than the original problem, as the input string no longer needs to be considered explicitly. In fact, we investigate this problem below for an arbitrary probabilistic CFG $\mathcal{G}$, which may or may not be the result of intersection.

The algorithm in Figure 23.8 is a special case of an algorithm by Knuth, which generalizes Dijkstra's algorithm to compute the shortest path in a weighted graph. It finds the probability $p_{\max}(A)$ of the

MOSTPROBABLEPARSE($\mathcal{G}$)

1: $\mathcal{E} \leftarrow \Sigma$
2: **repeat**
3:     $\mathcal{F} \leftarrow \{A \mid A \notin \mathcal{E} \wedge \exists A \rightarrow X_1 \cdots X_m [X_1, \ldots, X_m \in \mathcal{E}]\}$
4:     **if** $\mathcal{F} \leftarrow \emptyset$ **then**
5:         report failure and halt
6:     **for all** $A \in \mathcal{F}$ **do**
7:         $q(A) \leftarrow \displaystyle\max_{\substack{\pi=(A \rightarrow X_1 \cdots X_m): \\ X_1, \ldots, X_m \in \mathcal{E}}} p(\pi) \cdot p_{max}(X_1) \cdot \ldots \cdot p_{max}(X_m)$
8:     choose $A \in \mathcal{F}$ such that $q(A)$ is maximal
9:     $p_{max}(A) \leftarrow q(A)$
10:     $\mathcal{E} \leftarrow \mathcal{E} \cup \{A\}$
11: **until** $S \in \mathcal{E}$
12: output $p_{max}(S)$

FIGURE 23.8 Knuth's generalization of Dijkstra's algorithm, applied to finding the most probable parse in a probabilistic CFG $\mathcal{G}$.

most probable subparse with root labeled A. The value $p_{max}(S)$, where S is the start symbol, then gives us the probability of the most probable parse. The algorithm can be easily extended to return the parse itself. For notational convenience, we let $p_{max}(a) = 1$ for each terminal a. The set of terminals is here denoted as Σ.

In each iteration, the value of $p_{max}(A)$ is established for a nonterminal A; the set $\mathcal{E}$ contains all grammar symbols X for which $p_{max}(X)$ has already been established. Initially, this is Σ, as we let $p_{max}(a) = 1$ for each $a \in \Sigma$. The set $\mathcal{F}$ contains those nonterminals not yet in $\mathcal{E}$ that are candidates to be added next. Each nonterminal A in $\mathcal{F}$ is such that a subparse with root labeled A exists consisting of a rule $A \rightarrow X_1 \cdots X_m$ and subparses with roots labeled $X_1, \ldots, X_m$ matching the values of $p_{max}(X_1), \ldots, p_{max}(X_m)$ found earlier. The nonterminal A for which such a subparse has the highest probability is then added to $\mathcal{E}$.

In a practical implementation, $\mathcal{F}$ and q would not be constructed anew for each iteration. They would merely be revised every time a nonterminal A is added to $\mathcal{E}$. This revision consists in removing A from $\mathcal{F}$ and finding new rules whose right-hand side nonterminals are now all in $\mathcal{E}$. This allows adding new elements to $\mathcal{F}$ and/or updating q to assign higher values to elements in $\mathcal{F}$. Typically, $\mathcal{F}$ would be organized as a priority queue.

Knuth's algorithm can be combined with a construction of the intersection grammar. If the grammar has no cycles, several simplified algorithms exist. A particularly simple algorithm is an extended form of CYK parsing for grammars in Chomsky normal form, which computes values $p_{max}([i, A, j])$ after computing all values $p_{max}([i', A', j'])$ with $i < i' < j' \leq j$ or $i \leq i' < j' < j$.

The probability of an ambiguous string is defined as the sum of the probabilities of all parses of that string. In contrast to finding the most probable derivation, the problem of finding the most probable string is generally difficult. The decision version of this problem is NP-complete if there is a specified bound on the string length, and undecidable otherwise.

Probabilistic parsing is particularly effective for lexicalized grammars as it allows a fine-grained encoding of dependencies between lexical elements. For example, the rule S[training] $\rightarrow$ NP [company] VP[training] could be given a high probability, whereas S[training] $\rightarrow$ NP [problem] VP[training] is given a low probability.

Probabilistic CFGs for natural languages are normally induced on the basis of samples of language use, rather than explicitly written by people. The simplest algorithms of grammar induction rely on input consisting of a multiset of parses also known as a tree bank. Tree banks are often the result of manual or (partially) automated annotation of a number of texts, for example, newspaper articles.

The nonterminal label of a node in the tree bank together with the label of its children form a rule. The probability of such a rule, say $A \rightarrow \alpha$, can be estimated as

$$\frac{C(A \rightarrow \alpha)}{C(A)}, \tag{23.1}$$

where

$C(A)$ is the number of nodes in the input data with label A
$C(A \rightarrow \alpha)$ is the number of occurrences of rule $A \rightarrow \alpha$ in the tree bank

The probabilities of lexicalized rules are often computed as the product of probabilities of a number of features that together determine the rule. For example, the probability of $A[b] \rightarrow B[b]\ C[c]$ can be expressed as the probability of B given A and b, times the probability of C as second member in the right-hand side, given A, b, and B, times the probability of c given A, b, B, and C as second member. The last probability can be approximated as, for example, the probability of c given b and C. Such approximations lead to a reduced number of parameters, which can be estimated more accurately if the available tree bank is small.

23.6 Translation

Automatic translation between natural languages is one of the most challenging applications in NLP. State-of-the-art approaches to this task are based on syntactic models, usually enriched with statistical parameters. In this section we consider one such model, called synchronous context-free grammar (SCFG), which is a notational variant of the syntax-directed translation schemata originally developed in the theory of compilers (Aho and Ullman, 1972).

An SCFG consists of synchronous rules, each obtained by pairing two CFG rules with the same left-hand side. The right-hand sides of such a pair of CFG rules must consist of identical multisets of nonterminals, possibly ordered differently, and possibly combined with different terminal symbols. Furthermore, there is an explicit bijection that pairs occurrences of identical nonterminals in the two right-hand sides.

As an example, the synchronous rule $\langle$VP $\rightarrow$ VB$^{\boxed{1}}$ PP$^{\boxed{2}}$, VP $\rightarrow$ PP$^{\boxed{2}}$ VB$^{\boxed{1}}$ ga$\rangle$ states that an English verb phrase composed of the two constituents VB ("verb in base form") and PP ("prepositional phrase") can be translated into Japanese by swapping the order of the translations of these constituents and by inserting the word "ga" at the right. Note the use of integers within boxes as superscripts to indicate a bijection between nonterminal occurrences in the two context-free rules.

An SCFG can derive pairs of sentences as follows. Starting with the pair of nonterminals $\langle S^{\boxed{1}},\ S^{\boxed{1}}\rangle$, synchronous rules are applied to rewrite pairs of nonterminals that have the same index. At the application of a rule, the indices in the newly added nonterminals are consistently renamed, in order to avoid clashes with the indices introduced at previous rewriting steps. The rewriting stops when all nonterminals have been rewritten.

The following toy SCFG will be the running example of this section:

$$
\begin{array}{rllll}
s_1: & \langle & S \rightarrow A^{\boxed{1}}C^{\boxed{2}}, & S \rightarrow A^{\boxed{1}}C^{\boxed{2}} & \rangle \\
s_2: & \langle & C \rightarrow B^{\boxed{1}}S^{\boxed{2}}, & C \rightarrow B^{\boxed{1}}S^{\boxed{2}} & \rangle \\
s_3: & \langle & C \rightarrow B^{\boxed{1}}S^{\boxed{2}}, & C \rightarrow S^{\boxed{2}}B^{\boxed{1}} & \rangle \\
s_4: & \langle & C \rightarrow B^{\boxed{1}}, & C \rightarrow B^{\boxed{1}} & \rangle \\
s_5: & \langle & A \rightarrow a_1, & A \rightarrow a_2 & \rangle \\
s_6: & \langle & A \rightarrow a_1, & A \rightarrow \varepsilon & \rangle \\
s_7: & \langle & B \rightarrow b_1, & B \rightarrow b_2 & \rangle
\end{array}
$$

In the represented translation, nonterminals B and S can be optionally inverted (rule s_2 or rule s_3), and the terminal symbol a_2, which is the translation of a_1 (by rule s_5), can be optionally deleted (rule s_6).

An example derivation of the string pair $\langle a_1 b_1 a_1 b_1, \, a_2 b_2 b_2 \rangle$ by the above SCFG is

$$
\begin{aligned}
\langle S^{\boxed{1}}, S^{\boxed{1}} \rangle \;\Rightarrow^{s_1}\;& \langle A^{\boxed{2}} C^{\boxed{3}}, \, A^{\boxed{2}} C^{\boxed{3}} \rangle \\
\Rightarrow^{s_3}\;& \langle A^{\boxed{2}} B^{\boxed{4}} S^{\boxed{5}}, \, A^{\boxed{2}} S^{\boxed{5}} B^{\boxed{4}} \rangle \\
\Rightarrow^{s_1}\;& \langle A^{\boxed{2}} B^{\boxed{4}} A^{\boxed{6}} C^{\boxed{7}}, \, A^{\boxed{2}} A^{\boxed{6}} C^{\boxed{7}} B^{\boxed{4}} \rangle \\
\Rightarrow^{s_4}\;& \langle A^{\boxed{2}} B^{\boxed{4}} A^{\boxed{6}} B^{\boxed{8}}, \, A^{\boxed{2}} A^{\boxed{6}} B^{\boxed{8}} B^{\boxed{4}} \rangle \\
\Rightarrow^{s_5}\;& \langle a_1 B^{\boxed{4}} A^{\boxed{6}} B^{\boxed{8}}, \, a_2 A^{\boxed{6}} B^{\boxed{8}} B^{\boxed{4}} \rangle \\
\Rightarrow^{s_7}\;& \langle a_1 b_1 A^{\boxed{6}} B^{\boxed{8}}, \, a_2 A^{\boxed{6}} B^{\boxed{8}} b_2 \rangle \\
\Rightarrow^{s_6}\;& \langle a_1 b_1 a_1 B^{\boxed{8}}, \, a_2 B^{\boxed{8}} b_2 \rangle \\
\Rightarrow^{s_7}\;& \langle a_1 b_1 a_1 b_1, \, a_2 b_2 b_2 \rangle.
\end{aligned}
$$

In the same way as a derivation in a CFG can be associated with a parse tree, a derivation in an SCFG can be associated with a pair of parse trees. These trees differ only in the (terminal) labels of the leaf nodes and in the ordering of siblings, as illustrated by Figure 23.9. We will refer to the two trees in a pair as the input tree and the output tree.

Given an SCFG $\mathcal{G}$ and a string w, the expression $w \circ \mathcal{G}$ denotes the set of all pairs of parse trees associated with derivations in $\mathcal{G}$ whose input tree has a yield w. Note that all the strings that are translations of w under $\mathcal{G}$ can be easily enumerated if we can enumerate the elements of $w \circ \mathcal{G}$.

The set $w \circ \mathcal{G}$ can have size exponential in $|w|$, and the number of possible translations of w under $\mathcal{G}$ can likewise be exponential. (There may even be an infinite number of translations, if there are synchronous rules whose left components are epsilon rules or unit rules.) In Section 23.3, we discussed a compact representation of a large set of parse trees in the form of a CFG. Below, this is extended to a construction of an SCFG $\mathcal{G}'$ that represents $w \circ \mathcal{G}$ in a compact way. This construction is called left composition, and in full generality it can be applied to a finite automaton and an SCFG. As in Section 23.3, we simplify the discussion by assuming a special type of finite automaton that recognizes just one string $w = a_1 \cdots a_n$.

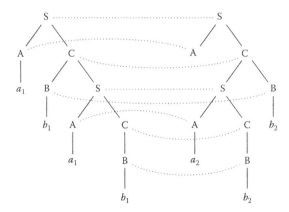

FIGURE 23.9 A pair of trees associated with a derivation in an SCFG. The dotted lines link pairs of nonterminal occurrences that had the same index during the rewriting process.

We further assume, without the loss of generality, that synchronous rules from $\mathcal{G}$ are either of the form $\langle A \to \alpha, \ A \to \alpha' \rangle$, where α and α' are nonempty strings of indexed nonterminals, or of the form $\langle A \to x, \ A \to y \rangle$, where x and y can each be a terminal symbol or the empty string. In the former case, we use a permutation π to denote the bijective relation pairing the nonterminal occurrences in α and α', and write the synchronous rule as $\left\langle A \to B_1^{\boxed{1}} \cdots B_m^{\boxed{m}}, \ A \to B_{\pi(1)}^{\boxed{\pi(1)}} \cdots B_{\pi(m)}^{\boxed{\pi(m)}} \right\rangle$.

The nonterminals of $\mathcal{G}'$ have the form $[i, A, j]$, where i and j denote input positions within w, and A is a nonterminal from $\mathcal{G}$.

LEFTCOMPOSITION$(w, \mathcal{G})$ {let $a_1 \cdots a_n = w$}
1:    $\mathcal{G}' \leftarrow$ SCFG with start nonterminal $[0, S, n]$ and empty set of synchronous rules
2:    **for all** $\left\langle A \to B_1^{\boxed{1}} \cdots B_m^{\boxed{m}}, \ A \to B_{\pi(1)}^{\boxed{\pi(1)}} \cdots B_{\pi(m)}^{\boxed{\pi(m)}} \right\rangle$ from $\mathcal{G}$ **do**
3:      **for all** $i_0, \ldots, i_m \ (0 \le i_0 \le \ldots \le i_m \le n)$ **do**
4:        add to $\mathcal{G}'$ the synchronous rule $\langle [i_0, A, i_m] \to [i_0, B_1, i_1]^{\boxed{1}} \cdots [i_{m-1}, B_m, i_m]^{\boxed{m}},$
         $[i_0, A, i_m] \to [i_{\pi(1)-1}, B_{\pi(1)}, i_{\pi(1)}]^{\boxed{\pi(1)}} \cdots [i_{\pi(m)-1}, B_{\pi(m)}, i_{\pi(m)}]^{\boxed{\pi(m)}} \rangle$
5:    **for all** $i \ (1 \le i \le n)$ and $\langle A \to a_i, \ A \to y \rangle$ from $\mathcal{G}$ **do**
6:      add to $\mathcal{G}'$ the synchronous rule $\langle [i-1, A, i] \to a_i, \ [i-1, A, i] \to y \rangle$
7:    **for all** $i \ (0 \le i \le n)$ and $\langle A \to \varepsilon, \ A \to y \rangle$ from $\mathcal{G}$ **do**
8:      add to $\mathcal{G}'$ the synchronous rule $\langle [i, A, i] \to \varepsilon, \ [i, A, i] \to y \rangle$
9:    **return** $\mathcal{G}'$

This construction may introduce many nonterminals into $\mathcal{G}'$ that are useless in the same way as algorithm INTERSECTION from Section 23.3 introduces useless nonterminals. Available techniques to eliminate useless nonterminals from $\mathcal{G}'$ are very similar to techniques discussed earlier.

If we remove the left components from synchronous rules of $\mathcal{G}'$, then we obtain a CFG $\mathcal{G}''$ that generates parse trees for all the possible translations of w under $\mathcal{G}$. These parse trees differ from the output trees in $w \circ \mathcal{G}$ only in the labels of nodes. In the former, there are labels of the form $[i, A, j]$ where as in the latter there are labels A.

Returning to the running example, consider the input string $w = a_1 b_1 a_1 b_1$. With the SCFG given earlier, this string can be translated into the five strings $a_2 b_2 a_2 b_2$, $a_2 a_2 b_2 b_2$, $a_2 b_2 b_2$, $b_2 a_2 b_2$, or $b_2 b_2$. There are eight pairs of trees in $w \circ \mathcal{G}$ as there are three derivations with output $a_2 b_2 b_2$ and two derivations with output $b_2 b_2$. After applying left composition and reduction of the grammar, we obtain

$$
\begin{aligned}
&\langle \ [0, S, 4] \to [0, A, 1]^{\boxed{1}} [1, C, 4]^{\boxed{2}}, && [0, S, 4] \to [0, A, 1]^{\boxed{1}} [1, C, 4]^{\boxed{2}} && \rangle \\
&\langle \ [1, C, 4] \to [1, B, 2]^{\boxed{1}} [2, S, 4]^{\boxed{2}}, && [1, C, 4] \to [1, B, 2]^{\boxed{1}} [2, S, 4]^{\boxed{2}} && \rangle \\
&\langle \ [1, C, 4] \to [1, B, 2]^{\boxed{1}} [2, S, 4]^{\boxed{2}}, && [1, C, 4] \to [2, S, 4]^{\boxed{2}} [1, B, 2]^{\boxed{1}} && \rangle \\
&\langle \ [2, S, 4] \to [2, A, 3]^{\boxed{1}} [3, C, 4]^{\boxed{2}}, && [2, S, 4] \to [2, A, 3]^{\boxed{1}} [3, C, 4]^{\boxed{2}} && \rangle \\
&\langle \ [3, C, 4] \to [3, B, 4]^{\boxed{1}}, && [3, C, 4] \to [3, B, 4]^{\boxed{1}} && \rangle \\
&\langle \ [0, A, 1] \to a_1, && [0, A, 1] \to a_2 && \rangle \\
&\langle \ [0, A, 1] \to a_1, && [0, A, 1] \to \varepsilon && \rangle \\
&\langle \ [1, B, 2] \to b_1, && [1, B, 2] \to b_2 && \rangle \\
&\langle \ [2, A, 3] \to a_1, && [2, A, 3] \to a_2 && \rangle \\
&\langle \ [2, A, 3] \to a_1, && [2, A, 3] \to \varepsilon && \rangle \\
&\langle \ [3, B, 4] \to b_1, && [3, B, 4] \to b_2 && \rangle
\end{aligned}
$$

The size of an SCFG $\mathcal{G}$, written as $|\mathcal{G}|$, is defined as the sum of the number of nonterminal and terminal occurrences in its synchronous rules. Let r be the length of the longest right-hand side of a context-free rule that is the input or output component of a synchronous rule. The time and space complexities of left composition are both $\mathcal{O}(|\mathcal{G}| \cdot n^{r+1})$, where n is the length of the input string.

In many practical applications, it is possible to factorize synchronous rules in such a way that the parameter r is reduced to a small integer. In the general case, however, an SCFG cannot be cast into an equivalent form with r bounded by some constant. This implies exponential behavior in the worst case.

Next to the left composition of SCFG $\mathcal{G}$ with input string w_1 there is the right composition of $\mathcal{G}$ with output string w_2. The definition of right composition is the natural mirror image of that of left composition, and it results in an SCFG $\mathcal{G}'$ that represents a set of tree pairs denoted by $\mathcal{G} \circ w_2$.

Left and right compositions may be combined, one after the other in either order, to construct an SCFG representing a set of tree pairs $w_1 \circ \mathcal{G} \circ w_2$, that is, the set of all derivations of $\mathcal{G}$ with input w_1 and output w_2. Important applications include inducing machine translation components from text that were manually translated.

23.7 Further Information

The formal properties of CFGs are discussed, among others, by Hopcroft and Ullman (1979) and Sippu and Soisalon-Soininen (1988). Interesting observations about the intersection of context-free languages and regular languages, as relating to the distinction between parsing and recognition, are due to Lang (1994), which offers a different perspective from that of, for example, Ruzzo (1979).

The algorithm GENERATINGINTERSECTION is close to a bottom-up variant of the parsing algorithm by Graham et al. (1980). The presentation of recognition algorithms as deduction systems is commonly identified with Shieber et al. (1995) and Sikkel (1997). The underlying idea, however, can be traced back to Cook (1970). The time complexity of deduction systems, implemented as dynamic programming algorithms, was discussed by McAllester (2002).

The algorithm in Figure 23.6 has been adapted from Eisner and Satta (1999).

The problem of finding the most probable derivation is discussed by Knuth (1977) and Nederhof (2003) in the general case, and by Jelinek et al. (1992) for grammars in Chomsky normal form. The problem of finding the most probable string is discussed by Paz (1971), Casacuberta and de la Higuera (2000), Sima'an (2002), and Blondel and Canterini (2003).

The result that general SCFGs cannot be cast in a normal form with a bound on rule length is from Aho and Ullman (1969). NP-hardness of problems relating to translation are discussed in Satta and Peserico (2005).

The running example from the beginning of this chapter is adapted from Manning and Schütze (1999), which is recommended as a good introduction to statistical natural language processing. A good general textbook on NLP is Jurafsky and Martin (2000).

The main journal of NLP is *Computational Linguistics*. Of at least equal importance are several annual and biennial conferences, among which are Annual Meeting of the Association for Computational Linguistics (ACL), European Chapter of the ACL (EACL), North American Chapter of the ACL (NAACL), International Conference on Computational Linguistics (COLING), Empirical Methods in Natural Language Processing (EMNLP) and Human Language Technology (HLT).

Defining Terms

Ambiguity: Existence of more than one interpretation of an element of language, for example, existence of several parses of one sentence, or several possible meanings of a word.

Disambiguation: The process of identifying one preferred interpretation from a set of interpretations of an ambiguous element of language.

Formal language: Language that is defined with mathematical rigor.

Grammar induction: Acquiring a grammar out of a sample of language.

Linguistics: The study of natural language.

Machine translation: Automated translation between natural languages.

Natural language: Language used in human communication that evolved without too many deliberate considerations.

Natural language processing: Automated analysis, generation or translation of language.

(Parse) item: Element stored in the table of a parsing or recognition algorithm representing existence of certain subparses.

Parse (tree): Structural interpretation of a sentence.

Parse forest: Structure containing a number of parses of one sentence.

Syntax: The study of the structure of sentences, as composed of words.

Tree bank: Multiset of parses representing an annotation of a sample of language use.

References

Aho, A. and Ullman, J. (1969). Properties of syntax directed translations. *Journal of Computer and System Sciences*, 3:319–334.

Aho, A. and Ullman, J. (1972). *Parsing*, volume 1 of *The Theory of Parsing, Translation and Compiling*. Prentice-Hall, Englewood Cliffs, NJ.

Bar-Hillel, Y., Perles, M., and Shamir, E. (1964). On formal properties of simple phrase structure grammars. In Bar-Hillel, Y., editor, *Language and Information: Selected Essays on their Theory and Application*, Chapter 9, pp. 116–150. Addison-Wesley, Reading, MA.

Blondel, V. and Canterini, C. (2003). Undecidable problems for probabilistic automata of fixed dimension. *Theory of Computing Systems*, 36:231–245.

Casacuberta, F. and de la Higuera, C. (2000). Computational complexity of problems on probabilistic grammars and transducers. In Oliveira, A., editor, *Grammatical Inference: Algorithms and Applications*, volume 1891 of *Lecture Notes in Artificial Intelligence*, pp. 15–24. Springer-Verlag, Berlin, Germany.

Cook, S. (1970). Path systems and language recognition. In *ACM Symposium on Theory of Computing*, pp. 70–72. Northampton, MA.

Eisner, J. and Satta, G. (1999). Efficient parsing for bilexical context-free grammars and head automaton grammars. In *37th Annual Meeting of the Association for Computational Linguistics, Proceedings of the Conference*, pp. 457–464. College Park, MD.

Graham, S., Harrison, M., and Ruzzo, W. (1980). An improved context-free recognizer. *ACM Transactions on Programming Languages and Systems*, 2:415–462.

Hopcroft, J. and Ullman, J. (1979). *Introduction to Automata Theory, Languages, and Computation*. Addison-Wesley, Reading, MA.

Jelinek, F., Lafferty, J., and Mercer, R. (1992). Basic methods of probabilistic context free grammars. In Laface, P. and De Mori, R., editors, *Speech Recognition and Understanding — Recent Advances, Trends and Applications*, pp. 345–360. Springer-Verlag, Berlin, Germany.

Jurafsky, D. and Martin, J. (2000). *Speech and Language Processing*. Prentice-Hall, Englewood Cliffs, NJ.

Knuth, D. (1977). A generalization of Dijkstra's algorithm. *Information Processing Letters*, 6(1):1–5.

Lang, B. (1994). Recognition can be harder than parsing. *Computational Intelligence*, 10(4):486–494.

Manning, C. and Schütze, H. (1999). *Foundations of Statistical Natural Language Processing*. MIT Press, Cambridge, MA.

McAllester, D. (2002). On the complexity analysis of static analyses. *Journal of the ACM*, 49(4):512–537.

Nederhof, M.-J. (2003). Weighted deductive parsing and Knuth's algorithm. *Computational Linguistics*, 29(1):135–143.

Paz, A. (1971). *Introduction to Probabilistic Automata*. Academic Press, New York.

Ruzzo, W. (1979). On the complexity of general context-free language parsing and recognition. In *Automata, Languages and Programming, Sixth Colloquium*, volume 71 of *Lecture Notes in Computer Science*, Graz. pp. 489–497. Springer-Verlag, Berlin, Germany.

Satta, G. and Peserico, E. (2005). Some computational complexity results for synchronous context-free grammars. In *Human Language Technology Conference and Conference on Empirical Methods in Natural Language Processing*, pp. 803–810. Vancouver, BC.

Shieber, S., Schabes, Y., and Pereira, F. (1995). Principles and implementation of deductive parsing. *Journal of Logic Programming*, 24:3–36.

Sikkel, K. (1997). *Parsing Schemata*. Springer-Verlag, New York.

Sima'an, K. (2002). Computational complexity of probabilistic disambiguation. *Grammars*, 5:125–151.

Sippu, S. and Soisalon-Soininen, E. (1988). *Parsing Theory, Vol. I: Languages and Parsing*, volume 15 of *EATCS Monographs on Theoretical Computer Science*. Springer-Verlag, Berlin, Germany.

Tomita, M. (1986). *Efficient Parsing for Natural Language*. Kluwer Academic Publishers, Dordrecht, the Netherlands.

24

Algorithmic Techniques for Regular Networks of Processors

Russ Miller
State University of New York at Buffalo

Quentin F. Stout
University of Michigan

24.1 Introduction

This chapter is concerned with designing algorithms for machines constructed from multiple processors. In particular, we discuss algorithms for machines in which the processors are connected to each other by some simple, systematic, interconnection patterns. For example, consider a chess board, where each square represents a processor (for example, a processor similar to one in a home computer) and every generic processor is connected to its four neighboring processors (those to the north, south, east, and west). This is an example of a mesh computer, a network of processors that is important for both theoretical and practical reasons.

The focus of this chapter is on algorithmic techniques. Initially, we define some basic terminology that is used to discuss parallel algorithms and parallel architectures. Following this introductory material, we define a variety of interconnection networks, including the mesh (chess board), which are used to allow processors to communicate with each other. We also define an abstract parallel model of computation, the PRAM, where processors are not connected to each other, but communicate directly with a global pool of memory that is shared amongst the processors. We then discuss several parallel programming paradigms, including the use of high-level data movement operations, divide-and-conquer, pipelining, and master–slave. Finally, we discuss the problem of mapping the structure of an inherently parallel problem onto a target parallel architecture. This mapping problem can arise in a variety of ways, and with a wide range of problem structures. In some cases, finding a good mapping is quite straightforward, but in other cases it is a computationally intractable NP-complete problem.

24.2 Terminology

In order to initiate our investigation, we first define some basic terminology that will be used throughout the remainder of this chapter.

24.2.1 Shared Memory versus Distributed Memory

In a shared memory machine, there is a single global image of memory that is available to all processors in the machine, typically through a common bus, set of busses, or switching network, as shown in Figure 24.1a. This model is similar to a blackboard, where any processor can read or write to any part of the board (memory), and where all communication is performed through messages placed on the board.

As shown in Figure 24.1b, each processor in a distributed memory machine has access only to its private (local) memory. In this model, processors communicate by sending messages to each other, with the messages being sent through some form of an interconnection network. This model is similar to that used by shipping services, such as the United States Postal Service, Federal Express, DHL, or UPS, to name a few. For example, suppose Tom in city X needs some information from Sue in city Y. Then Tom might send a letter requesting such information from Sue. However, the letter might get routed from city X to a facility (i.e., "post office") in city W, then to a facility in city Z, and finally to the facility in city Y before being delivered locally to Sue. Sue will now package up the information requested and go to a local shipping facility in city Y, which might route the package to a facility in city Q, then to a facility in city R, and finally to a facility in city X before being delivered locally to Tom. Note that there might be multiple paths between source

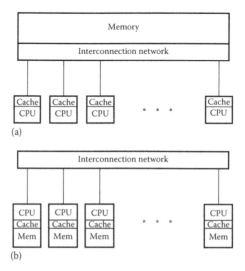

(a)

(b)

FIGURE 24.1 (a) Shared memory and (b) distributed memory machines.

and destination, that messages might move through different paths at different times between the same source and destination depending on congestion, availability of the communication path, and so forth. Also note that routing messages between processors that are closer to each other in terms of the interconnection network (fewer hops between processors) typically require less time than is required to route messages between pairs of processors that are farther apart (more hops between processors in terms of the interconnection network). In such message-passing systems, the overhead and delay can be significantly reduced if, for example, Sue sends the information to Tom without him first requesting the information. It is particularly useful if the data from Sue arrives before Tom needs to use it, for then Tom will not be delayed waiting for critical data. This analogy represents an important aspect of developing efficient programs for distributed memory machines, especially general-purpose machines in which communication can take place concurrently with calculation so that the communication time is effectively hidden.

For small shared memory systems, it may be that the network is such that each processor can access all memory cells in the same amount of time. For example, many symmetric multiprocessor (SMP) systems have this property. However, since memory takes space, systems with a large number of processors are typically constructed as modules (i.e., a processor/memory pair) that are connected to each other via an interconnection network. Thus, while memory may be logically shared in such a model, in terms of performance each processor acts as if it is distributed, with some memory being "close" (fast access) to the processor and some memory being "far" (slow access) from the processor. Notice the similarity to distributed memory machines, where there is a significant difference in speed between a processor accessing its own memory versus a processor accessing the memory of a distant processor. Such shared memory machines are called NUMA (nonuniform memory access) machines, and often the most efficient programs for NUMA machines are developed by using algorithms efficient for distributed memory architectures, rather than using ones optimized for uniform access shared memory architectures.

Efficient use of the interconnection network in a parallel computer is often an important consideration for developing and tuning parallel programs. For example, in either shared or distributed memory machines, communication will be delayed if a packet of information must pass through many communication links. Similarly, communication will be delayed by contention if many packets need to pass through the same link. As an example of contention at a link, in a distributed memory machine configured as a binary tree of processors, suppose that all leaf processors on one side of the machine need to exchange values with all leaf processors on the other side of the machine. Then a bottleneck occurs at the root since the passage of information proceeds in a sequential manner through the links in and out of the root. A similar bottleneck occurs in a system if the interconnect is merely a single Ethernet-based bus.

Both shared and distributed memory systems can also suffer from contention at the destinations. In a distributed memory system, too many processors may simultaneously send messages to the same processor, which causes a processing bottleneck. In a shared memory system, there may be memory contention, where too many processors try to simultaneously read or write from the same location.

Another common feature of both shared and distributed memory systems is that the programmer has to be sure that computations are properly synchronized, i.e., that they occur in the correct order. This tends to be easier in distributed memory systems, where each processor controls the access to its data, and the messages used to communicate data also have the side-effect of communicating the status of the sending processor. For example, suppose processor W is calculating a value, which will then be sent to processor R. If the program is constructed so that R does not proceed until the message from W arrives, then it is guaranteed of using the correct value in the calculations. In a shared memory system, the programmer needs to be more careful. For example, in the same scenario, W may write the new value to a memory location that R reads. However, if R reads before W has written, then it may proceed using the wrong value. This is known as a race condition, where the correctness of the calculation depends on the order of the operations. To avoid this, various locking or signaling protocols need to be enforced so that R does not read the location until after W has written to it. Race conditions are a common source of programming errors, and are often difficult to locate because they disappear when a deterministic, serial debugging approach is used.

24.2.2 Flynn's Taxonomy

In 1966, Michael Flynn classified computer architectures with respect to the instruction stream, that is, the sequence of operations performed by the computer, and the data stream, that is, the sequence of items operated on by the instructions (Flynn, 1966). While extensions and modifications to Flynn's taxonomy have appeared, Flynn's original taxonomy (Flynn, 1972) is still widely used. Flynn characterized an architecture as belonging to one of the following four classes.

- Single-instruction stream, single-data stream (SISD)
- Single-instruction stream, multiple-data stream (SIMD)
- Multiple-instruction stream, single-data stream (MISD)
- Multiple-instruction stream, multiple data stream (MIMD)

Standard serial computers fall into the SISD category, in which one instruction is executed per unit time. This is the so-called von Neumann model of computing, in which the stream of instructions and the stream of data can be viewed as being tightly coupled, so that one instruction is executed per unit time to produce one useful result. Modern "serial" computers include various forms of modest parallelism in their execution of instructions, but most of this is hidden from the programmer and only appears in the form of faster execution of a sequential program.

A SIMD machine typically consists of multiple processors, a control unit (controller), and an interconnection network, as shown in Figure 24.2. The control unit stores the program and broadcasts the instructions to all processors simultaneously. Active processors simultaneously execute the identical instruction on the contents of each active processor's own local memory. Through the use of a mask, processors may be in either an active or inactive state at any time during the execution of the program. Masks can be dynamically determined, based on local data or the processor's coordinates. Note that one side-effect of having a centralized controller is that the system is synchronous, so that no processor can execute a second instruction until all processors are finished with the first instruction. This is quite useful in algorithm design, as it eliminates many race conditions and makes it easier to reason about the status of processors and data.

MISD machines consist of two or more processors that simultaneously perform not necessarily identical instructions on the same data. This model is rarely implemented.

A MIMD machine typically consists of multiple processors and an interconnection network. In contrast to the single-instruction stream model, the multiple-instruction stream model allows each of the processors to store and execute its own program, providing multiple instruction streams. Each processor fetches its own data on which to operate. (Thus, there are multiple data streams, as in the SIMD model.) Often, all processors are executing the same program, but may be in different portions of the program at any given instant. This is the single-program multiple-data (SPMD) style of programming, and is an important mode of programming because it is rarely feasible to have a large number of different

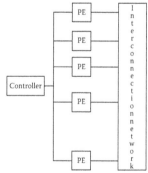

FIGURE 24.2 A SIMD machine. (PE is used to represent a processing element.)

programs for different processors. The SPMD style, like the SIMD architectures, also makes it somewhat simpler to reason about the status of data structures and processors.

MIMD machines have emerged as the most commonly used general-purpose parallel computers, and are available in a variety of configurations. Both shared and distributed memory machines are available, as are mixed architectures where small numbers of processors are grouped together as a shared memory SMP, and these SMPs are linked together in a distributed memory fashion.

24.2.3   Granularity

When discussing parallel architectures, the term granularity is often used to refer to the relative number and complexity of the processors. A fine-grained machine typically consists of a relatively large number of small, simple processors (in terms of local memory and computational power), while a coarse-grained machine typically consists of relatively few processors, each of which is large and powerful. Fine-grained machines typically fall into the SIMD category, where all processors operate in lockstep fashion (i.e., synchronously) on the contents of their own small, local, memory. Coarse-grained machines typically fall into the shared memory MIMD category, where processors operate asynchronously on the large, shared, memory. Medium-grained machines are typically built from commodity microprocessors, and are found in both distributed and shared memory models, almost always in MIMD designs.

For a variety of reasons, medium-grained machines currently dominate the parallel computer marketplace in terms of the number of installations. Such medium-grained machines typically utilize commodity processors and have the ability to efficiently perform as general-purpose (parallel) machines. Therefore, such medium-grained machines tend to have cost/performance advantages over systems utilizing special-purpose processors. In addition, they can also exploit much of the software written for their component processors. Fine-grained machines are difficult to use as general-purpose computers because it is often difficult to determine how to efficiently distribute the work to such simple processors. However, fine-grained machines can be quite effective in tasks such as image processing or pattern matching.

By analogy, one can also use the granularity terminology to describe data and algorithms. For example, a database is a coarse-grained view of data, while considering the individual records in the database is a fine-grained view of the same data.

24.3   Interconnection Networks

In this section, we discuss interconnection networks that are used for communication among processors in a distributed memory machine. In some cases, all communications are handled by processors sending messages to other processors that they have a direct connection to, where messages destined for processors farther away must be handled by a sequence of intermediate processors. In other cases the processors send messages into, and receive messages from, an interconnection network composed of specialized routers that pass the messages. Most large systems use the latter approach. We use the term node to represent the processors, in the former case, or the routers in the latter case, i.e., in any system, messages are passed from node to node in the interconnection network.

First, we define some terminology. The degree of node R is the number of other nodes that R is directly connected to via bi-directional communication links. (There are straightforward extensions to systems with unidirectional links.) The degree of the network is the maximum degree of any node in the network. The distance between two nodes is the number of communication links on a shortest path between the nodes. The communication diameter of the network is the maximum, over all pairs of nodes, distance between the nodes. The bisection bandwidth of the network corresponds to the minimum number of communication links that need to be removed (or cut) in order to partition the network into two pieces, each with the same number of nodes. Goals for interconnection networks include minimizing the degree of the nodes (to minimize the cost of building them), minimizing the communication diameter (to minimize the communication time for any single message), and maximizing the bisection bandwidth (to minimize contention when many messages are being sent concurrently). Unfortunately, these design goals are in conflict. Other important design goals include simplicity (to reduce the design costs for the hardware and software) and scalability (so that similar machines, with a range of sizes, can be produced). We informally call simple scalable interconnections regular networks. Regular networks make it easier for users to develop optimized code for a range of problem sizes.

Before defining some network models (i.e., distributed memory machines characterized by their interconnection networks, or the interconnection network used in a shared memory machine), we briefly discuss the parallel random access machine (PRAM), which is an idealized parallel model of computation, with a unit-time communication diameter. The PRAM is a shared memory machine that consists of a set of identical processors, where all processors have unit-time access to any memory location. The appeal of a PRAM is that one can ignore issues of communication when designing algorithms, focusing instead on obtaining the maximum parallelism possible in order to minimize the running time necessary to solve a given problem. The PRAM model typically assumes a SIMD strategy, so that operations are performed synchronously. If multiple processors try to simultaneously read or write from the same memory location, then a memory conflict occurs. There are several variations of the PRAM model targeted at handling these conflicts, ranging from the exclusive read exclusive write (EREW) model, which prohibits all such conflicts, to concurrent read concurrent write (CRCW) models, which have various ways of resolving the effects of simultaneous writes. One popular intermediate model is the concurrent read exclusive write (CREW) PRAM, in which concurrent reads to a memory location is permitted, but concurrent writes are not. For example, a classroom is usually conducted in a CREW manner. In the classroom, many students can read from the blackboard simultaneously (concurrent read), while if several students are writing simultaneously on the blackboard, they are doing so in different locations (exclusive write).

The unit-time memory access requirement for a PRAM is not scalable (i.e., it is not realistic for a large number of processors and memory). However, in creating parallel programs, it is sometimes useful to describe a PRAM algorithm and then either perform a stepwise simulation of every PRAM operation on the target machine, or perform a higher-level simulation by using global operations. In such settings, it is often useful to design the algorithm for a powerful CRCW PRAM model, since often the CRCW PRAM can solve a problem faster or more naturally than an EREW PRAM. Since one is not trying to construct an actual PRAM, objections to the difficulty of implementing CRCW are not relevant; rather, having a simpler and/or faster algorithm is the dominant consideration.

In the remainder of this section, several specific interconnection networks are defined. See Figure 24.3 for illustrations of these. The networks defined in this section are among the most commonly utilized networks. However, additional networks have appeared in both the literature and in real machines, and variations of the basic networks described here are numerous. For example, many small systems use only a bus as the interconnection network (where only one message at a time can be transmitted), reconfigurable meshes extend the capabilities of standard meshes by adding dynamic interconnection configuration (Li and Stout, 1991), and Clos networks have properties between those of completely connected crossbar systems and hypercubes.

24.3.1 Ring

In a *ring* network, the n nodes are connected in a circular fashion so that node R_i is directly connected to nodes R_{i-1} and R_{i+1} (the indices are computed modulo n, so that nodes R_0 and R_{n-1} are connected). While the degree of the network is only 2, the communication diameter is $\lfloor n/2 \rfloor$, which is quite high, and the bisection bandwidth is only 2, which is quite low.

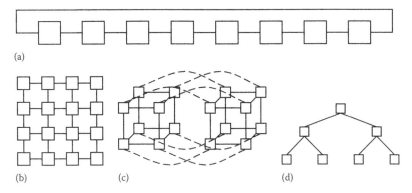

FIGURE 24.3 Sample interconnection networks (a through d): ring, mesh, hypercube, and tree.

24.3.2 Meshes and Tori

The n nodes of a two-dimensional square mesh network are configured so that an interior node $R_{i,j}$ is connected to its four neighbors, nodes $R_{i-1,j}$, $R_{i+1,j}$, $R_{i,j-1}$, and $R_{i,j+1}$. The four corner nodes are each connected to their two neighbors, while the remaining nodes that are on the edge of the mesh are each connected to three neighbors. So, by increasing the degree of the network to 4, as compared to the degree 2 of the ring, the communication diameter of the network is reduced to $2(\sqrt{n}-1)$, and the bisection bandwidth is increased to $\sqrt{n}$. The diameter is further reduced, to $2\lfloor\sqrt{n}/2\rfloor$, and the bisection bandwidth is increased, to $2\sqrt{n}$, in a two-dimensional torus, which has all the connections of the two-dimensional mesh plus connections between the first and the last nodes in each row and column. Meshes and tori of higher dimensions can be constructed, where the degree of a d-dimensional mesh or torus is $2d$, and, when n is a perfect dth power, the diameter is either $d(n^{1/d}-1)$ or $d\lfloor n^{1/d}/2\rfloor$, respectively, and the bisection bandwidth is either $n^{(d-1)/d}$ or $2n^{(d-1)/d}$, respectively. Notice that the ring is a one-dimensional torus.

For a two-dimensional mesh, and similarly for higher-dimensional meshes, the mesh can be rectangular, instead of square. This allows a great deal of flexibility in selecting the size of the mesh, and the same flexibility is available for tori as well.

24.3.3 Hypercube

A hypercube with n nodes, where n is an integral power of 2, has the nodes indexed by the integers $\{0,\ldots,n-1\}$. Viewing each integer in this range as a $(\log_2 n)$-bit string, two nodes are directly connected if and only if their indices differ by exactly one bit. Some advantages of a hypercube are that the communication diameter is only $\log_2 n$ and the bisection bandwidth is $n/2$. A disadvantage of the hypercube is that the number of communication links needed by each node grows as $\log_2 n$, unlike the fixed degree for nodes in ring and mesh networks. This makes it difficult to manufacture reasonably generic hypercube nodes that could scale to extremely large machines, though in practice this is not a concern because the cost of an extremely large machine would be prohibitive.

24.3.4 Tree

A complete binary tree of height k, $k \geq 0$ an integer, has $n = 2^{k+1}-1$ nodes. The root node is at level 0 and the 2^k leaves are at level k. Each node at level $1,\ldots,k-1$ has two children and one parent, the root node does not have a parent node, and the leaves at level k do not have children nodes. Notice that the degree of the network is 3 and that the communication diameter is $2k = 2\lfloor\log_2 n\rfloor$. One severe disadvantage of a tree is that when extensive communication occurs, all messages traveling from one side of the tree to the other must pass through the root, causing a bottleneck. This is because the bisection bandwidth is only 1. Fat trees, introduced by Leiserson (1985), alleviate this problem by increasing the bandwidth of the communication links near the root. This increase can come from changing the nature of the links, or, more easily, by using parallel communication links. Other generalizations of binary trees include complete t-ary trees of height k, where each node at level $0, \ldots, k-1$ has t children. There are $(t^{k+1} - 1)/(t - 1)$ nodes, the maximum degree is $t + 1$, and the diameter is $2k = 2\lfloor \log_t n \rfloor$.

24.4   Designing Algorithms

Viewed from the highest level, many parallel algorithms are purely sequential, with the same overall structure as an algorithm designed for a more standard "serial" computer. That is, there may be an input and initialization phase, then a computational phase, and then an output and termination phase. The differences, however, are manifested within each phase. For example, during the computational phase, an efficient parallel algorithm may be inherently different from its efficient sequential counterpart.

For each of the phases of a parallel computation, it is often useful to think of operating on an entire structure simultaneously. This is an SIMD-style approach, but the operations may be quite complex. For example, one may want to update all entries in a matrix, tree, or database, and view this as a single (complex) operation. For a fine-grained machine, this might be implemented by having a single (or few) data item per processor, and then using a purely parallel algorithm for the operation. For example, suppose an $n \times n$ array A is stored on an $n \times n$ two-dimensional torus, so that $A(i,j)$ is stored on processor $P_{i,j}$. Suppose one wants to replace each value $A(i,j)$ with the average of itself and the four neighbors $A(i-1,j)$, $A(i+1,j)$, $A(i,j-1)$, and $A(i,j+1)$, where the indices are computed modulo n (i.e., "neighbors" is in the torus sense). This average filtering can be accomplished by just shifting the array right, left, up, and down by one position in the torus, and having each processor average the four values received along with its initial value.

For a medium- or coarse-grained machine, operating on entire structures is most likely to be implemented by blending serial and parallel approaches. On such an architecture, each processor uses an efficient serial algorithm applied to the portion of the data in the processor, and communicates with other processors in order to exchange critical data. For example, suppose the $n \times n$ array of the previous paragraph is stored in a $p \times p$ torus, where p evenly divides n, so that $A(i,j)$ is stored in processor $P_{\lfloor ip/n \rfloor, \lfloor jp/n \rfloor}$. Then, to do the same average filtering on A, each processor $P_{k,l}$ still needs to communicate with its torus neighbors $P_{k\pm1,l}$, $P_{k,l\pm1}$, but now sends them either the leftmost or rightmost column of data, or the topmost or bottommost row. Once a processor receives its boundary set of data from its neighboring processors, it can then proceed serially through its subsquare of data and produce the desired results. To maximize efficiency, this can be performed by having each processor send the data needed by its neighbors, then perform the filtering on the part of the array that it contains that does not depend on data from the neighbors, and then finally perform the filtering on the elements that depend on the data from neighbors. Unfortunately, while this maximizes the possible overlap between communication and calculation, it also complicates the program because the order of computations within a processor needs to be rearranged.

24.4.1   Global Operations

To manipulate entire structures in one step, it is useful to have a collection of operations that perform such manipulations. These global operations may be very problem-dependent, but certain ones have been found to be widely useful. For example, the average filtering example above made use of shift operations to move an array around. Broadcast is another common global operation, used to send data from one processor to all other processors. Extensions of the broadcast operation include simultaneously performing a broadcast within every (predetermined and distinct) subset of processors. For example, suppose matrix A has been partitioned into submatrices allocated to different processors, and one needs to broadcast the first row of A so that if a processor contains any elements of column i, then it obtains the value of $A(1, i)$. In this situation, the more general form of a subset-based broadcast can be used.

Besides operating within subsets of processors, many global operations are defined in terms of a commutative, associative, semigroup operator $\otimes$. Examples of such semigroup operators include `minimum`, `maximum`, `or`, `and`, `sum`, and `product`. For example, suppose there is a set of values $V(i)$, $1 \leq i \leq n$, and the goal is to obtain the maximum of these values. Then $\otimes$ would represent maximum, and the operation of applying $\otimes$ to all n values is called reduction. If the value of the reduction is broadcast to all processors, then it is sometimes known as report. A more general form of the reduction operation involves labeled data items, i.e., each data item is embedded in a record that also contains a label, where at the end of the reduction operation the result of applying $\otimes$ to all values with the same label will be recorded in the record.

Global operations provide a useful way to describe major actions in parallel programs. Further, since several of these operations are widely useful, they are often made available in highly optimized implementations. The language APL provided a model for several of these operations, and some parallel versions of APL have appeared. Languages such as C* (Thinking Machines Corporation, 1991), UPC (El-Ghazawi et al., 2005), OpenMP (OpenMP Architecture Review Board, 2005), and FORTRAN 90 [Brainerd, Goldberg, and Adams, 1990] also provide for some forms of global operations, as do message-passing systems such as MPI (Snir et al., 1995). Reduction operations are so important that most parallelizing compilers detect them automatically, even if they have no explicit support for other global operations.

Besides broadcast, reduction, and shift, other important global operations include the following.

Sort: Let $X = \{x_0, x_1, \ldots, x_{n-1}\}$ be an ordered set such that $x_i < x_{i+1}$, for all $0 \leq i < n - 1$. (That is, X is a subset of a linearly ordered data type.) Given that the n elements of X are arbitrarily distributed among a set of p processors, the sort operation will (re)arrange the members of X so that they are ordered with respect to the processors. That is, after sorting, elements $x_0, \ldots, x_{\lfloor n/p \rfloor}$ will be in the first processor, elements $x_{\lfloor n/p \rfloor + 1}, \ldots, x_{\lfloor 2n/p \rfloor}$ will be in the second processor, and so forth. Note that this assumes an ordering on the processors, as well as on the elements.

Merge: Suppose that sets D_1 and D_2 are subsets of some linearly ordered data type, and D_1 and D_2 are each distributed in an ordered fashion among disjoint sets of processors $\mathcal{P}_1$ and $\mathcal{P}_2$, respectively. Then the merge operation combines D_1 and D_2 to yield a single sorted set stored in ordered fashion in the entire set of processors $\mathcal{P} = \mathcal{P}_1 \cup \mathcal{P}_2$.

Associative read/write: These operations start with a set of master records indexed by unique keys. In the associative read, each processor specifies a key and ends up with the data in the master record indexed by that key, if such a record exists, or else a flag indicating that there is no such record. In the associative write, each processor specifies a key and a value, and each master record is updated by applying $\otimes$ to all values sent to it. (Master records are generated for all keys written.)

These operations are extensions of the CRCW PRAM operations. They model a PRAM with associative memory and a powerful combining operation for concurrent writes. On most distributed memory machines the time to perform these more powerful operations is within a multiplicative constant of the time needed to simulate the usual concurrent read and concurrent write, and the use of the more powerful operations can result in significant algorithmic simplifications and speedups.

Compression: Compression moves data into a region of the machine where optimal inter-processor communication is possible. For example, compressing k items in a fine-grain two-dimensional mesh will move them to a $\sqrt{k} \times \sqrt{k}$ subsquare.

Scan (parallel prefix): Given a set of values a_i, $1 \leq i \leq n$, the scan computation determines $s_i = a_1 \otimes a_2 \otimes \cdots \otimes a_i$, for all i. This operation is available in APL. Note that the hardware feature known as "fetch-and-op" implements a variant of scan, where "op" is $\otimes$ and the ordering of the processors is not required to be deterministic.

All-to-all broadcast: Given data $D(i)$ in processor i, every processor j receives a copy of $D(i)$, for all $i \neq j$.

All-to-all personalized communication: Every processor P_i has a data item $D(i, j)$ that is sent to processor P_j, for all $i \neq j$.

Example: Maximal point problem

As an example of the use of global operations, consider the following problem from computational geometry. Let S be a finite set of planar (i.e., two-dimensional) points. A point $p = (p_x, p_y)$ in S is a maximal point of S if $p_x > q_x$ or $p_y > q_y$, for every point $(q_x, q_y) \neq p$ in S. The maximal point problem is to determine all maximal points of S (See Figure 24.4). The following parallel algorithm for the maximal point problem was apparently first noted by Atallah and Goodrich (1986).

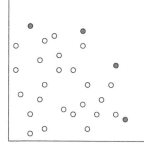

FIGURE 24.4 The maximal points of the set are shaded.

1. Sort the n planar points in reverse order by x-coordinate, with ties broken by reverse order by y-coordinate. Let (i_x, i_y) denote the coordinates of the ith point after the sort is complete. Therefore, after sorting, the points will be ordered so that if $i < j$ then either $i_x > j_x$ or $i_x = j_x$ and $i_y > j_y$.

2. Use a scan on the i_y values, where the operation $\otimes$ is taken to be maximum. The resulting values $\{L_i\}$ are such that L_i is the largest y-coordinate of any point with index less than i.

3. The point (i_x, i_y) is an extreme point if and only if $i_y > L_i$.

The running time $T(n)$ of this algorithm is given by

$$T(n) = Sort(n) + Scan(n) + O(1), \tag{24.1}$$

where $Sort(n)$ is the time to sort n items and $Scan(n)$ is the time to perform the scan. On all parallel architectures known to the authors, $Scan(n) = O(Sort(n))$, and hence on such machines the time of the algorithm is $\Theta(Sort(n))$. It is worth noting that for the sequential model (Kung et al., 1975), have shown that the problem of determining maximal points is as hard as sorting.

24.4.2 Divide-and-Conquer

Divide-and-conquer is a powerful algorithmic paradigm that exploits the repeated subdivision of problems and data into smaller, similar problems/data. It is quite useful in parallel computation because the logical subdivisions into subproblems can correspond to physical decomposition among processors, where eventually the problem is broken into subproblems that are each contained within a single processor. These small subproblems are typically solved by an efficient sequential algorithm within each processor.

As an example, consider the problem of labeling the figures of a black/white image, where the interpretation is that of black objects on a white background. Two black pixels are defined to be adjacent if they are vertical or horizontal neighbors, and connected if there is a path of adjacent black pixels between them. A figure (i.e., connected component) is defined to be a maximally connected set of black pixels in the image. The figures of an image are said to be labeled if every black pixel in the image has a label, with two black pixels having the same label if and only if they are in the same figure.

We utilize a generic parallel divide-and-conquer solution for this problem, given, for example, in (Miller and Stout, 1996, p. 30). Suppose that the $n \times n$ image has been divided into p subimages, as square as possible, and distributed one subimage per processor. Each processor labels the subimage it contains, using whatever serial algorithm is best and using labels that are unique to the processor (so that no two different figures can accidentally get the same label). For example, often the label used is a concatenation of the row and column coordinates of one of the pixels in the figure. Notice that, so, as long as the global row and column coordinates are used, the labels will be unique. After this step, the only figures that could have an incorrect global label are those that lie in two or more subimages, and any such figures must have a pixel on the border of each subimage it is in (see Figure 24.5). To resolve these labels, a record is prepared for each black pixel on the border of a subimage, where the record contains information about the pixel's position in the image, and its current label. There are far fewer such records than there are pixels

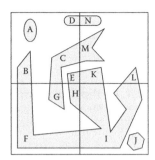

FIGURE 24.5 Divide-and-conquer for labeling figures. The 14 labels shown were generated after each quadrant performed its own, local, labeling algorithm. While the labels are unique, they need to be resolved globally. Notice that once the labels are resolved (not shown), the image will have only five unique labels, corresponding to the five figures.

in the original image, yet they contain all of the information needed to determine the proper global labels for figures crossing subimages. The problem of reconciling the local labels may itself be solved via divide-and-conquer, repeatedly merging results from adjacent regions, or may be solved via other approaches. Once these labels have been resolved, information is sent back to the processors generating the records, informing them of the proper final label.

One useful feature of many of the networks described in the section on Interconnection Networks is that they can be divided into similar subnetworks, in a manner that matches the divide-and-conquer paradigm. For example, if the component labeling algorithm just described were performed on a mesh computer, then each subregion of the image would correspond to a subsquare of the mesh. As another example, consider an implementation of quicksort on a hypercube. Suppose a pivot is chosen and that the data is partitioned into items smaller than the pivot and items larger than the pivot. Further, suppose that the hypercube is logically partitioned into two subcubes, where all of the small items are moved into one subcube and all of the large items are moved into the other subcube. Now, the quicksort routine may proceed recursively within each subcube. Because the recursive divide-and-conquer occurs within subcubes, all of the communication will occur within the subcubes and not cause contention with the other subcube.

24.4.3 Master–Slave

One algorithmic paradigm based on real-world organization paradigms is the master–slave (sometimes referred to as manager–worker) paradigm. In this approach, one processor acts as the master, directing all of the other slave processors. For example, many branch-and-bound approaches to optimization problems keep track of the best solution found so far, as well as a list of subproblems that need to be explored. In a master–slave implementation, the master maintains both of these items and is responsible for parceling out the subproblems to the slaves. The slaves are responsible for processing the subproblems and reporting the result to the master (which will determine if it is the current best solution), reporting new subproblems that need to be explored to the master, and notifying the master when it is free to work on a new subproblem. There are many variations on this theme, but the basic idea is that one processor is responsible for overall coordination, and the other processors are responsible for solving assigned subproblems. Note that this is a variant of the SPMD style of programming, in that there are two programs needed, rather than just one.

24.4.4 Pipelining and Systolic Algorithms

Another common parallel algorithmic technique is based on models that resemble an assembly line. A large problem, such as analyzing a number of images, may be broken into a sequence of steps that must be performed on each image (e.g., filtering, labeling, scene analysis). If one had three processors, and if each step takes about the same amount of time, one could start the first image on the first processor that does the filtering. Then the first image is passed on to the next processor for labeling, while the first processor starts filtering the second image. In the third time step, the initial image is at the third processor for scene analysis, the second image is at the second processor for labeling, and the third image is at the first processor for filtering. This form of processing is called pipelining, and it maps naturally to a parallel computer configured as a linear array (i.e., a one-dimensional mesh or, equivalently, a ring without the wraparound connection).

   This simple scenario can be extended in many ways. For example, as in a real assembly line, the processors need not all be identical, and may be optimized for their task. Also, if some task takes longer to perform than others, then more than one processor can be assigned to it. Finally, the flow may not be a simple line. For example, an automobile assembly process may have one line working on the chassis, while a different line is working on the engine, and eventually these two lines are merged. Such generalized pipelining is called systolic processing. For example, some matrix and image-processing operations can be performed in a two-dimensional systolic manner (see Ullman, 1984).

24.5 Mappings

Often, a problem has a natural structure to be exploited for parallelism, and this needs to be mapped onto a target machine. Several examples follow.

- The average filtering problem, discussed in the Section 24.4, has a natural array structure that can easily be mapped onto a mesh computer. If, however, one had the same problem, but a tree computer, then the mapping might be much more complicated.
- Some artificial intelligence paradigms exploit a blackboard-like communication mechanism that naturally maps onto a shared memory machine. However, a blackboard-like approach is more difficult to map onto a distributed-memory machine.
- Finite-element decompositions have a natural structure whereby calculations at each grid point depend only on values at adjacent points. A finite-element approach is frequently used to model automobiles, airplanes, and rocket exhaust, to name a few. However, the irregular (and perhaps dynamic) structure of such decompositions might need to be mapped onto a target-parallel architecture that bears little resemblance to the finite-element grid.
- A more traditional example consists of porting a parallel algorithm designed for one parallel architecture onto another parallel architecture.

   In all of these examples, one starts with a source structure that needs to be mapped onto a target machine. The goal is to map the source structure onto the target architecture so that calculation and communication steps on the source structure can be efficiently performed by the target architecture.

Usually, the most critical aspect is to map the calculations of the source structure onto the processors of the target machine, so that each processor performs the same amount of calculations. For example, if the source is an array, and each position of the array represents calculations that need to be performed, then one tries to map the array onto the machine so that all processors contain the same number of entries. If the source model is a shared-memory paradigm with agents reading from a blackboard, then one would map the agents to processors, trying to balance the computational work.

Besides trying to balance the computational load, one must also try to minimize the time spent on communication. The approaches used for these mappings depend on the source structure and target architecture, and some of the more widely used approaches are discussed in the following Sections.

24.5.1 Simulating Shared Memory

If the source structure is a shared memory model, and the target architecture is a distributed memory machine, then besides mapping the calculations of the source onto the processors of the target, one must also map the shared memory of the source onto the distributed memory of the target.

To map the memory onto the target machine, suppose that there are memory locations $0 \ldots n-1$ in the source structure, and p processors in the target. Typically one would map locations $0 \ldots \lfloor n/p-1 \rfloor$ to processor 0 of the target machine, locations $\lfloor n/p \rfloor \ldots \lfloor 2n/p - 1 \rfloor$ to processor 1, and so forth. Such a simple mapping balances the amount of memory being simulated by each target processor, and makes it easy to determine where data are located. For example, if a target processor needs to read from shared memory location i, it sends a message to target processor $\lfloor ip/n \rfloor$ asking for the contents of simulated shared memory location i.

Unfortunately, some shared memory algorithms utilize certain memory locations far more often than others, which can cause bottlenecks in terms of getting data in and out of processors holding the popular locations. If popular memory locations form contiguous blocks, then this congestion can be alleviated by stripping (mapping memory location i to processor $i \bmod p$) or pseudorandom mapping (Rau, 1991). Replication (having copies of frequently read locations in more than one processor) or adaptive mapping (dynamically moving simulated memory locations from heavily loaded processors to lightly loaded ones) are occasionally employed to relieve congestion, but such techniques are more complicated and involve additional overhead.

24.5.2 Simulating Distributed Memory

It is often useful to view distributed memory machines as graphs. Processors in the machine are represented by vertices of the graph, and communication links in the machine are represented by edges in the graph. Similarly, it is often convenient to view the structure of a problem as a graph, where vertices represent work that needs to be performed, and edges represent values that need to be communicated in order to perform the desired work. For example, in a finite-element decomposition, the vertices of a decomposition might represent calculations that need to be performed, while the edges correspond to the flow of data. That is, in a typical finite-element problem, if there is an edge from vertex p to vertex q, then the value of q at time t depends on the values of q and p at time $t - 1$. (Most finite-element decompositions are symmetric, so that p at time t would also depend on q at time $t - 1$.) Questions about mapping the structure of a problem onto a target architecture can then be answered by considering various operations on the related graphs.

An ideal situation for mapping a problem onto a target architecture is when the graph representing the structure of a problem is a subgraph of the graph representing the target architecture. For example, if the structure of a problem was represented as a connected string of p vertices and the target architecture was a ring of p processors, then the mapping of the problem onto the architecture would be straightforward and efficient. In graph terms, this is described through the notion of embedding. An embedding of an undirected graph $G = (V, E)$ (i.e., G has vertex set V and edges E) into an undirected graph $G' = (V', E')$ is a mapping ϕ of V into V' such that

- Every pair of distinct vertices $u, v \in V$, map to distinct vertices $\phi(u), \phi(v) \in V'$.
- For every edge $\{u, v\} \in E$, $\{\phi(u), \phi(v)\}$ is an edge in E'.

Let G represent the graph corresponding to the structure of a problem (*i.e.*, the source structure) and let G' represent the graph corresponding to the target architecture. Notice that if there is an embedding of G into G', then values that need to be communicated may be transmitted by a single communication step in the target architecture represented by G'. The fact that embeddings map distinct vertices of G to distinct vertices of G' ensures that a single calculation step for the problem can be simulated in a single calculation step of the target architecture.

One reason that hypercube computers were quite popular is that many graphs can be embedded into the hypercube (graph). An embedding of the one-dimensional ring of size 2^d into a d-dimensional hypercube is called a d-dimensional Gray code. In other words, if $\{0, 1\}^d$ denotes the set of all d-bit binary strings, then the d-dimensional Gray code G_d is a 1–1 map of $0 \dots 2^d - 1$ onto $\{0, 1\}^d$, such that $G_d(j)$ and $G_d((j + 1) \bmod 2^d)$ differ by a single bit, for $0 \leq j \leq 2^d - 1$. The most common Gray codes, called reflected binary Gray codes, are recursively defined as follows: G_d is a 1–1 mapping from $\{0, 1, \dots, 2^d - 1\}$ onto $\{0, 1\}^d$, given by $G_1(0) = 0$, $G_1(1) = 1$, and for $d \geq 2$,

$$G_d(x) = \begin{cases} 0 G_{d-1}(x) & 0 \leq x \leq 2^{d-1} - 1 \\ 1 G_{d-1}(2^d - 1 - x) & 2^{d-1} \leq x \leq 2^d - 1. \end{cases} \tag{24.2}$$

Alternatively, the same Gray code can be defined in a nonrecursive fashion as $G_d(x) = x \oplus \lfloor x/2 \rfloor$, where x and $\lfloor x/2 \rfloor$ are interpreted as d-bit strings. Further, the inverse of the reflected binary Gray code can be determined by

$$G_d^{-1}(y_0 \dots y_{d-1}) = x_0 \dots x_{d-1}, \tag{24.3}$$

where $x_{d-1} = y_{d-1}$, and $x_i = y_{d-1} \oplus \dots \oplus y_i$ for $0 \leq i < d - 1$.

Meshes can also be embedded into hypercubes. Let M be a d-dimensional mesh of size $m_1 \times m_2 \times \dots \times m_d$, and let $r = \sum_{i=1}^d \lceil \log_2 m_i \rceil$. Then M can be embedded into the hypercube of size 2^r. To see this, let $r_i = \lceil \log_2 m_i \rceil$, for $1 \leq i \leq d$. Let ϕ be the mapping of mesh node $(a_1, \dots, a_d)$ to the hypercube node which has as its label the concatenation $G_{r_1}(a_1) \dots G_{r_d}(a_d)$, where G_{r_i} denotes any r_i-bit Gray code. Then ϕ is an embedding. Wrapped dimensions can be handled using reflected Gray codes rather than arbitrary ones. (A mesh M is wrapped in dimension j if, in addition to the normal mesh adjacencies, vertices with indices of the form $(a_1, \dots, a_{j-1}, 0, a_{j+1}, \dots, a_d)$ and $(a_1, \dots, a_{j-1}, m_j - 1, a_{j+1}, \dots, a_d)$ are adjacent. A torus is a mesh wrapped in all dimensions.) If dimension j is wrapped and m_j is an integral power of 2, then the mapping ϕ suffices. If dimension j is wrapped and m_j is even, but not an integral power of 2, then to ensure that the first and last nodes in dimension j are mapped to adjacent hypercube nodes, use ϕ, but replace $G_{r_j}(a_j)$ with

$$\begin{cases} G_{r_j}(a_j) & \text{if } 0 \leq a_j \leq m_j/2 - 1 \\ G_{r_j}(a_j + 2^{r_j} - m_j) & \text{if } m_j/2 \leq a_j \leq m_j - 1, \end{cases} \tag{24.4}$$

where G_{r_j} is the r_j-bit reflected binary Gray code. This construction ensures that $G_{r_j}(m_j/2 - 1)$ and $G_{r_j}(2^{r_j} - m_j/2)$ differ by exactly one bit (the highest order one), which in turns ensures that the mapping takes mesh nodes neighboring in dimension j to hypercube neighbors.

Any tree T can be embedded into a $(|T| - 1)$-dimensional hypercube, where $|T|$ denotes the number of vertices in T, but this result is of little use since the target hypercube is exponentially larger than the source tree. Often one can map the tree into a more reasonably sized hypercube, but it is a difficult problem to determine the minimum dimension needed, and there are numerous papers on the subject.

In general, however, one cannot embed the source structure into the target architecture. For example, a complete binary tree of height two, which contains seven processors, cannot be embedded into a ring of any size. Therefore, one must consider weaker mappings, which allow for the possibility that the target machine has fewer processors than the source, and does not contain the communication links of the source. A weak embedding of a directed source graph $G = (V, E)$ into a directed target graph $G' = (V', E')$ consists of

- A map ϕ_v of V into V'.
- A map ϕ_e of E onto paths in G', such that if $(u, v) \in E$ then $\phi_e((u, v))$ is a path from $\phi_v(u)$ to $\phi_v(v)$.

(Note that if G is undirected, each edge becomes two directed edges that may be mapped to different paths in G'. Most machines that are based on meshes, tori, or hypercubes have the property that a message from processor P to processor Q may not necessarily follow the same path as a message sent from processor Q to processor P, if P and Q are not adjacent.) The map ϕ_v shows how computations are mapped from the source onto the target, and the map ϕ_e shows the communication paths that will be used in the target.

There are several measures that are often used to describe the quality of a weak embedding (ϕ_v, ϕ_e) of G into G', including the following.

Processor Load: The maximum, over all vertices $v' \in V'$, of the number of vertices in V mapped onto v' by ϕ_v. Note that if all vertices of the source structure represent the same amount of computation, then the processor load is the maximum computational load by any processor in the target machine. The goal is to make the processor load as close as possible to $|V|/|V'|$. If vertices do not all represent the same amount of work, then one should use labeled vertices, where the label represents the amount of work, and try to minimize the maximum, over all vertices $v' \in V'$, of the sum of the labels of the vertices mapped onto v'.

Link load (link congestion): The maximum, over all edges $(u', v') \in E'$, of the number of edges $(u, v) \in E$ such that (u', v') is part of the path $\phi_e((u, v))$. If all edges of the source structure represent the same amount of communication, then the link load represents the maximum amount of communication contending for a single communication link in the target architecture. As for processor load, if edges do not represent the same amount of communication, then weights should be balanced instead.

Dilation: The maximum, over all edges $(u, v) \in E$, of the path length of $\phi_e((u, v))$. The dilation represents the longest delay that would be needed to simulate a single communication step along an edge in the source, if that was the only communication being performed.

Expansion: The ratio of the number of vertices of G' divided by the number of vertices of G. As was noted in the example of trees embedding into hypercubes, large expansion is impractical. In practice, usually the real target machine has far fewer processors than the idealized source structure, so expansion is not a concern.

In some machines, dilation is an important measure of communication delay, but in most modern general-purpose machines it is far less important because each message has a relatively large start-up time that may be a few orders of magnitude larger than the time per link traversed. Link contention may still be a problem in such machines, but some solve this by increasing the bandwidth on links that would have heavy contention. For example, as noted earlier, fat-trees (Leiserson, 1985) add bandwidth near the root to avoid the bottlenecks inherent in a tree architecture. This increases the bisection bandwidth, which reduces the link contention for communication that poorly matches the basic tree structure.

For machines with very large message start-up times, often the number of messages needed becomes a dominant communication issue. In such a machine, one may merely try to balance calculation load and minimize the number of messages each processor needs to send, ignoring other communication effects. The number of messages that a processor needs to send can be easily determined by noting that processors p and q communicate if there are adjacent vertices u and v in the source structure such that ϕ_v maps u to p and v to q.

For many graphs that cannot be embedded into a hypercube, there are nonetheless useful weak embeddings. For example, keeping the expansion as close to 1 as is possible (given the restriction that a hypercube has a power of 2 processors), one can map the complete binary tree onto the hypercube with unit link congestion, dilation two, and unit processor contention. See, for example, (Leighton, 1992).

In general, however, finding an optimal weak embedding for a given source and target is an NP-complete problem. This problem, sometimes known as the mapping problem, is often solved by various heuristics. This is particularly true when the source structure is given by a finite-element decomposition or other approximation schemes for real entities, for in such cases the sources are often quite large and irregular. Fortunately, the fact that such sources often have an underlying geometric basis makes it easier to find fairly good mappings rather quickly.

For example, suppose the source structure is an irregular grid representing the surface of a three-dimensional airplane, and the target machine is a two-dimensional mesh. One might first project the airplane onto the x–y plane, ignoring the z-coordinates. Then one might locate a median x-coordinate, call it $\bar{x}$, where half of the plane's vertices lie to the left of $\bar{x}$ and half to the right. The vertices may then be mapped so that those that lie to the left of $\bar{x}$ are mapped onto the left half of the target machine, and those vertices that lie to the right of $\bar{x}$ are mapped to the right half of the target. In the left half of the target, one might locate the median y-coordinate, denoted $\bar{y}$, of the points mapped to that half, and map the points above $\bar{y}$ to the top-left quadrant of the target, and map points below $\bar{y}$ to the bottom-left. On the right half a similar operation would be performed for the points mapped to that side. Continuing in this recursive, divide-and-conquer manner, eventually the target machine would have been subdivided down into single processors, at which point the mapping would have been determined. This mapping is fairly straightforward, balances the processor load, and roughly keeps points adjacent in the grid near to each other in the target machine, and hence it does a reasonable approximation of minimizing communication time. This technique is known as recursive bisectioning, and is closely related to the serial data structure known as a K-d tree (Bentley, 1975).

An approach which typically results in less communication is to form a linear ordering of the vertices via their coordinates on a space-filling curve, and then divide the vertices into intervals of this ordering. Typically either Z-ordering (aka Morton-ordering) or Peano–Hilbert curves are used for this purpose. Peano–Hilbert curves are also used as orderings of the processors in meshes, where they are sometimes called proximity orderings (Miller and Stout, 1996, p. 150).

Neither recursive bisectioning nor space-filling curves minimize the number of messages sent by each processor, and hence if message start-up time is quite high it may be better to use recursive bisectioning where the plane is cut along only, say, the x-axis at each step. Each processor would end up with a cross-sectional slab, with all of the source vertices in the given range of x-coordinates. If grid edges are not longer than the width of such a slab, then each processor would have to send messages to only two processors, namely the processor with the slab to the left and the processor with the slab to the right.

Other complications can arise because the nodes or edges of such sources may not all represent the same amount of computation or calculation, respectively, in which case weighted mappings are appropriate. A variety of programs are available that perform such mappings, and over time the quality of the mapping achieved, and the time to achieve it, has significantly improved. For irregular source structures, such packages are generally superior to what one would achieve without considerable effort.

A more serious complication is that the natural source structure may be dynamic, adding nodes or edges over time. In such situations one often needs to dynamically adjust the mapping to keep the computational load balanced and keep communication minimal. This introduces additional overhead, which one must weigh against the costs of not adjusting the imbalance. Often the dynamical remappings are made incrementally, moving only a little of the data to correct the worst imbalances. Deciding how often to check for imbalance, and how much to move, typically depends quite heavily on the problem being solved.

24.6   Research Issues and Summary

The development of parallel algorithms and efficient parallel programs lags significantly behind that of algorithms and programs for standard serial computers. This makes sense due to the fact that commercially available serial machines have been available for approximately twice as long as have commercially available parallel machines. Parallel computing, including distributed computing, cluster computing, and grid computing, is in a rapidly growing phase, with important research and development still needed in almost all areas. Extensive theoretical and practical work continues in discovering parallel programming paradigms, in developing a wide range of efficient parallel algorithms, in developing ways to describe and manage parallelism through new languages or extensions of current ones, in developing techniques to automatically detect parallelism, and in developing libraries of parallel routines.

Another factor that has hindered parallel algorithm development is the fact that there are many different parallel computing models. As noted earlier, architectural differences can significantly affect the efficiency of an algorithm, and hence parallel algorithms have traditionally been tied to specific parallel models. One advance is that various hardware and software approaches are being developed to help hide some of the architectural differences. Thus, one may have, say, a distributed memory machine, but have a software system that allows the programmer to view it as a shared memory machine. While it is true that a programmer will usually only be able to achieve the highest performance by directly optimizing the code for a target machine, in many cases acceptable performance can be achieved without tying the code to excessive details of an architecture. This then allows the code to be ported to a variety of machines, encouraging code development. In the past, extensive code revision was needed every time the code was ported to a new parallel machine, strongly discouraging many users who did not want to plan for an unending parade of changes.

Another factor that has limited parallel algorithm development is that most computer scientists were not trained in parallel computing and have a limited knowledge of domain-specific areas (chemistry, biology, mechanical engineering, civil engineering, physics, and architecture, to name but a few). As the field matures, more courses will incorporate parallel computing and the situation will improve. There are some technological factors that argue for the need for rapid improvement in the training of programmers to exploit parallelism. The history of exponential growth in the clock rate of processors has come to a close, with only slow advances predicted for the future, so users can no longer expect to solve ever more complex problems merely through improvements in serial processors. Meanwhile, cluster computers, which are distributed memory systems where the individual nodes are commodity boards containing serial or small shared memory units, have become common throughout industry and academia. These are low-cost systems with significant computing power, but unfortunately, due to the dearth of parallel programmers, many of these systems are used only to run concurrent serial programs (known as embarrassingly parallel processing), or to run turnkey parallel programs (such as databases).

Parallelism is also becoming the dominant improvement in the capabilities of individual chips. Some graphics processing units (GPUs) already have over a hundred simple computational units that are vector processing systems, which can be interpreted as implementing SIMD operations. There is interest in exploiting these in areas such as data mining and numeric vector computing, but so far this has primarily been achieved for proof of concept demonstrations. Most importantly, standard serial processors are all becoming many-core chips with parallel computing possibilities, where the number of cores per chip is predicted to have exponential growth. Unfortunately it is very difficult to exploit their potential, and they are almost never used as parallel computers. Improving this situation has become an urgent problem in computer science and its application to problems in the disciplinary fields that require large multi-processor systems.

24.7 Further Information

A good introduction to parallel computing at the undergraduate level is *Parallel Computing: Theory and Practice* by Michael J. Quinn (1994). This book provides a nice introduction to parallel computing, including parallel algorithms, parallel architectures, and parallel programming languages. *Parallel Algorithms for Regular Architectures: Meshes and Pyramids* by Russ Miller and Quentin F. Stout focuses on fundamental algorithms and paradigms for fine-grained machines. It advocates an approach of designing algorithms in terms of fundamental data movement operations, including sorting, concurrent read, and concurrent write. Such an approach allows one to port algorithms in an efficient manner between architectures. *Introduction to Parallel Algorithms and Architectures: Arrays, Trees, Hypercubes* is a comprehensive book by F. Thomson Leighton (1992) that also focuses on fine-grained algorithms for several traditional interconnection networks. For the reader interested in algorithms for the PRAM, *An Introduction to Parallel Algorithms* by J. JáJá covers fundamental algorithms in geometry, graph theory, and string matching. It also includes a chapter on randomized algorithms. Finally, a new approach is used in *Algorithms Sequential & Parallel* by Miller and Boxer (2005), which presents a unified approach to sequential and parallel algorithms, focusing on the RAM, PRAM, Mesh, Hypercube, and Pyramid. This book focuses on paradigms and efficient implementations across a variety of platforms in order to provide efficient solutions to fundamental problems.

There are several professional societies that sponsor conferences, publish books, and publish journals in the area of parallel algorithms. These include the *Association for Computing Machinery (ACM)*, which can be found at http://www.acm.org, *The Institute for Electrical and Electronics Engineers, Inc. (IEEE)*, which can be found at http://www.ieee.org, and the *Society for Industrial and Applied Mathematics (SIAM)*, which can be found at http://www.siam.org.

Since parallel computing has become so pervasive, most computer science journals cover work concerned with parallel and distributed systems. For example, one would expect a journal on programming languages to publish articles on languages for shared-memory machines, distributed-memory machines, networks of workstations, and so forth. For several journals, however, the primary focus is on parallel algorithms. These journals include the *Journal for Parallel and Distributed Computing*, published by Academic Press (http://www.apnet.com), the *IEEE Transactions on Parallel and Distributed Systems* (http://computer.org/pubs/tpds), and for results that can be expressed in a condensed form, *Parallel Processing Letters*, published by World Scientific. Finally, several comprehensive journals should be mentioned that publish a fair number of articles on parallel algorithms. These include the *IEEE Transactions on Computers, Journal of the ACM*, and *SIAM Journal on Computing*.

Unfortunately, due to very long delays from submission to publication, most results that appear in journals (with the exception of *Parallel Processing Letters*) are actually quite old. (A delay of 3–5 years from submission to publication is not uncommon.) Recent results appear in a timely fashion in conferences, most of which are either peer or panel reviewed. The first conference devoted primarily to parallel computing was the *International Conference on Parallel Processing (ICPP)*, which had its inaugural conference in 1972. Many landmark papers were presented at ICPP, especially during the 1970s and 1980s. This conference merged with the *International Parallel Processing Symposium (IPPS)* (http://www.ippsxx.org), resulting in the *International Parallel and Distributed Symposium (IPDPS)* (http://www.ipdps.org). IPDPS is quite comprehensive in that in addition to the conference, it offers a wide variety of workshops and tutorials.

A conference that tends to include more theoretical algorithms is the *ACM Symposium on Parallelism in Algorithms and Architectures (SPAA)* (http://www.spaa-conference.org). This conference is an offshoot of the premier theoretical conferences in computer science, *ACM Symposium on Theory of Computing (STOC)* and *IEEE Symposium on Foundations of Computer Science (FOCS)*. A conference which focuses on very large parallel systems is *SC 'XY* (http://www.supercomp.org), where XY represents the last two digits of the year. This conferences includes the presentation of the Gordon Bell Prize for best parallelization. Awards are given in various categories, such as highest sustained performance and best price/performance. Other relevant conferences include the *International Supercomputing Conference* (http://www.supercomp.de) and the *IEEE International Conference on High Performance Computing* (http://www.hipc.org).

Finally, the IEEE Distributed Systems Online site, http://dsonline.computer.org, contains links to conferences, journals, people in the field, bibliographies on parallel processing, online course material, books, and so forth.

Defining Terms

Distributed memory: Each processor has access to only its own private (local) memory, and communicates with other processors via messages.

Divide-and-conquer: A programming paradigm whereby large problems are solved by decomposing them into smaller, yet similar, problems.

Global operations: Parallel operations that affect system-wide data structures.

Interconnection network: The communication system that links together all of the processors and memory of a parallel machine.

Master–slave (manager–worker): A parallel programming paradigm whereby a problem is broken into a collection of smaller problems, with a master processor keeping track of the subproblems and assigning them to the slave processors.

Parallel random access machine (PRAM): A theoretical shared-memory model, where typically the processors all execute the same instruction synchronously, and access to any memory location occurs in unit time.

Pipelining: A parallel programming paradigm that abstracts the notion of an assembly line. A task is broken into a sequence of fixed subtasks corresponding to the stations of an assembly line. A series of similar tasks are solved by starting one task through the subtask sequence, then starting the next task through as soon as the previous task has finished its first subtask. At any point in time, several tasks are in various stages of completion.

Shared memory: All processors have the same global image of (and access to) all of the memory.

Single-program multiple-data (SPMD): The dominant style of parallel programming, where all of the processors utilize the same program, though each has its own data.

References

Atallah, M.J. and Goodrich, M.T. 1986. Efficient parallel solutions to geometric problems, *Journal of Parallel and Distributed Computing*, 3: 492–507.

Bentley, J. 1975. Multidimensional binary search trees used for associative searching, *Communications of the ACM*, 18(9): 509–517.

Brainerd, W.S., Goldberg, C., and Adams, J.C. 1990. *Programmers Guide to FORTRAN 90*, McGraw-Hill, INC., New York.

El-Ghazawi, T., Carlson, W., Sterling, T., and Yellick, K. 2005. *UPC: Distributed Shared Memory Programming*, John Wiley & Sons, New York.

Flynn, M.J. 1966. Very high-speed computing systems, *Proceedings of the IEEE*, 54(12): 1901–1909.

Flynn, M.J. 1972. Some computer organizations and their effectiveness, *IEEE Transactions on Computers*, C-21:948–960.

JáJá, J. 1992. *An Introduction to Parallel Algorithms*, Addison-Wesley, Reading, MA.

Kung, H.T., Luccio, F., and Preparata, F.P. 1975. On finding the maxima of a set of vectors, *Journal of the ACM*, 22(4): 469–476.

Leighton, F.T. 1992. *Introduction to Parallel Algorithms and Architectures: Arrays, Trees, Hypercubes*, Morgan Kaufmann Publishers, San Mateo, CA.

Leiserson, C.E. 1985. Fat-trees: Universal networks for hardware-efficient supercomputing, *IEEE Transactions on Computers*, C-34(10):892–901.

Li, H. and Stout, Q.F. 1991. Reconfigurable SIMD parallel processors, *Proceedings of the IEEE*, 79:429–443.

Miller, R. and Boxer, L. 2005. Algorithms Sequential and Parallel: A unified Approach, second edn, Charles River Media Inc., Hingham, MA.

Miller, R. and Stout, Q.F. 1996. *Parallel Algorithms for Regular Architectures: Meshes and Pyramids*, MIT Press, Cambridge, MA.

OpenMP Architectural Review Board, 2005. *OpenMP Application Program Interface*, Eugene, OR.

Quinn, M.J. 1994. *Parallel Computing Theory and Practice*, McGraw-Hill, Inc., New York.

Rau, B.R. 1991. Pseudo-randomly interleaved memory. *Proceedings of the 18th annual international symposium on Computer Architecture*, ACM, New York, pp. 74–83.

Snir, M., Otto, S.W., Huss-Lederman, S., Walker, D.W., and Dongarra, J. 1995. *MPI: The Complete Reference*, MIT Press, Cambridge, MA.

Thinking Machines Corporation. 1991. *C* Programming Guide*, Version 6.0.2, Cambridge, MA.

Ullman, J.D. 1984. *Computational Aspects of VLSI*, Computer Science Press, Rockville, MD.

25

Parallel Algorithms

Guy E. Blelloch
Carnegie Mellon University

Bruce M. Maggs
Duke University and Akamai Technologies

25.1   Introduction

The subject of this chapter is the design and analysis of parallel algorithms. Most of today's algorithms are sequential, that is, they specify a sequence of steps in which each step consists of a single operation. These algorithms are well suited to today's computers, which basically perform operations in a sequential fashion. Although the speed at which sequential computers operate has been improving at an exponential rate for many years, the improvement is now coming at greater and greater cost. As a consequence, researchers have sought more cost-effective improvements by building "parallel" computers—computers that perform multiple operations in a single step. In order to solve a problem efficiently on a parallel computer, it is usually necessary to design an algorithm that specifies multiple operations on each step, i.e., a parallel algorithm.

As an example, consider the problem of computing the sum of a sequence A of n numbers. The standard algorithm computes the sum by making a single pass through the sequence, keeping a running sum of the numbers seen so far. It is not difficult however, to devise an algorithm for computing the sum that performs many operations in parallel. For example, suppose that, in parallel, each element of A with an even index is paired and summed with the next element of A, which has an odd index, i.e., $A[0]$ is paired with $A[1]$, $A[2]$ with $A[3]$, and so on. The result is a new sequence of $\lceil n/2 \rceil$ numbers that sum to the same value as the sum that we wish to compute. This pairing and summing step can be repeated until, after $\lceil \log_2 n \rceil$ steps, a sequence consisting of a single value is produced, and this value is equal to the final sum.

The parallelism in an algorithm can yield improved performance on many different kinds of computers. For example, on a parallel computer, the operations in a parallel algorithm can be performed simultaneously by different processors. Furthermore, even on a single-processor computer the parallelism in an algorithm can be exploited by using multiple functional units, pipelined functional units, or pipelined memory systems. Thus, it is important to make a distinction between the parallelism in an algorithm and the ability of any particular computer to perform multiple operations in parallel. Of course, in order for a parallel algorithm to run efficiently on any type of computer, the algorithm must contain at least as much parallelism as the computer, for otherwise resources would be left idle. Unfortunately, the converse does not always hold: some parallel computers cannot efficiently execute all algorithms, even if the algorithms contain a great deal of parallelism. Experience has shown that it is more difficult to build a general-purpose parallel computer than a general-purpose sequential computer.

The remainder of this chapter consists of nine sections. We begin in Section 25.2 with a discussion of how to model parallel computers. Next, in Section 25.3 we cover some general techniques that have proven useful in the design of parallel algorithms. Sections 25.4 through 25.8 present algorithms for solving problems from different domains. We conclude in Section 25.9 with a discussion of current research topics, a collection of defining terms, and finally sources for further information.

Throughout this chapter, we assume that the reader has some familiarity with sequential algorithms and asymptotic notation and analysis.

25.2   Modeling Parallel Computations

The designer of a sequential algorithm typically formulates the algorithm using an abstract model of computation called the random-access machine (RAM) model [2], Chapter 1. In this model, the machine consists of a single processor connected to a memory system. Each basic CPU operation, including arithmetic operations, logical operations, and memory accesses, requires one time step. The designer's goal is to develop an algorithm with modest time and memory requirements. The RAM model allows the algorithm designer to ignore many of the details of the computer on which the algorithm will ultimately be executed, but captures enough detail that the designer can predict with reasonable accuracy how the algorithm will perform.

Modeling parallel computations is more complicated than modeling sequential computations because in practice parallel computers tend to vary more in organization than do sequential computers. As a consequence, a large portion of the research on parallel algorithms has gone into the question of modeling, and many debates have raged over what the "right" model is, or about how practical various models are. Although there has been no consensus on the right model, this research has yielded a better understanding of the relationship between the models. Any discussion of parallel algorithms requires some understanding of the various models and the relationships among them.

In this chapter we divide parallel models into two classes: **multiprocessor models** and **work-depth models**. In the remainder of this section we discuss these two classes and how they are related.

25.2.1 Multiprocessor Models

A multiprocessor model is a generalization of the sequential RAM model in which there is more than one processor. Multiprocessor models can be classified into three basic types: local-memory machine models, modular memory machine models, and parallel random-access machine (PRAM) models. Figure 25.1 illustrates the structure of these machine models. A local-memory machine model consists of a set of *n* processors each with its own local-memory. These processors are attached to a common communication network. A modular memory machine model consists of *m* memory modules and *n* processors all attached to a common network. An *n*-processor PRAM model consists of a set of *n* processors all connected to a common shared memory [32,37,38,77].

The three types of multiprocessors differ in the way that memory can be accessed. In a local-memory machine model, each processor can access its own local memory directly, but can access the memory in another processor only by sending a memory request through the network. As in the RAM model, all local operations, including local memory accesses, take unit time. The time taken to access the memory in another processor, however, will depend on both the capabilities of the communication network and the pattern of memory accesses made by other processors, since these other accesses could congest the network. In a modular memory machine model, a processor accesses the memory in a memory module by sending a memory request through the network. Typically the processors and memory modules are arranged so that the time for any processor to access any memory module is roughly uniform. As in a local-memory machine model, the exact amount of

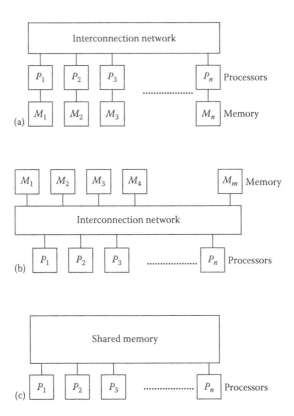

FIGURE 25.1 The three types of multiprocessor machine models: (a) a local-memory machine model; (b) a modular memory machine model; and (c) a PRAM model.

time depends on the communication network and the memory access pattern. In a **PRAM model**, a processor can access any word of memory in a single step. Furthermore, these accesses can occur in parallel, i.e., in a single step, every processor can access the shared memory.

The PRAM models are controversial because no real machine lives up to its ideal of unit-time access to shared memory. It is worth noting, however, that the ultimate purpose of an abstract model is not to directly model a real machine, but to help the algorithm designer produce efficient algorithms. Thus, if an algorithm designed for a PRAM model (or any other model) can be translated to an algorithm that runs efficiently on a real computer, then the model has succeeded. In Section 25.2.4 we show how an algorithm designed for one parallel machine model can be translated so that it executes efficiently on another model.

The three types of multiprocessor models that we have defined are broad and allow for many variations. The local-memory machine models and modular memory machine models may differ according to their network topologies. Furthermore, in all three types of models, there may be differences in the operations that the processors and networks are allowed to perform. In the remainder of this section we discuss some of the possibilities.

25.2.1.1 Network Topology

A network is a collection of switches connected by communication channels. A processor or memory module has one or more communication ports that are connected to these switches by communication channels. The pattern of interconnection of the switches is called the network topology. The topology of a network has a large influence on the performance and also on the cost and difficulty of constructing the network. Figure 25.2 illustrates several different topologies.

The simplest network topology is a bus. This network can be used in both local-memory machine models and modular memory machine models. In either case, all processors and memory modules are typically connected to a single bus. In each step, at most one piece of data can be written onto the bus. This data might be a request from a processor to read or write a memory value, or it might be the response from the processor or memory module that holds the value. In practice, the advantage of using a bus is that it is simple to build and, because all processors and memory modules can observe

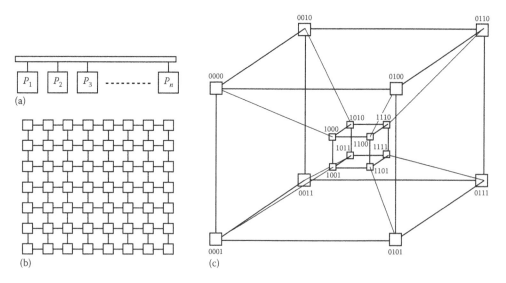

FIGURE 25.2 (a) Bus, (b) two-dimensional mesh, and (c) hypercube network topologies.

the traffic on the bus, it is relatively easy to develop protocols that allow processors to cache memory values locally. The disadvantage of using a bus is that the processors have to take turns accessing the bus. Hence, as more processors are added to a bus, the average time to perform a memory access grows proportionately.

A two-dimensional mesh is a network that can be laid out in a rectangular fashion. Each switch in a mesh has a distinct label (x, y) where $0 \leq x \leq X - 1$ and $0 \leq y \leq Y - 1$. The values X and Y determine the length of the sides of the mesh. The number of switches in a mesh is thus $X \cdot Y$. Every switch, except those on the sides of the mesh, is connected to four neighbors: one to the north, one to the south, one to the east, and one to the west. Thus, a switch labeled (x, y), where $0 < x < X - 1$ and $0 < y < Y - 1$, is connected to switches $(x, y + 1)$, $(x, y - 1)$, $(x + 1, y)$, and $(x - 1, y)$. This network typically appears in a local-memory machine model, i.e., a processor along with its local memory is connected to each switch, and remote memory accesses are made by routing messages through the mesh. Figure 25.2b shows an example of an 8×8 mesh.

Several variations on meshes are also popular, including three-dimensional meshes, toruses, and hypercubes. A torus is a mesh in which the switches on the sides have connections to the switches on the opposite sides. Thus, every switch (x, y) is connected to four other switches: $(x, y + 1 \bmod Y)$, $(x, y - 1 \bmod Y)$, $(x + 1 \bmod X, y)$, and $(x - 1 \bmod X, y)$. A hypercube is a network with 2^n switches in which each switch has a distinct n-bit label. Two switches are connected by a communication channel in a hypercube if and only if the labels of the switches differ in precisely one bit position. A hypercube with 16 switches is shown in Figure 25.2c.

A multistage network is used to connect one set of switches called the input switches to another set called the output switches through a sequence of stages of switches. Such networks were originally designed for telephone networks [15]. The stages of a multistage network are numbered 1 through L, where L is the depth of the network. The switches on stage 1 are the input switches, and those on stage L are the output switches. In most multistage networks, it is possible to send a message from any input switch to any output switch along a path that traverses the stages of the network in order from 1 to L. Multistage networks are frequently used in modular memory computers; typically processors are attached to input switches, and memory modules are attached to output switches. A processor accesses a word of memory by injecting a memory access request message into the network. This message then travels through the network to the appropriate memory module. If the request is to read a word of memory, then the memory module sends the data back through the network to the requesting processor. There are many different multistage network topologies. Figure 25.3a, for example, shows a depth-2 network that connects 4 processors to 16 memory modules. Each switch in this network has two channels at the bottom and four channels at the top. The ratio of processors to memory modules in this example is chosen to reflect the fact that, in practice, a processor is capable of generating memory access requests faster than a memory module is capable of servicing them.

A fat-tree is a network structured like a tree [56]. Each edge of the tree, however, may represent many communication channels, and each node may represent many network switches (hence the name "fat"). Figure 25.3b shows a fat-tree with the overall structure of a binary tree. Typically the

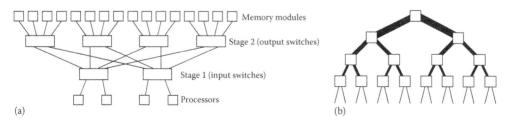

(a)
(b)

FIGURE 25.3 (a) 2-level multistage network and (b) fat-tree network topologies.

capacities of the edges near the root of the tree are much larger than the capacities near the leaves. For example, in this tree the two edges incident on the root represent 8 channels each, while the edges incident on the leaves represent only 1 channel each. A natural way to construct a local-memory machine model is to connect a processor along with its local memory to each leaf of the fat-tree. In this scheme, a message from one processor to another first travels up the tree to the least common-ancestor of the two processors, and then down the tree.

Many algorithms have been designed to run efficiently on particular network topologies such as the mesh or the hypercube. For extensive treatment such algorithms, see [55,67,73,80]. Although this approach can lead to very fine-tuned algorithms, it has some disadvantages. First, algorithms designed for one network may not perform well on other networks. Hence, in order to solve a problem on a new machine, it may be necessary to design a new algorithm from scratch. Second, algorithms that take advantage of a particular network tend to be more complicated than algorithms designed for more abstract models like the PRAM models, because they must incorporate some of the details of the network. Nevertheless, there are some operations that are performed so frequently by a parallel machine that it makes sense to design a fine-tuned network-specific algorithm. For example, the algorithm that routes messages or memory access requests through the network should exploit the network topology. Other examples include algorithms for broadcasting a message from one processor to many other processors, for collecting the results computed in many processors in a single processor, and for synchronizing processors.

An alternative to modeling the topology of a network is to summarize its routing capabilities in terms of two parameters, its latency and bandwidth. The latency, L, of a network is the time it takes for a message to traverse the network. In actual networks this will depend on the topology of the network, which particular ports the message is passing between, and the congestion of messages in the network. The latency is often modeled by considering the worst-case time assuming that the network is not heavily congested. The bandwidth at each port of the network is the rate at which a processor can inject data into the network. In actual networks this will depend on the topology of the network, the bandwidths of the network's individual communication channels and, again, the congestion of messages in the network. The bandwidth often can be usefully modeled as the maximum rate at which processors can inject messages into the network without causing it to become heavily congested, assuming a uniform distribution of message destinations. In this case, the bandwidth can be expressed as the minimum gap g between successive injections of messages into the network.

Three models that characterize a network in terms of its latency and bandwidth are the Postal model [14], the Bulk-Synchronous Parallel (BSP) model [85], and the LogP model [29]. In the Postal model, a network is described by a single parameter L, its latency. The Bulk-Synchronous Parallel model adds a second parameter g, the minimum ratio of computation steps to communication steps, i.e., the gap. The LogP model includes both of these parameters, and adds a third parameter o, the overhead, or wasted time, incurred by a processor upon sending or receiving a message.

25.2.1.2 Primitive Operations

A machine model must also specify the types of operations that the processors and network are permitted to perform. We assume that all processors are allowed to perform the same local instructions as the single processor in the standard sequential RAM model. In addition, processors may have special instructions for issuing nonlocal memory requests, for sending messages to other processors, and for executing various global operations, such as synchronization. There may also be restrictions on when processors can simultaneously issue instructions involving nonlocal operations. For example a model might not allow two processors to write to the same memory location at the same time. These restrictions might make it impossible to execute an algorithm on a particular model, or make the cost of executing the algorithm prohibitively expensive. It is therefore important to understand what instructions are supported before one can design or analyze a parallel algorithm. In this section we consider three classes of instructions that perform nonlocal operations: (1) instructions that perform concurrent accesses to the same shared memory location, (2) instructions for synchronization, and (3) instructions that perform global operations on data.

When multiple processors simultaneously make a request to read or write to the same resource—such as a processor, memory module, or memory location—there are several possible outcomes. Some machine models simply forbid such operations, declaring that it is an error if two or more processes try to access a resource simultaneously. In this case we say that the model allows only exclusive access to the resource. For example, a PRAM model might only allow exclusive read or write access to each memory location. A PRAM model of this type is called an exclusive-read exclusive-write (**EREW**) PRAM model. Other machine models may allow unlimited access to a shared resource. In this case we say that the model allows concurrent access to the resource. For example, a concurrent-read concurrent-write (**CRCW**) PRAM model allows both concurrent read and write access to memory locations, and a CREW PRAM model allows concurrent reads but only exclusive writes. When making a concurrent write to a resource such as a memory location there are many ways to resolve the conflict. The possibilities include choosing an arbitrary value from those written (arbitrary concurrent write), choosing the value from the processor with lowest index (priority concurrent write), and taking the logical or of the values written. A final choice is to allow for queued access. In this case concurrent access is permitted but the time for a step is proportional to the maximum number of accesses to any resource. A queue-read queue-write (QRQW) PRAM model allows for such accesses [36].

In addition to reads and writes to nonlocal memory or other processors, there are other important primitives that a model might supply. One class of such primitives support synchronization. There are a variety of different types of synchronization operations and the costs of these operations vary from model to model. In a PRAM model, for example, it is assumed that all processors operate in lock step, which provides implicit synchronization. In a local-memory machine model the cost of synchronization may be a function of the particular network topology. A related operation, broadcast, allows one processor to send a common message to all of the other processors. Some machine models supply more powerful primitives that combine arithmetic operations with communication. Such operations include the prefix and **multiprefix** operations, which are defined in Sections 25.4.2 and 25.4.3.

25.2.2 Work-Depth Models

Because there are so many different ways to organize parallel computers, and hence to model them, it is difficult to select one multiprocessor model that is appropriate for all machines. The alternative to focusing on the machine is to focus on the algorithm. In this section we present a class of models called work-depth models. In a work-depth model, the cost of an algorithm is determined by examining the total number of operations that it performs, and the dependencies among those operations. An algorithm's work W is the total number of operations that it performs; its depth D is the longest chain of dependencies among its operations. We call the ratio $\mathcal{P} = W/D$ the parallelism of the algorithm.

The **work-depth models** are more abstract than the multiprocessor models. As we shall see however, algorithms that are efficient in work-depth models can often be translated to algorithms that are efficient in the multiprocessor models, and from there to real parallel computers. The advantage of using a work-depth model is that there are no machine-dependent details to complicate the design and analysis of algorithms. Here we consider three classes of work-depth models: circuit models, vector machine models, and language-based models. We will be using a language-based model in this chapter, so we will return to these models in Section 25.2.5. The most abstract work-depth model is the circuit model. A circuit consists of nodes and directed arcs. A node represents a basic operation, such as adding two values. Each input value for an operation arrives at the corresponding

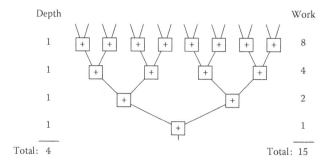

FIGURE 25.4 Summing 16 numbers on a tree. The total depth (longest chain of dependencies) is 4 and the total work (number of operations) is 15.

node via an incoming arc. The result of the operation is then carried out of the node via one or more outgoing arcs. These outgoing arcs may provide inputs to other nodes. The number of incoming arcs to a node is referred to as the fan-in of the node and the number of outgoing arcs is referred to as the fan-out. There are two special classes of arcs. A set of input arcs provide input values to the circuit as a whole. These arcs do not originate at nodes. The output arcs return the final output values produced by the circuit. These arcs do not terminate at nodes. By definition, a circuit is not permitted to contain a directed cycle. In this model, an algorithm is modeled as a family of directed acyclic circuits. There is a circuit for each possible size of the input.

Figure 25.4 shows a circuit for adding 16 numbers. In this figure all arcs are directed toward the bottom. The input arcs are at the top of the figure. Each + node adds the two values that arrive on its two incoming arcs, and places the result on its outgoing arc. The sum of all of the inputs to the circuit is returned on the single output arc at the bottom.

The **work** and **depth** of a circuit are measured as follows. The work is the total number of nodes. The work in Figure 25.4, for example, is 15. (The work is also called the size of the circuit.) The depth is the number of nodes on the longest directed path from an input arc and an output arc. In Figure 25.4, the depth is 4. For a family of circuits, the work and depth are typically parameterized in terms of the number of inputs. For example, the circuit in Figure 25.4 can be easily generalized to add n input values for any n that is a power of two. The work and depth for this family of circuits is $W(n) = n - 1$ and $D(n) = \log_2 n$.

Circuit models have been used for many years to study various theoretical aspects of parallelism, for example to prove that certain problems are difficult to solve in parallel. See [48] for an overview.

In a vector model an algorithm is expressed as a sequence of steps, each of which performs an operation on a vector (i.e., sequence) of input values, and produces a vector result [19,69]. The work of each step is equal to the length of its input (or output) vector. The work of an algorithm is the sum of the work of its steps. The depth of an algorithm is the number of vector steps.

In a language model, a work-depth cost is associated with each programming language construct [20,22]. For example, the work for calling two functions in parallel is equal to the sum of the work of the two calls. The depth, in this case, is equal to the maximum of the depth of the two calls.

25.2.3 Assigning Costs to Algorithms

In the work-depth models, the cost of an algorithm is determined by its work and by its depth. The notions of work and depth can also be defined for the multiprocessor models. The work W performed by an algorithm is equal to the number of processors multiplied by the time required for the algorithm to complete execution. The depth D is equal to the total time required to execute the algorithm.

The depth of an algorithm is important, because there are some applications for which the time to perform a computation is crucial. For example, the results of a weather forecasting program are useful only if the program completes execution before the weather does!

Generally, however, the most important measure of the cost of an algorithm is the work. This can be argued as follows. The cost of a computer is roughly proportional to the number of processors in the computer. The cost for purchasing time on a computer is proportional to the cost of the computer multiplied by the amount of time used. The total cost of performing a computation, therefore, is roughly proportional to the number of processors in the computer multiplied by the amount of time, i.e., the work.

In many instances, the cost of running a computation on a parallel computer may be slightly larger than the cost of running the same computation on a sequential computer. If the time to completion is sufficiently improved, however, this extra cost can often be justified. As we shall see, however, there is often a tradeoff between time-to-completion and total work performed. To quantify when parallel algorithms are efficient in terms of cost, we say that a parallel algorithm is **work-efficient** if asymptotically (as the problem size grows) it requires at most a constant factor more work than the best sequential algorithm known.

25.2.4 Emulations among Models

Although it may appear that a different algorithm must be designed for each of the many parallel models, there are often automatic and efficient techniques for translating algorithms designed for one model into algorithms designed for another. These translations are **work-preserving** in the sense that the work performed by both algorithms is the same, to within a constant factor. For example, the following theorem, known as Brent's Theorem [24], shows that an algorithm designed for the circuit model can be translated in a work-preserving fashion to a PRAM model algorithm.

THEOREM 25.1 *[Brent's Theorem] Any algorithm that can be expressed as a circuit of size (i.e., work) W and depth D and with constant fan-in nodes in the circuit model can be executed in $O(W/P + D)$ steps in the CREW PRAM model.*

PROOF   The basic idea is to have the PRAM emulate the computation specified by the circuit in a level-by-level fashion. The level of a node is defined as follows. A node is on level 1 if all of its inputs are also inputs to the circuit. Inductively, the level of any other node is one greater than the maximum of the level of the nodes that provide its inputs. Let l_i denote the number of nodes on level i. Then, by assigning $\lceil l_i/P \rceil$ operations to each of the P processors in the PRAM, the operations for level i can be performed in $O(\lceil l_i/P \rceil)$ steps. Concurrent reads might be required since many operations on one level might read the same result from a previous level. Summing the time over all D levels, we have

$$T_{PRAM}(W, D, P) = O\left(\sum_{i=1}^{D} \left\lceil \frac{l_i}{P} \right\rceil\right)$$

$$= O\left(\sum_{i=1}^{D} \left(\frac{l_i}{P} + 1\right)\right)$$

$$= O\left(\frac{1}{P}\left(\sum_{i=1}^{D} l_i\right) + D\right)$$

$$= O(W/P + D).$$

The last step is derived by observing that $W = \sum_{i=1}^{D} l_i$, i.e., that the work is equal to the total number of nodes on all of the levels of the circuit.

The total work performed by the PRAM, i.e., the processor-time product, is $O(W + PD)$. This emulation is work-preserving to within a constant factor when the parallelism ($\mathcal{P} = W/D$) is at least as large as the number of processors P, for in this case the work is $O(W)$. The requirement that the parallelism exceed the number of processors is typical of work-preserving emulations.

Brent's Theorem shows that an algorithm designed for one of the work-depth models can be translated in a work-preserving fashion to a multiprocessor model. Another important class of work-preserving translations are those that translate between different multiprocessor models. The translation we consider here is the work-preserving translation of algorithms written for the PRAM model to algorithms for a modular memory machine model that incorporates the feature of network topology. In particular we consider a butterfly machine [55, Chapter 3.6] model in which P processors are attached through a butterfly network of depth $\log P$ to P memory banks. We assume that, in constant time, a processor can hash a virtual memory address to a physical memory bank and an address within that bank using a sufficiently powerful hash function. This scheme was first proposed by Karlin and Upfal [47] for the EREW PRAM model. Ranade [72] later presented a more general approach that allowed the butterfly to efficiently emulate CRCW algorithms.

THEOREM 25.2   *Any algorithm that takes time T on a P-processor PRAM model can be translated into an algorithm that takes time $O(T(P/P' + \log P'))$, with high probability, on a P'-processor butterfly machine model.*

Sketch of proof: Each of the P' processors in the butterfly emulates a set of P/P' PRAM processors. The butterfly emulates the PRAM in a step-by-step fashion. First, each butterfly processor emulates one step of each of its P/P' PRAM processors. Some of the PRAM processors may wish to perform memory accesses. For each memory access, the butterfly processor hashes the memory address to a physical memory bank and an address within the bank, and then routes a message through the network to that bank. These messages are pipelined so that a butterfly processor can have multiple outstanding requests. Ranade proved that if each processor in the P-processor butterfly sends at most P/P' messages, and if the destinations of the messages are determined by a sufficiently powerful random hash function, then the network can deliver all of the messages, along with responses, in $O(P/P' + \log P')$ time. The $\log P'$ term accounts for the latency of the network, and for the fact that there will be some congestion at memory banks, even if each processor sends only a single message.

This theorem implies that the emulation is work preserving when $P \geq P' \log P'$, i.e., when the number of processors employed by the PRAM algorithm exceeds the number of processors in the butterfly by a factor of at least $\log P'$. When translating algorithms from one multiprocessor model (e.g., the PRAM model), which we call the guest model, to another multiprocessor model (e.g., the butterfly machine model), which we call the host model, it is not uncommon to require that the number of guest processors exceed the number of host processors by a factor proportional to the latency of the host. Indeed, the latency of the host can often be hidden by giving it a larger guest to emulate. If the bandwidth of the host is smaller than the bandwidth of a comparably sized guest, however, it is usually much more difficult for the host to perform a work-preserving emulation of the guest.

For more information on PRAM emulations, the reader is referred to [43,86].

25.2.5 Model Used in This Chapter

Because there are so many work-preserving translations between different parallel models of computation, we have the luxury of choosing the model that we feel most clearly illustrates the basic ideas behind the algorithms, a work-depth language model. Here we define the model that we use in this chapter in terms of a set of language constructs and a set of rules for assigning costs to the constructs. The description here is somewhat informal, but should suffice for the purpose of this chapter. The language and costs can be properly formalized using a profiling semantics [22].

Most of the syntax that we use should be familiar to readers who have programmed in Algol-like languages, such as Pascal and C. The constructs for expressing parallelism, however, may be unfamiliar. We will be using two parallel constructs—a parallel apply-to-each construct and a parallel-do construct—and a small set of parallel primitives on sequences (one-dimensional arrays). Our language constructs, syntax and cost rules are loosely based on the NESL language [20].

The apply-to-each construct is used to apply an expression over a sequence of values in parallel. It uses a set-like notation. For example, the expression

$$\{a * a : a \in [3, -4, -9, 5]\}$$

squares each element of the sequence $[3, -4, -9, 5]$ returning the sequence $[9, 16, 81, 25]$. This can be read: "in parallel, for each a in the sequence $[3, -4, -9, 5]$, square a." The apply-to-each construct also provides the ability to subselect elements of a sequence based on a filter. For example

$$\{a * a : a \in [3, -4, -9, 5] | a > 0\}$$

can be read: "in parallel, for each a in the sequence $[3, -4, -9, 5]$ such that a is greater than 0, square a." It returns the sequence $[9,25]$. The elements that remain maintain the relative order from the original sequence.

The parallel-do construct is used to evaluate multiple statements in parallel. It is expressed by listing the set of statements after the keywords **in parallel do**. For example, the following fragment of code calls FUNCTION1 on (X) and assigns the result to A and in parallel calls FUNCTION2 on (Y) and assigns the result to B.

> **in parallel do**
> $\quad A := \text{FUNCTION1}(X)$
> $\quad B := \text{FUNCTION2}(Y)$

The parallel-do completes when all of the parallel subcalls complete.

Work and depth are assigned to our language constructs as follows. The work and depth of a scalar primitive operation is one. For example, the work and depth for evaluating an expression such as $3 + 4$ is one. The work for applying a function to every element in a sequence is equal to the sum of the work for each of the individual applications of the function. For example, the work for evaluating the expression

$$\{a * a : a \in [0..n)\},$$

which creates an n-element sequence consisting of the squares of 0 through $n - 1$, is n. The depth for applying a function to every element in a sequence is equal to the maximum of the depths of the individual applications of the function. Hence, the depth of the previous example is one. The work for a parallel-do construct is equal to the sum of the work for each of its statements. The depth is equal to the maximum depth of its statements. In all other cases, the work and depth for a sequence of operations are equal to the sums of the work and depth for the individual operations.

In addition to the parallelism supplied by apply-to-each, we use four built-in functions on sequences, distribute, ++ (append), flatten, and ← (write,) each of which can be implemented in parallel. The function distribute creates a sequence of identical elements. For example, the expression

$$distribute(3, 5)$$

creates the sequence

$$[3, 3, 3, 3, 3].$$

The ++ function appends two sequences. For example $[2, 1]$++$[5, 0, 3]$ create the sequence $[2, 1, 5, 0, 3]$. The flatten function converts a nested sequence (a sequence in which each element is itself a sequence) into a flat sequence. For example,

$$flatten([[3, 5], [3, 2], [1, 5], [4, 6]])$$

creates the sequence

$$[3, 5, 3, 2, 1, 5, 4, 6].$$

The ← function is used to write multiple elements into a sequence in parallel. It takes two arguments. The first argument is the sequence to modify and the second is a sequence of integer-value pairs that specify what to modify. For each pair (i, v) the value v is inserted into position i of the destination sequence. For example

$$[0, 0, 0, 0, 0, 0, 0, 0] \leftarrow [(4, -2), (2, 5), (5, 9)]$$

inserts the -2, 5 and 9 into the sequence at locations 4, 2, and 5, respectively, returning

$$[0, 0, 5, 0, -2, 9, 0, 0].$$

As in the PRAM model, the issue of concurrent writes arises if an index is repeated. Rather than choosing a single policy for resolving concurrent writes, we will explain the policy used for the individual algorithms. All of these functions have depth one and work n, where n is the size of the sequence(s) involved. In the case of ←, the work is proportional to the length of the sequence of integer-value pairs, not the modified sequence, which might be much longer. In the case of ++, the work is proportional to the length of the second sequence.

We will use a few shorthand notations for specifying sequences. The expression $[-2..1]$ specifies the same sequence as the expression $[-2, -1, 0, 1]$. Changing the left or right bracket surrounding a sequence to a parenthesis omits the first or last elements, e.g., $[-2..1)$ denotes the sequence $[-2, -1, 0]$. The notation $A[i..j]$ denotes the subsequence consisting of elements $A[i]$ through $A[j]$. Similarly, $A[i, j)$ denotes the subsequence $A[i]$ through $A[j-1]$. We will assume that sequence indices are zero based, i.e., $A[0]$ extracts the first element of the sequence A.

Throughout this chapter our algorithms make use of random numbers. These numbers are generated using the functions rand_bit(), which returns a random bit, and rand_int(h), which returns a random integer in the range $[0, h - 1]$.

25.3 Parallel Algorithmic Techniques

As in sequential algorithm design, in parallel algorithm design there are many general techniques that can be used across a variety of problem areas. Some of these are variants of standard sequential techniques, while others are new to parallel algorithms. In this section we introduce some of these techniques, including parallel divide-and-conquer, randomization, and parallel pointer manipulation. We will make use of these techniques in later sections.

25.3.1 Divide-and-Conquer

A divide-and-conquer algorithm first splits the problem to be solved into subproblems that are easier to solve than the original problem, and then solves the subproblems, often recursively. Typically the subproblems can be solved independently. Finally, the algorithm merges the solutions to the subproblems to construct a solution to the original problem.

The divide-and-conquer paradigm improves program modularity, and often leads to simple and efficient algorithms. It has therefore proved to be a powerful tool for sequential algorithm designers. Divide-and-conquer plays an even more prominent role in parallel algorithm design. Because the subproblems created in the first step are typically independent, they can be solved in parallel. Often the subproblems are solved recursively and thus the next divide step yields even more subproblems to be solved in parallel. As a consequence, even divide-and-conquer algorithms that were designed for sequential machines typically have some inherent parallelism. Note however, that in order for divide-and-conquer to yield a highly parallel algorithm, it is often necessary to parallelize the divide step and the merge step. It is also common in parallel algorithms to divide the original problem into as many subproblems as possible, so that they can all be solved in parallel.

As an example of parallel divide-and-conquer, consider the sequential mergesort algorithm. Mergesort takes a set of n keys as input and returns the keys in sorted order. It works by splitting the keys into two sets of $n/2$ keys, recursively sorting each set, and then merging the two sorted sequences of $n/2$ keys into a sorted sequence of n keys. To analyze the sequential running time of mergesort we note that two sorted sequences of $n/2$ keys can be merged in $O(n)$ time. Hence the running time can be specified by the recurrence

$$T(n) = \begin{cases} 2T(n/2) + O(n) & n > 1 \\ O(1) & n = 1 \end{cases} \tag{25.1}$$

which has the solution $T(n) = O(n \log n)$. Although not designed as a parallel algorithm, mergesort has some inherent parallelism since the two recursive calls are independent, thus allowing them to be made in parallel. The parallel calls can be expressed as

ALGORITHM: MERGESORT(A)
1   **if** ($|A| = 1$) **then return** A
2   **else**
3   **in parallel do**
4           $L :=$ MERGESORT($A[0..|A|/2)$)
5           $R :=$ MERGESORT($A[|A|/2..|A|)$)
6   **return** MERGE(L, R)

Recall that in our work-depth model we can analyze the depth of an algorithm that makes parallel calls by taking the maximum depth of the two calls, and the work by taking the sum of the work of the two calls. We assume that the merging remains sequential so that the work and depth to merge two sorted sequences of $n/2$ keys is $O(n)$. Thus for mergesort the work and depth are given by the recurrences

$$W(n) = 2W(n/2) + O(n) \tag{25.2}$$

$$D(n) = \max(D(n/2), D(n/2)) + O(n) \tag{25.3}$$

$$= D(n/2) + O(n) \tag{25.4}$$

As expected, the solution for the work is $W(n) = O(n \log n)$, i.e., the same as the time for the sequential algorithm. For the depth, however, the solution is $D(n) = O(n)$, which is smaller than the work. Recall that we defined the parallelism of an algorithm as the ratio of the work to the depth. Hence, the parallelism of this algorithm is $O(\log n)$ (not very much). The problem here is that the merge step remains sequential, and is the bottleneck.

As mentioned earlier, the parallelism in a divide-and-conquer algorithm can often be enhanced by parallelizing the divide step and/or the merge step. Using a parallel merge [52], two sorted sequences of $n/2$ keys can be merged with work $O(n)$ and depth $O(\log \log n)$. Using this merge algorithm, the recurrence for the depth of mergesort becomes

$$D(n) = D(n/2) + O(\log \log n) \qquad (25.5)$$

which has solution $D(n) = O(\log n \log \log n)$. Using a technique called **pipelined divide-and-conquer** the depth of mergesort can be further reduced to $O(\log n)$ [26]. The idea is to start the merge at the top level before the recursive calls complete.

Divide-and-conquer has proven to be one of the most powerful techniques for solving problems in parallel. In this chapter, we will use it to solve problems from computational geometry, for sorting, and for performing fast Fourier transforms (FFT). Other applications range from solving linear systems to factoring large numbers to performing n-body simulations.

25.3.2 Randomization

Random numbers are used in parallel algorithms to ensure that processors can make local decisions that, with high probability, add up to good global decisions. Here we consider three uses of randomness.

25.3.2.1 Sampling

One use of randomness is to select a representative sample from a set of elements. Often, a problem can be solved by selecting a sample, solving the problem on that sample, and then using the solution for the sample to guide the solution for the original set. For example, suppose we want to sort a collection of integer keys. This can be accomplished by partitioning the keys into buckets and then sorting within each bucket. For this to work well, the buckets must represent nonoverlapping intervals of integer values, and each bucket must contain approximately the same number of keys. **Random sampling** is used to determine the boundaries of the intervals. First each processor selects a random sample of its keys. Next all of the selected keys are sorted together. Finally these keys are used as the boundaries. Such random sampling is also used in many parallel computational geometry, graph, and string matching algorithms.

25.3.2.2 Symmetry Breaking

Another use of randomness is in **symmetry breaking**. For example, consider the problem of selecting a large independent set of vertices in a graph in parallel. (A set of vertices is independent if no two are neighbors.) Imagine that each vertex must decide, in parallel with all other vertices, whether to join the set or not. Hence, if one vertex chooses to join the set, then all of its neighbors must choose not to join the set. The choice is difficult to make simultaneously for each vertex if the local structure at each vertex is the same, for example if each vertex has the same number of neighbors. As it turns out, the impasse can be resolved by using randomness to break the symmetry between the vertices [58].

25.3.2.3 Load Balancing

A third use of randomness is load balancing. One way to quickly partition a large number of data items into a collection of approximately evenly sized subsets is to randomly assign each element to a subset. This technique works best when the average size of a subset is at least logarithmic in the size of the original set.

25.3.3 Parallel Pointer Techniques

Many of the traditional sequential techniques for manipulating lists, trees, and graphs do not translate easily into parallel techniques. For example, techniques such as traversing the elements of a linked list, visiting the nodes of a tree in postorder, or performing a depth-first traversal of a graph appear to be inherently sequential. Fortunately these techniques can often be replaced by parallel techniques with roughly the same power.

25.3.3.1 Pointer Jumping

One of the oldest parallel pointer techniques is **pointer jumping** [88]. This technique can be applied to either lists or trees. In each pointer jumping step, each node in parallel replaces its pointer with that of its successor (or parent). For example, one way to label each node of an n-node list (or tree) with the label of the last node (or root) is to use pointer jumping. After at most $\lceil \log n \rceil$ steps, every node points to the same node, the end of the list (or root of the tree). This is described in more detail in Section 25.4.4.

25.3.3.2 Euler Tour Technique

An Euler tour of a directed graph is a path through the graph in which every edge is traversed exactly once. In an undirected graph each edge is typically replaced by two oppositely directed edges. The Euler tour of an undirected tree follows the perimeter of the tree visiting each edge twice, once on the way down and once on the way up. By keeping a linked structure that represents the Euler tour of a tree it is possible to compute many functions on the tree, such as the size of each subtree [83]. This technique uses linear work, and parallel depth that is independent of the depth of the tree. The Euler tour technique can often be used to replace a standard traversal of a tree, such as a depth-first traversal.

25.3.3.3 Graph Contraction

Graph contraction is an operation in which a graph is reduced in size while maintaining some of its original structure. Typically, after performing a graph contraction operation, the problem is solved recursively on the contracted graph. The solution to the problem on the contracted graph is then used to form the final solution. For example, one way to partition a graph into its connected components is to first contract the graph by merging some of the vertices with neighboring vertices, then find the connected components of the contracted graph, and finally undo the contraction operation. Many problems can be solved by contracting trees [64,65], in which case the technique is called **tree contraction**. More examples of graph contraction can be found in Section 25.5.

25.3.3.4 Ear Decomposition

An ear decomposition of a graph is a partition of its edges into an ordered collection of paths. The first path is a cycle, and the others are called ears. The end-points of each ear are anchored on previous paths. Once an ear decomposition of a graph is found, it is not difficult to determine if two edges lie on a common cycle. This information can be used in algorithms for determining biconnectivity, triconnectivity, 4-connectivity, and planarity [60,63]. An ear decomposition can be found in parallel using linear work and logarithmic depth, independent of the structure of the graph. Hence, this technique can be used to replace the standard sequential technique for solving these problems, depth-first search.

25.3.4 Other Techniques

Many other techniques have proven to be useful in the design of parallel algorithms. Finding small graph separators is useful for partitioning data among processors to reduce communication [75, Chapter 14]. Hashing is useful for load balancing and mapping addresses to memory [47,87]. Iterative techniques are useful as a replacement for direct methods for solving linear systems [18].

25.4 Basic Operations on Sequences, Lists, and Trees

We begin our presentation of parallel algorithms with a collection of algorithms for performing basic operations on sequences, lists, and trees. These operations will be used as subroutines in the algorithms that follow in later sections.

25.4.1 Sums

As explained near the beginning of this chapter, there is a simple recursive algorithm for computing the sum of the elements in an array.

ALGORITHM: SUM(A)
1   **if** $|A| = 1$ **then return** $A[0]$
2   **else return** SUM($\{A[2i] + A[2i + 1] : i \in [0..|A|/2)\}$)

The work and depth for this algorithm are given by the recurrences

$$W(n) = W(n/2) + O(n) \tag{25.6}$$

$$D(n) = D(n/2) + O(1) \tag{25.7}$$

which have solutions $W(n) = O(n)$ and $D(n) = O(\log n)$. This algorithm can also be expressed without recursion (using a **while** loop), but the recursive version foreshadows the recursive algorithm for the SCAN function.

As written, the algorithm only works on sequences that have lengths equal to powers of 2. Removing this restriction is not difficult by checking if the sequence is of odd length and separately adding the last element in if it is. This algorithm can also easily be modified to compute the "sum" using any other binary associative operator in place of $+$. For example the use of max would return the maximum value in the sequence.

25.4.2 Scans

The plus-scan operation (also called all-prefix-sums) takes a sequence of values and returns a sequence of equal length for which each element is the sum of all previous elements in the original sequence. For example, executing a plus-scan on the sequence $[3, 5, 3, 1, 6]$ returns $[0, 3, 8, 11, 12]$. An algorithm for performing the scan operation [81] is shown below.

ALGORITHM: SCAN(A)
1   **if** $|A| = 1$ **then return** $[0]$
2   **else**
3       $S = $ SCAN($\{A[2i] + A[2i + 1] : i \in [0..|A|/2)\}$)
4       $R = \{$**if** $(i \bmod 2) = 0$ **then** $S[i/2]$ **else** $S[(i - 1)/2] + A[i - 1] : i \in [0..|A|)\}$
5   **return** R

The algorithm works by element-wise adding the even-indexed elements of A to the odd-indexed elements of A, and then recursively solving the problem on the resulting sequence (Line 3). The result S of the recursive call gives the plus-scan values for the even positions in the output sequence R. The value for each of the odd positions in R is simply the value for the preceding even position in R plus the value of the preceding position from A.

The asymptotic work and depth costs of this algorithm are the same as for the SUM operation, $W(n) = O(n)$ and $D(n) = O(\log n)$. Also, as with the SUM operation, any binary associative operator can be used in place of the $+$. In fact the algorithm described can be used more generally to solve various recurrences, such as the first-order linear recurrences $x_i = (x_{i-1} \otimes a_{i-1}) \oplus b_{i-1}, 0 < i < n$, where $\otimes$ and $\oplus$ are both binary associative operators [51].

Scans have proven so useful in the design of parallel algorithms that some parallel machines provide support for scan operations in hardware.

25.4.3 Multiprefix and Fetch-and-Add

The multiprefix operation is a generalization of the scan operation in which multiple independent scans are performed. The input to the multiprefix operation is a sequence A of n pairs (k, a), where k specifies a key and a specifies an integer data value. For each key value, the multiprefix operation performs an independent scan. The output is a sequence B of n integers containing the results of each of the scans such that if $A[i] = (k, a)$ then

$$B[i] = sum(\{b : (t, b) \in A[0..i) \mid t = k\})$$

In other words, each position receives the sum of all previous elements that have the same key. As an example,

$$\text{MULTIPREFIX}([(1, 5), (0, 2), (0, 3), (1, 4), (0, 1), (2, 2)])$$

returns the sequence

$$[0, 0, 2, 5, 5, 0]$$

The fetch-and-add operation is a weaker version of the multiprefix operation, in which the order of the input elements for each scan is not necessarily the same as the order in the input sequence A. In this chapter we do not present an algorithm for the multiprefix operation, but it can be solved by a function that requires work $O(n)$ and depth $O(\log n)$ using concurrent writes [61].

25.4.4 Pointer Jumping

Pointer jumping is a technique that can be applied to both linked lists and trees [88]. The basic pointer jumping operation is simple. Each node i replaces its pointer $P[i]$ with the pointer of the node that it points to, $P[P[i]]$. By repeating this operation, it is possible to compute, for each node in a list or tree, a pointer to the end of the list or root of the tree. Given a set P of pointers that represent a tree (i.e., pointers from children to parents), the following code will generate a pointer from each node to the root of the tree. We assume that the root points to itself.

ALGORITHM: POINT_TO_ROOT(P)
1   for j from 1 to $\lceil \log |P| \rceil$
2      $P := \{P[P[i]] : i \in [0..|P|)\}$

The idea behind this algorithm is that in each loop iteration the distance spanned by each pointer, with respect to the original tree, will double, until it points to the root. Since a tree constructed from $n = |P|$ pointers has depth at most $n - 1$, after $\lceil \log n \rceil$ iterations each pointer will point to the

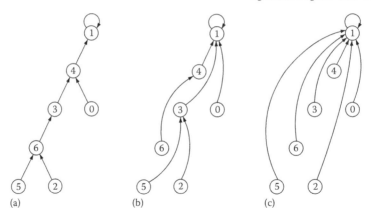

FIGURE 25.5 The effect of two iterations of algorithm POINT_TO_ROOT. (a) The input tree $P = [4, 1, 6, 4, 1, 6, 3]$. (b) The tree $P = [1, 1, 3, 1, 1, 3, 4]$ after one iteration of the algorithm. (c) The final tree $P = [1, 1, 1, 1, 1, 1, 1]$.

root. Because each iteration has constant depth and performs $\Theta(n)$ work, the algorithm has depth $\Theta(\log n)$ and work $\Theta(n \log n)$.

Figure 25.5 illustrates algorithm POINT_TO_ROOT applied to a tree consisting of seven nodes.

25.4.5 List Ranking

The problem of computing the distance from each node to the end of a linked list is called list ranking. Function POINT_TO_ROOT can be easily modified to compute these distances, as shown below.

ALGORITHM: LIST_RANK(P)
1    $V = \{\textbf{if } P[i] = i \textbf{ then } 0 \textbf{ else } 1 : i \in [0..|P|)\}$
2   **for** j **from** 1 **to** $\lceil \log |P| \rceil$
3        $V := \{V[i] + V[P[i]] : i \in [0..|P|)\}$
4        $P := \{P[P[i]] : i \in [0..|P|)\}$
5   **return** V

In this function, $V[i]$ can be thought of as the distance spanned by pointer $P[i]$ with respect to the original list. Line 1 initializes V by setting $V[i]$ to 0 if i is the last node (i.e., points to itself), and 1 otherwise. In each iteration, Line 3 calculates the new length of $P[i]$. The function has depth $\Theta(\log n)$ and work $\Theta(n \log n)$.

It is worth noting that there are simple sequential algorithms that perform the same tasks as both functions POINT_TO_ROOT and LIST_RANK using only $O(n)$ work. For example, the list ranking problem can be solved by making two passes through the list. The goal of the first pass is simply to count the number of elements in the list. The elements can then be numbered with their positions from the end of the list in a second pass. Thus, neither function POINT_TO_ROOT nor LIST_RANK are work-efficient, since both require $\Theta(n \log n)$ work in the worst case. There are, however, several work-efficient parallel solutions to both of these problems.

The following parallel algorithm uses the technique of random sampling to construct a pointer from each node to the end of a list of n nodes in a work-efficient fashion [74]. The algorithm is easily generalized to solve the list-ranking problem.

1. Pick m list nodes at random and call them the start nodes.
2. From each start node u, follow the list until reaching the next start node v. Call the list nodes between u and v the sublist of u.
3. Form a shorter list consisting only of the start nodes and the final node on the list by making each start node point to the next start node on the list.
4. Using pointer jumping on the shorter list, for each start node create a pointer to the last node in the list.
5. For each start node u, distribute the pointer to the end of the list to all of the nodes in the sublist of u.

The key to analyzing the work and depth of this algorithm is to bound the length of the longest sublist. Using elementary probability theory, it is not difficult to prove that the expected length of the longest sublist is at most $O((n \log m)/m)$. The work and depth for each step of the algorithm are thus computed as follows.

1. $W(n, m) = O(m)$ and $D(n, m) = O(1)$
2. $W(n, m) = O(n)$ and $D(n, m) = O((n \log m)/m)$
3. $W(n, m) = O(m)$ and $D(n, m) = O(1)$
4. $W(n, m) = O(m \log m)$ and $D(n, m) = O(\log m)$
5. $W(n, m) = O(n)$ and $D(n, m) = O((n \log m)/m)$

Thus, the work for the entire algorithm is $W(m, n) = O(n + m \log m)$, and the depth is $O((n \log m)/m)$. If we set $m = n/\log n$, these reduce to $W(n) = O(n)$ and $D(n) = O(\log^2 n)$.

Using a technique called contraction, it is possible to design a list ranking algorithm that runs in $O(n)$ work and $O(\log n)$ depth [8,9]. This technique can also be applied to trees [64,65].

25.4.6 Removing Duplicates

This section presents several algorithms for removing the duplicate items that appear in a sequence. Thus, the input to each algorithm is a sequence, and the output is a new sequence containing exactly one copy of every item that appears in the input sequence. It is assumed that the order of the items in the output sequence does not matter. Such an algorithm is useful when a sequence is used to represent an unordered set of items. Two sets can be merged, for example, by first appending their corresponding sequences, and then removing the duplicate items.

25.4.6.1 Approach 1: Using an Array of Flags

If the items are all nonnegative integers drawn from a small range, we can use a technique similar to bucket sort to remove the duplicates. We begin by creating an array equal in size to the range, and initializing all of its elements to 0. Next, using concurrent writes we set a flag in the array for each number that appears in the input list. Finally, we extract those numbers with flags that have been set. This algorithm is expressed as follows.

ALGORITHM: REM_DUPLICATES(V)
```
1   RANGE := 1 + MAX(V)
2   FLAGS := distribute (0,RANGE) ← {(i, 1) : i ∈ V}
3   return {j : j ∈ [0..RANGE]| FLAGS[j] = 1}
```

This algorithm has depth $O(1)$ and performs work $O(|V| + \text{MAX}(V))$. Its obvious disadvantage is that it explodes when given a large range of numbers, both in memory and in work.

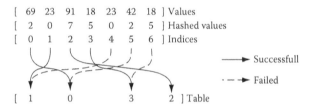

FIGURE 25.6 Each key attempts to write its index into a hash table entry.

25.4.6.2 Approach 2: Hashing

A more general approach is to use a hash table. The algorithm has the following outline. The algorithm first creates a hash table that contains a prime number of entries, where the prime is approximately twice as large as the number of items in the set V. A prime size is best, because it makes designing a good hash function easier. The size must also be large enough that the chance of collisions in the hash table is not too great. Let m denote the size of the hash table. Next, the algorithm computes a hash value $hash(V[j], m)$ for each item $V[j] \in V$, and attempts to write the index j into the hash table entry $hash(V[j], m)$. For example, Figure 25.6 describes a particular hash function applied to the sequence [69,23,91,18,42,23,18]. We assume that if multiple values are simultaneously written into the same memory location, one of the values will be correctly written (the arbitrary concurrent write model). An item $V[j]$ is called a winner if the index j is successfully written into the hash table. In our example, the winners are $V[0]$, $V[1]$, $V[2]$, and $V[3]$, i.e., 69, 23, 91, and 18. The winners are added to the duplicate-free sequence that is being constructed, and then set aside. Among the losers, we must distinguish between two types of items, those that were defeated by an item with the same value, and those that were defeated by an item with a different value. In our example, $V[5]$ and $V[6]$ (23 and 18) were defeated by items with the same value, and $V[4]$ (42) was defeated by an item with a different value. Items of the first type are set aside because they are duplicates. Items of the second type are retained, and the algorithm repeats the entire process on them using a different hash function. In general, it may take several iterations before all of the items have been set aside, and in each iteration the algorithm must use a different hash function.

The code for removing duplicates using hashing is shown below.

ALGORITHM: REMOVE_DUPLICATES(V)
```
1   m := NEXT_PRIME(2 * |V|)
2   TABLE := distribute(-1, m)
3   i := 0
4   RESULT := {}
5   while |V| > 0
6       TABLE := TABLE ← {(hash(V[j], m, i), j) : j ∈ [0..|V|)}
7       WINNERS := {V[j] : j ∈ [0..|V|) | TABLE[hash(V[j], m, i)] = j}
8       RESULT := RESULT ++ WINNERS
9       TABLE := TABLE ← {(hash(k, m, i), k) : k ∈ WINNERS}
10      V := {k ∈ V | TABLE[hash(k, m, i)] ≠ k}
11      i := i + 1
12  return RESULT
```

The first four lines of function REMOVE_DUPLICATES initialize several variables. Line 1 finds the first prime number larger than $2 * |V|$ using the built-in function NEXT_PRIME. Line 2 creates the hash table, and initializes its entries with an arbitrary value (-1). Line 3 initializes i, a variable that simply counts iterations of the while loop. Line 4 initializes the sequence RESULT to be empty. Ultimately, RESULT will contain a single copy of each distinct item from the sequence V.

The bulk of the work in function REMOVE_DUPLICATES is performed by the while loop. While there are items remaining to be processed, the code performs the following steps. In Line 6, each item $V[j]$ attempts to write its index j into the table entry given by the hash function $hash(V[j], m, i)$. Note that the hash function takes the iteration i as an argument, so that a different hash function is used in each iteration. Concurrent writes are used so that if several items attempt to write to the same entry, precisely one will win. Line 7 determines which items successfully wrote indices in Line 6, and stores the values of these items in an array called WINNERS. The winners are added to the RESULT in Line 8. The purpose of Lines 9 and 10 is to remove all of the items that are either winners or duplicates of winners. These lines reuse the hash table. In Line 9, each winner writes its value, rather than its index, into the hash table. In this step there are no concurrent writes. Finally, in Line 10, an item is retained only if it is not a winner, and the item that defeated it has a different value.

It is not difficult to prove that, provided that the hash values $hash(V[j], m, i)$ are random and sufficiently independent, both between iterations and within an iteration, each iteration reduces the number of items remaining by some constant fraction until the number of items remaining is small. As a consequence, $D(n) = O(\log n)$ and $W(n) = O(n)$.

The remove-duplicates algorithm is frequently used for set operations. For instance, given the code for REMOVE_DUPLICATES, it is easy to write the code for the set union operation.

25.5 Graphs

Graph problems are often difficult to parallelize since many standard sequential graph techniques, such as depth-first or priority-first search, do not parallelize well. For some problems, such as minimum-spanning tree and biconnected components, new techniques have been developed to generate efficient parallel algorithms. For other problems, such as single-source shortest paths, there are no known efficient parallel algorithms, at least not for the general case.

We have already outlined some of the parallel graph techniques in Section 25.3. In this section we describe algorithms for breadth-first search (BFS), for finding connected components, and for finding minimum spanning trees. These algorithms use some of the general techniques. In particular, randomization and graph contraction will play an important role in the algorithms. In this chapter we limit ourselves to algorithms on sparse undirected graphs. We suggest the following sources for further information on parallel graph algorithms [75, Chapters 2–8], [45, Chapter 5], [35, Chapter 2].

25.5.1 Graphs and Graph Representations

A graph $G = (V, E)$ consists of a set of vertices V and a set of edges E in which each edge connects two vertices. In a directed graph each edge is directed from one vertex to another, while in an undirected graph each edge is symmetric, i.e., goes in both directions. A weighted graph is a graph in which each edge $e \in E$ has a weight $w(e)$ associated with it. In this chapter we use the convention that $n = |V|$ and $m = |E|$. Qualitatively, a graph is considered sparse if m is much less than n^2 and dense otherwise. The diameter of a graph, denoted $D(G)$, is the maximum, over all pairs of vertices (u, v), of the minimum number of edges that must be traversed to get from u to v.

There are three standard representations of graphs used in sequential algorithms: edge lists, adjacency lists, and adjacency matrices. An edge list consists of a list of edges, each of which is a pair of vertices. The list directly represents the set E. An adjacency list is an array of lists. Each array element corresponds to one vertex and contains a linked list of the indices of the neighboring vertices, i.e., the linked list for a vertex v contains the indices of the vertices $\{u|(v, u) \in E\}$). An adjacency matrix is an $n \times n$ array A such that A_{ij} is 1 if $(i, j) \in E$ and 0 otherwise. The adjacency

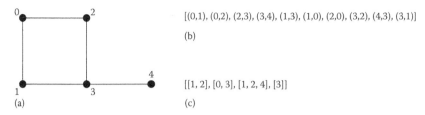

[(0,1), (0,2), (2,3), (3,4), (1,3), (1,0), (2,0), (3,2), (4,3), (3,1)]

(b)

[[1, 2], [0, 3], [1, 2, 4], [3]]

(c)

FIGURE 25.7 Representations of an undirected graph: (a) a graph G with 5 vertices and 5 edges; (b) an edge-list representation of G; and (c) the adjacency-list representation of G. Values between square brackets are elements of an array, and values between parentheses are elements of a pair.

matrix representation is typically used only when the graph is dense since it requires $\Theta(n^2)$ space, as opposed to $\Theta(n + m)$ space for the other two representations. Each of these representations can be used to represent either directed or undirected graphs.

For parallel algorithms we use similar representations for graphs. The main change we make is to replace the linked lists with arrays. In particular the edge-list is represented as an array of edges and the adjacency-list is represented as an array of arrays. Using arrays instead of lists makes it easier to process the graph in parallel. In particular, they make it easy to grab a set of elements in parallel, rather than having to follow a list. Figure 25.7 shows an example of our representations for an undirected graph. Note that for the edge-list representation of the undirected graph each edge appears twice, once in each direction (this property is important for some of the algorithms described in this chapter[*]). To represent a directed graph we simply only store the edge once in the desired direction. In the text we will refer to the left element of an edge pair as the source vertex and the right element as the destination vertex.

In designing algorithms, it is sometimes more efficient to use an edge list and sometimes more efficient to use an adjacency list. It is therefore important to be able to convert between the two representations. To convert from an adjacency list to an edge list (representation (c) to representation (b) in Figure 25.7) is straightforward. The following code will do it with linear work and constant depth:

$$flatten(\{\{(i, j) : j \in G[i]\} : i \in [0..|G|]\})$$

where G is the graph in the adjacency list representation. For each vertex i this code pairs up i with each of i's neighbors. The flatten is used since the nested apply-to-each will return a sequence of sequences that needs to be flattened into a single sequence.

To convert from an edge list to an adjacency list is somewhat more involved, but still requires only linear work. The basic idea is to sort the edges based on the source vertex. This places edges from a particular vertex in consecutive positions in the resulting array. This array can then be partitioned into blocks based on the source vertices. It turns out that since the sorting is on integers in the range $[0..|V|]$, a radix sort can be used (see Section 25.6.2), which requires linear work. The depth of the radix sort depends on the depth of the multiprefix operation (see Section 25.4.3).

25.5.2 Breadth-First Search

The first algorithm we consider is parallel BFS. BFS can be used to solve problems such as determining if a graph is connected or generating a spanning tree of a graph. Parallel BFS is similar to the sequential version, which starts with a source vertex s and visits levels of the graph one after the other using a

---

[*] If space is of serious concern, the algorithms can be easily modified to work with edges stored in just one direction.

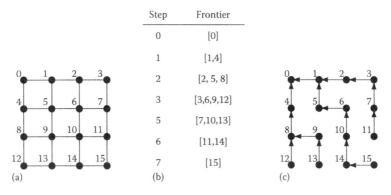

FIGURE 25.8 Example of parallel BFS: (a) a graph G; (b) the frontier at each step of the BFS of G with $s = 0$; and (c) a BFS tree.

queue to keep track of vertices that have not yet been visited. The main difference is that each level is going to be visited in parallel and no queue is required. As with the sequential algorithm each vertex will only be visited once and each edge at most twice, once in each direction. The work is therefore linear in the size of the graph, $O(n + m)$. For a graph with diameter D, the number of levels visited by the algorithm will be at least $D/2$ and at most D, depending on where the search is initiated. We will show that each level can be visited in constant depth, assuming a concurrent write model, so that the total depth of parallel BFS is $O(D)$.

The main idea of parallel BFS is to maintain a set of frontier vertices that represent the current level being visited, and to produce a new frontier on each step. The set of frontier vertices is initialized with the singleton s (the source vertex). A new frontier is generated by collecting all of the neighbors of the current frontier vertices in parallel and removing any that have already been visited. This is not sufficient on its own, however, since multiple vertices might collect the same unvisited vertex. For example, consider the graph in Figure 25.8. On step 2 vertices 5 and 8 will both collect vertex 9. The vertex will therefore appear twice in the new frontier. If the duplicate vertices are not removed the algorithm can generate an exponential number of vertices in the frontier. This problem does not occur in the sequential BFS because vertices are visited one at a time. The parallel version therefore requires an extra step to remove duplicates.

The following function performs a parallel BFS. It takes as input a source vertex s and a graph G represented as an adjacency-array, and returns as its result a BFS tree of G. In a BFS tree each vertex visited at level i points to one of its neighbors visited at level $i - 1$ (see Figure 25.8c). The source s is the root of the tree.

ALGORITHM: BFS(s, G)
```
1   FRONT := [s]
2   TREE := distribute(−1, |G|)
3   TREE[s] := s
4   while (|FRONT| ≠ 0)
5       E := flatten({{(u, v) : u ∈ G[v]} : v ∈ FRONT})
6       E′ := {(u, v) ∈ E | TREE[u] = −1}
7       TREE := TREE ← E′
8       FRONT := {u : (u, v) ∈ E′ | v = TREE[u]}
9   return TREE
```

In this code FRONT is the set of frontier vertices, and TREE is the current BFS tree, represented as an array of indices (pointers). The pointers (indices) in TREE are all initialized to -1, except for the source s which is initialized to point to itself. Each vertex in TREE is set to point to its parent in the BFS tree when it is visited. The algorithm assumes the arbitrary concurrent write model.

We now consider each iteration of the algorithm. The iterations terminate when there are no more vertices in the frontier (Line 4). The new frontier is generated by first collecting into an edge-array the set of edges from current frontier vertices to the neighbors of these vertices (Line 5). An edge from v to u is kept as the pair (u, v) (this is backwards from the standard edge representation and is used below to write from v to u). Next, the algorithm subselects the edges that lead to unvisited vertices (Line 6). Now for each remaining edge (u, v) the algorithm writes the source index v into the BFS tree entry for the destination vertex u (Line 7). In the case that more than one edge has the same destination, one of the source indices will be written arbitrarily—this is the only place that the algorithm uses a concurrent write. These indices become the parent pointers for the BFS tree, and are also used to remove duplicates for the next frontier set. In particular, the algorithm checks whether each edge succeeded in writing its source by reading back from the destination. If an edge reads the value it wrote, its destination is included in the new frontier (Line 8). Since only one edge that points to a given destination vertex will read back the same value, no duplicates will appear in the new frontier.

The algorithm requires only constant depth per iteration of the while loop. Since each vertex and its associated edges are visited only once, the total work is $O(m + n)$. An interesting aspect of this parallel BFS is that it can generate BFS trees that cannot be generated by a sequential BFS, even allowing for any order of visiting neighbors in sequential BFS. We leave the generation of an example as an exercise. We note, however, that if the algorithm used a priority concurrent write (see Section 25.2.5) on Line 7, then it would generate the same tree as a sequential BFS.

25.5.3   Connected Components

We now consider the problem of labeling the connected components of an undirected graph. The problem is to label all the vertices in a graph G such that two vertices u and v have the same label if and only if there is a path between the two vertices. Sequentially the connected components of a graph can easily be labeled using either depth-first or breadth-first search. We have seen how to perform a BFS, but the technique requires depth proportional to the diameter of a graph. This is fine for graphs with small diameter, but does not work well in the general case. Unfortunately, in terms of work, even the most efficient polylogarithmic-depth parallel algorithms for depth-first search and BFS are very inefficient. Hence, the efficient algorithms for solving the connected components problem use different techniques.

The two algorithms we consider are based on graph contraction. Graph contraction works by contracting the vertices of a connected subgraph into a single vertex. The techniques we use allow the algorithms to make many such contractions in parallel across the graph. The algorithms therefore proceed in a sequence of steps, each of which contracts a set of subgraphs, and forms a smaller graph in which each subgraph has been converted into a vertex. If each such step of the algorithm contracts the size of the graph by a constant fraction, then each component will contract down to a single vertex in $O(\log n)$ steps. By running the contraction in reverse, the algorithms can label all the vertices in the components. The two algorithms we consider differ in how they select subgraphs for contraction. The first uses randomization and the second is deterministic. Neither algorithm is work efficient because they require $O((n + m) \log n)$ work for worst-case graphs, but we briefly discuss how they can be made work efficient in Section 25.5.3.3. Both algorithms require the concurrent write model.

25.5.3.1 Random-Mate Graph Contraction

The random-mate technique for graph contraction is based on forming a set of star subgraphs and contracting the stars. A star is a tree of depth one—it consists of a root and an arbitrary number of children. The random-mate algorithm finds a set of nonoverlapping stars in a graph, and then contracts each star into a single vertex by merging each child into its parent. The technique used to form the stars uses randomization. For each vertex the algorithm flips a coin to decide if that vertex is a parent or a child. We assume the coin is unbiased so that every vertex has a 50% probability of being a parent. Now for each child vertex the algorithm selects a neighboring parent vertex and makes that parent the child's root. If the child has no neighboring parent, it has no root. The parents are now the roots of a set of stars, each with zero or more children. The children either belong to one of these stars or are left on their own (if they had no neighboring parents). The algorithm now contracts these stars. When contracting, the algorithm updates any edge that points to a child of a star to point to the root. Figure 25.9 illustrates a full contraction step. This contraction step is repeated until all components are of size 1.

To analyze the number of contraction steps required to complete we need to know how many vertices the algorithm removes on each contraction step. First we note that a contraction step is only going to remove children, and only if they have a neighboring parent. The probability that a vertex will be deleted is therefore the probability that a vertex is a child multiplied by the probability that at least one of its neighbors is a parent. The probability that it is a child is 1/2 and the probability that at least one neighbor is a parent is at least 1/2 (every vertex that is not fully contracted has one or more neighbors). The algorithm is therefore expected to remove at least 1/4 of the remaining vertices at each step, and since this is a constant fraction, it is expected to complete in $O(\log n)$ steps. The full probabilistic analysis is somewhat involved since it is possible to have a streak of bad flips, but it is not too hard to show that the algorithm is very unlikely to require more than $O(\log n)$ contraction steps.

The following algorithm uses random-mate contraction for labeling the connected components of a graph. The algorithm works by contracting until each component is a single vertex and then expanding so that it can label all vertices in that component with the same label. The input to the algorithm is a graph G in the edge-list representation (note that this is a different representation than used in BFS), along with the labels of the vertices. The labels of the vertices are assumed to be initialized to be unique indices in the range $0..|V| - 1$. The output of the algorithm is a label for each vertex such that two vertices will have the same label if and only if they belong to the same component. In fact, the label of each vertex will be the original label of one of the vertices in the component.

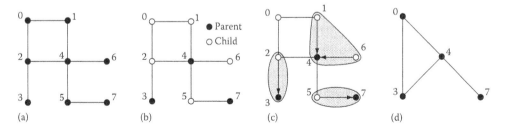

FIGURE 25.9 Example of one step of random-mate graph contraction. (a) The original graph G. (b) G after selecting the parents randomly. (c) The stars formed after each child is assigned a neighboring parent as its root (each star is shaded). (d) The graph after contracting each star and relabeling the edges to point to the roots. Children with no neighboring parent remain in the graph.

ALGORITHM: CC_RANDOM_MATE(LABELS,E)
1   **if** ($|E| = 0$) **then return** LABELS
2   **else**
3   CHILD := $\{rand\_bit() : v \in [1..n]\}$
4   HOOKS := $\{(u, v) \in E \mid \text{CHILD}[u] \textbf{ and } \wedge\neg \text{ CHILD}[v]\}$
5   LABELS := LABELS $\leftarrow$ HOOKS
6         $E' := \{(\text{LABELS}[u], \text{LABELS}[v]) : (u, v) \in E \mid \text{LABELS}[u] \neq \text{LABELS}[v]\}$
7   LABELS$'$:= CC_RANDOM_MATE(LABELS,E')
8   LABELS$'$:= LABELS$'$ $\leftarrow \{(u, \text{LABELS}' [v]) : (u, v) \in \text{HOOKS}\}$
9   **return** LABELS$'$

The algorithm works recursively by (a) executing one random-mate contraction step, (b) recursively applying itself to the contracted graph, and (c) expanding the graph by passing the labels from each root of a contracted star [from step (a)] to its children. The graph is therefore contracted while going down the recursion and expanded while coming back up. The termination condition is that there are no remaining edges (Line 1). To form stars for the contraction step the algorithm flips a coin for each vertex (Line 3) and subselects all edges HOOKS that go from a child to a parent (Line 4). We call these edges *hook edges* and they represent a superset of the star edges (each child can have multiple hook edges, but only one root in a star). For each hook edge the algorithm writes the parent's label into the child's label (Line 5). If a child has multiple neighboring parents, then one of the parent's labels is written arbitrarily—we assume an arbitrary concurrent write. At this point each child is labeled with one of its neighboring parents, if it has one. The algorithm now updates each edge by reading the labels from its two endpoints and using these as its new endpoints (Line 6). In the same step, the algorithm removes any edges that are within the same star. This gives a new sequence of edges E'. The algorithm has now completed the contraction step, and calls itself recursively on the contracted graph (Line 7). The LABELS$'$ returned by the recursive call are passed on to the children of the stars, effectively expanding the graph (Line 8). The same hooks that were used for contraction can be used for this update.

Two things should be noted about this algorithm. First, the algorithm flips coins on all of the vertices on each step even though many have already been contracted (there are no more edges that point to them). It turns out that this will not affect our worst-case asymptotic work or depth bounds, but it is not difficult to flip coins only on active vertices by keeping track of them—just keep an array of the labels of the active vertices. Second, if there are cycles in the graph, then the algorithm will create redundant edges in the contracted subgraphs. Again, keeping these edges is not a problem for the correctness or cost bounds, but they could be removed using hashing as discussed in Section 25.4.6.

To analyze the full work and depth of the algorithm we note that each step only requires constant depth and $O(n + m)$ work. Since the number of steps is $O(\log n)$ with high probability, as mentioned earlier, the total depth is $O(\log n)$ and the work is $O((n + m) \log n)$, both with high probability. One might expect that the work would be linear since the algorithm reduces the number of vertices on each step by a constant fraction. We have no guarantee, however, that the number of edges is also going to contract geometrically, and in fact for certain graphs they will not. In Section 25.5.3.3 we discuss how to improve the algorithm so that it is work-efficient.

25.5.3.2 Deterministic Graph Contraction

Our second algorithm for graph contraction is deterministic [41]. It is based on forming a set of disjoint subgraphs, each of which is tree, and then using the POINT_TO_ROOT routine (Section 25.4.4) to contract each subgraph to a single vertex. To generate the trees, the algorithm hooks each vertex into a neighbor with a smaller label (by hooking a into b we mean generating a directed edge from a to b). Vertices with no smaller-labeled neighbors are left unhooked. The result of the hooking is a set of disjoint trees since hooking only from larger to smaller guarantees there are no cycles, and

every node is hooked into at most one parent. Figure 25.10 shows an
example of a set of trees created by hooking. Since a vertex can have
more than one neighbor with a smaller label, a given graph can have
many different hookings. For example, in Figure 25.10 vertex 2 could
have hooked into vertex 1, rather than vertex 0.

The following algorithm performs the tree-based graph contraction.
We assume that the labels are initialized to the indices of the vertices.

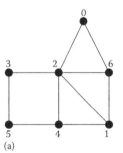

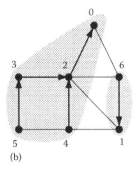

(a)

(b)

ALGORITHM: CC_TREE_CONTRACT (LABELS,E)
1   **if** $(|E| = 0)$
2   **then return** LABELS
3   **else**
4   HOOKS $:= \{(u, v) \in E \mid u > v\}$
5   LABELS $:=$ LABELS $\leftarrow$ HOOKS
6   LABELS $:=$ POINT_TO_ROOT(LABELS)
7       $E' := \{(\text{LABELS}[u],\text{LABELS}[v]) : (u, v) \in E \mid \text{LABELS}[u] \neq \text{LABELS}[v]\}$
8   **return** CC_TREE_CONTRACT(LABELS, E')

FIGURE 25.10 Tree-based
graph contraction. (a) A graph
G and (b) the hook edges
induced by hooking larger to
smaller vertices and the subgraphs induced by the trees.

The structure of the algorithm is similar to the random-mate graph
contraction algorithm. The main differences are how the hooks are
selected (Line 4), the pointer jumping step to contract the trees (Line 6),
and the fact that no relabeling is required when returning from the
recursive call. The hooking step simply selects edges that point from
larger numbered vertices to smaller numbered vertices. This is called
a *conditional hook*. The pointer jumping step uses the algorithm in
Section 25.4.4. This labels every vertex in the tree with the root of the
tree. The edge relabeling is the same as in the random-mate algorithm.
The reason the contraction algorithm does not need to relabel the vertices after the recursive call is
that the pointer jumping step does the relabeling.

Although the basic algorithm we have described so far works well in practice, in the worst case it
can take $n - 1$ steps. Consider the graph in Figure 25.11a. After hooking and contracting only one
vertex has been removed. This could be repeated up to $n - 1$ times. This worst-case behavior can be
avoided by trying to hook in both directions (from larger to smaller and from smaller to larger) and
picking the hooking that hooks more vertices. We make use of the following lemma.

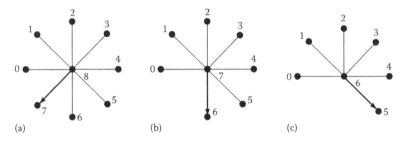

(a)   (b)   (c)

FIGURE 25.11 A worst-case graph: (a) a star graph G with the maximum index at the root of the star; (b) G after one
step of contraction; (c) G after two steps of contraction.

LEMMA 25.1   Let $G = (V, E)$ be an undirected graph in which each vertex has at least one neighbor. Then either $|\{u \mid (u, v) \in E, u < v\}| \geq |V|/2$ or $|\{u \mid (u, v) \in E, u > v\}| > |V|/2$.

PROOF   Every vertex must either have a neighbor with a lesser index or a neighbor with a greater index. This means that if we consider the set of vertices with a lesser neighbor and the set of vertices with a greater neighbor, then one of those sets must consist of at least one half the vertices.

This lemma will guarantee that if we try hooking in both directions and pick the better one we will remove at least 1/2 of the vertices on each step, so that the number of steps is bounded by $\lceil \log_2 n \rceil$.

We now consider the total cost of the algorithm. The hooking and relabeling of edges on each step takes $O(m)$ work and constant depth. The tree contraction using pointer jumping on each step requires $O(n \log n)$ work and $O(\log n)$ depth, in the worst case. Since there are $O(\log n)$ steps, in the worst case the total work is $O((m + n \log n) \log n)$ and depth $O(\log^2 n)$. However, if we keep track of the active vertices (the roots) and only pointer jump on active vertices, then the work is reduced to $O((m + n) \log n)$, since the number of vertices decreases geometrically in each step. This requires the algorithm to expand the graph on the way back up the recursion as done for the random-mate algorithm. The total work with this modification is the same work as the randomized technique, although the depth has increased.

25.5.3.3 Improved Versions of Connected Components

There are many improvements to the two basic connected component algorithms we described. Here we mention some of them.

The deterministic algorithm can be improved to run in $O(\log n)$ depth with the same work bounds [13,79]. The basic idea is to interleave the hooking steps with the pointer jumping steps. The one tricky aspect is that we must always hook in the same direction (e.g., from smaller to larger), so as not to create cycles. Our previous technique to solve the star-graph problem therefore does not work. Instead each vertex checks if it belongs to any tree after hooking. If it does not then it can hook to any neighbor, even if it has a larger index. This is called an unconditional hook.

The randomized algorithm can be improved to run in optimal work, $O(n + m)$ [33]. The basic idea is to not use all of the edges for hooking on each step, and instead use a sample of the edges. This technique, first applied to parallel algorithms, has since been used to improve some sequential algorithms, such as deriving the first linear-work algorithm for finding a minimum spanning tree [46].

Another improvement is to use the EREW model instead of requiring concurrent reads and writes [42]. However this comes at the cost of greatly complicating the algorithm. The basic idea is to keep circular linked lists of the neighbors of each vertex, and then to splice these lists when merging vertices.

25.5.3.4 Extensions to Spanning Trees and Minimum Spanning Trees

The connected-component algorithms can be extended to finding a spanning tree of a graph or minimum spanning tree of a weighted graph. In both cases we assume the graphs are undirected.

A spanning tree of a connected graph $G = (V, E)$ is a connected graph $T = (V, E')$ such that $E' \subseteq E$ and $|E'| = |V| - 1$. Because of the bound on the number of edges, the graph T cannot have any cycles and therefore forms a tree. Any given graph can have many different spanning trees.

It is not hard to extend the connected-component algorithms to return the spanning tree. In particular, whenever components are hooked together the algorithm can keep track of which edges were used for hooking. Since each edge will hook together two components that are not connected yet, and only one edge will succeed in hooking the components, the collection of these edges across all steps will form a spanning tree (they will connect all vertices and have no cycles). To determine which edges were used for contraction, each edge checks if it successfully hooked after the attempted hook.

A minimum spanning tree of a connected weighted graph $G = (V, E)$ with weights $w(e)$ for $e \in E$ is a spanning tree $T = (V, E')$ of G such that

$$w(T) = \sum_{e \in E'} w(e) \tag{25.8}$$

is minimized. The connected-component algorithms can also be extended to find a minimum spanning tree. Here we will briefly consider an extension of the random-mate technique. Let us assume, without loss of generality, that all of the edge weights are distinct. If this is not the case, then lexicographical information can be added to the edges weights to break ties. It is well known that if the edge weights are distinct, then there is a unique minimum spanning tree. Furthermore, given any $W \subset V$, the minimum weight edge from W to $V - W$ must be in the minimum spanning tree. As a consequence, the minimum edge incident on a vertex will be in the minimum spanning tree. This will be true even after we contract subgraphs into vertices, since each subgraph is a subset of V.

For the minimum-spanning-tree algorithm, we modify the random-mate technique so that each child u instead of picking an arbitrary parent to hook into, finds the incident edge (u, v) with minimum weight and hooks into v if it is a parent. If v is not a parent, then the child u does nothing (it is left as an orphan). Figure 25.12 illustrates the algorithm. As with the spanning-tree algorithm, we keep track of the hook edges and add them to a set E'. This new rule will still remove 1/4 of the vertices on each step on average since a vertex has 1/2 probability of being a child, and there is 1/2 probability that the vertex at the other end of the minimum edge is a parent. The one complication in this minimum spanning-tree algorithm is finding for each child the incident edge with minimum weight. Since we are keeping an edge list, this is not trivial to compute. If the algorithm used an adjacency list, then it would be easy, but since the algorithm needs to update the endpoints of the edges, it is not easy to maintain the adjacency list. One way to solve this problem is to use a priority concurrent write. In such a write, if multiple values are written to the same location, the one coming from the leftmost position will be written. With such a scheme the minimum edge can be found by presorting the edges by weight so the lowest weighted edge will always win when executing a concurrent write. Assuming a priority write, this minimum-spanning-tree algorithm has the same work and depth as the random-mate connected-components algorithm.

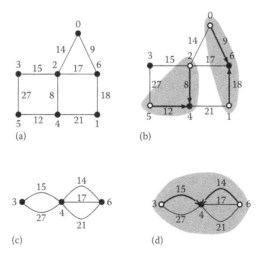

FIGURE 25.12 Example of the minimum-spanning-tree algorithm. (a) The original weighted graph G. (b) Each child (light) hooks across its minimum weighted edge to a parent (dark), if the edge is incident on a parent. (c) The graph after one step of contraction. (d) The second step in which children hook across minimum weighted edges to parents.

There is also a linear-work logarithmic-depth randomized algorithm for finding a minimum-spanning tree [27], but it is somewhat more complicated than the linear-work algorithms for finding connected components.

25.6 Sorting

Sorting is a problem that admits a variety of parallel solutions. In this section we limit our discussion to two parallel sorting algorithms, QuickSort and radix sort. Both of these algorithms are easy to program, and both work well in practice. Many more sorting algorithms can be found in the literature. The interested reader is referred to [3,45,55] for more complete coverage.

25.6.1 QuickSort

We begin our discussion of sorting with a parallel version of QuickSort. This algorithm is one of the simplest to code.

ALGORITHM: QUICKSORT(A)
1   **if** $|A| = 1$ **then return** A
2   $i := rand\_int(|A|)$
3   $p := A[i]$
4   **in parallel do**
5       $L :=$ QUICKSORT($\{a : a \in A \mid a < p\}$)
6       $E := \{a : a \in A \mid a = p\}$
7       $G :=$ QUICKSORT($\{a : a \in A \mid a > p\}$)
8   **return** $L{+}{+}E{+}{+}G$

We can make an optimistic estimate of the work and depth of this algorithm by assuming that each time a partition element p is selected, it divides the set A so that neither L nor H has more than half of the elements. In this case, the work and depth are given by the recurrences

$$W(n) = 2W(n/2) + O(n) \qquad\qquad (25.9)$$

$$D(n) = D(n/2) + 1 \qquad\qquad (25.10)$$

which have solutions $W(n) = O(n \log n)$ and $D(n) = O(\log n)$. A more sophisticated analysis [50] shows that the expected work and depth are indeed $W(n) = O(n \log n)$ and $D(n) = O(\log n)$, independent of the values in the input sequence A.

In practice, the performance of parallel QuickSort can be improved by selecting more than one partition element. In particular, on a machine with P processors, choosing $P - 1$ partition elements divides the keys into P sets, each of which can be sorted by a different processor using a fast sequential sorting algorithm. Since the algorithm does not finish until the last processor finishes, it is important to assign approximately the same number of keys to each processor. Simply choosing $p - 1$ partition elements at random is unlikely to yield a good partition. The partition can be improved, however, by choosing a larger number, sp, of candidate partition elements at random, sorting the candidates (perhaps using some other sorting algorithm), and then choosing the candidates with ranks $s, 2s, \ldots, (p-1)s$ to be the partition elements. The ratio s of candidates to partition elements is called the oversampling ratio. As s increases, the quality of the partition increases, but so does the time to sort the sp candidates. Hence there is an optimum value of s, typically larger than one, that minimizes the total time. The sorting algorithm that selects partition elements in this fashion is called sample sort [23,76,89].

25.6.2 Radix Sort

Our next sorting algorithm is radix sort, an algorithm that performs well in practice. Unlike Quick-Sort, radix sort is not a comparison sort, meaning that it does not compare keys directly in order to determine the relative ordering of keys. Instead, it relies on the representation of keys as b-bit integers.

The basic radix sort algorithm (whether serial or parallel) examines the keys to be sorted one "digit" position at a time, starting with the least significant digit in each key. Of fundamental importance is that this intermediate sort on digits be stable: the output ordering must preserve the input order of any two keys with identical digit values in the position being examined.

The most common implementation of the intermediate sort is as a counting sort. A counting sort first counts to determine the rank of each key—its position in the output order—and then permutes the keys by moving each key to the location indicated by its rank. The following algorithm performs radix sort assuming one-bit digits.

ALGORITHM: RADIX_SORT(A, b)
1   **for** i **from** 0 **to** $b - 1$
2   FLAGS $:= \{(a >> i) \bmod 2 : a \in A\}$
3   NOTFLAGS $:= \{1 - b : b \in FLAGS\}$
4        $R_0 :=$ SCAN(NOTFLAGS)
5        $s_0 :=$ SUM(NOTFLAGS)
6        $R_1 :=$ SCAN(FLAGS)
7        $R := \{$**if** FLAGS$[j] = 0$ **then** $R_0[j]$ **else** $R_1[j] + s_0 : j \in [0..|A|)\}$
8        $A := A \leftarrow \{(R[j], A[j]) : j \in [0..|A|)\}$
9   **return** A

For keys with b bits, the algorithm consists of b sequential iterations of a for loop, each iteration sorting according to one of the bits. Lines 2 and 3 compute the value and inverse value of the bit in the current position for each key. The notation $a >> i$ denotes the operation of shifting a to the right by i bit positions. Line 4 computes the rank of each key that has bit value 0. Computing the ranks of the keys with bit value 1 is a little more complicated, since these keys follow the keys with bit value 0. Line 5 computes the number of keys with bit value 0, which serves as the rank of the first key that has bit value 1. Line 6 computes the relative order of the keys with bit value 1. Line 7 merges the ranks of the even keys with those of the odd keys. Finally, Line 8 permutes the keys according to rank.

The work and depth of RADIX_SORT are computed as follows. There are b iterations of the for loop. In each iteration, the depths of Lines 2, 3, 7, 8, and 9 are constant, and the depths of Lines 4, 5, and 6 are $O(\log n)$. Hence the depth of the algorithm is $O(b \log n)$. The work performed by each of Lines 2 through 9 is $O(n)$. Hence, the work of the algorithm is $O(bn)$.

The radix sort algorithm can be generalized so that each b-bit key is viewed as b/r blocks of r bits each, rather than as b individual bits. In the generalized algorithm, there are b/r iterations of the for loop, each of which invokes the SCAN function 2^r times. When r is large, a multiprefix operation can be used for generating the ranks instead of executing a SCAN for each possible value [23]. In this case, and assuming the multiprefix operation runs in linear work, it is not hard to show that as long as $b = O(\log n)$ (and $r = \log n$, so that there are only $O(1)$ iterations), the total work for the radix sort is $O(n)$, and the depth is the same order as the depth of the multiprefix operation.

Floating-point numbers can also be sorted using radix sort. With a few simple bit manipulations, floating-point keys can be converted to integer keys with the same ordering and key size. For example, IEEE double-precision floating-point numbers can be sorted by first inverting the mantissa and exponent bits if the sign bit is 1, and then inverting the sign bit. The keys are then sorted as if they were integers.

25.7 Computational Geometry

Problems in computational geometry involve calculating properties of sets of objects in k-dimensional space. Some standard problems include finding the minimum distance between any two points in a set of points (closest-pair), finding the smallest convex region that encloses a set of points (convex-hull), and finding line or polygon intersections. Efficient parallel algorithms have been developed for most standard problems in computational geometry. Many of the sequential algorithms are based on divide-and-conquer and lead in a relatively straightforward manner to efficient parallel algorithms. Some others are based on a technique called plane sweeping, which does not parallelize well, but for which an analogous parallel technique, the plane sweep tree has been developed [1,10]. In this section we describe parallel algorithms for two problems in two dimensions—closest pair and convex hull. For convex hull we describe two algorithms. These algorithms are good examples of how sequential algorithms can be parallelized in a straightforward manner.

We suggest the following sources for further information on parallel algorithms for computational geometry: [6,39], [45, Chapter 6], and [75, Chapters 9 and 11].

25.7.1 Closest Pair

The closest-pair problem takes a set of points in k dimensions and returns the two points that are closest to each other. The distance is usually defined as Euclidean distance. Here we describe a closest-pair algorithm for two-dimensional space, also called the planar closest-pair problem. The algorithm is a parallel version of a standard sequential algorithm [16,17], and for n points, it requires the same work as the sequential versions, $O(n \log n)$, and has depth $O(\log^2 n)$. The work is optimal.

The algorithm uses divide-and-conquer based on splitting the points along lines parallel to the y-axis, and is expressed as follows.

ALGORITHM: CLOSEST_PAIR(P)
1   **if** $(|P| < 2)$ **then return** (P, ∞)
2   $x_m :=$ MEDIAN$(\{x : (x, y) \in P\})$
3   $L := \{(x, y) \in P \mid x < x_m\}$
4   $R := \{(x, y) \in P \mid x \geq x_m\}$
5   **in parallel do**
6       $(L', \delta_L) :=$ CLOSEST_PAIR(L)
7       $(R', \delta_R) :=$ CLOSEST_PAIR(R)
8   $P' :=$ MERGE_BY_Y(L', R')
9   $\delta_P :=$ BOUNDARY_MERGE$(P', \delta_L, \delta_R, x_m)$
10 **return** (P', δ_P)

This function takes a set of points P in the plane and returns both the original points sorted along the y-axis, and the distance between the closest two points. The sorted points are needed to help merge the results from recursive calls, and can be thrown away at the end. It would not be difficult to modify the routine to return the closest pair of points in addition to the distance between them. The function works by dividing the points in half based on the median x value, recursively solving the problem on each half and then merging the results. The MERGE_BY_Y function merges L' and R' along the y-axis and can use a standard parallel merge routine. The interesting aspect of the algorithm is the BOUNDARY_MERGE routine, which works on the principle described by Bentley and Shamos [16,17], and can be computed with $O(\log n)$ depth and $O(n)$ work. We first review the principle and then show how it can be performed in parallel.

The inputs to BOUNDARY_MERGE are the original points P sorted along the y-axis, the closest distance within L and R, and the median point x_m. The closest distance in P must be either the distance δ_L,

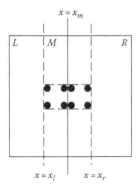

FIGURE 25.13 Merging two rectangles to determine the closest pair. Only eight points can fit in the $2\delta \times \delta$ dashed rectangle.

the distance δ_R, or a distance between a point in L and a point in R. For this distance to be less than δ_L or δ_R, the two points must lie within $\delta = \min(\delta_L, \delta_R)$ of the line $x = x_m$. Thus, the two vertical lines at $x_l = x_m - \delta$ and $x_r = x_m + \delta$ define the borders of a region M in which the points must lie (see Figure 25.13). If we could find the closest distance in M, call it δ_M, then the closest overall distance would be $\delta_P = \min(\delta_L, \delta_R, \delta_M)$.

To find δ_M we take advantage of the fact that not many points can be packed close together within M since all points within L or R must be separated by at least δ. Figure 25.13 shows the tightest possible packing of points in a $2\delta \times \delta$ rectangle within M. This packing implies that if the points in M are sorted along the y-axis, each point can determine the minimum distance to another point in M by looking at a fixed number of neighbors in the sorted order, at most 7 in each direction. To see this consider one of the points along the top of the $2\delta \times \delta$ rectangle. To determine if there are any points below it that are closer than δ we need only to consider the points within the rectangle (points below the rectangle must be further than δ away). As Figure 25.13 illustrates, there can be at most 7 other points within the rectangle. Given this property, the following function performs the border merge.

ALGORITHM: BOUNDARY_MERGE($P, \delta_L, \delta_R, x_m$)
1 $\delta := \min(\delta_L, \delta_R)$
2 $M := \{(x,y) \in P \mid (x \geq x_m - \delta) \wedge (x \leq x_m + \delta)\}$
3 $\delta_M := \min(\{\min(\{distance(M[i], M[i+j]) : j \in [1..7]\})$
4               $: i \in [0..|P| - 7)\}$
5 **return** $\min(\delta, \delta_M)$

For each point in M this function considers the seven points following it in the sorted order and determines the distance to each of these points. It then takes the minimum overall distances. Since the distance relationship is symmetric, there is no need to consider points appearing before a point in the sorted order.

The work of BOUNDARY_MERGE is $O(n)$ and the depth is dominated by the taking the minimum, which has $O(\log n)$ depth.* The work of the merge and median steps in CLOSEST_PAIR are also $O(n)$, and the depth of both are bounded by $O(\log n)$. The total work and depth of the algorithm can therefore be expressed by the recurrences

$$W(n) = 2W(n/2) + O(n) \tag{25.11}$$

$$D(n) = D(n/2) + O(\log n) \tag{25.12}$$

which have solutions $W(n) = O(n \log n)$ and $D(n) = O(\log^2 n)$.

25.7.2 Planar Convex Hull

The convex-hull problem takes a set of points in k dimensions and returns the smallest convex region that contains all the points. In two dimensions the problem is called the planar convex-hull problem, and it returns the set of points that form the corners of the region. These points are a subset of the

---

* The depth of finding the minimum or maximum of a set of numbers can actually be improved to $O(\log \log n)$ with concurrent reads [78].

original points. We will describe two parallel algorithms for the planar convex-hull problem. They are both based on divide-and-conquer, but one does most of the work before the divide step, and the other does most of the work after.

25.7.2.1   QuickHull

The parallel QuickHull algorithm is based on the sequential version [71], so named because of its similarity to the QuickSort algorithm. As with QuickSort, the strategy is to pick a "pivot" element, split the data into two sets based on the pivot, and recurse on each of the sets. Also as with QuickSort, the pivot element is not guaranteed to split the data into equal-sized sets, and in the worst case the algorithm requires $O(n^2)$ work. In practice, however, the algorithm is often very efficient, probably the most practical of the convex hull algorithms. At the end of the section we briefly describe how the splits can be made so that the work is guaranteed to be bounded by $O(n \log n)$.

The QuickHull algorithm is based on the recursive function SUBHULL, which is expressed as follows.

ALGORITHM: SUBHULL(P, p_1, p_2)
1   $P' := \{p \in P \mid \text{LEFT\_OF?}(p, (p_1, p_2))\}$
2   **if** $(|P'| < 2)$
3   **then return** $[p_1] + +P'$
4   **else**
5     $i := \text{MAX\_INDEX}(\{\text{DISTANCE}(p, (p_1, p_2)) : p \in P'\})$
6     $p_m := P'[i]$
7   **in parallel do**
8       $H_l := \text{SUBHULL}(P', p_1, p_m)$
9       $H_r := \text{SUBHULL}(P', p_m, p_2)$
10   **return** $H_l + +H_r$

This function takes a set of points P in the plane and two points p_1 and p_2 that are known to lie on the convex hull, and returns all the points that lie on the hull clockwise from p_1 to p_2, inclusive of p_1, but not of p_2. For example in Figure 25.14 SUBHULL$([A, B, C, \ldots, P], A, P)$ would return the sequence $[A, B, J, O]$.

The function SUBHULL works as follows. Line 1 removes all the elements that cannot be on the hull because they lie to the right of the line from p_1 to p_2. Determining which side of a line a point lies on can easily be calculated with a few arithmetic operations. If the remaining set P' is either empty or has just one element, the algorithm is done. Otherwise the algorithm finds the point p_m farthest from the line (p_1, p_2). The point p_m must be on the hull since as a line at infinity parallel to (p_1, p_2) moves toward (p_1, p_2), it must first hit p_m. In Line 5 the function MAX_INDEX returns the index of the maximum value of a sequence, which is then used to extract the point p_m. Once p_m is found, SUBHULL is called twice recursively to find the hulls from p_1 to p_m, and from p_m to p_2. When the recursive calls return, the results are appended.

The following function uses SUBHULL to find the full convex hull.

ALGORITHM: QUICKHULL(P)
1   $X := \{x : (x, y) \in P\}$
2   $x_{\min} := P[\text{MIN\_INDEX}(X)]$
3   $x_{\max} := P[\text{MAX\_INDEX}(X)]$
4   **return** SUBHULL$(P, x_{\min}, x_{\max}) + +$ SUBHULL$(P, x_{\max}, x_{\min})$

We now consider the cost of the parallel QuickHull, and in particular the SUBHULL routine, which does all the work. The call to MAX_INDEX

[$A B C D E F G H I J K L M N O P$]
A [$B D F G H J K M O$] P [$C E I L N$]
A [$B F$] J [O] $P N$ [$C E$]
$A B J O P N C$

FIGURE 25.14   An example of the QuickHull algorithm.

FIGURE 25.15 Contrived set of points for worst-case QuickHull.

uses $O(n)$ work and $O(\log n)$ depth. Hence, the cost of everything other than the recursive calls is $O(n)$ work and $O(\log n)$ depth. If the recursive calls are balanced so that neither recursive call gets much more than half of the data then the number of levels of recursion will be $O(\log n)$. This will lead to the algorithm running in $O(\log^2 n)$ depth. Since the sum of the sizes of the recursive calls can be less than n (e.g., the points within the triangle AJP will be thrown out when making the recursive calls to find the hulls between A and J and between J and P), the work can be as little as $O(n)$, and often is in practice. As with QuickSort, however, when the recursive calls are badly partitioned the number of levels of recursion can be as bad as $O(n)$ with work $O(n^2)$. For example, consider the case when all the points lie on a circle and have the following unlikely distribution. The points x_{min} and x_{max} appear on opposite sides of the circle. There is one point that appears half way between x_{min} and x_{max} on the sphere and this point becomes the new x_{max}. The remaining points are defined recursively. That is, the points become arbitrarily close to x_{min} (see Figure 25.15).

Kirkpatrick and Seidel [49] have shown that it is possible to modify QuickHull so that it makes provably good partitions. Although the technique is shown for a sequential algorithm, it is easy to parallelize. A simplification of the technique is given by Chan et al. [25]. Their algorithm admits even more parallelism and leads to an $O(\log^2 n)$-depth algorithm with $O(n \log h)$ work where h is the number of points on the convex hull.

25.7.2.2 MergeHull

The MergeHull algorithm [68] is another divide-and-conquer algorithm for solving the planar convex hull problem. Unlike QuickHull, however, it does most of its work after returning from the recursive calls. The function is expressed as follows.

ALGORITHM: MERGEHULL(P)
1   **if** ($|P| < 3$) **then return** P
2   **else**
3   **in parallel do**
4           $H_L = $ MERGEHULL($P[0..|P|/2)$)
5           $H_R = $ MERGEHULL($P[|P|/2..|P|)$)
6   **return** JOIN_HULLS(H_L, H_R)

This function assumes the input P is presorted according to the x coordinates of the points. Since the points are presorted, H_L is a convex hull on the left and H_R is a convex hull on the right. The JOIN_HULLS routine is the interesting part of the algorithm. It takes the two hulls and merges them into one. To do this it needs to find lower and upper points l_1 and u_1 on H_L and l_2 and u_2 on H_R such that l_1, l_2 and u_1, u_2 are successive points on H (see Figure 25.16). The lines b_1 and b_2 joining these upper and lower points are called the upper and lower bridges, respectively. All the points between l_1 and u_1 and between u_2 and l_2 on the "outer" sides of H_L and H_R are on the final convex hull, while the points on the "inner" sides are not on the convex hull. Without loss of generality we only consider how to find the upper bridge b_1. Finding the lower bridge b_2 is analogous.

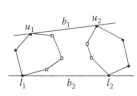

FIGURE 25.16 Merging two convex hulls.

To find the upper bridge one might consider taking the points with the maximum y values on H_L and H_R. This approach does not work in general, however, since u_1 can lie as far down as the point with the

minimum x or maximum x value (see Figure 25.17). Instead there is a
nice solution due to Overmars and van Leeuwen [68] based on a dual
binary search. Assume that the points on the convex hulls are given in
order (e.g., clockwise). At each step the binary-search algorithm will
eliminate half of the remaining points from consideration in either H_L
or H_R or both. After at most $\log |H_L| + \log |H_R|$ steps the search will be
left with only one point in each hull, and these will be the desired points
u_1 and u_2. Figure 25.18 illustrates the rules for eliminating part of H_L
or H_R on each step.

FIGURE 25.17 A bridge
that is far from the top of the
convex hull.

We now consider the cost of the algorithm. Each step of the binary
search requires only constant work and depth since we need only to
consider the two middle points M_1 and M_2, which can be found in
constant time if the hull is kept sorted. The cost of the full binary search to find the upper bridge is
therefore bounded by $D(n) = W(n) = O(\log n)$. Once we have found the upper and lower bridges
we need to remove the points on H_L and H_R that are not on H and append the remaining convex hull

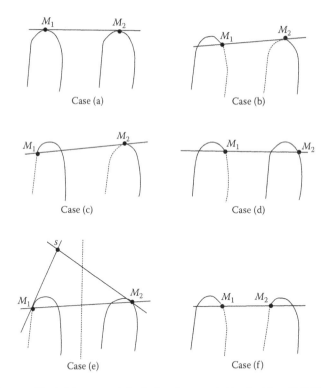

FIGURE 25.18 Cases used in the binary search for finding the upper bridge for MergeHull. The points M_1 and M_2
mark the middle of the remaining hulls. In case (a), all of H_L and H_R lie below the line through M_1 and M_2. In this
case, the line segment between M_1 and M_2 is the bridge. In the remaining cases, dotted lines represent the parts of the
hulls H_L and H_R that can be eliminated from consideration. In cases (b) and (c), all of H_R lies below the line through
M_1 and M_2, and either the left half of H_L or the right half of H_L lies below the line. In cases (d) through (f), neither H_L
nor H_R lies entirely below the line. In case (e), the region to eliminate depends on which side of a line separating H_L
and H_R the intersection of the tangents appears. The mirror images of cases (b) through (e) are also used. Case (f) is
actually an instance of case (d) and its mirror image.

points. This requires linear work and constant depth. The overall costs of MERGEHULL are therefore

$$D(n) = D(n/2) + O(\log n) = O(\log^2 n) \qquad (25.13)$$

$$W(n) = 2W(n/2) + O(n) = O(n \log n) \qquad (25.14)$$

This algorithm can be improved to run in $O(\log n)$ depth using one of two techniques. The first involves modifying the search for the bridge points so that it runs in constant depth with linear work [12]. This involves sampling every $\sqrt{n}$th point on each hull and comparing all pairs of these two samples to narrow the search down to regions of size $\sqrt{n}$ in constant depth. The regions can then be finished in constant depth by comparing all pairs between the two regions. The second technique [1,11] uses divide-and-conquer to separate the point set into $\sqrt{n}$ regions, solves the convex hull on each region recursively, and then merges all pairs of these regions using the binary-search method. Since there are $\sqrt{n}$ regions and each of the searches takes $O(\log n)$ work, the total work for merging is $O\left((\sqrt{n})^2 \log n\right) = O(n \log n)$ and the depth is $O(\log n)$. This leads to an overall algorithm that runs in $O(n \log n)$ work and $O(\log n)$ depth. The algorithms above require concurrent reads (CREW). The same bounds can be achieved with exclusive reads (EREW) [66].

25.8 Numerical Algorithms

There has been an immense amount of work on parallel algorithms for numerical problems. Here we briefly discuss some of the problems and results. We suggest the following sources for further information on parallel numerical algorithms [75, Chapters 12–14], [45, Chapter 8], [53, Chapters 5,10, and 11] and [18].

25.8.1 Matrix Operations

Matrix operations form the core of many numerical algorithms and led to some of the earliest work on parallel algorithms. The most basic matrix operation is matrix multiplication. The standard triply nested loop for multiplying two dense matrices is highly parallel since each of the loops can be parallelized:

ALGORITHM: MATRIX_MULTIPLY(A, B)
1   $(l, m) := dimensions(A)$
2   $(m, n) := dimensions(B)$
3   **in parallel for** $i \in [0..l)$ **do**
4   **in parallel for** $j \in [0..n)$ **do**
5       $R_{ij} := sum(\{A_{ik} * B_{kj} : k \in [0..m)\})$
6   **return** R

If $l = m = n$, this routine does $O(n^3)$ work and has depth $O(\log n)$, due to the depth of the summation. This has much more parallelism than is typically needed, and most of the research on parallel matrix multiplication has concentrated on what subset of the parallelism to use so that communication costs can be minimized. Sequentially it is known that matrix multiplication can be performed using less than $O(n^3)$ work. Strassen's algorithm [82], for example, requires only $O(n^{2.81})$ work. Most of these more efficient algorithms are also easy to parallelize because they are recursive in nature (Strassen's algorithm has $O(\log n)$ depth using a simple parallelization).

Another basic matrix operation is to invert matrices. Inverting dense matrices has proven to be more difficult to parallelize than multiplying dense matrices, but the problem still supplies plenty of parallelism for most practical purposes. When using Gauss–Jordan elimination, two of the three nested loops can be parallelized, leading to an algorithm that runs with $O(n^3)$ work and $O(n)$ depth. A recursive block-based method using matrix multiplication leads to the same depth, although the work can be reduced by using a more efficient matrix-multiplication algorithm. There are also more sophisticated, but less practical, work-efficient algorithms with depth $O(\log^2 n)$ [28,70].

Parallel algorithms for many other matrix operations have been studied, and there has also been significant work on algorithms for various special forms of matrices, such as tridiagonal, triangular, and sparse matrices. Iterative methods for solving sparse linear systems has been an area of significant activity.

25.8.2 Fourier Transform

Another problem for which there is a long history of parallel algorithms is the discrete Fourier transform (DFT). The FFT algorithm for solving the DFT is quite easy to parallelize and, as with matrix multiply, much of the research has gone into reducing communication costs. In fact the butterfly network topology is sometimes called the FFT network, since the FFT has the same communication pattern as the network [55, Section 3.7]. A parallel FFT over complex numbers can be expressed as follows:

ALGORITHM: FFT(A)
1   $n := |A|$
2   **if** ($n = 1$) **then return** A
3   **else**
4   **in parallel do**
5   EVEN $:=$ FFT($\{A[2i] : i \in [0..n/2)\}$)
6   ODD $:=$ FFT($\{A[2i + 1] : i \in [0..n/2)\}$)
7   **return** $\{\text{EVEN}[j] + \text{ODD}[j]e^{2\pi ij/n} : j \in [0..n/2)\} ++ \{\text{EVEN}[j] - \text{ODD}[j]e^{2\pi ij/n} : j \in [0..n/2)\}$

The algorithm simply calls itself recursively on the odd and even elements and then puts the results together. This algorithm does $O(n \log n)$ work, as does the sequential version, and has a depth of $O(\log n)$.

25.9   Research Issues and Summary

Recent work on parallel algorithms has focused on solving problems from domains such as pattern matching, data structures, sorting, computational geometry, combinatorial optimization, linear algebra, and linear and integer programming. For pointers to this work, see Section 25.10.

Algorithms have also been designed specifically for the types of parallel computers that are available today. Particular attention has been paid to machines with limited communication bandwidth. For example, there is a growing library of software developed for the BSP model [40,62,85].

The parallel computer industry has been through a period of financial turbulence, with several manufacturers failing or discontinuing sales of parallel machines. In the past few years, however, a large number of inexpensive small-scale parallel machines have been sold. These machines typically consist of 4–8 commodity processors connected by a bus to a shared-memory system. As these machines reach the limit in size imposed by the bus architecture, manufacturers have reintroduced parallel machines based on the hypercube network topology (e.g., [54]).

25.10   Further Information

In a chapter of this length, it is not possible to provide comprehensive coverage of the subject of parallel algorithms. Fortunately, there are several excellent textbooks and surveys on parallel algorithms including [4–6,18,30,31,35,45,48,53,55,67,75,80].

There are many technical conferences devoted to the subjects of parallel computing and computer algorithms, so keeping abreast of the latest research developments is challenging. Some of the best work in parallel algorithms can be found in conferences such as the ACM Symposium on Parallel Algorithms and Architectures, the IEEE International Parallel and Distributed Processing Symposium, the International Conference on Parallel Processing, the International Symposium on Parallel Architectures, Algorithms, and Networks, the ACM Symposium on the Theory of Computing, the IEEE Symposium on Foundations of Computer Science, the ACM–SIAM Symposium on Discrete Algorithms, and the ACM Symposium on Computational Geometry.

In addition to parallel algorithms, this chapter has also touched on several related subjects, including the modeling of parallel computations, parallel computer architecture, and parallel programming languages. More information on these subjects can be found in [7,43,44,86], and [21,34], respectively. Other topics likely to interest the reader of this chapter include distributed algorithms [59] and VLSI layout theory and computation [57,84].

Defining Terms

CRCW: A shared memory model that allows for concurrent reads and concurrent writes to the memory.

CREW: This refers to a shared memory model that allows for concurrent reads but only EWs to the memory.

Depth: The longest chain of sequential dependencies in a computation.

EREW: A shared memory model that allows for only ERs and EWs to the memory.

Graph contraction: Contracting a graph by removing a subset of the vertices.

List contraction: Contracting a list by removing a subset of the nodes.

Multiprefix: A generalization of the scan (**prefix sums**) operation in which the partial sums are grouped by keys.

Multiprocessor model: A model of parallel computation based on a set of communicating sequential processors.

Pipelined divide-and-conquer: A divide-and-conquer paradigm in which partial results from recursive calls can be used before the calls complete. The technique is often useful for reducing the depth of an algorithm.

Pointer jumping: In a linked structure, replacing a pointer with the pointer it points to. Used for various algorithms on lists and trees.

PRAM model: A multiprocessor model in which all processors can access a shared memory for reading or writing with uniform cost.

Prefix sums: A parallel operation in which each element in an array or linked-list receives the sum of the previous elements.

Random sampling: Using a randomly selected sample of the data to help solve a problem on the whole data.

Recursive doubling: The same as pointer jumping.

Scan: A parallel operation in which each element in an array receives the sum of all the previous elements.

Tree contraction: Contracting a tree by removing a subset of the nodes.

Symmetry breaking: A technique to break the symmetry in a structure such as a graph which can locally look the same to all the vertices. Usually implemented with randomization.

Work: The total number of operations taken by a computation.

Work-depth model: A model of parallel computation in which one keeps track of the total work and depth of a computation without worrying about how it maps onto a machine.

Work-efficient: A parallel algorithm is work-efficient if asymptotically (as the problem size grows) it requires at most a constant factor more work than the best known sequential algorithm (or the optimal work).

Work-preserving: A translation of an algorithm from one model to another is work-preserving if the work is the same in both models, to within a constant factor.

References

1. Aggarwal, A., Chazelle, B., Guibas, L., Dú, C.O., and Yap, C., Parallel computational geometry, *Algorithmica*, 3(3), 293–327, 1988.
2. Aho, A.V., Hopcroft, J.E., and Ullman, J.D., *The Design and Analysis of Computer Algorithms*, Addison-Wesley, Reading, MA, 1974.
3. Akl, S.G., *Parallel Sorting Algorithms*, Academic Press, Orlando, FL, 1985.
4. Akl, S.G., *The Design and Analysis of Parallel Algorithms*, Prentice Hall, Englewood Cliffs, NJ, 1989.
5. Akl, S.G., *Parallel Computation: Models and Methods*, Prentice Hall, Englewood Cliffs, NJ, 1997.
6. Akl, S.G. and Lyons, K.A., *Parallel Computational Geometry*, Prentice Hall, Englewood Cliffs, NJ, 1993.
7. Almasi, G.S. and Gottlieb, A., *Highly Parallel Computing*, Benjamin/Cummings, Redwood City, CA, 1989.
8. Anderson, R.J. and Miller, G.L., A simple randomized parallel algorithm for list-ranking, *Information Processing Letters*, 33(5), 269–273, January 1990.
9. Anderson, R.J. and Miller, G.L., Deterministic parallel list ranking, *Algorithmica*, 6(6), 859–868, 1991.
10. Atallah, M.J., Cole, R., and Goodrich, M.T., Cascading divide-and-conquer: A technique for designing parallel algorithms, *SIAM Journal of Computing*, 18(3), 499–532, June 1989.
11. Atallah, M.J. and Goodrich, M.T., Efficient parallel solutions to some geometric problems, *Journal of Parallel and Distributed Computing*, 3(4), 492–507, December 1986.
12. Atallah, M.J. and Goodrich, M.T., Parallel algorithms for some functions of two convex polygons, *Algorithmica*, 3(4), 535–548, 1988.
13. Awerbuch, B. and Shiloach, Y., New connectivity and MSF algorithms for shuffle-exchange network and PRAM, *IEEE Transactions on Computers*, C–36(10), 1258–1263, October 1987.
14. Bar-Noy, A. and Kipnis, S., Designing broadcasting algorithms in the postal model for message-passing systems, *Mathematical Systems Theory*, 27(5), 341–452, September/October 1994.
15. Beneš, V.E., *Mathematical Theory of Connecting Networks and Telephone Traffic*, Academic Press, New York, 1965.
16. Bentley, J.L., Multidimensional divide-and-conquer, *Communications of the Association for Computing Machinery*, 23(4), 214–229, April 1980.
17. Bentley, J.L. and Shamos, M.I., Divide-and-conquer in multidimensional space, in *Conference Record of the Eighth Annual ACM Symposium on Theory of Computing*, 220–230, New York, May 1976.
18. Bertsekas, D.P. and Tsitsiklis, J.N., *Parallel and Distributed Computation: Numerical Methods*, Prentice-Hall, Englewood Cliffs, NJ, 1989.

19. Blelloch, G.E., *Vector Models for Data-Parallel Computing*, MIT Press, Cambridge, MA, 1990.
20. Blelloch, G.E., Programming parallel algorithms, *Communications of the ACM*, 39(3), 85–97, March 1996.
21. Blelloch, G.E., Chandy, K.M., and Jagannathan, S., Eds., *Specification of Parallel Algorithms. Volume 18 of DIMACS Series in Discrete Mathematics and Theoretical Computer Science*, American Mathematical Society, Providence, RI, 1994.
22. Blelloch, G.E. and Greiner, J., Parallelism in sequential functional languages, in *Proceedings of the Symposium on Functional Programming and Computer Architecture*, 226–237, La Jolla, CA, June 1995.
23. Blelloch, G.E., Leiserson, C.E., Maggs, B.M., Plaxton, C.G., Smith, S.J., and Zagha, M., An experimental analysis of parallel sorting algorithms, *Theory of Computing Systems*, 31(2), 135–167, March/April 1998.
24. Brent, R.P., The parallel evaluation of general arithmetic expressions, *Journal of the Association for Computing Machinery*, 21(2), 201–206, April 1974.
25. Chan, T.M., Snoeyink, J., and Yap, C.-K., Output-sensitive construction of polytopes in four dimensions and clipped Voronoi diagrams in three, in *Proceedings of the Sixth Annual ACM–SIAM Symposium on Discrete Algorithms*, 282–291, San Francisco, CA, January 1995.
26. Cole, R., Parallel merge sort, *SIAM Journal of Computing*, 17(4), 770–785, August 1988.
27. Cole, R., Klein, P.N., and Tarjan, R.E., Finding minimum spanning forests in logarithmic time and linear work, in *Proceedings of the Eighth Annual ACM Symposium on Parallel Algorithms and Architectures*, 243–250, Padua, Italy, June 1996.
28. Csanky, L., Fast parallel matrix inversion algorithms, *SIAM Journal on Computing*, 5(4), 618–623, April 1976.
29. Culler, D.E., Karp, R.M., Patterson, D., Sahay, A., Santos, E.E., Schauser, K.E., Subramonian, R., and von Eicken, T., LogP: A practical model of parallel computation, *Communications of the Association for Computing Machinery*, 39(11), 78–85, November 1996.
30. Cypher, R. and Sanz, J.L.C., *The SIMD Model of Parallel Computation*, Springer-Verlag, New York, 1994.
31. Eppstein, D. and Galil, Z., Parallel algorithmic techniques for combinatorial computation, *Annual Review of Computer Science*, 3, 233–83, 1988.
32. Fortune, S. and Wyllie, J., Parallelism in random access machines, in *Conference Record of the Tenth Annual ACM Symposium on Theory of Computing*, 114–118, San Diego, CA, May 1978.
33. Gazit, H., An optimal randomized parallel algorithm for finding connected components in a graph, *SIAM Journal on Computing*, 20(6), 1046–1067, December 1991.
34. Gelernter, D., Nicolau, A., and Padua, D., Eds., *Languages and Compilers for Parallel Computing. Research Monographs in Parallel and Distributed*, MIT Press, Cambridge, MA, 1990.
35. Gibbons, A. and Rytter, W., *Efficient Parallel Algorithms*, Cambridge University Press, Cambridge, U.K., 1988.
36. Gibbons, P.B., Matias, Y., and Ramachandran, V., The QRQW PRAM: Accounting for contention in parallel algorithms, in *Proceedings of the Fifth Annual ACM–SIAM Symposium on Discrete Algorithms*, 638–648, Arlington, VA, January 1994.
37. Goldschlager, L.M., A unified approach to models of synchronous parallel machines, in *Conference Record of the Tenth Annual ACM Symposium on Theory of Computing*, 89–94, San Diego, CA, May 1978.
38. Goldschlager, L.M., A universal interconnection pattern for parallel computers, *Journal of the Association for Computing Machinery*, 29(3), 1073–1086, October 1982.
39. Goodrich, M.T., Parallel algorithms in geometry, in *CRC Handbook of Discrete and Computational Geometry*, J.E. Goodman and J.O'Rourke (Eds.), 669–682. CRC Press, Boca Raton, FL, 1997.
40. Goudreau, M., Lang, K., Rao, S., Suel, T., and Tsantilas. T., Towards efficiency and portability: Programming with the BSP model, in *Proceedings of the Eighth Annual ACM Symposium on Parallel Algorithms and Architectures*, 1–12, Padua, Italy, June 1996.

41. Greiner, J. and Blelloch, J.E., Connected components algorithms, in *High Performance Computing: Problem Solving with Parallel and Vector Architectures*, G.W. Sabot (Ed.), Addison-Wesley, Reading, MA, 1995.

42. Halperin, S. and Zwick, U., An optimal randomised logarithmic time connectivity algorithm for the EREW PRAM, *Journal of Computer and Systems Sciences*, 53(3), 395–416, December 1996.

43. Harris, T.J., A survey of PRAM simulation techniques, *ACM Computing Surveys*, 26(2), 187–206, June 1994.

44. Hennessy, J.L. and Patterson, D.A., *Computer Architecture: A Quantitative Approach*, 2nd edn., Morgan Kaufmann, San Francisco, CA, 1996.

45. JáJá, J., *An Introduction to Parallel Algorithms*, Addison Wesley, Reading, MA, 1992.

46. Karger, D.R., Klein, P.N., and Tarjan, R.E., A randomized linear-time algorithm to find minimum spanning trees, *Journal of the Association for Computing Machinery*, 42(2), 321–328, March 1995.

47. Karlin, A.R. and Upfal, E., Parallel hashing: An efficient implementation of shared memory, *Journal of the Association for Computing Machinery*, 35(5), 876–892, October 1988.

48. Karp, R.M. and Ramachandran, V., Parallel algorithms for shared-memory machines, in *Handbook of Theoretical Computer Science*, J. van Leeuwen (Ed.), Vol. A: *Algorithms and Complexity*, 869–941. Elsevier Science Publishers, Amsterdam, the Netherlands, 1990.

49. Kirkpatrick, D.G. and Seidel, R., The ultimate planar convex hull algorithm? *SIAM Journal on Computing*, 15(1), 287–299, February 1986.

50. Knuth, D.E., *Sorting and Searching*, Vol. 3 of *The Art of Computer Programming*, Addison-Wesley, Reading, MA, 1973.

51. Kogge, P.M. and Stone, H.S., A parallel algorithm for the efficient solution of a general class of recurrence equations, *IEEE Transactions on Computers*, C–22(8), 786–793, August 1973.

52. Kruskal, C.P., Searching, merging, and sorting in parallel computation, *IEEE Transactions on Computers*, C–32(10), 942–946, October 1983.

53. Kumar, V., Grama, A., Gupta, A., and Karypis, G., *Introduction to Parallel Computing: Design and Analysis of Algorithms*, Benjamin/Cummings, Redwood City, CA, 1994.

54. Laudon, J. and Lenoski, D., The SGI Origin: A ccNUMA highly scalable server, in *Proceedings of the 24th Annual International Symposium on Computer Architecture*, 241–251, Denver, CO, June 1997.

55. Leighton, F.T., *Introduction to Parallel Algorithms and Architectures: Arrays, Trees, Hypercubes*, Morgan Kaufmann, San Mateo, CA, 1992.

56. Leiserson, C.E., Fat-trees: Universal networks for hardware-efficient supercomputing, *IEEE Transactions on Computers*, C–34(10), 892–901, October 1985.

57. Lengauer, T., VLSI theory, in *Handbook of Theoretical Computer Science*, J.van Leeuwen (Ed.), Vol. A: *Algorithms and Complexity*, 837–868. Elsevier Science Publishers, Amsterdam, the Netherlands, 1990.

58. Luby, M., A simple parallel algorithm for the maximal independent set problem, *SIAM Journal on Computing*, 15(4), 1036–1053, November 1986.

59. Lynch, N.A., *Distributed Algorithms*, Morgan Kaufmann, San Francisco, CA, 1996.

60. Maon, Y., Schieber, B., and Vishkin, U., Parallel ear decomposition search (EDS) and st-numbering in graphs, *Theoretical Computer Science*, 47, 277–298, 1986.

61. Matias, Y. and Vishkin, U., On parallel hashing and integer sorting, *Journal of Algorithms*, 12(4), 573–606, December 1991.

62. McColl, W.F., BSP programming, in *Specification of Parallel Algorithms*, Vol. 18: *DIMACS Series in Discrete Mathematics and Theoretical Computer Science*, G.E. Blelloch, K.M. Chandy, and S. Jagannathan (Eds.), 25–35. American Mathematical Society, Providence, RI, May 1994.

63. Miller, G.L. and Ramachandran, V., A new graph triconnectivity algorithm and its parallelization, *Combinatorica*, 12(1), 53–76, 1992.

64. Miller, G.L. and Reif, J.H., Parallel tree contraction part 1: Fundamentals, in *Randomness and Computation*, Vol. 5: *Advances in Computing Research*, S. Micali (Ed.), 47–72. JAI Press, Greenwich, CT, 1989.

65. Miller, G.L. and Reif, J.H., Parallel tree contraction part 2: Further applications, *SIAM Journal of Computing*, 20(6), 1128–1147, December 1991.

66. Miller, R. and Stout, Q.F., Efficient parallel convex hull algorithms, *IEEE Transactions on Computers*, 37(12), 1605–1619, December 1988.

67. Miller, R. and Stout, Q.F., *Parallel Algorithms for Regular Architectures*, MIT Press, Cambridge, MA, 1996.

68. Overmars, M.H. and van Leeuwen, J., Maintenance of configurations in the plane, *Journal of Computer and System Sciences*, 23(2), 166–204, October 1981.

69. Pratt, V.R. and Stockmeyer, L.J., A characterization of the power of vector machines, *Journal of Computer and System Sciences*, 12(2), 198–221, April 1976.

70. Preparata, F. and Sarwate, D., An improved parallel processor bound in fast matrix inversion, *Information Processing Letters*, 7(3), 148–150, April 1978.

71. Preparata, F.P. and Shamos, M.I., *Computational Geometry—An Introduction*, Springer-Verlag, New York, 1985.

72. Ranade, A.G., How to emulate shared memory, *Journal of Computer and System Sciences*, 42(3), 307–326, June 1991.

73. Ranka, S. and Sahni, S., *Hypercube Algorithms: With Applications to Image Processing and Pattern Recognition*, Springer-Verlag, New York, 1990.

74. Reid-Miller, M., List ranking and list scan on the Cray C90, *Journal of Computer and Systems Sciences*, 53(3), 344–356, December 1996.

75. Reif, J.H., Ed., *Synthesis of Parallel Algorithms*, Morgan Kaufmann, San Mateo, CA, 1993.

76. Reif, J.H. and Valiant, L.G., A logarithmic time sort for linear size networks, *Journal of the Association for Computing Machinery*, 34(1), 60–76, January 1987.

77. Savitch, W.J. and Stimson, M., Time bounded random access machines with parallel processing, *Journal of the Association for Computing Machinery*, 26(1), 103–118, January 1979.

78. Shiloach, Y. and Vishkin, U., Finding the maximum, merging and sorting in a parallel computation model, *Journal of Algorithms*, 2(1), 88–102, March 1981.

79. Shiloach, Y. and Vishkin, U., An $O(\log n)$ parallel connectivity algorithm, *Journal of Algorithms*, 3(1), 57–67, March 1982.

80. Siegel, H.J., *Interconnection Networks for Large-Scale Parallel Processing: Theory and Case Studies*, 2nd edn., McGraw–Hill, New York, 1990.

81. Stone, H.S., Parallel tridiagonal equation solvers, *ACM Transactions on Mathematical Software*, 1(4), 289–307, December 1975.

82. Strassen, V., Gaussian elimination is not optimal, *Numerische Mathematik*, 14(3), 354–356, 1969.

83. Tarjan, R.E. and Vishkin, V., An efficient parallel biconnectivity algorithm, *SIAM Journal of Computing*, 14(4), 862–874, November 1985.

84. Ullman, J.D., *Computational Aspects of VLSI*, Computer Science Press, Rockville, MD, 1984.

85. Valiant, L.G., A bridging model for parallel computation, *Communications of the ACM*, 33(8), 103–111, August 1990.

86. Valiant, L.G., General purpose parallel architectures, in *Handbook of Theoretical Computer Science*, J.van Leeuwen (Ed.), Vol. A: *Algorithms and Complexity*, 943–971. Elsevier Science Publishers, Amsterdam, the Netherlands, 1990.

87. Vishkin, U., Parallel-design distributed-implementation (PDDI) general purpose computer, *Theoretical Computer Science*, 32(1–2), 157–172, July 1984.

88. Wyllie, J.C., The complexity of parallel computations, Technical Report TR-79-387, Department of Computer Science, Cornell University, Ithaca, NY, August 1979.

89. Yang, M.C.K., Huang, J.S., and Chow, Y., Optimal parallel sorting scheme by order statistics, *SIAM Journal on Computing*, 16(6), 990–1003, December 1987.

26

Self-Stabilizing Algorithms

Sébastien Tixeuil
Université Pierre et Marie Curie

26.1 Introduction

The study of distributed systems and algorithms helps in understanding the specific features of these systems compared to classic centralized systems: information is local (each element of the system only holds a fraction of the information, and must obtain more by communicating with other elements) and time is local (the elements of the system can run their instructions at different speeds). These two factors result in nondeterministic behaviors, as two consecutive executions of the same distributed system are likely to be different. The fact that certain elements of the system can become faulty increases even further this nondeterminism and the difficulty of predicting the overall system's behavior.

When the number of components in a distributed system is increased, the possibility for one or several of these components to become faulty also increases. When the production costs of these components are reduced to achieve economies of scale, the rate of potential defects again increases. Finally, when the components of the systems are deployed in an environment that is not necessarily controlled, the risks of faults occurring become impossible to overlook.

26.1.1 Fault Taxonomy in Distributed Systems

A first criterion for classifying faults in distributed systems is localization in time. Usually, three types of possible faults are distinguished:

1. *Transient faults*: Faults that are arbitrary in nature can strike the system, but there is a point in the execution beyond which these faults no longer occur.

2. *Permanent faults*: Faults that are arbitrary in nature can strike the system, but there is a point in the execution beyond which these faults always occur.

3. *Intermittent faults*: Faults that are arbitrary in nature can strike the system at any moment in the execution.

Transient faults and permanent faults are, of course, specific cases of intermittent faults. However, with a system in which intermittent faults rarely occur, a system that tolerates transient faults can be useful, because the useful life span can be long enough.

A second criterion is the nature of the faults. An element of the distributed system can be represented by an automaton, whose states represent the possible values of the element's variables, and whose transitions represent the code run by the element. We can then distinguish the following faults depending on whether they involve the state or the code of the element:

1. *State-related faults*: Changes in an element's variables may be caused by disturbances in the environment (e.g., electromagnetic waves), attacks (e.g., buffer overflow) or simply faults on the part of the equipment used. For example, it is possible for some variables to have values that they are not supposed to have if the system is running normally.

2. *Code-related faults*: An arbitrary change in an element's code is most often the result of an attack (e.g., the replacement of an element by a malicious opponent), but certain less serious types correspond to bugs or a difficulty in handling the load. There are several different subcategories of code-related faults:

    a. *Crashes*: At a given moment during the execution, an element stops its execution permanently and no longer performs any action.

    b. *Omissions*: At different moments during the execution, an element may omit to communicating with the other elements of the system, either in transmission or in reception.

    c. *Duplications*: At different moments during the execution, an element may perform an action several times, even though its code states that this execution must be performed only once.

    d. *Desequencing*: At different moments during the execution, an element may perform the right actions, but in the wrong order.

    e. *Byzantine faults*: These simply correspond to an arbitrary type of fault, and are, therefore, the faults that cause the most harm.

Crashes are included in omissions (an element that no longer communicates is perceived by the rest of the system as an element that has ended its execution). Omissions are trivially included in Byzantine faults. Duplications and desequencing are also included in Byzantine faults, but are generally regarded as behaviors strictly related to communication capabilities.

A third criterion is the extent (or span) of the faults, that is, how many of the individual system components can be hit by faults or attacks.

26.1.2 Fault-Tolerant Algorithm Categories

When faults occur on one or several of the elements that comprise a distributed system, it is essential to be able to deal with them. If a system tolerates no fault whatsoever, the failure of a single one of its elements can compromise the execution of the entire system: This is the case for a system in which an entity has a central role (such as the DNS). In order to preserve the system's useful life span, several ad hoc methods have been developed, which are usually specific to a particular type of fault that is likely to occur in the system in question. However, these solutions can be categorized depending on whether the effect is visible or not to an observer (e.g., a user). A masking solution hides the occurrence of faults to the observer, whereas a nonmasking solution does not present this characteristic: If The effect of faults is visible over a certain period of time, then the system resumes behaving properly.

A masking approach may seem preferable at first, since it applies to a greater number of applications. Using a non-masking approach to regulate air traffic would make collisions possible following the occurrence of faults. However, a masking solution is usually more costly (in resources and in time) than a nonblocking solution, and can only tolerate faults so long as they have been anticipated. For problems such as routing, where being unable to transport information for a few moments will not have catastrophic consequences, a non-masking approach is perfectly well suited. Two major categories for fault-tolerant algorithms can be distinguished:

1. *Robust algorithms*: These use redundancy on several levels of information, of communications, or of the system's nodes, in order to overlap to the extent that the rest of the code can safely be executed. They usually rely on the hypothesis that a limited number of faults will strike the system, so as to preserve at least a majority of correct elements (sometimes more if the faults are more severe). Typically, these are masking algorithms.
2. *Self-stabilizing algorithms*: These rely on the hypothesis that the faults are transient (in other words, limited in time), but do not set constraints regarding the extent of the faults (which may involve all of the system's elements). An algorithm is self-stabilizing [19,20] if it manages, in a finite time, to present an appropriate behavior independent from the initial state of its elements, meaning that the variables of the elements may exist in a state that is arbitrary (and impossible to achieve by running the application normally). Typically, self-stabilizing algorithms are non-masking, because between the moment when the faults cease and the moment when the system has stabilized to an appropriate behavior, the execution may turn out to be somewhat erratic.

Robust algorithms are quite close to what we conceive intuitively as fault-tolerance. If an element is susceptible to faults, then each element is replaced with three identical elements, and each time an action is undertaken, the action is performed three times by each of the elements, and the action actually undertaken is the one that corresponds to the majority of the three individual actions. Self-stabilization would seem to be related more to the concept of convergence in mathematics or control theory, where the objective is to reach a fixed point regardless of the initial position; the fixed point corresponds here to an appropriate execution. Being capable of starting with an arbitrary state may seem odd (since it would seem that the initial states of the elements are always well known), but studies [70] have shown that if a distributed system is subjected to stopping and restarting-type node failures (which correspond to a definite failure followed by a reinitialization), and communications cannot be totally reliable (some communications may be lost, duplicated or, desequenced), then an arbitrary state of the system can actually be achieved. Even if the probability of the execution that leads to this arbitrary state is negligible in normal conditions, it is not impossible for an attack on the system to attempt to reproduce such an execution. In any case, and regardless of the nature of what led the system to this arbitrary state, a self-stabilizing algorithm is capable of providing an appropriate behavior in a finite amount of time. In fact, self-stabilizing distributed algorithms are found in a number of protocols used in computer networks [46,64].

Figure 26.1 sums up the relative capabilities of self-stabilizing and robust algorithms, respectively. The three axes take into account the three possibilities to classify faults in distributed systems that are described in Section 26.1.1. With a self-stabilizing algorithm, an external user may experience erratic behavior (the stabilizing phase) after the faults have actually ceased, while a robust algorithm will always appear as behaving properly. In contrast, a self-stabilizing algorithm makes no assumption about the extent or the nature of the faults, while robust systems will generally put constraints on those. The rest of the chapter is organized as follows: Section 26.2 presents the most common

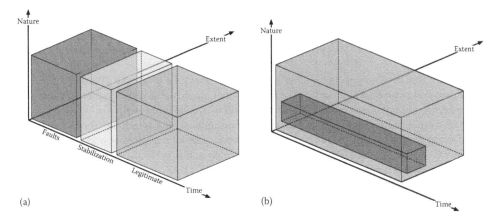

FIGURE 26.1 (a) Self-stabilization vs. (b) robustness.

hypotheses that are made in the self-stabilizing literature. Section 26.3 gives several examples of self-stabilizing algorithms, using various kinds of problems, hypotheses, and proof techniques. Section 26.4 presents the main variants of self-stabilization and concludes the chapter.

26.2   Models

Traditionally, a distributed system is usually represented by a graph, in which the nodes are the system's machines and the edges represent the ability of two machines to communicate. Thus, two machines are connected if they are capable of communicating information to one another (e.g., using a network connection). In some cases, the edges of the graph are oriented so as to represent the fact that the communication can only take place one way (e.g., wireless communication from a satellite to an antenna on the ground). From now on, we will indiscriminately use the words machine, node, or process depending on the context.

26.2.1   System Hypotheses

In the context of self-stabilization, the hypotheses made for the system generally do not include, as with robust algorithms, conditions on the completeness or the globality of the communications. Many algorithms run on systems with nodes that only communicate locally. However, several hypotheses may be crucial for the algorithm to run properly, and involve the hypotheses made regarding the scheduling of the system:

1. *Atomicity of the communications*: most of the self-stabilizing algorithms discussed in the literature use communication primitives with a high level of atomicity. At least three historic models are found in the literature:

   a. *The state model* (or shared memory model [19]): In one atomic step, a node can read the state of each of the neighboring nodes, and update its own state.
   b. *The shared register model* [24]: In one atomic step, a node can read the state of one of its neighboring nodes, or update its own state, but not both simultaneously.
   c. *The message passing model* [1,25,50]: This is the classic model for distributed algorithms, for which in one atomic step, a node sends a message to one of the neighboring nodes, or receives a message from one of the neighboring nodes, but not both simultaneously.

With the recent study of the self-stabilization property in wireless and ad hoc sensor networks, several models for local diffusion with potential collisions have appeared. In the model that presents the highest degree of atomicity [53], a node can, in one atomic step, read its own state and partially write the state of each of the neighboring nodes. If two nodes simultaneously write the state of a common neighbor, a collision occurs and none of the information is written. A more realistic model [43] consists of defining two distinct and atomic actions for local diffusion on one hand and the reception of a locally diffused message on the other.

In the case of bidirectional communications, it is possible to simulate a model using another model. For example, [20] shows how to transform the shared memory model into a message passing model. In the models that are specific to wireless networks, [52] shows how to transform the local diffusion model with collisions into a shared memory model; in a similar fashion [41] shows that the model in [43] can be transformed into a shared memory model. There are two problems with these transformations:

a. the transformation uses up resources (time, memory, energy in the case of sensors), which could be avoided using a direct solution in the model closest to the considered system;

b. the transformation is only possible in systems with bidirectional communications: this is due to the fact that acknowledgments have to be sent regularly to ensure that the highest level model is properly simulated.

2. *Spatial scheduling*: Historically, self-stabilizing algorithms rely on the hypothesis that two neighboring nodes cannot execute their codes simultaneously. This makes it possible to break symmetry in certain configurations [4,65]. Usually, three main possibilities are distinguished for scheduling, depending on which constraints are wanted:

a. *Central scheduling*: At a given moment, only one of the system's nodes can run its code;

b. *Global (or synchronous) scheduling*: At a given moment, all of the system's nodes run their codes;

c. *Distributed scheduling*: At a given moment, an arbitrary subset of the system's nodes runs its code. This type of spatial scheduling is the most realistic.

Other variations are also possible (e.g., a k-locally central scheduling [60]: at a given moment, in each neighborhood of node at distance at most k, only one of the nodes executes its code), but in practice, they are equivalent to one of the three models mentioned above (see [67,69]). The more constrained the spatial scheduling model is, the easier it is to solve problems. For example, [4] shows that it is impossible to color an arbitrary graph in a distributed and deterministic fashion. On the other hand, [36] shows that if the spatial scheduling is locally central, then such a solution is possible. Some algorithms, which rely on the hypothesis of one of these models, can be run in another model, as before, at the price of a greater consumption of resources. Because the most general model is the distributed model, it may be transformed into a more constrained model using a mutual exclusion algorithm [14,19,38] (for the central model), or a synchronization algorithm [2] (for the global model).

3. *Temporal scheduling*: The first self-stabilizing [19] algorithms were independent of the concept of time, that is, they were written in a purely asynchronous model, where no hypothesis is stated regarding the relative speeds of the system's nodes. Later on, scheduling models with heavier constraints began to appear, particularly for the description of real systems. Schedulers are usually divided into three main types:

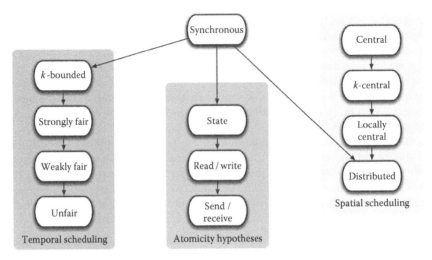

FIGURE 26.2 Taxonomy of system hypotheses in self-stabilization.

  a. *Arbitrary (aka unfair, adversarial) scheduling*: no hypothesis is made regarding the respective execution properties of the system's nodes, other than the simple progression (at each moment, at least one node executes some actions);

  b. *Fair scheduling*: each node runs local actions infinitely often;

  c. *Bounded scheduling*: between the executions of two actions for the same system node, each node executes a bounded number of actions.

Other variations are possible (e.g., [48] makes a distinction between weak fairness and strong fairness), but as before, the more constrained the temporal scheduling model is, the easier it is to solve problems. Bounded scheduling can be constrained further in order to obtain synchronous (or global) scheduling. As with the variations on the previous models, there are algorithms for transforming the execution from one model to another. For example, alternators [35,45,54] can be used to construct a bounded model based on a fair or arbitrary model. On the other hand, because of its unbounded nature, the strict fair model cannot be (strictly) constructed from the arbitrary model.

The general taxonomy of common system hypotheses made in self-stabilizing literature is presented in Figure 26.2. An arrow from a hypothesis a to a hypothesis b implies that a is stronger than b, that is, an algorithm p_a assuming a is weaker than an algorithm p_b assuming b. That is, p_b will work under hypotheses b and a, but p_a will only work with hypothesis a. Since the kinds of scheduling presented in this section are quite different from the notion of scheduling used, for example, in parallel algorithms, the most used term to refer to those hypotheses in the self-stabilizing literature is the daemon. The two terms are used interchangeably in the remaining of the text.

26.2.2 Program Model

For the formal description of our program, we use simplified UNITY notation [33]. A program consists of a set of processes. A process contains a set of constants that it can read but not update. A process maintains a set of variables. Each variable ranges over a fixed domain of values. We use small case letters to denote singleton variables, and capital ones to denote sets. Some variables are persistent from one activation of the process to the next, and are called state variables, while some other are wiped out between two activations of a process, and are called local variables. When there is no ambiguity, the generic term variable refers to a state variable.

An action has the form ⟨*name*⟩ : ⟨*guard*⟩ ⟶ ⟨*command*⟩. A guard is a Boolean predicate. In the shared memory model, this predicate is over the variables of the process and its communication neighbors. In the shared register model, this predicate is over the variables of the process and a single input register. In the message passing model, this predicate is over the variables of the process and the **receive**(*m*) primitive. In this context, **receive**(*m*) evaluates to *true* if a message matching *m* has been received, and to *false* otherwise. A command is a sequence of statements assigning new values to the variables of the process (in all communication models) and sending messages (using the **send**(*m*) primitive) in the message passing model. Besides assignments, conditionals and loops are also available (with the **if**-**fi** and **do**-**od** constructions, respectively). We refer to a variable *var* and an action *ac* of process *p* as *var.p* and *ac.p*, respectively. We make use of two operators: [] denotes alternation (that is, *a*[]*b* denotes the fact that either *a* or *b* is executed, but not both, and the choice is nondeterministic) while *[] denotes iteration (that is, *[*a*] denotes the fact that *a* is repetitively evaluated). A parameter is used to define a set of actions as one parametrized action. For example, let *j* be a parameter ranging over values 2, 5, and 9; then a parametrized action *ac.j* defines the set of actions: *ac.(j* := 2) [] *ac.(j* := 5) [] *ac.(j* := 9).

A configuration of the system is the assignment of a value to every variable of each process from the variable's corresponding domain (and possibly a similar assignment of message contents in channels between processes in case of the message passing model). Each process contains a set of actions. An action is enabled in some configuration if its guard is *true* at this state. A computation is a maximal sequence of configurations such that for each configuration s_i, the next configuration s_{i+1} is obtained by executing the command of an action that is enabled in s_i. Maximality of a computation means that the computation is infinite or it terminates in a state where none of the actions are enabled. Such state is a fix-point.

A configuration conforms to a predicate if this predicate is **true** in this configuration; otherwise the state violates the predicate. By this definition, every state conforms to predicate **true** and none conforms to **false**. Let R and S be predicates over the configuration of the system. Predicate R is closed with respect to the program actions if every configuration of the computation that starts in a configuration conforming to R also conforms to R. Predicate R converges to S if R and S are closed and any computation starting from a configuration conforming to R contains a configuration conforming to S. The program stabilizes to R if **true** converges to R.

DEFINITION 26.1 (Self-stabilization) Starting from an arbitrary initial configuration, any execution of a *self-stabilizing* algorithm contains a subsequent configuration from which every execution satisfies the specification.

26.3 Designing Self-Stabilizing Algorithms

In the context of self-stabilization, depending on the problem that we wish to solve, the minimum time required for going back to a correct configuration varies significantly. Problems are generally divided into two categories:

1. *Static problems*: We wish to perform a task that consists of calculating a function that depends on the system in which it is assessed. For example, it can consist of coloring the nodes of a network so as to never have two adjacent nodes with the same color; another example is the calculation of the shortest paths to a destination [21].

2. *Dynamic problems*: We wish to perform a task that performs a service for upper layer algorithms. The model transformation algorithms such as token passing fall into this category.

The example designs that are featured in this section are Maximal Matching (Section 26.3.1), Generalized Diners (Section 26.3.2), Census (Section 26.3.3), and Token Passing (Section 26.3.4). The first two are written for the state model (aka. shared memory model) and the last two for the message passing model. Maximal Matching and Census are instances of static problems, while Generalized Diners and Token Passing are instances of dynamic ones. The proof techniques used to show the self-stabilization properties of the presented algorithms include attractors, potential functions, and Markov chains. The examples that are presented in this section previously appeared in [12,16,27,58].

26.3.1 Maximal Matching

Algorithm. In the following, we present and motivate our algorithm for computing a maximal matching. The algorithm is self-stabilizing and does not make any assumption on the network topology. A set of edges $M \subseteq E$ is *matching* if and only if $x, y \in M$ implies that x and y do not share a common end point. A matching M is *maximal* if no proper superset of M is also a matching. Note that a maximal matching differs from a *maximum* matching, that is required to have maximum cardinality among all possible maximal matchings. In the remaining of the section, n denotes the number of nodes and m denotes the number of edges, respectively.

Each process i has a variable *points_to.i* pointing to one of its neighbors or to *null*. We say that processes i and j *are married* to each other if and only if i and j are neighbors and their *points_to*-values point to each other. In this case we will also refer to i as being married without specifying j. However, we note that in this case j is unique. A process which is not married is *unmarried*. Figure 26.3 depicts two possible configurations of our algorithm: In Figure 26.3a, some nodes point to some other without the converse being true (leaving those nodes unmatched) while Figure 26.3b depicts a maximal matching configuration.

We also use a variable *matched.i* to let neighboring processes of i know if process i is married or not. To determine the value of *matched.i* we use a predicate *PRmarried(i)* which evaluates to true if and only if i is married. Thus predicate *PRmarried(i)* allows process i to know if it is currently married and the variable *matched.i* allows neighbors of i to know if i is married. Note that the value of *matched.i* is not necessarily equal to *PRmarried(i)*.

Our self-stabilizing scheme is given in Algorithm 2. It is composed of four mutually exclusive guarded rules as described below.

The *update* rule updates the value of *matched.i* if it is necessary, while the three other rules can only be executed if the value of *matched.i* is correct. In the *marriage* rule, an unmarried process that is currently being pointed to by a neighbor j tries to marry j by setting *points_to.i* $= j$. In

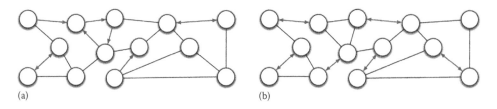

(a) (b)

FIGURE 26.3 Possible configurations of Algorithm 2. (a) A possible arbitrary initial configuration and (b) a legitimate maximal matching configuration.

the *seduction* rule, an unmarried process that is not being pointed to by any neighbor, points to an unmarried neighbor with the objective of marriage. Note that the identifier of the chosen neighbor has to be larger than that of the current process. This is enforced to avoid the creation of cycles of pointer values. In the *abandonment* rule, a process i resets its *points_to.i* value to *null*. This is done if the process j which it is pointing to does not point back at i and if either (1) j is married, or (2) j has a lower identifier than i. Condition (1) allows a process to stop waiting for an already married process while the purpose of Condition (2) is to break a possible initial cycle of *points_to*-values.

We note that if *PRmarried(i)* holds at some point of time then from then on it will remain true throughout the execution of the algorithm. Moreover, the algorithm will never actively create a cycle of pointing values since the *seduction* rule enforces that $j > i$ before process i will point to process j. Moreover, all initial cycles are eventually broken since the guard of the *abandonment* rule requires that $j \leq i$.

Algorithm 2 $\mathcal{MM}$: A self-stabilizing maximal matching algorithm

---

    process i
    const
        N: communication neighbors of i
    parameter
        $r : N$
    var
        matched.i : {**true, false**}
        points_to.i : {**null**} $\cup N$
    predicate
        $PRmarried(i) \equiv (points\_to.i = r \wedge points\_to.r = i)$

           *[
update: $matched.i \neq PRmarried(i) \longrightarrow$
                        $matched.i := PRmarried(i)$

        []
marriage: $matched.i = PRmarried(i) \wedge points\_to.i = \textbf{null} \wedge points\_to.r = i \longrightarrow$
                        $points\_to.i := r$

        []
seduction: $matched.i = PRmarried(i) \wedge points\_to.i = \textbf{null} \wedge \forall k \in N : points\_to.k \neq i$
                        $\wedge(points\_to.r = \textbf{null} \wedge r > i \wedge \neg matched.r) \longrightarrow$
                        $points\_to.i := Max\{j \in N : (points\_to.j = \textbf{null} \wedge j > i \wedge \neg matched.j)\}$

        []
abandonment: $matched.i = PRmarried(i) \wedge points\_to.i = j \wedge points\_to.j \neq i$
                        $\wedge(matched.j \vee j \leq i) \longrightarrow$
                        $points\_to.i := \textbf{null}$

        ]

Proof of Correctness. The proof of correctness presented in this section demonstrates how to cope with an adversarial daemon. Intuitively, we can only assume progress of the computation (i.e., at least one node that is activatable makes a move), so the main goal of the proof is to show that whatever the

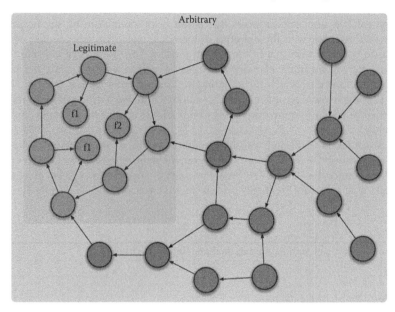

FIGURE 26.4 Proving self-stabilization with unfair scheduling.

choices of the daemon are, only a finite number of moves can be made by each node before reaching a legitimate configuration, as there exists no chain of consecutive configurations that form a loop in the set of illegitimate configurations (see Figure 26.4). For the maximal matching problem, the situation is facilitated by the fact that a final configuration is eventually reached, in which a maximal matching is constructed.

In the following, we will first show that when Algorithm 2 has reached a stable configuration it also defines a maximal matching. We will then bound the number of steps the algorithm needs to stabilize for the adversarial daemon. We now proceed to show that if Algorithm 2 reaches a stable configuration then the *points_to* and *matched*-values will define a maximal matching M where $(i, j) \in M$ if and only if $(i, j) \in E$, $points\_to.i = j$, and $points\_to.j = i$ while both *matched.i* and *matched.j* are true. A configuration is stable (or terminal) if no processes are eligible to execute a move. In order to perform the proof, we define the following five mutually exclusive predicates:

$PRmarried(i) \quad \equiv \exists j \in N(i) : (points\_to.i = j \text{ and } points\_to.j = i)$
$PRwaiting(i) \quad \equiv \exists j \in N(i) : (points\_to.i = j \text{ and } points\_to.j \neq i \text{ and } \neg PRmarried(j))$
$PRcondemned(i) \equiv \exists j \in N(i) : (points\_to.i = j \text{ and } points\_to.j \neq i \text{ and } PRmarried(j))$
$PRdead(i) \quad \equiv (points\_to.i = null) \text{ and } (\forall j \in N(i) : PRmarried(j))$
$PRfree(i) \quad \equiv (points\_to.i = null) \text{ and } (\exists j \in N(i) : \neg PRmarried(j))$

Note first that each process will evaluate exactly one of these predicates to true. Moreover, also note that *PRmarried(i)* is the same as in Algorithm 2. We now show that in a stable configuration each process i evaluates either *PRmarried(i)* or *PRdead(i)* to true, and when this is the case, the *points_to*-values define a maximal matching. To do so, we first note that in any stable configuration the *matched*-values reflects the current status of the process.

LEMMA 26.1   In a stable configuration we have *matched.i = PRmarried(i)* for each $i \in V$.

PROOF   This follows directly because if *matched.i ≠ PRmarried(i)* then *i* is eligible to execute the *update* rule.

We show in the following three lemmas that no process will evaluate either *PRwaiting(i)*, *PRcondemned(i)*, or *PRfree(i)* to true in a stable configuration.

LEMMA 26.2   In a stable configuration *PRcondemned(i)* is false for each $i \in V$.

PROOF   If there exists at least one process *i* in the current configuration such that *PRcondemned(i)* is true then *points_to.i* is pointing to a process $j \in N(i)$ that is married to a process *k* where $k \neq i$. From Lemma 26.1 it follows that in a stable configuration we have *matched.i = PRmarried(i)* and *matched.j = PRmarried(j)*. Thus in a stable configuration the predicate (*matched.i = PRmarried(i)* ∧ *points_to.i = j* ∧ *points_to.j ≠ i* ∧ *matched.j*) evaluates to true. But then process *i* is eligible to execute the *abandonment* rule contradicting that the current configuration is stable.

LEMMA 26.3   In a stable configuration *PRwaiting(i)* is false for each $i \in V$.

PROOF   Assume that the current configuration is stable and that there exists at least one process *i* such that *PRwaiting(i)* is true. Then it follows that *points_to.i* is pointing to a process $j \in N(i)$ such that *points_to.j ≠ i* and *j* is unmarried. Note first that if *points_to.j = null* then process *j* is eligible to execute a *marriage* move. Moreover, if $j < i$ then process *i* can execute an *abandonment* move. Assume, therefore, that *points_to.j ≠ null* and that $j > i$. It then follows from Lemma 26.2 that ¬*PRcondemned(j)* is true and since *j* is not married we also have ¬*PRmarried(j)*. Thus *PRwaiting(j)* must be true. Repeating the same argument for *j* as we just did for *i*, it follows that if both *i* and *j* are ineligible for a move then there must exist a process *k* such that *points_to.j = k*, $k > j$, and *PRwaiting(k)* also evaluate to true. This sequence of processes cannot be extended indefinitely since each process must have a higher identifier than the preceding one. Thus there must exist some process in *V* that is eligible for a move and the assumption that the current configuration is stable is incorrect.

LEMMA 26.4   In a stable configuration *PRfree(i)* is false for each $i \in V$.

PROOF   Assume that the current configuration is stable and that there exists at least one process *i* such that *PRfree(i)* is true. Then it follows that *points_to.i = null* and that there exists at least one process $j \in N(i)$ such that *j* is not married.

Next, we look at the value of the different predicates for the process *j*. Since *j* is not married it follows that *PRmarried(j)* evaluates to false. Moreover, from Lemmas 26.2 and 26.3 we have that both *PRwaiting(j)* and *PRcondemned(j)* must evaluate to false. Finally, since *i* is not married we cannot have *PRdead(j)*. Thus we must have *PRfree(j)*. But then the process with the smaller identifier of *i* and *j* is eligible to propose to the other, contradicting the fact that the current configuration is stable.

From Lemmas 26.2 through 26.4 we immediately get the following corollary.

COROLLARY 26.1   In a stable configuration either *PRmarried*(*i*) or *PRdead*(*i*) holds for every *i* ∈ *V*.

We can now show that a stable configuration also defines a maximal matching.

THEOREM 26.1   *In any stable configuration the matched and points_to-values define a maximal matching.*

PROOF   From Corollary 26.1 we know that in a stable configuration either *PRmarried*(*i*) or *PRdead*(*i*) holds for every *i* ∈ *V*. Moreover, from Lemma 26.1 it follows that *matched.i* is true if and only if *i* is married. It is then straightforward to see that the *points_to*-values define a matching.

To see that this matching is maximal assume to the contrary that it is possible to add one more edge (*i*, *j*) to the matching so that it still remains a legal matching. To be able to do so we must have *points_to.i* = *null* and *points_to.j* = *null*. Thus we have ¬*PRmarried*(*i*) and ¬*PRmarried*(*j*) which again implies that both *PRdead*(*i*) and *PRdead*(*j*) evaluates to true. But according to the *PRdead* predicate, two adjacent processes cannot be dead at the same time. It follows that the current matching is maximal.

In the following we will show that Algorithm 2 will reach a stable configuration after at most $3 \cdot n + 2 \cdot m$ steps under the distributed adversarial daemon.

First we note that as soon as two processes are married they will remain so for the rest of the execution of the algorithm.

LEMMA 26.5   If processes *i* and *j* are married in a configuration *C*, that is, *points_to.i* = *j* and *points_to.j* = *i*, then they will remain married in any ensuing configuration *C′*.

PROOF   Assume that *points_to.i* = *j* and *points_to.j* = *i* in some configuration *C*. Then process *i* can neither execute the *marriage* nor the *seduction* rule since these require that *points_to.i* = *null*. Similarly, *i* cannot execute the *abandonment* rule since this requires that *points_to.j* ≠ *i*. The exact same argument for process *j* shows that *j* also cannot execute any of the three rules: *marriage*, *seduction*, and *abandonment*. Thus the only rule that processes *i* and *j* can execute is *update* but this will not change the values of *points_to.i* or *points_to.j*.

A process discovers that it is married by executing the *update* rule. Thus this is the last rule a married process will execute in the algorithm. This is reflected in the following.

COROLLARY 26.2   If a process *i* executes an *update* move and sets *matched.i* = true then *i* will not move again.

PROOF   From the predicate of the *update* rule it follows that when process *i* sets *matched.i* = true there must exist a process *j* ∈ *N*(*i*) such that *points_to.i* = *j* and *points_to.j* = *i*. Thus from Lemma 26.5 the only subsequent move *i* can make is an *update* move. But since the value of *matched.i* is correct and *points_to.i* and *points_to.j* will not change again, this will not happen.

Since a married process cannot become "unmarried" we also have the following restriction on the number of times the *update* rule can be executed by any process.

COROLLARY 26.3   Any process executes at most two *update* moves.

We will now bound the number of moves from the set {*marriage, seduction, abandonment*}. Each such move is performed by a process i in relation to one of its neighbors j. We denote any such move made by either i or j with respect to the other as an i, j-*move*.

LEMMA 26.6   For any edge $(i, j) \in E$, there can at most be three steps in which an i, j-move is performed.

PROOF   Let $(i, j) \in E$ be an edge such that $i < j$. We then consider four different cases depending on the initial values of *points_to.i* and *points_to.j*. Note from Algorithm 2 that the only values that *points_to.i* and *points_to.j* can take on are *points_to.i* $\in$ {*null*} $\cup N(i)$ and *points_to.j* $\in$ {*null*} $\cup N(j)$. For each case, we will show that there can at most be three steps in which i, j-moves occur.

1. *points_to.i* $\neq j$ and *points_to.j* $\neq i$. Since $i < j$ the first i, j-move cannot be process j executing a *seduction* move. Moreover, as long as *points_to.i* $\neq j$, process j cannot execute a *marriage* move. Thus process j cannot execute an i, j-move until after process i has first made an i, j-move. It follows that the first possible i, j-move is that i executes a *seduction* move simultaneously as j makes no i, j-move. Note that at the starting configuration of this move, we must have ¬*matched.j*. If the next i, j-move is performed by j simultaneously as i performs no move then this must be a *marriage* move which results in *points_to.i* $= j$ and *points_to.j* $= i$. Then by Lemma 26.5 there will be no more i, j-moves. If process i makes the next i, j-move (independently of what process j does) then this must be an *abandonment* move. But this requires that the value of *matched.j* has changed from false to true. Then by Corollary 26.2 process j will not make any more i, j-moves and since *points_to.j* $\neq null$ and *points_to.j* $\neq i$ for the rest of the algorithm, it follows that process i cannot execute any future i, j-move. Thus there can at most be two steps in which i, j-moves are performed.

2. *points_to.i* $= j$ and *points_to.j* $\neq i$. If the first i, j-move only involves process j then this must be a *marriage* move resulting in *points_to.i* $= j$ and *points_to.j* $= i$ and from Lemma 26.5, neither i nor j will make any future i, j-moves. If the first i, j-move involves process i then this must be an *abandonment* move. Thus in the configuration prior to this move we must have *matched.j* $=$ true. It follows that either *matched.j* $\neq$ *PRmarried(j)* or *points_to.j* $\neq null$. In both cases process j cannot make an i, j-move simultaneously as i makes its move. Thus following the *abandonment* move by process i we are at Case (1) and there can at most be two more i, j-moves. Hence, there can at most be a total of three steps with i, j-moves.

3. *points_to.i* $\neq j$ and *points_to.j* $= i$. If the first i, j-move only involves process i then this must be a *marriage* move resulting in *points_to.i* $= j$ and *points_to.j* $= i$ and from Lemma 26.5 neither i nor j will make any future i, j-moves. If the first i, j-move involves process j then this must be an *abandonment* move. If process i does not make a simultaneous i, j-move then this will result in configuration (1) and there can at most be two more steps with i, j-moves for a total of three steps containing i, j-moves. If process i does make a simultaneous i, j-move with process j executing an abandonment move, then this must be a *marriage* move. We are now at a similar configuration as Case (2) but with ¬*matched.j*. If the second i, j-move involves process i then this must be an *abandonment* move implying that *matched.j* has changed to true. It then follows from Corollary 26.2

that process j (and therefore also process i) will not make any future i, j-move leaving a total of two steps containing i, j-moves. If the second i, j-move does not involve i then this must be a *marriage* move performed by process j resulting in $points\_to.i = j$ and $points\_to.j = i$ and from Lemma 26.5 neither i nor j will make any future i, j-moves.

4. $points\_to.i = j$ and $points\_to.j = i$. In this case it follows from Lemma 26.5 that neither process i nor process j will make any future i, j-moves.

It should be noted in the proof of Lemma 26.6 that only an edge (i, j) where we initially have either $points\_to.i = j$ or $points\_to.j = i$ (but not both) can result in three i, j-moves, otherwise the limit is two i, j-moves per edge. When we have three (i, j)-moves across an edge (i, j) we can charge these moves to the process that was initially pointing to the other. In this way each process will at most be incident on one edge for which it is charged three moves. From this observation we can now give the following bound on the total number of steps needed to obtain a stable solution.

THEOREM 26.2   *Algorithm 2 will stabilize after at most $3 \cdot n + 2 \cdot m$ steps under the distributed adversarial daemon.*

PROOF   From Corollary 26.2 we know that there can be at most $2n$ *update* moves, each of which can occur in a separate step. From Lemma 26.6 it follows that there can at most be three i, j-moves per edge. But as observed, there is at most one such edge incident on each process i for which process i is charged for, otherwise the limit is two i, j-moves. Thus the total number of i, j-moves is at most $n + 2 \cdot m$ and the result follows.

From Theorem 26.2 it follows that Algorithm 2 will use $O(m)$ moves on any connected system when assuming a distributed daemon. Since the distributed daemon encompasses the sequential daemon this result also holds for the sequential daemon. To see that this is a tight bound for the stabilization time, consider a complete graph in which each process $i_n, i_{n-1}, \ldots, i_1$ has a unique identifier such that $i_n > i_{n-1} > \cdots > i_1$. We will now show that there exists an initial configuration and a sequence of moves such that $\Omega(m)$ moves are executed before the system reaches a stable configuration. Consider that initially every process is unmarried and not pointing to anyone. Then the processes $i_{n-1}, \ldots, i_1$ will be eligible to execute *seduction* moves and point to i_n. Following this, i_n may now execute a *marriage* move, and become married to i_{n-1}. Thus the processes $i_{n-2}, \ldots, i_1$ are now eligible to execute *abandonment* moves. Observe that following these moves, two moves have been executed for every edge incident to i_n, and the processes $i_{n-2}, \ldots, i_1$ are once again not pointing to any other process. Furthermore, by Lemma 26.5 we know that neither i_n nor i_{n-1} will execute any further *nonupdate* moves (note that no moves were executed for any edge incident on i_{n-1}, with the exception of the edge (i_n, i_{n-1})). In the same manner, we can now reason that in addition to the above, two moves are executed for every edge incident on i_{n-2} (with the exception of those incident on i_n or i_{n-1}). Repeating this argument gives that $\Omega(m)$ *nonupdate* moves are executed before the system reaches a stable configuration.

26.3.2 Generalized Diners

Algorithm. An instance of the generalized diners problem defines for each process p a set of communication neighbors $N.p$ and a set of conflict neighbors $M.p$. Both relations are symmetric. That is, for any two processes p and q if $p \in N.q$ then $q \in N.p$. The same applies to $M.p$. Throughout the computation each process requests critical section (CS for short) access an arbitrary number of times: from zero to infinity. A program that solves the generalized diners satisfies the following two properties for each process p:

safety—if the action that executes the CS is enabled in p, it is disabled in all processes of $M.p$;

liveness—if p wishes to execute the CS, it is eventually allowed to do so.

In this section it is assumed that in any computation, the action execution is weakly fair. That is, if an action is enabled in all but finitely many states of an infinite computation then this action is executed infinitely often.

K-hop diners is a restriction of generalized diners. In k-hop diners, for each process p, $M.p$ contains all processes whose distance to p in the graph formed by the communication topology is no more than k.

The main idea of the algorithm is to coordinate CS request notifications between multiple conflict neighbors of the same process. We assume that for each process p there is a tree that spans $M.p$. This tree is rooted in p. A stabilizing breadth-first construction of a spanning tree is a relatively simple task [20].

The processes in this tree propagate CS request of its root. The request reflects from the leaves and informs the root that its conflict neighbors are notified. This mechanism resembles information propagation with feedback [9].

The access to the CS is granted on the basis of the priority of the requesting process. Each process has an identifier that is unique throughout the system. A process with lower identifier has higher priority. To ensure liveness, when executing the CS, each process p records the identifiers of its lower priority conflict neighbors that also request the CS. Before requesting it again, p then waits until all these processes access the CS.

Each process p has access to a number of constants. The set of identifiers of its communication neighbors is N, and its conflict neighbors is M. For each of its conflict neighbors r, p knows the appropriate spanning tree information: the parent identifier—$dad.p.r$, and a set of ids of its children—$KIDS.p.r$.

Process p stores its own request state in variable $state.p.p$ and the state of each of its conflict neighbors in $state.p.r$. Notice that p's own state can be only **idle** or **req**, while for its conflict neighbors p also has **rep**. To simplify the description, depending on the state, we refer to the process as being idle, requesting, or replying. In *YIELD*, process p maintains the ids of its lower priority conflict neighbors that should be allowed to enter the CS before p requests it again. Variable *needcs* is an external Boolean variable that indicates if CS access is desired. Notice that CS entry is guaranteed only if *needcs* remains **true** until p requests the CS.

There are five actions in the algorithm. The first two: *join* and *enter* manage CS entry of p itself. The remaining three—*forward*, *back*, and *stop*—propagate CS request information along the tree. Notice that the latter three actions are parametrized over the set of p's conflict neighbors.

Action *join* states that p requests the CS when the application variable *needcs* is **true**, p itself, as well as its children in its own spanning tree, is idle and there are no lower priority conflict neighbors to wait for. As action *enter* describes, p enters the CS when its children reply and the higher priority processes do not request the CS themselves. To simplify the presentation, we describe the CS execution as a single action.

Action *forward* describes the propagation of a request of a conflict neighbor r of p along r's tree. Process p propagates the request when p's parent $dad.p.r$ is requesting and p's children are idle. Similarly, *back* describes the propagation of a reply back to r. Process p propagates the reply either if its parent is requesting and p is the leaf in r's tree or all p's children are replying. The second disjunct of *back* is to expedite the stabilization of Algorithm 3. Action *stop* resets the state of p in r's tree to idle when its parent is idle. This action removes r from the set of lower priority processes to await before initiating another request.

The operation of Algorithm 3 in legitimate states is illustrated in Figure 26.5. We focus on the conflict neighborhood $M.a$ of a certain node a. We consider representative nodes in the spanning

---

Algorithm 3 $\mathcal{KDP}$: A self-stabilizing k-hops diners algorithm

---

process p
const
  M: k-hop conflict neighbors of p
  N: communication neighbors of p
  $(\forall q : q \in M : dad.p.q \in N, KIDS.p.q \subset N)$
    parent id and set of children ids for each k-hop neighbor
parameter
  $r : M$
var
  $state.p.p$: {**idle**, **req**},
  $(\forall q : q \in M : state.p.q$: {**idle**, **req**, **rep**}),
  $YIELD$: $\{\forall q : q \in M : q > p\}$ lower priority processes to wait for
  $needcs$: **boolean**, application variable to request the CS

   $*[$
join: $needcs \wedge state.p.p = $ **idle** $\wedge YIELD = \varnothing \wedge$
    $(\forall q : q \in KIDS.p.p : state.q.p = $ **idle**$) \longrightarrow$
     $state.p.p := $ **req**

   $[]$
enter: $state.p.p = $ **req** $\wedge$
    $(\forall q : q \in KIDS.p.p : state.q.p = $ **rep**$) \wedge$
    $(\forall q : q \in M \wedge q < p : state.p.q = $ **idle**$) \longrightarrow$
     /* Critical Section */
     $YIELD := \{\forall q : q \in M \wedge q > p : state.p.q = $ **rep**$\}$,
     $state.p.p := $ **idle**

   $[]$
forward: $state.p.r = $ **idle** $\wedge state.(dad.p.r).r = $ **req** $\wedge$
    $((KIDS.p.r = \varnothing) \vee (\forall q : q \in KIDS.p.r : state.q.r = $ **idle**$)) \longrightarrow$
     $state.p.r := $ **req**

   $[]$
back: $state.p.r = $ **req** $\wedge state.(dad.p.r).r = $ **req** $\wedge$
    $((KIDS.p.r = \varnothing) \vee (\forall q : q \in KIDS.p.r : state.q.r = $ **rep**$)) \vee$
    $state.p.r \neq $ **rep** $\wedge state.(dad.p.r).r = $ **rep** $\longrightarrow$
     $state.p.r := $ **rep**

   $[]$
stop: $(state.p.r \neq $ **idle** $\vee r \in YIELD) \wedge$
    $state.(dad.p.r).r = $ **idle** $\longrightarrow$
     $YIELD := YIELD \setminus \{r\}$,
     $state.p.r := $ **idle**

   $]$

---

tree of $M.a$. Specifically, we consider one of a's children—e, a descendant—b, b's parent—c and one of b's children—d.

Initially, the states of all processes in $M.a$ are idle. Then, a executes *join* and sets $state.a.a$ to **req** (see Figure 26.5a). This request propagates to process b, which executes *forward* and sets $state.b.a$ to **req** as well (Figure 26.5b). The request reaches the leaves and bounces back as the leaves change their state to **rep**. Process b then executes *back* and changes its state to **rep** as well (Figure 26.5c). After the

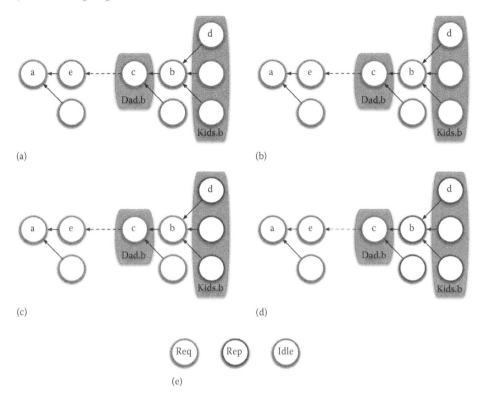

FIGURE 26.5 Phases of Algorithm 3 operation. (a) *a* initiates request, (b) *b* propagates request, (c) *b* propagates reply, (d) *b* resets, and (e) color code.

reply reaches *a* and if none of the higher priority processes are requesting the CS, *a* executes *enter*. This action resets *state.a.a* to **idle**. This reset propagates to *b* which executes *stop* and also changes *state.b.a* to **idle** (Figure 26.5d).

Proof of Correctness. The proof of correctness that is presented in this section demonstrates the use of attractors (see Figure 26.6). An attractor *a* is a predicate on configurations that is closed for any subsequent execution, and that attracts any execution starting from any arbitrary initial state. In other words, any execution of the system has a configuration that satisfies *a*. Usually, attractors are nested as presented in Figure 26.6, meaning that each nested attractor is a refinement of the previous one. Figure 26.6a presents the notion of nested attractors that do not require a fairness condition: starting from any state, only a finite number of steps may be executed before a particular intermediate attractor is satisfied. In contrast, Figure 26.6b presents some paths that are cycling, that is, there exists an execution such that the attractor is never reached (e.g., the cycle of light gray nodes with *f* label). However, for every configuration in a cyclic path, there exists another path that leads to a configuration that satisfies the attractor. Assuming weak fairness (i.e., in all but finitely many configurations, if it is possible that a particular action makes the configuration satisfy the attractor, then this action is eventually executed), we are able to prove that the attractor is eventually satisfied. Of course, there can be such nested attractors as well.

We present Algorithm 3 correctness proof as follows. We first state a predicate we call *InvK* and demonstrate that Algorithm 3 stabilizes to it in Theorem 26.3. We then proceed to show that if *InvK*

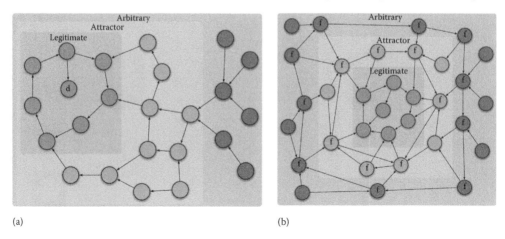

(a)   (b)

FIGURE 26.6 Proving self-stabilization with attractors. (a) Without fairness condition and (b) with fairness condition.

holds, then Algorithm 3 satisfies the safety and liveness properties of the *k*-hop diners in Theorems 26.4 and 26.5, respectively.

Throughout this section, unless otherwise specified, we consider the conflict neighbors of a certain node *a* (see Figure 26.5). That is, we implicitly assume that *a* is universally quantified over all processes in the system. We focus on the following nodes: $e \in KIDS.a.a$, $b \in M.a$, $c \equiv dad.b.a$ and $d \in KIDS.b.a$.

Since we discuss the states of *e*, *b*, *c*, and *d* in the spanning tree of *a*, it is clear from the context, we omit the specifier of the conflict neighborhood. For example, we use *state.b* for *state.b.a*. Moreover, for clarity, we attach the identifier of the process to the actions it contains. For example, *forward.b* is the *forward* action of process *b*.

Our global predicate consists of the following predicates that constrain the states of each individual process and the states of its communication neighbors. The predicate below relates the states of the root of the tree *a* to the states of its children.

$$(state.a = \textbf{idle}) \Rightarrow (\forall e : e \in KIDS.a : state.e \neq \textbf{req}) \qquad (Inv.a)$$

The following sequence of predicates relates the state of *b* to the state of its neighbors.

$$state.b = \textbf{idle} \wedge state.c \neq \textbf{rep} \wedge (\forall d : d \in KIDS.b : state.d \neq \textbf{req}) \qquad (I.b.a)$$

$$state.b = \textbf{req} \wedge state.c = \textbf{req} \qquad (R.b.a)$$

$$state.b = \textbf{rep} \wedge \qquad\qquad (\forall d : d \in KIDS.b : state.d = \textbf{rep}) \qquad (P.b.a)$$

We denote the disjunction of the above three predicates as follows:

$$I.b.a \vee R.b.a \vee P.b.a \qquad (Inv.b.a)$$

The following predicate relates the states of all processes in *M.a*.

$$(\forall a :: Inv.a \wedge (\forall b : b \in M.a : Inv.b.a)) \qquad (InvK)$$

To aid in exposition, we mapped the states and transitions for individual processes in Figure 26.7. Note that to simplify the picture, for the intermediate process *b* we only show the states and transitions if *Inv.f.a* holds for each ancestor *f* of *b*. For *b*, the *I.b*, *R.b*, and *P.b* denote the states conforming to the

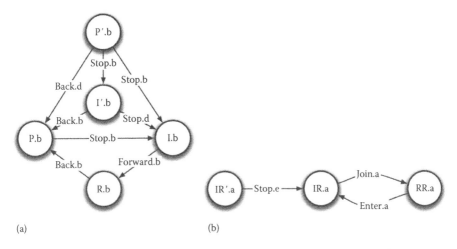

FIGURE 26.7 State transitions for an individual process. (a) Intermediate process *b* if *Inv* holds for ancestors and (b) root process *a*.

respective predicates, while the primed versions $I'.b$ and $P'.b$ signify the states where *b* is respectively idle and replying but *Inv.b.a* does not hold. Notice that if *Inv.c.a* holds for *b*'s parent *c*, the primed version of *R* does not exist. Indeed, to violate *R*, *b* should be requesting while *c* is either idle or replying. However, if *Inv.c.a* holds and *c* is in either of these two states, *b* cannot be requesting.

For *a*, *IR.a* and *RR.a* denote the states where *a* is respectively idle and requesting while *Inv.a* holds. In states *IR'.a*, *a* is idle while *Inv.a* does not hold. Notice that since *state* = **req** falsifies the antecedent of *Inv.a*, the predicate always holds if *a* is requesting. The state transitions in Figure 26.7 are labeled by actions whose execution effects them. Loopback transitions are not shown.

THEOREM 26.3 *Algorithm 3 stabilizes to InvK.*

PROOF By the definition of stabilization, *InvK* should be closed with respect to the execution of the actions of Algorithm 3, and Algorithm 3 should converge to *InvK*. We prove the closure first.

Closure. To aid in the subsequent convergence proof, we show a property that is stronger than just the closure of *InvK*. We demonstrate the closure of the following conjunction of predicates: *Inv.a* and *Inv.b.a* for a set of descendants of *a* up to a certain depth of the tree. To put it another way, in showing the closure of *Inv.b.a* for *b* we assume that the appropriate predicates hold for all its ancestors. Naturally, the closure of *InvK* follows. By definition of predicate closure, we need to demonstrate that if the predicate holds in a certain state, the execution of any action in this state does not violate the predicate.

Let us consider *Inv.a* and the root process *a* first. Notice that the only two actions that can potentially violate *Inv.a* are *enter.a* and *forward.e*. Let us examine each action individually. If *enter.a* is enabled, each child of *a* is replying. Hence, when it is executed and it changes the state of *a* to **idle**, *Inv.a* holds. If *forward.e* is enabled, *a* is requesting. Thus, executing the action and setting the state of *e* to **req** does not violate *Inv.a*.

Let us now consider *Inv.b.a* for an intermediate process $b \in M.a$. We examine the effect of the actions of *b*, *b*'s parent—*c*, and one of *b*'s children—*d* in this sequence.

We start with the actions of *b*. If *I.b* holds, *forward.b* is the only action that can be enabled. If it is enabled, *c* is requesting. Thus, if it is executed, *R.b* holds and *Inv.b.a* is not violated. If *R.b* holds then *back.b* is the only action that can be enabled. However, if *back.b* is enabled and *R.b* holds, then all children of *b* are replying. If *back.b* is executed, the resultant state conforms to *P.b*. If *P.b* holds, then *stop.b* can exclusively be enabled. If *P.b* holds and *stop.b* is enabled, then *c* is idle and all children of *b* are replying. The execution of *back.b* sets the state of *b* to **idle**. The resulting state conforms to *I.b* and *Inv.b.a* is not violated.

Let us examine the actions of *c*. Recall that we are assuming that *Inv.c* and the respective invariants of all of *b*'s ancestors hold. If *I.b* holds, *forward.c* and *join.c* (in case *b* is a child of *a*) are the actions that can possibly be enabled. If either is enabled, *b* is idle. The execution of either action changes the state of *c* to **req**. *I.b* and *Inv.b.a* still hold. If *R.b* holds, none of the actions of *c* are enabled. Indeed, actions *forward.c*, *back.c*, *join.c* and *enter.c* are disabled. Moreover, if *R.b* holds, *c* is requesting: Since *Inv.c* holds, *c* must be in *R.c* which means that *c*'s parent is not idle. Hence, *stop.c* is also disabled. Since *P.b* does not mention the state of *c*, the execution of *c*'s actions does not affect the validity of *P.b*.

Let us now examine the actions of *d*. If *I.b* holds, the only possibly enabled action is *stop.d*. The execution of this action changes the state of *d* to **idle**, which does not violate *I.b*. *R.b* does not mention the state of *d*. Hence, its action execution does not affect *R.b*. If *P.b* holds, all actions of *d* are disabled. This concludes the closure proof of *InvK*.

Convergence. We prove convergence by induction on the depth of the tree rooted in *a*. Let us show convergence of *a*. The only illegitimate set of states is *IR'.a*. When *a* conforms to *IR'.a*, *a* is idle and at least one child *e* is requesting. In such a state, all actions of *a* that affect its state are disabled. Moreover, for every child of *a* that is idle, all relevant actions are disabled as well. For the child of *a* that is not idle, the only enabled action is *stop.e*. After this action is executed, *e* is idle. Thus, eventually *IR.a* holds.

Let *a* conform to *Inv.a*. Moreover, let every descendant process *f* of *a* up to depth *i* conform to *Inv.f.a*. Let the distance from *a* to *b* be *i* + 1. We shall show that *Inv.b.a* eventually holds. Notice that according to the preceding closure proof, the conjunction of *Inv.a* and *Inv.f.a* for each process *f* in the distance no more than *i* is closed.

Note that according to Figure 26.7, there is no loop in the state transitions containing primed states. Hence, to prove that *b* eventually satisfies *Inv.b.a* we need to show that *b* does not remain in a single primed set of states indefinitely. Process *b* can satisfy either *I'.b* or *P'.b*. Let us examine these cases individually.

Let *b* ∈ *I'.b*. Since *Inv.c.a* holds, if *b* is idle, *c* cannot satisfy *P.c*. Thus, for *b* to satisfy *I'.b*, at least one child *d* of *b* must be requesting. However, if *b* is idle then *stop.d* is enabled. Notice that when *b* is idle, none of its non-requesting children can start to request. Thus, when this *stop* is executed for every requesting child of *b*, *b* leaves *I'.b*.

Suppose *b* ∈ *P'.b*. This means that there exists at least one child *d* of *b* that is not replying. However, for every such process *d*, *back.d* is enabled. Notice that when *b* is replying, none of its replying children can change state. Thus, when *back* is executed for every non-replying child of *b*, *b* leaves *P'.b*.

Hence, Algorithm 3 converges to *InvK*.

THEOREM 26.4 *If InvK holds and enter.a is enabled, then for every process b ∈ M.a, enter.b is disabled.*

PROOF   If *enter.a* is enabled, every child of *a* is replying. Due to *InvK*, this means that every descendant of *a* is also replying. Thus, for every process *x* ∈ *M.a* whose priority is lower than *a*'s priority, *enter.x* is disabled. Note also, that since *enter.a* is enabled, for every process *y* ∈ *M.a* whose priority is higher than *a*'s, *state.a.y* is **idle**. According to *InvK*, none of the ancestors of *a* in *y*'s tree, including *y*'s children, are replying. Thus, *enter.y* is disabled. In short, when *enter.a* is enabled, neither higher nor lower priority processes of *M.a* have *enter* enabled. The theorem follows.

LEMMA 26.7 If *InvK* holds and some process *a* is requesting, then eventually either *a* stops requesting or none of its descendants are idle.

PROOF Notice that the lemma trivially holds if *a* stops requesting. Thus, we focus on proving the second claim of the Lemma. We prove it by induction on the depth of *a*'s tree. Process *a* is requesting and so it is not idle. By the assumption of the Lemma, *a* will not be idle. Now let us assume that this lemma holds for all its descendants up to distance *i*. Let *b* be a descendant of *a* whose distance from *a* is *i* + 1 and let *b* be idle.

By inductive assumption, *b*'s parent *c* is not idle. Due to *InvK*, if *b* is idle, *c* is not replying. Hence, *c* is requesting. If there exists a child *d* of *b* that is not idle, then *stop.d* is enabled at *d*. When *stop.d* is executed, *d* is idle. Notice that when *b* and *d* are idle, all actions of *d* are disabled. Thus, *d* continues to be idle. When all children of *b* are idle and its parent is requesting, *forward.b* is enabled. When it is executed, *b* is not idle. Notice, that the only way for *b* to become idle again is to execute *stop.b*. However, by inductive assumption *c* is not idle. This means that *stop.b* is disabled. The lemma follows.

LEMMA 26.8 If *InvK* holds and some process *a* is requesting, then eventually all its children in *M.a* are replying.

PROOF Notice that when *a* is requesting, the conditions of Lemma 26.7 are satisfied. Thus, eventually, none of the descendants of *a* are idle. Notice that if a process is replying, it does not start requesting without being idle first (see Figure 26.7). Thus, we have to prove that each individual process is eventually replying. We prove it by induction on the height of *a*'s tree.

If a leaf node *b* is requesting and its parent is not idle, *back.b* is enabled. When it is executed, *b* is replying. Assume that each node whose longest distance to a leaf of *a*'s tree is *i* is replying. Let *b*'s longest distance to a leaf be *i* + 1. By assumption, all its children are replying. Due to Lemma 26.7, its parent is not idle. In this case *back.b* is enabled. After it is executed, *b* is replying. By induction, the lemma holds.

LEMMA 26.9 If *InvK* holds and the computation contains infinitely many states where *a* is idle, then for every descendant in *a*'s tree there are infinitely many states where it is idle as well.

PROOF We first consider the case where the computation contains a suffix where *a* is idle in every state. In this case we prove the lemma by induction on the depth of *a*'s tree with *a* itself as a base case. Assume that there is a suffix where all descendants of *a* up to depth *i* are idle. Let us consider process *b* whose distance to *a* is *i* + 1 and this suffix. Notice that this means that *c* remains idle in every state of this suffix. If *b* is not idle, *stop.b* is enabled. Once it is executed, no relevant actions are enabled at *b* and it remains idle afterward. By induction, the lemma holds.

Let us now consider the case where no computation suffix of continuously idle *a* exists. Yet, there are infinitely many states where *a* is idle. Thus, *a* leaves the idle state and returns to it infinitely often. We prove by induction on the depth of the tree that every descendant of *a* behaves similarly. Assume that this claim holds for the descendants up to depth *i*. Let *b*'s distance to *a* be *i* + 1.

When *InvK* holds, the only way for *b*'s parent *c* to leave **idle** is to execute *forward.c* (see Figure 26.7). Similarly, the only way for *c* to return to **idle** is to execute *stop.c* while *c* is replying.* However, *forward.c* is enabled only when *b* is idle. Moreover, according to *InvK* when *c* is requesting, *b* is not idle. Thus, *b* leaves **idle** and returns to it infinitely many times as well. By induction, the lemma follows.

LEMMA 26.10   If *InvK* holds and process *a* is requesting such that and *a*'s priority is the highest among the processes that ever request the CS in *M.a*, then *a* eventually executes the CS.

PROOF   If *a* is requesting, then, by Lemma 26.8, all its children are eventually replying. Therefore, the first and second conjuncts of the guard of *enter.a* are **true**. If *a*'s priority is the highest among all the requesting processes in *M.a*, then each process *z*, whose priority is higher than that of *a* is idle. According to Lemma 26.9, *state.a.z* is eventually **idle**. Thus, the third and last conjunct of *enter.a* is enabled. This allows *a* to execute the CS.

LEMMA 26.11   If *InvK* holds and process *a* is requesting, *a* eventually executes the CS.

PROOF   Notice that by Lemma 26.8, for every requesting process, the children are eventually replying. According to *InvK*, this implies that all the descendants of the requesting process are also replying. For the remainder of the proof we assume that this condition holds.

We prove this lemma by induction on the priority of the requesting processes. According to Lemma 26.10, the requesting process with the highest priority eventually executes the CS. Thus, if process *a* is requesting and there is no higher priority process $b \in M.a$ which is also requesting then, by Lemma 26.10, *a* eventually enters the CS.

Suppose, on the contrary, that there exists a requesting process $b \in M.a$ whose priority is higher than *a*'s. If every such process *b* enters the CS finitely many times, then, by the repeated application of Lemma 26.10, there is a suffix of the computation where all processes with priority higher than *a*'s are idle. Then, by Lemma 26.10, *a* enters the CS. Suppose there exists a higher priority process *b* that enters the CS infinitely often. Since *a* is requesting, *state.b.a* = **rep**. When *b* executes the CS, it enters *a* into *YIELD.b*. We assume that *b* enters the CS infinitely often. However, *b* can request the CS again only if *YIELD.b* is empty. The only action that takes *a* out of *YIELD.b* is *stop.b*. However, this action is enabled if *state.b.a* is **idle**. Notice that, if *InvK* holds, the only way for the descendants of *a* to move from replying to idle is if *a* itself moves from requesting to idle. That is *a* executes the CS. Thus, each process *a* requesting the CS eventually executes it.

LEMMA 26.12   If *InvK* holds and process *a* wishes to enter the CS, *a* eventually requests.

PROOF   We show that *a* wishing to enter the CS eventually executes *join.a*. We assume that *a* is idle and *needcs.a* is **true**. Then, *join.a* is enabled if *YIELD.a* is empty. Note that *a* adds a process to *YIELD* only when it executes the CS. Thus, as *a* remains idle, processes can only be removed from *YIELD.a*.

Let us consider a process $b \in YIELD.a$. If *b* executes the CS finitely many times, then there is a suffix of the computation where *b* is idle. According to Lemma 26.9, for all descendants of *b*, including *a*, *state.a.b* is idle. If this is the case *stop.a* is enabled. When it is executed *b* is removed from *YIELD.a*.

---

* The argument is slightly different for $c = a$ as it executes *join.a* and *enter.a* instead.

Let us consider the case, where b executes the CS infinitely often. In this case, b enters and leaves **idle** infinitely often. According to Lemma 26.9, *state.a.b* is idle infinitely often. Moreover, a moves to idle by executing *stop.a*, which removes b from *YIELD.a*. The lemma follows.

The Theorem in the following text follows from Lemmas 26.11 and 26.12.

THEOREM 26.5   *If InvK holds, a process wishing to enter the CS is eventually allowed to do so.*

We draw the following corollary from Theorems 26.3 through 26.5.

COROLLARY 26.4   Algorithm 3 is a self-stabilizing solution to the k-hop diners problem.

26.3.3 Census

Algorithm. Each node i has a unique identifier and is aware of its input degree $\delta^-.i$ (the number of its incident arcs), which is also placed in noncorruptible memory. A node i arbitrarily numbers its incident arcs using the first $\delta^-.i$ natural numbers. When receiving a message, the node i knows the number of the corresponding incoming link (that varies from 1 to $\delta^-.i$).

Each node maintains a local memory. The local memory of i is represented by a list denoted by $(i_1; i_2; \ldots; i_k)$. Each i_α is a nonempty list of pairs $\langle identifier, colors\rangle$, where *identifier* is a node identifier, and where *colors* is an array of Booleans of size $\delta-.i$. Each Boolean in the *colors* array is either *true* (denoted by ●) or *false* (denoted by ○). We assume that natural operations on Boolean arrays, such as unary *not*(denoted by ¬), binary *and* (denoted by ∧), and binary *or* (denoted by ∨) are available.

The goal of the Census algorithm is to guarantee that the local memory of each node contains the list of lists of identifiers (whatever the *colors* value in each pair $\langle identifier, colors\rangle$) that are predecessors of i in the communication graph. Each predecessor of i is present only once in the list. For the Census task to be satisfied, we must ensure that the local memory of each node i can contain as many lists of pairs as necessary. We assume that a minimum of

$$(n-1) \times \left(\log_2(k) + \delta^-.i\right)$$

bits space is available at each node i, where n is the number of nodes in the system and k is the number of possible identifiers in the system.

For example,

$$((j, [●\ ○\ ○]; q, [○\ ●\ ○]; t, [○\ ○\ ●])(z, [●\ ●\ ●]))$$

is a possible local memory for node i, assuming that $\delta^-.i$ equals 3. From the local memory of node i, it is possible to deduce the knowledge that node i has about its ancestors. With the previous example, node j is a direct ancestor of i (it is in the first list of the local memory of i) and this information was carried through incoming channel number 1 (only the first position of the *colors* array relative to node j is *true*). Similarly, nodes q and t are direct ancestors of i and this information was obtained through incoming links 2 and 3, respectively. Then, node z is at distance 2 from i, and this information was received through incoming links numbered 1, 2, and 3.

Each node sends and receives messages. The contents of a message is represented by a list denoted by $(i_1; i_2; \ldots; i_k)$. Each i_α is a nonempty list of *identifiers*.

For example,

$$((i)(j; q; t)(z))$$

is a possible contents of a message. It means that i sent the message (since it appears first in the message), that i believes that j, q, and t are the direct ancestors of i, and that z is an ancestor at distance 2 of i.

The distance from i to j is denoted by $d.i.j$, which is the minimal number of arcs from i to j. We assume that the graph is strongly connected, so the distance from i to j is always defined. Yet, since the graph may not be bidirectional, $d.i.j$ may be different from $d.j.i$. The *age* of i, denoted by $\chi.i$, is the greatest distance $d.j.i$ for any j in the graph. The network diameter is then equal to

$$\max_i \chi.i = D$$

Our algorithm can be seen as a knowledge collector on the network. The local memory of a node then represents the current knowledge of this node about the whole network. The only certain knowledge a node may have about the network is local: its identifier, its incoming degree, and the respective numbers of its incoming channels. This is the only information that is stored in non-corruptible memory. In a nutshell, Figure 26.8 provides a possible corrupted initial configuration (in Figure 26.8a, where some information is missing and some information is incorrect, for example, nodes x, y, and z are nonexistent) and a legitimate configuration (in Figure 26.8b, where the link contents are not displayed).

The algorithm for each node consists in updating in a coherent way its knowledge upon receipt of other process' messages, and communicating this knowledge to other processes after adding its constant information about the network. More precisely, each information placed in a local memory is related to the local name of the incoming channel that transmitted this information. For example, node i would only emit messages starting with singleton list (i) and then not containing i since it is trivially an ancestor of i at distance 0. Coherent update consists in three kinds of actions: the first two being trivial coherence checks on messages and local memory, respectively.

Check Message Coherence Since all nodes have the same behavior, when a node receives a message that does not start with a singleton list, the message is trivially considered as erroneous and is ignored. For example, messages of the form $((j; q; t)(k)(m; y)(p; z))$ are ignored.

Check Local Coherence Regularly and at each message receiving, a node checks for local coherence. We only check here for trivial inconsistencies (see the *problem()* helper function later): a node is incoherent if there exist at least one pair $\langle identifier, colors \rangle$ such that $colors = [\circ \cdots \circ]$ (which means that some information in the local memory was not obtained from any of the input channels). If a problem is detected upon time-out, then the local memory is reinitialized, else if a problem is detected upon a message receipt, the local memory is completely replaced by the information contained in the message.

Trust Most Recent Information When a node receives a message through an incoming channel, this message is expected to contain more recent and thus more reliable information. The node removes all previous information obtained through this channel from its local memory. Then it integrates new information and only keeps old information (from its other incoming channels) that does not clash with new information.

For example, assume that a message $mess = ((j)(k; l)(m)(p; q; r; i))$ is received by node i through its incoming link 1 and that $\delta_i^- = 2$. The following informations can be deduced

1. j is a direct ancestor of i (it appears first in the message),
2. k and l are ancestors at distance 2 of i and may transmit messages through node j,
3. m is an ancestor at distance 3 of i,
4. p, q, and r are ancestors at distance 4 of i, j obtained this information through m.

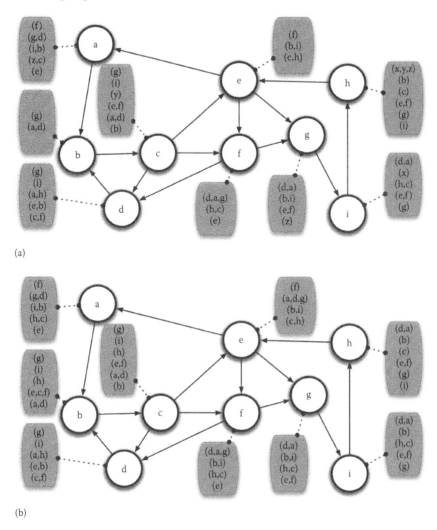

(a)

(b)

FIGURE 26.8 Two possible configurations for Algorithm 4. (a) A possible corrupted initial configuration and (b) a legitimate configuration (messages not displayed).

These informations are compatible with a local memory of i such as:

$$((j, [\bullet\circ]; q, [\circ\bullet])(k, [\bullet\circ]; e, [\circ\bullet]; w, [\circ\bullet])(m, [\circ\bullet]; y, [\bullet\bullet])(p, [\bullet\circ]; z, [\circ\bullet]; h, [\bullet\circ]))$$

Upon the receipt of message *mess* at i, the following operations take place: (1) the local memory of i is cleared from previous information coming from link 1, (2) the incoming message is "colored" by the number of the link (here each identifier α in the message becomes a pair $\alpha, [\bullet\circ]$ since it is received by link number 1 and *not* by link number 2), and (3) the local memory is enriched as in the following (where "$\leftarrow$" denotes information that was acquired upon receipt of a message, and where

"→" denotes information that is to be forwarded to the node output links):

$$
\begin{array}{llll}
(& (q,[\circ\bullet]) & (e,[\circ\bullet];w,[\circ\bullet]) & (m,[\circ\bullet];y,[\circ\bullet]) & (z,[\circ\bullet]) &) \\
\leftarrow (& (j,[\bullet\circ]) & (k,[\bullet\circ];l,[\bullet\circ]) & (m,[\bullet\circ]) & \left(\begin{array}{l} p,[\bullet\circ];q,[\bullet\circ]; \\ r,[\bullet\circ];i,[\bullet\circ] \end{array}\right) &) \\
\end{array}
$$

$$
\rightarrow (\quad (j,[\bullet\circ];q,[\circ\bullet]) \quad \left(\begin{array}{l} k,[\bullet\circ];e,[\circ\bullet]; \\ w,[\circ\bullet];l,[\bullet\circ] \end{array}\right) \quad (m,[\bullet\bullet];y,[\circ\bullet]) \quad \left(\begin{array}{l} p,[\bullet\circ];z,[\circ\bullet]; \\ q,[\bullet\circ];r,[\bullet\circ] \end{array}\right) \quad)
$$

This message enabled the modification of the local memory of node i in the following way: l is a new ancestor at distance 2. This was acquired through incoming link number 1 (thus through node j). Nodes m and y are confirmed to be ancestors at distance 3, but *mess* sends information *via* nodes j and q, while y only transmits its informations *via* node q. Moreover, q and r are part of the new knowledge of ancestors at distance 4. Finally, although i had information about h ($h,[\bullet\circ]$) before receiving *mess*, it knows now that the information about h is obsolete.

The property of resilience to intermittent link failures of our algorithm is mainly due to the fact that each message is self-contained and independently moves toward a complete correct knowledge about the network. More specifically:

1. The fair loss of messages is compensated by the infinite spontaneous retransmission by each process of their current knowledge.
2. The finite duplication tolerance is due to the fact that our algorithm is *idempotent* in the following sense: If a process receives the same message twice from the same incoming link, the second message does not modify the knowledge of the node.
3. The desequencing can be considered as a change in the relative speeds of two messages toward a complete knowledge about the network. Each message independently gets more accurate and complete, so that their relative order is insignificant. A formal treatment of this last and most important part can be found in the later part of this section.

We now describe helper functions that will enhance the readability of our algorithm. Those functions operate on lists, integers, and pairs $\langle identifier,colors \rangle$. The specifications of those functions use the following notations: l denotes a list of identifiers, p denotes an integer, lc denotes a list of pair $\langle identifier,colors \rangle$, Ll denotes a list of lists of identifiers, and Llc denotes a list of lists of pairs $\langle identifier,colors \rangle$.

We assume that classical operations on generic lists are available: \ denotes the binary operator "minus" (and returns the first list from which the elements of the second have been removed), ∪ denotes the binary operator "union" (and returns the list without duplicates of elements of both lists), + denotes the binary operator "concatenate" (and returns the list resulting from concatenation of both lists), ♯ denotes the unary operator that returns the number of elements contained in the list, and [] takes an integer parameter p so that $l[p]$ returns a reference to the pth element of the list l if $p \le ♯l$ (in order that it can be used on the left part of an assignment operator ":="), or expand l with $p - ♯l$ empty lists and returns a reference to the pth element of the updated list if $p > ♯l$.

$colors(p) \rightarrow$ **array of Booleans** returns the array of Booleans that correspond to the pth incoming link, that is, the array that is such that $[\underbrace{\circ \cdots \circ}_{p-1 \text{times}} \bullet \underbrace{\circ \cdots \circ}_{\delta^-.i-p \text{times}}]$.

$clean(lc,p) \rightarrow$ **list of couples** returns the empty list if lc is empty and a list of pairs lc_2 such that for each $\langle identifier_{lc}, colors_{lc} \rangle \in lc$, if $colors_{lc} \wedge \neg colors(p) \ne [\circ \cdots \circ]$, then $\langle identifier_{lc}, colors_{lc} \wedge \neg colors(p) \rangle$ is in lc_2, else $\langle identifier_{lc}, * \rangle$ is not in lc_2.

$emit(i,Llc)$ sends the message resulting from $(i) + identifiers(Llc)$ on every outgoing link of i.

identifiers(*Llc*) → **list of list of identifiers** returns the empty list if *Llc* is empty and returns a
list *Ll* of list of identifiers (such that each pair ⟨*identifier,colors*⟩ in *Llc* becomes *identifier*
in *Ll*) otherwise.

merge(*lc, l, p*) → **list of couples** returns the empty list if *lc* and *l* are both empty and

$$\bigcup_{\substack{\langle i,c \rangle \in lc \\ i \in l}} (\langle i, c \vee colors(p) \rangle) \cup \bigcup_{\substack{\langle i,* \rangle \notin lc \\ i \in l}} (\langle i, colors(p) \rangle)$$

otherwise.

new(*lc, l*) → **list of couples** returns the empty list if *lc* is empty and the list of pairs ⟨*identifier,
colors*⟩ whose *identifier* is in *lc* but not in *l* otherwise.

problem(*Llc*) → **Boolean** returns *true* if there exist two integers *p* and *q* such that $p \leq \sharp(Llc)$
and $q \leq \sharp(Llc[p])$ and $Llc[p][q]$ is of the form ⟨*identifier, colors*⟩ and all Booleans in *colors*
are *false* (∘). Otherwise, this function returns *false*.

In addition to its local memory, each node makes use of the following local variables when
processing messages: α is the current index in the local memory and message main list, *i_ pertinent* is
a Boolean that is *true* if the αth element of the local memory main list contains pertinent information,
m_ pertinent is a Boolean that is *true* if the αth element of the message main list contains pertinent
information, *known* is the list of all identifiers found in the local memory and message found up
to index α, *temp* is a temporary list that stores the updated αth element of the local memory
main list.

We are now ready to present our Census Algorithm (noted Algorithm 4 in the remaining of
the paper). This algorithm is message driven: processes execute their code when they receive an
incoming message. In order to perform correctly in configurations where no messages are present,
Algorithm 4 also uses a spontaneous action that will emit a message.

Proof of Correctness. The intuitive reason for self-stabilization of the protocol is as follows: anytime a node sends a message, it adds up its identifier and pushes further in the list of lists included in
every message potential fake identifiers. As the network is strongly connected, the message eventually
goes through every possible cycle in the network, and every node on every such cycle removes its
identifier from a list at a certain index (corresponding to the length of the cycle). So, after a message
has visited all nodes, there exists at least one empty list in the list of lists of the message, and all fake
identifiers are all included beyond this empty list. Since every outgoing message is truncated from
the empty list onward, no fake identifier remains in the system forever.

The proof of correctness that is presented in this section demonstrates the use of potential func-
tions (see Figure 26.9, where each configuration $c_1, c_2, |dots$ is associated with a number $f_1, f_2, |dots$
obtained through the potential function for this configuration). Intuitively, a potential function
maps configurations into a finite set endowed with a total order (typically the set of natural integers),
and exhibits the following property: When an action of the distributed system is executed to move
from one configuration to the other, the potential function decreases. Since the set is finite and
the order well founded, the lowest element is eventually reached. Now if the lowest element of the
potential function matches the set of legitimate configurations of the problem specification, this
implies stabilization. Of course, the difficult part of the proof lies in finding a suitable such potential
function.

In this section, we show that Algorithm 4 is a self-stabilizing Census algorithm. In more details,
independently of the initial configuration of network channels (noninfinitely full) and of the initial
configuration of local memories of nodes, every node ends up with a local memory that reflect
the contents of the network, even if unreliable communication media is used for the underlying
communication between nodes.

Algorithm 4 $\mathcal{CA}$: A self-stabilizing census algorithm

process *i*
const
    $\delta^-.i$: input degree of *i*
parameter
    message: list of lists of identifiers
var
    local_memory.i: list of lists of pairs
lvars
    α: integer
    i_pertinent, m_pertinent: boolean
    known, temp: list of identifiers
        *[

problem: *problem(local_memory.i)* $\longrightarrow$
                local_memory.i := ()
                emit(i, local_memory.i)

        []

update: **receive**(*message*) **from link** *p* $\longrightarrow$
                i_pertinent:= $\neg$ *problem(local_memory.i)*
                m_pertinent:= ($\sharp$(*message*[1]) = 1)
                α := 0;*known*:= (*i*);
                do *m_pertinent* $\vee$ *i_pertinent* $\longrightarrow$
                    α := α + 1;*temp*:= ()
                    local_memory.i[α] := *clean(local_memory.i*[α], *p*)
                    if *i_pertinent* $\longrightarrow$
                        temp:= *new(local_memory.i*[α],*known*)
                        temp= () $\longrightarrow$ *i_pertinent*:=*false*
                  fi
                  if *m_pertinent* $\longrightarrow$
                    if *message*[α]*known*= () $\longrightarrow$
                        m_pertinent:=*false*
                    []   *temp*:= *merge(temp,message*[α]*known, p*)
                    fi
                  fi
                  if *temp*$\neq$ () $\longrightarrow$
                    local_memory.i[α] := *temp*
                    known:=*known*$\cup$ *identifiers(temp)*
                fi
                od
                local_memory.i := (*local_memory.i*[1], . . . , *local_memory.i*[α])

        []

resend: *true* $\longrightarrow$
                emit(i, local_memory.i)

        ]

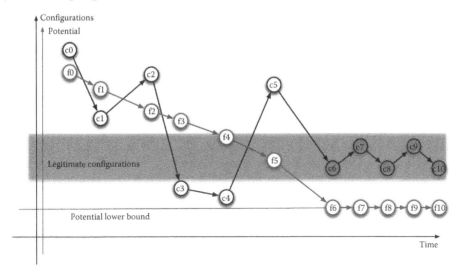

FIGURE 26.9 Proving self-stabilization with potential functions.

First, we define a formal measure on messages that circulate in the network and on local memories of the nodes. This measure is either the distance between the current form of the message and its canonical form (that denotes optimal knowledge about the network), or between the current value of the local memories and their canonical form (when a node has a perfect knowledge about the network). We use this measure to compute the potential function result of a configuration (in the sequel, the result of this potential function is called the configuration weight).

Then, we show that after a set of emissions and receptions of messages, the weight of a configuration decreases. An induction shows that this phenomenon continue to appear and that the weight of a configuration reaches 0, that is, a configuration where each message is correct and where each node has an optimal knowledge about the network. We also show that such a configuration (whose weight is 0) is stable when a message is emitted or received. According to the previous definitions, a configuration of weight 0 is a legitimate configuration after finite time.

These two parts prove, respectively, the convergence and closure of our algorithm, and establish its self-stabilizing property.

As the Census problem is static and deterministic, when we consider only node local memories, there is a single legitimate configuration. This legitimate configuration is when each node has a global correct knowledge about the network. It is also the stable configuration the system would reach had it been started from a zero knowledge configuration (where the local memory of each node is null, and where no messages are in transit in the system).

In this legitimate configuration, all circulating messages are of the same kind. Moreover, on every particular link, all messages have the same contents. The canonical form of a message circulating on a link between nodes j and i is the list of lists starting with the singleton list (j) followed by the $\chi.j$ lists of ancestors of j at distance between 1 and $\chi.j$. The canonical form of node i's local memory is the list of lists of pairs Llc of the $\chi.j$ lists of pairs $\langle$ *identifier,colors* $\rangle$ such that: (1) *identifiers*(Llc) is the list of the $\chi.i$ lists of ancestors of i at distance 1 to $\chi.i$, and (2) if a shortest path from node j to node i passes through the pth input channel of i, then the Boolean array *colors* associated to node j in Llc has *colors*$[p] = \bullet$. For the sake of simplicity, we will also call the αth list of a canonical message or a canonical local memory a canonical list.

PROPOSITION 26.1   The canonical form of node i's local memory and that of its incoming and outgoing channels are coherent.

PROOF   If node i's local memory is in canonical form, then the emit action trivially produces a canonical message.

Conversely, upon receipt by node i of a canonical message through incoming link j, the local memory of i is replaced by a new identical canonical memory. Indeed, clean first removes from the αth list of i's local memory all pairs $\langle identifier, colors\rangle$ such that $colors= colors(p)$, yet by the definition of canonical memory, each such *identifier* is that of a node such that the shortest path from *identifier* to i is of length α and passes through j. Moreover, the list l used by merge is the list of nodes at distance $\alpha - 1$ of node i, so for any *identifier* appearing in l, two cases may occur:

1. There exists a path from *identifier* to i that is of length $< \alpha$, then $identifiers\in known$ and it does not appear in the new list of length α,
2. There exists a shorter path from *identifier* to i through j of length α, then $\langle identifier, colors(p)\rangle$ is one of the elements that were removed by clean and this information is put back into node i's local memory.

COROLLARY 26.5   The set of legitimate configurations is closed.

PROOF   Starting from a configuration where every message and every local memory is canonical, none of the local memories is modified, and none of the emitted message is non-canonical.

We define a weight on configurations as a function on system configurations that returns a positive integer. As configurations of weight zero are legitimate, the weight of a configuration c denotes the "distance" from c toward a legitimate configuration.

In order to evaluate the weight of configurations, we define a measure on messages and local memory of nodes as an integer written using $D + 2$ digits in base 3 (where D denotes the graph diameter). The weight of a configuration is then the pair of the maximum weight of local memories, and the maximum weight of circulating messages. For sake of clarity, a single integer will denote the weight of the configuration when both values are equal. Let m be a circulating message on a communication link whose canonical message is denoted by $\tilde{m}$. Note that since a canonical message is of size $\leq D + 1$, we have $\tilde{m}[D + 2] = ()$. The weight of m is the integer written using $D + 2$ base 3 digits and whose αth digit is: (1) 0, if $m[\alpha] = \tilde{m}[\alpha]$, (2) 1, if $m[\alpha] \leftarrow \tilde{m}[\alpha]$, and (3) 2, if $m[\alpha] \Uparrow \tilde{m}[\alpha]$. Then, $3^{D+2} - 1$ is the biggest weight for a message, and corresponds to a message that is totally erroneous. At the opposite, 0 is the smallest weight for a message, and corresponds to a canonical message, or to a message that begins with a canonical message.

Let m be the local memory of a node i whose canonical local memory is $\tilde{m}$. The weight of m is the integer written using $D + 1$ digits (in base 3) and whose αth digit is: (1) 0, if $m[\alpha] = \tilde{m}[\alpha]$, (2) 1, if $m[\alpha] \neq \tilde{m}[\alpha]$ and $identifiers(m[\alpha]) \subseteq identifiers(\tilde{m}[\alpha])$ and for any $\langle identifier, colors_1 \rangle$ of $m[\alpha]$, the associated $\langle identifier, colors_2\rangle$ in $\tilde{m}[\alpha]$ satisfies: $(colors_1 \wedge colors_2) = colors_1$, and (3) 2, otherwise. Then $3^{D+1} - 1$ is the biggest weight of a local memory and denotes a totally erroneous local memory. At the opposite, 0 is the smallest weight and denotes a canonical local memory.

Let us notice that in both cases (weight of circulating messages and of nodes local memories), the αth digit 0 associated to the αth list denotes that this particular list is in its final form (the canonical form). The αth digit 1 means that the αth list is coherent with the αth canonical list, but still lacks some information. On the contrary, the αth digit 2 signals that the related αth position contains informations that shall not persist and that are thus unreliable. The weight of a message indicates how much of the information it contains is pertinent. After defining message weight and, by extension, configuration weights, we first prove that starting from an arbitrary initial configuration, only messages of weight lower or equal to $3^{D+1} - 1$ are emitted, which stands for the base case for our induction proof.

LEMMA 26.13   In any configuration, only messages of weight lower than 3^{D+1} may be emitted.

PROOF   Any message that is emitted from a node i on a link from i to j is by function *emit*. This function ensures that this message starts with the singleton list (i). This singleton list is also the first element of the canonical message for this channel. Consequently, the biggest number that may be associated to a message emitted by node i starts with a 0 and is followed by $D + 1$ digits equal to 2. Its overall weight is at most $3^{D+1} - 1$.

LEMMA 26.14   Assume $\alpha \geq 1$. The set of configurations whose weight is strictly lower than $3^{\alpha-1}$ is an attractor for the set of configuration whose weight is strictly lower than 3^{α}.

PROOF   A local memory of weight strictly lower that 3^{α} contains at most α erroneous lists, and it is granted that it starts with $D + 2 - \alpha$ canonical lists.

By definition of the *emit* function, each node i that owns a local memory of weight strictly below 3^{α} shall emit the singleton list (i) followed by $D + 2 - \alpha$ canonical lists. Since canonical messages sent by a node and its canonical local memory are coherent, it must emit messages that contain at least $D + 2 - \alpha + 1$ canonical lists, which means at worst $\alpha - 1$ erroneous lists. The weight of any message emitted in such a configuration is then strictly lower than $3^{\alpha-1}$.

It follows that messages of weight exactly 3^{α} which remain are those from the initially considered configuration. Hence they are in finite number. Such messages are either lost or received by some node in a finite time. The first configuration that immediately follows the receiving or loss of those initial messages is of weight (3^{α} (local memory), $3^{\alpha-1}$ (messages)).

The receiving by each node of at least one message from any incoming channel occurs in finite time. By the time each node receives a message, and according to the local memory maintenance algorithm, each node would have been updated. Indeed, the receiving of a message from an input channel implies the cleaning of all previous information obtained from this channel. Consequently, in the considered configuration, all lists in the local memory result from corresponding lists in the latest messages sent through each channel. Yet, all these latest messages have a weight strictly lower than $3^{\alpha-1}$ and by the coherence property on canonical forms, they present information that are compatible with the node canonical local memory, up to index $D + 3 - \alpha$. By the same property, and since all input channels contribute to this information, it is complete. In the new configuration, each node i maintains a local memory whose first $D + 3 - \alpha$ lists are canonical, and thus the weight of its local memory is $3^{\alpha-1}$. Such a configuration is reached within finite time and its weight is ($3^{\alpha-1}$ (local memory), $3^{\alpha-1}$ (messages)).

PROPOSITION 26.2   The set of configurations whose weight is 0 is an attractor for the set of all possible configurations.

PROOF   By induction on the maximum degree of the weight on configurations, the base case is proved by Lemma 26.13, and the induction step is proved by Lemma 26.14. Starting from any initial configuration whose weight is greater that 1, a configuration whose weight is strictly inferior is eventually reached. Since the weight of a configuration is positive or zero, and that the order defined on configurations weights is total, eventually a configuration whose weight is zero is eventually reached. By definition, this configuration is legitimate.

THEOREM 26.6   *Algorithm 4 is self-stabilizing.*

PROOF   Consider a message m of weight 0. Two cases may occur: (1) m is canonical, or (2) m starts with a canonical message, followed by at least one empty list, (possibly) followed by several erroneous lists. Assume that m is *not* canonical, then it is impossible that m was emitted, since the **truncate** part of Algorithm 4 ensures that no message having an empty list can be emitted; then m is an erroneous message that was present in the initial configuration.

Similarly, the only local memories that may contain an empty list are those initially present (e.g., due to a transient failure).

As a consequence, after the receipt of a message by each node and after the receipt of all initial messages, all configurations of weight 0 are legitimate (they only contain canonical messages and canonical local memories).

By Proposition 26.2, the set of legitimate configurations is an attractor for the set of all possible configurations, and Corollary 26.5 proves closure of the set of legitimate configurations. Therefore, Algorithm 4 is self-stabilizing.

In the convergence part of the proof, we only assumed that computations were maximal, and that message loss, duplication, and desequencing could occur. In order to provide an upper bound on the stabilization time for our algorithm, we assume strong synchrony between nodes and a reliable communication medium between nodes. Note that these assumptions are used for complexity results only, since our algorithm was proven correct even in the case of asynchronous unfair computations with link intermittent failures. In the following D denotes the network diameter.

LEMMA 26.15   Assuming a synchronous reliable system S, the stabilization time of Algorithm 4 is $O(D)$.

PROOF   Since the network is synchronous, we consider system steps as: (1) each node receives all messages that are located at each of its incoming links and updates its local memory according to the received information, and (2) each node sends as many messages as received on each of its outgoing links. Intuitively, within one system step, each message is received by one process and sent back. Within one system step, all messages are received, and messages of weight strictly inferior to that of the previous step are emitted (see the proof of Lemma 26.14). In the same time, when a process has received messages from each of its incoming links, its weights are bounded by $3^{D+1-\alpha}$, where D is the network diameter, and α is the number of the system step (see the proof of Lemma 26.14). Since the maximal initial weight of a message and of a local memory is 3^{D+2}, after $O(D)$ system steps, the weight of each message and of each local memory is 0, and the system has stabilized.

26.3.4 Token Passing

Algorithm. The "benchmark" self-stabilizing algorithm for dynamic problems (and the first published algorithm [19]) is the mutual exclusion algorithm on a unidirectional ring. Mutual exclusion is ensured by circulating a "token" (a local predicate at a given node) fairly in the system. In a self-stabilizing context, it is necessary to recover both from configuration where there is no initial token and where there exists several superfluous ones (see e.g., Figure 26.10a for a possible inital configuration with several tokens, and Figure 26.10b for a legitimate configuration with a single

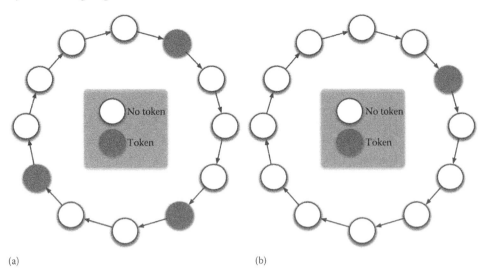

FIGURE 26.10 Possible configurations of a self-stabilizing token passing algorithm. (a) A possible arbitrary initial configuration, and (b) a possible legitimate configuration.

token). In practice, several criteria need to be taken into account: the stabilization time, the service time (maximum time between two token passings on a given node), the memory used (in bits), and the transparency with respect to underlying communication algorithms.

Due to impossibility results in uniform networks (unidirectional rings where nodes cannot be distinguished from one another), probabilistic self-stabilization was introduced [39]. A probabilistic self-stabilizing algorithm has the property of reaching a legitimate configuration in finite time with probability 1. The algorithm of [39] can intuitively be viewed as follows: each process i maintains a local predicate that states whether it has a token, and executes the following code: at each pulse, if i is currently holding a token (i.e., the token predicate is true), it transmits it to its successor with probability p, and keeps it with probability $1 - p$. The original paper [39] performs in synchronous odd-sized networks and uses a combinatorial trick to guarantee the presence of at least one token (this trick was further refined to handle the case of arbitrary-sized networks [6] and arbitrary minimal number of tokens [37]).

While the original version of the algorithm is written for the shared memory model, we present a version of the algorithm that is written for the message passing model. So, we have to rely on time-outs to recover from configurations with no tokens. We consider that the protocol has a parameter k that is used as a time-out value; k should be big enough so that the time-out is not triggered if a token is already present, but given the probabilistic nature of the token propagation, this is guaranteed only with high probability. Our protocol is presented as Algorithm 4.

Proof of Correctness. To show stabilization, it is sufficient to prove that starting from a configuration with several tokens, every execution ends up in a configuration with exactly one token. The proof is by showing that in any configuration with two tokens (or more) there is a positive probability that the two tokens merge (the "last" one always get a "transmit" to the successor, while the "first" one always get a "keep" at the current node). When two tokens merge, they stay so. Overall, a configuration with m tokens is always reachable from a configuration with $m + 1$ tokens with a strictly

---

Algorithm 5 $\mathcal{TP}$: A probabilistically self-stabilizing token passing algorithm

---

    process i
    const
        $p.i$: predecessor of i on the ring
        $s.i$: successor of i on the ring
    parameter
        k: integer
        p: probability to pass the token
    var
        token.i: {**null**, **token**}
        timeout.i: integer
    function
        random(x): draws a random number z in $[0..1]$, returns true if $z \leq x$, false otherwise.
    macro
        send_token $\equiv$
                    if *token.i* = **token** $\wedge$ *random*(p) $\longrightarrow$
                        send(**token**) **to** *s.i*
                        token.i := **null**
                    fi

                *[
    token: *timeout.i* > 0 $\wedge$ **receive**(**token**) **from** *p.i* $\longrightarrow$
                          token.i := **token**
                          timeout.i := k
                          send_token

                []
    elapse: *timeout.i* > 0 $\longrightarrow$
                          timeout.i := *timeout.i* $-$ 1
                          send_token

                []
    timeout: *timeout.i* = 0 $\longrightarrow$
                          token.i := **token**

                ]

---

positive probability, while a configuration with $m+1$ tokens is never reachable from a configuration with m tokens. As a result, a configuration with exactly one token is eventually reached.

    The intuitive proof argument being settled, we focus here on the complexity of the stabilization time. Since the scheduler is synchronous, there is no nondeterministic choice other than probabilistic in every execution, and Markov chains are an attractive tool to compute expected bounds for the stabilization time. We follow the terminology of [63] about Markov Chains. The classical hypothesis can be used since the network has a synchronous behavior; for an asynchronous setting, see [28]. Let $P_{n \times n}$ be a stochastic matrix, that is the sum of every line is equal to 1. A discrete Markov Chain, denoted by $(X_t)_{t \leq 0}$ on a set of states X is a sequence of random variables $X_0, X_1, \ldots$ with $X_i \in X$ and so that X_{i+1} only depends on X_i and $\Pr(X_{i+1} = y | X_i = x) = p_{x,y}$. The matrix P is called the transition probability matrix. A node x leads to a node y if $\exists j \geq i, \Pr(X_j = y | X_i = x) > 0$. A state y is an absorbing state if y does not lead to any other state. The expected hitting time or hitting time $\mathbb{E}_x^y$ is the average number of steps starting from node x to reach node y for the first time. We will make use of the following theorem for Markov chains:

THEOREM 26.7 *The vector of hitting times* $\mathbb{E}^t = (\mathbb{E}_x^t : x \in V)$ *is the minimal nonnegative solution of the following system of linear equations:*

$$\begin{cases} \mathbb{E}_t^t = 0 \\ \mathbb{E}_x^t = 1 + \sum_{y \neq t} p_{x,y} \mathbb{E}_y^t \text{ for } x \in V \end{cases}$$

Applying Theorem 26.7 to a specific Markov chain, we obtain a useful Lemma for the analysis of Algorithm 4:

LEMMA 26.16 Let C_d be a chain of $d+1$ states $0, 1, \ldots, d$ and $q \in]0, 1/2]$. If state 0 is absorbing and the transition matrix is of the form:

$$\begin{cases} p_{i,i-1} = p_{i,i+1} = q \text{ for } 1 \leq i \leq d-1 \\ p_{i,i} = 1 - 2q \text{ for } 1 \leq i \leq d-1 \\ p_{d,d} = 1 - q \end{cases}$$

then, the hitting time to state 0 starting from state i is $\mathbb{E}_i^0 = \frac{i}{2q}(2d - i + 1)$.

PROOF We make a use of Theorem 26.7 for the computation of $\mathbb{E}_i^0$. We have

$$\begin{cases} \mathbb{E}_1^0 = 1 + (1 - 2q)\mathbb{E}_1^0 + q\mathbb{E}_2^0 \\ \mathbb{E}_i^0 = 1 + q\mathbb{E}_{i-1}^0 + (1 - 2q)\mathbb{E}_i^0 + q\mathbb{E}_{i+1}^0 \text{ for } 2 \leq i \leq d-1 \\ \mathbb{E}_d^0 = 1 + (1 - q)\mathbb{E}_d^0 + q\mathbb{E}_{d-1}^0 \end{cases}$$

Noting that $\mathbb{E}_i^0 = \sum_{j=1}^i \mathbb{E}_j^{j-1}$, we are interested by $\mathbb{E}_j^{j-1}$ for $1 \leq j \leq d$. Therefore, $\mathbb{E}_d^{d-1} = 1 + (1 - q)\mathbb{E}_d^{d-1} = 1/q$ and

$$\begin{aligned} \mathbb{E}_j^{j-1} &= 1 + (1 - 2q)\mathbb{E}_j^{j-1} + q\mathbb{E}_{j+1}^{j-1} \\ &= 1 + (1 - 2q)\mathbb{E}_j^{j-1} + q\left(\mathbb{E}_{j+1}^j + \mathbb{E}_j^{j-1}\right) \\ &= 1/q + \mathbb{E}_{j+1}^j \\ &= \frac{d-j}{q} \end{aligned}$$

This implies that $\mathbb{E}_i^0 = \sum_{j=1}^i (d - j)/q = \frac{1}{q}\left(di - \frac{i(i-1)}{2}\right)$.

THEOREM 26.8 *In a unidirectional n-sized ring containing an arbitrary number k of tokens ($k \geq 2$), the stabilization time of Algorithm 4 is $\frac{n^2}{8p(1-p)} \leq \mathbb{E}(T) \leq \frac{n^2}{2p(1-p)}\left(\frac{\pi^2}{6} - 1\right) + \frac{n\log n}{p(1-p)}$. For constant p, $\mathbb{E}(T) = \Theta(n^2)$.*

PROOF For any $k \geq 2$, the evolution of the ring with exactly k tokens under Algorithm 26.3.4 can be described by a Markov chain $\mathcal{S}_k$ whose state space is the set of k-tuples of positive integers whose sum is equal to n (these integers represent the distances between successive tokens on the ring), with an additional state $\delta = (0, \ldots, 0)$ to represent transitions to a configuration with fewer than k tokens. To prove the upper bound of the Theorem, we will prove an upper bound on the hitting time of this state δ, independently of the initial state. Consider two successive tokens on the ring. On any given round, each will move forward, independently of the other, with probability p, and stay in place with probability $1 - p$. Thus, with probability $p(1 - p)$, the distance between them will decrease by 1; with the same probability, it will increase by 1; and, with probability $1 - 2p(1 - p)$, the distance will remain the same. Thus, locally, the distance between consecutive tokens follows the same evolution rule as that of the chain C_n of Lemma 26.16.

What follows is a formal proof, using the technique of couplings of Markov chains, that the expected time it takes for two tokens among k to collide is no longer than the expected time for $C_{n/k}$ to reach state 0.

For any state $\mathbf{x} = (x^1, \ldots, x^k)$ of S_k, let $m(\mathbf{x}) = \min_i x^i$ denote the minimum distance between two successive tokens, and let $i(\mathbf{x}) = \min\{j : x^j = m(\mathbf{x})\}$ denote the smallest index where this minimum is realized. Let $(X_t)_{t \geq 0}$ denote a realization of the Markov chain S_k. We define a coupling $(X_t, Y_t)_{t \geq 0}$ of the Markov chains S_k and C_d, where $d = \lfloor n/k \rfloor$ and $q = p(1 - p)$, as follows :

- $Y_0 = m(X_0)$;
- $Y_{t+1} = \min \left\{ d, Y_t + \left(X_{t+1}^{i(X_t)} - X_t^{i(X_t)} \right) \right\}$

In other words, the evolution of Y_t is determined by selecting two tokens that are separated by the minimum distance in X_t, and making the change in Y_t reflect the change in distance between these two tokens (while capping Y_t at d).

A trivial induction on t shows that $Y_t \geq m(X_t)$ holds for all t, so that (X_t) will reach state δ no later than (Y_t) reaches 0. Thus, the time for S_k to reach δ (i.e., the time during which the ring has exactly k tokens) is stochastically dominated by the time for C_d to reach 0. By Lemma 26.16, the expectation of this time is no longer than

$$\frac{d(d+1)}{2q} \leq \frac{1}{2q} \left(\frac{n^2}{k^2} + \frac{n}{k} \right)$$

Summing over all values of k from 2 to n, we get, for the expected stabilization time T,

$$
\begin{aligned}
\mathbb{E}(T) & \leq \frac{1}{2p(1-p)} \sum_{k=2}^{n} \left[\frac{n^2}{k^2} + \frac{n}{k} \right] \\
& \leq \frac{1}{2p(1-p)} \left(\left(\frac{\pi^2}{6} - 1 \right) n^2 + n \ln(n) \right)
\end{aligned}
$$

The lower bound comes from the fact that, when $k = 2$, the expected time for $C_{n/2}$ to reach state 0 from state $n/2$ is at least $\frac{n^2}{8p(1-p)}$.

REMARK 26.1   Our upper bound on the expected convergence time is minimal for $p = 1/2$. The precise study of Algorithm 4 show that the convergence time hardly depends on the initial number of tokens: for n high enough and $p = 1/2$, $\mathbb{E}(T) > n^2/2$ for two initial tokens at distance $n/2$, and $\mathbb{E}(T) < 1.3n^2$ for n tokens.

26.4   Research Issues and Summary

In its core formulation, self-stabilization is a useful paradigm for forward recovery in distributed systems and networks. Because self-stabilization only considers the effect of faults, there is no assumption about the nature or the extent of faults. In practice, when a faulty component is diagnosed in a self-stabilizing network (e.g., because it exhibits erratic behavior), it is sufficient to remove this

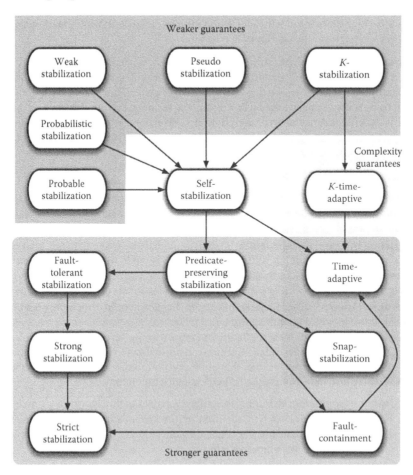

FIGURE 26.11 Taxonomy of self-stabilization main variants.

component from the system to recover proper behavior automatically [64], as a self-stabilizing system does not require any kind of initialization.

On the negative side, "eventually" does not give any complexity guarantee on the stabilization time, and in some cases, it is possible that a single hazard triggers a correction wave in the whole network. Moreover, the fact that a system participant is not able to detect stabilization (arbitrary memory corruption could lead the participants believe that the system is stabilized when it is not) prevents safety-related specifications from having self-stabilizing solutions. Those limitations led to defining new forms of self-stabilization, that are presented in Figure 26.11 and detailed in the remaining of the section. In Figure 26.11, each arrow denotes the (transitive) relation "provides weaker guarantees than" between two variants of self-stabilization.

26.4.1 Weaker than Self-Stabilization

The guarantees that are given are weaker than that of self-stabilization, and this permits in general to solve strictly more difficult specifications than those than can be solved by a self-stabilizing system. Thus problems that are provably intractable in a strict self-stabilizing setting may be come solvable.

This permits to widen the scope of self-stabilization to new applications, while maintaining attractive fault-tolerance properties to the developed applications.

1. **Restricting the nature of the faults**. This approach, that we denote by probable stabilization consists of considering that truly arbitrary memory corruptions are highly unlikely. Probabilistic arguments are used to establish that, in general, memory corruptions that result from faults can be detected using traditional techniques from information theory, such as data redundancy or error detection and correction codes. In particular, in [42], error detection codes are used to determine that memory corruption has occurred, with a high probability. If the article only considers the case where a single corruption arises (in other words, only one node in the system is affected by this corruption), it makes it possible to return to a normal behavior in a single correction step. For a system, even a large-scale one, where memory corruptions localized in each neighborhood and are not malicious (i.e., they can be detected using techniques such a cyclic redundancy checks), this approach is well indicated.

2. **Restricting the extent of the faults**. If we assume that the faults that can occur only ever concern a very small part of the network, it is possible to design algorithms that converge more quickly than traditional self-stabilizing algorithms. In order to have a formal framework, we consider that the distance to a legitimate configuration is equal to the number of nodes whose memories have to be changed in order to achieve a legitimate configuration (as with a Hamming distance). Of course, it is possible that even if we are at a distance k from a legitimate configuration, more than k nodes have, in fact, corrupted memories. From the perspective of returning to a normal state, only the closest legitimate configuration is considered. Studies that attempt to minimize the stabilization time in context where few faults occur usually divide stabilization into two levels [55]:

   a. *"visible" stabilization*: Here, only the output variables of the algorithm are involved. The output variables are typically used by the system's user. For example, if we consider a tree construction algorithm, only the pointer oriented toward the parent node is included in the output variables.

   b. *"internal" stabilization*: Here, all of the algorithm's variables are involved. This type of stabilization corresponds to the traditional concept of self-stabilization.

   In many studies, only the "visible" stabilization is performed quickly (i.e., in time relative to the number of faults that strike the system, rather than in time relative to the size of said system), while the "internal" stabilization most of the time remains proportional to the size of the network. Algorithms that present this constraint are not capable of tolerating a high frequency of faults. Consider an algorithm whose time for visible stabilization depends on k (the number of faults) and whose internal stabilization depends on n (the size of the system). Now, if a new fault occurs while the visible stabilization is being performed, but not during the internal stabilization, this can lead to a global state containing a number of faults greater than k, and there is no longer any guarantee on the new time for visible stabilization. The notion of k-stabilization [7] is defined as self-stabilization, when restricting the starting configurations to those configurations that are at a distance of k or less from a legitimate configuration. Because of the less hostile environment, it is possible to solve problems that are impossible in the case of general self-stabilization [44,69], and to offer reduced visible stabilization times. A particular instance of k-stabilization is *node*-stabilization [68]: in the initial configuration, links do not contain any message, yet the memory of the nodes may be arbitrarily corrupted, as this would happen if the networking equipments of a system were shut down but not the computing devices.

3. **Restricting the stabilization guarantees**.

   a. Section 26.3.4 already presented the concept of probabilistic stabilization. Probabilistic stabilization weakens the convergence requirement by only requiring expected finite time convergence with probability 1. In [13,14], weak and strong variants of probabilistic stabilization are presented: a weak probabilistic algorithm only guarantees correctness with probability 1, while a strong probabilistic algorithm guarantees certain correctness (i.e., when a legitimate configuration is reached, all subsequet executions from this configuration conform to the specification). For example, the algorithm presented in Section 26.3.4 falls in the category of weak probabilistic algorithms [27].

   b. The notion of pseudo-stabilization [11] removes the guarantee of reaching a configuration from which the behavior satisfies the specification. Instead, pseudo-stabilization guarantees that every execution has a suffix that satisfies the specification. The main weakening here is that there is no guarantee that the system ever stabilizes (as it may never reach a legitimate configuration). However, for an external observer, the behavior of the system is eventually correct.

   c. Weak stabilization [34] breaks the requirement that every execution reaches a legitimate configuration. Instead, weak stabilization guarantees that from every possible initial configuration c, there exists at least one execution starting from c that reaches a legitimate configuration. Recently, a strong connection between weak stabilization and probabilistic stabilization was demonstrated [18]: essentially a weak stabilizing protocol can be turned into a probabilistic stabilizing algorithm that operates under a probabilistic daemon (i.e., a distributed scheduler whose choices are probabilistic).

26.4.2 Stronger than Self-Stabilization

The guarantees that are given are stronger than that of self-stabilization, and this permits in general to solve strictly less difficult specifications than those than can be solved by a self-stabilizing system. That is, the set of problems than can be solved is strictly smaller than the set of problems that can be solved by strictly self-stabilizing algorithms, yet the guarantees are stronger and may even match those of robust algorithms.

1. **Stronger safety guarantees**.

   a. Predicate-preserving stabilization refers to the fact that in addition to being self-stabilizing, the algorithm also preserves some distributed predicate on configurations, either in the stabilizing phase or in the stabilized phase, in spite of the occurrence of new faults (of limited nature). One such instance is the algorithm described in Section 26.3.3, where message losses, duplications, and desequencings are tolerated both in the stabilizing and stabilized phases. Another instance is that of route-preserving stabilization [47]: the algorithm maintains a shortest path spanning tree in a self-stabilizing way, and has an additional property of path preservation, meaning that if a tree is initially constructed toward a destination, any message transmitted toward that destination reaches it in a finite time, even if the cost of every edge in the system continuously changes. Super-stabilization [22,40,49] is a special instance of predicate-preserving stabilization. This property states that a super-stabilizing algorithm is self-stabilizing on one hand and, on the other, preserves a predicate (typically a safety predicate) when changes in topology occur in a legitimate configuration. Thus, changes in topology are limited: if these changes occur during the stabilization phase, the system can never stabilize. On the other hand, if they occur only after a correct global state is achieved, the system remains stable.

b. Fault-tolerant self-stabilization is characteristic of algorithms that aim at providing tolerance to both arbitrary transient faults (self-stabilization) and permanent ones (robustness), the two main trends in distributed fault-tolerance (see Section 27.1). While mostly impossibility results were obtained in this context [3,8] except for some particular problems such as time synchronization [26], recent results [17] hint that the ultimate properties found in fault-tolerant algorithms are more related to pseudo-stabilization than to self-stabilization.

c. Strict stabilization [62] refers to a different scheme to tolerate both transient and permanent Byzantine faults. The Byzantine contamination radius is defined as the maximum distance from which the effect of Byzantine nodes can be felt. This contamination radius must obviously be as small as possible. A problem is r-restrictive if its specification prohibits combinations of states in a configuration for nodes at a distance of r at the most. For example, the problem of coloring nodes in a network is 1-restrictive, since two neighboring nodes cannot have the same color. On the other hand, the tree construction problem is r-restrictive (for any r between 1 and $n - 1$) because the correction implies that all of the parents that are chosen must form a tree. The main Theorem in [62] states that if a problem is r-restrictive, the best contamination radius that can be obtained is r. The follow-up paper [61] provides a *Byzantine-insensitive* link coloring algorithm: the subset of correct nodes, once stabilized, cannot be influenced again by Byzantine processes.

d. For problems such as tree construction that have r-restrictive specifications for some arbitrary r, the weaker notion of strong stabilization has been introduced [59]. While strict stabilization contains the action of Byzantine processes in space, strong stabilization contains the action of Byzantine processes in time: even if a Byzantine process may execute an infinite number of malicious actions in a infinite execution, the correct processes may only be impacted a finite number of times. However, and similarly to pseudo-stabilization, there is no bound on the time after which Byzantine actions are harmless (a Byzantine process may execute only correct actions over a long-unbounded-period of time, and then arbitrarily behave for some time, making correct processes execute a finite (bounded) number of corrective actions).

2. **Stronger complexity guarantees**.

a. For certain problems, a memory corruption can cause a cascade of corrections in the entire system [5], yet it would be natural for the stabilization to be quicker when the number of failures that strike the system is smaller. This is the principle behind time-adaptive self-stabilization [10,23,56], also known as scalable stabilization [32] and fault local stabilization [57]. Note that time adaptive protocols have the property of fault containment [31] in the sense that a visible fault cannot spread on the whole network: it is contained near to its initial location before it disappears. It is possible to arrange self-stabilizing, k-stabilizing, and time-adaptive algorithms into classes, depending on the difficulty in solving problems that can be solved in each case. For example, if it is possible to solve a problem in a self-stabilizing way, it is also possible to solve it in a k-stabilizing way (if you can do more, you can do less). Likewise, if it is possible to solve a problem in a time-adaptive way, it is also possible to solve it without trying to constrain the visible stabilization time. Thus, the class of problems that can be solved in a time-adaptive way is a subset of the class of problems that can be solved in a self-stabilizing way, which is itself a subset of the class of problems that have k-stabilizing solutions. These inclusions are strict: Some problems can be solved in a k-stabilizing way, but not in a self-stabilizing [69], others can be solved in a self-stabilizing way, but not in a time-adaptive way [29].

b. Given a problem specification, a snap-stabilizing system [9] is guaranteed to perform according to this specification regardless of the initial state. That is, a snap-stabilizing system has a stabilization time of 0. It is important to note that a snap-stabilizing protocol does not guarantee that the system never works in a fuzzy manner. Actually, the main idea behind the snap-stabilization is the following: the protocol is seen as a function and the function ensures two properties despite the arbitrary initial configuration of the system:

 i. Upon an external (w.r.t. the protocol) request at a process p, the process p (called the initiator) starts a computation of the function in finite time using special actions called starting actions;

 ii. If the process p starts an computation, then the computation performs an expected task.

With such properties, the protocol always satisfies its specifications. Indeed, when the protocol receives a request, this means that an external application (or a user) requests the computation of a specific task provided by the protocol. In this case, a snap-stabilizing protocol guarantees that the requested task is executed as expected. On the contrary, when there is no request, there is nothing to guarantee. Due to the "start" and "correctness" properties it has to ensure, snap-stabilization requires specifications based on a sequence of actions ("request," "start,"...) rather than a particular subset of configurations (e.g., the legitimate configurations). Most of the literature on snap-stabilization deals with the shared memory model only, with the notable recent exception of [15]. Due to the 0 stabilization time complexity, snap-stabilization actually also guarantees stronger safety properties (e.g., mutual exclusion in [15]) than those of self-stabilization.

There remains the special case of k-time adaptive stabilization. In our classification, it is both a weakening of self-stabilization in the sense that fewer faults are allowed, and a strengthening of self-stabilization in the sense that the stabilization time complexity guarantee is proportional to the number of faults that hit the network.

26.5 Further Information

Advances on all aspects of self-stabilization are reported in the annual Symposium on Stabilization, Safety, and Security of distributed systems (SSS). Theoretical aspects of self-stabilization are also covered by theoretical distributed computing conferences such as Principles of Distributed Computing (PODC), Distributed Computing (DISC), On Principles of Distributed Systems (OPODIS), and International Colloquium on Structural Information and Communication Complexity (Sirocco). Practical aspects of self-stabilization are covered by International Conference on Distributed Computing Systems (ICDCS), Dependable Systems and Networks (DSN), Symposium on Reliable Distributed Systems (SRDS), and the many conferences dedicated to sensor networks and autonomic computing.

A book [20] published in 2000 is dedicated to self-stabilization, and [30,51,66] all include a chapter on self-stabilization. Ref. [46] surveys self-stabilization in network protocols, while Ref. [69] describes self-stabilization with respect to scalability properties.

Defining Terms

Configuration: Global state of the system at a particular time.

Daemon: Predicate on executions, used to abstract system hypotheses.

Execution: Maximal sequence of configurations.

Self-stabilization: Property of a distributed system to eventually converge to a configuration from which every execution assuming a particular daemon conforms to a particular specification.

Specification: Predicate on executions.

References

1. Y. Afek and G. M. Brown. Self-stabilization over unreliable communication media. *Distributed Computing*, 7(1):27–34, 1993.
2. L. O. Alima, J. Beauquier, A. K. Datta, and S. Tixeuil. Self-stabilization with global rooted synchronizers. In *Proceedings of the 18th International Conference on Distributed Computing Systems*, pp. 102–109, Amsterdam, the Netherlands, IEEE Press, May 1998.
3. E. Anagncstou and V. Hadzilacos. Tolerating transient and permanent failures (extended abstract). In A. Schiper, editor, *WDAG*, volume 725 of *Lecture Notes in Computer Science*, pp. 174–188, Lausanne, Switzerland, Springer, 1993.
4. D. Angluin. Local and global properties in networks of processors (extended abstract). In *STOC '80: Proceedings of the Twelfth Annual ACM Symposium on Theory of Computing*, pp. 82–93, New York, ACM Press, 1980.
5. B. Awerbuch, B. Patt-Shamir, G. Varghese, and S. Dolev. Self-stabilization by local checking and global reset (extended abstract). In G. Tel and P. M. B. Vitányi, editors, *Distributed Algorithms, 8th International Workshop, WDAG '94*, volume 857 of *Lecture Notes in Computer Science*, pp. 326–339, Terschelling, the Netherlands, Springer, 1994.
6. J. Beauquier, S. Cordier, and S. Delaët. Optimum probabilistic self-stabilization on uniform rings. In *Proceedings on the Workshop on Self-Stabilizing Systems*, pp. 15.1–15.15, Las Vegas, NV, 1995.
7. J. Beauquier, C. Genolini, and S. Kutten. Optimal reactive k-stabilization: The case of mutual exclusion. In *Proceedings of the Eighteenth Annual ACM Symposium on Principles of Distributed Computing*, pp. 209–218, New York, 1999.
8. J. Beauquier and S. Kekkonen-Moneta. Fault-tolerance and self-stabilization: Impossibility results and solutions using self-stabilizing failure detectors. *International Journal of Systems Science*, 28(11):1177–1187, 1997.
9. A. Bui, A. K. Datta, F. Petit, and V. Villain. Snap-stabilization and pif in tree networks. *Distributed Computing*, 20(1):3–19, 2007.
10. J. Burman, T. Herman, S. Kutten, and B. Patt-Shamir. Asynchronous and fully self-stabilizing time-adaptive majority consensus. In J. H. Anderson, G. Prencipe, and R. Wattenhofer, editors, *OPODIS*, volume 3974 of *Lecture Notes in Computer Science*, pp. 146–160, Pisa, Italy, Springer, 2005.
11. J. E. Burns, M. G. Gouda, and R. E. Miller. Stabilization and pseudo-stabilization. *Distributed Computing*, 7(1):35–42, 1993.
12. P. Danturi, M. Nesterenko, and S. Tixeuil. Self-stabilizing philosophers with generic conflicts. In A. K. Datta and M. Gradinariu, editors, *Eighth International Symposium on Stabilization, Safety, and*

Security on Distributed Systems (SSS 2006), Lecture Notes in Computer Science, pp. 214–230, Dallas, Texas, Springer Verlag, November 2006.

13. A. K. Datta, M. Gradinariu, and S. Tixeuil. Self-stabilizing mutual exclusion using unfair distributed scheduler. In *IEEE International Parallel and Distributed Processing Symposium (IPDPS'2000)*, pp. 465–470, Cancun, Mexico, IEEE Press, May 2000.

14. A. K. Datta, M. Gradinariu, and S. Tixeuil. Self-stabilizing mutual exclusion with arbitrary scheduler. *The Computer Journal*, 47(3):289–298, 2004.

15. S. Delaët, S. Devismes, M. Nesterenko, and S. Tixeuil. Snap-stabilization in message-passing systems. In *International Conference on Distributed Systems and Networks (ICDCN 2009)*, number 5404 in *LNCS*, pp. 281–286, Hyderabad, India, January 2009. Also a Brief Anouncement in PODC 2008.

16. S. Delaët and S. Tixeuil. Tolerating transient and intermittent failures. *Journal of Parallel and Distributed Computing*, 62(5):961–981, May 2002.

17. C. Delporte-Gallet, S. Devismes, and H. Fauconnier. Robust stabilizing leader election. In T. Masuzawa and S. Tixeuil, editors, *SSS*, volume 4838 of *Lecture Notes in Computer Science*, pp. 219–233, Paris, France, Springer, 2007.

18. S. Devismes, S. Tixeuil, and M. Yamashita. Weak vs. self vs. probabilistic stabilization. In *Proceedings of the IEEE International Conference on Distributed Computing Systems (ICDCS 2008)*, Beijing, China, June 2008.

19. E. W. Dijkstra. Self-stabilizing systems in spite of distributed control. *Communications of the ACM*, 17(11):643–644, 1974.

20. S. Dolev. *Self-Stabilization*. MIT Press, Cambridge, MA, March 2000.

21. S. Dolev, M. G. Gouda, and M. Schneider. Memory requirements for silent stabilization. *Acta Informatica*, 36(6):447–462, 1999.

22. S. Dolev and T. Herman. Superstabilizing protocols for dynamic distributed systems. *Chicago Journal of Theoritcal Computer Science*, 4:1–40, 1997.

23. S. Dolev and T. Herman. Parallel composition for time-to-fault adaptive stabilization. *Distributed Computing*, 20(1):29–38, 2007.

24. S. Dolev, A. Israeli, and S. Moran. Self-stabilization of dynamic systems assuming only read/write atomicity. *Distributed Computing*, 7(1):3–16, 1993.

25. S. Dolev, A. Israeli, and S. Moran. Resource bounds for self-stabilizing message-driven protocols. *SIAM Journal of Computing*, 26(1):273–290, 1997.

26. S. Dolev and J. L. Welch. Self-stabilizing clock synchronization in the presence of byzantine faults. *Journal of the ACM*, 51(5):780–799, 2004.

27. P. Duchon, N. Hanusse, and S. Tixeuil. Optimal randomized self-stabilizing mutual exclusion in synchronous rings. In *Proceedings of the 18th Symposium on Distributed Computing (DISC 2004)*, number 3274 in *Lecture Notes in Computer Science*, pp. 216–229, Amsterdam, the Netherlands, Springer Verlag, October 2004.

28. M. Duflot, L. Fribourg, and C. Picaronny. Randomized finite-state distributed algorithms as markov chains. In J. L. Welch, editor, *DISC*, volume 2180 of *Lecture Notes in Computer Science*, pp. 240–254, Lisbon, Portugal, Springer, 2001.

29. C. Genolini and S. Tixeuil. A lower bound on *k*-stabilization in asynchronous systems. In *Proceedings of IEEE 21st Symposium on Reliable Distributed Systems (SRDS'2002)*, Osaka, Japan, October 2002.

30. S. Ghosh. *Distributed Systems (Computer and Information Sciences)*. Boca Raton, FL, Chapman & Hall/CRC, 2006.

31. S. Ghosh, A. Gupta, T. Herman, and S. V. Pemmaraju. Fault-containing self-stabilizing distributed protocols. *Distributed Computing*, 20(1):53–73, 2007.

32. S. Ghosh and X. He. Scalable self-stabilization. *Journal of Parallel and Distributed Computing*, 62(5):945–960, 2002.

33. M. G. Gouda. *Elements of Network Protocol Design*. John Wiley & Sons, Inc., New York, 1998.

34. M. G. Gouda. The theory of weak stabilization. In A. K. Datta and T. Herman, editors, *WSS*, volume 2194 of *Lecture Notes in Computer Science*, pp. 114–123, Lisbon, Portugal, Springer, 2001.

35. M. G. Gouda and F. Furman Haddix. The alternator. *Distributed Computing*, 20(1):21–28, 2007.

36. M. Gradinariu and S. Tixeuil. Self-stabilizing vertex coloring of arbitrary graphs. In *International Conference on Principles of Distributed Systems (OPODIS'2000)*, pp. 55–70, Paris, France, December 2000.

37. M. Gradinariu and S. Tixeuil. Tight space uniform self-stabilizing l-mutual exclusion. In *IEEE International Conference on Distributed Computing Systems (ICDCS'01)*, pp. 83–90, Phoenix, Arizona, IEEE Press, May 2001.

38. M. Gradinariu and S. Tixeuil. Conflict managers for self-stabilization without fairness assumption. In *Proceedings of the International Conference on Distributed Computing Systems (ICDCS 2007)*, p. 46, Toronto, Ontario, Canada, IEEE, June 2007.

39. T. Herman. Probabilistic self-stabilization. *Information Processing Letters*, 35(2):63–67, 1990.

40. T. Herman. Superstabilizing mutual exclusion. *Distributed Computing*, 13(1):1–17, 2000.

41. T. Herman. Models of self-stabilization and sensor networks. In S. R. Das and S. K. Das, editors, *Distributed Computing - IWDC 2003, 5th International Workshop*, volume 2918 of *Lecture Notes in Computer Science*, pp. 205–214, Kolkata, India, Springer, 2003.

42. T. Herman and S. V. Pemmaraju. Error-detecting codes and fault-containing self-stabilization. *Information Processing Letters*, 73(1–2):41–46, 2000.

43. T. Herman and S. Tixeuil. A distributed tdma slot assignment algorithm for wireless sensor networks. In *Proceedings of the First Workshop on Algorithmic Aspects of Wireless Sensor Networks (AlgoSensors'2004)*, number 3121 in *Lecture Notes in Computer Science*, pp. 45–58, Turku, Finland, Springer-Verlag, July 2004.

44. A. Israeli and M. Jalfon. Token management schemes and random walks yield self-stabilizing mutual exclusion. In *Proceedings of the Ninth Annual ACM Symposium on Principles of Distributed Computing*, pp. 119–131, Quebec, Canada, 1990.

45. C. Johnen, L. O. Alima, A. K. Datta, and S. Tixeuil. Optimal snap-stabilizing neighborhood synchronizer in tree networks. *Parallel Processing Letters*, 12(3–4):327–340, 2002.

46. C. Johnen, F. Petit, and S. Tixeuil. Auto-stabilisation et protocoles réseaux. *Technique et Science Informatiques*, 23(8):1027–1056, 2004.

47. C. Johnen and S. Tixeuil. Route preserving stabilization. In *Proceedings of the Sixth Symposium on Self-stabilizing Systems (SSS'03)*, Lecture Notes in Computer Science, San Francisco, CA, Springer Verlag, June 2003. Also in the Proceedings of DSN'03 as a one page abstract.

48. M. H. Karaata. Self-stabilizing strong fairness under weak fairness. *IEEE Transactions on Parallel and Distributed Systems*, 12(4):337–345, 2001.

49. Y. Katayama, E. Ueda, H. Fujiwara, and T. Masuzawa. A latency optimal superstabilizing mutual exclusion protocol in unidirectional rings. *Journal of Parallel and Distributed Computing*, 62(5):865–884, 2002.

50. S. Katz and K. J. Perry. Self-stabilizing extensions for message-passing systems. *Distributed Computing*, 7(1):17–26, 1993.

51. A. D. Kshemkalyani and M. Singhal. *Distributed Computing: Principles, Algorithms, and Systems*. Cambridge University Press, Cambridge, U.K., 2008.

52. S. S. Kulkarni and M. Arumugam. Transformations for write-all-with-collision model. *Computer Communications*, 29(2):183–199, 2006.

53. S. S. Kulkarni and U. Arumugam. Collision-free communication in sensor networks. In S.-T. Huang and T. Herman, editors, *Self-Stabilizing Systems, 6th International Symposium, SSS 2003*, volume 2704 of *Lecture Notes in Computer Science*, pp. 17–31, San Francisco, CA, Springer, 2003.

54. S. S. Kulkarni, C. Bolen, J. Oleszkiewicz, and A. Robinson. Alternators in read/write atomicity. *Information Processing Letters*, 93(5):207–215, 2005.

55. S. Kutten and T. Masuzawa. Output stability versus time till output. In A. Pelc, editor, *DISC*, volume 4731 of *Lecture Notes in Computer Science*, pp. 343–357, Lemesos, Cyprus, Springer, 2007.

56. S. Kutten and B. Patt-Shamir. Stabilizing time-adaptive protocols. *Theoritical Computer Science*, 220(1):93–111, 1999.

57. S. Kutten and D. Peleg. Fault-local distributed mending. *Journal of Algorithms*, 30(1):144–165, 1999.

58. F. Manne, M. Mjelde, L. Pilard, and S. Tixeuil. A new self-stabilizing maximal matching algorithm. In *Proceedings of the 14th International Colloquium on Structural Information and Communication Complexity (Sirocco 2007)*, volume 4474, pp. 96–108, Castiglioncello (LI), Italy, Springer Verlag, June 2007.

59. T. Masuzawa and S. Tixeuil. Bounding the impact of unbounded attacks in stabilization. In A. K. Datta and M. Gradinariu, editors, *Eighth International Symposium on Stabilization, Safety, and Security on Distributed Systems (SSS 2006), Lecture Notes in Computer Science*, pp. 440–453, Dallas, Texas, Springer Verlag, November 2006.

60. T. Masuzawa and S. Tixeuil. On bootstrapping topology knowledge in anonymous networks. In A. K. Datta and M. Gradinariu, editors, *Eighth International Symposium on Stabilization, Safety, and Security on Distributed Systems (SSS 2006), Lecture Notes in Computer Science*, pp. 454–468, Dallas, Texas, Springer Verlag, November 2006.

61. T. Masuzawa and S. Tixeuil. Stabilizing link-coloration of arbitrary networks with unbounded byzantine faults. *International Journal of Principles and Applications of Information Science and Technology (PAIST)*, 1(1):1–13, December 2007.

62. M. Nesterenko and A. Arora. Tolerance to unbounded byzantine faults. In *21st Symposium on Reliable Distributed Systems (SRDS 2002)*, p. 22, Osaka, Japan, IEEE Computer Society, 2002.

63. J. R. Norris. *Markov Chains* (Cambridge Series in Statistical and Probabilistic Mathematics). Cambridge University Press, New York, first edition, 1997.

64. R. Perlman. *Interconnexion Networks*. Addison-Wesley, Reading, MA, 2000.

65. N. Sakamoto. Comparison of initial conditions for distributed algorithms on anonymous networks. In *PODC*, pp. 173–179, Atlanta, GA, 1999.

66. G. Tel. *Introduction to Distributed Algorithms*. Cambridge University Press, New York, 2001.

67. S. Tixeuil. *Auto-stabilisation Efficace*. PhD thesis, University of Paris Sud XI, Orsay, France, January 2000.

68. S. Tixeuil. On a space-optimal distributed traversal algorithm. In Springer Verlag, editor, *Fifth Workshop on Self-stabilizing Systems (WSS'2001)*, volume *LNCS* 2194, pp. 216–228, Lisbon, Portugal, October 2001.

69. S. Tixeuil. *Wireless Ad Hoc and Sensor Networks*, Chapter 10 Fault-tolerant distributed algorithms for scalable systems. Washington, DC, ISTE, ISBN: 978 1 905209 86, October 2007.

70. G. Varghese and M. Jayaram. The fault span of crash failures. *Journal of the ACM*, 47(2):244–293, 2000.

27

Theory of Communication Networks

Gopal Pandurangan
Purdue University

Maleq Khan
Purdue University

27.1 Introduction

Communication networks have become ubiquitous today. The Internet, the global computer communication network that interconnects millions of computers, has become an indispensable part of our everyday life. This chapter discusses theoretical and algorithmic underpinnings of distributed communication networks focusing mainly on themes motivated by the Internet. The Internet is a distributed wide area communication network that connects a variety of end systems or *hosts* by a network of communication links and packet switches (e.g., routers). A packet switch takes a packet arriving on one of its incoming communication links and forwards that packet on one of its outgoing communication links. From the sending host to the receiving host, the sequence of communication links and packet switches is known as a **route** or a path through the network. Throughout we will use the term node to denote a host (processor) or a packet switch.

The Internet is a complex system, but fortunately it has a layered architecture which is extremely helpful in understanding and analyzing its functionality. This is called the network protocol stack and is organized as follows. The understanding of different layers of the stack allows us to tie the theoretical and algorithmic results that will be discussed to specific functions and protocols in the Internet.

Application layer: The application layer is closest to the end user. This layer interacts with software applications that implement a communicating component. The layered architecture allows one to create a variety of distributed application protocols running over multiple hosts. The application in one host uses the protocol to exchange packets of data with the application in another host. Some examples of application layer implementations include Telnet, File Transfer Protocol (FTP), Simple Mail Transfer Protocol (SMTP), and the Hypertext Transfer Protocol (HTTP). An application architecture determines how a network application is structured over the various hosts. The traditional application architecture paradigm has been the client–server paradigm. In a client–server architecture, there is an always-on host, called the server, which services requests from many other hosts, called clients. For example, all the above applications—Web, FTP, Telnet, and e-mail—are client-server based. The last few years has seen the emergence of a new paradigm called Peer-to-Peer (P2P) architecture. In P2P architecture, there is no concept of a dedicated, always-on, server. Instead hosts, called peers, communicate directly and a peer can act both as a client (while requesting information) or as a server (when servicing requests for other peers). Many of today's most popular and traffic-intensive applications, such as file distribution (e.g., BitTorrent), file searching (e.g., Gnutella/LimeWire), and Internet telephony (e.g., Skype) are P2P based. A key application of P2P is the decentralized searching and sharing of data and resources. The P2P paradigm is inherently scalable as the peers serve the dual role of clients and servers.

Transport layer: The transport layer segments the data for transport across the network. Generally, the transport layer is responsible for making sure that the data is delivered error free and in the proper sequence. Flow control (i.e., sender/receiver speed matching) and congestion control (source throttles its transmission rate as a response to network congestion) occur at the transport layer. Transmission Control Protocol (TCP) and User Datagram Protocol (UDP) are the two transport protocols of the Internet, with only the former providing flow and congestion controls.

Network layer: The network layer is responsible for breaking data into packets (known as datagrams) and moving the packets from the source to the destination. The network layer defines the network address. It includes the Internet **Protocol** (IP), which defines network addresses called **IP addresses**. Because this layer defines the logical network layout, routers can use this layer to determine how to forward packets. Internet's routing protocols are a key part of this layer. The routing protocols help in configuring the forwarding tables of the routers, which indicates to which of the neighbors a packet is to be forwarded based on its destination address. There are two types of protocols based on whether the routing is made within an autonomous system (intra-AS routing) or between autonomous systems (inter-AS routing). **Shortest path routing** is typically used for intra-AS routing. A protocol called Border Gateway Protocol (BGP) is used for inter-AS routing.

Link layer: Link layer's functionality is to move data from one node to an adjacent node over a single link on the path from the source to the destination host. Services offered by link layer include link access, reliable delivery, error detection, error correction, and flow control. A key protocol of link access called medium access control addresses the **multiple access problem**: how multiple nodes that share a single broadcast link can coordinate their transmissions so as to avoid collisions. Examples of link layer protocols include Ethernet, the main wired local area network (LAN) technology, and 802.11 wireless LAN, and token ring.

Physical layer: The physical layer is responsible for moving individual bits of data from one node to the next and its protocols are dependent on the actual transmission medium of the link (copper wire, fiber-optic cables, etc.)

27.1.1 Overview

In Section 27.3, we discuss routing algorithms that are a key part of the network layer's functionality. In Section 27.4, we discuss the basics of queuing theory, which is needed for modeling network delay and understanding the performance of queuing strategies. We consider the traditional stochastic queuing theory as well as the adversarial queuing theory. We will then discuss the theory of contention resolution protocols in Section 27.5 that addresses the multiple access problem, a problem handled by the link layer. In Section 27.6, we will address theoretical issues behind congestion/flow control and resource allocation, an important functionality of the transport layer, in particular, the TCP protocol. In Section 27.7, we will discuss P2P networks, the new emerging paradigm at the application layer.

An underlying theme of this chapter is the emphasis on distributed or decentralized algorithms and protocols. These require only local information (as opposed to global information) which is typically the only information available to the nodes to begin with. Also distributed algorithms are more robust since they do not rely on a central node that might fail. Distributed algorithms can react rapidly to a local change at the point of change. This is especially very useful for problems, such as routing. Distributed algorithms are inherently scalable and this is crucial for deployment in a large-scale communication network. Section 27.2 gives a brief introduction to distributed computing model and complexity measures that will be used later.

27.2 Distributed Computing Model

We will focus on the message-passing model of distributed computing where the network is modeled as a graph with nodes (vertices) connected by communication links (edges). There can be weights associated with the links that might denote the delay, or the capacity (bandwidth) of the link. Each edge e supports message passing between the two nodes associated with the endpoints of e. Messages can be sent in both directions (we will assume that the graph is undirected unless otherwise stated). For example, at the network layer of the Internet, nodes correspond to hosts and packet switches (routers), each identified by a unique identifier called the IP address, and the messages exchanged between them are basic data transmission units called packets. Alternatively, at the application layer (as in a P2P network), nodes could correspond to hosts (computers) and edges to TCP/IP connections that are maintained between pairs of nodes.

In the message-passing model, information is communicated in the network by exchanging messages. We assume that each node has a unique identity number (e.g., IP address) and at the beginning of computation, each vertex v accepts as input its own identity number and the weights of the edges adjacent to v. Thus, a node has only local knowledge limited to itself and its neighbors. We assume that each processor knows its neighbors, in the network and that it communicates directly only with its neighbors.

Two important models can be distinguished based on processor synchronization. In a synchronous model, each processor has an internal clock and the clocks of all processors are synchronized. The processor speeds are uniform and each processor takes the same amount of time to perform the same operation. Thus communication is synchronous and occurs in discrete clock "ticks" (time steps). In one time step, each node v can send an arbitrary message of size $O(\log n)$ (n is the number of nodes in the network) through each edge $e = (v, u)$ that is adjacent to v, and the message arrives at u by the end of this time step. (Note that a $O(\log n)$-size address is needed to uniquely address all nodes.) We will assume that the weights of the edges are at most polynomial in the number of vertices n, and therefore, the weight of a single edge can be communicated in one time step. This model of the distributed computation is called the (synchronous) $\mathcal{CONGEST}(\log n)$ model or simply the $\mathcal{CONGEST}$ model [39]. The $\mathcal{CONGEST}$ model is not very realistic for the Internet. However, it has been widely used as model to study distributed algorithms and captures the notion that there is a bound on the amount of messages that can be sent in a unit time. At the other extreme is the $\mathcal{LOCAL}$ model [39], where there is no such bound. We will adopt the $\mathcal{CONGEST}$ model here.

In an asynchronous model, no assumptions are made about any internal clocks or on the speeds of the processors. The steps in an asynchronous algorithm are determined by conditions or events and not by clock ticks. However, we do make two reasonable timing assumptions. First, we assume that messages do arrive (eventually) and in the same order they are sent (i.e., there is first-in first-out (FIFO) queuing). Second, we assume that if a processor has an event that requires it to perform a task, then it will eventually perform the task. Between the two extreme models, we can define "intermediate" models that are partially synchronous, where the processors have some partial information about timing (e.g., almost synchronized clocks or approximate bounds on message delivery time, etc.), not the complete information as they do in the synchronous model. Although intermediate models can provide a more realistic model of real networks, such as the Internet, we will restrict our attention to synchronous and asynchronous models in this chapter. Algorithms designed for the synchronous model can often be translated to work for the asynchronous model (see below), and algorithms for the latter will work for an intermediate model as well.

There are two important complexity measures for comparing distributed algorithms. The first is the time complexity, or the time needed for the distributed algorithm to converge to the solution. In the synchronous model, time is measured by the number of clock ticks called rounds (processors compute in "lock step"). For an asynchronous algorithm, this definition is meaningless since a single message from a node to a neighbor can take a long time to arrive. Therefore the following definition is used: time complexity in an asynchronous model is the time units from start to finish, assuming that each message incurs a delay of at most one time unit [39]. Note that this definition is used only for performance evaluation and has no effect on correctness issues. The second important measure is message complexity, which measures the total number of messages (each of size $O(\log n)$) that are sent between all pairs of nodes during the computation.

We will assume synchronous ($\mathcal{CONGEST}$) model unless otherwise stated, since it is simpler and easier to design algorithms for this model. Using a tool called synchronizers, one can transform a synchronous algorithm to work in an asynchronous model with no increase in time complexity and at the cost of some increase in the message complexity [39].

27.3 Routing Algorithms

We discuss three basic routing modes that are used in the Internet: unicast, broadcast, and multicast. In unicast, a node sends a packet to another specific node; in broadcast, a node sends a packet to every node in the network; in multicast, a node sends a packet to a subset of nodes. We focus on fundamental distributed network algorithms that arise in these routing modes.

27.3.1 Unicast Routing

Unicast routing (or just simply routing) is the process of determining a "good" path or route to send data from the source to the destination. Typically, a good path is one that has the least cost. Consider a weighted network $G = (V, E, c)$ with positive real-valued edge (link) costs given by the cost function c. The cost of a path $p = (e_1, \ldots, e_k)$ is defined as $c(p) = \sum_{i=1}^{k} c(e_i)$. For a source–destination pair $(s, t) \in (V \times V)$, the goal is to find a **least-cost** (or **shortest**) **path**, i.e., a path from s to t in G that has minimum cost. If all edges in the graph have cost 1, the shortest path is the path with the smallest number of links between the source and the destination. The Internet uses least-cost path routing. One way to classify routing algorithms is according to whether they are global or local:

- A global routing algorithm computes the least-cost path between a source and a destination using complete, global knowledge of the network. In practice, it happens to be referred to as a link state (LS) algorithm.

- A local routing algorithm computes the least-cost path in an iterative, distributed fashion. No node has complete information about all the edge costs. In practice, it happens to be referred to as a **distance vector** (DV) algorithm.

The LS and DV algorithms are essentially the only routing algorithms used in the Internet today.

27.3.1.1 A Link State Routing Algorithm

A LS algorithm knows the global network topology and all link costs. One way to accomplish this is by having each node broadcasts its identity number and costs of its incident edges to all other nodes in the network using a broadcasting algorithm, e.g., flooding. (Broadcast algorithms are described in Section 27.3.3.) Each node can then run the (centralized) link state algorithm and compute the same set of shortest paths as any other node. A well-known LS algorithm is the Dijkstra's algorithm for computing least-cost paths. This algorithm takes as input a weighted graph $G = (V, E, c)$, a source vertex $s \in V$ and computes shortest paths (and their values) from s to all nodes in V. Dijkstra's algorithm and its run-time analysis is given in Chapter 6. Note that this run-time is the time needed to run the algorithm in a single node. The message complexity and the time complexity of the algorithm is determined by the broadcast algorithm. For example, if broadcast is done by flooding (cf. Section 27.3.3.1) then the message complexity is $O(|E|^2)$, since $O(|E|)$ messages have to be broadcast, each of which causes $O(|E|)$ messages to be sent by flooding. The time complexity is $O(|E|D)$, where D is the **diameter** of the network. Internet's Open Shortest Path First (OSPF) protocol uses a LS routing algorithm as mentioned earlier. Since the LS algorithm is centralized, it may not scale well when the networks become larger.

27.3.1.2 A Distance Vector Algorithm

The DV algorithm is distributed and asynchronous. It is distributed because information is exchanged only between neighbors. Nodes then perform local computation and distribute the results back to its neighbors. This process continues till no more information is exchanged between neighbors. The algorithm is asynchronous in that it does not require all of the nodes to operate in lock step with each other.

The DV algorithm that is described below is called the distributed Bellman–Ford (DBF) algorithm. It is used in the Routing Information Protocol (RIP) and the Border Gateway Protocol (BGP) of the Internet.

We will describe the basic idea behind the DBF algorithm [2,15]. Suppose we want to compute the shortest (least-cost) path between s and all other nodes in a given undirected graph $G = (V, E, c)$ with real-valued positive edge weights. During the algorithm, each node x maintains a distance label $a(x)$ which is the current known shortest distance from s to x, and a variable $p(x)$ that contains the identity of the previous node on the current known shortest path from s to x. Initially, $a(s) = 0$, $a(x) = \infty$, and $p(x)$ are undefined for all $x \neq s$. When the algorithm terminates, $a(x) = d(s, x)$, where $d(s, x)$ is the shortest path distance between s and x, and $p(x)$ holds the neighbor of x on the shortest path from x to s. Thus, if node x wants to route to s along the shortest path it has to forward its data to $p(x)$.

The DBF consists of two basic rules: the update rule and the send rule. The update rule determines how to update the current label according to a message from a neighboring node. The send rule determines what values to send to its neighbors and is applied whenever a node adopts a new label.

Update rule: Suppose x with a label $a(x)$ receives $a(z)$ from a neighbor z. If $a(z) + c(z, x) < a(x)$, then it updates $a(x)$ to $a(z) + c(z, x)$ and sets $p(x)$ to be z. Otherwise $a(x)$ and $p(x)$ are not changed.

Send rule: Let $a(x)$ be a new label adopted by x. Then x sends $a(x)$ to all its neighbors.

27.3.1.2.1 Correctness and Analysis of the DBF Algorithm

We assume a synchronous setting where computation takes place in discrete rounds, i.e., all nodes simultaneously receive messages from their neighbors, perform the update rule (if necessary), and send the message to their neighbors (if update was performed). The following theorem gives the correctness and the complexity of the DBF algorithm.

THEOREM 27.1 *If all nodes work in a synchronous way, then the DBF algorithm terminates after at most n rounds. When it terminates, $a(x) = d(s, x)$ for all nodes x. The message complexity is $O(n|E|)$.*

PROOF Fix a vertex $x \in V$, we prove that the algorithm computes a shortest path from s to x.

Let $P = v_0, v_1, \ldots, v_k$, where $v_0 = s$ and $v_k = x$ be a shortest path from s to x. Note that $k < n$.

We prove by induction on i that after the ith round, the algorithm has computed the shortest path from s to v_i, i.e., $a(v_i) = d(s, v_i)$.

The hypothesis holds for $v_0 = s$ in round zero.

Assume that it holds for $j \leq i - 1$. After the ith iteration

$$a[v_i] \leq a[v_{i-1}] + c(v_{i-1}, v_i) \tag{27.1}$$

which is the shortest path from s to v_i, since P is the shortest path from s to v_i, and the right-hand side is the distance between s to v_i on that path.

Since, in each round $O(|E|)$ messages are exchanged, the total message complexity is $O(n|E|)$. $\quad\square$

From the above proof, it can be seen that the algorithm will compute the correct values even if the nodes operate asynchronously. An important observation is that the algorithm is "self-terminating"—there is no signal that the computation should stop; it just stops. The earlier rules can easily be generalized to compute the shortest path between all pairs of nodes by maintaining in each node x a distance label $a_y(x)$ for every $y \in V$. This is called the distance vector of x. Each node stores its own distance vector and the distance vectors of each of its neighbors. Whenever something changes, say the weight of any of its incident edges or its distance vector, the node will send its distance vector to all of its neighbors. The receiving nodes then update their own distance vectors according to the update rule.

A drawback of the DBF algorithm is that convergence time can be made arbitrarily large, if the initial distance labels are not correct. Consider the following simple network consisting of 4 nodes and 3 links shown in Figure 27.1. The goal is to compute the shortest paths to D. Initially each link has weight 1 and each node had calculated the shortest path distance to D. Thus the distances of A, B, and C to D are 3, 2, and 1, respectively. Now suppose the weight of edge (C, D) changes from 1 to a large positive number, say L. Assuming a synchronous behavior, in the subsequent iteration, C will set its distance label to D as 3, since B supposedly has a path of length 2 to D. In the following iteration, B will change its distance estimate to 4, since the best path it has is through C. This will

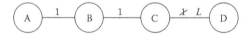

FIGURE 27.1 An example network for count-to-infinity problem.

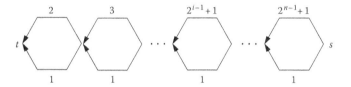

FIGURE 27.2 An example where the number of messages in the DBF algorithm is $\Omega(2^n)$.

continue until C distance label reaches L. Thus the number of iterations taken to converge to the correct shortest path is proportional to L. This problem is referred to as the "**count-to-infinity**" **problem** and shows that the algorithm can be very slow in reacting to a change in an edge cost. Some heuristics have been proposed to alleviate this problem; we refer to [5,27] for details.

27.3.1.2.2 *An Approximate Distributed Bellman–Ford Algorithm*

The DBF algorithm can suffer from exponential message complexity in an asynchronous setting. For example, consider the graph shown in Figure 27.2. There are 2^n distinct paths from s to t, each one with a distinct length. It is possible, in an asynchronous environment, to create an execution instance of DBF such that t will receive $\Omega(2^n)$ messages [2]. Awerbuch et al. [2] proposed a simple modification to the DBF algorithm that will result in a polynomial message complexity. However, the modified algorithm may not compute a shortest path, but will instead compute an approximate shortest path, in particular, the paths computed can be worse than the shortest path by at most a constant factor. Thus this can be called a distributed approximation algorithm. The only difference between the modified algorithm and the DBF algorithm is in the update rule.

Multiplicative update rule: Let $\alpha = 1 + 1/n$. Suppose x, with a label $a(x)$ receives $a(z)$ from a neighbor z. If $a(z) + c(x, z) < a(z)/\alpha$, then $a(x)$ is updated to $a(z) + c(x, z)$ and $p(x)$ is set to z.

Theorem 27.2 gives the performance of the modified DBF. For a proof we refer to [2].

THEOREM 27.2   *The length of the computed path between s and x (for any node x) at the end of the execution of the algorithm is at most $e.d(s, x)$, where e is the base of the natural logarithm. The number of messages sent is bounded by $O(|E|n \log(n\Delta))$, where Δ is the largest edge cost.*

27.3.2 Multicommodity Flow-Based Routing

A drawback of shortest path routing is that each source and destination is selected independent of other paths. It is quite possible that least-cost paths can cause a high congestion, i.e., many paths may go through the same edge. Alternate to least-cost paths routing, a broad class of routing algorithms is based on viewing packet traffic as flows between sources and destinations in a network. In this approach, the routing problem is formulated as a constrained optimization problem known as a multicommodity network flow problem with the goal of optimizing some appropriate global objective. In a multicommodity flow problem, we are given a directed network $G = (V, E)$ with capacity $C(i, j)$ for each directed edge (i, j). Let $r_i(j)$ be the traffic rate (or demand) from source i to destination j. While routing traffic from source to destination, we obtain a flow, i.e., traffic can be split and routed along multiple paths. Let $f_{ik}(j)$ denote the traffic flow destined for j that is sent on edge (i, k). The goal is to find paths between source–destination pairs and to determine the amount of flow on these paths under a set of constraints and objectives. Two types of constraints are standard:

Capacity constraint: The flow through an edge cannot exceed its capacity, i.e.,

$$\sum_j f_{ik}(j) \le C(i,k). \tag{27.2}$$

Flow conservation constraint: The amount of flow for j coming into node i equals the amount of traffic for j leaving node i, i.e.,

$$r_i(j) + \sum_k f_{ki}(j) = \sum_k f_{ik}(j). \tag{27.3}$$

We have the following different problems based on different objective functions:

Maximum multicommodity flow: The objective is to maximize the sum of the traffic rates, i.e., total throughput, that can be delivered in the network. Formally, we want to

$$\text{maximize} \sum_{i,j} r_i(j),$$

subject to capacity and flow conservation constraints.

Maximum concurrent flow problem: The objective is to maximize the fraction of traffic routed for each source–destination pair. This can be considered a "fairer" objective compared to the maximum multicommodity flow. Formally, the objective is to maximize the fraction μ subject to capacity and flow constraints. The latter may be rewritten as

$$\mu r_i(j) + \sum_k f_{ki}(j) = \sum_k f_{ik}(j). \tag{27.4}$$

Both the above problems can be solved using linear programming formulations. These algorithms are complex and are not easy to implement distributively. Awerbuch and Leighton [3] gave a local control algorithm for solving the above problems. It gives an $(1 + \epsilon)$-approximation for a desired $\epsilon > 0$. See Chapter 7 for a description of this algorithm. This algorithm can be implemented in a distributed manner.

Minimum delay routing problem (MDRP): The objective is to minimize the average delay subject to capacity and flow conservation constraints. The average delay on a link is assumed to be an increasing (convex) function of the amount of traffic on the link. Thus the average delay per packet on link (i,k) is given by $D_{ik}(f_{ik})$, where D_{ik} is a continuous and increasing (convex) function and f_{ik} is the total traffic on link (i,k). ($D_{ik}(f_{ik})$ tends to infinity as f_{ik} approaches C_{ik}.) The objective is to minimize the total average delay given by

$$D_T = \sum_{i,k} D_{ik}(f_{ik}). \tag{27.5}$$

27.3.2.1 A Distributed Algorithm for Minimum Delay Routing

The problem is complicated since the objective function is not linear and the challenge is to come up with a distributed algorithm. We briefly describe a distributed algorithm due to Gallager [17].

Let $t_i(j)$ be the total flow (traffic) at node i destined for j; thus $t_i(j) = r_i(j) + \sum_k f_{ki}(j)$.

Let $\phi_{ik}(j)$ be the fraction of the node flow $t_i(j)$ that is routed over edge (i,k). Since node flow $t_i(j)$ at node i is the sum of the input traffic and the traffic routed to i from other nodes

$$t_i(j) = r_i(j) + \sum_\ell t_\ell(j)\phi_{\ell i}(j) \quad \forall i,j. \tag{27.6}$$

The above equation implicitly expresses the flow conservation at each node. Now we can express f_{ik}, the traffic on link (i, k) as

$$f_{ik} = \sum_j t_i(j)\phi_{ik}(j). \tag{27.7}$$

The variable set ϕ is called the routing variable and it satisfies the following conditions:

1. $\phi_{ik}(j) = 0$, if (i, k) is not an edge or if $i = j$.
2. $\sum_k \phi_{ik}(j) = 1$ for all j.

Note that the traffic flow set $t = \{t_i(j)\}$ and link flow set $F = \{f_{ik}\}$ can be obtained from $r = \{r_i(j)\}$ and $\phi = \{\phi_{ik}(j)\}$. Therefore D_T can be expressed as a function of r and ϕ using Equations 27.6 and 27.7. The MDRP can be restated as follows:

For a given network $G = (V, E)$, and input traffic flow set $r_i(j)$, and delay function $D_{ik}(f_{ik})$ for each link (i, k), find the variable set ϕ that minimizes the average delay subject to the above conditions on ϕ, and the capacity and flow constraints.

We will assume that for each i, j $(i \neq j)$ there is a routing path from i to j, thus there is a sequence of nodes $i, k, \ell, \ldots, m$, such that $\phi_{ik}(j) > 0, \phi_{k\ell}(j) > 0, \ldots \phi_{mj}(j) > 0$. Gallager shows that the above conditions guarantee that the set of Equation 27.6 has a unique solution for t.

27.3.2.1.1 Necessary and Sufficient Conditions for Optimality

Gallager derived the necessary and sufficient conditions that must be satisfied to solve MDRP. These conditions are summarized in Theorem 27.3.

THEOREM 27.3 [17] *The necessary condition for a minimum of D_T with respect to ϕ for all $i \neq j$ and $(i, k) \in E$ is*

$$\frac{\partial D_T}{\partial \phi_{ik}(j)} \begin{cases} = \lambda_{ij} & : & \phi_{ik}(j) > 0 \\ \geq \lambda_{ij} & : & \phi_{ik}(j) = 0 \end{cases} \tag{27.8}$$

where λ_{ij} is some positive number.

The sufficient condition to minimize D_T with respect to ϕ for all $i \neq j$ and $(i, k) \in E$ is

$$D'_{ik}(f_{ik}) + \frac{\partial D_T}{\partial r_k(j)} \begin{cases} = \dfrac{\partial D_T}{\partial r_i(j)} & : & \phi_{ik}(j) > 0 \\[2mm] \geq \dfrac{\partial D_T}{\partial r_k(j)} & : & \phi_{ik}(j) = 0 \end{cases} \tag{27.9}$$

where
$\frac{\partial D_T}{\partial r_k(j)}$ *is called the marginal distance from i to j*
$\frac{\partial D_T}{\partial \phi_{ik}(j)}$ *is called the marginal delay*
$D'_{ik}(f_{ik})$ *is the marginal link delay.*

Thus Gallager's optimality conditions mandate two properties:

1. For a given node i and a destination j, the marginal link delays should be same for all links (i, k) for which there is nonzero flow, i.e., $\phi_{ik}(j) > 0$. Furthermore, this marginal delay must be less than or equal to the marginal delay on the links on which the traffic flow is zero. This is the necessary condition.
2. Under optimal routing at node i with respect to a particular destination j, the marginal distance through a neighbor k plus the marginal link delay should be equal to the marginal distance from node i if traffic is forwarded through k. It should be greater if no traffic is forwarded through k. This is the sufficiency condition.

Gallager's algorithm for MDRP: Gallager's algorithm iteratively computes the routing variables $\phi_{ik}(j)$ in a distributed manner. The idea is to progress incrementally toward the optimality conditions. Each node i incrementally decreases those routing variables $\phi_{ik}(j)$ for which the marginal delay $D'_{ik}(f_{ik}) + \frac{\partial D_T}{\partial r_k(j)}$ is large and increases those for which it is small. The decrease is by a small quantity ϵ; the excess fraction is moved over to the link (i, m) that has the smallest marginal delay (the current best link).

Gallager proves that for suitably small ϵ, the above iterative algorithm converges to the minimum delay routing solution. The choice of ϵ is critical in determining the convergence and its rate. A small ϵ may take a large time to converge, while a large ϵ may cause the system to diverge or oscillate around the minimum. Gallager's algorithm uses a blocking technique to make sure that the routes are loop-free at every instant. Loop-freedom ensures that the traffic that is routed along a path does not come back along a cycle. The algorithm ensures that if the routing variables are loop-free to begin with, then they will be loop-free during the execution of the algorithm. Thus, one way to start the algorithm is to assign the routing variables based on shortest path routing, i.e., they are set to 1 for the edges along the shortest paths, and 0 otherwise. To change the routing variables, each node needs to calculate marginal delays. This can be done in a distributed manner by propagating the value from the destination (its marginal delay is zero) as follows. For each destination j, each node i waits until it has received the value $\frac{\partial D_T}{\partial r_k(j)}$ from each of its downstream neighbors $k \neq j$ (i.e., neighbors that are closer to j on a routing path from i to j). The node i then calculates $\frac{\partial D_T}{\partial r_i(j)}$ (additional information needed are $\phi_{ik}(j)$ and $D'_{ik}(f_{ik})$, which are known locally) and broadcasts to all of its neighbors. Loop-freedom is essential to guarantee that this procedure is free from deadlocks [17].

27.3.2.2 Confluent Flow Routing

Most flows in today's communication networks, especially the Internet, are **confluent**. A flow is said to be confluent if all the flows arriving at a node destined for a particular destination depart from the node along a single edge. Internet flows are confluent because Internet routing is destination based. (Recall from our discussion of shortest-path routing, the forwarding tables of the routers are set to have one entry per destination.) Destination-based routing makes forwarding simple and the forwarding table size linear in the number of the nodes in the network. Both shortest path routing and BGP routing are destination based. One can study multicommodity flow problems with the condition that flows must be confluent. The primary goal is to find the confluent flows that minimize maximum congestion in the network. The **congestion** of an edge (node) is defined as the total amount of flow going through the edge (node). Formally, we consider the following problem studied in [11].

Minimum congestion ratio routing: We are given a directed graph $G = (V, E)$ with n nodes, m edges, and capacities $C(i, j)$ for each $(i, j) \in E$. Represent the traffic associated with a particular destination as a separate commodity. Let there be a total of k commodities. Let commodity i be associated with destination t_i and a set $S_i \subset V$ sources. The commodity demand, given by a function $d : \{1, \ldots, k\} \times V \to \mathcal{R}^+$, specifies the demand of each commodity for each vertex. A flow $f : \{1, \ldots, k\} \times V \times V \to \mathcal{R}$ (specifies the amount of flow of each commodity type through each pair of vertices) is confluent if for any commodity i, there is at most one outgoing flow at any node v. (It is easy to see that for any commodity i, flow f induces a set of **arborescences**, each of which is rooted at a destination t_i.) Given a flow f, the congestion ratio at an edge (u, v) with $c(u, v) > 0$, denoted by $r(u, v)$ is the ratio between the total flow $\sum_{i \in \{1, \ldots, k\}} f(i, u, v)$ on this edge and the capacity of the edge $C(u, v)$. The congestion ratio of flow f is the maximum congestion ratio among all edges. The minimum congestion ratio problem is to find a confluent flow to satisfy all the demands with minimum congestion ratios.

The above problem is NP-hard to solve optimally [12]. Hence the focus is to look for an approximately optimal solution that can be implemented in a distributed fashion. We present an algorithm due to Chen et al. [11] called the locally independent rounding algorithm (LIRA). The main idea is to cast the above problem as an integer linear program and then solve its relaxation. The relaxation is the standard multicommodity flow problem where confluence constraints need not be satisfied. Let $D_i = \sum_{v \in V} d(i, v)$, i.e., the total demand for commodity i. Let $x : \{1, \ldots, k\} \times V \times V \to [0, 1]$ and $\rho \in \mathcal{R}^+$. $x(i, u, v)$ will denote the fraction of the total commodity i that enters node u that leaves for v. If the flow is confluent, $x(i, u, v)$ will be 0 or 1 for all $i \in \{1, \ldots, k\}$ and $u, v \in V$. If $x(i, u, v)$ is allowed to be fractional, we get a **splittable flow** (need not be confluent). The following linear program computes a splittable flow with minimum congestion ratio:

Minimize r

subject to

$$\sum_{v \in V} f(i, u, v) = d(i, u), \quad \forall u \neq t_i, \forall i \tag{27.10}$$

$$\sum_{i \in \{1, \ldots, k\}} f(i, u, v) \leq rc(u, v), \quad \forall u, v \in V \tag{27.11}$$

$$0 \leq x(i, u, v) \leq 1, \forall u, v \in V, \quad \forall i \in \{1, \ldots, k\} \tag{27.12}$$

$$\sum_{w \in V} x(i, v, w) = 1, \forall v \in V, \quad \forall i \in \{1, \ldots, k\} \tag{27.13}$$

$$0 \leq f(i, u, v) \leq D_i x(i, u, v) \forall u, v \in V, \quad \forall i \in \{1, \ldots, k\}. \tag{27.14}$$

Equation 27.10 captures flow conservation and Inequality 27.11 captures capacity constraints. The above multicommodity problem can be solved approximately in a distributed manner by using the local control algorithm of Awerbuch and Leighton [3] (see Section 27.3.2). This gives a $1 + \epsilon$ approximation. The splittable flow computed by this algorithm is then made confluent by using the technique of randomized rounding (see Chapter 33) as follows. Let f be the flow computed by the algorithm in [3]. Then each node chooses for each commodity a unique outgoing edge independently at random. Node u chooses, for commodity i, edge (u, v) with probability

$$p(i, u, v) = \frac{f(i, u, v)}{\sum_{(u,v') \in E} f(i, u, v')}. \tag{27.15}$$

This rounding algorithm can be easily implemented in a distributed manner since each node makes its own choice based on local information.

The following theorem can be shown [11].

THEOREM 27.4 *Given a splittable flow f with congestion ratio C, the above rounding algorithm produces a confluent flow ϕ with $O(max(C, D/c_{min} \log n))$ congestion ratio with high probability, where $D = max_i D_i$ and c_{min} is the minimum edge capacity.*

Note that if $C = \Omega(D/c_{min} \log n)$, then LIRA is a constant factor approximation algorithm for the multicommodity flow problem. If $C = \Omega(D/c_{min})$, then LIRA achieves a logarithmic approximation. These performance guarantees do not hold, if these conditions are not true. Note that $\Omega(D/c_{max}\delta)$ is a lower bound on C, where δ is the maximum degree of a node. If capacities are more or less uniform and the maximum degree is not large then these conditions are reasonable. Stronger bounds on various special cases of the confluent flow problem have been established. We refer to [10,12] for details.

27.3.3 Broadcast Routing

Broadcasting is another important communication mode: sending a message from a source node to all other nodes of the network. We will consider two basic broadcasting approaches: flooding and spanning tree-based routing.

27.3.3.1 Flooding

Flooding is a natural and basic algorithm for broadcast. Suppose a source node s wants to send a message to all nodes in the network. s simply forwards the message over all its edges. Any vertex $v \neq s$, upon receiving the message for the first time (over an edge e) forwards it on every other edge. Upon receiving the message again it does nothing. It is easy to check that this yields a correct broadcast algorithm. The complexity of flooding can be summarized by Theorem 27.5.

THEOREM 27.5   *The message complexity of flooding is $\Theta(|E|)$ and the time complexity is $\Theta(Diam(G))$ in both the synchronous and asynchronous models.*

PROOF   The message complexity follows from the fact that each edge delivers the message at least once and at most twice (one in each direction). To show the time complexity, we use induction on t to show that after t time units, the message has already reached every vertex at a distance of t or less from the source.   □

27.3.3.2 Minimum Spanning Tree Algorithms

The **minimum spanning tree** (MST) problem is an important and commonly occurring primitive in the design and operation of communication networks. The formal definition of the problem, some properties of MST, and several sequential algorithms are given in Chapter 6. Of particular interest here, is that an MST can be used naturally for broadcast. Any node that wishes to broadcast simply sends messages along the spanning tree [27]. The advantage of this method over flooding is that redundant messages are avoided. The message complexity is $O(n)$, which is optimal.

Here we focus on distributed algorithms for this problem and few more properties of MST. The first distributed algorithm for the MST problem was given by Gallager et al. [18] in 1983. This algorithm is known as GHS algorithm and will work in an asynchronous model also.

27.3.3.2.1   GHS Algorithm

GHS algorithm assumes that the edge weights are distinct. If all the edges of a connected graph have distinct weights, then the MST is unique. Suppose, to the contrary, there are two different MSTs, T and T'. Let e be the minimum-weight edge that is in T but not in T'. The graph $\{e\} \cup T'$ must contain a cycle, and at least one edge in this cycle, say e', is not in T, as T contains no cycles. Since the edge weights are all distinct and e' is in one but not both of the trees, the weight of e is strictly less than the weight of e'. Thus $\{e\} \cup T' - \{e'\}$ is a spanning tree of smaller weight than T'; this is a contradiction.

We are given an undirected graph $G = (V, E)$. Let T be the MST on G. A fragment F of T is defined as a connected subgraph of T, that is, F is a subtree of T. An outgoing edge of a fragment is an edge in E where one adjacent node to the edge is in the fragment and the other is not. The minimum-weight outgoing edge (MOE) of a fragment F is the edge with minimum weight among all outgoing edges of F. As an immediate consequence of the blue rule for MST (see Chapter 6 in [47]), the MOE of a fragment $F = (V_F, E_F)$ is an edge of the MST. Consider a cut $(V_F, V - V_F)$ of G. The MOE of F is the minimum-weight edge in the cut $(V_F, V - V_F)$, and therefore the MOE is an edge of the MST. Thus adding the MOE of F to F along with the node at the other end of MOE yields another fragment of the MST. The algorithm starts with each individual node as a fragment by itself and ends with one fragment—the MST. That is, at the beginning there are $|V|$ fragments and at the end there is a single fragment, which is the MST. All fragments find their MOE simultaneously in parallel. Initially, each node (a singleton fragment) is a core node; subsequently each fragment will have one core node (determined as explained below). To find the MOE of a fragment, the core node in the fragment broadcasts a message to all nodes in the fragment using the edges in the fragment. Each node in the fragment, after receiving the message, finds the minimum outgoing edge adjacent to it and reports to the core node. Once the MOE of the fragment is found, the fragment attempts to combine with the fragment at the other end of the edge. Each fragment has a level. A fragment containing only a single node is at level 0. Let the fragment F be at level L, the edge e be the MOE of F, and the fragment F' be at the other end of e. Let L' be the level of F'. We have the following rules for combining fragments:

1. If $L < L'$, fragments F and F' are combined into a new fragment at level L', and the core node of F' is the core node of the new fragment.
2. If $L = L'$ and fragments F and F' have the same minimum-weight outgoing edge, F and F' are combined into a new fragment at level $L + 1$. The node incident on the combining edge with the higher identity number is the core node of the new fragment.
3. Otherwise, fragment F waits until fragment F' reaches a level high enough to apply any of the above two rules of combining.

The waiting of the fragments in the above procedure cannot cause a deadlock. The waiting is done to reduce the communication cost (number of messages) required for a fragment to find its MOE. The communication cost is proportional to the fragment size, and thus communication is reduced by small fragments joining into large ones rather than vice versa. The maximum level a fragment can reach is $\log n$, where $n = |V|$. The algorithm takes $O(n \log n + |E|)$ messages and $O(n \log n)$ time. It can be shown that any distributed algorithm for MST problem requires $\Omega(n \log n + |E|)$ messages [48]. Thus the communication (message) complexity of GHS algorithm is optimal. However, its time complexity is not optimal.

27.3.3.2.2 Kutten and Peleg's Algorithm

Kutten and Peleg's [28] distributed MST algorithm runs in $O(D + \sqrt{n} \log^* n)$ time, where D is the diameter of the graph G. The algorithm consists of two parts. In the first part, similar to GHS algorithm, this algorithm begins with each node as a singleton fragment. Initially, all of these singleton fragments are active fragments. Each active fragment finds its MOE and merges with another fragment by adding MOE to the MST. Once the depth of a fragment reaches $\sqrt{n}$, it becomes a terminated fragment. A terminated fragment stops merging onto other fragments. However, an active fragment can merge onto a terminated fragment. The depth of a fragment is measured by broadcasting a message from the root of the fragment up to the depth of at most $\sqrt{n}$ and by using a counter in the message. At the end of the first part, the size (number of nodes) of a fragment is at least $\sqrt{n}$, and there are at most $\sqrt{n}$ fragments. The first part of the algorithm takes $O(\sqrt{n} \log^* n)$ time (see [28] for details).

In the second part of the algorithm, a breadth-first tree B is built on G. Then the following pipeline algorithm is executed. Let $r(B)$ be the root of B. Using the edges in B, $r(B)$ collects weights of the interfragment edges, computes the MST, T', of the fragments by considering each fragment as a super node. It then broadcasts the edges in T' to the other nodes using the breadth-first tree B.

This pipeline algorithm follows the principle used by Kruskal's algorithm. Each node v, except the root r, maintains two lists of interfragment edges, Q and U. Initially, Q contains only the interfragment edges adjacent to v, and U is empty. At each pulse, v sends the minimum-weight edge in Q that does not create a cycle with the edges in U to its parent and moves this edge from Q to U. If Q is empty, v sends a terminate message to its parent. The parent after receiving an edge from a child, adds the edge in its Q list. A leaf node starts sending edges upward at pulse 0. An intermediate node starts sending at the first pulse after it has received at least one message from each of its children.

Observe that the edges reported by each node to its parent in the tree are sent in nondecreasing weight order, and each node sends at most $\sqrt{n}$ edges upward to its parent. Using the basic principle of Kruskal's algorithm, it can be easily shown that the root $r(B)$ receives all the interfragment edges required to compute T' correctly. To build B, it takes $O(D)$ time. Since the depth of B is D and each node sends at most $\sqrt{n}$ edges upward, the pipeline algorithm takes $O(D + \sqrt{n})$ time. Thus the time complexity of Kutten and Peleg's algorithm is $O(D + \sqrt{n} \log^* n)$. The communication complexity of this algorithm is $O(|E| + n^{1.5})$.

Peleg and Rabinovich [40] showed that $\tilde{\Omega}(D + \sqrt{n})$ time * is required for the distributed construction of MST. Elkin [16] showed that even finding an approximate MST in distributed time on graphs of small diameter (e.g., $O(\log n)$) within a ratio H requires $\Omega\left(\sqrt{\frac{n}{H \log n}}\right)$ time.

27.3.3.2.3 Khan and Pandurangan's Approximation Algorithm

Khan and Pandurangan [24] gave an algorithm for $O(\log n)$-approximate MST that runs in $\tilde{O}(D+L)$ time. L is called local shortest path diameter. L can be small in many classes of graphs, especially those of practical interest such as wireless networks. In the worst case, L can be as large as $n - 1$.

Khan and Pandurangan's algorithm [24] is based on simple scheme called the Nearest Neighbor Tree (NNT) scheme. The tree constructed using this scheme is called nearest neighbor tree, which is an $O(\log n)$-approximation to MST. The NNT scheme is as follows: (1) each node chooses a unique rank from a totally ordered set, and (2) each node, except the one with the highest rank, connects via the shortest path to the nearest node of higher rank, that is, add the edges in the shortest path to the NNT. The added edges can create cycle with the existing (previously added) edges of NNT. Cycle formation is avoided by removing some existing edges.

The following notations and definitions are used to describe the algorithm:

- $|Q(u, v)|$ or simply $|Q|$—the number of edges in path Q from u to v.
- $w(Q)$—the weight of the path Q, which is defined as the sum of the weights of the edges in path Q.
- $P(u, v)$—a shortest path (in the weighted sense) from u to v.
- $d(u, v)$—the (weighted) distance between u and v, that is, $d(u, v) = w(P(u, v))$.
- $W(v)$—the weight of the largest edge adjacent to v. $W(v) = \max_{(v,x) \in E} w(v, x)$.
- $l(u, v)$—the number of the edges in the shortest path from u to v. If there are more than one shortest path from u to v, $l(u, v)$ is the number of edges of the shortest path having the least number of edges, i.e., $l(u, v) = \min\{|P(u, v)| \mid P(u, v)$ is a shortest path from u to $v\}$.

ρ-neighborhood. ρ-neighborhood of a node v, denoted by $\Gamma_\rho(v)$, is the set of the nodes that are within distance ρ from v. $\Gamma_\rho(v) = \{u \mid d(u, v) \le \rho\}$.

(ρ, λ)-neighborhood. (ρ, λ)-neighborhood of a node v, denoted by $\Gamma_{\rho,\lambda}(v)$, is the set of all nodes u such that there is a path $Q(v, u)$ such that $w(Q) \le \rho$ and $|Q| \le \lambda$. Clearly, $\Gamma_{\rho,\lambda}(v) \subseteq \Gamma_\rho(v)$.

Local Shortest Path Diameter (LSPD). LSPD is denoted by $L(G, w)$ (or L for short) and defined by $L = \max_{v \in V} L(v)$, where $L(v) = \max_{u \in \Gamma_{W(v)}(v)} l(u, v)$.

27.3.3.2.3.1 Rank Selection.

The nodes select unique ranks as follows. First a leader is elected by a leader election algorithm (e.g., see [35]). (Or assume that there is one such node, say which initiates the algorithm.) Let s be the leader node. The leader picks a number $p(s)$ from the range $[b - 1, b]$,

---

* The $\tilde{\Omega}$ notation hides logarithmic factors.

where b is a (real) number arbitrarily chosen by s, and sends this number $p(s)$ along with its ID (identity number) to all of its neighbors. As soon as a node u receives the first message from a neighbor v, it picks a number $p(u)$ from $[p(v) - 1, p(v))$ so that it is smaller than $p(v)$, and sends $p(u)$ and ID(u) to its neighbors. If u receives another message later from another neighbor v', u simply stores $p(v')$ and ID(v'), and does nothing else. $p(u)$ and ID(u) constitute u's rank $r(u)$ as follows. For any two nodes u and v, $r(u) < r(v)$ iff (1) $p(u) < p(v)$, or (2) $p(u) = p(v)$ and ID$(u) <$ ID(v). At the end of execution of the above procedure of rank selection (1) each node knows the ranks of all of its neighbors, (2) the leader s has the highest rank among all nodes in the graph, and (3) each node v, except the leader, has one neighbor u, i.e., $(u, v) \in E$, such that $r(u) > r(v)$.

27.3.3.2.3.2 Connecting to a Higher-Ranked Node.   Each node v, except the leader s, executes the following algorithm simultaneously to find the nearest node of higher rank and connect to it. Each node v needs to explore only the nodes in $\Gamma_{W(v)}(v)$ to find a node of higher rank. This is because at the end of rank selection procedure, v has at least one neighbor u such that $r(u) > r(v)$, and if u is a neighbor of v, then $d(u, v) \leq W(v)$.

Each node v executes the algorithm in phases. In the first phase, v sets $\rho = 1$. In the subsequent phases, it doubles the value of ρ; that is, in the ith phase, $\rho = 2^{i-1}$. In a phase of the algorithm, v explores the nodes in $\Gamma_\rho(v)$ to find a node u (if any) such that $r(u) > r(v)$. If such a node with higher rank is not found, v continues to the next phase with ρ doubled. Node v needs to increase ρ to at most $W(v)$. Each phase of the algorithm consists of one or more rounds. In the first round, v sets $\lambda = 1$. In the subsequent rounds, values for λ are doubled. In a particular round, v explores all nodes in $\Gamma_{\rho,\lambda}(v)$. At the end of each round, v counts the number of nodes it has explored. If the number of nodes remains the same in two successive rounds of the same phase (i.e., v already explored all nodes in $\Gamma_\rho(v)$), v doubles ρ and starts the next phase. If at any point of time v finds a node of higher rank, it terminates its exploration.

Since all of the nodes explore their neighborhoods simultaneously, many nodes may have overlapping ρ-neighborhoods. This might create a congestion of the messages in some edges that may result in increased running time of the algorithm, in some cases by a factor of $\Theta(n)$. Consider the network given in Figure 27.3. If $r(v) < r(u_i)$ for all i, when $\rho \geq 2$ and $\lambda \geq 2$, an exploration message sent to v by any u_i will be forwarded to all other u_is. Note that values for ρ and λ for all u_is will not necessarily be the same at a particular time. Thus congestion at any edge (v, u_i) can be as much as the number of such nodes u_i, which can be, in fact, $\Theta(n)$ in some graphs. However, to improve the running time of the algorithm, the congestion on all edges is controlled to be $O(1)$ by sacrificing the quality of the NNT, by a constant factor, as explained below.

If at any time step, a node v receives more than one, say $k > 1$, messages from different originators u_i, $1 \leq i \leq k$, v forwards only one message and replies back to the other originators as follows. Let u_i be in phase ρ_i. For any pair of originators u_i and u_j, if $r(u_i) < r(u_j)$ and $\rho_j \leq \rho_i$, v sends back a *found* message to u_i telling that u_i can connect to u_j (note that the weight of the connecting path $w(Q(u_i, u_j)) \leq 2\rho_i$) instead of forwarding u_i's message. Now, there are at least one u_i left to which v did not send the *found* message back. If there is exactly one such u_i, v forwards its message; otherwise, v takes the following actions. Let u_s be the node with lowest rank among the rest of the u_is (i.e., those u_is which were not sent a *found* message by v), and u_t, with $t \neq s$, be an arbitrary node among the rest of u_is. Now, it must be the case that $\rho_s < \rho_t$ (otherwise, v would send a *found* message to u_s), i.e., u_s is in an earlier phase than u_t. This can happen if in some previous phase, u_t exhausted its ρ-value with smaller λ-value leading to a smaller

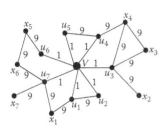

FIGURE 27.3   A network with possible congestion in the edges adjacent to v. Weight of the edges (v, u_i) is 1 for all i, and 9 for the rest of the edges. Assume $r(v) < r(u_i)$ for all i.

number of rounds in that phase and a quick transition to the next phase. Node v forwards *explore* message of u_s only and sends back *wait* messages to all u_t ($t \neq s$). A node after receiving a *wait* message simply repeats its exploration phase with the same ρ and λ.

Suppose a node u_i found a higher ranked node u_j via the path $Q(u_i, u_j)$. If u_i's nearest node of higher rank is u', then $w(Q) \leq 4d(u_i, u')$. Assume that u_j is found when u_i explored the (ρ, λ)-neighborhood for some ρ and λ. Then $d(u_i, u') > \rho/2$, otherwise, u_i would find u' as a node of higher rank in the previous phase and would not explore the ρ-neighborhood. Now, u_j could be found by u_i in two ways: (1) The *explore* message originated by u_i reaches u_j and u_j sends back a *found* message. In this case, $w(Q) \leq \rho$. (2) Some node v receives two *explore* messages originated by u_i and u_j via the paths $R(u_i, v)$ and $S(u_j, v)$ respectively, where $r(u_i) < r(u_j)$ and $w(S) \leq \rho$; and v (on behalf of u_j) sent a *found* message to u_i. In this case, $w(Q) = w(R) + w(S) \leq 2\rho$, since $w(R) \leq \rho$. Thus, in both cases, we have $w(Q) \leq 4d(u_i, u')$.

The cost of the NNT produced by this algorithm is at most $4\lceil \log n \rceil * c(\mathrm{MST})$. The time complexity of the algorithm is $O(D + L \log n)$ and the message complexity is $O(|E| \log L \log n)$. For more details of this algorithm, the proof of correctness, and the analysis of approximation ratio, time complexity, and message complexity, readers are referred to [24].

27.3.4 Multicast Routing

We can view multicasting as a generalization of broadcasting, where the message has to be sent to only a (required) subset of nodes. This formulation leads to the minimum cost Steiner tree problem, a generalization of the MST problem (see also Chapter 7). The Steiner tree connects the required set of nodes and can possibly use other nodes in the graph (these are called Steiner nodes). Finding the minimum cost Steiner tree in a arbitrary weighted graph is NP-hard. The following approximation algorithm is well known (e.g., [49]). Transform the given graph $G = (V, E, w)$ as a *complete metric graph* $G' = (V, E', w')$ where the pairwise edge costs in E' are the corresponding pairwise (weighted) shortest path distances in G, i.e., $w'(u, v) = \mathrm{dist}(u, v)$, where $\mathrm{dist}(u, v)$ is the shortest path distance between u and v in G. It can be shown that an MST T' on the required set of nodes in G' is a 2-approximation to the optimal Steiner tree in G. To get a 2-approximate Steiner tree in G, we replace the edges in T' by the corresponding shortest paths in G. This may possibly create cycles, so some edges may have to be deleted.

Thus, distributed algorithms for minimum-cost Steiner tree are at least as involved as those for the MST problem. Distributed approximation algorithms have been recently proposed for this problem. Based on the MST approximation described above, an $O(n \log n)$-time 2-approximate distributed algorithm has been proposed in [9]. The NNT approach of [24] described in Section 27.3.3.2 yields a faster $O(\log n)$-approximation with the same time bounds. Another advantage of the NNT approach is that this algorithm works with virtually no change. Indeed the basic algorithm is as follows: every required node chooses a unique rank and connects to its closest required node of higher rank.

27.4 Queuing

Queuing theory is the primary methodological framework for analyzing network delay. We study two different approaches here, the first characterized by the stochastic modeling of arrival and service times (Section 27.4.1), and the second by a more general adversarial model (Section 27.4.3).

27.4.1 Stochastic Queuing Models

In a typical queuing system, scenario is characterized by a variable set of customer(s) who contend for (limited) resources served by server(s). A customer leaves the system once it receives the service.

We note that integral to queuing situations is the idea of uncertainty in, for example, interarrival times and service times. Hence, stochastic models are used to model queuing systems.

Our study of queuing is basically motivated by its use in the study of communication systems and computer networks. Nodes, especially, routers and switches in a network may be modeled as individual queues. The whole system may itself be modeled as a queuing network providing the required service to the data that need to be carried. Queuing theory provides the theoretical framework for the design and the study of such networks.

For a single queuing system where customers are identified based on their arrival order, we denote the following:

$$N(t) = \text{number of customers in the queue at time } t$$
$$\alpha(t) = \text{number of customers who arrived in } [0, t]$$
$$T_i = \text{time spent in the system by customer } i$$

Assume that the following three limits exist:

$$N = \lim_{t \to \infty} \frac{1}{t} \int_0^t N(t) \tag{27.16}$$

$$\lambda = \lim_{t \to \infty} \frac{\alpha(t)}{t} \tag{27.17}$$

$$T = \lim_{t \to \infty} \frac{\sum_{i=1}^{\alpha(t)} T_i}{\alpha(t)} \tag{27.18}$$

where
N is called the steady-state time average of the number of customers in the system
λ is the steady-state arrival rate of customers
T the steady-state time average customer delay.

These three quantities are related by the following basic equation known as Little's Lemma holds: $N = \lambda T$ (see [45] for a proof).

27.4.1.1 The *M/M/1* System

The name *M/M/1* reflects the standard queuing theory nomenclature: (1) The first letter indicates the nature of the arrival process. For example, *M* stands for memoryless, i.e., a Poisson arrival process, *G* stands for a general distribution of interarrival times, *D* stands for deterministic interarrival times. (2) The second letter indicates the nature of the probability of the server process. For example, *M* stands for memoryless, i.e., an exponential service time model; *G* for general, and *D* for deterministic (3) The last number indicates the number of servers (can range from 1 to ∞). The queue discipline that we will assume (unless otherwise stated) is FIFO. This specifies how to determine which customer to serve next from among the set of customers waiting for service. We will assume a Poisson arrival process throughout; this makes the models amenable to precise analysis.

The *M/M/1* system is a simple and basic queuing system where there is a single queue with a single server. Customers arrive according to a Poisson process with rate λ. The probability distribution of the service time is exponential with mean $\frac{1}{\mu}$ s. The successive interarrival times and service times are statistically independent of each other.

Using a Markov chain analysis, the following properties of the *M/M/1* queuing system can be shown (see for e.g., [5]):

1. The expected number of customers in the system at steady state is $N = \frac{\lambda}{\mu - \lambda}$.
2. The average delay per customer is (by Little's Lemma) $T = N/\lambda = \frac{1}{\mu - \lambda}$.

3. The average time spent by the packet waiting in the queue (not including the transmission time) is $W = T - 1/\mu$.

4. The average number of customers in the queue is $N_Q = \lambda W = \frac{\rho^2}{1-\rho}$, where $\rho = \lambda/\mu$ is called the utilization factor.

We look at an application of $M/M/1$ queues to communication networks. We assume that packets arrive according to a Poisson process to an edge and the packet transmission rate is exponentially distributed. Interarrival and service times are independent, i.e., the length of the arriving packet does not affect the arrival time of the next packet (this is not true in reality when you have a network of queues, see [5]).

Statistical multiplexing verses time-division multiplexing: These are two basic schemes of multiplexing traffic in a communication link. In statistical multiplexing, the packets from different traffic streams are merged into a single queue and transmitted on a first-come first-serve basis. In time division multiplexing (TDM), with m traffic streams, the edge capacity is essentially subdivided into m parts and each part is allocated to one traffic stream.

Given m identical Poisson processes each with arrival rate λ/m, which scheme gives better delay results? We assume that the average packet service time is $1/\mu$. Then statistical multiplexing gives a delay corresponding to an arrival rate λ and a service rate μ, i.e., $T = \frac{1}{\mu-\lambda}$. On the other hand, TDM, i.e., transmitting through m separate channels, gives a much bigger per packet delay $T' = \frac{m}{\mu-\lambda}$ since the service rate per channel now reduces to $\frac{\mu}{m}$. This is the reason why in **packet-switched networks**, such as the Internet, where traffic is mostly irregular and bursty, statistical multiplexing is used. In **circuit-switched networks**, such as telephone networks, where each traffic stream is more or less "regular" (as opposed to Poisson), TDM is preferred. The reason is that there will be no waiting time in the queue, if such a stream is transmitted on a dedicated line.

27.4.1.2 $M/G/1$ System

In this system, we have a single server, an infinite waiting room, exponentially distributed interarrival times (with parameter λ), and an arbitrary service time distribution for which at least the mean value and the standard deviation are known. The service discipline is FIFO. Let X_i be the service time of the ith customer. X_i's are identically distributed, mutually independent, and independent of interarrival times.

$E[X] = 1/\mu = $ average service time.
$E[X^2] = $ second moment service time.

The expected waiting time in queue for a customer in this system is given by the Pollaczek–Khintchine formula (P–K formula).

P–K formula: The expected waiting time in queue in a stable $M/G/1$ system is

$$W = \frac{\lambda E[X^2]}{2(1 - \rho)}, \tag{27.19}$$

where $\rho = \lambda/\mu$. See [5] for a proof. Using Little's law, the expected number of customers in the queue is

$$N_Q = \lambda W = \frac{\lambda^2 E[X^2]}{2(1 - \rho)}. \tag{27.20}$$

27.4.1.3 $M/G/\infty$ System

In $M/G/\infty$ system, arrivals are Poisson with rate λ; however, service times are assumed to be independent with distribution G, i.e., the service of a customer is independent of the services of the other customers and of the arrival processes.

Let $X(t)$ denote the number of customers arrived in the system till time t (some or all of them might have already left the system). Conditioning $\Pr(N(t) = j)$ over $X(t)$ we get

$$\Pr(N(t) = j) = \sum_{n=0}^{\infty} \Pr(N(t) = j | X(t) = n) e^{-\lambda t} \frac{\lambda t}{n!}, \tag{27.21}$$

because $\Pr(X(t) = n) = e^{-\lambda t} \frac{\lambda t}{n!}$ due to Poisson arrival with rate λ.

The probability that a customer who arrives at time x will still be present at t is $1 - G(t - x)$.

Let p be the probability that an arbitrary customer who arrives during the interval $[0, t]$ is still in the network at time t. We know, for a Poisson distribution, given that an arrival happens at time interval $[0, t]$, then the arrival is uniform in this interval. Therefore, $\frac{dx}{t}$ is the probability that an arbitrary customer arrives in the interval dx. Thus,

$$p = \int_0^t (1 - G(t - x)) \frac{dx}{t} = \int_0^t (1 - G(x)) \frac{dx}{t}, \tag{27.22}$$

independently of the others. Then,

$$P(N(t) = j | X(t) = n) = \begin{cases} \binom{n}{j} p^j (1 - p)^{n-j} & : \quad j = 0, 1, \ldots, n \\ 0 & : \quad j > n \end{cases}. \tag{27.23}$$

Thus,

$$\Pr(N(t) = j) = \sum_{n=j}^{\infty} \binom{n}{j} p^j (1 - p)^{n-j} e^{-\lambda t} \frac{\lambda t}{n!} \tag{27.24}$$

$$= e^{-\lambda t} \frac{(\lambda t p)^j}{j!} \sum_{n=j}^{\infty} \frac{(\lambda t (1 - p))^{n-j}}{(n - j)!} \tag{27.25}$$

$$= e^{-\lambda t p} \frac{(\lambda t p)^j}{j!}. \tag{27.26}$$

Hence, we find that $N(t)$ is Poisson distributed with mean $\lambda \int_0^t (1 - G(x)) dx$.

27.4.1.4 Application to P2P Networks

A P2P network (cf. Section 27.7) can be modeled as an $M/M/\infty$ system.

The arrival of new nodes is Poisson distributed with rate λ. The duration of time a node stays connected to the network is independently and exponentially distributed with parameter μ. Without the loss of generality, let $\lambda = 1$ and let $N = 1/\mu$.

THEOREM 27.6 [37] *If $\frac{t}{N} \to \infty$ then with high probability, $N(t)| = N \pm o(N)$.*

PROOF The number of nodes in the network at time t, $N(t)$, is Poisson distributed with mean $\lambda \int_0^t (1 - G(x)) dx$, which is $N(1 - e^{-t/N})$.

If $t/N \to \infty$, $E[N(t)] = N + o(N)$.

Now, using a Chernoff bound for the Poisson distribution [1], for $t = \Omega(N)$, and for some constants b and $c > 1$:

$$\Pr \left(N(t) - E[N(t)]| \leq \sqrt{bN \log N} \right) \geq 1 - 1/N^c. \tag{27.27}$$

$\square$

27.4.2   Communication Networks Modeled as Network of Queues

So far we have only looked at a single stand-alone queueing system. However, most real systems are better represented as a network of queues. An obvious example is the Internet, where we can model each outgoing link of each router as a single queueing system, and where an end-to-end path traverses a multitude of intermediate routers. In a queuing network, a customer finishing service in a service facility is immediately proceeding to another service facility or is leaving the system.

When we look at a network of queues, the analysis of the system becomes much more complex. The main reasons are

- Even if the input queue is an $M/M/1$ queue, this is not true for any internal node.
- The interarrival times in the downstream queues are highly correlated with the service times of the upstream queues.
- The service times of the same packet in different queues are not independent.

To illustrate with a simple example, consider two edges of equal capacity in tandem connecting two queues, 1 and 2, with the first queue feeding the second. Assume that packet arrives according to a Poisson process with independent exponentially distributed packet lengths. Then the first queue behaves as an $M/M/1$, but the second queue does not. This is because the interarrival times at the second queue are strongly correlated with the packet lengths: longer packets will wait less at the second queue than shorter packets, since their transmission time at the first queue takes longer, thereby giving more time for the second queue to empty out.

To deal with this situation, Kleinrock [25] introduced the following independence assumption:

Adopt an $M/M/1$ model for each communication link regardless of the interaction of the traffic on this link with traffic on other links.

Although, the Kleinrock independence assumption is typically violated in real networks, the cost function that is based on the Kleinrock assumption represents a useful measure of performance in practice. This turns out to be a reasonable approximation for networks involving Poisson stream arrivals at the entry points, packet lengths that are nearly exponentially distributed, a densely connected network, and moderate-to-heavy traffic loads.

One basic classification of queueing networks is the distinction between open and closed queuing networks. In an open network, new customers may arrive from outside the system (coming from a conceptually infinite population) and later on leave the system. In a closed queuing network, the number of customers is fixed and no customer enters or leaves the system. An example for an open queuing network may be the Internet, where new packets arrive from outside the system (in fact, from the users).

27.4.2.1   Jackson's Theorem

Jackson' theorem is an important result in the analysis of a network of queues, which shows that under certain conditions a network of queues behaves as a collection of $M/M/1$ queues.

Assume a system of K FIFO queues.

Let $n_i(t)$ be the number of customers at queue i at time t.

Let $n(t) = (n_1(t), \ldots, n_K(t))$ be the states of the system at time t.

The steady-state probability of the system being in state $(n_1, \ldots, n_K)$ is denoted as $P(n)$. From this steady state probability, we can derive the marginal probability $P_i(n_i)$ that the node i contains exactly n_i customers. In some cases, it is possible to represent the state probabilities as follows:

$$P(n) = \frac{1}{G(n)} P_1(n_1) \times \cdots \times P_K(n_K), \tag{27.28}$$

where $G(n)$ is the so-called normalization constant (it depends on the number of customers in the system). In the case of an open queuing network we have always $G(n) = 1$, in the case of a closed queuing network $G(n)$ must be chosen such that the normalization condition $\sum P(n) = 1$ holds. Equation 27.28 represents a product form solution. A nice property of this equation is that we can decompose the system and look at every service center separately.

The theorem of Jackson specifies the conditions under which a product form solution in open queuing networks exist. These conditions are the following:

- The number of customers in the network is not limited.
- New customers arrive at queue i according to a Poisson process with rate r_i (some or all r_i's may be 0).
- Service at queue i is exponential with rate μ_i.
- A customer can leave the system from any node (or a subset).
- When a customer is served at queue i it proceeds with probability P_{ij} to queue j, and with probability $1 - \sum_j P_{ij}$ it leaves the system.
- All choices are independent of previous choices.
- In every node the service discipline is FIFO.

Let λ_i be the arrival rate to queue i

$$\lambda_i = r_i + \sum_{j=1}^{K} \lambda_j P_{ji}, \quad i = 1, \ldots, K. \tag{27.29}$$

Assume that the above system has a unique solution and let $\rho_i = \lambda_i/\mu_i$. Then Jackson's theorem can be stated as

THEOREM 27.7 *If in an open network the condition $\rho_i < 1$ holds for $i = 1, \ldots, K$, then under the above-mentioned conditions the system has a steady state $(n_1, \ldots, n_K)$ and the steady-state probability of the network can be expressed as the product of the state probabilities of the individual nodes:* $\lim_{t \to \infty} P_t(n) = P(n) = P_1(n_1) \times \cdots \times P_K(n_K)$ *and* $P_j(n_j) = \rho_j^{n_j}(1 - \rho_j)$ $n_j \geq 0$.

The theorem implies that the system operates as a collection of independent *M/M/1* queues, though the arrival process to each of the queues is not necessarily Poisson.

An application of Jackson's theorem can be seen in routing on the butterfly network (Chapter 3.6 in [30]) with Poisson arrivals to the inputs, and exponential transition time through the edges.

Assume that the arrival rate of packets to each of the inputs is λ.

Then the arrival rate to each of the internal nodes is λ.

Let $\mu > \lambda$, i.e., $\rho < 1$. Then the system is stable. Since for an n input butterfly network there are $\log n$ links (queues) from any input node to a output node, the expected time in the system for a customer (packet) is $\leq \frac{\log n}{\mu - \lambda}$.

27.4.3 Adversarial Queuing

Adversarial queuing theory [8] addresses some of the restrictions inherent in probabilistic analysis and queuing theory based on time-invariant stochastic generation. This theory is aimed at the systemic study of queuing with little or no probabilistic assumption. Adversarial queuing is based on an adversarial generation of packets in dynamic packet routing, where packets are injected continuously into the network.

27.4.3.1 The Adversarial Model

A **routing network** is a directed graph. Time proceeds in discrete time steps. A packet must travel along a path from its source to its destination. When a packet reaches its destination, it is absorbed. During each step, a packet may be sent from its current node along one of the outgoing edges from that node. At most one packet may travel along any edge of the network in a step. A packet that wishes to travel along any edge at a particular time step but is not sent can wait in a queue for that edge. The delay of a packet is the number of steps, which the packet spends waiting in queues. At each step, an adversary injects a set of packets and specifies the route for each packet. The route of each packet is fixed at the time of injection (nonadaptive routing). The following load condition is imposed: Let τ be any sequence of w consecutive time steps (where w is some positive integer) and $N(\tau, e)$ the number of packets injected by the adversary during time interval τ that traverse edge e. Then, for every sequence τ and for every edge e, a (w, ρ)-adversary injects new packets with $N(\tau, e)/w \leq \rho$. ρ is called the rate of injection with window size w. Clearly, if $\rho > 1$, an adversary would congest some edge and the network would be unstable.

If there are more than one packet waiting in the queue of an edge e, a scheduling policy determines which packet is to be moved along edge e in each time step. A **scheduling policy** is greedy if it moves a packet along an edge whenever there is a packet in the queue. A network system is stable if there is a finite number M such that the number of packets in any queue is bounded by M. The following theorem shows a general stability result for adversarial queuing.

THEOREM 27.8   *In a directed acyclic graph, with any greedy scheduling and any adversarial packet generation with rate $\rho = 1$, a network system is stable.*

PROOF   Let $Q_t(e)$ be the number of packets in the queue of edge e at time t, and $A_t(e)$ the number of packets that have been injected into the system by time t, are still in the system (not absorbed yet), and are destined to cross edge e. Clearly $Q_t(e) \leq A_t(e)$. Since the injection rate is 1, there exists some window size w such that for any window of time $(t - w, t]$ and for any edge e, the adversary can inject at most w packets. Then for any t' such that $t - w < t' < t$

$$A_{t'}(e) \leq A_{t-w}(e) + w. \tag{27.30}$$

For an edge e, let $e_1, e_2, \ldots, e_k$ be the edges entering the tail of e. A function ψ is defined as follows:

$$\psi(e) = \max\{2w, Q_0(e)\} + \sum_{i=1}^{k} \psi(e_i). \tag{27.31}$$

The theorem is proved by showing that for all $t = lw \geq 0 \ (l \geq 1)$ and for all edges e

$$A_t(e) \leq \psi(e) \tag{27.32}$$

Since the graph is directed acyclic, it is easy to see that $\psi(e)$ is finite and is a function of the number of nodes in the network; but it does not depend on time t. Thus, as $Q_t(e) \leq A_t(e)$, Inequality 27.32 gives the stability of the network.

Inequality 27.32 is proved by induction on l. The claim holds when $l = 0$. Now assume $t = lw$ for $l \geq 1$. Consider the following two cases.

Case 1. $A_{t-w}(e) \leq w + \sum_{i=1}^{k} \psi(e_i)$:

$$A_t(e) \leq A_{t-w}(e) + w \leq 2w + \sum_{i=1}^{k} \psi(e_i) \leq \psi(e). \tag{27.33}$$

Case 2. $A_{t-w}(e) > w + \sum_{i=1}^{k} \psi(e_i)$:

By induction hypothesis, $A_{t-w}(e_i) \leq \psi(e_i)$. Further, notice that $A_{t-w}(e)$ is at most $Q_{t-w}(e) + \sum_{i=1}^{k} A_{t-w}(e_i)$. Thus,

$$Q_{t-w}(e) \geq A_{t-w}(e) - \sum_{i=1}^{k} A_{t-w}(e_i) \tag{27.34}$$

$$\geq w + \sum_{i=1}^{k} \psi(e_i) - \sum_{i=1}^{k} A_{t-w}(e_i) \tag{27.35}$$

$$\geq w. \tag{27.36}$$

That is, there are at least w packets in the queue of edge e at the beginning of time step $t - w$. Thus, by a greedy scheduling, w packets cross e in the next w time steps; but the adversary can inject at most w packets into the system for edge e. Therefore, $A_t(e) \leq A_{t-w}(e)$. By induction hypothesis, $A_{t-w}(e) \leq \psi(e)$. Thus, $A_t(e) \leq \psi(e)$. □

If the network contains cycles, the system may not be stable for some scheduling policies, such as FIFO and LIS (Longest in System, where priority is given to the packets that have been in the network for the longest amount of time). However, some scheduling policies, such as FTG (Furthest to go, where priority is given to the packet that has the largest number of edges still to be traversed), are stable in a directed network with cycles and adversarial packet generation with rate 1. For proofs of these claims and further details, readers are referred to [8].

Adversarial queuing theory also deals with stochastic adversaries. A (w, ρ) stochastic adversary generates packets following some probability distribution such that $E[N(\tau, e)]/w \leq \rho$. The adversary can follow different probability distributions at different time steps. If $\rho = 1$, an adversary can inject zero packets with probability $1/2$ and two packets with probability $1/2$, which is analogous to a random walk on line $[0, \infty)$, and can make the system unstable; after t steps, the expected queue size is approximately $\sqrt{t}$. Hence it is assumed that $\rho < 1$. Further constraint is imposed on the pth moment for $p > 2$: there exist constants $p > 2$ and V such that for all sequence τ of w consecutive time steps and all edge e, $E[N^p(\tau, e)] \leq V$. A network system with arbitrary directed acyclic graph, an arbitrary greedy scheduling, and stochastic adversaries with rate $\rho < 1$ is stable [8].

27.5 Multiple Access Protocols

Multiple access channels provide a simple and efficient means of communication in distributed networks, such as the Ethernet and wireless networks. Ethernet is a broadcast LAN, i.e., when a station (node) wants to transmit a message to one or more stations, all stations connected to the Ethernet LAN will receive it. Since several nodes share the same broadcast channel and only one message can be broadcast at each step, a multiple access protocol (or a contention resolution protocol) is needed to coordinate the transmissions and avoid collisions. Such a protocol decides when to transmit a message and what to do in the event of a collision.

We assume that one message can be sent at a given time. If more than one message is sent at a given time, no message is received (i.e., a collision occurs). A station can check if the "channel" is free. Note that this does not eliminate collisions. Unfortunately, collisions can still occur, since there is a noneligible delay between the time when a station begins to transmit and the other stations detect the transmission. Hence, if two or more stations attempt to transmit within this window of time, a collision will occur. A station can detect if the message it sent was broadcast successfully. A station's protocol uses only that station's history of success and failure transitions (such a protocol is called acknowledgment-based); it has no knowledge about other stations, or even the number of other active stations. In the case of a collision, the messages (that were not sent) are queued at their respective stations for retransmission at some point in the future. It is not a good idea to retransmit immediately, since this would result in another collision. Rather, messages are retransmitted after some delay, which is chosen probabilistically. Such a strategy is called a backoff strategy defined formally below.

Backoff protocol:
  Each station follows the following protocol:

  1. The backoff counter b is initially set to 0.
  2. While the station queue is not empty do

        a. with probability $1/f(b)$ try to broadcast.

        b. If the broadcast succeeded $b = 0$, else $b = b + 1$.

  In principle, f can be an arbitrary function. We have different classes of backoff protocols depending on f.
  Exponential backoff: $f(b) = 2^b$ (The Ethernet uses $2^{\min[b,10]}$).
  Linear backoff: $f(b) = b + 1$.
  Polynomial backoff: $f(b) = (b + 1)^\alpha$ (for some constant $\alpha > 1$).

27.5.1 Analysis Model

The key theoretical question is: Under what condition, backoff protocols have good performance? For example, is exponential backoff better than polynomial backoff? To answer this rigorously we need a model and a performance measure. We consider the following model [21].

  Time is partitioned into steps. At the beginning of a step, each station receives a new message with probability λ_i, for $1 \leq i \leq N$, where N is the number of stations in the system. The arrival of new messages is assumed to be independent over time and among stations. (Note that no assumptions are made about the independence of the state of the system from one time step to the next, or between stations.) The overall arrival rate is defined to be $\lambda = \sum_{i=1}^{N} \lambda_i$. Arriving messages are added to the end of the queues located at each station. Queues can be of infinite size. The system starts with empty queues. The acknowledgment of a collision or successful transmission is taking place within one step. Message lengths are assumed to be smaller than the duration of a time step.

  The main performance measure is stability. Let $W_{avg}(t)$ be the average waiting time of messages (packets) that entered the system at time t. Let $L(t)$ be the number of messages in queues at the end of step t. We say that a system is *stable* if $E[W_{avg}(t)]$ and $E[L(t)]$ are both bounded with respect to t. (By Little's Lemma $E[L(t)] = \lambda E[W_{avg}(t)]$ if both exist and are finite).

  There are important differences between the above model and real life, e.g., the Ethernet [21]. For example, in the Ethernet, the backoff counter is never allowed to exceed a specified value $b_{max} = 10$. A problem with placing such a bound is that any protocol becomes unstable for any fixed λ and large enough N [21]. Another difference is that in the Ethernet these messages are discarded if they are not delivered within a specified amount of time. Discarding messages contributes to stability but at the cost of losing a nonzero fraction of the messages. Also assumptions regarding message arrival distribution, message length, and synchronization may not be fully realistic.

  We next state two key results of [21]. The first states that the exponential backoff protocol is unstable for any finite $N > 2$ when the arrival rate at each station is λ/N, for $\lambda \geq \lambda_0 \approx 0.6$. Previous results showed similar instability results but assumed infinite number of stations (which renders the result somewhat less realistic). The second states that polynomial backoff protocol is stable for any $\alpha > 1$ and any $\lambda < 1$. This was surprising because if the number of station is infinite, it can be shown that polynomial backoff is unstable [22]. The main approach in showing the above results is by analyzing the behavior of the associated Markov chain. For backoff protocols, with finite number of stations, one can associate every possible configuration of backoff counters and queues $(\mathbf{b}, \mathbf{q}) = \{(b_1, \ldots, b_N, q_1, \ldots, q_N) | b_i \geq 0, q_i \geq 0 \text{ for } 1 \leq i \leq N\}$ with a unique state of the Markov chain. The initial state is identified with $(0, \ldots, 0)$. A potential function method is then used to analyze the evolution of the chain. For showing instability results, for example, the potential function is chosen such that the expected change in potential is at least a fixed positive constant (thus the function grows linearly with time) and this is then shown to imply that the protocol is unstable. For example, the following potential function is used to show that the exponential backoff function is unstable.

Define a potential function of the system at time t:

$$\Phi(t) = C \sum_{i=1}^{N} q_i + \sum_{i=1}^{N} 2^{b_i} - N, \tag{27.37}$$

where

$C = 2N - 1$
$q_i = $ number of packets queued at station i at time t
$b_i = $ the value of the backoff counter at station i at time t.

The main thrust of the proof is to show that $E[\Phi(t)]$ increases by a constant δ in each step. This will then imply that the system is unstable because of Lemma 27.1.

LEMMA 27.1   If there is a constant $\delta > 0$ (independent of t) such that for all $t > 0$

$$E[\Delta\Phi(t)] = E[\Phi(t) - \Phi(t-1)] \geq \delta, \tag{27.38}$$

then the system is unstable.

PROOF   A message enters the queue of station i at time t with probability $\lambda_i = \lambda/N$.
   The expected wait time of this message is at least $q_i + 2^{b_i}$.
   For vectors $\bar{q}$ and $\bar{b}$, let $P(t, \bar{q}, \bar{b})$ be the probability that at time t, and for some station $i = 1, \ldots, N$, the queue of station i has q_i items and the counter of i is b_i.

$$E[W_{\text{avg}}(t)] \geq \sum_{q,b} \sum_{i=1}^{N} \lambda_i (q_i + 2^{b_i}) P(t, \bar{q}, \bar{b}). \tag{27.39}$$

$$E[\Phi(t)] = \sum_{q,b} \sum_{i=1}^{N} (Cq_i + 2^{b_i} - 1) P(t, \bar{q}, \bar{b}). \tag{27.40}$$

Thus,

$$E[W_{\text{avg}}(t)] \geq \frac{\lambda}{CN} \Phi(t) \geq t\delta \frac{\lambda}{CN} \Phi(0). \tag{27.41}$$

$$E[W_{\text{avg}}(t)] \to \infty \text{ as } t \to \infty. \tag{27.42}$$

□

Once a protocol is determined to be stable (for a given arrival rate), the next step is to compute the expected delay that a message incurs, i.e., $E[W_{avg}]$, which is an important performance measure, especially for high-speed communication networks. The work of Hastad et al. [21] also showed that long delays are unavoidable in backoff protocols: they showed that the delay of any stable exponential or polynomial backoff protocol is at least polynomial in N.

We next consider a protocol due to Raghavan and Upfal [41] that gives a protocol that guarantees a delay of $O(\log N)$, provided the total arrival rate λ is less than a fixed constant $\lambda' \approx 1/10$. The main feature of this protocol is that each sender has a **transmission buffer** of size $O(\log N)$ and a *queue*. (The protocol assumes that each station knows N.) Messages awaiting transmission are stored in the buffer or in the queue. Throughout the execution of the protocol a sender is in one of the two states: a normal state or a reset state. We need the following definitions. Let μ, α, γ be suitably chosen fixed constants (these are fixed in the proof) [41]. *Count_attempts(s)* keeps a count of the number of times station s tries to transmit a message from its buffer in the $4\mu N \log N$ most recent steps. *Failure_counts(s)* stores the failure rates in transmission attempts of packets from the buffer of s in the most recent $\mu \log N$ attempts. *Random_number()* is a function that returns a random number uniformly chosen in the range $[0, 1]$ independent of the outcomes of previous calls to the function.

Communication protocol for station s:
   While in the normal state repeat:

1. Place new messages in the buffer.
2. Let X denote the number of packets in the buffer.
   if *Random_number()* $\leq X/8\alpha \log N$ then

   (a) Try to transmit a random message from the buffer.
   (b) Update *Count_attempts(s)* and *Failure_counts(s)*.

   else
   if *Random_number()* $\geq 1 - 1/N^2$ the transmit the message at the head of the queue.
3. If (*Count_attempts(s)* $\geq \mu \log n$ and *Failure_counts(s)* $> 5/8$) or if ($X > 2\alpha \log N$) then

   (a) Move all messages in the buffer to the end of the queue.
   (b) Switch to the reset state for $4\mu N^2 \log N + \gamma \log N$ steps.

   While in the reset state repeat:

1. Append any new messages to the queue.
2. If *Random_number()* $\leq 1/N^2$ then transmit the message at the head of the queue.

The main intuition behind the protocol is that most of the time, things behave normally, i.e., there is not too many collisions and the buffer size is $O(\log N)$ and the delay is $O(\log N)$. However, there can be bad events (e.g., every station generates a packet in every one of N^{10} consecutive steps), but these happen with very low probabilities. The reset state and the queue are then used as an emergency mechanism to handle such cases. Note that the delay for sending messages in the queue is polynomial in N, but this happens with a very low probability so that the overall expectation is still $O(\log N)$.

Raghavan and Upfal also show that for a class of protocols that include backoff protocols, if the total arrival rate λ is less than some fixed constant $\lambda_1 < 1$, then the delay must be $\Omega(N)$. Using a more sophisticated protocol, Goldberg et al. [20] are able to show a constant expected delay for arrival rate up to $1/e$. This protocol is practical, if one assumes that the clocks of the stations are synchronized. If not, it is shown that $O(1)$ delay can still be achieved, but the constant is too large to be practical.

27.6 Resource Allocation and Flow Control

Congestion and flow controls key ingredients of modern communication networks (typically implemented in the transport layer), which allows many users to share the network without causing congestion failure. Of course, congestion control can be trivially enforced if users do not send any packets! Thus, in addition to congestion control, we would like to maximize network throughput. It is possible to devise schemes that can maximize network throughput, for example, while denying access to some users. We would like to devise congestion control schemes that operate with high network utilization, while at the same time ensuring that all users get a fair share of resources. This can be cast as an optimization problem (defined formally below) and can be solved in a centralized fashion. However, to be useful in a network, we need to devise schemes for solving this *distributively*. The additive increase/multiplicative decrease scheme of Internet's flow control (TCP) (see, e.g., [27]) is an attempt to solve the above resource allocation problem. The framework we study next provides a provably good distributed algorithm for optimal resource allocation and flow control, and can be viewed as a rigorous generalization of the simple flow control scheme of TCP.

27.6.1 Optimization Problem

Consider a network that consists of a set $L = \{1, \ldots, L\}$ links of capacity c_l, $l \in L$. The network is shared by a set $S = \{1, \ldots, S\}$ of sources (users). The goal is to share the network resources (i.e., link bandwidths) in a fair manner. Fairness is captured as follows. Each user s has an associated utility function $U_s(x_s)$ (an increasing and strictly concave function)—the utility of the source as a function of its transmission rate x_s. We assume that a *fair* resource allocation tries to maximize the *sum* of the utilities of all the users in the network. We assume that the route of each source is fixed, i.e., $L(s) \subseteq L$ is a set of links that source s uses. For each link l, let $S(l) = \{s \in S | l \cdot \in L(s)\}$ be the set of sources that use link l. Note that $l \in L(s)$ if and only if $s \in S(l)$.

The resource allocation problem can be formulated as the following nonlinear program (e.g., [23,32]). The objective is to choose source rates $x = \{x_s, s \in S\}$ so as to

$$\max_{x_s \geq 0} \sum_s U_s(x_s)$$

subject to

$$\sum_{s \in S(l)} x_s \leq c_l, l = 1, \ldots, L. \tag{27.43}$$

The constraint says that the aggregate source rate at any link l does not exceed the capacity. If the utility functions are strictly concave, then the above nonlinear program has a unique optimal solution. To solve this optimization problem directly, one should have the knowledge of the utility functions and the routes of all sources in the network. This requires coordination among all sources the which is not practical in large networks such as the Internet. Our goal is a distributed solution, where each source adapts its transmission rate based only on local information. We next describe a distributed solution due to Low and Lapsley [32] that works on the dual of the above nonlinear problem (Equation 27.43).

27.6.2 Dual Problem

To set up the dual, we define the Lagrangian

$$L(x, p) = \sum_s U_s(x_s) - \sum_l p_l \left(\sum_{s \in S(l)} x_s - c_l \right) \tag{27.44}$$

$$= \sum_s \left(U_s(x_s) - x_s \sum_{l \in L(s)} p_l \right) + \sum_l p_l c_l. \tag{27.45}$$

$p = \{p_1, \ldots, p_L\}$ are the Lagrange multipliers. The first term is separable in x_s, and hence

$$\max_{x_s} \sum_s \left(U_s(x_s) - x_s \sum_{l \in L(s)} p_l \right) = \sum_s \max_{x_s} \left(U_s(x_s) - x_s \sum_{l \in L(s)} p_l \right). \tag{27.46}$$

The objective function of the dual problem is thus (e.g., see [6])

$$\min_{p_l \geq 0} D(p) := \sum_s \max_{x_s \geq 0} \left(U_s(x_s) - x_s \sum_{l \in L(s)p_l} \right) + \sum_l p_l c_l. \tag{27.47}$$

The first term of the dual-objective function $D(p)$ is decomposable into S separable subproblems, one for each source. Furthermore, a key observation that emerges from the dual formulation is that each source can individually compute its optimal source rate without the need to coordinate with other sources, provided it knows the optimal Lagrange multipliers. One can naturally interpret the Langrange multiplier, p_l, as the price per unit bandwidth at link l. Then $\sum_{l \in L(s)} p_l$ is the total price per unit bandwidth for all links in the path of s. Thus, the optimal rate of a source depends only on the aggregate price on its route and can be computed by solving a simple maximization problem:

$$x_s(p) = U_s'^{-1} \left(\sum_{l \in L(s)} p_l \right), \tag{27.48}$$

where $U_s'^{-1}(.)$ denotes the inverse of the derivative of U_s. Of course, to compute x_s we first need to compute the minimal prices that we focus on next.

The dual problem of computing the optimal prices is solved by using the gradient projection method (see e.g., [6]). This yields the following adjustment rule that progressively converges to the optimal:

$$p_l(t + 1) = \max \left(p_l(t) + (\gamma \sum_{s \in S(l)} x_s(p(t)) - \gamma c_l), 0 \right), \tag{27.49}$$

where $\gamma > 0$ is the stepsize parameter (has to be sufficiently small for convergence [33]). The above formula has nice interpretation in the language of economics: if the demand $\sum_{s \in S(l)} x_s(p(t))$ for bandwidth at link l exceeds the supply c_l, then raise the price $p_l(t)$; otherwise, reduce price $p_l(t)$. Another key observation is that, given the aggregate source rate $\sum_{s \in S(l)} x_s(p(t))$ that goes through link l, the adjustment rule is completely distributed and can be implemented by individual links (or the routers) using only local information. Equations 27.48 and 27.49 suggest the following iterative (and synchronous) distributed algorithm:

1. Link l's algorithm: In step $t \geq 1$, a link l receives rates $x_s(t)$ from all sources $s \in S(l)$ ($x_s(0)$ can be zero) that go through link l. It then computes a new price $p_l(t+1)$ ($p_l(0) = 0$) based on Equation 27.49 and communicates the new price to all sources $s \in S(l)$ that use link l.

2. Source s's algorithm: In step $t \geq 1$, a source s receives from the network the aggregate price $\sum_{l \in L(s)} p_l$ of links on its path. It then chooses a new transmission rate $x_s(p(t+1))$ computed by Equation 27.48 and communicates this new rate to links $l \in L(s)$ in its path.

Low and Lapsley [32] show the convergence of the algorithm when the stepsize parameter is appropriately chosen. They further show that the algorithm can be extended to work in an asynchronous setting where the sources and the links need not compute in lockstep.

27.7   Peer-to-Peer Networks

P2P network is a highly **dynamic network:** nodes (peers) and edges (currently established connections) appear and disappear over time. In contrast to a static network, a P2P network has no fixed infrastructure with respect to the participating peers, although the underlying network (Internet) can be static. Thus a P2P network can be considered as an **overlay network** over an underlying network, where the communication between adjacent peers in the P2P network may in fact go through one or more intermediate peers in the underlying network. Peers communicate using only local information since a peer does not have global knowledge of the current topology, or even the identities (e.g., IP addresses) of other peers in the current network.

In a dynamic network, even maintaining a basic property such as connectivity becomes nontrivial. Consider search—a common and pervasive P2P application. Clearly, connectivity is important for reachability of search queries. Having a network of low diameter is also important. Consider P2P systems running the Gnutella protocol [50]. They use a flooding-like routing algorithm: queries fan-out from the source node and the search radius is limited by a "Time to Live (TTL)" parameter (this is set to an initial value at the source node, and is decreased in each hop). Thus a low diameter is not only helpful in enlarging the scope of search but also in reducing the associated network traffic. At the same time, it is desirable to have nodes with bounded degree; this will also help the traffic explosion problem in Gnutella-like networks by reducing the fan-out in each stage.

27.7.1   A Distributed Protocol for Building a P2P Network

A fundamental challenge is to design distributed protocols that maintain network connectivity and low diameter, and which operates without having global knowledge. We next discuss the protocol due to Pandurangan, Raghavan, and Upfal [37], which is a distributed protocol for constructing and maintaining connected, low-diameter, constant-degree P2P networks. The protocol specifies what nodes a newly arriving node should connect, and when and how to replace lost connections (edges) when an existing node leaves. The model assumes that an incoming node knows the identities (IP addresses) of a small number of existing nodes (at least one such address is needed to get access to the network).

We assume that there is a central element called the host server which, at all times, maintains a cache containing a list of nodes (i.e., their IP addresses). In particular, assume that the host server maintains K nodes at all times, where K is a constant. The host server is reachable by all nodes at all times; however, it need not know the topology of the network at any time, or even the identities of all nodes currently on the network. We only expect that (1) when the host server is contacted on its IP address it responds and (2) any node on the P2P network can send messages to its neighbors.

Before we state the protocol, we need some terminology. When a node is in the cache, we refer to it as a cache node. It accepts connections from all other nodes. A node is *new* when it joins the network, otherwise it is *old*. The protocol ensures that the degree (number of neighbors) of all nodes will be in the interval $[D, C + 1]$, for two constants D and C.

A new node first contacts the host server, which gives it D random nodes from the current cache to connect to. The new node connects to these, and becomes a d-node; it remains a d-node until it subsequently either enters the cache or leaves the network. The degree of a d-node is always D. At some point the protocol may put a d-node into the cache. It stays in the cache until it acquires a total of C connections, at which point it leaves the cache as a c-node. A c-node might lose connections after it leaves the cache, but its degree is always at least D. A c-node has always one preferred connection made precise below. The protocol is summarized below as a set of rules applicable to various situations that a node may find itself in.

27.7.1.1 P2P Protocol for Node v

1. *On joining the network*: Connect to D cache nodes, which are chosen uniformly at random from the current cache. Note that $D < K$.

2. *Reconnect rule*: If a neighbor of v leaves the network, and that connection was not a preferred connection, connect to a random node in the cache with probability $D/d(v)$, where $d(v)$ is the degree of v before losing the neighbor.

3. *Cache replacement rule*: When a cache node v reaches degree C while in the cache (or if v drops out of the network), it is replaced in the cache by a d-node from the network. Let $r_0(v) = v$, and let $r_k(v)$ be the node replaced by $r_{k-1}(v)$ in the cache. The replacement d-node is found by the following rule:

   $k = 0$;
   while (a d-node is not found) **do**
         search neighbors of $r_k(v)$ for a d-node;
         $k = k + 1$;
   endwhile

4. *Preferred node rule*: When v leaves the cache as a c-node, it maintains a preferred connection to the d-node that replaced it in the cache. If v is not already connected to that node, this adds another connection to v. Thus the maximum degree of a node is $C + 1$. Also, a c-node can follow a chain of preferred connections to reach a cache node.

5. *Preferred reconnect rule*: If v is a c-node and its preferred connection is lost, then v reconnects to a random node in the cache and this becomes its new preferred connection.

Finally, note that the degree of a d-node is always D. Moreover, every d-node connects to a c-node. A c-node may lose connections after it leaves the cache, but its degree is always at least D. A c-node has always one preferred connection to another c-node.

27.7.2 Protocol Analysis

To analyze the protocol, we assume the $M/M/\infty$ model (see Section 27.4). In other words, we have a "stochastic adversary," which deletes and inserts nodes. (A worst-case adversary is trivial to analyze this setting, since such an adversary can make sure that the network is always disconnected.) This also turns out to be a reasonable model (it approximates real-life P2P networks quite well [43]). Under this model the steady-state size of the network depends only on the ratio $\lambda/\mu = N$ (see Section 27.4.1). Let G_t be the network at time t (G_0 has no vertices).

27.7.3 Connectivity

Theorem 27.9 shows that the protocol keeps the network connected with large probability at any instant of time (after an initial time period of $\Omega(\log N)$).

THEOREM 27.9 *There is a constant a such that at any given time $t > a \log N$,*
$$Pr(G_t \text{ is connected}) \geq 1 - O\left(\frac{\log^2 N}{N}\right).$$

The main idea in the proof of the above theorem is using the preferred connections as a "backbone": we can show that at all times, each node is connected to some cache node directly or through a path in the network. The backbone is kept connected with large probability by incoming nodes: here the "randomness" in choosing the cache node by an incoming node proves crucial. Also, an important consequence from the proof is that the above theorem does not depend on the state of the network at time $t - c \log N$; therefore it can be shown to recover rapidly from fragmentation—a nice "self-correcting" property of the protocol. More precisely we have the following corollary:

THEOREM 27.10 *There is a constant c such that if the network is disconnected at time t,*
$$Pr(G_{t+c\log N} \text{ is connected}) \geq 1 - O\left(\frac{\log^2 N}{N}\right).$$
Also, if the network is not connected then it has a connected component of size $N(1 - o(1))$.

The theorem also shows that if the network becomes disconnected, then a very large fraction of the network remains connected with high probability.

27.7.4 Diameter

The key result of [37] is that of the diameter of G_t: it is only logarithmic in the size of the network at any time t with large probability.

THEOREM 27.11 *At any any time t (after the network has converged, i.e., $t/N \to \infty$), with high probability, the largest connected component of G_t has diameter $O(\log N)$. In particular, if the network is connected $\left(\text{which has probability } 1 - O\left(\frac{\log^2 N}{N}\right)\right)$ then, with high probability, its diameter is $O(\log N)$.*

The high-level idea of the proof is to show that the network at any time resembles a random graph, i.e., the classical *Erdos–Renyi or $G(n, p)$* random graph (n is the number of vertices and p is the probability of having an edge between any two vertices) [7]. The intuition underlying the analysis is that connections are chosen randomly, and thus the topology should resemble a $G(n, p)$ random graph. In a $G(n, p)$ random graph, the connectivity threshold is $\log n/n$. That is, if p (probability that an edge occurs between two nodes) is greater than $(1 + \epsilon) \log n/n$ then almost surely the graph is connected with $O(\log n)$ diameter; if it is less than $(1 - \epsilon)$ then almost surely the graph will not be connected (for every constant $\epsilon > 0$) [7,13].

First, it can be shown that there is a very small ($\approx 1/N$) probability that any two c-nodes are connected to each other (note that d-nodes are connected to c-nodes); however, this alone is not sufficient to guarantee low diameter since this is below the threshold needed for connectivity. We appeal to a key idea from the theory of $G(n, p)$ graphs: if $p = c/n$, for a suitably large constant c, then almost surely, there is a *giant connected component* in the graph whose size is polynomial in n and has $O(\log n)$ diameter [7,13]. The protocol of Pandurangan et al. [37] is designed in such a way that a fraction of nodes have a so-called preferred connection to a set of random nodes and these connections are always maintained. Using these preferred connections one can (indirectly) show that the giant component that emerges from the random process is indeed the whole graph. A number of works have subsequently used random graphs to efficiently design P2P with good properties, see e.g., [14,19,29].

In a dynamic network, there is the additional challenge of quantifying the work done by the algorithm to maintain the desired properties. An important advantage of the aforementioned protocol is that it takes $O(\log N)$ (N is the steady-state network size) work per insertion/deletion. In a subsequent paper, using the same stochastic model, Liben-Nowell, Balakrishnan, and Karger [31] show that $\Omega(\log N)$ work is also required to maintain connectivity. A drawback of the protocol of [37] is that it is not fully decentralized—the entering node needs to connect to a small set of random nodes, which is assumed to be available in a centralized fashion. A number of subsequent papers have addressed this and devised fully decentralized protocols, which guarantee similar properties [14,19,29,31,36].

Before we end our discussion on P2P, we briefly mention another important design approach referred to as a Distributed Hash Table (DHT) (see e.g., [36,42,46]). A DHT does two things: (1) creates a fully decentralized index that maps file identifiers to peers and (2) allows a peer to search for a file very efficiently (typically logarithmically in the size of the network) without flooding. Such systems have been called structured P2P networks, unlike Gnutella, for example, which are unstructured. In unstructured networks, there is no relation between the file identifier and the peer where it resides. Many of the commercially deployed P2P networks are unstructured.

27.8 Research Issues and Summary

This chapter has focussed on theoretical and algorithmic underpinnings of today's communication networks, especially the Internet. The topics covered are routing algorithms (including algorithms for broadcast and multicast), queuing theory (both stochastic and adversarial), multiple access problem, resource allocation/flow control, and P2P network protocols. We have given only a brief overview of many of these, which illustrates the basic theoretical problems involved and the approaches taken to solve them. Tools and techniques from a variety of areas are needed—distributed algorithms, probabilistic models and analysis, graph theoretic algorithms, and nonlinear optimization to name a few. "Further Information" gives pointers to gain background on these tools and techniques, and further reading on these topics.

There are many research challenges. Designing efficient distributed approximation algorithms for fundamental network optimization problems, such as minimum Steiner tree and minimum delay routing, is an important goal. Designing provably efficient P2P networks remains a key problem at the application layer. The goal is to design provably fast protocols for building and maintaining P2P networks with good topological properties, while at the same time guaranteeing efficient and robust services, such as search and file sharing. For these we need efficient distributed algorithm that works well on dynamic networks. Another problem is understanding confluent flows and its impact on routing in the Internet. Developing realistic network models will be a key step in analysis.

27.9 Further Information

A good introduction to computer networks and the Internet is [27]. Bertsekas and Gallager's book [5] provides an introduction to communication networks blending theory and practice. Kleinrock's books [26] are standard references for Queuing theory. Further information on adversarial queuing theory can be found in [8]. References for distributed algorithms include [35,48]. The book by Peleg [39] focuses on the locality aspects of distributed computing. A survey of various P2P schemes can be found in [34]. For more details on resource allocation and congestion control, we refer to the survey of Low and Srikant [33].

Some topics that we have not covered here include topology models for the Internet, the traffic modeling, and the quality of service. For a survey on these issues we refer to [38]. There is large body of work concerning packet routing in tightly coupled interconnection networks (parallel architectures, such as trees, meshes, and hypercubes). Leighton's book [30] is a good reference for this. Scheideler's book [44] deals with universal routing algorithms, i.e., algorithms that can be applied to arbitrary topologies.

The theory of communication networks is one of the most active areas of research with explosive growth in publications in the last couple of decades. Research gets published in many different conferences. Outlets for the theoretical aspects of distributed computing and networks include *ACM Symposium on Principles of Distributed Computing (PODC)*, *International Symposium on Distributed Computing (DISC)*, *ACM Symposium on Parallel Algorithms and Architectures*, and *International Conference on Distributed Computing Systems (ICDCS)*. Traditional theory conferences, such as *ACM Symposium on Theory of Computing (STOC)*, and *IEEE Symposium on Foundation of Computer Science (FOCS)*, and *ACM-SIAM Symposium on Discrete Algorithms (SODA)*, also contain papers devoted to theoretical aspects of communication networks. The *IEEE Conference on Computer Communications (INFOCOM)* conference covers all the aspects of networking, including theory.

Defining Terms

Arborescence: An arborescence in a directed graph is a subgraph such that there is a unique path from a given root to any other node in the subgraph.

Circuit-switched networks: A communication network where a physical path (channel) is obtained for and dedicated to a single connection between two end-points in the network for the duration of the connection.

Confluent flow: A flow is said to be confluent if all the flows arriving at a node destined for a particular destination departs from the node along a single edge.

Congestion: The congestion of an edge is defined as the total amount of flow going through the edge.

Count-to-infinity problem: A problem that shows that distance vector routing algorithm can be slow in converging to the correct values in case of a change in link costs.

Diameter: The diameter of a network is the number of links in a (unweighted) shortest path between the furthest pair of nodes.

Distance vector: In some routing algorithms, each node maintains a vector, where the ith component of the vector is an estimate of its distance to the ith node in the network.

Dynamic network: A network in which the topology changes from time to time.

IP address: An IP address (Internet Protocol address) is a unique address that the network devices, such as routers, host computers, and printers, use in order to identify and communicate with each other on a network utilizing the Internet Protocol.

Minimum spanning tree: A minimum spanning tree is a tree spanning all nodes in the networks that has the minimum cost among all possible such trees. The cost of a tree is the sum of the costs of all edges in the tree.

Multiple access problem: The problem of coordinating the simultaneous transmissions of multiple nodes in a common link.

Overlay network: A structured virtual topology over a physical network. A link in the overlay network is a virtual or logical link, which corresponds to a path, perhaps through many physical links, in the underlying network.

Packet-switched networks: A communication network where messages are divided into packets and each packet is then transmitted individually. The packets can follow different routes to its destination. Once all the packets forming a message arrive at the destination, they are combined into the original message.

Route: A route of a packet is a sequence of links through which the packet travels in the network.

Routing network: A routing network is a directed graph where the data packets flow along the edges and in the directions of the edges.

Scheduling policy: A scheduling policy is a method of selecting a packet to be forwarded next along an edge from the packets waiting in the queue of that edge.

Shortest path or least-cost path: The path between source and destination that has the minimum cost. The cost of a path is the sum of the costs of the links in the path.

Shortest path routing: Finding a shortest path between the source and the destination, and routing the packets through the links along the shortest path.

Splittable flow: A flow that need not be confluent.

Transmission buffer: A buffer in a node to temporarily store the packets that are ready for transmission.

References

1. Alon, N. and Spencer, J., *The Probabilistic Method*, 2nd edition, John Wiley, Hoboken, NJ, 2000.
2. Awerbuch, B., Bar-Noy, A., and Gopal, M., Approximate distributed Bellman-Ford algorithms, *IEEE Transactions on Communications*, **42**(8), 1994, 2515–2517.
3. Awerbuch, B. and Leighton, F., A simple local-control approximation algorithm for multicommodity flow, *Proceedings of the 34th IEEE Symposium on Foundations of Computer Science (FOCS)*, Palo Alto, CA, 1993, pp. 459–468.
4. Awerbuch, B. and Peleg, D., Network synchronization with polylogarithmic overhead, *31st IEEE Symposium on Foundations of Computer Science (FOCS)*, St. Louis, MO, 1990, pp. 514–522.
5. Bertsekas, D. and Gallager, R., *Data Networks*, 2nd edition, Prentice Hall, Englewood Cliffs, NJ, 1992.
6. Bertsekas, D. and Tsitsiklis, J., *Parallel and Distributed Computation*, Prentice Hall, Englewood Cliffs, NJ, 1989.
7. Bollobas, B., *Random Graphs*, Cambridge University Press, Cambridge, U.K., 2001.
8. Borodin, A., Kleinberg, J., Raghavan, P., Sudan, M., and Williamson, D., Adversarial queuing theory, *Journal of the ACM*, **48**(1), Jan. 2001, 13–38.
9. Chalermsook, P. and Fakcharoenphol, J., Simple distributed algorithms for approximating minimum Steiner trees, *Proceedings of International Computing and Combinatorics Conference (COCOON)*, Kunming, China, 2005, pp. 380–389.
10. Chen, J., Kleinberg, R.D., Lovasz, L., Rajaraman, R., Sundaram, R., and Vetta, A., (Almost) tight bounds and existence theorems for confluent flows, *Proceedings of STOC*, Chicago, IL, 2004, pp. 529–538.
11. Chen, J., Marathe, M., Rajmohan, R., and Sundaram, R., The confluent capacity of the Internet: Congestion vs. dilation, *Proceedings of the 26th IEEE International Conference on Distributed Computing Systems (ICDCS)*, Lisbon, Portugal, 2006, pp. 5–5.
12. Chen, J., Rajmohan, R., and Sundaram, R., Meet and merge: Approximation algorithms for confluent flows, *Journal of Computer and System Sciences*, **72**, 2006, 468–489.

13. Chung, F. and Lu, L., Diameter of random sparse graphs, *Advances in Applied Mathematics*, **26**, 2001, 257–279.

14. Cooper, C., Dyer, M., and Greenhill, C., Sampling regular graphs and a peer-to-peer network, *Proceedings of the 16th Annual ACM-SIAM Symposium on Discrete Algorithms (SODA)*, New York–Philadelphia, 2005, pp. 980–988.

15. Cormen, T., Leiserson, C., Rivest, R., and Stein, C., *Introduction to Algorithms*, MIT Press, Cambridge, MA, and McGraw Hill, New York 2001.

16. Elkin, M., Unconditional lower bounds on the time-approximation tradeoffs for the distributed minimum spanning tree problem, *Proceedings of the ACM Symposium on Theory of Computing*, Chicago, IL, 2004, pp. 331–340.

17. Gallager, R., A minimum delay routing algorithm using distributed computation, *IEEE Transactions on Communications*, **25**(1), 1997, 73–85.

18. Gallager, R., Humblet, P., and Spira, P., A distributed algorithm for minimum-weight spanning trees, *ACM Transactions on Programming Languages and Systems*, **5**(1), 1983, 66–77.

19. Gkantsidis, C., Mihail, M., and Saberi, A., On the random walk method in peer-to-peer networks, *Proceedings of INFOCOM*, Hong Kong, China, 2004, pp. 130–140.

20. Goldberg, L., Mackenzie, P., Paterson, M., and Srinivasan, A., Contention resolution with constant expected delay, *JACM*, **47**(6), 2000, 1048–1096.

21. Hastad, J., Leighton, T., and Rogoff, B., Analysis of backoff protocols for multiple access channels, *SIAM Journal on Computing*, **25**(4), 1996, 740–774.

22. Kelley, F., Stochastic models of computer communication systems, *Journal of Royal Statistical Society, Series B*, **47**, 1985, 379–395.

23. Kelly, F., Maulloo, A., and Tan, D., Rate control for communication networks: Shadow prices, proportional fairness and stability, *Journal of the Operational Research Society*, **49**(3), 1998, 237–252.

24. Khan, M. and Pandurangan, G., A Fast distributed approximation algorithm for minimum spanning trees, *Distributed Computing*, **20**, 2008, 391–402. Conference version: *Proceedings of the 20th International Symposium on Distributed Computing (DISC)*, Stockholm, Sweden, 2006, *LNCS* 4167, Springer-Verlag, Berlin, Heidelberg, 2006, pp. 355–369.

25. Kleinrock, L., *Communication Nets: Stochastic Message Flow and Delay*, McGraw-Hill, New York, 1964.

26. Kleinrock, L., *Queueing Systems, Vols. 1 and 2*, Wiley, New York, 1975 and 1976.

27. Kurose, J.F. and Ross, K.W., *Computer Networking: A Top-Down Approach*, 4th edition, Addison Wesley, Reading, MA, 2008.

28. Kutten, S. and Peleg D., Fast distributed construction of k-dominating sets and applications, *Journal of Algorithms*, **28**, 1998, 40–66.

29. Law, C. and Siu, K., Distributed construction of random expander networks, *Proceedings of the 22nd IEEE Conference on Computer Communications (INFOCOM)*, San Francisco, CA, 2003, pp. 2133–2143.

30. Leighton, F., *Introduction to Parallel Algorithms and Architectures*, Morgan Kaufmann, San Mateo, CA, 1992.

31. Liben-Nowell, D., Balakrishnan, H., and Karger, D., Analysis of the evolution of peer-to-peer systems, *ACM Conference on Principles of Distributed Computing (PODC)*, Monterey, CA, July 2002, pp. 233–242.

32. Low, S. and Lapsley, D., Optimization flow control-I: Basic algorithm and convergence, *IEEE/ACM Transactions on Networking*, **7**(6), 1999, 861–874.

33. Low, S. and Srikant, R., A Mathematical framework for designing a low-loss, low-delay Internet, *Networks and Spatial Economics*, **4**(1), 2004, 75–101.

34. Lua, E., Crowcroft, J., Pias, M., Sharma, R., and Lim, S., A survey and comparison of peer-to-peer overlay network schemes, *IEEE Communications Surveys and Tutorials*, **7**(2), 2005, 72–93.

35. Lynch, N., *Distributed Algorithms*, Morgan Kaufamnn Publishers, San Mateo, CA, 1996.

36. Malkhi, D., Naor, M., and Ratajczak, D., Viceroy: A scalable and dynamic emulation of the butterfly, *Proceedings of the 21st Annual Symposium on Principles of Distributed Computing (PODC)*, Monterey, CA, 2002, pp. 183–192.

37. Pandurangan, G., Raghavan, P., and Upfal, E., Building low diameter peer-to-peer networks, *IEEE Journal on Selected Areas in Communications*, **21**(6), Aug. 2003, 995–1002. Conference version in *Proceedings of the 42nd Annual IEEE Symposium on the Foundations of Computer Science (FOCS)*, Las Vegas, NV, 2001, pp. 492–499.

38. Park, K., The Internet as a complex system, in *The Internet as a Large-Scale Complex System*, K. Park and W. Willinger (Eds.), Sante Fe Institute, Studies in the Sciences of Complexity, Oxford University Press, 2005.

39. Peleg, D., *Distributed Computing: A Locality Sensitive Approach*, SIAM, Philadelphia, PA, 2000.

40. Peleg, D. and Rabinovich, V., A near-tight lower bound on the time complexity of distributed MST construction, *Proceedings of the 40th IEEE Symposium on Foundations of Computer Science*, New York, 1999, pp. 253–261.

41. Raghavan, P. and Upfal, E., Stochastic contention resolution with short delays, *SIAM Journal on Computing*, **28**(2), 1998, 709–719.

42. Ratnasamy, S., Francis, P., Handley, M., Karp, R., and Shenker, S., A scalable content-addressable network, *Proceedings of the 2001 Conference on Applications, Technologies, Architectures, and Protocols for Computer Communications*, San Diego, CA, 2001, pp. 161–172.

43. Saroiu, S., Gummadi, P.K., and Gribble, S.D., A measurement study of peer-to-peer file sharing systems, *Proceedings of Multimedia Computing and Networking (MMCN)*, San Jose, CA, 2002.

44. Scheideler, C., *Universal Routing Strategies for Interconnection Networks, LNCS* 1390, Springer-Verlag, Heidelberg, Germany, 1998.

45. Stidham, S., A last word on $L = \lambda W$, *Operations Research*, **22**, 1974, 417–421.

46. Stoica, I., Morris, R., Karger, D., Kaashoek, M., and Balakrishnan, H., Chord: A scalable peer-to-peer lookup service for Internet applications, *Proceedings of ACM SIGCOMM*, San Diego, CA, 2001, pp. 149–160.

47. Tarjan, R., *Data Structures and Network Algorithms*, SIAM, Philadelphia, PA, 1987.

48. Tel, G., *Introduction to Distributed Algorithms*, 2nd edition, Cambridge University Press, Cambridge, U.K., 2000.

49. Vazirani, V., *Approximation Algorithms*, Springer-Verlag, Berlin, 2001.

50. The Gnutella Protocol Specification v0.4. http://www9.limewire.com/developer/gnutella_protocol_0.4.pdf

28

Network Algorithmics*

George Varghese
University of California

Network Algorithmics studies techniques for efficient network protocol implementations. It is primarily about two things: a set of fundamental networking performance bottlenecks, and a set of interdisciplinary techniques to address these bottlenecks.

Network Algorithmics is an interdisciplinary approach because it encompasses such fields as architecture and operating systems (for speeding up servers), hardware design (for speeding up network devices such as routers), and algorithm design (for designing scalable algorithms). Network Algorithmics is also a systems approach because it exploits the fact that routers and servers are systems, in which efficiencies can be gained by moving functions in time and space between subsystems.

Bottlenecks arise from the simultaneous desire to make networks easy to use while at the same time realizing the performance of the raw hardware. Ease of use comes from the use of powerful network abstractions such as socket interfaces and prefix-based forwarding. Such abstractions can exact a large performance penalty compared to the raw transmission capacity of optical links. The bottlenecks are different for the two fundamental categories of networking devices, endnodes and routers.

Endnodes are the endpoints of the network and include personal computers, workstations, and large servers. To run an arbitrary code, personal endnodes typically have an operating system that mediates between applications and the network. To ease software development, most operating systems are structured as layered software; to protect the operating system from other applications, operating systems implement a set of protection mechanisms; finally, core operating system routines such as schedulers and allocators are written using general mechanisms that target as wide a class of applications as possible. Unfortunately, the combination of layered software, protection mechanisms, and excessive generality can slow down networking software greatly even with the fastest processors.

Unlike endnodes, routers are special-purpose devices devoted to networking. Thus there is very little structural overhead within a "router," with only the use of a very lightweight operating system and a clearly separated forwarding path that often is completely implemented in hardware. Instead

---

of structure, the fundamental problems faced by routers are caused by scale and services. Optical links keep getting faster, going from 1 to 40 Gbps links. Besides bandwidth scaling, there is also population scaling as endpoints get added to the Internet as more enterprises go online. While scaling has historically driven the design of routers, recently it has become important to instrument routers to provide network guarantees—delay in times of congestion, protection during attacks, and availability when failures occur. Finding ways to implement these new services at high speeds will be a major challenge for router vendors in the next decade.

In the rest of the chapter, we make these concepts more specific by providing examples from Router Algorithmics (Section 28.2) and Endnode Algorithmics (Section 28.3). Before we do so, we provide a brief survey of the measures we use and a model of memory technologies. This is because optimizing memory accesses is crucial for both endnode and router algorithmics.

28.1 Measures and Memories

The goodness of the algorithms in this chapter is measured by space (the amount of memory) and speed (the number of memory accesses) for various operations. In conventional algorithms, speed is measured by the amount of processing steps (e.g., instructions). However, to a first approximation we will ignore such processing steps. This is reasonable in hardware design because complex processing steps can often be done in a single clock cycle. This is also reasonable in software designs because instruction cycle times are fast compared to memory access times.

Thus the main measures are the memory required for an algorithm and the number of memory accesses for each operation of the algorithm. Unfortunately, this requires more qualification because there are different types of memories. More precisely, the main space measure is the amount of fast memory (cache in software, SRAM in hardware). If a design only uses slow memory (main memory in software, DRAM in hardware) then the amount of memory used is often irrelevant because such memory is typically cheap and plentiful. Similarly, if a design uses both fast and slow memory, the main speed measure is the number of slow memory accesses (because fast memory accesses can be ignored in comparison). If a design uses only fast memory, then the speed measure is the number of fast memory accesses.

The distinction between slow (dynamic random access memory, DRAM) and fast (static RAM, SRAM) memories is worth internalizing before we proceed further. A SRAM contains N registers addressed by $\log N$ address bits A. SRAM is so named because the underlying transistors storing a bit refresh themselves and so are "static." Besides transistors to store the bits, an SRAM also needs a decoder that decodes A into a unary value used to select the right register. Accessing an SRAM on-chip is only slightly slower than accessing a register because of the added decode delay. It is possible to obtain on-chip SRAMs with 0.5 ns access times. Access times of 1–2 ns for on-chip SRAM and 5–10 ns for off-chip SRAM are common. On-chip SRAM is limited to around 64 Mbits today. The Level 1 and Level 2 caches in modern processors are built from SRAM.

In order to refresh itself, an SRAM bit cell requires at least five transistors. Thus SRAM is always less dense or more expensive than memory technology based on dynamic RAM (DRAM). A DRAM cell uses only a single transistor connected to an output capacitance that can be manufactured to take much less space than the transistors in an SRAM. Loss due to leakage via the capacitance is fixed by refreshing the DRAM cell externally within a few ms. Main memory in modern computers is often DRAM. DRAM chips appear to quadruple in capacity every 3 years [FPCe97] and are heading toward 1 Gigabit on a single chip.

It seems clear that DRAM will always be denser but slower than SRAM. There are, however, a number of interesting combinations of DRAMs and SRAMs that are useful in network algorithmics. *Interleaved DRAMs:* While memory latency is critical for computation speed, memory throughput (often called bandwidth) is also important for many network applications. Suppose a DRAM has a word size of 32 bits and a cycle time of 100 ns. Then the throughput using a single copy of the DRAM is limited to 32 bits every 100 ns. Clearly, throughput can be improved using accesses to multiple DRAMs. Multiple DRAMs (called banks) can be strung together on a single bus. While using multiple memory banks is a very old idea, it is only in the last 10 years that memory designers have integrated several banks into a single memory chip where the address and data lines for all banks are multiplexed using a common high-speed network called a bus. Prominent examples include SDRAM with two banks, and RDRAM with 16 banks. Memory interleaving is often used with pipelining any data structure such as a tree can be split into separate banks of the interleaved memory such as RAMBUS. While the first lookup has moved on to the second level of the tree in say Bank 2, a second lookup can be scheduled to the first level (i.e., the root level) of the tree in say Bank 1.

Wide Word Parallelism: A common theme in many networking designs is to use wide memory words that can be processed in parallel. This can be implemented using DRAM and exploiting page mode, or by using SRAM and making each memory word wider. Note that in software designs wide words can be exploited by aligning data structures to the cache line size (the width of bits loaded on every access); in hardware designs, one can choose the width of memory (up to say 5000 bits or so after which electrical issues may be problematic) to fit the problem.

28.2 Router Algorithmics

We model a router in Figure 28.1 as a box with a set of input links on the left, and a set of output links on the right; the task of the router is to switch a packet from an input link to the appropriate output link based on the packet destination address. A router has three main performance bottlenecks: lookup, switching, and output queuing.

Lookup: Each packet arriving on an input link contains a 32 bits IP address. The router has a forwarding information database (FIB) that contains prefixes with their associated output links. Prefixes are used as a form of variable length "area code" to compress the FIB. Even core routers store only several hundred thousand prefixes compared to the several hundred million assigned Internet addresses. A prefix like 01*, where the * denotes the usual "don't care" symbol, matches IP addresses that start with 01.

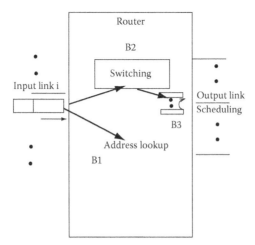

FIGURE 28.1 A model of a router labeled with the three main bottlenecks in the forwarding path: address lookup (B1), switching (B2), and output scheduling (B3).

Suppose that the FIB contains only two prefixes 100* with associated output link 6, and prefix 1* has output link 2. Thus a packet whose destination address starts with 100 matches both prefix 100* and 1*. The disambiguating rule that IP routers use is to match an address to the longest matching prefix. Assuming no longer matching prefixes, this packet should be forwarded to output link 6.

Many routers today also offer a more complex lookup called packet classification where the lookup takes as input the destination address as well as source address and TCP ports that we also briefly describe.

Switching: In Figure 28.1, after address lookup determines the output link, the packet must be switched to the corresponding output link. While early routers used a single bus for switching, for performance reasons, most routers use a crossbar switch that uses several *parallel* buses, one for each input and one for each output. An input and an output are connected by turning on transistors connecting the corresponding input bus and output bus. While the data path is conceptually easy, a challenging problem is to match available inputs and outputs every packet arrival time, finding a maximal match that avoids conflicts.

Output Scheduling: Once the packet in Figure 28.1 has been looked up and switched to output link 6, output link 6 may be congested and thus the packet may have to be placed in a queue for output link 6. Many older routers simply place the packet in a first-in first-out (FIFO) transmission queue. However, some routers employ more sophisticated packet scheduling to provide fair bandwidth allocation and delay guarantees.

Besides the major tasks of lookups, switching and queuing, there are also a number of other tasks that are less time critical such as validating header fields and checksums, computing routes to populate the forwarding table (often done in software by separate route processors), management protocols such as Simple Network Management Protocol (SNMP) and Internet Control Message Protocol (ICNP). ICNP is basically a protocol for sending error messages such as "time-to-live exceeded."

28.2.1 Lookups

Figure 28.1 described address lookups as a key function of a router. For most routers, address lookup is really the longest matching prefix lookup. In this section we will describe a series of techniques for prefix lookups starting with unibit tries (used in early routers) to simple multibit tries (used in several routers) to compressed multibit tries (probably the best current technique, and used in possibly the fastest router today, Cisco's CRS-1).

The simplest technique is the unibit trie. Consider the sample prefix database of Figure 28.2. This database will be used to illustrate the algorithmic solutions in this section. It contains 9 prefixes called P1 to P9, with the bit strings shown in the figure. In the figure, an address D that starts with 1 followed by a string of 31 zeroes will match P4, P6, P7, and P8. The longest match is P7.

Figure 28.3 shows a unibit trie for the sample database of Figure 28.2. A unibit trie is a tree in which each node is an array containing a 0-pointer and a 1-pointer. At the root all prefixes that start with 0 are stored in the subtrie pointed to by the 0-pointer, and all prefixes that start with a 1 are stored in the subtrie pointed to by the 1-pointer. Each subtrie is then constructed recursively in a similar fashion using the remaining bits of the prefixes allocated to the subtrie.

There are two other fine points to note. In some cases, a prefix may be a substring of another prefix. For example, P4 = 1* is a substring of P2 = 111*. In that case, the smaller string P4 is stored inside a trie node on the path to the longer string. For example, P4 is stored at the right child to the root; note that the path to this right child is the string 1, which is the same as P4.

Finally, in the case of a prefix like P3 = 11001, after we follow the first three bits, we might naively expect to find a string of nodes corresponding to the last two bits. However, since no other prefixes share more than the first 3 bits with P3, these nodes

P1 = 101*
P2 = 111*
P3 = 11001*
P4 = 1*
P5 = 0*
P6 = 1000*
P7 = 100000*
P8 = 100*
P9 = 110*

FIGURE 28.2

Sample prefix database used for lookup algorithms.

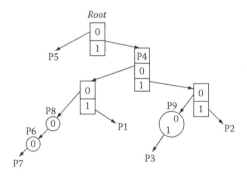

FIGURE 28.3 The one bit trie for the sample database of Figure 28.2.

would only contain 1-pointer a piece. Such a string of trie nodes with only 1-pointer each is a called a one-way branch.

Clearly one-way branches can greatly increase wasted storage by using whole nodes (containing at least two pointers) when only a single bit suffices. One-way branches are best compressed as in Figure 28.3, using a text string (i.e., "01") to represent the pointers that would have been followed in the one-way branch.

To search for the longest matching prefix of a destination address D, the bits of D are used to trace a path through the trie. The path starts with the root and continues until search fails by ending at an empty pointer, or at a text string that does not completely match. While following the path, the algorithm keeps track of the last prefix encountered at a node in the path. When search fails, this is the longest matching prefix that is returned.

For example, if D begins with 1110, the algorithm starts by following the 1-pointer at the root to arrive at the node containing P4. The algorithm remembers P4, and uses the next bit of D (a 1) to follow the 1-pointer to the next node. At this node, the algorithm follows the 1-pointer to arrive at P2. When the algorithm arrives at P2, it overwrites the previously stored value (P4) by the newer prefix found (P2). At this point, search terminates because P2 has no outgoing pointers.

The literature on tries [Knu73] does not use text strings to compress one-way branches as in Figure 28.3. Instead, the classical scheme, called a Patricia trie, uses a skip count. This count records the number of bits in the corresponding text string, not the bits themselves. For example, the text string node "01" in our example would be replaced with the skip count "2" in a Patricia trie. Unfortunately, this works very badly with prefix matching, an application that Patricia tries were not designed to handle in the first place. When search fails, a search in a Patricia trie has to backtrack, and go back up the trie searching for a possible shorter match. Unfortunately, the BSD implementation of IP forwarding [WS95] decided to use Patricia tries as a basis for best matching prefix. Text strings are a much better idea than using skip counts as in Patricia tries.

Multibit Tries: Most large memories use DRAM. DRAM has a large latency (around 60 ns) compared to register access times (2–5 ns). Since a unibit trie may have to make 32 accesses for a 32 bits prefix, the worst-case search time of a unibit trie is at least $32*80 = 2.56\,\mu s$. Now this disadvantage can be worked around by pipelining the 32 memory accesses as described at the end of Section 28.1, as was done in early high-speed routers. However, the hardware complexity increases with the number of pipeline levels (32 is high). Thus in either case, it is worth reducing the height of the trie. This clearly motivates multibit trie search.

To search a trie in strides of 4 bits, the main problem is dealing with prefixes like 10101* (length 5) whose lengths are not a multiple of the chosen stride length 4. If we search 4 bits at a time, how can we ensure that we do not miss prefixes like 10101*?

Controlled prefix expansion solves this problem by transforming an existing prefix database into a new database with less prefix lengths, but with potentially more prefixes. By eliminating all lengths that are not multiples of the chosen stride length, expansion allows faster multibit trie search at the cost of an increased database size.

For example, removing odd prefix lengths reduces the number of prefix lengths from 32 to 16, and would allow trie search two bits at a time. To remove a prefix like 101* of length 3, observe that 101* represents addresses that begin with 101, which in turn represents addresses that begin with 1010* or 1011*. Thus 101* (of length 3) can be replaced by two prefixes of length 4 (1010* and 1011*), both of which inherit the forwarding information associated with 101*. However, the expanded prefixes may collide with an existing prefix at the new length. In that case, the expanded prefix is removed. The existing prefix is given priority because it was originally of longer length.

Figure 28.4 shows a trie for the database of Figure 28.2 using expanded tries with a fixed stride length of 3. Thus each trie node uses 3 bits. The replicated entries within trie nodes in Figure 28.4 correspond exactly to expanded prefixes. For example, notice that P6 in Figure 28.2 has three expansions (100001, 100010, 100011).

These three expanded prefixes are pointed to by the 100-pointer in the root node of Figure 28.4 (because all three expanded prefixes start with 100) and are stored in the 001, 010, and 011 entries of the right child of the root node. Notice also that the entry 100 in the root node also has a stored prefix P8 (besides the pointer pointing to P6's expansions), because P8 = 100* is itself an expanded prefix.

Thus each trie node element is a record containing two entries: a stored prefix and a pointer. Trie search proceeds 3 bits at a time. Each time a pointer is followed, the algorithm remembers the stored prefix (if any). When search terminates at an empty pointer, the last stored prefix in the path is returned.

For example, if address D begins with 1110, search for D starts at the 111 entry at the root node, which has no outgoing pointer but a stored prefix (P2). Thus search for D terminates with P2. A search for an address that starts with 100000 follows the 100 pointer in the root (and remembers P8). This leads to the node on the lower right, where the 000 entry has no outgoing pointer but a stored prefix (P7). The search terminates with result P7. Both the pointer and stored prefix can be retrieved in one memory access using wide memories.

Lulea Compressed Tries: Unfortunately, the use of larger strides can increase speed but considerably waste memory. For example, a router that wishes to do lookups in 4 memory accesses may use strides

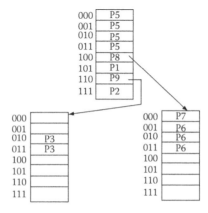

FIGURE 28.4 Expanded trie (that has two strides of 3 bits each) corresponding to prefix database of Figure 28.2.

of 8, which in turn will result in 256 element trie nodes which can be very wasteful of memory. This is particularly a problem if the lookup is to be implemented using SRAM (e.g., on chip memory). The Lulea technique is a way to have one's cake and eat it: the use of large strides for faster access and yet a way to compress the lookup structure using bitmap compression to considerably reduce storage.

We know that a string with repetitions (e.g., AAAABBAAACCCCC) can be compressed using a bitmap denoting repetition points (i.e., 10001010010000) together with a compressed sequence (i.e., ABAC). Similarly, the root node of Figure 28.4 contains a repeated sequence (P5, P5, P5, P5) caused by expansion.

The Lulea scheme [DBCP97] avoids this obvious waste by compressing repeated information using a bitmap and a compressed sequence without paying a high penalty in search time. This allows the entire database to fit into expensive SRAM or on-chip memory. It does, however, pay a high price in insertion times.

Some expanded trie entries (e.g., the 110 entry at the root of Figure 28.4) have two values, a pointer and a prefix. To make compression easier, the algorithm starts by making each entry have exactly one value by pushing forwarding information associated with prefixes down to the trie leaves. Since the leaves do not have a pointer, we only have forwarding information at leaves, and only pointers at non-leaf nodes. This process is called leaf pushing.

For example, to avoid the extra stored prefix in the 110 entry of the root node of Figure 28.4, the P9 stored prefix is pushed to all the entries in the leftmost trie node with the exception of the 010 and 011 entries (both of which continue to contain P3). Similarly, the P8 stored prefix in the 100 root node entry is pushed down to the 100, 101, 110, and 111 entries of the rightmost trie node. Once this is done, each node entry contains either a stored prefix or a pointer, but not both. More precisely, by stored prefix we mean some representation of the stored prefix and its associated fowarding information.

The Lulea scheme starts with a conceptual leaf-pushed expanded trie and replaces consecutive identical elements with a single value. A node bitmap (with 0's corresponding to removed positions) is used to allow fast indexing on the compressed nodes.

Consider the root node in Figure 28.4. After leaf pushing the root has the sequence P5, P5, P5, P5, ptr1, P1, ptr2, P2 (ptr1 is a pointer to the trie node containing P6 and P7, and ptr2 is a pointer to the node containing P3). After replacing consecutive values by the first value we get P5, -, -, -, ptr1, P1, ptr2, P2 as shown in the middle frame of Figure 28.5. The rightmost frame shows the final result with a bitmap indicating removed positions (10001111) and a compressed list (P5, ptr1, P1, ptr2, P1).

If there are N original prefixes and pointers within an original (unexpanded) trie node, the number of entries within the compressed node can be shown to be never more than $2N + 1$. Intuitively, this is because N prefixes partition the address space into at most $2N + 1$ disjoint subranges and each subrange requires at most one compressed node entry.

Search uses the number of bits specified by the stride to index into the current trie node, starting with the root and continuing until a null pointer is encountered. However, while following pointers, an uncompressed index must be mapped to an index into the compressed node. This mapping is accomplished by counting bits within the node bitmap.

Consider the data structure on the right of Figure 28.5. Consider a search for an address that starts with 100111. If we were dealing with just the uncompressed node on the left of Figure 28.5, we could use 100 to index into the fifth array element to get ptr1. However, we must now obtain the same information from the compressed node representation on the right of Figure 28.5.

Instead, we use the first three bits (100) to index into the root node bitmap. Since this is the second bit set (the algorithm needs to count the bits set before a given bit), the algorithm indexes into the second element of the compressed node. This produces a pointer ptr1 to the rightmost trie node. Next, imagine the rightmost leaf node of Figure 28.4 (after leaf pushing) also compressed in the same

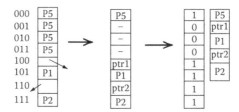

FIGURE 28.5 Compressing the root node of Figure 28.4 (after leaf pushing) using the Lulea bitmap compression scheme.

way. The node contains the sequence P7, P6, P6, P6, P8, P8, P8, P8. Thus the corresponding bitmap is 11001000, and the compressed sequence is P7, P6, P8.

Thus in the rightmost leaf node, the algorithm uses the next 3 bits (111) of the destination address to index into bit 8. Since there is no pointer to follow in the equivalent uncompressed node, the search terminates at this point. However, to retrieve the best matching prefix (if any) stored at this node, the algorithm must find any prefix stored before this entry.

This would be trivial with expansion as the value P8 would have been expanded into the 111 entry, but since the expanded sequence of P8 values has been replaced by a single P8 value in the compressed version, the algorithm has to work harder. Thus the Lulea algorithm *counts* the number of bits set before position 8 (which happens to be 3), and then indexes into the 3rd element of the compressed sequence. This gives the correct result, P8.

The Lulea paper [DBCP97] describes a trie that uses fixed strides of 16, 8, and 8. But how can the algorithm efficiently count the bits set in a large bitmap? For example, a 16 bits stride uses 64 kbits.

To speed up counting set bits, the algorithm accompanies each bitmap with a summary array that contains a cumulative count of the number of set bits associated with fixed size "chunks" of the bitmap. More precisely, the summary array $S[j]$ for $j \geq 1$ contains the number of bits set in the bitmap from positions 0 to jc, where c is the chosen chunk size. Using 64 bits chunks, the summary array takes at most 25% beyond the storage required for the bitmap.

Using the summary array, counting the bits set up to position i now takes two steps. First, access the summary array at position j, where $j = i \div c$ (i.e., j is the number of the chunk containing bit i). Then access chunk j and count the bits set in chunk j up to position i. The sum of the two values gives the count of bits set up to position i.

Despite compact storage, the Lulea scheme has two disadvantages. First, counting bits requires at least one extra memory reference per node. Second, leaf pushing makes worst-case insertion times large. A prefix added to a root node can cause information to be pushed to thousands of leaves. The full Tree bitmap scheme, which we study next, overcomes these problem by abandoning leaf pushing and using two bitmaps per node.

Tree Bitmap: The Tree bitmap [EDV04] scheme starts with the goal of achieving the same storage and speed as the Lulea scheme, but adds the goal of fast insertions. While we have argued that fast insertions are not as important as fast lookups, they clearly are desirable. Also, if the only way to handle an insertion or deletion is to rebuild the Lulea compressed trie, then a router must keep two copies of its routing database, one that is being built and one that is being used for lookups. This can potentially double the storage cost from 32 to 64 bits per prefix. This in turn can halve the number of prefixes that can be supported by a chip which places the entire database in on-chip SRAM.

To obtain fast insertions and hence avoid the need for two copies of the database, the first problem in Lulea that must be handled is the use of leaf-pushing. When a prefix of small length is inserted, leaf pushing can result in pushing down the prefix to a large number of leaves, making insertion slow.

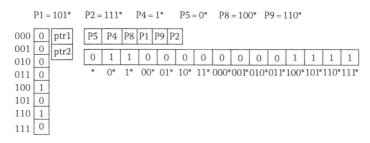

FIGURE 28.6 The Tree bitmap scheme allows the compression of Lulea without sacrificing fast insertions by using two bitmaps per node. The first bitmap describes valid versus null pointers, and the second bitmap describes internally stored prefixes.

The first and main idea in the Tree bitmap scheme is that there are two bitmaps per trie node, one for all the internally stored prefixes and one for the external pointers. Figure 28.6 shows the Tree bitmap version of the root node in Figure 28.5.

Recall that in Figure 28.5, the prefixes P8 = 100* and P9 = 110* in the original database are missing from the picture on the left side of Figure 28.5 because they have been pushed down to the leaves to accommodate the two pointers (*ptr1* that points to nodes containing longer prefixes such as P6 = 1000*, and *ptr2* that points to nodes containing longer prefixes such as P3 = 11001*). This results in the basic Lulea trie node in which each element contains either a pointer or a prefix, but not both. This allows the use of a single bitmap to compress a Lulea node, as shown on the extreme right of Figure 28.5.

By contrast, the same trie node in Figure 28.6 is split into two compressed arrays, each with its own bitmap. The first array, shown vertically, is a pointer array which contains a bitmap denoting the (two) positions where non-null pointers exist, and a compressed array containing the non-null pointers *ptr1* and *ptr2*.

The second array, shown horizontally, is the internal prefix array which contains a list of all the prefixes contained within the first 3 bits. The bitmap used for this array is very different from the Lulea encoding, and has 1 bit set for every possible prefix stored within this node. Possible prefixes are listed lexicographically from starting from ∗, followed by 0∗ and 1∗, and then on to the length two prefixes (00*, 01*, 10*, 11*), and finally the length three prefixes. Bits are set when the corresponding prefixes occur within the trie node.

Thus in Figure 28.6, the prefixes P8 and P9 which were leaf pushed in Figure 28.5 have been resurrected, and now correspond to bits 12 and 14 in the internal prefix bitmap. In general, for an r bit trie node, there are $2^{r+1} - 1$ possible prefixes of lengths $\leq r$, which requires the use of a $2^{r+1} - 1$ bitmap. The scheme gets its name because the internal prefix bitmap represents a trie in a linearized format: each row of the trie is captured top-down from left to right.

The second idea in the Tree bitmap scheme is to keep the trie nodes as small as possible to reduce the required memory access size for a given stride. Thus a trie node is of fixed size, and only contains a pointer bitmap, an internal prefix bitmap, and child pointers. But what about the forwarding information (i.e., the next router to which packets destined to a prefix should be sent) associated with stored prefixes?

The trick is to store the forwarding information associated with the internal prefixes stored within each trie node in a separate array associated with this trie node. Putting forwarding information in a separate result array potentially requires two memory accesses per trie node (one for the trie node, and one to fetch the result node for stored prefixes).

However, a simple lazy evaluation strategy is to not access the result nodes till search terminates. Upon termination, the algorithm makes a final access to the correct result node. This is the result node that corresponds to the last trie node encountered in the path that contained a valid prefix. This adds only a single memory reference at the end, in addition to the one memory reference required per trie node.

The third idea is to use only one memory access per node, unlike Lulea which uses at least two memory accesses. Lulea needs two memory accesses per node because it uses large strides of 8 or 16 bits. This increases the bitmap size so much that the only feasible way to count bits is to use an additional chunk array that must be accessed separately. The Tree bitmap scheme gets around this by simply using smaller stride nodes, say of 4 bits. This makes the bitmaps small enough that the entire node can be accessed by a single wide access (we are exploiting locality). Combinatorial logic can be used to count the bits.

Tree Bitmap Search Algorithm: The search algorithm starts with the root node and uses the first r bits of the destination address (corresponding to the stride of the root node, 3 in our example) to index into the pointer bitmap at the root node at position P. If there is a 1 in this position, there is a valid child pointer. The algorithm counts the number of 1s to the left of this 1 (including this 1) and denote this count by I. Since the pointer to the start position of the child pointer block (say y) is known and so is the size of each trie node (say S), the pointer to the child node can be calculated as $y + (I * S)$.

Before moving on to the child, the algorithm must also check the internal bitmap to see if there are one or more stored prefixes corresponding to the path through the multibit node to position P. For example, suppose P is 101 and a 3 bits stride is used at the root node bitmap as in Figure 28.6. The algorithm first checks to see whether there is a stored internal prefix 101*. Since 101* corresponds to the 13th bit position in the internal prefix bitmap, the algorithm can check if there is a 1 in that position (there is one in the example). If there was no 1 in this position, the algorithm would back up to check whether there is an internal prefix corresponding to 10*. Finally, the algorithm checks for the prefix 1*.

This search algorithm appears to require a number of iterations proportional to the logarithm of the internal bitmap length. However, for bitmaps of up to 512 bits or so in hardware, this is just a matter of simple combinatorial logic. Intuitively, such logic performs all iterations in parallel, and uses a priority encoder to return the longest matching stored prefix.

Once it knows there is a matching stored prefix within a trie node, the algorithm does not immediately retrieve the corresponding forwarding information from the result node associated with the trie node. Instead, the algorithm moves to the child node while remembering the stored prefix position and the corresponding parent trie node. The intent is to remember the last trie node T in the search path that contained a stored prefix, and the corresponding prefix position.

Search terminates when it encounters a trie node with a 0 set in the corresponding position of the extending bitmap. At this point, the algorithm makes a final access to the result array corresponding to T to read off the forwarding information. Further tricks to reduce memory access width are described in Eatherton' MS Thesis [Eat95], which describes a number of other useful ideas.

Intuitively, insertions in Tree bitmap are very similar to insertions in a simple multibit trie without leaf pushing. A prefix insertion may cause a trie node to be changed completely; a new copy of the node is created and linked in atomically to the existing trie. Compression results in [EDV04] show that Tree bitmap has all the features of the Lulea scheme in terms of compression and speed, along with fast insertions. Tree bitmap also has the ability to be tuned for hardware implementations ranging from the use of RAMBUS-like memories to on-chip SRAM.

28.2.2 Switching

Early routers used a simple bus to switch packets between input and output links. A bus is a wire that allows only one input to send to one output at a time. Today, however, almost every core router uses an internal crossbar switch that allows disjoint link pairs to communicate in parallel, to increase effective throughput. Each router must match input links (that wish to send packets) to output links (that can accept packets from the corresponding links) in the duration of reception of a single packet. This is 8 ns for a 40 bytes packet at 40 Gbps.

Now caches cannot be relied upon to finesse lookups because of the rarity of large trains of packets to the same destination. Similarly, it is unlikely that two consecutive packets at a switch input port are destined to the same output port. This makes it hard to amortize the switching overhead over multiple packets.

Thus to operate at wire-speed the switching system must decide which input and output links should be matched in a minimum packet arrival time. This makes the control portion of an Internet switch (that sets up connections) much harder to build than say a telephone switch. A second important difference between telephone switches and packet switches is the need for packet switches to support multicast connections. Multicast complicates the scheduling problem even further because some inputs require sending to multiple outputs.

To simplify the problem, most routers internally segment variable size packets into fixed size cells before sending to the switch fabric. Mathematically, the switching component of a router reduces to solving a bipartite matching problem: the router must match as many input links as possible (to as many output links as possible) in a fixed cell arrival time. While optimal algorithms for bipartite matching are well known to run in milliseconds, solving the same problem every 8 ns at 40 Gbps requires some systems thinking. For example, the solutions described in this section will trade accuracy for time, use hardware parallelism and randomization, and exploit the fact that typical switches have 32–64 ports to build fast priority queue operations using bitmaps.

Simpler Approaches: Before describing bus and crossbar-based switches, it is helpful to consider one of the simplest switch implementations based on shared memory. Packets are read into a memory from the input links and are read out of memory to the appropriate output links. Such designs have been used as part of Time Slot Interchange switches in telephony for years. They also work well for networking for small size switches.

The main problem is memory bandwidth. If the chip takes in eight input links and has eight output links, the chip must read and write each packet or cell once. Thus the memory has to run at 16 times the speed of each link. Up to a point, this can be solved by using a wide memory access width. The idea is that the bits come in serially on an input link and are accumulated into an input shift register. When a whole cell has been accumulated, the cell can be loaded into the cell-wide memory. Later they can be read out into the output shift register of the corresponding link and be shifted out onto the output link.

A second simpler approach used in early routers was to use a bus. Buses have speed limitations because of the number of different sources and destinations that a single shared bus has to handle. These sources and destinations add extra electrical loading that slows down signal rise times and ultimately the speed of sending bits on the bus. Other electrical effects include the effect of multiple connectors (from each line card) and reflections on the line.

Crossbar Switches: The classical way to get around the bottleneck of a single bus is to use a crossbar switch. A crossbar switch essentially has a set of $2N$ parallel buses, one bus per source line card, and one bus per destination line card. If one thinks of the source buses as being horizontal and the destination buses as being vertical, the matrix of buses forms what is called a crossbar.

Potentially, this provides an N-fold speedup over a single bus, because in the best case all N buses will be used in parallel at the same time to transfer data instead of a single bus. Of course, to get this speedup requires finding N disjoint source–destination pairs at each time slot. Trying to get close to this bound is the major scheduling problem studied in this section.

Head-of-line Blocking: The simplest way to schedule a crossbar is to use a single queue at each input link and try to match the packet at the head of every input queue to any available output. This can easily be done by a simple centralized scheduler using fairly simple logic or by more elegant distributed algorithms such as the take-a-ticket scheduler described in [SKO$^+$94].

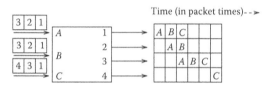

FIGURE 28.7 Example of head of line blocking caused by schemes like take-a-ticket. For each output port, a horizontal time scale is drawn labeled with the input port that sent a packet to that output port during the corresponding time period, and the blank mark if there is none. Note the large number of blanks showing potentially wasted opportunities that limit parallelism.

Any such scheduler will miss transmission opportunities as shown in Figure 28.7. The figure shows a switch with three input ports A, B, and C and four output ports $1, 2, 3$, and 4. There is a queue of packets at each input, and each input packet is labeled with the output port to which it is destined. Thus input port A has a packet for output port 1 at the head of its queue, followed by a packet destined to port 2, and lastly a packet destined to port 3.

At the right of the picture is a time–space picture showing transmissions. Assume fixed size packets or cells. Note that in the first time slot (first column) only one packet can be scheduled because the heads of all input queues are destined for output port 1. Because output port 1 can only accept 1 packet (say from input port A), the other input ports B and C are starved and cannot transmit. In the next time slot, output port 1 grants to say B and A can send its second packet to output port 2. However, input port C is starved. Continuing in this fashion (always being fair in round-robin fashion when multiple inputs contend for the same output) we take six cell times to send all packets.

Note that that there were nine potential transmission opportunities in three iterations because there are three input ports and three iterations. However, after the end of three cell times, there is one packet in B's queue and two in C's queue. Thus only six of potentially nine packets have been sent, thus taking limited advantage of parallelism.

It is easy to see visually from the right of Figure 28.7 that only roughly half of the transmission opportunities (more precisely, 9 out of 24) are used. Now, of course, no algorithm can do better for certain scenarios. However, other algorithms such as iSLIP [Mea97] (described next) can extract more parallel opportunities, and finish the same nine packets in four iterations instead of six.

In the first iteration of Figure 28.7, all inputs have packets waiting for output 1. Since only one (i.e., A) can send a packet to output 1 at a time, the entire queue at B (and C) is stuck waiting for A to complete. Since the entire queue is held hostage by the progress of the head of the queue or line, this is called head of line (HOL) blocking. iSLIP and PIM (both described below) get around this limitation by allowing packets behind a blocked packet to make progress (for example, the packet destined to output port 2 in the input queue at B can be sent to output port 2 in iteration 1 of Figure 28.7) at the cost of a more complex scheduling algorithm.

The loss of throughput caused by HOL blocking can be analytically captured using a simple traffic models. A classic analysis [KHM87] shows that the utilization with HOL is close to 58%. Given that we can lose roughly half the throughput of an expensive switch because of head of line blocking, it is worthwhile finding a way to avoid it and get 100% throughput as long as the extra hardware cost for the new mechanism is small.

Simulated Output Queuing: A natural response to HOL is to use output queuing in place of input queuing. Suppose that packets can somehow be sent to an output port without any queuing at the input. Then it is impossible for a packet P destined to a busy output port to block another packet behind it. This is because packet P is sent off to the queue at the output port where it can only block packets sent to the same output port.

The simplest way to do this would be to design the switch to be N times faster than the input links. Then even if all N inputs send to the same output in a given cell time, all N cells can be sent through the switch to be queued at the output. Thus pure output queuing requires an N-fold speedup within the fabric, which is expensive or infeasible.

A practical implementation of output queuing was provided by the Knockout [YHA87] switch design. The idea was to optimize for the common case where $k << N$ cells arrive at the same time at an input port so the the the fabric runs only k times as fast as an input link. This can be done using k parallel buses. The main technical challenge is to fairly handle the case where the expected case is violated and $M > k$ cells get sent to the same output at the same time.

A simple tournament tree (implemented using a tree of 2 by 2 switching elements that randomly chooses a winner) can fairly pick a winner. However, picking the other $k - 1$ choices from the same tree actually results in unfairness. Instead k tournament trees are used with the losers of all previous tournaments fed into subsequent tournaments. Speed is regained by pipelining so that later tournament trees do have to wait till previous tournaments have completed.

While the Knockout switch is important to understand because of the techniques it introduced, it is complex to implement and makes assumptions about traffic distributions. These assumptions are untrue for real topologies in which more than k clients frequently gang up to concurrently send to a popular server. More importantly, researchers devised relatively simple ways to combat HOL blocking without going to output queuing.

Avoiding HOL Blocking using Virtual Output Queues: The main idea is to retrofit input queuing to avoid head-of-line blocking. PIM does so by allowing an input port to schedule not just the head of its input queue, but also other cells that can make progress when the head is blocked. At first glance, this looks very hard. There could be a hundred thousand cells in each queue; attempting to maintain even 1 bit of scheduling state for each cell will take too much memory to store and process.

However, the first significant observation is that cells in each input port queue can only be destined to N possible output ports. If cell $P1$ is before cell $P2$ in input queue X, and both $P1$ and $P2$ are destined for the same output queue Y, then $P1$ must be sent before $P2$. We need not schedule any cells other than the first cell sent to every distinct output port. Thus we decompose the single input queue at each port into a separate input queue per output at each input port. These are called virtual output queues (VOQs).

The second significant observation is that communicating the scheduling needs of any input port takes only a bitmap of N bits, where N is the size of the switch. In the bitmap, a 1 in position i implies that there is at least one cell destined to output port i. Thus if each of N input ports communicates an N bits vector describing its scheduling needs, the scheduler only needs to process N^2 bits. For small $N \leq 32$, this is not a great deal of bits to communicate via control buses or to store in control memories.

The first algorithm to introduce VOQs was parallel iterative matching or PIM [AOST93]. PIM is a very simple matching algorithm. In the first iteration, every input sends requests to every output it has a packet destined to. If an output port gets multiple requests, it randomly picks one input port to send a grant to. If an input port gets multiple grants, it randomly chooses one grant to accept and a match is made between that input and that output in the crossbar. While the worst-case time to reach a maximal match for N inputs is N iterations, a simple argument shows that expected number of iterations is closer to $\log N$. The DEC AN-2 implementation [AOST93] used three iterations for a 30-port switch.

While PIM was a seminal scheme because it introduced the idea that HOL could be avoided at reasonable hardware cost using VOQs, it has two problems. First, it uses randomization and second (more importantly) it uses a logarithmic number of iterations.

iSLIP [Mea97] is a very popular and influential scheme that essentially "derandomizes" PIM and also achieves very close to maximal matches after just one or two iterations. The base idea is extremely simple. Instead of using random selection to pick fairly among multiple contenders, iSLIP uses a round robin pointer (per input and output port) to rotate the priority fairly among ports. While the round-robin pointers can be initially synchronized and cause something akin to head-of-line blocking, they tend to break free and result in maximal matches over the long run, at least as measured in simulation. Thus the subtlety in iSLIP is not the use of round-robin pointers, but the apparent lack of long-term synchronization among N such pointers running concurrently.

More precisely, each output (respectively input) maintains a pointer g initially set to the first input (respectively output) port. When an output has to choose between multiple input requests, it chooses the lowest input number $\geq g$. Similarly, when an input port has to choose between multiple output port requests, it chooses the lowest output port number that is $\geq a$, where a is the pointer of the input port. If an output port is matched to an input port X, then the output port pointer is incremented to the first port number greater than X in circular order (i.e., g becomes $X + 1$ unless X was the last port, in which case g wraps around to the first port number).

This simple device of a "rotating priority" allows each resource (output port, input port) to be shared reasonably fairly across all contenders at the cost of $2N$ extra pointers of size $\log_2 N$ bits each. This is, of course, in addition to the N^2 bits of scheduling state needed on every iteration.

Figure 28.8 shows the first few iterations of the same scenario as Figure 28.7 using a two-iteration iSLIP. Since each row is an iteration of a match, each match is shown using two rows. Thus the three rows of Figure 28.8 show the first two iterations of the first match (called a round in the figure) and the first iteration of the second match.

The upper left picture starts with each input port sending requests to each output port for which it has a cell destined. Each output port has a so-called grant pointer g, which is initialized for all outputs to be A. Similarly, each input has a so-called accept pointer called a, which is initialized for all inputs to 1.

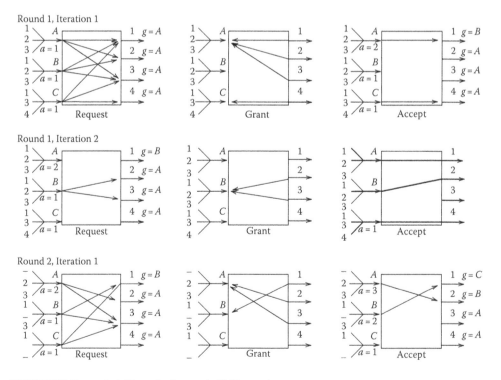

FIGURE 28.8 One and a half rounds of a sample iSLIP scenario.

When output 1 receives requests from all three input ports, it grants to A because A is the smallest input greater than or equal to $g_1 = A$. By contrast, in say PIM, port 1 randomly chooses say input port B. At this stage the determinism of iSLIP seems a real disadvantage because A has sent requests to output ports 3 and 4 as well. Because 3 and 4 also have grant pointers $g_3 = g_4 = A$, ports 3 and 4 grant to A as well, ignoring the claims of B and C. As before, since C is the lone requester for port 4, C gets the grant from 4.

When the popular A gets three grants back from ports 1, 2, and 3, A accepts 1. This is because Port 1 is the first output $\geq A$'s accept pointer a_A which was equal to 1. Similarly C chooses 4. Having done so, A increments a_A to 2 and C increments a_C to 1 (one greater than 4 in circular order is 1). Only at this stage does output 1 increment its grant pointer g_1 to B (one greater than the last successful grant) and port 4 similarly increments to A (one greater than C in circular order).

Note that although ports 2 and 3 gave grants to A, they do not increment their grant pointers because A spurned their grants. If they did, it would be possible to construct a scenario where output ports keep incrementing their grant pointer beyond some input port I after unsuccessful grants, thereby continually starving input port I. Note also that the match is only of size 2; thus this iSLIP scenario can be improved by a second iteration shown in the second row of Figure 28.8. Notice that at the end of the first iteration the matched inputs and outputs are not connected by solid lines (denoting data transfer). This data transfer will await the end of the final (in this case second) iteration.

The second iteration (middle row of Figure 28.8) starts with only inputs unmatched on previous iterations (i.e, B) requesting, and only to hitherto unmatched outputs. Thus B requests to 2 and 3 (and not to 1 though B has a cell destined to 1 as well). Both 2 and 3 grant B, and B chooses 2 (lowest greater than or equal to its accept pointer of 1). One might think that B should increment its accept pointer to 3 (1 plus the last accepted which was 2). However, to avoid starvation iSLIP does not increment pointers on iterations other than the first, for reasons that will be explained.

Thus even after B is connected to 2, 2's grant pointer remains at A and B's accept pointer remains at 1. Since this is the final iteration, all matched pairs, including pairs such as $A, 1$ matched in prior iterations, are all connected and data transfer (solid lines) occurs.

The third row provides some insight into how the initial synchronization of grant and accept pointers gets broken. Because only one output port has granted to A, that port (i.e., 1) gets to move on and this time provide priority to ports beyond A. Thus even if A had a second packet destined to 1 (which it does not in this example), 1 would still grant to B.

Note also that iSLIP may look worse than PIM because it required two iterations per match for iSLIP to achieve the same match sizes that PIM can do using one iteration per match. However, this is more illustrative of the startup penalty that iSLIP pays rather than a long-term penalty. In practice, as soon as the iSLIP pointers desynchronize, iSLIP does very well with just one iteration, and some commercial implementations use just one iteration: iSLIP is extremely popular. The Cisco GSR, one of the most popular routers in the market, uses iSLIP.

One might summarize iSLIP as PIM with the randomization replaced by round-robin scheduling of input and output pointers. However, this characterization misses out on two subtle aspects of iSLIP. These are

- *Grant pointers are incremented only in the third phase after a grant is accepted*: Intuitively, if O grants to an input port I, there is no guarantee that I will accept. Thus if O were to increment its grant pointer beyond O, it can cause traffic from I to O to be persistently starved. Even worse, [Mea97] shows that this simplistic round-robin scheme reduces the throughput to just 63% (for Bernoulli arrivals) because the pointers tend to synchronize and move in lock-step.

- *All pointers are incremented only after the first iteration accept is granted*: Once again, this rule prevents starvation but the scenario is more subtle.

iSLIP can also be extended to handle priorities and multicast traffic.

28.2.3 Packet Classification

Routers have evolved from traditional destination-based forwarding devices to what are called packet classification routers. In modern routers, the route and resources allocated to a packet are determined by the destination address as well as other header fields of the packet such as the source address and TCP/UDP port numbers. Packet classification unifies the forwarding functions required by firewalls, resource reservations, QoS routing, unicast routing, and multicast routing. In classification, the forwarding database of a router consists of a potentially large number of rules on key header fields. A given packet header can match multiple rules. So each rule is given a cost, and the packet is forwarded using the least cost matching rule.

Several variants of packet classification have already established themselves in the Internet. First, many routers implement firewalls [CB95] at trust boundaries, such as the entry and exit points of a corporate network. A firewall database consists of a series of packet rules that implement security policies. A typical policy may be to allow remote login from within the corporation, but to disallow it from outside the corporation.

Second, the need for predictable and guaranteed service has led to proposals for reservation protocols like DiffServ [SWG] that reserve bandwidth between a source and a destination. Third, the cries for routing based on traffic type have become more strident recently—for instance, the need to route Web traffic between Site 1 and Site 2 on say Route A and other traffic on say Route B.

Formalization of Packet Classification: Assume that the information relevant to a lookup is contained in K distinct header fields in each message. These header fields are denoted $H[1], H[2], \ldots,$ $H[K]$, where each field is a string of bits. For instance, the relevant fields for an IPv4 packet could be the Destination Address (32 bits), the Source Address (32 bits), the Protocol Field (8 bits), the Destination Port (16 bits), the Source Port (16 bits), and TCP flags (8 bits). Thus, the combination $(D, S, \text{TCP-ACK}, 63, 125)$, denotes the header of an IP packet with destination D, source S, protocol TCP, destination port 63, source port 125, and the ACK bit set.

A classifier or rule database router consists of a finite set of rules, $R_1, R_2, \ldots, R_N$. Each rule is a combination of K values, one for each header field. Each field in a rule is allowed three kinds of matches: exact match, prefix match, or range match. In an exact match, the header field of the packet should exactly match the rule field—for instance, this is useful for protocol and flag fields. In a prefix match, the rule field should be a prefix of the header field—this could be useful for blocking access from a certain subnetwork. In a range match, the header values should lie in the range specified by the rule—this can be useful for specifying port number ranges.

Each rule R_i has an associated directive $disp_i$, which specifies how to forward the packet matching this rule. The directive specifies if the packet should be blocked. If the packet is to be forwarded, the directive specifies the outgoing link to which the packet is sent, and perhaps also a queue within that link if the message belongs to a flow with bandwidth guarantees.

A packet P is said to match a rule R if each field of P matches the corresponding field of R— the match type is implicit in the specification of the field. For instance, if the destination field is specified as 1010∗, then it requires a prefix match; if the protocol field is UDP, then it requires an exact match; if the port field is a range, such as 1024–1100, then it requires a range match. For instance, let $R = (1010∗, ∗, TCP, 1024–1080, ∗)$ be a rule, with $disp = block$. Then, a packet with header $(10101\ldots111, 11110\ldots000, TCP, 1050, 3)$ matches R and is therefore blocked. The packet $(10110\ldots000, 11110\ldots000, TCP, 80, 3)$, on the other hand, does not match R.

Since a packet may match multiple rules in the database, each rule R in the database is associated with a non-negative number, $cost(R)$. Ambiguity is avoided by returning the least cost rule matching the packet's header. The cost function generalizes the implicit precedence rules that are used in practice to choose between multiple matching rules.

As an example of a rule database, consider the topology and firewall database [CB95] shown in Figure 28.9, where a screened subnet configuration interposes between a company subnetwork

Destination	Source	Destination Port	Source Port	Protocol and Flags	Disp	Comments
M	*	25	*	*	Allow	Inbound e-mail
M	*	53	*	UDP	Allow	DNS Access
M	S	53	*	*	Allow	Secondary DNS
M	*	23	*	*	Allow	Incoming SSH
TI	TO	123	123	UDP	Allow	NTP
*	Net	*	*	*	Allow	Outgoing packets
Net	*	*	*	TCP ack	Allow	Return packs OK
*	*	*	*	*	Block	Drop all else

FIGURE 28.9 The top half of the figure shows the topology of a small company, and the bottom half shows a sample firewall database for this company as described in the book by Cheswick and Bellovin [CB95]. The *block* flags are not shown in the figure; the first 7 rules have *block = false* (i.e., allow) and the last rule has *block = true* (i.e., block). We assume that all the addresses within the company subnetwork (shown on top left) start with the prefix *Net* including *M* and *TI*.

(shown on top left) and the rest of the Internet (including hackers). There is a so-called bastion host *M* within the company that mediates all access to and from the external world. *M* serves as the mail gateway and also provides external name server access. *TI*, *TO* are Network Time Protocol (NTP) sources, where *TI* is internal to the company and *TO* is external. *S* is the address of the secondary name server which is external to the company.

Clearly, the site manager wishes to allow communication from within the network to *TO* and *S*, and yet wishes to block hackers. The database of rules shown on the bottom of Figure 28.9 implements this intention. Assume that all addresses of machines within the company's network start with the prefix *Net*. Thus *M* and *TI* both match the prefix *Net*. All packets matching any of the first seven rules are allowed; the remaining (last rule) are dropped by the screening router. As an example, consider a packet sent to *M* from *S* with UDP destination port equal to 53. This packet matches Rules 2, 3, and 8, but must be allowed through because the first matching rule is Rule 2.

Note that the description above uses *N* for the number of rules and *K* for the number of packet fields. *K* is sometimes called the number of dimensions for reasons that will become clearer.

Before we describe algorithmic solutions to the problem of packet classification, we first describe a conceptually simple but hardware-intensive solution called ternary CAMs.

CAMs: A ternary content addressable memory (TCAM) is a content addressable memory where the first cell that matches a data item will be returned using a parallel lookup in hardware. A ternary CAM allows each stored bit of data to be either a 0, a 1, or a wildcard. Thus clearly ternary CAMs can be used for rule matching as well as for prefix matching. However, the CAMs must provide wide lengths—for example, the combination of the IPv4 destination, source, and two port fields is 96 bits.

However, CAMs have smaller density and consume larger power than algorithmic solutions using SRAMs. For packet classification, CAMs also have the problem of rule multiplication caused by ranges. In CAM solutions, each range has to be replaced by a potentially large number of prefixes, thus causing extra entries. Some algorithmic solutions can handle ranges in rules without converting ranges to rules. While better CAM cell designs that reduce density and power requirements may emerge, it is still important to understand the corresponding advantages and disadvantages of algorithmic solutions. The remainder of this section is devoted to this topic.

Geometric View of Classification: A geometric view of classification was introduced by Lakshman and Stidialis [LS98]. Observe that we can view a 32 bits prefix like 00* as a range of addresses from 000...00 to 001...11 on the number line from 0 to $2^{32} - 1$. If prefixes correspond to line segments geometrically, two-dimensional rules correspond to rectangles, three-dimensional rules to cubes, and so on. A given packet header is a point. The problem of packet classification reduces to finding the lowest cost box that contains the given point.

The first advantage of the geometric view is that it enables the application of algorithms from computational geometry. For example, Lakshman and Stiliadis [LS98] adapt a technique from computational geometry known as fractional cascading to do binary search for two field rules matching in $O(\log N)$ time, where N is the number of rules. In other words, two-dimensional rule matching is asymptotically as fast as one-dimensional rule matching using binary search. Unfortunately, the constants for fractional cascading are quite high. Perhaps this suggests that adapting existing geometric algorithms may actually not result in the most efficient algorithms.

A second advantage of the geometric viewpoint is that it is suggestive and useful. For example, a very useful concept that the geometric view provides is the number of disjoint (i.e., nonintersecting) classification regions. Since rules can overlap, this is not the number of rules. For example in two dimensions, with N rules one can create N^2 classification regions by having $N/2$ rules that correspond geometrically to horizontal strips together with $N/2$ rules that correspond geometrically to vertical strips. The intersection of the $N/2$ horizontal strips with the $N/2$ vertical strips creates $O(N^2)$ disjoint classification regions. Similar constructions can be used to generate $O(N^K)$ regions for K-dimensional rules.

Using the geometric viewpoint just described, it is easy to adapt a lower bound from computational geometry. Thus, it is known that general multidimensional range searching over N ranges in k dimensions requires $\Omega((\log N)^{K-1})$ worst-case time if the memory is limited to about linear size [Cha9a,Cha9b] or requires $O(N^K)$ size memory. While $\log N$ could be reasonable (say 10 memory accesses), $\log^4 N$ will be very large (say 10,000 memory accesses). Notice that this lower bound is consistent with solutions for the two-dimensional case that take linear storage but are as fast as $O(\log N)$.

The lower bound implies that for perfectly general rule sets, algorithmic approaches to classification either require a large amount of memory or a large amount of time. Unfortunately, classification at high speeds, especially for core routers, requires the use of limited and expensive SRAM. Thus the lower bound seems to imply that CAM are required for reasonably sized classifiers (say 10,000 rules) that must be searched at high speeds (e.g., 40 Gbps).

Bit Vector Schemes: A surprisingly simple practical scheme is one due tp Lakshman and Staliadis [LS98]. The database of Figure 28.9 is first sliced into columns, one for each field i. All possible values of field i are stored in a Table corresponding to field i. Along with each possible value M of field i, the algorithm stores the set of rules $S(M)$ that match M in field i as a bit vector. When a packet arrives, each of its header fields is examined in Table i to find the longest match. Thus if field i matches $M(i)$, we know the set $S(M(i))$ of rules that match the packet in field i. By intersecting the sets $S(M(i))$ for all i, we get the rules that match in all the fields.

For example, Figure 28.10 shows the sliced database together with bit vectors for each sliced field value. The bit vector has 8 bits, one corresponding to each of the eight possible rules in Figure 28.9. Bit j is set for value M in field i if value M matches Rule j in field i.

Despite the elegance, note that since there are N rules, the bitmaps are N bits long. Hence, computing the AND requires $O(N)$ operations. So the algorithm is effectively doing linear search. The only reason it is practical for databases of thousands of rules is that we can use wide word memory parallelism to process a large bitmap of 1000 bits in one cycle.

Decision Tree Approaches: The idea is to build a decision tree where each node in the tree tests bits on some field. Depending on the values of the test, a new branch is taken to another decision tree node because certain sets of rules have been eliminated by each test. We now describe a Decision Tree approach called HiCuts [GM99a].

Figure 28.11 shows a fragment of a HiCuts decision tree on the database of Figure 28.9. The nodes contain range comparisons on values of any specified fields, and the edges are labeled with *True* or *False*. Thus the root node tests whether the destination port field is less than 50. The fragment only follows the case when this test is false. Looking at Figure 28.9 notice that this branch eliminates $R1$ (i.e., rule 1) and $R4$, because these rules contain port numbers 25 and 23, respectively.

Destination Prefixes		Source Prefixes		DstPort Prefixes		SrcPort Prefixes		Flags Prefixes	
M	11110111	S	11110011	25	10000111	123	11111111	UDP	11111101
T1	00001111	TO	11011011	53	01100111	*	11110111	TCP	10110111
Net	00000111	Net	11010111	23	00010111			*	10110101
*	00000101	*	11010011	123	00001111				
				*	00000111				

FIGURE 28.10 The database of Figure 28.9 together with bit vectors for every possible sliced value. The bit vector has 8 bits, one corresponding to each of the eight possible rules in Figure 28.9. Bit j is set for value M in field i if value M matches Rule j in field i. The last slice (flags prefixes) corresponds to the protocol and flags column in Figure 28.9.

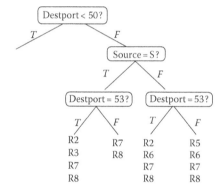

FIGURE 28.11 The HiCuts data structure is essentially a range tree which has pointers corresponding to some ranges of some dimension variable with linear search at end.

The next test checks whether the source address is equal to that of the secondary name server S in Figure 28.9. If this test evaluates to true, then $R5$ is eliminated (because it contains TO) and so is $R6$ (because it contains Net and S does not belong to the internal prefix Net). This leads to a second test on the destination port field. If the value is not 53, the only possible rules that can match are $R7$ and $R8$.

Thus on a packet header in which the destination port is 123 and the source is S, the search algorithm takes the right branch at the root, the left branch at the next node, and a right branch at the final node. At this point the packet header is compared to rules $R7$ and $R8$ using linear search.

The algorithm needs an efficient decision tree building algorithm that minimizes the worst-case height, and yet has reasonable storage. Rather than consider the general optimization problem, which is NP-complete, HiCuts [GM99a] uses the following more restricted heuristic based on the repeated use of the following greedy strategy.

Pick a field: The HiCuts paper suggests first picking a field to cut on at each stage based on the number of distinct field values in that field. For example, in Figure 28.11, this heuristic would pick the destination port field.

Pick the number of cuts: For each field, rather than just pick one range check as in Figure 28.11, one can pick k ranges or cuts. Of course, these can be implemented as separate range checks as in Figure 28.11. To choose k, the algorithm suggested in [GM99b] is to keep doubling k and to stop when the storage caused by the k cuts exceeds a prespecified threshold.

Several details are needed to actually implement this somewhat general framework. Assuming the cuts or ranges are equally spaced, the storage cost of k cuts on a field is estimated by counting the sum of the rules assigned to each of the k cuts. Clearly, cuts that cause rule replication will have a large storage estimate. The threshold that defines acceptable storage is a constant (called spfac for space factor) times the number of rules at the node. The intent is to keep the storage linear in the number of rules up to a tunable constant factor.

Finally, the process stops when all decision tree leaves have no more than binth (bin threshold) rules. binth controls the amount of linear searching at the end of tree search. The current state of the art appears to be a generalization of HiCuts called HyperCuts [BSV03], which takes the decision tree approach a step further by allowing the use of several cuts in a single step using multidimensional array indexing. Because each cut is now a general hypercube, the scheme is called HyperCuts. HyperCuts appears to work significantly faster than HiCuts on many real databases [BSV03].

28.2.4 Output Scheduling

Returning to our model of a router (Figure 28.1), recall that once the packet arrives at the output port, the packet could be placed in a FIFO queue. If congestion occurs and the output link buffers fill up, packets arriving at the tail of the queue are dropped. Many routers use such a default output link scheduling mechanism that is often referred to as FIFO with tail-drop.

However, there are other options. First, we could place packets in multiple queues based on packet headers and schedule these output queues according to some scheduling policy. Second, even if we had a single queue we need not always drop from the tail when buffers overflow; we can even, surprisingly, drop a packet when the packet buffer is not full. These options are useful to provide router support for congestion and to providing QoS guarantees to flows. Minimally, such QoS guarantees involve sharing bandwidth and ideally requires guaranteeing delays to say video and audio flows. We now describe some of these scheduling options for routers.

Random Early Detection (RED): RED is a packet scheduling algorithm implemented in most modern routers even at the highest speeds, and has become a de facto standard. In a nutshell, a RED router monitors the average output queue length; when it goes beyond a threshold, it randomly drops arriving packets with a certain probability even though there may be space to buffer the packet. The dropped packet acts as a signal to the source to slow down early, preventing a large number of dropped packets later.

The implementation of RED is more complex than it seems. For example, we need to calculate the output queue size using a weighted average with weight w. The weight w is used to bias the average toward older samples. The weighted average calculation can be made efficient by only allowing the w to be a reciprocal of a power of two so that multiplications reduce to easy bit shifting. There is a certain amount of interpolation to be done as well, which can also be done easily using shifts if the parameters are powers of 2.

Token Bucket Policing: So far with RED we assumed that all packets are placed in a single output queue; the RED drop decision is taken at the input of this queue. In addition, many ISPs require limiting the rate of traffic for a flow. More specifically, an ISP may want to limit NEWS traffic in its network to no more than 1 Mbps. Such examples of bandwidth limiting can be easily accomplished by a technique called token bucket policing which uses only a single queue per flow and a single counter per flow.

Conceptually there is a bucket per flow that fills with "tokens" at the specified average rate of R per sec The bucket has size limited to the specified burst size of B tokens. Thus when the bucket is full all incoming tokens are dropped. When packets arrive for a flow, they are allowed out only if the bucket contains a number of tokens equal to the size of packet in bits. If not, the packet is queued until sufficient tokens arrive. Since there can be at most B tokens, a burst is limited to at most B bits followed by the more steady rate of R bits per second. This can be easily implemented using a counter and a timer per flow.

Unfortunately, token bucket shaping requires a separate queue for each flow, which is expensive. If one wishes to limit oneself to a single queue, a simpler technique is to use a token bucket policer. The idea would be to simply drop any packet that arrives to find the token bucket empty. In other words, a policer is a shaper without the buffer. A policer only needs a counter and a timer per flow, which is simple to implement at high speeds using the efficient timer implementations.

Providing Bandwidth Guarantees: Another requirement for customers is to provide predictable bandwidth guarantees to some streams, say voice over IP (VoIP) traffic. An ideal scheme called bit-by-bit round robin [DKS89] sends one bit out from each queue in round-robin order. Since bits cannot be split within packets, simulated bit-by-bit round robin [DKS89] suggests doing a discrete time simulation of the perfectly fair bit-by-bit system, and then sending packets out in the same order they would have in the bit-by-bit system. It was also discovered that this algorithm could also provide good delay bounds for say video traffic. Unfortunately, the sorting required is hard to implement at gigabit speeds.

While bit-by-bit round robin provides both bandwidth guarantees and delay bounds, many applications can benefit from just bandwidth guarantees. Now, the simplest scheduler to implement is packet-by-packet round robin where the scheduler serves one packet from each active queue in each round. Unfortunately, packet-by-packet round robin (unlike the more complex bit-by-bit round robin) is unfair. If Queue 1 only has large packets and Queue 2 only has small packets, then Queue 1 will get more bandwidth in terms of bits per second than Queue 2.

By relaxing the specification to focus only on bandwidth guarantees, it is possible to fix the unfairness of packet-by-packet round robin without simulating bit-by-bit round robin. Each queue is given a quantum which is like a periodic salary that gets credited to the queue's bank account on every round-robin cycle. A queue cannot spend (i.e., send packets of the corresponding size) more than is contained in its account. However, the balance remains in the account for possible spending in the next round-robin cycle.

More precisely, for each queue i, the Deficit Round Robin (DRR) algorithm keeps a quantum size Q_i and a deficit counter D_i. The larger the quantum size assigned to a queue, the larger the share of the bandwidth it receives. On each round-robin scan, the algorithm will service as many packets as possible for queue i with size $< Q_i + D_i$. If packets remain in queue i's queue, the algorithm stores the "deficit" or remainder in D_i for the next opportunity. It is easy to prove that the DRR algorithm is fair in the long term for any combination of packet sizes, and that it takes only a few more instructions to implement than packet-by-packet round robin.

Delay Guarantees: Bandwidth guarantees are good for throughput but may not suffice for latency-critical traffic like voice or video. The problem is that if an occasional voice packet arrives it needs to be scheduled before a round-robin scheduler finally moves its pointer to this queue. First, let us consider what an ideal delay bound should be in Figure 28.12. The left figure shows three video flows that traverse a common output link; the flows have reserved five, eight, and two bandwidth units, respectively, of a 15-unit output link. The right figure shows the ideal "view" of Video 2 if it had its own dedicated router with an output link of eight units. Thus the ideal delay bound is the delay that a flow would have received in isolation assuming an output link bandwidth equal to its own reservation.

In theory, the simulated bit-by-bit round-robin algorithm [DKS89] guarantees isolation and delay bounds. Thus it was used as the basis for the IntServ proposal as a scheduler that could integrate video, voice, and data. However, bit-by-bit round robin is expensive to implement.

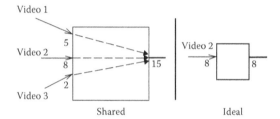

FIGURE 28.12 Defining what an ideal delay bound should be.

Strict bit-by-bit round robin takes $O(n)$ time per packet, where n is the number of concurrent flows. Faster approximations that take $O(log(n))$ time were introduced by Stiliadis and Verma [SV96] and Bennett and Zhang [BZ96]. We will only provide some intuition behind another variant of these algorithms called Virtual Clock [Zha91].

Suppose the rate of flow F is r. What is the departure time of a packet p of flow F arriving at a router dedicated to F that always transmits at r bits per second? Well, if p arrives before the previous packet from flow F (say *prev*) is transmitted, then p has to wait for *prev* to depart; otherwise p gets transmitted right away. Thus in an ideal system, packet p will depart by: Maximum (Arrival Time(p), Departure Time (*prev*)) + Length(p)/r. This is a recursive equation and can easily be solved if we know the arrival times of all packets in flow F up to and including packet p. We now use p's departure time in the ideal system as a deadline in the real system.

The packet deadlines translate into a scheduling algorithm using a classic real-time scheduler called Earliest Deadline First: the scheduler sorts the deadlines of all packets at the head of each flow queue. The scheduler then sends the packet with the earliest deadline first. This earliest deadline packet scheduler is called Virtual Clock [Zha91]. Finally, note that sorting can be done (at some expense) using hardware heaps [BL00]. In practice, however, most routers implement DRR but do not implement algorithms that guarantee delay bounds.

28.3   Endnode Algorithmics

From router algorithms, we now turn to endnode algorithmics. Our purpose is two fold. First, having learned about some of the important techniques for building a high-performance router, it is worth understanding some of the basic bottlenecks and techniques for building a high performance server, say a Web server. Second, the principles used in endnode algorithms are often different from router algorithmics. There is less of a focus on hardware or even clever algorithms. Instead, there is more of a focus on systems issues and on combating structural overheads caused by existing operating systems and architectures.

We now proceed to examine two major endnode bottlenecks: copying and control overhead.

28.3.1   Avoiding Redundant Copying

Figure 28.13 shows the sequence of data transfers involved in reading file data from the disk to the sending of the corresponding segments via the network adaptor. The file is read from disk into the application buffer via say a read() system call. The combination of the HTTP response and the application buffer is then sent to the network over the TCP connection to the client by say a write() system call. The TCP code in the network subsystem of the kernel breaks up the response data into network-sized segments, and transmits them to the network adaptor after adding a TCP checksum to each segment.

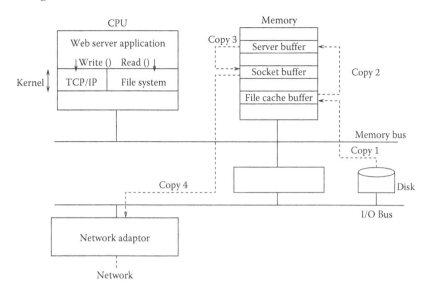

FIGURE 28.13 Redundant copies involved in handling a GET request at a server.

The details are more complex. First, the file is read into a piece of kernel memory called the file cache (Copy 1). This is useful to serve subsequent requests to a popular file. The file is then copied from the file cache into the Web server application buffer (Copy 2, see Figure 28.13).

The Web server then does a write() to the corresponding socket. Since the application can freely reuse its buffer (or even deallocate it) at any time after the write(), the network subsystem in the kernel cannot simply transmit out of the application buffer. Thus UNIX (and many other operating systems) provide what is known as copy semantics. The application buffer specified in the write() call is copied to a socket buffer (another buffer within the kernel, at a different address in memory than either the file cache or the application buffer). This is called Copy 3 in Figure 28.13. Finally, each segment is sent out to the network (after IP and link headers have been pasted) by copying the data from the socket buffer to memory within the network adaptor. This is called Copy 4.

In between, the TCP software in the kernel must make a pass over the data to compute the TCP checksum. The TCP checksum essentially computes the sum of 16 bits words in each TCP segment's data.

Each of the four copies and the checksum consume resources. All four copies and the checksum calculation consume bandwidth on the memory bus. The copies between memory locations (Copies 2 and 3) are actually worse than the others because they require one READ and one WRITE across the bus for every word of memory transferred. The TCP checksum requires only one READ for every word, and a single WRITE to append the final checksum. Finally, Copies 1 and 4 can be as expensive as Copies 2 and 3 if the CPU orchestrates the copy (so-called programmed I/O); however, if the devices themselves copy (so-called direct memory access or DMA), the cost is only a single READ or WRITE per word across the bus.

The copies also consume I/O bus bandwidth and ultimately memory bandwidth itself. A memory that supplies a word of size W bits every x nanoseconds has a fundamental limit on throughput of W/x bits per nanosecond. For example, even assuming DMA, these copies ensure that the memory bus is used 7 times for each word in the file sent out by the server. Thus the Web server throughput cannot exceed $T/7$ where T is the smaller of the speed of the memory and the memory bus.

Secondly, and more basically, the extra copies consume memory. The same file (Figure 28.13) could be stored in the file cache, the application buffer, and the socket buffer. While memory seems to be cheap and plentiful (especially when buying a PC!), it does have some limits, and Web servers would like to use as much as possible for the file cache to avoid slow disk I/O. Thus triply replicating a file reduces the file cache by a factor of three, which in turn reduces the cache hit rate and server performance.

In summary, redundant copies hurt performance in two fundamental and orthogonal ways. First, by using more bus and memory bandwidth than strictly necessary, the Web server runs slower than bus speeds even when serving documents that are in memory. Second, by using more memory than it should the Web server will have to read an unduly large fraction of files from disk instead of from the file cache.

Avoiding Copies using Shared memory: Modern UNIX variants [Ste98] provide a convenient system call known as mmap() to allow an application like a server to map a file into its virtual memory address space. Other operating systems provide equivalent functions. Conceptually, when a file is mapped into an application's address space, it is as if the application has cached a copy of the file in its memory. This seems very redundant because the file system also maintains cached files. However, using the magic of virtual memory the cached file is really only a set of mappings so that other applications and the file server cache can gain common access to one set of physical pages for the file.

Avoiding File System Copies by I/O Splicing: While some have tried to move applications within the kernel to avoid copies, a much more maintainable idea keeps the server application in user space. This is done using a simple idea called I/O splicing to eliminate all the copying in Figure 28.13. I/O splicing is shown in Figure 28.14 and was first introduced in [FP93]. The idea is to introduce a new system call that *combines* the old call to read a file with the old call to send a message to the network. By allowing the kernel to splice together these two hitherto separate system calls, all redundant copies can be avoided. Many systems have system calls such as SendFile() which are now used by several commercial vendors.

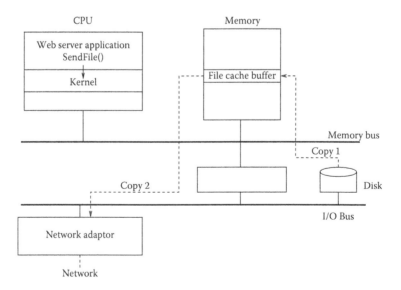

FIGURE 28.14 In I/O splicing, all the indirection caused by copying to and from user space buffers is removed by a single system call that "splices" together the I/O stream from the disk with the I/O stream to the network. As always, Copy 1 can be removed for files in the cache.

28.3.2 Reducing Control Overhead

Figure 28.13 reviewed the copying overhead involved in a Web server by showing the potential copies involved in responding to a GET (a request to get data) at a server. By contrast, Figure 28.15 shows the potential control overhead involved in a large Web server that handles many clients.

For the purposes of understanding the possible control overhead involved in serving a GET request, the relevant aspects are slightly different from that in the earlier section on copy overhead. First, assume the client has sent a TCP SYN request to the server that arrives at the adaptor from which it is placed in memory. The kernel is then informed of this arrival via an interrupt. The kernel notifies the Web server via the unblocking of an earlier system call; the Web server application will accept this connection if it has sufficient resources.

In the second step of processing, some server process parses the Web request. For example, assume the request is GET File 1. In the third step, the server needs to locate where the file is on disk, for example by navigating directory structures that may also be stored on disk. Once the file is located, in the fourth step, the server process initiates a read to the file system (another system call). If the file is in the file cache, the read request can be satisfied quickly; failing a cache hit, the file subsystem initiates a disk seek to read the data from disk. Finally, after the file is in an application buffer, the server sends out the HTTP response by writing to the corresponding connection (another system call).

So far the only control overhead appears to be that of system calls and interrupts. However, that is because we have not examined closely the structure of the networking and application code.

First, if the networking code is structured naively, with a single process per layer in the stack, then the process scheduling overhead (in the order of 100s of microseconds) of processing a packet can easily be much larger than a single packet arrival time. This potential scheduling overhead is shown in Figure 28.15 with a dashed line to the TCP/IP code in the kernel. Fortunately, most networking code is structured more monolithically with minimal control overhead, although there are some clever techniques that can do even better.

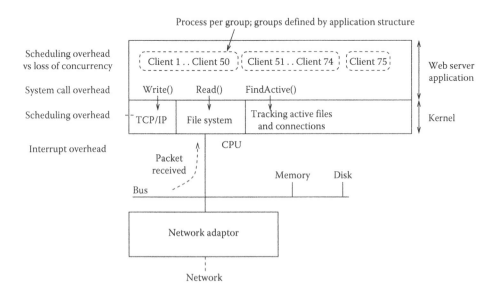

FIGURE 28.15 Control overhead involved in handling a GET request at a server.

Second, our description of Web processing has focused on a single client. Since we are assuming a large Web server that is working concurrently on behalf of thousands of clients, it is unclear how the Web server should be structured. At one extreme, if each client is a separate process (or thread) running the Web server code, concurrency is maximized (because when client 1 is waiting for a disk read, client 2 could be sending out network packets) at the cost of high process scheduling overhead.

On the other hand, if all clients are handled by a single event-driven process, then context switching overhead is minimized, but the single process must internally schedule the clients to maximize concurrency. In particular, it must know when file reads have completed and when network data has arrived.

Many operating systems provide a system call for this purpose that we have generically called FindActive(). For example, in UNIX the specific name for this generic routine is the select() system call. While even an empty system call is expensive because of the kernel-to-application boundary crossing, an inefficient select() implementation can be even more expensive.

Thus there are challenging questions as to how to structure both the networking and server code in order to minimize scheduling overhead and maximize concurrency. For this reason, Figure 28.15 shows the clients partitioned into groups, each of which is implemented in a single process or thread. Note that placing all clients in a single group yields the event-driven approach, while placing each client in a separate group yields the process (or thread) per client approach.

Thus an unoptimized implementation can incur considerable process switching overhead (100s of microseconds) if the application and networking code is poorly structured. Even if process structuring overhead is removed, system calls can cost 10s of microseconds, and interrupts can cost microseconds. To put these numbers in perspective, observe that on a 10-Gigabit Ethernet link, a 40-byte packet can arrive at a PC every 3.2 μs.

We now begin attacking the bottlenecks described in Figure 28.15.

Event Driven Schedulers: In terms of programming, the simplest way to implement a Web server is to structure the processing of each client as a separate process. Thus the web server application does not need to perform scheduling between clients; the operating system does this automatically on the application's behalf. However, process context switching and restoring is expensive. To avoid such process scheduling costs, the application must implement its own internal scheduler that juggles the state of each client.

For example, the application may have to implement a state machine that remembers that Client 1 is in Stage 2 (HTTP processing), while Client 2 is in Stage 3 (waiting for disk I/O), and Client 3 is Stage 4 (waiting for a socket buffer to clear up to send the next part of the response).

However, the kernel has an advantage over an application program because the kernel sees all I/O completion events. For example, if Client 1 is blocked waiting for I/O, in a per thread implementation, when the disk controller interrupts the CPU to say that the data is now in memory, the kernel can now attempt to schedule the Client 1 thread.

Thus if the Web server application is to do its own scheduling between clients, the kernel must pass information across the API to allow a single threaded application to view the completion of all I/O that it has initiated. Many operating systems provide such a facility that we generically called FindActive() in Figure 28.15. For example, Windows NT 3.5 has a I/O Completion Port (IOCP) mechanism, and UNIX provides the select() system call.

The main idea is that the application stays in a loop invoking the FindActive() call. Assuming there is always some work to do on behalf of some client, the call will return with a list of I/O descriptors (e.g., file 1 data is now in memory, connection 5 has received data) with pending work. When the Web server processes these active descriptors, it loops back to making another FindActive() call.

If there is always some client that needs attention (typically true for a busy server), there is no need to sleep and invoke the costs of context switching (e.g., scheduler overhead, Translation Lookaside Buffer misses) when scheduling between clients. Application-specific internal scheduling is more

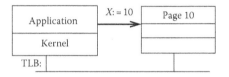

FIGURE 28.16 Reading and writing to memory is not mediated by the kernel.

efficient than invoking the general purpose, external CPU scheduler, because the application knows the minimum set of context that must be saved when moving from client to client.

Avoiding System Calls: We now turn our attention to one more bottleneck in Figure 28.15. When an application wants to send data, it must somehow tell the adaptor where the data is. Similarly, when the application wants to receive data, it must specify buffers where the received packet data should be written to. Today, in UNIX this is typically done using system calls where the application tells the kernel about data it wishes to send, and buffers it wishes to receive to. Even if we implement the protocol in user-space the kernel must service these system calls for every packet sent and received.

This appears to be required because there can be several applications sending and receiving data from a common adaptor; since the adaptor is a shared resource, it seems unthinkable for an application to write directly to the device registers of a network adaptor without kernel mediation to check for malicious or erroneous use. Or is it?

A simple analogy suggests that alternatives may be possible. In Figure 28.16 we see that when an application wants to set the value of a variable X equal to 10, it does not actually make a call to the kernel. If this were the case, every read and write in a program would be slowed down very badly. Instead, the hardware determines the virtual page of X, translates it to a physical page (say 10) via a cache called the Translation Lookaside Buffer (TLB), and then allows direct access as long as the application has Page 10 mapped into its virtual memory.

If Page 10 is not mapped into the applications virtual memory, the hardware generates an exception and causes the kernel to intervene to determine why there is a page access violation. Notice that the kernel was involved in setting up the virtual memory for the application (only the kernel should be allowed to do so for reasons of security), and may be involved if the application violates its page accesses that the kernel set up. However, the kernel is not involved in every access. Could we hope for a similar approach for application access to adaptor memory to avoid wasted system calls?

To see if this is possible we need to examine more carefully what information an application sends and receives from an adaptor. Clearly, we must prevent incorrect or malicious applications from damaging other applications or the kernel itself. Figure 28.17 shows an application that wishes to receive data directly from the adaptor. Typically, an application that does so must queue a descriptor. A descriptor is a small piece of information that describes the buffer in main memory where the data

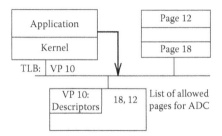

FIGURE 28.17 Application device channels.

for the next packet (for this application) should be written to. Thus we should consider carefully and separately both descriptor memory as well as the actual buffer memory.

We can deal with descriptor memory quite easily by recalling that the adaptor memory is memory mapped. Suppose the adaptor has 10,000 bytes of memory which is considered memory on the bus, and the physical page size of the system is 1000 bytes. This means that the adaptor has 10 physical pages. Suppose we allocate two physical pages to each of five high-performance applications (e.g., Web, FTP) that want to use the adaptor to transfer data. Suppose the Web application gets 2 physical pages 9 and 10, and FTP gets pages 3 and 4. Then the kernel maps the physical pages 9 and 10 into the Web application's page table, and physical pages 3 and 4 into the FTP application's page table.

Now the Web application can write directly to physical pages 9 and 10 without any danger; if it tries to write into pages 3 and 4, the virtual memory hardware will generate an exception. Thus we are exploiting existing hardware in the form of the TLB (recall this is the Translation Lookaside Buffer) to protect access to pages. So now let us assume that page 10 is a sequence of free buffer descriptors written by the Web application; each buffer descriptor describes a page of main memory (assume this can be done using just 32 bits) that will be used to receive the next packet described for the Web application.

For example, Page 10 could contain the sequence 18, 12 (see Figure 28.17). This means that the Web application has currently queued physical pages 18 and 12 for the next incoming packet and its successor. We assume that pages 18 and 12 are in main memory and are physically locked pages that are assigned to the Web application by the kernel when the Web application first started.

When a new packet arrives for the Web application, the adaptor will demultiplex the packet to the descriptor page 10 using a packet filter, and then it will write the data of the packet (using DMA) to page 18. When it is done, the adaptor will write the descriptor 18 to a page of written page descriptors, say Page 9, that the Web application is authorized to read. It is up to the Web application to finish processing written pages and periodically queue new free buffer descriptors to the adaptor.

This sounds fine but there is a serious security flaw. Suppose the Web application, through malice or error, writes the sequence 155, 120 to its descriptor page (which it can do). Suppose further that Page 155 is in main memory and is where the kernel stores its data structures. When the adaptor gets the next packet for the Web application it will write it to page 155, overwriting the kernel data structures. This causes a serious problem, at least causing the machine to crash.

The reader may wonder why virtual memory hardware cannot detect this problem. The reason is that virtual memory hardware (observe the position of the TLB in Figure 28.16) only protects against unauthorized access by processes running on the CPU. This is because the TLB intercepts every READ (or WRITE) access done by the CPU and can do checks. However, devices like adaptors that do DMA bypass the virtual memory system and directly access memory.

This is not a problem in practice because applications cannot program the devices (like disks, adaptors) to read or write to specific places at the application's command. Instead, access is always mediated by the kernel. If we are getting rid of the kernel, then we have to ensure that everything the application can instruct the adaptor to do is carefully scrutinized.

The solution used in the Application Device Channel (ADC) [DDP94] solution is to have the kernel pass the adaptor a list of valid physical pages that each application using the adaptor directly can access. This can be done once the application first starts and before data transfer begins. In other words, the time-consuming computation involved in authorizing pages is shifted in time from the data transfer phase to application initialization. For example, when the Web application first starts it can ask the kernel for two physical pages, say 18 and 12, and then ask the kernel to authorize the use of these pages to the adaptor.

The kernel is then bypassed for normal data operation. However, if now the Web application queues the descriptor 155, and a new packet arrives, the adaptor will first check the number 155 against its authorized list for the application (i.e., 18, 12). Since it is not in the list, the adaptor will not overwrite the kernel data structures (phew!).

In summary, ADCs are based on shifting protection functions in space from the kernel to the adaptor, using some precomputed information (list of allowed physical pages, passed from the kernel to the adaptor augmented with the normal virtual memory hardware.

VIA [CIC97] stands for virtual interface architecture, and is a commercial standard proposed by Microsoft and Intel that advocates the ideas in ADCs. The term virtual interface makes sense because one can think of an application device channel as providing each application with their own virtual interface that they can manipulate without kernel intervention.

28.4 Conclusions

In this chapter we have described a set of bottlenecks for endnodes and routers and a set of the most effective known techniques for overcoming these bottlenecks. While the techniques seem ad hoc, the following is an attempt at a systematic classification based on 15 principles that can be seen to underlie the specific techniques described earlier.

- *P1, Avoid obvious waste in common situations*: In a system, there may be wasted resources in special sequences of operations. If these patterns occur commonly, it may be worth eliminating the waste. An example is making multiple copies of a packet between operating system and user buffers. Notice that each operation (single packet copy) considered by itself has no obvious waste. It is the sequence of operations that has obvious waste.

- *P2, Shift computation in time*: Systems have an aspect in space and time. Many efficiencies can be gained by shifting computation in time. Two generic methods that fall under time-shifting are

- *P2a, Precomputation* refers to computing quantities before they are actually used to save time at the point of use. The Lulea and Tree BitMap algorithms precompute information when prefixes are inserted.

- *P2b, Lazy evaluation* refers to postponing expensive operations at critical times. A famous example of lazy evaluation is Copy-on-write [Wik89]. Suppose we have to copy a virtual address space *A* to another space *B* for process migration. Copy-on-write makes page table entries in *B*'s virtual address space point to the corresponding page in *A*. When a process using *B* writes to a location, then a separate copy of the corresponding page in *A* is made for *B*, and the write is performed. Since we expect the number of pages that are written in *B* to be small compared to the total number of pages, this avoids unnecessary copying.

- *P3, Relax specifications*: When a system is first designed top-down, functions are partitioned among subsystems. When implementation difficulties arise, the basic system structure may have to be redone.

  Two techniques that arise from this principle are

  3a, Trading Certainty for Time: A simple example is stochastic fair queuing [McK91] that reduces the memory for DRR even further by hashing flows into buckets and doing DRR only on the buckets.

  3b, Shifting Computation in Space: This refers to moving computation from one subsystem to another as shifting computation in space. In networking, for example, the need for routers to fragment packets has recently been avoided by having end systems calculate a packet size that will pass all routers.

- *P4, Leverage other system components*: For example, algorithms are designed to fit the features offered by the hardware. Some techniques that fall under this principle are

  P4a, Exploit local access costs: The Tree Bitmap algorithm, for example, crucially relies on the ability of hardware to do wide memory accesses.

P4b, Trade memory for speed: The obvious technique is to use more memory such as lookup tables to save processing time. A less obvious technique is to compress a data structure to make it more likely to fit into cache because cache accesses are cheaper than memory accesses as in the Lulea IP lookup algorithm.

- *P5, Add hardware to improve performance*: The following list of specific hardware techniques are often used in networking hardware.

5a, Use memory interleaving and pipelining: All the IP lookup algorithms described in this chapter can be sped up by assigning each level of the tree to a separate bank in memory such as RAMBUS and pipelining.

5b, Use wide word parallelism: A common theme in many networking designs such as the bit vector scheme for packet classification described above is to use wide memory words that can be processed in parallel.

5c, Combine DRAM and SRAM: While the use of SRAM as a cache for DRAM databases is classical, a more creative idea is described in [RV03,SIPM02] where a large number of high-speed counters are implemented by storing only the low-order bits of each counter in SRAM, while the entire counter is backed up in DRAM.

- *P6, Replace inefficient general-purpose routines*: It can pay to design an optimized and specialized routine such as the SendFile() call described earlier that can splice data from the file system to the network without going through user space.

- *P7, Avoid unnecessary generality*: For example, RED calculations in routers can be speeded up by restricting parameters to powers of 2.

- *P8, Do not confuse specification and implementation*: Implementors are free to change the reference implementation as long as the two implementations have the same external effects. The use of ADCs, for example, is an unusual structuring technique.

- *P9, Pass information like hints in module interfaces*: A hint is information passed from a client to a service which, if correct, can avoid expensive computation by the service. A complex example of such a hint in networking is a so-called packet filter by which users pass to the kernel a specification of which packets they wish to receive.

- *P10, Pass information in protocol headers*: For distributed systems, the logical extension to prior Principle *P9* is to pass information such as hints in message headers. Two classic examples in networking including tag switching and multi protocol label switching (in which lookup hints are passed between routers) and differentiated services (where type of service information is passed from the edge routers to core routers).

- *P11, Optimize the expected case*: It pays to make common behaviors efficient, even at the cost of making uncommon behaviors more expensive. This is more often used for endnode bottlenecks. A classic example in networking is so-called header prediction where the cost of processing a TCP packet can be greatly reduced by assuming that the next packet received is closely related to the last packet processed.

- *P12, Add or exploit state to gain speed*: If an operation is expensive, consider maintaining additional but redundant state to speed up the operation. For example, the Lucent Bit vector scheme adds extra bitmaps to the basic structure.

- *P13, Exploit degrees of freedom*: It helps to be aware of the variables that are under one's control, and the evaluation criteria used to determine good performance. For example, a major breakthrough in switching occurred with the realization that the use of per-output/input queues and a matching algorithm could finesse the problem of head-of-line blocking.

- *P14, Use special techniques for finite universes such as integers*: When dealing with small universes such as moderately sized integers, techniques like bucket sorting, array lookup, and using bitmaps are often more efficient than general-purpose sorting and searching algorithms. Once again, the switch scheduling algorithm iSLIP relies on the fact that computing the first bit set in a bitmap is feasible for reasonable bit widths (up to 64 bits) which works fine for small switches of up to 64 ports.

- *P15, Use algorithmic techniques*: Even where there are major bottlenecks such as virtual address translation, systems designers finesse the need for clever algorithms by passing hints, using caches, and performing table lookup. In many cases, Principles 1–14 need to be applied before any algorithmic issues become bottlenecks. Even when this is done the real breakthroughs may arise from applying algorithmic thinking as opposed to merely reusing existing algorithms. For example, Patricia tries with skip counts are not the lookup algorithm of choice compared to say Tree bitmap.

Finally, networking continues to change and new bottlenecks are beginning to emerge in security, measurement, and application-aware processing. While the specific techniques and bottlenecks described in this chapter may change or become less important as time passes, we hope that the ways of thinking embodied in the 15 principles will continue to remain relevant. More details of this approach and a description of other bottlenecks can be found in the textbook [Var05]. The author is grateful to Rick Adams of Elsevier for permission to use material and figures from [Var05].

References

[AOST93] T. Anderson, S. Owicki, J. Saxe, and C. Thacker. High speed switch scheduling for local area networks. *ACM Transactions on Computer Systems*, 11(4):319–352, 1993.

[BL00] R. Bhagwan and W. Lin. Fast and scalable priority queue architecture for high-speed network switches. In *IEEE INFOCOM*, pp. 538–547, IEEE, Tel Aviv, Israel, 2000.

[BSV03] F. Baboescu, S. Singh, and G. Varghese. Packet classification for core routers: Is there an alternative to CAMs? In *Proceedings of the IEEE INFOCOM*, Los Alamitos, CA, 2003.

[BZ96] J. Bennett and H. Zhang. Hierarchical packet fair queuing algorithms. In *Proceedings of the SIGCOMM*, Stanford, CA, 1996.

[CB95] W. Cheswick and S. Bellovin. *Firewalls and Internet Security*. Addison-Wesley, Reading, MA, 1995.

[Cha9a] B. Chazelle. Lower bounds for orthogonal range searching, I: The reporting case. *Journal of the ACM*, 37:200–212, 1990.

[Cha9b] B. Chazelle. Lower bounds for orthogonal range searching, II: The arithmetic model. *Journal of the ACM*, 37:439–463, 1990.

[CIC97] Compaq, Intel, and Microsoft Corporations. Virtual Interface Architecture Specification. In *http://www.viaarch.org*, 1997.

[DBCP97] M. Degermark, A. Brodnik, S. Carlsson, and S. Pink. Small forwarding tables for fast routing lookups. In *Proceedings of the ACM SIGCOMM*, pp. 3–14, ACM, Cannes, France 1997.

[DDP94] P. Druschel, B. Davie, and L. Peterson. Experiences with a high-speed network adapter: A software perspective. In *Proceedings of the ACM SIGCOMM*, London, U.K., September 1994.

[DKS89] A. Demers, S. Keshav, and S. Shenker. Analysis and simulation of a fair queueing algorithm. *Proceedings of the Sigcomm '89 Symposium on Communications Architectures and Protocols*, 19(4):1–12, September 1989. Part of ACM Sigcomm Computer Communication Review.

[Eat95] W. Eatherton. Hardware-based internet protocol prefix lookups. MS thesis. Washington University Electrical Engineering Department, Washington, D.C., 1995.

[EDV04] W. Eatherton, Z. Dittia, and G. Varghese. Tree bitmap: Hardware software IP lookups with incremental updates. *ACM Computer Communications Review*, 34(2):97–123, April 2004.

[FP93]   K. Fall and J. Pasquale. Exploiting In-kernel data paths to improve I/O throughput and CPU availability. In *USENIX Winter*, pp. 327–334, Usenix Association, San Diego, CA, 1993.

[FPCe97]   R. Fromm, S. Perissakis, N. Cardwell et al. The energy efficiency of IRAM architectures. In *International Symposium on Computer Architecture (ISCA 97)*, Denver, CO, June 1997.

[GM99a]   P. Gupta and N. McKeown. Designing and implementing a fast crossbar scheduler. *IEEE Micro*, 19(1):20–28, February 1999.

[GM99b]   P. Gupta and N. McKeown. Packet classification on multiple fields. In *Proceedings of the ACM SIGCOMM*, pp. 147–160, ACM, Cambridge, MA, 1999.

[KHM87]   M. Karol, M. Hluchyj, and S. Morgan. Input versus output queuing on a space division switch. *IEEE Transactions on Communications*, 35(12): 13477–1356, December 1987.

[Knu73]   D. Knuth. *Fundamental Algorithms*, vol 3. *Sorting and Searching*. Addison-Wesley, Reading, MA, 1973.

[LS98]   T. V. Lakshman and D. Stidialis. High speed policy-based packet forwarding using efficient multi-dimensional range matching. In *Proceedings of the ACM SIGCOMM*, Vancouver, British Columbia, Canada, September 1998.

[McK91]   P. McKenney. Stochastic fairness queueing. In *Internetworking: Research and Experience*, Wiley, Chichester, vol. 2, 113–131, January 1991.

[Mea97]   N. McKeown et al. The Tiny Tera: A packet switch core. In *IEEE Micro*, 17(1):26–33, January 1997.

[RV03]   S. Ramabhadran and G. Varghese. Efficient implementation of a statistics counter architecture. In *Proceedings ACM SIGMETRICS*, San Diego, CA, 2003.

[SIPM02]   D. Shah, S. Iyer, B. Prabhakar, and N. McKeown. Maintaining statistics counters in router line cards. In *IEEE Micro*, 76–81, January 2002.

[SKO+94]   R. Souza, P. Krishnakumar, C. Ozveren, R. Simcoe, B. Spinney, R. Thomas, and R. Walsh. GIGAswitch: A high-performance packet switching platform. In *Digital Technical Journal* 6(1):9–22, *Winter*, 1994.

[Ste98]   W.R. Stevens. *UNIX Network Programming*. Prentice-Hall, Upper Saddle River, NJ, 1998.

[SV96]   D. Stiliadis and A. Varma. Frame-based fair queueing: A new traffic scheduling algorithm for packet-switched networks. In *Proceedings of the ACM SIGMETRICS*, Philadelplia, PA, 1996.

[SWG]   Differentiated Services Working Group. Differentiated Services (diffserv) Charter. In *http://www.ietf.org/html.charters/diffserv-charter.html*.

[Var05]   G. Varghese. *Network Algorithmics*. Morgan Kaufman/Elsevier, San Francisco, CA, 2005.

[Wik89]   Wikipedia. In *Copy-on-write*, 1989.

[WS95]   G.R. Wright and W.R. Stevens. *TCP/IP Illustrated Volume 2*. Addison-Wesley, Reading, MA, 1995.

[YHA87]   Y. Yeh, M. Hluchyj, and A. Acampora. The Knockout Switch: A simple modular architecture for high-performance packet switching. *IEEE Journal on Selected Areas in Communication*, 5:1426–1435, October 1987.

[Zha91]   L. Zhang. Virtual clock: A new traffic control algorithm for packet-switched networks. *ACM Transactions on Computer Systems*, 9(2):101–124 1991.

29

Algorithmic Issues in Grid Computing

Yves Robert
Ecole Normale Supérieure de Lyon

Frédéric Vivien
INRIA

29.1 Introduction

A Grid is a virtual computer created through the interconnection of geographically dispersed, and usually **heterogeneous**, computing and data resources. In this chapter, we focus on the algorithmic issues to be addressed when trying to efficiently deploy one or several applications on a Grid (or, more generally, on a distributed and heterogeneous collection of computing resources).

The most obvious problem with Grids is their heterogeneity. For instance, any Grid will contain processors of different characteristics and the running time of any application will thus depend on processors that have been enrolled to execute it. The obvious pitfall for a parallel computation using processors of different speeds would be to slow down its execution at the pace of the least efficient processor. An intuitive approach to circumvent this problem would be to pick only fast processors or, at least, to pick processors of equivalent processing power. As shown in Section 29.2.2, this intuition can lead to very poor performance. Indeed, in some cases, it is not computing power that matters, but communication capabilities. In a Grid, of course, the communication capabilities of the different entities will be also heterogeneous. Communications in Grids can be even more a problem than in traditional parallel computers because of the irregularity of the interconnection network and also because of longer distance, and thus slower, communication links. In Grids, resources are abundant, but the resources used to solve a problem must be carefully chosen: **resource selection** is key to performance.

Grid resources are not only heterogeneous, they are also irregular. In general, a Grid has no reason to have a nice hierarchical design pattern: it is built through a time-evolving aggregation of time-evolving platforms bought and assembled by independent and not necessarily cooperating organizations. To efficiently deploy on such a platform any parallel application that is not embarrassingly parallel, we need some accurate platform model* and some accurate application model.

Problem:

- Very accurate models are intractable. Using such models forbids to have any significant insight of the problem characteristics and thus of what efficient solutions should be.
- Tractable models are often over-simplistic and do not enable to predict the actual behavior of practical applications.

These classical issues of platform modeling and application modeling may seem even more difficult in a Grid computing framework, as Grid platforms are likely to be of very large scale. However, because Grids are complex and difficult to efficiently exploit, their typical use is restricted to run very large and time-consuming applications. For such applications, we can relax classical optimization problems, such as the absolute minimization of the total execution time (or **makespan**), and, for instance, concentrate on throughput maximization. The latter objective is much simpler, while it can be shown to provide an asymptotically optimal solution for the execution time. In other words, it is sometimes less difficult to efficiently **schedule** 10,000 tasks on a heterogeneous Grid than to schedule 10 tasks on a homogeneous cluster.

Section 29.2 aims at assessing the importance of three key ingredients of Grid algorithmic techniques: (1) static approaches; (2) application models; and (3) platform models. In Section 29.3 we cover in detail an interesting case study, that of matrix product under memory constraints. Matrix product is the archetype of parallel algorithms; we revisit this well-known computational kernel in a client–server framework: we deal with the problem of multiplying two very large matrices, which initially reside on the disk of a master processor. The master aims at enrolling adequately chosen worker processors in order to perform the computation. Because the matrices would not fit in the main memory of the workers, communication overhead will be the main bottleneck. We present a theoretical study of this problem and, from its conclusions, we derive practical algorithms. We conclude this chapter by stating some research directions in Section 29.4, by defining terms in Section 29.5, and by pointing to additional sources of information in Section 29.6.

29.2   Underlying Principles

In this section, we illustrate the impact of three key components of algorithm design and scheduling techniques for Grid applications. In Section 29.2.1, we emphasize the importance of static approaches, which are likely to achieve much better performance than their dynamic counterparts in the presence of heterogeneous resources, as they can predict, and thus avoid, the pitfalls due to the heterogeneity of resources. Then, in Section 29.2.2, we revisit an application model and show the benefits of a steady-state approach aimed at throughput optimization (as opposed to a traditional approach targeted to the search of a minimal makespan). Along the way, we stress the impact of resource selection on application performance: a careful resource selection may enable to reduce execution time while sparing resources. Finally, in Section 29.2.3, we stress the need to carefully define platform models, showing that slightly different models can lead to greatly different complexity results and to solutions of different practical importance.

---

* We may also need such a model to deploy any embarrassingly parallel application that requires large volumes of input data: these data will induce large communications which can suffer from potential communication bottlenecks.

29.2.1 The Importance of Static Approaches

In this section, we show that dynamic strategies may be the victims of the obvious pitfalls of platform heterogeneity: applications are run at the pace of the least efficient processor. Hence the importance of static approach or semi-static approach.

To illustrate this point, we deal first with the simple problem of distributing independent chunks of computation to heterogeneous processors. We then use the result of this first study to design optimal allocation schemes for finite-difference computations over a tiled iteration space.

29.2.1.1 Distributing Independent Chunks

Given B independent chunks of computations, each of equal size (i.e., each requiring the same amount of work), how can we assign these chunks to p physical processors P_1, P_2, ..., P_p of respective execution times w_1, w_2, ..., w_p, so that the workload is best balanced? Here the execution time is understood as the number of time units needed to perform one chunk of computation, that is, processor P_i executes any computation chunk in w_i time units. Obviously, the model is crude, as communication costs are ignored. Still, how should we distribute chunks to processors? The intuition is that the load of P_i, that is, the number n_i of chunks assigned to it, should be inversely proportional to w_i. The processing time for P_i will be $n_i w_i$, and we must ensure that $\sum_{i=1}^{p} n_i = B$. The goal is to minimize the total execution time:

$$T_{\text{exe}} = \max_{\sum_{i=1}^{p} n_i = B} (n_i \times w_i).$$

Consider a toy example with three processors of (relative) execution times $w_1 = 3$, $w_2 = 5$, and $w_3 = 8$. In Table 29.1, we report the optimal allocations for $B = 1$ up to $B = 10$. These allocations are computed incrementally, using a simple dynamic programming algorithm. Let $C = (n_1, n_2, \ldots, n_p)$ with $\sum_{i=1}^{p} n_i = B$ denote the optimal solution for B chunks. The execution time is $\max_{1 \leq i \leq p} n_i w_i$. The entry "selected processor" denotes the index of the processor chosen to receive the next chunk. We select index j such that

$$(n_j + 1)w_j = \min_{1 \leq i \leq p} (n_i + 1)w_i,$$

and P_j is assigned the $(B + 1)$th chunk.

For instance at step 4, i.e., to allocate the fourth chunk, we start from the solution for three chunks, i.e., $(n_1, n_2, n_3) = (2, 1, 0)$. Which processor P_i should receive the fourth chunk, i.e., which n_i should be increased? There are three possibilities $(n_1 + 1, n_2, n_3) = (3, 1, 0)$, $(n_1, n_2 + 1, n_3) = (2, 2, 0)$, and $(n_1, n_2, n_3 + 1) = (2, 1, 1)$ of execution times 9 (P_1 is the slowest), 10 (P_2 is the slowest), and 8 (P_3 is the slowest), respectively. Hence we select $j = 3$ and we retain the solution $(n_1, n_2, n_3) = (2, 1, 1)$.

TABLE 29.1 Running the Dynamic Programming Algorithm with Three Processors: $w_1 = 3$, $w_2 = 5$, and $w_3 = 8$

Number of Chunks B	n_1	n_2	n_3	T_{exe}	Selected Processor
0	0	0	0		1
1	1	0	0	3	2
2	1	1	0	5	1
3	2	1	0	6	3
4	2	1	1	8	1
5	3	1	1	9	2
6	3	2	1	10	1
7	4	2	1	12	1
8	5	2	1	15	2
9	5	3	1	15	3
10	5	3	2	16	

The complexity of the dynamic programming algorithm is $O(pB)$ (it can be easily improved to $O(\log(p)B)$).

Even though we have determined the optimal allocation, we point out that the load is not perfectly balanced among processors. For instance with $B = 10$, the load of P_1 and P_2 is 15 time-units while that of P_3 is 16. In fact, it is easy to see that a perfect load-balancing can be achieved only for values of B which are multiples of $B_{\text{opt}} = L\left(\sum_{i=1}^{p}\frac{1}{w_i}\right)$, where $L = \text{lcm}(w_1, w_2, \ldots, w_p)$. In the example, $L = 120$ and $B_{\text{opt}} = 79$. Out of these 79 chunks, 40 go to P_1, 24 to P_2, and 15 to P_3, so that the execution time of each processor is 120.

Finally, we have to acknowledge that statically computing the optimal allocation may have been somewhat useless! Indeed, we always have the possibility to use a fully dynamic scheme, where processors pick up a new chunk as soon as they are finished with their current one. This dynamic scheme will self regulate and automatically converge to the optimal distribution. The next example shows that dynamic strategies may well achieve much poorer performance than their static counterparts.

29.2.1.2 Finite-Difference Computations

Consider the two-dimensional rectangular iteration space represented in Figure 29.1. Tiles are rectangular, and their edges are parallel to the axes. All tiles have the same fixed size. The size of the tiled iteration space is $N_1 \times N_2$. Dependencies between tiles are summarized by the vector pair $\left\{\left(\begin{smallmatrix}1\\0\end{smallmatrix}\right), \left(\begin{smallmatrix}0\\1\end{smallmatrix}\right)\right\}$. In other words, the computation of a tile cannot be started before both its left and lower neighbor tiles have been executed. Such computations are the representative of finite-difference methods (Wolfe (1996)).

Assume, as before, that there are p different-speed processors, processor P_i, $1 \leq i \leq p$, with an execution time w_i. Assume also that the columns of tiles are assigned to processors, so as to increase both the locality and the granularity of the computations. What is the best distribution of tile columns to processors so that the execution time is minimized? As before, we use a crude model and neglect all communication costs.[*]

When targeting homogeneous processors, a natural approach is to use a pure cyclic allocation of tile columns to processors (Ohta et al., 1995; Andonov and Rajopadhye, 1997; Högstedt et al., 1997), or again to use a dynamic approach where processors request a new tile column as soon as

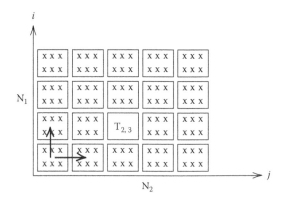

FIGURE 29.1 A tiled iteration space with horizontal and vertical dependencies.

---

[*] The following results and conclusions can be extended to include communication costs, assuming for instance that processors are arranged along a (virtual) ring (Boulet et al., 1999).

⋮

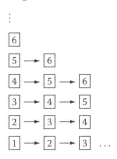

FIGURE 29.2 Stepwise execution with homogeneous processors (numbers in tiles denote execution steps).

they are finished with their current one. Both approaches lead to the same execution pattern where the execution progresses up along each column (see Figure 29.2). However, we will see that such a cyclic allocation leads to very poor performance with heterogeneous processors.

Indeed, the best execution time that we can hope for is

$$T_{opt} \approx \frac{N_1 \times N_2}{\sum_{i=1}^{p} \frac{1}{w_i}}.$$

Here, T_{opt} is the total number of tiles divided by the cumulated speed of the computing platform, and it is a lower bound on the actual optimal execution time (even if the load is perfectly balanced, there will be some idle time at the beginning and the end of the execution). However, using a cyclic allocation, the execution progresses at the pace of the slowest processor, and we are far from reaching T_{opt}. Because of dependencies, the dynamic approach (that allocates the next column to the next available processor) reduces to the cyclic allocation: hence it becomes very inefficient in a heterogeneous framework.

Consider again the example with three processors, where $w_1 = 3$, $w_2 = 5$, and $w_3 = 8$. As illustrated in Figure 29.3, the execution time will be

$$T_{exe} \approx \frac{8}{3} N_1 N_2 \approx 2.67\, N_1 N_2 \quad \text{instead of} \quad T_{opt} \approx \frac{120}{79} N_1 N_2 \approx 1.52\, N_1 N_2.$$

Instead of using a cyclic allocation, we better assign blocks of B consecutive columns to processors. If we take $B = 10$ and run the dynamic programming algorithm of Table 29.1, we obtain $T_{exe} \approx 1.6\, N_1 N_2$ as illustrated in Figure 29.4. Because the load is not perfectly balanced when using a block size $B = 10$, we do not achieve an execution time equal to T_{opt}, but we are much closer! Had we used $B = 79$, we would have obtained the latter value.

For the general case with p processors P_i of execution time w_i, we (asymptotically) reach T_{opt} using the following algorithm:

FIGURE 29.3 Cyclic allocation with three processors progressing at the pace of the slowest one.

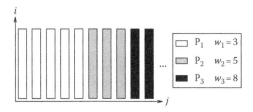

FIGURE 29.4 Block allocation with three processors achieving a good load balancing.

1. Compute $L = \text{lcm}(w_1, w_2, \ldots, w_p)$ (in the example, $L = \text{lcm}(3, 5, 8) = 120$).
2. Assign first $n_1 = L/w_1$ columns to P_1, next $n_2 = L/w_2$ columns to P_2, and so on.
3. Set the block size, or the period of the allocation, to $B = n_1 + n_2 + \cdots + n_p$ (in the example, $B = n_1 + n_2 + n_3 = 40 + 24 + 15 = 79$).
4. Change schedule:

   - Sort processors so that $n_1 w_1 \leq n_2 w_2 \leq \cdots \leq n_p w_p$.
   - Process tiles horizontally within blocks.

We point out that each processor must now progress horizontally across its columns, while in the previous version it was progressing vertically along each column and finishing the execution of that column before starting the execution of the next one.

To summarize, the static approach consists in allocating panels of B tile columns to the p processors in a periodic fashion. Inside each panel, processors receive an amount of columns inversely proportional to their speed. Within each panel, the work is well-balanced, and dependencies do not slow down the execution.

The heterogeneity of the platform had a deep impact on the form of the optimal solution: the shapes of the optimal solutions for the homogeneous and heterogeneous cases are quite different. Furthermore, the intuitive dynamic approach is very inefficient: only a static approach can avoid falling in the trap laid off by the heterogeneity of processors.

29.2.2 The Importance of Application Models

We illustrate here the well-known fact that application models have a deep impact on our capacity to analyze and solve problems. This is especially true in the case of Grid computing as we must reconsider some assumptions and approaches that we have been long accustomed to, but which may now be no longer meaningful. This is because of the particularities of Grid platforms and of the applications to be executed on such platforms. To make this point, we consider an application made of a large number of same-size independent jobs to be executed on a tree-shape platform; studying this example in the steady-state framework enables us to find, in polynomial time, asymptotically optimal solutions for a problem that is NP-complete in the classical framework.

Along the way, we will show that resource selection can have a deep impact on performance: a careful resource selection will enable us to reduce execution time while sparing resources. Furthermore, the rule for the selection of resources, in our example, is rather counter-intuitive: processing speeds do not matter!

29.2.2.1 Motivation

The traditional objective of scheduling algorithms is to minimize the total execution time, or makespan: given a task graph and a set of computing resources, find a mapping of the tasks onto the processors, and order the execution of the tasks so that: (1) task precedence constraints are satisfied; (2) resource constraints are satisfied; and (3) a minimum schedule length is provided. However, makespan minimization turns out to be NP-hard in most practical situations (Shirazi et al., 1995; Ausiello et al., 1999). The advent of more heterogeneous architectural platforms is likely to even increase the computational complexity of the process of mapping applications to machines.

An idea to circumvent the difficulty of makespan minimization is to lower the ambition of the scheduling objective. Instead of aiming at the absolute minimization of the execution time, why not consider asymptotic optimality? After all, the number of tasks to be executed on the computing platform is expected to be very large: otherwise why deploy the corresponding application on a computational Grid? To state this informally: if there is a nice (meaning, polynomial) way to derive, say, a schedule whose length is two hours and three minutes, as opposed to an optimal schedule that would run for only two hours, we would be satisfied.

This approach has been pioneered by Bertsimas and Gamarnik (1999). Steady-state scheduling allows one to relax the scheduling problem in many ways. The costs of the initialization and clean-up phases are neglected. The initial integer formulation is replaced by a continuous or rational formulation. The precise ordering and allocation of tasks and messages are not required, at least in the first step. The main idea is to characterize the activity of each resource during each time unit: which (rational) fraction of time is spent computing, which is spent receiving or sending to which neighbor. Such activity variables are gathered into a linear program, which includes conservation laws that characterize the global behavior of the system. The actual schedule then arises naturally from these quantities.

29.2.2.2 Master–Worker Tasking

In this section, we deal with the master–worker paradigm on a heterogeneous platform, where resources have different speeds of computation and communication. The target platform is composed of one specific node M, referred to as the master, together with p workers P_1 to P_p (see Figure 29.5). The master initially holds (or generates the data for) a large collection of independent, identical tasks to be allocated to the workers. The question for the master is to decide on the number of tasks (i.e., task files) to forward to each of the workers. Due to heterogeneity, these workers may receive different amounts of work (maybe none for some of them).

We enforce the following rules of execution:

- The master sends tasks to workers sequentially, and without preemption.
- There is a full computation/communication overlap for each worker.
- Worker P_i receives a task in c_i time-units.
- Worker P_i processes a task in w_i time-units.
- The master M does not compute any task (but a master with computation time w_m can be simulated as a worker with same computation time and zero communication time).

We are using quite a realistic model of execution, which is called the one-port model and corresponds to the mode of operation of non-threaded MPI (Message Passing Interface) libraries. Because the number of tasks to be executed on the computing platform is expected to be large, we target steady-state optimization problems rather than standard makespan minimization problems. The optimal steady-state scheduling can be characterized very efficiently, with low-degree polynomial complexity. It is defined as follows: for each worker processor, determine the fraction of time spent computing, and the fraction of time spent receiving tasks; for the master, determine the fraction of time spent communicating along each communication link. The objective is that the (averaged) overall number of tasks processed at each time-step is maximum.

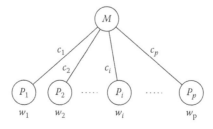

FIGURE 29.5 The heterogeneous star platform.

Formally, after a start-up phase, we want the resources to operate in a periodic mode, where worker P_i executes α_i tasks per time-unit. We point out that α_i is a rational number, not an integer, so that there will remain some work to reconstruct a true schedule with an integer number of tasks. First we express the constraints for computations: P_i needs $\alpha_i \cdot w_i$ to compute α_i tasks, and it computes these within one time-unit, so necessarily

$$\alpha_i \cdot w_i \leq 1.$$

As for communications, the master M sends tasks sequentially to the workers, and it takes $\alpha_i \cdot c_i$ time-units to send α_i tasks along the link to P_i, so necessarily

$$\sum_{i=1}^{p} \alpha_i \cdot c_i \leq 1. \tag{29.1}$$

Finally, the objective is to maximize throughput, namely

$$\rho = \sum_i \alpha_i.$$

Altogether, we have a linear programming problem in rational numbers, we obtain rational values for all variables in polynomial time (polynomial in p, the size of the heterogeneous platform). When we have the optimal solution, we take the least common multiple of the denominators, and multiply all variables by this quantity, thereby deriving an integer period T for the steady-state operation, together with an integer number of messages going through each link. Because it arises from the linear program, $\log T$ is indeed a number polynomial in the problem size, but T itself is not. Hence, describing what happens at every time-step during the period would be exponential in the problem size. Instead, we have a more compact description of the schedule: we only need the duration of the p time intervals during which the master is sending tasks to each worker (some possibly zero), and the duration of the p time intervals during which each worker is computing (again, some possibly zero).

It turns out that the linear program is so simple that it can be solved analytically:

1. Sort the workers by increasing communication times. Re-number them so that $c_1 \leq c_2 \leq \cdots \leq c_k$.
2. Let q be the largest index so that $\sum_{i=1}^{q} \frac{c_i}{w_i} \leq 1$. If $q < p$ let $\varepsilon = 1 - \sum_{i=1}^{q} \frac{c_i}{w_i}$, otherwise let $\varepsilon = 0$.
3. Then $\rho = \sum_{i=1}^{q} \frac{1}{w_i} + \frac{\varepsilon}{c_{q+1}}$.

When $q = p$ the result is expected: it basically says that workers can be fed with tasks fast enough so that they are all kept computing steadily. However, if $q < p$ the result is surprising: in the situation when the communication bandwidth is limited, some workers will partially starve; in the optimal solution, these are those with slow communication rates, whatever their processing speeds. In other words, a slow processor with a fast communication link is to be preferred to a fast processor with a slow communication link. The optimal strategy is bandwidth-centric as it delegates work to the fastest communicating workers, and it does not matter if these workers are computing slowly (but slow workers will not contribute much to the overall throughput).

Consider the example shown in Figure 29.6. Note that in steady-state mode, communications for the next tasks occur in parallel with the computations for the current tasks. Figure 29.7 shows that 11 tasks are computed every 18 time-units, so that $\rho = 11/18 \approx 0.6$. If we had used a purely demand-driven strategy (with workers posting requests and the master serving them in FIFO order), we would have computed only 5 tasks every 36 time-units, therefore achieving a throughput

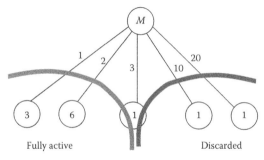

FIGURE 29.6 An example for the bandwidth-centric strategy.

Tasks	Communication	Computation
6 Tasks to P_1	$6c_1 = 6$	$6w_1 = 18$
3 Tasks to P_2	$3c_2 = 6$	$3w_2 = 18$
2 Tasks to P_3	$2c_3 = 6$	$2w_3 = 2$

FIGURE 29.7 Achieved throughput for the bandwidth-centric strategy.

of only $\rho = 5/36 \approx 0.14$. In fact, we see in this example that the demand-driven strategy end up with a round-robin distribution of tasks to all workers, regardless of the heterogeneity of the communication links (and we could make the comparison arbitrarily worse, simply by increasing the value of c_5). The conclusion is that even if resources are cheap and abundant, resource selection is key to performance.

29.2.2.3 Extension to Tree-Shaped Platforms

In fact, minimizing the total execution time to schedule a bag of independent tasks on a heterogeneous star platform is a problem of polynomial complexity (Beaumont et al., 2002). But still, we may prefer to use the regular periodic schedule provided by the steady-state approach than the irregular schedule returned by the makespan minimization algorithm of Beaumont et al. (2002). In any case, we have no choice when moving to general tree-shaped platforms: the makespan minimization problem becomes NP-hard (Dutot, 2003), while the steady-state approach can be smoothly extended from star to tree platforms. We now outline this extension.

For a node P_i in the tree, let w_i denote the time it needs to compute a task, and α_i the (fractional) number of tasks that it executes every time unit. For any edge $P_i \rightarrow P_j$ in the tree, let $c_{i,j}$ denote the time needed to send a task from P_i to P_j, and let sent$(P_i \rightarrow P_j)$ denote the (fractional) number of tasks sent by P_i to P_j every time unit. The master $M = P_0$ is the root of the tree. If $P_{i_1}, P_{i_2}, \ldots, P_{i_k}$ denote its children in the tree, we have an equation quite similar to Equation 29.1:

$$\sum_{j=1}^{k} \text{sent}(P_0 \rightarrow P_{i_j})c_{0,i_j} \leq 1.$$

If the master computes tasks as well, then we have:

$$\alpha_0 w_0 \leq 1.$$

Consider now an internal node P_i. Let P_{i_0} denote its parent in the tree, and let $P_{i_1}, P_{i_2}, \ldots, P_{i_k}$ denote its children in the tree. As before, there are equations to constrain the computations and the communications of P_i:

$$\alpha_i w_i \leq 1 \qquad \text{and} \qquad \sum_{j=1}^{k} \text{sent}(P_i \rightarrow P_{i_j}) c_{i,i_j} \leq 1.$$

But there is another constraint to write, namely a conservation law: the number of tasks received by P_i is equal to the sum of the number of tasks that it executes itself plus the number of tasks that it sends to its children:

$$\text{sent}(P_{i_0} \rightarrow P_i) = \alpha_i + \sum_{j=1}^{k} \text{sent}(P_i \rightarrow P_{i_j}).$$

It is important to understand that the conservation law applies to the steady-state operation: overall, the flow of incoming tasks equals the flow of tasks consumed in place plus the flow of outgoing tasks. The law links tasks that arrive during different periods: incoming tasks arrive during the current period while consumed and outgoing tasks had arrived during the previous period.

Finally, if P_i is a tree leaf whose parent is P_{i_0}, we easily derive that

$$\alpha_i w_i \leq 1 \qquad \text{and} \qquad \text{sent}(P_{i_0} \rightarrow P_i) = \alpha_i.$$

The total throughput to optimize is given by

$$\rho = \sum_{\text{all} P_i} \alpha_i.$$

We have built a linear program, and we can follow the same steps as for star-shaped platforms to define the period and reconstruct the final schedule.

29.2.2.4 Asymptotic Optimality

We keep the previous notations to prove that the solution to the steady-state scheduling problem is asymptotically optimal. Given a tree platform and a time bound T, let $nb_{\text{opt}}(T)$ be the optimal number of tasks that can be computed by any schedule, be it periodic or not, within T time units. We have the following result:

PROPOSITION 29.1 $\quad nb_{\text{opt}}(T) \leq \rho \times T.$

PROOF Consider an optimal schedule. Consider any node P_i, with parent P_{i_0} and children $P_{i_1}, P_{i_2}, \ldots, P_{i_k}$. Let $t_i(T)$ be the total number of tasks that are executed by P_i within the T time units. Similarly, for each edge $e_{i,i_j} : P_i \rightarrow P_{i_j}$ in the tree, let $t_{i,i_j}(T)$ be the total number of tasks that have been forwarded by P_i to P_{i_j} within the T time units. The following equations hold true:

- $t_i(T) \cdot w_i \leq T$ (time for P_i to process its tasks)
- $\sum_{j=1}^{k} t_{i,i_j}(T) \cdot c_{i,i_j} \leq T$ (time for P_i to forward outgoing tasks in the one-port model)
- $t_{i_0,i}(T) = t_i(T) + \sum_{j=1}^{k} t_{i,i_j}(T)$ (conservation equation)

Let $\alpha_i = \frac{w_i t_i(T)}{T}$, and $\text{sent}(P_i \rightarrow P_{i_j}) = \frac{c_{i,i_j} t_{i,i_j}(T)}{T}$. All the equations of the previous linear program hold, hence $\sum_i \frac{\alpha_i}{w_i} \leq \rho$, the optimal value. Going back to the original variables, we derive

$$nb_{\text{opt}}(T) = \sum_i t_i(T) \leq \rho \times T$$

Essentially Proposition 29.1 says that no schedule can execute more tasks than the optimal steady-state. There remains to bound the potential loss due to the initialization and the clean-up phase. Consider the following algorithm (assume T is large enough):

- Solve the linear program: compute the maximal throughput ρ, compute all the values α_i and sent$(P_i \rightarrow P_j)$, and determine T_{period}. For each processor P_i, determine per_i, the total number of tasks that it receives per period. Note that all these quantities are independent of T: they depend only upon w_i and c_{ij}, characteristics of the platform.
- Initialization: the master sends per_i tasks to each processor P_j. This requires I units of time, where I is a constant independent of T.
- Let J be the maximum time for each processor to consume per_i tasks ($J = \max_i\{per_i \cdot w_i\}$). Again, J is a constant independent of T.
- Let $r = \lfloor \frac{T-I-J}{T_{\text{period}}} \rfloor$.
- Steady-state scheduling: during r periods of duration T_{period}, operate the platform in steady state.
- Clean-up: do not forward any task but consume in place all remaining tasks. This requires at most J time units. Do nothing during the very last units ($T - I - J$ may not be evenly divisible by T_{period}).

The number of tasks processed by the above algorithm within T time units is equal to $nb(T) = (r+1) \times T_{\text{period}} \times \rho$.

PROPOSITION 29.2   The previous scheduling algorithm based upon steady-state operation is asymptotically optimal:

$$\lim_{T \to +\infty} \frac{nb(T)}{nb_{\text{opt}}(T)} = 1.$$

PROOF   Using Proposition 29.1, $nb_{\text{opt}}(T) \leq \rho T$. From the description of the algorithm, we have $nb(T) = ((r+1)T_{\text{period}}) \cdot \rho \geq (T - I - J) \cdot \rho$. Hence the result as I, J, T_{period}, and ρ are constants independent of T.

29.2.2.5   Why Steady-State Scheduling?

In addition to its simplicity, which enables one to tackle more complex problems, steady-state scheduling enjoys two main advantages over traditional scheduling:

- *Efficiency:* Steady-state scheduling provides, by definition, a periodic schedule, which is described in compact form and is thus possible to implement efficiently in practice. This is to be contrasted with the need for classical schedules to list the starting dates and target processors for each task in the application.
- *Adaptability:* Because the schedule is periodic, it is possible to dynamically record the observed performance during the current period, and to inject this information into the algorithm that will compute the optimal schedule for the next period. This makes it possible to react on the fly to resource availability variations, which is the common case for example on non-dedicated Grid platforms.

29.2.2.6 Conclusion

We have seen that when we target Grid platforms, we have to reconsider our usual assumptions on application modeling. Indeed, relaxing some classical hypotheses can enable to solve optimization problems, which would otherwise be too difficult, while the cost of the relaxation itself may be negligible. Among the classical hypotheses and methods, one should reconsider the choice of the objective function: rather than optimizing the makespan, the minimization of the applications' flow-times, slowdowns, or other quality of service metrics, or the maximization of the platform throughput may be more relevant.

29.2.3 The Importance of Platform Models

We illustrate here the well-known fact that platform models have a deep impact on our capacity to analyze problems and to find solutions of practical importance. As anyone would have guessed, we will see that the more accurate and complex the platform model, the more difficult is the optimization problem. We will also see that using a more accurate platform model—which may forbid us to derive optimal solutions in polynomial-time—strategies that are.

To show the impact and the importance of platform models, we focus on the mapping of iterative algorithms. Such algorithms typically operate on a large collection of application data, which is partitioned over the processors. At each iteration, some independent calculations are carried out in parallel, and then some communications take place. This scheme is very general, and encompasses a broad spectrum of scientific computations, from mesh based solvers (e.g., elliptic PDE solvers) to signal processing (e.g., recursive convolution), and image processing algorithms (e.g., mask-based algorithms such as thinning). Our aim here is twofold. The first one is to show the impact of the network model on problem complexity. The second one is to stress the importance of an accurate model.

An abstract view of the problem is the following: an iterative algorithm repeatedly operates on a large rectangular matrix of data samples. This data matrix is split into vertical slices that are allocated to the computing resources (processors). At each step of the algorithm, the slices are updated locally, and then boundary information is exchanged between consecutive slices. This (virtual) geometrical constraint advocates that processors be organized as a virtual ring. Then each processor will only communicate twice, once with its (virtual) predecessor in the ring, and once with its (virtual) successor. Note that there is no reason a priori to restrict to a uni-dimensional partitioning of the data, and to map it onto a uni-dimensional ring of processors: more general data partitionings, such as two-dimensional, recursive, or even arbitrary slicings into rectangles, could be considered. But uni-dimensional partitionings are very natural for most applications, and, as will be shown, the problem to find the optimal one is already difficult.

The target architecture is a fully heterogeneous platform, composed of different-speed processors that communicate through links of potentially different bandwidths. On the architecture side, the problem is twofold: (1) select the processors that will participate in the solution and decide for their ordering that will represent the arrangement into a ring; (2) assign communication routes from each participating processor to its successor in the ring. One major difficulty of this ring embedding process is that some of the communication routes may have to share some physical communication links: indeed, the communication networks of heterogeneous platforms typically are sparse, i.e., far from being fully connected. If two or more routes share the same physical link, we have to decide which portion of the link bandwidth is to be assigned to each route. Once the ring and the routing have been decided, there remains to determine the best partitioning of the application data. Clearly, the quality of the final solution depends on many application and architecture parameters, and we should expect the optimization problem to be very difficult to solve.

29.2.3.1 Framework

29.2.3.1.1 Computing Costs

The target computing platform is modeled as a directed graph $G = (P, E)$. Each node P_i in the graph, $1 \leq i \leq |P| = p$, models a computing resource, and is weighted by its relative execution time w_i: P_i requires w_i time-steps to process a unit-size task.

29.2.3.1.2 Communication Costs

Graph edges represent communication links and are labeled with available bandwidths. If there is an oriented link $e \in E$ from P_i to P_j, let b_e denote the bandwidth of the link. It will take D_c/b_e time-units to transfer a single message of size D_c from P_i to P_j using link e. Note that all messages have same size D_c, since only boundary information is exchanged between consecutive processors.

29.2.3.1.3 Application Parameters: Computations

Let D_w be the total size of the work to be performed at each step of the algorithm. Processor P_i will accomplish a share $\alpha_i \cdot D_w$ of this total work, where $\alpha_i \geq 0$ for $1 \leq i \leq p$ and $\sum_{i=1}^{p} \alpha_i = 1$. Note that we allow $\alpha_j = 0$ for some index j, meaning that processor P_j does not participate in the computation. Indeed, there is no reason a priori for all resources to be involved, especially when the total work is not very large: the extra communications incurred by adding more processors may slow down the whole process, despite the increased cumulated speed. Also note that we allow $\alpha_i \cdot D_w$ to be rational: we assume the total work to be a divisible load (Bharadwaj et al., 1996; Gallet et al., 2009).

29.2.3.1.4 Application Parameters: Communications in the Ring

We arrange participating processors along a ring (yet to be determined). After updating its data slice of size $\alpha_i \cdot D_w$, each active processor P_i sends a message of fixed length D_c (typically some boundary data) to its successor and one to its predecessor. We assume that the successor of the last processor is the first processor, and that the predecessor of the first processor is the last processor. To illustrate the relationship between D_w and D_c, we can view the original data matrix as a large rectangle composed of D_w columns of height D_c, so that one single column is exchanged between any pair of consecutive processors in the ring. Let $\text{succ}(i)$ and $\text{pred}(i)$ denote the successor and the predecessor of P_i in the virtual ring. The time needed to transfer a message of size D_c from P_i to P_j is $D_c \cdot c_{i,j}$, where the communication delay $c_{i,j}$ will be instantiated in the following text according to whether links are assumed to be homogeneous or heterogeneous, to be shared or dedicated.

29.2.3.1.5 Objective Function

The total cost of a single step in the iterative algorithm is the maximum, over all participating processors, of the time spent computing and communicating:

$$T_{\text{step}} = \max_{1 \leq i \leq p} \mathbb{I}\{i\}[\alpha_i \cdot D_w \cdot w_i + D_c \cdot (c_{i,\text{pred}(i)} + c_{i,\text{succ}(i)})] \tag{29.2}$$

where $\mathbb{I}\{i\}[x] = x$ if P_i is involved in the computation, and 0 otherwise. Here we use a model where each processor sequentially sends messages to its two neighbors, and we implicitly assume asynchronous receptions. (Other architectural models could be worth investigating.)

In summary, our goal is to determine the best way to select q processors out of the p processors available, to assign them computational workloads, to arrange them along a ring and, if needed, to share the network bandwidth so that the total execution time per step is minimized. We denote our optimization problem as BUILDRING(p, w_i, E, b_e, D_w, D_c): knowing the number of processors and their respective execution times, the existing communication links and their respective bandwidths, the overall amount of work and the size of the border that two successive processors are exchanging, what is the shortest achievable iteration step T_{step}? The decision problem associated to the BUILDRING optimization problem is BuildRingDec(p, w_i, E, b_e, D_w, D_c, K): is there a solution to BuildRing, such that T_{step} is no greater than K?

We look at the complexity of our problem with respect to the model used for representing the interconnection network. We start from a simplistic model that we gradually make more accurate and thus more complicated. Along the way, we study how the problem complexity evolves.

29.2.3.2 Fully Connected Platforms with Homogeneous Communication Links

Here, we assume that all communication links have the same characteristics and that each pair of processors is connected by a direct (physical) network link. This corresponds, for example, to a cluster where processors are linked through a switch. Under these assumptions, we do not have to care about communication routing, or whether two distinct communications may use the same physical link.

By hypothesis, we have $c_{i,j} = c$ for all i and j, where c is a constant (it is the inverse of the bandwidth of any link). We have only two cases to consider: (1) only one processor is active; (2) all processors are involved. Indeed, as soon as a single communication occurs, we can have several ones happening in parallel for the same cost, and the best is then to divide the computing load among all resources. In the former case, the only active processor is obviously the fastest one and $T_{\text{step}} = D_w \cdot \min_{1 \leq i \leq p} w_i$. In the latter case, the load is most balanced when the execution time is the same for all processors: otherwise, removing a small fraction of the load of the processor with largest execution time, and giving it to a processor finishing earlier, would decrease the maximum computation time. This leads to $\alpha_i.w_i = Constant$ for all i, with $\sum_{i=1}^{p} \alpha_i = 1$. We derive that $T_{\text{step}} = D_w \cdot w_{\text{cumul}} + 2D_c \cdot c$, where $w_{\text{cumul}} = \frac{1}{\sum_{i=1}^{p} \frac{1}{w_i}}$. If the platform is given, there is an application-dependent threshold determining whether only the fastest computing resource, as opposed to all the resources, should be involved. Given D_c, the fastest processor will do all the job for small values of D_w, namely, when $D_w \leq D_c \cdot \frac{2c}{\min_{1 \leq i \leq p} w_i - w_{\text{cumul}}}$. For larger values of D_w, all processors should be involved.

29.2.3.3 Fully Connected Platforms with Heterogeneous Communication Links

We are now assuming a more general model, where communication links can have different characteristics but where the interconnection network is still assumed to be complete. As the interconnection network is still complete, we assume that if a communication must occur from a processor P_i to a processor P_j, it happens along the direct physical link connecting these two processors. Therefore, as previously, we do not have to care about communication routing, or whether two distinct communications may use the same physical link.

Before formally stating the complexity of our problem on such a platform, let us consider the special case where all processors are involved in an optimal solution. All the p processors require the same amount of time to compute and communicate: otherwise, we would once again slightly decrease the computing load of the processor completing its assignment last (computations followed by communications) and assign extra work to another one. Hence (see Figure 29.8 for an illustration) we have

$$T_{\text{step}} = \alpha_i \cdot D_w \cdot w_i + D_c \cdot (c_{i,\text{pred}(i)} + c_{i,\text{succ}(i)})$$

for all i. Since $\sum_{i=1}^{p} \alpha_i = 1$, and defining $w_{\text{cumul}} = \frac{1}{\sum_{i=1}^{p} \frac{1}{w_i}}$ as before, we obtain

$$\frac{T_{\text{step}}}{D_w \cdot w_{\text{cumul}}} = 1 + \frac{D_c}{D_w} \sum_{i=1}^{p} \frac{c_{i,\text{pred}(i)} + c_{i,\text{succ}(i)}}{w_i}. \tag{29.3}$$

Therefore, T_{step} will be minimal when $\sum_{i=1}^{p} \frac{c_{i,\text{pred}(i)} + c_{i,\text{succ}(i)}}{w_i}$ is minimal. This will be achieved by the ring that corresponds to the shortest Hamiltonian cycle in the graph $G = (P, E)$, where each edge

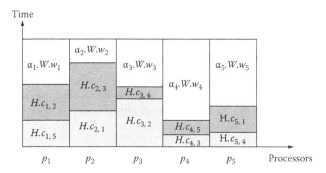

Time

p_1 p_2 p_3 p_4 p_5 Processors

FIGURE 29.8 Summary of computation and communication times with $q = 5$ processors.

$e_{i,j}$ is given the weight $d_{i,j} = \frac{c_{i,j}}{w_i} + \frac{c_{j,i}}{w_j}$. Therefore, if all processors are involved, our optimization problem is equivalent to a traveling salesman problem. This hints at the NP-completeness of the problem. Formally, the following theorem states that, as soon as the communication links become heterogeneous, the optimization problem becomes NP-complete (see [Legrand et al., 2004] for a detailed proof):

THEOREM 29.1   BUILDRINGDEC$(p, w_i, E = \{(P_j, P_k)\}_{1 \le j,k \le p, j \ne k}, b_e, D_w, D_c, K)$ is NP-complete.

29.2.3.4 Heterogeneous Platforms

We are now considering general (sparse) interconnection networks. The problem becomes even harder. Formally, the decision problem associated to the BUILDRING optimization problem turns out to be NP-complete. To really show the complexity induced by the fact that the interconnection network is no longer complete, we consider the toy example presented in Figure 29.9. This example contains five processors and five bidirectional communication links. For the sake of simplicity, processors P_1 to P_4 are labeled in the order that they appear in the 4-processor ring that we construct leaving out the other processor, R. Also, links are labeled with letters from a to e instead of indices; we use b_x to denote the bandwidth of link x.

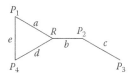

FIGURE 29.9   A small-size cluster.

29.2.3.4.1 Routing

There may no longer exist a direct route between any two processors. We then assume that we can freely decide how to route messages from one processor to another. We thus have to define the communication path S_i (S stands for successor) from P_i to $P_{\text{succ}(i)}$. Similarly, we have to define the communication path $\mathcal{P}_i$ ($\mathcal{P}$ stands for predecessor) from P_i to $P_{\text{pred}(i)}$. In our example, for S_1, we arbitrarily choose to use links a and b so that the communication path from P_1 to its successor P_2 is $S_1 = \{a, b\}$. But for the path $\mathcal{P}_2$ we may use links b, d, and e so that the communication path from P_2 to its predecessor P_1 is $\mathcal{P}_2 = \{b, d, e\}$. Here is the complete and the arbitrary list of paths (note that many other choices could have been made):

- From P_1: to P_2, $S_1 = \{a, b\}$ and to P_4, $\mathcal{P}_1 = \{e\}$
- From P_2: to P_3, $S_2 = \{c\}$ and to P_1, $\mathcal{P}_2 = \{b, d, e\}$
- From P_3: to P_4, $S_3 = \{c, b, d\}$ and to P_2, $\mathcal{P}_3 = \{c\}$
- From P_4: to P_1, $S_4 = \{e\}$ and to P_3, $\mathcal{P}_4 = \{d, b, c\}$

29.2.3.4.2 Communication Costs

It takes D_c/b_e time-units to transfer a single message of size D_c through link e. When a communication uses several links, the overall speed is defined by the link with the smallest available bandwidth. When there are several messages sharing the link, each of them receives a portion (to be determined) of the available bandwidth. For instance, if there are two messages sharing link e, and if the first message is allocated two-thirds of the bandwidth, i.e., $2b_e/3$, then the second message cannot use more than $b_e/3$. We assume that the portions of the bandwidth allocated to the messages can be freely determined by the user, the only rule is that the sum of all these portions cannot exceed the total link bandwidth. Let $s_{i,m}$ be the portion of the bandwidth b_{e_m} of the physical link e_m that has been allocated to the path $\mathcal{S}_i$. Of course if a link e_r is not used in the path, then $s_{i,r} = 0$. Let $c_{i,\mathrm{succ}(i)} = \frac{1}{\min_{e_m \in S_i} s_{i,m}}$: then P_i requires $D_c \cdot c_{i,\mathrm{succ}(i)}$ time-units to send its message of size D_c to its successor $P_{\mathrm{succ}(i)}$. Similarly, let $p_{i,m}$ be the portion of the bandwidth b_{e_m} of the physical link e_m that has been allocated to the path $\mathcal{P}_i$, and $c_{i,\mathrm{pred}(i)} = \frac{1}{\min_{e_m \in \mathcal{P}_i} p_{i,m}}$. Then P_i requires $D_c \cdot c_{i,\mathrm{pred}(i)}$ time-units to send its message of size D_c to its predecessor $P_{\mathrm{pred}(i)}$. Because of the resource constraints on the link bandwidths, the allocated bandwidth portions must satisfy the following set of inequations:

$$
\begin{array}{ll}
\text{Link a: } s_{1,a} \leq b_a & \text{Link b: } s_{1,b} + s_{3,b} + p_{2,b} + p_{4,b} \leq b_b \\
\text{Link c: } s_{2,c} + s_{3,c} + p_{3,c} + p_{4,c} \leq b_c & \text{Link d: } s_{3,d} + p_{2,d} + p_{4,d} \leq b_d \\
\text{Link e: } s_{4,e} + p_{1,e} + p_{2,e} \leq b_e &
\end{array}
$$

Finally, we write the communication costs of the communication paths. For P_1, because $\mathcal{S}_1 = \{a, b\}$, we get $c_{1,2} = \frac{1}{\min(s_{1,a}, s_{1,b})}$; and because $\mathcal{P}_1 = \{e\}$, we get $c_{1,4} = \frac{1}{p_{1,e}}$. We proceed likewise for P_2 to P_4.

Now that we have all the constraints, we can (try to) compute the α_i, $s_{i,j}$, and $p_{i,j}$ minimizing the objective function T_{step}. Equation 29.4 summarizes the whole system of (in)equations, which is quadratic in the unknowns α_i, $s_{i,j}$, $c_{i,j}$, and $p_{i,j}$.[*] For instance, the constraint $s_{1,a} \cdot c_{1,2} \geq 1$ comes from the condition $c_{1,2} = \frac{1}{\min(s_{1,a}, s_{1,b})}$. The fact that this system is quadratic is easily explained by the fact that we are allocating bandwidths, while we are trying to optimize time, which is the inverse of bandwidths.

$$
\textsc{Minimize} \quad \max_{1 \leq i \leq 4} \left(\alpha_i \cdot D_w \cdot w_i + D_c \cdot (c_{i,i-1} + c_{i,i+1})\right) \quad \text{subject to}
$$

$$
\begin{array}{lll}
\sum_{i=1}^{4} \alpha_i = 1 & & \\
s_{1,a} \leq b_a & s_{1,b} + s_{3,b} + p_{2,b} + p_{4,b} \leq b_b & s_{2,c} + s_{3,c} + p_{3,c} + p_{4,c} \leq b_c \\
s_{3,d} + p_{2,d} + p_{4,d} \leq b_d & s_{4,e} + p_{1,e} + p_{2,e} \leq b_e & \\
s_{1,a} \cdot c_{1,2} \geq 1 & s_{1,b} \cdot c_{1,2} \geq 1 & p_{1,e} \cdot c_{1,4} \geq 1 \\
s_{2,c} \cdot c_{2,3} \geq 1 & p_{2,b} \cdot c_{2,1} \geq 1 & p_{2,d} \cdot c_{2,1} \geq 1 \\
p_{2,e} \cdot c_{2,1} \geq 1 & s_{3,c} \cdot c_{3,4} \geq 1 & s_{3,b} \cdot c_{3,4} \geq 1 \\
s_{3,d} \cdot c_{3,4} \geq 1 & p_{3,c} \cdot c_{3,2} \geq 1 & s_{4,e} \cdot c_{4,1} \geq 1 \\
p_{4,d} \cdot c_{4,3} \geq 1 & p_{4,b} \cdot c_{4,3} \geq 1 & p_{4,c} \cdot c_{4,3} \geq 1
\end{array}
$$

$$
(29.4)
$$

29.2.3.4.3 Conclusion

To build up System 29.4, we have used arbitrary communication paths. There are of course many other paths to be tried. Worse, there are many other rings to be tried, made with the same processors arranged differently or using other processors. The number of processors q must be varied too. In other words, if we assume that (1) the processor selection was performed; (2) the selected processors

---

[*] We did not express in Equation 29.4 the inequations stating that all the unknowns are nonnegative.

were arranged into a ring; and that (3) the communication paths were defined, then the remaining load (and bandwidth) distribution problem is solved by a closed-form expression in the case of complete graphs and by a quadratic system in the general case.

Having a more realistic network model, first assuming that the network is no more homogeneous, and then that it is no more complete, each time makes our problem far more complicated. The question now is whether we gain anything by considering more accurate models that forbid us to derive optimal solutions in polynomial-time.

29.2.3.5 Heuristics

We use Equation 29.3 as a basis for a greedy algorithm to grow a solution ring iteratively. The greedy heuristic starts by selecting the best pair of processors. Then, it iteratively includes a new node in the current solution ring. Assume that we have already selected a ring of r processors. For each remaining processor P_k, we search where to insert it in the current ring: for each pair of successive processors (P_i, P_j) in the ring, we evaluate with Equation 29.3 the quality of the ring obtained by inserting P_k between P_i and P_j. Together with the ring, we have $2r$ communicating paths, and a certain portion of the initial bandwidth has been allocated to these paths. To build the two new paths involving P_k, we reason on the graph $G = (V, E, b)$ where each edge is labeled with the remaining available bandwidth: $b(e_m)$ is no longer the initial bandwidth of edge e_m, but what has been left by the $2r$ paths. We then build the required communication paths, allocate them bandwidths using the max–min fairness heuristic (Bertsekas and Gallager, 1987), and then obtain a value for T_{step}. We retain the processor and the pair that minimize T_{step}. Finally, among the solution that only uses the fastest processor and the rings built for each ring size in $[2, p]$, we pick the solution having the smallest T_{step}. It is important to try all the values of the ring size as T_{step} may not vary monotonically with the ring size.

A concise way to describe the heuristic is the following: we greedily grow a ring by peeling off the bandwidths to insert new processors. When we are done with the heuristic, we have a q-processor ring, q workloads, $2q$ communicating paths, bandwidth portions, and communication costs for these paths, and a feasible value of T_{step}.

To assess the importance of having an accurate communication model, we compare the above heuristic run on the exact platform graph, to the very same heuristic run on the clique version of the same platform: the platform graph is then a complete graph of same vertices where each (virtual) edge is labeled by the bandwidth of the shortest-path (in terms of bandwidth) of the corresponding computing nodes of the actual platform. Clearly, the value of T_{step} achieved by the second heuristic may well not be feasible as the actual network is not fully connected. Therefore, to compute the actual performance of the solution built, we keep the ring, and the communicating paths between adjacent processors in the ring, and we compute feasible bandwidth values using the max–min fairness heuristic.

We only report the conclusion of the experiments (Legrand et al., 2004). When both heuristics are compared on actual platforms, the first one, that takes possible link contentions into account, clearly outperforms the second when communication matters, i.e., when the communication-to-computation ratio is high.

29.2.3.6 Conclusion

Here we have studied the complexity of a particular problem, namely, the load-balancing of iterative computations with respect to the model of the interconnection network. As expected, the more accurate and complex the network model, the more difficult the optimization problem. Then, one may wonder whether using an accurate network model—which forbids us to derive optimal solutions in polynomial-time—is more of a loss than a gain. In other words, we may wonder whether a solution, optimal for a simpler model, and roughly adapted to the general case, would not have better performance than a solution directly designed for the general case, but relying on a bunch of non-guaranteed heuristics. Using models of actual platforms, one can evaluate the heuristics designed for two different models of interconnection networks. The comparison is meaningful as the heuristic for the more general model is equivalent to the other heuristic when used on the simpler model. The evaluation clearly shows that, in this context, we must use accurate platform models in order to design better heuristics and to better harness the processing power of Grid platforms.

29.3 Case Study: Matrix Product under Memory Constraints

29.3.1 Introduction

Matrix product is a key computational kernel in many scientific applications, and it has been extensively studied on parallel architectures. Two well-known parallel versions are Cannon's algorithm (Cannon, 1969) and the ScaLAPACK outer product algorithm (Blackford et al., 1997). Typically, parallel implementations work well on 2D processor arrays, because the input matrices are sliced horizontally and vertically into square blocks that are mapped one-to-one onto the physical resources; several communications can take place in parallel, both horizontally and vertically. Even better, most of these communications can be overlapped with (independent) computations. All these characteristics render the matrix product kernel quite amenable to an efficient parallel implementation on 2D processor arrays. On a Grid, however, the computing resources are interconnected by a sparse network: there are no direct links between any pair of processors and assuming the interconnection network to be a 2D grid would lead to communication contentions and performance degradations. A new parallelization approach should thus be undertaken.

Furthermore, as the Grid may contain long-distance, and thus slow-communicating, network links, it becomes necessary to include the cost of both the initial distribution of the matrices to the processors and of collecting back the results. These input/output operations have always been neglected in the analysis of the conventional algorithms. This is because only $\Theta(n^2)$ coefficients need to be distributed in the beginning, and gathered at the end, as opposed to the $\Theta(n^3)$ computations* to be performed (where n is the problem size). The assumption that these communications can be ignored could have made sense on dedicated parallel machines like the Intel Paragon, but it is no longer reasonable on heterogeneous platforms. Furthermore, when processors cannot store all the matrices in their memory, the total volume of communication required can be larger than $\Theta(n^2)$ as the same matrix element may have to be sent several times to the same processor.

We therefore adopt an application scenario where input files are read from a fixed repository (a disk on a data server). Computations will be delegated to available resources in the target architecture, and results will be returned to the repository. This calls for a master–worker paradigm, or more precisely for a computational scheme where the master (the processor holding the input data) assigns computations to other resources, the workers. In this centralized approach, all matrix files originate from, and must be returned to, the master. The master distributes both data and computations to the workers. Finally, because we investigate the parallelization of large problems, we cannot assume that full matrix panels can be stored in worker memories and be reused for subsequent updates (e.g., as in ScaLAPACK).

To summarize, the target platform is composed of several workers with different computing powers, different bandwidth links to/from the master, and different, limited, memory capacities. The first problem is resource selection. Which workers should be enrolled in the execution? All of them, or maybe only the faster computing ones, or else only the faster-communicating ones?

---

* Of course, there are $\Theta(n^3)$ computations if we only consider algorithms that uses the standard way of multiplying matrices; this excludes Strassen's and Winograd's algorithms (Cormen et al., 1990).

Once participating resources have been selected, there remain several scheduling decisions to take: How to minimize the number of communications? In which order workers should receive input data and return results? What amount of communications can be overlapped with (independent) computations?

The rest of this section surveys recent work by Dongarra et al. (2008a,b). In Section 29.3.2, we state the scheduling problem precisely, and we introduce some notations. Next, in Section 29.3.3, we proceed with the analysis of the total communication volume that is needed in the presence of memory constraints. We show how to improve a well-known bound by Toledo (1999) and Ironya et al. (2004), and we outline an algorithm (Dongarra et al., 2008a) almost achieving this bound on platforms with a single worker. We deal with homogeneous platforms in Section 29.3.4, and with heterogeneous ones in Section 29.3.5. We report on some MPI experiments in Section 29.3.6, and we conclude in Section 29.3.7.

29.3.2 Framework and Notations

Here, we formally state the hypotheses on the application and on the target platform.

29.3.2.1 Application

We deal with the computational kernel $C \leftarrow C + A \cdot B$. We partition the three matrices A, B, and C as illustrated in Figure 29.10. More precisely

- We use a block-oriented approach. The atomic elements that we manipulate are not matrix coefficients but instead square blocks of size $q \times q$ (hence with q^2 coefficients). This is to harness the power of Level 3 BLAS routines (Blackford et al., 1996). Typically, $q = 80$ or 100 when using ATLAS-generated routines (Whaley and Dongarra, 1998; Clint Whaley et al., 2001).
- The input matrix A is of size $n_A \times n_{AB}$:
  - We split A into r horizontal stripes A_i, $1 \le i \le r$, where $r = n_A/q$.
  - We split each stripe A_i into t square $q \times q$ blocks $A_{i,k}$, $1 \le k \le t$, where $t = n_{AB}/q$.

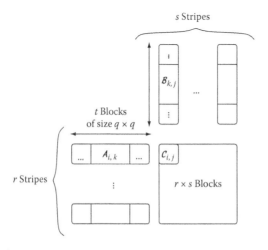

FIGURE 29.10 Partition of the three matrices.

- The input matrix $\mathcal{B}$ is of size $n_{\mathcal{AB}} \times n_{\mathcal{B}}$:
    - We split $\mathcal{B}$ into s vertical stripes $\mathcal{B}_j$, $1 \leq j \leq s$, where $s = n_{\mathcal{B}}/q$.
    - We split each stripe $\mathcal{B}_j$ into t square $q \times q$ blocks $\mathcal{B}_{k,j}$, $1 \leq k \leq t$.
- We compute $\mathcal{C} = \mathcal{C} + \mathcal{A} \cdot \mathcal{B}$. Matrix $\mathcal{C}$ is accessed (both for input and output) by square $q \times q$ blocks $\mathcal{C}_{i,j}$, $1 \leq i \leq r$, $1 \leq j \leq s$; there are $r \times s$ such blocks.

We point out that, with such a decomposition, all stripes and blocks have same size. This will greatly simplify the analysis of communication costs.

29.3.2.2 Platform

We target a star network $\mathcal{S} = \{P_0, P_1, P_2, \ldots, P_p\}$, composed of a master P_0 and of p workers P_i, $1 \leq i \leq p$ (see Figure 29.11). Because we manipulate large data blocks, we adopt a linear cost model, both for computations and communications (i.e., we neglect start-up overheads). We have the following notations:

- It takes $X.w_i$ time-units to execute a task of size X on P_i.
- It takes $X.c_i$ time units for the master P_0 to send a message of size X to P_i or to receive a message of size X from P_i.

Our star platforms are thus fully heterogeneous, both in terms of computations and of communications. Without the loss of generality, we assume that the master has no processing capability (otherwise, add a fictitious extra worker paying no communication cost to simulate computation at the master).

For the communication model, we once again use the strict one-port model, which is defined as follows:

- The master can only send data to, and receive data from, a single worker at a given time-step, and it cannot be enrolled in more than one communication at any time-step.
- A given worker cannot start an execution before it has terminated to receive the needed data from the master; similarly, it cannot start sending the results back to the master before finishing the computation.

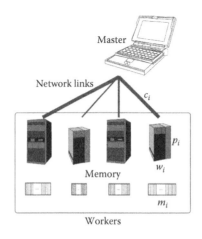

FIGURE 29.11 Fully heterogeneous platform.

Our final assumption is related to memory capacity; we assume that a worker P_i can only store m_i blocks (either from A, B, and/or C). For large problems, this memory limitation will considerably impact the design of the algorithms, as data reuse will be greatly dependent on the amount of available buffers.

29.3.3 Minimization of the Communication Volume

The initial assumption is that communication costs dominate the problem. Therefore, we do not directly target makespan minimization, but communication volume minimization. Experiments, in Section 29.3.6, will show that the communication-volume minimization approach effectively leads to algorithms with shorter makespans.

We thus want to derive a lower bound on the total number of communications (sent from, or received by, the master) that are needed to execute any matrix multiplication algorithm. As we are only interested in minimizing the total communication volume, we can simulate any parallel algorithm on a single worker and we only need to consider the one-worker case.

We deal with the following formulation of the problem:

- The master sends blocks A_{ik}, B_{kj}, and C_{ij}.
- The master retrieves final values of blocks C_{ij}.
- We enforce limited memory on the worker; only m buffers are available, which means that at most m blocks of A, B, and/or C can simultaneously be stored in the worker.

First, we improve a lower bound on the communication volume established by Toledo (1999) and Ironya et al. (2004). Then, we describe an algorithm that aims at reusing C blocks as much as possible after they have been loaded, and we assess its performance.

29.3.3.1 Lower Bound on the Communication Volume

To derive the lower bound, the idea is to estimate the number of computations made, thanks to m consecutive communication steps (once again, the unit here is a matrix block). Using Loomis–Whitney inequality (Ironya et al., 2004), one can then show that a lower bound for the communication-to-computation ratio is

$$\text{CCR}_{\text{opt}} \geq \sqrt{\frac{27}{8m}}.$$

29.3.3.2 The Maximum Reuse Algorithm

The aforementioned lower bound on the communication volume is obtained when the three matrices A, B, and C are equally accessed during a sequence of communications. This may suggest to allocate one-third of the memory to each of these matrices. In fact, Toledo (1999) uses this memory layout. A closer look to the problem shows that the multiplied matrices, A and B, have the same behavior that differs from the behavior of the result matrix C. Indeed, if an element of C is no longer used, it cannot be simply discarded from the memory as the elements of A and B are, but it must be sent back to the master. Intuitively, sending an element of C to a worker also costs the communication needed to retrieve it from the worker, and is thus twice as expensive as sending an element of A or B. Hence the motivation to design an algorithm, which reuses as much as possible the elements of C.

Cannon's algorithm (Cannon, 1969) and the ScaLAPACK outer product algorithm (Blackford et al., 1997) both distribute square blocks of C to the processors. Intuitively, squares are better than elongated rectangles because their perimeter (which is proportional to the communication volume)

1	μ				μ²															
A	B	B	B	B	C	C	C	C	C	C	C	C	C	C	C	C	C	C	C	C

FIGURE 29.12 Memory layout for the maximum reuse algorithm when $m = 21$: $\mu = 4$; one block is used for $\mathcal{A}$, μ for $\mathcal{B}$, and μ^2 for $\mathcal{C}$.

is smaller for the same area. We use the same approach here, even if there are no optimality results to justify it.

The maximum reuse algorithm uses the memory layout illustrated in Figure 29.12. Four consecutive execution steps are shown in Figure 29.13. Assume that there are m available buffers. First we find μ as the largest integer, such that $1 + \mu + \mu^2 \leq m$. The idea is to use one buffer to store $\mathcal{A}$ blocks, μ buffers to store $\mathcal{B}$ blocks, and μ^2 buffers to store $\mathcal{C}$ blocks. In the outer loop of the algorithm, a $\mu \times \mu$ square of $\mathcal{C}$ blocks is loaded. Once these μ^2 blocks have been loaded, they are repeatedly updated in the inner loop of the algorithm until their final value is computed. Then the blocks are returned to the master, and μ^2 new $\mathcal{C}$ blocks are sent by the master and stored by the worker. As illustrated in Figure 29.12, we need μ buffers to store a row of $\mathcal{B}$ blocks, but only one buffer for $\mathcal{A}$ blocks: $\mathcal{A}$ blocks are sent in sequence, each of them is used in combination with a row of μ $\mathcal{B}$ blocks to update the corresponding row of $\mathcal{C}$ blocks. This leads to the following sketch of the algorithm:

Outer loop: while there remain $\mathcal{C}$ blocks to be computed

- Store μ^2 blocks of $\mathcal{C}$ in worker's memory: $\{\mathcal{C}_{i,j} \mid i_0 \leq i < i_0 + \mu, \ j_0 \leq j < j_0 + \mu\}$.
- **Inner loop**: For each k from 1 to t

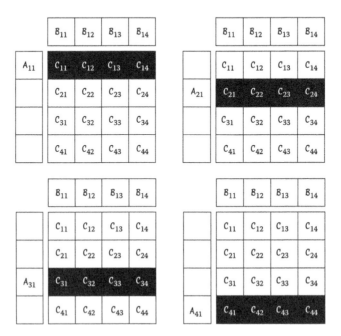

FIGURE 29.13 Four steps of the maximum reuse algorithm, with $m = 21$ and $\mu = 4$. The elements of $\mathcal{C}$ updated are displayed on white on black.

1. Send a row of μ elements $\{\mathcal{B}_{k,j} \mid j_0 \le j < j_0 + \mu\}$.
2. Sequentially send μ elements of column $\{\mathcal{A}_{i,k} \mid i_0 \le i < i_0 + \mu\}$. For each $A_{i,k}$, update μ elements of $\mathcal{C}$.

- Return results to master.

The performance of one iteration of the outer loop of the maximum reuse algorithm can be readily determined:

- We need $2\mu^2$ communications to send and retrieve $\mathcal{C}$ blocks.
- For each value of t
  - We need μ elements of $\mathcal{A}$ and μ elements of $\mathcal{B}$.
  - We update μ^2 blocks.

In terms of block operations, the communication-to-computation ratio achieved by the algorithm is thus

$$\text{CCR} = \frac{2\mu^2 + 2\mu t}{\mu^2 t} = \frac{2}{t} + \frac{2}{\mu}.$$

For large problems, i.e., large values of t, we see that CCR is asymptotically close to the value $\text{CCR}_\infty = \frac{2}{\sqrt{m}}$. We point out that, in terms of data elements, the communication-to-computation ratio is divided by a factor q. Indeed, a block consists of q^2 coefficients but an update requires q^3 floating-point operations. Also, the ratio CCR_∞ achieved by the maximum reuse algorithm is lower by a factor $\sqrt{3}$ than the ratio achieved by the blocked matrix-multiply algorithm of Toledo (1999). Finally, we remark that the performance of the maximum reuse algorithm is quite close to the lower bound derived earlier:

$$\text{CCR}_\infty = \frac{2}{\sqrt{m}} = \sqrt{\frac{32}{8m}}.$$

29.3.4 Algorithms for Homogeneous Platforms

We now adapt the maximum reuse algorithm to fully homogeneous platforms. We have a limitation on the memory capacity. So we must first decide which part of the memory will be used to store which part of the original matrices, in order to maximize the total number of computations per time unit.

We load into the memory of each worker μ blocks of $\mathcal{A}$ and μ blocks of $\mathcal{B}$ to compute μ^2 blocks of $\mathcal{C}$ (in other words, we waste some memory in order to decrease the number of communications and the synchronization effects). In addition, we need 2μ extra buffers, split into μ buffers for $\mathcal{A}$ and μ for $\mathcal{B}$, in order to overlap computation and communication steps. In fact, μ buffers for $\mathcal{A}$ and μ for $\mathcal{B}$ would suffice for each update, but we need to prepare for the next update while computing. Overall, the number μ^2 of $\mathcal{C}$ blocks that we can simultaneously load into memory is defined by the largest integer μ such that: $\mu^2 + 4\mu \le m$.

We have to determine the number of participating workers $\mathfrak{P}$. On the communication side, we know that in a round (computing a $\mathcal{C}$ block entirely), the master exchanges with each worker $2\mu^2$ blocks of $\mathcal{C}$ (μ^2 sent and μ^2 received), and sends μt blocks of $\mathcal{A}$ and μt blocks of $\mathcal{B}$. Also during this round, on the computation side, each worker computes $\mu^2 t$ block updates. If we enroll too many processors, the communication capacity of the master will be exceeded. There is a limit on the number of blocks sent per time unit, hence on the maximal processor number $\mathfrak{P}$, which we compute as follows: $\mathfrak{P}$ is the smallest integer, such that $2\mu tc \times \mathfrak{P} \ge \mu^2 tw$. Indeed, this is the smallest value to saturate the communication capacity of the master required to sustain the corresponding computations. Finally, we cannot use more processors than are available and, in the context of matrix multiplication, we have $c = q^2 \tau_c$ and $w = q^3 \tau_a$, where τ_c and τ_a, respectively, represent the elementary communication and computation times. Hence we obtain the formula:

$$\mathfrak{P} = \min \left\{ p, \left\lceil \frac{\mu q}{2} \frac{\tau_a}{\tau_c} \right\rceil \right\}.$$

Finally, the participating workers receive data in a round-robin fashion.

29.3.5 Algorithms for Heterogeneous Platforms

We now consider the general problem, i.e., when processors are heterogeneous in terms of memory size as well as computation or communication time. As in the previous section, m_i is the number of $q \times q$ blocks that fit in the memory of worker P_i, and we need to load into the memory of P_i $2\mu_i$ blocks of $\mathcal{A}$, $2\mu_i$ blocks of $\mathcal{B}$, and μ_i^2 blocks of $\mathcal{C}$. This number of blocks loaded into the memory changes from worker to worker, because it depends upon their memory capacities. We first compute all the different values of μ_i, μ_i being the largest integer, such that: $\mu_i^2 + 4\mu_i \leq m_i$. To adapt the maximum reuse algorithm to heterogeneous platforms, the first idea would be to adopt a steady-state-like approach. The problem with such a solution, however, is that workers may not have enough memory to execute it! Therefore, this solution cannot always be realized in practice and, to avoid such memory problems, resource selection will be performed through a step-by-step simulation. However, we point out that a steady-state solution can be seen as an upper bound of the performance that can be achieved.

The different memory capacities of the workers imply that we assign them chunks of different sizes. This requirement complicates the global partitioning of the $\mathcal{C}$ matrix among the workers. To take this into account, while simplifying the implementation, the algorithm only assigns full matrix column blocks. This is done in a two-phase approach.

In the first phase, the allocation of blocks to processors is pre-computed using a processor selection algorithm later described. We start as if we had a huge matrix of size $\infty \times \sum_{i=1}^{p} \mu_i$. Each time a processor P_i is chosen by the processor selection algorithm, it is assigned a square chunk of μ_i^2 $\mathcal{C}$ blocks. As soon as some processor P_i has enough blocks to fill up μ_i block columns of the initial matrix, we decide that P_i will indeed execute these columns during the parallel execution. Therefore we maintain a panel of $\sum_{i=1}^{p} \mu_i$ block columns and fill them out by assigning blocks to processors. We stop this phase as soon as all the $r \times s$ blocks of the initial matrix have been allocated columnwise by this process. Note that worker P_i will be assigned a block column after it has been selected $\lceil \frac{r}{\mu_i} \rceil$ times by the algorithm.

In the second phase we perform the actual execution. Messages will be sent to workers according to the previous selection process. The first time a processor P_i is selected, it receives a square chunk of μ_i^2 $\mathcal{C}$ blocks, which initializes its repeated pattern of operation; the following t times, P_i receives μ_i $\mathcal{A}$ and μ_i $\mathcal{B}$ blocks, which requires $2\mu_i c_i$ time-units.

To decide which processor to select at each step of the first phase, one can imagine two variants of an incremental algorithm, a global one that aims at optimizing the overall communication-to-computation ratio, and a local one that selects the best processor for the next stage.

29.3.5.1 Global Selection Algorithm

The intuitive idea for this algorithm is to select the processor that maximizes the ratio of the total work achieved so far (in terms of block updates) over the completion time of the last communication. The latter represents the time spent by the master so far, either sending data to workers or staying idle waiting for the workers to finish their current computations. Estimating computations is easy: P_i executes μ_i^2 block updates per assignment. Communications are slightly more complicated to deal with; we cannot just use the communication time $2\mu_i c_i$ of P_i for the $\mathcal{A}$ and $\mathcal{B}$ blocks because we need to take its ready time into account. Indeed, if P_i is currently busy executing work, it cannot receive additional data too much in advance because its memory is limited. Please see Dongarra et al. (2008b) for further details.

29.3.5.2 Local Selection Algorithm

The global selection algorithm picks, as the next processor, the one that maximizes the ratio of the total amount of work assigned over the time needed to send all the required data. Instead, the local selection algorithm chooses, as destination of the ith communication, the processor that maximizes the ratio of the amount of work assigned by this communication over the time during which the communication link is used to performed this communication (i.e., the elapsed time between the end of $(i-1)$th communication and the end of the ith communication). As previously, if processor P_j is the target of the ith communication, the ith communication is the sending of μ_j blocks of $\mathcal{A}$ and μ_j blocks of $\mathcal{B}$ to processor P_j, which enables it to perform μ_j^2 updates.

29.3.6 MPI Experiments

An experimental evaluation enables to assess the practical importance of the algorithm designed above.

29.3.6.1 Platforms

We used a heterogeneous cluster composed of twenty-seven processors. It is composed of four different homogeneous sets of machines. The different sets are composed of (1) 8 SuperMicro servers 5013-GM, with processors P4 2.4 GHz; (2) 5 SuperMicro servers 6013PI, with processors P4 Xeon 2.4 GHz; (3) 7 SuperMicro servers 5013SI, with processors P4 Xeon 2.6 GHz; (4) 7 SuperMicro servers IDE250W, with processors P4 2.8 GHz.

All nodes have 1 GB of memory and are running the Linux operating system. The nodes are connected with a switched 10 Mbps Fast Ethernet network. As this platform may not be as heterogeneous as we would like, we sometimes artificially modify its heterogeneity. In order to artificially slow down a communication link, we send the same message several times to one worker. The same idea works for processor speeds: we ask a worker to compute a given matrix-product several times in order to slow down its computation capability. In all experiments, except the last batch, we used nine processors: one master and eight workers.

29.3.6.2 Algorithms

We chose four different algorithms from the literature that we compare our algorithms to. The closest work addressing our problem is Toledo's out-of-core algorithm (Toledo, 1999). Hence, this work serves as the baseline reference. Then we studied hybrid algorithms, i.e., algorithms which use our memory layout and are based on classical principles, such as round-robin, min–min (Maheswaran et al., 1999), or a dynamic demand-driven approach. Here, we only report on the latter as the round-robin and mini-min based algorithms had far worse performance. The first two algorithms below use our memory allocation, the only difference between them is the order in which the master sends blocks to workers.

Heterogeneous algorithm (**Het**) is our algorithm.
Overlapped Demand-Driven, Optimized Memory Layout (**ODDOML**) is a demand-driven algorithm. In our memory layout, two buffers of size μ_i are reserved for matrix $\mathcal{A}$, and two for matrix $\mathcal{B}$. In order to use the two available extra buffers (the second for $\mathcal{A}$ and the second for $\mathcal{B}$), one sends the next block to the first worker, which can receive it. This would be a dynamic version of our algorithm, if it took worker selection into account.

Block Matrix Multiply (**BMM**) is Toledo's algorithm (Toledo, 1999). It splits each worker memory equally into three parts and allocates one slot for a square block of $\mathcal{A}$, another for a square block of $\mathcal{B}$, and the last one for a square block of $\mathcal{C}$, with the square blocks having the same size. It sends blocks to the workers in a demand-driven fashion. First a worker receives a block of $\mathcal{C}$, then it receives corresponding blocks of $\mathcal{A}$ and $\mathcal{B}$ in order to update $\mathcal{C}$, until $\mathcal{C}$ is fully computed.

Note that the two algorithms using our optimized memory layout are considering matrices as composed of square blocks of size $q \times q = 80 \times 80$, while **BMM** loads three panels, each of size one-third of the available memory, for $\mathcal{A}$, $\mathcal{B}$, and $\mathcal{C}$.

When launching an algorithm on the platform, the very first step we do is to determine the platform's parameters. For that purpose, we launch a benchmark on it, in order to get the memory size, the communication speed, and the computation speed. The different speeds are determined by sending and computing a square block of size $q \times q$ ten times on each worker, and computing the median of the times obtained. This step represents at most 2% of the total time of execution. The times reported also take into account the decision process of the algorithms.

29.3.6.3 Evaluation Methodology

In the first three sets of experiments, we have only one parameter of heterogeneity, either the amount of memory, the communication speed, or the computation speed. We test the algorithms on such platforms with five matrices of increasing sizes. As we do not want to change several parameters at a time, we only change the value of parameter s (rather than, for instance, always consider square matrices). Matrix $\mathcal{A}$ is of size 8000×8000 whereas $\mathcal{B}$ is of increasing sizes 8000×64000, 8000×80000, 8000×96000, 8000×112000, and 8000×128000. For the other 14 experiments, which correspond to fully heterogeneous platforms, $\mathcal{A}$ is of size 8000×8000 and $\mathcal{B}$ is of size 8000×80000. The heterogeneous workers have different memory capacities, which implies that each algorithm, even **BMM**, assigns them chunks of different sizes. In order to simplify the global partitioning of matrix $\mathcal{C}$, we decide to only assign workers full matrix column blocks.

As we want to assess the performance of the studied algorithms independently of matrix sizes, we look at the relative cost of the algorithms rather than at their absolute execution times. The relative cost of a given algorithm on a particular instance is equal to the execution time, or makespan, achieved on that instance by the algorithm, divided by the minimum over all studied algorithms of the makespan achieved on that instance. Using relative cost also enables us to build statistics on the performance of algorithms.

Besides relative cost, we take into account the number of processors used. To assess the efficiency of a given algorithm, we look at its relative work, which is equal, for a given instance, to its makespan times the number of enrolled processors, divided by the minimum of this value over all studied algorithms.

29.3.6.4 Experimental Results

Figure 29.14 summarizes all our MPI experiments. Figure 29.14a and b respectively presents the relative cost and the relative work obtained for each experiment by our heterogeneous algorithm (**Het**), by Toledo's algorithm (**BMM**), and by the best of the dynamic heuristics using our memory layout (**ODDOML**). The results show the superiority of our memory allocation. Furthermore, if we add the resource selection of **Het**, not only do we achieve, most of the time, the best makespan, but also the best relative work as we also spare resources. Using our memory layout (**ODDOML**) rather than Toledo's (**BMM**) enables us to gain 19% of execution time on average. When this is combined with resource selection, this enables us to gain additionally 10%, which is 27% against Toledo's running time. We achieve this significant gain while sparing resources. Our **Het** algorithm

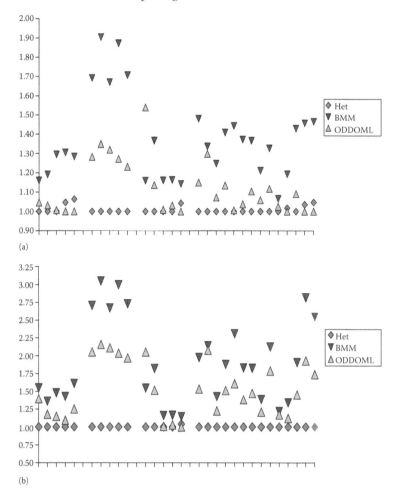

FIGURE 29.14 Summary of experiments. (a) Relative cost and (b) relative work.

is on average 1% away from the best achieved makespan. At worst **Het** is 14% away from the best makespan, **ODDOML** 61%, and **BMM** 128%. Altogether, we have thus been able to design an efficient, thrifty, and reliable algorithm.

29.3.7 Conclusion

Through a careful study of necessary communications, we have been able to design an algorithm minimizing the overall amount of communications. Experiments have then shown that the overall execution time of this algorithm was minimal when compared to existing approaches, stressing once again the importance of communications. This case study illustrates two facts that we had already emphasized: (1) a static approach can be more effective than classical dynamic approaches; and (2) resource selection enables to reduce execution time while limiting the number of participating resources.

29.4 Research Issues and Summary

In this chapter, we have discussed how to extend classic algorithm design and scheduling strategies to deal with Grid computing platforms. In the context of heterogeneous computations, static strategies have several drawbacks. Indeed, the network may be non-dedicated, and possible variations in the available computing power of processors have to be taken into account. Nevertheless, as we have shown, the best solution may not involve fully dynamic strategies as these can easily fall in the pitfalls laid by the heterogeneity of resources. One should rather consider mixed strategies: problems due to variations in processor speeds can be addressed by remapping data and computations from time to time between well identified static phases. We have shown that both accurate communication models and sophisticated resource selection strategies play a key role in the design of efficient solutions.

We believe that the design of efficient data allocation techniques for regular problems, similar to the ScaLAPACK library on homogeneous platforms, relies both on the design of efficient static distribution schemes (to cope with heterogeneity) and on the design of efficient remapping strategies (to cope with speed variations during the computations). However, the difficulty of algorithmic issues is not to be underestimated. Data decomposition, scheduling heuristics, and load balancing were already known to be difficult problems in the context of classical parallel architectures. They become extremely difficult in the context of heterogeneous clusters, not to mention Grid platforms.

Future research directions include scheduling structured applications on cluster and Grid platforms. Several solutions already exist, but most are over-simplistic (e.g., using very crude platform models or not taking communication costs into account) or target not so-pertinent objectives (e.g., trying to schedule applications one-by-one and independently when several are competing for the same resources). There is, thus, much room for improvement. This is especially true when one wants to schedule a collection of identical task-graphs concurrently submitted (by one or several users) to a Grid platform. (Such a work pattern is sometimes called a workflow.) This case can occur when the same application is run on different inputs, possibly when different users use the same reference application. A bag of identical tasks as dealt with in Section 29.2.2 is a very simple instance of this problem. The first extension of this case would be to move from independent tasks to more complex direct acyclic graphs (DAGs), such as collections of pipelines, fork-joins, or series-parallel graphs. The second extension is to consider several collections that execute concurrently and that compete for CPU and network resources. In both cases, finding the best trade-offs between performance-oriented criteria (throughput, response time, slowdown, or stretch) and maybe other criteria, such as reliability or energy minimization, is quite a difficult algorithmic challenge.

Another difficult challenge comes from the many sources of uncertainties in the definition of the algorithmic problem: the many application and platform parameters are only approximatively known, and may evolve over time. These uncertainties may require the design and the use of robust solutions able to cope with them, that is, whose performance do not significantly suffer from these uncertainties. One may also have to look for fault tolerance solutions, as faults may become an important problem in platforms containing a very large number of resources.

29.5 Further Information

Steady-state scheduling was pioneered by Bertsimas and Gamarnik (1999). Beaumont et al. (2005) have written a survey of steady-state scheduling on heterogeneous clusters. The other platform model used in this chapter, the divisible load model, is surveyed in Gallet et al. (2009). More references on this model can be found in Robertazzi (2003). About platform modeling, we refer the interested reader to Casanova (2005) specifically for network modeling issues, and more generally to Cappello et al. (2005).

There is a very abundant literature on scheduling and load balancing strategies for parallel and/or distributed systems. The survey (Shirazi et al., 1995) is a good starting point to this work. The most famous heuristic to schedule a task graph on an heterogeneous platform is HEFT (Topcuoglu et al., 2002) but, as we have already stated, there is still plenty of room to design far more powerful solutions to that problem.

Finally, Section 29.2.2 is mainly based on Beaumont et al. (2005), Section 29.2.3 on Legrand et al. (2004), and Section 29.3 on Dongarra et al. (2008a,b).

Defining Terms

Heterogeneity: resources are said to be heterogeneous when they do not all have exactly the same characteristics. For instance, the set of processors in a Grid is heterogeneous if they do not all have the same processing capabilities or the same amount of memory.

Makespan: the makespan of an application is the time at which the application's execution is completed.

Resource selection: is the problem of deciding which, among all available resources, should be used to solve a given problem.

Scheduling: is the problem of deciding at what time (when), and on which resource (where), to execute each of the (atomic) tasks that must be executed on the given platform.

References

Andonov, R. and Rajopadhye, S. 1997. Optimal orthogonal tiling of two-dimensional iterations. *Journal of Parallel and Distributed Computing*, 45(2):159–165.

Ausiello, G., Crescenzi, P., Gambosi, G., Kann, V., Marchetti-Spaccamela, A., and Protasi, M. 1999. *Complexity and Approximation*. Springer, Berlin, Germany.

Beaumont, O., Legrand, A., Marchal, L., and Robert, Y. 2005. Steady-state scheduling on heterogeneous clusters. *International Journal of Foundations of Computer Science*, 16(2):163–194.

Beaumont, O., Legrand, A., and Robert, Y. 2002. A polynomial-time algorithm for allocating independent tasks on heterogeneous fork-graphs. In: *ISCIS XVII, 17th International Symposium on Computer and Information Sciences*. CRC Press, Orlando, FL, pages 115–119.

Bertsekas, D. and Gallager, R. 1987. *Data Networks*. Prentice Hall, Englewood Cliffs, NJ.

Bertsimas, D. and Gamarnik, D. 1999. Asymptotically optimal algorithms for job shop scheduling and packet routing. *Journal of Algorithms*, 33(2):296–318.

Bharadwaj, V., Ghose, D., Mani, V., and Robertazzi, T. 1996. *Scheduling Divisible Loads in Parallel and Distributed Systems*. IEEE Computer Society Press, Los Alamitos, CA.

Blackford, L., Choi, J., Cleary, A., Demmel, J., Dhillon, I., Dongarra, J., Hammarling, S., Henry, G., Petitet, A., Stanley, K., Walker, D., and Whaley, R. C. 1996. ScaLAPACK: A portable linear algebra library for distributed-memory computers—Design issues and performance. In: *Supercomputing '96*. IEEE Computer Society Press, Los Alamitos, CA.

Blackford, L. S., Choi, J., Cleary, A., D'Azevedo, E., Demmel, J., Dhillon, I., Dongarra, J., Hammarling, S., Henry, G., Petitet, A., Stanley, K., Walker, D., and Whaley, R. C. 1997. *ScaLAPACK Users' Guide*. SIAM, Philadelphia, PA.

Boulet, P., Dongarra, J., Robert, Y., and Vivien, F. 1999. Static tiling for heterogeneous computing platforms. *Parallel Computing*, 25:547–568.

Cannon, L. E. 1969. A cellular computer to implement the Kalman filter algorithm. PhD thesis, Montana State University, Bozeman, MT.

Cappello, F., Fraigniaud, P., Mans, B., and Rosenberg, A. L. 2005. An algorithmic model for heterogeneous hyper-clusters: Rationale and experience. *International Journal of Foundations of Computer Science*, 16(2):195–216.

Casanova, H. 2005. Network modeling issues for grid application scheduling. *International Journal of Foundations of Computer Science*, 16(2):145–162.

Clint Whaley, R., Petitet, A., and Dongarra, J. J. 2001. Automated empirical optimizations of software and the atlas project. *Parallel Computing*, 27(1–2):3–35.

Cormen, T. H., Leiserson, C. E., and Rivest, R. L. 1990. *Introduction to Algorithms*. The MIT Press, Cambridge, MA.

Dongarra, J., Pineau, J.-F., Robert, Y., Shi, Z., and Vivien, F. 2008a. Revisiting matrix product on master-worker platforms. *International Journal of Foundations of Computer Science*, 19(6):1317–1336.

Dongarra, J., Pineau, J.-F., Robert, Y., and Vivien, F. 2008b. Matrix product on heterogeneous master-worker platforms. In: *13th ACM SIGPLAN Symposium on Principles and Practice of Parallel Programming*, Salt Lake City, UT.

Dutot, P. 2003. Master–slave tasking on heterogeneous processors. In: *International Parallel and Distributed Processing Symposium IPDPS'2003*. IEEE Computer Society Press, Nice, France.

Gallet, M., Robert, Y., and Vivien, F. 2009. Divisible load scheduling. In: *Introduction to Scheduling*. CRC Press, Boca Raton, FL.

Högstedt, K., Carter, L., and Ferrante, J. 1997. Determining the idle time of a tiling. In: *Principles of Programming Languages*. ACM Press, Paris, France, pages 160–173. Extended version available as Technical Report UCSD-CS96-489.

Ironya, D., Toledo, S., and Tiskin, A. 2004. Communication lower bounds for distributed-memory matrix multiplication. *Journal of Parallel Distributed Computing*, 64(9):1017–1026.

Legrand, A., Renard, H., Robert, Y., and Vivien, F. 2004. Mapping and load-balancing iterative computations on heterogeneous clusters with shared links. *IEEE Transactions on Parallel Distributed Systems*, 15(6):546–558.

Maheswaran, M., Ali, S., Siegel, H., Hensgen, D., and Freund, R. 1999. Dynamic matching and scheduling of a class of independent tasks onto heterogeneous computing systems. In: *Eight Heterogeneous Computing Workshop*. IEEE Computer Society Press, Los Alamitos, CA, pages 30–44.

Ohta, H., Saito, Y., Kainaga, M., and Ono, H. 1995. Optimal tile size adjustment in compiling general DOACROSS loop nests. In: *1995 International Conference on Supercomputing*. ACM Press, New York, pages 270–279.

Robertazzi, T. 2003. Ten reasons to use divisible load theory. *IEEE Computer*, 36(5):63–68.

Shirazi, B. A., Hurson, A. R., and Kavi, K. M. 1995. *Scheduling and Load Balancing in Parallel and Distributed Systems*. IEEE Computer Science Press, Los Alamitos, CA.

Toledo, S. 1999. A survey of out-of-core algorithms in numerical linear algebra. In: *External Memory Algorithms and Visualization*. American Mathematical Society Press, Providence, RI, pages 161–180.

Topcuoglu, H., Hariri, S., and Wu, M. Y. 2002. Performance-effective and low-complexity task scheduling for heterogeneous computing. *IEEE Transactions on Parallel Distributed Systems*, 13(3):260–274.

Whaley, R. C. and Dongarra, J. 1998. Automatically tuned linear algebra software. In: *Proceedings of the ACM/IEEE Symposium on Supercomputing (SC'98)*. IEEE Computer Society Press, Washington, DC.

Wolfe, M. 1996. *High Performance Compilers for Parallel Computing*. Addison-Wesley, Redwood City, CA.

30

Uncheatable Grid Computing

Wenliang Du
Syracuse University

Mummoorthy Murugesan
Syracuse University

Jing Jia
Syracuse University

30.1 Introduction

The increasing needs for conducting complex computations have motivated computational outsourcing. Many users conduct complex computations only occasionally, and buying expensive computing equipment only for these occasional tasks is a waste of resources. In the situation like this, outsourcing their computational tasks to a third party (or a group of third parties) is a more viable solution. Moreover, the increasing reach and speed of the Internet have made computational outsourcing much convenient.

Although computational outsourcing frees users from conducting the computation by themselves, it introduces two challenging security issues. The first is privacy. Computational outsourcing usually involves sending input data to a third party. These data might contain private information. For example, if the task is to conduct data mining on a medical database to discover useful knowledge, the database often needs to be sent to the party that conducts the computation. This causes a major threat to privacy, because the patients' private information is in the database. A number of studies have been focusing on this privacy issue [1–4].

The other issue is integrity, i.e., to guarantee that computation tasks are carried out properly. Once computing power becomes a commodity, there will be incentives for the third parties to cheat, namely, they might not conduct the actual computation, but still claim that they have done so.

It is important to guarantee the correctness of those computations. This chapter focuses on the integrity issue.

A promising computational outsourcing paradigm is a computational grid, which is a novel, evolving infrastructure that provides unified, coordinated access to computing resources such as processor cycles, storage, etc. Wide varieties of systems, from small workstations to supercomputers, are linked to form a grid, thus presenting what is essentially a powerful virtual computer. All the complexities involved in managing resources of a grid are hidden from the clients, providing seamless access to computing resources. As a great advancement towards cost reduction, computational grids can be used as a replacement for supercomputers that are presently used in many computationally intensive scientific problems [5,6].

The class of problems dealt by grid computing is that which involves tremendous computations and can be broken down into independent tasks. A general grid computing environment includes a supervisor and a group of participants who allow the idle cycles of their processors to be used for computations. The participants need know the tasks assigned to others, nor do they need to communicate with other participants. After the participants complete their tasks, they report back the results to the supervisor.

The past few years have seen a tremendous growth in grid computing with its effect being felt in the biotechnology industry (e.g., DNA sequencing and drug discovery), entertainment industry, financial industry, etc. The success of the projects like SETI@home [7], IBM smallpox research [8], GIMPS [9] has made the potential of grid computing visible.

For instance, IBM's smallpox research [8] uses grid computing to find potential drugs to counter the smallpox virus. Its main task is to screen hundreds of thousands of molecules, a task that can take years even with supercomputers. Using grid computing, any computer can be invited to participate and assist in this important effort. By downloading and running the software, participants can add their CPUs to the global grid. Every time their computers are idle, the computing resources can be contributed to the grid, accelerating the screening process while dramatically reducing the cost of the project. The result is that rather than spending years, it will be possible to screen hundreds of millions of molecules in just months. Another highly profiled grid computing project is SETI@home[7], which is a scientific experiment that uses Internet-connected computers in the Search for Extraterrestrial Intelligence (SETI). SETI@home has more than 4.5 million users (according to 2004 statistics) contributing their computers' unused processing power, to form a 15 teraflops grid, faster than IBM's powerful supercomputer *ASCI White* (12 teraflops). Also the cost of the SETI grid is only 500K dollars, whereas ASCI White costs 110 million dollars [7].

On December 2, 2003, Michael Shafer a volunteer in the GIMPS project discovered the largest known prime number, $2^{20996011} - 1$. He used a 2 GHz Pentium 4 DEll Dimension PC for 19 days to prove the number prime. The new number, expressed as 2 to the 20,996,011th power minus 1, has 6,320,430 decimal digits and was discovered on November 17. It is more than two million digits larger than the previous largest known prime number. Tens of thousands of people have volunteered the use of their PCs in this worldwide project that has harnessed the power of 211,000 computers, in effect creating a supercomputer capable of performing 9 trillion calculations per second. This enabled GIMPS to find the prime in just 2 years instead of the 25,000 years a single PC would have required.

However, the untrusted environments in which the computations are performed tend to cast suspicion on the veracity of the results returned by the participants. The ease with which anyone can become a part of the grid poses a big threat to the computation being performed. The true identity of the participant may not be known to the supervisor. The supervisor does not have any control over the participants' machines and cannot prevent them from manipulating the codes provided. Thus, the participant may not have performed the necessary computations but claim to have done so. This cheating behavior, if undetected, may render the results useless. Project managers from SETI@home have reportedly uncovered attempts by some users "to forge the amount of time they have donated in order to move up on the Web listings of top contributors" [10]. Yet SETI participants are volunteers who do not get paid for the cycles they contribute. When participants are paid for their contribution, they have strong incentives to cheat for maximizing their gain.

Therefore, we need methods to detect the cheating behavior in grid computing. This objective can be formulated as the following uncheatable grid computing problem:

PROBLEM 30.1 (*Uncheatable Grid Computing*). A participant is assigned a task to compute $f(x)$ for all the inputs $x \in D = \{x_1, \ldots, x_n\}$, where $n = |D|$; the participant needs to return the results of interest to the supervisor. A dishonest participant might compute $f(x)$ for only $x \in D'$, where D' is a subset of D, but claim to have computed f for all the inputs. How does the supervisor efficiently detect whether the participant is telling the truth or a lie?

A straightforward solution is to double check every result. The supervisor can assign the same task to more than one participant and compare the results. This simple scheme leads to the wastage of processor cycles that are precious resources in grid computing. Moreover, it introduces $O(n)$ communication cost for each participant. Note that in grid computing, the supervisor only needs the participant to return the results of interest, which is usually a very small number compared to n. Therefore, $O(n)$ overhead is substantial.

An improved solution is to use sampling techniques. The supervisor randomly selects a small number of inputs from D (we call these randomly selected input samples or sample inputs); it only double checks the results of these sample inputs. If the dishonest participant computes only one half of the inputs, the probability that he or she can successfully cheat the supervisor is one out of 2^m, where m is the number of samples. If we make m large enough, e.g., $m = 50$, cheating is almost impossible. This solution has a very small computational overhead ($O(m)$), because $m \ll n$. However, this scheme still suffers from the $O(n)$ communication cost because it requires the participant to send all the results back to the supervisor, including those that are of no interest to the supervisor.

To reduce the communication cost, a commitment-based sampling (CBS) scheme was developed [11]. The scheme reduces the communication overhead to $O(m \log n)$. Because n is usually large (e.g., $n = 2^{40}$), this result is a substantial improvement.

The organization of the chapter is as follows: We will discuss related work in Section 30.2. In Section 30.3, we formally model grid computing and its threat, and define uncheatable grid computing. In Section 30.4, we describe a naive solution. In Section 30.5, we present the CBS scheme for uncheatable grid computing. Section 30.6 further improves the CBS scheme by getting rid of the unappealing interaction. In Section 30.7, we present an implementation of the CBS scheme on two docking programs: Autodock and FTDock. Finally, Section 30.8 draws the conclusion and lays out the future work.

30.2 Related Work

30.2.1 Ringer Scheme

To defeat cheating in grid computing, Golle and Mironov proposed a ringer scheme [12]. The ringer scheme is designed for a class of distributed computations called inversion of a one-way function (IOWF),[*] and the solution can secure such computations against cheating. The IOWF computation is defined as the following:

---

[*] A function f is a one-way function, if for a general value of y in the domain of function f, it is impossible to find a value x such that $f(x) = y$.

Let $f : D \mapsto T$ be a one-way function, and $y = f(x)$ for some $x \in D$. Given f and y only, the objective of the computation is to discover x by exhaustive search of the domain D.

In the basic ringer scheme, during the initialization stage, the supervisor randomly selects several inputs $x_i \in D$. These inputs are called ringers, and they are kept secret with the supervisor. The supervisor then computes ringer images $y_i = f(x_i)$ for each x_i. In addition to assigning the grid computing task (inputs D, the function f, and the screener function S), the supervisor also assigns the ringer images to the participant.

Not only does the participant need to compute f on x for all $x \in D$ and return the results of interest, but also has to return all the ringers corresponding to the precomputed ringer images. The supervisor then verifies whether the participant has found all the ringers assigned to it or not. If yes, then the supervisor is assured with high probability that the participant has indeed conducted all the computations. Because participants cannot distinguish ringers from nonringers, their cheating can be caught if they cheat on ringers. By choosing the appropriate number of ringers, the probability for catching cheaters is quite high.

Szajda et al. extend the ringer scheme to deal with other general classes of computations, including optimization and Monte Carlo simulations [13]. They propose effective ways to choose ringers for those computations. It is still unknown whether the schemes proposed in [13] can be extended further to generic computations.

30.2.1.1 Limitations of the Ringer Scheme

The ringer scheme is limited to a specific computation, the IOWF computation. Therefore, the function f has to be a one-way function; otherwise the participant can calculate f^{-1} on the ringer images and find their corresponding secret ringers, without conducting the computations on all inputs.

However, if we can convert any generic function f into a one-way function, we can make the ringer scheme work for generic computations. There is a quite straightforward way to achieve this, as shown in the following:

$$\hat{f}(x) = h(x, f(x)),$$

where function h is a one-way function.

Because h is a one-way function, there is no easy way for a cheating participant to construct $\hat{f}^{-1}$, even if it knows f and f^{-1}. Namely, given $\hat{f}(x)$, to find x is no easier than the brute-force approach, i.e., trying all possible values of x. After this conversion, any generic function f is converted to a one-way function $\hat{f}$, and we can use the ringer scheme on the new function $\hat{f}$.

30.2.1.2 New Attack When |T| Is Small

The above improvement lifts the limitation of the original ringer scheme however, it does come with a cost: the new scheme is now subject to the following attack when the size (represented as $|T|$) of the output domain T is small:

1. For $\forall x \in D$ and $\forall y \in T$, the cheating participant computes $h(x, y)$.
2. If $h(x, y)$ matches with a ringer, output x.

The above attack uses brute-force method to try all the possible combinations of x and y. Because the ringers are unique, the participant can use this method to identify them without computing f at all. Let h_c be the cost of computing $h(x, y)$, and f_c be the cost of computing $f(x)$. The total cost for the brute-force attack is $|D| \times |T| \times h_c$. Since the cost of conducting $f(x)$ for all $x \in D$ is $|D| \times f_c$, in order to make the brute-force attack more expensive than the honest computing, the following inequality must hold:

$$|T| \cdot h_c \geq f_c.$$

It appears that to defeat the above brute-force attack, we just need to increase the cost of $h(x, y)$. We can easily achieve this by conducting many rounds of hash function. Unfortunately, increasing h_c increases the computational overhead on the participant side because

$$\text{Overall computation cost} = (f_c + h_c) \cdot |D|.$$

An alternative to defeat the attack is to increase $|T|$. This can be achieved by also treating the intermediate results in the computation as outputs. The cost of the brute-force attack can exceed the cost of the task, if we include enough intermediate results in outputs. However, introducing intermediate results makes the scheme application dependent, whether there exists a generic scheme is still unknown.

30.2.2 Redundancy Scheme

Another approach for uncheatable grid computing is to double check computation results by simply having other participants redo them entirely. This is the method used by SETI@home to deal with cheating users. Golle and Stubblebine [14] proposed an efficient scheme based on probabilistic redundant execution. In this scheme, for each task, the supervisor draws a random number n from a predetermined probability distribution; then the supervisor randomly picks n participants and assigns a copy of this task to each participant. The verification succeeds, if all the participants return the same results; otherwise, the same process will be repeated until the verification succeeds.

30.2.3 The Input-Chaff Injection Scheme

Du and Goodrich proposed schemes for efficiently performing uncheatable grid computing to search for high-value rare events [15]. The schemes are intended to prevent the participants from not reporting the results, if they have found the high-value rare events in grid computing. The technique is called chaff injection, which involves introducing elements to task inputs or outputs that provide rare-event obfuscation. Output-chaff injection applies to contexts where rare events are not immediately identifiable by clients. It involves the use of one-way hash functions applied to computation outcomes in a way that defends against coalitions of hoarding clients. Input-chaff injection applies to contexts where rare events are easily identified by clients (as in the SETI@home application for finding patterns of intelligence in extraterrestrial signals). It involves the injection of a number of obfuscated inputs that will test positive as rare events.

30.2.4 Other Related Work

There are also some alternative ways to defeat cheating, which have not been specifically addressed in the literature. For example, one method would be to use tamper-resistant software. In this case, a code obfuscater would be used to convert task programs to equivalent programs that are much harder to understand and reverse-engineer. Thus, it would become hard for malicious attackers to modify the program to accomplish what he wants. Nevertheless, the tamper-resistant approach is only heuristically secure, and many tamper-resistant schemes cannot withstand attacks from really determined attackers [16], including groups of colluding cheaters.

The problem addressed in this chapter is close to another body of literature: the problem of malicious hosts in the study of mobile agents [17,18]. Several practical solutions have been proposed, which include remote auditing [16,19], code obfuscation with timing constraints [20], computing with encrypted functions [21], replication, and voting [22]. The major difference between the mobile-agent setting and the grid computing framework is the threat model. The mobile-agent work assumes a malicious cheating model, i.e., a malicious host can do whatever it takes to cheat; this includes spending more CPU cycles than the honest behavior. However, this chapter studies a different model, in which it is irrational for a participant to cheat with a cost more expensive than the honest behavior.

Cryptographic protocol, such as private information retrieval (PIR) [23] and probabilistically checkable proofs (PCP) [24], can also be used to achieve uncheatable grid computing. However, their expensive computation cost makes them an inappropriate choice for grid computing in practice.

30.3 Problem Definition

30.3.1 Model of Grid Computing

We consider a grid computing in which untrusted participants are taking part. The computation is organized by a supervisor. Formally, such computations are defined by the following elements [12]:

- *A function $f : X \mapsto T$ defined on a finite domain X.* The goal of the computation is to evaluate f for all $x \in X$. For the purpose of distributing the computation, the supervisor partitions X into subsets. The evaluation of f on subset X_i is assigned to participant i.
- *A screener S.* The screener is a program that takes as input a pair of the form $(x; f(x))$ for $x \in X$, and returns a string $s = S(x; f(x))$. S is intended to screen for "valuable" outputs of f that are reported to the supervisor by means of the string s. We assume that the run time of S is of negligible cost compared to the evaluation of f.

30.3.2 Models of Cheating

A participant can choose to cheat for a variety of reasons. These reasons can be categorized into two models. We assume that the participant is given a domain $D \subset X$, and its task is to compute $f(x)$ for all $x \in D$. From now on, we use D as the domain of f for the participant.

30.3.2.1 Semi-Honest Cheating Model

In this model, the participant follows the supervisor's computations with one exception: for $x \in \check{D} \subset D$, the participant uses $\check{f}(x)$ as the result of $f(x)$. Function $\check{f}$ is usually much less expensive than function f; for instance, $\check{f}$ can be a random guess. In other words, the participant does not compute the required function f on inputs $x \in \check{D}$. The goal of the cheating participant in this model is to reduce the amount of computations, such that it can maximize its gain by performing more tasks during the same period of time. If the participants are getting paid, the semi-honest participant might be guided by the lure of money.

30.3.2.2 Malicious Cheating Model

In this model, the behavior of the participant can be arbitrary. For example, a malicious participant might have calculated function f for all $x \in D$, but when it computes the screener function S, instead of computing $S(x; f(x))$, it might compute $S(x; z)$, where z is random number. In other words, the participant intentionally returns wrong results to the supervisor, for the purpose of disrupting the computations. A malicious cheater may be a competitor, or a nonserious participant playing pranks.

To maximize their gains, rational cheaters tend to use minimal cost to falsify the contributions they have never made. Their behaviors fall into the semi-honest cheating model. Therefore, in this paper, we focus on the semi-honest cheating model.

30.3.3 Definition of Uncheatable Grid Computing

Assume that a participant is assigned the task that consists of computing $f(x)$ for all $x \in D$, where $D = \{x_1, \ldots, x_n\}$. If a participant computes the function f only on $x \in D'$, where $D' \subseteq D$, we define the honesty ratio r as the value of $\frac{|D'|}{|D|}$. Honesty ratio is the fraction of the overall computation actually performed by the participant. When the participant is fully honest, the honesty ratio is $r = 1$; when it only conducts 50% of the required computations, its honesty ratio is $r = 0.5$.

DEFINITION 30.1 *(Uncheatable Grid Computing).* Let $Pr(r)$ be the probability that a participant with honesty ratio r can cheat without being detected by the supervisor. Let $C_{cheating}$ be the expected cost of successful cheating, and C_{task} be the overall computation cost of the required task. We say grid computing is uncheatable, if one of the following or both inequalities are true:

$$Pr(r) < \varepsilon \quad \text{for a given } \varepsilon \ (0 < \varepsilon \leq 1)$$

$$\text{or} \quad C_{cheating} > C_{task}$$

In other words, uncheatable grid computing either makes the probability of successful cheating negligible or makes the cost of successful cheating even more expensive than the cost of honestly conducting the entire task. Since honestly performing the tasks is now less time-consuming, a semi-honest cheater is more likely to avoid short cuts to complete his task.

30.4 Naive Sampling Schemes

30.4.1 Parallel Tasks

Assume that the task of a participant is divided into n independent subtasks $T_1, \ldots, T_n$, with I_i being the input for T_i and O_i being the output for T_i, for $i = 1, \ldots, n$. Our goal is to verify whether the participant has conducted these subtasks.

One way to verify is to use probabilistic sampling, namely after the computation, the participant sends all the results $O_1, \ldots, O_n$ of the subtasks to the supervisor. (If each result O_i contains a large amount of data, the participant can send $hash(O_i)$ to the supervisor, instead of O_i. This way, the communication cost can be significantly reduced). The supervisor then randomly selects m of them to verify. If any of these subtask verifications fails to reproduce the same results as sent by the participant, the supervisor concludes that the participant is cheating. If the participant has conducted all the subtasks, it can pass the verification. However, if the participant has only conducted r portion of the total n tasks (we call $0 \leq r \leq 1$ the honesty ratio), the probability that the supervisor fails to detect the cheating is the following:

$$Pr(\text{cheating succeeds}) = (r + (1 - r)q)^m, \tag{30.1}$$

where q is the probability for the participant to guess the correct answer without conducting the actual computation.

Since $r + (1 - r)q$ is less than 1 when q is less than 1, the above probability can be arbitrarily small if we choose m that is large enough.

30.4.2 Sequential Tasks

The above scheme assumes that the n subtasks are independent to each other, they can be verified independently. In a sequential task, task T_i actually depends on the results from T_j, where $j = 1, \ldots, i-1$. For the semisequential computation model, we show that we can verify the computation using the same way as that for the parallel computation.

30.4.2.1 Semisequential Model

Figure 30.1.a depicts our scheme. The participant also sends $O_1, \ldots, O_n$ to the supervisor. The supervisor treats each T_i as an independent subtask (although in reality all the subtasks are dependent), with the inputs being O_{i-1} and I_i, and the output being O_i. The supervisor can use the sampling scheme to verify whether some selected subtasks are correctly computed from the given inputs.

Although one of the input, O_{i-1}, for each subtask T_i cannot be verified (its verification requires the verification of $T_{i-1}, T_{i-2}, \ldots, T_1$), the other input, I_i, can be verified, because they are independent of the output of each task and can usually be derived from the original input that is known to both the supervisor and the participants. It is important that at least one of the inputs should be verified, otherwise, the reusing attack is possible. We will describe such an attack next in the purely sequential model. Here, we assume that the cost of computing T_i from I_i and O_{i-1} is approximately the same regardless of what the value of O_{i-1} is.

It should be noted that for this model, the participant can no longer send $hash(O_i)$; it must send O_i because O_i is needed as an input when we verify the task T_{i+1}. Therefore, if the size of O_i is large and the number of subtasks is large, this sampling scheme might not be efficient because of the overwhelming communication overhead.

30.4.2.2 Purely-Sequential Model

It looks like that we can use a similar scheme to conduct the sampling for the purely sequential model. The scheme is depicted in Figure 30.1.b. However, this scheme suffers from a reusing attack.

As shown in Figure 30.2, the participant can break the chain at any of the links and replace the second part of the chain with a new chain. As long as all the computations on this new chain are correct, the sampling scheme cannot detect any misbehaviors, unless the link at the breaking point is selected as one of the samples. However, the chance of selecting the breaking point is low because only one single breaking point is needed to launch this attack.

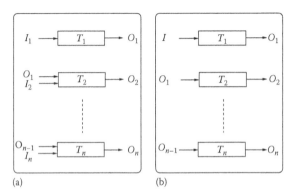

(a)   (b)

FIGURE 30.1 Verifying sequential computation model: (a) semisequential model and (b) purely sequential model.

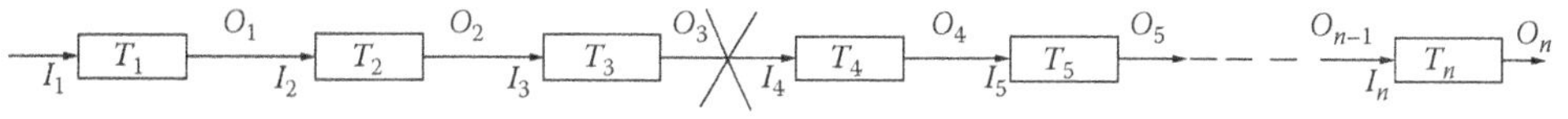

Link broken by the participant

FIGURE 30.2 Attack for sequential tasks.

Therefore, in order to cheat, the participant can reuse existing (correct) computations, and use them for the second part of the chain. The participant can get away from the cheating as long as the subtasks before and after the breaking point are not sampled simultaneously.

30.4.3 Limitation of the Naive Sampling Scheme

The major problem with the naive sampling scheme is the communication cost, which is $O(n)$, where n is the total number of subtasks. Although the communication cost might not cause a problem at the participant side, it can cause a problem at the supervisor side. For example, if the task of grid computing is to break a 64-bit password using the brute-force method, the total communication cost at the supervisor side is $O(2^{64})$, which is about 16 million terabytes. Very few networks can handle such a heavy network load.

To solve this problem, we describe a CBS scheme, which reduces the communication overhead from $O(n)$ to $O(m \log n)$. Because $m \ll n$ and n can be large (e.g., $n = 2^{30}$), this result is a substantial improvement.

30.5 Commitment-Based Sampling Scheme

To reduce communication costs, we cannot ask each participant to send back all the outputs; it is desirable if we can ask the participants to send back the results for those sample inputs. However, doing this is nontrivial, because we have to prevent the participant from computing the results for the sample inputs after it learns which inputs are samples. For example, if x_k is selected as a sample, the supervisor needs $f(x_k)$ from the participant to check whether the participant has correctly calculated $f(x_k)$. However, without a proper security measure, the participant who has not computed $f(x_k)$ can always compute it after learning x_k is a sample. This defeats the purpose of sampling. Therefore, a security measure must guarantee that the value of $f(x_k)$ sent by the participant is calculated before it knows that x_k is a sample.

One way to solve the above problem is to use commitment. Before the participant knows that x_k is a sample, it needs to send the commitment for $f(x_k)$ to the supervisor. Once the participant commits, it cannot change $f(x_k)$ without being caught. The supervisor then tells x_k to the participant, which has to reply with the original value of $f(x_k)$ that was committed. Since any input has an equal probability to become a sample, the participant has to commit all the results for those n inputs. Obviously, the participant cannot afford to send the commitment for each single input, because the $O(n)$ communication cost makes it no better than the naive sampling scheme. The participant cannot hash all these n results together to form one single commitment either: although this method achieves the commitment for all results, it makes verifying a single result difficult, because the supervisor needs to know all the other $n - 1$ results.

In summary, we need a commitment scheme that (1) allows all the n results to be committed efficiently, and (2) allows the verification of each single result to be performed efficiently. We use the Merkle tree [25] to achieve these goals.

30.5.1 Commitment-Based Sampling Scheme

The Merkle tree (also called hash tree) is a complete binary tree equipped with a function *hash* and an assignment Φ, which maps a set of nodes to a set of a fixed-size strings. In a Merkle tree, the leaves of the tree contain the data, and the Φ value of an internal tree node is the hash value of the concatenation of the Φ values of its two children.

To build a Merkle tree for our problem, the participant constructs n leaves $L_1, \ldots, L_n$. Then it builds a complete binary tree with these leaves. The Φ value of each node is defined as the following (we use V to denote an internal tree node, and V_{left} and V_{right} to denote V's two children):

$$\Phi(L_i) = f(x_i) \text{ for } i = 1, \ldots, n,$$
$$\Phi(V) = hash(\Phi(V_{\text{left}})||\Phi(V_{\text{right}})), \tag{30.2}$$

where "$||$" represents the concatenation of two strings, and the function *hash* is a one-way hash function such as MD5(128 bit) or SHA1(160 bit). To make a commitment on all the data on the leaves, the participant just needs to send $\Phi(R)$ to the supervisor, where R is the root of the Merkle tree. Figure 30.3 depicts an example of the Merkle tree built for our purpose.

After receiving the commitment, the supervisor randomly selects a number of samples and sends them to the participant. The participant needs to provide the evidence to show that before making the commitment, it has already computed f for those samples. Let x be a sample, and L be x's corresponding leaf node in the tree. Let λ denote the path from L to the root (not including the root), and let H represent the length of the path. In order to prove its honesty regarding $f(x)$, the participant sends $f(x)$ to the supervisor; in addition, for each node $v \in \lambda$, the participant also sends $\Phi(v$'s sibling) to the supervisor. We use $\lambda_1, \ldots, \lambda_H$ to represent these Φ values.

To verify the participant's honesty on sample x, the supervisor first verifies the correctness of $f(x)$. If $f(x)$ sent by the participant is incorrect, the participant is caught cheating immediately. Even if $f(x)$ from the participant is correct, it cannot prove the participant's honesty because the participant who did not compute $f(x)$ can compute the correct $f(x)$ after knowing that x is the sample. The supervisor uses the commitment $\Phi(R)$ made by the participant to ensure that the correct $f(x)$ is

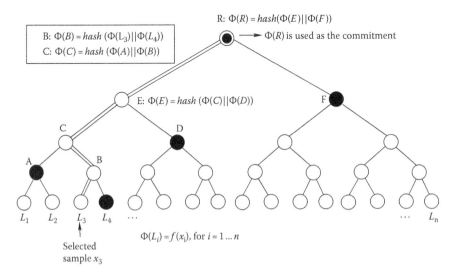

FIGURE 30.3 CBS scheme: the Merkle tree and the verification.

used at the time of building the Merkle tree. To achieve this, the supervisor uses $f(x)$ (correct) and $\lambda_1, \ldots, \lambda_H$ to reconstruct the root of the Merkle tree R', thus getting $\Phi(R')$. Only if $\Phi(R') = \Phi(R)$ will the supervisor trust that the participant has correctly computed $f(x)$ before building the Merkle tree. The communication cost of this process is proportional to the height of the tree. Because the Merkle tree is a complete binary tree with n leaves, its height is $O(\log n)$, where $n = |D|$.

We demonstrate how the verification works using an example depicted in Figure 30.3. Assume that x_3 is selected as a sample, whose corresponding leaf node in the tree is L_3. The participant finds the path from L_3 to the root (depicted by the double lines). Then the participant sends to the supervisor $f(x_3)$ and all the Φ values of the sibling nodes (L_4, A, D, and F) along the path. The sibling nodes are depicted by the black nodes in the figure. To verify whether, before committing $\Phi(R)$, the participant has computed $f(x_3)$ or not, the supervisor first makes sure if $f(x_3)$ is correct. Then, the supervisor reconstructs the root R' from $f(x_3)$, $\Phi(L_4)$, $\Phi(A)$, $\Phi(D)$, and $\Phi(F)$.* If $\Phi(R') = \Phi(R)$, we can say that the participant knows $f(x_3)$ before building the Merkle tree.

The scheme described above is called the CBS scheme. Its steps are described in the following:

Step 1: Building Merkle Tree. Using Equation 30.2, the participant builds a Merkle tree with leaf nodes $L_1, \ldots, L_n$, and $\Phi(L_i) = f(x_i)$, for $i = 1, \ldots, n$. The participant then sends $\Phi(R)$ to the supervisor.

Step 2: Sample Selection. The supervisor randomly generates m numbers ($i_1, \ldots, i_m$) in the domain $[1, n]$, and sends these m numbers to the participant. These numbers are the sample inputs.

Step 3: Participant's Proof of Honesty. For each $i \in \{i_1, \ldots, i_m\}$, the participant finds the path λ from the leaf node L_i to the root R; then, for each node $v \in \lambda$, the participant sends to the supervisor $\Phi(v$'s sibling). These Φ values are denoted by $\lambda_1, \ldots, \lambda_H$. The participant also sends $f(x_i)$ to the supervisor.

Step 4: Supervisor's Verification. For each $i \in \{i_1, \ldots, i_m\}$, the supervisor verifies whether $f(x_i)$ from the participant is correct.

1. If $f(x_i)$ is incorrect, the verification stops and the participant is caught cheating.
2. If $f(x_i)$ is correct, using the recursive procedure defined in Equation 30.2, the supervisor reconstructs the root $\Phi(R')$ of the hash tree from $f(x_i)$ and $\lambda_1, \ldots, \lambda_H$. If $\Phi(R) \neq \Phi(R')$, the verification stops and the participant is caught cheating. If $\Phi(R) = \Phi(R')$, the verification succeeds for the sample i.

If the above verification succeeds for all $i \in \{i_1, \ldots, i_m\}$, the supervisor is convinced with a high probability that the participant has not cheated.

To verify whether $f(x_i)$ is correct does not necessarily mean that the supervisor has to recompute $f(x_i)$. There are many computations whose verification is much less expensive than the computations themselves. For example, factoring large numbers is an expensive computation, but verifying the factoring results is trivial.

Regarding the communication cost, for each sample, the participant needs to send $O(\log n)$ data to the supervisor. Therefore, the total communication overhead for m samples is $O(m \log n)$. Actually, due to path overlaps, the actual communication cost could be less than $m \log n$, so $m \log n$ is the upper bound on the communication cost.

---

* The reconstruction of R' can be conducted using the following procedure: with $f(x_3)$ and $\Phi(L_4)$, we can compute $\Phi(B)$; then, with $\Phi(A)$, we can compute $\Phi(C)$; then, with $\Phi(D)$, we can compute $\Phi(E)$; finally, we can compute $\Phi(R')$ from $\Phi(E)$ and $\Phi(F)$.

30.5.2 Security Analysis

In the following theorem, we use L to denote input x's corresponding leaf node. We use T to denote the Merkle tree built by the participant, and we use R to denote the root of the tree.

Let λ be the path from the leaf L to the root R, and let $\lambda_1, \ldots, \lambda_H$ represent the Φ values of the sibling nodes along the path λ. According to the property of the Merkle tree, $\Phi(R)$ can be computed using $\Phi(L)$ and $\lambda_1, \ldots, \lambda_H$. We use $\Lambda(\Phi(L), \lambda_1, \ldots, \lambda_H) = \Phi(R)$ to represent this calculation, where $\Phi(R)$ is already committed to the supervisor by the participant.

THEOREM 30.1 (Soundness). *If the participant has computed $f(x)$ at the time of building the Merkle tree, it will succeed in proving its honesty on x within the CBS algorithm.*

PROOF   If the participant is indeed honest, according to the CBS scheme, when building the Merkle tree, we have $\Phi(L) = f(x)$. Therefore, during the verification, the supervisor gets

$$\Phi(R') = \Lambda(f(x), \lambda_1, \ldots, \lambda_H) = \Lambda(\Phi(L), \lambda_1, \ldots, \lambda_H) = \Phi(R).$$

Therefore, according to the CBS scheme, the participant succeeds in proving its honesty on x.

THEOREM 30.2 (Uncheatability). *In the CBS scheme, it is computationally infeasible to convince the supervisor if the participant did not compute $f(x)$ at the time of Merkle tree construction, i.e., for an input x, a value other than $f(x)$ is used while constructing the Merkle tree.*

PROOF   According to the CBS scheme, the participant sends $f(x)$ and $\lambda'_1, \ldots, \lambda'_H$ to the supervisor. After verifying the correctness of $f(x)$, the supervisor uses $\Phi(R') = \Lambda(f(x), \lambda'_1, \ldots, \lambda'_H)$ to reconstruct the root (denoted by R') of the tree. The supervisor believes that the participant is honest on $f(x)$ only if $\Phi(R') = \Phi(R)$.

If the participant is dishonest and $\Phi(L) \neq f(x)$, to cheat successfully, the participant must find $\lambda'_1, \ldots, \lambda'_H$, such that

$$\Lambda(f(x), \lambda'_1, \ldots, \lambda'_H) = \Lambda(\Phi(L), \lambda_1, \ldots, \lambda_H) = \Phi(R).$$

Because Λ consists of a series of one-way hash functions, given $\Phi(R)$, when $\Phi(L) \neq f(x)$, according to [25], as long as the underlying one-way hash function is strong (e.g., SHA-2 family), it is computationally infeasible to find $\lambda'_1, \ldots, \lambda'_H$ that satisfy the above equation. This proves that it is computationally infeasible for the dishonest participant to convince the supervisor that it knows $f(x)$ at the time of building the Merkle tree.

In the following theorem, let q be the probability that the participant can guess the correct result of $f(x)$, i.e., $Pr_{\text{guess}}(\Phi(L) = f(x)) = q$. Let D' be the set of inputs that are computed honestly by the participant, so honesty ratio is $r = \frac{|D'|}{|D|}$.

THEOREM 30.3   *When m samples are used in the CBS scheme, the probability that a participant with honesty ratio r can cheat successfully is*

$$Pr(\text{cheating succeeds}) = (r + (1 - r)q)^m. \tag{30.3}$$

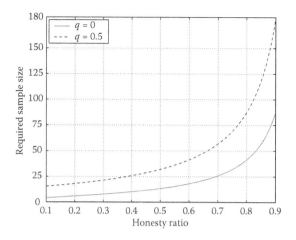

FIGURE 30.4 Required sample size versus cheating efforts ($\varepsilon = 0.0001$).

PROOF   Since each sample is uniformly randomly selected, the probability that a sample x belongs to D' is r. When $x \in D'$, i.e., the participant has indeed computed $f(x)$, according to Theorem 30.1, the participant should be able to convince the supervisor of its honesty on sample x. When $x \in D - D'$, i.e., the participant did not compute $f(x)$ when building the tree, according to Theorem 30.2, it is computationally infeasible for the participant to cheat unless $\Phi(L)$ happens to equal $f(x)$. Since $Pr_{\text{guess}}(\Phi(L) = f(x)) = q$, when $x \in D - D'$, the probability to cheat successfully is q.

Combining both cases of $x \in D'$ and $x \in D - D'$, for one sample x, the probability that the participant can prove its honesty on sample x is $(r + (1 - r)q)$. Therefore, the probability that the participant can prove its honesty on all m samples is $(r + (1 - r)q)^m$.

Therefore, if we want to keep the probability of successful cheating below a small threshold ε, i.e., $(r + (1 - r)q)^m \leq \varepsilon$, we just need to find the sample size m, such that

$$m \geq \frac{\log \varepsilon}{\log (r + (1 - r)q)}. \tag{30.4}$$

Figure 30.4 shows how large m should be for different honesty ratios r, given $\varepsilon = 0.0001$. For example, let us consider a situation where the participant has conducted only one half of the task, which means only one half of the leaf nodes in the Merkle tree contain the actually computed results, and the other half contain guessed results. When the probability of guessing the correct results is 0.5 (i.e., $q = 0.5$), we need at least 33 samples to ensure the probability of successful cheating to be below $\varepsilon = 0.0001$. When $q \approx 0$ (i.e., it is almost impossible to make a correct guess on $f(x)$ without computing it), we only need 14 samples.

30.5.3 Storage Usage Improvement

It should be noted that the CBS scheme requires the participant to store the entire Merkle tree in its memory or hard disk, and the amount of space required is $O(|D|)$. Today's hard disk technologies make it possible for a participant to accept tasks with $|D|$ as large as 2^{30} (by using giga bytes of storage); however, the storage might not be enough when $|D|$ is much larger than 2^{30}. For example, as mentioned by [12], it is possible to assign a task of size 2^{40} to a participant. If we use 128-bit hash functions, 2^{40} inputs require 16,000 GB (16 Terabytes) to store only the leaf nodes and 32,000 GB to store the complete Merkle tree, which is unrealistic nowadays.

We noticed that if a task is as large as 2^{40}, then computing $f(x)$ must be very fast; otherwise, it might take the participant unreasonably long time to finish the task. Based on this observation, we can make a trade-off between time and storage in the following way: Assume that the height of the entire Merkle tree is $H = \log |D|$, and the root is at level 0. Instead of storing the entire Merkle tree, the participant only stores the tree up to level $H - \ell$, where $0 < \ell < H$. Figure 30.5a depicts the part of the tree that needs to be stored. The total amount of storage required is $O\left(\frac{|D|}{2^\ell}\right)$, a decrease of 2^ℓ-folds.

To prove that it has computed $f(x)$ (in Step 3 of the CBS scheme), the participant must find the path from the sample x's corresponding leaf node to the root, and then send to the supervisor the Φ values of the sibling nodes along this path. Unfortunately because the lower part of the Merkle tree is not saved, the sibling nodes in the lower part of the tree cannot be obtained from the storage. The shading area in Figure 30.5a represents the subtree that contains the sample x but not saved in the storage. Figure 30.5b depicts an example of the unsaved subtree. From the figure, we can see that the nodes V_1, V_2, and V_3 are also the sibling nodes along the path, but their Φ values are not saved. To provide their Φ values to the supervisor, the participant must recompute them. From Figure 30.5b, it is not difficult to see that recomputing those Φ values requires the rebuilding of the whole subtree depicted in the shading area. For each sample, this cost includes computing f for 2^ℓ times and computing the hash function for $2^\ell - 1$ times. Therefore, the cost of the rebuilding is $m \cdot (2^\ell f_c + (2^\ell - 1)h_c)$, where f_c is the cost of computing $f(x)$ for one input, and h_c is the cost of

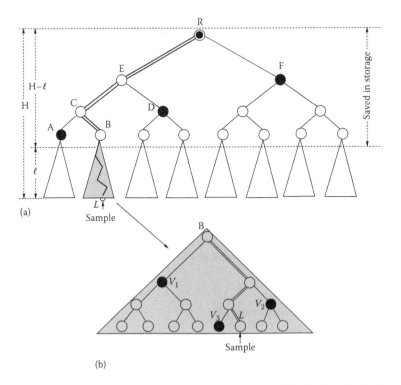

(a)

(b)

FIGURE 30.5 Storage usage improvement: (a) Storing partial merkle tree; and (b) details in the shading area.

computing the one-way hash function. Because 2^ℓ is significantly larger than 1, we simply take the cost to be $m \cdot 2^\ell \cdot (f_c + h_c)$.

We use the relative computation overhead (rco) to indicate how this computation overhead impacts the entire task. The *rco* is defined as the ratio of the total computation overhead for m samples to the cost of computing the entire Merkle tree based on D (the cost is approximate to $|D| \cdot (f_c + h_c)$). Let $S = 2^{H-\ell+1}$ represent the amount of space for storing the partial tree. Hence, we have the following formula:

$$rco = \frac{m \cdot 2^\ell \cdot (f_c + h_c)}{|D| \cdot (f_c + h_c)} = \frac{m \cdot 2^\ell}{2^H} = \frac{m}{2^{(H-\ell)}} = \frac{2m}{S}.$$

The above equation indicates that *rco* is only affected by m and S, and not by the amount of inputs in D. The more storage a participant uses for storing the tree, the lower is the relative computation overhead. For example, when $m = 64$, if we use 4G (2^{32}) hard disk space to store the partial Merkle tree, we have $rco = 2^{-25}$. This means that, regardless of how large a task is, compared to the cost of the task, the computation overhead at the participant side is negligible when we use 4G disk space. Therefore, even for a task of size 2^{40}, using 4G disk space provides a feasible solution both storage-wise and computation-wise.

30.5.4 Comparison with the Extended Ringer Scheme

The performance of the CBS and the extended ringer scheme is similar in general, but the CBS scheme has more advantage under the following circumstances:

1. In the CBS scheme, the supervisor needs to verify whether $f(x)$ sent by the participant is correct or not. For many computations, the verification of $f(x)$ can be much more efficient than computing $f(x)$ itself. For example, if $f(x)$ is to find a Hamiltonian cycle in a graph x, the computation is expensive because the Hamiltonian-cycle problem is NP-complete [26]. However, verifying whether a cycle is a Hamiltonian cycle in x can be achieved in polynomial time. Therefore, the supervisor does not need to recompute $f(x)$ for such computations in CBS scheme. This is not true for the ringer scheme. Since the ringer scheme needs to precompute $f(x)$, the recomputation is inevitable. Therefore, for the computations whose verification is much more efficient than the computations themselves (such as NP-complete problems), the CBS scheme can achieve significant savings for the supervisor.

2. When $f(x)$ is a nondeterministic function, the ringer scheme will not work. Given the same inputs, a nondeterministic function may return different results each time it is executed. For example, a nondeterministic Hamiltonian-cycle computation might output different Hamiltonian cycles in the same graph each time they are executed. The ringer scheme does not work for the nondeterministic function, because the participant might not be able to regenerate the same results as those precomputed ringer images. However, the CBS scheme still works as long as the nondeterministic function is verifiable. Unfortunately, when a nondeterministic function is difficult to verify, the CBS scheme also becomes inapplicable.

3. As we mentioned before, the extended ringer scheme does not work well when $|T|$ is small; we have to introduce the intermediate results to make $|T|$ significantly large. However, introducing the intermediate results can be application-dependent. The CBS scheme does not have such a problem, and it works well even when $|T|$ is small.

30.6 Noninteractive CBS Scheme

The CBS scheme has an extra round of interaction between the supervisor and the participant. This interaction involves the participant sending the commitment and the supervisor sending the samples. The interaction ensures that the supervisor sends the samples only after it receives the commitment. Although the communication cost of this extra round of interaction is not a concern, the interaction is often found less appealing because of the implementation issues involved in grid computing.

In many grid computing architectures, the supervisor might not be able to directly interact with the participants. For example, in the grid architecture for computational economy (GRACE) architecture [27], which represents a futuristic paradigm of a service-oriented computing industry, there exists a grid resource broker (GRB), which acts as a mediator between the supervisor and the participant. The GRB provides middleware services in the form of resource management and assignment of tasks. The GRB is responsible for finding more resources (participants) and scheduling of tasks among the resources depending on their availability and capability. GRACE provides services that help resource owners and user-agents maximize their objective functions. The resource providers can contribute their resource to the grid and charge for services. They can use GRACE mechanism to define their charging and access policies and the GRB works according to those policies.

In the GRACE architecture, the supervisor assigns a big bulk of tasks to GRB, and relies on GRB to interact with, and assign tasks to, the participants. The supervisor does not even know which participant is conducting what tasks. If the supervisor wants to verify the participant's honesty on its own using the CBS scheme, it will be difficult because GRB hides the participants from the supervisor.

It will be more desirable if we can conduct CBS without that extra round of interaction. One way to achieve this is to let the participant generate the sample choices. Obviously, if the participant is to select the samples, the sample selection must satisfy the following properties:

1. The samples are selected after the Merkle tree is built: If the samples are selected before building the Merkle tree, the verification procedure will never detect cheating. It will verify only those computations which have actually been performed by the participant.

2. The samples must be hard to predict: If the samples can be predicted by the participant, the participant can just perform the tasks on the samples beforehand and renders the verification process useless.

When the supervisor selects the samples, the above two requirements are easily enforced because the supervisor does not tell the participant the sample choices until the participant sends the commitment. How can we enforce these requirements when we rely on the participant to generate the sample choices?

30.6.1 Noninteractive CBS Scheme

The CBS scheme can be modified, such that the sample choices are generated by the participant. This improved scheme is called the noninteractive (NI-CBS) scheme. Since the only different part of this improved scheme is the sample selection step, we will not repeat the steps that are the same as in the CBS scheme.

Step 1: Building Merkle Tree. This step is exactly like the CBS scheme. At the end, the participant sends $\Phi(R)$ to the supervisor.

Step 2: Sample Selection. Let g be a one-way hash function. Assume that $D = \{x_1, \ldots, x_n\}$ is assigned to the participant. The participant uses the following method to generate m numbers $\{i_1, \ldots, i_m\}$ in the domain $[1, n]$:

$$i_k = (g^k(\Phi(R)) \bmod n) + 1 \quad \text{for} \ k = 1, \ldots, m, \tag{30.5}$$

where

$$g^k(\Phi(R)) = \begin{cases} g(\Phi(R)), & \text{for k=1,} \\ g(g^{k-1}(\Phi(R))), & \text{for k=2}\ldots \text{m.} \end{cases}$$

Inputs x_i, for $i \in \{i_1, \ldots, i_m\}$, are the selected samples. In other words, the kth sample is the result of applying the one-way hash function g on R for k times.

Step 3: Participant's Proof of Honesty. This step is also exactly like the CBS scheme.

Step 4: Supervisor's Verification. Using Equation 30.5, the supervisor regenerates the sample choices $\{i_1, \ldots, i_m\}$ from $\Phi(R)$. It then uses Step 4 of the CBS scheme to verify the participant's results.

30.6.2 Security Analysis

Assume the participant has conducted the computations only for the inputs in D', where $D' \subset D$, and the honesty ratio is $r = \frac{|D'|}{|D|} < 1$. Also assume that the sample choices generated by the participant are $S_1, \ldots, S_m$. To guarantee a successful cheating, the participant needs to make sure that all the S_i for $i = 1, \ldots, m$ fall into D'. Assuming the perfect randomness of the one-way hash values, the probability that all these m sample choices are in the set of D' is r^m. Namely when building the Merkle tree, the participant can use whatever values to replace $f(x)$ for $x \in D - D'$, the probability to produce the sample choices that are all in set D' is r^m.

The one-way hash function acts as an unbiased random-bit generator for the sample generation. For the participant to cheat, it has to be able to predict the output of the one-way function. However, there is no way for the participant to force the one-way function to produce certain values or to guess which values it will produce. It is also computationally infeasible for the participant to work in the reverse way, i.e., the participant cannot select the samples first, and then build a Merkle tree that generates these selected samples. The one-way function is, in effect, the supervisor's surrogate in the NI-CBS scheme.

Unfortunately, the noninteractive feature brings up a potential attack. In the CBS scheme, the participant has only one chance to cheat. If the supervisor chooses to use $m = 10$ samples, the participant with honesty ratio $r = 0.5$ has a probability of 1 in 2^{10} to cheat successfully. If one cheating attempt fails, the supervisor will not give the participant more chances to cheat. The probability of 1 in 2^{10} tends to be small enough for the interactive scheme, but it is still too large for the noninteractive scheme. The participant can use the following strategy to cheat:

1. Build the Merkle tree, using random numbers as the results of $f(x)$ for the inputs x that are not in D' (i.e., $x \in D - D'$).
2. Compute the sample selections from the root of the Merkle tree. If they are all within D', cheating is successful; otherwise pick other random numbers as $f(x)$ for $x \in D - D'$.
3. Revise the Merkle tree based on the newly selected values, and repeat Step 2 until the cheating becomes successful.

The participant can use the above strategy to repeatedly make many cheating attempts until it finds out that all the m generated samples are in D'. Since the process is noninteractive, the supervisor knows nothing about these attempts.

There are two ways to defeat this strategy: one way is to increase the number of samples. For example, we can use 128 samples, because making 2^{128} attempts is a computationally infeasible task. However, this also increases the computation cost for the verification at the supervisor side, because the supervisor now has to verify 128 computations, much more than it needs to be done in the CBS scheme.

Another way to defeat the cheating strategy is to let the participant pay for all those cheating attempts. If the cost of conducting the cheating becomes more expensive than the cost of conducting all the required computations (i.e., computing $f(x)$ for all $x \in D$), the cheating brings no benefit. To achieve this, we increase the cost of sample generation. Let C_g be the cost of the one-way function g, and C_f be the cost of function f. Because the probability that each attempt being successful is r^m, the expected number of attempts the participant needs to make is $\frac{1}{r^m}$. Therefore, to make the expected cost of the total cheating attempts more expensive than the total cost of the task, we need the following inequality:

$$\frac{1}{r^m} \cdot m \cdot C_g \geq n \cdot C_f. \tag{30.6}$$

To achieve the above inequality, we can increase either m or C_g. Increasing m is straightforward; increasing C_g needs some further explanation. Most of the one-way functions are very fast. To find a one-way function g such that C_g satisfies the above inequality, we just need to let $g \equiv (h)^k$, namely applying the one-way hash function h for k times, where k is a number that makes C_g expensive enough. If we let the left side of the inequality to be just slightly greater than the right side, this extra cost of g does not bring significant overhead to the supervisor or the honest participant because the ratio between the cost of the sample generation $(m \cdot C_g)$ and the cost of the entire task $(n \cdot C_f)$ is about r^m, which can be significantly small when we choose a proper value for m.

30.7 Experimental Study

Drug discovery process has long been acknowledged as an expensive and computationally intensive procedure. This process starts with identifying a protein that causes a disease. Subsequently, millions of molecules are tested for finding lead molecules that will bind with the protein to modulate the disease. Instead of doing laborious experiments in laboratories, which normally takes years of analyzing all the lead molecules, molecular modeling is used to do the simulations of the same computations in few hours. Researchers have investigated methods to accelerate this matching process (also known as docking), and the most recent one involves the PC grid computing.

Drug discovery is one of the most successful grid computing projects, which makes use of the idle time of computers connected to the Internet. Millions of volunteers donate their otherwise unused computer time for the projects such as FightAIDS@Home [28], Smallpox drug discovery [8], Cancer drug discovery, and bioterrorism [29].

To determine the effectiveness of the CBS scheme, the CBS scheme is applied to some popular drug discovery programs, in particular, to two molecular docking programs, AutoDock [30] and FTDock [31]. AutoDock is based on simulated annealing method for docking flexible molecules. FightAIDS@Home [28], a PC grid computing project, uses this software for searching drugs for AIDS. FTDock finds the best matching configuration for two molecules by searching and scoring all the possible translations and orientations. In FTDock, Fast Fourier Transform (FFT) is used to score each configuration as a surface correlational value.

In the experimental analysis, the following are conducted:

- Software Modules for CBS have been designed and developed in C++ and C in UNIX platform.
- CBS has been integrated with AutoDock and FTDock software.
- Time analysis is performed for AutoDock and FTDock to compare the time taken by the CBS versus the time taken for the actual task.

TABLE 30.1 AutoDock-CBS

Total Steps	Total Time (s)	SA Time (s)	CBS Time (s)	%
100,000	115.56	102.34	6.9	6.7
200,000	221.85	207.69	13.63	6.6
300,000	311.8	297.74	20.52	6.8
500,000	512.61	498.7	34.29	6.8
1,000,000	1022.7	1008.71	68.19	6.7

30.7.1 AutoDock

AutoDock is developed and distributed as a free software by The Scripps Research Institute, California, USA. AutoDock is implemented in C++, and is available for a variety of platforms. For the implementation, a Sun Enterprise 450 with four UltraSPARC-II 296 MHz processors running SunOS 5.8 is used. The Autodock version 3.0 consists of simulated annealing and genetic algorithm methods of docking; in the experimental study, only simulated annealing is used.

The simulated annealing function writes out the energy value of each randomly generated configuration. The CBS module, which is integrated to the simulated annealing, accumulates each of these energy values. At the end of the annealing procedure, a Merkle tree is constructed by the CBS tree function.

Table 30.1 shows timings of docking runs and the time taken by the CBS algorithms. The total steps are the number of leaves in the Merkle tree. The total time (without using CBS algorithm) is split into setup time and the simulated annealing (SA) time. The time taken by the CBS is listed under the CBS column. The percentage columns list the percentage of the task time (i.e., SA time) that is taken by the CBS algorithms. It is found that the overhead for the CBS algorithm is less than 7% compared to the actual SA time. This test was performed using 1hvr as protein and xk2A as inhibitor.

30.7.2 FTDock

FTDock is developed and distributed by the Biomolecular Modeling Laboratory of Cancer Research, London, UK. Implemented in C, it uses a Fast Fourier Transform for finding the surface complementarity of two molecules for all possible translations and rotations. For calculating the Fourier Transform, it uses the Fastest Fourier Transform in the West (FFTW) library. FFTW is a C subroutine library for computing the discrete Fourier transform (DFT) and is available at [32].

At each run of DFT, it generates N^3 (N is the size of the grid) scores. The CBS module constructs the Merkle tree for all of these scores. In FTDock, it is possible to define how many scores are ranked. In that case, CBS works on only the selected data values. After accumulating all the scores, CBS tree routine creates the root hash and also builds the Merkle tree.

As shown in Table 30.2, the actual computation time (FFT time) is proportional to the total number of scores. For the experimental study, KALLIKREIN A and TRYPSIN INHIBITOR (CRYSTAL FORM II) [33] are used as protein and inhibitor, respectively. All these experiments are conducted on Sun Enterprise 450 with four UltraSPARC-II 296 MHz processors running SunOS 5.8. From the

TABLE 30.2 FTDock-CBS

Total Scores	Total Time (s)	FFT Time (s)	CBS Time (s)	%
1,000	1718.07	100.96	0.04	0.04
2,000	1810.09	201.38	0.1	0.05
5,000	2115.93	505.85	0.21	0.04
10,000	2646.24	1036.74	0.44	0.04
15,000	3093.07	1500.34	0.63	0.04

experiments, it is observed that the CBS calculations take less than 0.05% of the time taken for the actual task, FFT.

30.8 Conclusion and Future Work

This chapter describes the CBS scheme to detect cheating participants in grid computing. To prevent the participant from changing the computation results after learning the samples, the CBS scheme uses the Merkle tree for the participant to commit its results before learning the sample selections. The CBS scheme can be used for generic computations in grid computing. It is efficient in communication as well as in computation. Based on the CBS scheme, two important issues are addressed: (1) how to reduce the storage requirement, and (2) how to convert the CBS scheme from an interactive scheme to a noninteractive scheme. The experimental study shows that CBS scheme takes only a small portion of the actual task for docking computations used in drug discovery projects.

One assumption made in the CBS scheme and the ringer scheme is that $|D|$ should be significantly large. When each participant is assigned a task with very few inputs, the sampling scheme and the ringer scheme does not work well. For example, when $|D| = 1$, i.e., each task consists of only one input, the cost of generating the precomputed results (for the ringer scheme) or the cost of verifying a sample (for the CBS scheme) is as expensive as conducting the task. Therefore, these two schemes are no better than the naive double-check-every-result scheme. Developing efficient schemes for a situation when $|D|$ is small is a good topic for future work.

References

1. M. J. Atallah and J. Li, Secure outsourcing of sequence comparisons, *International Journal of Information Security Archive*, 4(4), October 2005, 277–287.
2. M. J. Atallah, K. N. Pantazopoulos, J. R. Rice, and E. H. Spafford, Secure outsourcing of scientific computations, *Advances in Computers*, 54, 2001, 216–272.
3. K. LeFevre, D. J. DeWitt, and R. Ramakrishnan, Incognito: Efficient full-domain k-anonymity, in *Proceedings of the 2005 ACM SIGMOD*, Baltimore, MD, June 12–June 16, 2005.
4. R. J. Bayardo and R. Agrawal, Data privacy through optimal k-anonymization, in *Proceedings of the 21st IEEE International Conference on Data Engineering (ICDE)*, Tokyo, Japan, April 2005.
5. I. Foster and C. Kesselman, Eds., *The Grid: Blueprint for a New Computing Infrastructure*. Morgan-Kaufmann, San Francisco, CA, 1999.
6. IBM Grid computing. Available: http://www-1.ibm.com/grid/about_grid/what_is.shtml.
7. SETI@Home: The search for extraterrestrial intelligence project. University of California, Berkeley. Available: http://setiathome.berkeley.edu/.
8. The Smallpox Research Grid. Available: http://www-3.ibm.com/solutions/lifesciences/research/smallpox.
9. The Great Internet Mersenne Prime Search. Available: http://www.mersenne.org/prime.htm.
10. D. Bedell, Search for extraterrestrials–or extra cash, in *The Dallas Morning News*, Dallas, TX, December 2, 1999, also available at: http://www.dallasnews.com/technology/1202ptech9pcs.htm.
11. W. Du, J. Jia, M. Mangal, and M. Murugesan, Uncheatable grid computing, in *Proceedings of the 24th International Conference on Distributed Computing Systems (ICDCS)*, Tokyo, Japan, March 23–26, 2004, pp. 4–11.
12. P. Golle and I. Mironov, Uncheatable distributed computations, in *CT-RSA 2001: Proceedings of the 2001 Conference on Topics in Cryptology*, San Francisco, CA, April 8–12, 2001, pp. 425–440.
13. D. Szajda, B. Lawson, and J. Owen, Hardening functions for large scale distributed computations, *IEEE Symposium on Security and Privacy*, Berkeley, CA, 2003.

14. P. Golle and S. Stubblebine, Secure distributed computing in a commercial environment, in *Proceedings of Financial Crypto 2001*, ser. *Lecture Notes in Computer Science*, P. Syverson, Ed., vol. 2339. Springer, Berlin, Germany, 2001, pp. 289–304.

15. W. Du and M. T. Goodrich, Searching for high-value rare events with uncheatable grid computing, in *Proceedings of the Applied Cryptography and Network Security Conference*, New York, June 7–10, 2005, pp. 122–137.

16. F. Monrose, P. Wykoff, and A. D. Rubin, Distributed execution with remote audit, in *Proceedings of ISOC Symposium on Network and Distributed System Security*, San Diego, CA, February 1999, pp. 103–113.

17. G. Vigna, Ed., *Mobile Agents and Security*, ser. *Lecture Notes in Computer Science*, vol. 1419. Springer, Berlin, Germany, 1998.

18. B. S. Yee, A sanctuary for mobile agents, in *Secure Internet Programming: Security Issues for Mobile and Distributed Objects, Lecture Notes in Computer Science*, J. Vitek and C. Jensen, Eds., Vol. 1603, Springer-Verlag, Berlin, 1999, pp. 261–273.

19. G. Vigna, Protecting mobile agents through tracing, in *Proceedings of the 3rd Workshop on Mobile Object Systems*, Jyvaskyla, Finland, June 1997.

20. F. Hohl, Time limited blackbox security: Protecting mobile agents from malicious hosts, *Mobile Agents and Security, Lecture Notes in Computer Science*, vol. 1419. Springer, Berlin, Germany, 1998, pp. 92–113.

21. T. Sander and C. F. Tschudin, Protecting mobile agents against malicious hosts, *Lecture Notes in Computer Science*, vol. 1419. Springer-Verlag, New York, 1998, pp. 44–60.

22. F. S. Y. Minsky, R. van Renesse, and S. D. Stoller, Cryptographic support for fault-tolerant distributed computing, in *Proceedings of Seventh ACM SIGOPS European Workshop, System Support for Worldwide Applications*, Connemara, Ireland, September 1996, pp. 109–114.

23. C. Cachin, S. Micali, and M. Stadler, Computationally private information retrieval with polylogarithmic communication, in *Proceedings of Advances in Cryptology—EUROCRYPT*, Berlin, Germany, 1999, pp. 402–414.

24. R. O. W. Aiello, S. Bhatt, and S. Rajagopalan, Fast verification of any remote procedure call: Short witness-indistinguishable one-round proofs for np, in *Proceedings of the 27th International Colloquium on Automata, Languages and Programming*, Geneva, Switzerland, July 2000, pp. 463–474.

25. R. C. Merkle, Protocols for public key cryptography, in *IEEE Symposium on Security and Privacy*, Oakland, CA, 1980, pp. 122–134.

26. T. H. Cormen, C. E. Leiserson, and R. L. Rivest, *Introduction to Algorithms*. The MIT Press, Cambridge, MA, 1989.

27. R. Buyya, Economic based distributed resource management and scheduling for grid computing, PhD dissertation, Monash University, Melbourne, Australia, April 2002.

28. The FightAIDS@Home Research Grid. Available:http://fightaidsathome.scripps.edu.

29. The PatriotGrid: Anthrax Research Project. Available:http://www.grid.org/projects/patriot.htm.

30. AutoDock. Available:http://www.scripps.edu/pub/olson-web/doc/autodock/.

31. FTDock. Available: http://www.bmm.icnet.uk/docking/.

32. FFTW (Fastest Fourier Transform in the West). Available: http://www.fftw.org/.

33. FTDock Test System. Available : http://www.bmm.icnet.uk/docking/systems/2KAI.html.

31

DNA Computing: A Research Snapshot

Lila Kari
University of Western Ontario

Kalpana Mahalingam
Indian Institute of Technology

31.1 Introduction

During the last decades, we have witnessed exciting new developments in computation theory and practice, from entirely novel perspectives. DNA Computing, known also under the names of molecular computing, biocomputing, or biomolecular computing, is an emergent field lying at the crossroads of computer science and molecular biology. It is based on the idea that data can be encoded as DNA strands, and molecular biology tools can be used to perform arithmetic and logic operations.

This chapter intends to give the reader a basic understanding of the tools and methods used in DNA computing, and a snapshot of theoretical and experimental research in this field. It includes descriptions of a representative DNA computing experiment, of the design of autonomous and programmable molecular computers, and of models of DNA memories. It describes research into computation in and by living cells, as well as into DNA self-assembly, a process that can be used for computation or can produce either static DNA nanostructures or dynamic DNA nanomachines. This chapter is intended to offer the reader a glimpse of the astonishing world of DNA computing, rather than being an exhaustive review of the research in the field.

The chapter is organized as follows.* Section 31.2 introduces basic molecular biology notions about DNA as an information-encoding medium, and the bio-operations that can be performed on DNA strands. As a representative example of a DNA bio-algorithm, in which a computational problem is solved exclusively by bio-operations performed on DNA strands in test tubes, Section 31.3 describes an experiment demonstrated by Adleman's team, that solved a 20-variable instance of the 3-SAT problem. Section 31.4 exemplifies the research on molecular autonomous programmable DNA computers by the description of Benenson automata and their potential applications to smart drug design. Section 31.5 presents research into DNA memory technology such as nested primer molecular memory (Section 31.5.1), and organic DNA memory (Section 31.5.2), as well as explores the topic of optimal DNA sequence design for DNA computing purposes (Section 31.5.3). Section 31.6 explores the fascinating possibilities offered by *in vivo* computing, that is, computation in and by living cells, as well as the potential insights it could offer into understanding the computational processes in nature. Section 31.7 describes one of the most important directions of research in DNA computing, with significant implications for the rapidly evolving field of nanotechnology: DNA computing by self-assembly. It includes descriptions of experimental computation by self-assembly (Section 31.7.1), of static intricate DNA nanostructures (Section 31.7.2), and of impressive DNA nanomachines (Section 31.7.3). Finally, Section 31.8 offers conclusions and brief comments on the latest developments in the field of DNA computing.

31.2   Molecular Biology Basics

This chapter was written with the expectation that the reader is a computer scientist with a limited knowledge of biology. This section provides the basic notions of molecular biology (DNA structure and DNA bio-operations) necessary for understanding the text.

DNA (*deoxyribonucleic acid*) is found in every cellular organism as the storage medium for genetic information, [39]. DNA is a polymer whose monomer units are *nucleotides*, and the polymer is known as a "polynucleotide." More precisely, DNA is a linear chain made up of four different types of nucleotides, each consisting of a base (Adenine, Cytosine, Guanine, or Thymine) and a sugar-phosphate unit. The sugar-phosphate units are linked together by covalent bonds to form the backbone of the DNA single strand. Since nucleotides may differ only by their bases, a DNA strand can be viewed as simply a word over the four-letter alphabet {A, C, G, T}. A DNA single strand has an orientation, with one end known as the 5′ end, and the other as the 3′ end, based on their chemical properties. By convention, a word over the DNA alphabet represents the corresponding DNA single strand in the 5′ ⟶ 3′ orientation, that is, GGTTTTT stands for the DNA single strand 5′-GGTTTTT-3′. A short single-stranded polynucleotide chain, usually less than 20 nucleotides long, is called an *oligonucleotide*. A DNA strand is sometimes called an *n-mer*, where *n* signifies its length, that is, the number of its nucleotide monomers.

A crucial feature of DNA single strands is their *Watson–Crick complementarity*, [91]: A is complementary to T, and G is complementary to C. Two complementary DNA single strands with opposite orientation will bind to each other by hydrogen bonds between their individual bases to form a stable DNA double strand with the backbones at the outside and the bound pairs of bases lying inside. In a DNA double strand, two complementary bases situated on opposite strands, and bound together by hydrogen bonds, form a *base-pair (bp)*.

---

* The general audience for whom this paper is intended, our fields of expertize, and the space available for this chapter, affected both the depth and the breadth of our exposition. In particular, the list of research topics presented here is by no means exhaustive, and it is only meant to give a sample of DNA computing research. Moreover, the space devoted to various fields and topics was influenced by several factors and, as such, has no relation to the respective importance of the field or the relative size of the body of research in that field.

As an example, the DNA single strand 5'- AAAAACC - 3' will bind to the DNA single strand 5'-GGTTTTT-3' to form the 7 base-pair-long (7bp) double strand

$$5' - AAAAACC - 3'$$
$$3' - TTTTTGG - 5'.$$

Formally, the Watson–Crick complement of a DNA single strand w will be denoted by WK(w) or $\overleftarrow{w}$. Using our convention on directionality of strands, note that WK(AAAAACC) = GGTTTTT.

Another nucleic acid that has been used for computations is RNA, [23]. *Ribonucleic acid* or *RNA* is a nucleic acid that is similar to DNA, but differs from it in three main aspects: RNA is typically single-stranded while DNA is usually double-stranded, RNA nucleotides contain the sugar ribose, while DNA nucleotides contain the sugar deoxyribose, and in RNA the nucleotide Uracil, U, substitutes for Thymine, which is present in DNA.

The genome consists of DNA sequences, some of which are genes that can be transcribed into *messenger RNA (mRNA)*, and then translated into proteins according to the *genetic code* that maps each 3-letter RNA segment (called *codon*) into an amino acid. Several designated triplets, called start (stop) codons, signal the initiation (termination) of a translation. A protein is a sequence over the 20-letter alphabet of amino acids.

There are many possible *DNA bio-operations* that can be used for computations, [2,38,64], such as: hybridization by Watson–Crick complementarity, cut-and-paste operations achievable by enzymes, the synthesis of desired DNA strands up to a certain length, making exponentially many copies of a DNA strand, the separation of strands by length, the extraction of DNA strands that contain a certain subsequence, and reading-out a DNA strand. These bio-operations, some briefly explained later, have all been used to control DNA computations and DNA robotic operations.

DNA single strands with opposite orientation will join together to form a double helix in a process called *base-pairing, annealing*, or *hybridization*. The reverse process—a double-stranded helix coming apart to yield its two constituent single strands—is called *melting*. As the name suggests, melting is achieved by raising the temperature, and annealing by lowering it.

One class of enzymes, called *restriction endonucleases*, each recognize a specific short sequence of DNA, known as a *restriction site*. Any double-stranded DNA that contains the restriction site within its sequence is cut by the enzyme at that location in a specific pattern. Depending on the enzyme, the cutting operation leaves either two "blunt-ended" DNA double strands or, more often, two DNA strands that are double-stranded but have single-stranded overhangs known as "sticky-ends." Another enzyme, called *DNA ligase*, can link together two partially double-stranded DNA strands with complementary sticky-ends, by sealing their backbones with covalent bonds. The process is called *ligation*.

The separation of DNA strands by length is possible by using a technique called *gel electrophoresis*. The negatively charged DNA molecules are placed at the top of a wet gel, to which an electric field is applied, drawing them to the bottom. Larger molecules travel more slowly through the gel. After a period, the molecules spread out into distinct bands according to their size.

The extraction of DNA single strands that contain a specific subsequence v, from a heterogeneous solution of DNA single strands, can be accomplished by *affinity purification*. After synthesizing strands Watson–Crick complementary to v, and attaching them to magnetic beads, the heterogeneous solution is passed over the beads. Those strands containing v anneal to its Watson–Crick complementary sequence and are retained. Strands not containing v pass through without being retained.

The *DNA polymerase* enzymes perform several functions including the replication of DNA by a process called *Polymerase Chain Reaction*, or *PCR*. The PCR replication reaction requires a guiding DNA single strand called *template*, and an oligonucleotide called *primer*, that is annealed to the template. The DNA polymerase enzyme then catalyzes DNA synthesis by successively adding nucleotides to one end of the primer. The primer is thus extended at its $3'$ end, in the direction $5'$ to $3'$ only, until the desired strand is obtained that starts with the primer and is complementary to the template. If two primers are used, the result is the exponential multiplication of the subsequence of the template strand that is flanked by the two primers, in a process called *amplification*, schematically explained in the following. For the purpose of this explanation, if x is a string of letters over the DNA alphabet {A, C, G, T}, then $\bar{x}$ will denote its simple complement, for example,

$$\overline{\text{AACCTTGG}} = \text{TTGGAACC}.$$

Let us assume now that one desires to amplify the subsequence between x and y from the DNA double strand $\dfrac{5' - \alpha x \beta y \delta - 3'}{3' - \bar{\alpha}\bar{x}\bar{\beta}\bar{y}\bar{\delta} - 5'}$, where $\alpha, x, \beta, y, \delta$ are DNA segments. Then, one uses as primers the strand x and the Watson–Crick complement of y. After heating the solution and thus melting the double-stranded DNA into its two constituent strands, the solution is cooled and the Watson–Crick complement of y anneals to the "top" strand, while x anneals to the "bottom" strand. The polymerase enzyme extends the $3'$ ends of both primers into the $5'$ to $3'$ direction, producing partially double-stranded strands $\dfrac{5' - \alpha x \beta y \delta - 3'}{3' - \bar{\alpha}\bar{x}\bar{\beta}\bar{y} \ - 5'}$ and $\dfrac{5' - \ x \beta y \delta - 3'}{3' - \bar{\alpha}\bar{x}\bar{\beta}\bar{y}\bar{\delta} - 5'}$. In a similar fashion, the next heating–cooling cycle will result in the production of the additional strands $5' - x\beta y - 3'$ and $3' - \bar{x}\bar{\beta}\bar{y} - 5'$. These strands are Watson–Crick complementary and will, from now on, be produced in excess of the other strands, since both are replicated during each cycle. At the end, an order of 2^n copies of the desired subsequences flanked by x and y will be present in the solution, where n is the number of the heating–cooling cycles.

The earlier and other bio-operations have all been harnessed in biocomputing. To encode information using DNA, one can choose an encoding scheme mapping the original alphabet onto strings over {A, C, G, T}, and proceed to synthesize the information-encoding strings as DNA single strands. A biocomputation consists of a succession of bio-operations, [20], such as the ones described in this section. The DNA strands representing the output of the biocomputation can then be read out (using a sequencer) and decoded.

31.3 Adleman's 3-SAT Experiment

The idea of using DNA molecules as the working elements of a computer goes back to 1994, [1], when Adleman solved a 7-node instance of the Hamiltonian Path Problem by exclusively using DNA strands to encode information, and molecular biology techniques as algorithmic instructions. This section describes another, more complex, DNA computing experiment that solves a large instance of the 3-SAT problem, [15].

The 3-SAT problem is an NP-complete computational problem for which the fastest known sequential algorithms require exponential time. The problem became a testbed for the performance of DNA computers after Lipton, [54], demonstrated that it was well suited to take advantage of the parallelism afforded by molecular computation. In 2002, Adleman and his group solved a 20-variable 3-SAT problem, [15]. Unlike the initial proof-of-concept DNA computing experiment, this was the first experiment that demonstrated that DNA computing devices can exceed the computational power of an unaided human. Indeed, the answer to the problem was found after an exhaustive search of more than 1 million (2^{20}) possible solution candidates.

The input to a 3-SAT problem is a Boolean formula in 3CNF, that is, in conjunctive normal form where each conjunct has only at most three literals. This formula is called *satisfiable* if there exists a truth value assignment to its variables that satisfies it, that is, that makes the whole formula true. Thus, the output to the 3-SAT problem is "yes" if such a satisfying truth value assignment exists, and "no" otherwise.

The input formula of Adleman's experiment was the 20-variable, 24-clause, 3CNF formula:

$\Phi = (x_3 \lor \overline{x_{16}} \lor x_{18}) \land (x_5 \lor x_{12} \lor \overline{x_9}) \land (\overline{x_{13}} \lor \overline{x_2} \lor x_{20}) \land (x_{12} \lor \overline{x_9} \lor \overline{x_5}) \land (x_{19} \lor \overline{x_4} \lor x_6) \land (x_9 \lor x_{12} \lor \overline{x_5}) \land (\overline{x_1} \lor x_4 \lor \overline{x_{11}}) \land (x_{13} \lor \overline{x_2} \lor \overline{x_{19}}) \land (x_5 \lor x_{17} \lor x_9) \land (x_{15} \lor x_9 \lor \overline{x_{17}}) \land (\overline{x_5} \lor \overline{x_9} \lor \overline{x_{12}}) \land (x_6 \lor x_{11} \lor x_4) \land (\overline{x_{15}} \lor \overline{x_{17}} \lor x_7) \land (\overline{x_6} \lor x_{19} \lor x_{13}) \land (\overline{x_{12}} \lor \overline{x_9} \lor x_5) \land (x_{12} \lor x_1 \lor x_{14}) \land (x_{20} \lor x_3 \lor x_2) \land (x_{10} \lor \overline{x_7} \lor \overline{x_8}) \land (\overline{x_5} \lor x_9 \lor \overline{x_{12}}) \land (x_{18} \lor \overline{x_{20}} \lor x_3) \land (\overline{x_{10}} \lor x_{18} \lor \overline{x_{16}}) \land (x_1 \lor \overline{x_{11}} \lor \overline{x_{14}}) \land (x_8 \lor \overline{x_7} \lor \overline{x_{15}}) \land (\overline{x_8} \lor x_{16} \lor \overline{x_{10}}),$

where $\overline{x_i}$ denotes the negation of x_i, $1 \leq i \leq 20$. This formula Φ was designed so as to have a unique satisfying truth assignment. The unique solution is: $x_1 = F, x_2 = T, x_3 = F, x_4 = F, x_5 = F, x_6 = F, x_7 = T, x_8 = T, x_9 = F, x_{10} = T, x_{11} = T, x_{12} = T, x_{13} = F, x_{14} = F, x_{15} = T, x_{16} = T, x_{17} = T, x_{18} = F, x_{19} = F, x_{20} = F.$

Adleman's solution was based on the following nondeterministic algorithm.

Input: A Boolean formula Φ in 3CNF.

Step 1: Generate the set of all possible truth value assignments.

Step 2: Remove the set of all truth value assignments that make the first clause false.

Step 3: Repeat Step 2 for all the clauses of the input formula.

Output: The remaining (if any) truth value assignments.

To implement this algorithm, one needs to first encode the input data as DNA strands. This was achieved as follows. Every variable x_k, $k = 1, \ldots, 20$, was associated with two distinct 15-mer DNA single strands called "value sequences." One of them, denoted by X_k^T, represented true (T), while the second, denoted by X_k^F, represented false (F).

The following are some examples of the particular 15-mer sequences—none of which contained the nucleotide G—synthesized and used in the experiment:

$$X_1^T = \text{TTACACCAATCTCTT}, \quad X_1^F = \text{CTCCTACAATTCCTA},$$

$$X_{20}^T = \text{ACACAAATACACATC}, \quad X_{20}^F = \text{CAACCAAACATAAAC}.$$

Each of the possible 2^{20} truth assignments was represented by a 300-mer "library strand" consisting of the ordered catenation of one 15-mer value sequence for each variable, that is, by a strand

$$\alpha_1 \alpha_2 \ldots \alpha_{20}, \text{ where } \alpha_i \in \{X_i^T, X_i^F\}, 1 \leq i \leq 20.$$

To obtain these "library strands," the 40 individual 15-mer sequences (each present in multiple copies in solution) were assembled using the mix-and-match combinatorial synthesis technique of [23].

The biocomputation wetware essentially consisted of a glass "library module" filled with a gel containing the library, as well as one glass "clause module" for each of the 24 clauses of the formula. Each clause module was filled with gel containing probes (immobilized DNA single strands) designed to bind only library strands encoding truth assignments statisfying that clause.

The strands were moved between the modules with the aid of gel electrophoresis, that is, by applying an electric current that resulted in the migration of the negatively charged DNA strands through the gel.

The protocol started with the library passing through the first "clause module," wherein library strands containing the truth assignments satisfying the 1st clause were captured by the immobilized probes, while library strands that did not satisfy the first clause continued into a buffer reservoir. The captured strands were then released by raising the temperature, and used as input to the second clause module, and so on. At the end, only the strand representing the truth assignment that satisfied all the 24 clauses remained.

The output strand was PCR amplified with primer pairs corresponding to all four possible true–false combinations of assignments for the first and last variable x_1 and x_{20}. None except the primer pair $(X_1^F, WK(X_{20}^F))$ showed any bands, thus indicating two truth values of the satisfying assignment, namely $x_1 = F$ and $x_{20} = F$. The process was repeated for each of the variable pairs (x_1, x_k), $2 \leq k \leq 19$, and, based on the lengths of the bands observed, value assignments were given to the variables. These experimentally derived values corresponded to the unique satisfying assignment for the formula Φ, thus concluding the experiment.

One of the remarkable features of this DNA computing experiment was that simple bio-operations could carry on a complex computation such as the one needed for solving a relatively large instance of 3-SAT.

31.4 Benenson Automata

Following Adleman's, [1], proof-of-concept demonstration of DNA computing, several research teams embarked on the quest for a DNA implementation of a Universal Turing Machine. In the process, they demonstrated various molecular scale autonomous programmable computers ([10–12,56,74]) allowing both input and output information to be in molecular form. In this section we illustrate one such programmable computer, [10], known as the "Benenson automaton."

In [10] the authors construct a simple two-state automaton, Figure 31.1, over a two-letter alphabet set, by using double-stranded DNA molecules and restriction enzymes.

Any particular two-state automaton will have a subset of the eight possible transition rules, and a subset of the two-state set as final states. The automaton implemented by Benenson et al. takes strings of symbols over the two-letter alphabet $\{a, b\}$ as input, and accepts only those strings that contain an even number of letters b (see Figure 31.1).

The main engine of the Benenson automaton is the *FokI* enzyme, a member of an unusual class of restriction enzymes that recognize a specific DNA sequence and cleave nonspecifically a short distance away from that sequence. *FokI* binds the sequence

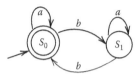

FIGURE 31.1 A finite automaton with two states S_0 and S_1 over the alphabet set $\{a, b\}$, that accepts only words with an even number of letters b.

$$5' - GGATG - 3'$$
$$3' - CCTAC - 5''$$

and cleaves DNA (regardless of the sequence composition) 9 bp away on the "top" strand, and 13 bp away on the "bottom" strand, leaving thus four-letter-long sticky ends.

The input symbols, a, b, and the terminator t, are encoded respectively as the 6bp sequences:

$$a: \begin{array}{c} 5' - CTGGCT - 3' \\ 3' - GACCGA - 5' \end{array} \quad b: \begin{array}{c} 5' - CGCAGC - 3' \\ 3' - GCGTCG - 5' \end{array} \quad t: \begin{array}{c} 5' - TGTCGC - 3' \\ 3' - ACAGCG - 5' \end{array}$$

An example of the encoding of an input string can be seen in the first line of Figure 31.3. The input ab is encoded as a DNA double strand that contains a restriction site for *FokI*, followed closely by the catenation of the encodings for the symbol string abt.

Every state/symbol pair is encoded as a 4-mer DNA strand in the following way. The state/symbol pair $S_0 a$ is encoded as $5' - GGCT - 3'$ (the 4-mer suffix of the encoding for the symbol a), while $S_1 a$ is encoded as $5' - CTGG - 3'$ (the 4-mer prefix of the encoding for a). The encodings are chosen in a similar way for the state/symbol pairs involving the input symbol b, and the terminator symbol t. This method permits the encoding of a symbol to be interpreted in two ways: If the 4-mer suffix of the encoded symbol is detected, then the symbol is interpreted as being read in state S_0; and if the 4-mer prefix of the encoded symbol is detected, then the symbol is being interpreted as being read in state S_1.

There are two output detection molecules. S_0-D is a 161-mer DNA double strand with an overhang $3' - AGCG - 5'$, which "detects" the last state of the computation as being S_0. S_1-D is a 251-mer DNA

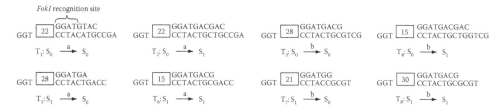

FIGURE 31.2 Transition molecules encoding the set of all possible transition rules of a two-state automaton. The boxes represent DNA double-strands of length equal to the number inside the box. (Adapted from Benenson, Y. et al., *Nature*, 414, 430, 2001.)

double strand with an overhang $3' - ACAG - 5'$, which detects the last state of the computation as being S_1. The two output detection molecules have different lengths, so that they can be easily differentiated by gel electrophoresis.

The eight possible transition molecules of a two-state Benenson automaton are shown in Figure 31.2. Each has a four-letter overhang, for example, $3' - CCGA - 5'$ for T_1, that can selectively bind to DNA molecules encoding the current state/symbol pair, as detailed in the following example.

The automaton processes the encoding for the input string ab as shown in Figure 31.3. First, the *FokI* enzyme recognizes its restriction site and cleaves the input encoding the symbols abt, thereby exposing the 4-nucleotide sticky-end $5' - GGCT - 3'$ that encodes the state/symbol pair S_0a.

The transition molecule T_1 encoding the automaton rule $S_0a \longrightarrow S_0$ "detects" this state/symbol, by binding exactly to the cleaved input molecule and forming a fully double-stranded DNA molecule with the aid of the enzyme ligase.

The transition molecule T_1, now incorporated in the current molecule, contains a *FokI* recognition site. Moreover, the 3bp "spacer" sequence that follows the *FokI* restriction site in T_1 ensures that the next cleaving of *FokI* will expose a suffix of the encoding of the next input symbol b, which will be correctly interpreted as S_0b.*

The overhang of the current DNA molecule is now $5' - CAGC - 3'$, which is interpreted as S_0b. The sticky-end of this sequence fits the transition rule T_4, encoding the automaton rule $S_0b \longrightarrow S_1$. Thus, the combination of the current DNA strand with T_4 and the enzyme ligase leads to another fully double-stranded DNA strand.

A last use of *FokI* exposes the overhang $5' - TGTC - 3'$ of the current DNA molecule. This is a prefix of the terminator, which is interpreted as S_1t. The overhang is complementary to the sticky-end $3' - ACAG - 5'$ of the detector molecule S_1-D, corresponding to the last state of the computation being S_1. The state S_1 is not final, and thus the outcome of the computation is that the input ab is not accepted by this automaton.

Note that any two-state two-symbol automaton could be built using the aforementioned method. The automaton could be made to perform a different task by choosing a different set of transition molecules.

As an exciting application, Benenson et al. [12] demonstrated how the aforementioned method can be used for medical diagnosis and treatment. As a proof of principle, Benenson et al. [12] programmed an automaton to identify and analyze the mRNA of disease-related genes associated with models of small-cell lung cancer and prostate cancer, and to produce a single-stranded DNA molecule modeled after an anticancer drug.

---

* Note that, if the automaton would have had instead the transition rule T_2, encoding $S_0a \longrightarrow S_1$, then the length of the "spacer" in T_2 would have been 5bp, placing the position of the cut 2bp more to the left. This would have exposed a prefix of the encoding for b, which would have been correctly interpreted as S_1b.

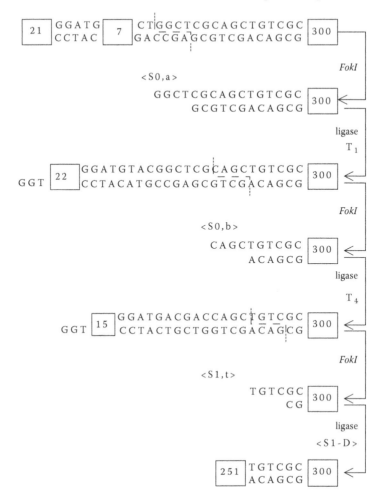

FIGURE 31.3 Example of the computation of a Benenson automaton corresponding to Figure 31.1, for the input *ab*. The numbers in the boxes indicate the lengths of the corresponding double-stranded DNA sequences. (Adapted from Benenson, Y. et al., *Nature*, 414, 430, 2001.)

The computational process of this automaton was formalized in [86]. The authors of [86] study the computational power of Benenson automata and show that they are capable of computing arbitrary Boolean functions.

31.5 DNA Memory

This section discusses various aspects of DNA-encoded information. A brief description of several models for a DNA memory is followed by a more detailed description of the Nested Primer Molecular Memory (NPMM) (Section 31.5.1), and *in vivo* Organic DNA Memory (Section 31.5.2). In addition, Section 31.5.3 provides a discussion on the various approaches to the optimal design of information-encoding DNA sequences for DNA computing, including software simulations, algorithmic methods, and theoretical studies.

There are several reasons to consider DNA memory as an alternative to all the currently available implementations of memories. The first is the extraordinary information-encoding density that can be achieved by using DNA strands. Indeed, the possibility of storing vast amounts of data in a small space is probably one of the most alluring features of DNA computing. In [71], Reif reported a calculation based on the facts that a mole contains 6.02×10^{23} DNA base monomers, and the mean molecular weight of a monomer is approximately 350 g/mole. Thus, 1 g of DNA contains 2.1×10^{21} DNA bases. Since there are 4 DNA bases that can encode 2 bits, it follows that 1 gram of DNA can store approximately 4.2×10^{21} bits. Thus, DNA has the potential of storing data on the order of 10^{10} more compactly than conventional storage technologies. In addition, the robustness of DNA data ensures the maintenance of the archived information over extensive periods of time [4,18,85].

The idea of a content-addressable DNA memory, able to store binary words of fixed length, was first proposed by Baum in [8]. In his model, each word in the memory would be a DNA single strand constructed as follows. Two distinct DNA sequences would be assigned to each component, the first encoding a "1," and the other encoding a "0." DNA molecules encoding a particular binary word would then be obtained by catenating the appropriate DNA sequences corresponding to its bits, in any order. This DNA memory would be associative, or content addressable. Given a subset of the component values, one could retrieve any words consistent with these values from the DNA memory as follows. For each component, one could synthesize the complement of the corresponding subsequence of encoding DNA, and affix it to a solid support, for example, a magnetic bead. This complement would then bond to DNA memory molecules that contain that subsequence, that is, code for that component value. These molecules could then be extracted magnetically. After iterative extractions based on each component, one would end up with DNA molecules matching the constraints exactly. As calculated in [8], the storage that is in principle possible using these techniques is astonishing, making it possible to imagine DNA memories vastly larger than the brain.

Several other authors, [16,46,60,71,88], have proposed various DNA memory models. Almost all of this research consists of preliminary experiments on a very small scale. Recently, Yamamoto et al. proposed a DNA memory with over 10 million addresses [99]. Their proposed DNA memory is addressable by using nested PCR, was named NPMM, and will be discussed in more detail in the Section 31.5.1.

31.5.1 Nested Primer Molecular Memory

NPMM, [99], is a pool of DNA strands wherein each strand codes both data information and its corresponding address information. The data information is a DNA sequence that uses a suitable encoding to express the information as a sequence over the DNA alphabet {A, C, G, T}. The address information consists of several components expressed as DNA sequences flanking the data. For example, data could be stored as

$$[CL_i, BL_j, AL_k, \text{DATA}, AR_q, BR_r, CR_s],$$

where $i, j, k, q, r, s \in \{0, \dots, 15\}$, and each of the components, for example, CL_0, represents a 20-mer DNA sequence. To retrieve the data, one uses nested PCR consisting of three steps. The first step is to use PCR with the primer pair $(CL_i, WK(CR_s))$. As a result, one can extract from the memory solution the set of DNA molecules starting with CL_i and ending in CR_s. This is because PCR results in a significant difference in concentration between the amplified molecules (starting with CL_i and ending with CR_s), and nonamplified molecules. The latter can be disregarded for all practical purposes. The second PCR is performed with primer pair $(BL_j, WK(BR_r))$. At this point we possess DNA-encoded data that was flanked by CL_iBL_j and BR_rCR_s. The third step is PCR using primer pair

$(AL_k, WK(AR_q))$. This will result in the target DNA-encoding data with "left address" $CL_iBL_jAL_k$ and "right address" $CR_qBR_rAR_s$. The molecules can then be sequenced and decoded to allow the retrieval of the target DNA-encoded data.

The aforementioned NPMM model can realize both enormous address space and high specificity. The hierarchical address structure enables a few DNA sequences to express very large address spaces. One of the main disadvantages of NPMM is that during PCR mutations can occur. This can be avoided by proper selection of DNA sequences, which will be discussed in detail in Section 31.5.3. The authors discuss, in [99], the limitation of scaling up the proposed DNA memory by using a theoretical model based on combinatorial optimization with some experimental restrictions. The results reveal that the size of the address space of this model of DNA memory is close to the theoretical limit.

31.5.2 Organic DNA Memory

The development of a DNA memory technology utilizing living organisms has a much greater potential than any of the existing counterparts to render long life expectancy of data, [97,98]. Since a naked DNA molecule is easily destroyed in any open environment, a living organism can act as a host that protects the information-encoding DNA sequence. In [97], the authors propose a candidate for a living host for DNA memory sequences, that tolerates the addition of artificial gene sequences and survives extreme environmental conditions.

The experiment had several key stages, [97]. In the first stage, in the process of identifying candidates to carry the embedded DNA molecules, the authors considered several microorganisms, and chose two well-understood bacteria, *Escherichia coli*, and *Deinococcus radiodurans*. In particular, *Deinococcus* can survive extreme conditions including cold, dehydration, vacuum, acid, and radiation, and is therefore, known as a polyextremophile.

During the *information encoding* stage, a certain encoding scheme was chosen, that assigned 3-mer sequences to various symbols. For example, AAA encoded the digit "0," AAC the digit "1," and AGG the letter "A" of the English alphabet. Each of the encoding DNA sequences contained only three of the four DNA nucleotides. Using this encoding, any English text could be codified as a DNA sequence, and the text chosen for this experiment was "And the oceans are wide."

In addition, several DNA sequences were chosen to act as "sentinels" and tag the beginning and the end of the encoded messages, for later identification and retrieval. These sequences were chosen by searching the genomes of both *E. coli* and *Deinococcus*, and identifying a set of fixed-size DNA sequences that do not exist in either genome, yet satisfy all the genomic constraints and restrictions. Twenty five such DNA sequences, 20 base-pairs each, were selected, in such a way so as not to cause unnecessary mutation or damage to the bacteria. All sequences contained multiple stop codons TAA, TGA, and TAG as subsequences. Without their presence, the bacterium could misinterpret the memory strands, transcribe them into mRNA, and then translate the mRNA into artificial proteins that could destroy the integrity of the embedded message, or even kill the bacterium.

A 46bp DNA sequence was then created, consisting of two different 20bp selected "sentinel" sequences, connected by a 6bp sequence that was the recognition site of a particular enzyme. This double-stranded DNA molecule was cloned into a recombinant plasmid (a circular double-stranded DNA sequence)—see Figure 31.4.

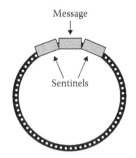

FIGURE 31.4 A recombinant plasmid with two DNA fragments as sentinels protecting the DNA-encoded message in between. (Reproduced from Wong, P. et al., *Commun. ACM*, 46, 95, 2003.)

The embedded DNA was then inserted into cloning vectors (circular DNA molecules that can self-replicate within a bacterial host). The resultant vectors were then transferred into *E.coli* by high-voltage shocks, allowing the vector to multiply.

The vector and the encoded DNA were then incorporated into the genome of *Deinococcus* for permanent information storage and retrieval. Using prior knowledge of the sequences at both borders, the message was retrieved by PCR, read-out, and decoded into the original English text it encoded.

The proposed organic DNA memory has an enormous potential capacity for storing information, especially considering that 1mL of liquid can contain up to 10^9 bacteria. Potential disadvantages are random mutations, but these are unlikely given the well developed natural mechanisms that exist in cells for detecting and correcting errors.

31.5.3 Design of DNA Sequences

In most DNA-based computations, there are three basic stages: (1) encoding the input data using single- or double-stranded DNA molecules, (2) performing the biocomputation using bio-operations, and finally (3) decoding the result. One of the main problems in DNA computing is associated with step (1), and concerns the design of optimal information-encoding oligonucleotides to ensure that mismatched pairing due to the Watson–Crick complementarity is minimized, and that undesirable bonds between DNA strands are avoided.

Indeed, in laboratory experiments, the complementarity of the bases may pose potential problems if some DNA strands form nonspecific hybridizations and partially anneal to strands that are not their exact complements, or if they stick to themselves or to each other in unintended ways. There are several approaches to addressing this so-called *sequence design problem*. In this chapter, we discuss the software simulation approach, the algorithmic approach, and the theoretical approach to the design of DNA strands optimal for DNA computing.

31.5.3.1 Software Simulation

Software simulation tools verify the computation protocol correctness before it is carried out in a laboratory experiment. Several software packages [24,25,32,33] written for DNA computing purposes are available. For example, the software *Edna* simulates biochemical processes and reactions that can occur during a laboratory experiment.

Edna, [28], is a simulation tool that uses a cluster of PCs and simulates the processes that could happen in test tubes. *Edna* can be used to determine if a particular choice of encoding strategy is appropriate, to test a proposed protocol and estimate its performance and reliability, and even to help assess the complexity of the protocols. Test tube operations are assigned a cost that takes into account many of the reaction conditions. The measure of complexity used by *Edna* is the sum of the costs added up over all operations in a protocol. Other features offered by the software allow the prediction of DNA melting temperature (the temperature at which a DNA double strand "melts" into its two constituent single strands), taking into account various reaction conditions. All molecular interactions simulated by the software are local and reflect the randomness inherent in biomolecular processes.

31.5.3.2 Algorithmic Method

In most DNA-based computations there is an assumption that a strand will bind only to its perfect Watson–Crick complement. For example, the results of DNA computations are retrieved from the test tubes by using strands that are complementary to the ones used for computation.

In practice, it is, however, possible that a DNA molecule will bind to another molecule, which differs from its perfect complementary molecule by one or even several nucleotides. One way to avoid this is to ensure that two molecules in the solution differ in more than one location. This property can be formalized in terms of the Hamming distance.

Given two words w_1, w_2 of equal length, the Hamming distance $H(w_1, w_2)$ is defined as the number of locations in which the words w_1 and w_2 are distinct. For a set of words, the Hamming distance constraint requires that any two words w_1 and w_2 in the set have $H(w_1, w_2) \geq d$. If we are dealing with DNA words, another constraint necessary to avoid mishybridizations is $H(w_1, WK(w_2)) \geq d$, where $WK(w)$ denotes the Watson–Crick complement of w. Yet another consideration is that, when retrieving the results from the solution, hybridization should occur simultaneously for all molecules in the solution. This implies that the respective melting temperatures should be comparable for all hybridization reactions that are taking place. This is the third main constraint that the set of words under consideration needs to adhere to.

To address the problem of designing DNA code words according to these three constraints, an algorithm based on a stochastic local search method was proposed in [89]. The melting temperature constraint is simplified to the constraint requiring that, for each strand, the number of C and G nucleotides amounts to 50% of its total nucleotide count. The algorithm produces a set of DNA sequences of equal length that satisfy the Hamming distance and the temperature constraints. The algorithm is based on the following:

Input: Number k of words needed to produce, and the word length n.

Step 1: Produce a random set of k words of length n.

Step 2: Modify the set so that the set satisfies the first constraint.

Step 3: Repeat Step 2 for all the given constraints.

Output: The set of words (if one can be found).

More specifically, for *Step 2* two words w_1 and w_2 are chosen from the set that violate at least one of the constraints. With a probability $1 - \theta$, θ being the noise parameter, one of these words is altered by randomly substituting one base in a way that maximally decreases the number of constraint violations. The algorithm terminates either when there are no more constraint violations in the set of words or when the number of loop iterations has exceeded some maximum threshold. Empirical results prove this technique to be effective and the noise parameter θ is empirically determined to be optimal as 0.2, regardless of the problem instance.

31.5.3.3 Theoretical Studies

In this section, we discuss the formal language theoretical approach to the problem of designing code words, introduced in [35]. We begin by reviewing some basic notions and notations. An alphabet is a finite, nonempty set of symbols. Let Σ be such an arbitrary alphabet. Then Σ^* denotes the set of all words over this alphabet, including the empty word λ. Σ^+ is the set of all nonempty words over Σ. Σ^i is the set of all words over Σ of length i. We also denote by $\text{Sub}_k(L)$, the set of all subwords of length k of words from the set L. For more details of formal language theory, theory of codes, and combinatorics on words the reader is referred to [14,34,55,75,83,102].

Every biomolecular protocol involving DNA generates molecules whose sequences of nucleotides form a language over the four-letter alphabet $\Delta = \{A, G, C, T\}$. The Watson–Crick complementarity of the nucleotides defines a natural involution θ, $A \mapsto T$ and $G \mapsto C$ which is an antimorphism on Δ^*. An involution θ is a mapping such that θ^2 is the identity. An antimorphism θ is such that $\theta(uv) = \theta(v)\theta(u)$ for all words u, v from Δ^*. For example, if $u = AGACT$, then the Watson–Crick complement $\theta(u)$ is AGTCT.

Given a DNA language, that is, a language over the DNA alphabet Δ, undesirable Watson–Crick bonds between its words can be avoided if the language satisfies certain properties. There are two types of unwanted hybridizations: intramolecular and intermolecular. The intramolecular

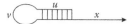

FIGURE 31.5 Intramolecular hybridization (hybridization within the same molecule). A DNA sequence of the type $uv\overleftarrow{u}x$, where $\overleftarrow{u}$ denotes the Watson–Crick complement of u, will form a secondary structure called a hairpin. Such hairpin structures are avoided by hairpin-free languages and by θ-k-m-subword codes.

hybridization happens when two sequences, one being the reverse complement of the other, appear within the same DNA strand (see Figure 31.5). In this case the DNA strand forms a hairpin.

Suppose we want to avoid the type of hybridization shown in Figure 31.5. A language L is called (Jonoska et al., [36]) a θ-k-m-subword code if for all words $u \in \Sigma^k$ we have $\Sigma^* u \Sigma^m \theta(u) \Sigma^* \cap L = \emptyset$. An analogous definition of hairpin-free languages was given in [41]. In this definition, Σ is an arbitrary alphabet and θ is any antimorphic involution. If Σ is taken to be the DNA alphabet Δ, then this property is essentially saying that the DNA sequences from the set L do not form the hairpin structures illustrated in Figure 31.5.

DNA strand sets that avoid all types of unwanted intermolecular bindings (See Figure 31.6) were introduced in [36], and called θ-k-codes. A language L is said to be a θ-k-code if $\theta(x) \neq y$ for all $x, y \in \mathrm{Sub}_k(L)$. The relationship $\theta(x) = y$ indicates that the molecules corresponding to x and y can form chemical bonds between them as shown in Figure 31.6. For a suitable k, a θ-k-code avoids all kinds of unwanted intermolecular hybridizations.

The θ-k-code property is meant to ensure that DNA strands do not form unwanted hybridizations during DNA computations. Sets theoretically designed to have this property have been successfully tested in practical wet-lab experiments [36]. In [44], the concept of θ-k-code has been extended to the Hamming bond-free property: $H(\theta(x), y) > d$ for any subwords $x, y \in \mathrm{Sub}_k(L)$, where d is an empirically chosen positive integer parameter.

Suppose we use codes that have the language properties we have described, what may happen during the course of computation is that the properties initially present deteriorate over time. This leads to another study, which investigates how bio-operations such as cutting, pasting, splicing, contextual insertion, and deletion affect the various bond-free properties of DNA languages. Invariance under these bio-operations has been studied in [36,37,40]. Bounds on the sizes of some other codes with desirable properties that can be constructed are explored by Marathe et al. [57].

The approach to DNA-encoded information and its properties described in this section has been meaningful to both DNA computing experimental research, and to DNA computing theoretical research by establishing a mathematical framework for describing and reasoning about DNA-encoded information. In addition, the notions defined and investigated in this context proved to be fruitful from the purely computer science point of view. Indeed, it turns out that some of these notions generalize classical concepts in coding theory and combinatorics of words such as primitive,

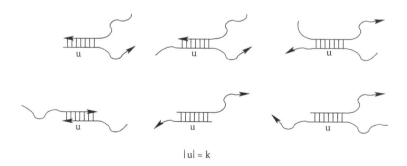

$|u| = k$

FIGURE 31.6 Intermolecular hybridizations (hybridizations between different molecules): If one DNA molecule contains a subword u of length k, and the other one contains the subword $\leftarrow u$, then the molecules will bind to each other, as shown in this figure. Such hybridizations are avoided by θ-k-subword codes.

commutative, conjugate, and palindromic words, as well as prefix, suffix, infix, and comma-free codes (see, e.g., [19,35,40,42,43]).

31.6   Computation in Living Cells

An example of the attempts to understand nature as computation is the research on computation in and by living cells. This is also sometimes called cellular computing, or *in vivo* computing, and one particular area of study in this field concerns the computational capabilities of gene assembly in unicellular organisms called ciliates [22,50].

Ciliates, unicellular protozoa named for their wisp-like cover of cilia, possess two types of nuclei: an active macronucleus containing the functional genes, and a functionally inert micronucleus that contributes only to genetic information exchange. In the process of conjugation, after two ciliates exchange genetic information and form new micronuclei, they have to assemble in real-time new macronuclei necessary for their survival. This is accomplished by a process that involves reordering some fragments of DNA from the micronuclear DNA (permutations and some inversions) and deleting other fragments. The process of obtaining the macronuclear DNA from the micronuclear DNA, by removing the noncoding sequences and permuting the coding fragments to obtain the correct order, is called *gene rearrangement* [66] (see Figure 31.7). The function of the various noncoding eliminated sequences is unknown and they represent a large portion (up to 98%) of the micronuclear sequences. As an example, the micronuclear *Actin I* gene in *Oxytricha nova* is composed of 9 coding segments present in the permuted order 3-4-6-5-7-9-2-1-8, and separated by 8 noncoding sequences. Instructions for the gene rearrangement (also called gene unscrambling) are apparently carried in the micronuclear DNA itself: pointer sequences present at the junction between coding and noncoding sequences, as well as certain RNA "templates," permit reassembly of the functional macronuclear gene.

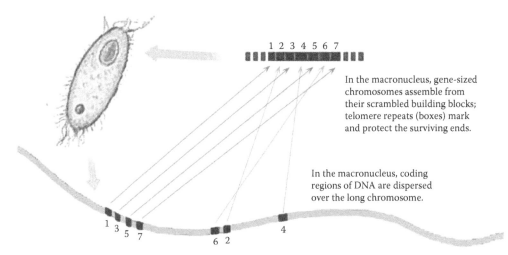

1 2 3 4 5 6 7

In the macronucleus, gene-sized chromosomes assemble from their scrambled building blocks; telomere repeats (boxes) mark and protect the surviving ends.

In the macronucleus, coding regions of DNA are dispersed over the long chromosome.

1 3 5 7   6 2   4

FIGURE 31.7 Overview of gene rearrangement in ciliates. Dispersed micronuclear coding fragments 1–7 (bottom) reassemble during macronuclear development to form the functional gene copy (top). (From Landweber, L. and Kari, L., The evolution of cellular computing: Nature's solution to a computational problem, *Proceedings of the DNA 4*, University of Pennsylvania, Philadelphia, PA (Reproduced from L. Kari, H. Rubin, and D. Wood, Eds.); *Biosystems*, 52, 3, 1999.)

The process of gene assembly is interesting from both the biological and the computational point of view. From the computational point of view, this study led to many novel and challenging research themes, [22]. Among others, it was proved that various models of gene assembly have full Turing machine capabilities, [50]. From the biological point of view, the joint effort of computer scientists and biologists led to a plausible hypothesis (supported already by some experimental data) about the "bioware" that implements the process of gene assembly, which is based on the new concept of template-guided recombination, [3,63,67].

Other approaches to cellular computing include developing of an *in vivo* programmable and autonomous finite-state automaton within *E. coli*, [59], and designing and constructing *in vivo* cellular logic gates and genetic circuits that harness the cell's existing biochemical processes [49,92,93].

31.7 DNA Self-Assembly

One of the most important achievements of DNA computing has been its contribution to the extensive body of research in nanosciences, by providing computational insights into a number of fundamental issues. Perhaps, the most prominent is its contribution to the understanding of self-assembly, which is among the key concepts in nanosciences (see [70] for an overview of self-assembly written for computer scientists). Self-assembly originates in the formal two-dimensional model of tiling defined and studied by Wang, [90]. Tiles are squares with "coloured edges" that, when placed on the plane, form a valid tiling only if the colours at their abutting edges are equal. Tiling systems can be used to simulate the computation of a universal Turing Machine (see, e.g., [13,90]). Winfree et al. [94,95], were the first to implement the concepts of computational tiling by using self-assembling "DNA tiles." This section describes several models of self-assembly, as well as applications of self-assembly to computation, (Section 31.7.1), to the creation of static DNA shapes and patterns (Section 31.7.2), as well as to engineering dynamic DNA nanomachines (Section 31.7.3).

31.7.1 DNA Self-Assembly for Computation

Molecular self-assembly is an automatic process in which molecules assemble into covalently bonded, well-defined stable structures without guidance or management from an outside source.

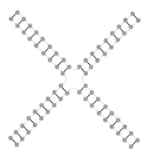

FIGURE 31.8 DNA Holiday junction. Four DNA single strands can be designed so as to form a branched structure in which each single strand participates in two consecutive double-stranded "branches."

There are two types of self-assembly, namely, intramolecular self-assembly and intermolecular self-assembly. Most often the term molecular self-assembly refers to intermolecular self-assembly, while the intramolecular analog is more commonly called folding.

One of the important features of DNA used in self-assembly is that DNA has many rigid, well-characterized forms that are not a linear double helix. For example, one can build DNA structures called Holiday junctions, [77], wherein four DNA single strands self-assemble and create a branched structure, as seen in Figure 31.8. This can be accomplished by designing each of the four DNA single strands to be partially complementary to one of the others, so that each strand participates in two consecutive double-stranded "branches."

This type of complex self-assembled DNA structures have been used by Winfree et al. who introduced a two-dimensional self-assembly model for DNA computation, [95]. In this model, using techniques similar to the ones described earlier (demonstrated, e.g., in [27]), rectangular DNA structures are built, with four sticky single-stranded ends at their corners. These DNA structures behave as "tiles" and can be designed to perform two-dimensional assembly as follows. Depending on their composition, the sticky ends of a tile (which are single-stranded DNA strands) will attach only to Watson–Crick complementary corresponding ends from other tiles. This results in spontaneous self-assemblies of the tiles into essentially planar conformations that can be designed so as to perform any deterministic or nondeterministic computation.

Yokomori et al., [101], introduced another self-assembly model that is Turing-universal. Other computations that have been achieved by the self-assembly of complex DNA nanostructures include bit-wise cumulative XOR, [56], and binary counters [5]. As another example, Seelig et al. [76] reported the design and experimental implementation of DNA-based digital logic circuits. The authors demonstrate AND, OR, and NOT gates, signal restoration, amplification, feedback, and cascading.

Regarding intramolecular self-assembly, Hagiya et al. [31] and Sakamoto et al. [74], proposed a new method of DNA computing that involves a self-acting DNA molecule containing, on the same strand, the input, program, and working memory. The computation, also called Whiplash PCR, proceeds as follows (see Figure 31.9). The single DNA strand contains at its 5′ end state transitions of the type $A \longrightarrow B$, each encoded as a DNA rule block "$\overleftarrow{B} - \overleftarrow{A}$ *−stopper sequence*." The 3′ end of the DNA sequence contains, at any given time, the encoding of the "current state," say A. In Step (1), cooling the solution will lead the 3′ end of the DNA strand, A, to attach to its complement in the corresponding rule block, namely to $\overleftarrow{A}$. In Step (2), PCR is used to extend the now-attached end A by the encoding of the new state B, and the extension process is stopped by the stopper sequence. Then, in Step (3), by raising the temperature, the new current state B detaches and becomes loose, and the next state transition cycle can begin.

The main motivation for using this model is that self-acting molecules can compute in parallel, in a single-tube reaction, allowing for a multiple program, multiple input architecture. Hagiya et al., [31], showed how to theoretically learn µ-formulas using this model. Subsequently, Winfree [96] showed how to solve several NP-complete problems in $O(1)$ biosteps using whiplash PCR.

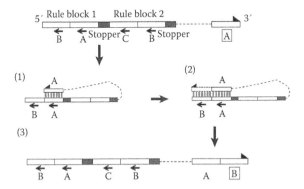

FIGURE 31.9 Whiplash PCR ($\overleftarrow{A}$ denotes the Watson–Crick complement of A). The 5′ end of the DNA strand contains rule blocks "$\overleftarrow{B} - \overleftarrow{A}$ − stopper sequence" coding for state transitions of the type $A \rightarrow B$. The 3′ end contains the encoding of the current state A. In Step (1), cooling the solution leads the 3′ end A to attach to the corresponding part of the rule block. In Step (2), PCR extends the now-attached end encoding the state A, by the new state B, stopping at the stopper sequence. Raising the temperature results, in Step (3), in the loose end encoding the new state B becoming detached and ready for the next state transition cycle.

31.7.2 DNA Nanoscale Shapes

DNA nanotechnology (see [78] for a comprehensive survey), uses the unique molecular recognition properties of DNA to create intricate structures such as self-assembling branched DNA complexes with useful properties.

DNA is a building block ideally suited for nanostructures, as it combines self-assembly properties with programmability. In nanotechnology, DNA is used as a structural material rather than as a carrier of biological information. Using DNA, impressive nanostructures have been created, including two- and three-dimensional structures. Examples include DNA nanostructures such as Sierpinski triangles [73], and cubes [17,78].

Another design theme was introduced in [81]. The authors report that a single strand of DNA (1669 nucleotides long) was folded into a highly rigid nanoscale octahedron structure, of a diameter of about 22 nanometers. The long strand of DNA was designed so as to have a number of self-complementary regions, which would induce the strand to fold back on itself to form a sturdy octahedron. Folding the DNA into the octahedral structure simply required the heating and the cooling of the solution containing the DNA, magnesium ions, and five 40-mer accessory DNA sequences. Moreover, this design permitted the assembly of the first clonable DNA three-dimensional nanostructure.

Rothemund, [72], presented a simple model that uses several short strands of DNA to direct the folding of a long single strand of DNA into desired shapes. The author demonstrates the generality of the method, called "scaffolded DNA origami," that permits the fabrication (out of DNA) of any two-dimensional shape roughly 100 nm in diameter. The techniques used are similar to the ones used in the design of the Holiday junction (Section 31.7), in that the constituent DNA strands form complex structures by their design, which makes it possible for some single strands to participate in two DNA helices—they wind along one helix, then switch to another.

The design process of a DNA origami involves several steps. The first step is to build an approximate geometric model of the desired shape. The shape is approximated by cylinders that are models of the DNA double-helices that will ultimately be used for the construction. The second step is to fill the shape by folding a single long "scaffold strand" back and forth in a raster pattern such that, at each moment, the scaffold strand represents either the "main" strand or the "complement" strand of a double helix. (The process is analogous to drawing out a shape using a single line, and without taking the pencil off the paper.) Once the geometric model and the folding path have been designed, the third step is to use a computer program to generate a set of "staple strands" that provide Watson–Crick complements for the scaffold. The staple strands are designed to bind to portions of the scaffold strand, holding it thus together into the desired shape. The staple strands are then fine-tuned to minimize strain in the construction and optimize the binding specificity and binding energy.

To test this method, Rothemund [72] used as a scaffold circular genomic DNA (7249 nucleotide long) from the virus *M13mp18*. Approximately 250 short staple strands were synthesized and mixed with the scaffold, in 100-fold excess to it. The strands annealed in less than 2 h and AFM (Atomic Force Microscopy) imaging showed that indeed the desired shape was realized. The generality of the method was proved by assembling six different shapes such as squares, triangles, five-pointed stars, and smiley faces (Figure 31.10).

This method not only provides access to structures that approximate the outline of any desired shape, but also enables the creation of structures with arbitrary-shaped holes or surface patterns. In addition, it seems likely that scaffolded DNA origami can be adapted to design three-dimensional structures.

31.7.3 DNA Nanomachines

In addition to static DNA nanoscale shapes and patterns, an impressive array of ingenious DNA nanomachines (see [6,53] for comprehensive reviews), were designed with potential uses to

(a)   (b)   (c)

FIGURE 31.10 (a) A folding path that creates a "DNA smiley." An even number of double helices are filled into the desired shape. (b), (c) A long single-stranded DNA self-assembles into the desired shape (Atomic Force Microscopy images). (Reproduced from Rothemund, P., *Nature*, 440, 297, 2006.)

nanofabrication, engineering, and computation. DNA nanomachines, that is, DNA-based nano-devices that convert static DNA structures into machines that can move or change conformation, have been developed rapidly since one of the earliest examples was reported in 1998, [100]. They include molecular switches that can be driven between two conformations [52], DNA "tweezers" [103], DNA "walkers" that can be moved along a track [80,82] and autonomous molecular motors [7,30,69].

We illustrate some of the construction principles used with the construction of a device referred to as "molecular tweezers," [103]. The DNA tweezers consist of two partially double-stranded DNA arms connected by a short single DNA strand acting as a flexible hinge (Figure 31.11). The resulting structure is similar in form to a pair of open tweezers. A so-called set strand is designed in such a way as to be complementary to both single stranded "tails" at the end of the arms. The addition of this set strand to the mixture will result in its annealing to both tails of the arms, bringing thus the arms of the tweezers together in a "closed" configuration. A short region of the set strand remains single stranded even after it hybridizes to the arms. This region is used as a toehold that allows a new "reset strand" to strip the set strand from the arms, by hybridizing with the set strand itself. The tweezers are thus returned to the "open" configuration. A number of variations on the tweezers have also been reported [21,26,58,84].

Considerable efforts have been expended on realizing "walking devices." Several molecular machines, which can walk along a track have been proposed [65,79,80,82,87]. Shin and Pierce [82],

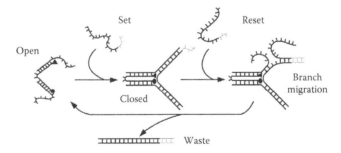

FIGURE 31.11 A DNA nanomachine, "DNA tweezers," driven by repeated sequential additions of "set" and "reset" strands. In the open state, the DNA tweezer consists of two double helical arms (with short single-stranded tails), connected by a single stranded hinge. Hybridization of the tails with the set strand closes the arms. The set strand can be removed via branch migration, by which the fully complementary "reset strand" strips the set strand from the tweezer tails, returning it to the open state. (Reproduced from Liedl, T. et al., *Nanotoday*, 2, 36, 2007.)

introduced a DNA device with two distinguishable feet that "walks" directionally on a linear DNA track with four distinct single strands periodically protruding from it and acting as anchors. The walker is double-stranded and has two single-stranded extensions acting as "legs." Specific attachment strands bind the legs to the single-stranded extensions anchors periodically placed along the double-stranded track. Each step requires the sequential addition of two strands: the first lifts the back foot from the track, by strand displacement—a process by which an invading single DNA strand can "displace" one of the constituent strands of a double-strand, by replacing it with itself, provided the newly formed structure is more stable. The second strand places the released foot ahead of the stationary foot. Brownian motion provides movement, and the order of adding the "instruction" strands ensures directionality. A similar walking device, based on the movement pattern of inchworms, was developed by Sherman and Seeman, [80]: here the front foot steps forward and the back foot catches up.

The walkers described earlier need "fuel DNA," and cannot walk autonomously. Recently, [79] suggested a three-leg molecular walking machine that can walk autonomously in two or three dimensions on a designed route. It uses an enzyme as a source of power, and a track of DNA equipped with many single-stranded DNA anchors arranged in a certain pattern.

We end this section by mentioning research on DNA nanodevices in conjunction with the cell's genetic mechanisms. Cellular organisms exhibit complex biochemical interaction networks comprising genes, RNA and proteins. These include gene activation and inhibition, gene transcription into RNA, and the translation of RNA into proteins. By adapting parts of these genetic mechanisms, one can engineer novel molecular machinery *in vitro* and potentially *in vivo*, [53]. Several attempts have been made to use genetic mechanisms to control DNA-based nanodevices, [47,48,61,62]. For example, pioneering work by Noireaux, Bar-Ziv and Libchaber, [61,62], demonstrates the principles of "cell-free genetic circuit assembly," wherein the term cell-free refers to the fact that the experiments were performed *in vitro*, not *in vivo*. Overall, the successful combination of DNA-based nanodevices with genetic machinery points to promising possible applications to biotechnology, bioengineering, and medicine.

31.8 Conclusion

The excitement that the first DNA computing experiment created in 1994 was primarily due to the fact that computing with DNA offered a completely new way of both looking at and performing computations: by cutting and pasting DNA strands using enzymes, by using the Watson–Crick complementarity of DNA strands, by selecting DNA strands containing a certain pattern, and so on. This novel way of viewing computation has the potential to change the very meaning of the word "compute," and DNA computing has already made significant contributions to the field of computer science. At the same time, research into the computational abilities of cellular organisms has the potential to uncover the laws governing biological information and to enable us to harness the computational power of cells.

This chapter was intended to give the reader a snapshot of DNA computing research by addressing both theoretical and experimental aspects, both "classical" and very recent trends, the impact of molecular computation on the theory and technology of computation, as well as on understanding of the basic mechanisms of bioprocesses.

DNA computing is a rapidly developing field and the recent years already saw some interesting new developments. Experimental advances include self-assembly of cylinders and Möbius strips [29], and new designs for DNA nanomachines, fueled by light, [51]. Progress in the theory of DNA computing include studies of pseudoknot-free DNA/RNA words, [45], of the time-optimal self-assembly of three-dimensional shapes [9], and the proposal of a simple scalable DNA-gate motif for the purpose of synthesizing large-scale circuits [68].

It is envisaged that research into *in vitro* and *in vivo* DNA computing constitutes just a preliminary step that may ultimately lead to molecular computing becoming a viable complementary tool for computation, as well as providing more insight into the computational nature of bioprocesses taking place in living organisms.

Acknowledgments

This research was supported by Natural Sciences and Engineering Research Council of Canada (NSERC) Discovery Grant and Canada Research Chair Award to L. K. We thank Bo Cui, Elena Czeizler, Eugen Czeizler, Zachary Kincaid, Shinnosuke Seki, and the anonymous referee for their suggestions and comments, and Yan Zeng for Figures 31.9 and 31.8.

References

1. L. Adleman, Molecular computation of solutions to combinatorial problems, *Science* 266 (1994) 583–585.
2. M. Amos, *Theoretical and Experimental DNA Computation*, Springer Verlag, Berlin, *Natural Computing Series* 2005.
3. A. Angeleska, N. Jonoska, M. Saito, L. Landweber, RNA-guided DNA assembly, *Journal of Theoretical Biology* 248 (2007) 706–720.
4. C. Bancroft, T. Bowler, B. Bloom, C. Clelland, Long-term storage of information in DNA, *Science* 293 (2001) 1763–1765.
5. R. Barish, P. Rothemund, E. Winfree, Two computational primitives for algorithmic self-assembly: Copying and counting, *Nanoletters* 5 (2005) 2586–2592.
6. J. Bath, A. Turberfield, DNA nanomachines, *Nature Nanotechnology* 2 (2007) 275–284.
7. J. Bath, S. Green, A. Turberfield, A free-running DNA motor powered by a nicking enzyme, *Angewandte Chemie International Edition* 44 (2005) 4358–4361.
8. E. Baum, How to build an associative memory vastly larger than the brain, *Science* 268 (1995) 583–585.
9. F. Becker, E. Remila, N. Schabanel, Time optimal self-assembling of 2D and 3D shapes: The case of squares and cubes, *Prelim. Proc. of DNA 14* (*DNA n* denotes the main international conference in the field of DNA Computing, *The nth International Meeting on DNA Computing.*), Prague, Czech Republic (A. Goel, F. Simmel, P. Sosik, Eds.), June (2008) pp. 78–87.
10. Y. Benenson, T. Paz-Elizur, R. Adar, E. Keinan, Z. Livneh, E. Shapiro, Programmable and autonomous computing machine made of biomolecules, *Nature* 414 (2001) 430–434.
11. Y. Benenson, R. Adar, T. Paz-Elizur, Z. Livneh, E. Shapiro, DNA molecule provides a computing machine with both data and fuel, *Proceedings of the National Academy of Sciences U S A* 100 (2003) 2191–2196.
12. Y. Benenson, B. Gil, U. Ben-Dor, R. Adar, E. Shapiro, An autonomous molecular computer for logical control of gene expression, *Nature* 429 (2004) 423–429.
13. R. Berger, *Undecidability of the domino problem*, *Memoirs of the American Mathematical Society* 66 (1966), 72 pp.
14. J. Berstel, D. Perrin, *Theory of Codes*, Academic Press Inc. Orlando, FL, 1985.
15. R. Braich, N. Chelyapov, C. Johnson, P. Rothemund, L. Adleman, Solution of a 20-variable 3-SAT problem on a DNA computer, *Science* 296 (2002) 499–502.
16. J. Chen, R. Deaton, Y. Wang, A DNA-based memory with in vitro learning and associative recall, *Natural Computing* 4 (2005) 83–101.
17. J. Chen, N. Seeman, Synthesis from DNA of a molecule with the connectivity of a cube, *Nature* 350 (1991) 631–633.

18. J. Cox, Long-term data storage in DNA, *Trends in Biotechnology* 19 (2001) 247–250.

19. E. Czeizler, L. Kari, S. Seki, On a special class of primitive words, *Proceedings of Mathematical Foundations of Computer Science (MFCS)*, Torun, Poland, *LNCS* 5162 (2008) pp. 265–277.

20. M. Daley, L. Kari, DNA computing: Models and implementations, *Comments on Theoretical Biology* 7 (2002) 177–198.

21. B. Ding, N. Seeman, Operation of a DNA robot arm inserted into a 2D DNA crystalline substrate, *Science* 314 (2006) 1583–1585.

22. A. Ehrenfeucht, T. Harju, I. Petre, D. Prescott, G. Rozenberg, *Computation in Living Cells: Gene Assembly in Ciliates*, Springer Verlag, Berlin, *Natural Computing Series*, 2004.

23. D. Faulhammer, A. Cukras, R. Lipton, L. Landweber. Molecular computation: RNA solutions to chess problems, *Proceedings of the National Academy of Sciences U S A* 97 (2000), 1385–1389.

24. U. Feldkamp, W. Banzhaf, H. Rauhe, A DNA sequence compiler, *Preliminary Proceedings of DNA 6*, Leiden, Netherlands (A. Condon, G. Rozenberg, Eds.), June (2000).

25. U. Feldkamp, S. Saghafi, W. Banzhaf, H. Rauhe, DNA sequence generator: A program for the construction of DNA sequences, *Proceedings of DNA 7*, Tampa, FL (N. Jonoska, N. Seeman, Eds.), *LNCS* 2340 (2002) pp. 23–32.

26. L. Feng, S. Park, J. Reif, H. Yan, A two state DNA lattice switched by DNA nanoactuator, *Angewandte Chemie International Edition* 42 (2003) 4342–4346.

27. T. Fu, N. Seeman, DNA double cross-over structures, *Biochemistry* 32 (1993) 3211–3220.

28. M. Garzon, C. Oehmen, Biomolecular computation in virtual test tubes, *Proceedings of DNA 7*, Tampa, FL (N. Jonoska, N. Seeman, Eds.), *LNCS* 2340 (2002) pp. 117–128.

29. M. Gopalkrishnan, N. Gopalkrishnan, L. Adleman, Self-assembly of cylinders and Möbius strips by DNA origami, *Preliminary Proceedings of DNA 14*, Prague, Czech Republic (A. Goel, F. Simmel, P. Sosik, Eds.), June (2008) 181.

30. S. Green, D. Lubrich, A. Turberfield, DNA hairpins: Fuel for autonomous DNA devices, *Biophysical Journal* 91 (2006) 2966–2975.

31. M. Hagiya, M. Arita, D. Kiga, K. Sakamoto, S. Yokoyama, Towards parallel evaluation and learning of boolean μ-formulas with molecules, *Proceedings of DNA 3*, Philadelphia, PA (H. Rubin, D. Wood, Eds.), *DIMACS* 48 (1997) pp. 57–72.

32. A. Hartemink, D. Gifford, Thermodynamic simulation of deoxyoligonucleotide hybridization for DNA computation, *Proceedings of DNA 3*, Philadelphia, PA (H. Rubin, D. Wood, Eds.), *DIMACS* 48 (1997) pp. 25–38.

33. A. Hartemink, D. Gifford, J. Khodor, Automated constraint-based nucleotide sequence selection for DNA computation, *Proc. of DNA 4*, Philadelphia, PA, (L. Kari, H. Rubin, D. Wood, Eds.), *Biosystems* 52 (1999) pp. 227–235.

34. J. Hopcroft, J. Ullman, R. Motwani, *Introduction to Automata Theory, Languages and Computation*, 2nd edition, Boston MA: Addison Wesley 2001.

35. S. Hussini, L. Kari, S. Konstantinidis, Coding properties of DNA languages, *Theoretical Computer Science* 290 (2003) 1557–1579.

36. N. Jonoska, K. Mahalingam, J. Chen, Involution codes: With application to DNA coded languages, *Natural Computing* 4 (2005) 141–162.

37. N. Jonoska, L. Kari, K. Mahalingam, Involution solid and join codes, *International Conference on Developments in Language Theory (DLT)*, Santa Barbara, CA, *LNCS* 4036 (2006) pp. 192–202.

38. L. Kari, *DNA* Computing: Arrival of biological mathematics, *The Mathematical Intelligencer* 19 (1997) 9–22.

39. L. Kari, R. Kitto, G. Gloor, A computer scientist's guide to molecular biology, *Soft Computing*, (G. Paun, T. Yokomori, Eds.), Vol. 5 (2001) pp. 95–101.

40. L. Kari, S. Konstantinidis, E. Losseva, G. Wozniak, Sticky-free and overhang-free DNA languages, *Acta Informatica* 40 (2003) 119–157.

41. L. Kari, S. Konstantinidis, E. Losseva, P. Sosik, G. Thierrin, Hairpin structures in DNA words, *Proceedings of DNA 11*, London, Ontario, Canada (A. Carbone, N. Pierce, Eds.), *LNCS* 3892 (2006) pp. 158–170.

42. L. Kari, K. Mahalingam, Watson-Crick conjugate and commutative words, *Proceedings of DNA 13*, Memphis, TN (M. Garzon, H. Yan, Eds.), *LNCS* 4848 (2008), pp. 273–283.

43. L. Kari, K. Mahalingam, Watson-Crick palindromes in DNA computing, *Natural Computing* (2009) DOI 10.1007/s11047-009-9131-2.

44. L. Kari, S. Konstantinidis, P. Sosik, Bond-free languages: Formalizations, maximality and construction methods, *International Journal of Foundations of Computer Science* 16 (2005) 1039–1070.

45. L. Kari, S. Seki, On pseudoknot-bordered words and their properties, *Journal of Computer and System Sciences*, 75 (2009) 113–121.

46. S. Kashiwamura, M. Yamamoto, A. Kameda, T. Shiba, A. Ohuchi, Potential for enlarging DNA memory: The validity of experimental operations of scaled-up nested primer molecular memory, *Biosystems* 80 (2005) 99–112.

47. J. Kim, J. Hopfield, E. Winfree, Neural network computation by in vitro transcriptional circuits, *Advances in Neural Information Processing Systems (NIPS)* 17 (2004) 681–688.

48. J. Kim, K. White, E. Winfree, Construction of an in vitro bistable circuit from synthetic transcriptional switches, *Molecular Systems Biology* 2 (2006) doi:10.1038/msb4100099.

49. T. Knight Jr., G. Sussman, Cellular gate technology, *Unconventional Models of Computation* (1998) 257–272.

50. L. Landweber, L. Kari, The evolution of cellular computing: nature's solution to a computational problem, *Proceedings of DNA 4*, Philadelphia, PA (L. Kari, H. Rubin, D. Wood, Eds.), *Biosystems* 52 (1999) 3–13.

51. X. Liang, H. Nishioka, N. Takenaka, H. Asanuma, Construction of photon-fuelled DNA nanomachines by tethering azobenzenes as engines, *Preliminary Proceedings of DNA 14*, Prague, Czech Republic (A. Goel, F. Simmel, P. Sosik, Eds.) June (2008) pp. 17–25.

52. T. Liedl, M. Olapinski, F. Simmel, A surface-bound DNA switch driven by a chemical oscillator, *Angewandte Chemie International Edition* 45 (2006) 5007–5010.

53. T. Liedl, T. Sobey, F. Simmel, DNA-based nanodevices, *Nanotoday* 2 (2007) 36–41.

54. R. Lipton, DNA solution of hard computational problems, *Science* 268 (1995) 542–545.

55. M. Lothaire, *Combinatorics on Words*, Cambridge University Press, Cambridge U. K. 1997.

56. C. Mao, T. LaBean, J. Reif, N. Seeman, Logical computation using algorithmic self-assembly of DNA triple-crossover molecules, *Nature* 407 (2000) 493–496.

57. A. Marathe, A. Condon, R. Corn, On combinatorial word design, *Journal of Computational Biology* 8 (2001) 201–219.

58. J. Mitchell, B. Yurke, DNA scissors, *Proceedings of DNA 7*, Tampa, FL (N. Jonoska, N. Seeman, Eds.), *LNCS* 2340 (2002), pp. 258–268.

59. H. Nakagawa, K. Sakamoto, Y. Sakakibara, Development of an *in vivo* computer based on *Escherichia Coli*, *Proceedings of DNA 11*, London, Ontario, Canada (A. Carbone, N. Pierce, Eds.), *LNCS* 3892 (2006) pp. 203–212.

60. A. Neel, M. Garzon, P. Penumatsa, Improving the quality of semantic retrieval in DNA-based memories with learning, *International Conference on Knowledge-Based Intelligent Information and Engineering Systems*, Wellington, New Zealand, *LNCS* 3213 (2004) pp. 18–24.

61. V. Noireaux, A. Libchaber, A vesicle bioreactor as a step toward an artificial cell assembly, *Proceedings of the National Academy of Sciences* 101 (2004) 17669–17674.

62. V. Noireaux, R. Bar-Ziv, A. Libchaber, Principles of cell-free genetic circuit assembly, *Proceedings of the National Academy of Sciences* 100 (2003) 12672–12677.

63. M. Nowacki, V. Vijayan, Y. Zhou, K. Schotanus, T. Doak, L. Landweber, RNA-mediated epigenetic programming of a genome-rearrangement pathway, *Nature* 451 (2008) 153–158.

64. G. Paun, G. Rozenberg, A. Salomaa, *DNA Computing—New Computing Paradigms*, Springer-Verlag, Berlin, 1998.

65. R. Pei, S. Taylor, D. Stefanovic, S. Rudchenko, T. Mitchell, M. Stojanovic, Behaviour of polycatalytic assemblies in a substrate displaying matrix, *Journal of American Chemical Society* 128 (2006) 12693–12699.

66. D. Prescott, The DNA of ciliated protozoa, *Microbiology and Molecular Biology Reviews* 58 (1994) 233–267.

67. D. Prescott, A. Ehrenfeucht, G. Rozenberg, Template- guided recombination for IES elimination and unscrambling of genes in stichotrichous ciliates, *Journal of Theoretical Biology* 222 (2003) 323–330.

68. L. Qian, E. Winfree, A simple DNA gate motif for synthesizing large-scale circuits, *Preliminary Proceedings of DNA 14*, Prague, Czech Republic (A. Goel, F. Simmel, P. Sosik, Eds.), June (2008) pp. 139–151.

69. J. Reif, The design of autonomous DNA nanomechanical devices: Walking and rolling DNA, *Proceedings of DNA 8*, Sapporo, Japan (M. Hagiya, A. Ohuchi, Eds.), *LNCS* 2568 (2003) pp. 22–37.

70. J. Reif, T. LaBean, Autonomous programmable biomolecular devices using self-assembled DNA nanostructures, *Communications of the ACM* 50 (2007) 46–53.

71. J. Reif, T. LaBean, M. Pirrung, V. Rana, B. Guo, C. Kingsford, G. Wickham, Experimental construction of very large scale DNA databases with associative search capability, *Proceedings of DNA 7*, Tampa, FL (N. Jonoska, N. Seeman, Eds.), *LNCS* 2340 (2002) pp. 231–247.

72. P. Rothemund, Folding DNA to create nanoscale shapes and patterns, *Nature* 440 (2006) 297–302.

73. P. Rothemund, N. Papadakis, E. Winfree, Algorithmic self-assembly of DNA Sierpinski triangles, *PLoS Biology* 2 (2004) 2041–2053.

74. K. Sakamoto, H. Gouzu, K. Komiya, D. Kiga, S. Yokoyama, T. Yokomori, M. Hagiya, Molecular computation by DNA hairpin formation, *Science* 288 (2000) 1223–1226.

75. A. Salomaa, *Formal Languages*, Academic Press, New York, 1973.

76. G. Seelig, D. Soloveichik, D. Zhang, E. Winfree, Enzyme-free nucleic acid logic circuits, *Science* 314 (2006) 1585–1588.

77. N. Seeman, DNA in a material world, *Nature* 421 (2003) 427–431.

78. N. Seeman, Nanotechnology and the double-helix, *Scientific American* 290 (2004) 64–75.

79. H. Sekiguchi, K. Komiya, D. Kiga, M. Yamamura, A realization of DNA molecular machine that walks autonomously by using a restriction enzyme, *Proceedings of DNA 13*, Memphis, TN (M. Garzon, H. Yan, Eds.), *LNCS* 4848 (2008) pp. 34–65.

80. W. Sherman, N. Seeman, A precisely controlled DNA biped walking device, *Nano Letters* 4 (2004) 1203–1207.

81. W. Shih, J. Quispe, G. Joyce, A 1.7-kilobase single-stranded DNA that folds into a nanoscale octahedron, *Nature* 427 (2004) 618–621.

82. J. Shin, N. Pierce, A synthetic DNA walker for molecular transport, *Journal of American Chemical Society* 126 (2004) 10834–10835.

83. H. J.Shyr, *Free Monoids and Languages*, Hon Min Book Company, Taichung, Taiwan, 2001.

84. F. Simmel, B. Yurke, Using DNA to construct and power a nanoactuator, *Physical Review E63* (2001) 041913.

85. G. Smith, C. Fiddes, J. Hawkins, J. Cox, Some possible codes for encrypting data in DNA, *Biotechnology Letters* 25 (2003) 1125–1130.

86. D. Soloveichik, E. Winfree, The computational power of Benenson automata, *Theoretical Computer Science* 344 (2005) 279–297.

87. Y. Tian, Y. He, Y. Chen, P. Yin, C. Mao, A DNAzyme that walks processively and autonomously along a one-dimensional track, *Angewandte Chemie International Edition* 44 (2005) 4355–4358.

88. Y. Tsuboi, Z. Ibrahim, O. Ono, DNA-based semantic memory with linear strands, *International Journal of Innovative Computing, Information and Control* 1 (2005) 755–766.

89. D. Tulpan, H. Hoos, A. Condon, Stochastic local search algorithms for DNA word design, *Proceedings of DNA 8*, Sapporo, Japan (M. Hagiya, A. Ohuchi, Eds.), *LNCS* 2568 (2003) pp. 229–241.

90. H. Wang, Proving theorems by pattern recognition—II, *Bell System Technical Journal* 40 (1961) 1–41.

91. J. Watson, N. Hopkins, J. Roberts, J. Steitz, A. Weiner, *Molecular Biology of the Gene*, Benjamin Cummings Menlo Park, CA, 1987.

92. R. Weiss, G. Homsy, T. Knight, Jr., Toward in-vivo digital circuits, *Evolution as Computation, Natural Computing Series*, Springer, Berlin, Heidelberg, New York (2002) 275–295.

93. R. Weiss, S. Basu, The device physics of cellular logic gates, *Proceedings of the First Workshop on Non-Silicon Computation*, Cambridge, MA (2002) 54–61.

94. E. Winfree, F. Liu, L. Wenzler, N. Seeman, Design and self-assembly of two-dimensional DNA crystals, *Nature* 394 (1998) 539–544.

95. E. Winfree, X. Yang, N. Seeman, Universal computation via self-assembly of DNA: some theory and experiments, *Proceedings of DNA 2*, Princeton, NJ (L. Landweber, E. Baum, Eds.), *DIMACS* 44 (1996) 191–213.

96. E. Winfree, Whiplash PCR for O(1) computing, Caltech Technical Report, CaltechCSTR:1998.23 (1998).

97. P. Wong, K. Wong, H. Foote, Organic data memory using the DNA approach, *Communications of the ACM* 46 (2003) 95–98.

98. N. Yachie, K. Sekiyama, J. Sugahara, Y. Ohashi, M. Tomita, Alignment based approach for durable data storage into living organisms, *Biotechnology Progress* 23 (2007) 501–505.

99. M. Yamamoto, S. Kashiwamura, A. Ohuchi, DNA memory with 16.8M addresses, *Proceedings of DNA 13*, Memphis, TN (M. Garzon, H. Yan, Eds.), *LNCS* 4848 (2008) pp. 99–108.

100. X. Yang, A. Vologodskii, B. Liu, B. Kemper, N. Seeman, Torsional control of double stranded DNA branch migration, *Biopolymers* 45 (1998) 69–83.

101. T. Yokomori, Yet another computation model of self-assembly, *Proceedings of DNA 5*, Cambridge, MA (E. Winfree, D. Gifford, Eds.), *DIMACS Series* 54 (1999) 153–167.

102. S. S. Yu, *Languages and Codes*, Lecture Notes, Department of Computer Science, National Chung-Hsing University, Taichung, Taiwan, 2005.

103. B. Yurke, A. Turberfield, A. Mills, F. Simmel, J. Neumann, A DNA-fuelled molecular machine made of DNA, *Nature* 406 (2000) 605–608.

32

Computational Systems Biology

T.M. Murali
*Virginia Polytechnic Institute
and State University*

Srinivas Aluru
*Iowa State University and Indian
Institute of Technology*

32.1 Introduction

The functioning of a living cell is governed by an intricate network of interactions among different types of molecules. A collection of long DNA molecules called chromosomes, that together constitute the genome of the organism, encode for much of the cellular molecular apparatus including various types of RNAs and proteins. Short DNA sequences that are part of chromosomal DNA, called genes, can be transcribed repeatedly to result in various types of RNAs. Some of these RNAs act directly, such as micro (miRNA), ribosomal (rRNA), small nuclear (snRNA), and transfer (tRNA) RNAs. Many genes result in messenger RNAs (mRNAs), which are translated to corresponding proteins, a diverse and important set of molecules critical for cellular processes. A plethora of small molecules that are outside the hereditarily derived genes–RNAs–proteins system, called metabolites, play a crucial role in biological processes as intermediary molecules that are both products and inputs to biochemical enzymatic reactions.

These complex interactions define, regulate, and even initiate and terminate biological processes, and also create the molecules that take part in them. They are pervasive in all aspects of cell function, including the transmission of external signals to the interior of the cell, controlling processes that result in protein synthesis, modifying protein activities and their locations in the cell, and driving biochemical reactions. Gene products coordinate to execute cellular processes—sometimes by acting together, such as multiple proteins forming a protein supercomplex (e.g., the ribosome), or by acting in a concerted way to create biochemical pathways and networks (e.g., metabolic pathways that break down food, and photosynthetic pathways that convert sunlight to energy in plants). It is the same gene products that also regulate the expression of genes, often through binding to cis-regulatory sequences upstream of genes, to calibrate gene expression for different processes and to even decide which pathways are appropriate to trigger based on external stimuli.

The genomic revolution of the past two decades provides the parts list for systems biology. Advances in high-throughput experimental techniques are enabling measurements of mRNA, protein, and metabolite levels, and the detection of molecular interactions on a massive scale. In parallel, automated parsing and manual curation have extracted information on molecular interactions that have been deposited in the scientific literature over decades of small-scale experiments. In combination, these efforts have provided us with large-scale publicly-available datasets of molecular interactions and measurements of molecular activity, especially for well-studied model organisms such as *Saccharomyces cerevisiae* (baker's yeast), *Caenorhabditis elegans* (a nematode), and *Drosophila melanogaster* (the fruitfly), pathogens such as *Plasmodium falciparum* (the microbe that causes malaria), and for *Homo sapiens* itself.

These advances are transforming molecular biology from a reductionist, hypothesis-driven experimental field into an increasingly data-driven science, focused on understanding the functioning of the living cell at a systems level. How do the molecules within the cell interact with each other over time and in response to external conditions? What higher-level modules do these interactions form? How have these modules evolved and how do they confer robustness to the cell? How does disease result from the disruption of normal cellular activities? Understanding the complex interactions between these diverse and large body of molecules at various levels, and inferring the complex pathways and intermediaries that govern each biological process, are some of the grand challenges that constitute the field of systems biology. The data deluge has resulted in an ever-increasing importance placed on the computational analysis of biological data and computationally-driven experimental design. Research in this area of computational systems biology (CSB) spans a continuum of approaches [IL03] that includes simulating systems of differential equations, Boolean networks, Bayesian analysis, and statistical data mining.

Computational systems biology is a young discipline in which the important directions are still in a state of flux and being defined. In this chapter, we focus primarily on introducing and formulating the most well-studied classes of algorithmic problems that arise in the phenomenological and data-driven analyses of large-scale information on the behavior of molecules in the cell. We focus on research where the problem formulations and algorithms developed have actually been applied to biological data sets. Where possible, we refer to theoretical results and tie the work in the CSB literature to research in the algorithms community. The breadth of topics in CSB and the diversity of the connections between CSB and theoretical computer science preclude an exhaustive coverage of topics and literature within the scope of this short chapter. We caution the reader that our treatment of the topics and their depth and citation to relevant literature are by no means exhaustive. Rather, we attempt to provide a self-contained and logically interconnected survey of some of the important problem areas within this discipline, and provide pointers to a reasonable body of literature for further exploration by the reader. By necessity, this chapter introduces a number of biological terms and concepts that a computer scientist may not be familiar with. A glossary at the end of the chapter provides an easy resource for cross-reference.

32.2 An Illustrative Example

To elucidate how a typical biological process may unfold, and help explain some of the models used in systems biology, consider a generic process by which a eukaryotic cell responds to an external signal, e.g., a growth factor. See Figure 32.1 for a specific illustration of such a process. The growth factor binds to a specific receptor protein on the cell surface. The receptor protein dimerizes (i.e., protein molecules with bound growth factors themselves bind to each other). The dimerized form of the receptor is active; it phosphorylates (adds phosphate groups to) other proteins in the cytoplasm

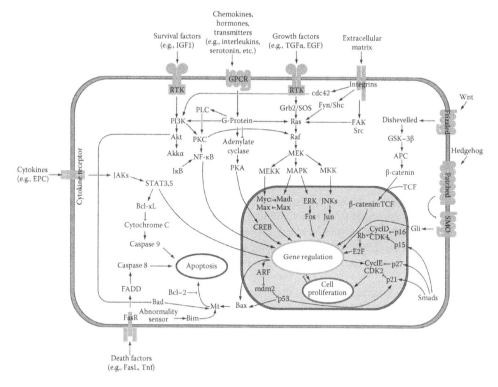

FIGURE 32.1 An illustration of a cellular signaling network (a Wikipedia image released by the author into the public domain).

of the cell, which in turn physically interact with or phosphorylate other proteins. Multiple such signaling cascades may be activated. The cascade culminates in the activation of a transcription factor (TF); the TF moves to the nucleus, where it binds to target sites on genomic DNA that recognize the TF. The TF recruits the cellular apparatus for transcription, resulting in the expression of numerous genes. These genes are converted to proteins in the ribosome of the cell, and then transported to various locations within the cell to perform their activities. Some proteins may be TFs themselves and cause the expression of other genes. Others may catalyze enzymatic reactions that produce or consume metabolites. Synthesized proteins may activate other signaling or reaction cascades. Ultimately, the initial binding of the growth factor with its receptor changes the levels and the activities of numerous genes, proteins, metabolites, and other compounds, and modulates global responses, such as cell migration, adhesion, and proliferation.

High-throughput experiments shed light on many of these interaction types. Interactions between signaling proteins (e.g., kinases and phosphatases) and their substrates constitute directed protein phosphorylation networks. Undirected protein–protein interaction (PPI) networks represent physical interactions between proteins. Directed transcriptional regulatory networks connect TFs to genes they regulate. Biochemical networks describe metabolic reactions with information on the enzymes that catalyze each reaction. Taken together, the known molecular interactions for an organism constitute its wiring diagram, a graph where each node is a molecule and each edge is a directed or undirected interaction between two molecules. As generally conceived, a wiring diagram contains molecules and interactions of many types.

Wiring diagrams usually contain direct interactions between molecules. Sometimes they are augmented with indirect interactions. A prominent example is a genetic interaction: two genes have a genetic interaction if the action of one gene is modified by the other. An extreme case of a genetic interaction is a synthetically lethal interaction, where knocking out each of two genes does not kill the cell but knocking out both genes results in cell death. For other examples of indirect or conceptual interactions, see functional linkage networks (FLNs) in Section 32.3 and reverse-engineered gene networks in Section 32.6.2.

32.3   Gene Function Prediction

The genomes of more than 600 organisms (including more than 70 eukaryotes) have been completely sequenced [LMTK08]. However, a fundamental roadblock to progress in systems biology is the poor state of knowledge about the biological functions of the genes in sequenced genomes [Kar04,RKKe04, Rob04]. Many genes of unknown function might support important cellular functions. Discovering the functions of these genes will provide critical insights into the biology of many organisms. In addition, discovering these functions will improve our ability to annotate genomes that are sequenced in the future.

The phrase "gene function" has a variety of meanings. It often refers to the molecular function of the protein the gene codes for, e.g., whether the protein catalyzes a reaction or binds DNA to regulate the expression of a target gene. More generally, the phrase refers to the context in which the protein acts in the cell, e.g., the component of the cell it is localized to; the pathway or biological process it is a member of; the cell type the gene is expressed in (in the case of multi-cellular organisms); or the developmental stage during which the gene is active. In this section, we will restrict our attention to three structured, controlled vocabularies (ontologies) developed by the Gene Ontology (GO) Consortium [ABBB00]: molecular function, cellular component, and biological process. Each ontology is a Directed Acyclic Graph (DAG) where each function is connected to parent functions by relationships such as "is_a" or "part_of". By design, a function represents a more specific biological concept than any of its parents. GO annotations follow the true path rule: if a gene is annotated with a function, then the gene must be annotated with all parents of that function.

A powerful method for predicting gene function relies on the evolutionary conservation of gene and protein sequences. Thus, if a gene in an organism has a nucleotide sequence, amino-acid sequence, or protein structure very similar to that of a gene with a known function [GJF07], then the function can be transferred to the first gene. These methods are primarily useful for determining the molecular function of a gene, which often depends directly on the structure of the protein encoded by the gene. Further, these methods do not provide annotations for the more than 40% of eukaryotic genes that do not have high sequence or structural similarity to genes in other organisms [EKO03].

A promising approach to gene function prediction starts by constructing an Functional Linkage Network (FLN) connecting genes of interest. In such a network, each node is a gene and each edge connects two genes that may share the same function based on some experimental or computational evidence. For instance, two genes may be linked if they have similar expression profiles in some experiment; if the proteins they code for interact physically or catalyze reactions involving the same metabolite; or, if knocking-out or silencing the expression of both genes produces the same phenotype. Constructing biologically meaningful FLNs from functional genomic data is an active area of research. Most existing methods proceed by estimating the probability that a given pair of genes should be functionally linked based on a given type of data and then integrating signals from multiple data sets [MT07,LLC$^+$08]. Many such FLNs are available in publicly accessible databases [BCS06].

In this section, we focus on the problem of predicting gene functions assuming that an FLN is given as input. We denote the FLN by an undirected graph $G = (V, E)$, where V is the set of nodes

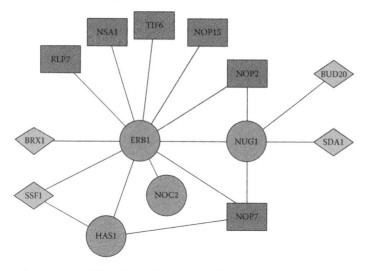

FIGURE 32.2 A subgraph of an FLN in *S. cerevisiae* for the biological process "ribosome biogenesis." Each node is a gene. Each edge corresponds to two genes whose protein products interact. Using the notation in the text, the rectangles are members of V_f^+ (where f is ribosome biogenesis), the circles are elements of V_f^0, and the diamonds are in V_f^-. To improve readability, we display only interactions involving genes in V_f^0. (Figure taken from Karaoz, U. et al., *Proc. Natl. Acad. Sci. U S A.*, 101, 2888, 2004.)

and E is the set of edges. For an edge $(u, v) \in E$, let $0 \leq w_{uv} \leq 1$ denote the weight of the edge in E between u and v; we interpret w_{uv} as a measure of confidence that u and v should be annotated with the same function. Note that the edge (u, v) suggests that u and v could perform the same function in the cell but does not specify what that function is. Let f be a function of interest in the GO. We cast the problem of predicting which nodes (genes) in G have the function f as a semi-supervised learning problem [CSZ06]. We partition V into three subsets V_f^+, V_f^0, and V_f^- corresponding to positive examples, unknown examples, and negative examples, respectively. A node v is in V_f^+ if v is annotated either with f or with a descendant of f in the GO DAG; v is in V_f^0 if $v \notin V_f^+$ and there is a function f' that is an ancestor of f that annotates v; and v is a member of V_f^- otherwise. See Figure 32.2 for an example.

For each gene in V_f^0, our goal is to predict whether that gene should be an element of V_f^+ or V_f^-. We formulate the problem in general terms as computing a (mathematical) function $r : V \rightarrow \mathbb{R}$ that is smooth over the nodes of G, i.e., for every edge $(u, v) \in E$, the larger w_{uv} is, the closer $r(u)$ and $r(v)$ are. After computing such a function, we predict every node $v \in V_f^0$ such that $r(v) \geq t$, for some input threshold t, as being annotated with f. Note that $r(v)$ is directly compared with the threshold, and not with the $r(u)$ values for each node u connected to v. This is because edge weights are used to ensure that r is smooth—i.e, for a highly weighted edge (u, v), it is likely both u and v are classified the same way due to the closeness of $r(u)$ and $r(v)$. There are a number of ways to ensure that r is smooth. A popular technique [ZGL03] is to fix $r(u) = 1$ for each $u \in V_f^+$, $r(u) = -1$ for every $u \in V_f^-$, and to compute r so that it minimizes the energy

$$E(G, r) = \sum_{\substack{(u,v) \in E \\ u \in V_f^0 \text{ or } v \in V_f^0}} w_{uv}(r(u) - r(v))^2. \tag{32.1}$$

For a node $v \in V$, let N_v denote the neighbors of v in G. Karaoz et al. [KMLe04] restrict $r(v)$ to be either -1 or 1; in this case, they can equivalently minimize

$$- \sum_{\substack{(u,v) \in E \\ u \in V_f^0 \text{ or } v \in V_f^0}} w_{uv} r(u) r(v).$$

An algorithm that iterates over all the nodes in V_f^0 and for each node v sets

$$r(v) = \text{sgn} \left(\sum_{u \in N_v} w_{uv} r(u) \right) \qquad (32.2)$$

until the $r(v)$ values converge will yield a value of energy that is at most half as large as the smallest value possible [KBDS93]. Nabieva et al. [NJAe05] and Murali et al. [MWK06] note that this problem can also be solved by computing minimum cuts in an appropriately transformed version of G. Nabieva et al. solve the problem as a special case of the NP-hard minimum multiway k-cut problem using integer linear programming. However, their approach allows a gene to have only one among a set of k functions. The approach adopted by Murali et al. transforms G into a flow network; they compute the minimum s–t cut in this graph using standard approaches [GT88].

Other approaches to gene function prediction based on FLNs include the use of frameworks such as Markov Random Fields, support vector machines (SVMs), and decision trees. We refer the reader to three recent surveys for discussions of these and other approaches [NBH07,SUS07,PCTMe08].

A thorny issue in gene function prediction is that biological experiments rarely report that a gene does not perform a particular function. Hence, the set V_f^- is hard to define accurately. A few approaches attempt to predict gene functions only from V_f^+ and V_f^0 [CX04,NJAe05]. How best to exploit the hierarchical dependence between functions in GO is an active research problem [BST06, PCTMe08]. There may be other types of dependencies between functions, e.g., genes annotated with function f_1 may have a surprisingly large number of edges in G to genes annotated with function f_2. What is the best way to detect such dependencies and utilize them to predict gene function? Finally, the question of systematically using FLNs to transfer function between organisms has received surprisingly little attention [NK07,SSKe05].

32.4 Gene Expression Analysis

Gene expression is the process by which a gene is first transcribed to a messenger RNA (mRNA) and then translated into a protein. The expression level of a gene is the number of copies of its mRNA that is present in a cell. Genome sequencing and the concomitant advent of DNA microarrays have allowed biologists to simultaneously measure the expression levels of all the genes in a sample of cells; the expression level so measured is an average over all the cells in the sample. DNA microarrays have revolutionized biological research since they capture a snapshot of the activity of all genes in the cells in the sample. A typical gene expression dataset usually consists of measurements from multiple samples under a particular experimental condition; the samples can correspond to multiple time-points after exposing cells to a particular treatment or stimulus or to multiple patients diagnosed with a particular disease.

32.4.1 Gene Expression Clustering

Let V denote the set of genes in an organism. The gene expression data set for a condition consists of a set of samples S, each with an expression level for each gene in V; we denote by g_S the vector of expression levels for gene g in the samples in S. Let $d = |S|$ and $n = |V|$. Typically, $d \ll n$.

A natural problem that arises now is to cluster the vectors $V_S = \{g_S \mid g \in V\}$.* Clustering allows the grouping of genes based on the similarity of their responses to the condition; since similarly expressed genes may perform the same function in the cell, such clusters can be the basis for constructing FLNs for gene function prediction (see Section 32.3). Two popular methods are k-means clustering and hierarchical clustering. In k-means clustering, the goal is to partition V into k sets such that the sum of squared distances from each gene to the centroid of the partition it belongs to is minimized. This problem is known to be NP-hard even for $k = 2$ when d and n are part of the input [DFKe04]. Feldman et al. developed a polynomial-time approximation scheme (PTAS) for this problem [FMS07]. Given parameters $\varepsilon, \lambda > 0$, their algorithm computes a $(1 + \varepsilon)$-approximate solution in time $O(nkd + d(k/\varepsilon)^{O(1)} + 2^{\tilde{O}(k/\varepsilon)})$ with a probability of at least $1 - \lambda$. In practice, most applications of k-means clustering use Lloyd's heuristic [Llo82]—start with a random set of k centers and repeatedly apply the following two steps until convergence: (1) associate each gene with the center closest to it and (2) move each center to the centroid of the genes associated with it. Har-Peled and Sadri [HPS05] prove that the number of iterations taken by the variants of this algorithm is polynomial in $|V|$, k, and the *spread* of V_S, which is defined as the diameter of V_S divided by the distance between the two closest genes. Denoting by $\Delta_k^2(V_S)$ the optimal solution to the k-means problem for V_S, Ostrovsky et al. [ORSS06] developed a linear-time constant-factor approximation algorithm and a PTAS that return a $(1 + \varepsilon)$-optimal solution with constant probability in time $O(2^{O(k(1+\omega^2)/\varepsilon)}dn)$, when V_S is ω-separated, i.e., if $\Delta_k^2(V_S)/\Delta_{k-1}^2(V_S) \leq \omega^2$.

The agglomerative version of hierarchical clustering is typically used to analyze gene expression data [ESBB98]. It starts by putting each gene in a separate cluster and repeatedly merging the closest pair of clusters. Typically, these algorithms continue until only one cluster remains. To specify this algorithm completely, it suffices to define the measure of distance between two genes and between two sets of genes. Let $\delta(a, b)$ denote the distance between two genes a and b, and let A and B be two sets of genes. In *single-linkage clustering*, we define the distance $\delta(A, B) = \min_{a \in A, b \in B} \delta(a, b)$; hierarchical clustering under this model is equivalent to computing the minimum spanning tree of the complete graph whose nodes are genes and an edge between two genes has weight equal to the distance between them. For the frequently used Pearson correlation metric, Seal et al. [SKA05] provide an $O(n \log n)$ algorithm for single-linkage clustering by exploiting the geometric transformation that the Pearson correlation coefficient between two gene expression vectors is equal to the cosine of the angle between the corresponding vectors. In *complete-linkage clustering*, $\delta(A, B) = \max_{a \in A, b \in B} \delta(a, b)$, whereas in *average-linkage clustering*, $\delta(A, B)$ is the distance between the centroids of A and B. Naive algorithms usually run in $O(dn^2)$ time and may require $O(n^2)$ space (for storing all pair-wise distances between genes). Krznaric and Levcopoulos [KL02] present an $O(n \log n)$ time and $O(n)$ space algorithm for complete linkage clustering under the L_1 and L_∞ metrics; for every other fixed L_t metric, their algorithm approximates the complete linkage clustering to an arbitrarily-small factor with the same bounds. Borodin et al. present sub-quadratic algorithms for approximate versions of the agglomerative clustering problem [BOR04].

Displaying a hierarchical clustering is problematic since the order in which the leaves of the underlying tree should be laid out is unclear. This issue is important since practitioners still use visualizations of the clustering to detect important or interesting patterns in the data. Bar-Joseph et al. [BJDGe03] compute an ordering of the leaves that minimizes the sum of the similarity between adjacent leaves in the ordering in $O(4^t n^3)$ time. They also present a method that allows up to t (a user-specified number) clusters to be merged at any step; the method runs in $O(n^3)$ time.

Given a hierarchical clustering of the genes, for every $k > 1$, it is possible to obtain an induced k-clustering of the genes, i.e., a partition of the genes into k clusters by stopping the clustering

---

* Clustering the vectors corresponding to the samples is also useful. For the sake of concreteness, we focus on clustering the vectors corresponding to the genes.

algorithm when only k clusters remain. Dasgupta and Long [DL05] consider whether there is a hierarchical clustering such that for every $k > 1$, there is an induced k-clustering that is close to the optimal k-clustering of the genes. Defining the cost of a clustering to be the largest radius of one of its clusters, they modify Gonzalez's approximation algorithm for the k-center problem [Gon85] to produce a hierarchical clustering such that for every k, the induced k-clustering has cost at most eight times the optimal k-clustering. They also present a randomized algorithm that achieves an approximation factor of $2e \approx 5.44$.

32.4.2   Gene Expression Biclustering

The clustering algorithms discussed in the previous section suffer from two primary drawbacks. First, they operate in the space spanned by all the samples; thus, they may not detect the patterns of clustering that are apparent only in a sub-space of $\mathbb{R}^d$. Second, since many algorithms partition the set of genes into clusters, they are unable to correctly deal with genes that perform multiple functions; such genes should participate in multiple clusters but will be placed in at most one cluster.

Biclustering (also known as projective or subspace clustering) has emerged as a powerful algorithmic tool for tackling these problems. A typical definition of a *bicluster* is a pair (U, T), where $U \subset V$ and $T \subset S$ such that the genes in U are clustered well in the samples in T but are not clustered well in the samples in $S - T$. In this formulation, a bicluster includes only a subset of genes and samples. Hence, algorithms that compute biclusters capture condition-specific patterns of co-expression. Biclustering algorithms allow a gene or a sample to participate in multiple biclusters, each of which may correspond to a different pathway or biological process. Different biclusters may contain different numbers of genes and/or samples. A number of different methods have emerged for computing biclusters in gene expression data; two papers provide excellent surveys [MO04,TSS06].

A powerful approach to computing biclusters rests on representing gene expression data as a bipartite graph connecting genes to samples. Algorithms use different criteria to decide which gene-sample pairs to connect in such a graph. A bicluster is usually modeled as a bipartite clique (biclique) (U, T), where $U \subset V$ and $T \subset S$. The goal is to compute one or more bicliques of large size in the graph, where the size of a biclique is usually defined as $|U||T|$. Finding the biclique with the largest number of edges in an unweighted bipartite graph is known to be NP-hard [Pee03]. Ambühl et al. [AMS07] extend results by Khot [Kho06] to prove that this problem is hard to approximate, i.e., it does not have a PTAS under the assumption that NP does not have randomized algorithms that run in sub-exponential time. Lonardi et al. [LSY06] propose a random-sampling algorithm to compute the largest bicluster (formalized as a biclique in a bipartite graph). We describe a related approach by Mishra et al. [MRS04] in more detail. These authors consider ε-*bicliques*: each gene in such a biclique is connected to at least a $(1 - \varepsilon)$ fraction of the samples in the biclique. They pose the problem of computing an ε-biclique that has at least a $(1 - 2b)$ fraction of the number of edges in the maximum biclique, for a small constant $b \geq 0$. They present a random-sampling algorithm that is efficient if the largest biclique has at least some fraction ρ_G of the genes and some fraction ρ_S of the samples. Under this assumption, their algorithm runs in time linear in d, logarithmic in n, quasi-polynomial in ρ_G and ρ_S, and exponential in poly$(1/\varepsilon)$.

Mishra et al. also propose a strategy for computing multiple bicliques. Simply computing the k largest bicliques for some value of k may be unsatisfactory: these bicliques may have considerable overlap in their edge sets, and highly overlapping bicliques may not capture the diversity of biclusters present in the data. To preclude this possibility, the authors introduce the notion of δ-domination: one biclique δ-*dominates* another if the number of edges in the second biclique that do not belong to the first is at most a δ fraction of the size of the union of the two edge sets. Next, they introduce the notion of when a collection $\mathcal{C}$ of k ε-bicliques is *diverse*: when for every pair (U', T') and (U'', T'') of bicliques in $\mathcal{C}$, neither δ-dominates the other. Finally, they introduce the notion of when a collection $\mathcal{C}$ of k ε-bicliques *swamps* a biclique (U', T'): either one of the k bicliques in $\mathcal{C}$ δ-dominates (U', T'), or (U', T') does not contain many more edges than any biclique in $\mathcal{C}$. Armed with these definitions, they pose the problem of computing a collection $\mathcal{C}$ of k ε-bicliques that are diverse and swamp every large biclique in the graph (a biclique is *large* if it contains at least some fraction ρ_G of the genes and some fraction ρ_S of the samples). Their algorithm runs in time linear in d, logarithmic in n, quasi-polynomial in k, ρ_G and ρ_S, and exponential in poly$(1/\varepsilon)$.

Tanay et al. [TSS02] construct a bipartite graph between genes and samples that represents a discretized version of the data. They assess edge weights in this graph based on a statistical model. They define a bicluster to be a bipartite clique (biclique) of large total edge weight. Under the assumption that each gene is connected to at most a constant number of samples, they simply enumerate all bipartite cliques in this graph. In practice, they supplement this approach with local searches to improve the weight of bipartite cliques and with hashing techniques to speed up the search. In a follow-up paper, they extend their formulation to integrate the analysis of different types of genome-wide data [TSKS04]. In this work, the bipartite graph connects genes to properties. Properties include the expression of a gene in a sample, the regulation of a gene by a transcription factor, and the response of a gene to a chemical treatment. In another follow-up study, Tanay et al. [TSKS05] extended these techniques to analyze gene expression data from a new study in the context of a large compendium of data from prior studies; they recast the new dataset in terms of biclusters computed from the other datasets and biclusters discovered only upon the addition of the new data.

Motivated by the approach proposed by Tanay et al., Tan [Tan08] considers the problem of finding the biclique of largest total weight in a weighted bipartite graph where edge weights are positive or negative integers. Under the assumption that the absolute value of the ratio of the smallest edge weight to the largest edge weight is in the range $\Omega(n^{\delta-1/2}) \cap O(n^{1/2-\delta})$,* where $\delta > 0$ is an arbitrarily-small constant, Tan shows that this problem is hard to approximate within a factor of ε for some $\varepsilon > 0$ unless RP = NP. Tan et al. [TCZZ07] prove that other formulations of biclustering [LO02,BDCKY02] are also hard to approximate.

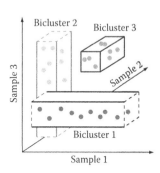

FIGURE 32.3 Example of orthogonal biclusters. Samples 2 and 3 belong to bicluster 1, samples 1 and 2 are elements of bicluster 2, and all three samples are in bicluster 3. A dashed face of a box indicates the dimension along which the box is unbounded.

The approaches discussed earlier are applicable when real-valued gene expression data are discretized. When such a discretization is not preferred, a geometric viewpoint may be more appropriate. From this perspective, the approaches discussed earlier compute orthogonal biclusters, i.e., each bicluster is a projection of a subset of the genes into an orthogonal subspace of $\mathbb{R}^d$ spanned by a subset of samples. See Figure 32.3 for examples of such biclusters. This image is based on the approach proposed by Procopiuc et al. [PJAM02]. In this model, a gene is an element of a bicluster if and only if the gene's expression levels in the samples in the bicluster span an interval of width at most w, where $w > 0$ is a parameter. They consider dense projective clusters, i.e., those that contain at least an α-fraction of the samples, $0 \leq \alpha \leq 1$. In addition, they introduce a condition that specifies the trade-off between the number of genes and the number of samples in a bicluster; this condition depends on a parameter β. Under this formulation, they present a Monte Carlo algorithm to compute a bicluster with the largest number of genes and with width at most $2w$ with probability of at least $1/2$, in $O(nd^{\log(2/\alpha)/\log(1/(2\beta))})$ time. Melkman and Shaham [MS04] describe a closely-related algorithm for the following model: for every pair of genes participating in a bicluster, the ratio of the

---

* We have rewritten Tan's condition assuming $n \geq d$.

expression levels of this pair of genes in each of the samples in the bicluster is a constant depending only on the two genes. They introduce the notion of sleeve-width to allow noise in this model.

Attention has also been paid to the problem of non-orthogonal projective clustering. The typical formulation of this problem seeks to approximate a set V of n points in $\mathbb{R}^d$ by a collection F of k shapes in $\mathbb{R}^d$. The shapes in F may be points, lines, j-dimensional subspaces ($j < d$), or non-linear shapes. By assigning each point in V to the closest subspace in F, we obtain a projective clustering of V. Algorithms attempt to minimize a value such as the sum of the distances from each point to the closest shape, the sum of the squares of these distances, or the largest of these, yielding the k-median, k-mean, or k-centre projective clustering problems, respectively. When k or d is part of the input, Megiddo and Tamir [MT83] have shown that many versions of these problems are NP-Hard. Feldman et al. [FFSS07] summarize known results for approximating the quality of the best clustering. They point out that all such approximations must be super-polynomial in k, unless P $=$ NP. Motivated by this observation, they consider (α, β) bi-criteria approximation algorithms that compute α j-dimensional flats whose quality is within a β factor of the best approximation by k such flats. Their algorithm achieves the performance guarantees of $\alpha(k, j, n) = \log n (jk \log \log n)^{O(j)}$ and $\beta(j) = 2^{O(j)}$ in time $dn(jk)^{O(j)}$ with probability of at least $1/2$. Their algorithm applies simultaneously to the median, mean, and centre versions of the problem.

We note that orthogonal projective clustering algorithms are more likely to be useful in practice, since they are easier to interpret: each bicluster is simply a set of genes and a set of samples. Moreover, existing orthogonal biclustering algorithms allow different computed biclusters to have differing numbers of genes and samples, a property essential to capturing the diversity of different biological processes.

32.5 Structure of the Wiring Diagram

In Section 32.2, we defined a wiring diagram as the network composed of the known molecular interactions for an organism. Specifically, the wiring diagram is a graph where each node is a molecule and each edge is a directed or undirected interaction between two molecules. A pair of molecules may be connected by multiple edges; each edge is usually annotated with information on the type of the interaction, e.g., physical interaction, phosphorylation, or regulation. As mentioned earlier, wiring diagrams are now available for a number of organisms. Some types of networks (e.g., PPI networks) have been experimentally studied on a much larger scale and for many more organisms than others (e.g., transcriptional regulatory networks [HGLR04]). Limitations in experimental techniques lead to considerable noise in available networks; they contain erroneous interactions (false positives) and miss many interactions (false negatives). Assessing error rates at the level of experiment types and individual interactions is an area of active research [SSR+06]. There are other types of uncertainties inherent in these data. For example, we may know that a set of proteins interact to form a protein complex, but we may not know precisely which pairs of proteins interact within the complex [BVH07].

Nevertheless, the computational studies of wiring diagrams (especially, PPI network and transcriptional regulatory networks) have yielded numerous insights into their structure and evolution. Preliminary studies of the PPI networks and metabolic networks suggested that their degree distributions follow the power law [AJB00,JTA+00]. More specifically, the fraction of nodes with degree $d \geq 1$ is proportional to $d^{-\gamma}$, with typical values of γ ranging between 2 and 3. More recent studies have cast doubts on these results arguing that power law distributions may arise from experimental biases and artifacts caused by sampling [HDBe05,SWM05].

What systems-level insights into cellular function can wiring diagrams reveal? One of the guiding principles of systems biology is that molecules within the cell organize themselves into modules [HHLM99]. A module may be loosely defined as a group of interacting molecules that act coherently in the cell. Modules can share both nodes and edges, especially since many genes and proteins are multi-functional. Modules can be hierarchical in the sense that one module may contain another. In a sense, such modules constitute building blocks of wiring diagrams. Examples of modules are densely interacting proteins (perhaps forming complexes), protein sub-networks that may be evolutionarily conserved in many organisms, biochemical pathways that synthesize a particular compound, and sets of genes that have co-evolved and are found in multiple genomes.

32.5.1 Network Decomposition

Graph clustering, or automatic decomposition of a network into modules or communities, has a rich history, with many problem formulations and techniques [CF06]. In the context of systems biology, a number of papers have studied the problem of decomposing the wiring diagram, mainly PPI networks, into modules [SI06]. These methods use various ad hoc heuristics, e.g., repeatedly removing the edge with the largest betweenness centrality [DDS05], local searches around multiple seeds [BH03,Bad03], and approaches akin to simulated annealing [SM03]. We describe a few approaches that find a partition or cover of an undirected graph into multiple modules using principled ideas.

Given an undirected graph $G = (V, E)$, Hartuv and Shamir [HS00] define an induced subgraph H of G to be highly connected if the minimum number of edges that must be removed in order to disconnect H is at least half the number of nodes in H. They present an output-sensitive recursive algorithm to compute all highly connected subgraphs of G. For each subgraph returned, the algorithm runs in time taken to compute the minimum cut in G.

Newman [New06] measures the quality of a partitioning of G into two subgraphs G_1 and G_2 using the notion of *modularity*, defined as

$$\frac{1}{4|E|} \sum_{(u,v)} \left(w_{uv} - \frac{d_u d_v}{|E|} \right) s_u s_v, \tag{32.3}$$

where the summation is over all pairs of nodes in V, $w_{uv} = 1$ if $(u, v) \in E$ and 0 otherwise, d_u is the degree of node $u \in V$, and $s_u = 1$ (respectively, -1) if $u \in G_1$ (respectively, G_2). He optimizes this quantity by computing the leading eigenvector of a symmetric matrix whose values are the elements within the summation. To find multiple modules, he recursively applies this algorithm, stopping when the largest eigenvalue for a subgraph is 0. Brandes et al. [BDGe06] prove that maximizing *modularity* is strongly NP-complete.

The $-d_u d_v / |E|$ term in Equation 32.3 arises from the fact that if the edges in G are rewired randomly while maintaining the degree of the nodes, then the probability that u and v are connected is $\frac{d_u d_v}{|E|}$. Intuitively, subtracting this quantity accounts for any modularity that a random graph with the same degree sequence as G may have. This notion arises repeatedly in CSB: what is the probability that an observed network module may arise in random data? A typical approach to answering this question empirically is to sample multiple times from the distribution of random networks with the same degree sequence as G, run the network decomposition algorithm on each sample, and use the distribution of module sizes thus obtained to estimate the desired probability. The Markov Chain Monte Carlo method is useful in this situation, but current algorithms have large run-times making them unsuitable for graphs with tens of thousands of nodes [GMZ03]. Given a degree sequence where the maximum degree $d_{\max} = O(|E|^{1/4 - \tau})$, where τ is any positive constant, Bayati et al. [BKS07] develop an algorithm that runs in $O(|E| d_{\max})$ time and generates any graph with the given degree sequence with probability within $1 \pm o(1)$ factor of uniform. They also use an approach called "sequential importance sampling" to convert this algorithm into a fully polynomial randomized approximation scheme. Specifically, for any $\varepsilon, \delta > 0$, with probability at least $1 - \delta$, the output of their algorithm is a graph with the given degree sequence; this graph is drawn from the set of all such graphs with uniform probability upto a multiplicative error of $1 \pm \varepsilon$. Their algorithm runs in time $O(|E|d_{\max}\varepsilon^{-2}\log(1/\delta))$.

32.5.2 Evolutionarily Conserved Modules

It has been observed that many PPIs are evolutionarily conserved between different species [YLLe04], i.e., if proteins a and b in one organism interact and if a (respectively, b) is orthologous to protein a' (respectively, b') in another organism, then a' and b' interact. It is natural to ask whether larger sets of interactions may be conserved and how such sets could be automatically computed from PPI networks for two different organisms. In the CSB community, a number of approaches have been developed that use evolutionary constraints to compute such Conserved Protein Interaction Modules (CPIMs) [SI06]. See Figure 32.4 for an illustration.

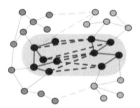

FIGURE 32.4 An illustration of a CPIM. Circles represent proteins. Solid lines connect interacting proteins. Dashed lines connect orthologous proteins. The figure contains two PPI networks, one on the left and the other on the right. The darker sub-networks and the pairs of orthologous proteins in those sub-networks (the nodes and the edges within the shaded oval) constitute a CPIM.

A number of these approaches [KYLe04,SIKe04,KKTS06,SSI05] share many common features. They combine the PPI networks of two species into a single alignment graph. A node in the alignment graph represents two orthologous proteins, one from each PPI network. An edge in the alignment graph represents an interaction that is conserved in both PPI networks. These methods add an edge to the alignment graph only if the proteins contributing to the nodes are connected through at most one intermediate protein in the respective PPI networks. The weight of an edge represents the likelihood that the corresponding interactions are conserved; this weight depends on the degree of orthology between the proteins and on assessed confidence estimates that the individual PPIs indeed take place in the cell. After constructing the alignment network, these authors find CPIMs by using various approaches to compute paths, complexes, and subgraphs of high weight in the alignment network and then expanding each such subgraph into the constituent PPIs. Sharan et al. [SSKe05] generalize this idea to more than two PPI networks. Liang et al. [LXTN06] propose a method where each node in the alignment graph is a pair of conserved PPIs. They develop criteria to connect two such nodes and reduce the problem of computing CPIMs to the problem of finding all maximal cliques in the alignment graph.

Narayanan and Karp have recently presented the Match-and-Split algorithm [NK07]. Like other methods, they define a pair of proteins to be similar if their sequence similarity is at least some threshold. They use combinatorial criteria to decide when the local neighborhoods of a pair of orthologs match. Under their model, they prove that a given pair of proteins can belong to at most one CPIM. This observation leads to a top-down partitioning algorithm that finds all maximal CPIMs in polynomial time.

32.5.3 Network Motifs

Milo et al. [MSOIe02] pioneered the study of network motifs and the bottom-up assembly of complex networks from such motifs. Informally, given two graphs G and H (where H is connected), we say that H *occurs* in G if H is isomorphic to a subgraph of G. We say that H is a *network motif* of G if the number of times that H occurs in G is surprisingly large (we make this notion precise below). Network motifs may play a key role in processing information in regulatory networks [SOMMA02].

A typical approach to computing network motifs (1) identifies the number of subgraphs of G that is isomorphic to a candidate network motif H, (2) determines the probability that H occurs at least this many times in random graphs with the same degree sequence as G, and (3) declares H to be a network motif if this probability is smaller than a user-specified threshold. These methods usually enumerate isomorphic subgraphs explicitly, which can be computationally expensive. Wernicke [Wer06] defines the concentration of a k-node subgraph H to be the ratio of the number of occurrences of H in G to the total number of occurrences of all connected k-node subgraphs in G. He presents a randomized algorithm that computes an unbiased estimator of the concentration of every connected k-node subgraph that occurs in G. He also shows how to adapt theorems by Bender and Canfield [Ben74, BC78] to estimate the expected concentration of a given k-node subgraph in random graphs with the same degree sequence as G without explicitly generating such random graphs.

Wiring diagrams contain the interactions of multiple types. Yeger-Lotem et al. [YLSKe04] and Zhang et al. [ZKWe05] consider the question of finding multi-colored network motifs. Researchers have also considered whether motifs might assemble into larger structures [GK07,ZKWe05] and how such relationships between consolidated subgraphs may reveal insights into the structure of the wiring diagram.

32.6 Condition-Specific Analysis of Wiring Diagrams

As described in the previous section, existing wiring diagrams are tremendous resources for systems biology, since they integrate information on multiple types of molecular interactions obtained from a variety of different experimental sources. However, such an experiment often does not yield information on when an interaction is activated within the cell. Therefore, the potential impact of wiring diagrams is diluted since they typically represent the universe of interactions that take place across diverse contexts in the cell. Another deficiency of existing wiring diagrams is that they are highly incomplete in spite of decades of small-scale experimentation and recent advances in high-throughput screening. For instance, a recent estimate [STdSe08] suggests that the human PPI network may contain 650,000 edges, about an order of magnitude greater than the number obtained by combining multiple existing databases [DMS08]. Note that this estimate is solely for PPIs. Our knowledge of other types of interactions (e.g., between transcription factors and their target genes, between small molecules and metabolites, or between recently-discovered regulatory molecules, such as microRNAs and their targets) is even more scarce than for PPIs. In this section, we focus on two classes of methods developed to address the two issues raised above.

32.6.1 Response Networks

Many algorithms have been developed to integrate the wiring diagram with gene expression measurements for a single condition (e.g., the time-course of response of a cell to a stress or data from patients diagnosed with a particular disease) in order to compute the sub-network of interactions that is perturbed in that condition. These approaches take a wiring diagram $G = (V, E)$ and a gene expression dataset $V_S = \{g_S \mid g \in V\}$ as input, where S is the set of samples. Their goal is to compute the subgraph G_S of G such that the genes in G_S show the most similar expression patterns over all subgraphs of G.

A common experimental design is to divide the set S of samples into two classes, a set corresponding to a treatment and a set corresponding to a control. In this situation, gene expression measurements are better represented by the estimates of differential expressions for each gene. It is possible to use a hypothesis testing framework to assess how different the expression levels of a gene in the treatment samples are from its expression levels in the control samples, e.g., by using the t-test. For each gene g, this computation yields a p-value $0 \le p_g \le 1$ representing the statistical significance of the difference between the two sets of expression levels of the gene. Given such a data set, Ideker et al. [IOSS02] apply an assumption that p_g arises from a normal distribution, and compute a z-score $z_g = N^{-1}(p_g)$, where N^{-1} is the inverse of the normal distribution function.

They define the *z*-score $z(G')$ of a subgraph G' of G to be the sum of the *z*-scores of the nodes in G' divided by the square root of the number of nodes in G'. Their goal is to compute the subgraph of G with the largest *z*-score. After showing that a version of this problem is NP-complete, they proceed to use simulated annealing to solve the problem.

Murali and Rivera [MR08] propose a method that is applicable when S contains enough samples to estimate the co-expression of any pair of genes. For every edge $e = (g, h)$ in E, they compute a weight w_e that is the absolute value of Pearson's correlation coefficient between g_S and h_S. They define the density of a subgraph G' of G as the total weight of the edges in G' divided by the number of nodes in G. Their goal is to compute the subgraph of G with largest density. This problem can be solved in polynomial time using parametric network flows [GGT89]. In practice, they use the greedy algorithm suggested by Charikar [Cha00], which computes a subgraph at least half as dense as the densest subgraph.

These two approaches have the drawback that they consider co-expression relationships only between pairs of genes that are adjacent in G. We have discussed earlier that G is incomplete for many organisms. In such situations, these approaches may ignore many co-expressed pairs of genes. Ulitsky and Shamir [US07] propose an innovative approach to mitigate this problem. They compute an undirected graph X_S where two genes are connected if they are highly co-expressed. In this graph, their goal is to find dense subgraphs under the constraint that each dense subgraph must induce a connected network in G. Thus, two genes may belong to a dense subgraph in X_S even if they are not directly connected in G. Ulitsky and Shamir develop a statistical model for this problem and propose a number of heuristics to compute multiple dense modules [US07].

In principle, these problems are related to the question of computing the largest clique in a graph, a problem well-known to be NP-complete [GJ79], and hard to approximate [Hås99]. Apart from the two papers mentioned above [Cha00,GGT89], theoretical studies of similar problems have usually dealt with unweighted graphs. Feige et al. [FKP01] compute the densest *k*-vertex subgraph of a given graph, namely, the subgraph with *k* vertices that contains the most edges among all *k*-vertex subgraphs. They develop an approximation algorithm for the problem, with approximation ratio $O(n^\delta)$, for some $\delta < 1/3$. Khot proves that this problem does not admit a PTAS [Kho06].

Holzapfel et al. [HKMT06] pose the γ-CLUSTER problem, where given an undirected graph G and a natural number k, they ask if G has a subgraph on k vertices whose average degree is at least $\gamma(k)$; they allow $\gamma : \mathbb{N} \to \mathbb{Q}_+$ to be any function that can be computed in polynomial time and satisfies $\gamma(k) \leq k - 1$, for all $k \in \mathbb{N}$. For $\gamma(k) = k - 1$, this problem is the clique problem. In contrast, for $\gamma(k) = 2$, the problem can be solved in polynomial time. They show that the problem remains NP-complete if $\gamma = 2 + \Omega(1/k^{1-\varepsilon})$ for some $\varepsilon > 0$ and has a polynomial-time algorithm for $\gamma = 2 + O(1/k)$.

The spectral radius of an undirected graph G is the largest eigenvalue of the adjacency matrix of the graph. It is well-known that the spectral radius of a graph is at least as large as its average degree. Andersen and Cioaba [AC07] pose the (k, λ)-spectral radius problem: does G have a subgraph on at most k vertices whose spectral radius is at least λ? When such a subgraph exists, they present an approximation algorithm that runs in $O(n\Delta k^2)$ time, where Δ is the maximum degree of the graph, that outputs a subgraph with spectral radius of at least $\lambda/4$ and with at most Δk^2 vertices.

32.6.2 Reverse-Engineering Gene Networks

The algorithms described in the previous section assume that a wiring diagram is available. However, as mentioned at the beginning of Section 32.6, existing wiring diagrams are incomplete. To surmount this difficulty, methods have been developed to reverse engineer interactions between genes from gene expression data. The primary assumption underlying these techniques is that if two genes are highly co-expressed, i.e., if their expression levels under one or more conditions have high correlations, then the genes may have a functional interaction. Based on this hypothesis, numerous methods have been developed to infer interactions between pairs of genes [BBAIB07,MS07]. Approaches investigated for gene network construction include gene relevance networks [BK99,DWFS98], Gaussian graphical models [dlFBHM04,SS05], mutual information (MI) based networks [BMSe05,BK00,ZAA08], and Bayesian networks [FLN00,Y⁺02].

In spite of the excitement surrounding these approaches, there is considerable debate about a number of issues. How should the co-expression between two genes be measured and what are the relative advantages of each measure? For example, Pearson's correlation coefficient can be computed in time linear in the number of samples and estimated with confidence even for relatively few samples, but it can only capture linear dependencies. More complex methods typically come with greater computational cost and/or the need for a large number of samples. Does the co-expression of two genes imply stable binding between the proteins that the genes code for, or a cause-and-effect relationship between the genes? These issues have been discussed in a number of papers in the last few years [MS07,BBAIB07,ZSA08,SBA07].

An important problem that arises in gene network construction is to find a sufficient number of samples relative to the network size to be inferred. One could limit the network size if the goal is to infer a subnetwork focused around a biological process that involves a subset of genes, but this requires knowing the genes in advance. In many cases, one would want to infer a gene network to precisely identify such subnetworks and discover unknown genes that may be a part of such networks. There are tens of thousands of genes in any complex organism, and in many cases it is impossible to find a reliable way to limit analysis to only a subset of them. The number of samples can be significantly increased by tapping into the public repositories of gene expression profiles resulting from microarray experiments carried out by many laboratories worldwide. Even so, the number of available samples falls short of what is ideally required by the underlying computational methods, with even the number of genes in an organism significantly outnumbering the number of available samples at present. In addition, the use of such large number of samples raises computational and statistical challenges. Gene expression data is inherently noisy and significantly influenced by many experiment-specific attributes. As a result, it is not meaningful to directly compare expression levels across multiple samples. Finally, little is known about general regulatory mechanisms (e.g., post-transcriptional effects) and thus no satisfactory models of genetic regulation are available.

To provide more insight into the construction of gene networks, we present MI based methods in greater detail. MI can capture non-linear dependencies, the underlying algorithms have polynomial complexity, and recent work demonstrates that these methods generate networks with good quality. Let n denote the number of genes and m denote the number of samples. In MI based methods, the expression level of gene g_i is taken to be random variable X_i, for which we have m recorded observations. The MI between a pair of genes g_i and g_j, denoted by $\mathcal{I}(X_i; X_j)$, is given by

$$\mathcal{I}(X_i; X_j) = \mathcal{H}(X_i) + \mathcal{H}(X_j) - \mathcal{H}(X_i, X_j)$$

where the entropy $\mathcal{H}(X)$ of a continuous variable X is given by

$$\mathcal{H}(X) = -\int p_X(\xi) \log p_X(\xi) d\xi,$$

and p_X is a probability density function for X. In this case p_{X_i}, p_{X_j}, and the joint probability density function p_{X_i, X_j} are unknown and have to be estimated based on available gene expression samples.

To reverse engineer gene networks using this approach, a method for computing the MI between a pair of genes and a criterion for assessing when the MI value is significant are needed. Several different techniques to estimate MI have been proposed [K⁺07], differing in precision and complexity. Simple histogram methods [BK00] are very fast but inaccurate, especially when the number of observations is small. Gaussian kernel estimators utilized by Margolin et al. [MNBe06] provide good precision but take $O(m^2)$ run-time. Daub et al. [DSSK04] propose a linear time method that is competitive with the

Gaussian kernel estimator. Their method is based on binning, which in its simplest form estimates the probability density of a random variable by dividing samples into fixed number of bins and counting samples per bin. Such a method is imprecise and sensitive to the selection of boundaries of bins [MRL95]. Daub et al. overcome this by using B-splines as a smoothing criterion: each observation belongs to k bins simultaneously with weights given by B-spline functions up to order k.

A standard way to assess the significance of MI values between g_i and g_j is to randomly permute the expression values of one of the genes, say g_i, and computing the MI again based on the permuted expression values of g_i and the unaltered expression values of g_j. A large number of such permutation tests are conducted, and $\mathcal{I}(X_i; X_j)$ is deemed significant if it is greater than the MI value of at least a fraction $1 - \epsilon$ of the permutations tested. Such testing is expensive, particularly when repeated for the $\binom{n}{2}$ gene pairs. Zola et al. [ZAA08] propose a method to make a permutation test applicable to all gene pairs, thereby reducing the complexity of permutation testing by a factor of $\Theta(n^2)$.

A particularly vexing issue in network inference is the difficulty of distinguishing indirect interactions from direct interactions. Consider three genes g_i, g_j, and g_k, where g_i directly interacts with g_j and g_j directly interacts with g_k, but g_i and g_k have no direct interaction. If all three genes are up-regulated in a condition, all three pairs of expression profiles will be correlated. Disentangling direct interactions from indirect ones is difficult, and becomes more complex with larger sets of genes that interact in more intricate ways. MI methods address this using Data Processing Inequality (DPI) [CT91], which states that in the case of the example mentioned above, $\mathcal{I}(X_i, X_k) \leq \mathcal{I}(X_i, X_j)$ and $\mathcal{I}(X_i, X_k) \leq \mathcal{I}(X_j, X_k)$. After computing the MI between all pairs of genes, the DPI is run in reverse by identifying such triplets of gene pairs and removing the one with the smallest MI value in each triplet [MNBe06,ZAA08]. Margolin et al. [MNBe06] prove that this algorithm correctly recovers the interaction network if the MI values can be estimated without errors, the network contains only pairwise interactions, and the network is a tree. They show that their algorithm can reconstruct networks with loops under certain other assumptions. ARACNe, the software implementation of their algorithm, runs in $O(n^3 + n^2 m^2)$ time. Zola et al. [ZAA08] developed a parallel method for MI based inference that combines Daub et al.'s $O(m)$ time B-spline MI estimator with a new method for reducing permutation testing complexity by $\Theta(n^2)$. Their software implementation TINGe scales to whole genome networks and much larger number of samples than previous approaches.

An illustration of a gene network inferred using the MI approach is shown in Figure 32.5. The network in Figure 32.5a shows interactions among 4,000 yeast genes with node size reflecting the degree of the node. A closer look at the connectivity among some of the genes involved in response to oxidative stress is shown in Figure 32.5b. Given the complexity of networks, the visualization and the navigation of these networks are of considerable importance to biologists. Shannon et al. developed a widely used software environment, termed Cytoscape, for this purpose [SMe03], which allows third party plugins for further enhancing the functionality of the environment as necessary. Another problem is that of comparing across multiple network inference methods, all the more important due to the diversity of approaches that are applied. While a few biologically validated networks exist, it is useful to have a wide array of benchmark data sets with different numbers of genes, samples, quality of samples, etc. It is common practice to validate network inference algorithms using synthetically generated networks by programs, such as SynTReN [dBLNe06] and COPASI [HSe06]. SynTReN takes a gene network as input (such as, a biologically validated network of yeast), and creates a synthetic benchmark network of a specified size with similar or desired topological properties, and the desired number of samples with user-specified noise. COPASI is capable of generating time series data—such data is particularly valuable in inferring the directionality of interactions. For example one may infer that gene g_i regulates gene g_j, if a time delay is observed between the rise in expression value of g_j when compared to the rise in expression value of g_i. Time series data can be especially valuable for gene network construction, but requires access to carefully timed experiments. Thus, undirected network inference using large public data repositories continues to be of value to gather sufficient number of samples to build robust gene networks.

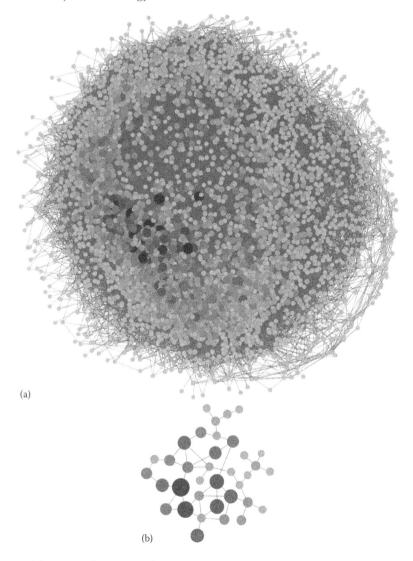

FIGURE 32.5 (a) Yeast regulatory network on 4000 genes inferred by TINGe software. The size of a node is proportional to its degree in the graph. (b) A closer look at the connectivity between some of the genes involved in response to oxidative stress.

32.7 Outlook and Resources for Further Study

Although the field is young, research in CSB is taking place at an intense level commensurate with the importance and the potential applications of this field. This is hardly surprising given the potential for broad impact in critical areas such as health and human disease, and the management of agricultural and animal resources. In this interdisciplinary field, apart from computer science and molecular biology, many other disciplines contribute to knowledge discovery including chemical engineering, physics, control theory, and statistics. In keeping with the scope and expected audience of this handbook, our coverage of topics is heavily influenced by areas with substantial contributions from computer science, and the role of algorithms and theory even within that. Not covered in detail here are many topics related to contributions with techniques from other areas of computer science, such as text mining for annotation and network extraction, heterogeneous data integration, models of network evolution, and data and graph mining. A number of techniques normally considered outside the realm of algorithms and theory have also been utilized in making substantial contributions to systems biology. Examples include metabolic flux analysis, physics driven models, and modeling of biological processes as complex control systems, although algorithmic questions naturally arise when these models eventually result in the need for computational solutions.

The field of systems biology is expected to grow quite rapidly, aided by both an increasing availability of experimental data, and a continued discovery and refinement of computational models and techniques. On the experimental side, high-throughput experimental techniques are continually improved and increasing amounts of such data are being generated, e.g., comprehensive gene expression profile measurements for various organisms. The predominant culture in the community of open data sharing through web portals is a significant driver of innovation, drawing in scientists from many fields with no interest in conducting the experiments per se, or in developing such expertise. New experimental techniques to measure the various aspects of cellular activity are expected to come on line, along with refined measurement capability for existing instrumentation. To match these experimental advances on the computational end, we need better graph theoretic models of biological processes, approaches to reason about networks, and techniques for modeling experiments and their results. In the future, we envision the routine use of computational models as a mechanism for suggesting useful experiments for biologists and the incorporation of the results of these experiments into more refined models of the cell.

For the reader interested in further forays into the field or keeping abreast of this rapidly growing field, a number of resources are available, some of which we mention here. The open access *PLoS Computational Biology* journal publishes a series on "Getting started in ...," which is a valuable resource for understanding many topics relevant to systems biology. The *Nature* Insights series provides an editorial and a compendium of commentaries on specific focus topics. *Nature* also launched a series of "Connections essays" to explore how large number of interacting components result in systems level behavior. Finally, *Nature Cell Biology* and *Nature Reviews Molecular Cell Biology* have partnered together to publish several review articles in various subfields of systems biology. We encourage readers of the chapter to tap into these and other resources for further study of this fascinating and emerging area of scientific discovery. Finally, computational biology is a vast research area and to permit a reasonable exposition within this chapter, we limited ourselves to the emerging important area of systems biology. Readers interested in a comprehensive introduction to the field of computational biology are referred to the handbook edited by one of the authors [Alu06].

Acknowledgment

Work on this book chapter is supported in part by the National Science Foundation under CCF-0811804.

Glossary

Molecules

Cis-element A short DNA sequence found in the upstream non-coding region of the gene that controls the transcriptional activity of a gene via transcription factors or other DNA-binding elements.

Enzyme A protein that catalyzes a biochemical reaction.

Gene A DNA sequence (a substring of a chromosome) that is involved in encoding one or more functional products, such as an RNA or a protein. The sequence includes coding regions that code for the functional product(s), and non-coding regions, such as introns.

Homolog A gene whose nucleotide sequence exhibits similarity to another gene or a set of genes.

Kinase A protein that adds a phosphate group to a protein, the substrate.

Ortholog A gene in one organism is orthologous to a gene in another organism if both genes have evolved from the same gene in a common ancestral organism. Orthology is usually established by comparing genetic sequences.

Phosphatase A protein that removes a phosphate group from a protein.

Promotor A regulatory sequence of DNA located in the upstream 5' non-coding region of the gene that controls the transcription of the gene.

Protein A chain of amino acids synthesized in a specific order from the transcription and the translation of a gene via the genetic code.

Transcription factor A protein that binds to a specific DNA sequence (cis-element) in the promoter region of a gene to control its transcription.

Interactions

Biochemical reaction A process where an enzyme interacts with one or more molecules (substrate) to produce a product.

Genetic interaction Interaction between a set of genes where the action of one gene is modified by another gene or a set of genes.

Phosphorylation Addition of a phosphate group to organic molecules by enzymes called kinases.

Protein–protein interaction An association between a set of protein molecules to form long-term protein complexes.

Transcriptional regulatory interaction An interaction between a transcription factor and a promotor of a gene to regulate gene expression.

Processes

Gene expression The conversion of a gene (DNA sequence) into functional gene product, such as RNA or protein.

Post-translational modification Changes made to a protein after translation, such as the addition of a chemical group, or the formation of a protein complex, or making structural changes to the protein.

Protein translation The process of converting messenger RNA, the product of gene expression, into a chain of amino acids using the genetic code.

Signal transduction A process by which signals are transmitted by proteins and other molecules from the outside of a cell to its interior or within a cell.

Gene Functions

Biological process A series of events accomplished by one or more ordered assemblies of molecular functions.

Cellular component A component of the cell.

Gene Ontology A standard nomenclature that is used to describe genes and gene product attributes across organisms.

Molecular function An activity, such as catalytic or binding activity, that occurs at the molecular level.

Phenotype An observable characteristic or trait of an organism.

Protein localization Positioning of a protein in an appropriate cellular area (e.g., an organelle, an interior membrane, etc.) where its activity is needed.

Organisms

Eukaryote An organism whose cells contain nuclei. Genomic DNA is contained within the nucleus of each cell.

Model organism An organism that is extensively studied with the expectation that knowledge gained here provides valuable insights into other related organisms.

Prokaryote An organism whose cells do not have nuclei.

References

[ABBB00]   M. Ashburner, C.A. Ball, J.A. Blake, D. Botstein et al. Gene ontology: Tool for the unification of biology. The Gene Ontology Consortium. *Nature Genetics*, 25(1):25–29, 2000.

[AC07]   R. Andersen and S.M. Cioaba. Spectral densest subgraph and independence number of a graph. *Journal of Universal Computer Science*, 13(11):1501–1513, 2007.

[AJB00]   R. Albert, H. Jeong, and A.L. Barabasi. Error and attack tolerance of complex networks. *Nature*, 406(6794):378–382, 2000.

[Alu06]   S. Aluru, editor. *Handbook of Computational Molecular Biology*. Chapman & Hall/CRC Computer and Information Science Series, Boca Raton, FL, 2006.

[AMS07]   C. Ambühl, M. Mastrolilli, and O. Svensson. Inapproximability results for sparsest cut, optimal linear arrangement, and precedence constrained scheduling. In *Proceedings of the 48th Annual IEEE Symposium on Foundations of Computer Science (FOCS)*, Providence, RI, pp. 329–337, 2007.

[Bad03]   J.S. Bader. Greedily building protein networks with confidence. *Bioinformatics*, 19(15): 1869–1874, 2003.

[BBAIB07]   M. Bansal, V. Belcastro, A. Ambesi-Impiombato, and D. Bernardo. How to infer gene networks from expression profiles. *Molecular Systems Biology*, 3:78, 2007.

[BC78]   E.A. Bender and E.R. Canfield. The asymptotic number of labeled graphs with given degree sequences. *Journal of Combinatorial Theory Series A*, 24(3):296–307, 1978.

[BCS06]   G.D. Bader, M.P. Cary, and C. Sander. Pathguide: a pathway resource list. *Nucleic Acids Research*, 34(Database issue):D504–D506, 2006.

[BDCKY02]   A. Ben-Dor, B. Chor, R.M. Karp, and Z. Yakhini. Discovering local structure in gene expression data: The order-preserving submatrix problem. In *Proceedings of the 6th International Conference on Computational Biology (RECOMB)*, Washington, D.C., pp. 49–57, 2002.

[BDGe06]   U. Brandes, D. Delling, M. Gaertler, R. Goerke et al. Maximizing modularity is hard. *Arxiv Preprint Physics/0608255*, 2006.

[Ben74]   E.A. Bender. The asymptotic number of non-negative integer matrices with given row and column sums. *Discrete Mathematics*, 10:217–223, 1974.

[BH03]   G.D. Bader and C.W. Hogue. An automated method for finding molecular complexes in large protein interaction networks. *BMC Bioinformatics*, 4(1):2, January 2003.

[BJDGe03]   Z. Bar-Joseph, E.D. Demaine, D.K. Gifford, N. Srebro et al. K-ary clustering with optimal leaf ordering for gene expression data. *Bioinformatics*, 19(9):1070–1078, 2003.

[BK99]   A.J. Butte and I.S. Kohane. Unsupervised knowledge discovery in medical databases using relevance networks. In *Proceeding of the American Medical Informatics Association Symposium*, Boston, MA, pp. 711–715, 1999.

[BK00]   A.J. Butte and I.S. Kohane. Mutual information relevance networks: Functional genomic clustering using pairwise entropy measurements. In *Proceedings of the Pacific Symposium on Biocomputing*, Honolulu, HI, pp. 418–429, 2000.

[BKS07]   M. Bayati, J.H. Kim, and A. Saberi. A sequential algorithm for generating random graphs. In *Approximation, Randomization, and Combinatorial Optimization. Algorithms and Techniques, 10th International Workshop, APPROX 2007, and 11th International Workshop, RANDOM 2007, Proceedings*, Princeton, NJ, *Lecture Notes in Computer Science*, volume 4627, pp. 326–340. Springer, Berlin, 2007.

[BMSe05]   K. Basso, A.A. Margolin, G. Stolovitzky, U. Klein et al. Reverse engineering of regulatory networks in human B cells. *Nature Genetics*, 37(4):382–390, 2005.

[BOR04]   A. Borodin, R. Ostrovsky, and Y. Rabani. Subquadratic approximation algorithms for clustering problems in high dimensional spaces. *Machine Learning*, 56(1):153–167, 2004.

[BST06]   Z. Barutcuoglu, R.E. Schapire, and O.G. Troyanskaya. Hierarchical multi-label prediction of gene function. *Bioinformatics*, 22(7):830–836, 2006.

[BVH07]   A. Bernard, D.S. Vaughn, and A.J. Hartemink. Reconstructing the topology of protein complexes. In *Proceedings of the 11th International Conference on Research in Computational Molecular Biology (RECOMB)*, Oakland, CA, pp. 32–46, 2007.

[CF06]   D. Chakrabarti and C. Faloutsos. Graph mining: laws, generators, and algorithms. *ACM Computing Surveys*, 38(1):Article 2, 2006.

[Cha00]   M. Charikar. Greedy approximation algorithms for finding dense components in a graph. In *Proceedings of the 3rd International Workshop on Approximation Algorithms for Combinatorial Optimization*, Stanford, CA, pp. 84–95, 2000.

[CSZ06]   O. Chapelle, B. Schölkopf, and A. Zien, editors. *Semi-Supervised Learning*. MIT Press, Cambridge, MA, 2006.

[CT91]   T.M. Cover and J.A. Thomas. *Elements of Information Theory*. Wiley-Interscience, New York, 1991.

[CX04]   Y. Chen and D. Xu. Global protein function annotation through mining genome-scale data in yeast *Saccharomyces cerevisiae*. *Nucleic Acids Research*, 32(21):6414–6424, 2004.

[dBLNe06]   T. Van den Bulcke, K. Van Leemput, B. Naudts, P. Van Remortel et al. SynTReN: A generator of synthetic gene expression data for design and analysis of structure learning algorithms. *BMC Bioinformatics*, 7:43, 2006.

[DDS05]   R. Dunn, F. Dudbridge, and C.M. Sanderson. The use of edge-betweenness clustering to investigate biological function in protein interaction networks. *BMC Bioinformatics*, 6(1):39, March 2005.

[DFKe04]   P. Drineas, A. Frieze, R. Kannan, S. Vempala et al. Clustering large graphs via the singular value decomposition. *Machine Learning*, 56(1):9–33, 2004.

[DL05]   S. Dasgupta and P.M. Long. Performance guarantees for hierarchical clustering. *Journal of Computer and System Sciences*, 70(4):555–569, 2005.

[dlFBHM04]   A. de la Fuente, N. Bing, I. Hoeschele, and P. Mendes. Discovery of meaningful associations in genomic data using partial correlation coefficients. *Bioinformatics*, 20(18):3565–3574, 2004.

[DMS08]   M.D. Dyer, T.M. Murali, and B.W. Sobral. The landscape of human proteins interacting with viruses and other pathogens. *PLoS Pathogens*, 4(2):e32, 2008.

[DSSK04]   C.O. Daub, R. Steuer, J. Selbig, and S. Kloska. Estimating mutual information using B-spline functions—an improved similarity measure for analysing gene expression data. *BMC Bioinformatics*, 5:118, 2004.

[DWFS98]   P. D'haeseleer, X. Wen, S. Fuhrman, and R. Somogyi. Mining the gene expression matrix: Inferring gene relationships from large scale gene expression data. In *Information Processing in Cells and Tissues*, Sheffiedd, U.K., pp. 203–212, 1998.

[EKO03]   A.J. Enright, V. Kunin, and C.A. Ouzounis. Protein families and tribes in genome sequence space. *Nucleic Acids Research*, 31(15):4632–4638, 2003.

[ESBB98]   M.B. Eisen, P.T. Spellman, P.O. Brown, and D. Botstein. Cluster analysis and display of genome-wide expression patterns. *Proceedings of the National Academy of Sciences U S A*, 95:14863–14868, 1998.

[FFSS07]   D. Feldman, A. Fiat, M. Sharir, and D. Segev. Bi-criteria linear-time approximations for generalized k-mean/median/center. In *Proceedings of the 23rd Annual Symposium on Computational Geometry*, Gycongju, South Korea, pp. 19–26, 2007.

[FKP01]   U. Feige, G. Kortsarz, and D. Peleg. The dense k-subgraph problem. *Algorithmica*, 29:410–421, 2001.

[FLN00]   N. Friedman, M. Linial, and I. Nachman. Using Bayesian networks to analyze expression data. *Journal of Computational Biology*, 7:601–620, 2000.

[FMS07]   D. Feldman, M. Monemizadeh, and C. Sohler. A PTAS for k-means clustering based on weak coresets. In *Proceedings of the 23rd Annual Symposium on Computational Geometry*, Gycongju, South Korea, pp. 11–18, 2007.

[GGT89]   G. Gallo, M.D. Grigoriadis, and R.E. Tarjan. A fast parametric maximum flow algorithm and applications. *SIAM Journal on Computing*, 18(1):30–55, 1989.

[GJ79]   M.R. Garey and D.S. Johnson. *Computers and Intractability: A Guide to the Theory of NP-Completeness*. W. H. Freeman, New York, 1979.

[GJF07]   A. Godzik, M. Jambon, and I. Friedberg. Computational protein function prediction: Are we making progress? *Cellular and Molecular Life Sciences*, 64(19–20):2505–2511, 2007.

[GK07]   J. Grochow and M. Kellis. Network motif discovery using subgraph enumeration and symmetry-breaking. In *Proceedings of the 11th Annual International Conference on Research in Computational Molecular Biology (RECOMB)*, Springer-Verlag Lecture Notes in Computer Science, volume 4453, Oakland, CA, pp. 92–106, 2007.

[GMZ03]   C. Gkantsidis, M. Mihail, and E. Zegura. The Markov chain simulation method for generating connected power law random graphs. In *Proceedings of the 5th Workshop on Algorithm Engineering and Experiments (ALENEX)*, Baltimore, CA, pp. 16–25, 2003.

[Gon85]   T. Gonzalez. Clustering to minimize the maximum intercluster distance. *Theoretical Computer Science*, 38:293–306, 1985.

[GT88]   A.V. Goldberg and R.E. Tarjan. A new approach to the maximum flow problem. *Journal of the Association for Computing Machinery*, 35:921–940, 1988.

[Hås99]   J. Håstad. Clique is hard to approximate within $1 - \varepsilon$. *Acta Mathematica*, 182(1):105–142, 1999.

[HDBe05]   J.-D.J. Han, D. Dupuy, N. Bertin, M.E. Cusick et al. Effect of sampling on topology predictions of protein–protein interaction networks. *Nature Biotechnology*, 23(7):839–844, 2005.

[HGLR04]   C.T. Harbison, D.B. Gordon, T.I. Lee, N.J. Rinaldi et al. Transcriptional regulatory code of a eukaryotic genome. *Nature*, 431(7004):99–104, 2004.

[HHLM99]   L.H. Hartwell, J.J. Hopfield, S. Leibler, and A.W. Murray. From molecular to modular cell biology. *Nature*, 402(Suppl 6761):C47–C52, 1999.

[HKMT06]   K. Holzapfel, S. Kosub, M.G. Maaß, and H. Täubig. The complexity of detecting fixed-density clusters. *Discrete Applied Mathematics*, 154(11):1547–1562, 2006.

[HPS05]   S. Har-Peled and B. Sadri. How fast is the k-means method? *Algorithmica*, 41(3):185–202, 2005.

[HS00]   E. Hartuv and R. Shamir. A clustering algorithm based on graph connectivity. *Information Processing Letters*, 76(4–6):175–181, 2000.

[HSe06]   S. Hoops, S. Sahle, R. Gauges et al. COPASI—a COmplex PAthway SImulator. *Bioinformatics*, 22:3067–3074, 2006.

[IL03]   T. Ideker and D. Lauffenburger. Building with a scaffold: Emerging strategies for high- to low-level cellular modeling. *Trends in Biotechnology*, 21(6):255–262, 2003.

[IOSS02]   T. Ideker, O. Ozier, B. Schwikowski, and A.F. Siegel. Discovering regulatory and signalling circuits in molecular interaction networks. *Bioinformatics*, 18 (Suppl 1):S233–S240, 2002.

[JTA$^+$00] H. Jeong, B. Tombor, R. Albert, Z. N. Oltvai, and A.L. Barabasi. The large-scale organization of metabolic networks. *Nature*, 407(6804):651–654, 2000.

[K$^+$07]   S. Khan et al. Relative performance of mutual information estimation methods for quantifying the dependence among short and noisy data. *Physical Review E*, 76(2 Pt 2):026209, 2007.

[Kar04]   P.D. Karp. Call for an enzyme genomics initiative. *Genome Biology*, 5(8):401–401.2, 2004.

[KBDS93]   S. Kasif, S. Banerjee, A.L. Delcher, and G. Sullivan. Some results on the complexity of symmetric connectionist networks. *Annals of Mathematics and Artificial Intelligence*, 9:327–344, 1993.

[Kho06]   S. Khot. Ruling out ptas for graph min-bisection, dense k-subgraph, and bipartite clique. *SIAM Journal on Computing*, 36(4):1025–1071, 2006.

[KKTS06]   M. Koyuturk, Y. Kim, U. Topkara, and S. Subramaniam. Pairwise alignment of protein interaction networks. *Journal of Computational Biology*, 13(2):182–199, 2006.

[KL02]   D. Krznaric and C. Levcopoulos. Optimal algorithms for complete linkage clustering in d dimensions. *Theoretical Computer Science*, 286(1):139–149, 2002.

[KMLe04]   U. Karaoz, T.M. Murali, S. Letovsky, Y. Zheng et al. Whole genome annotation using evidence integration in functional linkage networks. *Proceedings of the National Academy of Sciences U S A*, 101:2888–2893, 2004.

[KYLe04]   B.P. Kelley, B. Yuan, F. Lewitter, R. Sharan et al. PathBLAST: A tool for alignment of protein interaction networks. *Nucleic Acids Research*, 32(Web Server issue), 2004.

[LLC$^+$08] I. Lee, B. Lehner, C. Crombie, W. Wong, A.G. Fraser, and E.M. Marcotte. A single gene network accurately predicts phenotypic effects of gene perturbation in *Caenorhabditis elegans*. *Nature Genetics*, 40(2):181–188, January 2008.

[Llo82]   S.P. Lloyd. Least squares quantization in PCM. *IEEE Transactions on Information Theory*, 28:129–137, 1982. Special issue on quantization.

[LMTK08]   K. Liolios, K. Mavromatis, N. Tavernarakis, and N.C. Kyrpides. The Genomes On Line Database (GOLD) in 2007: Status of genomic and metagenomic projects and their associated metadata. *Nucleic Acids Research*, 36(Database issue):D475–D479, 2008.

[LO02]   L. Lazzeroni and A. Owen. Plaid models for gene expression data. *Statistica Sinica*, 12:61–86, 2002.

[LSY06]   S. Lonardi, W. Szpankowski, and Q. Yang. Finding biclusters by random projections. *Theoretical Computer Science*, 368(3):217–230, 2006.

[LXTN06]   Z. Liang, M. Xu, M. Teng, and L. Niu. Comparison of protein interaction networks reveals species conservation and divergence. *BMC Bioinformatics*, 7:457, 2006.

[MNBe06]   A.A. Margolin, T. Nemenman, K. Basso, C. Wiggins et al. ARACNE: An algorithm for the reconstruction of gene regulatory networks in a mammalian cellular context. *BMC Bioinformatics*, 7 (Suppl 1), 2006.

[MO04]   S.C. Madeira and A.L. Oliveira. Biclustering algorithms for biological data analysis: A survey. *IEEE/ACM Transactions on Computational Biology and Bioinformatics*, 1(1):24–45, 2004.

[MR08]   T.M. Murali and C.G. Rivera. Network legos: Building blocks of cellular wiring diagrams. *Journal of Computational Biology*, 15(7):829–844, 2008.

[MRL95]   Y. Moon, B. Rajagopalan, and U. Lall. Estimation of mutual information using kernel density estimators. *Physical Review, E*, 52(3):2318–2321, 1995.

[MRS04]   N. Mishra, D. Ron, and R. Swaminathan. A new conceptual clustering framework. *Machine Learning Journal*, 56(1–3):115–151, 2004.

[MS04]   A.A. Melkman and E. Shaham. Sleeved coclustering. In *Proceedings 10th ACM SIGKDD International Conference on Knowledge Discovery and Data Mining*, Seattle, WA, pp. 635–640, 2004.

[MS07]   F. Markowetz and R. Spang. Inferring cellular networks—a review. *BMC Bioinformatics*, 8(Suppl 6):S5, 2007.

[MSOIe02]   R. Milo, S. Shen-Orr, S. Itzkovitz, N. Kashtan et al. Network motifs: Simple building blocks of complex networks. *Science*, 298(5594):824–827, 2002.

[MT83]   N. Megiddo and A. Tamir. Finding least-distances lines. *SIAM Journal on Algebraic Discrete Methods*, 2:207–211, 1983.

[MT07]   F. Markowetz and O.G. Troyanskaya. Computational identification of cellular networks and pathways. *Molecular Biosystems*, 3(7):478–482, 2007.

[MWK06]   T.M. Murali, C.-J. Wu, and S. Kasif. The art of gene function prediction. *Nature Biotechnology*, 12:1474–1475, 2006.

[NBH07]   W.S. Noble and A. Ben-Hur. Integrating information for protein function prediction. In *Bioinformatics—from Genomes to Therapies*, volume 3, pp. 1297–1314. Wiley, Weinheim, Germany, 2007.

[New06]   M.E.J. Newman. Modularity and community structure in networks. *Proceedings of the National Academy of Sciences*, 103(23):8577–8582, 2006.

[NJAe05]   E. Nabieva, K. Jim, A. Agarwal, B. Chazelle et al. Whole-proteome prediction of protein function via graph-theoretic analysis of interaction maps. *Bioinformatics*, 21 (Suppl 1):i302–i310, 2005.

[NK07]   M. Narayanan and R.M. Karp. Comparing protein interaction networks via a graph match-and-split algorithm. *Journal of Computational Biology*, 14(7):892–907, 2007.

[ORSS06]   R. Ostrovsky, Y. Rabani, L.J. Schulman, and C. Swamy. The effectiveness of Lloyd-type methods for the k-means problem. In *Proceedings of the 47th Annual IEEE Symposium on Foundations of Computer Science (FOCS'06)*, pages 165–176, 2006.

[PCTMe08]   L. Pena-Castillo, M. Tasan, C.L. Myers, H. Lee et al. A critical assessment of *Mus musculus* gene function prediction using integrated genomic evidence. *Genome Biology*, 9 (Suppl 1): S2, 2008.

[Pee03]   R. Peeters. The maximum edge biclique problem is NP-complete. *Discrete Applied Mathematics*, 131(3):651–654, 2003.

[PJAM02]   C.M. Procopiuc, M.T. Jones, P.K. Agarwal, and T.M. Murali. A Monte-Carlo algorithm for fast projective clustering. In *Proceedings of the International Conference on Management of Data*, Berkeley, CA, pp. 418–427, 2002.

[RKKe04]   R.J. Roberts, P. Karp, S. Kasif, S. Linn et al. An experimental approach to genome annotation. Report of a workshop organised by the American Academy of Microbiology, 2004.

[Rob04]   R.J. Roberts. Identifying protein function—a call for community action. *PLoS Biology*, 2(3):E42, 2004.

[SBA07]   N. Soranzo, G. Bianconi, and C. Altafini. Comparing association network algorithms for reverse engineering of large-scale gene regulatory networks: Synthetic versus real data. *Bioinformatics*, 23(13):1640–1647, 2007.

[SI06]   R. Sharan and T. Ideker. Modeling cellular machinery through biological network comparison. *Nature Biotechnology*, 24(4):427–33, 2006.

[SIKe04]   R. Sharan, T. Ideker, B.P. Kelley, R. Shamir et al. Identification of protein complexes by comparative analysis of yeast and bacterial protein interaction data. In *Proceedings of the 8th Annual International Conference on Computational Molecular Biology*, San Diego, CA, pp. 282–289, 2004.

[SKA05]   S. Seal, S. Komarina, and S. Aluru. An optimal hierarchical clustering algorithm for gene expression data. *Information Processing Letters*, 93(3):143–147, 2005.

[SM03]   V. Spirin and L.A. Mirny. Protein complexes and functional modules in molecular networks. *Proceedings of the National Academy of Sciences U S A*, 100(21):12123–12128, 2003.

[SMe03]   P. Shannon, A. Markiel, O. Ozier et al. Cytoscape: a software environment for integrated models of biomolecular interaction networks. *Genome Research*, 13(11):2498–2504, 2003.

[SOMMA02] S.S. Shen-Orr, R. Milo, S. Mangan, and U. Alon. Network motifs in the transcriptional regulation network of *Escherichia coli*. *Nature Genetics*, 31(1):64–68, 2002.

[SS05]   J. Schafer and K. Strimmer. An empirical Bayes approach to inferring large-scale gene association networks. *Bioinformatics*, 21(6):754–764, 2005.

[SSI05]   S. Suthram, T. Sittler, and T. Ideker. The plasmodium protein network diverges from those of other eukaryotes. *Nature*, 438(7064):108–112, 2005.

[SSKe05]   R. Sharan, S. Suthram, R.M. Kelley, T. Kuhn et al. From the cover: conserved patterns of protein interaction in multiple species. *Proceedings of the National Academy of Sciences U S A*, 102(6):1974–1979, 2005.

[SSR⁺06]   S. Suthram, T. Shlomi, E. Ruppin, R. Sharan, and T. Ideker. A direct comparison of protein interaction confidence assignment schemes. *BMC Bioinformatics*, 7:360, 2006.

[STdSe08]   M.P. Stumpf, T. Thorne, E. de Silva, R. Stewart et al. Estimating the size of the human interactome. *Proceedings of the National Academy of Sciences U S A*, 105(19):6959–6964, 2008.

[SUS07]   R. Sharan, I. Ulitsky, and R. Shamir. Network-based prediction of protein function. *Molecular Systems Biology*, 3:88, 2007.

[SWM05]   M.P.H. Stumpf, C. Wiuf, and R.M. May. Subnets of scale-free networks are not scale-free: Sampling properties of networks. *Proceedings of the National Academy of Sciences U S A*, 102(12):4221–4224, 2005.

[Tan08]   J. Tan. Inapproximability of maximum weighted edge biclique and its applications. In *Proceedings of the 5th Annual Conference on Theory and Applications of Models of Computation, Springer-Verlag Lecture Notes in Computer Science*, volume 4978, San Diego, CA, pp. 282–293, 2008.

[TCZZ07]   J. Tan, K.S. Chua, L. Zhang, and S. Zhu. Algorithmic and complexity issues of three clustering methods in microarray data analysis. *Algorithmica*, 48(2):203–219, 2007.

[TSKS04]   A. Tanay, R. Sharan, M. Kupiec, and R. Shamir. Revealing modularity and organization in the yeast molecular network by integrated analysis of highly heterogeneous genomewide data. *Proceedings of the National Academy of Sciences U S A*, 101(9):2981–2986, 2004.

[TSKS05]   A. Tanay, I. Steinfeld, M. Kupiec, and R. Shamir. Integrative analysis of genome-wide experiments in the context of a large high-throughput data compendium. *Molecular Systems Biology*, 1(1):msb4100005-E1-msb4100005-E10, 2005.

[TSS02]   A. Tanay, R. Sharan, and R. Shamir. Discovering statistically significant biclusters in gene expression data. *Bioinformatics*, 18 (Suppl 1):S136–S144, 2002.

[TSS06]   A. Tanay, R. Sharan, and R. Shamir. Biclustering algorithms: a survey. In *Handbook of Computational Molecular Biology*, pages 26–1. Chapman & Hall/CRC Press Computer and Information Science Series, Boca Raton, FL, pp. 26-1–26-17, 2006.

[US07]   I. Ulitsky and R. Shamir. Identification of functional modules using network topology and high-throughput data. *BMC Systems Biology*, 1:8, 2007.

[Wer06]   S. Wernicke. Efficient detection of network motifs. *IEEE/ACM Transactions on Computational Biology and Bioinformatics*, 3(4):347–359, 2006.

[Y⁺02]   H. Yu et al. Using Bayesian network inference algorithms to recover molecular genetic regulatory networks. In *Proceedings of the International Conference on Systems Biology*, 2002.

[YLLe04]   H. Yu, N.M. Luscombe, H.X. Lu, X. Zhu et al. Annotation transfer between genomes: protein–protein interologs and protein-dna regulogs. *Genome Research*, 14(6):1107–1118, June 2004.

[YLSKe04]   E. Yeger-Lotem, S. Sattath, N. Kashtan, S. Itzkovitz et al. Network motifs in integrated cellular networks of transcription-regulation and protein–protein interaction. *Proceedings of the National Academy of Sciences U S A*, 101(16):5934–5939, 2004.

[ZAA08]   J. Zola, M. Aluru, and S. Aluru. Parallel information theory based construction of gene regulatory networks. In *Proceedings of the 15th International Conference on High Performance Computing (HiPC)*, Bangalore, India, *Springer-Verlag Lecture Notes in Computer Science*, volume 5374, pp. 336–349, 2008.

[ZGL03]   X. Zhu, Z. Ghahramani, and J. Lafferty. Semi-supervised learning using Gaussian fields and harmonic functions. In *Proceedings of the 20th International Conference on Machine Learning*, Washington, D.C., 2003.

[ZKWe05]   L.V. Zhang, O.D. King, S.L. Wong, D.S. Goldberg et al. Motifs, themes and thematic maps of an integrated *Saccharomyces cerevisiae* interaction network. *Journal of Biology*, 4(2):6, 2005.

[ZSA08]   M. Zampieri, N. Soranzo, and C. Altafini. Discerning static and causal interactions in genome-wide reverse engineering problems. *Bioinformatics*, 24(13):1510–1515, 2008.

33

Pricing Algorithms for Financial Derivatives

Ruppa K. Thulasiram
University of Manitoba

Parimala Thulasiraman
University of Manitoba

33.1 Introduction

Computational finance is a cross-disciplinary area that relies on mathematics, statistics, finance, and computational algorithms to make critical decisions. One of the core tasks in this area is to analyze and measure the risk component that a financial portfolio would create. Portfolio would generally comprise of stocks, bonds, and other instruments, such as derivatives. Derivatives are financial instruments that depend on some other assets, such as stocks. They are also referred commonly as options.

An option is a contract in which the buyer (holder) has the right but no obligation to buy (call) or sell (put) an underlying asset (for example, a stock) at a predetermined price (strike price) on or before a specified date (expiration date). The seller (also known as writer) has the obligation to honor the terms specified in the contract (option). The holder pays an option premium to the writer [44].

Option pricing is one of the fundamental problems in finance which has led to two Nobel prizes to be awarded. In 1998, Myron S. Scholes and Robert C. Merton received the Nobel prize for the Black–Scholes–Merton (BSM) model [6]. Recently, Charles Engle received a Nobel prize in 2003 for his autoregressive conditional heteroskedasticity (ARCH) model [32]. There are a multitude of financial instruments traded on the markets. The class of instruments known as American options are claims to payoffs at a time chosen by the holder of the security. These contracts give the holder the right to exercise the option prior to the expiration date in order to achieve maximum profit. Therefore, determining an optimum exercise policy is a key issue in pricing American securities. The solution for the optimal exercise policy must typically be performed numerically, and is usually a computationally intensive problem.

There are many numerical techniques proposed in the literature for option pricing, such as binomial lattice approach [68] for the asset price and then roll backward the tree checking at each node if early exercise is optimal. Binomial lattice approach is one of the traditionally used computing models [23] in the finance community due to its intuitiveness and the ease of implementation. Another traditional model is the Monte-Carlo (MC) [10] simulation of price movements. This is done by evaluating the option value from many simulated trajectories of the asset price and taking the average. The accuracy of the final price depends on the number of simulations performed. These trajectory calculations are independent of each other and hence lead to embarrassingly parallel situations (for example, see [66]). Some of the computational fluid dynamics techniques, such as finite-differencing [31,67] and finite-element methods [75], are emerging as important techniques for option pricing problem by solving the Black–Scholes and other models manifested as partial differential equations (PDE). Recently, fast Fourier transform (FFT) has been shown [16] to be suitable for option pricing problems. Also, it is shown [24] that FFT method can be used to study multifactor models for this problem and shown to be quite effective in terms of computational times when compared to other numerical techniques, such as MC and PDE approaches.

Economic and financial models used for evaluation and forecasting purposes typically lead to large dynamic, nonlinear problems that have to be solved for a certain time span. In particular, many of the problems in finance demand efficient algorithms and high-performance computing capabilities [2,74]. All the techniques mentioned earlier are suitable for parallel computing. However, parallel computing is not yet widely popular in the field of finance, though there is recognition for its need. The prices and the price dynamics of goods and financial securities the investor purchases determine the investor's overall budget. An investor has many assets to choose from over many periods and the market dynamics determine the price of each asset. Computational requirement under such conditions are large and when there are market frictions the computational requirements increase dramatically [41]. An important aspect of the portfolio optimization problem is the objective function that represents the investor's preferences. The need of investors to manage the risks associated with their portfolios, vis-à-vis, their liabilities also prompted the development of mathematical models for portfolio management. These models trace their roots to Markowitz's mean-variance optimization models [40]. Academics and practitioners are turning increasingly to mathematical models and computer simulations in order to understand the peculiarities of financial markets and to develop risk-management tools. One of the notable models is the highly celebrated Black–Scholes model for option pricing [6]. Immense computing power available now need to be harnessed in an effective manner to address the ever-increasing demand from the practitioners, for example, for solving the complex mathematical models for the risk management through portfolio adjustments. However, more and more models are developed these days for pricing the risks of equities, stocks, bonds, derivatives of these instruments without much attention to their solubility or tractability. The need to compute the solutions for such models dictates the design and development of efficient algorithms that not only solve the problem in hand but also try to utilize the supercomputing power of the present day. In other words, parallel algorithms are in great demand for financial problems, especially for long-dated options with many underlying assets. Many of the computational finance problems are shown to be well suited for solution on parallel architectures of many kinds [21,51]. However, the adoption of parallel computing to the modern day finance has been very modest [74]. Moreover, volatility of the market place demands the processing of information quickly and accurately in order to react in a very short time to avoid potential huge losses.

This chapter introduces the readers to the field of finance and algorithmic research opportunities therein. Though there are several problems in finance, such as portfolio optimization, calibration, and marking-to-the-market, this chapter focuses on option pricing problem, which forms the backbone of the above problems. In the next section, the basic idea of options is introduced followed by some definitions and explanations through numerical examples. Complex or exotic options and their solution techniques are introduced next. In Section 33.2 the design and the development of parallel algorithms for binomial lattice approach to solve the option pricing problem are discussed. In Section 33.3 the design and the development of parallel algorithms for quasi-Monte-Carlo (QMC) simulation to solve the option pricing problem are discussed. In Section 33.4 the design and the development of parallel algorithms for FFT technique to solve the option pricing problem are discussed. Some fundamental ideas of evolutionary algorithms for the option pricing problem are discussed in Section 33.5 followed by a list of keywords.

33.2 Binomial Lattice Algorithm for Pricing Options

This section discusses the design and the development of parallel algorithms for option pricing problem using the binomial lattice approach.

33.2.1 One-Step Binomial Method

One of the early models for option pricing is the Nobel prize winning BSM model [6] that is based on efficient market hypothesis. Black–Scholes developed a model to alleviate risk involved in financial investments. Merton [54] augmented it with stochastic calculus resulting in a stochastic PDE for the option. This model is valid for simple European options and a major assumption was that underlying stock would have constant volatility. A closed-form solution is available only for simplified BSM model with many assumptions, especially since numerical techniques for solving PDE were not popular to the finance community. This scenario changed in 1979 when discrete time approach was proposed by Cox–Ross–Rubenstein (CRR) [23]. The CRR binomial model is a simple and intuitive numerical technique developed for pricing options. The binomial option pricing model has proved over time to be the most flexible, and popular approach to price options. If constructed assuming the same initial conditions, binomial model agrees asymptotically with the BSM model. Moreover, binomial model can be used for pricing American style options. The standard binomial option pricing model assumes that the binomial tree is recombining with constant volatility, constant risk-less return, and constant payout return. However, these could be relaxed unlike the BSM Model.

Binomial model uses a binomial tree (Figure 33.1) structure to price options. The contract period between valuation (current) date and expiration date is divided into a certain number of time steps. Each node in the tree represents a possible price of the stock (underlying asset) at a particular time. In binomial method, knowing asset price is the basis for computing option value. The valuation of the binomial tree is iterative; it starts from the leaf nodes and works backwards toward the root node, which represents the valuation date.

In Figure 33.1, the interest is in valuing a European Call Option for current date whose underlying stock is priced at $20. At the end of the preset period, say 3 months, since the stock price is $30, and the strike price is $21, the value of the option is $9; whereas if the stock price at the end of 3 months had been $15 the value of the option is $0.

In deriving the one-step binomial model (refer [9,23,59]), a portfolio of stocks and options is set up in such a way that there is no uncertainty about the value of the portfolio at the end of the life of the derivative, say, three months. This amounts to assuming that the option is risk-free

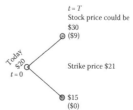

FIGURE **33.1** One-step binomial model.

and arbitrage-free. That is, since the portfolio has no risk, the return earned on it must equal the risk-free interest rate, which is the case with the BSM Model. This assumption would enable us to determine the option value as well as the cost of setting up the portfolio.

Consider a portfolio with a long position in Δ shares of stocks and a short position in one call option. First calculate the value of Δ that makes the portfolio risk-less. If the stock price goes up to $30 from the current price of $20, value of the share is 30Δ and value of the option is $9 (since strike price is $21). Therefore, the total value of the portfolio is $30\Delta - 9$. If the stock price goes down to $15 from the current price of $20, the value of the share is 15Δ and the value of the option is $0. Therefore, the total value of the portfolio is 15Δ. The portfolio will become risk-neutral if the value of Δ is chosen such that the value of the portfolio remains the same in both cases, i.e., $30\Delta - 9 = 15\Delta$. The risk-less portfolio, therefore, comprises of 0.6 shares of stock in long position and one short position of option. If the stock price goes up, the value of the portfolio is: $30 * 0.6 - 9 = 9$. If the stock price goes down, the value of the portfolio is: $15 * 0.6 = 9$. That is, regardless of the behavior of the stock price, value of the portfolio remains same at the end of the life of the derivative. This implies risk-neutral portfolios must, in the absence of arbitrage, earn the risk-free rate of interest. If the risk-free interest rate is 12% per annum, then the value of the portfolio today must be the present value (also known as discounted value) of 9. That is, using the growth rate formula [44], $9e^{-rt} = 9e^{-0.12*0.25} = 8.734$. Therefore, for a guaranteed return value of $9 at the end of 3 months, the investment today should be $8.734.

33.2.1.1 Option Value

If the stock price today is known to be $20, denoting the option price by f, the value of the portfolio today is, therefore, $20 * 0.6 - f = 8.734$ and $f = 3.266$. That is, when arbitration opportunity is excluded, the current value of the option must be $3.266. If the option value is larger than this amount, the portfolio would cost less than $8.734 to set up and would earn more than the risk-free interest rate.

33.2.1.2 Generalized One-Step Binomial Model

Generalizing these statements, the one-step binomial model of finding the current option value can be formulated as follows [23,73]:

Let the terms S and f represent the current stock price and derivative price. During the life of the derivative, stock price S can (1) either move up from S to a new level Su ($u > 1$); or (2) move down from S to a new level Sd ($d < 1$). The proportional increase in S when the up movement is $(u - 1)$ and the proportional decrease in S when the down movement is $(1 - d)$. When the stock price moves up from $S \longrightarrow Su$, we suppose that the pay-off from derivative is f_u and when it goes down from $S \longrightarrow Sd$, it is f_d (Figure 33.2).

First we find a closed form formula for the Δ defined before using these general terms. If the stock price goes up from $S \longrightarrow Su$, the value of the portfolio at the end of the life of the derivative is $Su\Delta - f_u$ and likewise for the other possibility it is $Sd\Delta - f_d$. For the risk-neutral portfolio, therefore, $Su\Delta - f_u = Sd\Delta - f_d$ or $S(u - d)\Delta = f_u - f_d$ or

$$\Delta = \frac{f_u - f_d}{S(u - d)} \qquad (33.1)$$

Emphasizing again, this would result in risk-neutral portfolio earning at least the risk-free interest rate. Equation 33.1 is the ratio of the change in the derivative price to the change in the stock price at time T. Let r be the risk-free

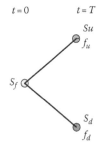

FIGURE 33.2 Generalized one-step binomial model.

interest rate. Using the growth rate formula, the present value of the portfolio must be $(Su\Delta - f_u)e^{-r\Delta t}$. Since the cost of setting up the portfolio is $S\Delta - f$, for a risk-less portfolio $S\Delta - f = (Su\Delta - f_u)e^{-r\Delta t}$. Substituting Equation 33.1 and manipulating one can get

$$f = e^{-r\Delta t}\left(pf_u + (1-p)f_d\right) \tag{33.2}$$

where

$$p = \frac{e^{r\Delta t} - d}{u - d} \tag{33.3}$$

Equations 33.2 and 33.3 comprise the one-step binomial model in pricing a derivative. The expression for p given in Equation 33.3 is interpreted as the probability of an up movement in the stock price. The expression $\left(pf_u + (1-p)f_d\right)$ in Equation 33.2 is referred to as the expected payoff from the derivative.

33.2.2 Two-Step Binomial Trees

Following the same steps as mentioned earlier, the two-step binomial model [23] can be described. At each step, let us assume for simplicity, the increase or the decrease of stock price is at the same rate, say 10% (Figure 33.3).

At each step the stock price either goes up by u times the initial value or goes down by d times the initial value. Applying Equation 33.2 repeatedly, we get the following relations:

$$f_u = e^{-r\Delta t}\left(pf_{uu} + (1-p)f_{ud}\right) \tag{33.4}$$

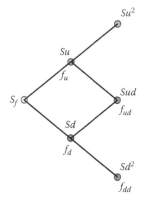

$$f_d = e^{-r\Delta t}\left(pf_{ud} + (1-p)f_{dd}\right) \tag{33.5}$$

$$f = e^{-r\Delta t}\left(pf_u + (1-p)f_d\right) \tag{33.6}$$

Substituting Equations 33.4 and 33.5 into Equation 33.6 one can get

FIGURE 33.3 General two-step binomial model.

$$f = e^{-2r\Delta t}\left(p^2 f_{uu} + 2p(1-p)f_{ud} + (1-p)^2 f_{dd}\right) \tag{33.7}$$

33.2.2.1 Numerical Example

Figure 33.4 describes the two-step binomial method in detail. Node A is current time. Nodes B and C represent all the possible prices of the stock down the road for 3 months. Similarly, nodes D, E, and F represent possible prices the stock would take, 3 months from B or C. The generation of a path consisting of possible prices in the future is known as downsizing operation. Tracking back to root node is known as upswing operations. The periods (Δt) are assumed to be equal, for simplicity. Also for simplicity, we assume that the up or the down movement of the stock price is at a constant rate, say 10% (i.e., $u = 1.1$ and $d = 0.9$). (These nodes could be located at different intervals. Similarly, the increase or decrease in the stock price need not be at the equal proportion as assumed for this example.)

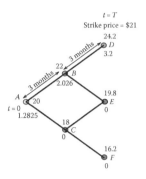

FIGURE 33.4 Numerical example of a two-step binomial model.

At D, E, and F (leaf nodes) the option prices are nothing but the local pay-off from the derivative. They can be directly calculated from the strike price and the speculated stock price. Therefore, at D, the *option price = stock price − strike price* = 24.2 − 21 = $3.2. At nodes E and F since the *stock price < strike price*, the option value at these nodes is $0. Also at C, the option price is $1. At B during the downswing, a first calculation suggests the option value to be $0 (since *stock price*($22) − *strike price*($21) = 1). Assuming a 10% increase or decrease (i.e., $u = 1.1$ and $d = 0.9$), at B, p can be calculated using Equation 33.3 to be $p = 0.6523$. Therefore, the option value at B can be calculated using Equation 33.2 as 2.026 in the upswing. If the assumption had been an American option, it would be undesirable to exercise the option at node B during the first pass, since the option value at node B is zero; whereas waiting for the second pass suggests a better option value. It is to be noted here that the block of nodes B, D, and E comprises what is required for a one-step binomial model, and hence the corresponding formula can be applied.

Having determined the stock prices and the option values at B and C, the option value at A can be calculated to be 1.2825 again by applying the one-step binomial formula given by Equation 33.2. Recall that the binomial tree is constructed in such a way that the up and down movements are at the same rate at each node and that the time steps are also of equal length. This assumption led to a constant p at each node in this example. These restrictions are relaxed in a realistic analysis of portfolios.

33.2.3 Binomial Lattice Pricing Algorithm

The recursive algorithm described earlier proceeds in two phases. First, lattice is constructed. Second, work backward through the lattice from time T in order to compute the present value of the option. (Mathematical convergence of the option price computed under the binomial method to the true option price is discussed in [30].) To compute the binomial lattice, first divide the time interval $[0, T]$ into L smaller intervals, each of length $\Delta t = \frac{T}{L}$. Over each subinterval, the asset price is assumed to move up from value S to Su, or down to Sd, with probabilities p and $1 − p$, respectively, as shown in Figure 33.3. It can be shown [73] that if $u = e^{\sigma\sqrt{\Delta t}}$, $d = e^{-\sigma\sqrt{\Delta t}}$, and $p = \frac{e^{r\Delta t} - d}{u - d}$ (where r is the current interest rate and σ is the volatility of the underlying asset) then over each time interval, the mean and the variance of asset price movements will match the mean and the variance of the continuous-time risk-neutralized asset price process. In the second phase of the algorithm, work backward from the leaf nodes at time T to compute the option price. At the expiration time T and leaf-node i, the value of the option is given by $F(S_i, T) = \max[0, K − S_i]$, K is the strike price. At time $T − 1$ and node j, the value of the option is given by $F(S_j, T − 1) = \max[K − S_j, (pF(S_j u, T) + (1 − p)F(S_j d, T))e^{-r\Delta t}]$, the greater of the values under immediate exercise of the option, $K - S_j$, or the expected value of holding the option for another period, $(pF(S_ju, T) + (1 - p)F(S_jd, T))e^{-r\Delta t}$.

33.2.4 Binomial Lattice Algorithm for Pricing Complex Options

The Modifications of algorithms described in the previous section to complex options, such as options on dividend-paying assets with focus only on the nature of these options, are discussed in this section. The computational complexity of the algorithms changes and the cause for this additional computational complexity of the algorithms is discussed.

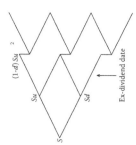

FIGURE 33.5 Binomial tree adjusted for a known discrete proportional dividend.

The occurrence of a dividend payment during the life of the option reduces the price of the asset on the ex-dividend date as shown in Figure 33.5. An ex-dividend date is the date set by the company (generally few days before the actual dividend date) to pay dividend to the share holders of the stock.

Consider three different situations as described briefly in the following.

33.2.4.1 Continuous Dividend Yield

If the underlying asset pays a continuous dividend yield at a rate of δ per unit time, then the data generating process in the Black–Scholes [44] model are given by

$$dS = (r - \delta)Sdt + \sigma SdB_t$$

where dB_t is the general Wiener process. As can be realized from this small change, one can price these options as the plain American puts and calls, with the difference in the formulas for the jump sizes and the probabilities. Hence, the class of options is a very simple variation of the vanilla options.

33.2.4.2 Known Discrete Proportional Dividend

If there is a one-time proportional dividend payment of δ during the life of the option that occurs at time τ, then the price at all the nodes of the binary tree preceding the dividend payment date will be $Su^j d^{i-j}$ (for $i\Delta t < \tau$) and $(1 - \delta)Su^j d^{i-j}$ for all times after the dividend payment date (i.e. for $i\Delta t \geq \tau$).

To price an option of this type, a binary tree with a one time jump of the tree structure is used. At the node i where $(i - 1)\Delta t \leq \tau < i\Delta t$ the input value is changed to $(1 - \delta)Su$ from Su to reflect the dividend date. Correspondingly, the formulas for the probabilities of up and down movements and the jump sizes are also modified. Hence, the natural load of computation before and after the ex-dividend date changes slightly from the earlier option.

33.2.4.3 Known Discrete Cash Dividend

It is a more plausible assumption that the dividend is a known dollar amount rather than a percentage yield. Therefore, the case of a known discrete cash dividend is considered. The problem here is that the cash dividend changes the structure of the binary tree. Suppose there is a one time dividend payment at time τ where $i\Delta t \leq \tau < (i + 1)\Delta t$. At time $(i + 1)\Delta t$, the price of the asset will decrease by the amount of the dividend payment, D. This will make the tree at the subsequent nodes non-recombining, what was originally recombining.

One way to cope with the non-recombining tree is to use the recursive algorithm. Another is to consider the asset price S as composed of two parts, certain (the discounted dividend payments) and uncertain $S*$, where $S* = S$ for $i\Delta t \geq \tau$ and $S* = S - De^{-r(\tau - i\Delta t)}$ for $i\Delta t \leq \tau$. One can model

$S*$ instead of S, and thus obtain a recombining tree. We can construct the tree for S from the tree for $S*$ and thus obtain the value of the option.

33.2.5 Parallel Pricing Algorithms for Single Underlying Assets

Pricing algorithms are described here for options with single asset and extension to multidimensional derivatives. These algorithms [68,69] are for the base CRR model [23] for option pricing. This discretized approach [23] approximates the BSM model of option pricing to a large extent by representing the asset price movement in a lattice and is easy to implement for experimentation.

33.2.5.1 Parallel Recursive (Non-Recombining) Algorithm

Let P be the number of processors. In the recursive algorithm, processor 0 is delegated as the master processor. Computations start at the first processor at the first level ($L = 0$), which corresponds to time $t = 0$. A non-recombining tree is built until the number of leaf nodes equals the number of available processors. Each leaf node is assigned to a distinct processor (one of them could be a master processor itself). Then each processor starts building its own tree. All processors are working independently on their subtrees without any information interchange. As soon as a processor creates a tree with a certain depth ($= L - \log_2 P$), it starts working backward and computes option price in every node. Each processor generates the tree and computes the option price recursively. At the end of the computation, each processor sends its final value to the master processor. The master processor upon receiving the values corresponding to all leaf nodes from the individual processors, proceeds with computing the option values recursively until the initial node at $t = 0$. The recursive algorithm utilizes a non-recombining binomial lattice, that is, $Sud = Sdu = S$ is not imposed. The sheer number of nodes created in a non-recombining lattice (2^L) restricts the usefulness of the recursive algorithm to short-dated options. When L is large (to accurately price options with long times to expiration (large T)), the non-recombining lattice algorithm is inefficient. Therefore, a recombining tree is created to reduce the number of nodes by utilizing $Sud = Sdu = S$. The recombining algorithm is explained in the following text.

33.2.5.2 Parallel Iterative (Recombining) Algorithm

This algorithm starts from level L. The number of levels in the tree and the number of processors are assumed to be power of two. The number of leaf nodes is always equal to the number of levels plus one ($L + 1$). All leaf nodes are evenly distributed among the processors but the last processor receives an additional node. Initially, the option prices at the leaf nodes are calculated by finding the difference between a possible stock price and strike (exercise) price, ($S_i - K$), which is the same as the local pay-off at these nodes. Every processor i, except the processor 0, sends the value of its boundary node to processor ($i - 1$). (Higher numbered processor sends data to the lower numbered neighbors.) In a given processor, for every pair of adjacent nodes at a certain level L_i, the processor computes option price for pair's parent node, which is at level ($L_i - 1$). The computation proceeds to the previous level L_{i-1} and option price computation is repeated with the additional exchange of boundary node values. Eventually the processor 0 computes the option price at level $L = 0$.

At each time step t, both algorithms operate in two modes simultaneously: communication and computation. In the communication mode, adjacent processors exchange data on option values. Processor 0 only receives data from processor 1. Processors 1 through ($P - 1$) receive data from the processor's higher numbered neighbor and send data to their lower-numbered neighbor. P only sends data to processor ($P - 1$). By forcing the lattice to recombine (imposing the condition that $Sud = Sdu = S$), the step of constructing the lattice is eliminated. The stock prices are instead calculated "on-the-fly" with proper indexing scheme. Computation at each node thus involves calculating the stock price and the option value. Based on the number of up and down movements from the initial node, asset price is calculated. This calculation only requires the node and time indices. For example, the stock price at node j and time L_i is calculated as $S_j^{L_i} = S_0 * u^{L_i - 2*(L_i - j)}$, for a recombining tree where $u = 1/d$.

33.2.6 Parallel Pricing Algorithm for Multiple Underlying Assets

The algorithms described earlier are for options with single underlying asset. As mentioned earlier, parallel algorithms are in great demand for financial problems, especially for long-dated options with many underlying assets. Financial derivatives with ten underlying assets are very common in the market place. One can easily extend the above algorithms for derivatives that have multiple assets as underlying components. This problem gives rise to immense effort in analyzing the effect of one of the underlying assets onto the other and the overall effect on the derivative itself. This leads to multi-dimensional analysis and in the simplest unoptimized form would require ten times more computing time.

As time periods and the number of assets increase, the interactions among these assets would modify the computational pattern and this would further increase the computational complexity for such problems. The problem becomes computationally intractable due to many real-world constraints to be satisfied in pricing such derivatives. One of the other effects is on managing the portfolio [41], for which the pricing of these derivatives form a fundamental problem that need to be solved accurately and expediently.

In experimental studies the number of assets is fixed for a given option, for simplicity. This helps us to keep the computation time step constant for all the assets once the number of periods L into which the total period $[0, T]$ is divided into is fixed. This results in having the same tree structure for all the assets. As explained in the single asset problem, each processor receives a leaf node from the master processor upon which the processor creates a subtree recursively. Similarly, for i assets, each processor is allocated i leaf nodes by the master processor. The processor expands on these leaf nodes and creates a subtree for each of the leaf nodes.

33.2.7 Theoretical Analysis

33.2.7.1 Recombining Algorithm

With an example binomial tree in Figure 33.6 with eight levels and four processors, the number of computations required for option pricing is analyzed in this section and then extended to larger trees with L levels and P processors.

33.2.7.1.1 Total Computations

The number of computations performed by each processor excluding the leaf nodes has to be determined. In the example, the longest path from any of the three leaf nodes in P_3 ends at level 6.

The number of nodes allocated to this processor initially is $L/P + 1$ (=3). Each of these nodes has a parent node at level 7. One of the two nodes at level 7 is boundary nodes. In the algorithm, the communication is assumed to take place from higher numbered processors to lower numbered neighboring processors. Therefore, the boundary node is created in P_3. Similarly, these two nodes at level 7 have parent node at level 6, which is also a boundary node local to the processor. Note that all these nodes form a triangle starting at level 6. These are the nodes computed by processor P_3 locally. The number of computation, therefore, is $1 + 2 + 3 + \cdots + L/P = (L/P) * (L/P + 1)/2$.

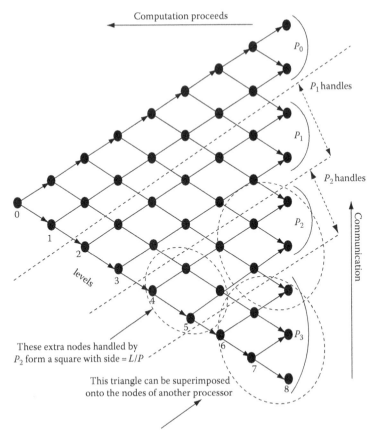

FIGURE 33.6 Computational complexity—example with eight levels and four processors.

The longest path from the leaf nodes of processor P_2 ends at level 4. Using the aforementioned technique of triangles the total number of computations performed by processor P_2 can be determined.

By superimposing the structure of local nodes of processor P_3 (represented as a triangle) (that is, the first node (counting from the bottom) at level 6 can be superimposed onto the third node at the same level). Most of the nodes handled by processor P_2 can be covered except for the square (actually a rhombus in the figure) at the bottom (consisting of first node at level 4, first and second nodes at level 5, and second node at level 6). The number of nodes on each side of this square is equal to L/P and hence, the total number of nodes in this square is $(L/P)^2$. Therefore, processor (P_2) has $(L/P) * (L/P + 1)/2 + (L/P)^2$ total nodes for computations.

Therefore, the number of total computations for P_i, $i : 0 \longrightarrow P - 1$ is $(L/P)(L/P + 1)/2 + (P - i - 1)(L/P)^2$.

33.2.7.1.2 Communication

Recall that at each level, the higher numbered processor sends a data value to the neighboring lower numbered processor. Therefore, P_0 performs no remote communications. Processor P_3 sends $(L/P + 1)$ data values to processor P_2 and processor P_2 in turn sends $(L/P + 1) + L/P$ data values to processor P_1, etc. In general, processor P_i communicates $1 + (L/P \times (P - i))$ data values to P_{i-1}.

Therefore, total number of communication is $\sum_i^{P-1} (1 + L/P \times (P - i)) = (P - 1)(1 + L/2)$

33.2.7.2 Non-Recombining Algorithm

In this algorithm, the master processor creates P leaves (for each asset in the case of the multi-asset derivative) and stores them in an array and distributes this array to distinct processors. The processors create their own subtree with the leaf nodes (of the master processor) as the root and perform the computations. At the end of the local computations, the processors send back the computed option values to the master processor in an array (for each of the asset tree in the case of the multi-asset derivative). Therefore, there are P sends and P receives by the master processor totaling $2P$ communications. The number of levels created by the master processor is $\log_2 P$. Therefore, to initially reach set number of levels L, each processor creates a subtree with $(L - \log_2 P)$ levels. The number of computations per processor is $2^{(L - \log_2 P + 1)} - 1$. Therefore, the total number of computations in the algorithm is $2^{L+1} - 1$.

33.3   Monte-Carlo Simulation

MC simulation is a widely used method for pricing financial options due to the absence of straightforward closed form solutions for many financial models. Since Boyle [7] introduced the MC simulation in pricing option in 1977, the literature in this area has grown very rapidly. For example, Hull and White [43] employed MC method in stochastic volatility application and obtained more accurate results than using BSM model [6,54]; the latter often overprices options about 10% and the error is exaggerated as the time to maturity increases. Schwartz and Torous [62] use MC method to simulate the stochastic process of prepayment behavior of mortgage holders and the results matched closely to that actually observed. Fu [34] gives introductory details concerning the use of MC simulation techniques for options pricing. In spite of the prevailing belief that American-style options cannot be valued efficiently in a simulation model, Tilley [71], Grant et al. [38], as well as Broadie and Glasserman [12] and some others have proposed MC methods for American-style options and obtained acceptable results. Examples about valuing exotic options can be found in [46]. The literature on MC methods in valuing options keeps growing. More examples can be found in [8,36]. The traditional MC methods have been shown to be a powerful and a flexible tool in computational finance [58].

While the traditional MC methods are widely applied in option pricing, however, their disadvantages are well known. In particular, for some complex problems that require a large number of replications to obtain precise results, a traditional MC method using pseudo-random numbers can be quite slow because its convergence rate is only $O(N^{-1/2})$, where N is the number of samples. Different variance reduction techniques have been developed for increasing the efficiency of the traditional MC simulation, such as control variates, antithetic variates, stratified sampling, Latin hypercube sampling, moment matching methods, and importance sampling. For detail about these techniques, please refer to [36]. Another technique for speeding up the MC methods and obtaining more accurate result is to use low discrepancy (LD) sequences instead of random sequences. The use of LD sequences in MC method leads to what is known as QMC method. The error bounds in QMC methods are in the order of $(\log N)^d \cdot N^{-1}$ where d is the problem dimension and N is the number of simulations.

Birge [5] reported how QMC method can be used in option pricing in 1994 and demonstrated improved estimates through both analytical and empirical evidences. In 1995, Paskov and Traub [57] performed tests about two LD algorithms (Sobol and Halton) and two randomized algorithms (classical MC and MC combined with antithetic variables) on Collateralized Mortgage Obligation (CMO). They obtained more accurate approximations with QMC methods than with traditional MC methods and concluded that for this CMO the Sobol sequence is superior to the other algorithms. Acworth et al. [1] compared some traditional MC methods and QMC sequences in option pricing and drew a similar conclusion. Boyle et al. [8] also found that QMC outperforms traditional MC and Sobol sequence outperforms other sequences. Galanti and Jung [35] used both pseudo-random sequences and LD sequences (Sobol, Halton and Faure) with MC simulations to value some complex options and demonstrated that LD sequences are a viable alternative to random sequences and the Sobol sequence exhibits better convergence properties than others. The MC and QMC methods for option pricing are discussed in this section.

33.3.1 Monte Carlo and Quasi-Monte Carlo Methods

In general, MC and QMC methods are applied to estimate the integral of function $f(x)$ over the $[0, 1]^d$ unit hypercube where d is the dimension of the hypercube,

$$I = \int_{[0,1]^d} f(x)\, dx \tag{33.8}$$

In MC methods, I is estimated by evaluating $f(x)$ at N independent points randomly chosen from a uniform random distribution over $[0, 1]^d$ and then evaluating the average

$$\hat{I} = \frac{1}{N} \sum_{i=1}^{N} f(x_i) \tag{33.9}$$

From the law of large numbers, $\hat{I} \to I$ as $N \to \infty$. The standard deviation is

$$\sqrt{\frac{1}{N-1} \sum_{i=1}^{N} (f(x_i) - I)^2} \tag{33.10}$$

Therefore, the error of MC methods is proportional to $1/\sqrt{N}$.

QMC methods compute the aforementioned integral based on LD sequences. The elements in a LD sequence are uniformly chosen from $[0, 1]^d$ rather than randomly. The discrepancy is a measure to evaluate the uniformity of points over $[0, 1]^d$. Let $\{q_n\}$ be a sequence in $[0, 1]^d$, the discrepancy D_N^* of q_n is defined as follows, using Niederreiter's notation [55].

$$D_N^*(q_n) = \sup_{B \in [0,1)^d} \left| \frac{A(B, q_n)}{N} - v_d(B) \right| \tag{33.11}$$

where B is a sub cube of $[0, 1)^d$ containing the origin, $A(B, q_n)$ is the number of points in q_n that falls into B, and $v_d(B)$ is the d-dimensional Lebesgue measure of B. The elements of q_n is said to be uniformly distributed, if its discrepancy $D_N^* \to 0$ as $N \to \infty$. From the theory of uniform distribution sequences [48], the estimate of the integral using a uniformly distributed sequence $\{q_n\}$ is $\hat{I} = \frac{1}{N} \sum_{n=1}^{N} f(q_n)$, as $N \to \infty$ then $\hat{I} \to I$. The integration error bound is given by the Koksman–Hlawka inequality:

$$\left| I - \frac{1}{N} \sum_{n=1}^{N} f(q_n) \right| \le V(f) D_N^*(q_n) \tag{33.12}$$

where $V(f)$ is the variation of the function in the sense of Hardy and Krause (see [48], which is assumed to be finite). The inequality suggests that a smaller error can be obtained by using sequences with smaller discrepancy. The discrepancy of many uniformly distributed sequences satisfies $O((\log N)^d/N)$. These sequences are called LD sequences [55]. Inequality 33.12 shows that the estimates using a LD sequence satisfy the deterministic error bound $O((\log N)^d/N)$.

33.3.2 Monte Carlo Simulations for Option Pricing

Under the risk-neutral measure, the price of a fairly valued European call option is the expectation of the payoff $E[e^{-rT}(S_T - K)^+]^*$. In order to compute the expectation, Black and Scholes [6,54] modeled the stochastic process generating the price of a non-dividend-paying stock as geometric Brownian motion:

$$dS_t = \mu S_t\, dt + \sigma\, S_t\, dW_t \tag{33.13}$$

where W is a standard Wiener Process, also known as Brownian motion. Under the risk-neutral measure, the drift μ is set to r, the risk-free interest rate.

To simulate the path followed by the asset price S, suppose the life of the option has been divided into n short intervals of length Δt ($\Delta t = T/n$), the updating of the stock price at $t+\Delta t$ from t is [44]:

$$S_{t+\Delta t} - S_t = rS_t\Delta t + \sigma S_t Z\sqrt{\Delta t} \tag{33.14}$$

where Z is a standard random variable, i.e., $Z \sim (0,1)$. This enables the value of $S_{\Delta t}$ to be calculated from initial value S_0 at time Δt, the value at time $2\Delta t$ to be calculated from $S_{\Delta t}$, and so on. Hence, a completed path for S has been constructed.

In practice, in order to avoid discretization errors, it is usual to simulate the logarithm of the asset price, $\ln S$, rather than S. From Itô's lemma (see [73]), the process followed by $\ln S$ of Equation 33.14 is

$$d \ln S = \left(r - \frac{\sigma^2}{2}\right) dt + \sigma\, dz \tag{33.15}$$

so that

$$\ln S_{t+\Delta t} - \ln S_t = \left(r - \frac{\sigma^2}{2}\right) dt + \sigma Z\sqrt{\Delta t} \tag{33.16}$$

or equivalently

$$S_{t+\Delta t} = S_t \exp[(r - \sigma^2/2)\Delta t + \sigma\sqrt{\Delta t}Z]. \tag{33.17}$$

Substituting independent samples $Z_1, \ldots, Z_n$ from the normal distribution into Equation 33.17 yields independent samples $S_T^{(i)}$, $i = 1, \ldots, n$, of the stock price at expiry time T. Hence, the option value is given by

$$C = \frac{1}{n}\sum_{i=1}^{n} C_i = \frac{1}{n}\sum_{i=1}^{n} e^{-rT} \max\left\{S_T^{(i)} - K, 0.0\right\}. \tag{33.18}$$

The QMC simulations follow the same steps as the MC simulations, except that the pseudo-random numbers are replaced by LD sequences. The basic LD sequences known in literature are Halton [42], Sobol [65], and Faure [33]. Niederreiter [55] proposed a general principle of generating LD sequences. In finance, several examples [1,8,35,57] have shown that the Sobol sequence is superior to others. For example, Galanti and Jung [35] observed that "the Sobol sequence outperforms the Faure sequence, and the Faure marginally outperforms the Halton sequence. At 15,000 simulations, the random sequence exhibits an error of 0.07%; the Halton and Faure sequences have errors of 0.1%; and the Sobol sequence has an error of 0.03%. These errors decrease as the number of simulations increase."

---

* $(S_T - K)^+$ represents 0 or positive value of the local pay-off.

33.3.3 Parallel QMC Method for Options

Due to the replicative nature, QMC simulation often consumes a large amount of computing time. Solution on a sequential computer will require hours and/or even days depending on the size of the problem [58]. MC simulations are well suited to parallel computing since it is an embarrassingly parallel problem (no communication between processors [66]). We can employ many processors to simulate various random walks, then average these values to produce the final result. In this scenario, the whole simulation time is minimized. There are three parallel techniques in using random sequence in literature, namely, Leapfrog, Blocking, and Parameterization. In recent times, the earlier three schemes have been proposed for parallelizing LD sequences (see, for example [13,52,56,61]). Srinivasan [66] compared the effectiveness of these three strategies in pricing financial derivatives and concluded that blocking is the most promising method if there are a large number of processors running at unequal speeds. However, the disadvantages of blocking scheme are well pronounced. First, if a processor consumes more LD elements than it was assigned, then the sub-sequences could overlap. Second, if a processor consumes less LD elements than it was assigned, then some elements will be wasted. The final result will be the same as the sequential computation that uses the same LD sequence with some gaps. Third, if processors run at unequal speeds, the fastest processor will finish its task first and wait for the slowest processor at synchronization point; then the overall computation time will be determined by the time elapsed on the slowest processor. It is wasteful to do this since the most powerful processors will idle most of the time. Hence, a good parallel algorithm should distribute computations based on the actual performance of processors at the moment of the execution of the program. The more powerful a processor is, the more tasks it will be assigned. That is, data, computations, and communications should be distributed unevenly among processors to provide the best execution performance. To achieve this, the algorithm must collect information of the entire computing space and compute the relative performances of actual processors in the run time; otherwise, the load of processors will be unbalanced resulting in poor performance. Currently, no parallel programming tool can implement such parallel algorithm except mpC [50]. As a new parallel programming tool, mpC offers a very convenient way to obtain the statistical information of the computing space and the power of each processor. By using mpC, a programmer can explicitly specify the uneven distribution of computations across parallel processors. In addition, mpC has its own mapping algorithms to ensure that each processor performs computations at the speed proportional to the volume of computation it performs. Hence, these two way mappings lead to a more balanced and faster parallel program. An adaptive, distributed QMC algorithm for option pricing [19,20] that takes into account the performances of both processors and communications is presented here.

Having known the power of each processors, the volumes of computation assigned to each processor can be computed as follows: Given N is the total tasks, p is the number of processors, $power_i$ is the ith processor's power. Then, the tasks assigned to the ith processor is

$$\text{task}_i = \left[N \times \frac{\text{power}_i}{\sum_{i=0}^{p} \text{power}_i} \right] \tag{33.19}$$

If there are tasks left, then assign it to host processor.

Suppose each simulation consumes q elements of the given LD sequence, and processor i has t_i tasks, then the whole number of elements will be consumed by processor i is $B = t_i \times q$. Note that B is not necessarily N/P where N is the number of points and P is the number of processors. Hence, the LD sequence is partitioned into uneven blocks. This partition of LD sequence is somewhat like the general blocking scheme in MPI (Message Passing Interface); however, it is superior to general blocking scheme, in which the LD sequence is partitioned into equal size or the burden is on the programmer to determine B. B is usually chosen to be greater than N/P to avoid the problems mentioned earlier in the general blocking scheme with N/P elements per processor.

That is, if a processor consumes more LD elements than it was assigned, then the sub-sequences could overlap. In general, it is unclear on the exact number of points of a sequence that a processor will consume. Because in QMC simulations, there are no safe stopping rules [36]. Without experimentation, it is difficulty to know the number of points of a LD sequence needed to achieve a desired accuracy. Unlike the traditional MC methods, one can use a standard error estimated from a number of simulations to determine if a desire precision is reached. Hence, the LD numbers must be generated more than needed. So the discrepancy of the contiguous points in a block might be different from the discrepancy of the points, which is enough for a sequential run. Therefore, there is a chance that the result produced from the blocking scheme of QMC is not the same as that of a sequential computing [66]. Further discussions are available in [19,20].

33.4 Fast Fourier Transform for Option Pricing

Many natural and artificial phenomena, such as business cycles, repeat in constant times. These phenomena can be approximated by Fourier series. The Fourier transform is used by engineers in a variety of ways including speech transmission, coding theory, and image processing. The financial engineers use Fourier inversion technique to price stock options with various models (for example [63]). Assuming that the characteristic function of the terminal asset price could be derived analytically, this study [63] uses the Fourier inversion technique to determine the risk-neutral probability of the profitable option.

Fundamental ideas on the use of FFT in finance are presented in [17,18]. Financial engineers use Fourier analysis to identify cyclic patterns in asset price movements. Such processes can be either described in the time domain h, which is a function of time $h(t)$, or in the frequency domain where the process is specified by giving its amplitude, H, as a function of frequency f, that is $H(f)$, with $-\infty < f < \infty$. One goes back and forth between the representations by means of the continuous Fourier transform equations, $h(t) = \int_{-\infty}^{\infty} H(f)e^{-2\pi ift}\, df$ called forward $(-i)$ Fourier transform and $H(f) = \int_{-\infty}^{\infty} h(t)e^{2\pi ift}\, dt$ called reverse (i) Fourier transform. Or their discretized forms given by: $H(f) = \frac{1}{N}\sum_{t=0}^{N-1} h(t)e^{2\pi ift/N}$ and $h(t) = \frac{1}{N}\sum_{t=0}^{N-1} H(f)e^{-2\pi ift/N}$

Since the call value is a function of strike price, by mapping call value and strike price to the earlier equations, Carr and Madan [16] applied the Fourier transform to the option pricing problem to get

$$C_T(k) = \frac{\exp(-\alpha k)}{\pi} \int_0^\infty e^{-ivk}\psi_T(v)\, dv \qquad (33.20)$$

where
$k = \log(K)$
K is strike price
α is a dampening factor introduced in the model
$\psi_T(v)$ is the Fourier transform of this call price given as

$$\psi_T(v) = \frac{e^{-rT}\phi_T(v - (\alpha + 1)i)}{\alpha^2 + \alpha - v^2 + i(2\alpha + 1)v} \qquad (33.21)$$

$\psi_T(v)$ is odd in its imaginary part and even in its real part, and ϕ_T is given by 33.26. Here k is the log of strike price K ($k = \log(K)$) and is identical to t in the discretized form for Fourier transform given earlier [70]. That is, the call value needs to be computed at various strike prices of the underlying assets in the option contract. Furthermore, v corresponds to f, $\psi_T(v)$ is the Fourier transform of the call price $C_T(k)$ and $q_T(s)$ is the risk-neutral density function of the pricing model. The integral in the right hand side of Equation 33.20 is a direct Fourier transform and lends itself to the application of the FFT in the form of summation given by the continuous and discretized forms for Fourier transform given earlier. This is called Carr–Madan (CM) model.

If $M = e^{-\alpha k}/\pi$ and $\omega = e^{-i}$ then

$$C_T(k) = M \int_0^\infty \omega^{vk}\psi_T(v)\, dv \tag{33.22}$$

If $v_j = \eta(j-1)$ and applying trapezoid rule for the integral on the right of Equation 33.22, $C_T(k)$ can be written as

$$C_T(k) \approx M \sum_{j=1}^N \psi_T(v_j)\omega^{v_jk}\eta, \quad k = 1,\ldots,N \tag{33.23}$$

where the effective upper limit of integration is $N\eta$ and v_j corresponds to various prices with η spacing.

To calculate the call values, Equation 33.22 has to be solved analytically. The discrete form of the earlier equation given as Equation 33.23 is not suitable to feed into the existing FFT algorithms, for example, Cooley–Tukey [22], Stockham auto sort [53], and Bailey [3]. Hence, the CM model in its current form cannot be used for faster pricing. This is a major drawback of using CM model for practical purposes. An explanation on the mathematical reasons for the drawback is presented in the next section, while proposing a solution for this drawback in the model.

To obtain numerical solution and to take advantage of parallel computing for real time pricing one needs to improve this mathematical model.

33.4.1 Mapping Mathematical Model to FFT Framework

Recent research on option valuation has successfully applied Fourier analysis to calculate option prices. As shown earlier in Equation 33.20, to obtain the analytically solvable Fourier transform, the call price function needs to be multiplied with an exponential factor, $e^{\alpha k}$ ($c_T(k) = e^{\alpha k}C_T(k)$), where α is a dampening coefficient [16]. The calculation of $\psi_T(v)$ in Equation 33.21 depends on the factor $\phi_T(u)$, where $u = v - (\alpha + 1)i$. The calculation of the intermediate function $\phi_T(u)$ requires the specification of the risk neutral density function, $q_T(s)$. The limits on the integral have to be selected in such a way as to generate real values for the FFT inputs. To generate the closed form expression of the integral, the integrands, especially the function $q_T(s)$, have to be selected appropriately. Without the loss of generality, one can use uniform distribution for $q_T(s)$. This implies the occurrence of a range of terminal log prices at equal probability, which could, of course, be relaxed and a normal or other distribution could be employed. Since the volatility is assumed constant (low), the variation in the drift is expected to cause a stiffness* in the system. However, by assuming uniform distribution for $q_T(s)$, variation in drift is eliminated and hence the stiffness is avoided. Therefore, the use of uniform distribution would make the integration easier.

For computation purposes, the upper limit of Equation 33.22 is assumed as a constant value and the lower limit is assumed as 0. The upper limit will be dictated based on the terminal spot price.

---

* Stiffness occurs when two processes controlling a physical phenomenon proceeds at two extremely different rates. It is common in scientific problems, such as chemical reactions and high temperature physics. When a system with such physical phenomenon is manifested in mathematics, such as differential or integral equation, the mathematical system is known to be stiff, where solution of such systems of equations would require special techniques to handle the stiffness. Drift and volatility in the finance systems act as two phenomena affecting the system away from equilibrium, hence may induce stiffness. The assumptions of uniform distribution for the density function to make the integration easier, in conjunction with assumed constant volatility, however, naturally avoids this issue.

In other words, to finish the call option in-the-money,* the upper limit will be smaller than the terminal asset price. Therefore, the equation is

$$\phi_T(u) = \int_0^\lambda (\cos(vk) + i\sin(vk))q_T(s)\,ds \qquad (33.24)$$

Without loss of generality, further modifications are required as derived in the following text. The purpose of these modifications is to generate feasible and tractable initial input conditions to the FFT algorithm from these equations. Moreover, these modifications make the implementation easier. From Equation 33.21

$$\psi_T(v) = \frac{e^{-rT}\phi_T(v - (\alpha + 1)i)(C - iD)}{(C^2 + D^2)} \qquad (33.25)$$

where $C = (\alpha^2 + \alpha - v^2)$; and $D = (2\alpha + 1)v$.
Now,

$$\phi_T(u) = \int_0^\lambda e^{ius}q_T(s)\,ds \qquad (33.26)$$

where λ is terminal spot price and integration is taken only in the positive axis.
To calculate $\phi_T(v - (\alpha + 1)i)$, $v - (\alpha + 1)i$ is substituted by u in Equation 33.26 that gives

$$\phi_T(v - (\alpha + 1)i) = \int_0^\lambda e^{(iv + \alpha + 1)s}q_T(s)\,ds \qquad (33.27)$$

Assuming $q_T(s)$ as an uniform distribution function of the terminal log price, Equation 33.27 can be shown as

$$\phi_T(v - (\alpha + 1)i) = q_T(s)\frac{e^{(iv + \alpha + 1)s}}{iv + \alpha + 1}\Big|_0^\lambda$$

$$= \frac{q_T(s)}{iv + \alpha + 1}\left(e^{iv\lambda}e^{(\alpha+1)\lambda} - 1\right)$$

$$= \frac{q_T(s)(\alpha + 1 - iv)}{(\alpha + 1)^2 + v^2}\left[\left(e^{(\alpha+1)\lambda}(\cos(\lambda v) + i\sin(\lambda v)) - 1\right)\right]$$

$$= \frac{q_T(s)}{(\alpha + 1)^2 + v^2}\left[\left[e^{(\alpha+1)\lambda}i\{(\alpha + 1)\cos(\lambda v) + v\sin(\lambda v)\}\right.\right.$$

$$\left.\left. - (\alpha + 1) + i\left[e^{(\alpha+1)\lambda}\{(\alpha + 1)\sin(\lambda v) - v\cos(\lambda v)\} + v\right]\right] \qquad (33.28)$$

If one sets $e^{(\alpha+1)\lambda}\{(\alpha + 1)\cos(\lambda v) + v\sin(\lambda v)\} - (\alpha + 1) = \Delta$ and $e^{(\alpha+1)\lambda}\{(\alpha + 1)\sin(\lambda v) - v\cos(\lambda v)\} + v = \Delta_x$ then Equation 33.28 can be simplified as

$$\phi_T(v - (\alpha + 1)i) = \frac{q_T(s)}{(\alpha + 1)^2 + v^2}(\Delta + i\Delta_x) \qquad (33.29)$$

---

* In-the-money call option is a situation where the underlying asset price of the option is larger than the strike price; at-the-money call means asset price equals the strike price; natural extension is for out-of-the-money call, which corresponds to a situation where the asset price is smaller than the strike price. These definitions are reversed for a put option.

Substituting Equation 33.29 in Equation 33.25 gives

$$\psi_T(v) = \frac{A}{B(C^2 + D^2)} \left[(C\Delta + D\Delta_x) + i(C\Delta_x - D\Delta) \right] \qquad (33.30)$$

where, $A = e^{-rT}q_T(s)$; $B = (\alpha + 1)^2 + v^2$; $C = \alpha^2 + \alpha - v^2$; $D = (2\alpha + 1)v$. This final expression can be used for parallel FFT algorithm to compute the call price function. The financial input data set for our parallel FFT algorithm is the calculated data points of $\psi_T(v)$ for different values of v.

One can then calculate call values for different strike price values v_j where j will range from 1 to N. The lower limit of strike price is 0 and the upper limit is $(N - 1)\eta$ where η is the spacing in the line of integration the smaller value of η gives fine grid integration and a smooth characteristics function of strike price and the corresponding calculated call value.

The value of k on the left side of Equation 33.23 represents the log of ratio of strike and terminal spot price. The implementation of FFT mathematical model returns N values of k with a spacing size of γ and these values are fed into a parallel algorithm to calculate N values of $C_T(k)$. Here focus is on the cases in the range of in-the-money to at-the-money call values. The value of k will be 0 for at-the-money call—that is strike price and exercise price are equal. The value of k will be negative when we are in-the-money and positive when we are out-of-the-money. If γ is the spacing in the k then the values for k can be obtained from the following equation:

$$k_u = -p + \gamma(u - 1) \quad \text{for } u = 1, \dots, N \qquad (33.31)$$

Hence, the log of the ratio of strike and exercise price will range from $-p$ to p where $p = \frac{N\gamma}{2}$. The substitution of Equation 33.31 in Equation 33.23 gives

$$C_T(k_u) \approx \frac{\exp(-\alpha k_u)}{\pi} \sum_{j=1}^{N} \left\{ e^{-iv_j(-p+\gamma(u-1))} \psi_T(v_j)\eta \right\} \quad \text{for } u = 1, \dots, N \qquad (33.32)$$

Replacing v_j with $(j - 1)\eta$ in Equation 33.32,

$$C_T(k_u) \approx \frac{\exp(-\alpha k_u)}{\pi} \sum_{j=1}^{N} \left\{ e^{-i\gamma\eta(j-1)(u-1)} e^{ipv_j} \psi_T(v_j)\eta \right\} \quad \text{for } u = 1, \dots, N \qquad (33.33)$$

The basic equation of FFT is

$$Y(k) = \sum_{j=1}^{N-1} e^{-i\frac{2\pi}{N}(j-1)(k-1)} x(j) \quad \text{for } k = 1, \dots, N. \qquad (33.34)$$

Comparing Equation 33.33 with the basic FFT equation 33.34 it is easy to note that $\gamma\eta = \frac{2\pi}{N}$. The smaller values of η will ensure fine grid for the integration. But, for call prices at relatively large strike spacings (γ), few strike prices will lie in the desired region near the stock price [16]. Furthermore, if the values of N are increased, one will get more intermediate points of the calculated call prices ($C_T(k_u)$) corresponding to different strike prices (v_j). This helps the investor to capture the call price movements of an option for different strike prices in the market. With N (=1024) number of calculated call values, assuming $\eta = 0.25$ with the intuition that it will ensure fine grid integration, γ is calculated as 0.02454. Similar to basic FFT equation, Equation 33.32 can also be parallelized. An efficient parallel FFT algorithm can compute fast and accurate solutions of equation 33.32. In the next section a data swapping technique that exploits data locality to reduce communication on a parallel computer is developed and effectively applied to the improved mathematical model. With the inputs derived from Equation 33.32, this algorithm can be implemented on a parallel machine.

33.4.2 Parallel FFT Algorithm

Figure 33.7 illustrates the classical Cooley–Tukey algorithm [22] (also known as FFT network) and the butterfly computation. There are N ($N = 2^m$) data elements and P ($P = 2^r$) processors where $N > P$. A butterfly computation is performed on each of the data points in every iteration where there are $\frac{N}{2}$ summations and $\frac{N}{2}$ differences.

The FFT is inherently a synchronous algorithm. In general, a parallel algorithm for FFT with blocked data distribution [37] where $\frac{N}{P}$ data is allocated to every processor involves communication for $\log P$ iterations and terminates after $\log N$ iterations. The input data points are bit reversed before feeding to the parallel FFT algorithm. With shuffled input data at the beginning, the first $\log N - \log P$ stages require no communication. That is, the data required for the butterfly computation resides in each local processor. Therefore, during the first ($\log N - \log P$) iterations, a sequential FFT algorithm can be used inside each processor (called local algorithm).

At the end of the ($\log N - \log P$)th iteration, the latest computed values for $\frac{N}{P}$ data points exist in each processor. The last $\log P$ stages require remote communications (called remote algorithm). The partners of each of the $\frac{N}{P}$ data points in processor P_i required to perform the actual butterfly computation at each iteration reside in a different processor P_j. In a blocked data distribution, therefore, $\frac{N}{P}$ amount of data is communicated by each processor for $\log P$ stages. The message size is $\frac{N}{P}$.

In Figure 33.7 a, one can note that calculating Y_0 in processor 0 requires two data points, one of which resides in the local processor (=0), and the other resides in processor 2, and hence requires one communication to calculate Y_0. Similarly, calculating Y_1, Y_2, and Y_3 needs three

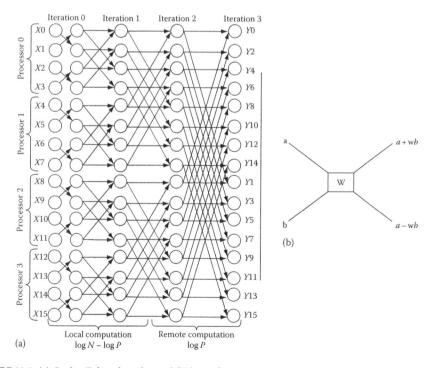

FIGURE 33.7 (a) Cooley-Tukey algorithm and (b) butterfly operation.

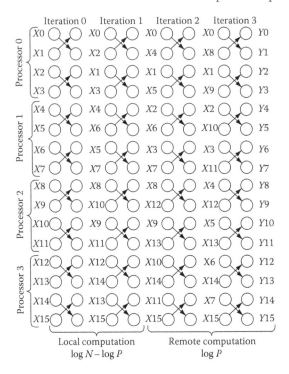

FIGURE 33.8 Data swapping algorithm.

more communications with processor 2. Each processor requires four communications to calculate four FFT output. In total, 16 communications are required.

In the data swapping algorithm depicted in Figure 33.8 [4] the same blocked data distribution is applied and the first $(\log N - \log P)$ stage requires no communication. However, in the last $\log P$ stage thats require communication, some of the data is swapped at each stage leaving the data to reside in the processor's local memory after swapping. Therefore, the identity of some of the data points in each processor changes at every stage of the $\log P$ stages.

In Figure 33.8, one can note that calculating the first two output data points in processor 0 needs two input data points with indices 0 and 8, and node with index 8 does not reside in the local processor. Therefore, one communication is needed to bring node 8 from processor 2. Similarly, calculating the next two output data points need one more communication. Therefore, in processor 0, two communications are needed to calculate four output data points. With the same arguments, each of the processors 1, 2, and 3 needs two communications. In total, eight communications are required to calculate FFT of 16 data points. Therefore, in the new parallel FFT algorithm, the number of communications is reduced by half.

It is interesting to note that communication between processors is point to point and swapping the data could be done by taking advantage of this point. However, in this case, only $\frac{N}{2P}$ amount of data (message size) is communicated by each processor at every stage. Also note that, data swapping between processors at each location allows both the upper and lower parts of the butterfly computations to be performed locally by each processor. This improvement enhances good data locality and thereby providing performance increase in the new FFT algorithm compared to the Cooley–Tukey algorithm.

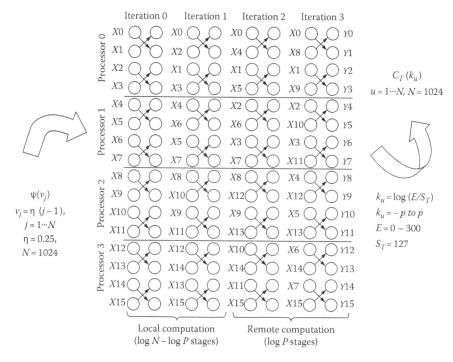

FIGURE 33.9 Input and output to the data swap algorithm.

Figure 33.9 shows how the data swap algorithm calculates the call values from the input data set generated from the Equation 33.30. As mentioned in Section 33.4.1, Equation 33.30 calculates $\psi_T(v_j)$ for different values of v_j (strike price), where $v_j = \eta(j-1)$ and $j = 1 \cdots N$. η is the strike price spacing and for a fixed strike price range (for example 0–300) larger values of N gives smaller values of η. With smaller strike price spacing, the data swap algorithm calculates more numbers of intermediate call values in the specified region of the strike price. This depicts the change in the call price for smaller change in the strike price. It is observed [4] the data swap algorithm performs better over the Cooley–Tukey algorithm when the data size (N) increases. Therefore, the data swap algorithm can capture the call prices for various strike prices faster than the Cooley–Tukey algorithm.

The data swap algorithm calculates the N number of call values ($C_T(k_u)$) where k_u is the log of the ratio of the strike price E and the terminal spot price S_T (market price of the underlying asset). The call values obtained from the data swap algorithm are normalized values with respect to the terminal spot price. In other words, call values for a given strike price, which is the log of the ratio of the strike and terminal spot prices, are normalized with terminal spot price as a base. When the *strike price < spot price* the call option is in-the-money (the value of k_u will be negative) and when strike and spot price are equal the call option is at-the-money (the value of k_u will be zero). Whereas the call option is out-of-the-money (the value of k_u will be positive), if *strike price > spot price*. When the call option is in-the-money, the investor would prefer to exercise the option (purchasing the option) at the strike price and immediately sell the asset in the market at the terminal spot price. Thus, the holder can profit. When the call option is at-the-money, the profit or loss is zero. But for the call option out-of-the-money, the investor would not prefer to exercise it as the spot price of the asset in the market is less than the exercise price (strike price) of the contract.

33.5　Summary and Further Research

There are various directions of research in computational finance based on commercial and/or algorithmic interests. This chapter has touched upon some of the techniques that are both of algorithmic value and commercial value. There are many other engineering techniques and various other problems in finance, such as portfolio optimization, that require good algorithms and faster solutions. We finish the chapter with a short discussion on an emerging algorithmic science for financial applications.

Nature inspired algorithms (also known as Swarm Intelligence or evolutionary algorithms) [26] have been used in many applications to solve many combinatorial optimization problems, including problems in telecommunications [14,15] and dynamic networks, such as mobile ad hoc networks [39]. Swarm intelligence is an artificial intelligence technique inspired by animal societies, such as bees [64,72], termites [60], and ants [27]. These small animals live in a hostile, decentralized environment and communicate with one another through a stigmergic method to accomplish their tasks, such as finding the food source. Ant Colony Optimization (ACO) is one such swarm intelligence technique inspired by real ants. The ants communicate with one another by depositing pheromone (scent) on the ground to attract their fellow ants to follow their trail, one of the stigmergic approaches in the animal world. As time progresses, the pheromone along longer path evaporates leaving the ants to converge along shorter paths. The objective of ACO in most applications is to find the shortest path.

These nature inspired algorithms have prospects in many areas of finance, such as to evolve trading rules, diagnosis of company's future, etc. (see [11]). A key advantage of using evolutionary algorithms to design a trading process is that these algorithms can simultaneously evolve good rules and good parameters for such rules [11].

For option pricing, the solution space consists of large number of price nodes. The ants have this large bounded space of price nodes to search through in deciding the best time (node) to exercise the option. The nodes within the search space are dispersed in many locations and are connected in some random manner. The goal of ACO in the current application is to find the best node to exercise the option. This can be achieved by directing a path to the node, thereby allowing other investors to quickly arrive at the node to exercise prices. This path can be created using a variety of techniques. However, these approaches can be used on structured graphs only. ACO is a probabilistic approach that allows the path to be created in a random way. That is, the graph is ad hoc and random. There is no real structure to the graph. Therefore, ACO allows flexibility by distributing the nodes randomly and thereby capturing the real market place.

Dorigo [25,29] proposed ACO that involves distributed computation, positive feedback, and a greedy heuristic to solve a given problem. Distributed computing helps in searching a wide area of the problem domain, positive feedback helps in finding good solutions and a greedy heuristic is needed to find solutions in early stages.

In finance, knowing the volatility of a financial instrument makes an investor powerful and could make precise decisions. However, predicting the volatility of an instrument, say stock price, is a daunting task and is a research area by itself. Researchers and practitioners, using historical data, have developed methods for predicting historical volatility. Historical volatility would only reflect past and the accuracy of stock prices predicted based on historic volatility is questionable. There have been some efforts in measuring volatility accurately. Implied volatility measures the intrinsic dependence of past stock prices not only with time but with other factors affecting the price over a period of time. Keber and Schuster [45] used generalized ant programming to derive analytical approximations to determine the implied volatility for American put options. Generalized ant programming is a new method inspired by genetic programming approach introduced by Koza [47] and Ant Colony System (ACS) [28]. They used experimental data and validation data sets for computing the implied volatility. Their results outperformed any other approximations.

The objective of ACO in most applications is to find the shortest path. However, in option pricing, the interest is not in finding the shortest path, but finding the best node that allows the investor to exercise the option. Therefore, the general ACO algorithm has to be improved to handle this problem. In [49], we have developed two bioinspired algorithms for option pricing problem using ACO.

Defining Terms

American option: An option that allows exercising anytime before the expiration date.

Arbitrage: An act of taking advantage of difference in prices in two different exchanges at a given time.

Bid: An initiation to buy a security, such as an option or futures contract, at a specified price at a future time.

Bid-ask spread: The difference between the offered price and the bid price.

Black–Scholes–Merton model: The Nobel prize winning model that calculates the price of an option, when the exercise price, time to exercise, the interest rate, and the expected volatility of the price of the underlying asset are given.

Call option: Right to buy a share or other asset at a fixed price (strike price) at some time (expiration date) in the future. Call buyer is known as Holder. The other party is known as Writer.

in-the-money-Call: When strike price is lower the current market price of the underlying asset, the call option is said to be in-the-money.

at-the-money-Call: When strike price is equal to the market price of the underlying asset, the call option is said to be at-the-money.

out-of-the-money-Call: When strike price is higher than the market price of the underlying asset, the call option is said to be out-of-the-money.

Commodity futures: Contracts to buy or sell commodities on a futures exchange at a price specified price and future date

Delta hedging (option's delta): The number of shares/stocks required to hedge against the price risk of holding an option.

Efficient market hypothesis: This hypothesis states that the process in financial markets reflect all publicly available information. One of the consequences is that, provided the conditions of an efficient market are met, individual investors will not achieve risk-adjusted returns above the market average in the long run.

European option: An option that allows exercising only at expiration date.

Exotic option: An option that has complex pricing mechanism that requires finding minimum, maximum, or averages of the asset prices; or that the expiration date could be floating; or may have a barrier price on the asset for the option to kick in, etc.

Futures: Futures are contracts to buy or sell a fixed quantity of an asset—goods or securities—at a fixed price and a fixed date, regardless of any intervening change in price or circumstances.

Mark-to-Market: When an investment or a liability is revalued to the current market price.

Options: An option gives the right, but not the obligation, to buy or sell something at a given price and time. A call option gives the right to buy while a put option gives the right to sell. A simple European or American option is known as Vanilla option.

Option pricing theory: Relates to the valuation of options or various other continent contracts where value depends on the underlying securities or assets.

Put option: Right to sell a share or other asset at a fixed price (strike price) at some time (expiration date) in the future.

Random walk: The convention that in efficient markets, successive price changes are random.

Risk: In investment risk is referred to as the uncertainty of return from any asset.

Risk-free asset: An asset providing a certain return over some holding period.

References

1. P. Acworth, M. Broadie, and P. Glasserman. A comparison of some Monte Carlo and quasi Monte Carlo methods for option pricing. In H. Niederreiter, P. Hellekalek, G. Larcher, and P. Zinterhof, editors, *Monte Carlo and Quasi-Monte Carlo Methods 1996*, pages 1–18. Springer, Berlin, 1998.
2. H. M. Amman. Supercomputing in economics: A new field of research? In J. L. Delhaye and E. Gelenbe, editors, High Performance Computing. Elsevier Science Publishers B.V., Amsterdam, 1989.
3. D. H. Bailey. FFTs in external or hierarchical memory Fourier. *The Journal of Supercomputing*, 4, 1990.
4. S. Barua, R. K. Thulasiram, and P. Thulasiraman. High performance computing for a financial application using fast Fourier transform. In *Proceedings of the European Parallel Computing Conference (EuroPar 2005)*, volume 3648, pages 1246–1253, Lisbon, Portugal, 2005.
5. J. R. Birge. Quasi-Monte Carlo Approaches to Option Pricing. Technical Report 94–19, Department of Industrial and Operations Engineering, University of Michigan, 1994.
6. F. Black and M. Scholes. The pricing of options and corporate liabilities. *Journal of Political Economy*, 81:637–654, January 1973.
7. P. Boyle. Options: A Monte Carlo approach. *Journal of Financial Economics*, 4:323–338, 1977.
8. P. Boyle, M. Broadie, and P. Glasserman. Monte Carlo methods for security pricing. *Journal of Economic Dynamics and Control*, 21:1267–1321, 1997.
9. P. P. Boyle. A lattice framework for option pricing with two-state variables. *Journal of Financial and Quantitative Analysis*, 23(1):1–12, March 1988.
10. P. P. Boyle. Options: A Monte Carlo approach. *Journal of Financial Economics*, 4:323–338, 1977.
11. A. Brabozan and M. O'Neill. *Biologically Inspired Algorithms for Financial Modelling*. Springer, New York, 2006.
12. M. Broadie and P. Glasserman. Pricing American-Style Securities Using Simulation. Technical Report 96–12, Columbia—Graduate School of Business, 1996. Available at http://ideas.repec.org/p/fth/colubu/96-12.html.
13. B. C. Bromley. Quasirandom number generators for parallel Monte Carlo algorithms. *Journal of Parallel and Distributed Computing*, 38(0132):101–104, 1996.
14. G. Di Caro, F. Ducatelle, and L. M. Gambardella. AntHocNet: An adaptive nature-inspired algorithm for routing in mobile ad hoc networks. *European Transactions on Telecommunications (ETT), Special Issue on Self Organization in Mobile Networking*, 16(5):443–455, 2005.
15. G. Di Caro and M. Dorigo. AntNet: Distributed stigmergetic control for communications networks. *Journal of Artificial Intelligence Research*, 9:317–365, 1998.
16. P. Carr and D. B. Madan. Option valuation using the Fast Fourier transform. *The Journal of Computational Finance*, 2(4):61–73, 1999.
17. A. Cerny. Introduction to fast Fourier transform in finance. *Journal of Derivatives*, 12(1): Fall 2004.
18. A. Cerny. *Mathematical Techniques in Finance—Tools for Incomplete Markets*—Chapter 7. Princeton University Press, Princeton, NJ, January 2004.
19. G. Chen. Distributed quasi Monte Carlo algorithms for option pricing on HNoWs using mpC. Master's thesis, Department of Computer Science, University of Manitoba, Winnipeg, MB, CA, 2006.

20. G. Chen, Parimala Thulasiraman, and Ruppa K. Thulasiram. Distributed quasi-Monte Carlo algorithm for option pricing on HNoWs using mpC. In *ANSS '06: Proceedings of the 39th Annual Symposium on Simulation*, pages 90–97, Washington, DC, USA, 2006. IEEE Computer Society.

21. I. J. Clark. Option pricing algorithms for the Cray T3D supercomputer. *Proceedings of the First National Conference on Computational and Quantitative Finance (Loose-Bound Volume—No Page Number)*, September 1998.

22. J. W. Cooley, P. A. Lewis, and P. D. Welch. Statistical methods for digital computers. In *The Fast Fourier Transform and Its Application to Time Series Analysis*. Wiley, New York, 1977.

23. J. C. Cox, S. A. Ross, and M. Rubinstein. Option pricing: A simplified approach. *Journal of Financial Economics*, 7:229–263, 1979.

24. M. A. H. Dempster and S. S. G. Hong. Spread Option Valuation and the Fast Fourier Transform. Technical Report WP 26/2000, Judge Institute of Management Studies, Cambridge, England, 2000.

25. M. Dorigo. Optimization, learning and natural algorithms. PhD thesis, DEI, Politecnico di Milano, Italy [in Italian], 1992.

26. M. Dorigo, E. Bonabeau, and G. Theraulaz. *Swarm Intelligence: From Natural to Artifical Systems*. Oxford University Press, New York, NY, 1999.

27. M. Dorigo and T. Stützle. *Ant Colony Optimization*. MIT Press, Cambridge, MA, May 2004.

28. M. Dorigo and Luca Maria Gambardella. Ant colony system: A cooperative learning approach to the traveling salesman problem. *IEEE Transactions on Evolutionary Computation*, 1(1):53–66, April 1997.

29. M. Dorigo, V. Maniezzo, and A. Colorni. The Ant System: Optimization by a colony of cooperating agents. *IEEE Transactions on Systems, Man, and Cybernetics Part B: Cybernetics*, 26(1):29–41, 1996.

30. D. Duffie. *Dynamic Asset Pricing Theory*. Princeton University Press, Princeton, NJ, 1996.

31. N. Ekvall. Two Finite Difference Schemes for Evaluation of Contingent Claims with Three Underlying State Variables. Technical Report 6520, Ekonomiska Forskningsinstitutet, Stockholm, Norway, 1993.

32. R. Engle. Autoregressive conditional heteroskedasticity with estimates of the variance of U.K. inflation. *Econometrica*, 50(9):987–1008, 1982.

33. H. Faure. Discrepance de suites associees a un systeme de numeration (en dimension s). *Acta Arithmetica*, 41:337–351, 1982.

34. M. C. Fu. Pricing of financial derivatives via simulation. In C. Alexopoulos, K. Kang, W. Lilegdon, and D. Goldsman, editors. In *Proceedings of the 1995 Winter Wimulation Conference*, pages 126–132. Institute of Electrical and Electronics Engineers, Piscataway, NJ, 1995.

35. S. Galanti and A. Jung. Low-discrepancy sequences: Monte Carlo simulation of option prices. *Journal of Derivatives*, 5(1):63–83, 1997.

36. P. Glasserman. *Monte Carlo Methods in Financial Engineering*. Springer, New York, 2004.

37. A. Grama, A. Gupta, V. Kumar, and G. Karypis. *Introduction to Parallel Computing*. Pearson Education Limited, Edinburgh Gate, Essex, 2nd edition, 2003.

38. D. Grant, G. Vora, and D. Weeks. Path-dependent options: Extending the Monte Carlo simulation approach. *Management Science*, 43(11):1589–1602, November 1997.

39. M. Gunes, U. Sorges, and I. Bouazzi. ARA—the ant-colony based routing algorithm for MANETs. In *Proceedings of the International Conference on Parallel Processing Workshops (ICPPW'02)*, pages 79–85, Vancouver, B.C., August 2002.

40. H. Markowitz. Portfolio selection. *Journal of Finance*, 7(1):77–91, 1952.

41. M. B. Haugh and A. W. Lo. Computational challenges in portfolio management—tomorrow's hardest problem. *Computing in Science and Engineering*, 3(3):54–59, May-June 2001.

42. J. H. Halton. On the efficiency of certain quasirandom sequences of points in evaluating multidimensional integrals. *Numerische Mathematik*, 2:84–90, 1960.

43. J. Hull and A. White. The pricing of options on assets with stochastic volatilities. *Journal of Finance*, 42:281–300, 1987.

44. J. C. Hull. *Options, Futures, and Other Derivative Securities.* Prentice Hall, Englewood-Cliffs, NJ, May 2006.

45. C. Keber and M. G. Schuster. Generalized ant programming in option pricing: Determining implied volatilities based on American put options. In *Proceedings of the IEEE International Conference on Computational Intelligence for Financial Engineering,* pages 123–130, Hong Kong Convention and Exhibition Centre, Hong Kong, March, 2003.

46. A. G. Z. Kemna and A. C. F. Vorst. A pricing method for options based on average asset values. *Journal of Banking and Finance,* 14:113–129, 1990.

47. J. R. Koza. *Genetic Programming: On the Programming of Computers by Means of Natural Selection.* MIT Press, Cambridge, MA, 1992.

48. L. Kuipers and H. Niederreiter. *Uniform Distribution of Sequences.* John Wiley & Sons, New York, 1974.

49. S. Kumar, R. Thulasiram, and P. Thulasiraman. A bioinspired algorithm for pricing options. In *ACM Canadian Conference on Computer Science and Software Engineering (C3S2E),* pages 11–22, Montreal, Canada, May 2008.

50. A. Lastovetsky. Adaptive parallel computing on heterogeneous networks with mpC. *Parallel Computing,* 28:1369–1407, 2002.

51. E. Laure and Han Moritsch. Portable parallel portfolio optimization in the AURORA financial management system. In Howard J. Siegel, editor, *Proceedings (Vol. 4528) of the SPIE International Symposium on Commercial Applications of High Performance Computing,* pages 193–204, Denver, CO, August 2001.

52. J. X. Li and G. L. Mullen. Parallel computing of a quasi-monte carlo algorithm for valuing derivatives. *Parallel Computing,* 26(5):641–653, 2000.

53. C. Van Loan. *Computational Frameworks for the Fast Fourier Transform* (Chapter 1.7). SIAM: Frontiers in Applied Mathematics, Philadelphia, PA, 1992.

54. R. C. Merton. Theory of rational option pricing. *Bell Journal of Economics,* 4:141–183, 1973.

55. H. Niederreiter. *Random Number Generation and Quasi-Monte Carlo Methods,* volume 63 of *CBMS-NSF Regional Conference Series in Applied Mathematics.* SIAM, Philadelphia, PA, 1992.

56. G. Okten and A. Srinivasan. Parallel quasi-Monte Carlo methods on a heterogeneous cluster. In H. Niederreiter et al., editor, *Proceedings of Fourth International Conference on Monte Carlo and Quasi-Monte Carlo,* pages 406–421, Hong Kong, 2000.

57. S. H. Paskov and J. F. Traub. Faster valuation of financial derivatives. *Journal of Portfolio Management,* 22(1):113–120, Fall 1995.

58. S. Rakhmayil, I. Shiller, and R. K. Thulasiram. Cost of option pricing errors associated with incorrect estimates of the underlying assets volatility: Parallel Monte Carlo simulation. *IMACS Journal of Mathematics and Computers in Simulation* (under review).

59. R. J. Rendleman and B. J. Bartter. Two-state option pricing. *Journal of Finance,* 34(5):1093–1109, December 1979.

60. M. Roth and S. Wicker. Termite: Ad-hoc networking with stigmergy. In *Proceedings of IEEE Global Telecommunications Conference (Globecom 2003),* pages 2937–2941, San Francisco, USA, 2003.

61. W. Schmid and A. Uhl. Techniques of parallel quasi-Monte Carlo integration with digital sequences and associated problems. *Mathematics and Computers in Simulation,* 55:249–257, 2000.

62. E. S. Schwartz and W. N. Torous. Prepayment and the valuation of mortgage-backed securities. *Journal of Finance,* 44(2):375–392, 1989.

63. L. O. Scott. Pricing stock options in a jump-diffusion model with stochastic volatility and interest rates. *Mathematical Finance,* 7(4):413–426, 1997.

64. T. D. Seeley. *The Wisdom of the Hive : The Social Physiology of Honey Bee Colonies.* Harvard University Press, Cambridge, MA, USA, 1995.

65. I. M. Sobol. On the distribution of points in a cube and the approximate evaluation of integers. *U.S.S.R. Computational Mathematics and Mathematical Physics,* 7(4):86–112, 1967.

66. A. Srinivasan. Parallel and distributed computing issues in pricing financial derivatives through quasi Monte Carlo. In *Proceedings (CD-ROM) of International Parallel and Distributed Processing Symposium (IPDPS02)*, Fort Lauderdale, FL, April 2002.

67. D. Tavalla and C. Randall. *Pricing Financial Instruments: The Finite Difference Method.* John Wiley & Sons, New York, NY, 2000.

68. R. K. Thulasiram, L. Litov, H. Nojumi, C. T. Downing, and G. R. Gao. Multithreaded algorithms for pricing a class of complex options. In *Proceedings (CD-ROM) of the International Parallel and Distributed Processing Symposium(IPDPS)*, San Francisco, CA, April 2001.

69. R. K. Thulasiram, C. T. Downing, and G. R. Gao. A multithreaded parallel algorithm for pricing American securities. In *Proceedings (CD-ROM) of the International Conference on Computational Finance 2000*, London, England, June 2000.

70. R. K. Thulasiram and P. Thulasiraman. Performance evaluation of a multithreaded Fast Fourier Transform algorithm for derivative pricing. *The Journal of Supercomputing*, 26(1):43–58, August 2003.

71. J. A. Tilley. Valuing American options in a path simulation model. *Transactions of the Society of Actuaries*, 45:83–104, 1993.

72. H. F. Wedde, M. Farooq, T. Pannenbaecker, B. Vogel, C. Mueller, J. Meth, and R. Jeruschkat. BeeAdHOC: An energy efficient routing algorithm for mobile ad hoc networks inspired by bee behavior. In *Proceedings of Genetic and Evolutionary Computation Conference*, pages 153–160, Washington, DC, June 2005.

73. P. Wilmott, S. Howison, and J. Dewynne. *The Mathematics of Financial Derivatives*. Cambridge University Press, Cambridge, UK, 1995.

74. S. A. Zenios. High-performance computing in finance: The last 10 years and the next. *Parallel Computing*, 25:2149–2075, December 1999.

75. R. Zvan, P. Forsyth, and K. R. Vetzal. A general finite element approach for PDE option pricing models. In *Proceedings of the First National Conference on Computational and Quantitative Finance (Loose-Bound volume—No Page Number)*, September 1998.

Index

O

P

Printed and bound by CPI Group (UK) Ltd, Croydon, CR0 4YY

23/10/2024

01778260-0001